P9-BJY-876

CHESS

5334 Problems, Combinations, and Games

CHESS

5334 Problems, Combinations, and Games

INRODUCTION BY BRUCE PANDOLFINI

László Polgár

Black Dog
& Leventhal
Publishers
New York

This paperback edition published by
Black Dog & Leventhal Publishers in 2006.

An earlier edition of this book was published in Germany by Könemann Verlagsgesellschaft mbH under the title *Polgár Chess, training in 5333+1 positions.*

Published by
Black Dog & Leventhal Publishers
151 West 19th Street
New York, NY 10011

Distributed by
Workman Publishing Company
225 Varick Street
New York, NY 10014

Printed and bound in the U.S.A.

ISBN-13: 978-1-57912-554-7

Library of Congress Cataloging-in-publication data

Polgár, László 1946–
Chess: 5334 problems, combinations and games/ Polgár, László ;
Introduction by Bruce Pandolfini
p. cm
Includes biographical references (p.)
ISBN 1-57912-554-9
1. Chess Problems 2. Chess—middle games 3, Chess—collection of games.
I. Title
GV141.P65 1995 95-22487
794.1'2-de20 CIP

l k j i h g f

CONTENTS

Foreword **9**

Introduction **11**

Key to symbols **13**

Mate in one (1–306) **53**

Mate in two (307–3718) **107**

White to move (307–450) **109**

Combinations – white to move (451–1254) **133**

Combinations – black to move (1255–1470) **267**

Compositions – white to move (1471–3514) **303**

Combinational compositions – white to move (3515–3718) **644**

Mate in three (3719–4462) **679**

Combinations – white to move (3719–4138) **681**

Combinations – black to move (4139–4462) **751**

600 miniature games (4463–5062) **807**

f3 (f6) combinations (4463–4562) **809**

g3 (g6) combinations (4563–4662) **837**

h3 (h6) combinations (4663–4762) **866**

f2 (f7) combinations (4763–4862) **895**

g2 (g7) combinations (4863–4962) **925**

h2 (h7) combinations (4963–5062) **955**

Simple endgames (5063–5206) **985**

White to move and draws (5063–5104) **987**

White to move and wins (5105–5206) **994**

Polgár sisters' tournament-game combinations (5207–5334) **1013**

White to move and wins (5207–5278) **1015**

Black to move and wins (5279–5333) **1027**

Mate in two – white to move (5334) **1037**

Solutions **1039**

Biography of László Polgár **1101**

Bibliography **1102**

FOREWORD

There are more than 5,000 problems in this book. Most of them can be solved within a few minutes. There are few books containing chess problems, games and endgames which are not only easy to solve but elegant and instructive as well. And there are even fewer that offer a sufficient number for regular and intensive training over a relatively long period. With this book, I would like to provide some help for those who are learning the game alone, and for parents and coaches too.

A word of guidance before we start. First, learn the names and the colors of the squares on the chessboard. Never move the pieces while solving a problem. Solving the problems in this book may result in a considerable improvement in your tactics, logic, concentration and imagination, and will also be a source of motivation – nothing is as encouraging as success and achievement. You can use your new knowledge in playing with partners and in tournaments too.

Studying the miniature games (maximum 25 moves) will help you to avoid errors in game opening, and it will improve your arsenal for attacking the king. Children are more interested in short games, and they can concentrate on them better than on long ones. The two-move miniature games (maximum number of pieces: 7) are not difficult, but they are enjoyable, and make it possible to proceed to more challenging chess problems. They may also inspire you to study the literature dealing with problems, or even to create your own chess problems.

This book draws on my private collection of five thousand chess books, and my archives of periodicals going back 150 years. I have also used the experience gained in the training of my three daughters – Olympic gold medallists, Oscar-award winners, holders of several Guinness records, champions of the world, for many years first and second in the world ranking – and during my fifteen year activity as a chess tutor in extracurricular classes at various schools.

In addition to studying the book, beginners who may wish to become tournament chess players should play every day – even against a chess computer if they do not have a regular partner. And the repeated problems in the book will repay repeated solving and study. The book aims to strengthen the positive skills and discipline needed for competition in chess (and any other field) – but above all it aims to give pleasure.

Apart from the foreword and the chapter "Key to symbols", the following 1000 pages are written in international chess terms, and thus the book can be read in any country by those who wish to improve their knowledge. The chapters containing problems and combinations might be useful for setting up problem-solving contests. The problems figuring in this collection cover the history of chess literature from 1140 to the present. There are problems by Pál Benkő, the international grandmaster, created especially for this collection, and there are 125 combinations from the tournament games of the Polgár sisters. This book is primarily intended for beginners and amateurs, from the basic level to master candidate (2,200 Elo points), and can be enjoyed by players of all ages.

I wish to thank Könemann Verlag for making it possible to publish a chess book of so wide a scope; Kulturtrade for editing the manuscript; my daughters for their help in professional matters; and my wife for her help with technical problems.

BUDAPEST, 1994
LÁSZLÓ POLGÁR

INTRODUCTION

By Bruce Pandolfini

Chess, as this László Polgár volume is titled, is perhaps the most comprehensive book ever conceived on the art of checkmate. It contains some 5,334 examples, primarily dealing with mate, mating attacks, or forced endgame wins. Mr. Polgár, the father and personal chess trainer of the renowned Polgár sisters of Hungary, has provided over 1,100 pages of brilliantly organized didactic and practical paradigms to equip the reader with the weapons needed to play power chess.

Here reconstructed is the path followed by Zsusza, Zsofia, and Judit Polgár in their meteoric climb to the top of the chess pantheon. It is said that each sister worked through and played out similar groups of their father's carefully chosen and arranged examples, until vital patterns and schemes became second nature, very like the way music students master finger exercises.

The format is beautifully logical, as with all grand strategies. There are sections of related problems, six large diagrams per page. Examples in a section are subdivided further, based on which player, White or Black, moves first. The text is minimal, but the presentation, is clear enough for even beginning students. Anyway, it's well known that the best chess instruction is largely visual.

Of the initial 4,462 problems here, 306 are mates in one move; 3,411 are mates in two moves (the attacker plays, the defender responds, the attacker mates); and 743 are mates in three moves (the attacker plays, the defender responds, the attacker plays a second move, the defender responds again, and the attacker gives mate). Some of these mates have been composed for instructional purposes, but others have been taken from real chess games. Each of these sections is large enough to constitute a book in itself, and the chapter on "mates in two" alone could enrich several years of study.

Following the mating problems is a phenomenal section with 600 miniature games (examples 4,463—5,062), all the moves of every game given in figurine algebraic notation for easy use. (A symbol of the moving piece appears instead of an abbreviating letter, so anyone can follow in any language.) Each game begins with the opening moves guiding the reader to a pivotal moment. A diagram prompts the reader to find the next move, suggested by the subsection in which the game appears. For instance, if a game is in the f3 section (examples 4,463—4,562), the winning move must be made to the square f3, if it is Black's move, or f6, if it is White's. What better clue could be given than an indication of the destination square?

Next is a chapter of basic endgame positions (5,063—5,206), each of just a few pieces and mainly concluded by promoting a pawn into a queen, important patterns because many games reduce to a situation where the extra queen assists checkmate.

From this Olympian preparation, the book moves into its final chapter (examples 5,207—5,333), consisting of short tactical games played by the Polgár sisters themselves. These contests all illustrate incisive attacks and winning stratagems utilized by Zsuzsa, Zsofia, or Judit in their confrontations with masters and world class players throughout their tournament and match careers. The solutions and pertinent factual information (place, date, and name of the Polgár opponent) are found in the concluding segment, containing all the answers for the book's 5,334 examples. The very last diagram, by the way, is a bonus problem created by the middle sister, Zsofia Polgár, the placement of the pieces tracing the outline of a rook! The tome ends with a short biographical sketch of the author and a helpful bibliography of sources of the problems not personally developed by László Polgár.

If you're new to the game, you can begin at page one and proceed step by step. *Chess* conveniently provides a pithy introduction to the moves and rules, for some reason omitted from most tactical chess books. Even experienced players might prefer reviewing this introduction before tackling the actual problems. Another approach, just as viable, is to focus on particular sections or themes for special study, since each grouping, though part of a logical whole, is also a self-contained unit.

You might decide not to work directly from the book but to set up problems on a chessboard. If so, first try to solve them without moving the pieces. This develops analytic vision and improves overall ability. Whatever your level of skill or interest, you'll find that László Polgár's *Chess* offers an abundance of instruction and pleasure and provides a superb foundation for any chess library.

KEY TO SYMBOLS

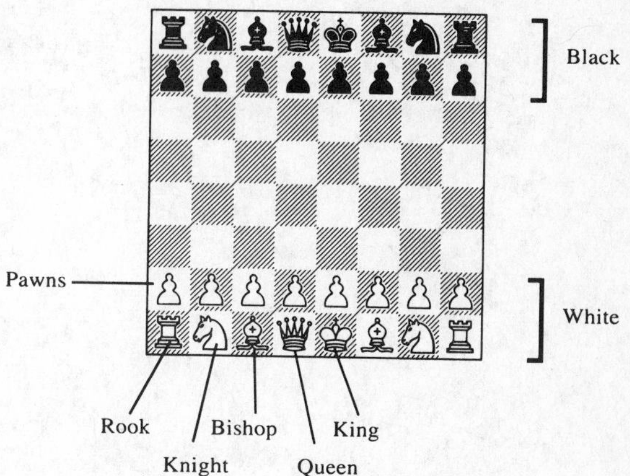

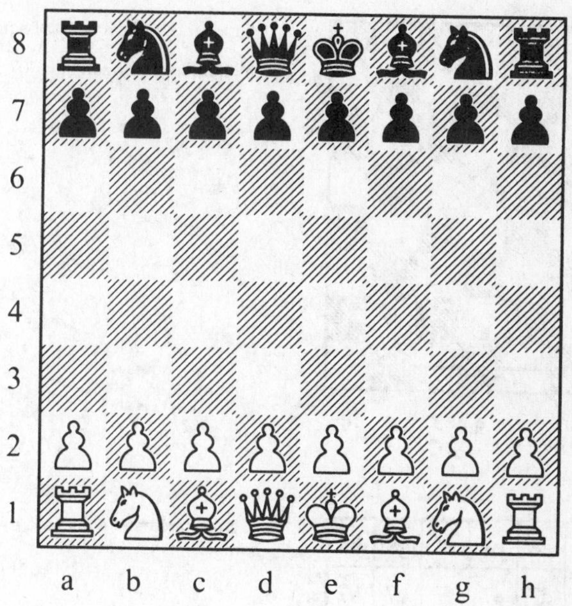

The starting position

Square a1 is always black.

Black is always at the top, white is always at the bottom.

THE STARTING POSTION

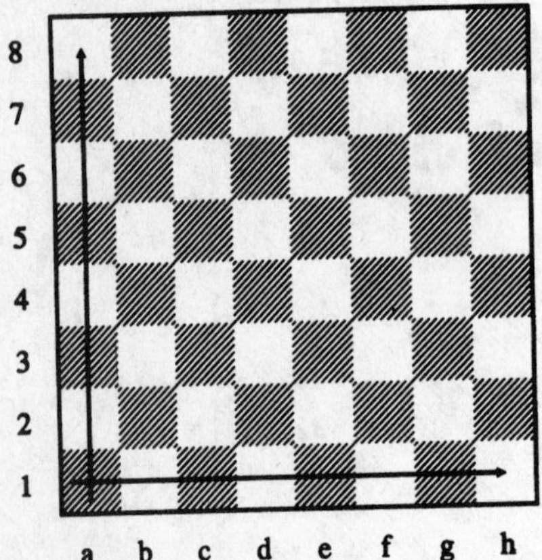

The letters: vertical rows of squares ↑.

The numbers: horizontal rows of squares →.

a8	b8	c8	d8	e8	f8	g8	h8
a7	b7	c7	d7	e7	f7	g7	h7
a6	b6	c6	d6	e6	f6	g6	h6
a5	b5	c5	d5	e5	f5	g5	h5
a4	b4	c4	d4	e4	f4	g4	h4
a3	b3	c3	d3	e3	f3	g3	h3
a2	b2	c2	d2	e2	f2	g2	h2
a1	b1	c1	d1	e1	f1	g1	h1

Each square can be denoted by a letter and a number.

Move number + symbol of the piece (no symbol for the pawn) + the square the piece travels to.

Here, 1. ♘c3.

= draw

a) +− white is winning

b) −+ black is winning

checkmate

THE KING

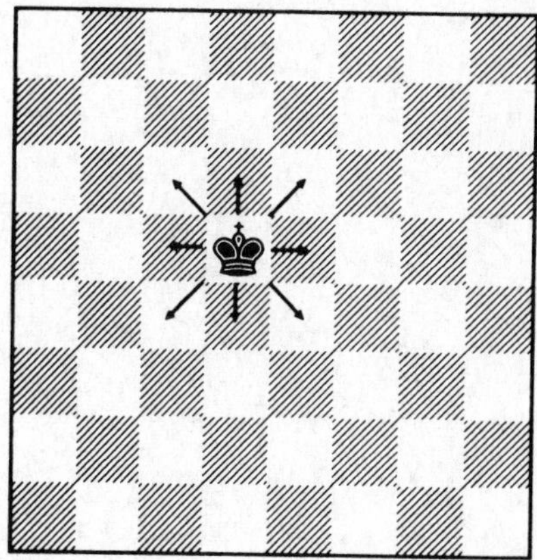

The king may move one square in any direction, so long as no piece is blocking its path.

The king can capture in any direction.

The king may not move onto a square
– occupied by one of his pieces,
– where it is checked by an enemy piece,
– adjacent to the enemy king.

THE QUEEN

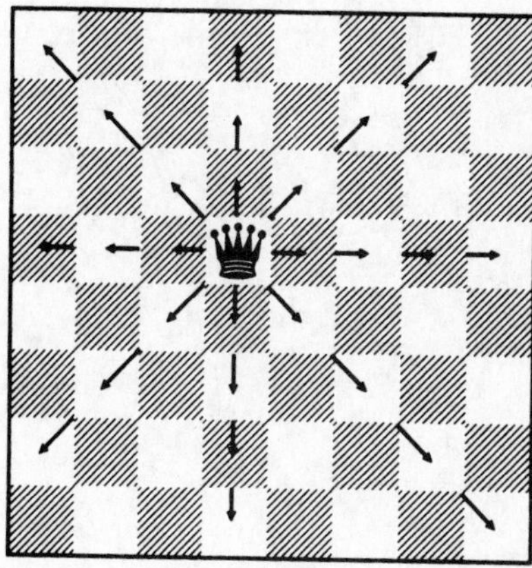

The queen may move any number of squares in any direction, so long as no piece is blocking its path. But it cannot move as a knight.

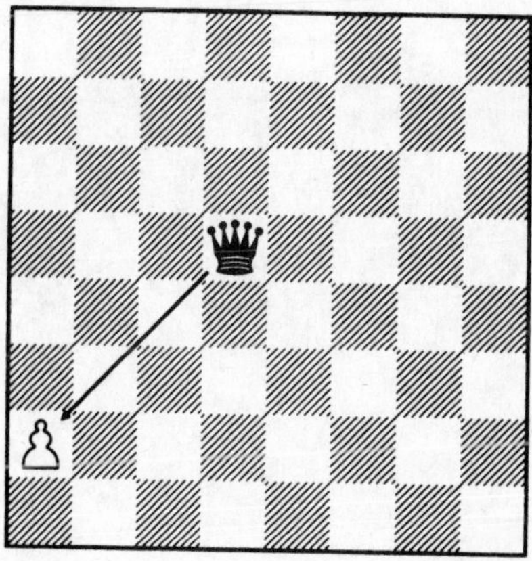

The queen can move here to capture the pawn.

THE ROOK

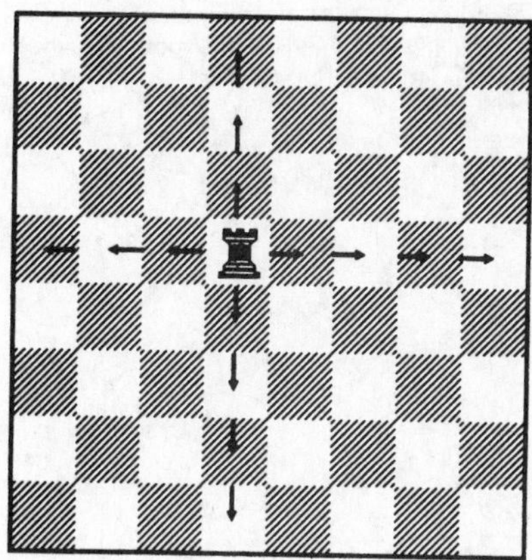

The rook may move any number of squares horizontally or vertically, so long as no piece is blocking its path.

The rook can move here to capture the pawn.

CASTLING

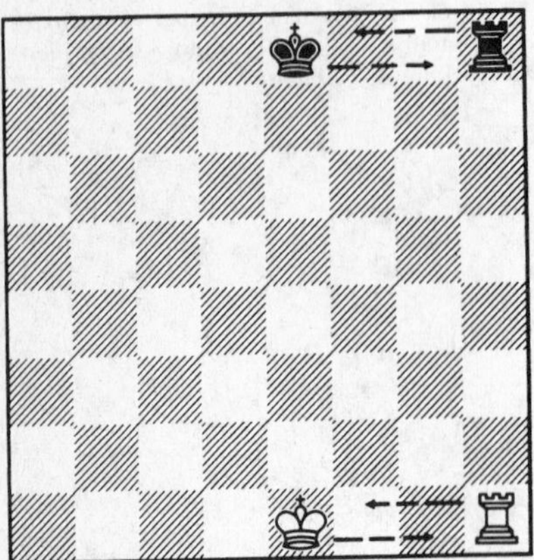

Castling king's side: the king moves two squares towards the rook, the rook leaps over the king. Recorded: 0-0.

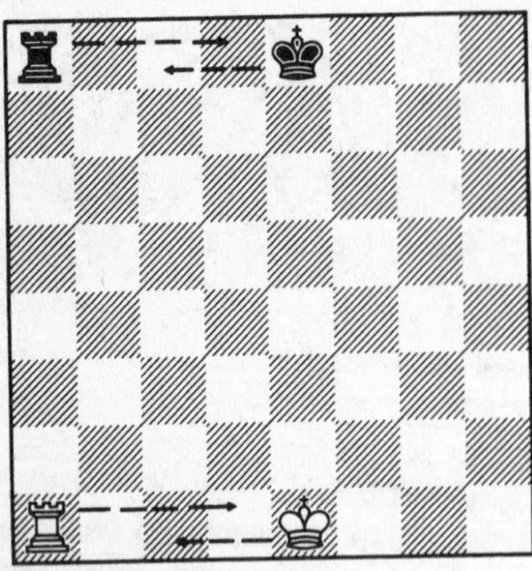

Castling queen's side: the king moves two squares towards the rook, the rook leaps over the king. Recorded: 0-0-0.

You cannot castle:

a) if the king is in check,
b) if there is a piece between the king and the rook,
c) if the king will be in check after castling,
d) if the square through which the king passes is under attack,
e) if the king or rook has already been moved.

THE KNIGHT

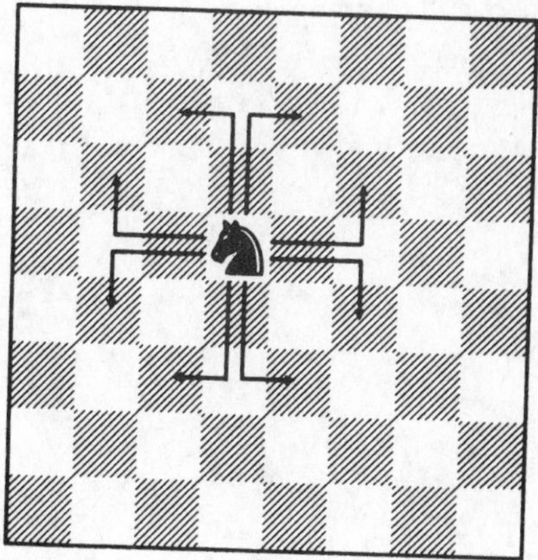

The knight may leap to any square in an "L" shape. It is the only piece that may jump over a piece in its way.

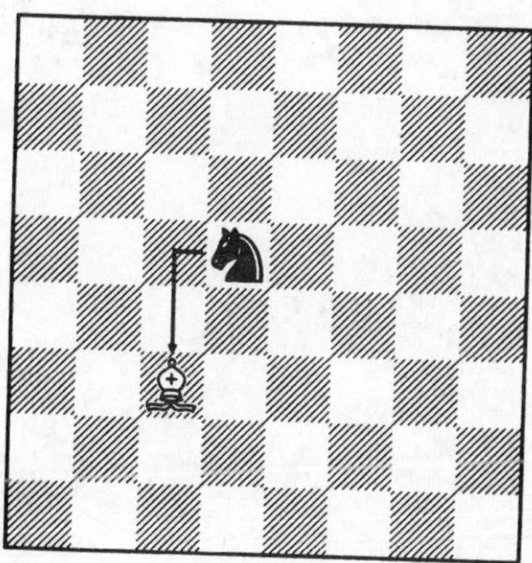

The knight can move here to capture the bishop.

THE BISHOP

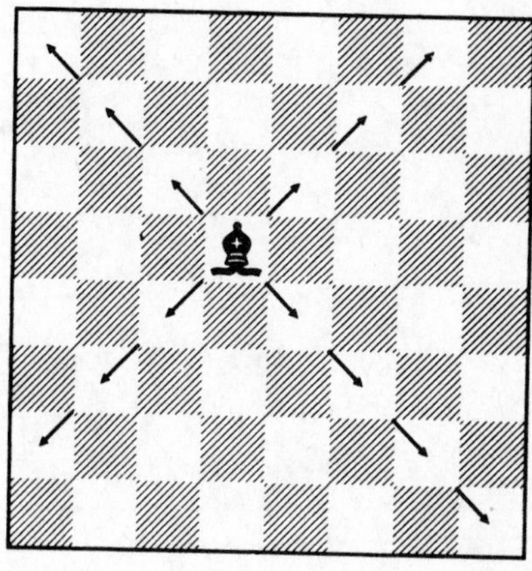

The bishop may move any number of squares diagonally, backwards or forwards, so long as no piece is blocking its path.

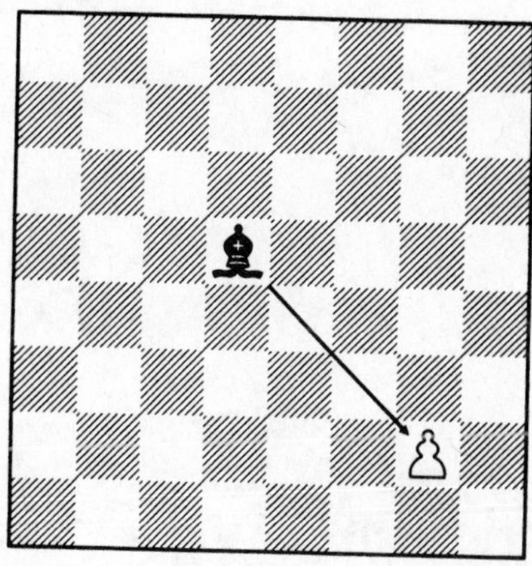

The bishop can move here to capture the pawn.

THE PAWN

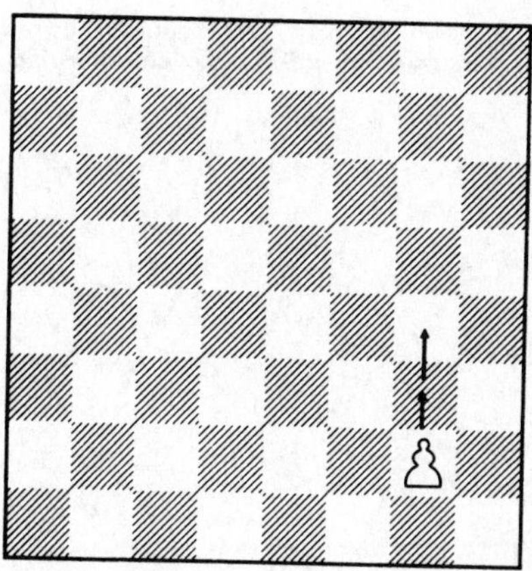

The pawn on its first move may move either one or two squares forwards.

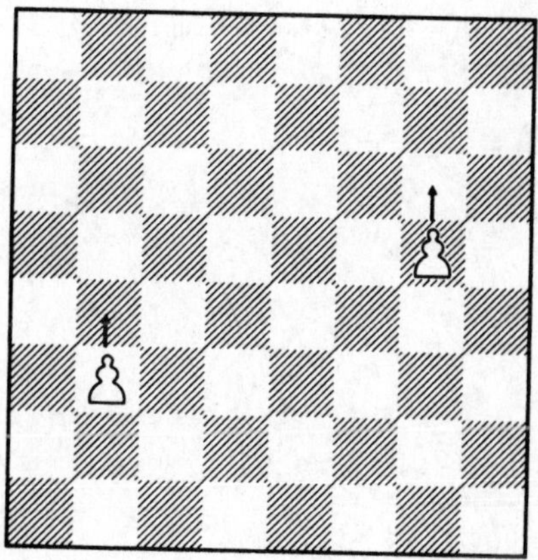

The pawn after the first move may only advance one square at a time.

The pawn captures by moving diagonally one square forwards, either to the left or the right.

The pawn may not move or capture backwards.

Neither pawn can move.

THE EN PASSANT PAWN CAPTURE

The possibility of en passant pawn capture arises only when a pawn takes advantage of its initial move of two steps forward.

If the pawn had moved only one square forward an enemy pawn could have captured it. But the en passant rule makes capture possible anyway.

The white pawn has been captured en passant!

PAWN PROMOTION

If a pawn reaches the 8th (or 1st) row of the board, it must be exchanged.

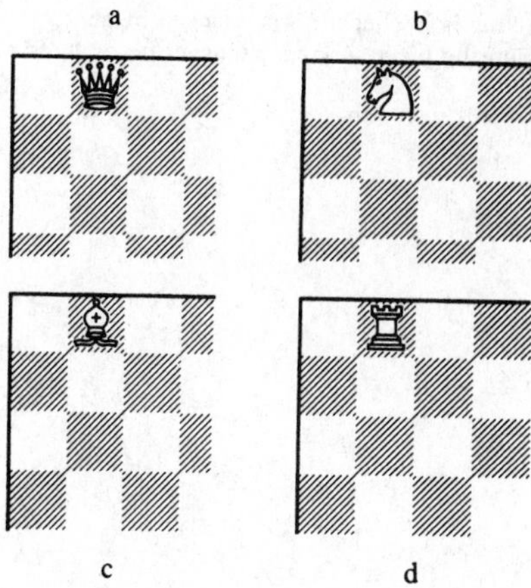

It can be exchanged for a) queen, b) knight, c) bishop or d) rook of its own color. But never for a king.

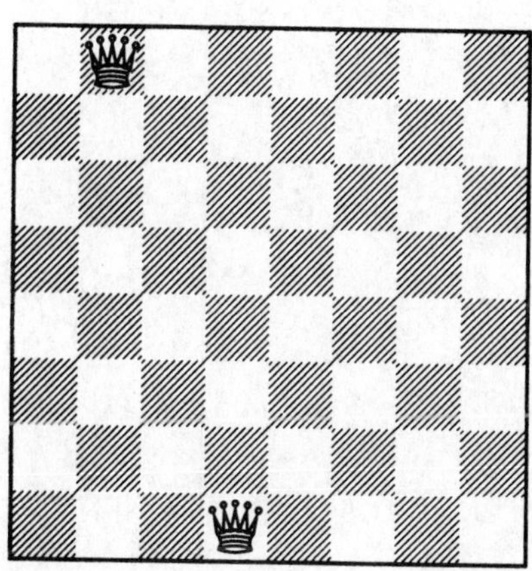

If you still have your original queen, you might have a new queen as well.

CHECK

A king is in check if it is attacked by an opposing piece. A king can never be captured.

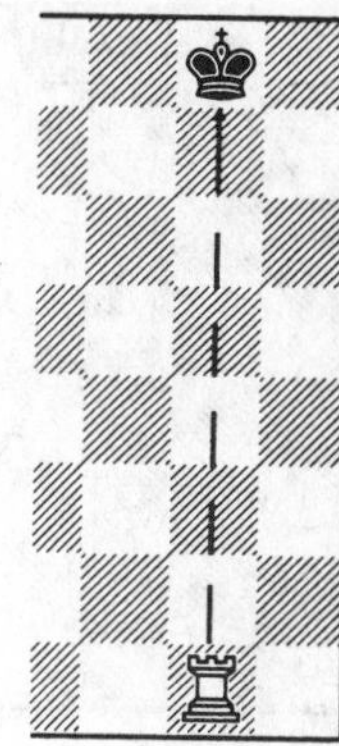

A king must be got out of check immediately
– by moving the king,
– or by capturing the opposing piece,
– or by interposing a friendly piece to block the check (impossible in the case of a knight check).

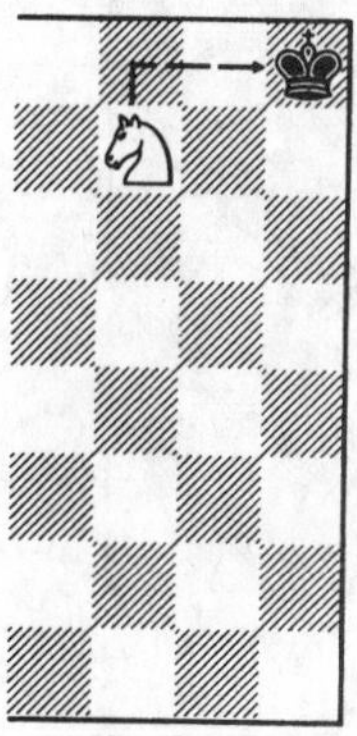

CHECKMATE

If the king cannot escape from check, this position is checkmate.

If the king cannot move and is not in checkmate, and the player whose turn it is cannot move any other piece, the game ends in a draw by stalemate. Here: in each case black to move.

Mate in one

(1–306)

1.

2.

3.

4.

5.

6.

7.

8.

9.

10.

11.

12.

13.

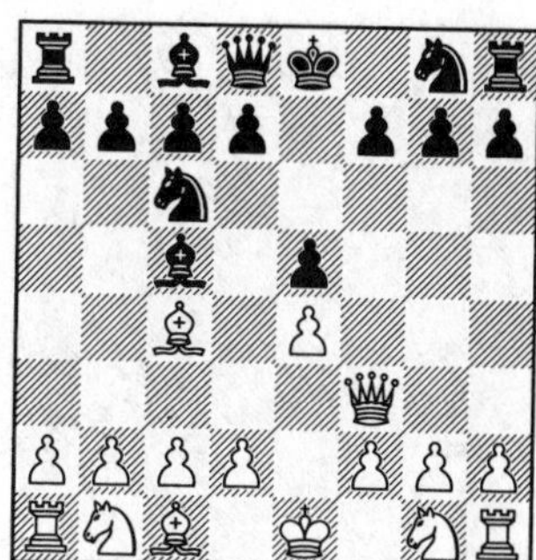

14.

15.

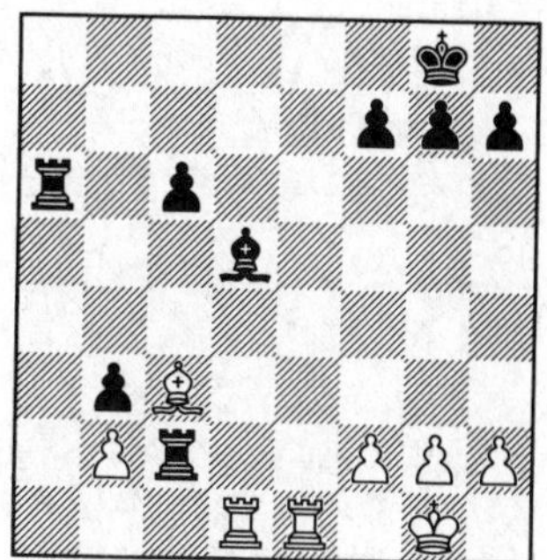

16.

17.

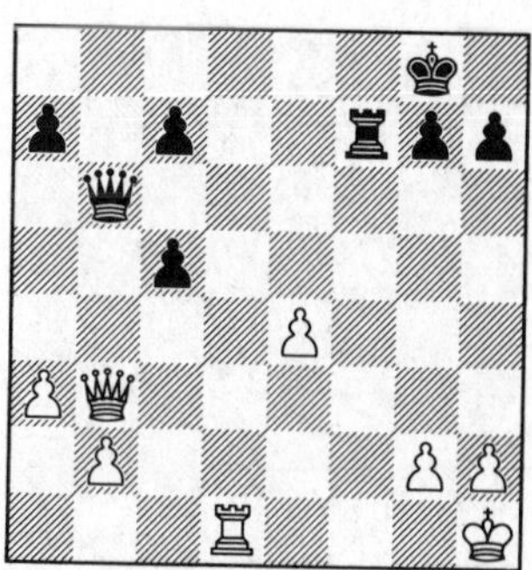

18.

19.

20.

21.

22.

23.

24.

25.

26.

27.

28.

29.

30.

31.

32.

33.

34.

35.

36.

37.

38.

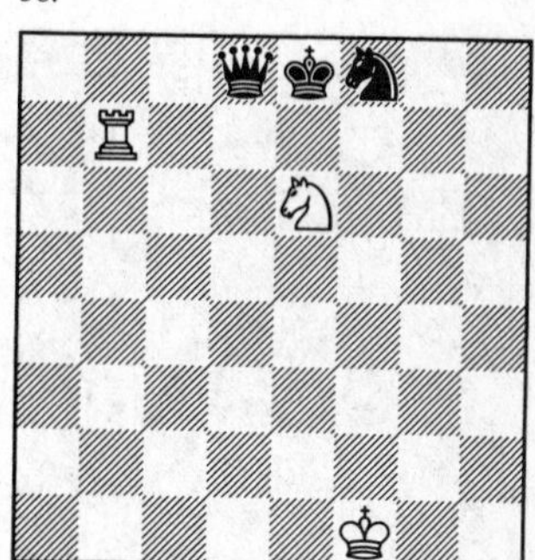

39.

40.

41.

42.

43.

44.

45.

46.

47.

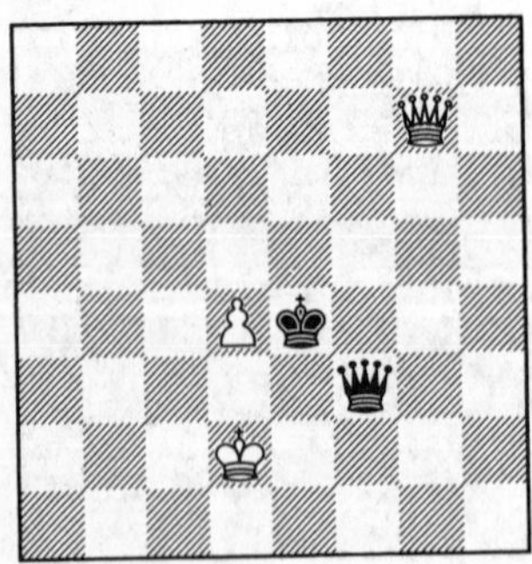

48.

49.

50.

51.

52.

53.

54.

55.

56.

57.

58.

59.

60.

61.

62.

63.

64.

65.

66.

67.

68.

69.

70.

71.

72.

73.

74.

75.

76.

77.

78.

79.

80.

81.

82.

83.

84.

85.

86.

87.

88.

89.

90.

91.

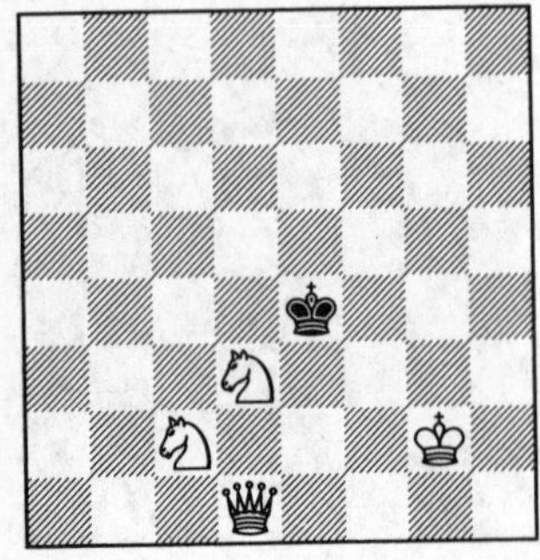

92.

93.

94.

95.

96.

97.

98.

99.

100.

101.

102.

103.

104.

105.

106.

107.

108.

109.

110.

111.

112.

113.

114.

115.

116.

117.

118.

119.

120.

121.

122.

123.

124.

125.

126.

127.

128.

129.

130.

131.

132.

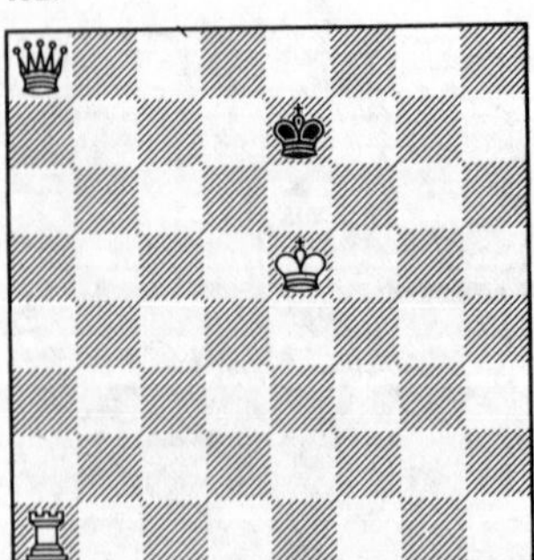

133.

134.

135.

136.

137.

138.

139.

140.

141.

142.

143.

144.

145.

146.

147.

148.

149.

150.

151.

152.

153.

154.

155.

156.

157.

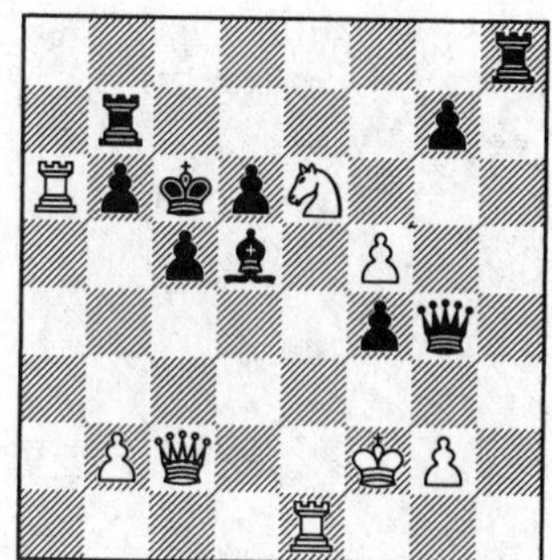

158.

159.

160.

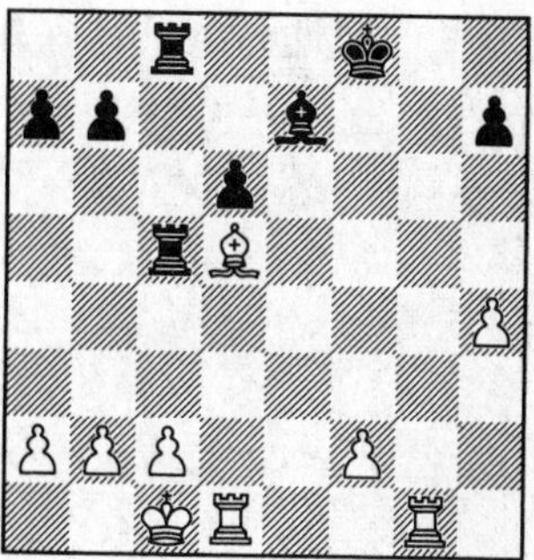

161.

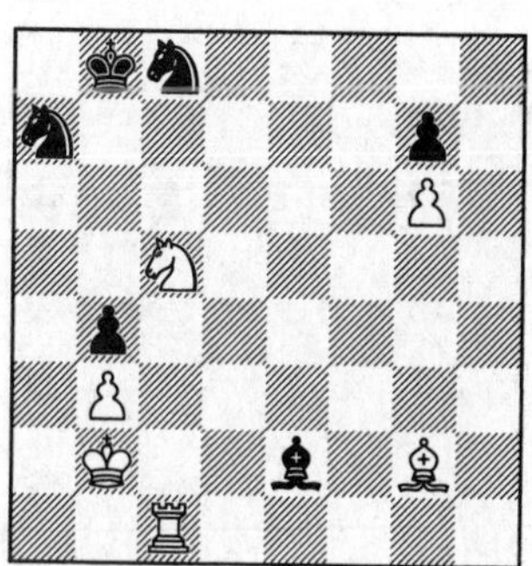

162.

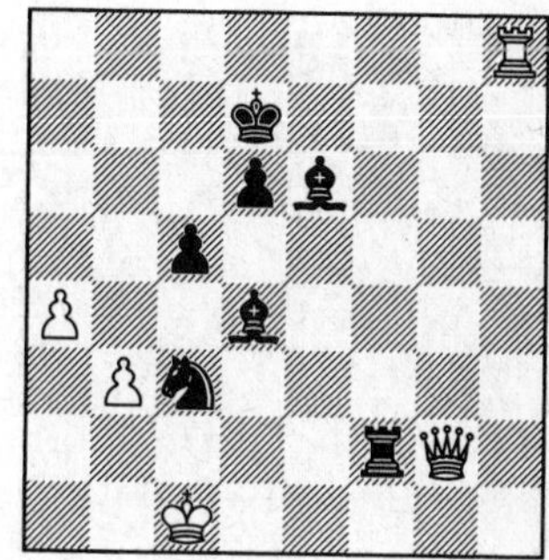

163.

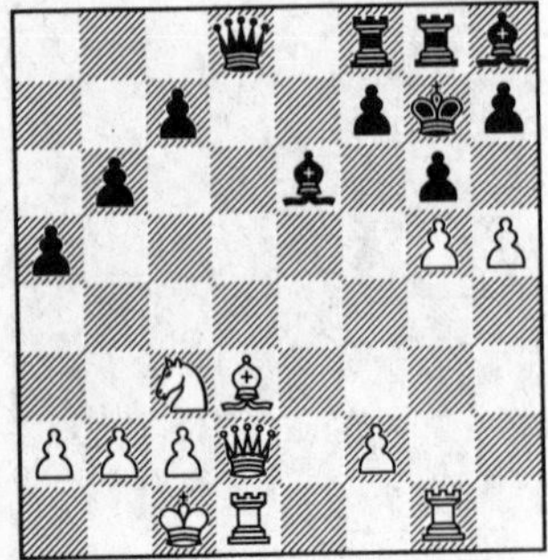

164.

165.

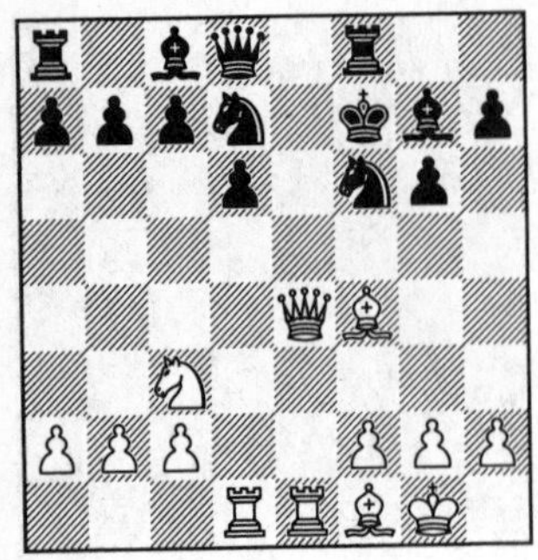

166.

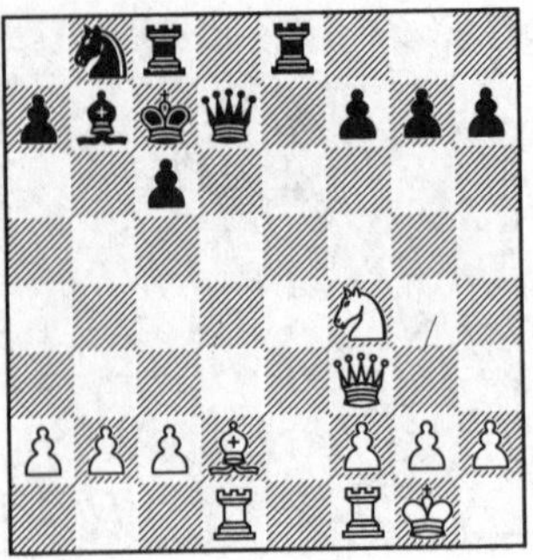

167.

168.

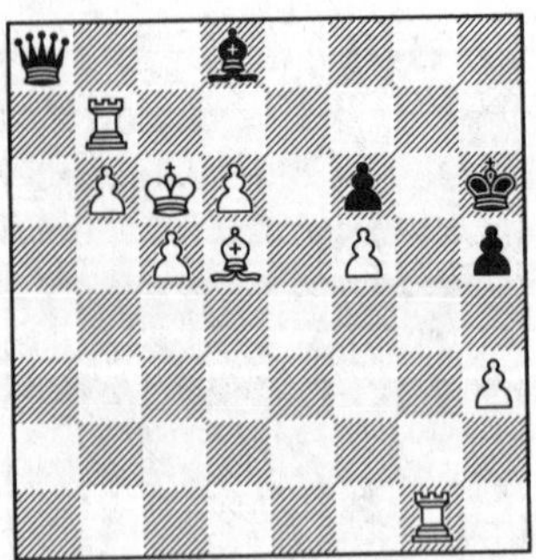

169.

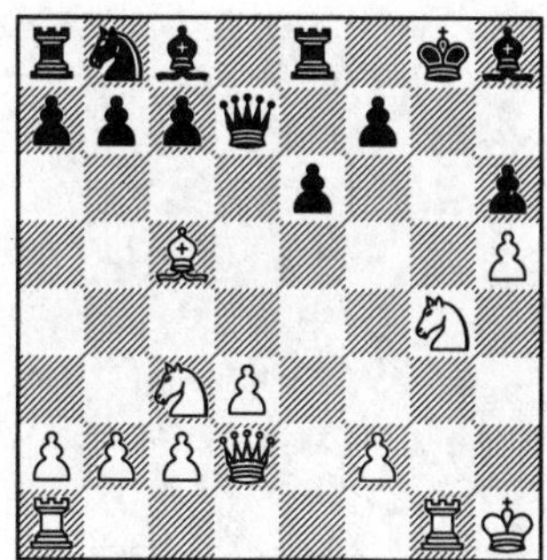

170.

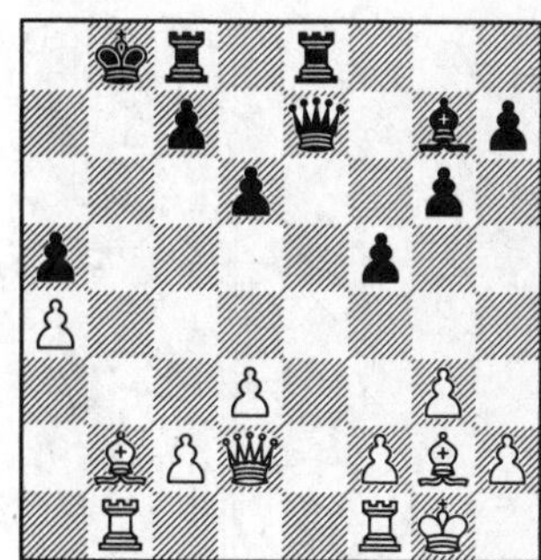

171.

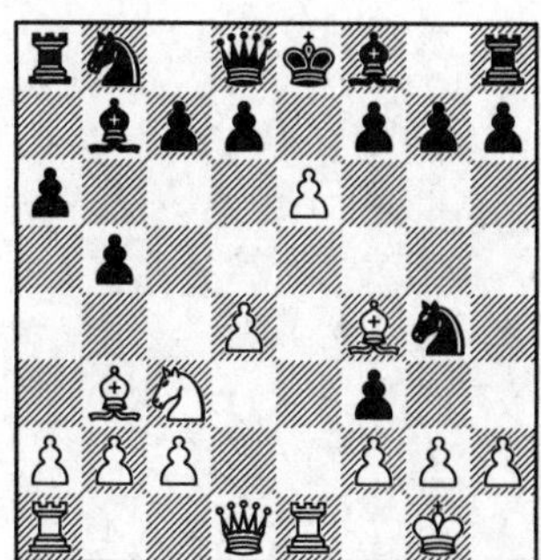

172.

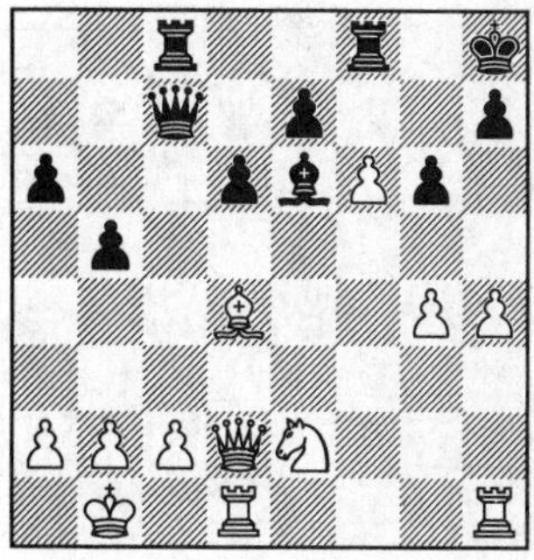

173.

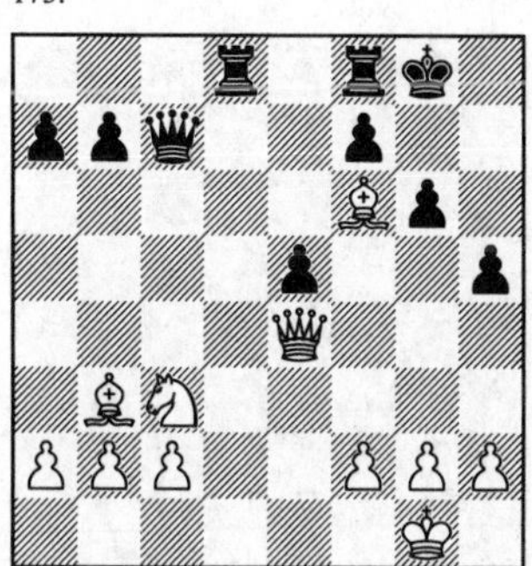

174.

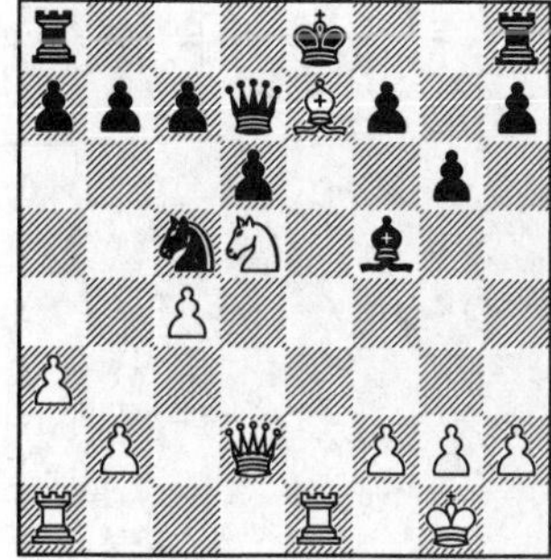

175.

176.

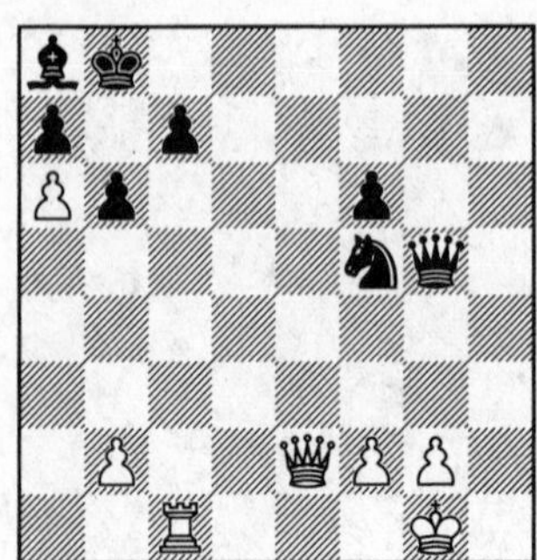

177.

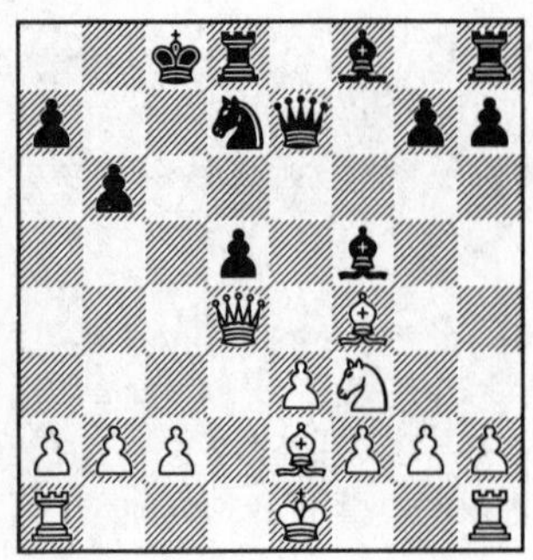

178.

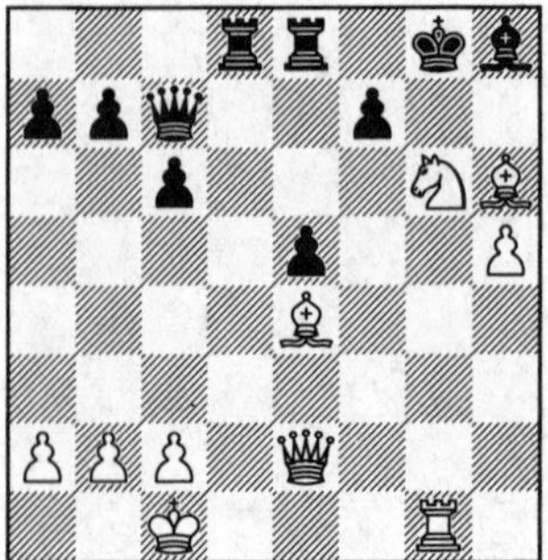

179.

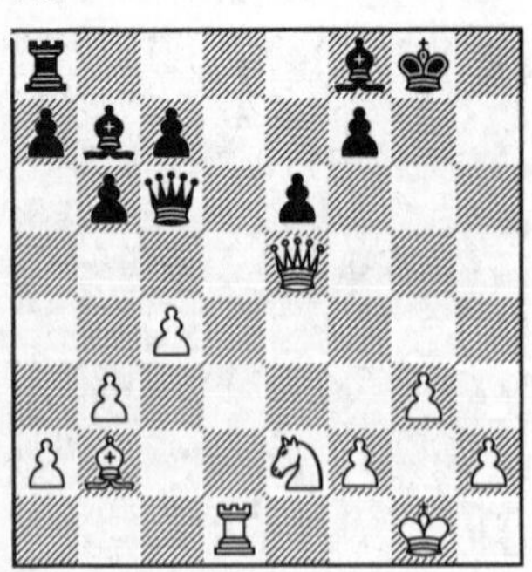

180.

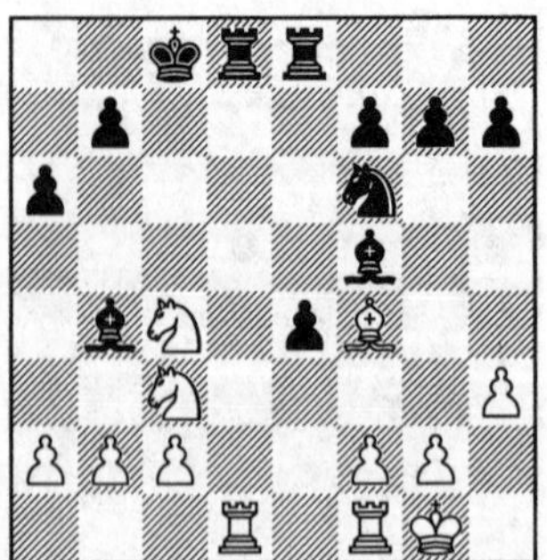

181.

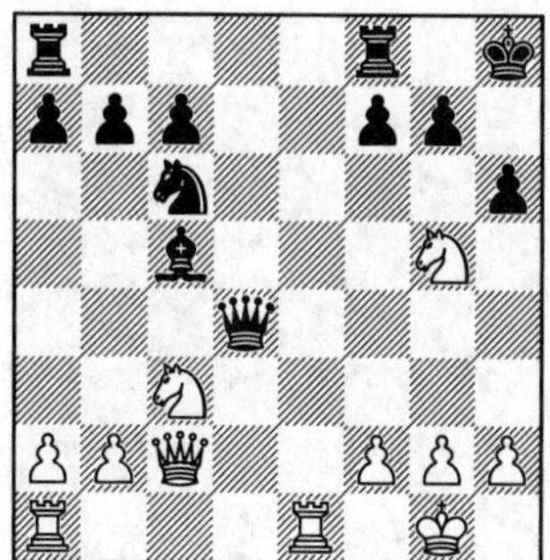

182.

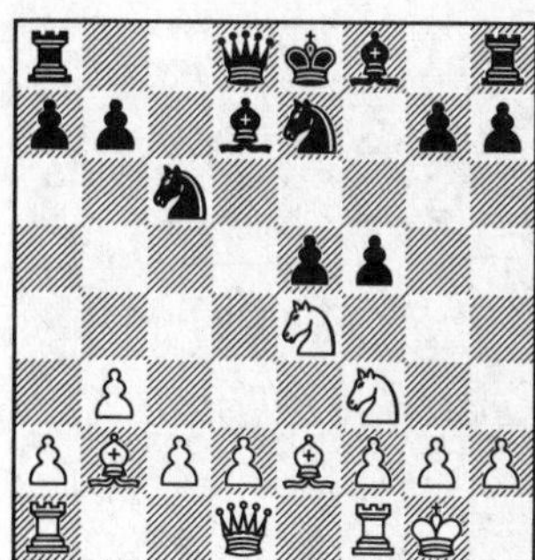

183.

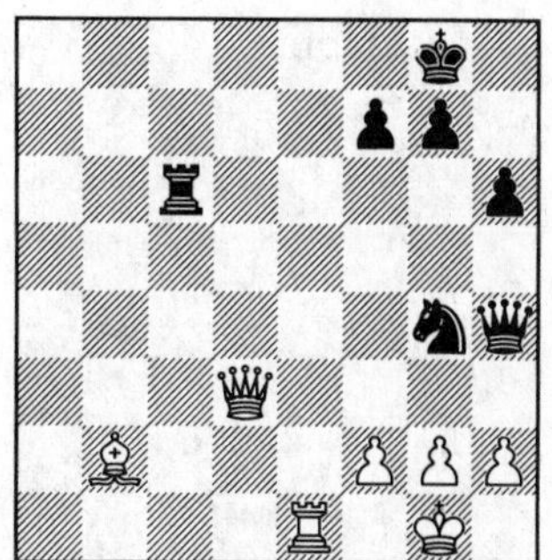

184.

185.

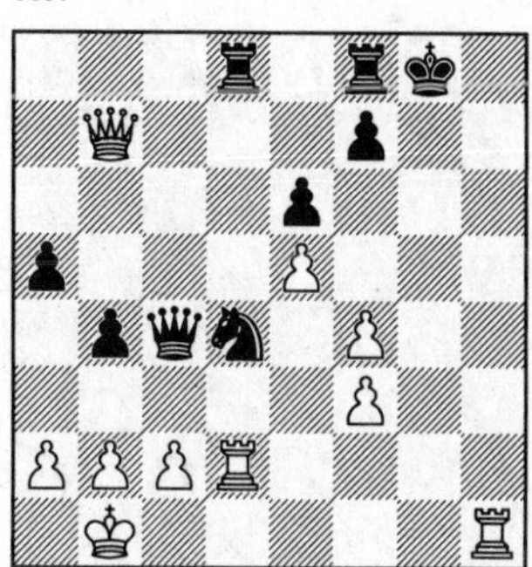

186.

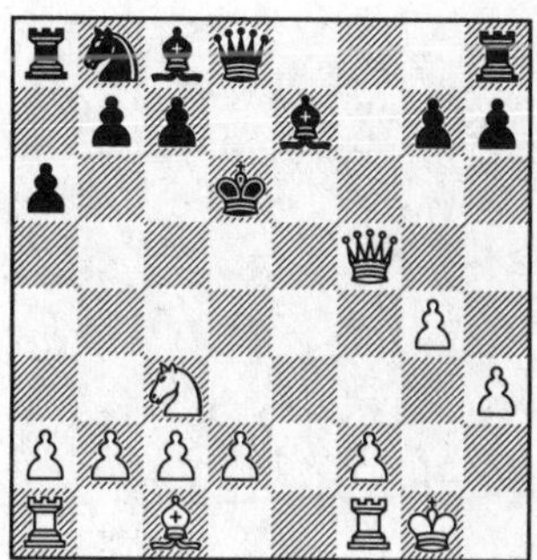

187.

188.

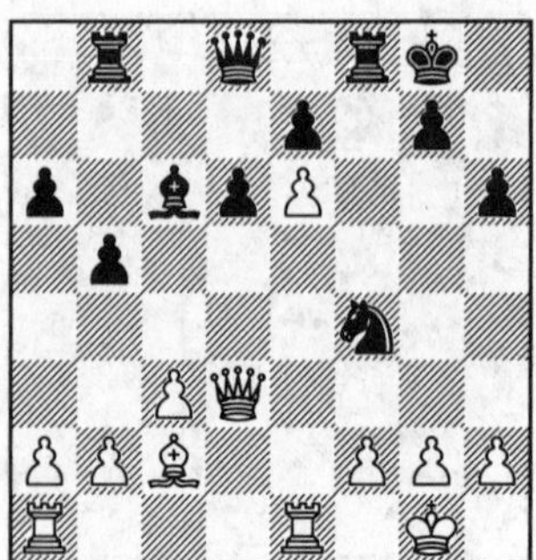

189.

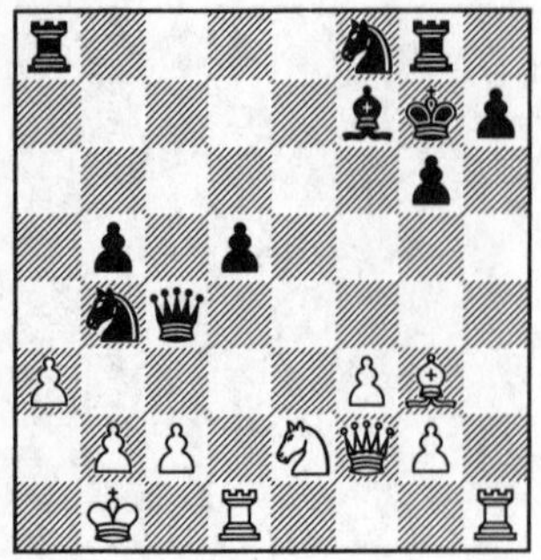

190.

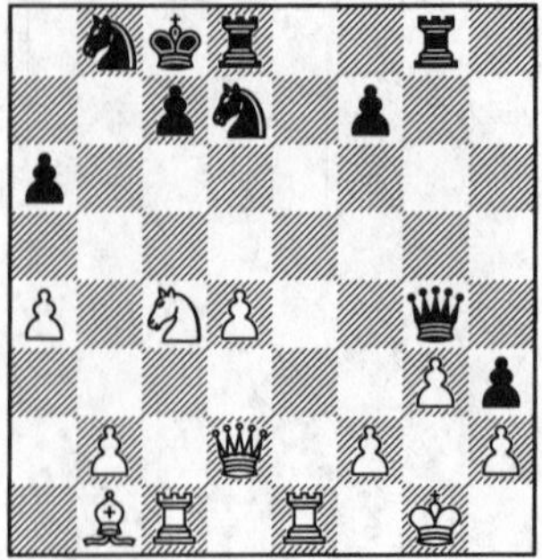

191.

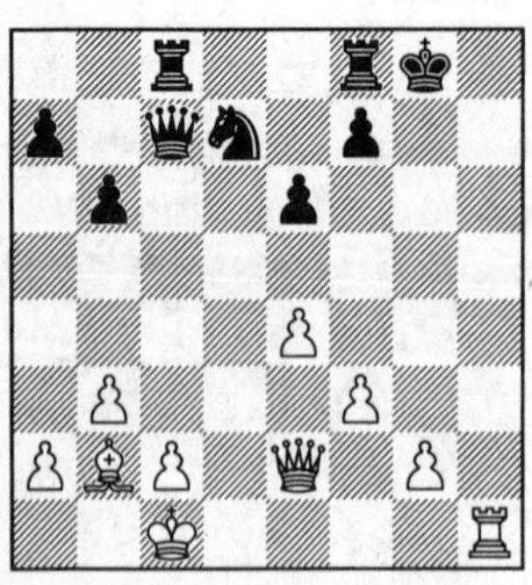

192.

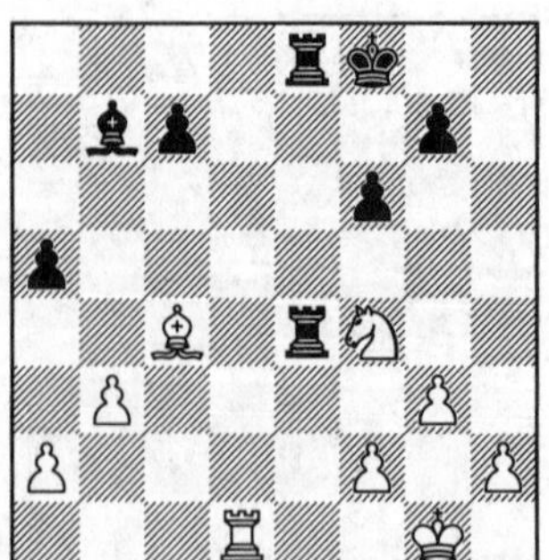

193.

194.

195.

196.

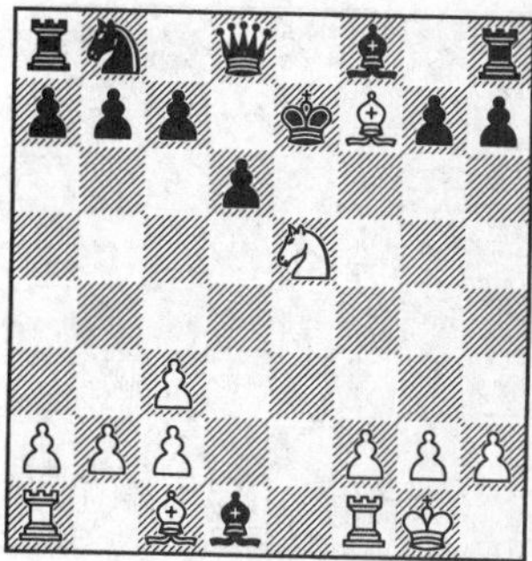

197.

198.

199.

200.

201.

202.

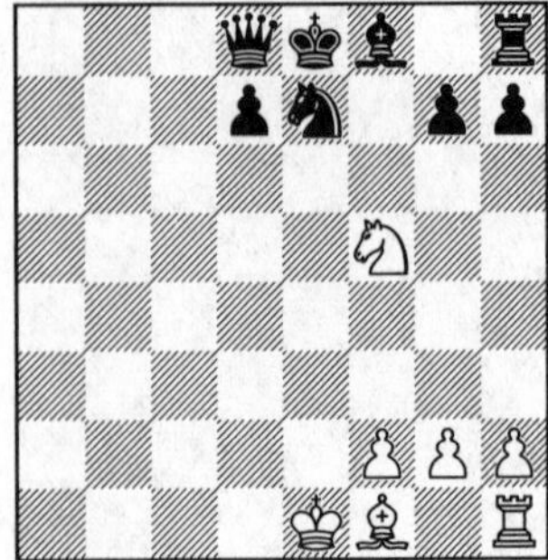

203.

204.

205.

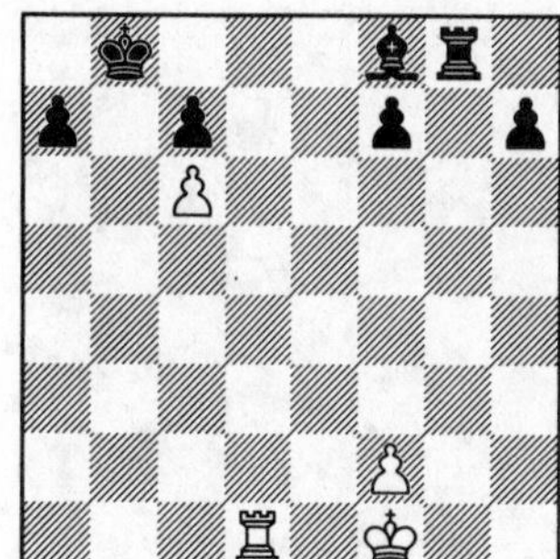

206.

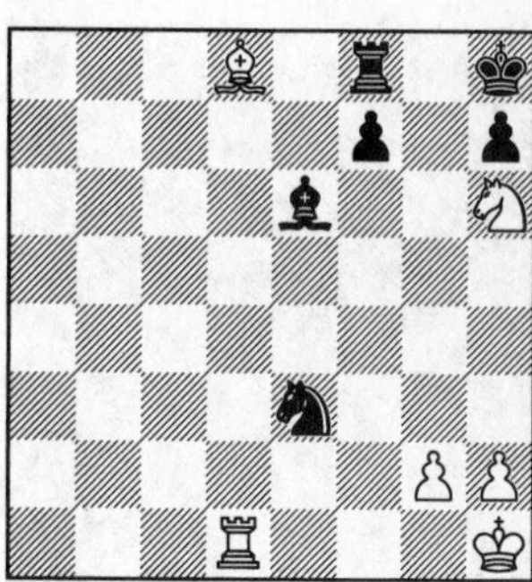

207.

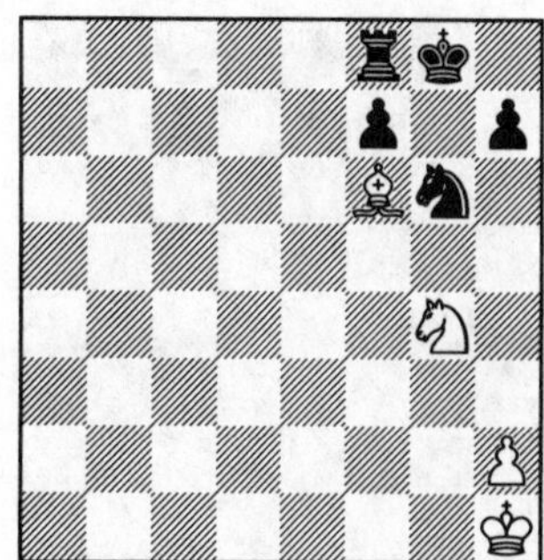

208.

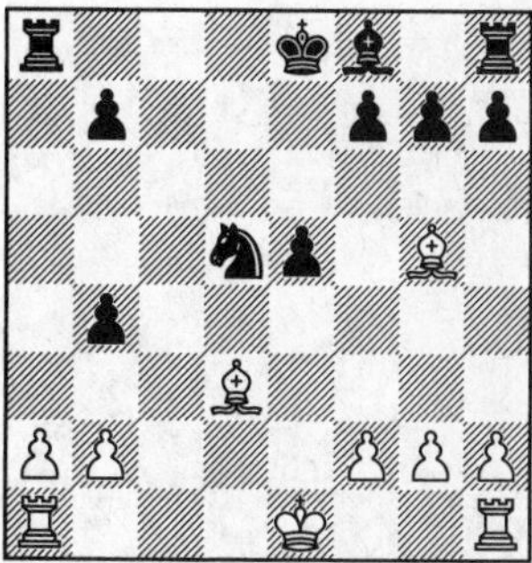

209.

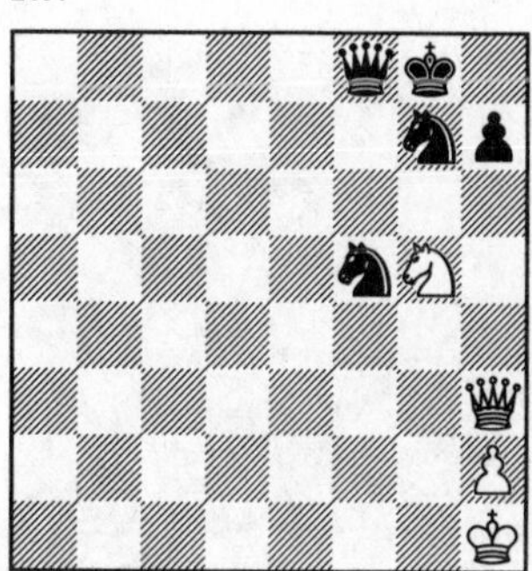

210.

211.

212.

213.

214.

215.

216.

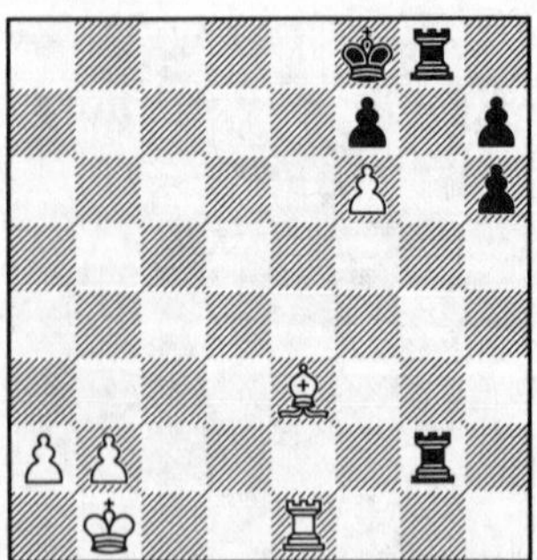

217.

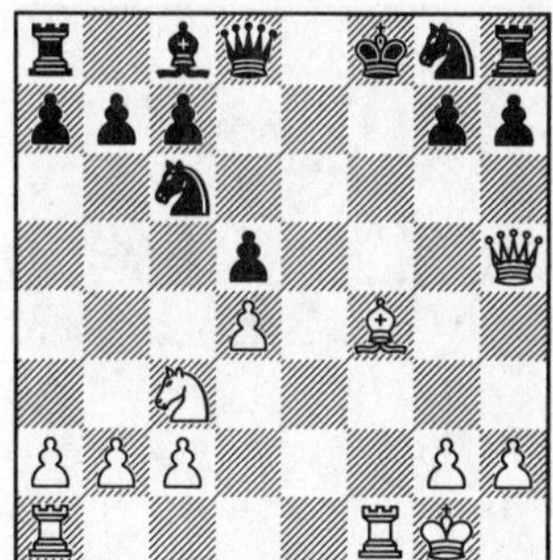

218.

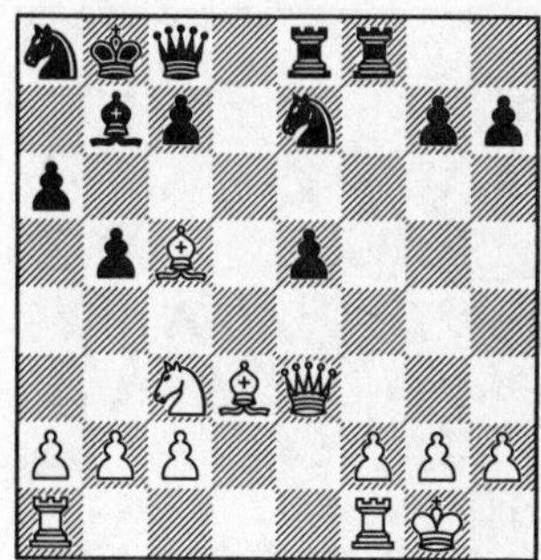

219.

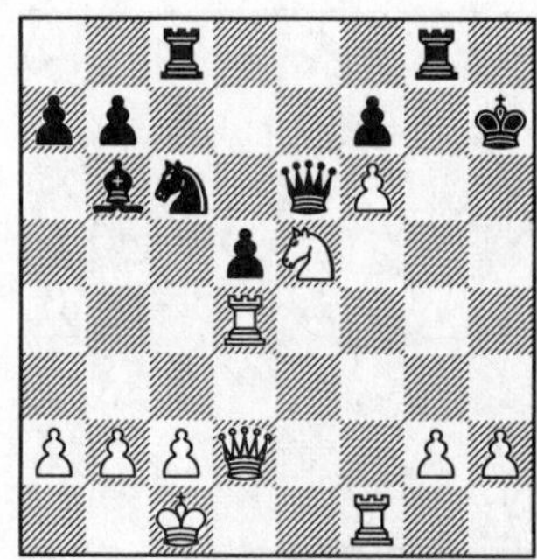

220.

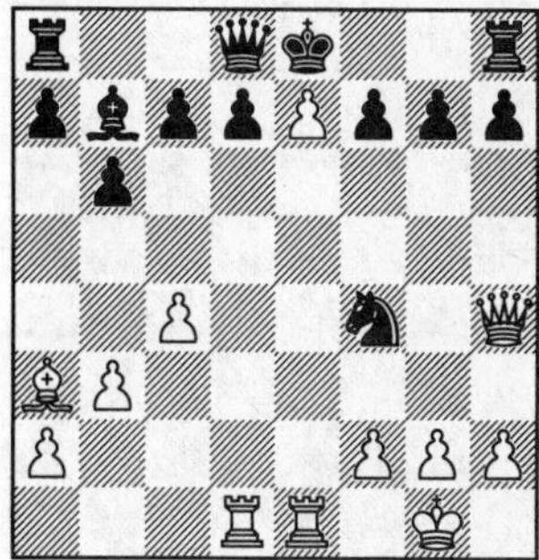

221.

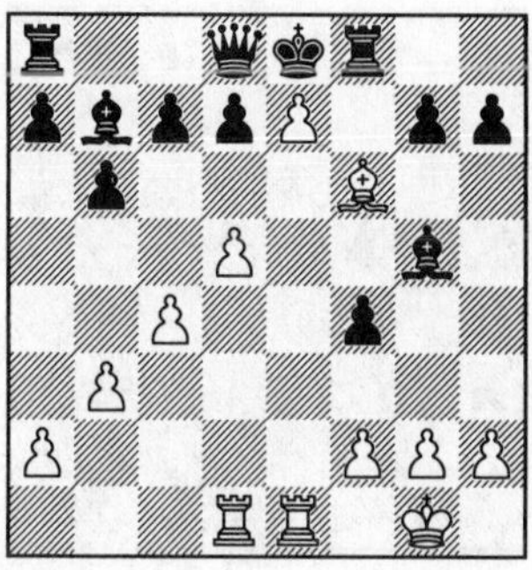

222.

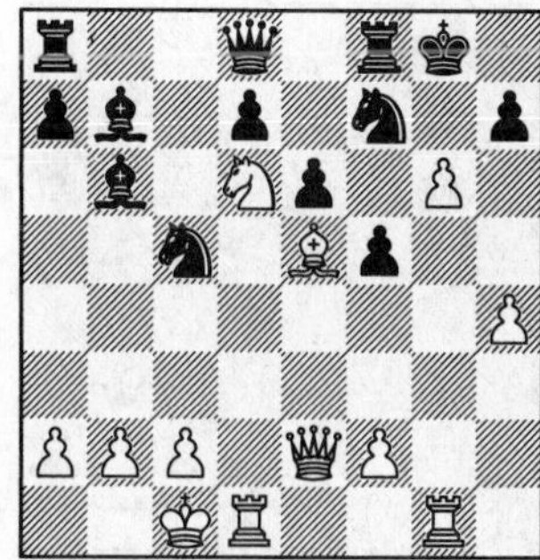

223.

224.

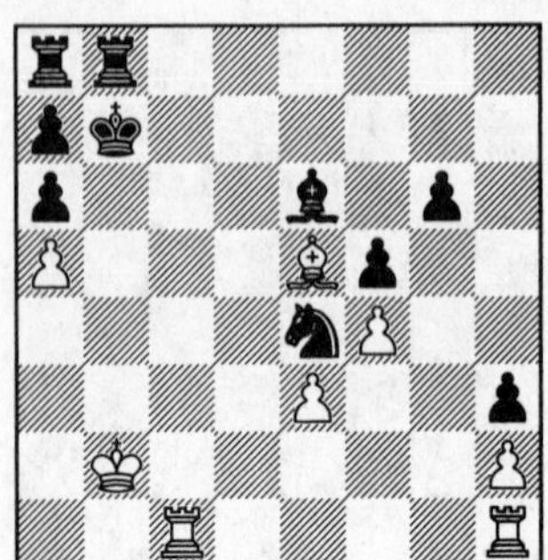

225.

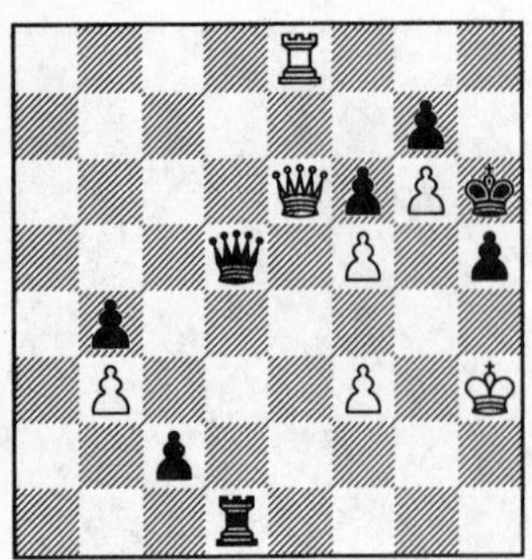

226.

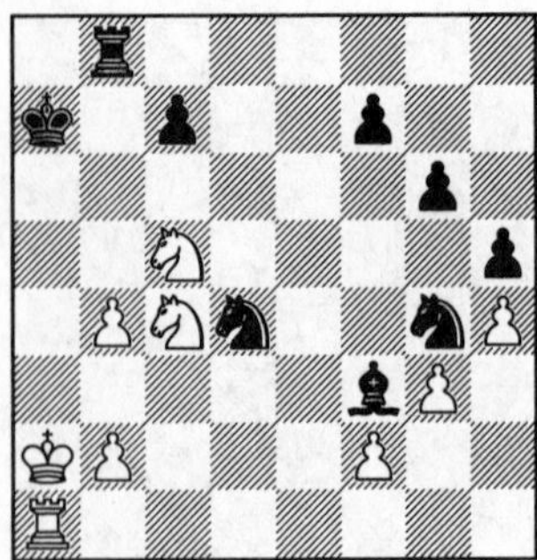

227.

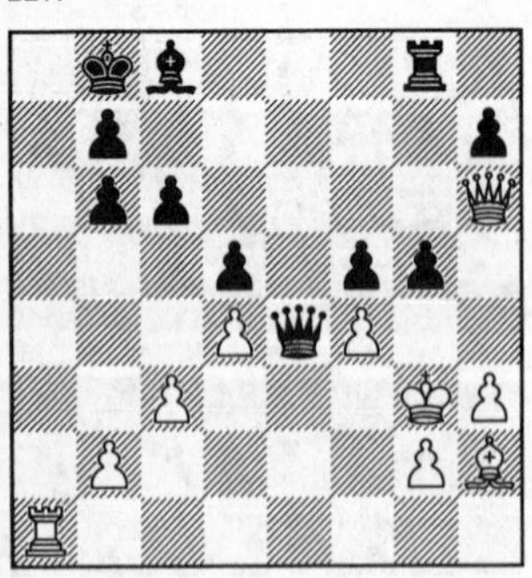

228.

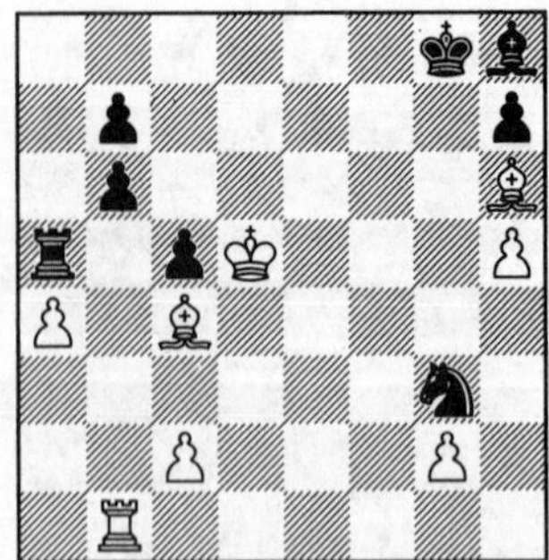

229.

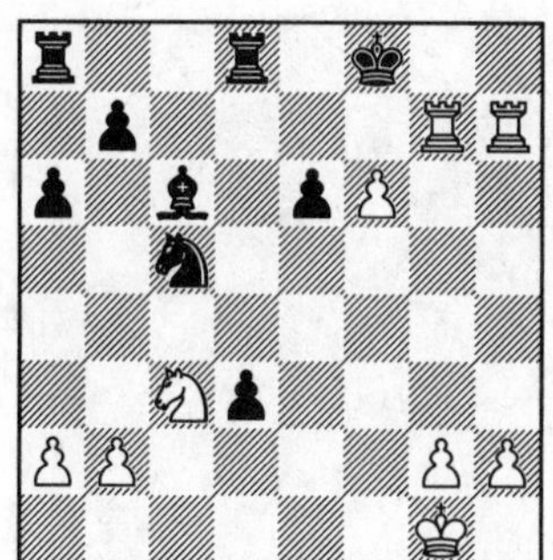

230.

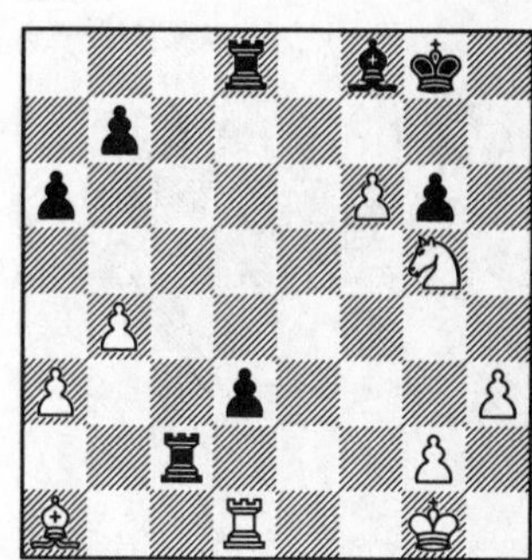

231.

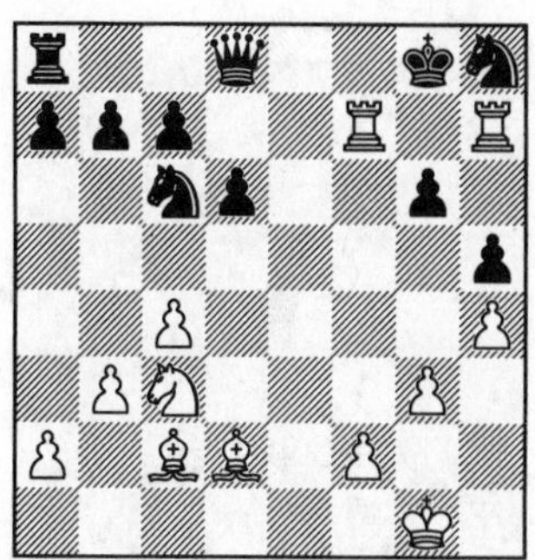

232.

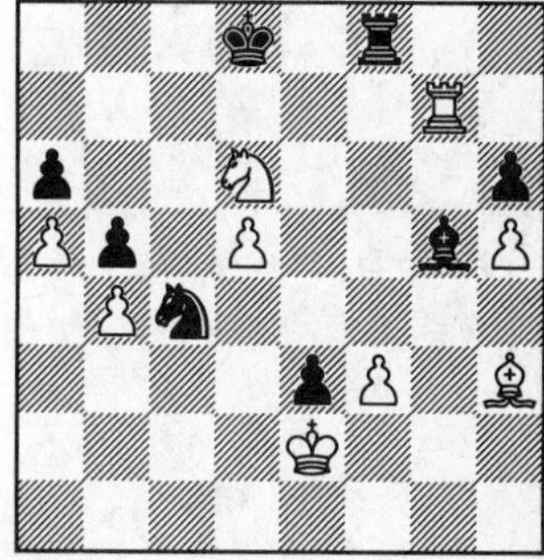

233.

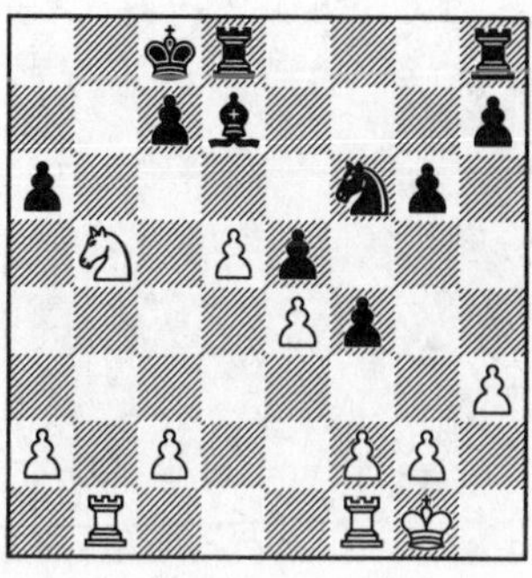

234.

235.

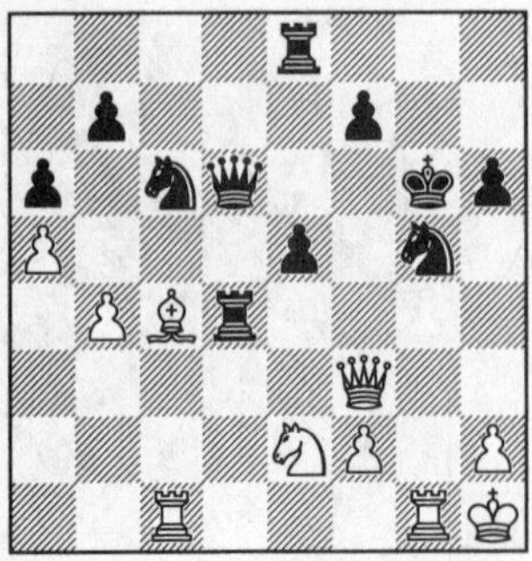

236.

237.

238.

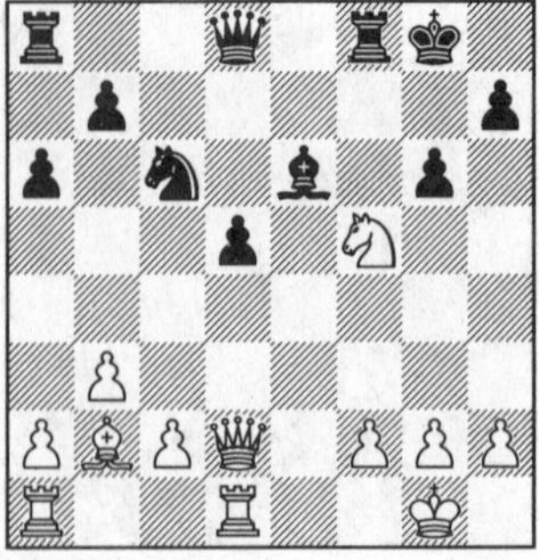

239.

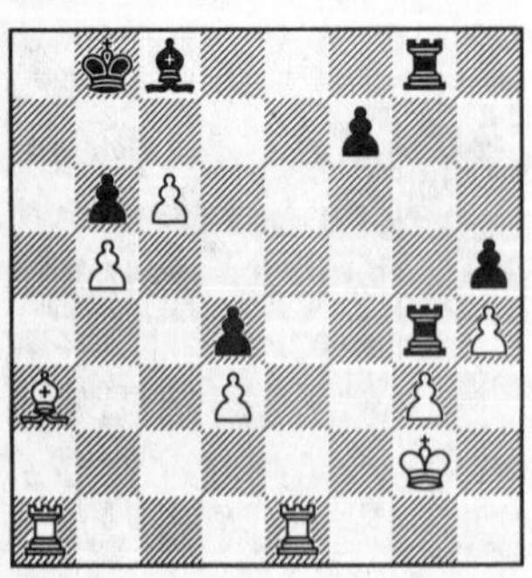

240.

241.

242.

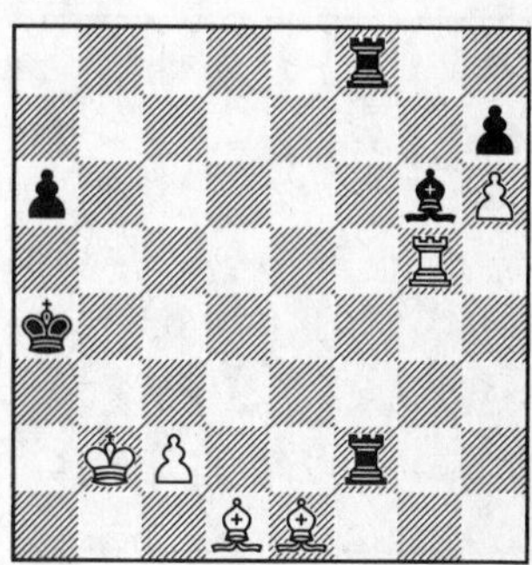

243.

244.

245.

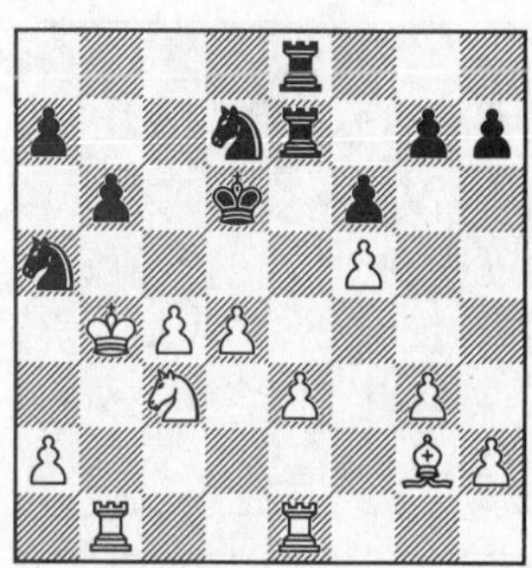

246.

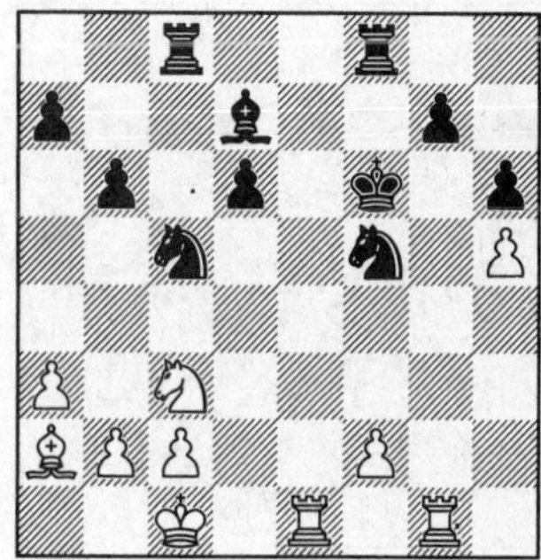

247.

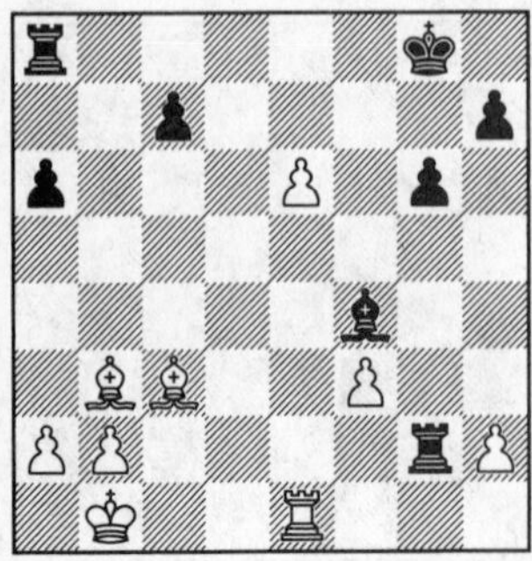

248.

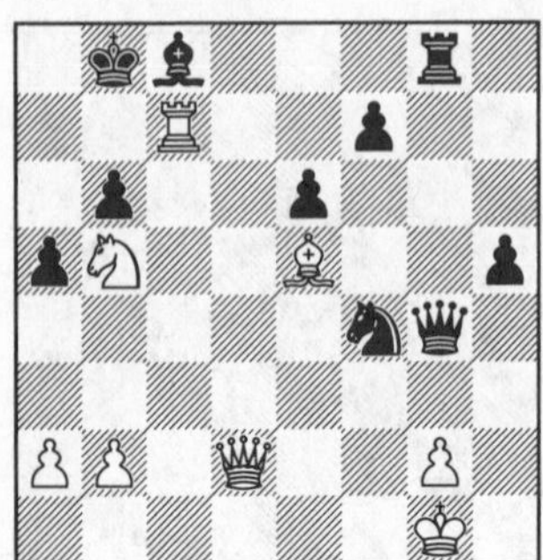

249.

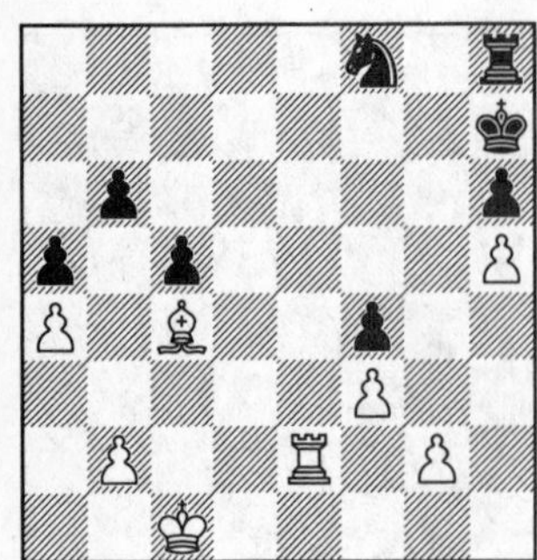

250.

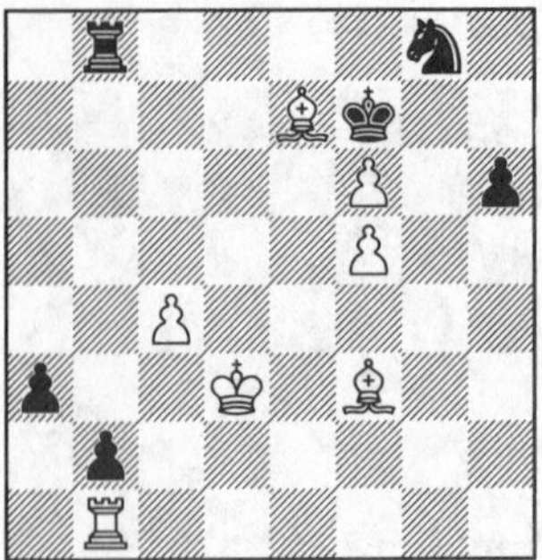

251.

252.

253.

254.

255.

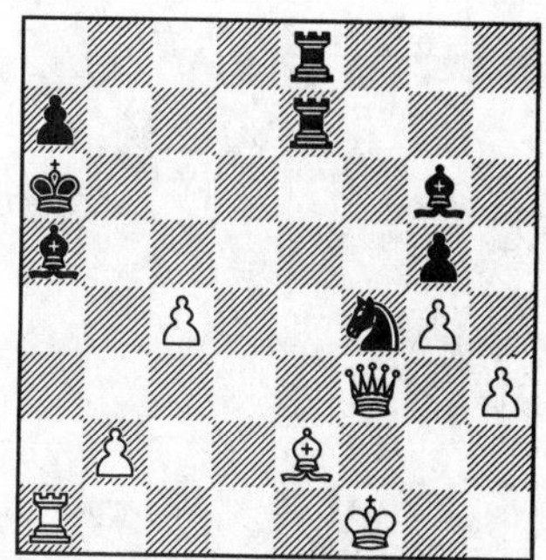

256.

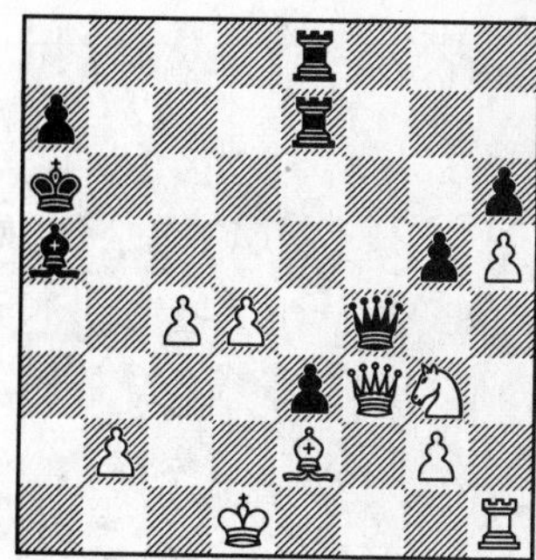

257.

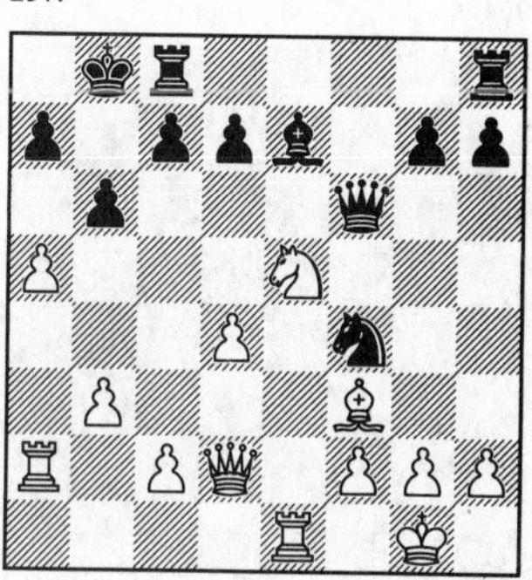

258.

259.

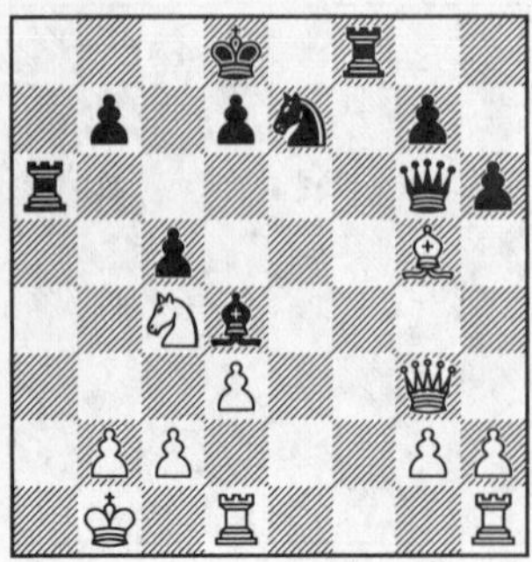

260.

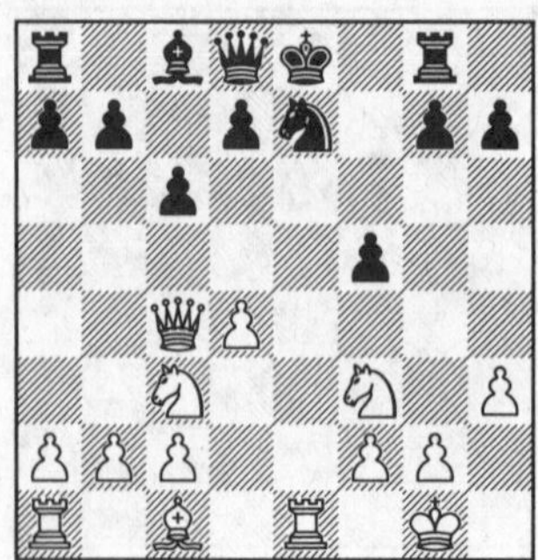

261.

262.

263.

264.

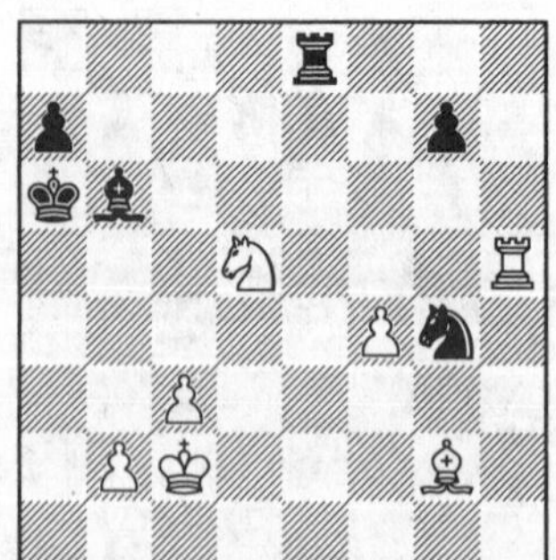

265.

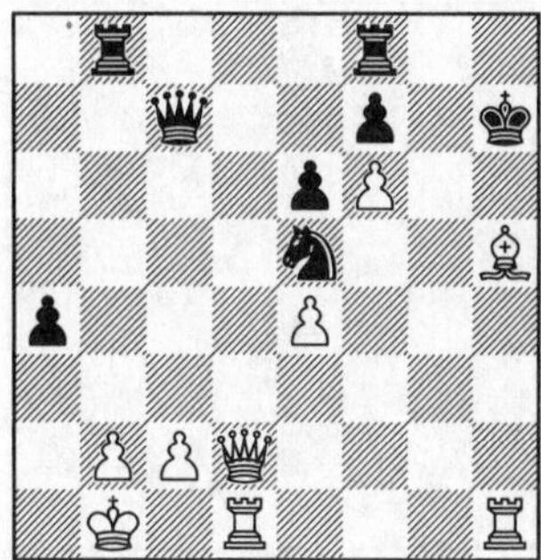

266.

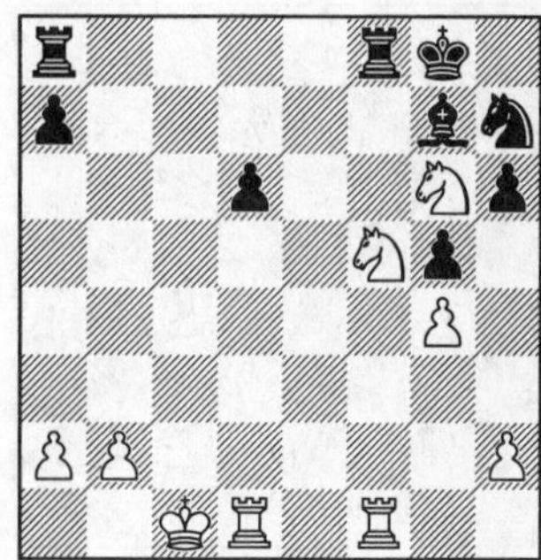

267.

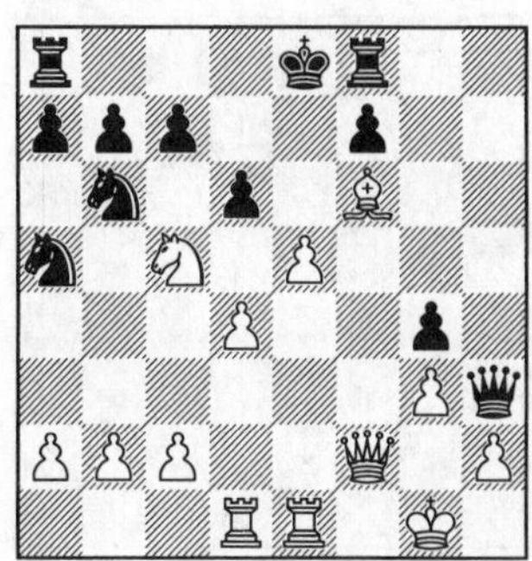

268.

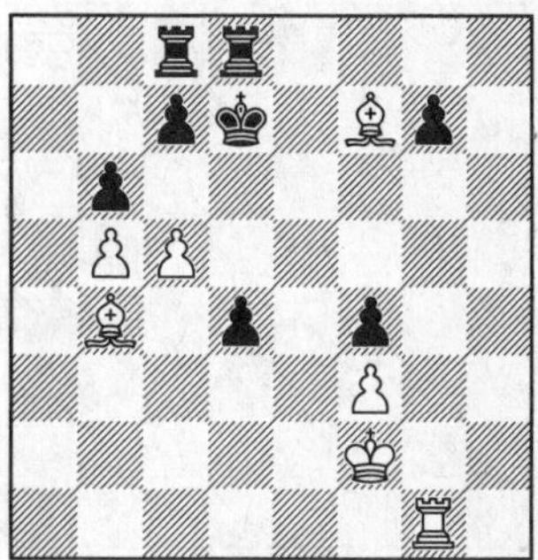

269.

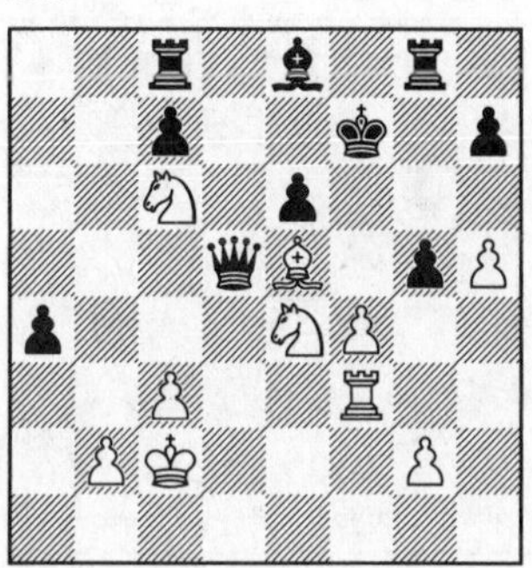

270.

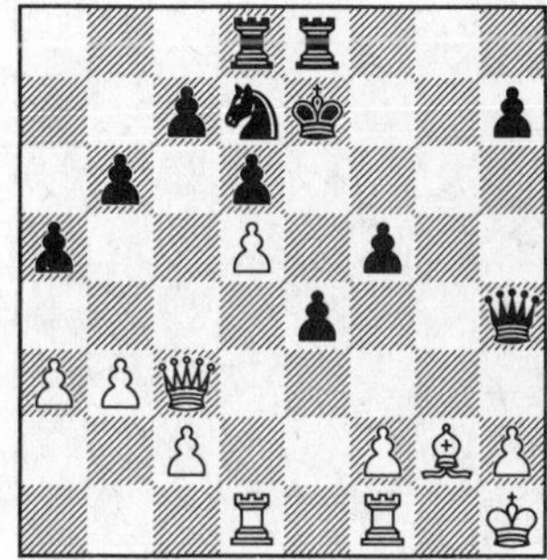

271.

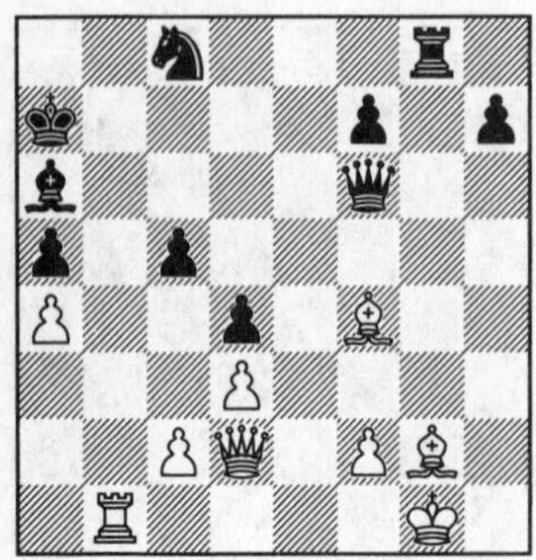

272.

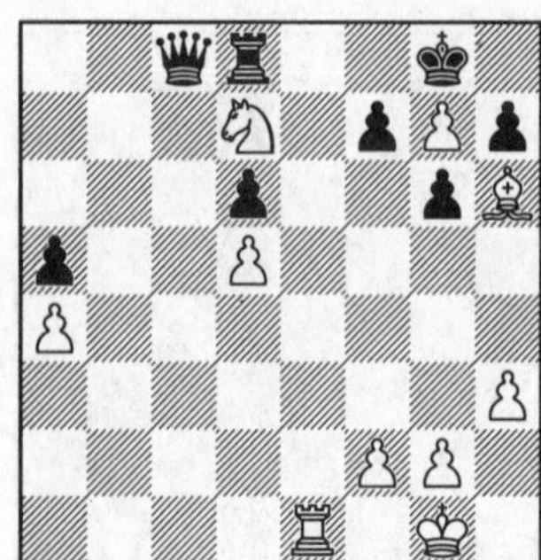

273.

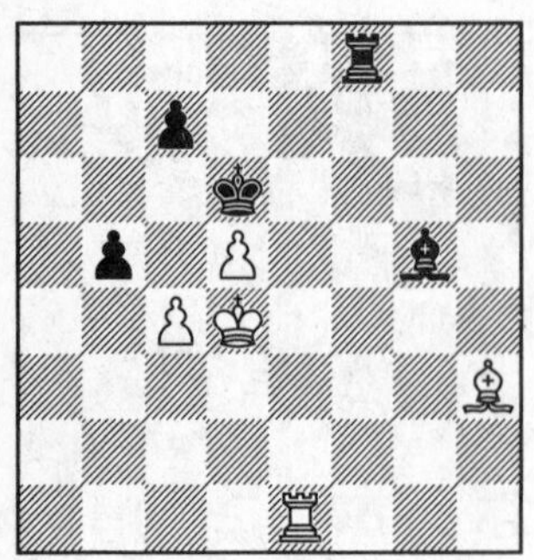

274.

275.

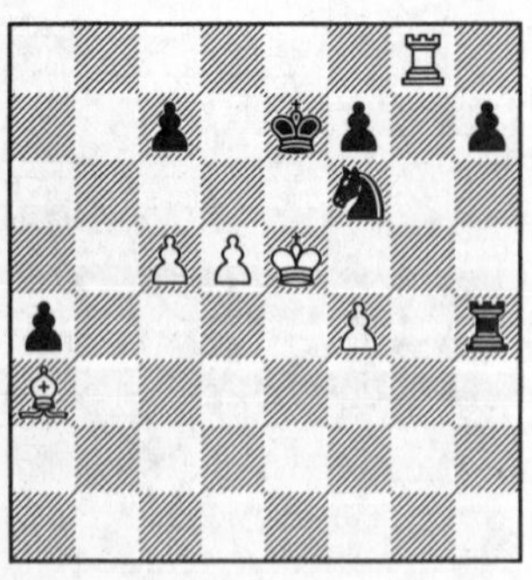

276.

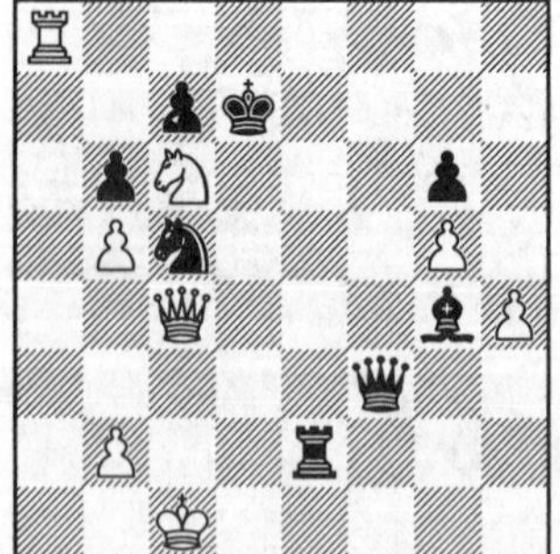

277.

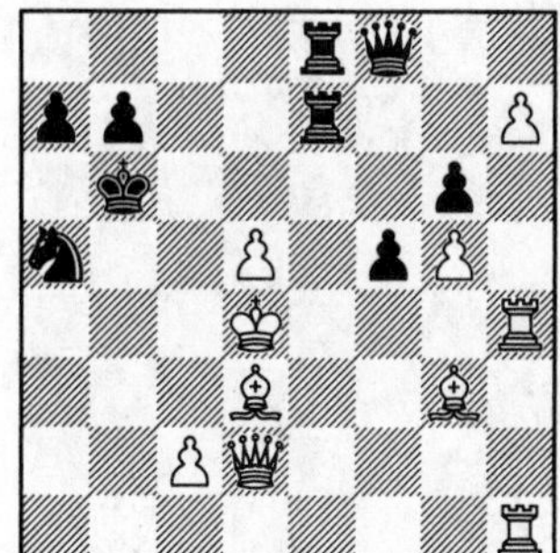

278.

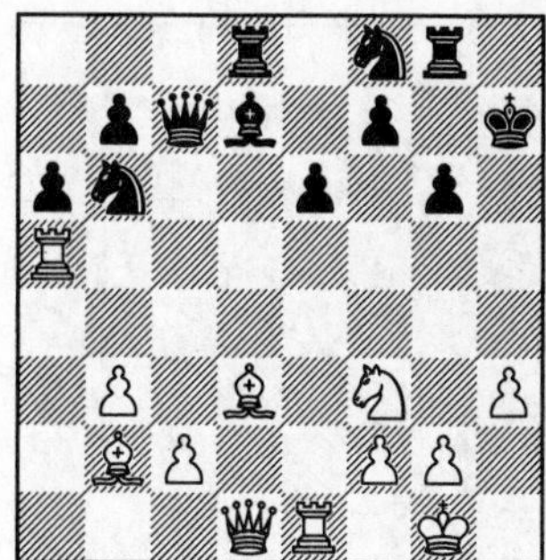

279.

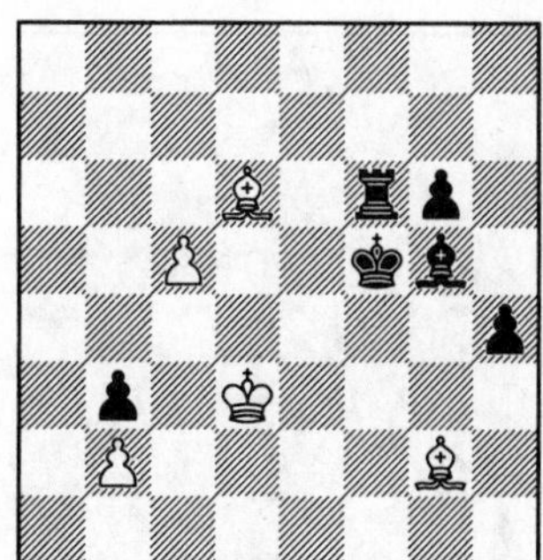

280.

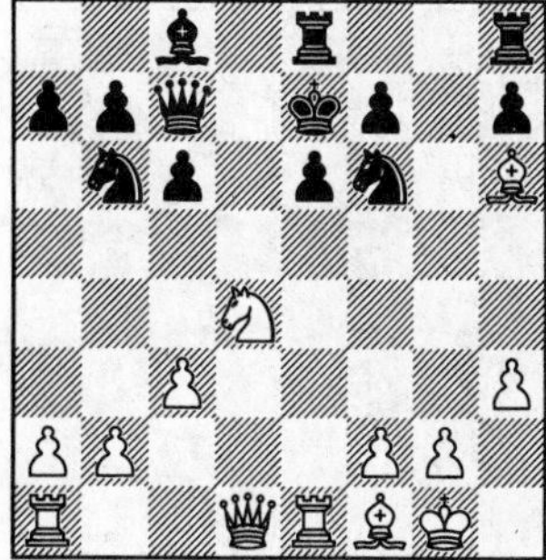

281.

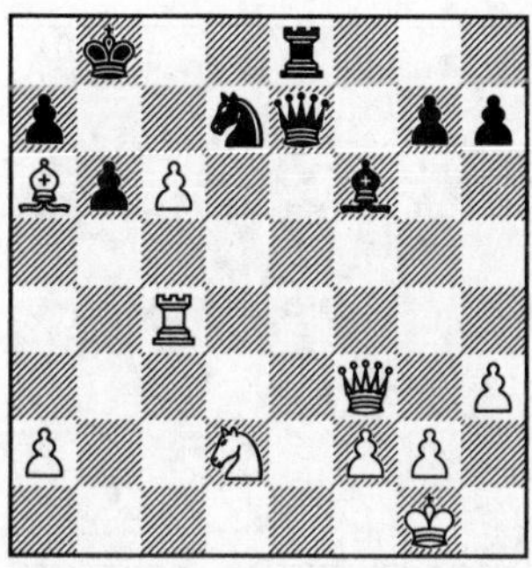

282.

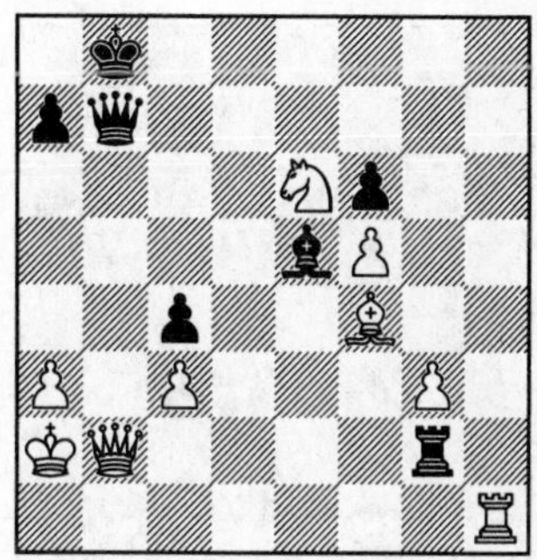

283.

284.

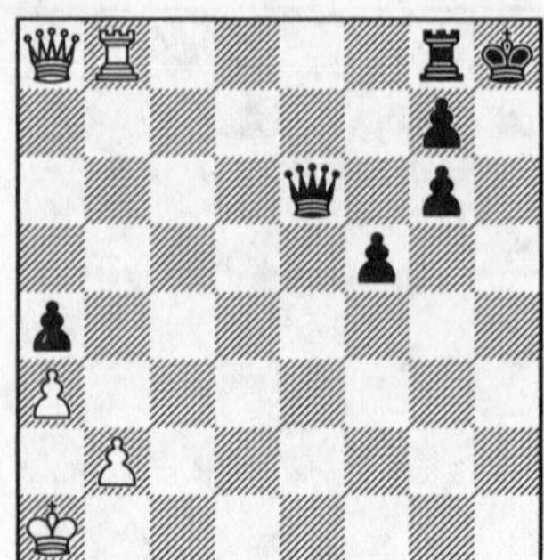

285.

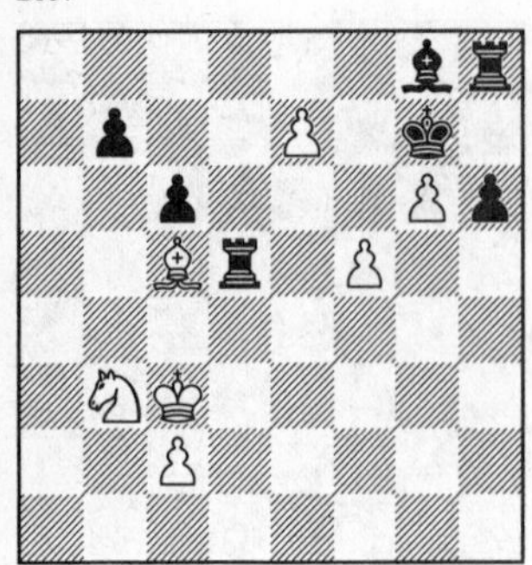

286.

287.

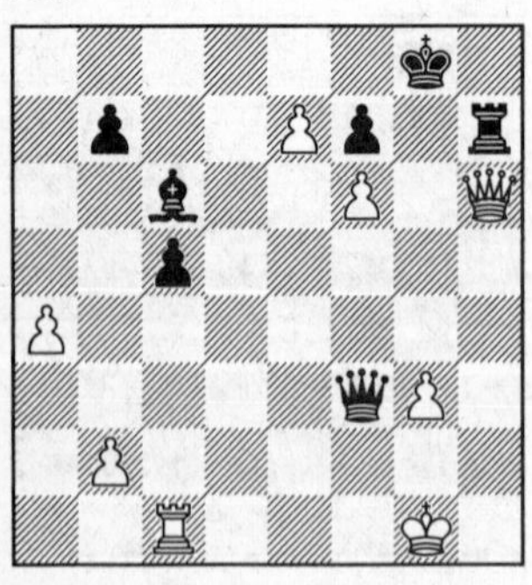

288.

289.

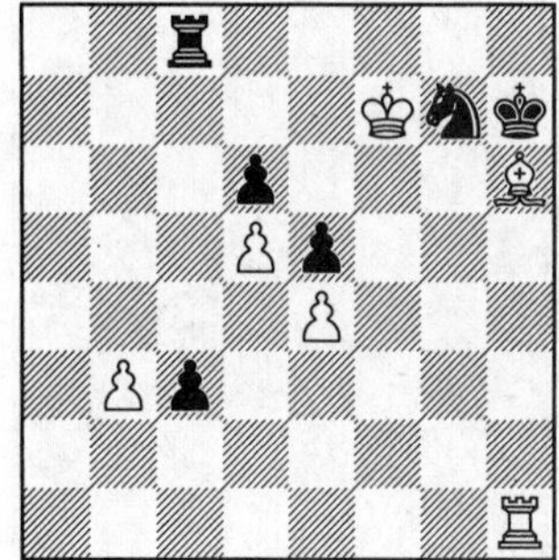

290.

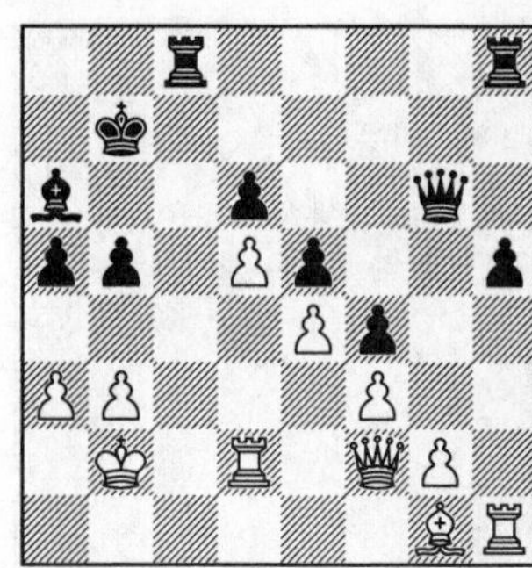

291.

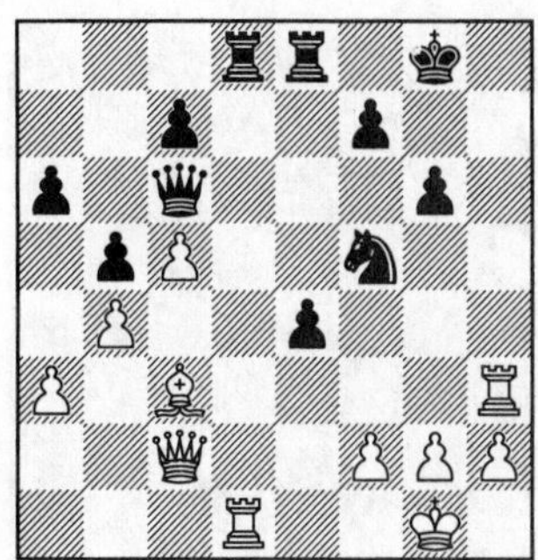

292.

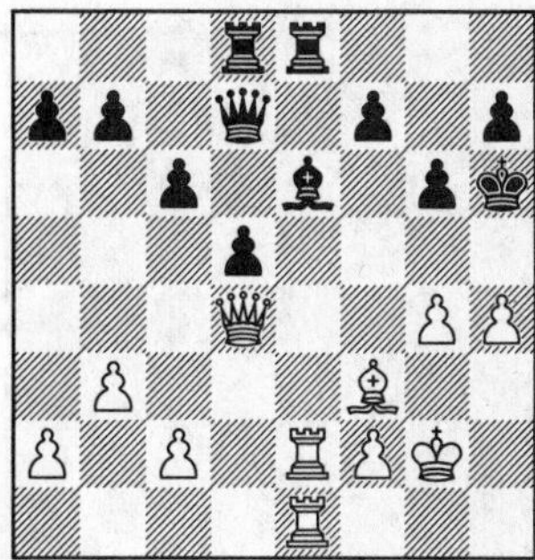

293.

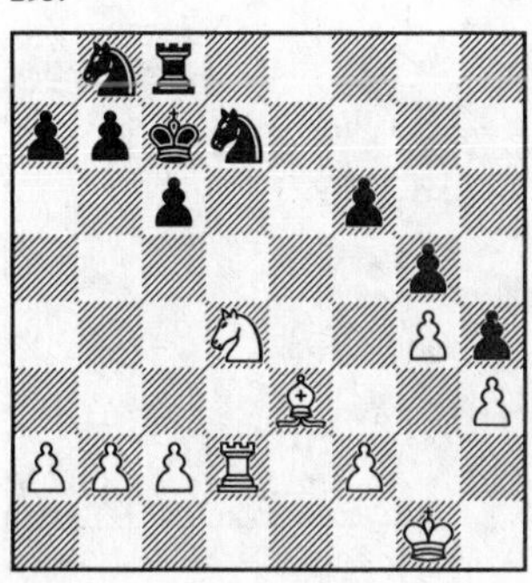

294.

295.

296.

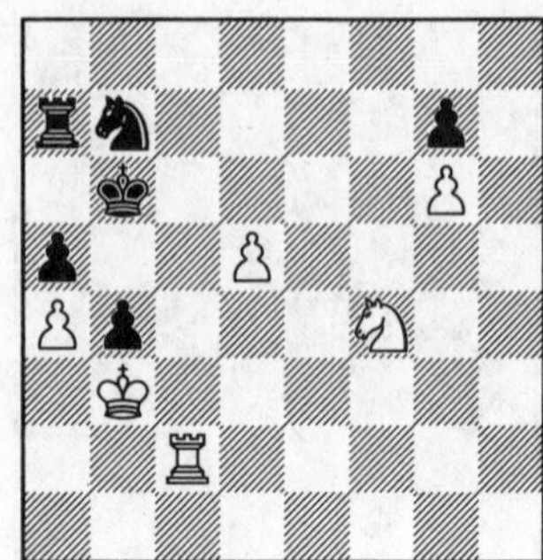

297.

298.

299.

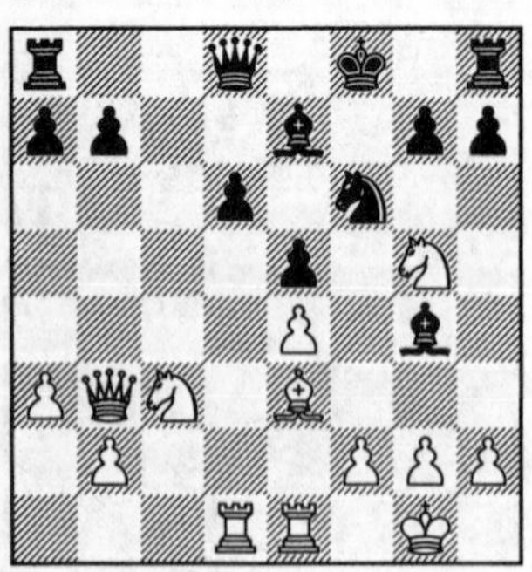

300.

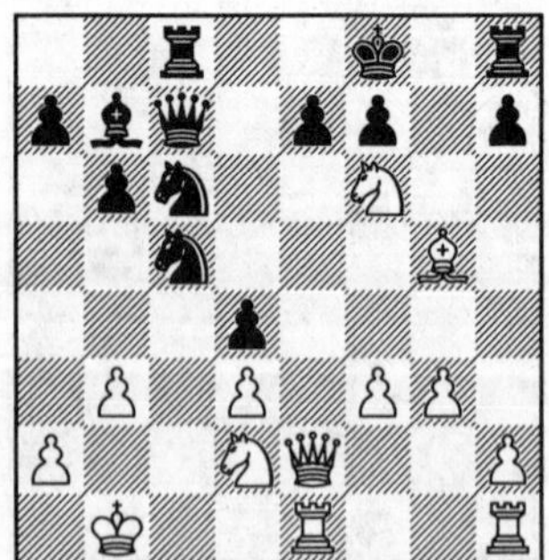

301.

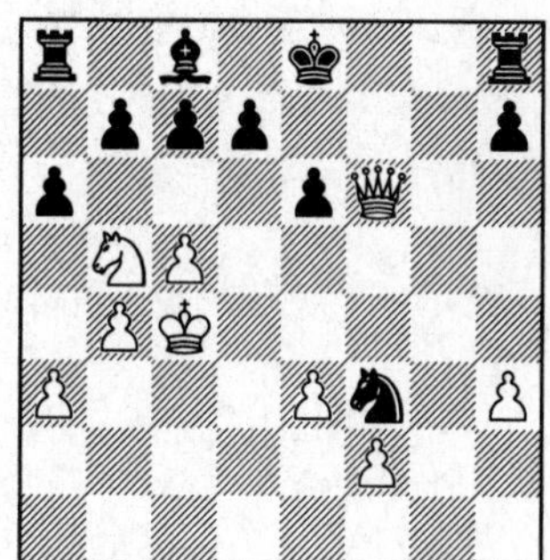

302.

303.

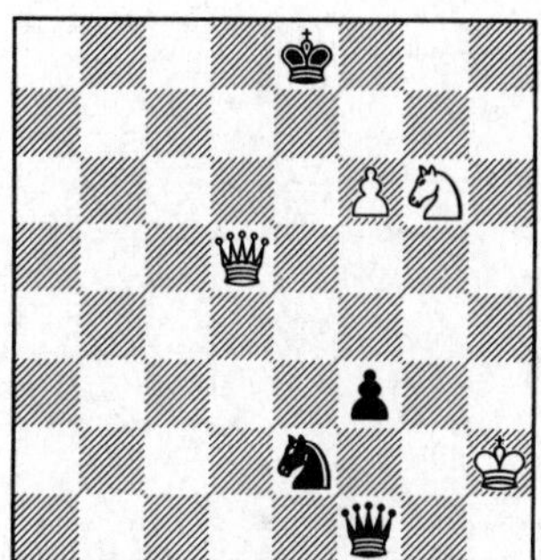

304.

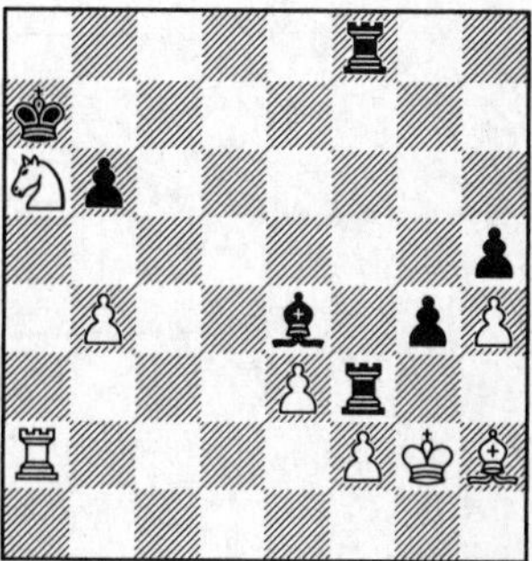

305.

306.

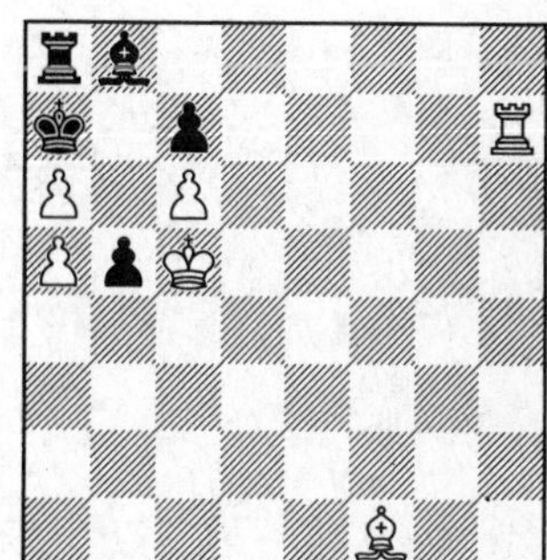

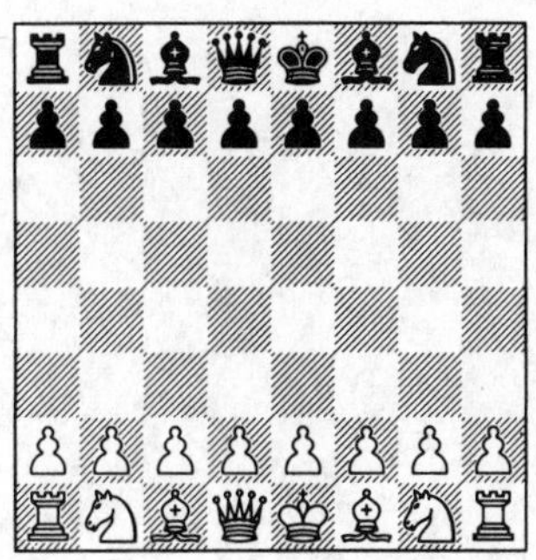

Mate in two

(307–3718)

307.

308.

309.

310.

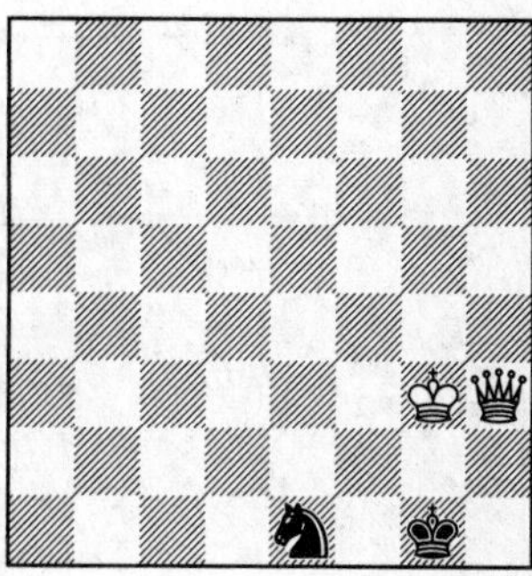

311.

312.

313.

314.

315.

316.

317.

318.

319.

320.

321.

322.

323.

324.

325.

326.

327.

328.

329.

330.

331.

332.

333.

334.

335.

336.

337.

338.

339.

340.

341.

342.

343.

344.

345.

346.

347.

348.

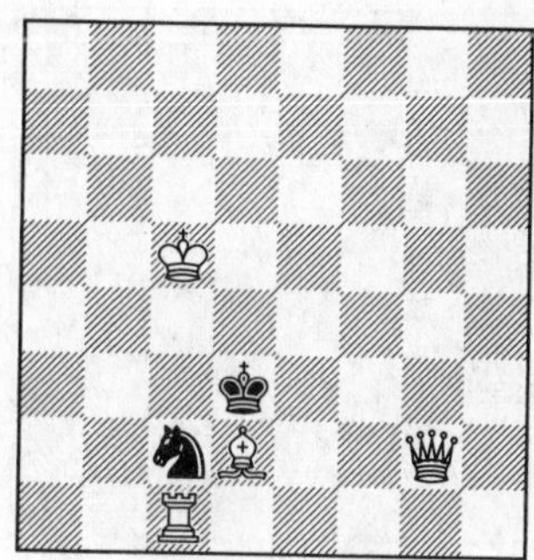

349.

350.

351.

352.

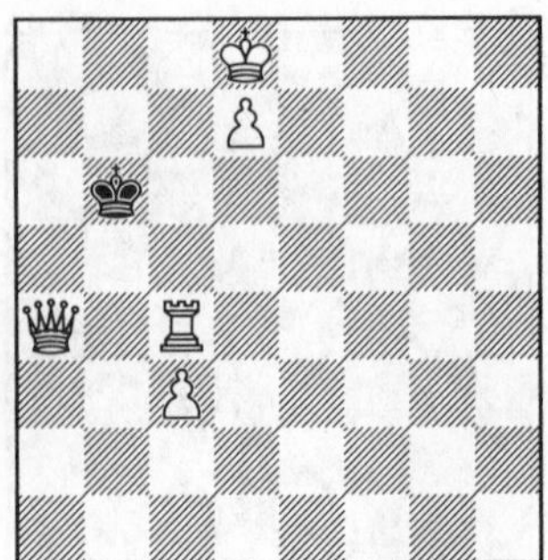

353.

354.

355.

356.

357.

358.

359.

360.

361.

362.

363.

364.

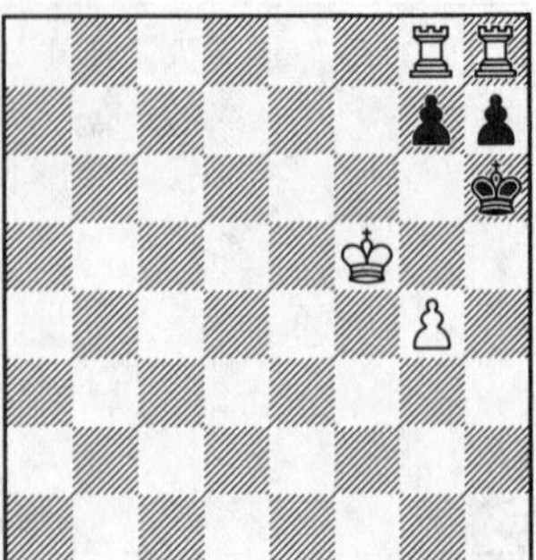

365.

366.

367.

368.

369.

370.

371.

372.

373.

374.

375.

376.

377.

378.

379.

380.

381.

382.

383.

384.

385.

386.

387.

388.

389.

390.

391.

392.

393.

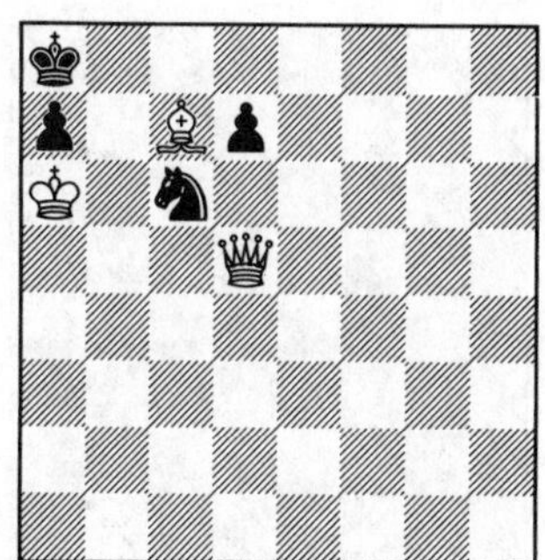

394.

395.

396.

397.

398.

399.

400.

401.

402.

403.

404.

405.

406.

407.

408.

409.

410.

411.

412.

413.

414.

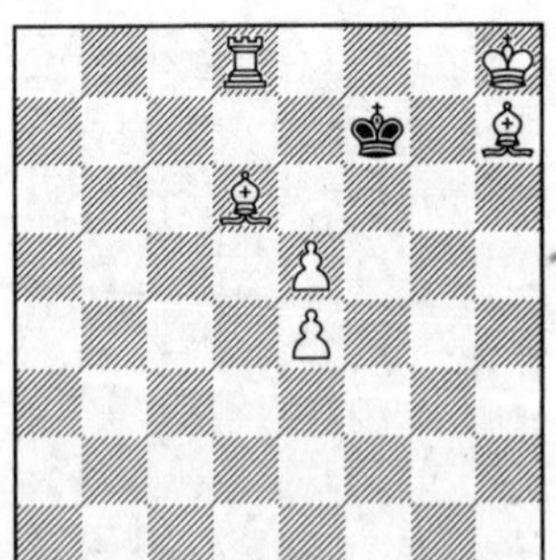

415.

416.

417.

418.

419.

420.

421.

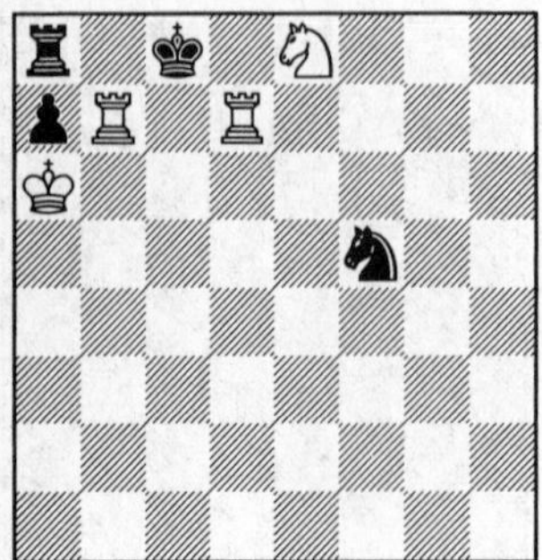

422.

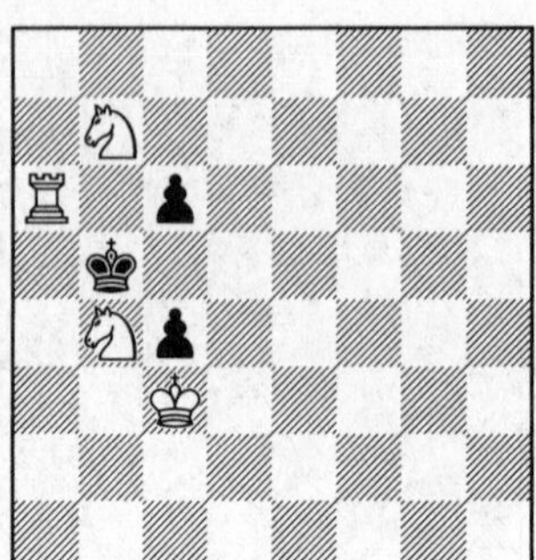

423.

424.

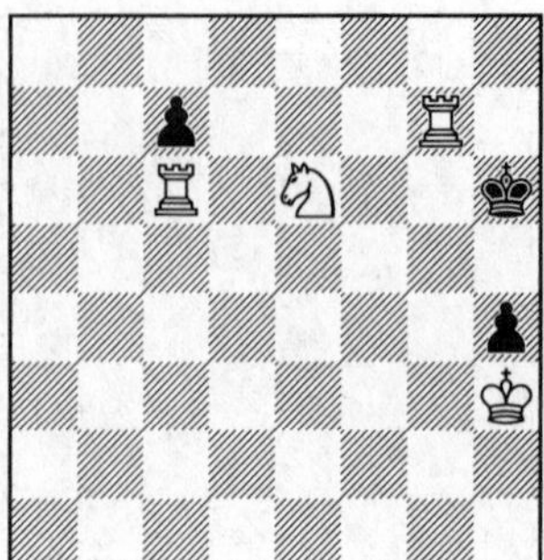

425.

426.

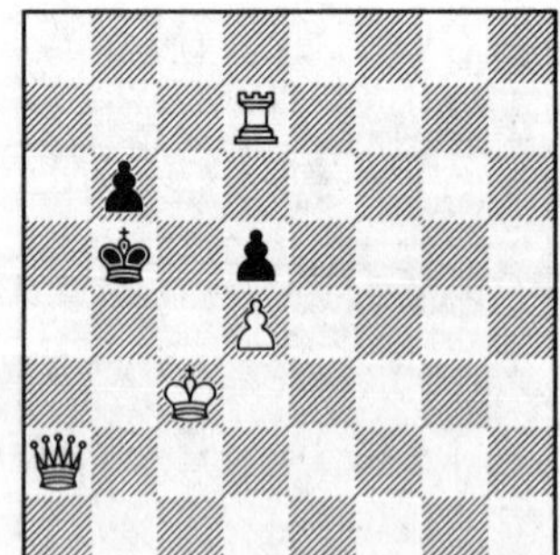

427.

428.

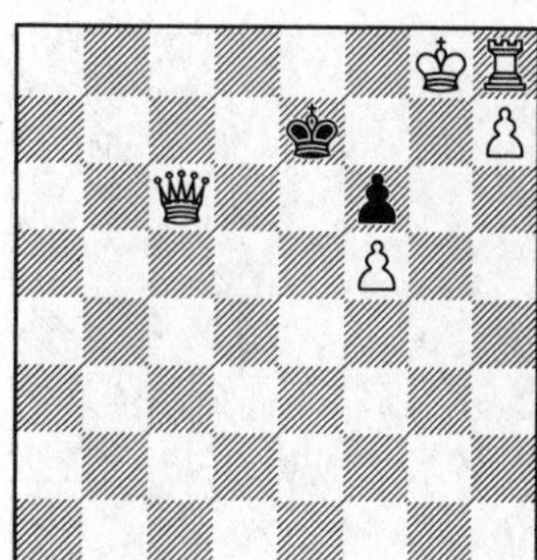

429.

430.

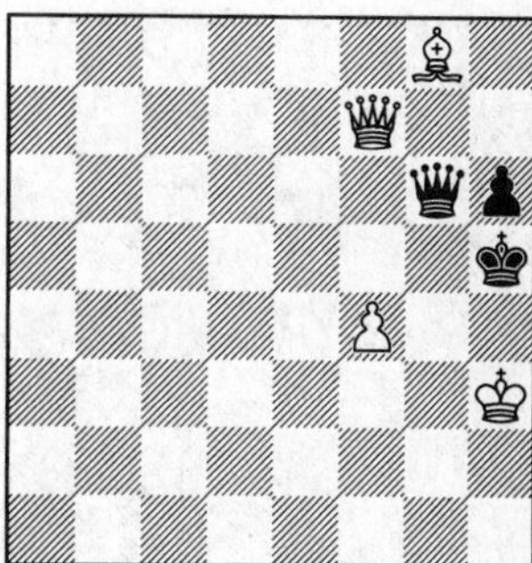

431.

432.

433.

434.

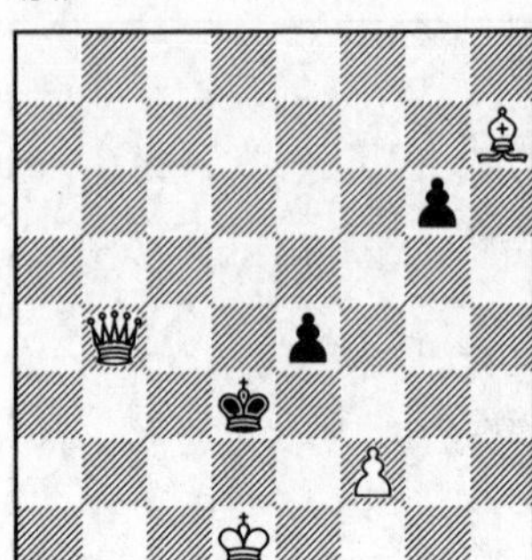

435.

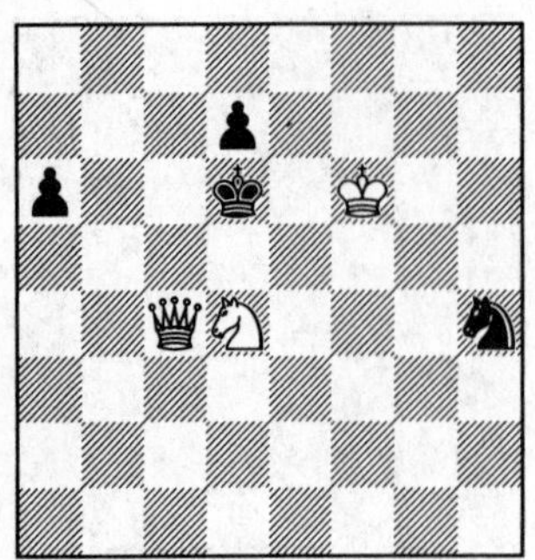

436.

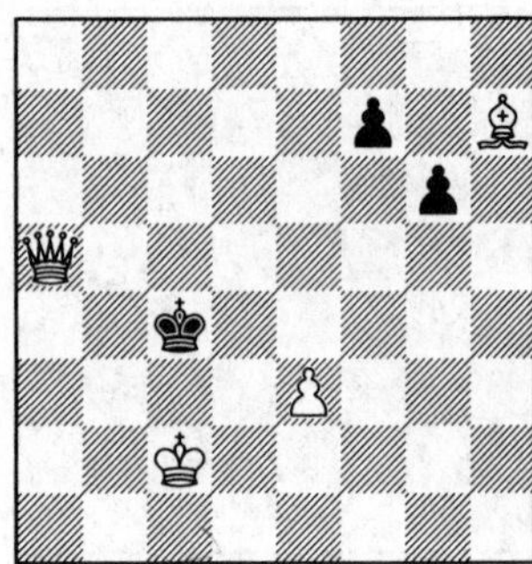

437.

438.

439.

440.

441.

442.

443.

444.

445.

446.

447.

448.

449.

450.

451.

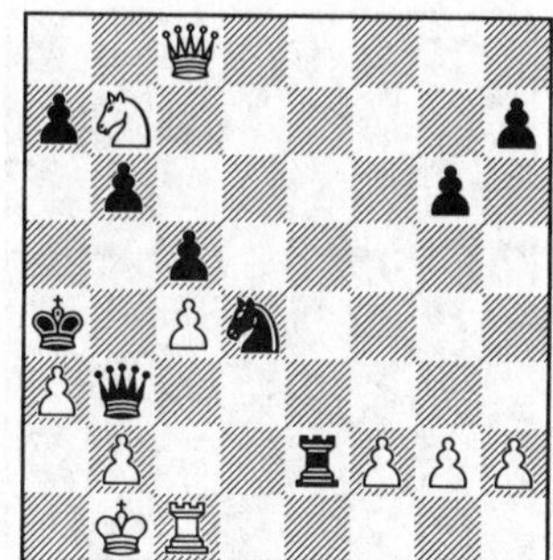

452.

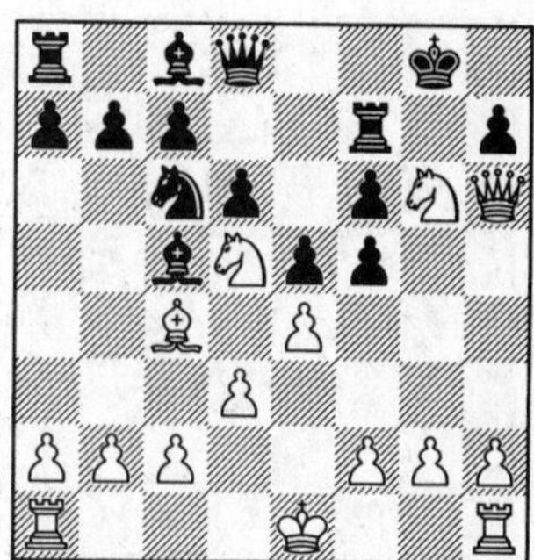

453.

454.

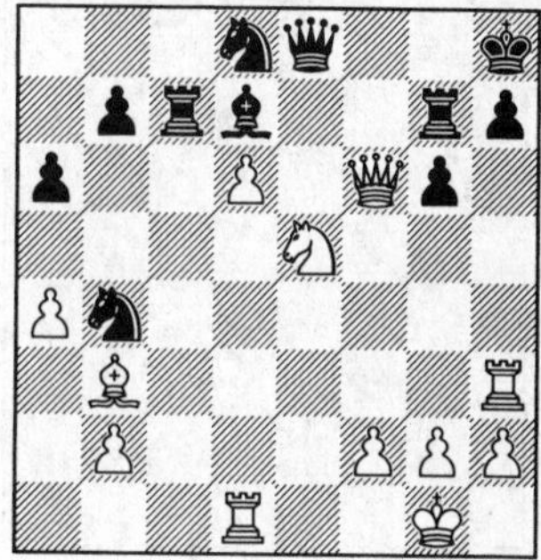

455.

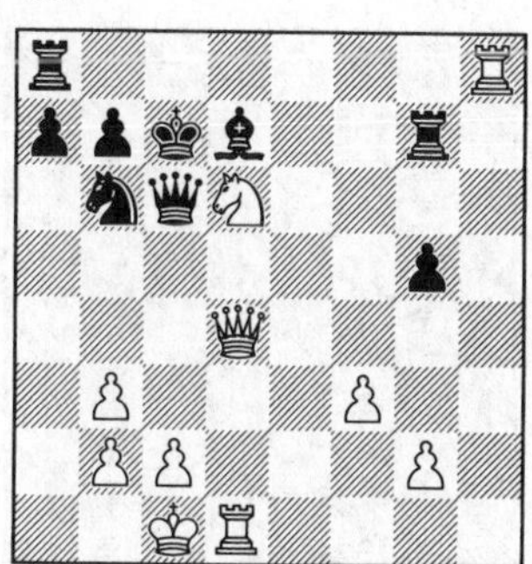

456.

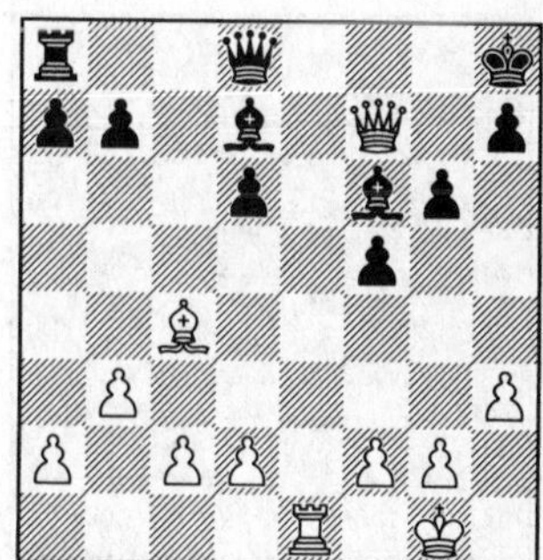

457.

458.

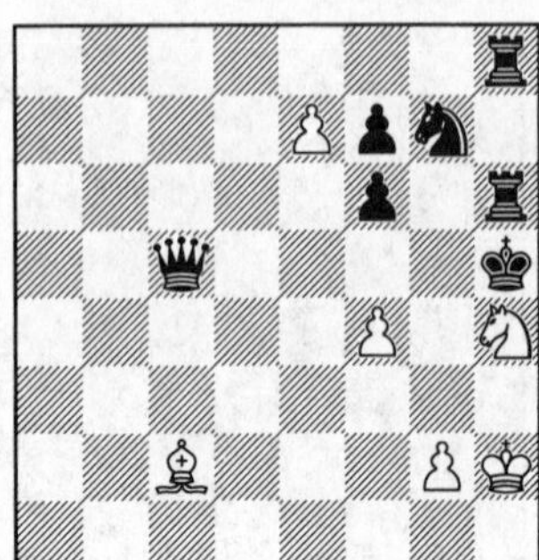

459.

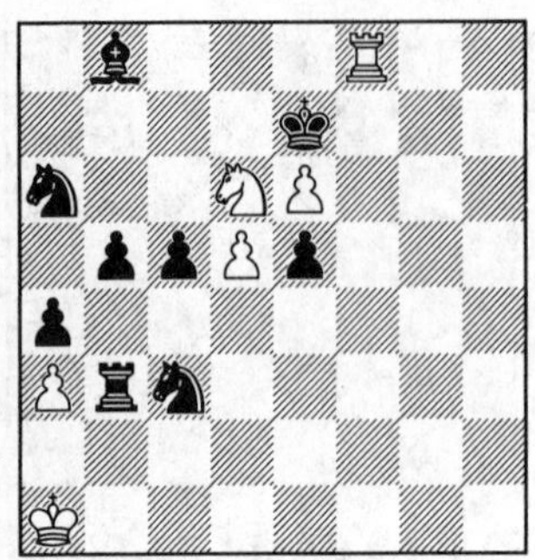

460.

461.

462.

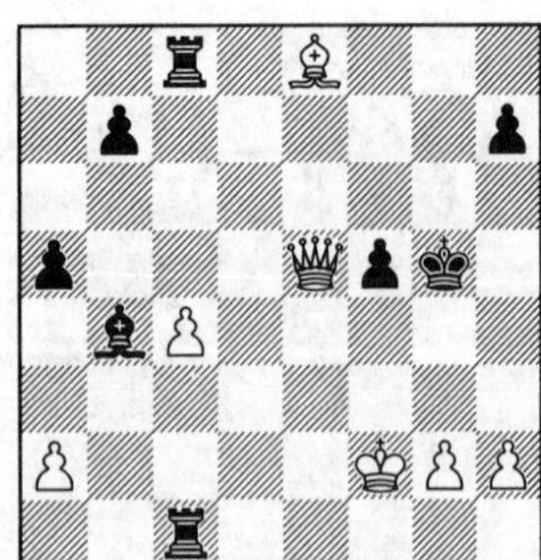

463.

464.

465.

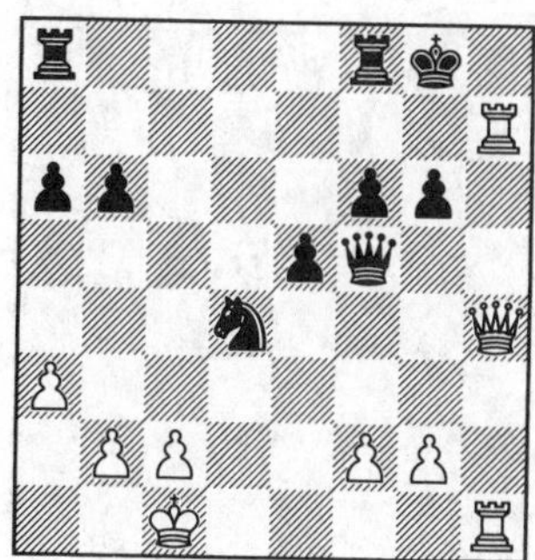

466.

467.

468.

469.

470.

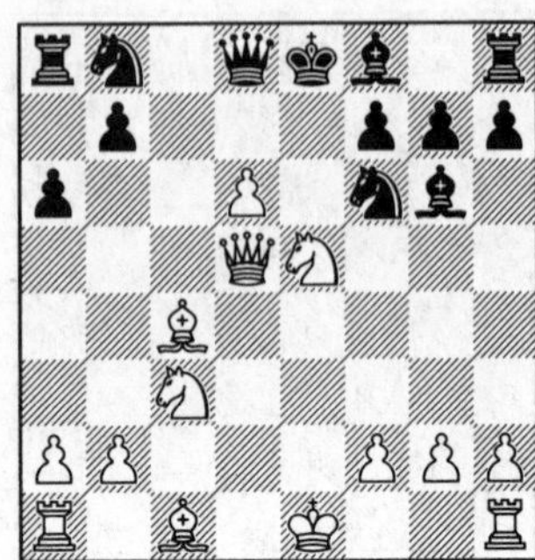

471.

472.

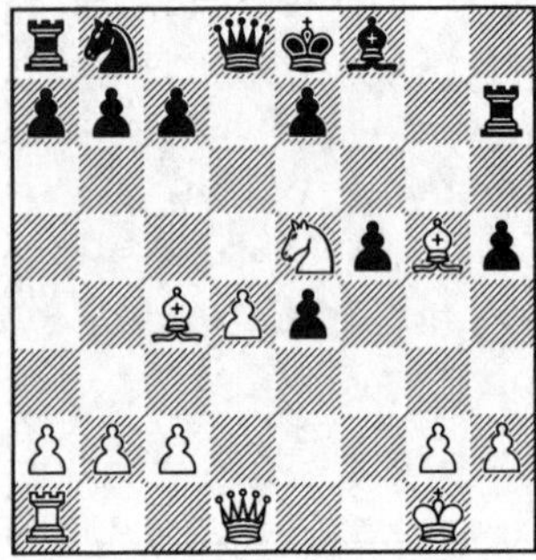

473.

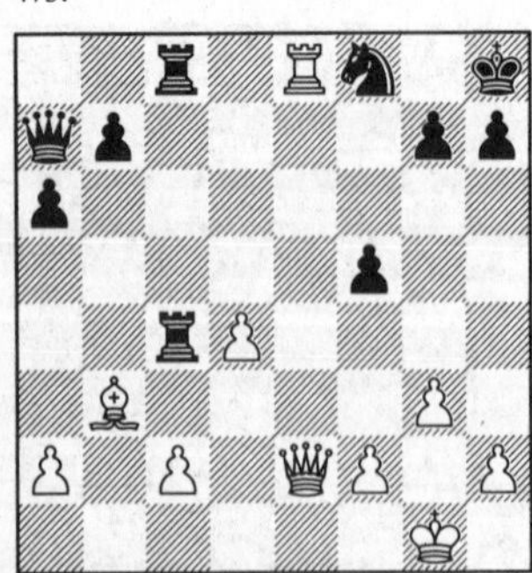

474.

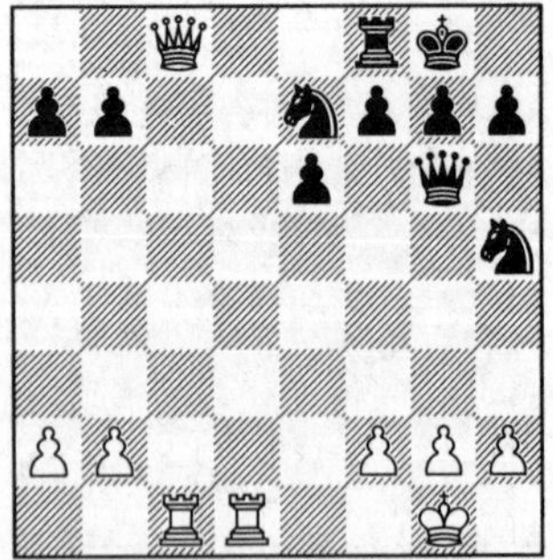

475.

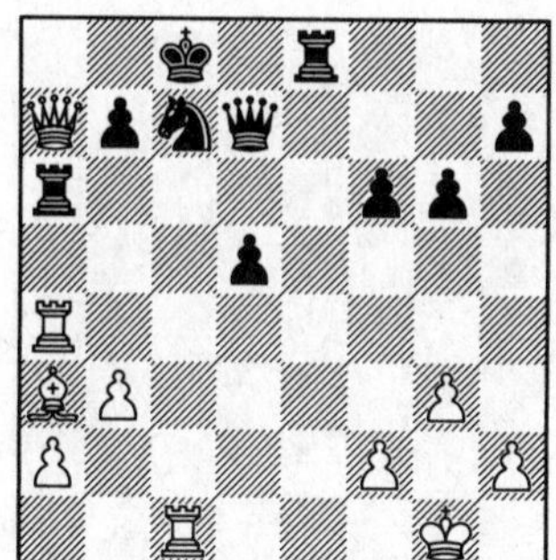

476.

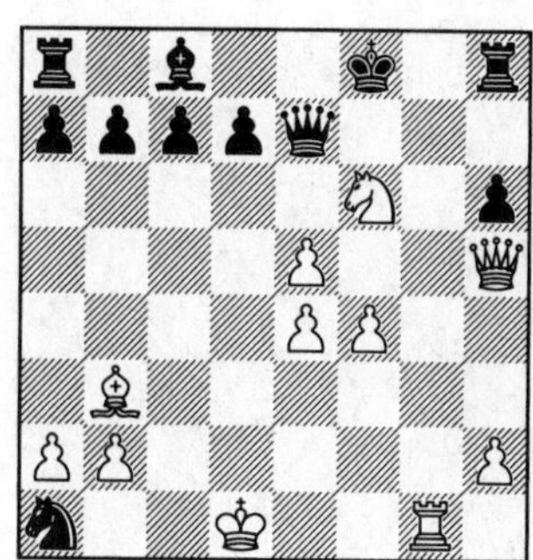

477.

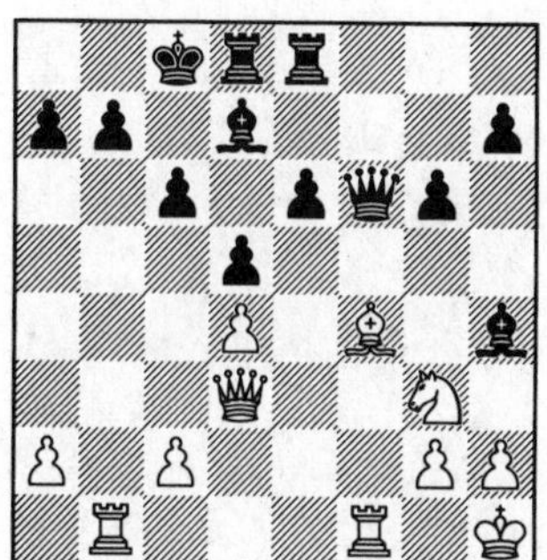

478.

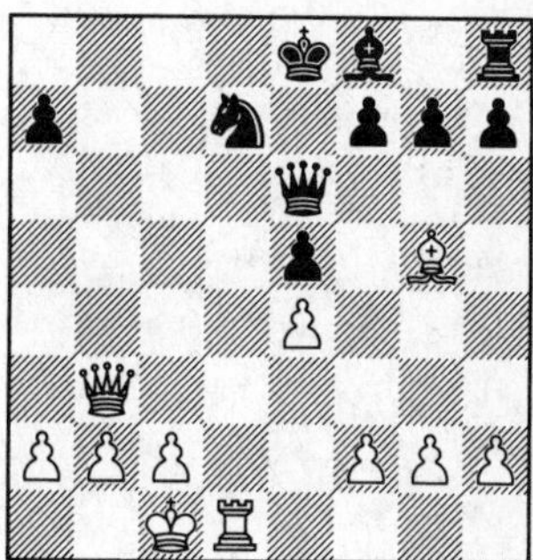

479.

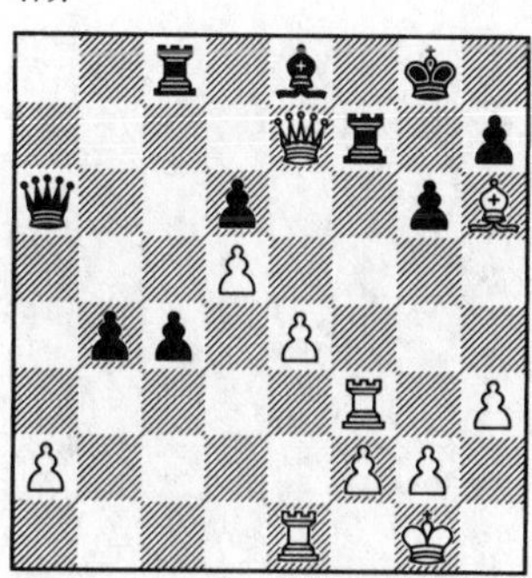

480.

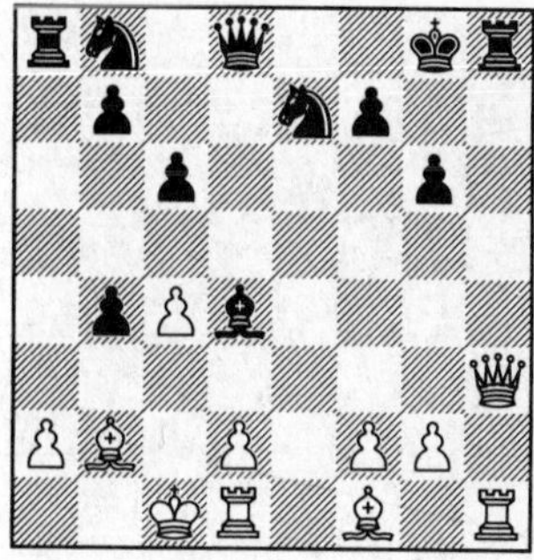

481.

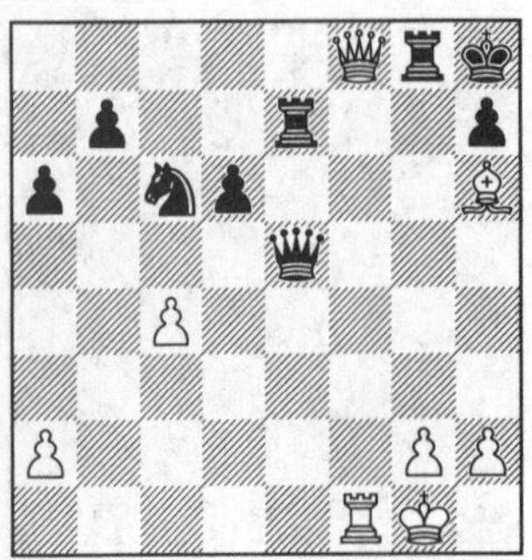

482.

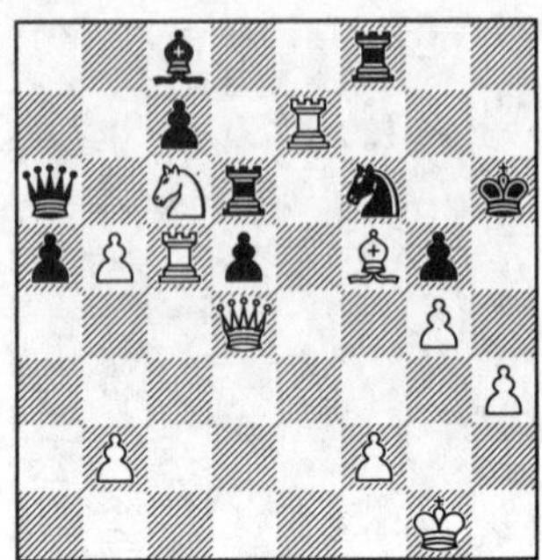

483.

484.

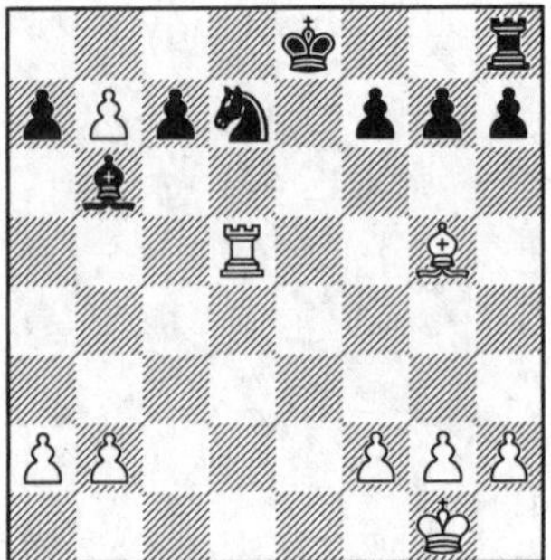

485.

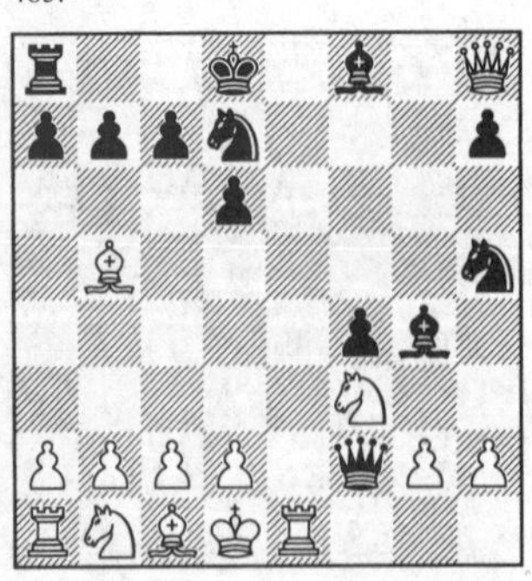

486.

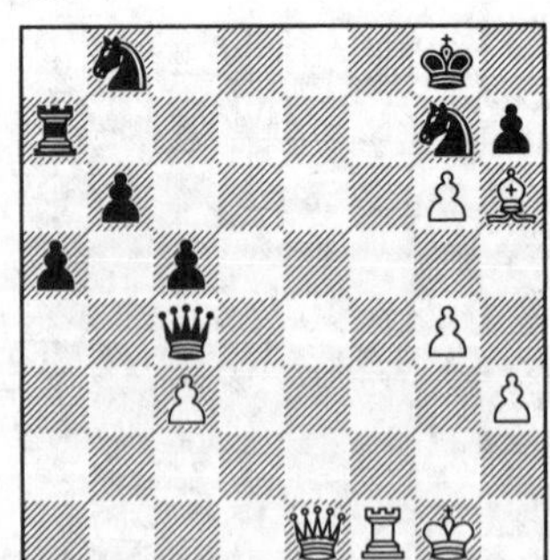

487.

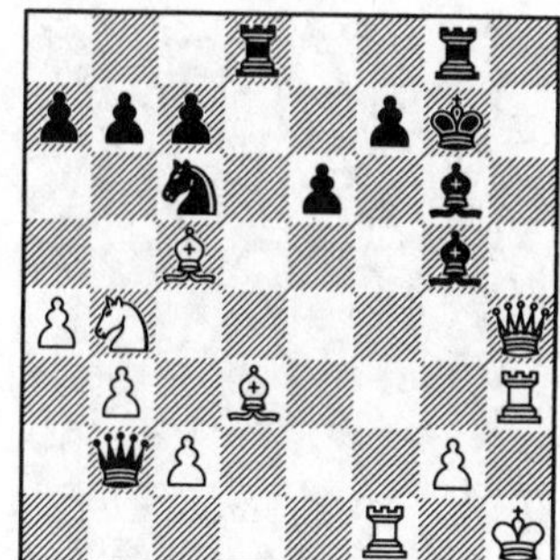

488.

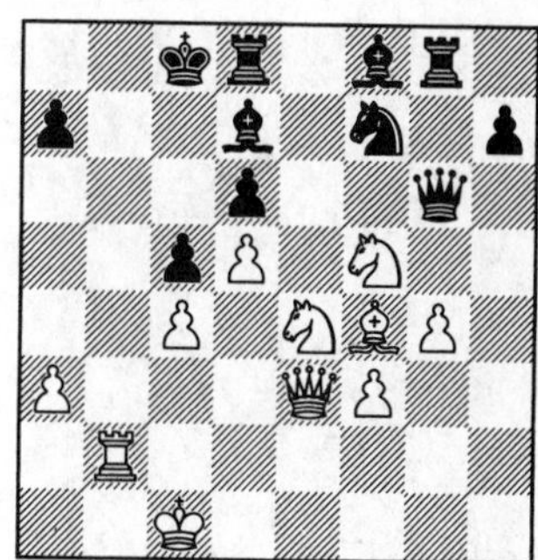

489.

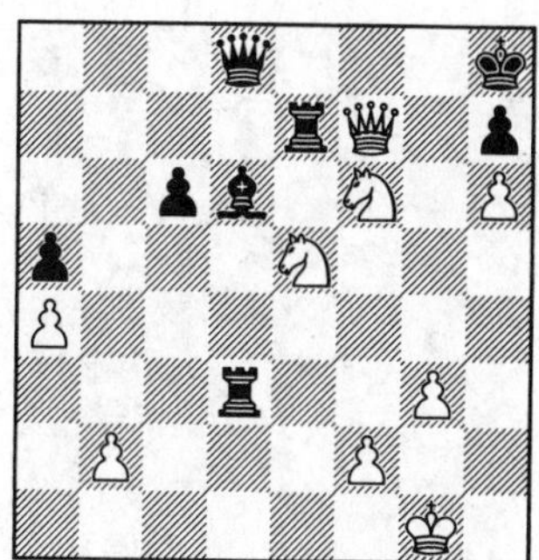

490.

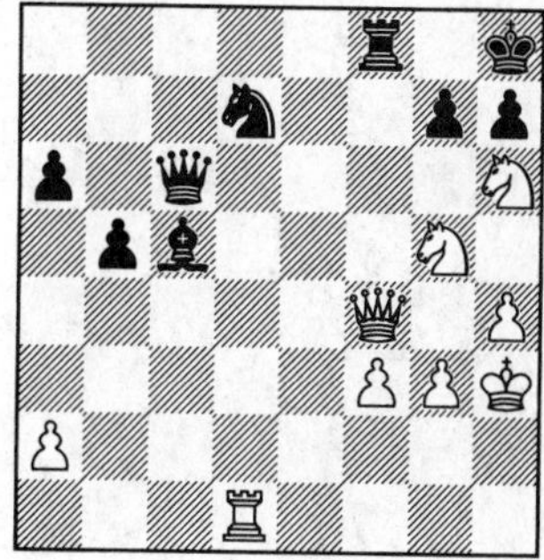

491.

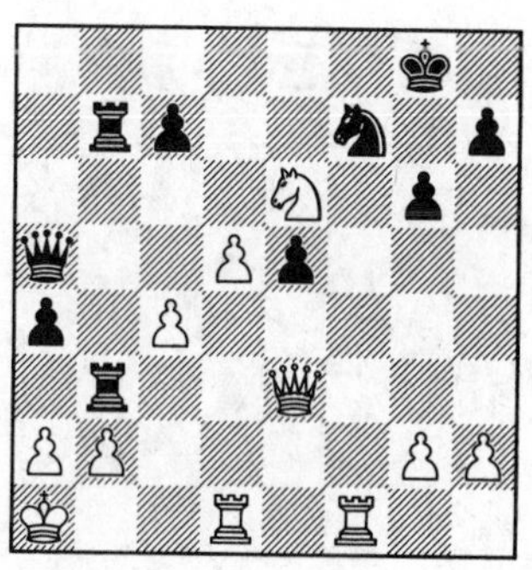

492.

493.

494.

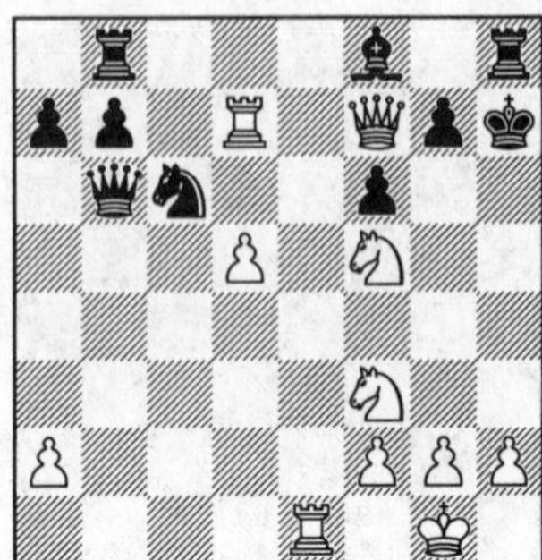

495.

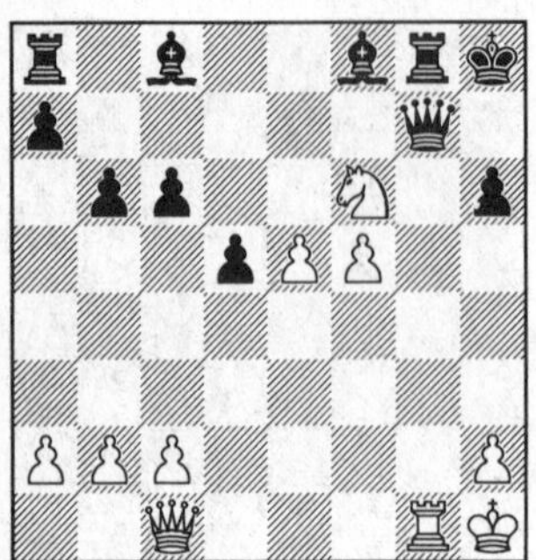

496.

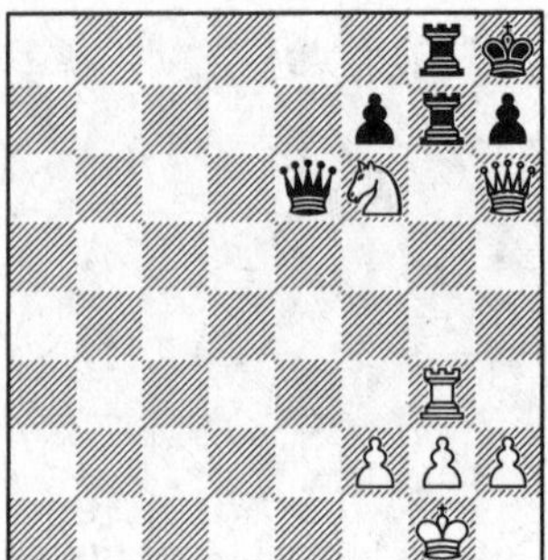

497.

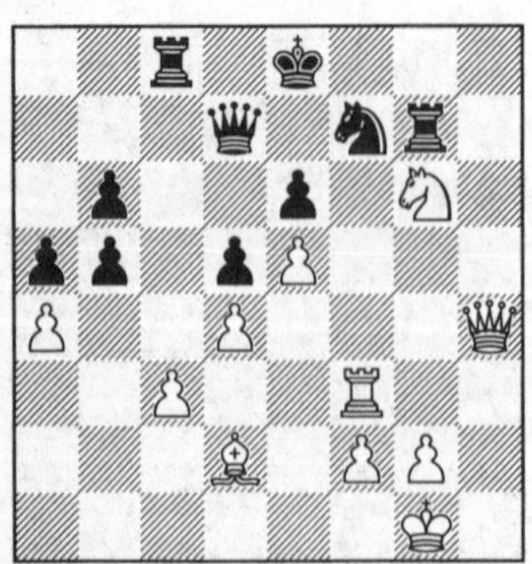

498.

499.

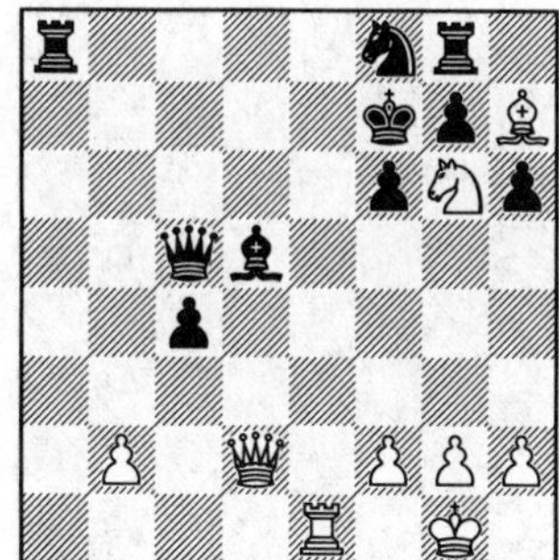

500.

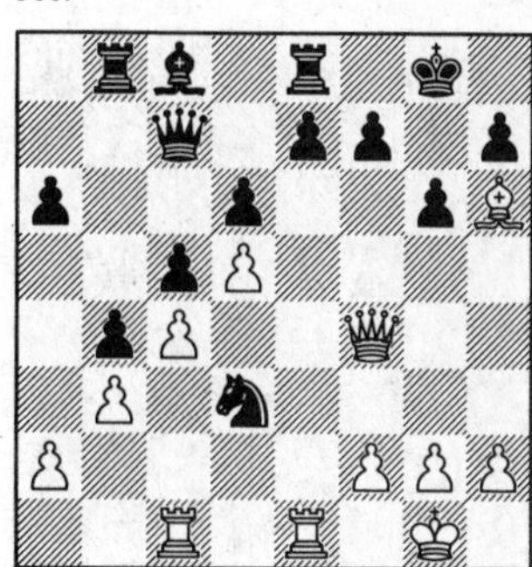

501.

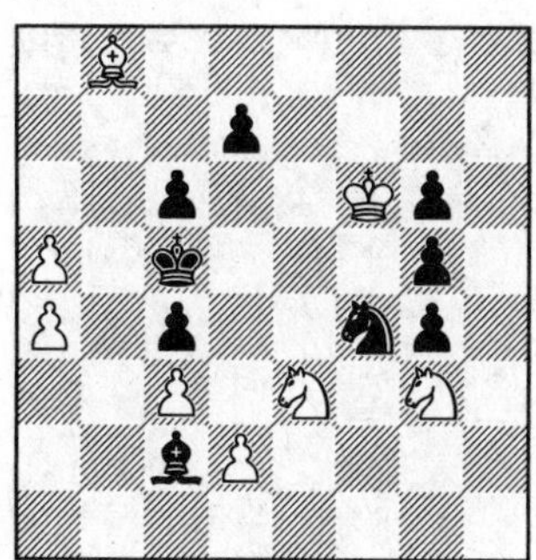

502.

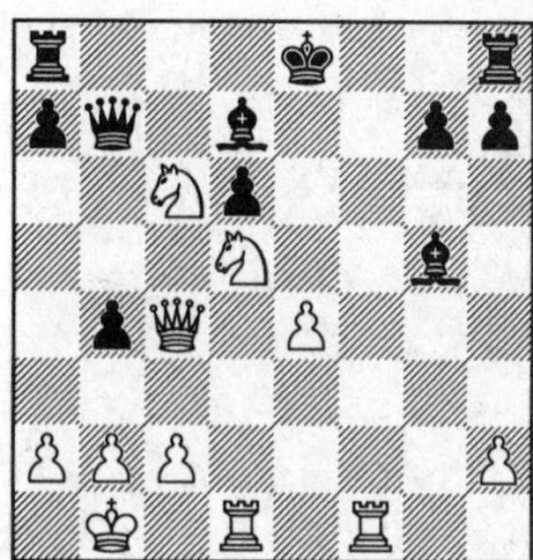

503.

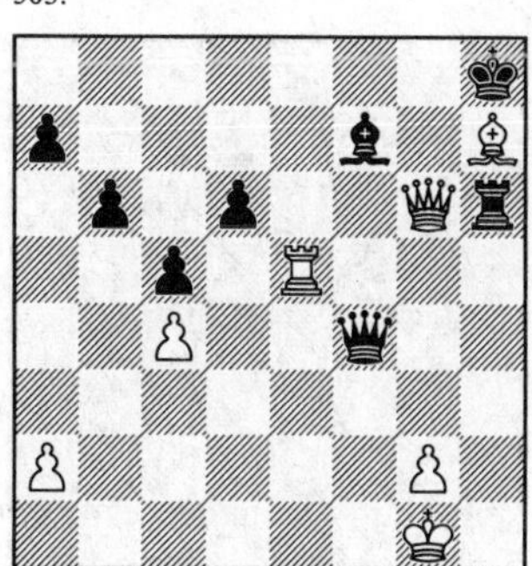

504.

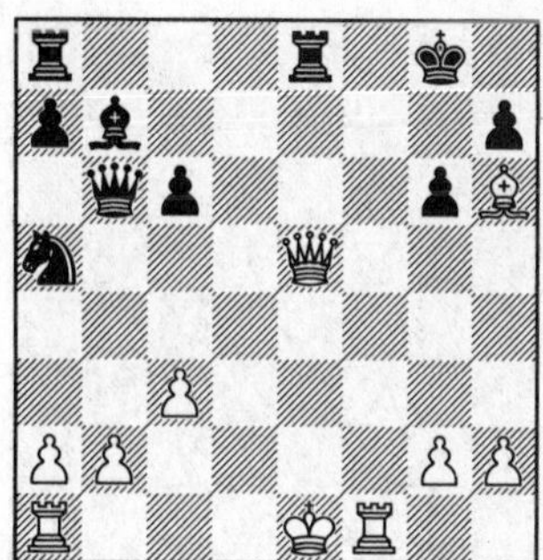

505.

506.

507.

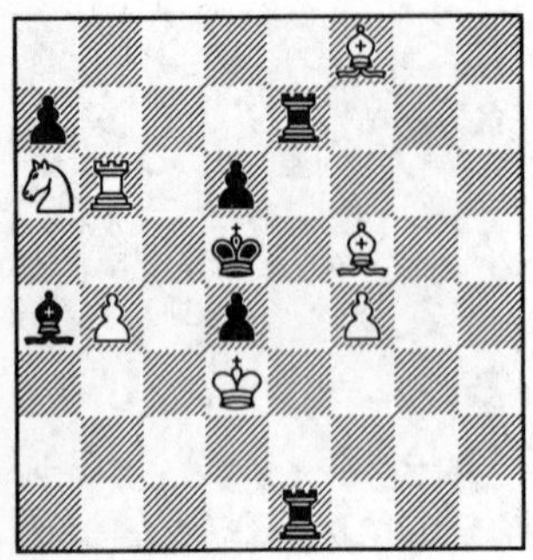

508.

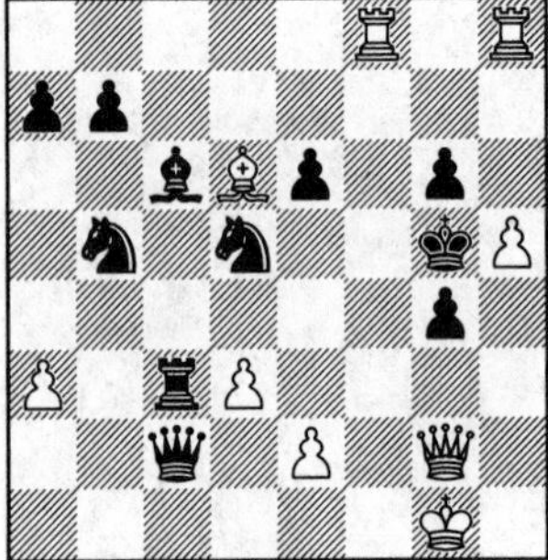

509.

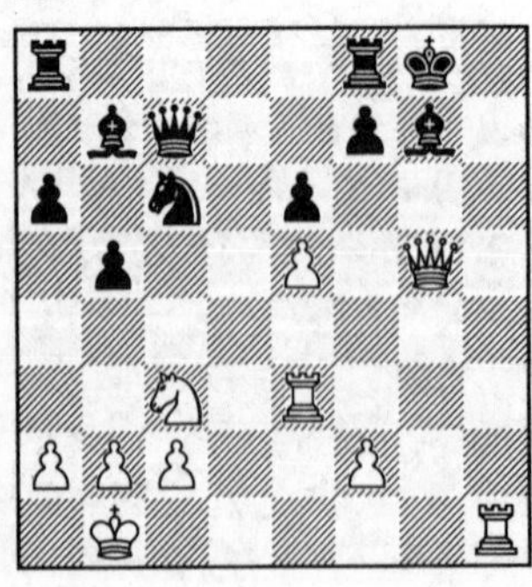

510.

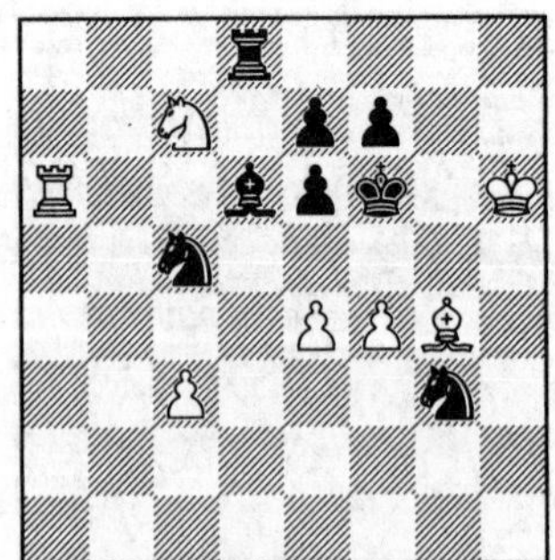

511.

512.

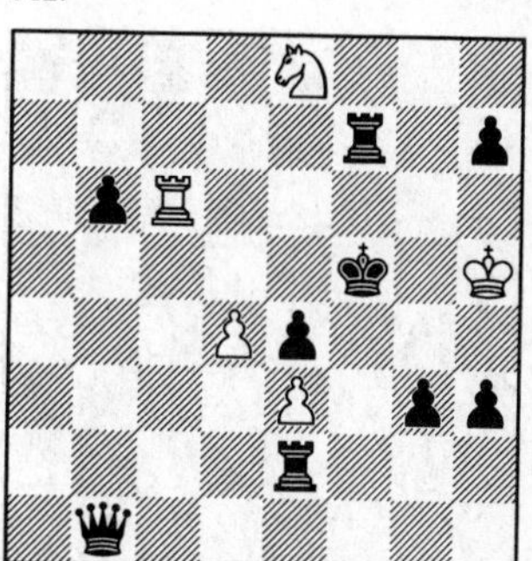

513.

514.

515.

516.

517.

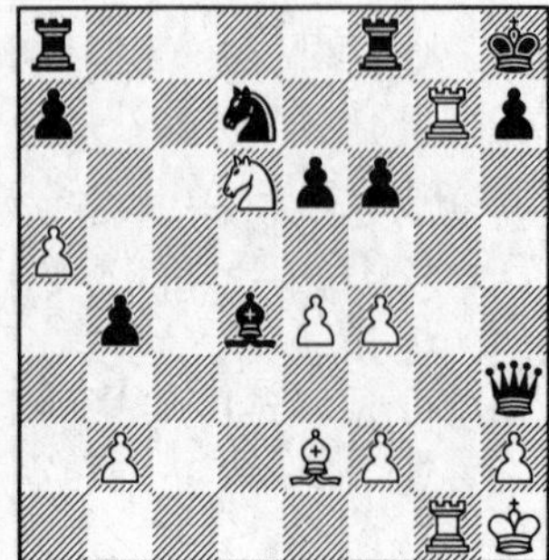

518.

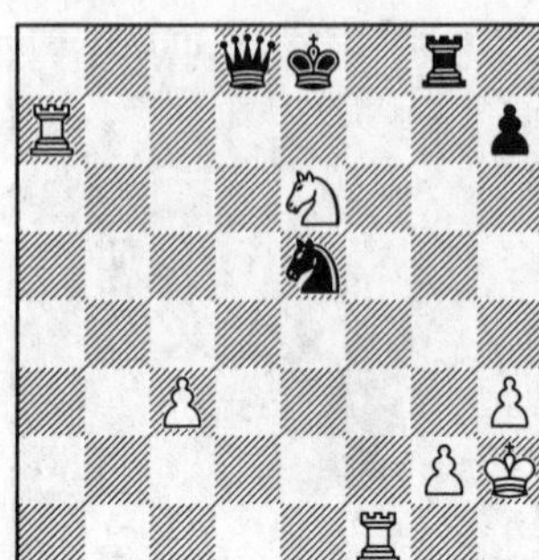

519.

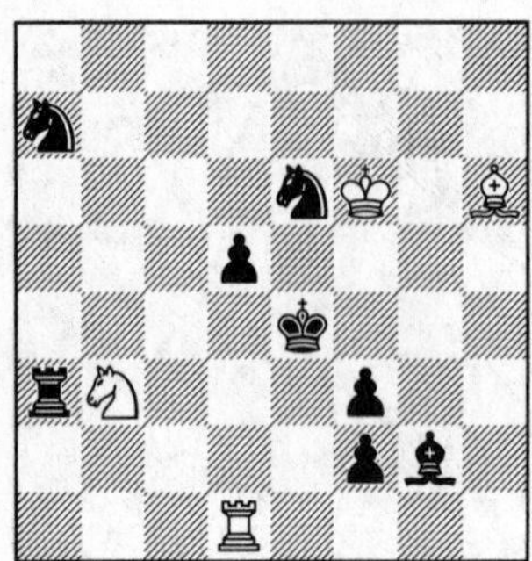

520.

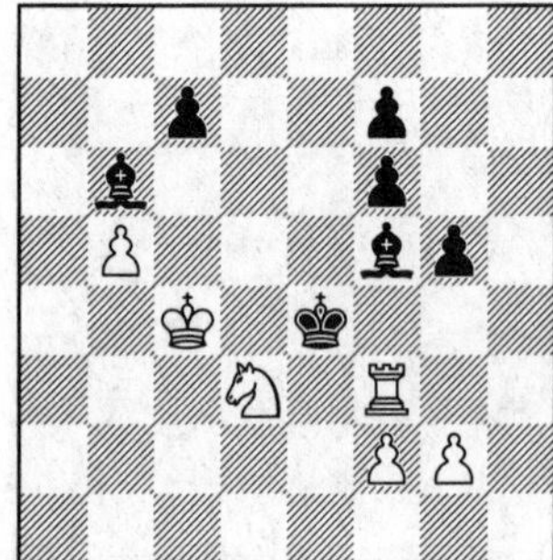

521.

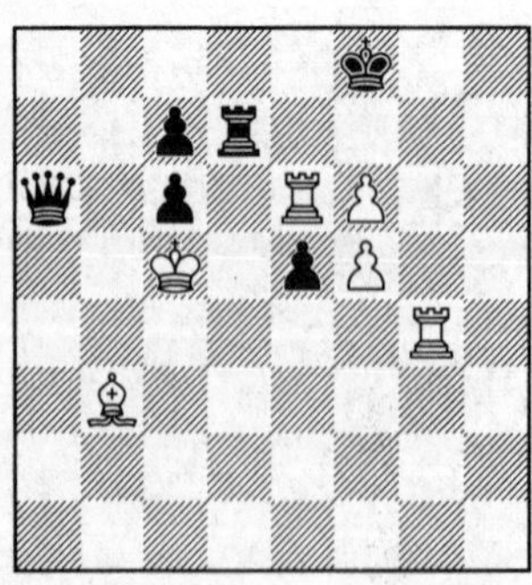

522.

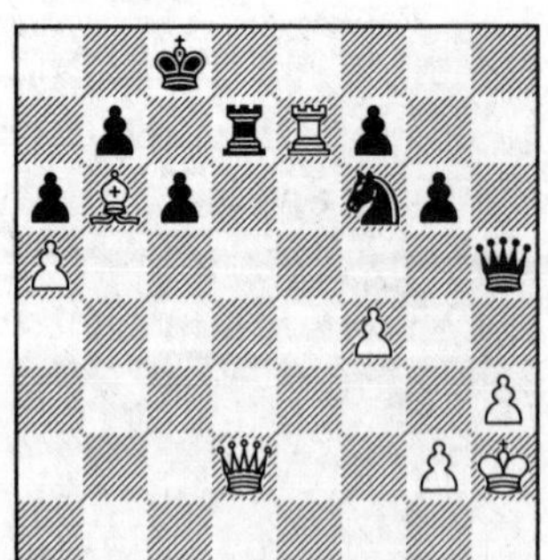

523.

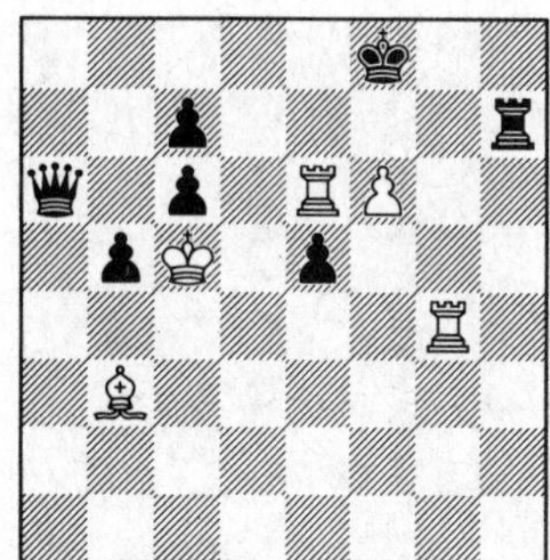

524.

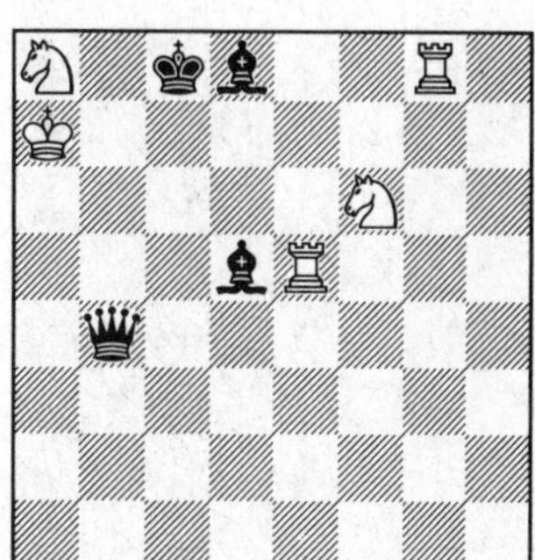

525.

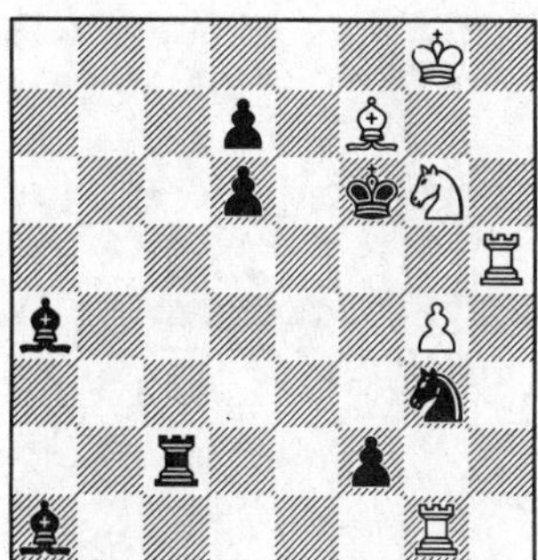

526.

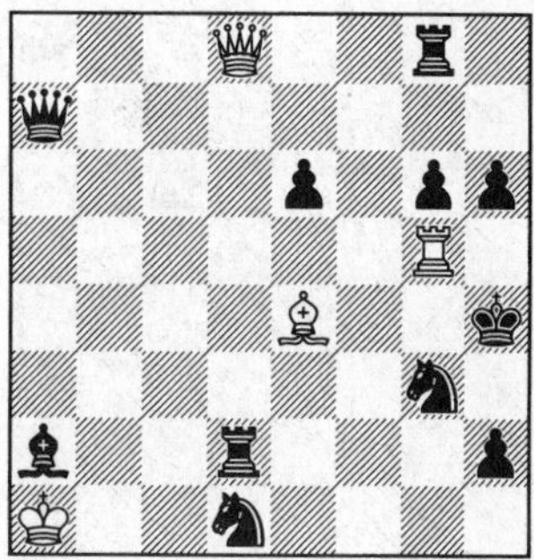

527.

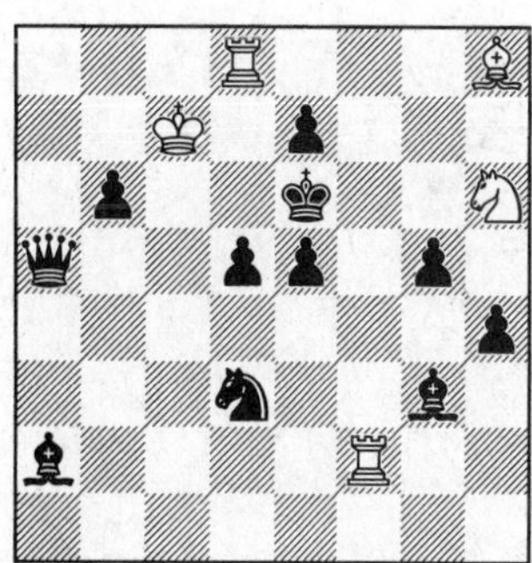

528.

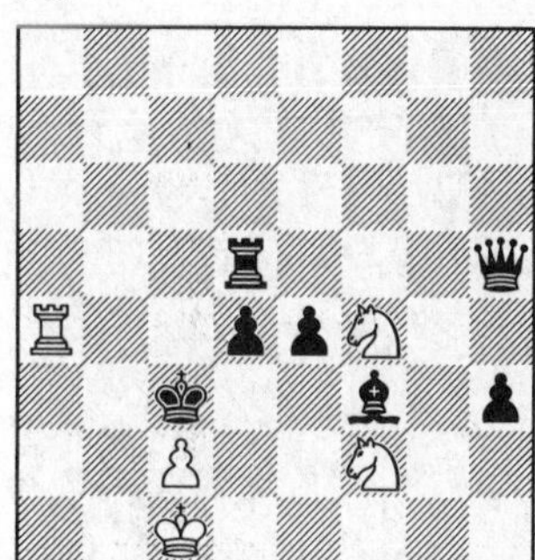

529.

530.

531.

532.

533.

534.

535.

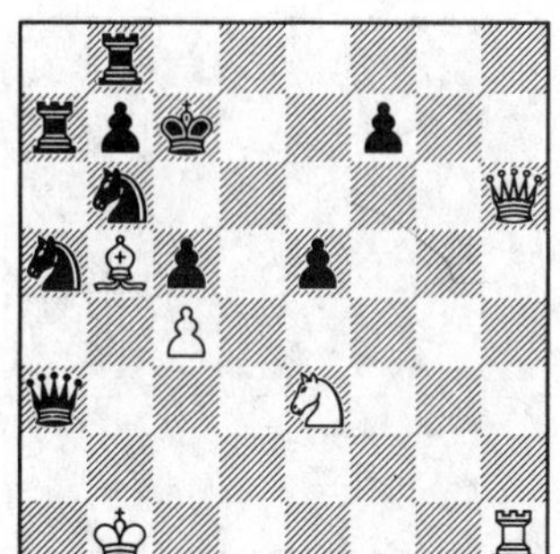

536.

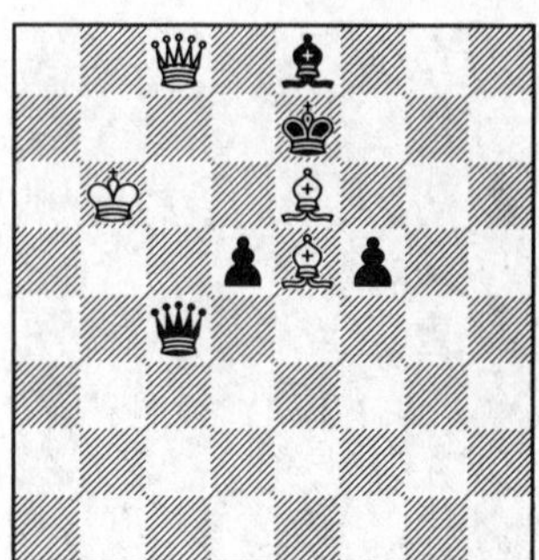

537.

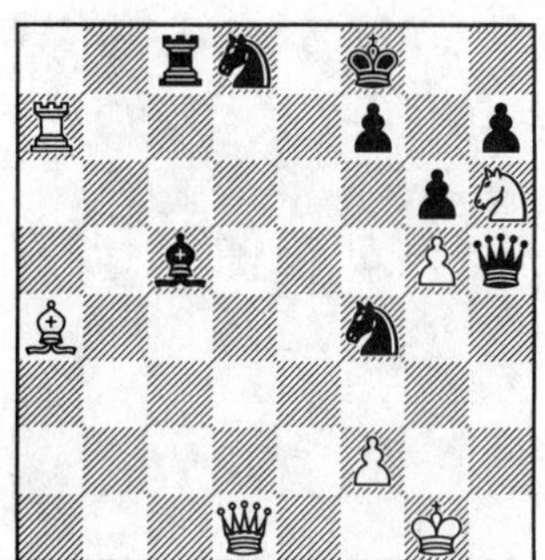

538.

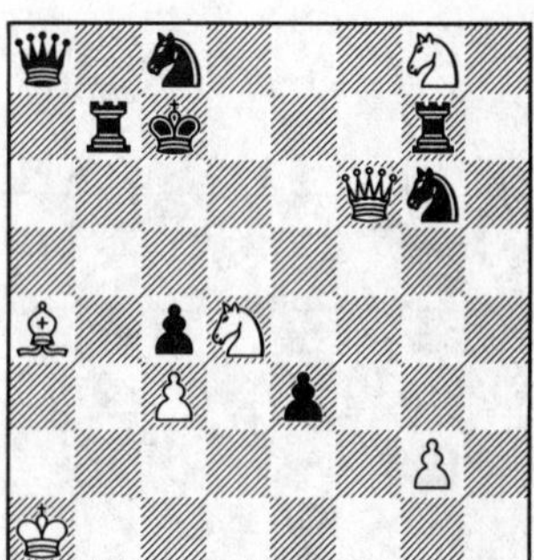

539.

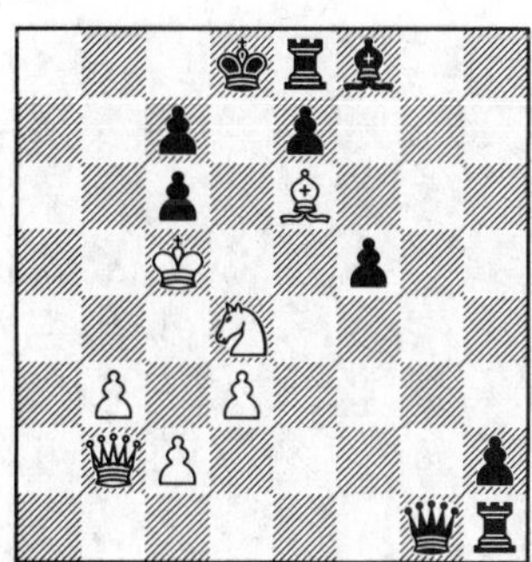

540.

541.

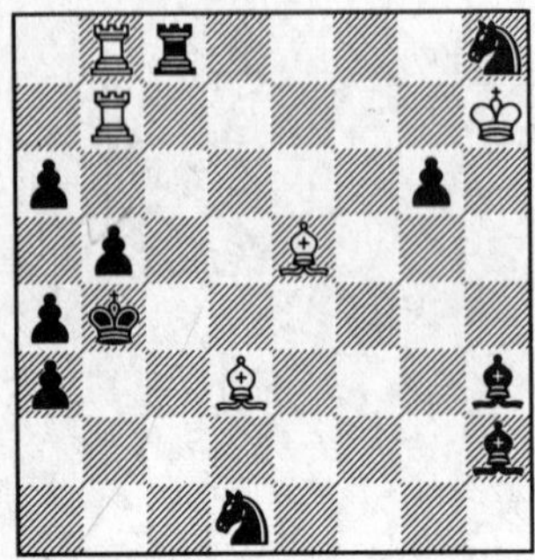

542.

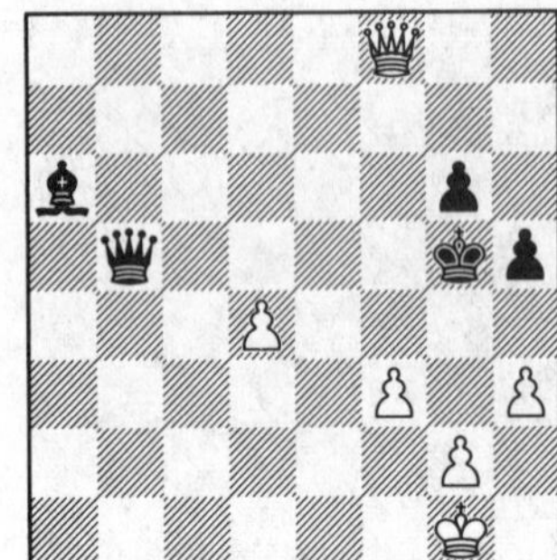

543.

544.

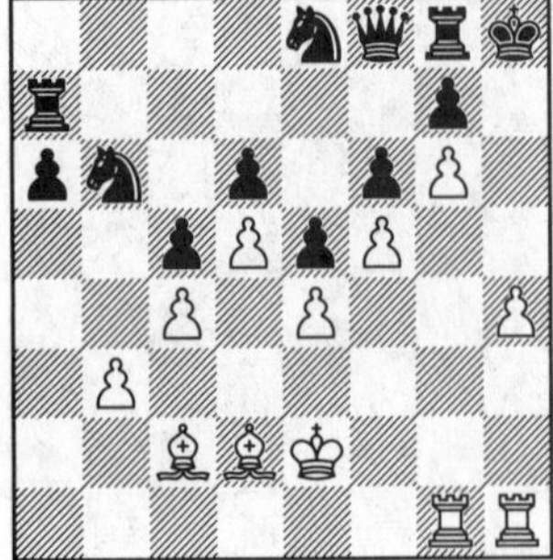

545.

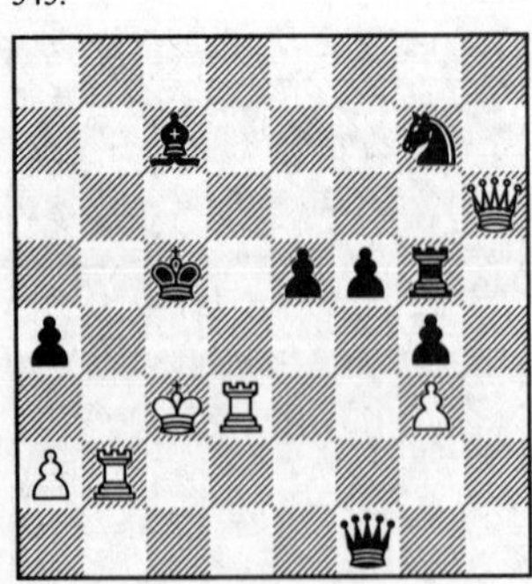

546.

547.

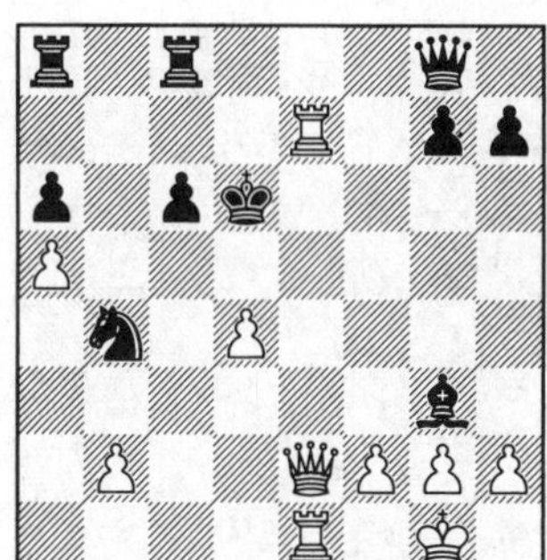

548.

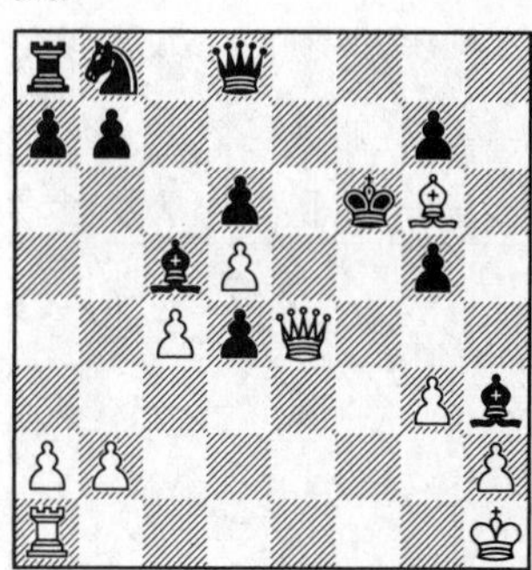

549.

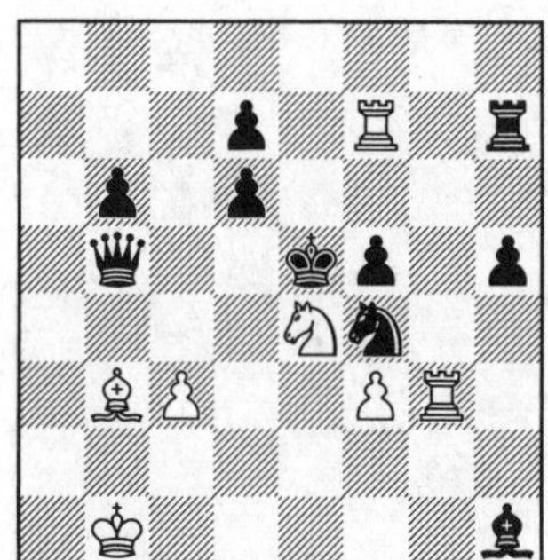

550.

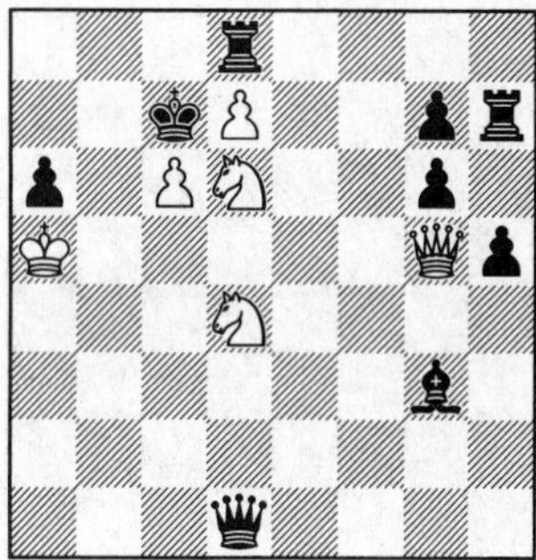

551.

552.

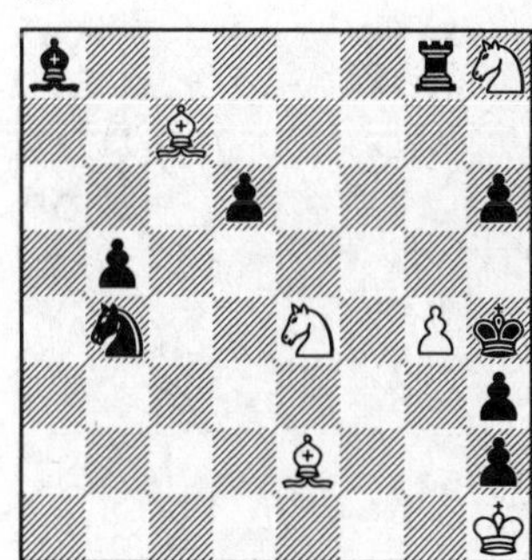

553.

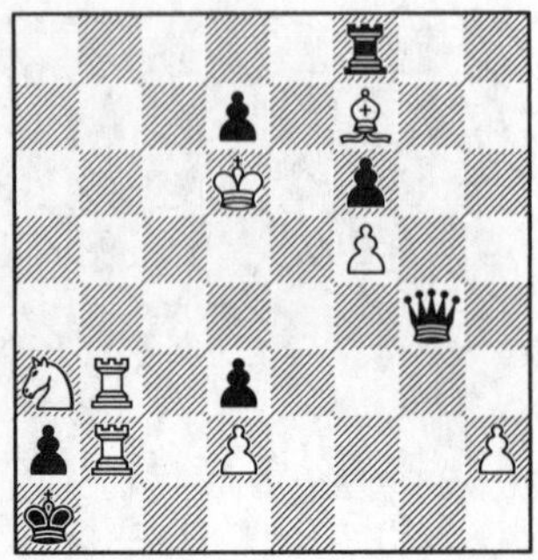

554.

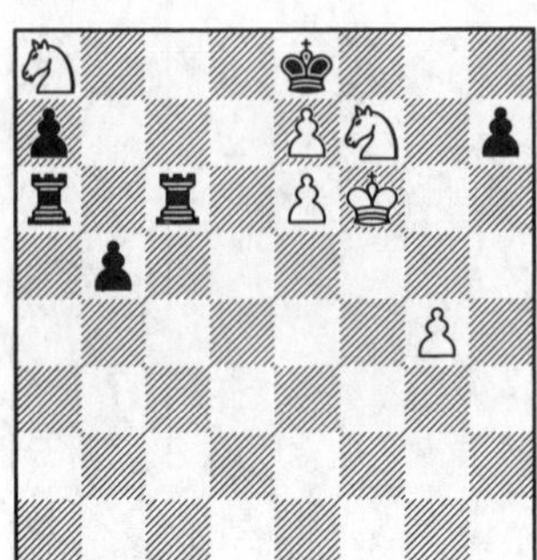

555.

556.

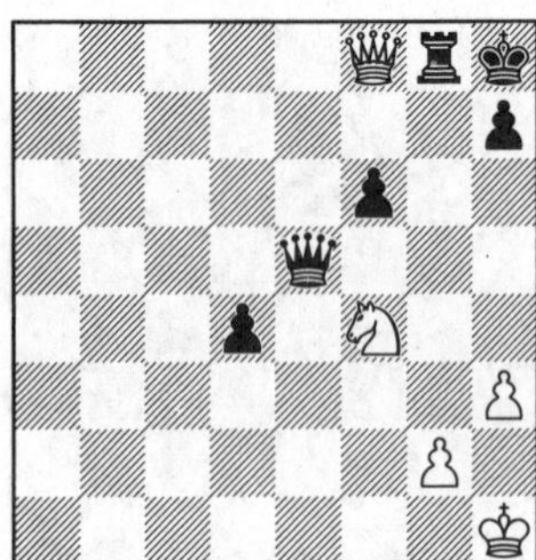

557.

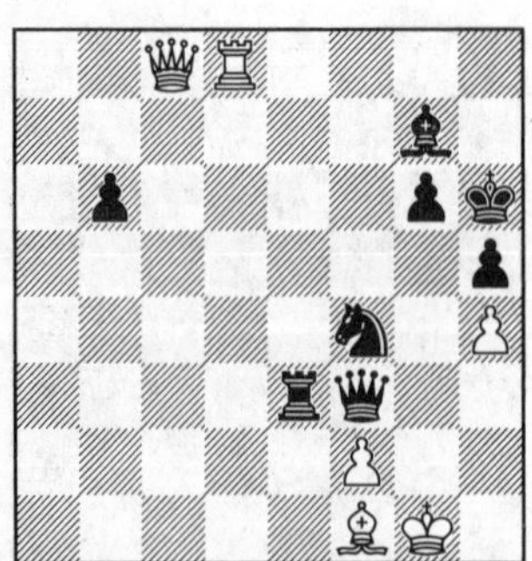

558.

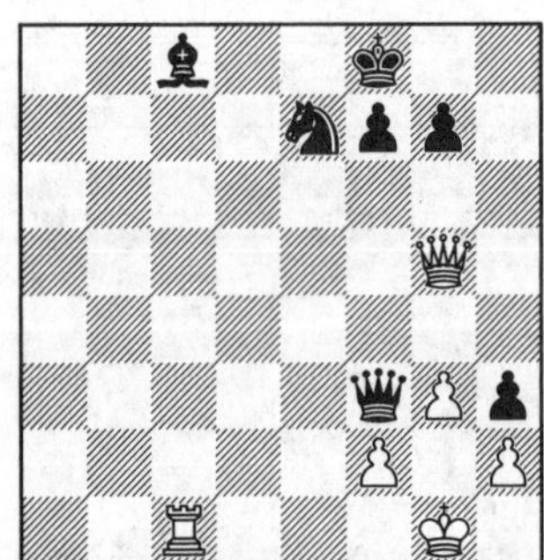

559.

560.

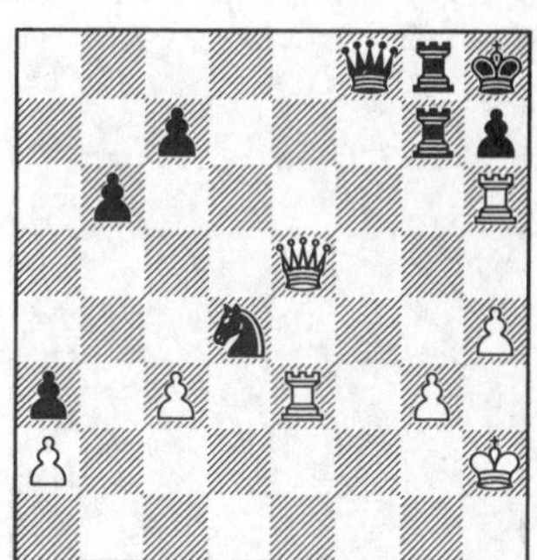

561.

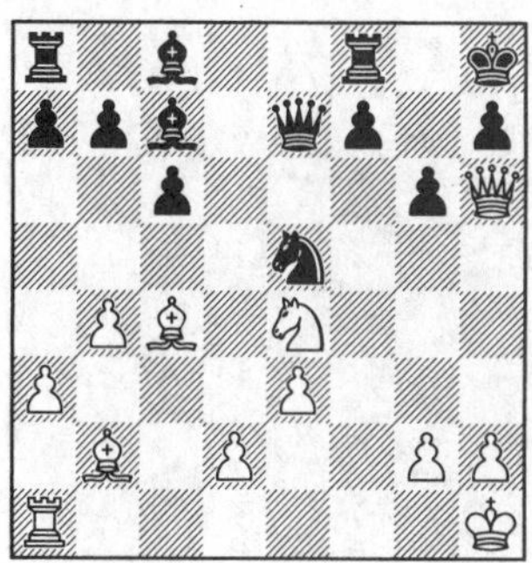

562.

563.

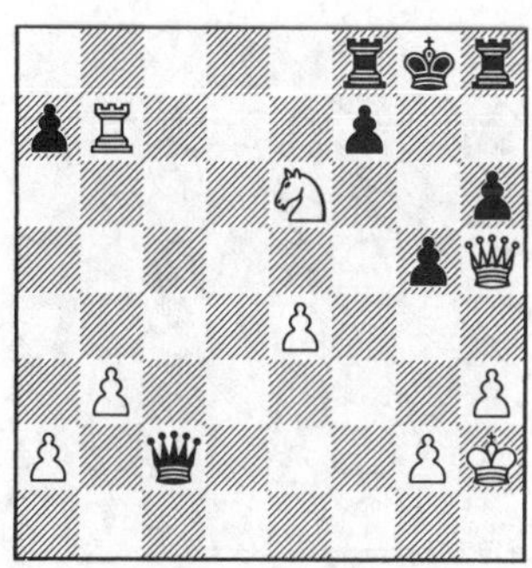

564.

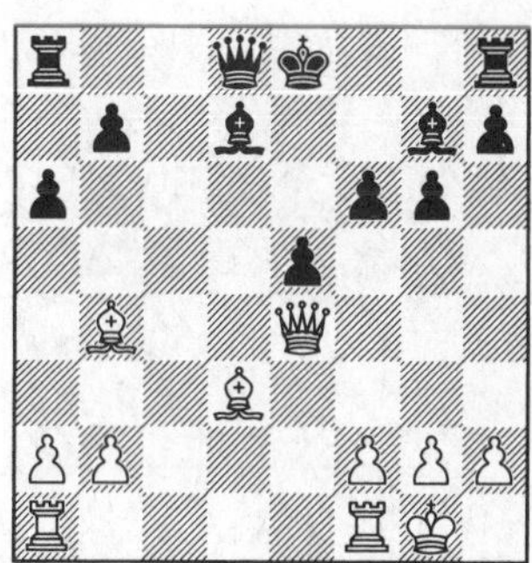

565.

566.

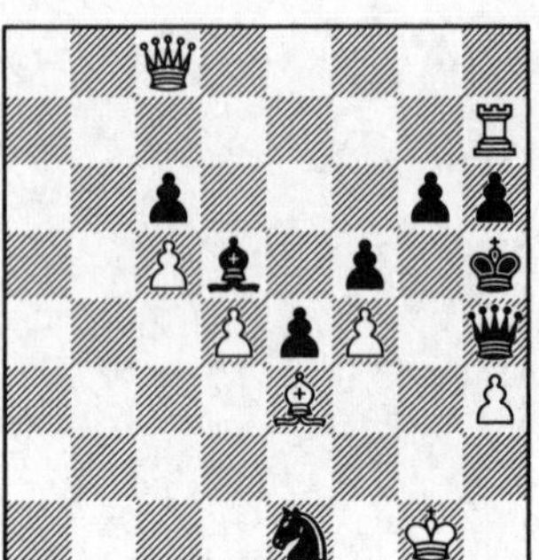

567.

568.

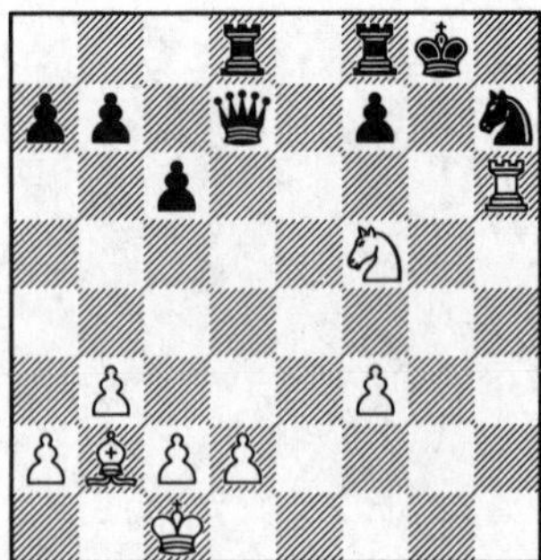

569.

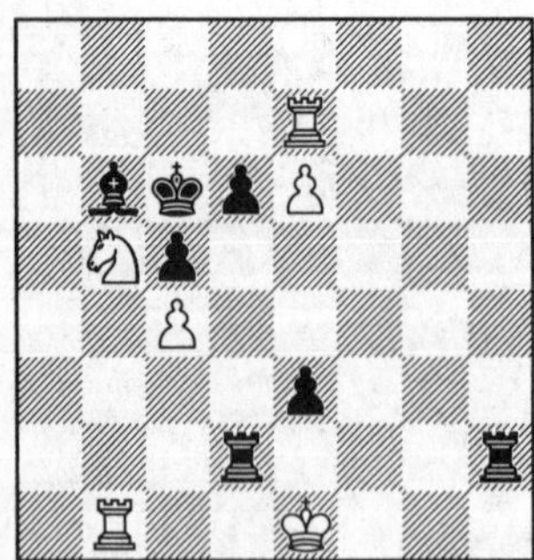

570.

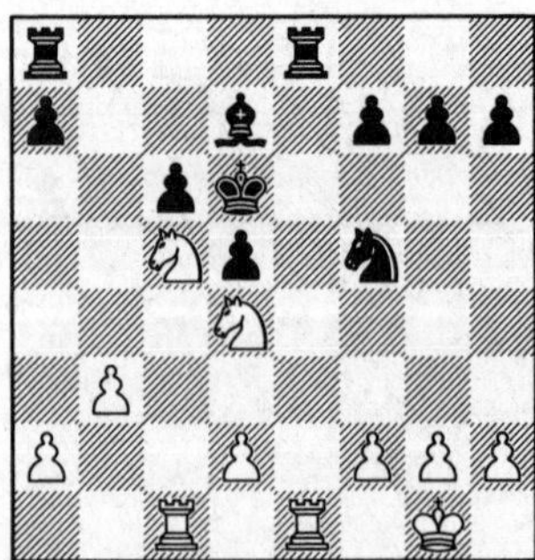

571.

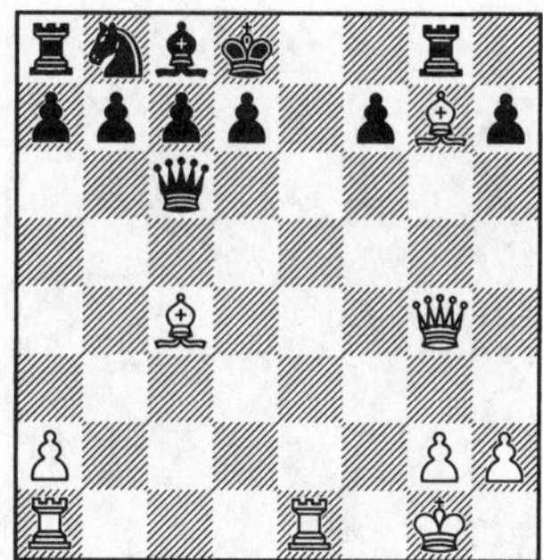

572.

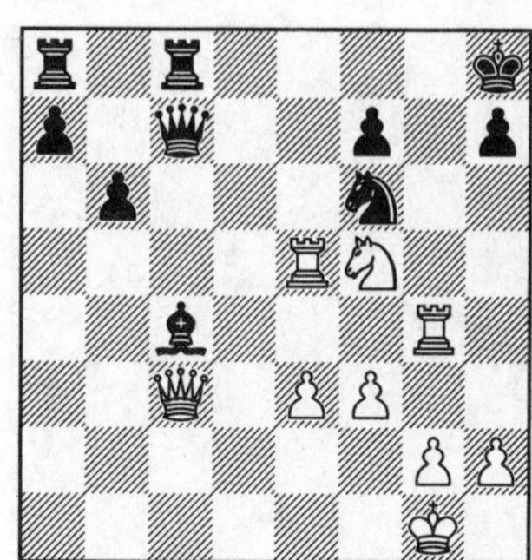

573.

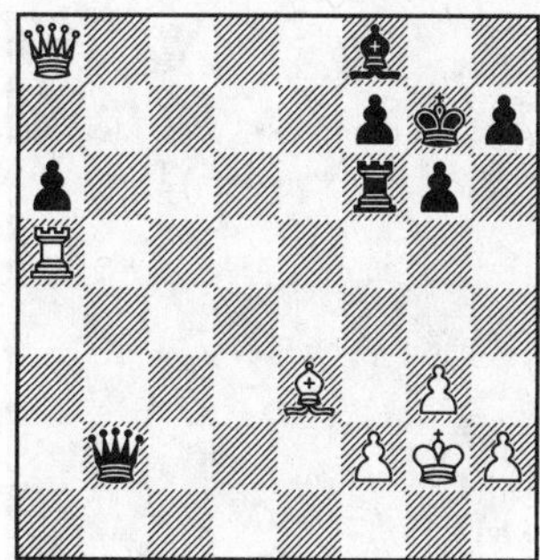

574.

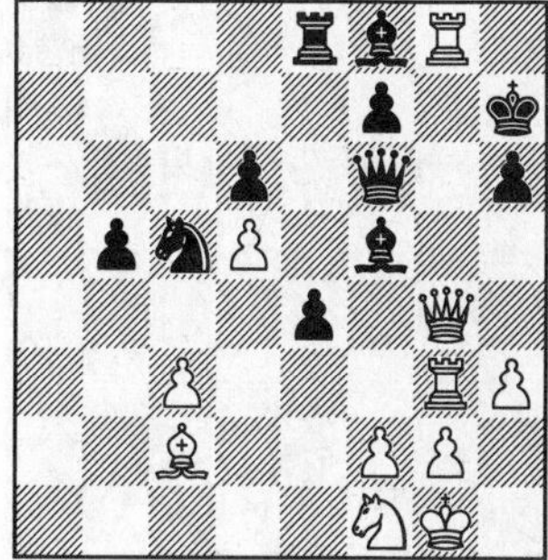

575.

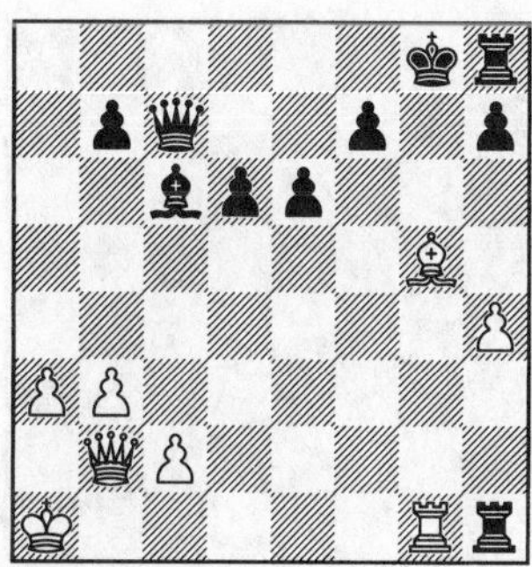

576.

577.

578.

579.

580.

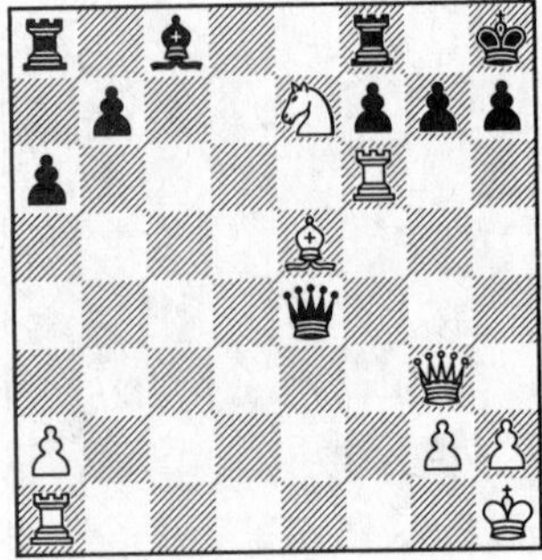

581.

582.

583.

584.

585.

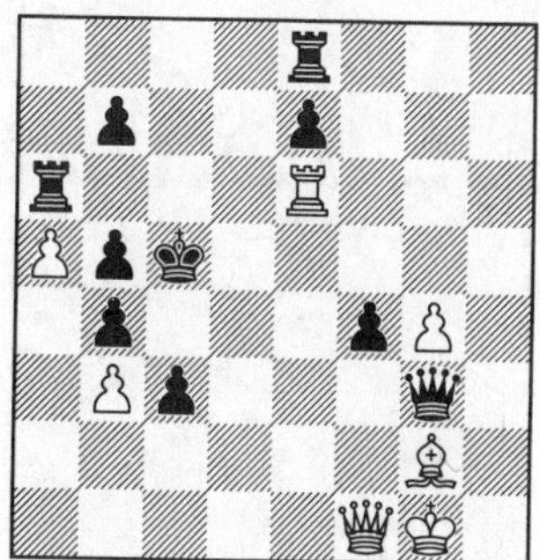

586.

587.

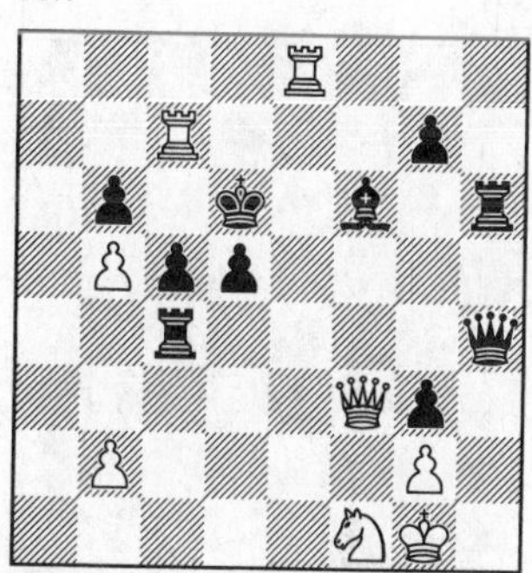

588.

589.

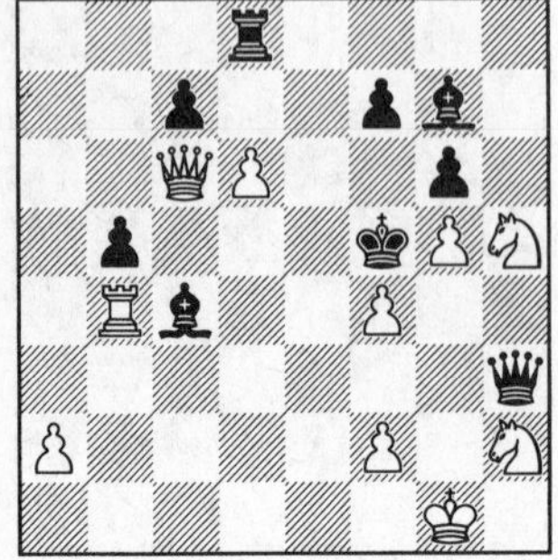

590.

591.

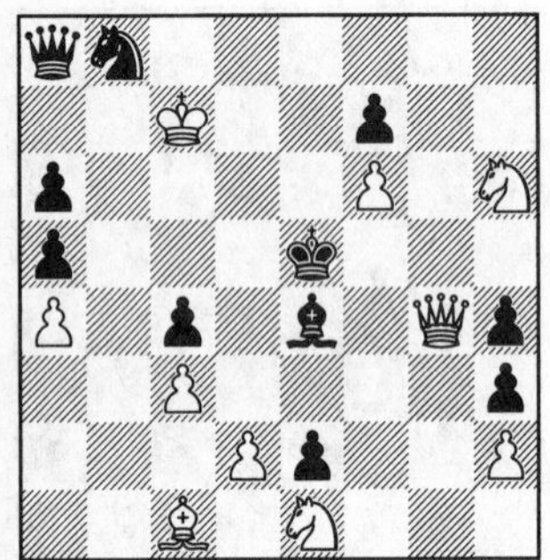

592.

593.

594.

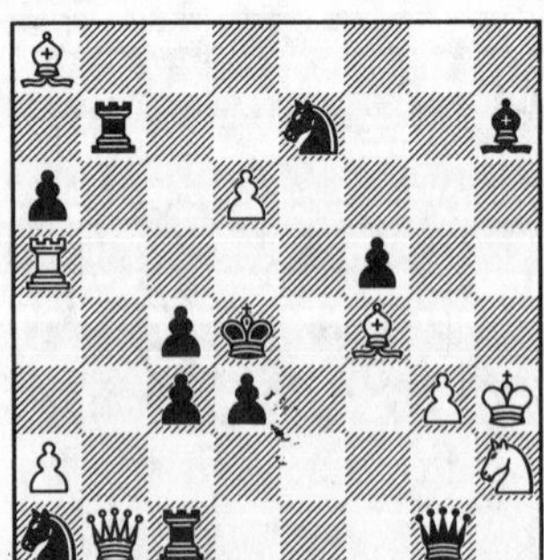

595.

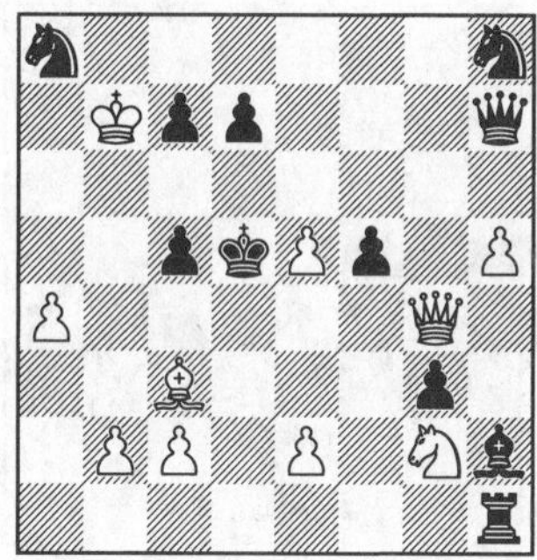

596.

597.

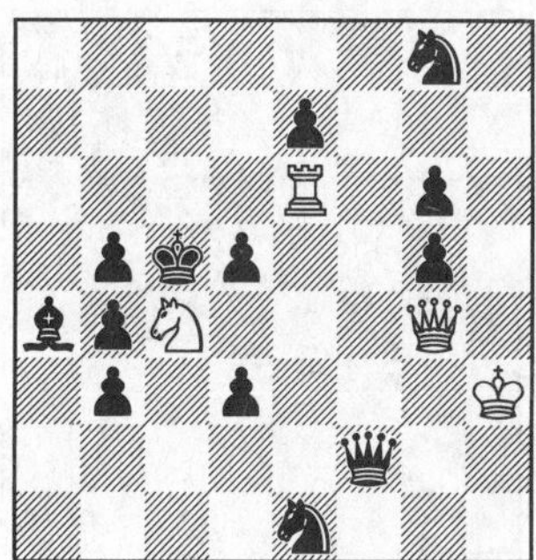

598.

599.

600.

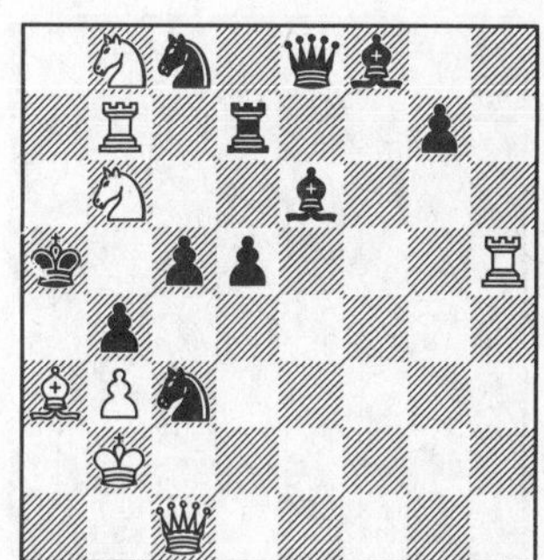

601.

602.

603.

604.

605.

606.

607.

608.

609.

610.

611.

612.

613.

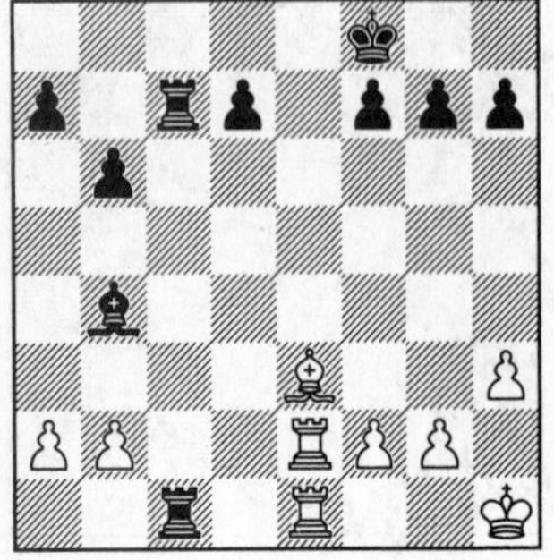

614.

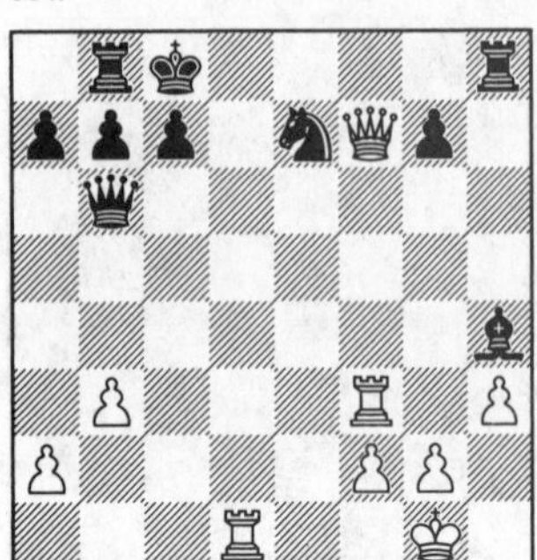

615.

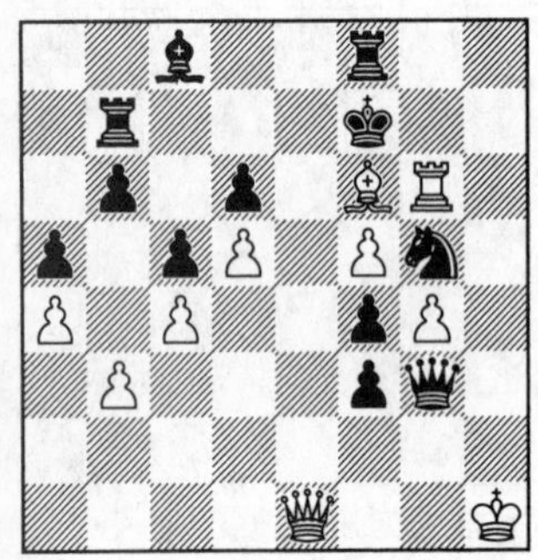

616.

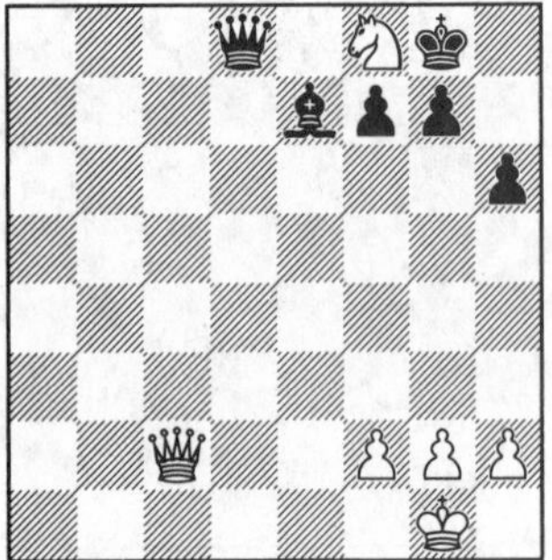

617.

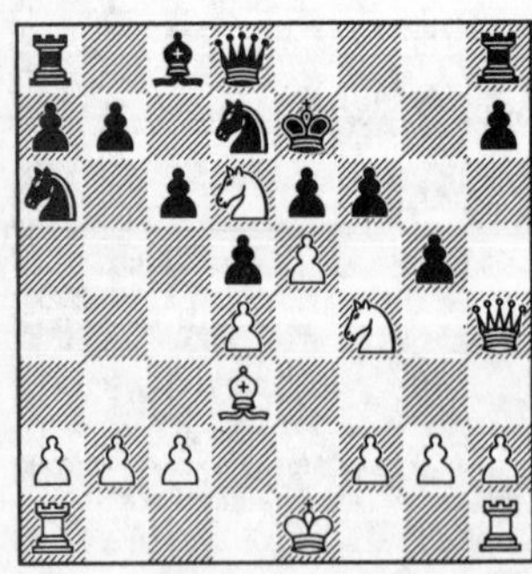

618.

619.

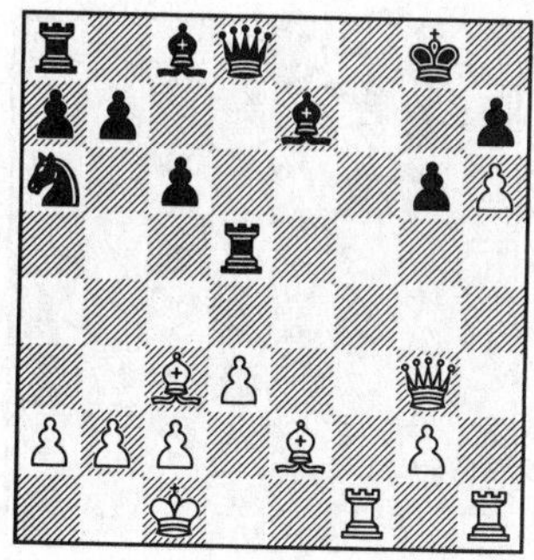

620.

621.

622.

623.

624.

625.

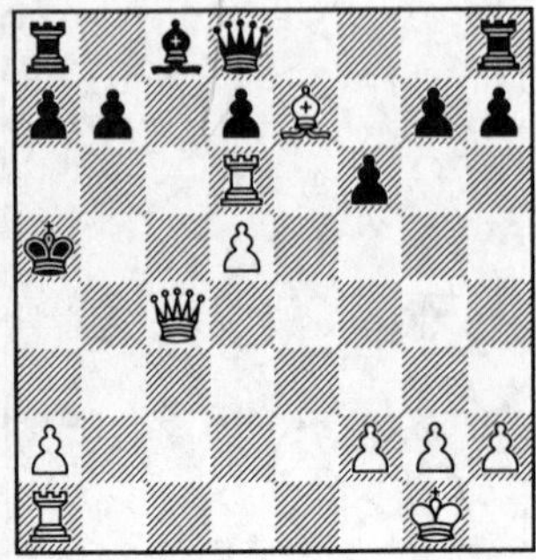

626.

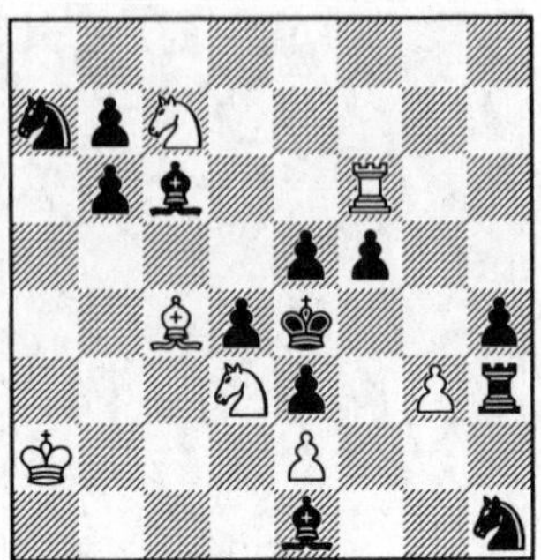

627.

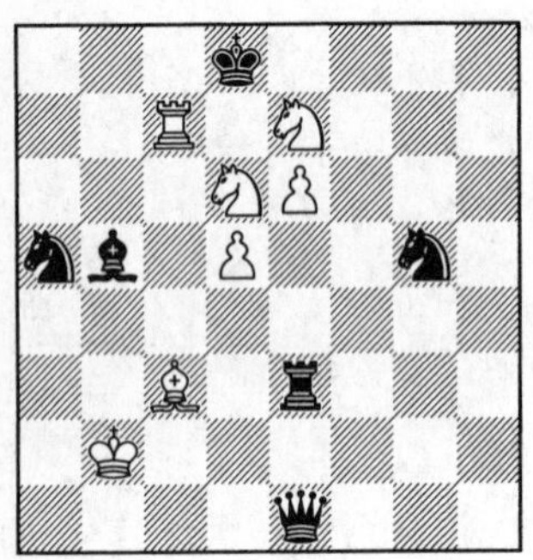

628.

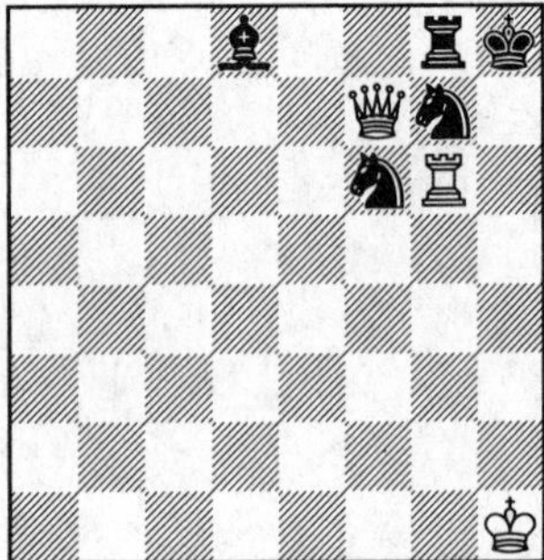

629.

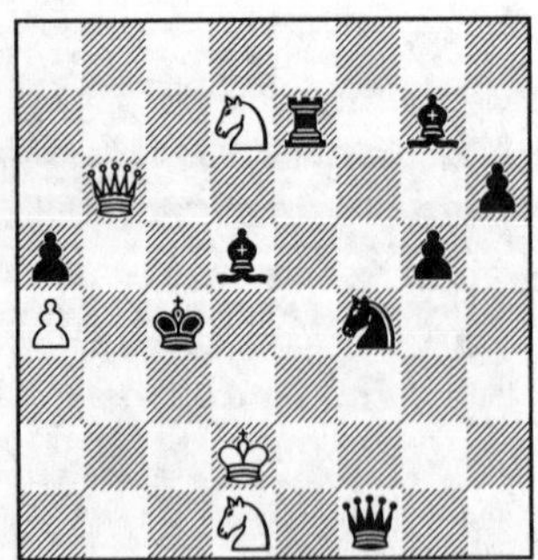

630.

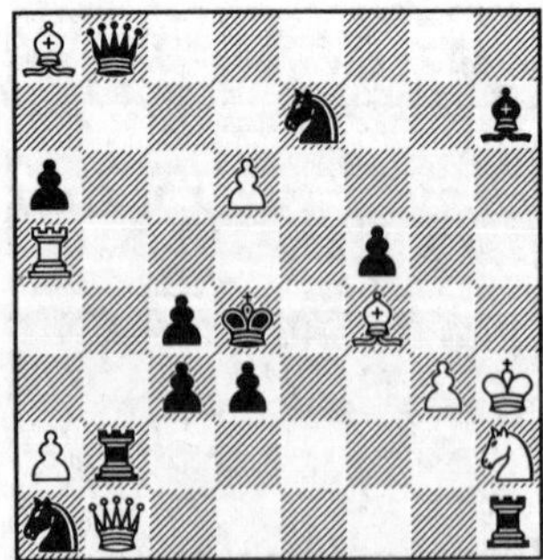

631.

632.

633.

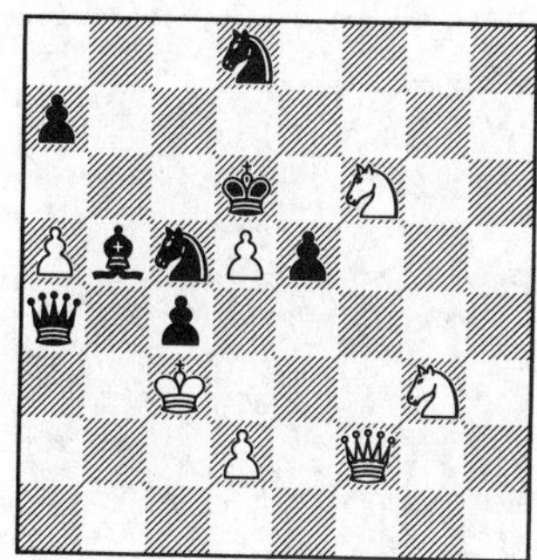

634.

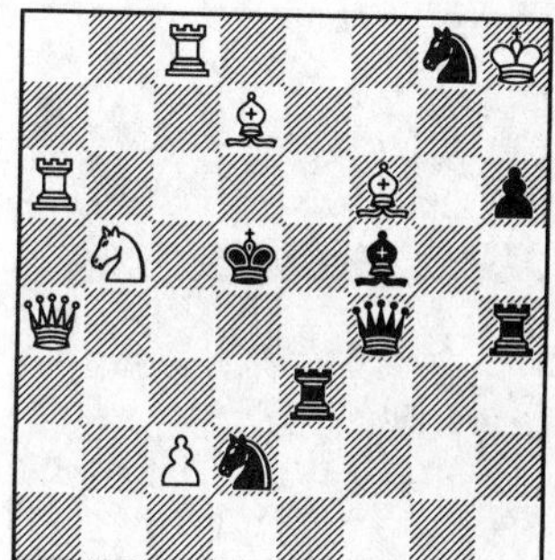

635.

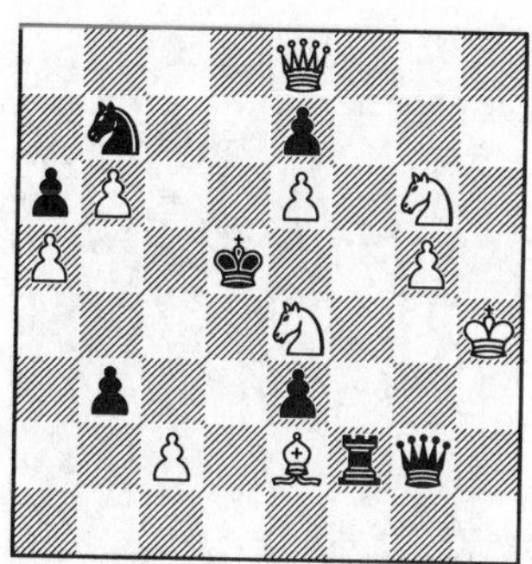

636.

637.

638.

639.

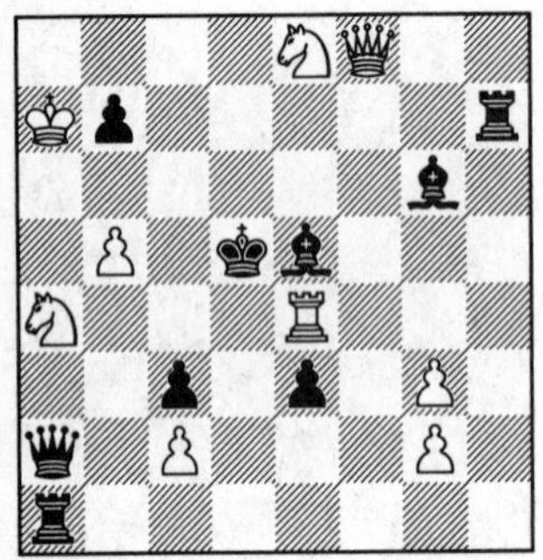

640.

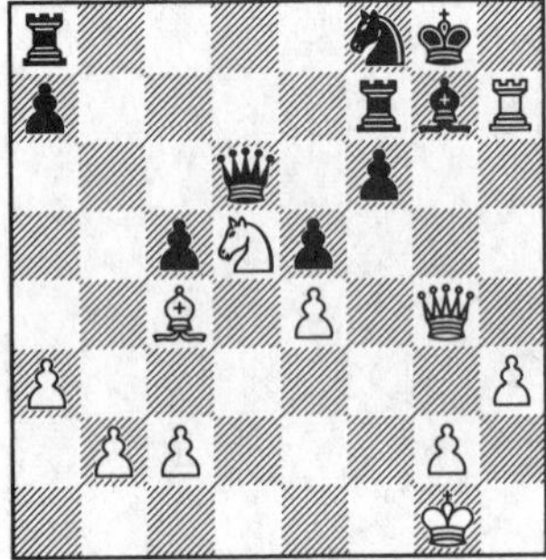

641.

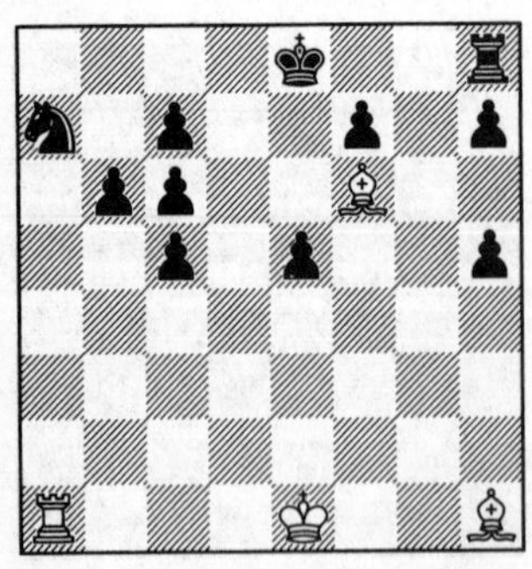

642.

643.

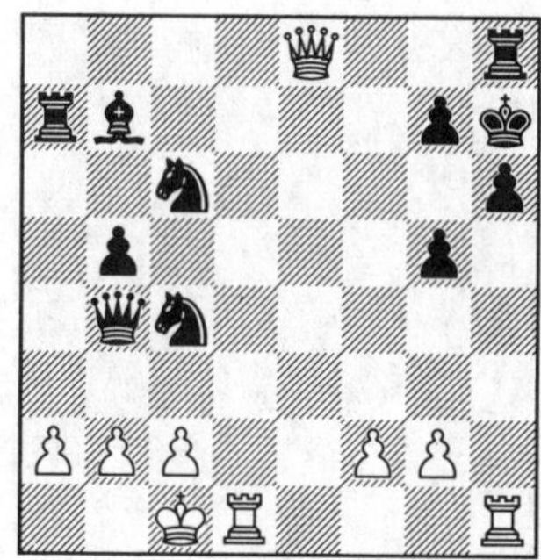

644.

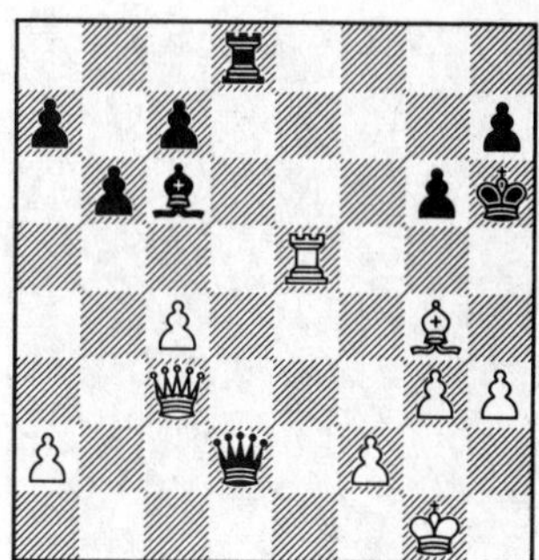

645.

646.

647.

648.

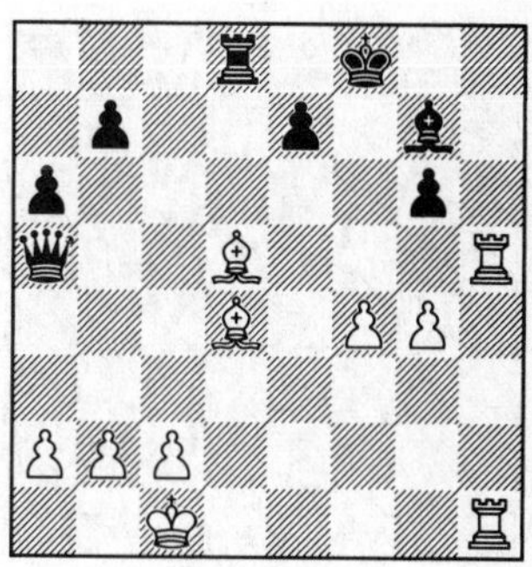

649.

650.

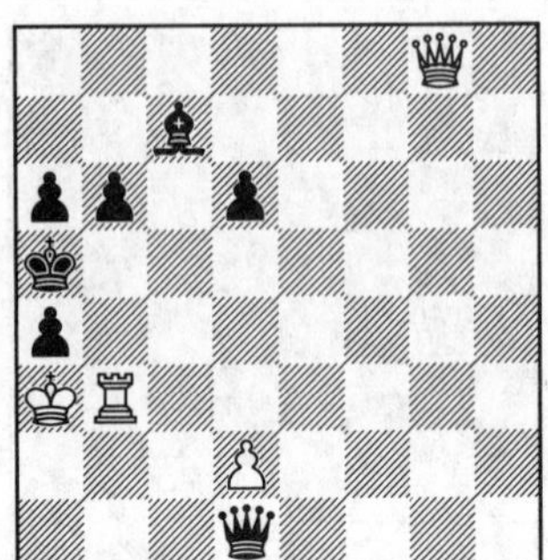

651.

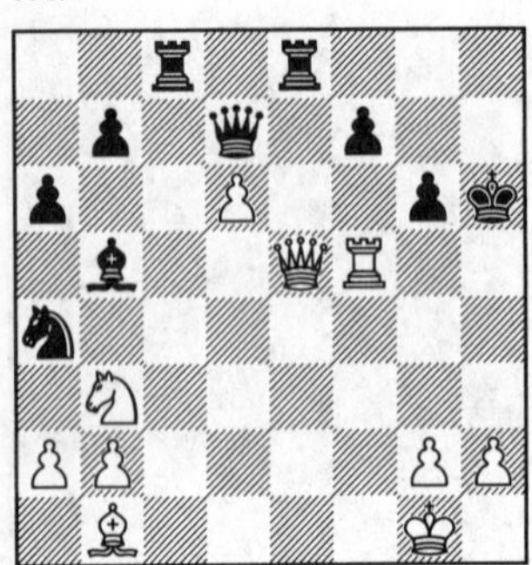

652.

653.

654.

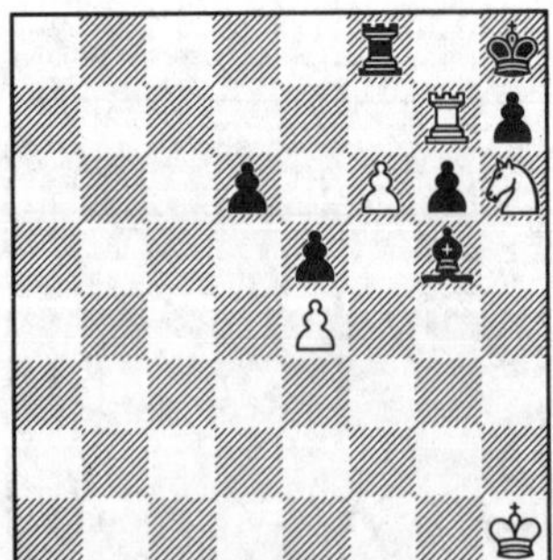

655.

656.

657.

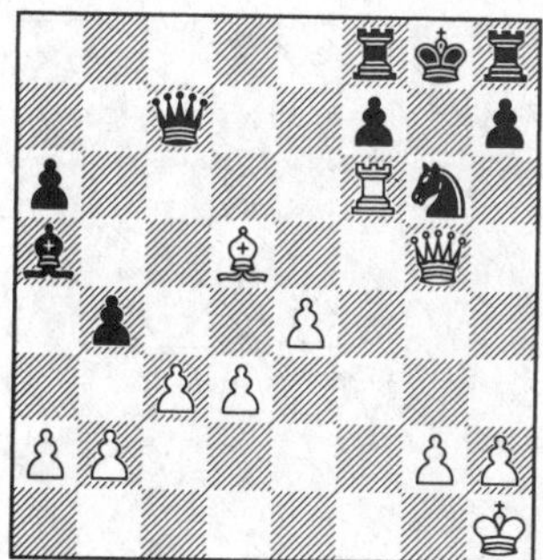

658.

659.

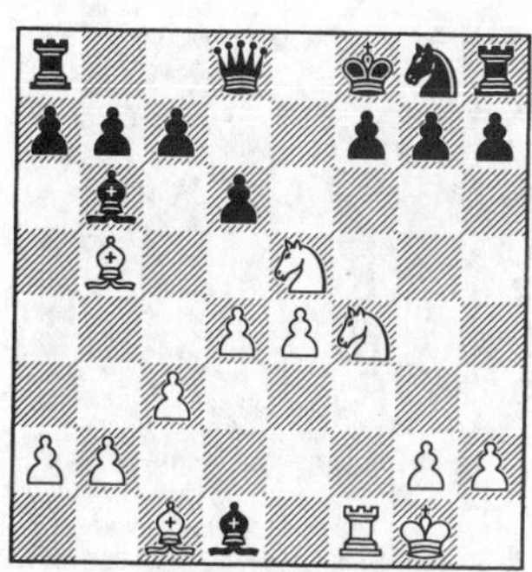

660.

661.

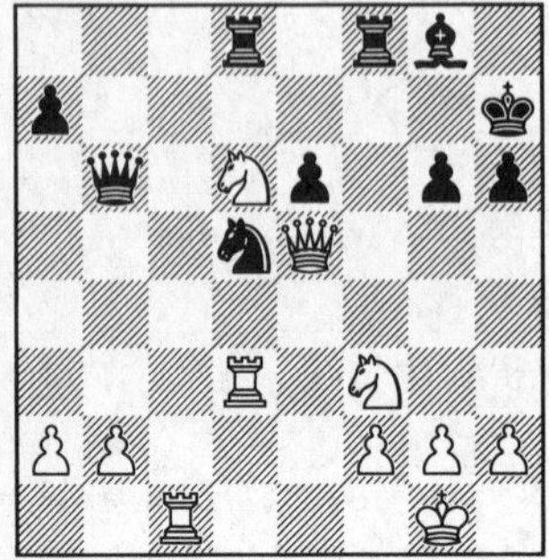

662.

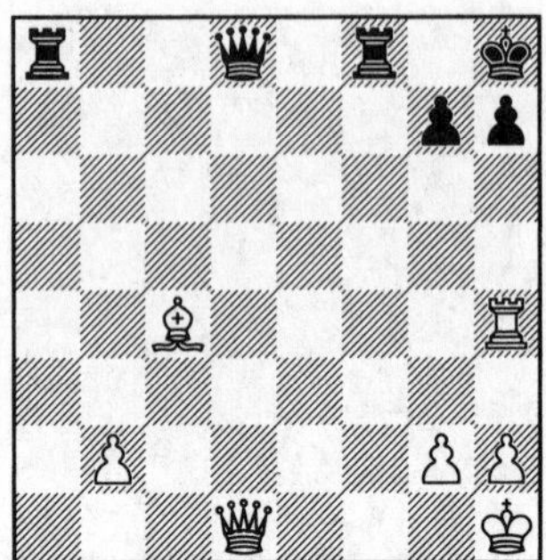

663.

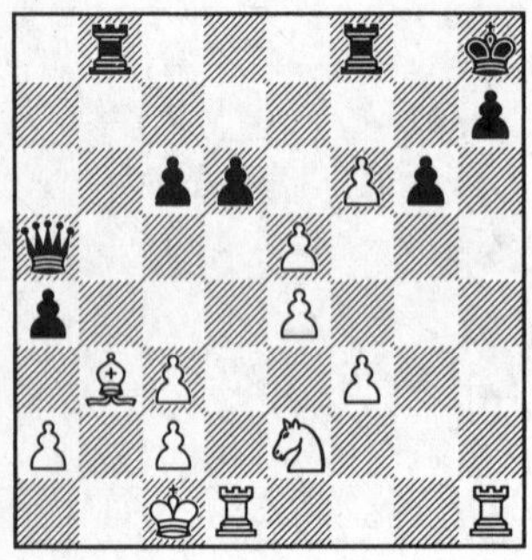

664.

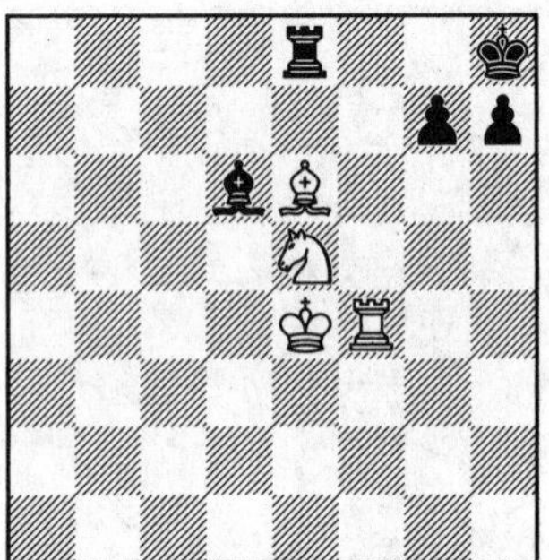

665.

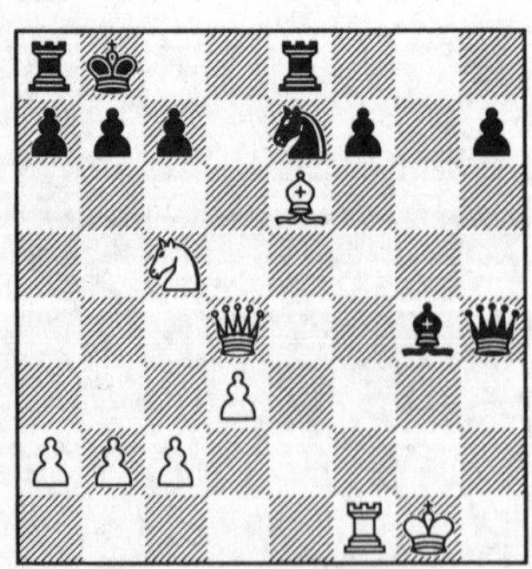

666.

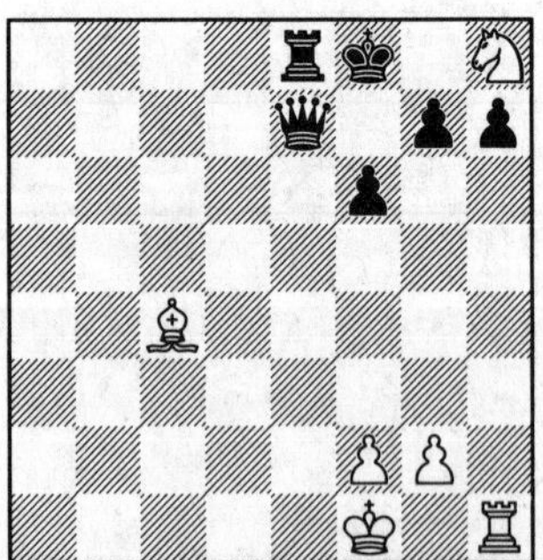

667.

668.

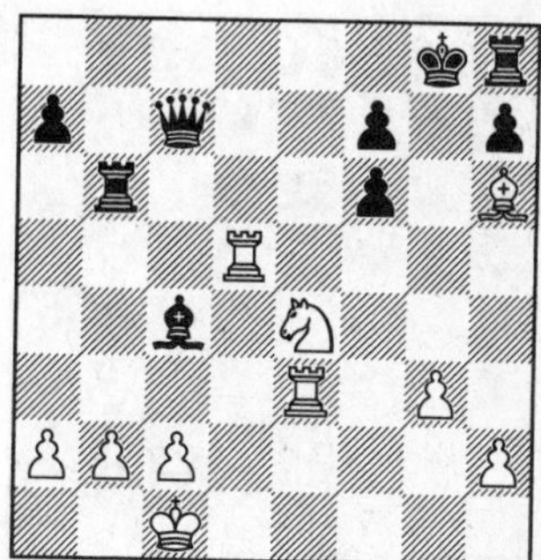

669.

670.

671.

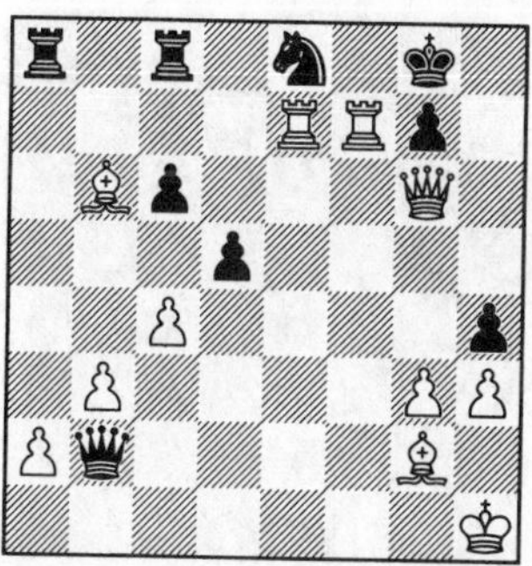

672.

673.

674.

675.

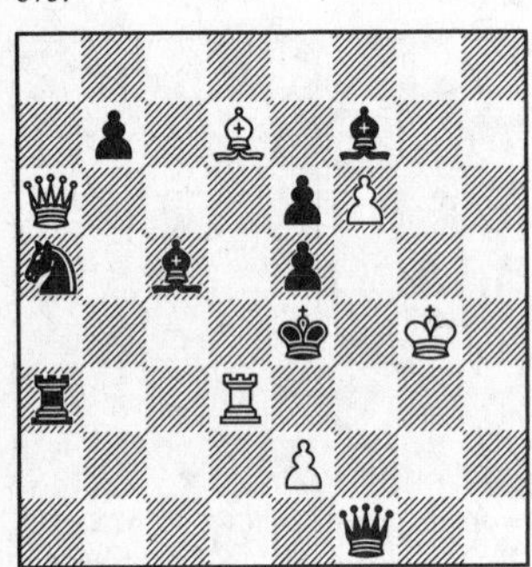

676.

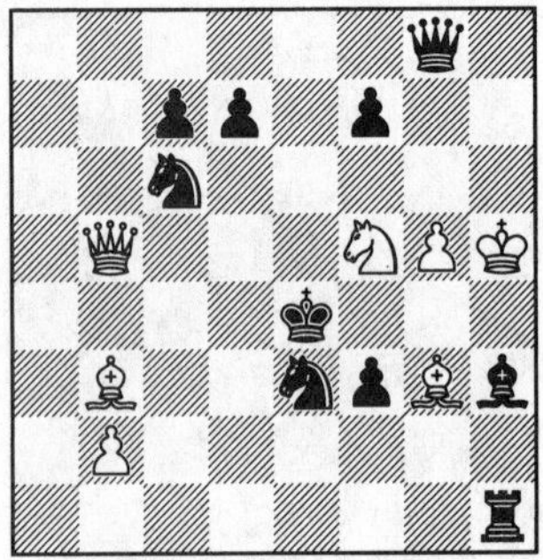

677.

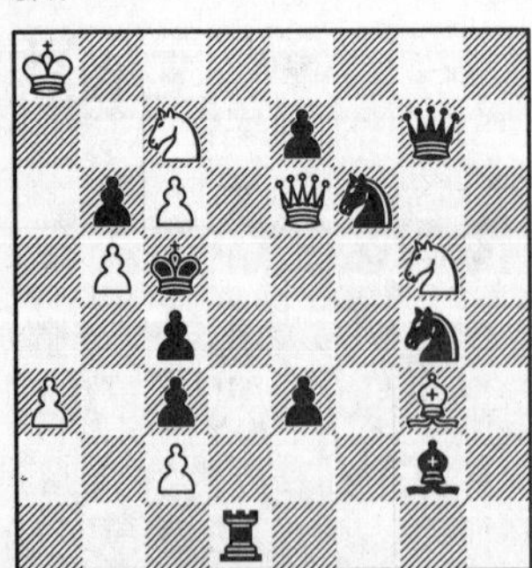

678.

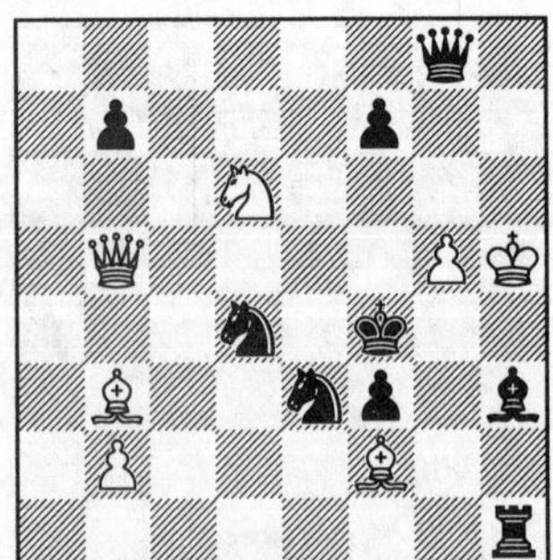

679.

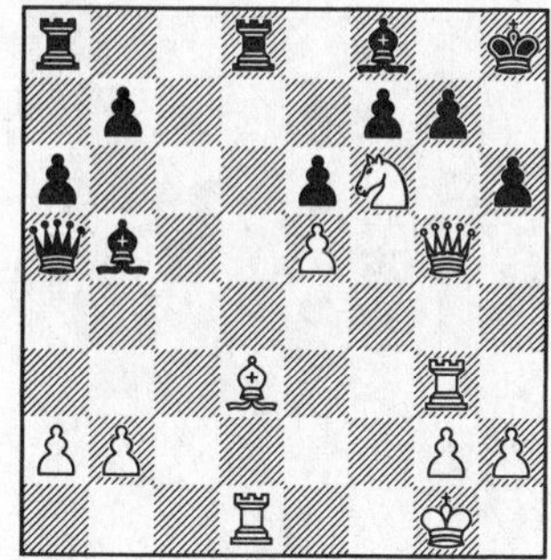

680.

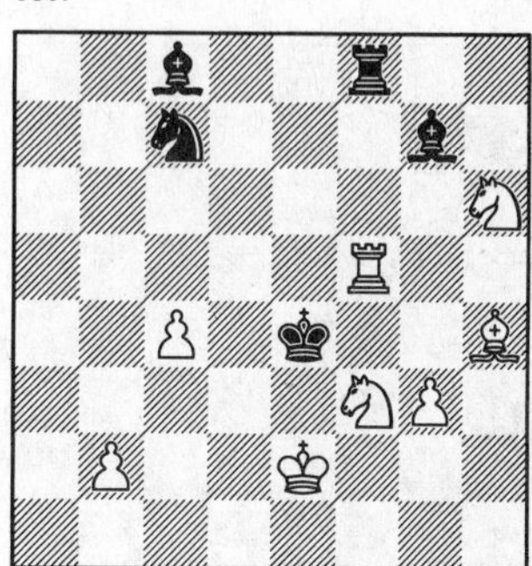

681.

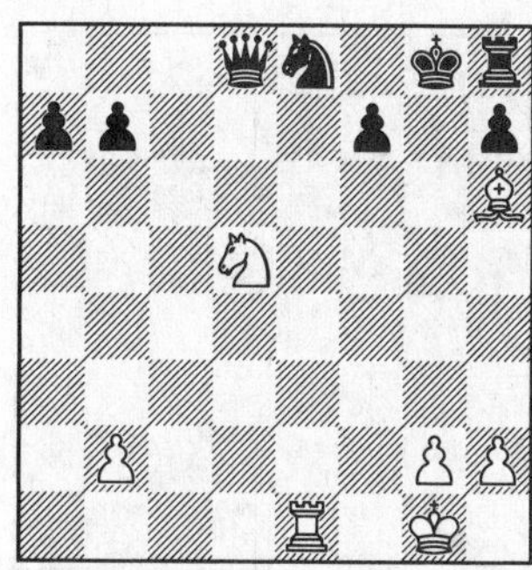

682.

683.

684.

685.

686.

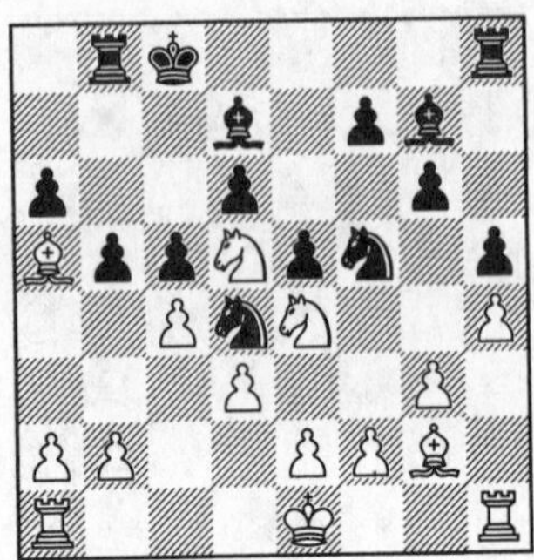

687.

688.

689.

690.

691.

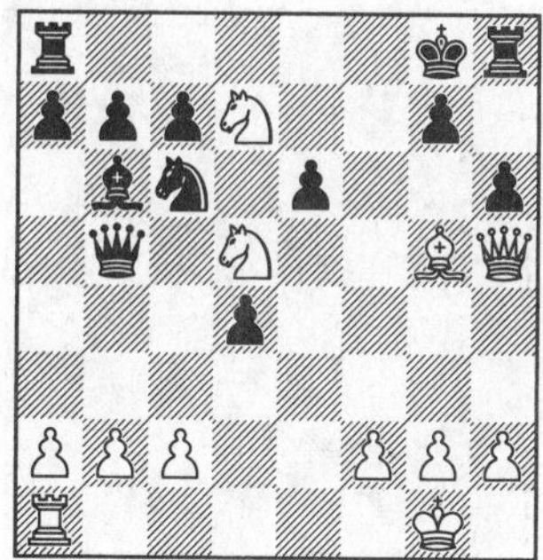

692.

693.

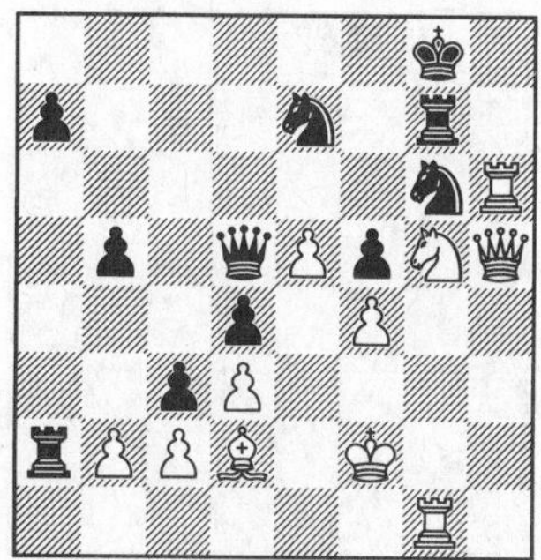

694.

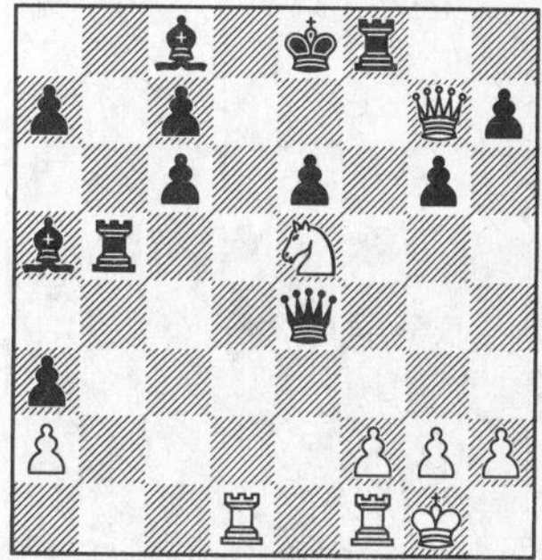

695.

696.

697.

698.

699.

700.

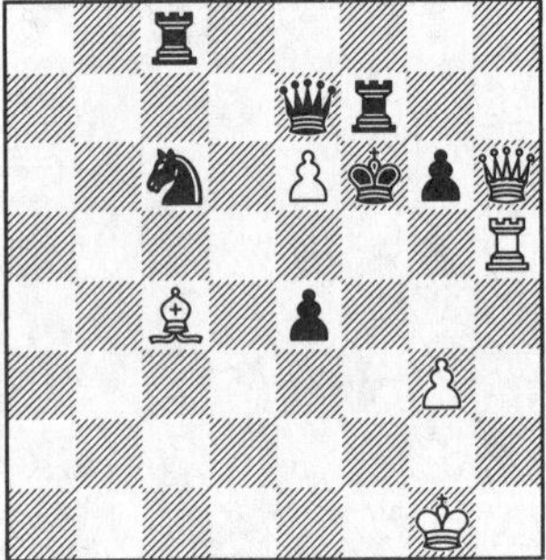

701.

702.

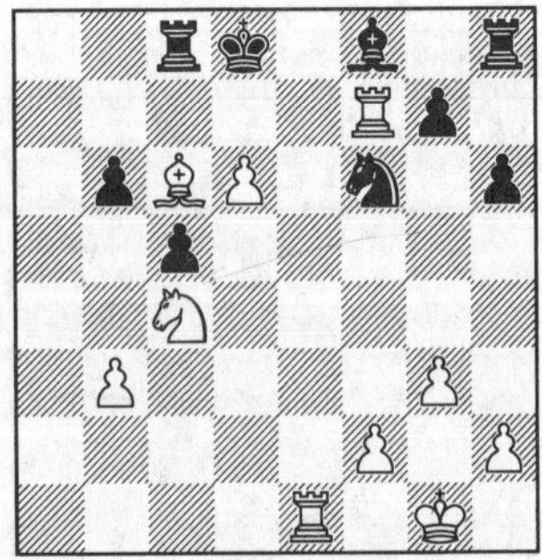

703.

704.

705.

706.

707.

708.

709.

710.

711.

712.

713.

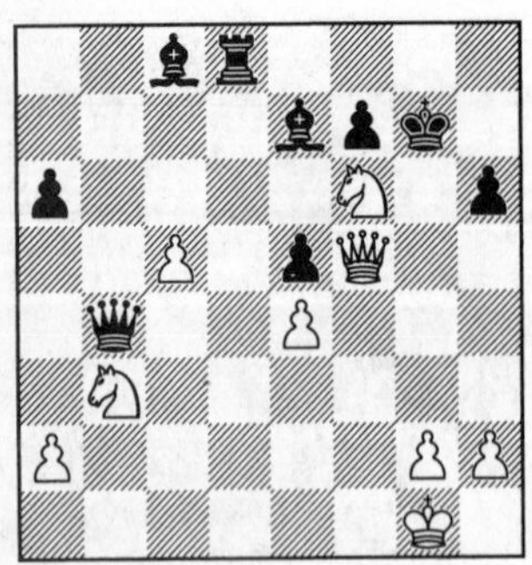

714.

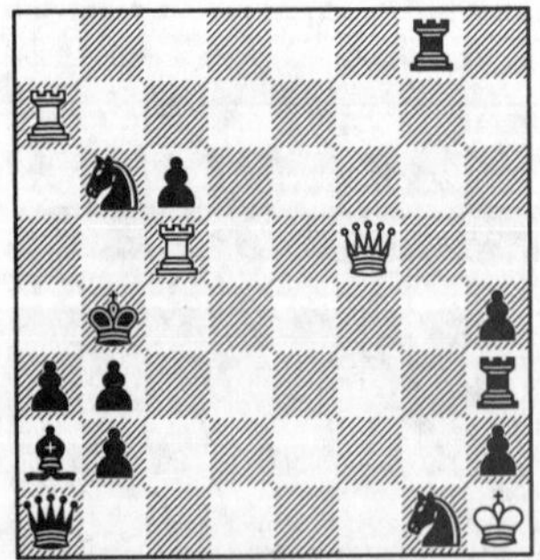

715.

716.

717.

718.

719.

720.

721.

722.

723.

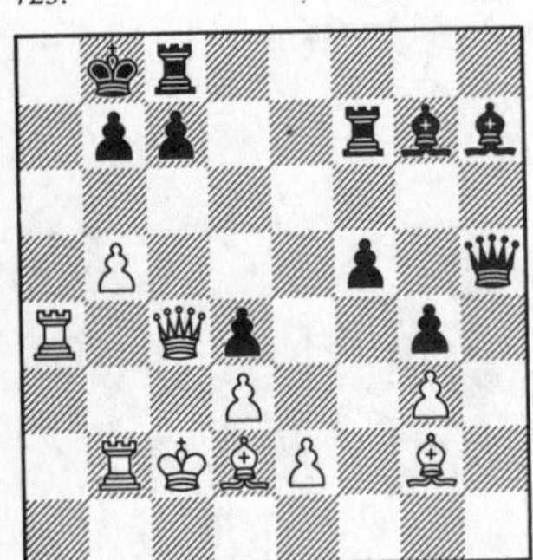

724.

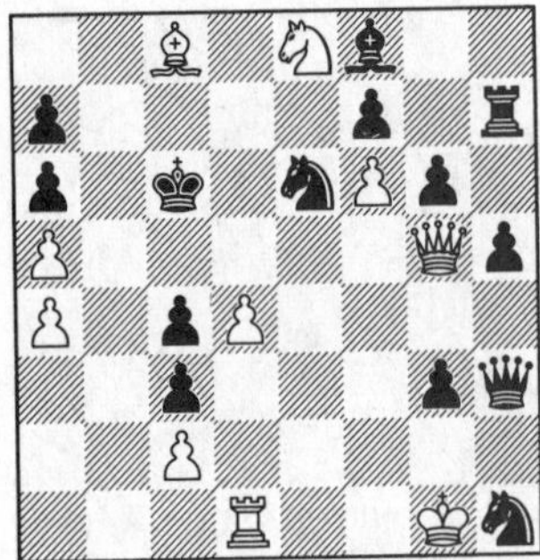

725.

726.

727.

728.

729.

730.

731.

732.

733.

734.

735.

736.

737.

AW
738.

739.

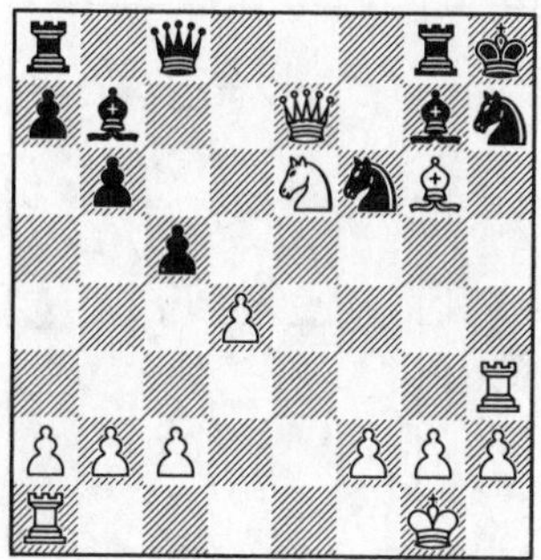

740.

741.

742.

743.

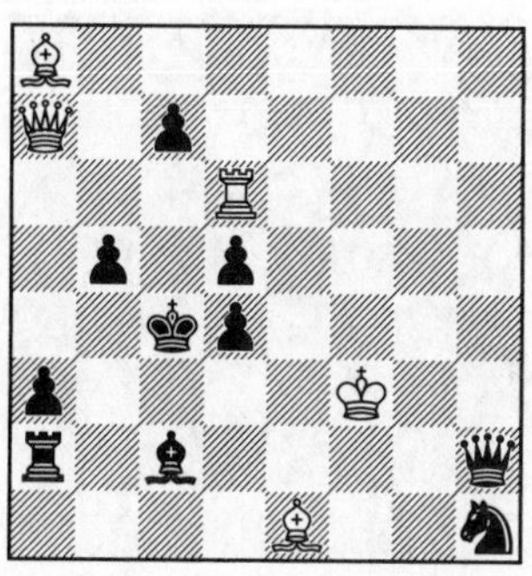

744.

745.

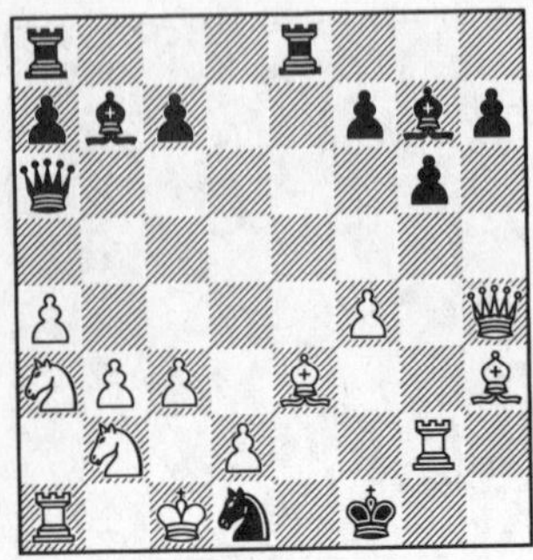

746.

747.

748.

749.

750.

751.

752.

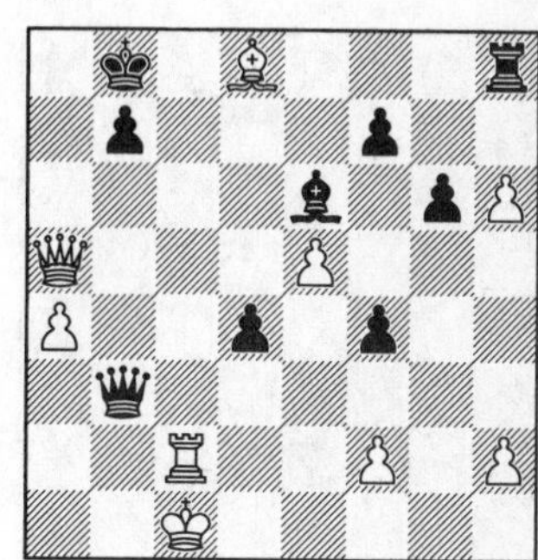

753.

754.

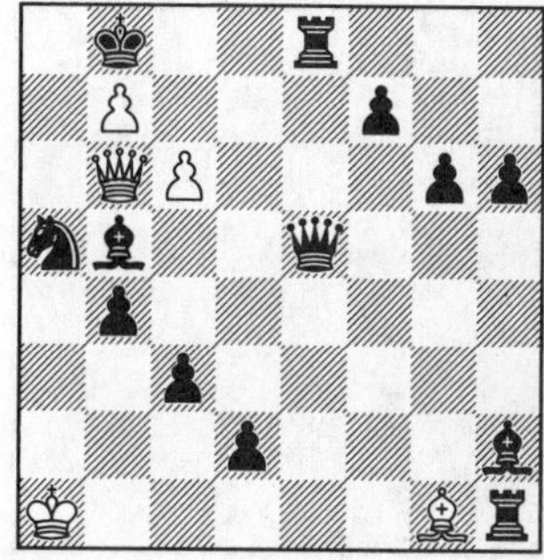

755.

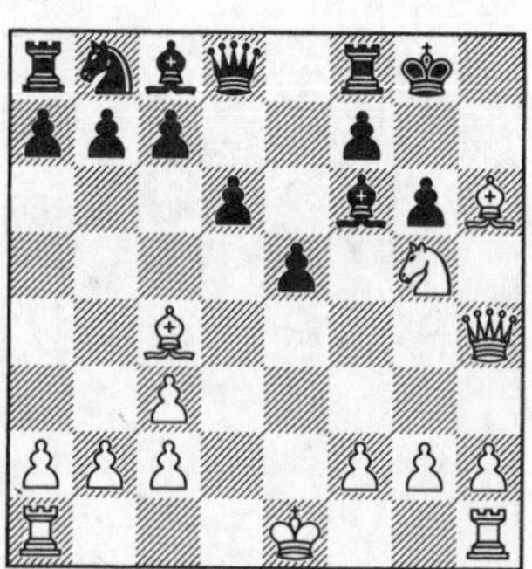

756.

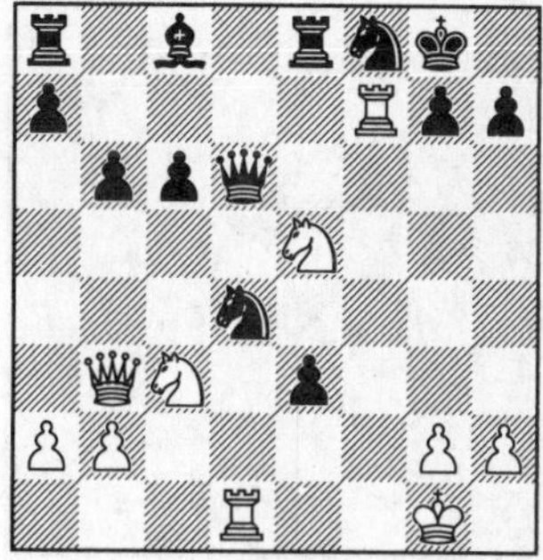

757.

758.

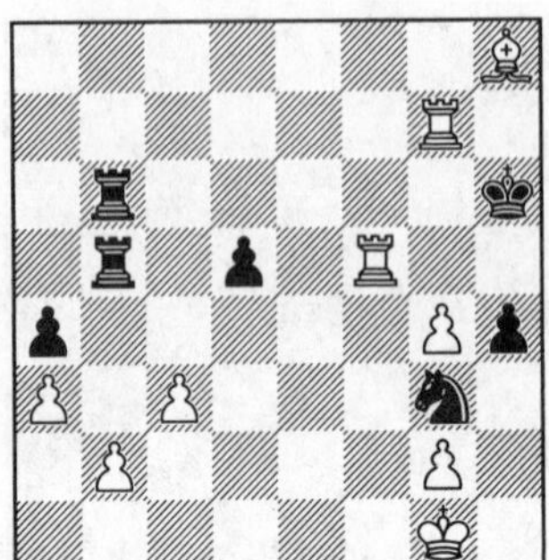

759.

760.

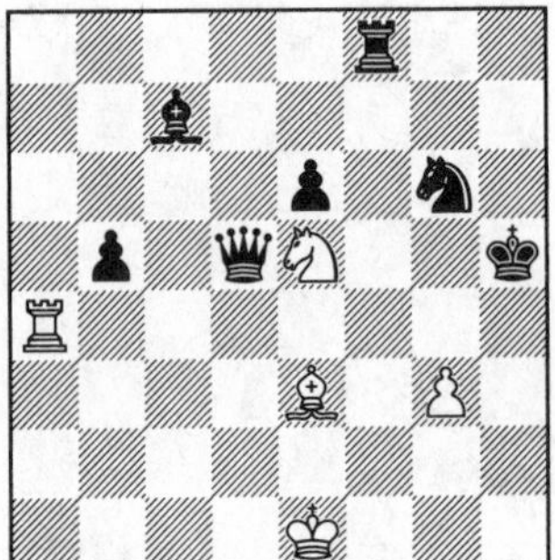

761.

762.

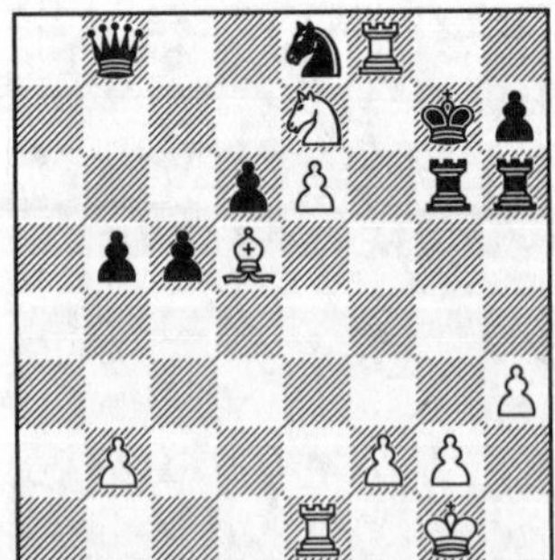

763.

764.

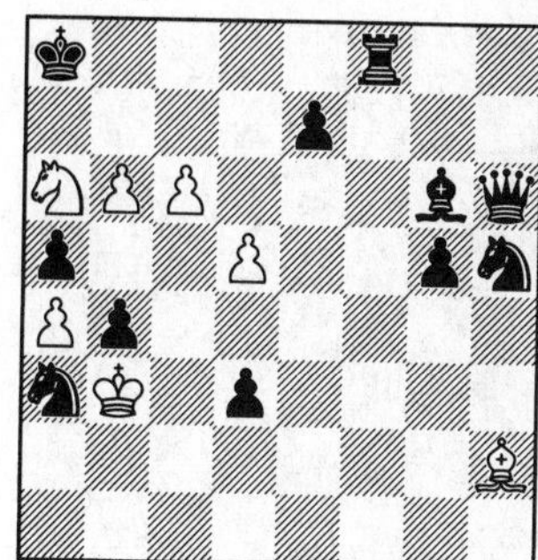

765.

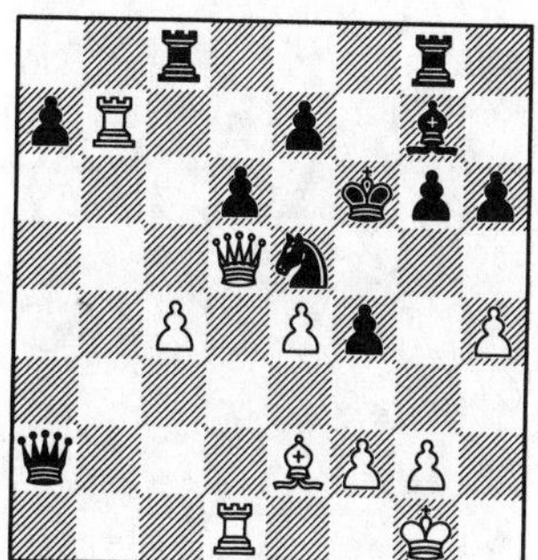

766.

767.

768.

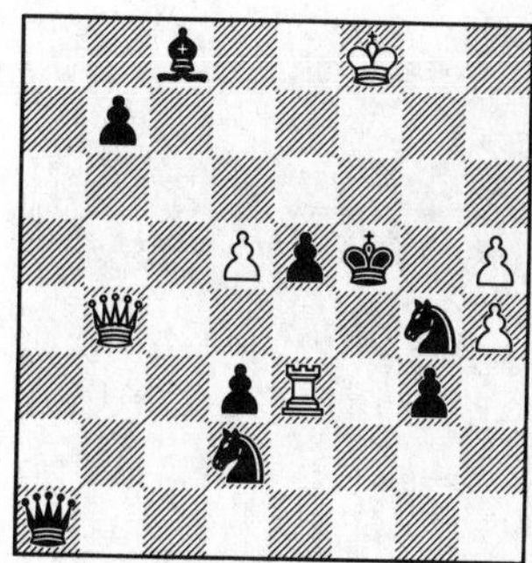

769.

770.

771.

772.

773.

774.

775.

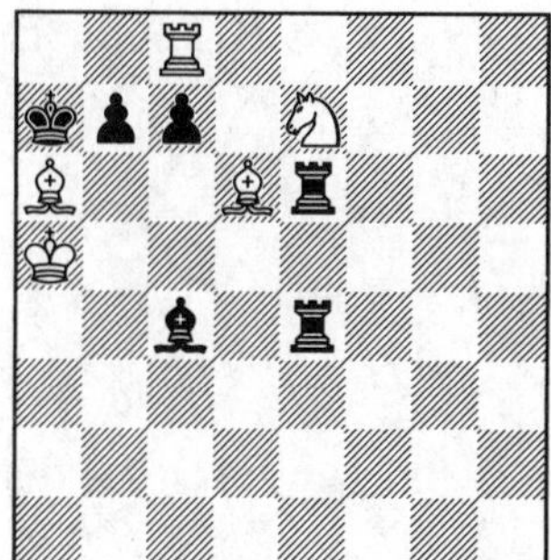

776.

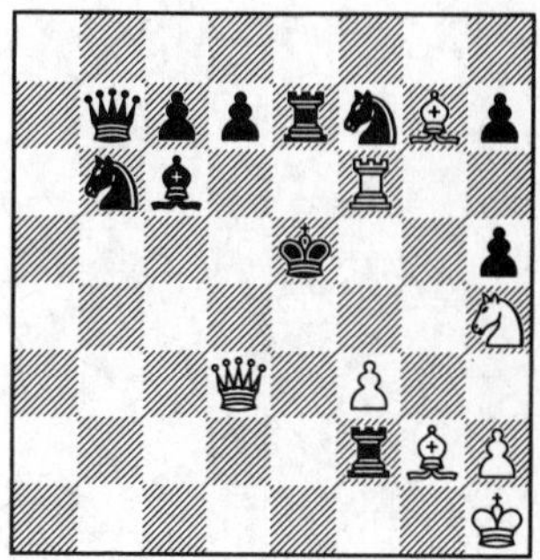

777.

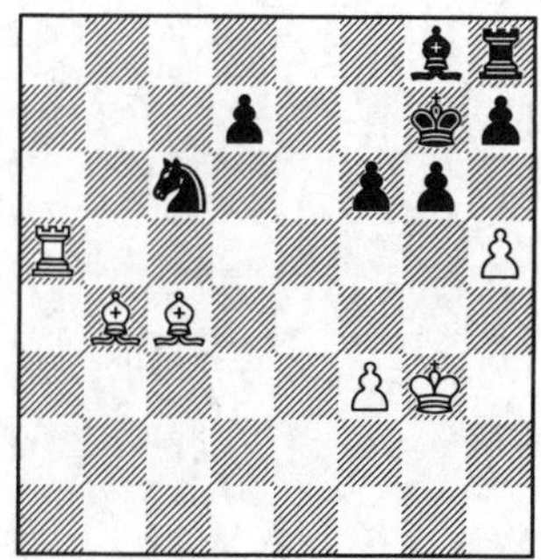

778.

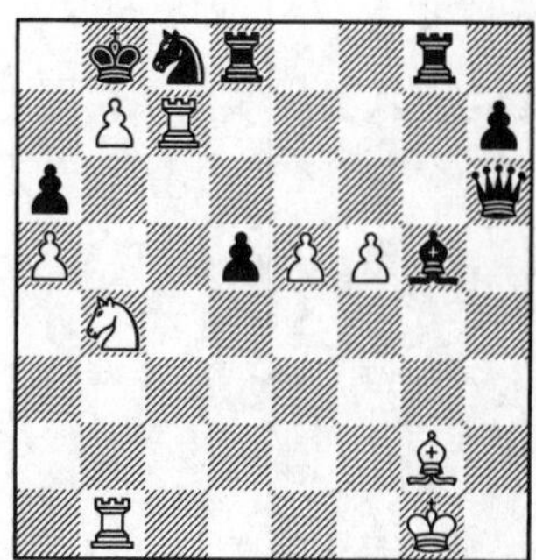

779.

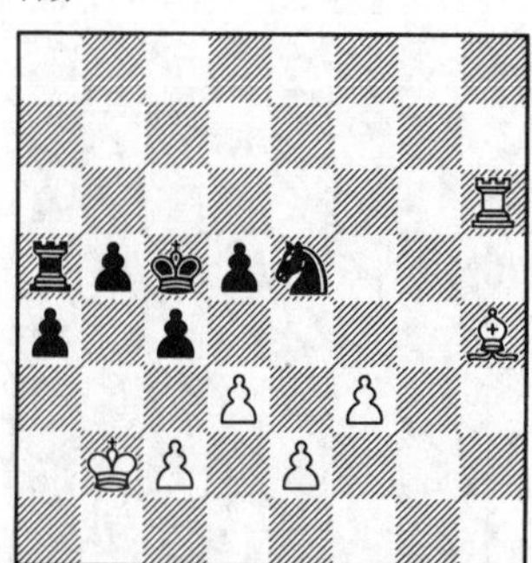

780.

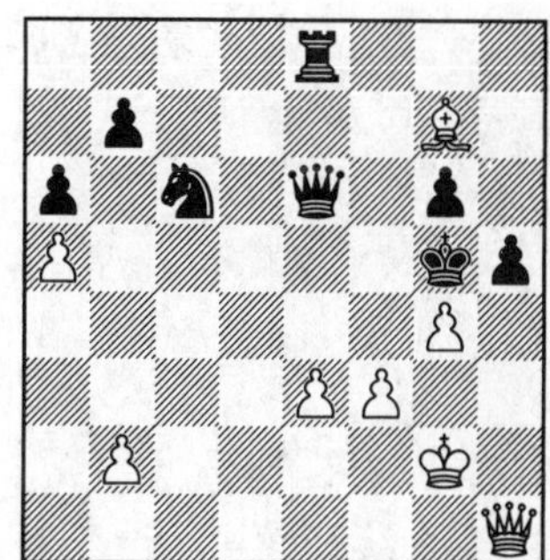

781.

782.

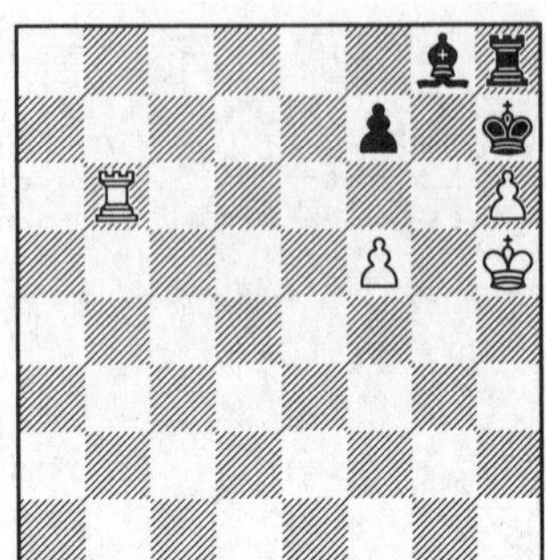

783.

784.

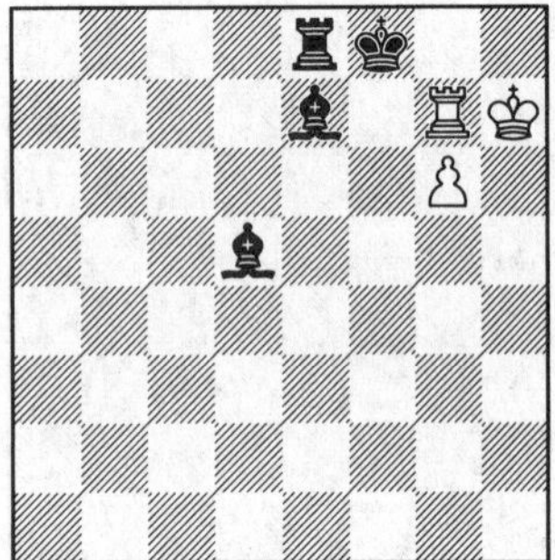

785.

786.

787.

788.

789.

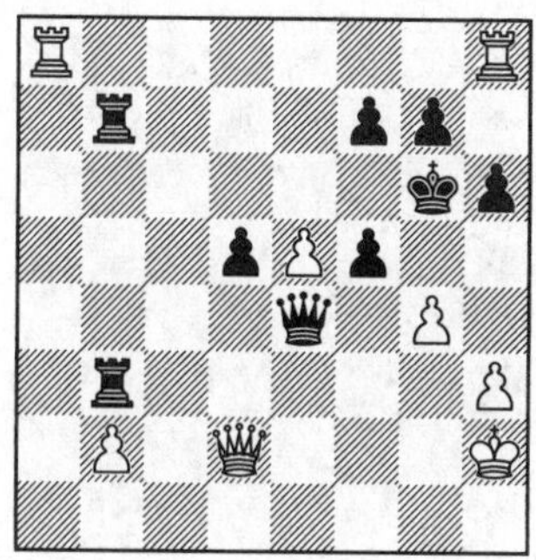

790.

791.

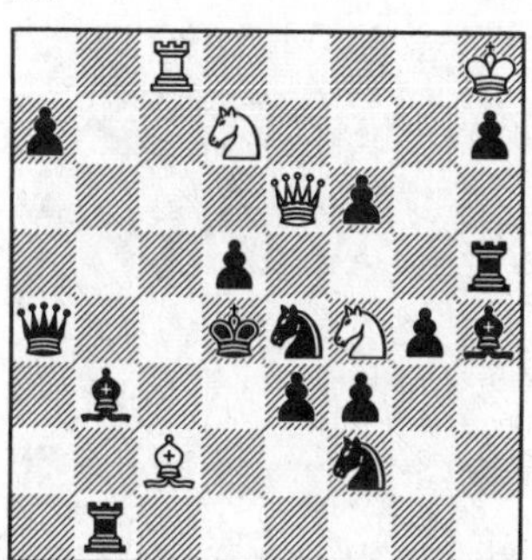

792.

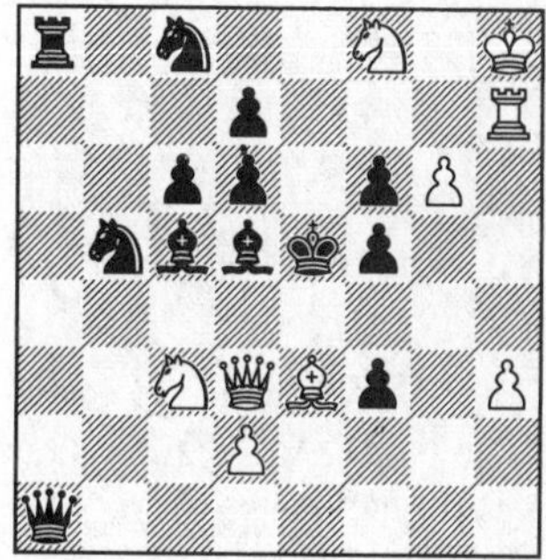

793.

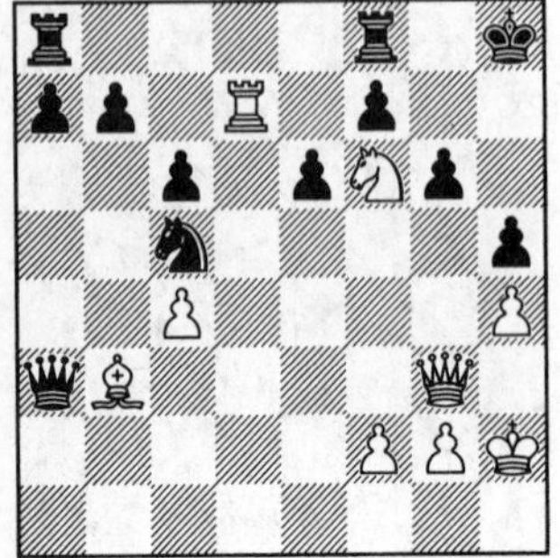

794.

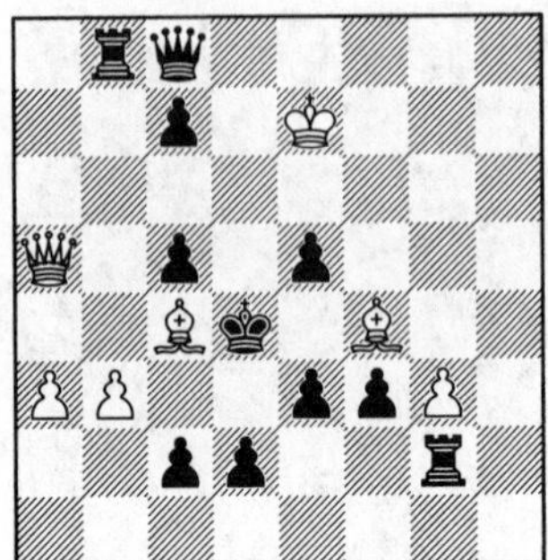

795.

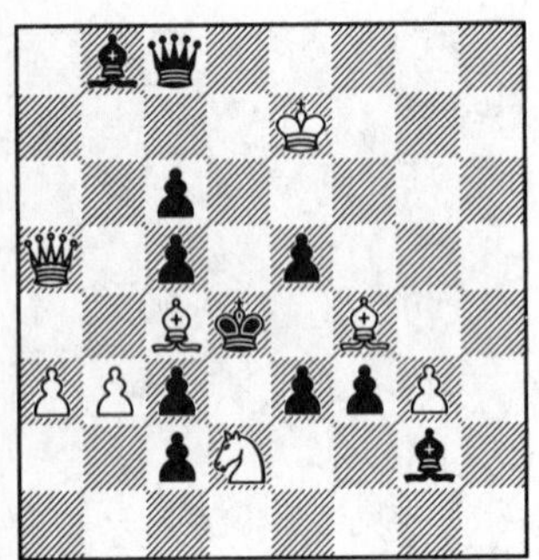

796.

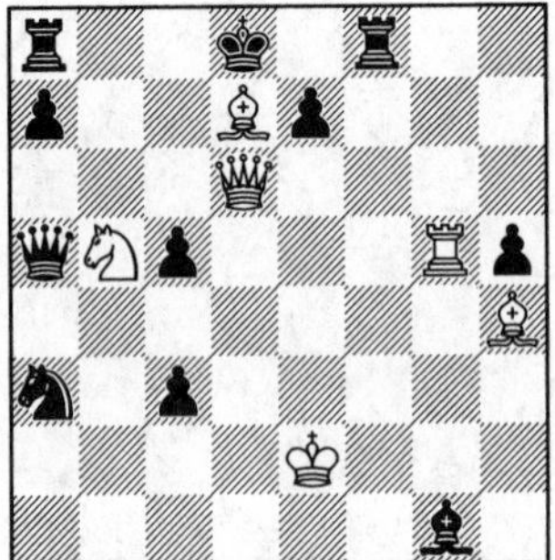

797.

798.

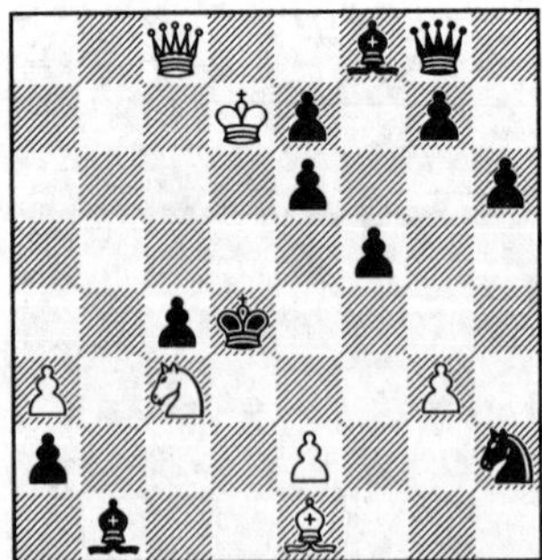

799.

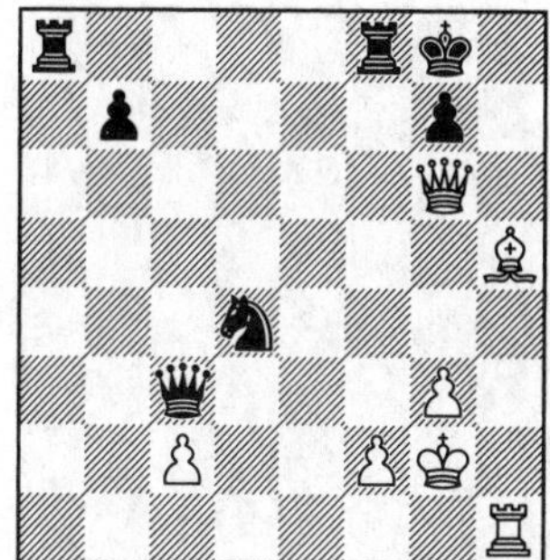

800.

801.

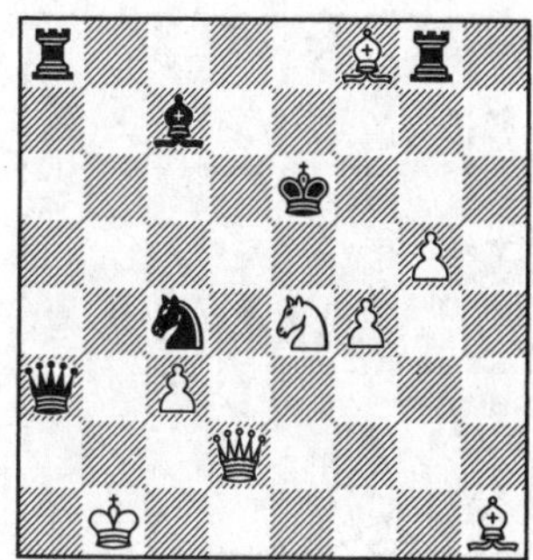

802.

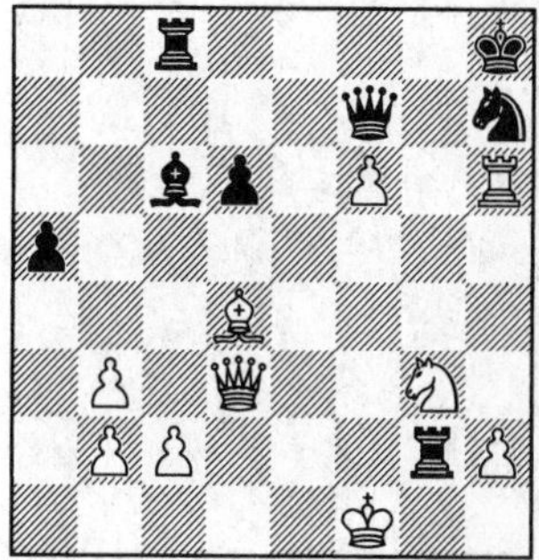

803.

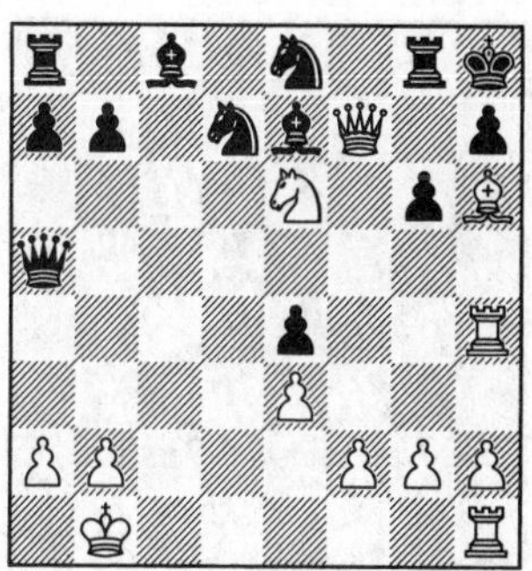

804.

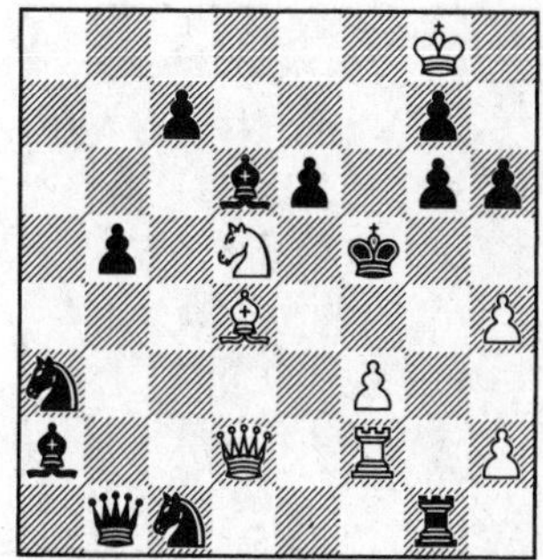

805.

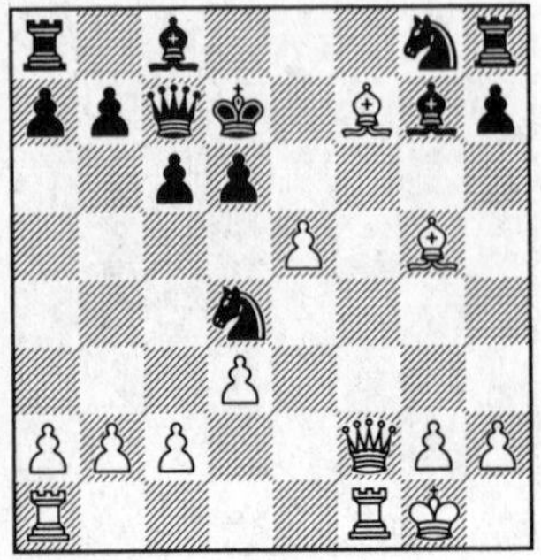

806.

807.

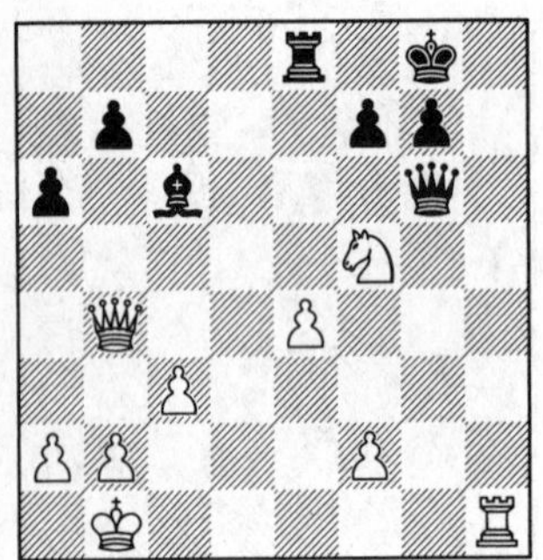

808.

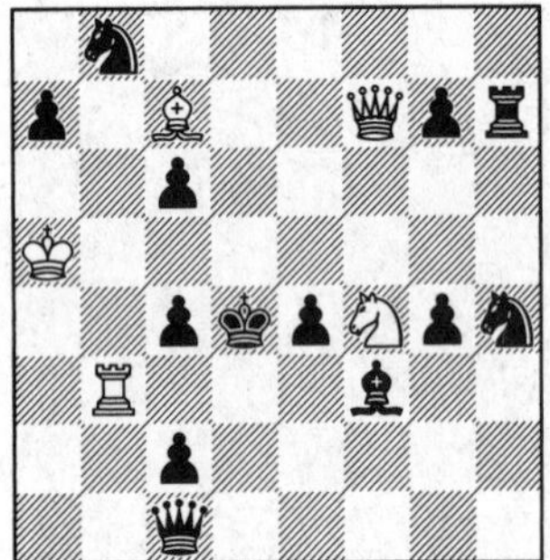

809.

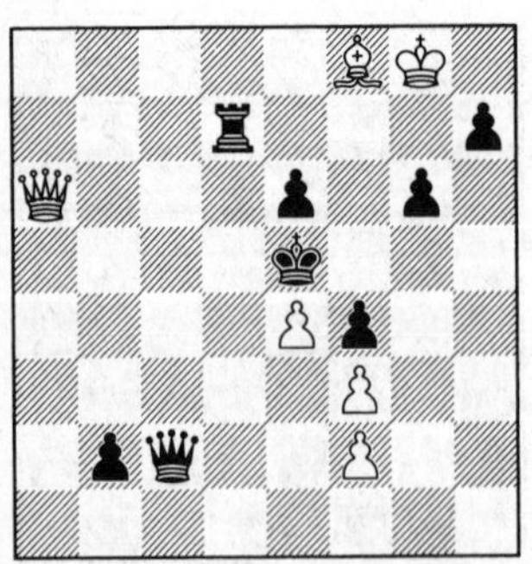

810.

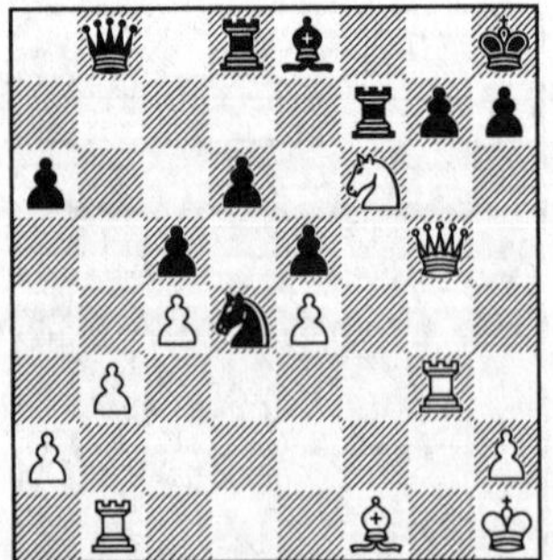

811.

812.

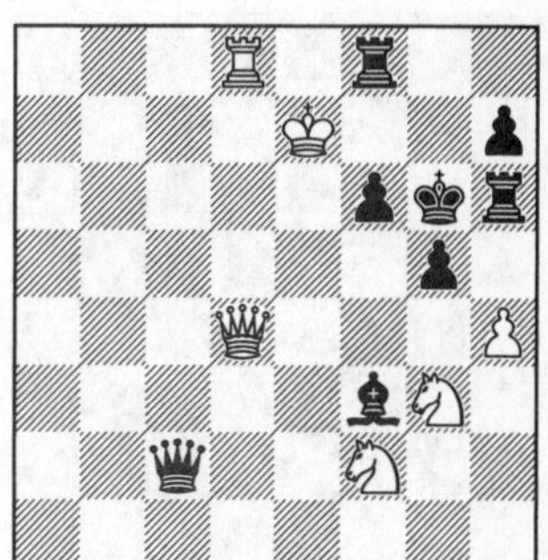

813.

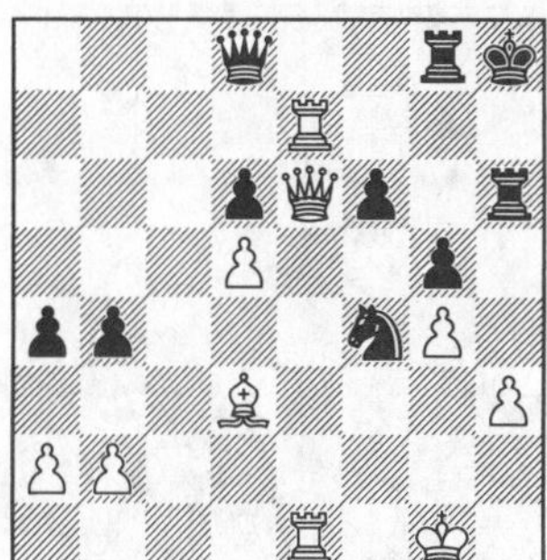

814.

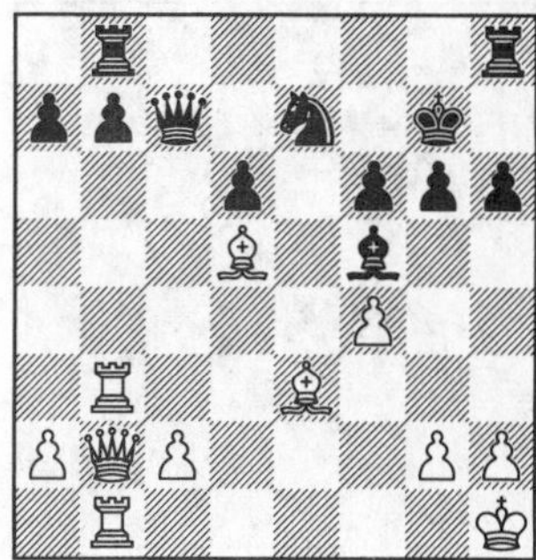

815.

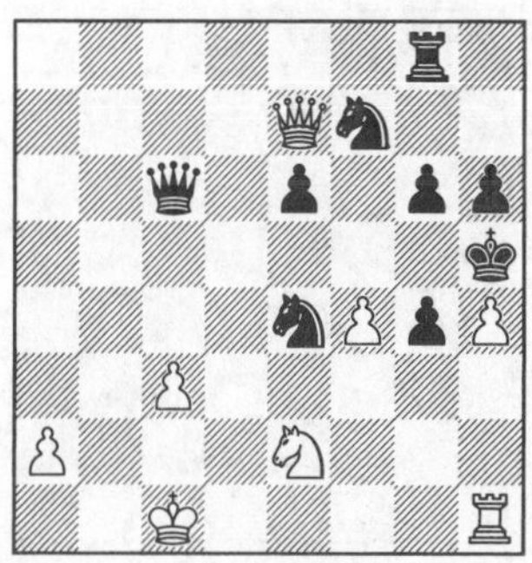

816.

817.

818.

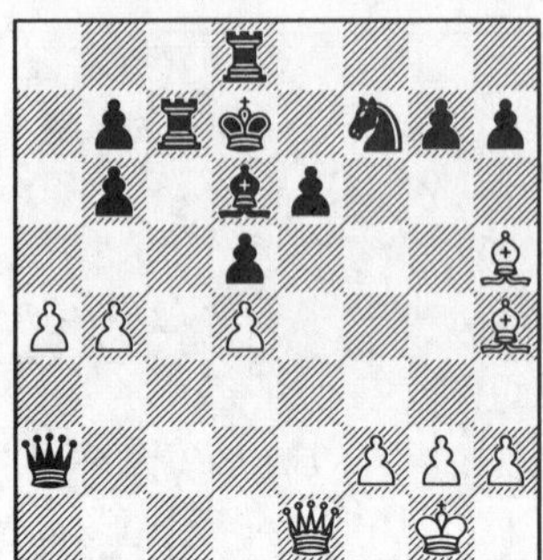

819.

820.

821.

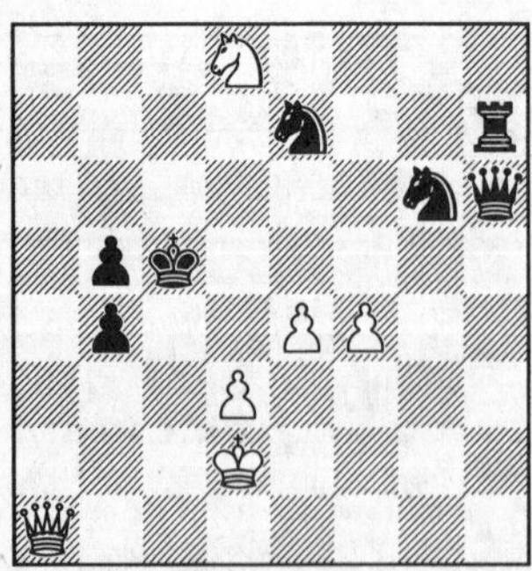

822.

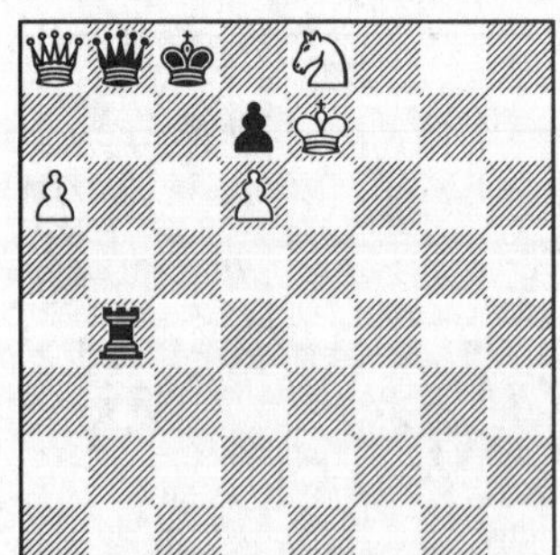

823.

824.

825.

826.

827.

828.

829.

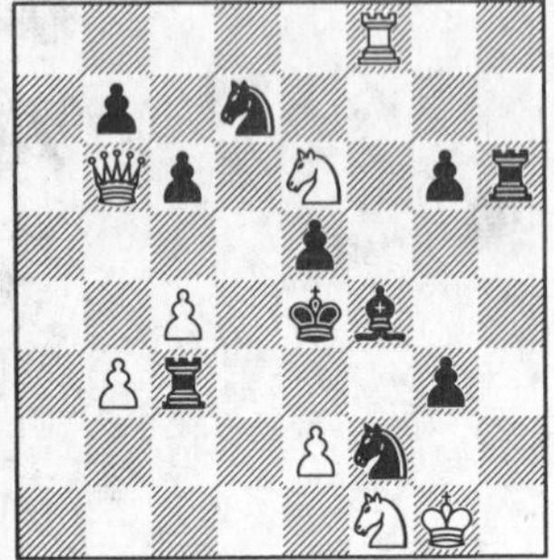

830.

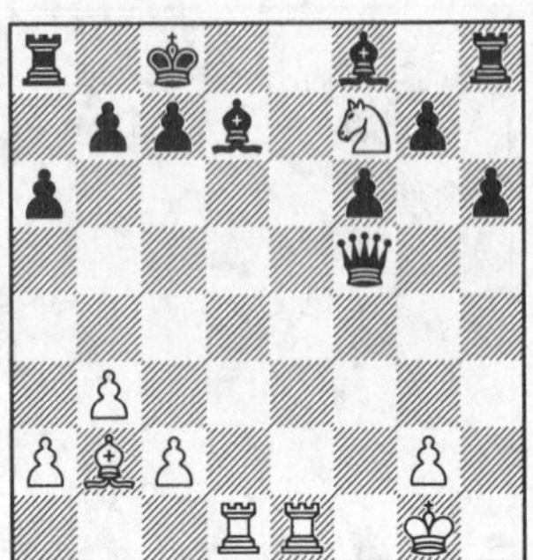

831.

832.

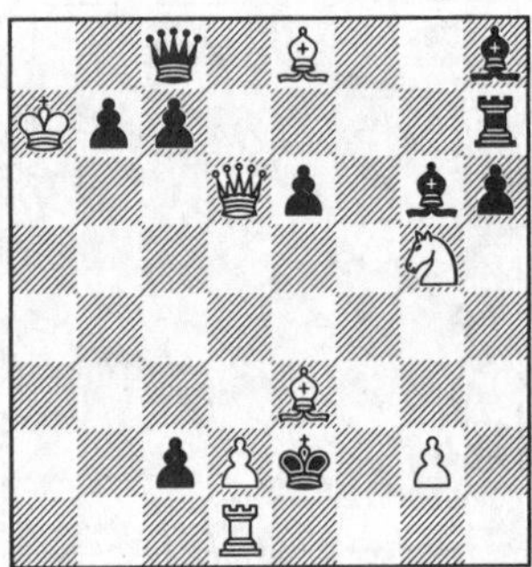

833.

834.

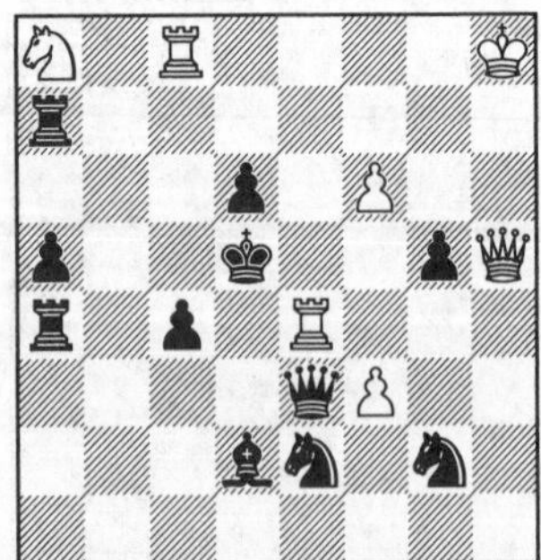

835.

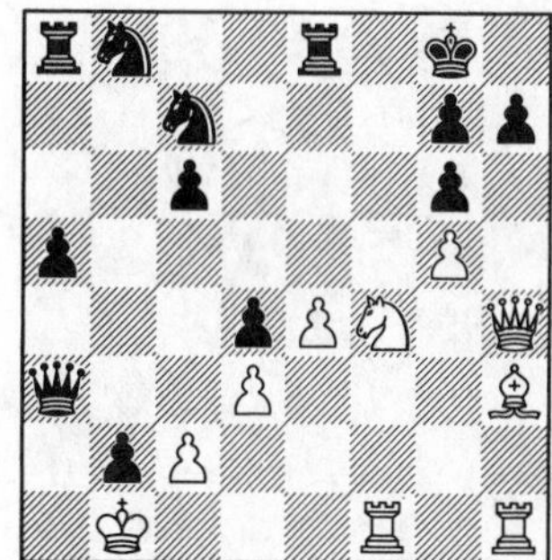

836.

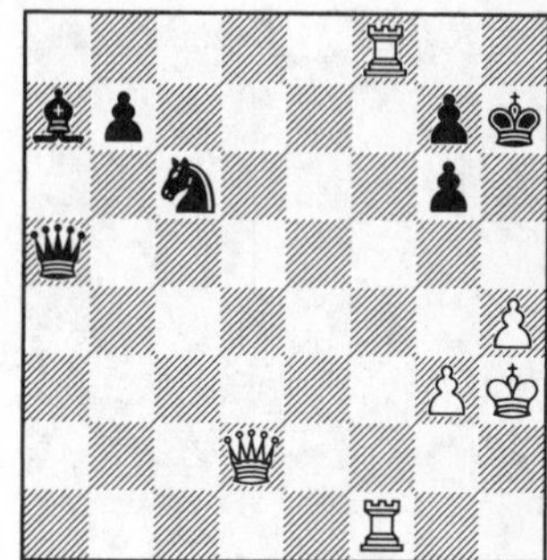

837.

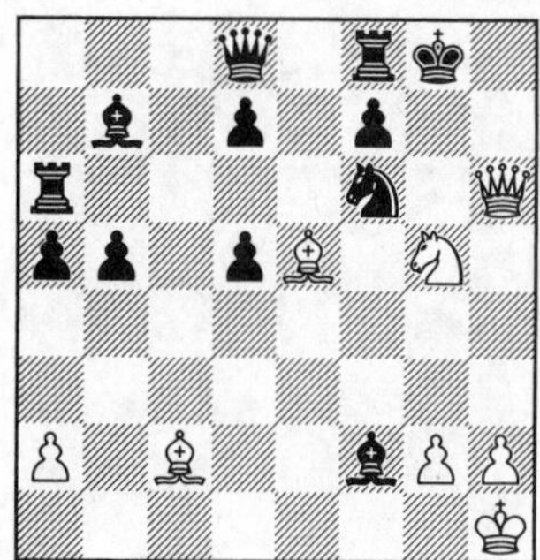

838.

839.

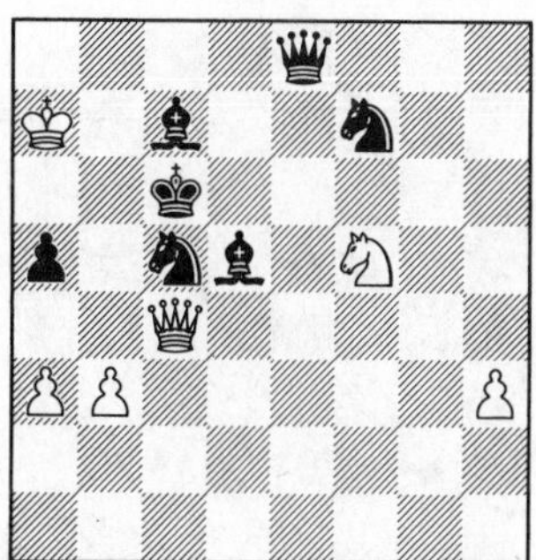

840.

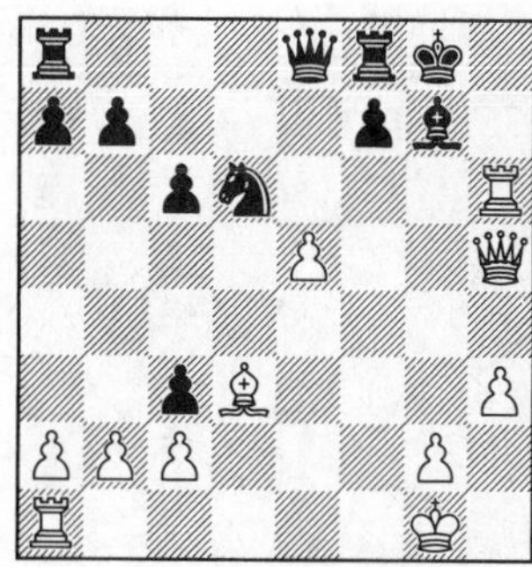

841.

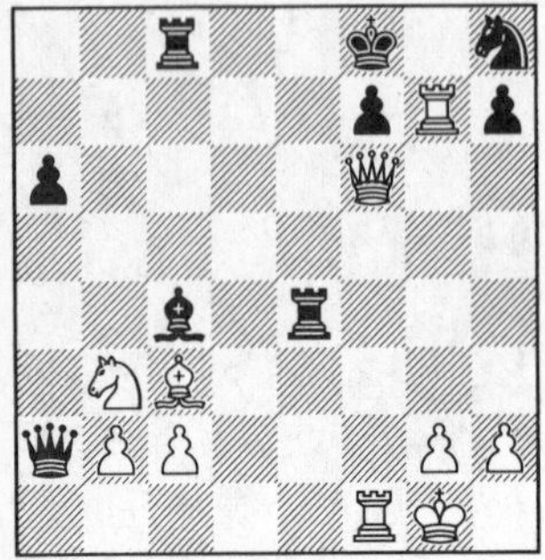

842.

843.

844.

845.

846.

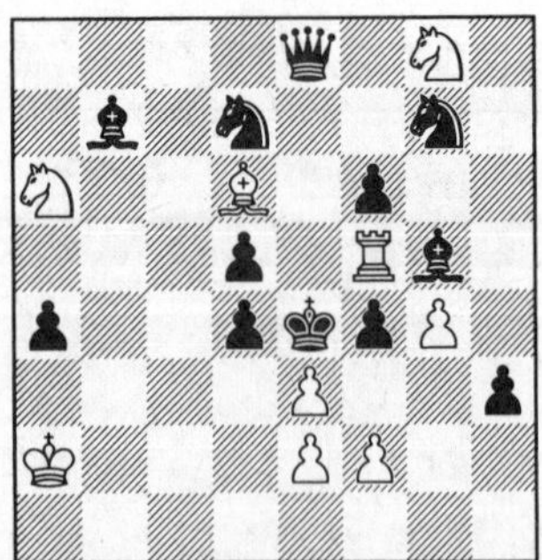

847.

848.

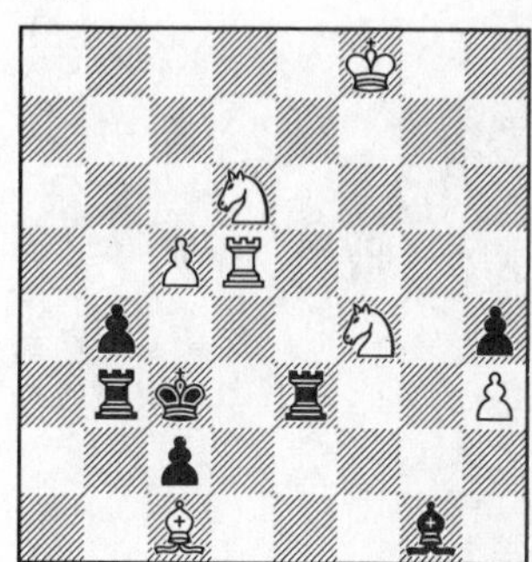

849.

850.

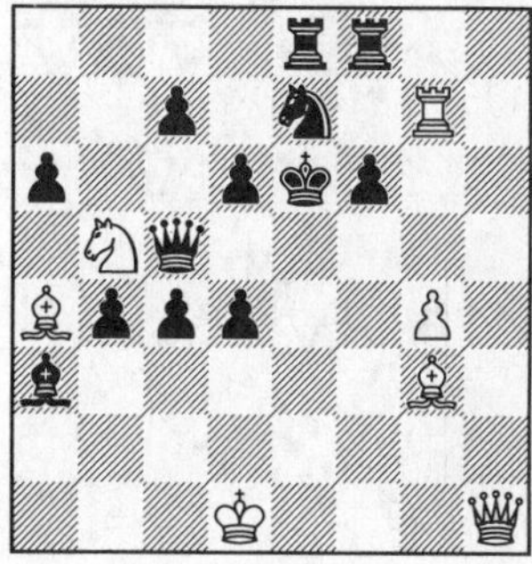

851.

852.

853.

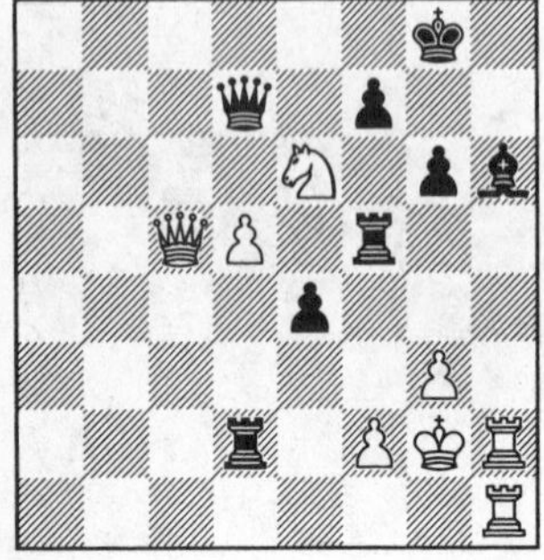

854.

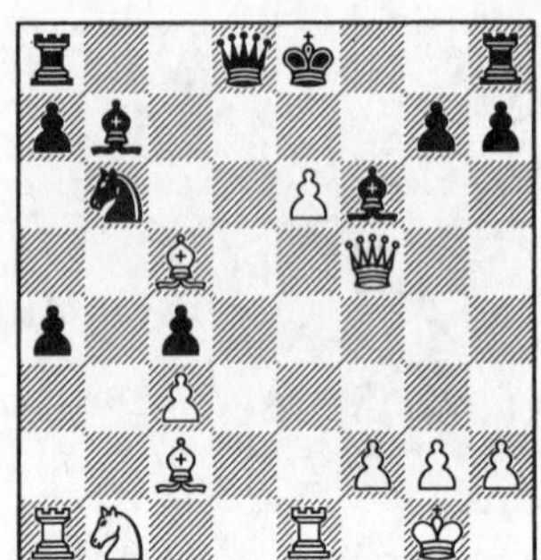

855.

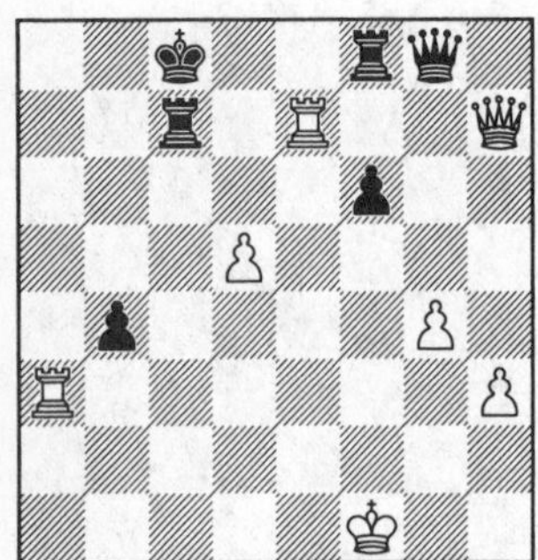

856.

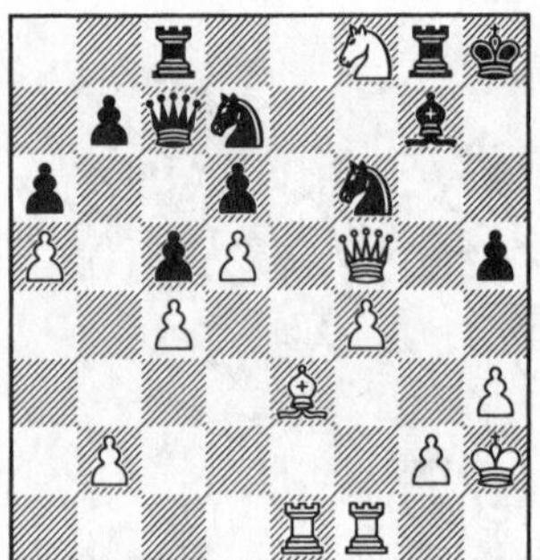

857.

858.

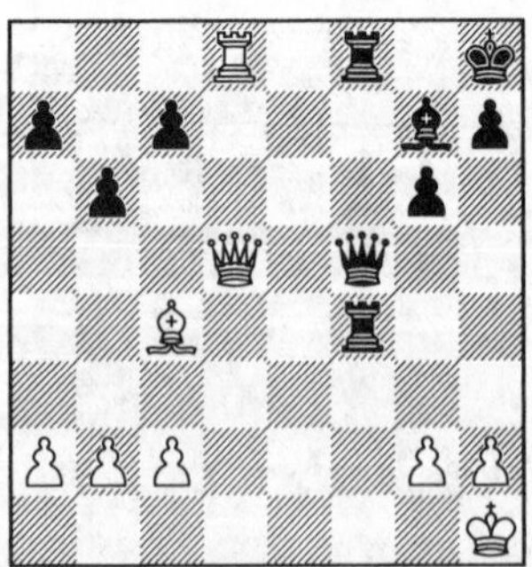

859.

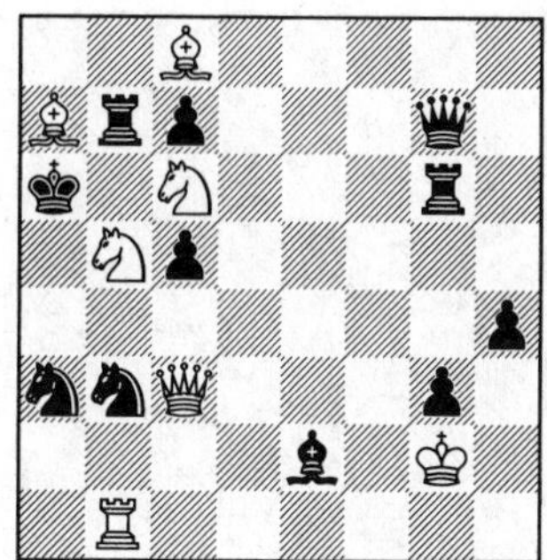

860.

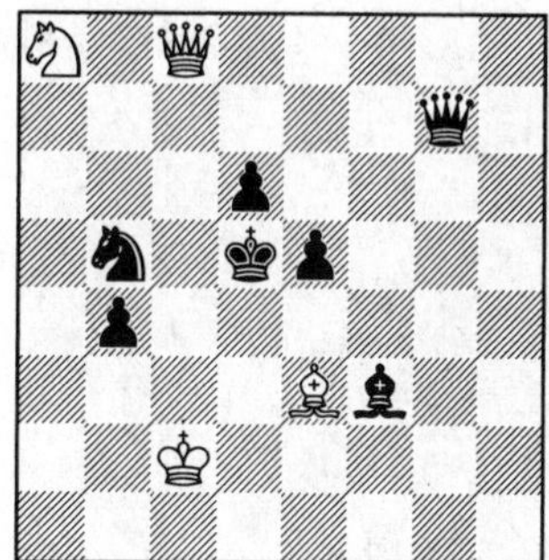

861.

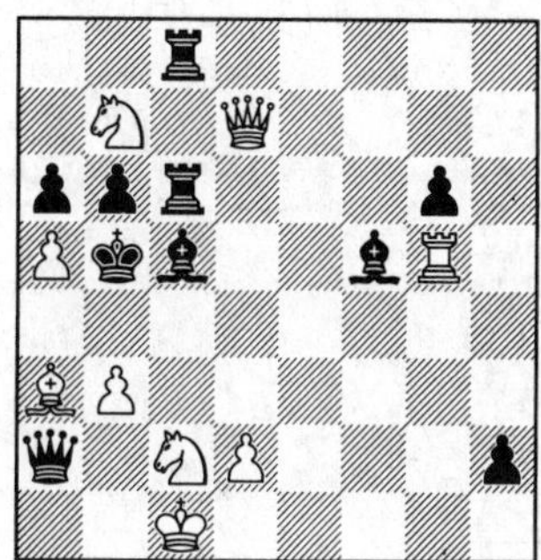

862.

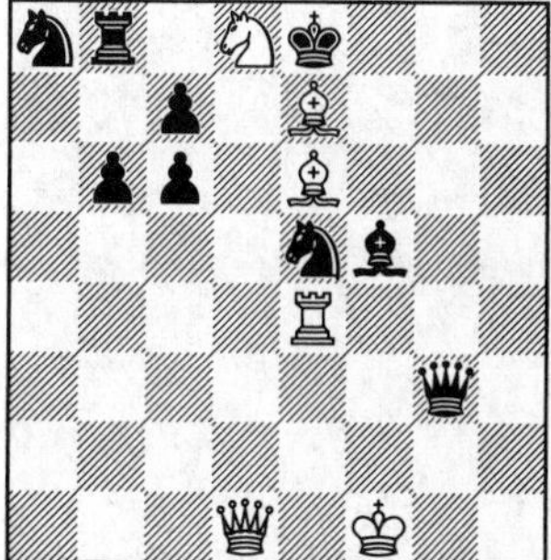

863.

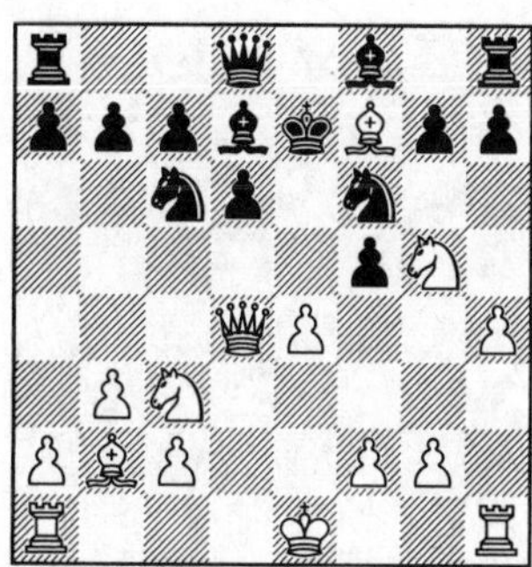

864.

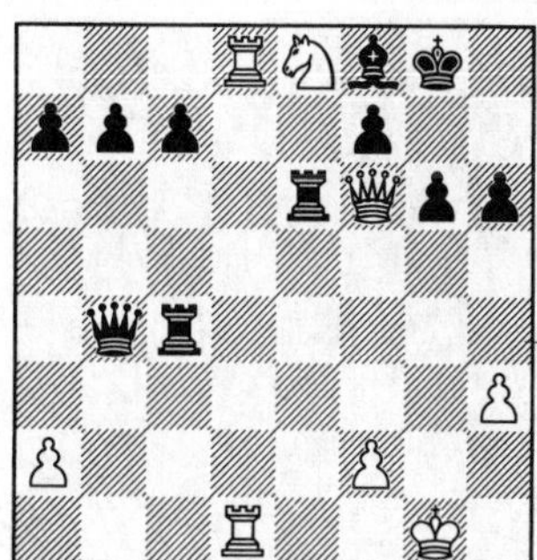

865.

866.

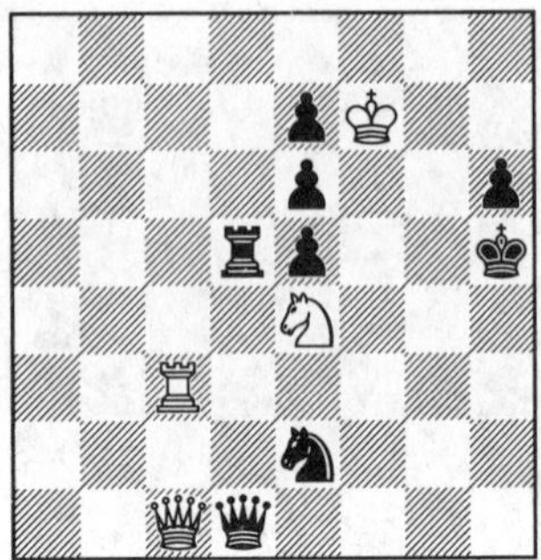

867.

868.

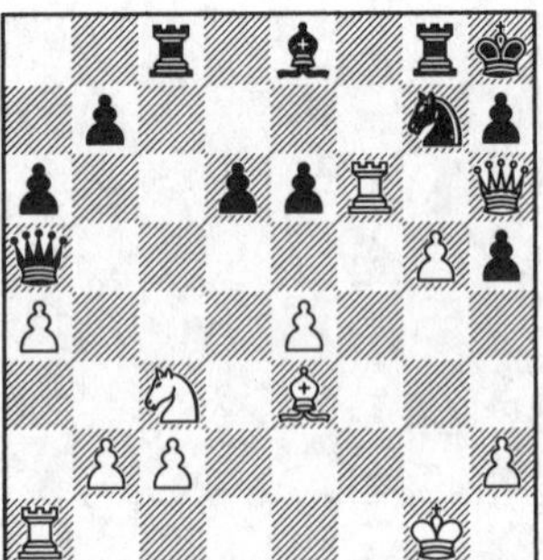

869.

870.

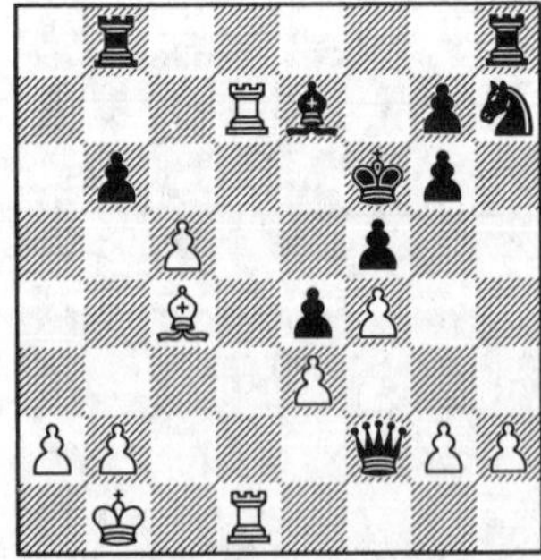

871.

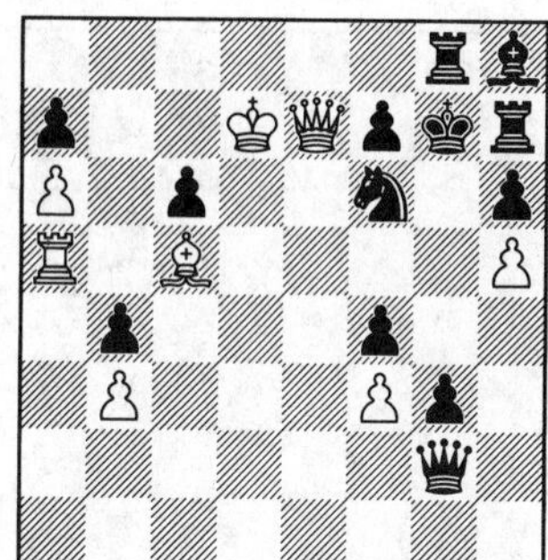

872.

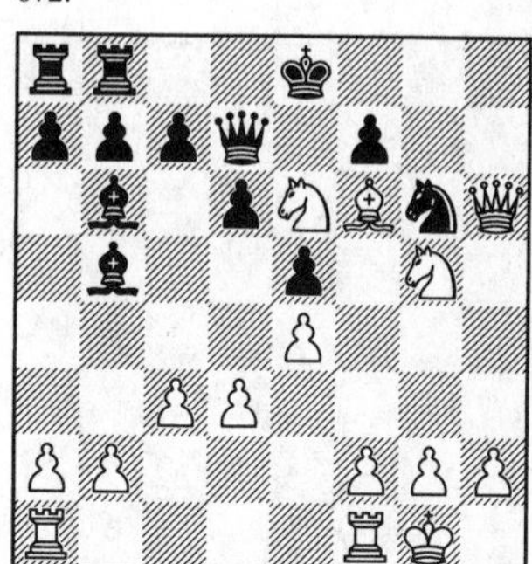

873.

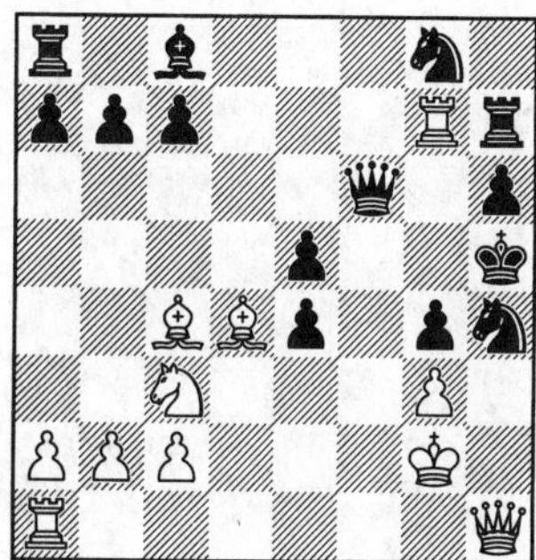

874.

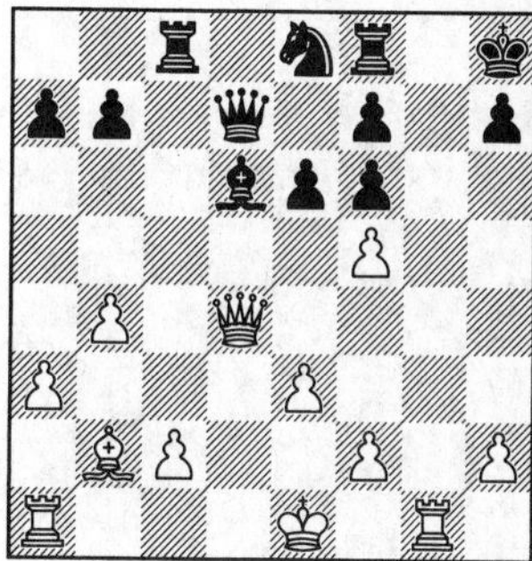

875.

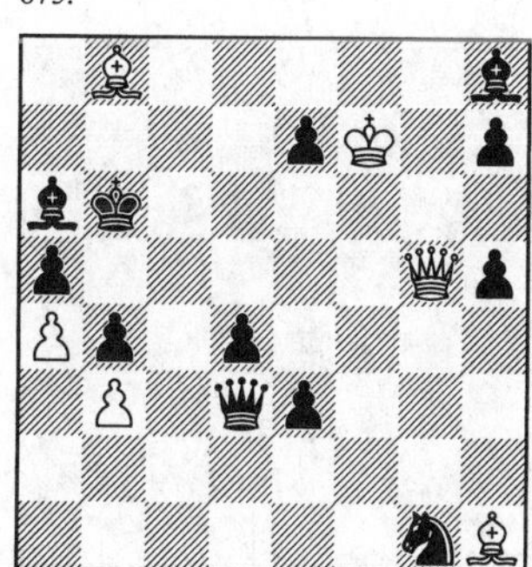

876.

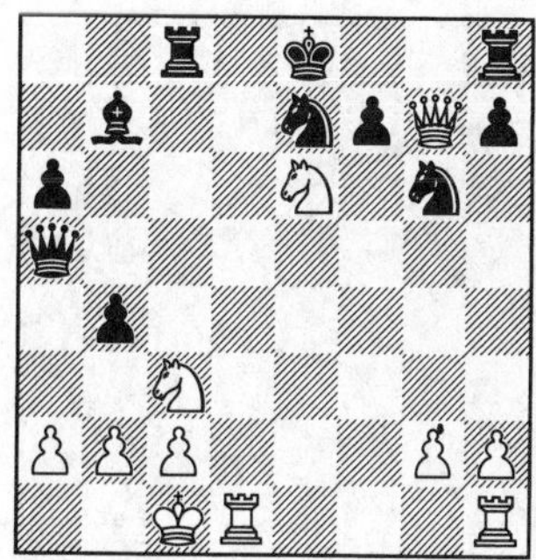

877.

878.

879.

880.

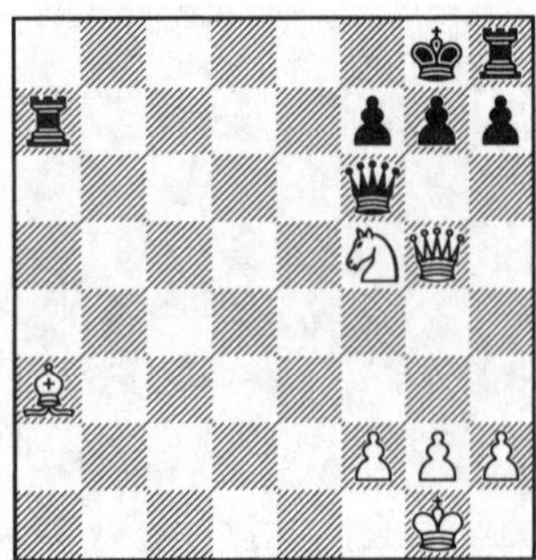

881.

882.

883.

884.

885.

886.

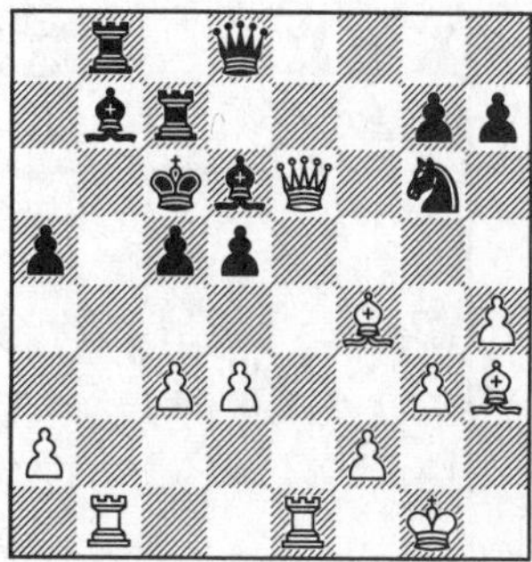

887.

888.

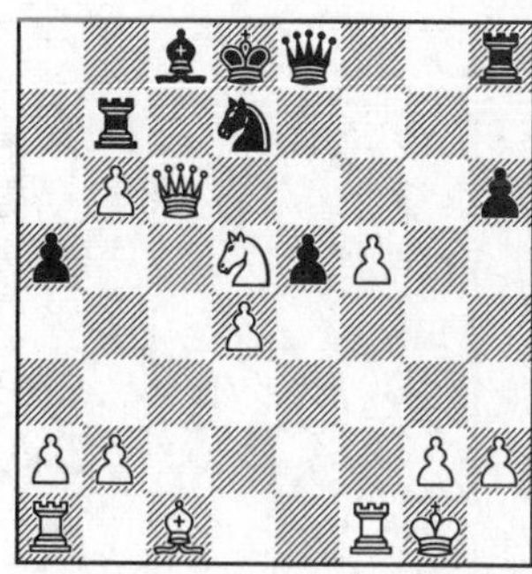

889.

890.

891.

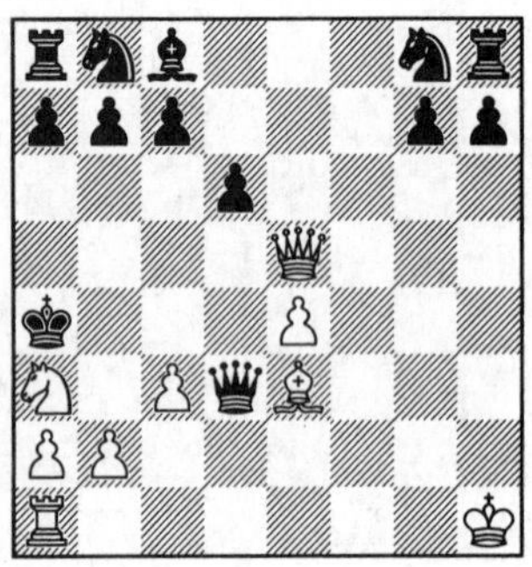

892.

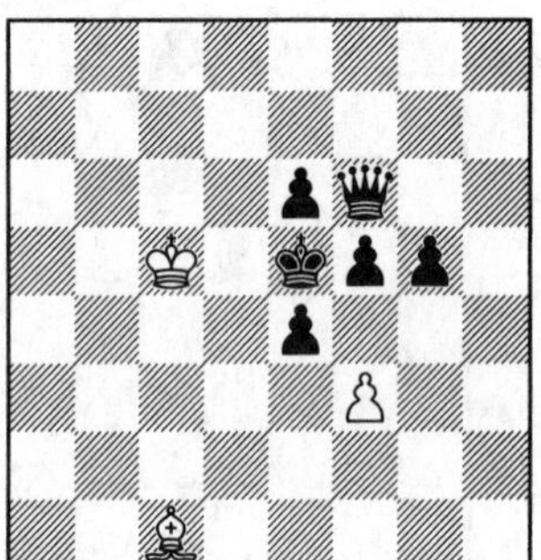

893.

894.

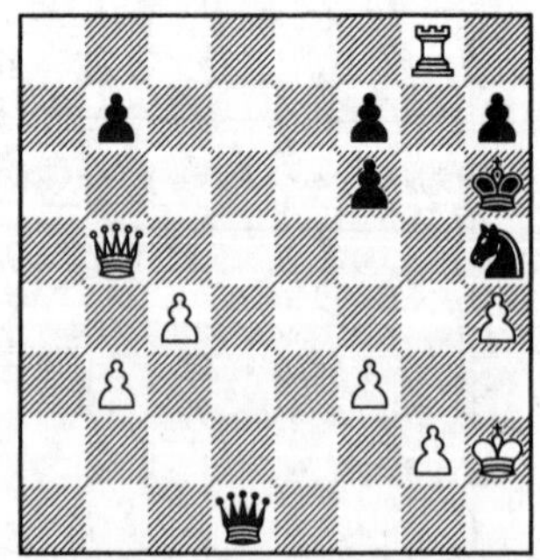

895.

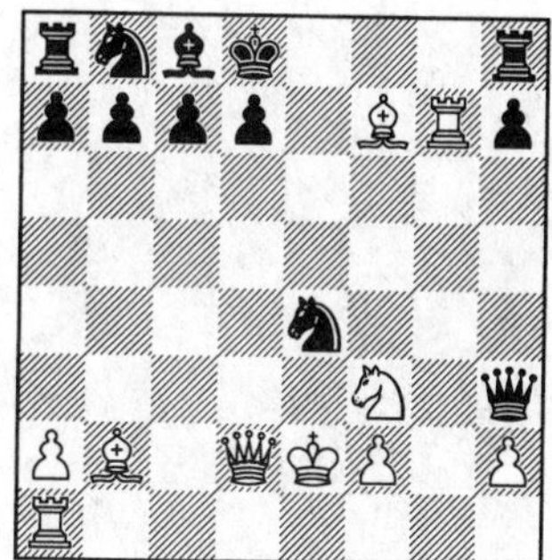

896.

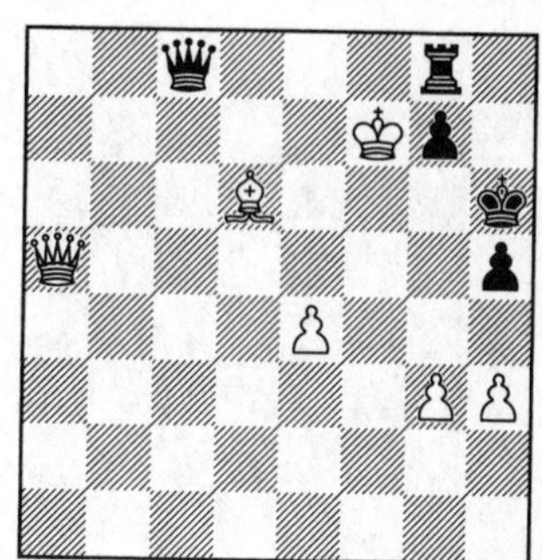

897.

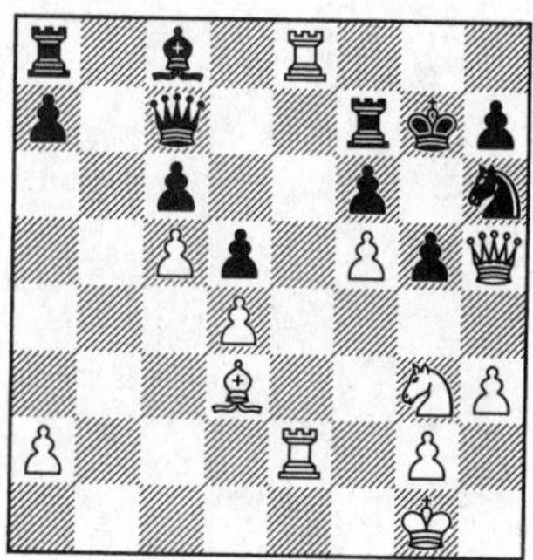

898.

899.

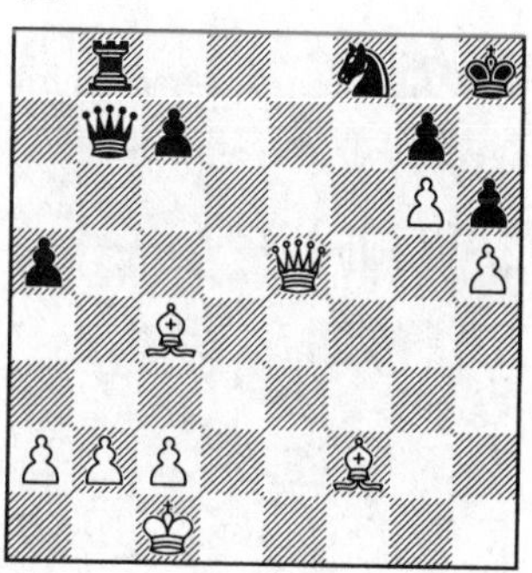

900.

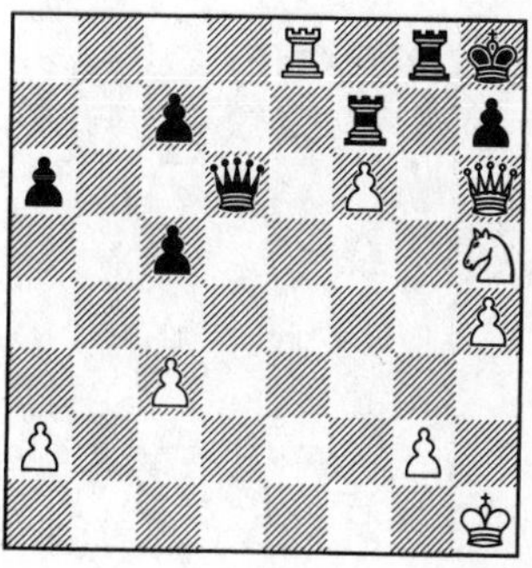

901.

902.

903.

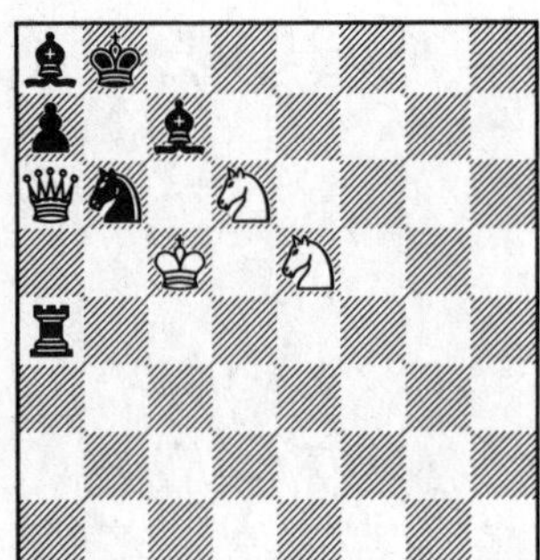

904.

905.

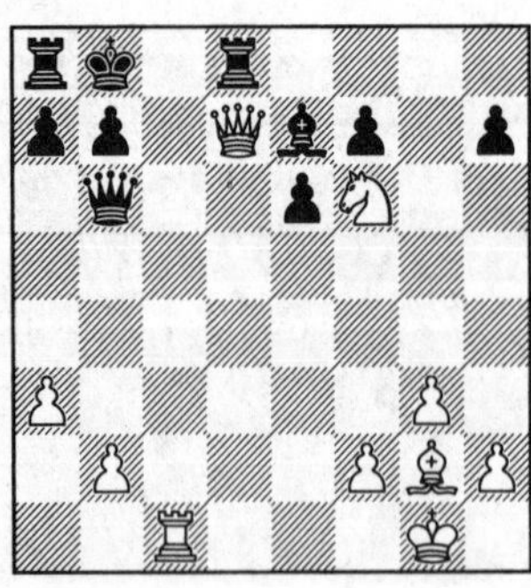

906.

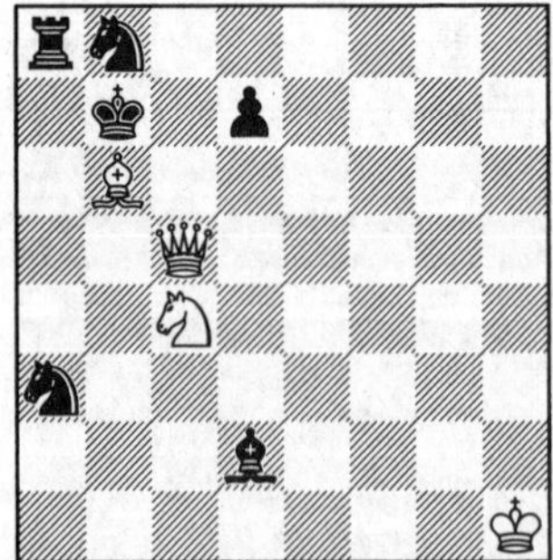

907.

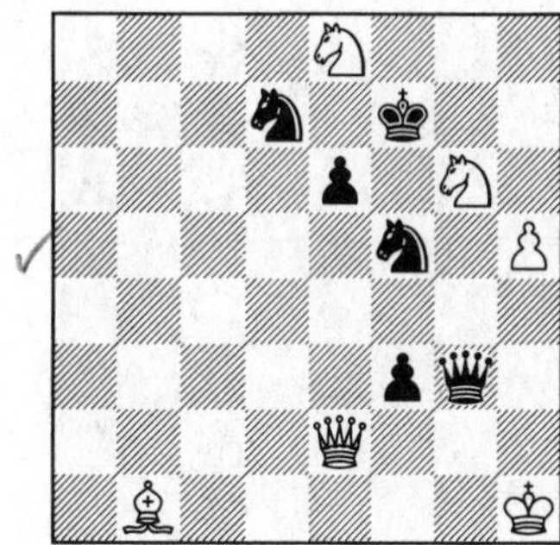

908.

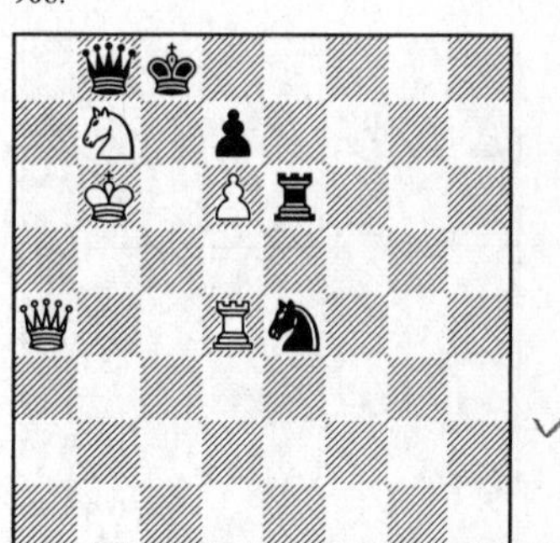

909.

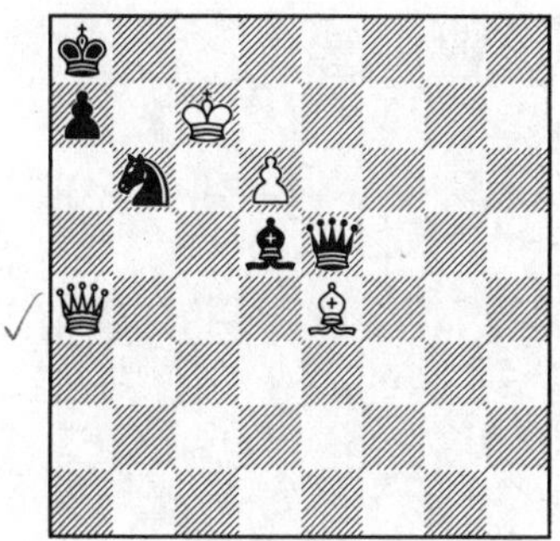

910.

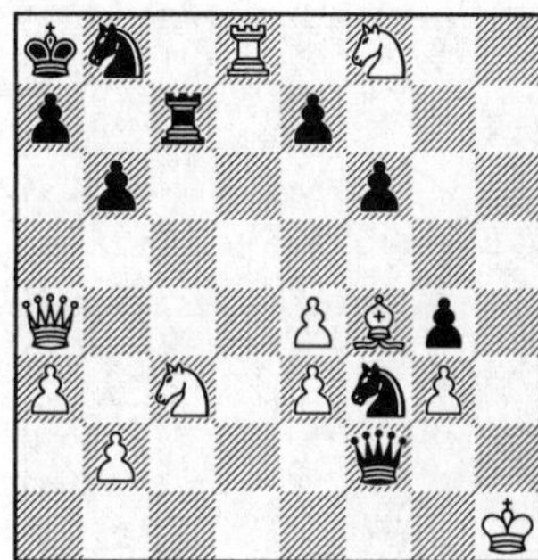

911.

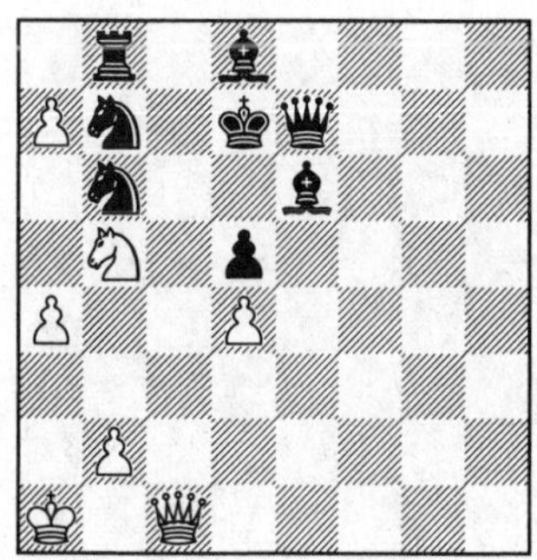

912.

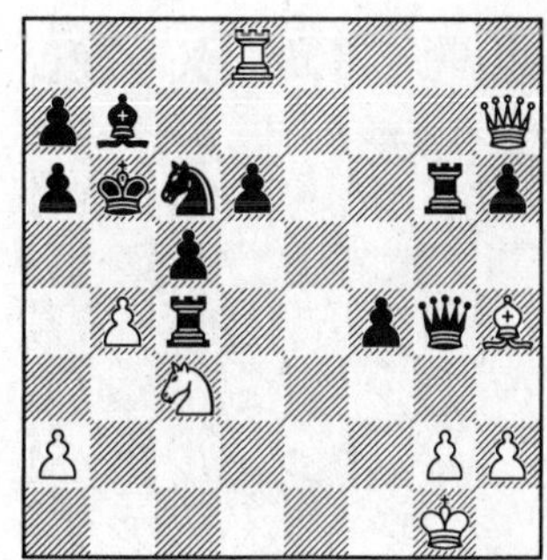

913.

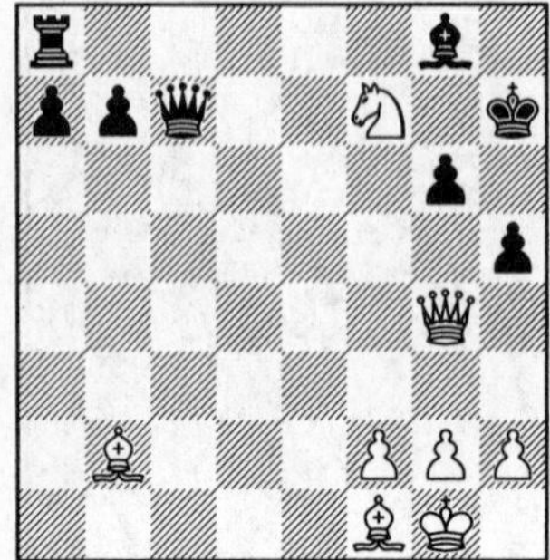

914.

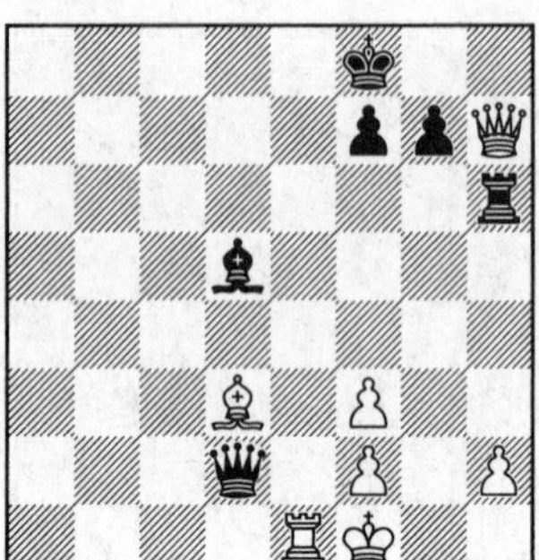

915.

916.

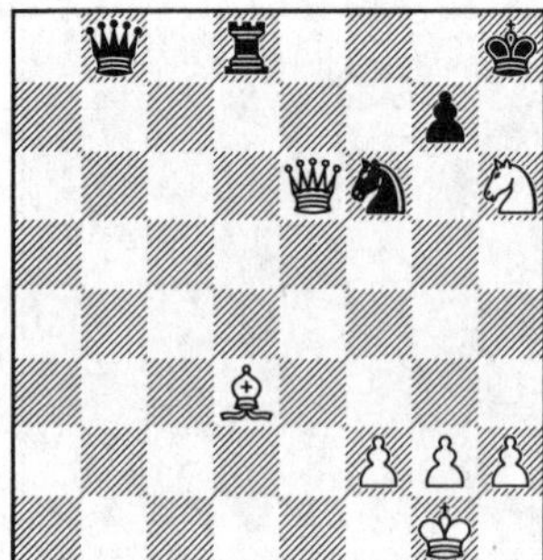

917.

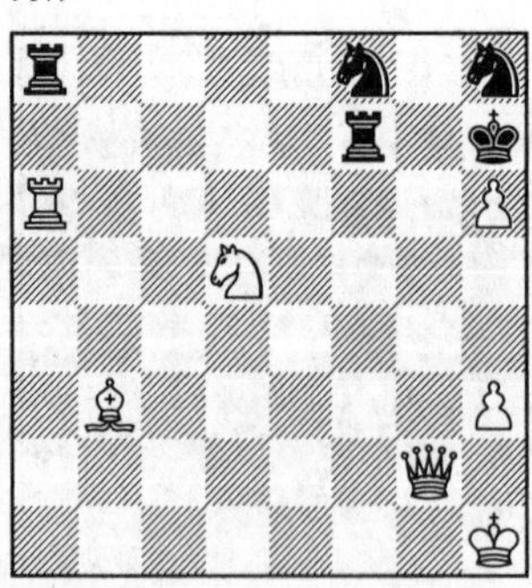

918.

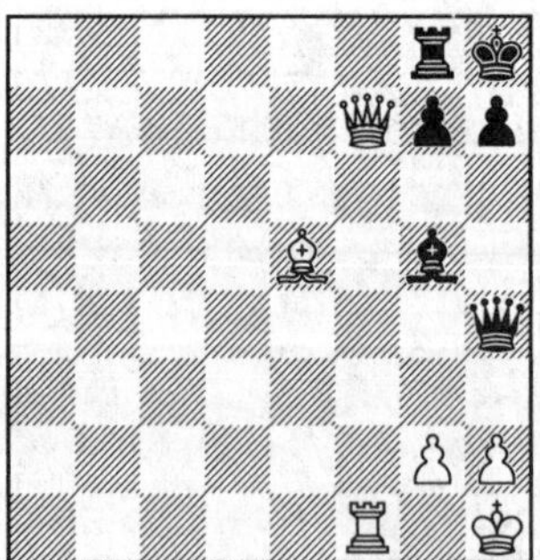

919.

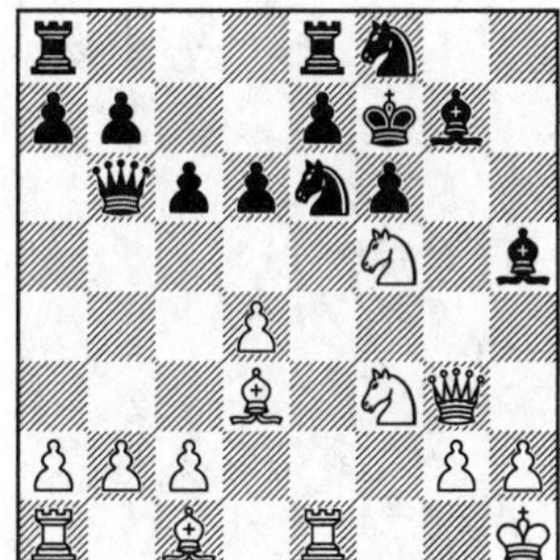

920.

921.

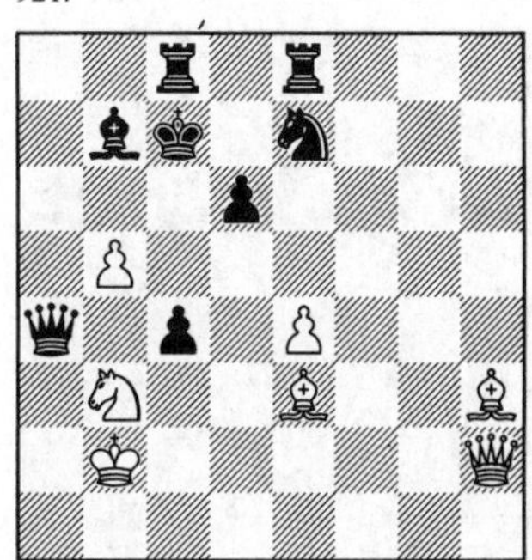

922.

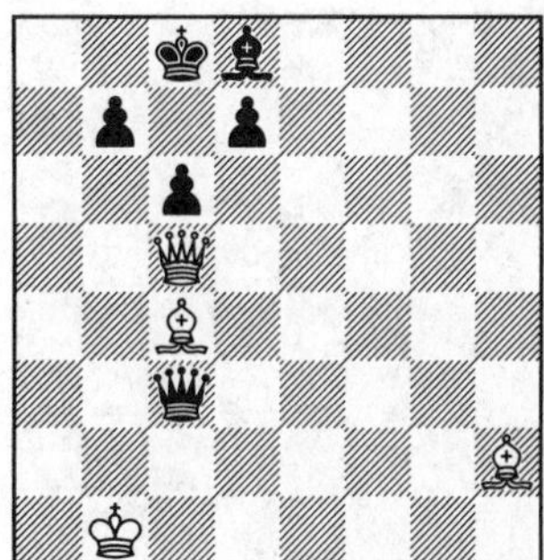

923.

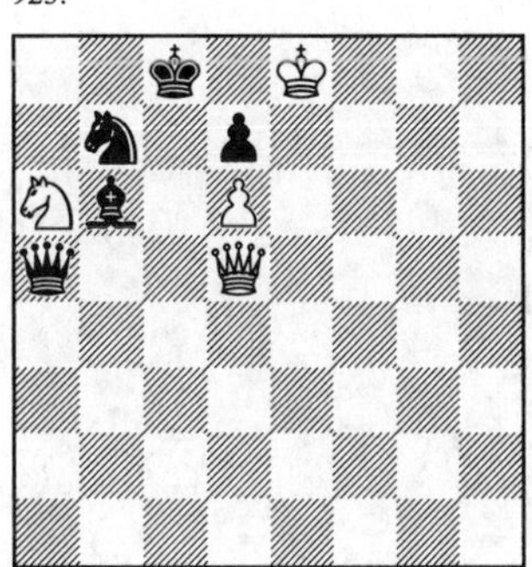

924.

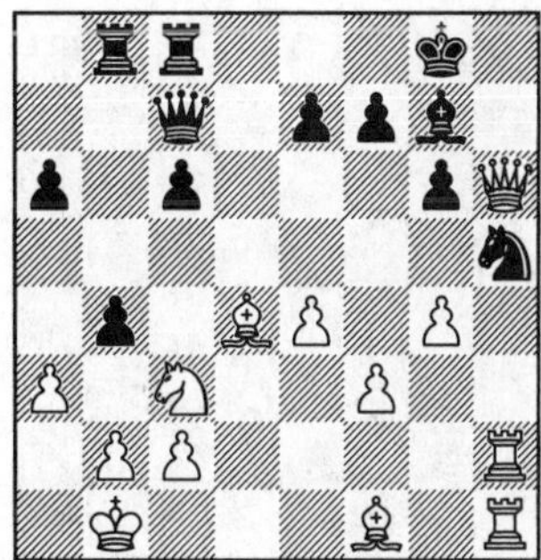

925.

926.

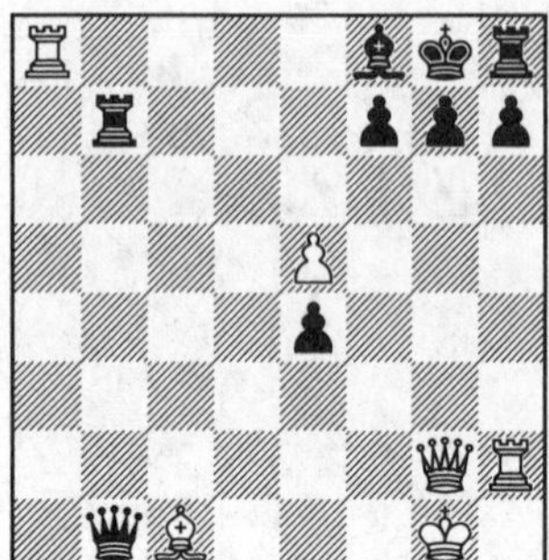

927.

928.

929.

930.

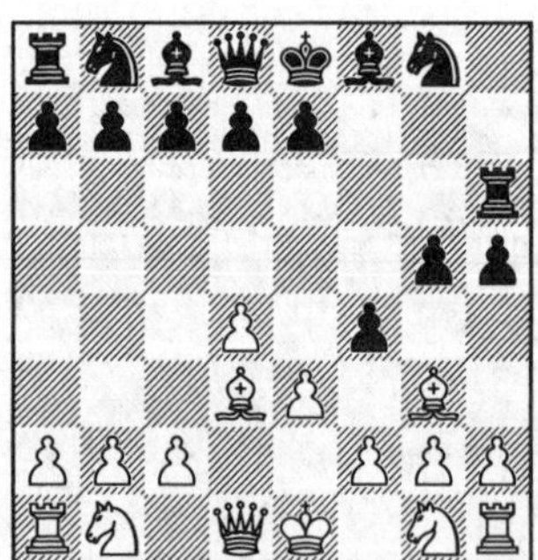

931.

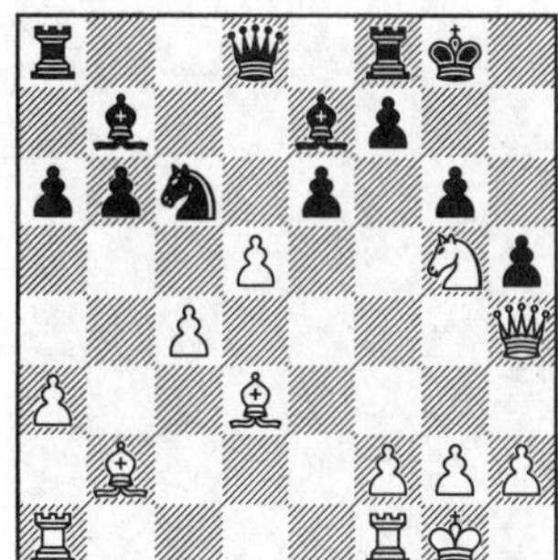

932.

933.

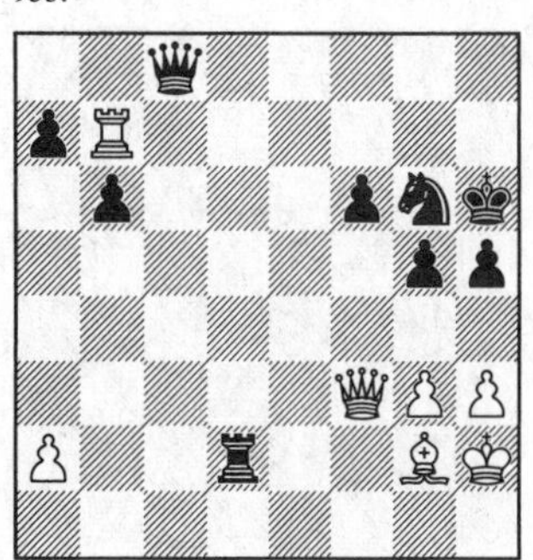

934.

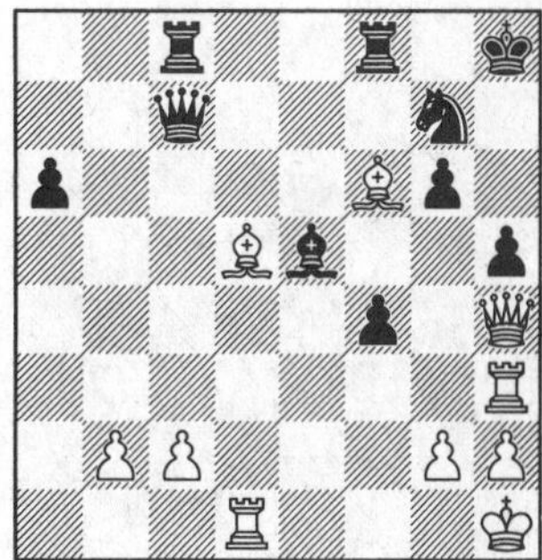

935.

936.

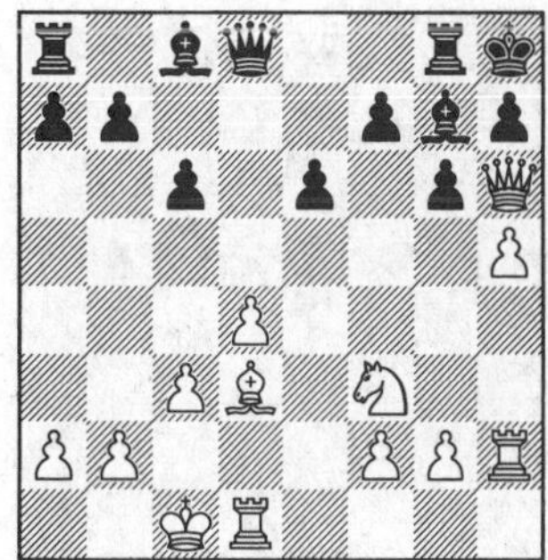

937.

938.

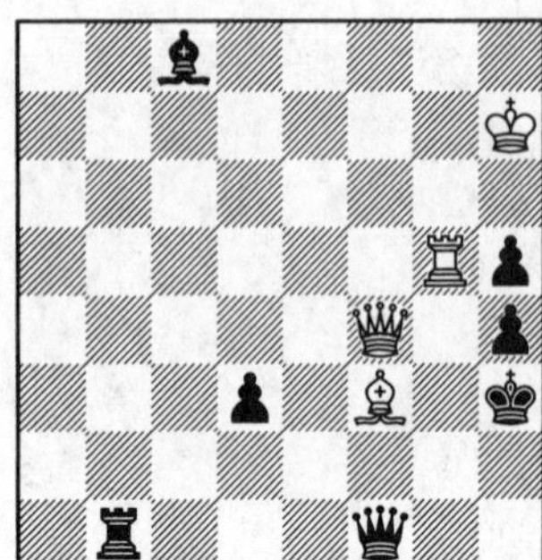

939.

940.

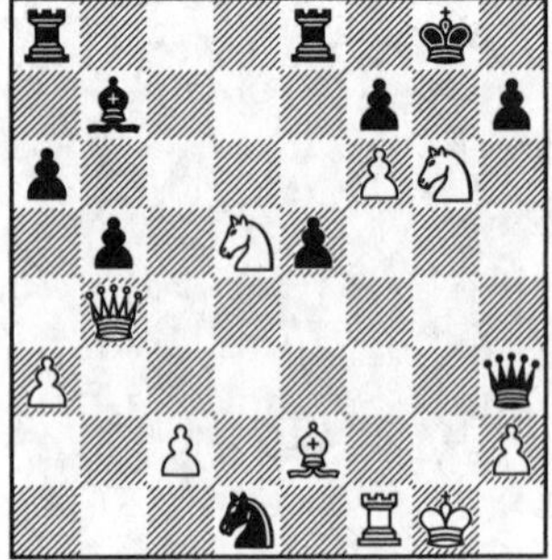

941.

942.

943.

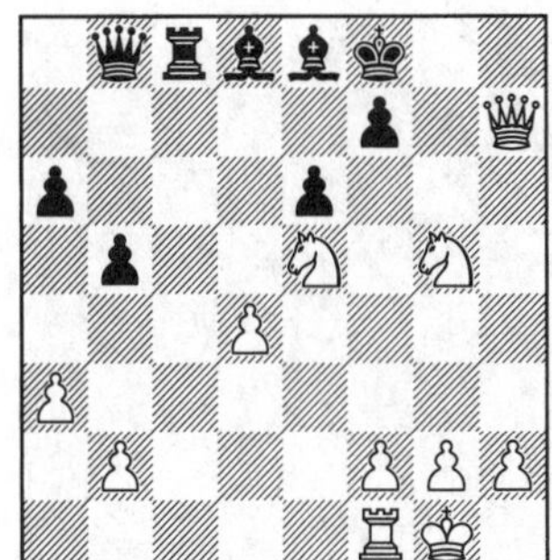

944.

945.

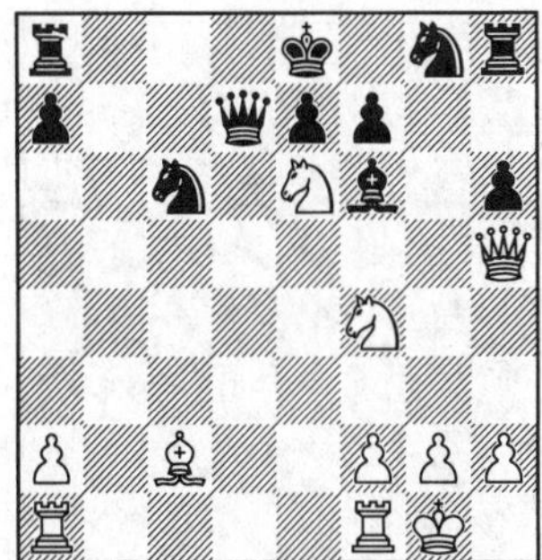

946.

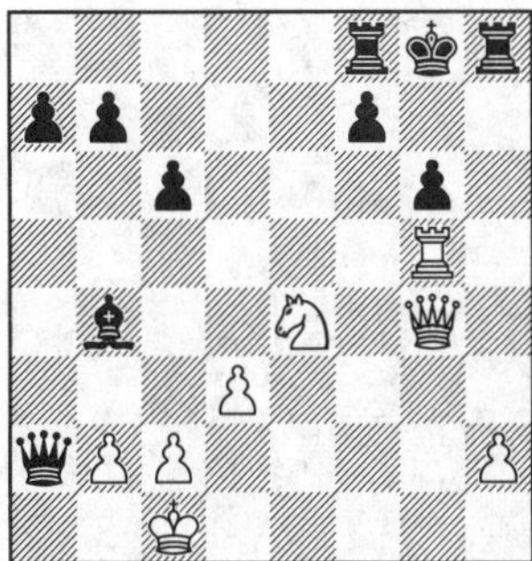

947.

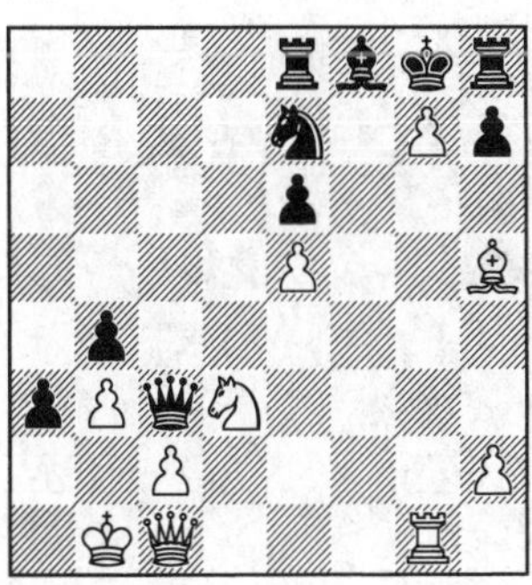

948.

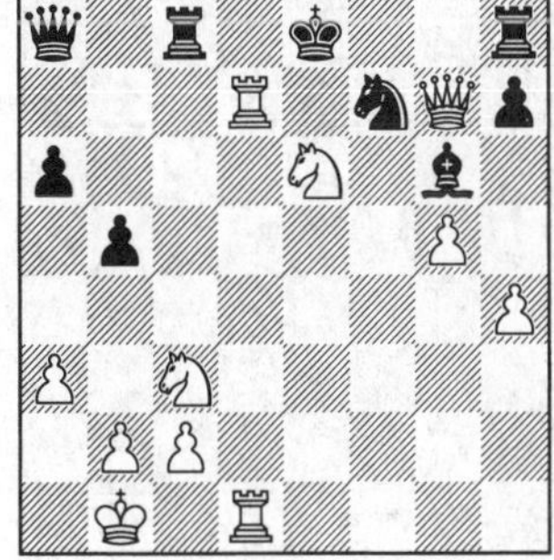

949.

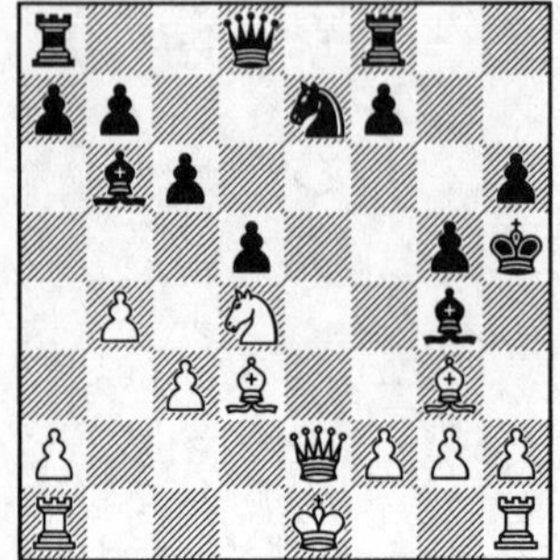

950.

951.

952.

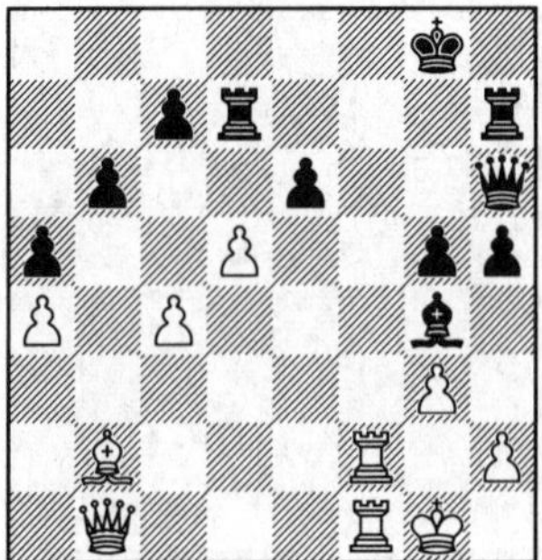

953.

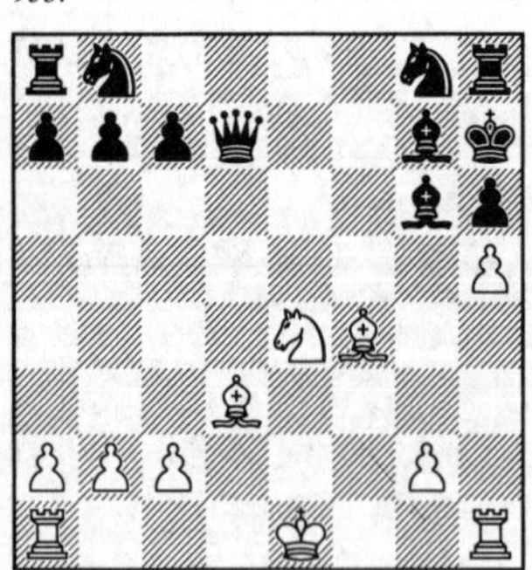

954.

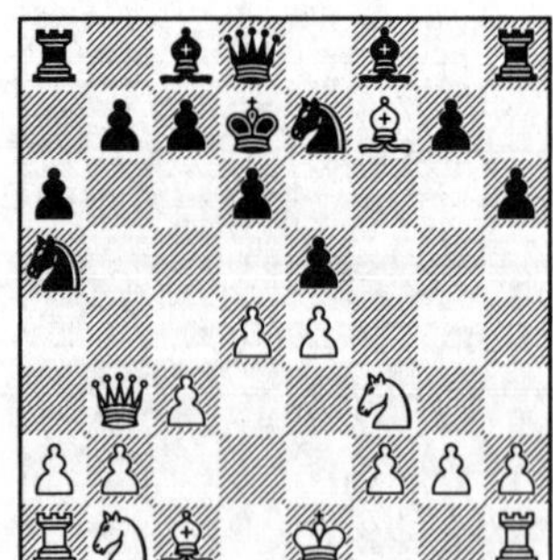

955.

956.

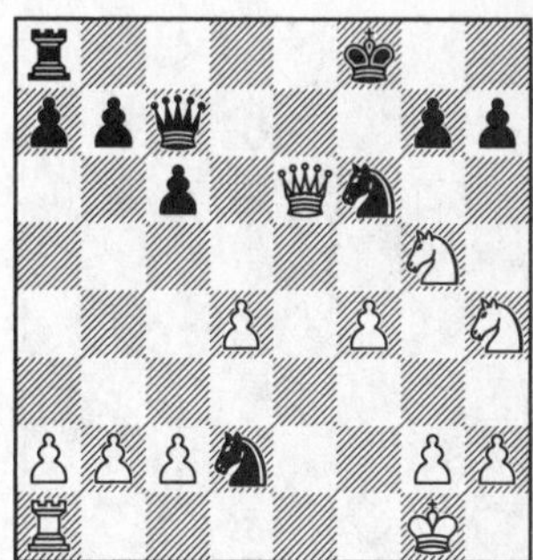

957.

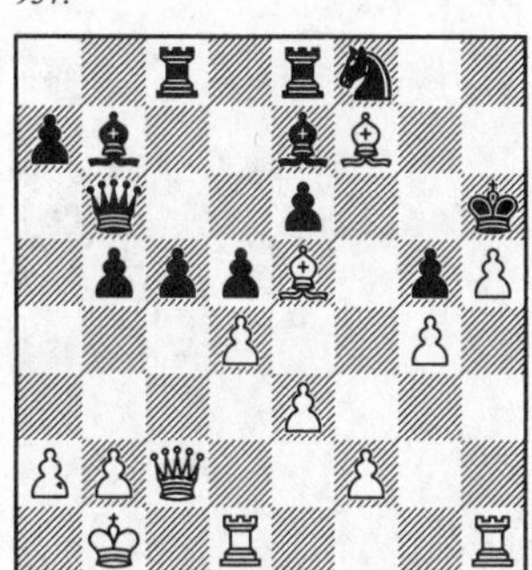

958.

959.

960.

961.

962.

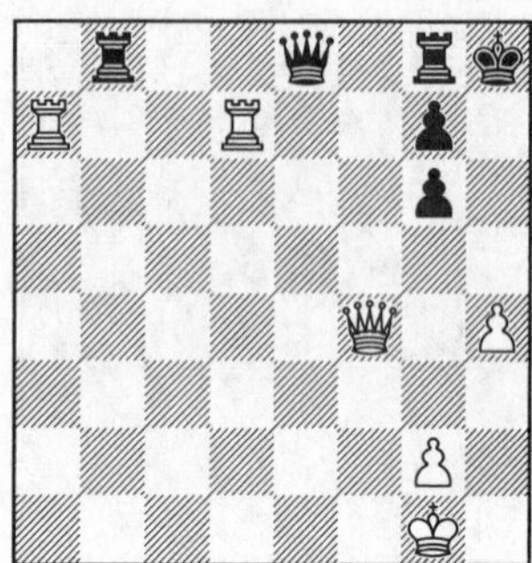

963.

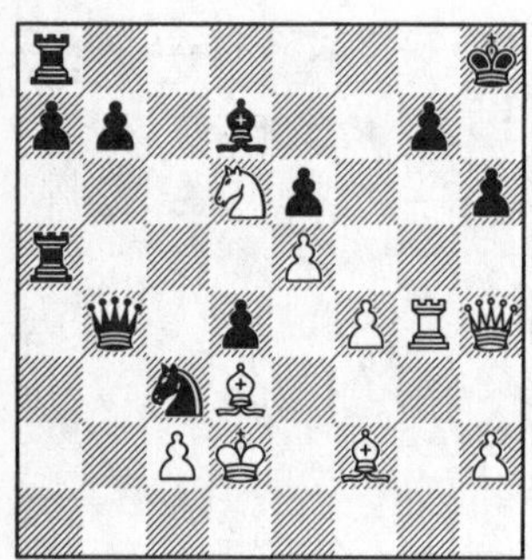

964.

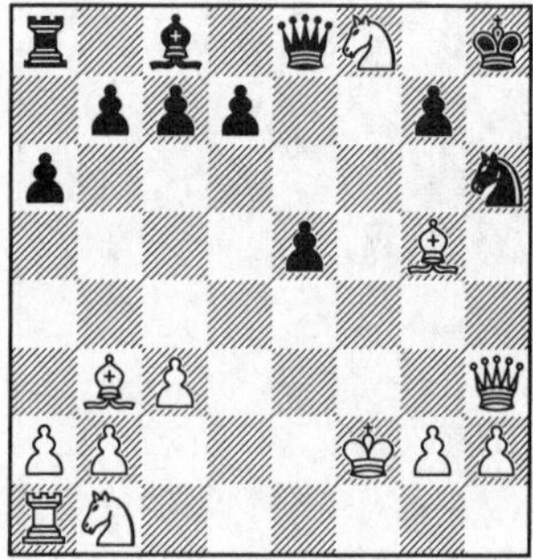

965.

966.

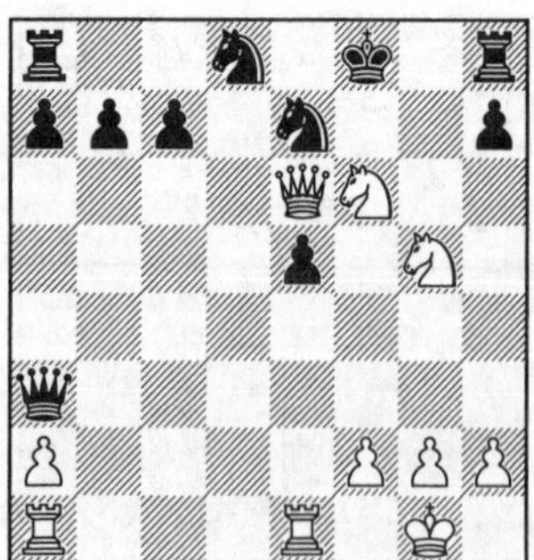

967.

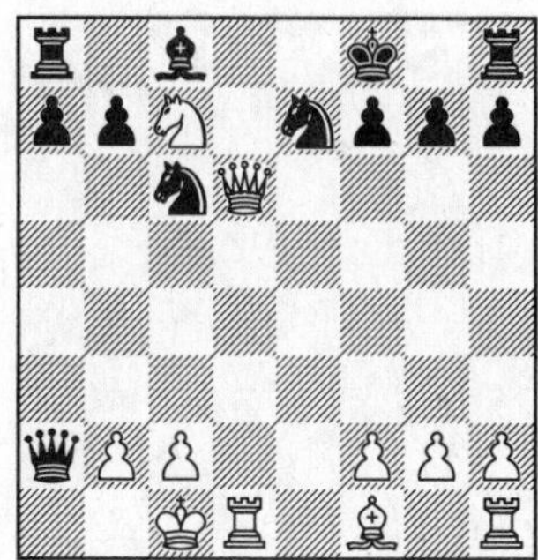

968.

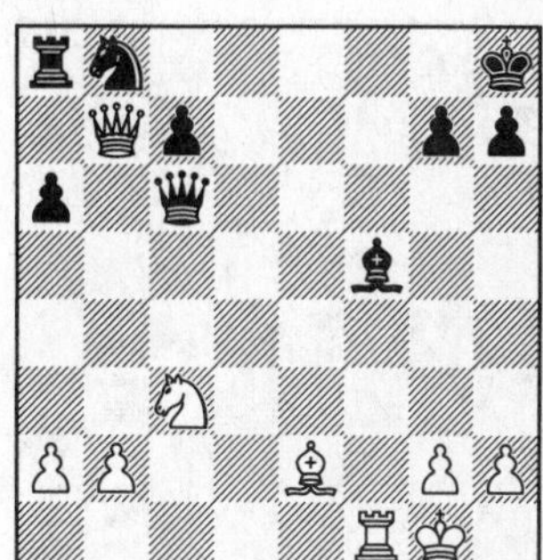

969.

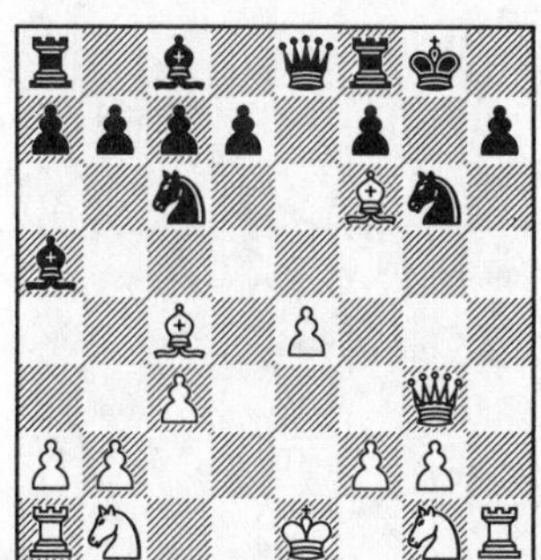

970.

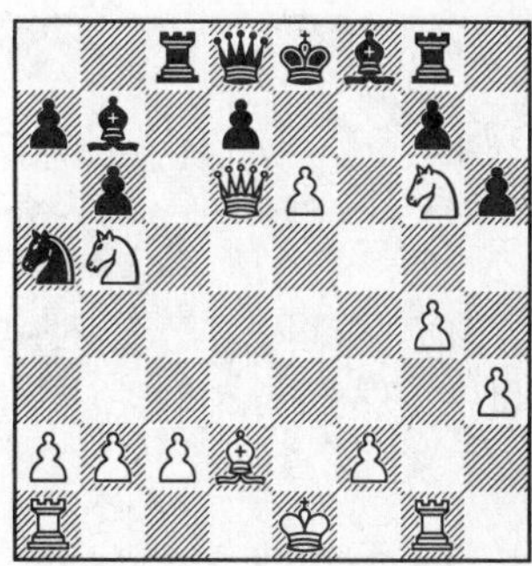

971.

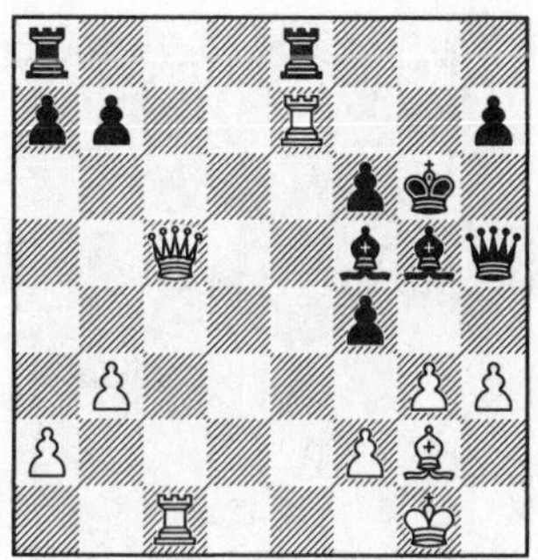

972.

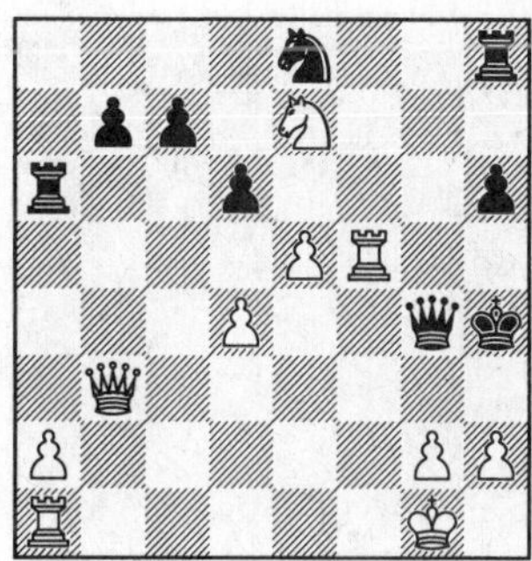

973.

974.

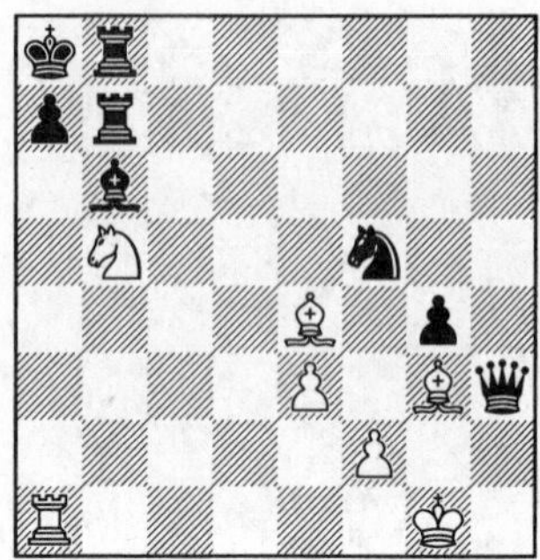

975.

976.

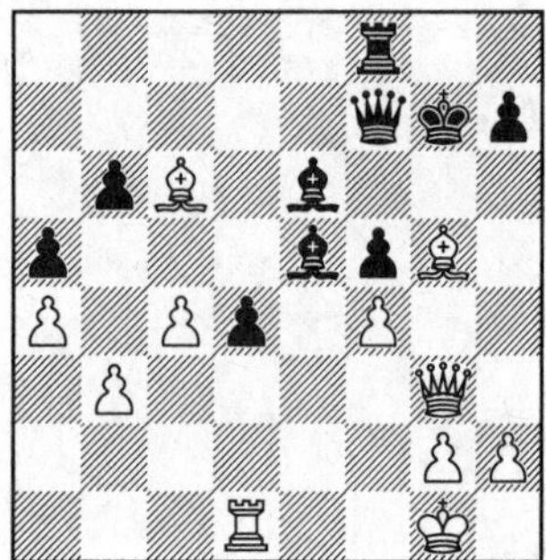

977.

978.

979.

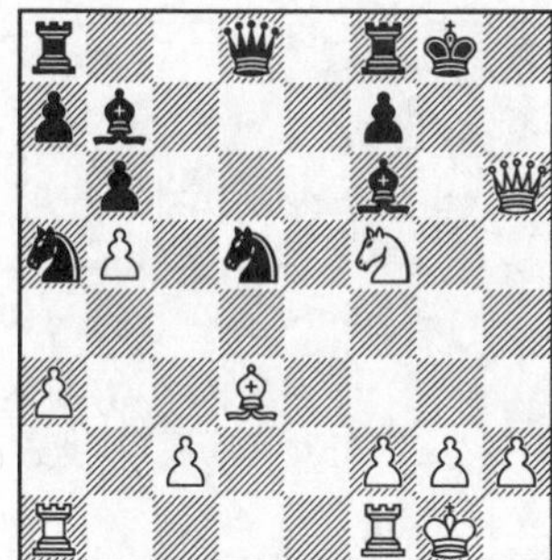

980.

981.

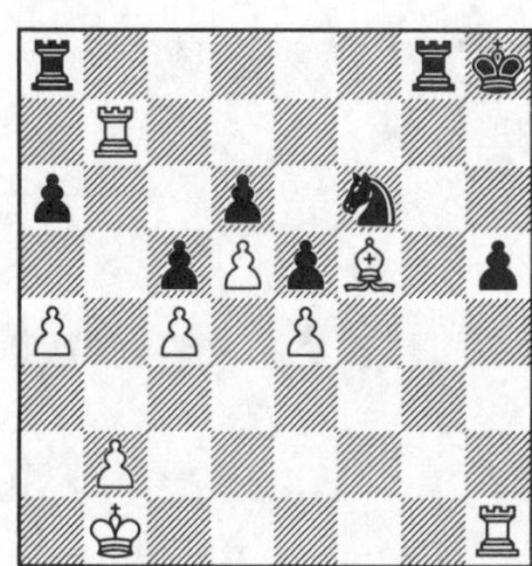

982.

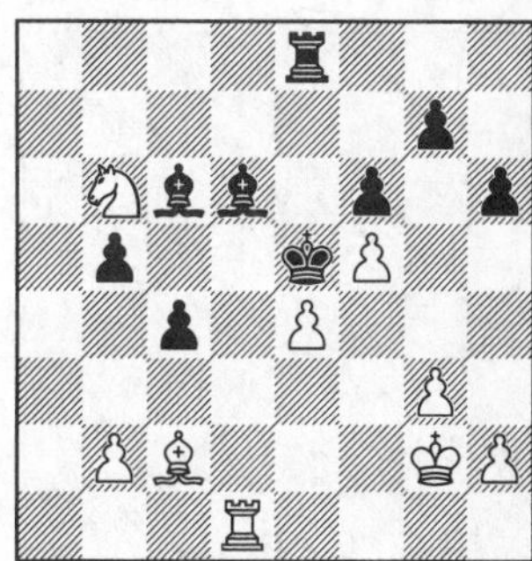

983.

984.

985.

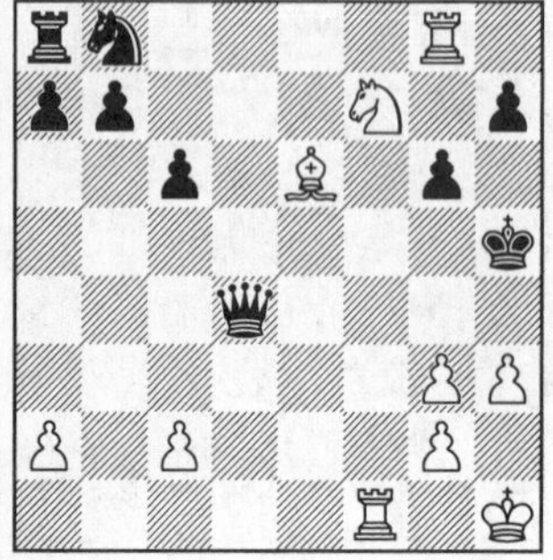

986.

987.

988.

989.

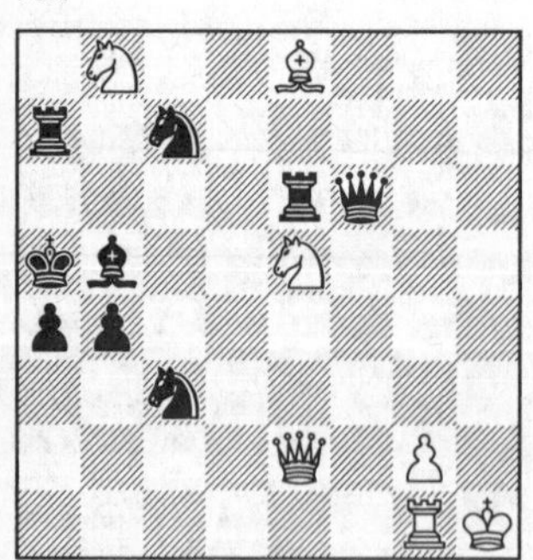

990.

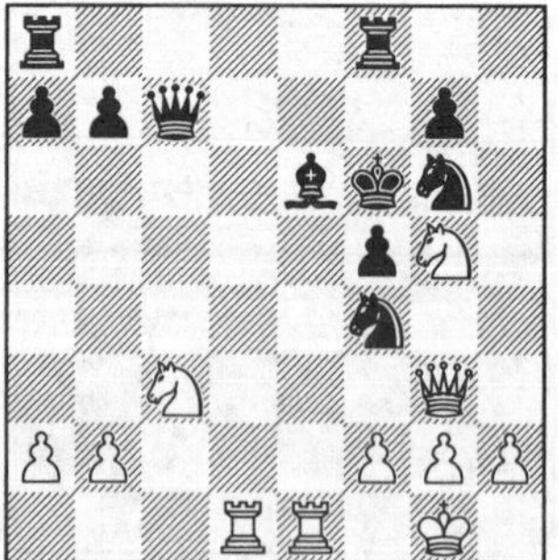

991.

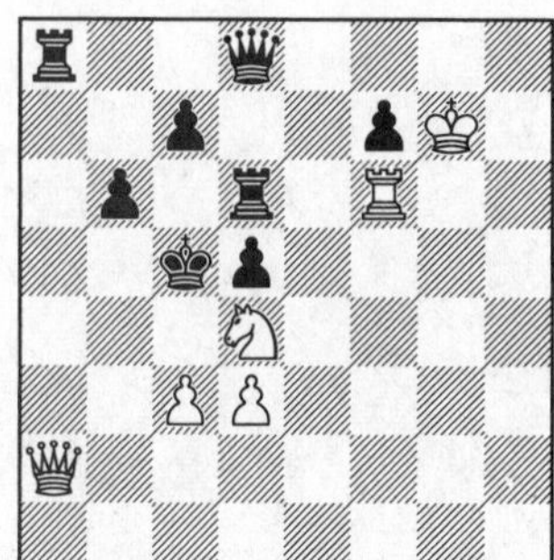

992.

993.

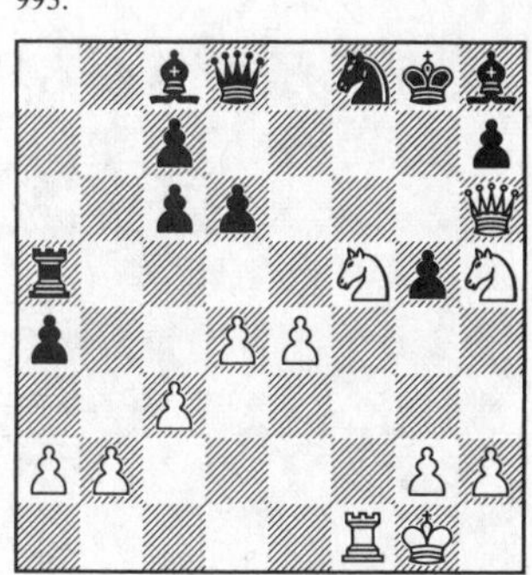

994.

995.

996.

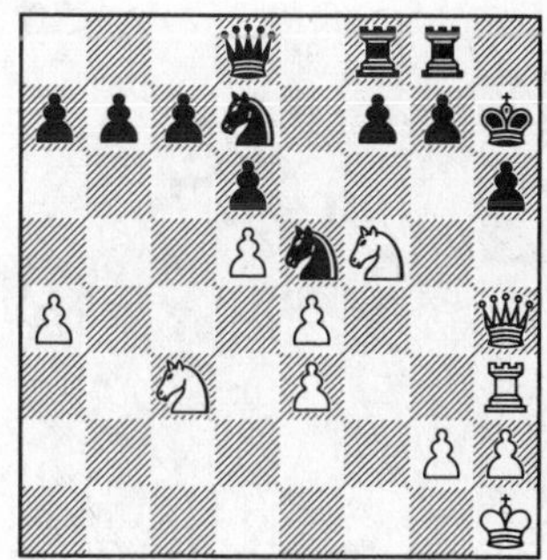

997.

998.

999.

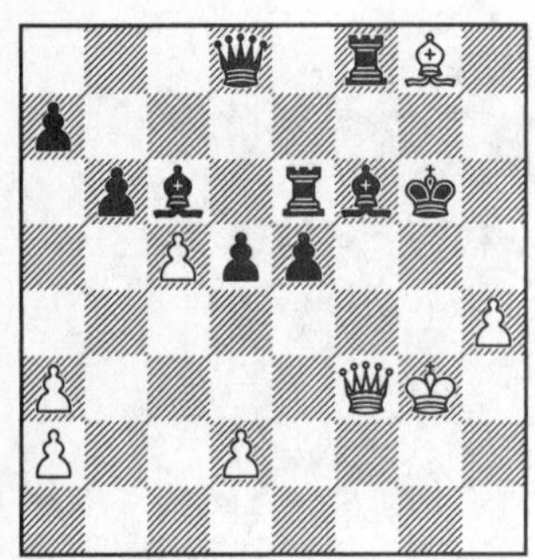

1000.

1001.

1002.

1003.

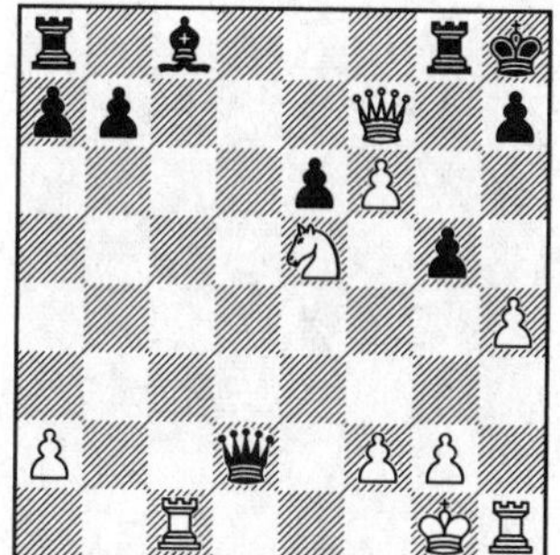

1004.

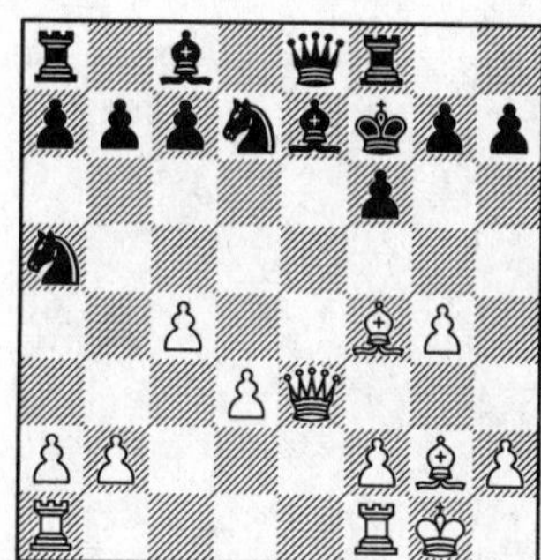

1005.

1006.

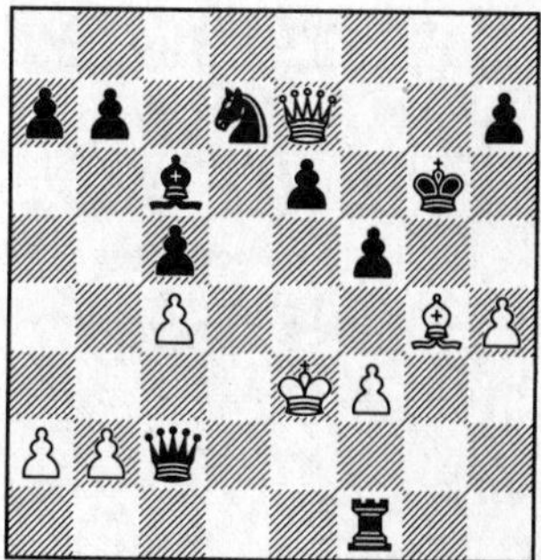

1007.

1008.

1009.

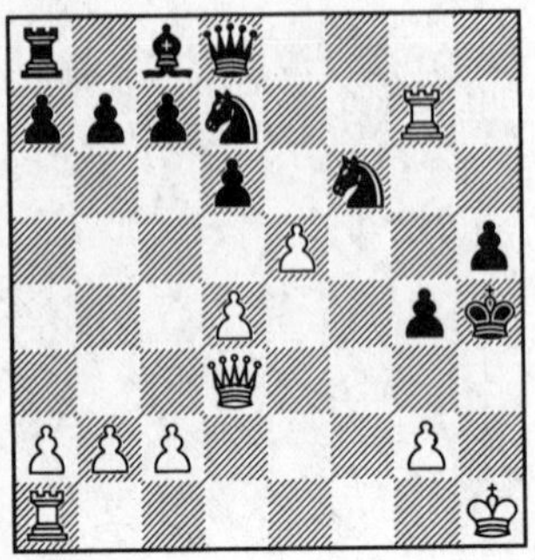

1010.

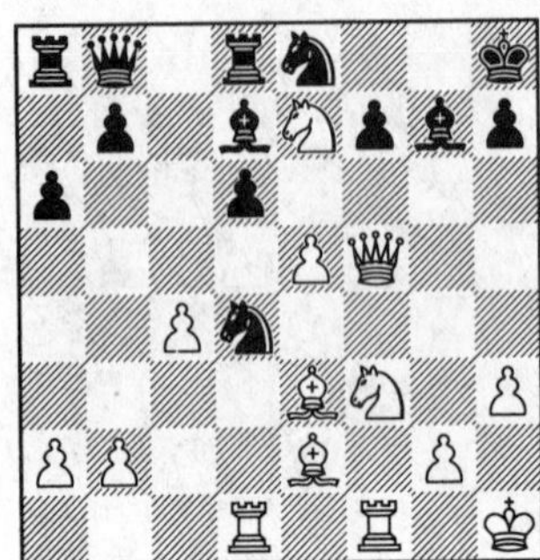

1011.

1012.

1013.

1014.

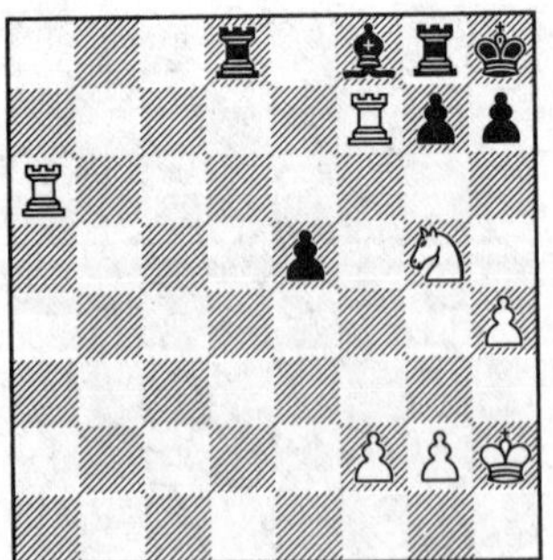

1015.

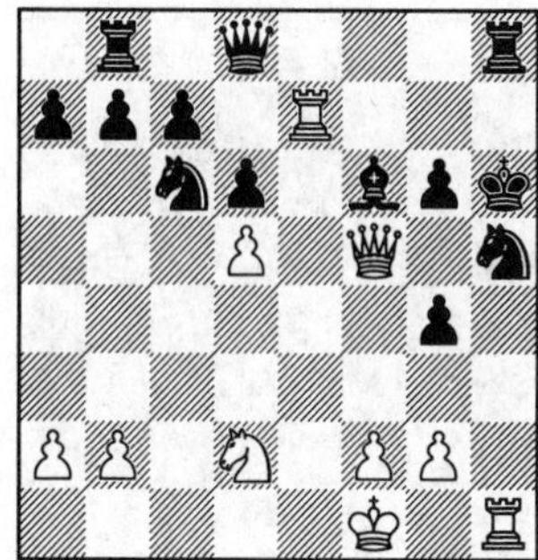

1016.

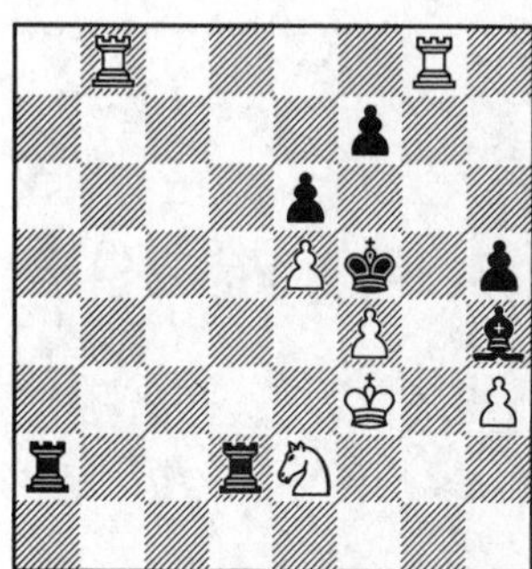

1017.

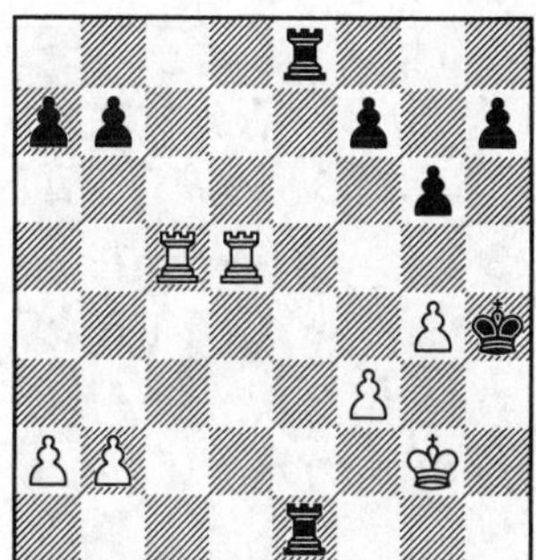

1018.

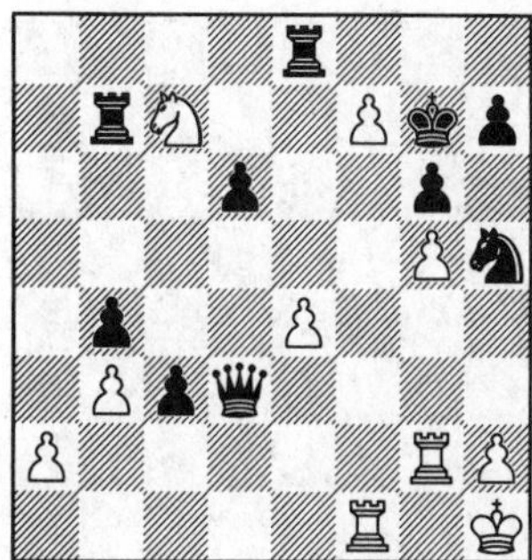

1019.

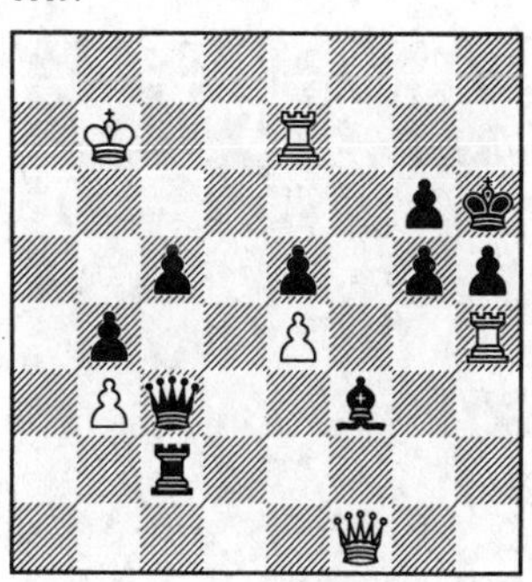

1020.

1021.

1022.

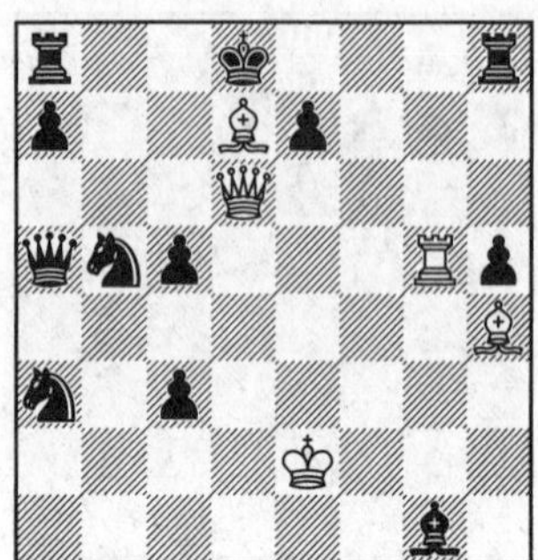

1023.

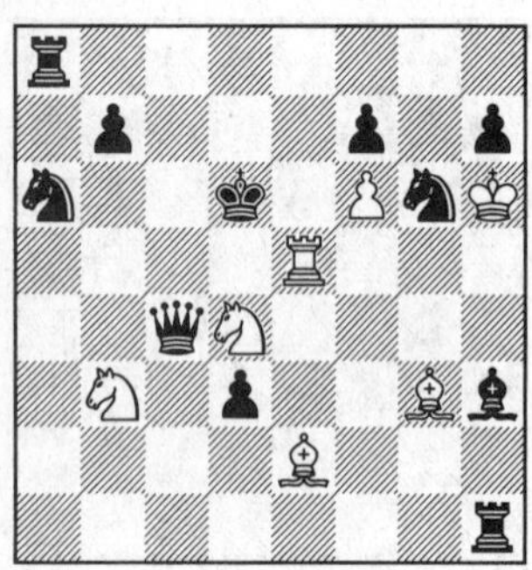

1024.

1025.

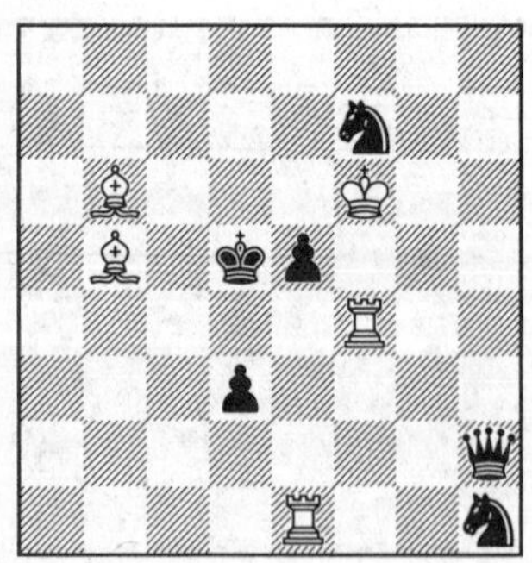

1026.

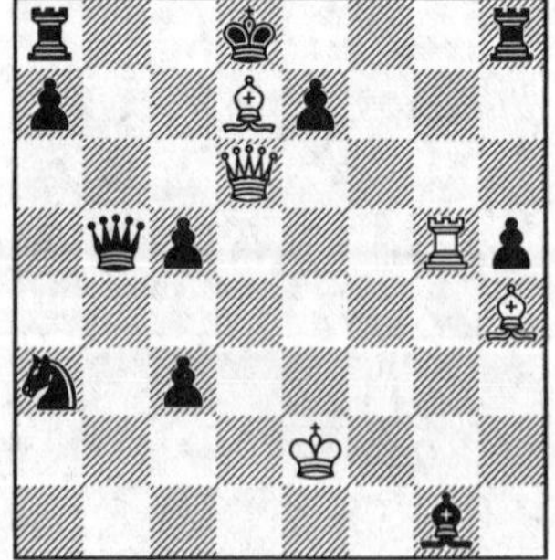

1027.

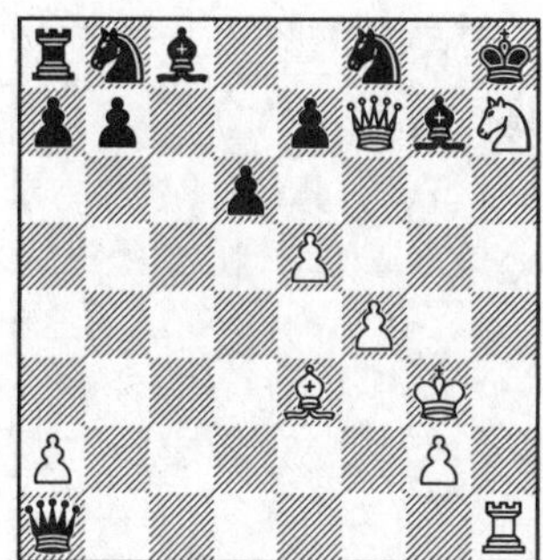

1028.

1029.

1030.

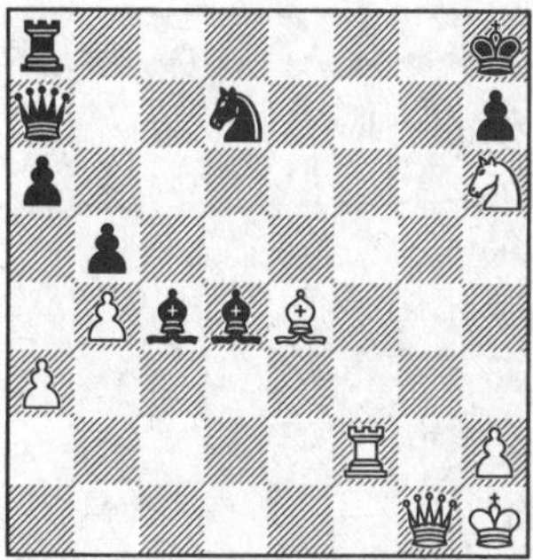

1031.

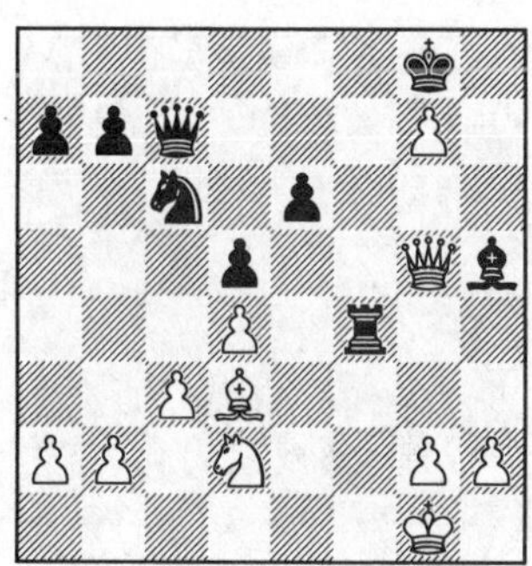

1032.

1033.

1034.

1035.

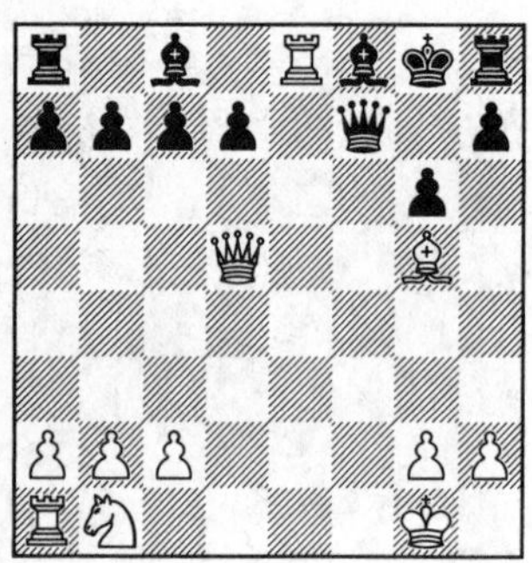

1036.

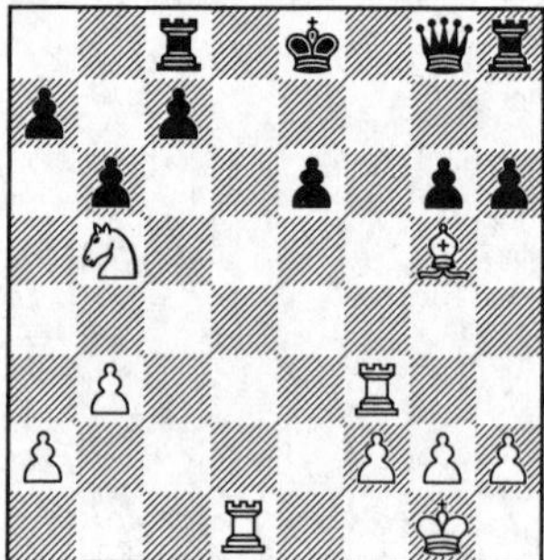

1037.

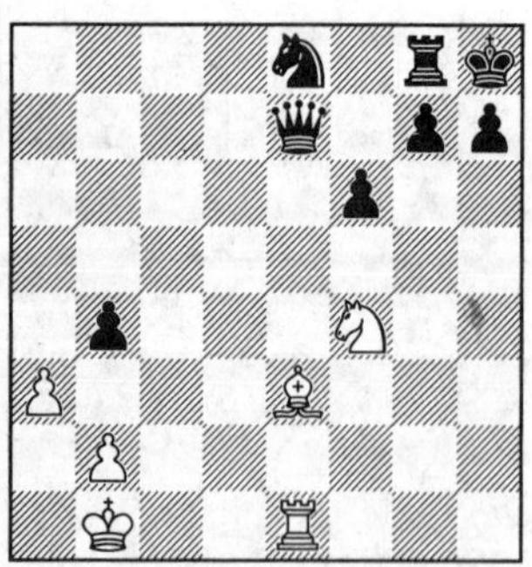

1038.

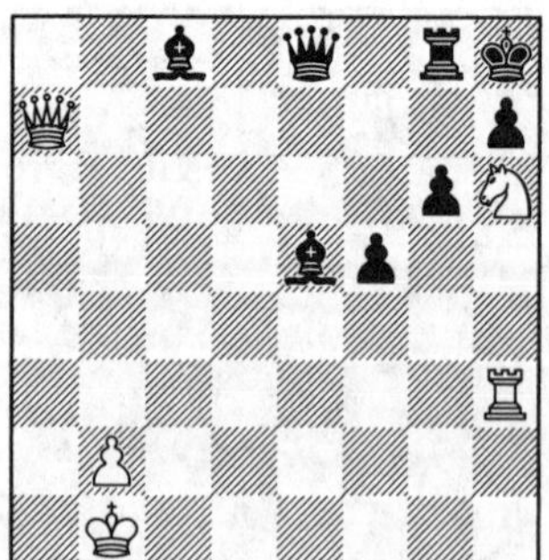

1039.

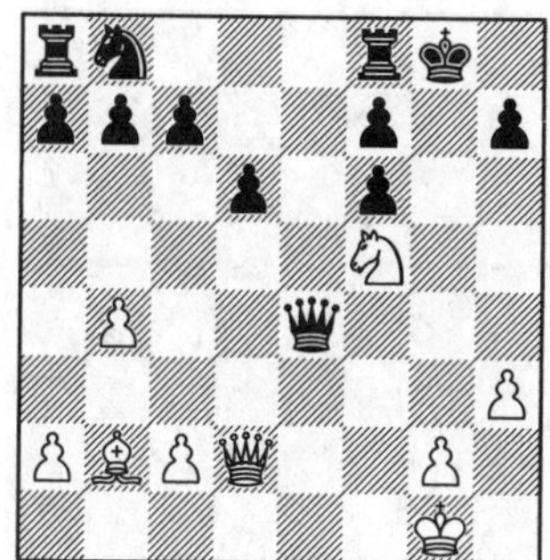

1040.

1041.

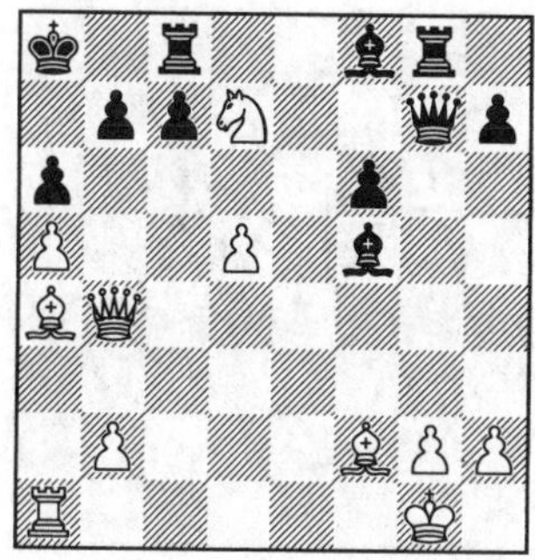

1042.

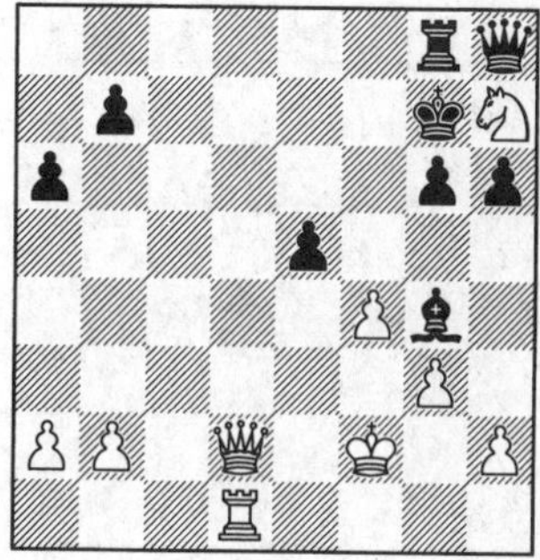

1043.

1044.

1045.

1046.

1047.

1048.

1049.

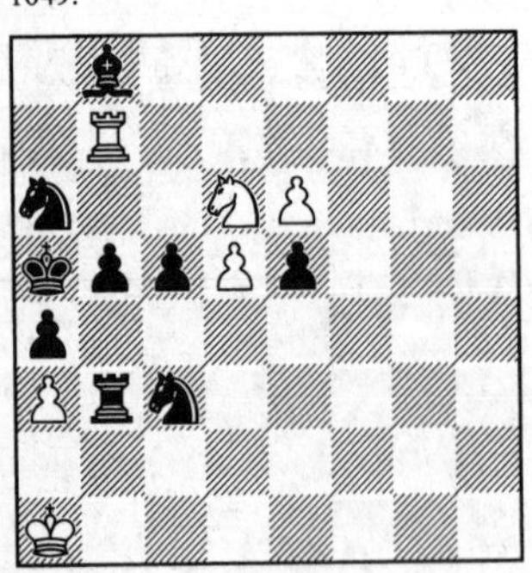

1050.

1051.

1052.

1053.

1054.

1055.

1056.

1057.

1058.

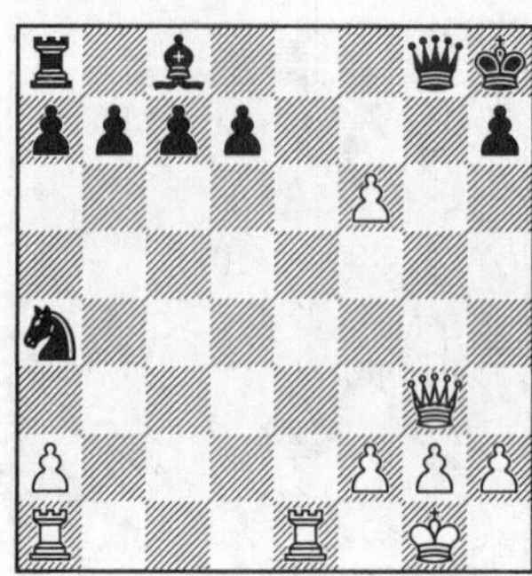

1059.

1060.

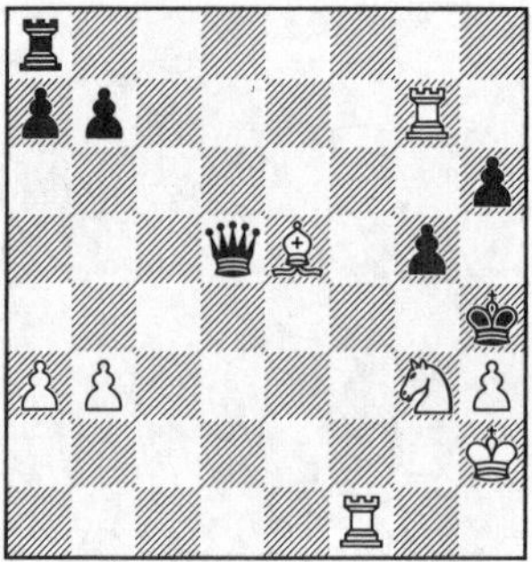

1061.

1062.

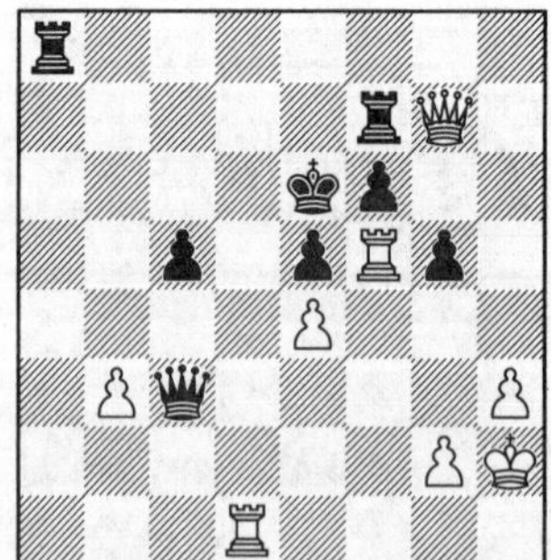

1063.

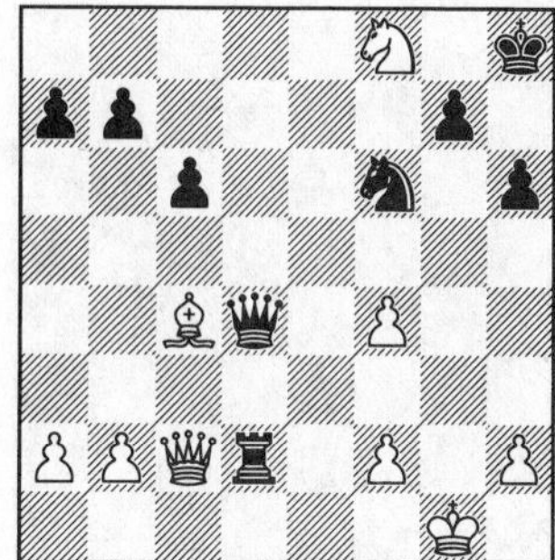

1064.

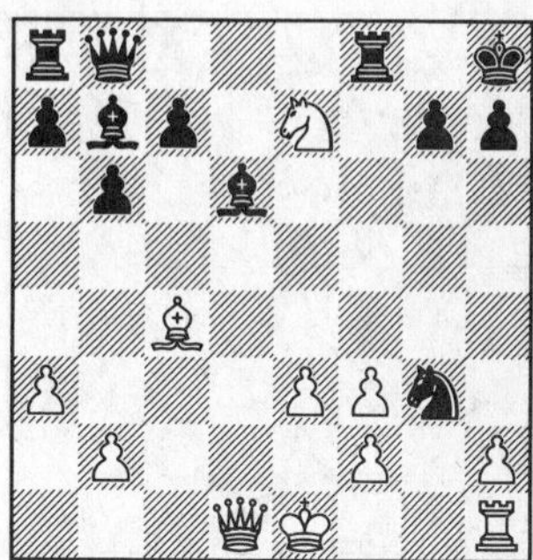

1065.

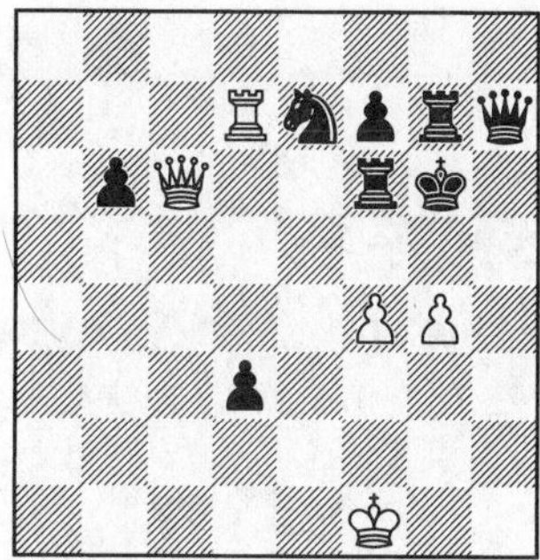

1066.

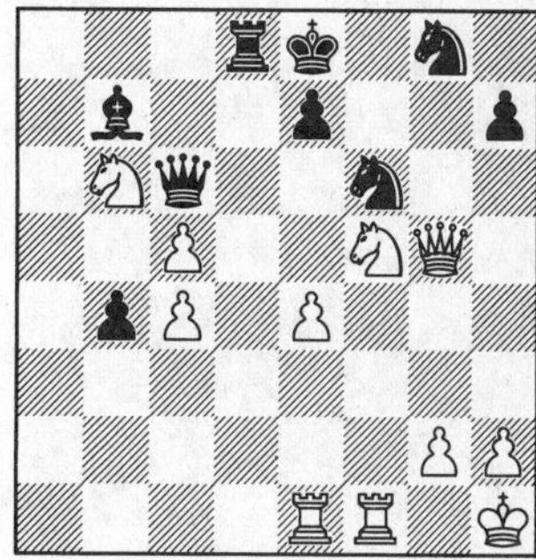

1067.

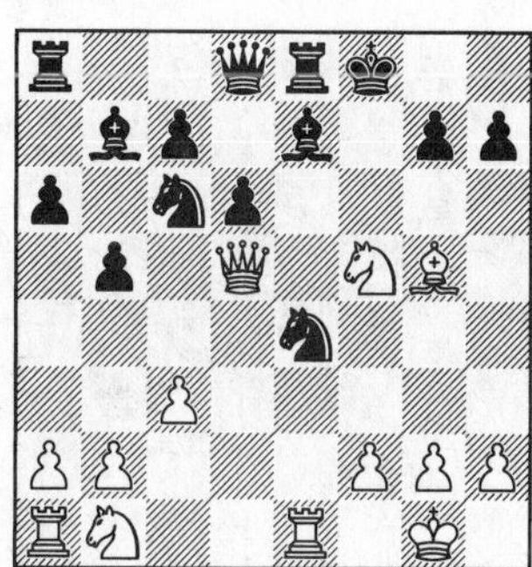

1068.

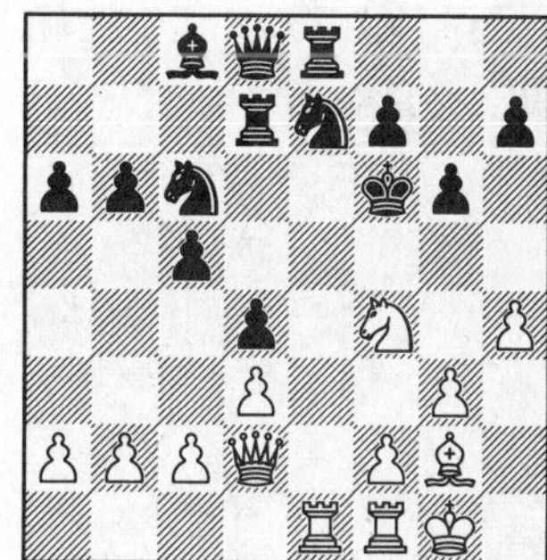

1069.

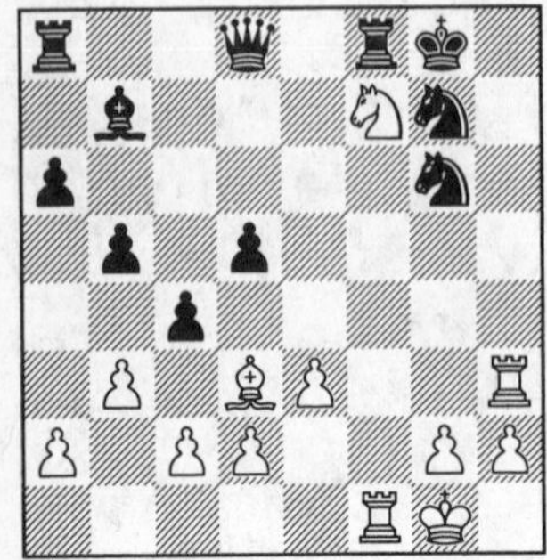

1070.

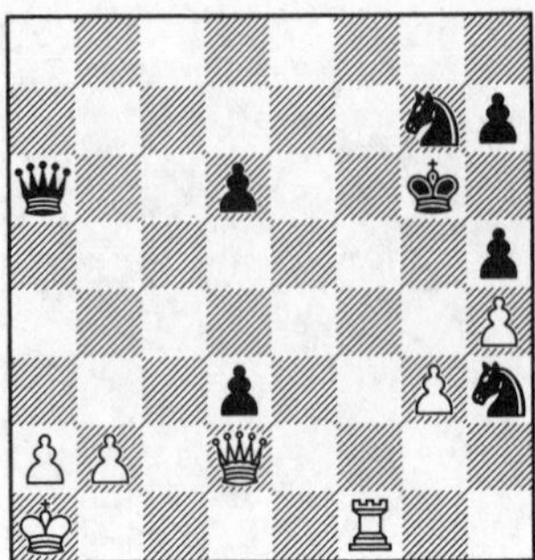

1071.

1072.

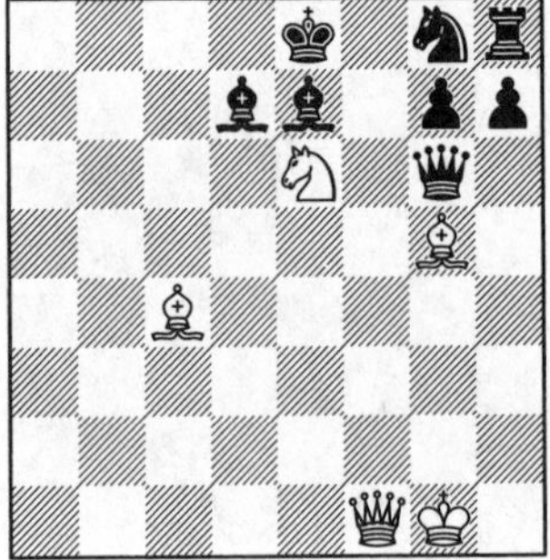

1073.

1074.

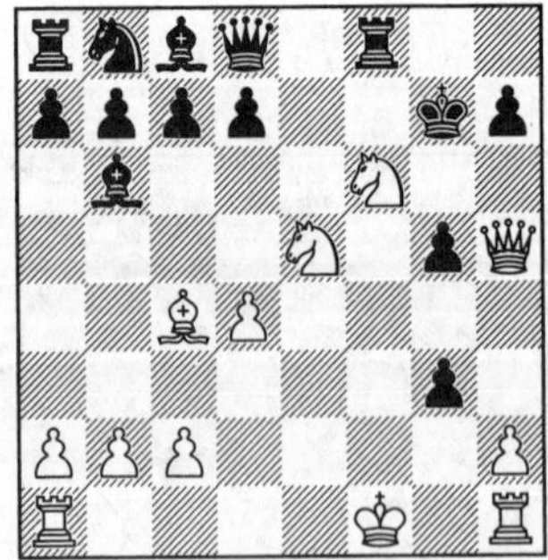

1075.

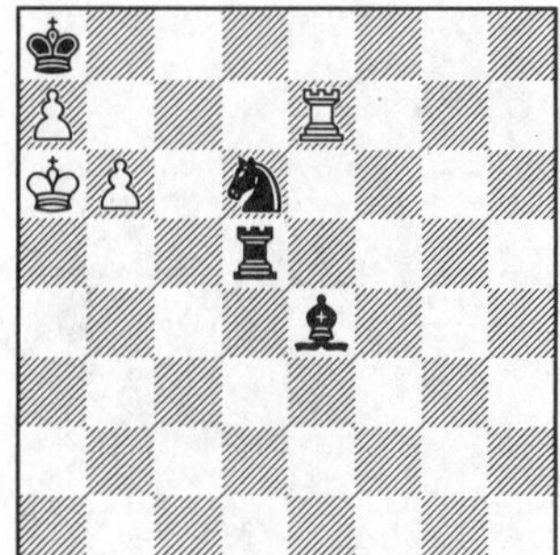

1076.

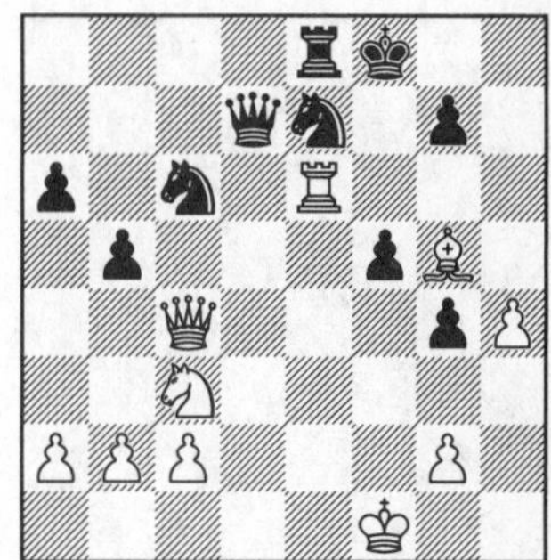

1077.

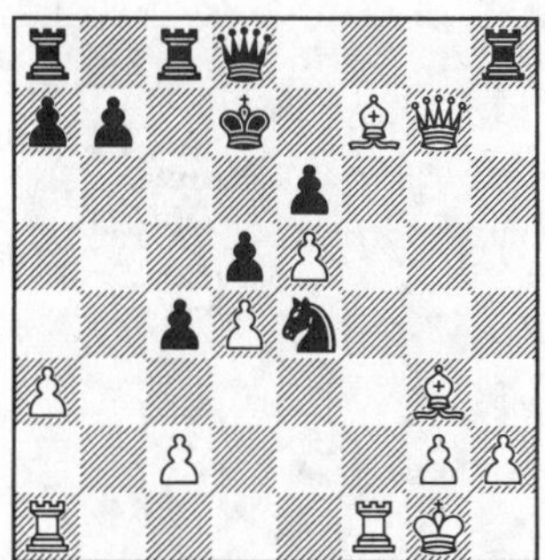

1078.

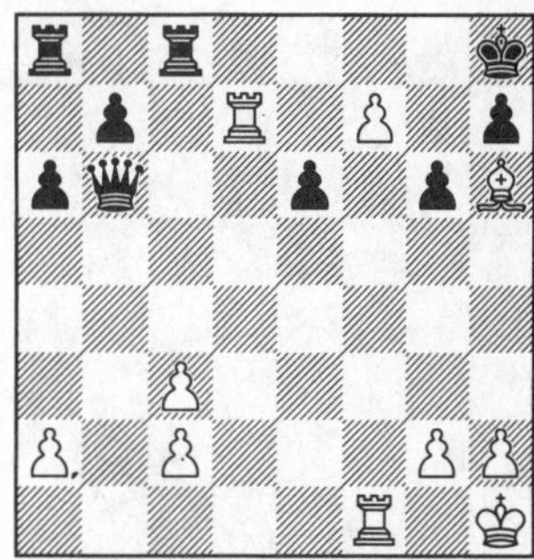

1079.

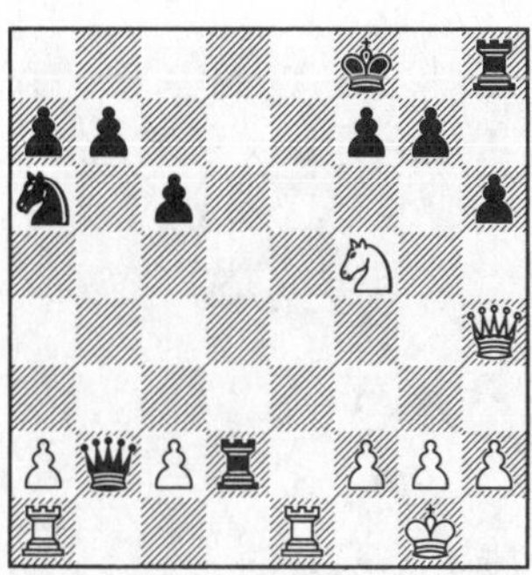

1080.

1081.

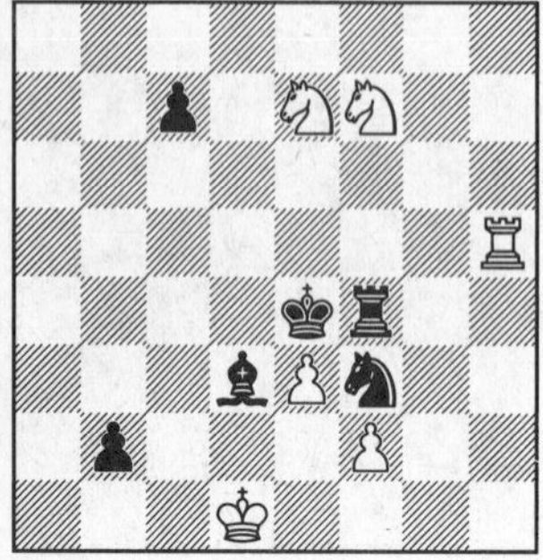

1082.

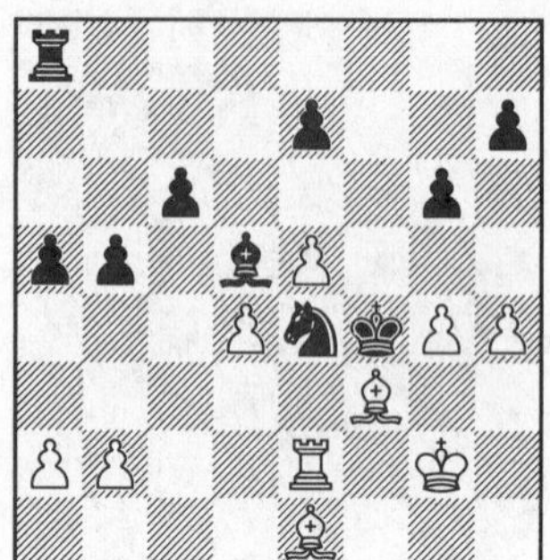

1083.

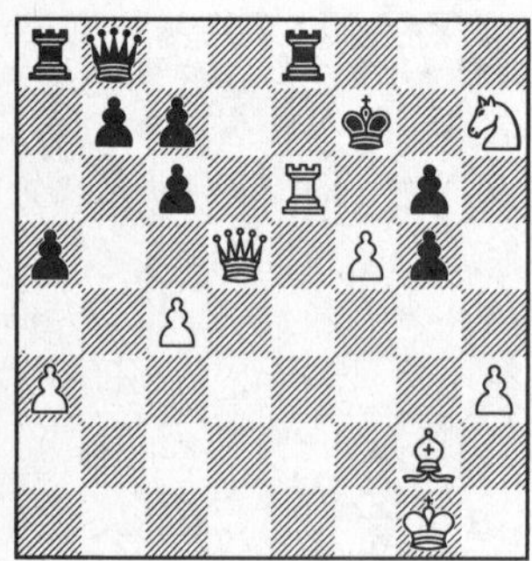

1084.

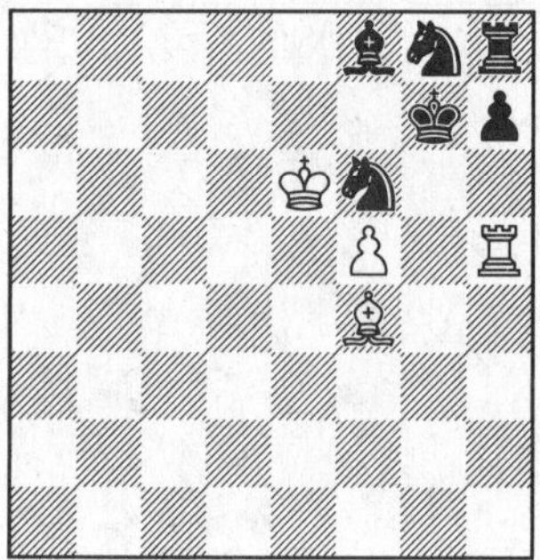

1085.

1086.

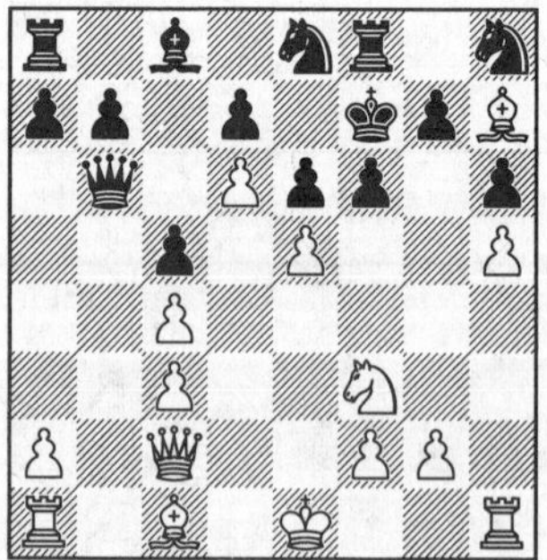

1087.

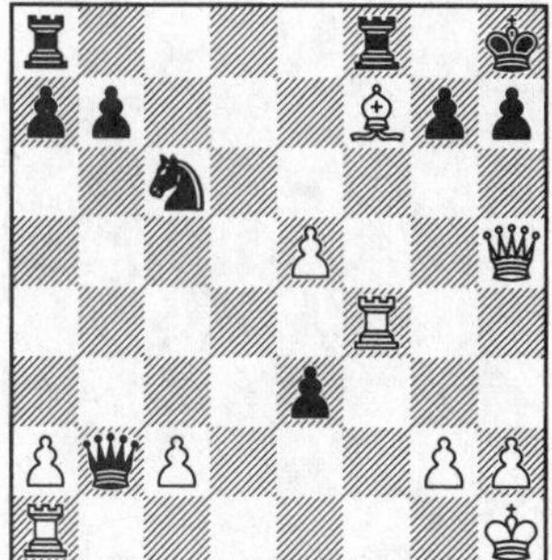

1088.

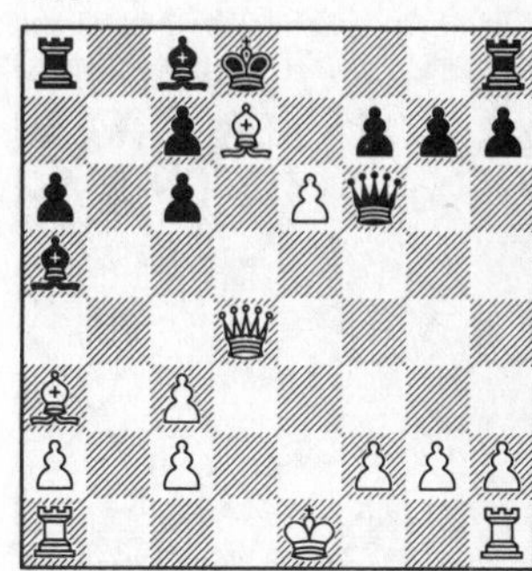

1089.

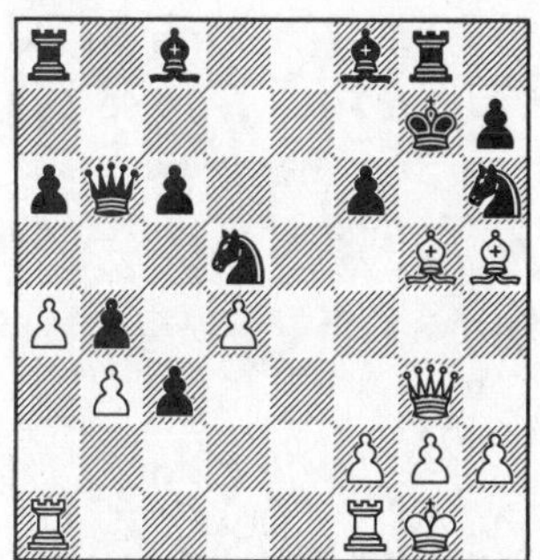

1090.

1091.

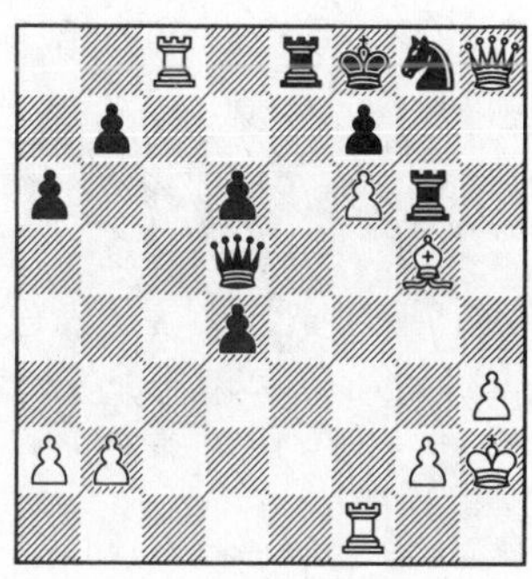

1092.

1093.

1094.

1095.

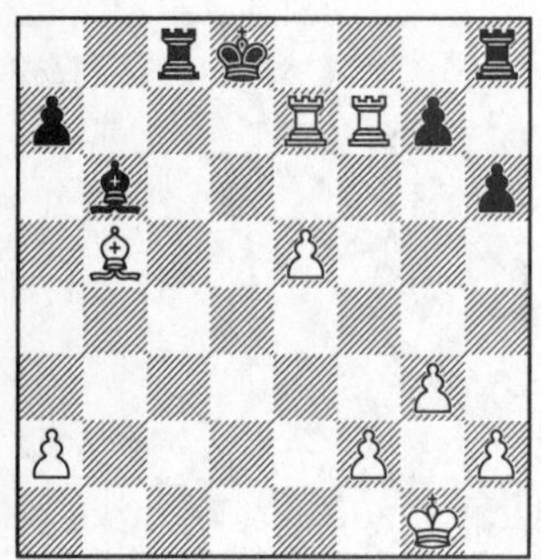

1096.

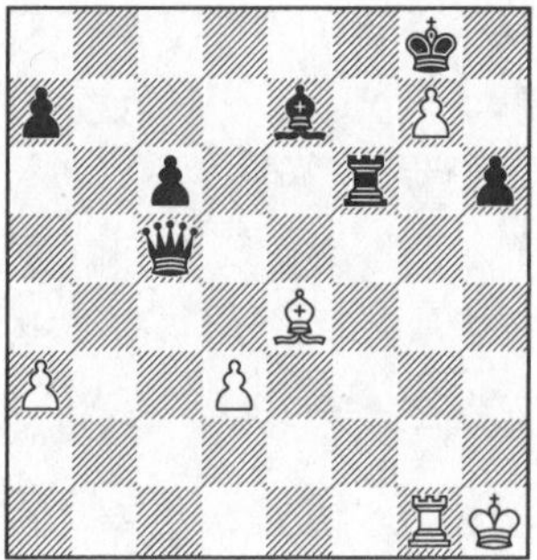

1097.

1098.

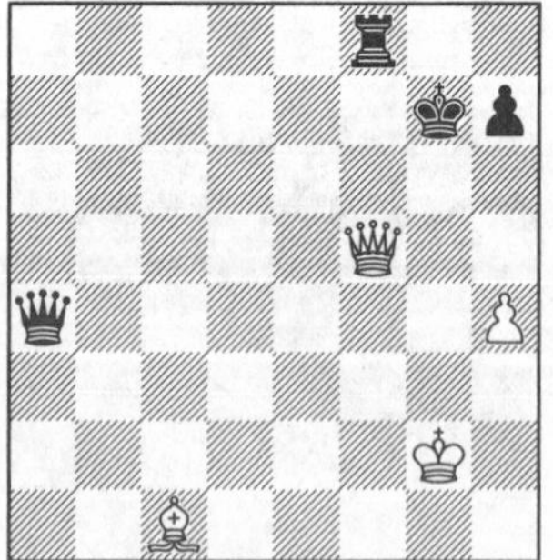

1099.

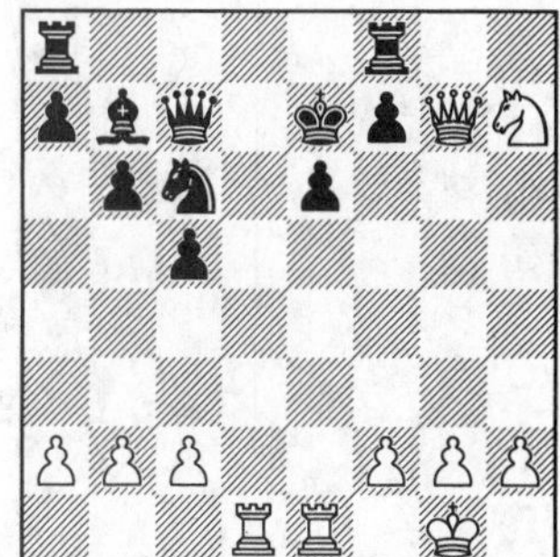

1100.

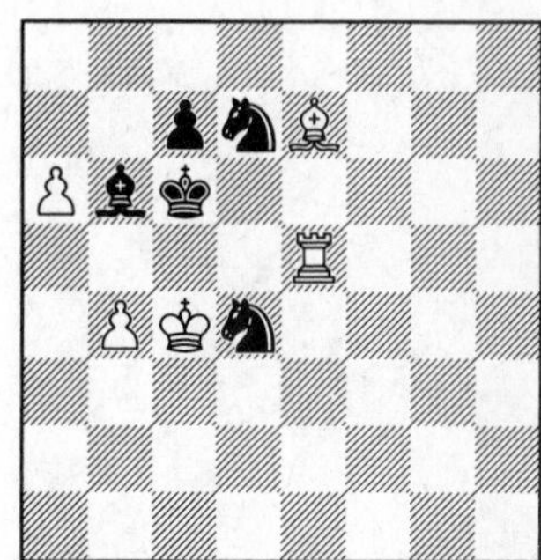

1101.

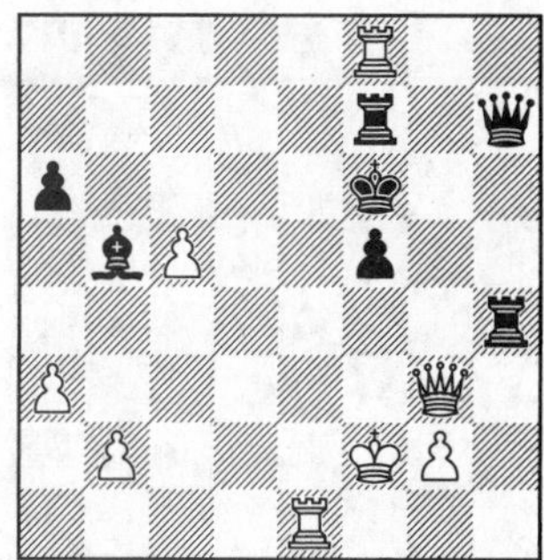

1102.

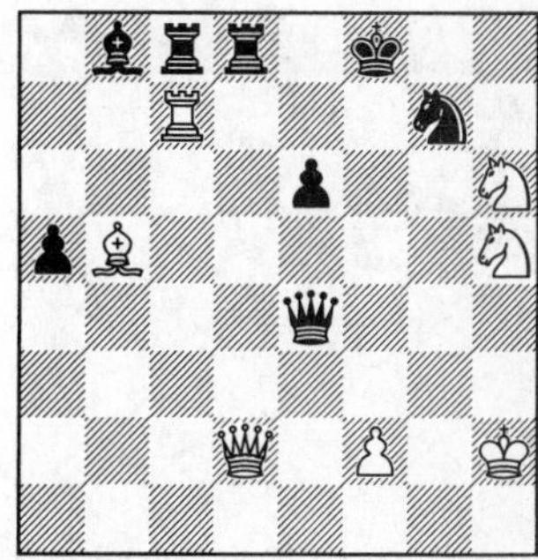

1103.

1104.

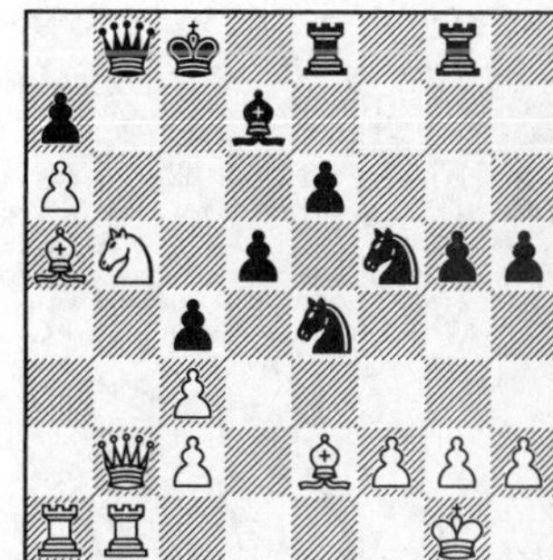

1105.

1106.

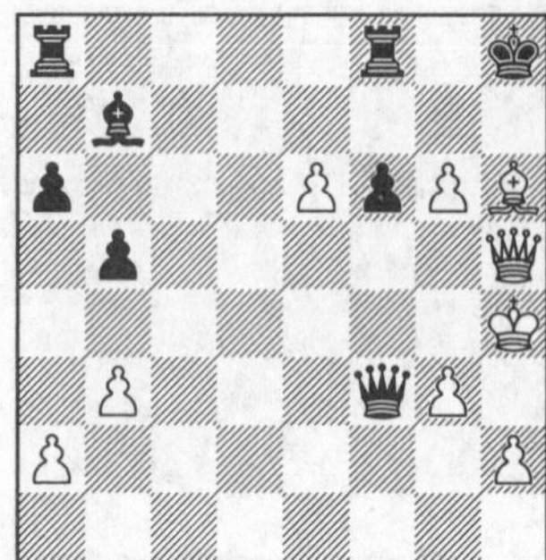

1107.

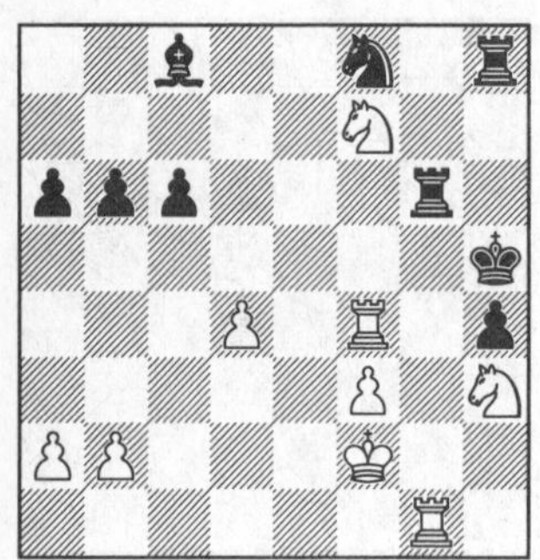

1108.

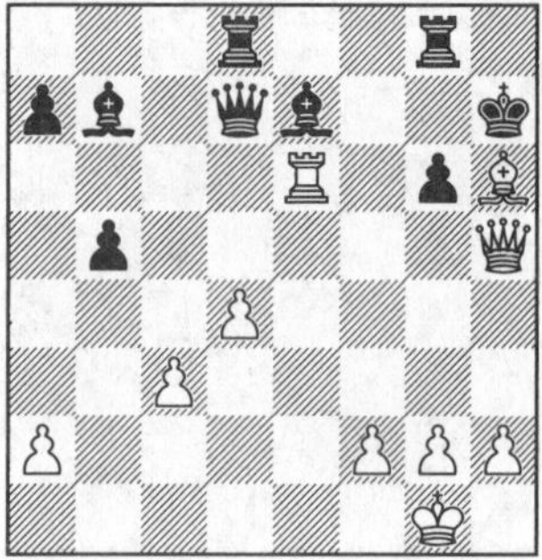

1109.

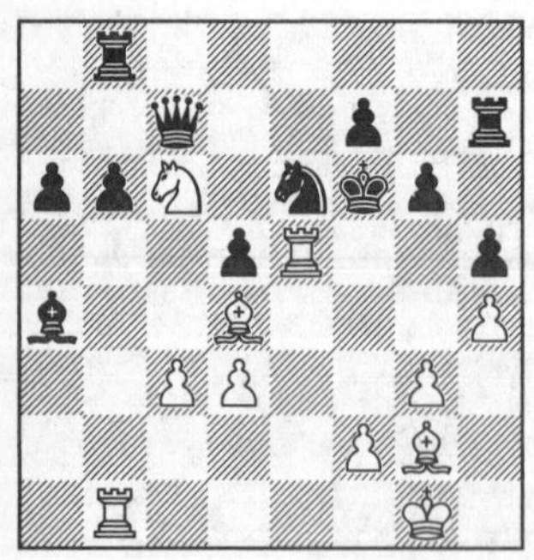

1110.

1111.

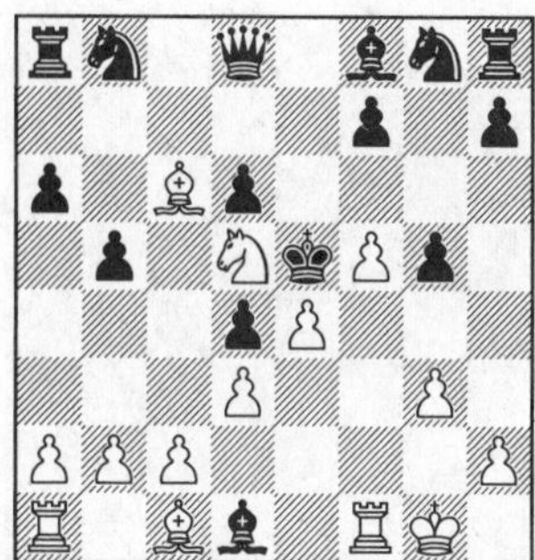

1112.

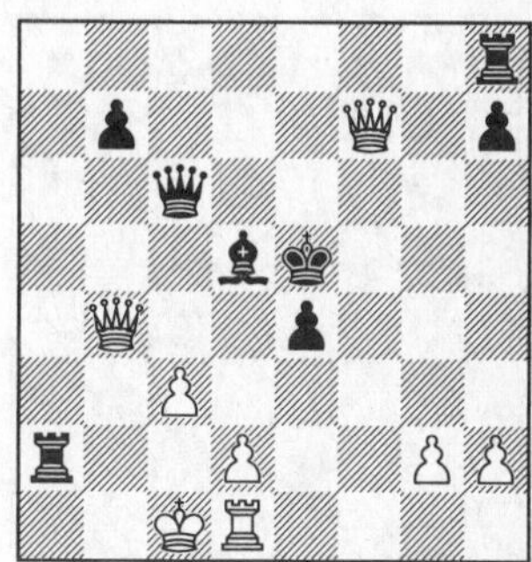

1113.

1114.

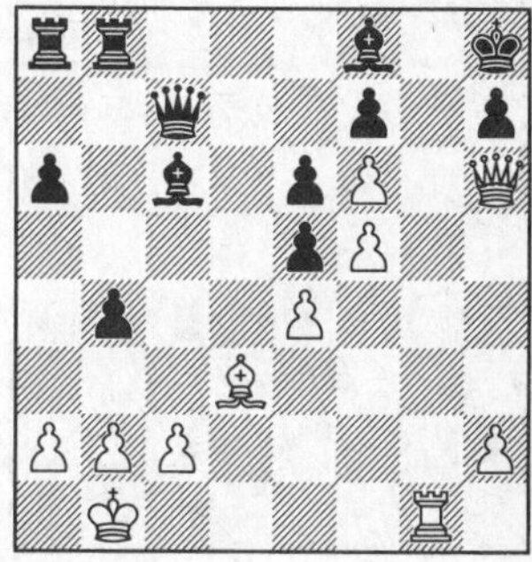

1115.

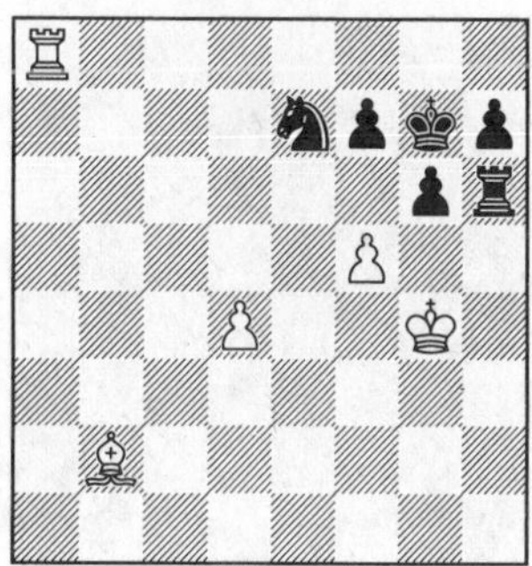

1116.

1117.

1118.

1119.

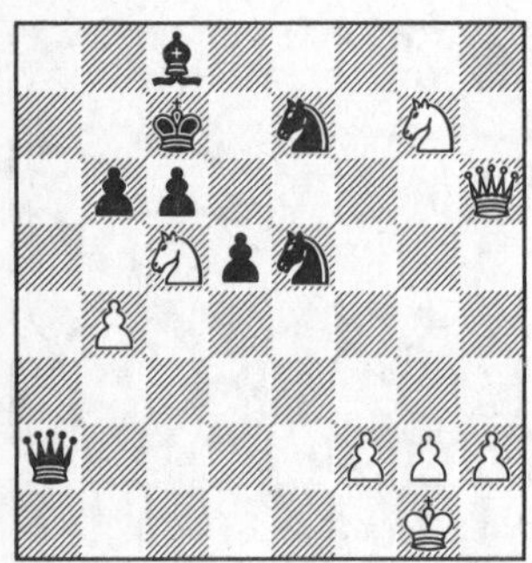

1120.

1121.

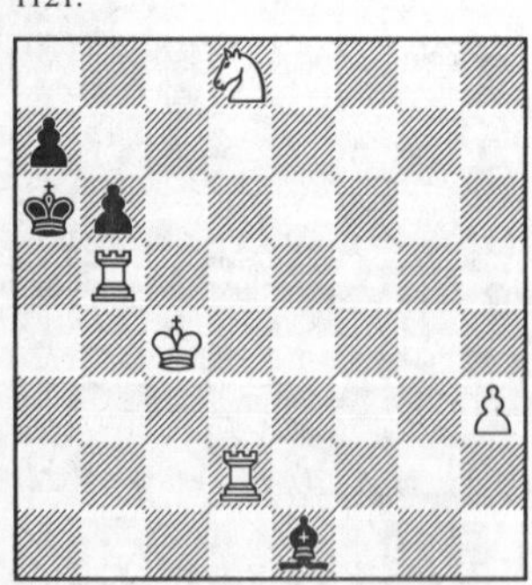

1122.

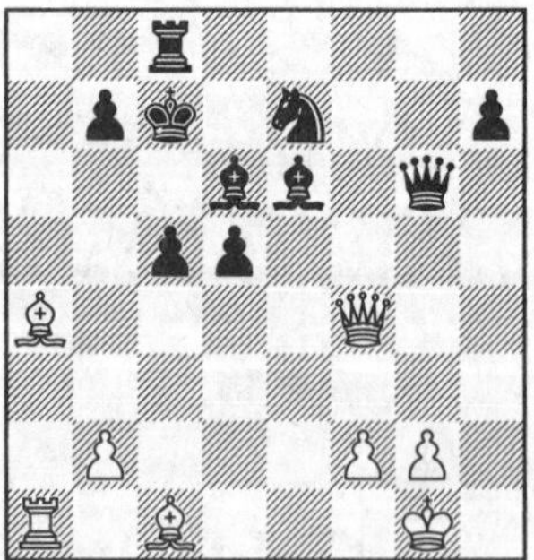

1123.

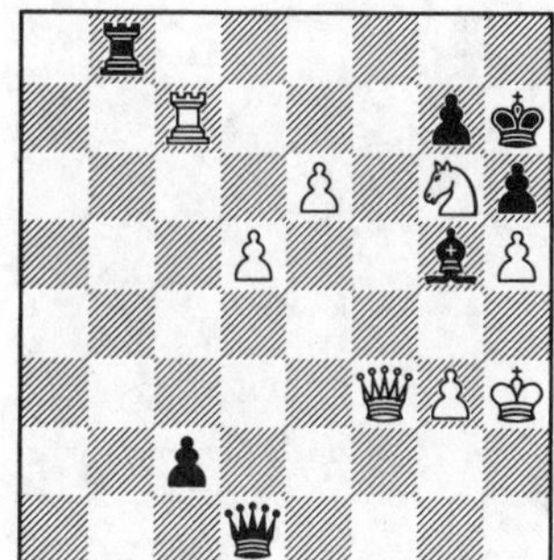

1124.

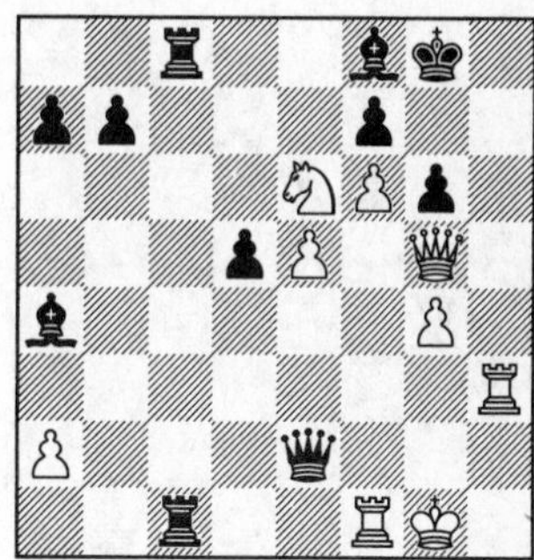

1125.

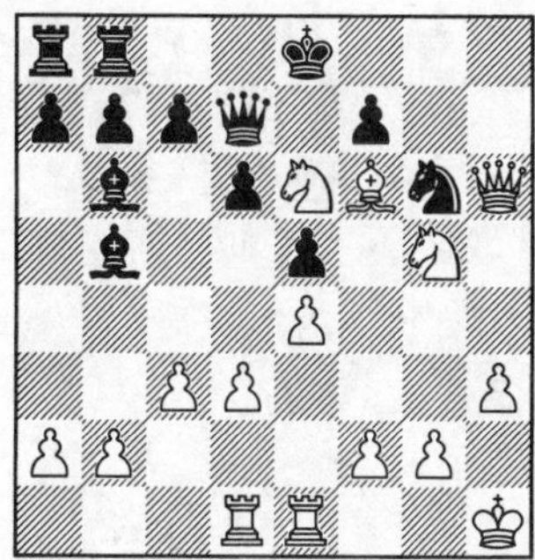

1126.

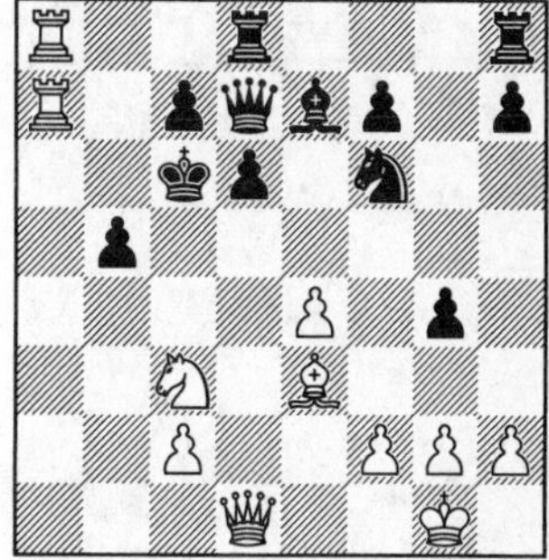

1127.

1128.

1129.

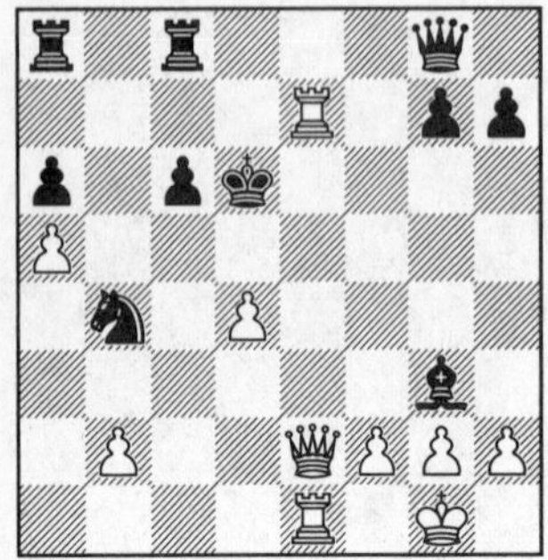

1130.

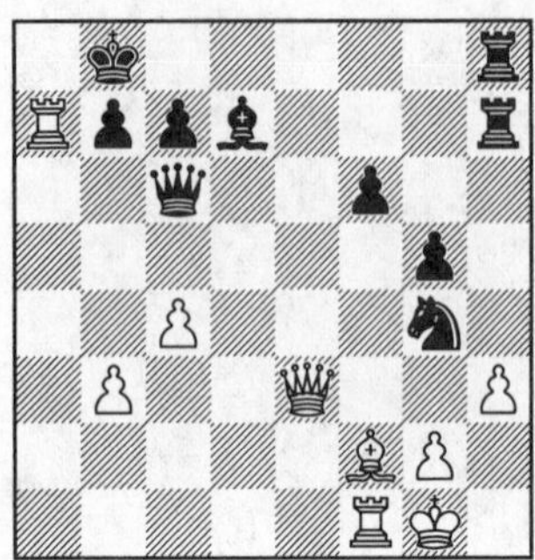

1131.

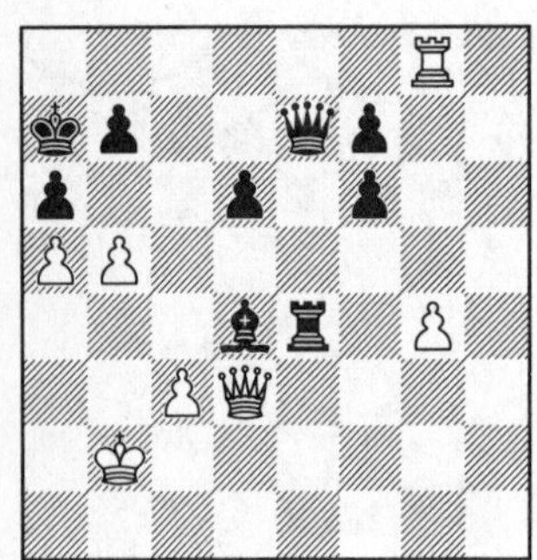

1132.

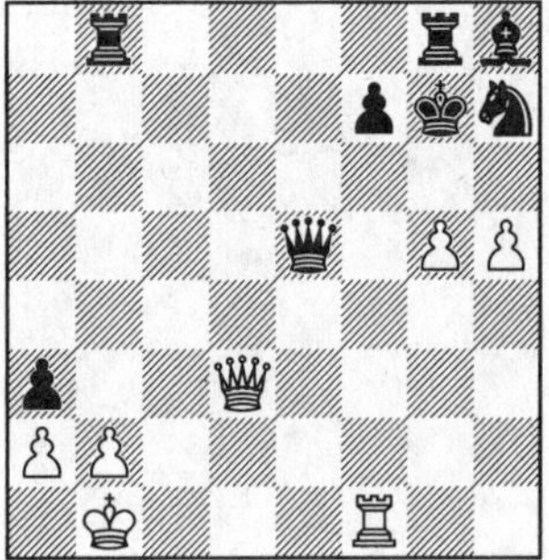

1133.

1134.

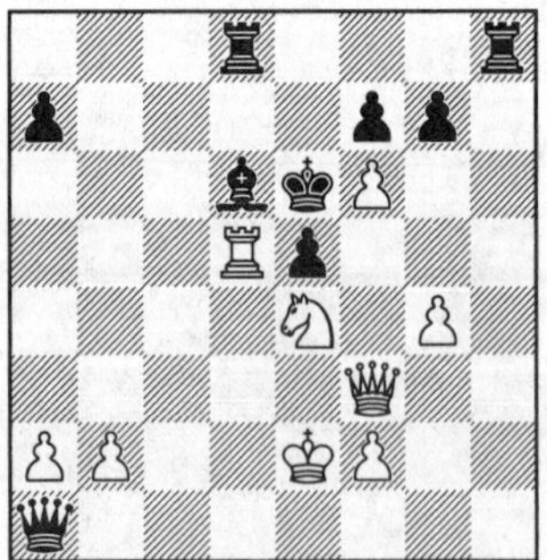

1135.

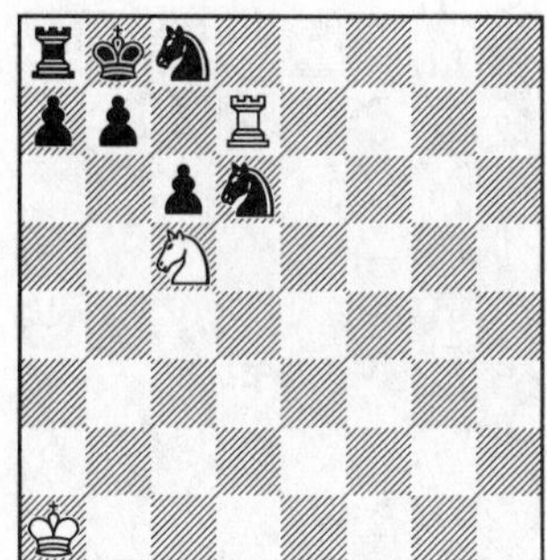

1136.

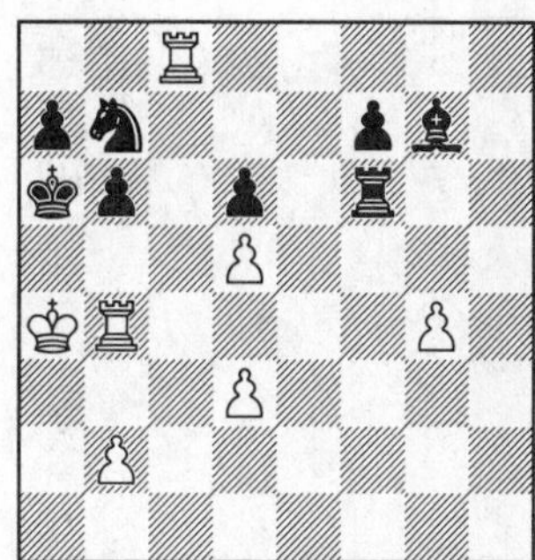

1137.

1138.

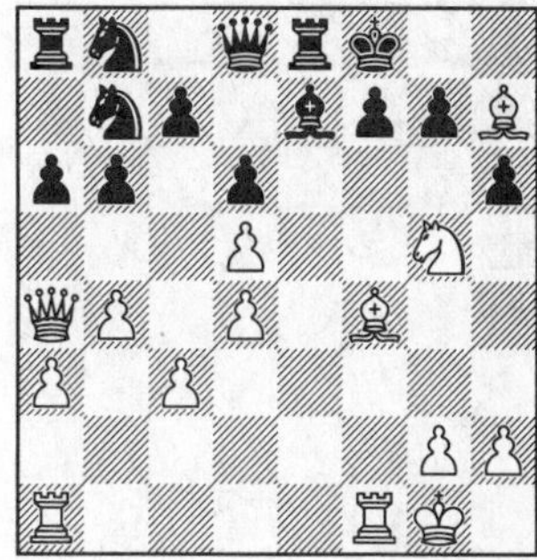

1139.

1140.

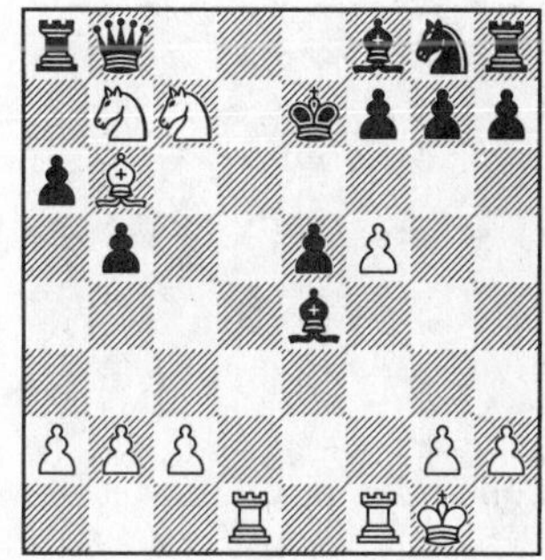

1141.

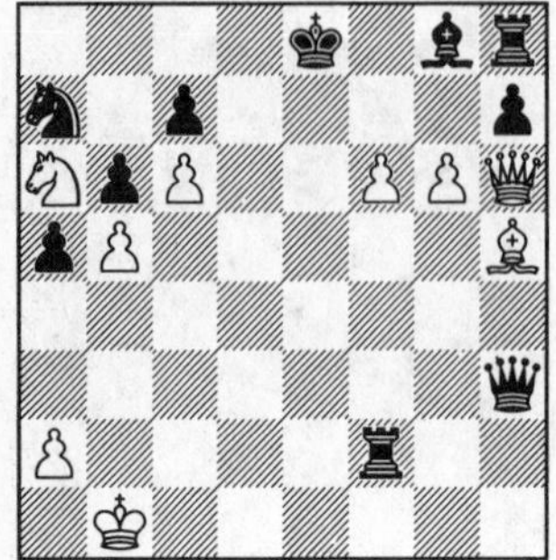

1142.

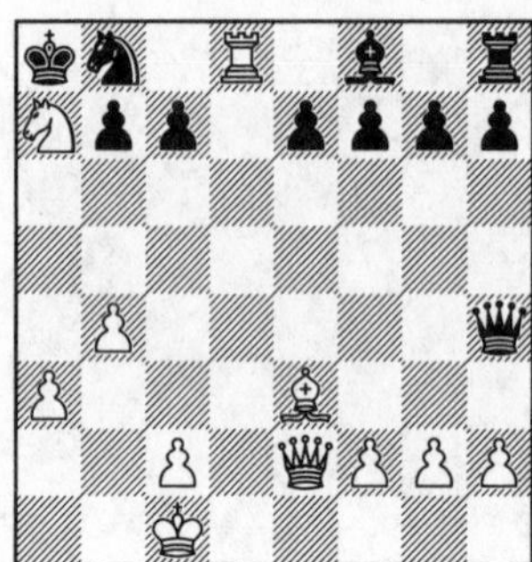

1143.

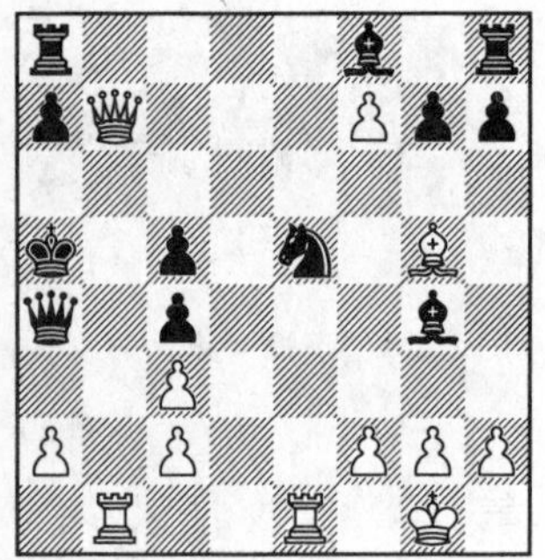

1144.

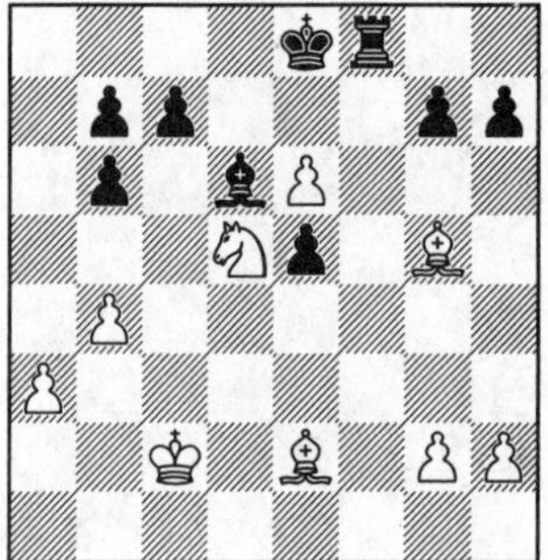

1145.

1146.

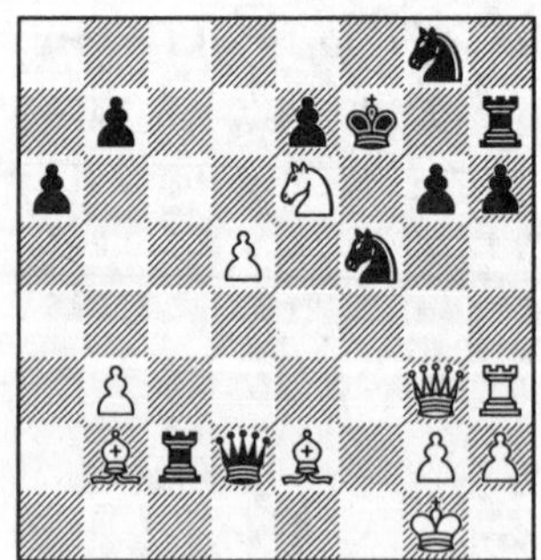

1147.

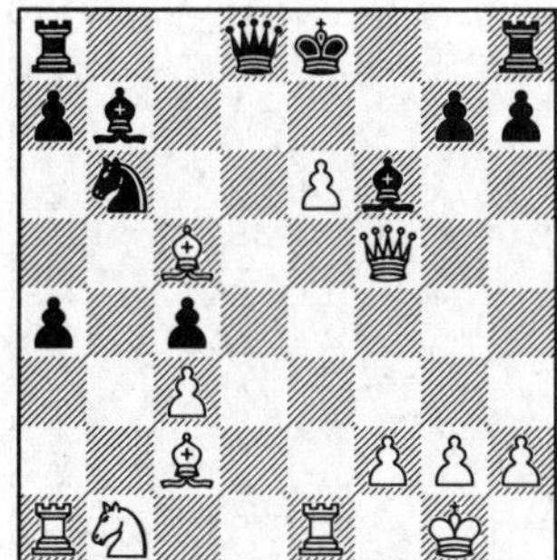

1148.

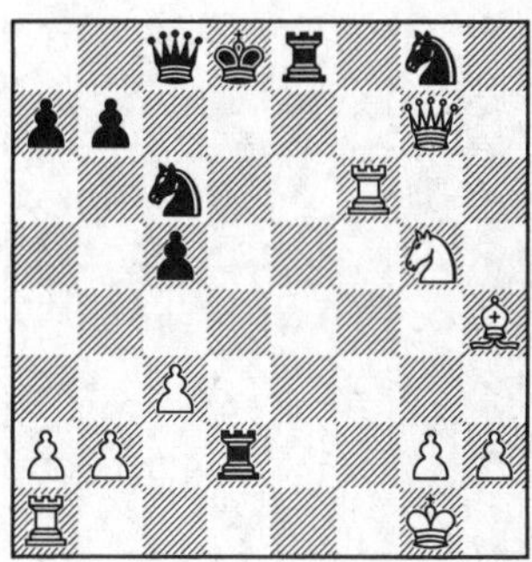

1149.

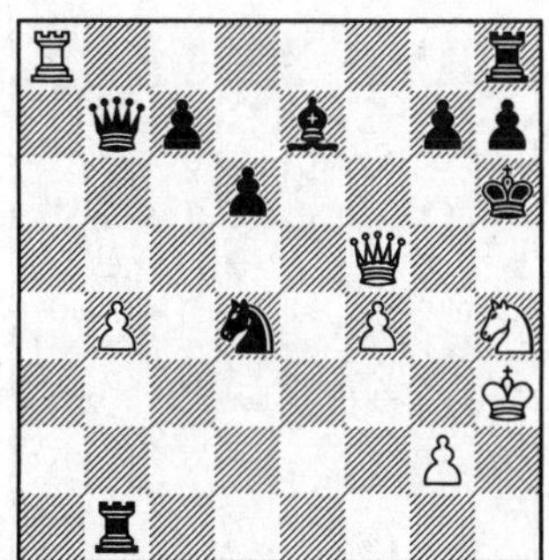

1150.

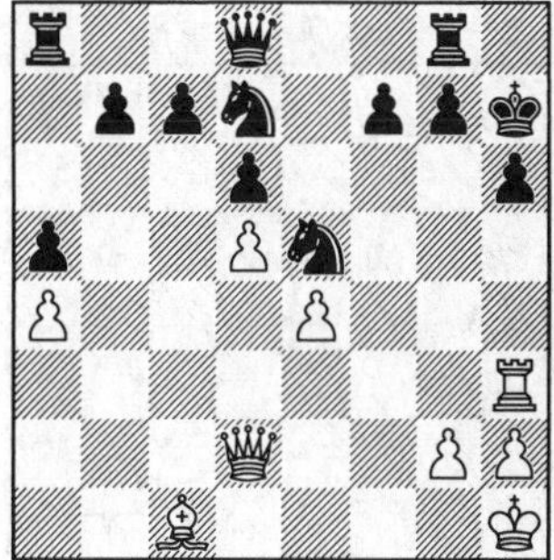

1151.

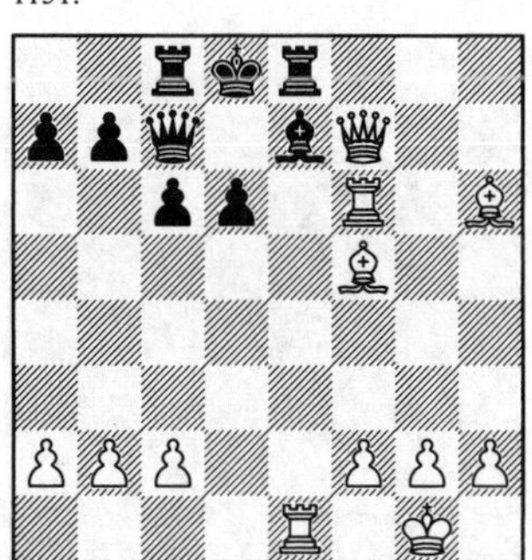

1152.

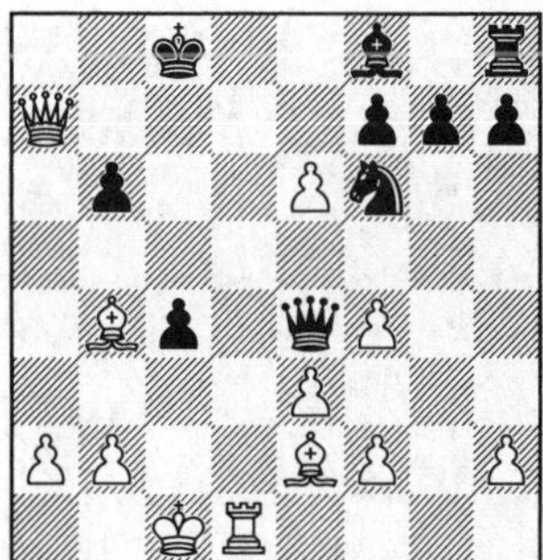

1153.

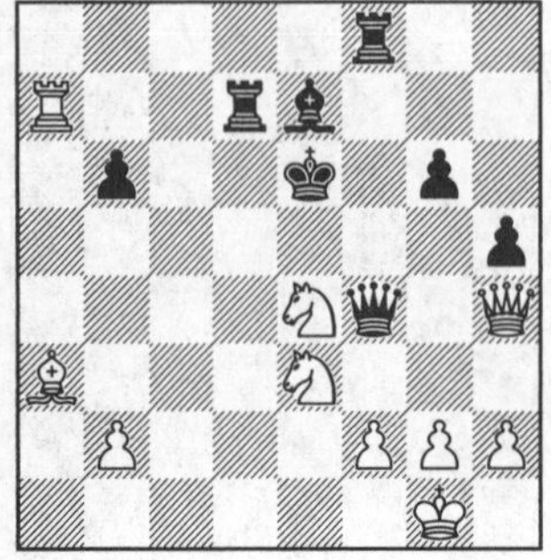

1154.

1155.

1156.

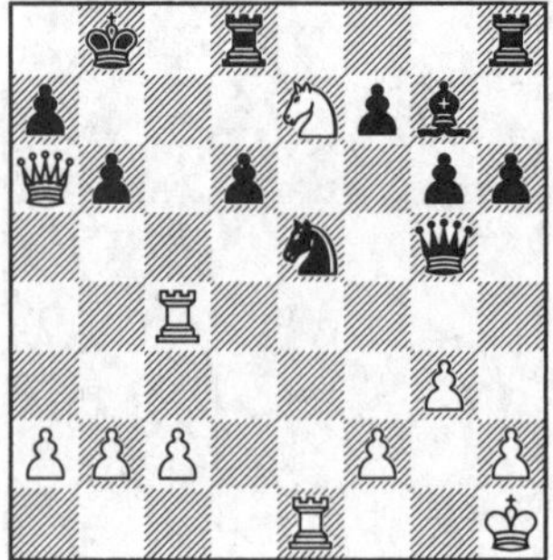

1157.

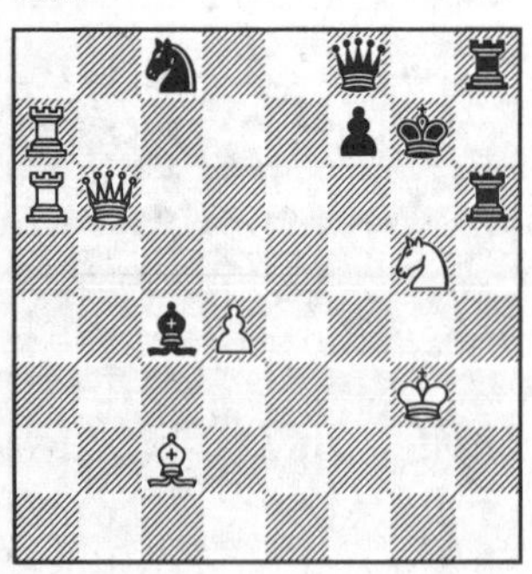

1158.

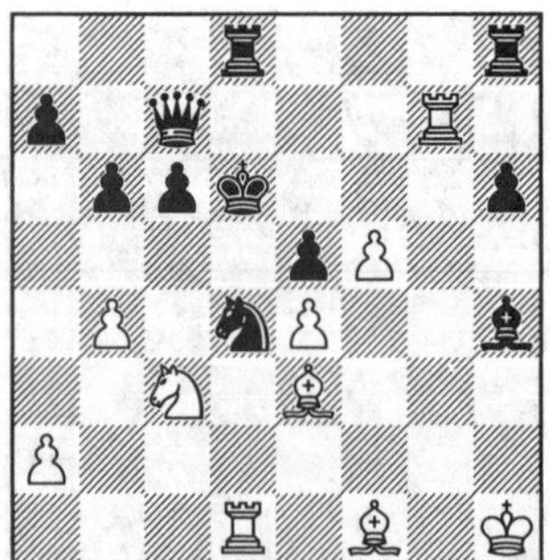

1159.

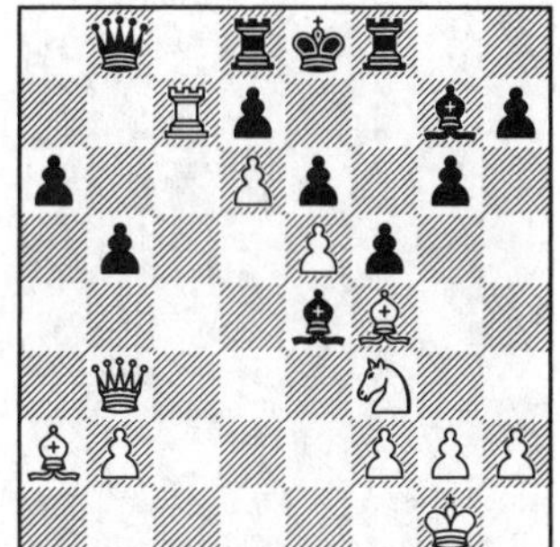

1160.

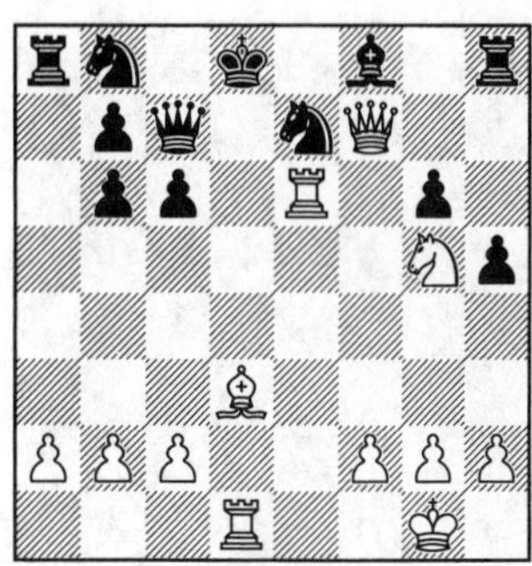

1161.

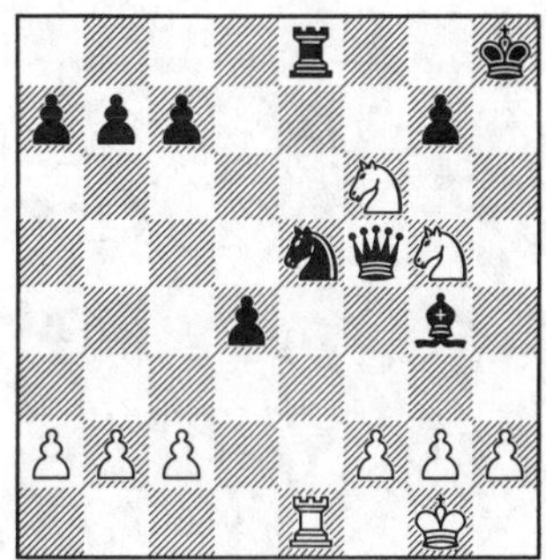

1162.

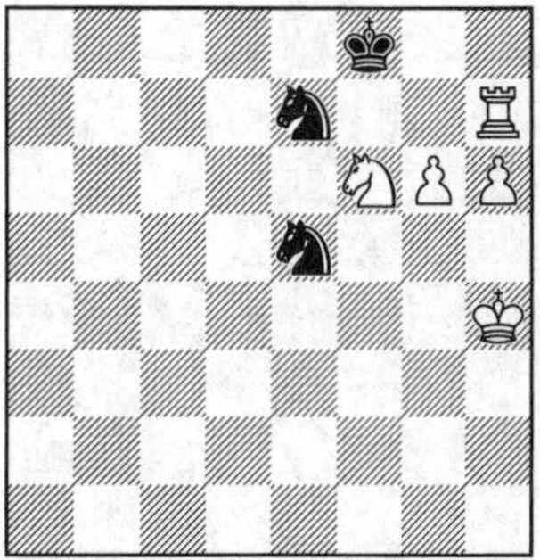

1163.

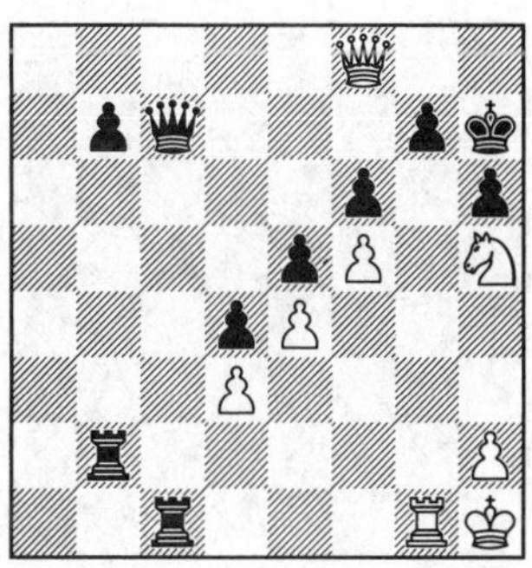

1164.

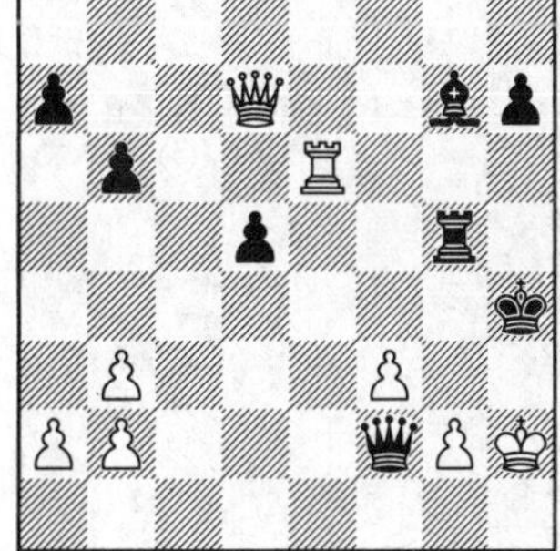

1165.

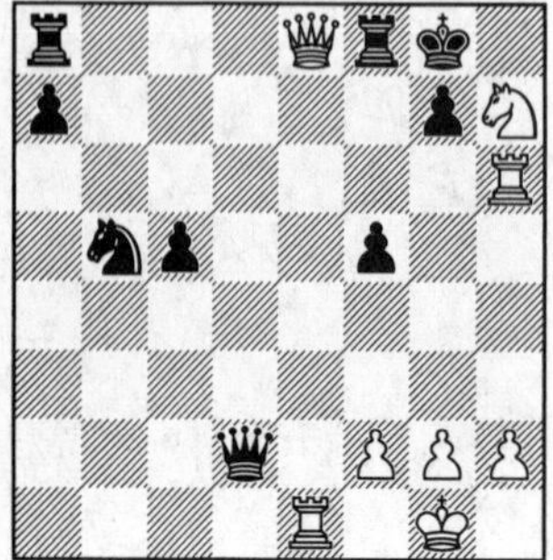

1166.

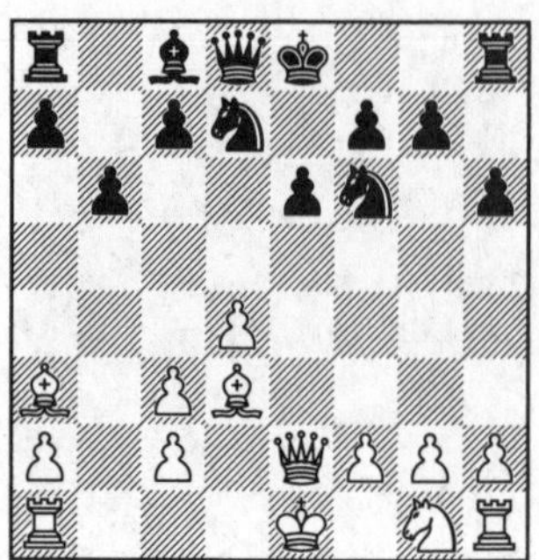

1167.

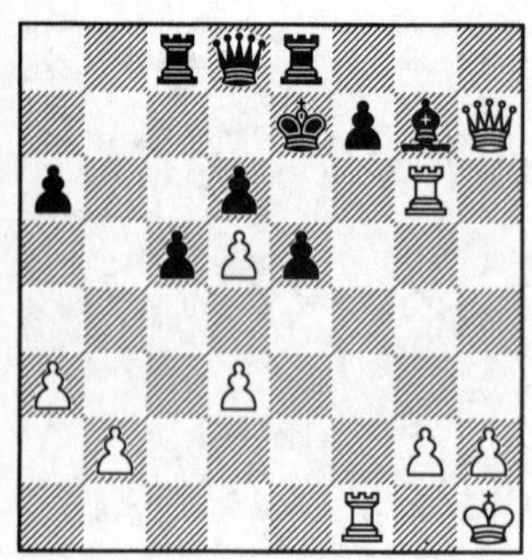

1168.

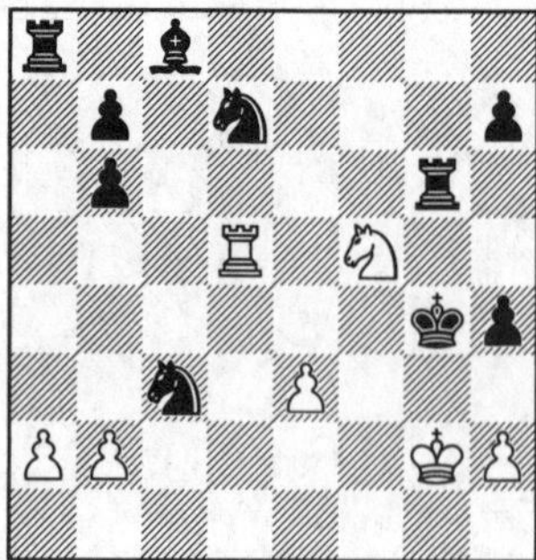

1169.

1170.

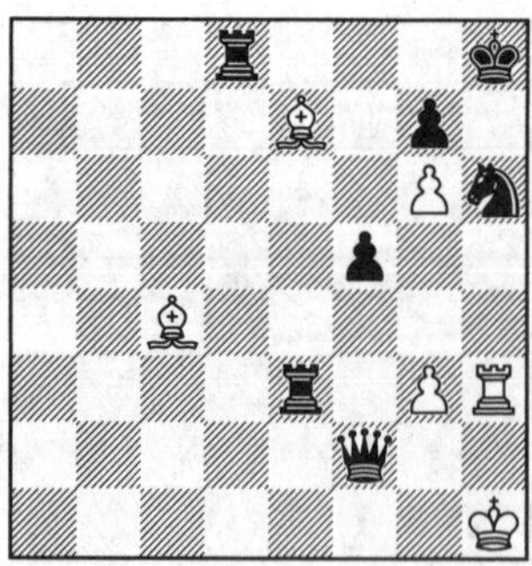

1171.

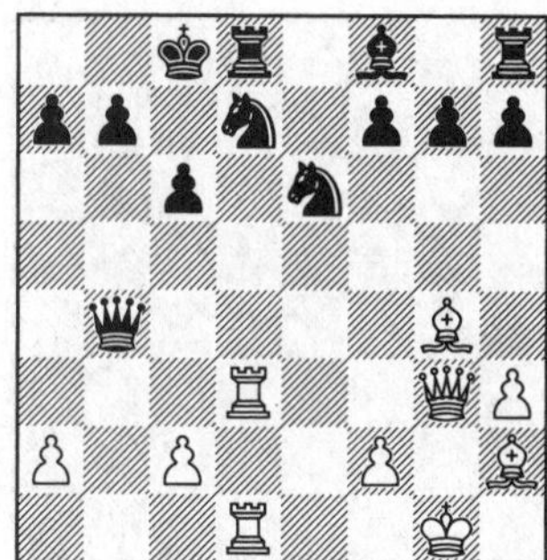

1172.

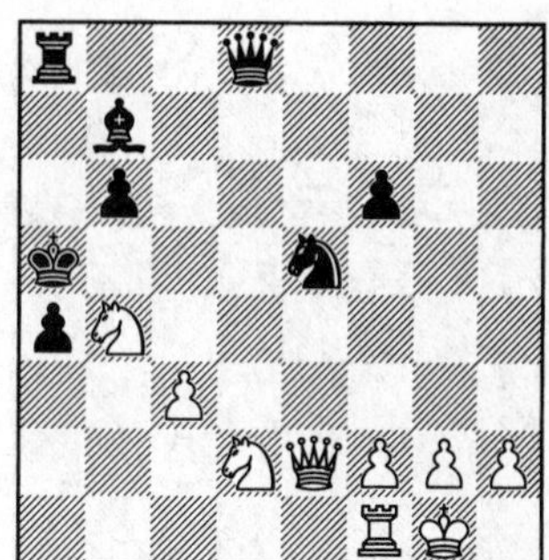

1173.

1174.

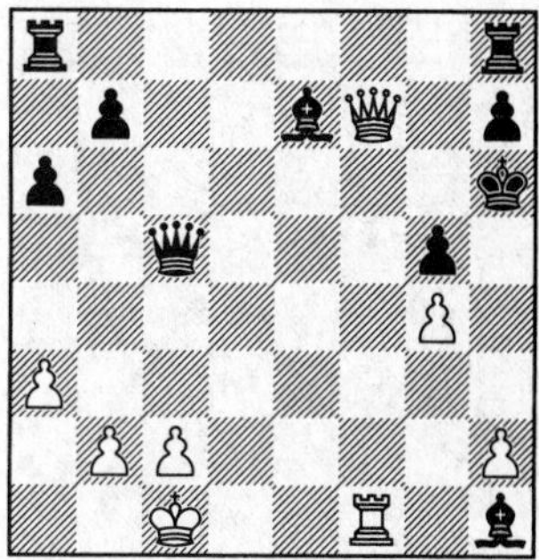

1175.

1176.

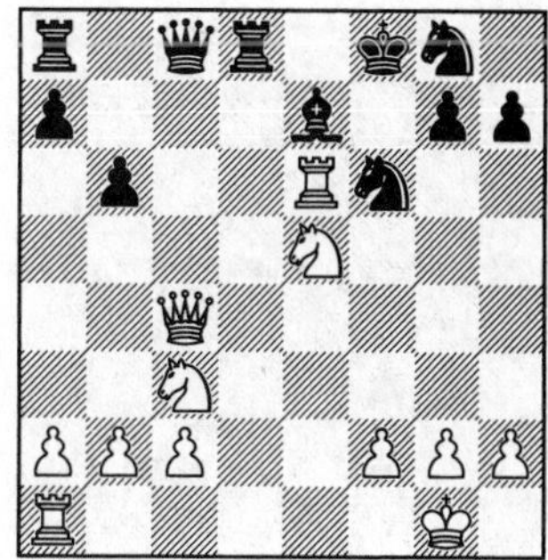

1177.

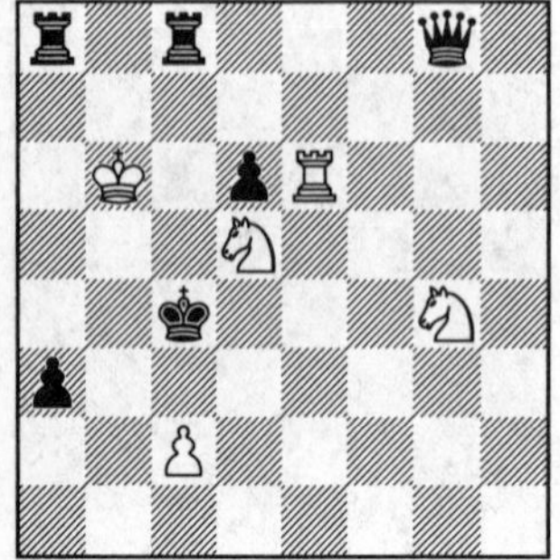

1178.

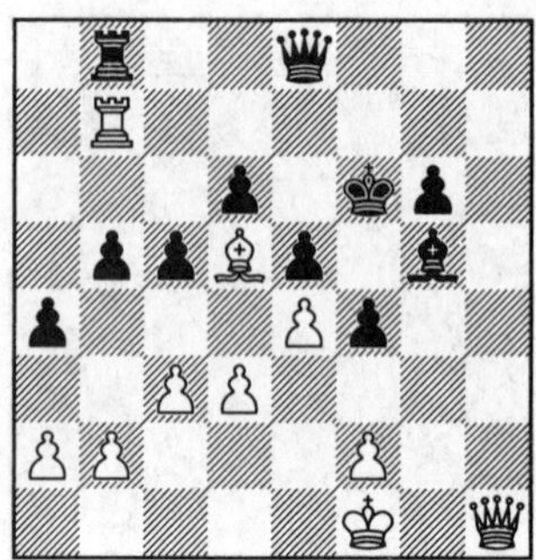

1179.

1180.

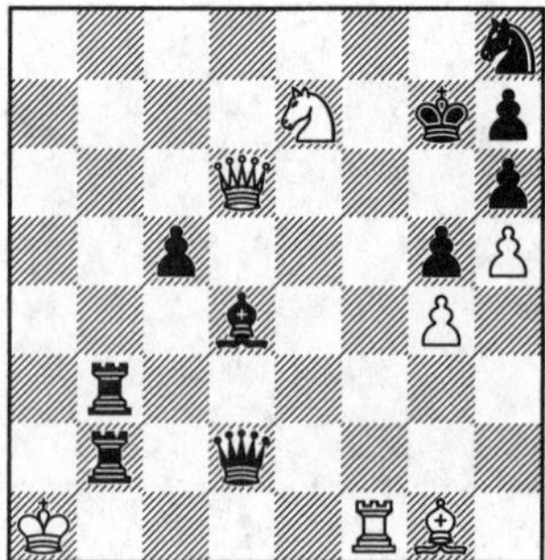

1181.

1182.

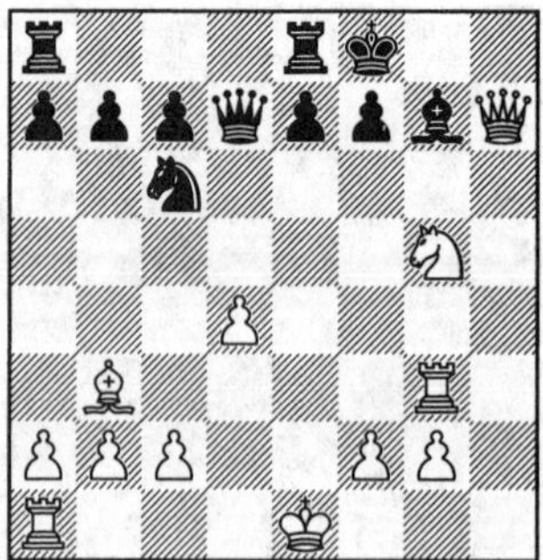

1183.

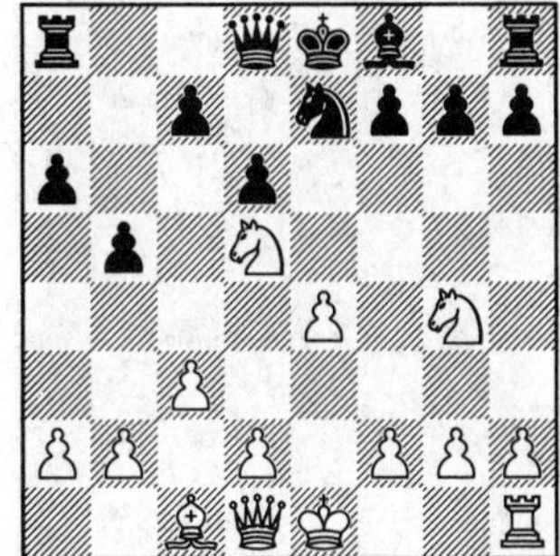

1184.

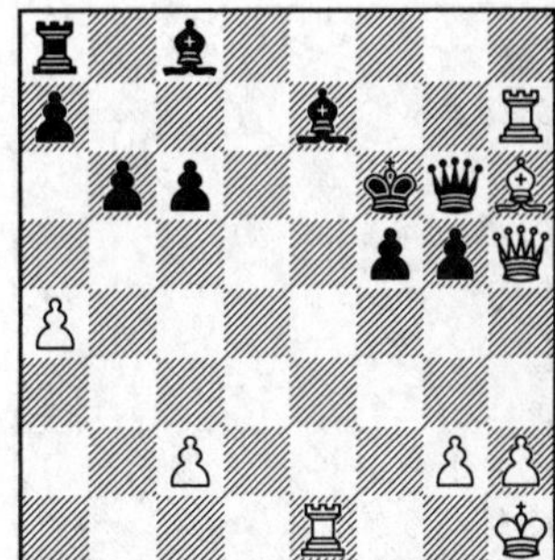

1185.

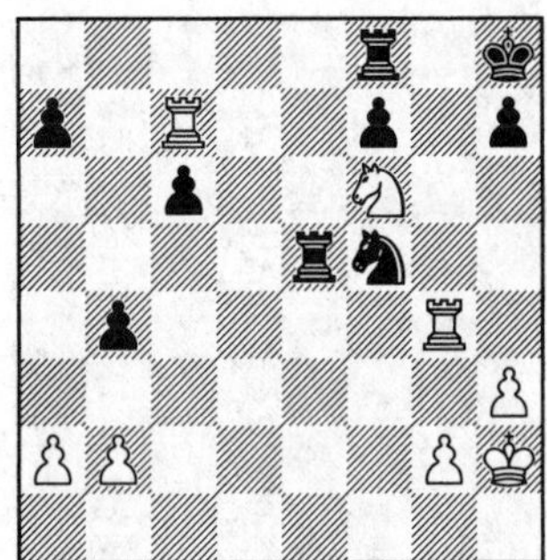

1186.

1187.

1188.

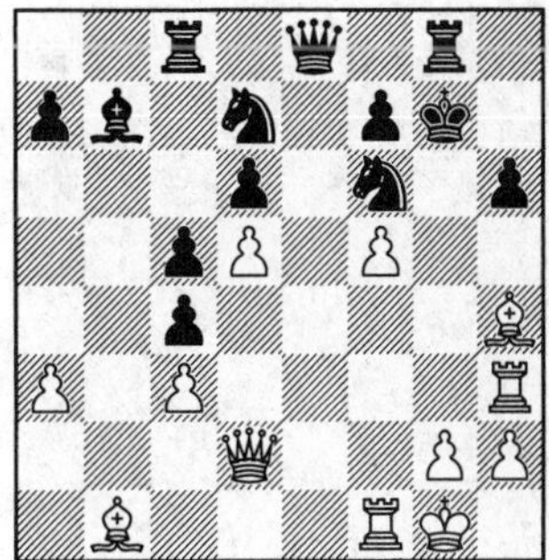

1189.

1190.

1191.

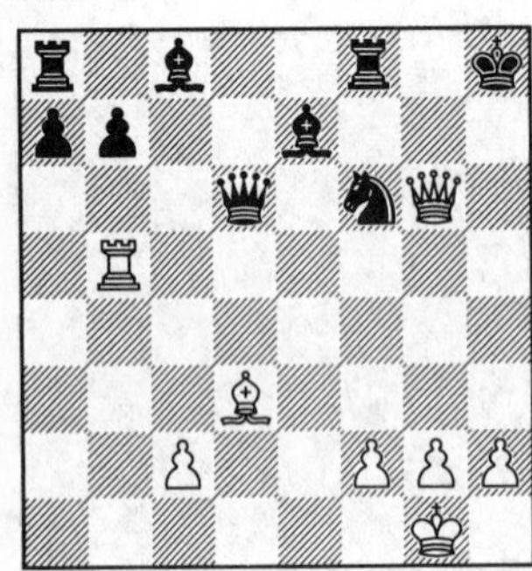

1192.

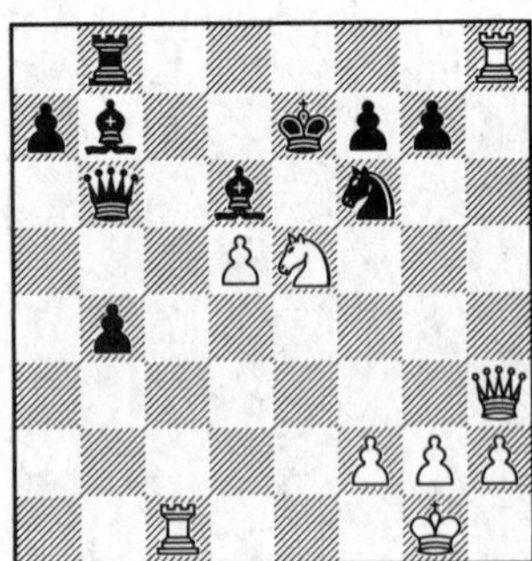

1193.

1194.

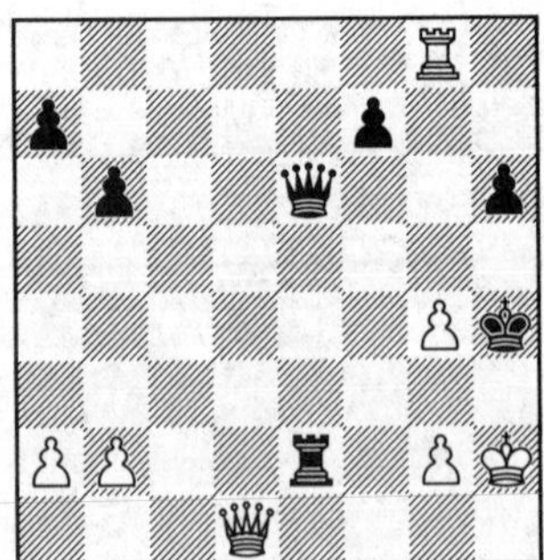

1195.

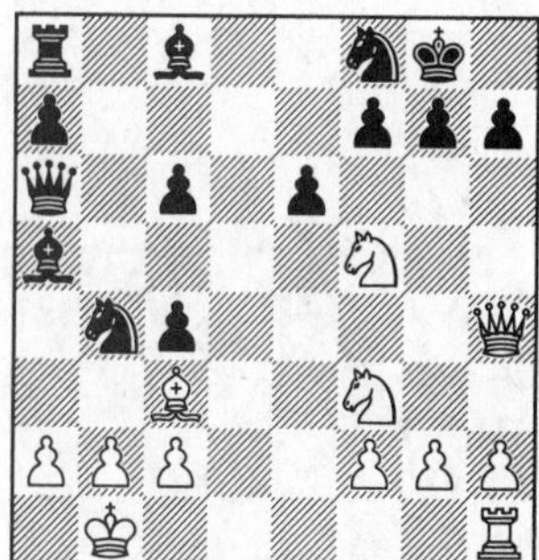

1196.

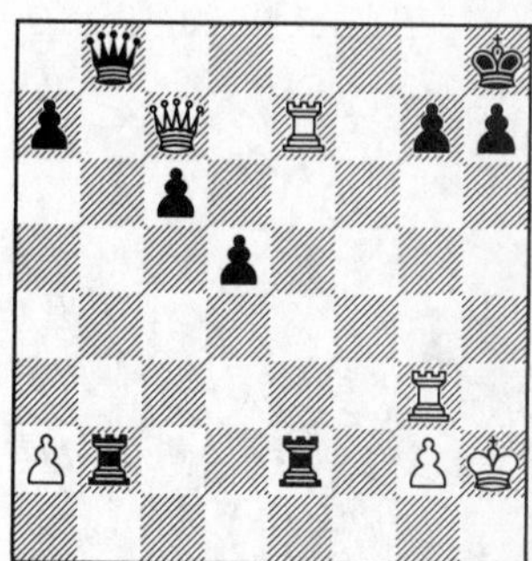

1197.

1198.

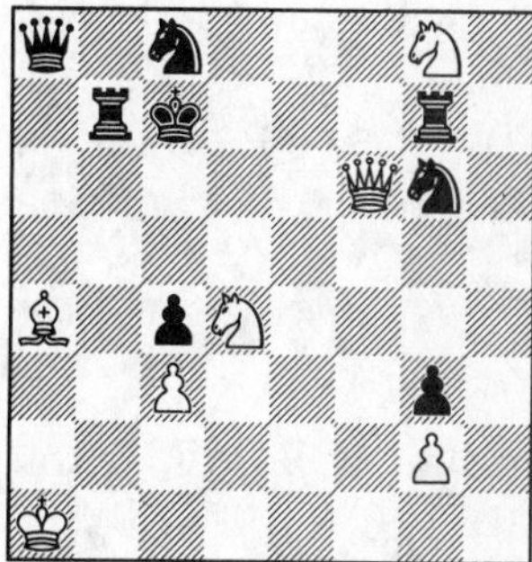

1199.

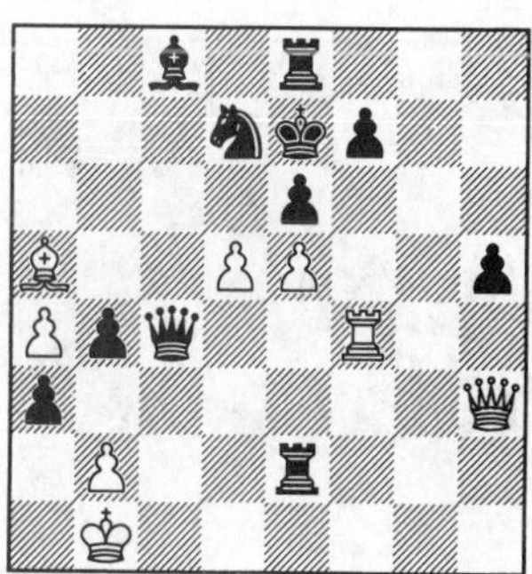

1200.

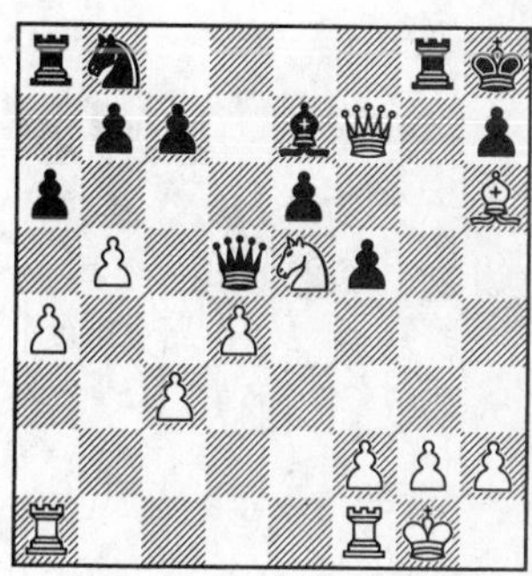

1201.

1202.

1203.

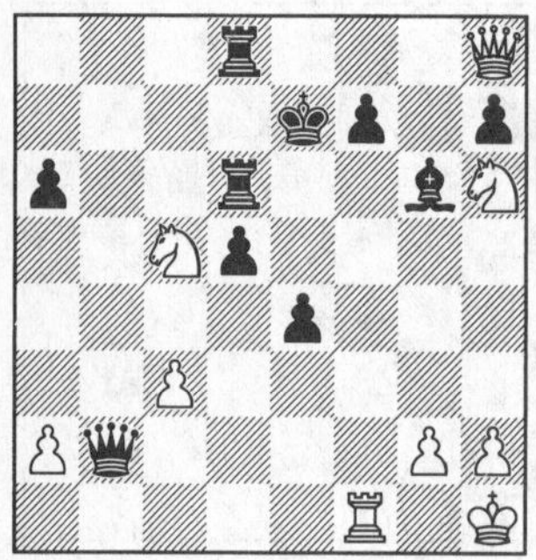

1204.

1205.

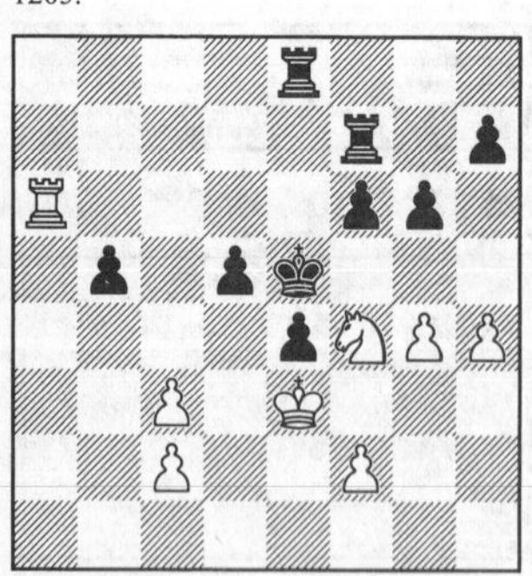

1206.

1207.

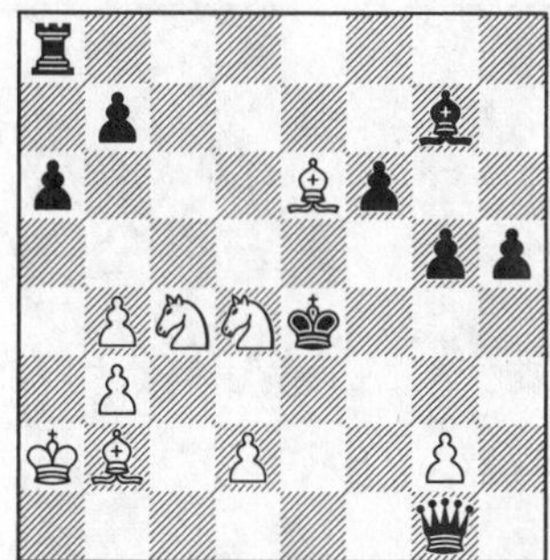

1208.

1209.

1210.

1211.

1212.

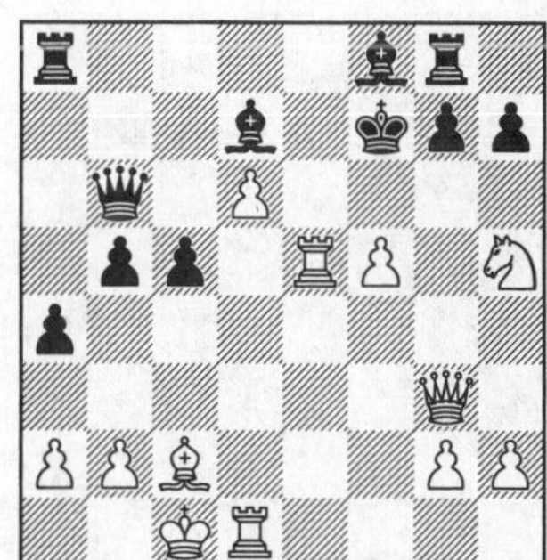

1213.

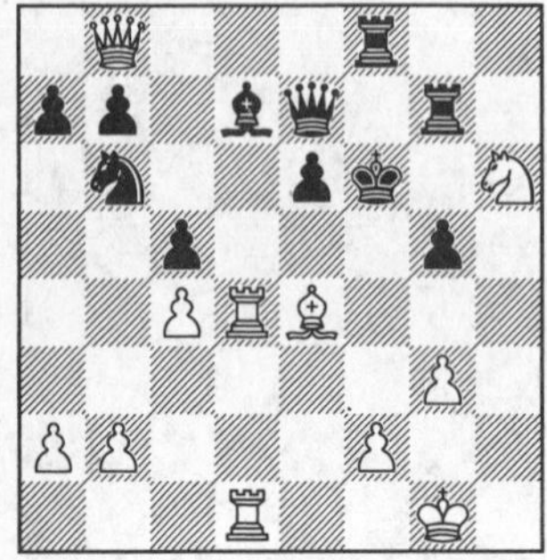

1214.

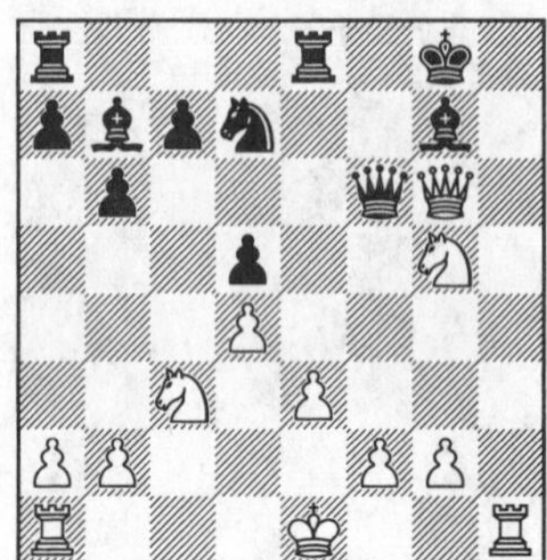

1215.

1216.

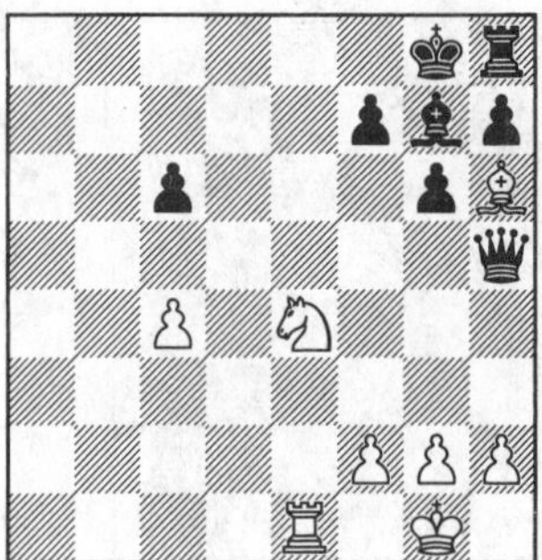

1217.

1218.

1219.

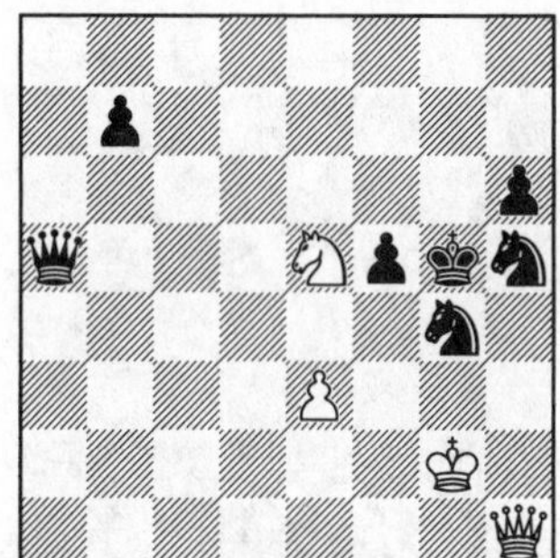

1220.

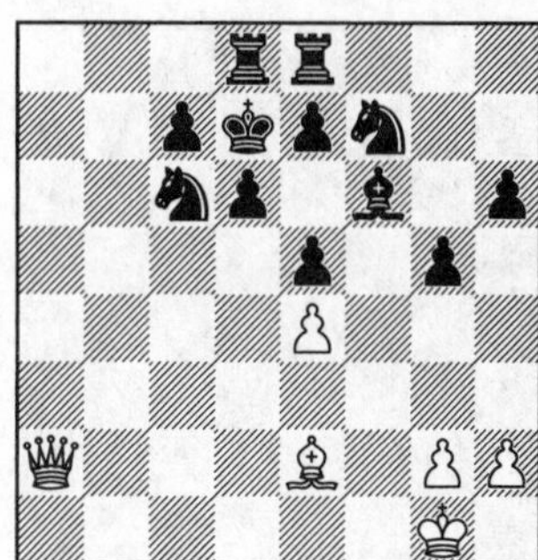

1221.

1222.

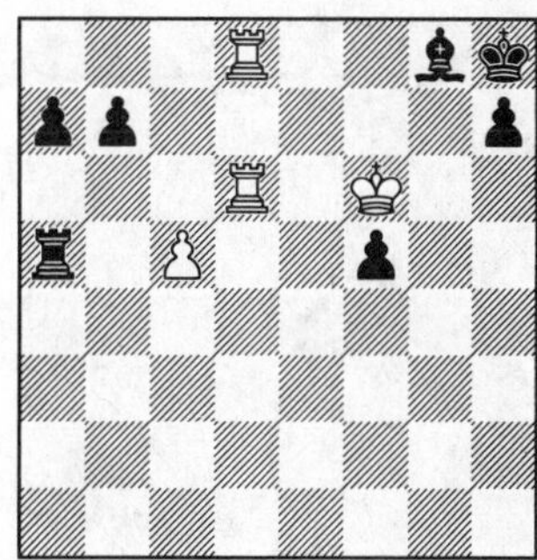

1223.

1224.

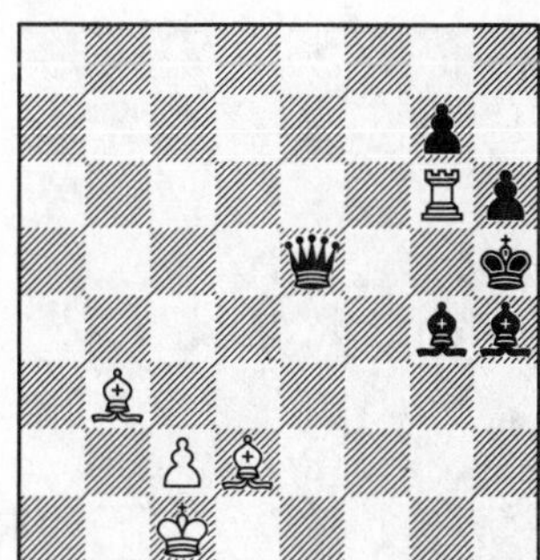

1225.

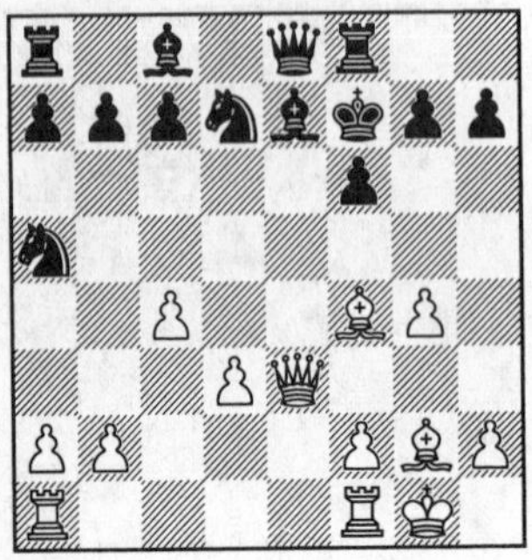

1226.

1227.

1228.

1229.

1230.

1231.

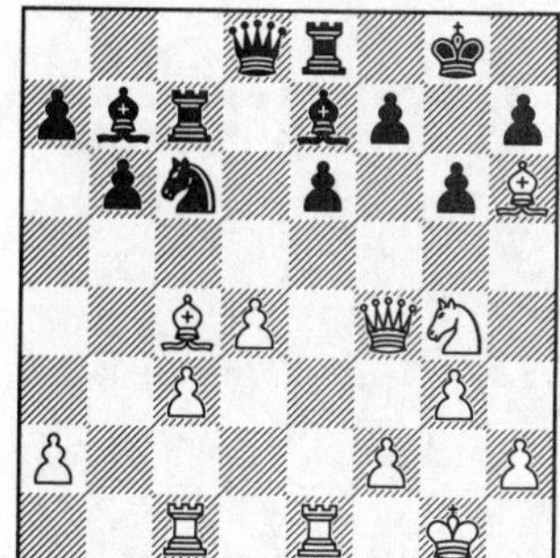

1232.

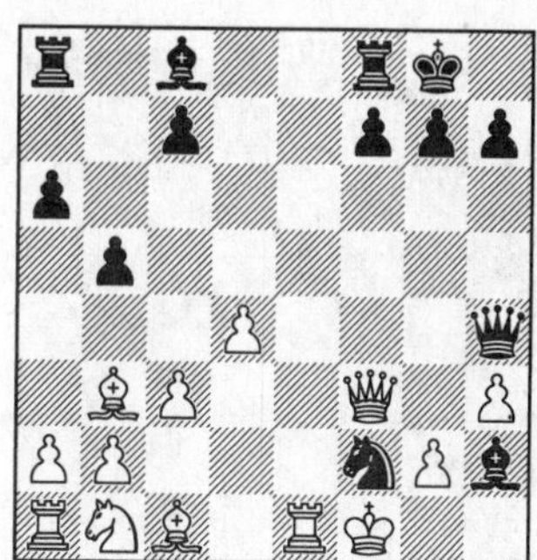

1233.

1234.

1235.

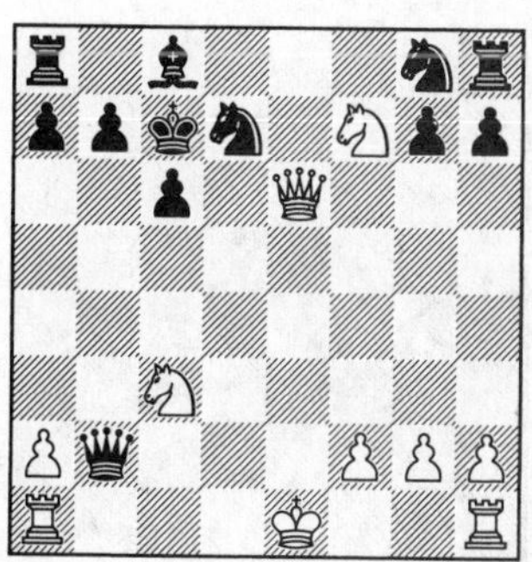

1236.

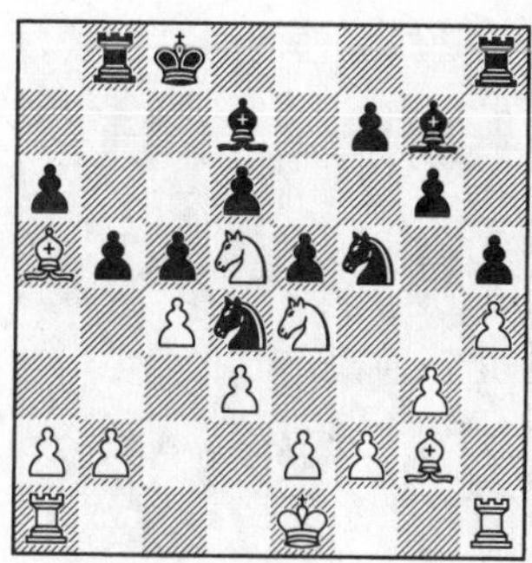

1237.

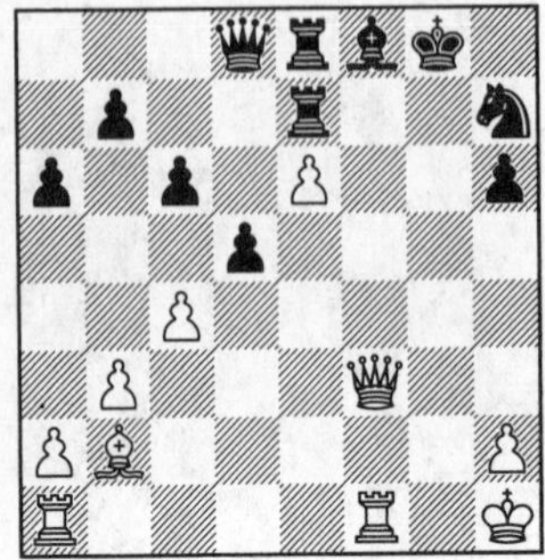

1238.

1239.

1240.

1241.

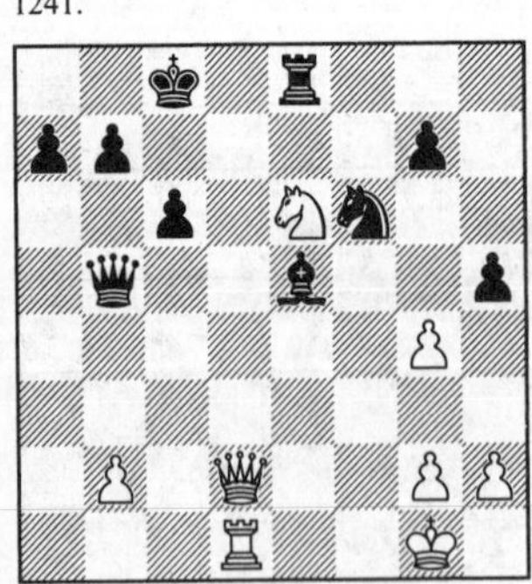

1242.

1243.

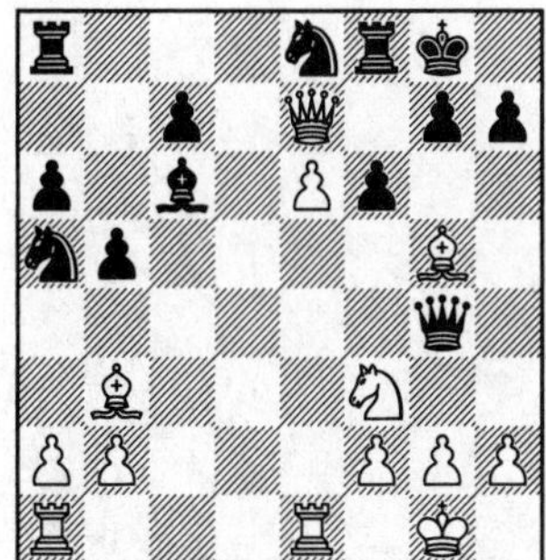

1244.

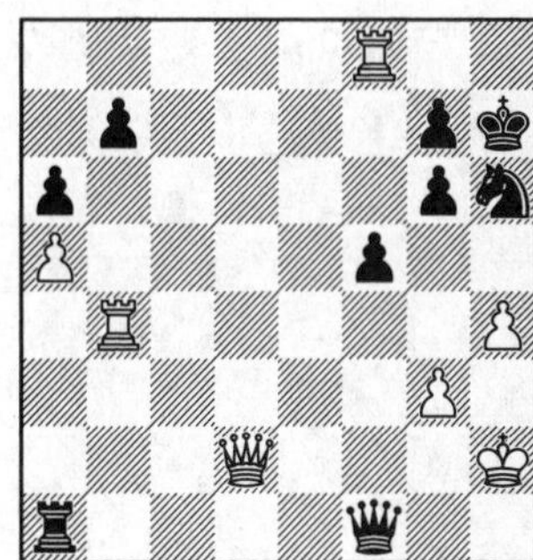

1245.

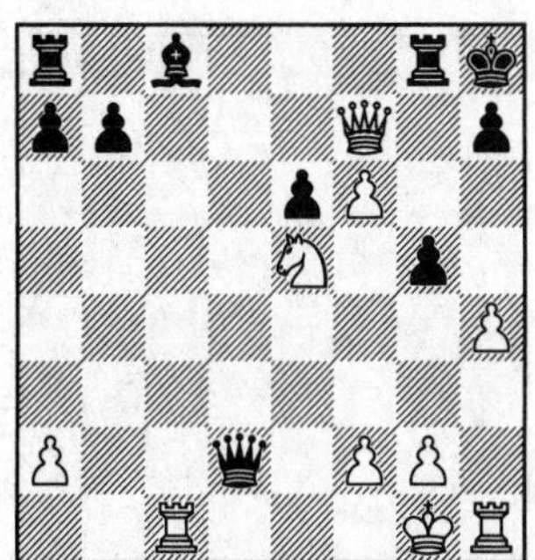

1246.

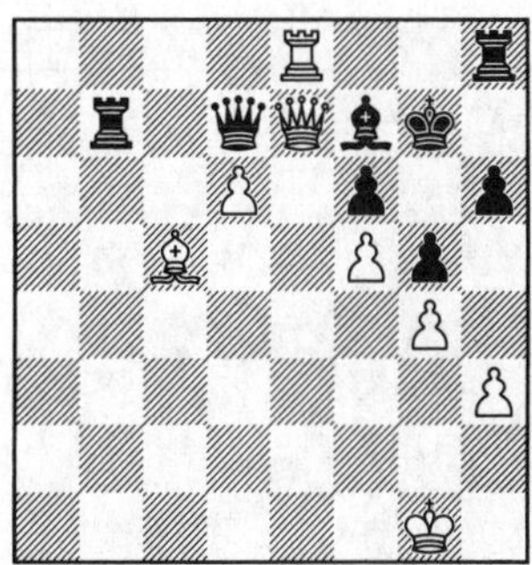

1247.

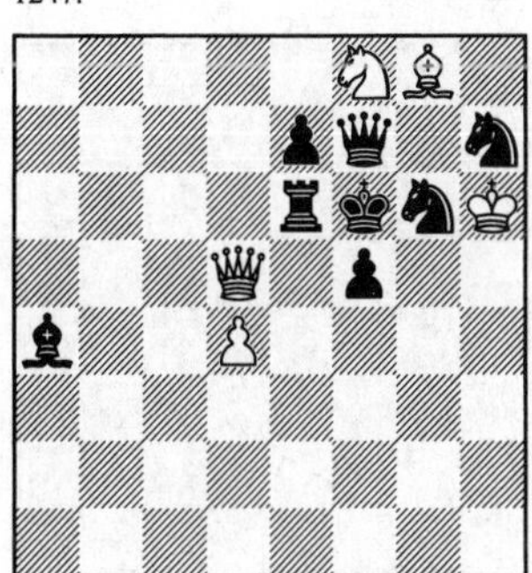

1248.

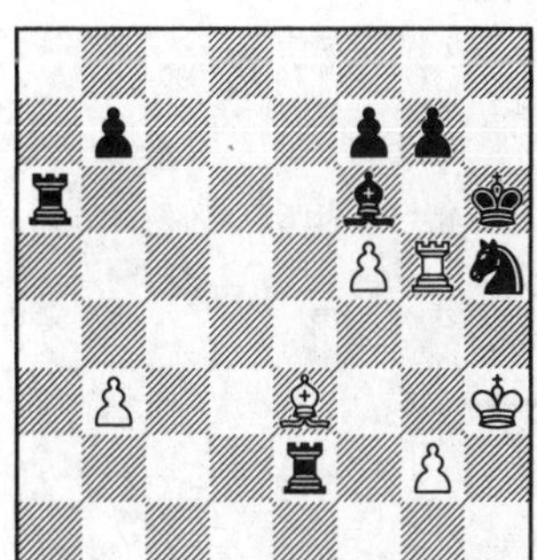

1249.

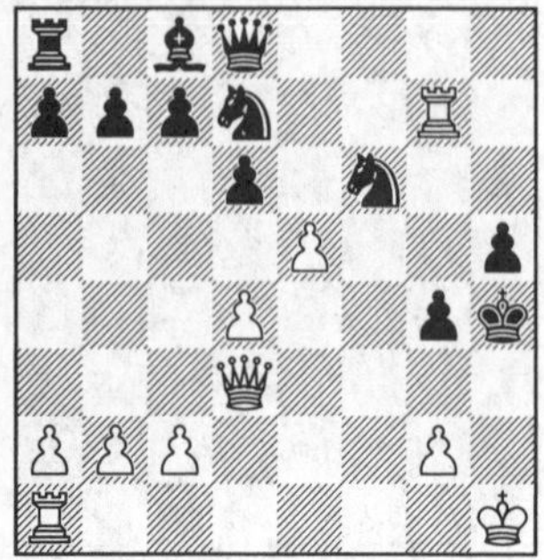

1250.

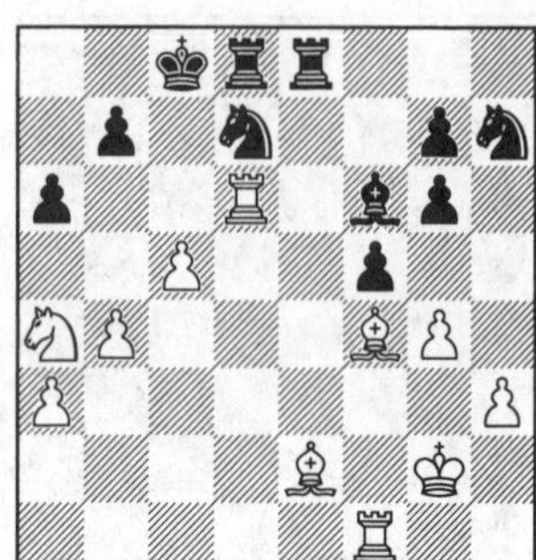

1251.

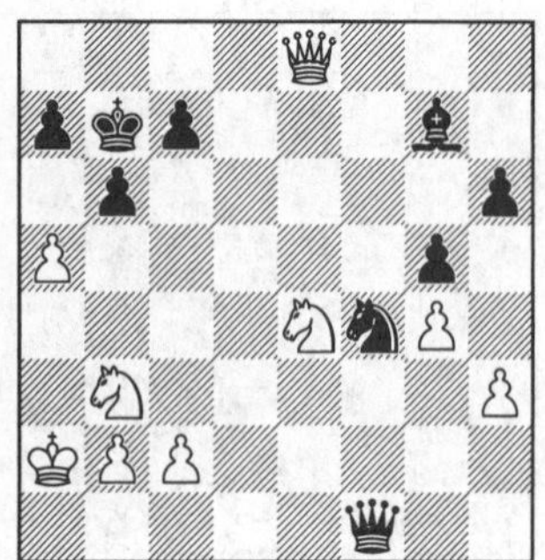

1252.

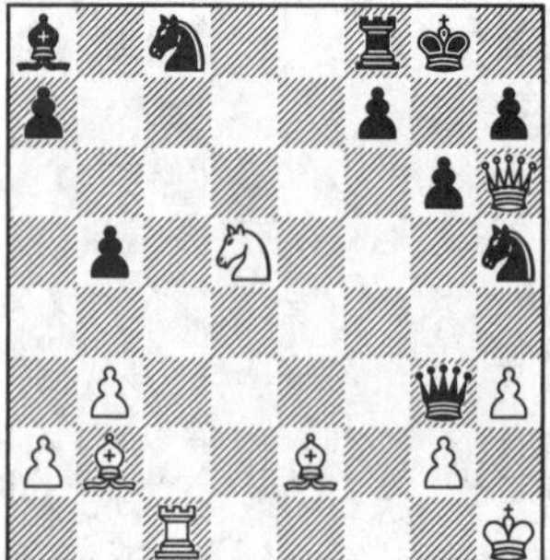

1253.

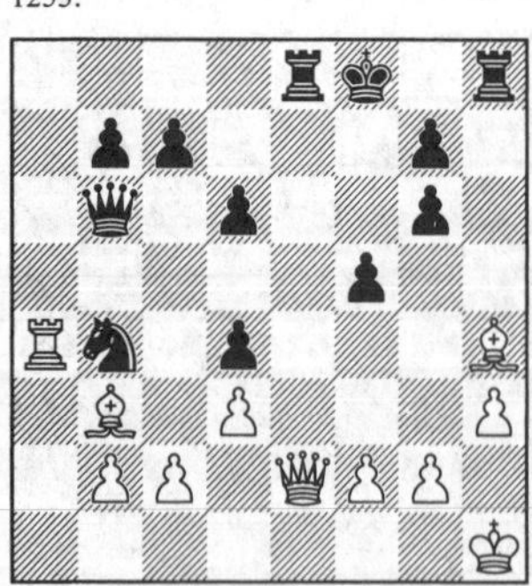

1254.

1255.

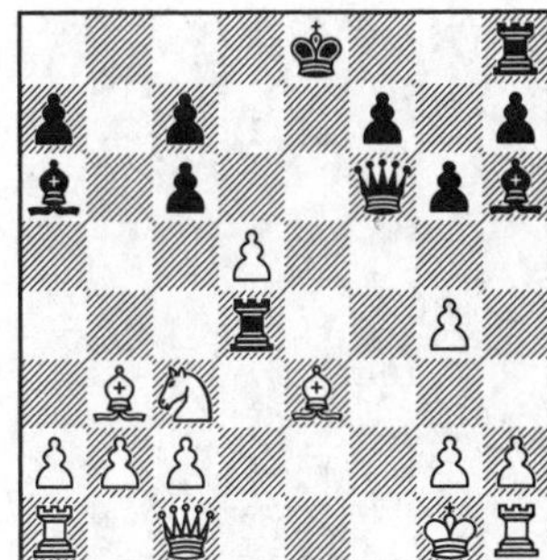

1256.

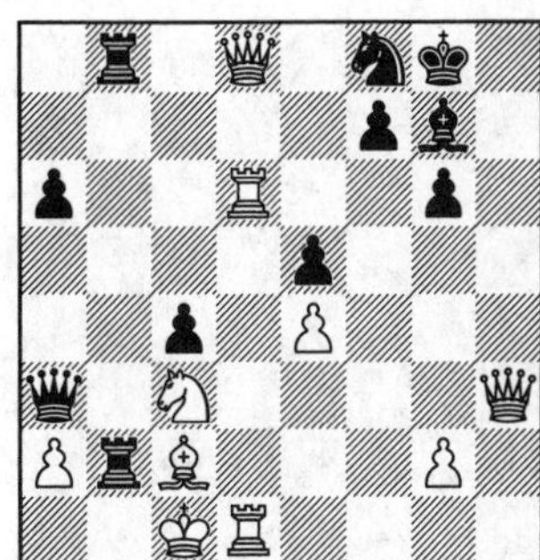

1257.

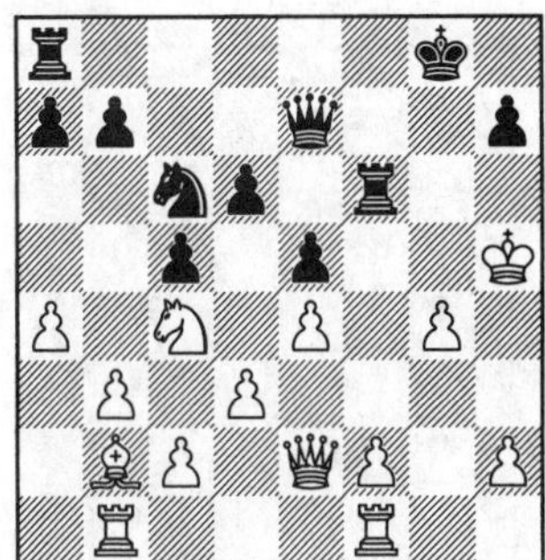

1258.

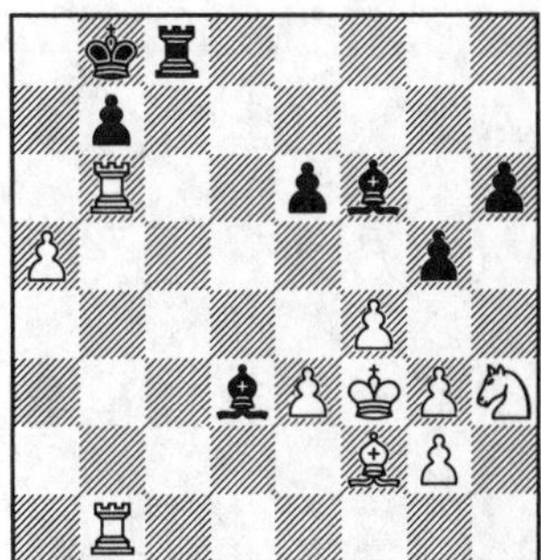

1259.

1260.

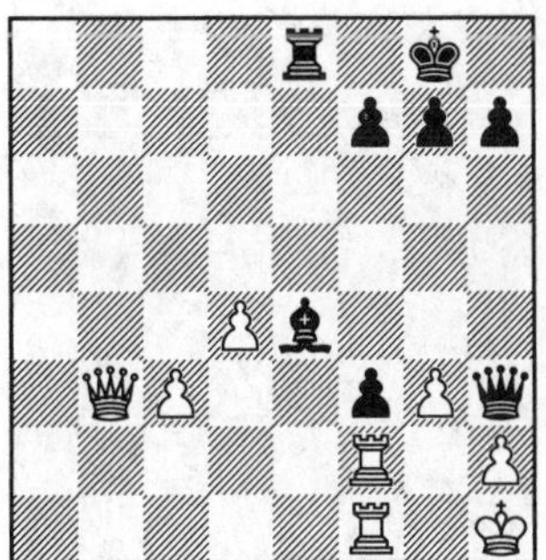

1261.

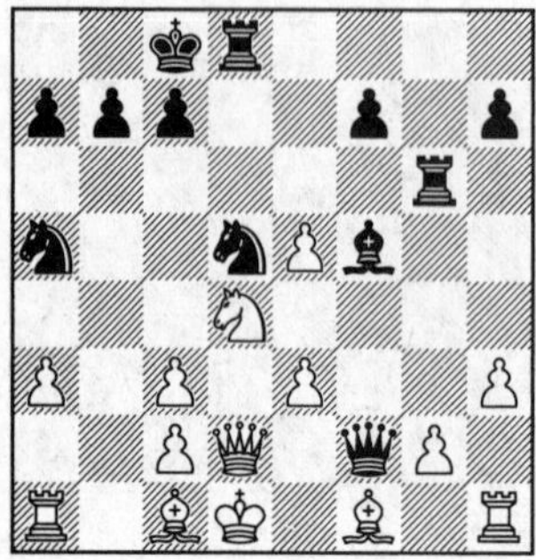

1262.

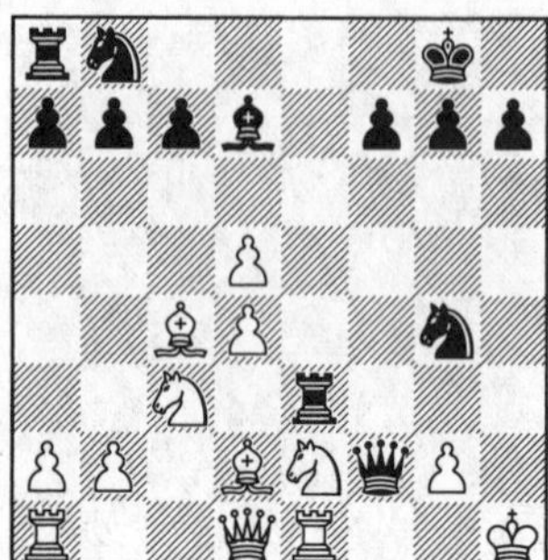

1263.

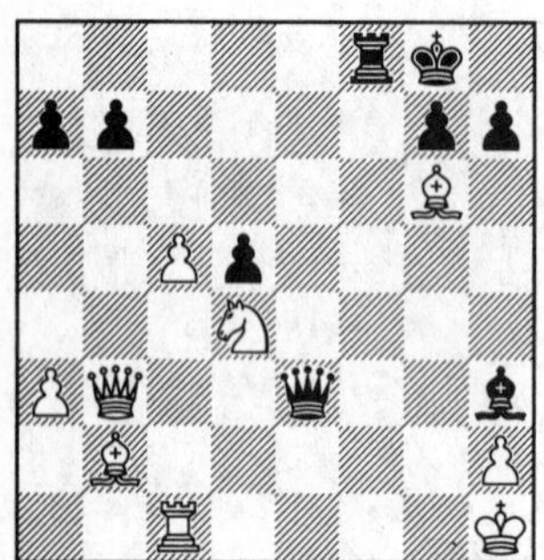

1264.

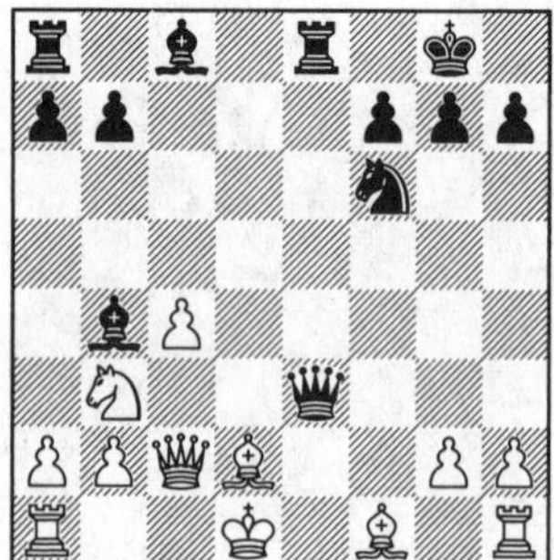

1265.

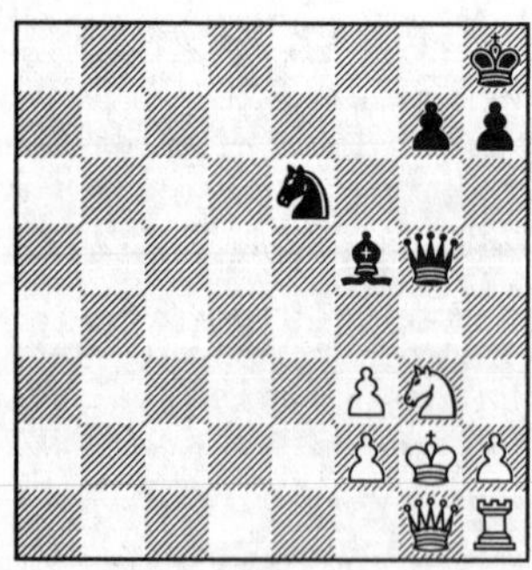

1266.

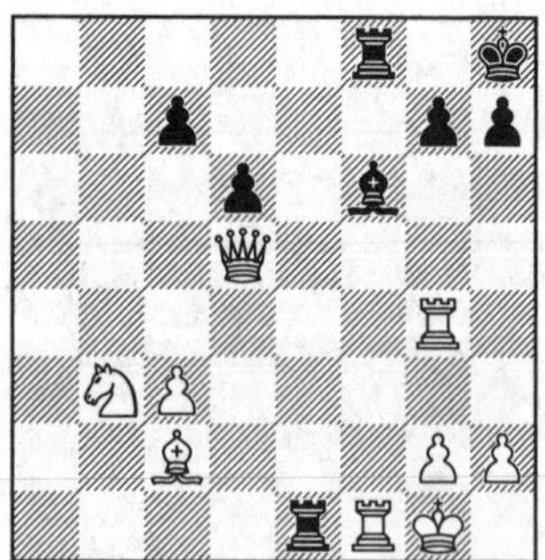

1267.

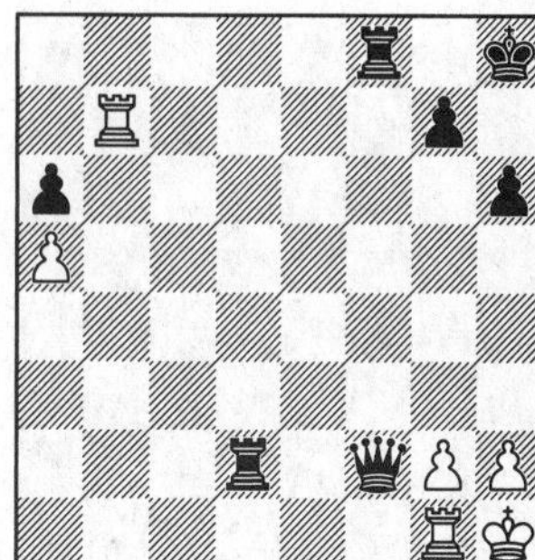

1268.

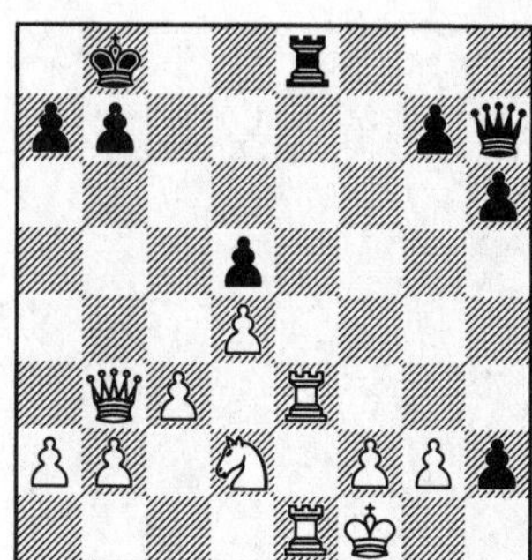

1269.

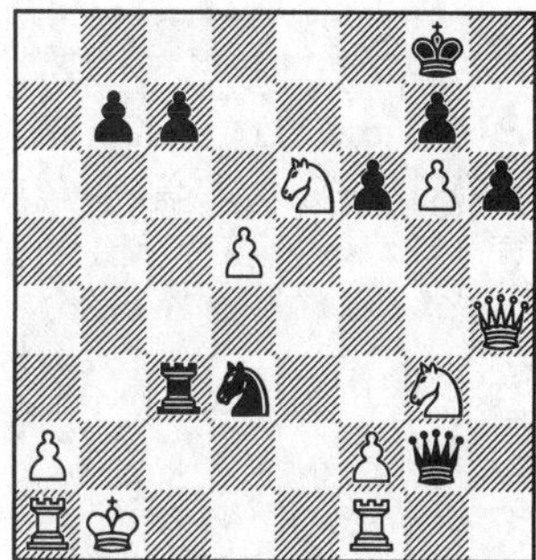

1270.

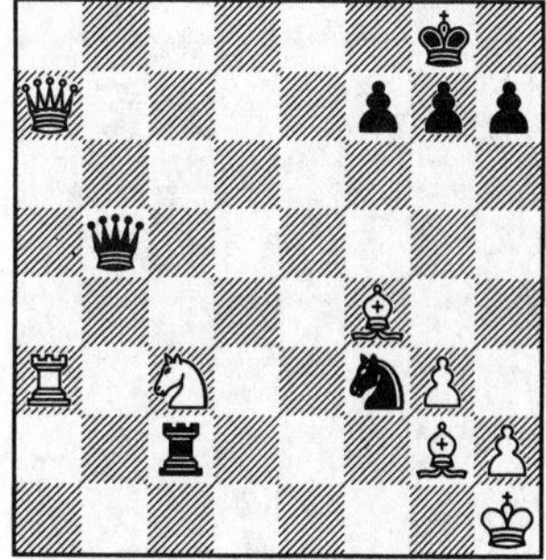

1271.

1272.

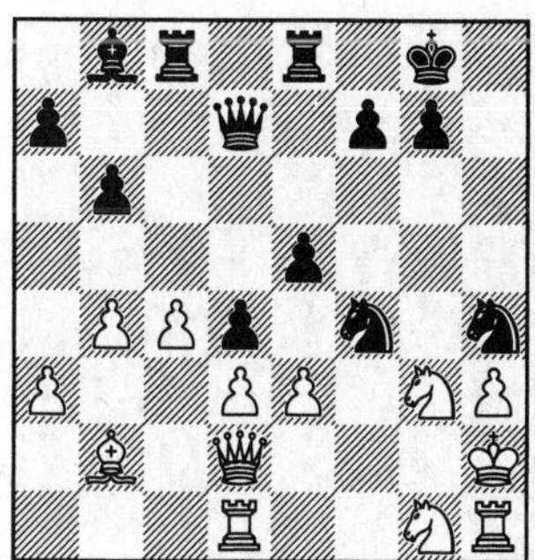

1273.

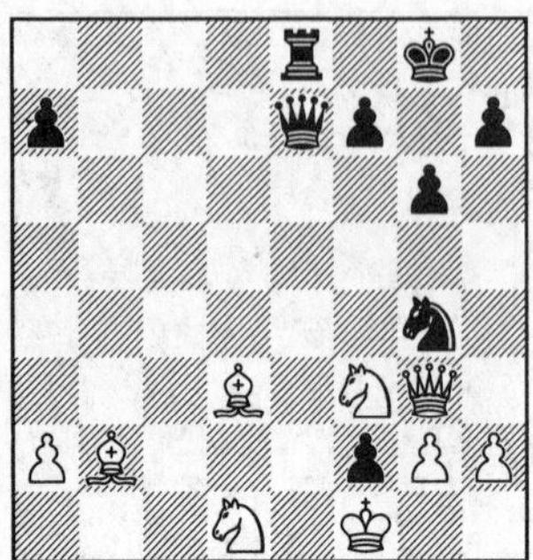

1274.

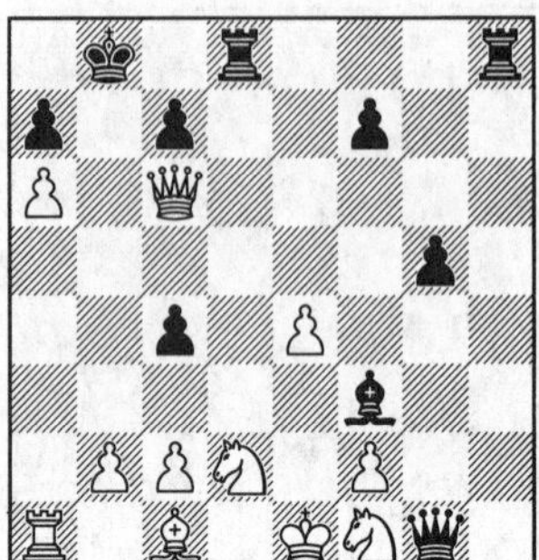

1275.

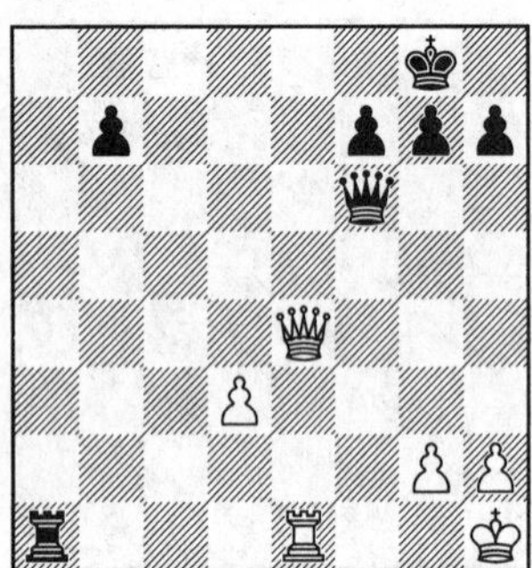

1276.

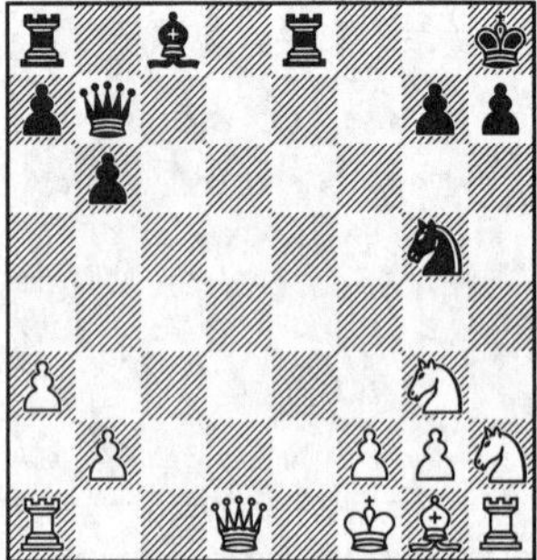

1277.

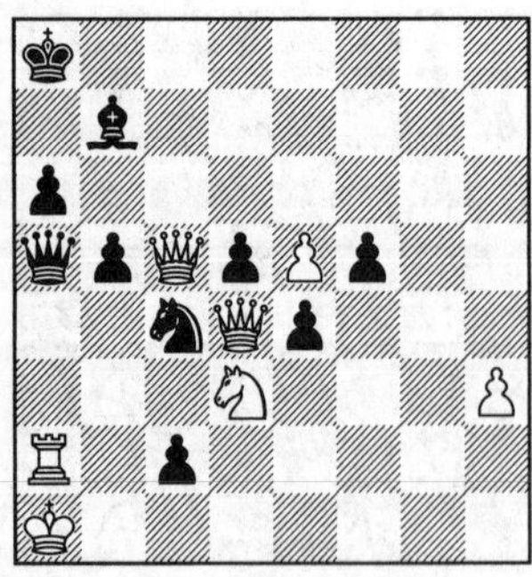

1278.

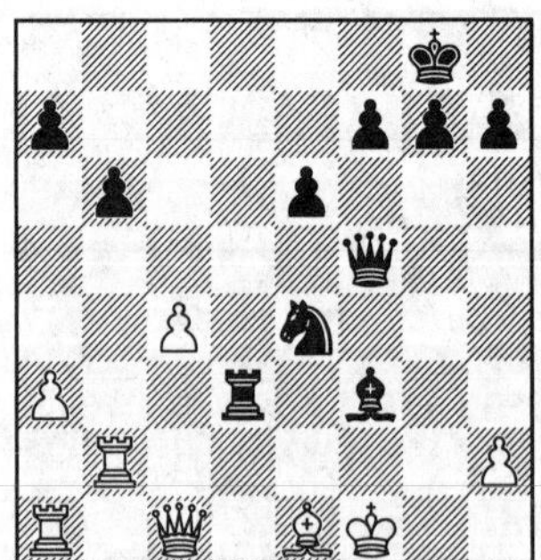

1279.

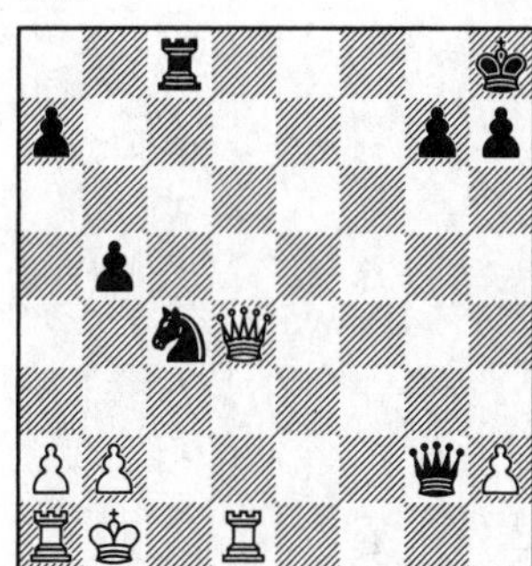

1280.

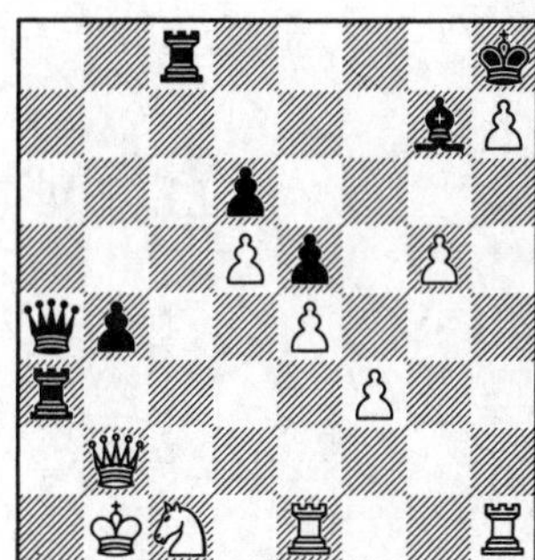

1281.

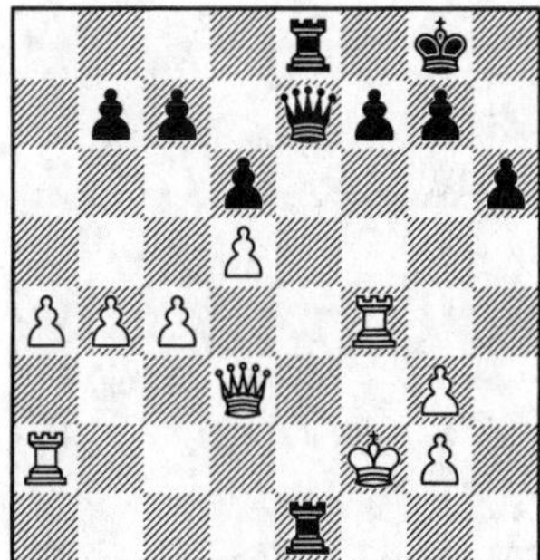

1282.

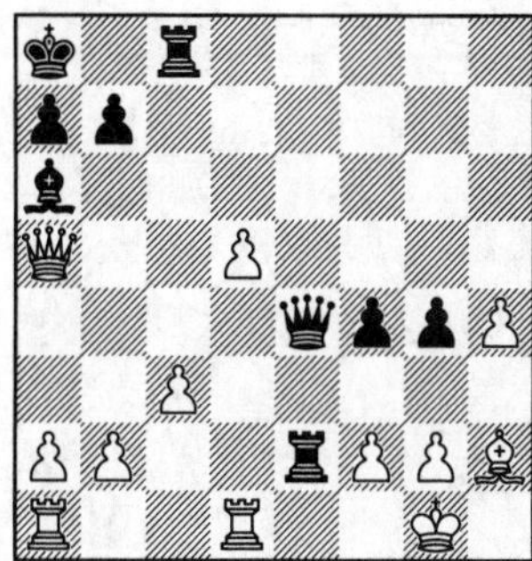

1283.

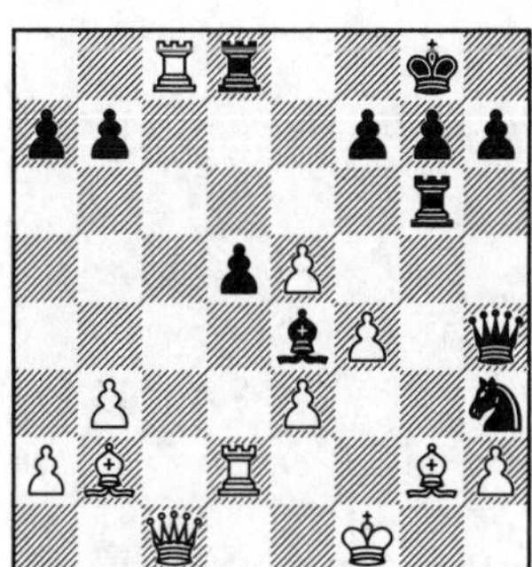

1284.

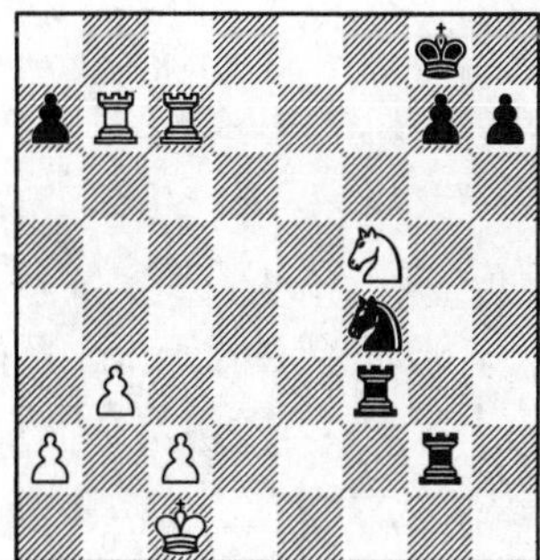

1285.

1286.

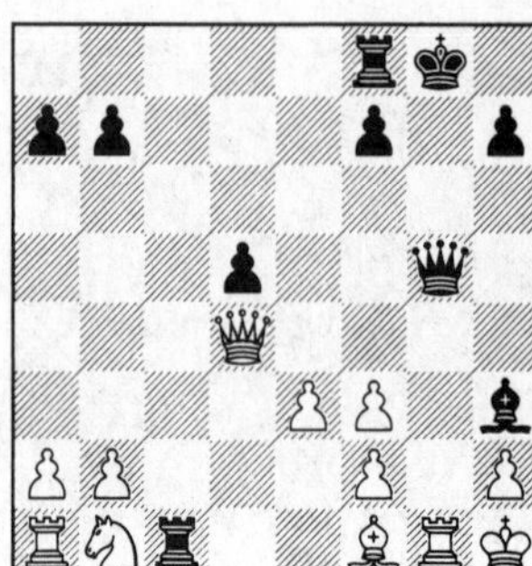

1287.

1288.

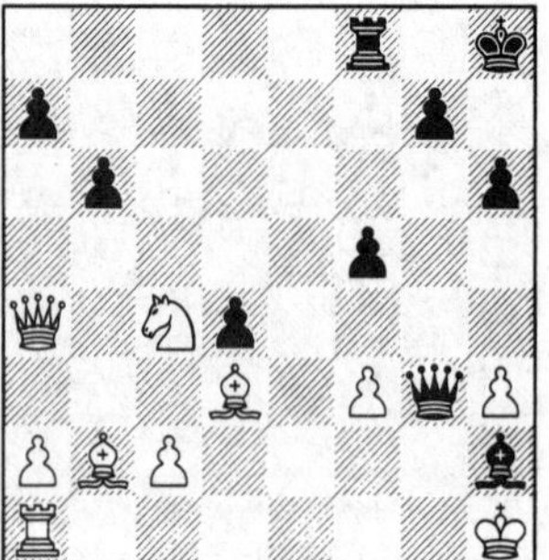

1289.

1290.

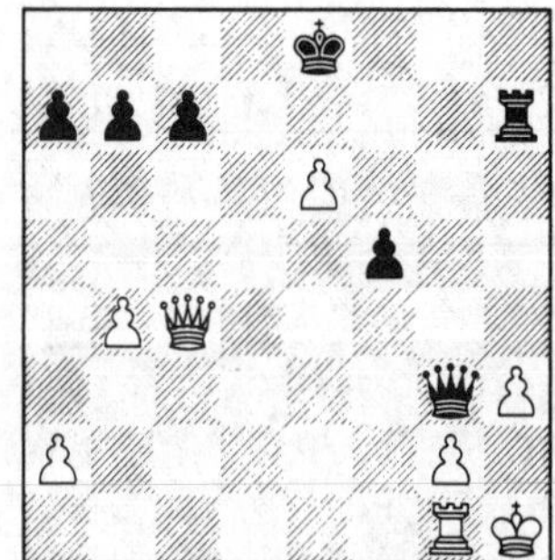

1291.

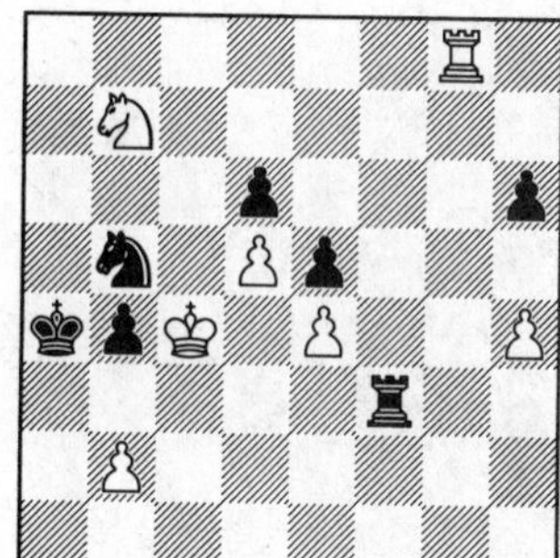

1292.

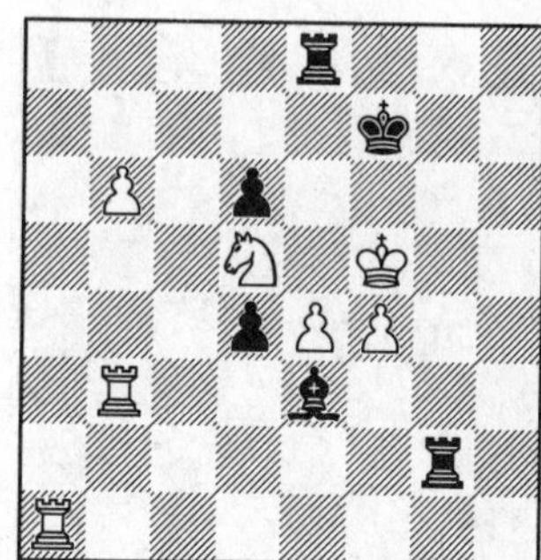

1293.

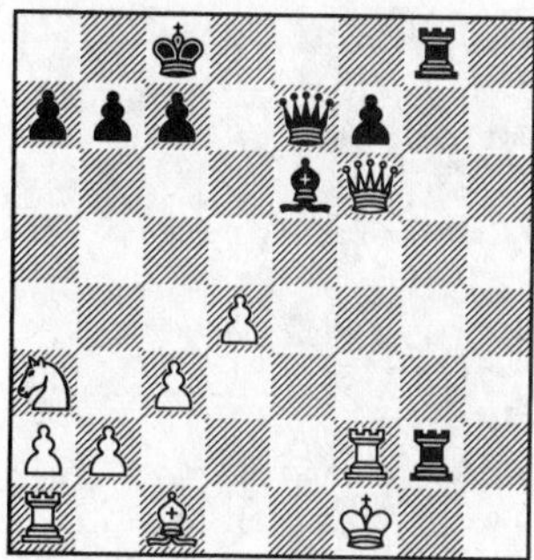

1294.

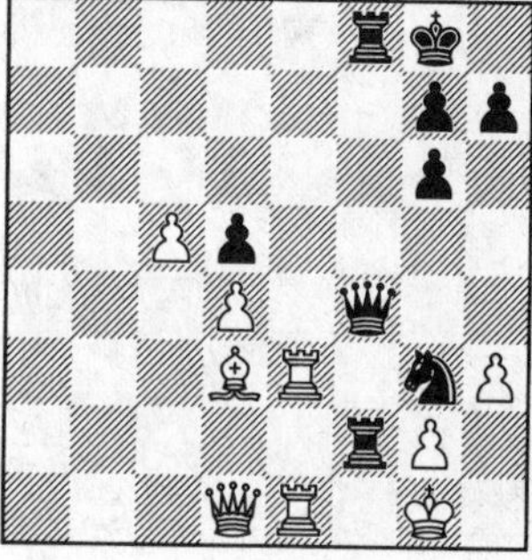

1295.

1296.

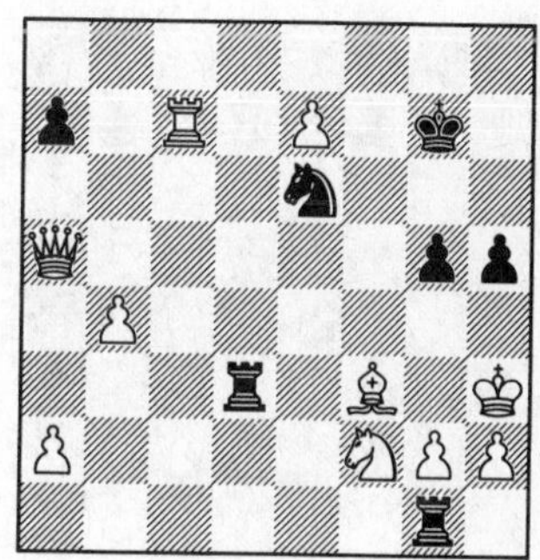

1297.

1298.

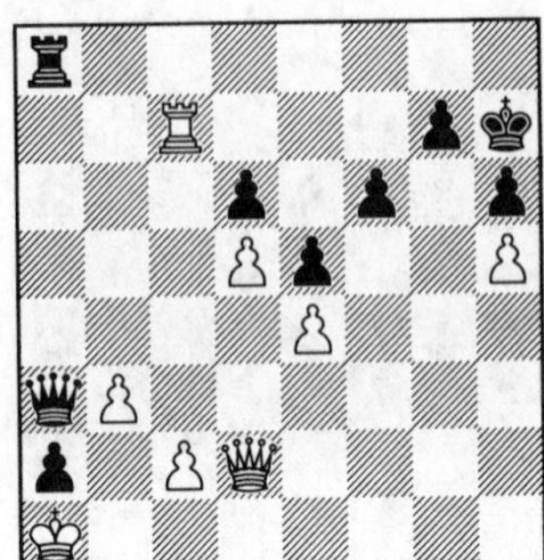

1299.

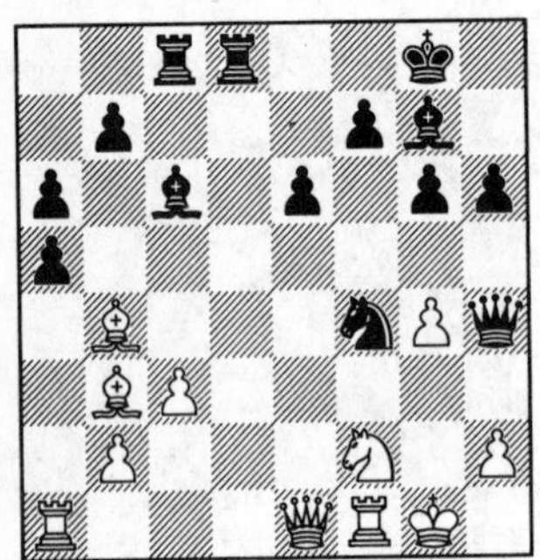

1300.

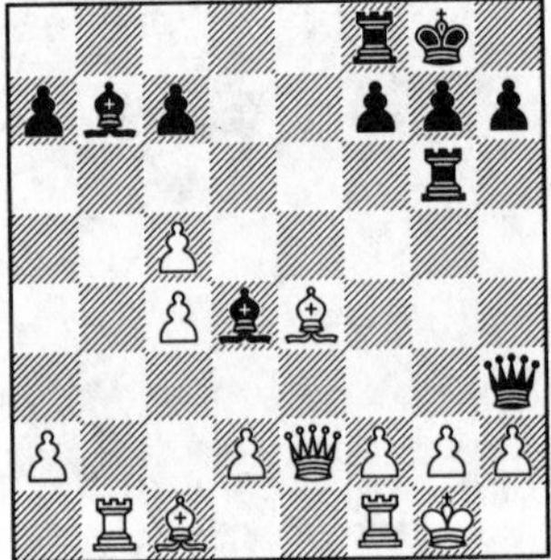

1301.

1302.

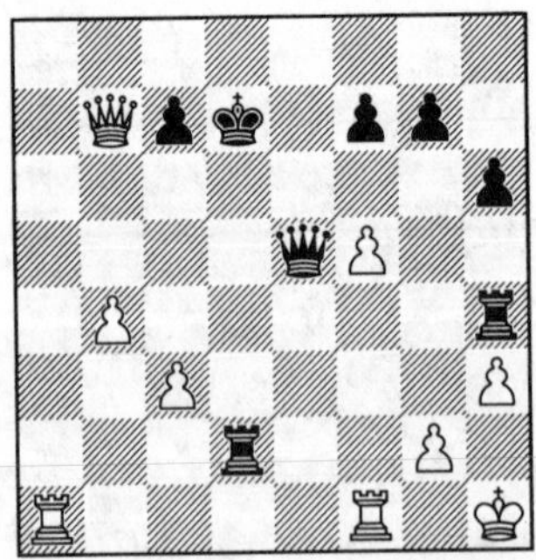

1303.

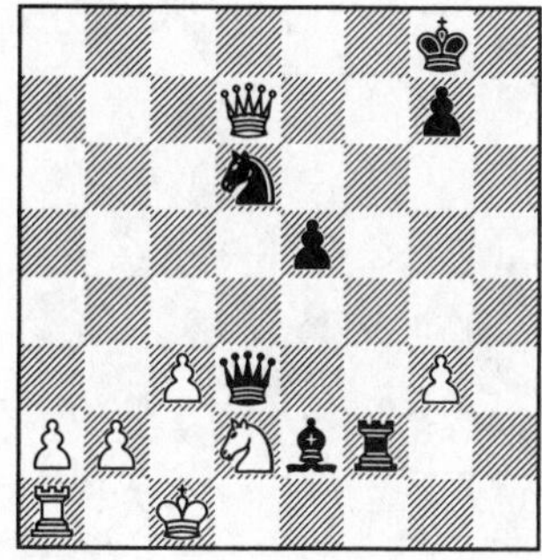

1304.

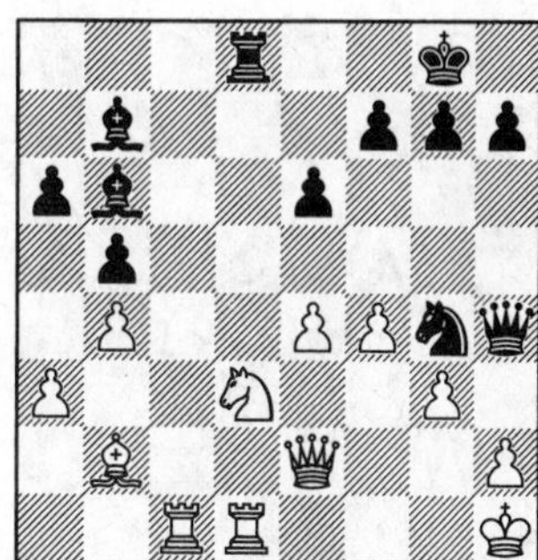

1305.

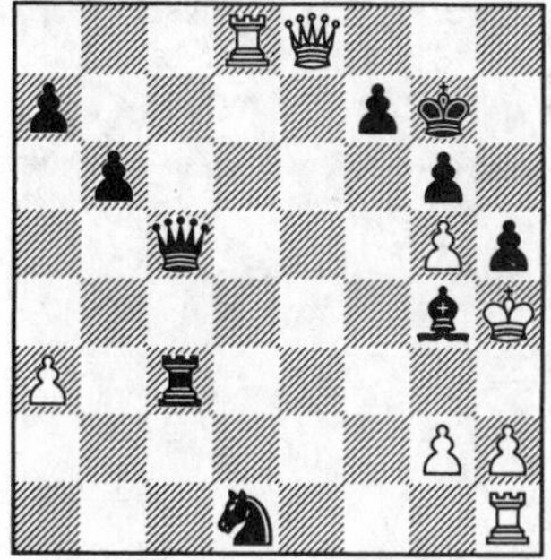

1306.

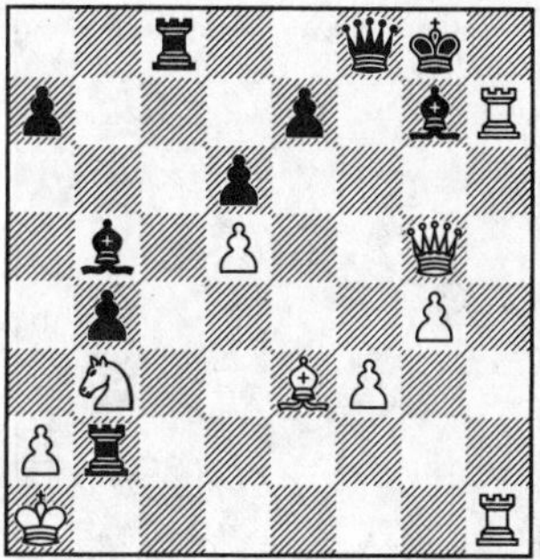

1307.

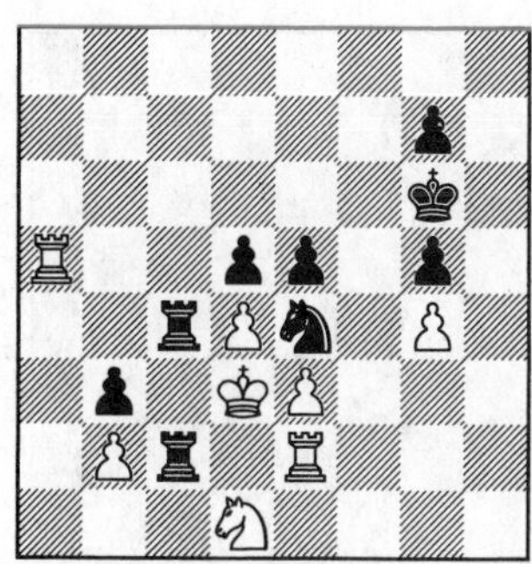

1308.

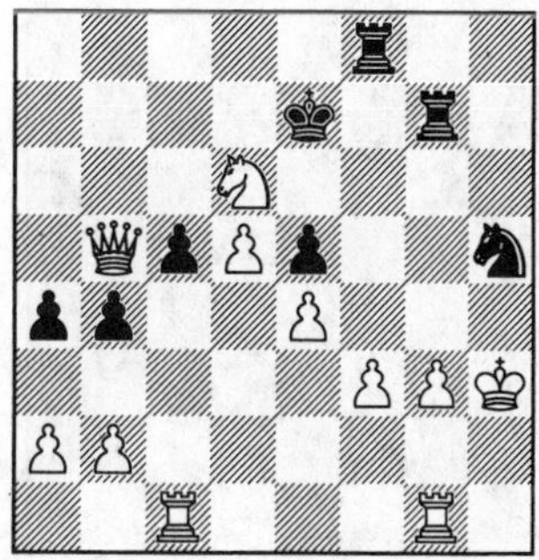

1309.

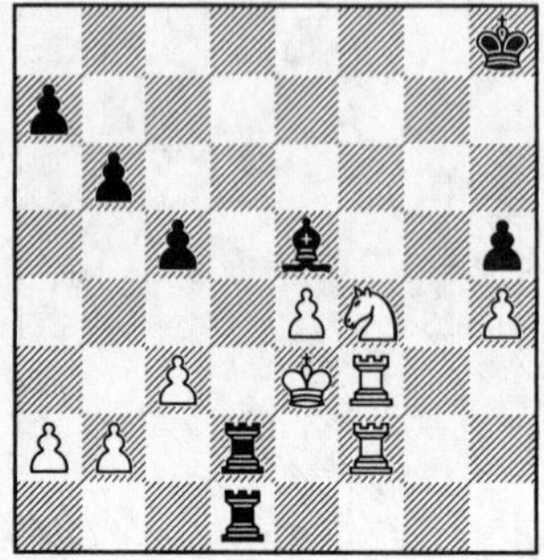

1310.

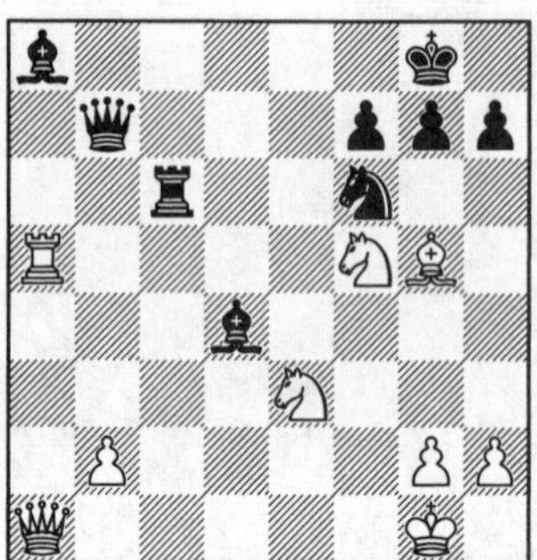

1311.

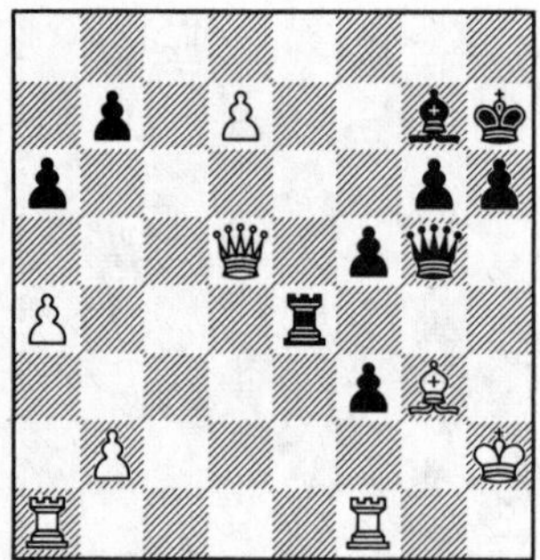

1312.

1313.

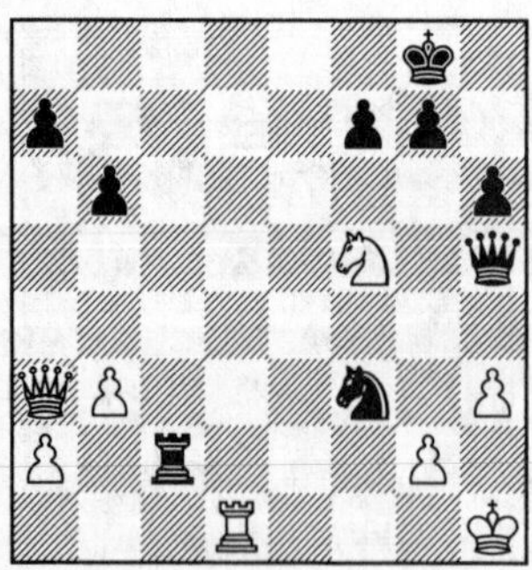

1314.

1315.

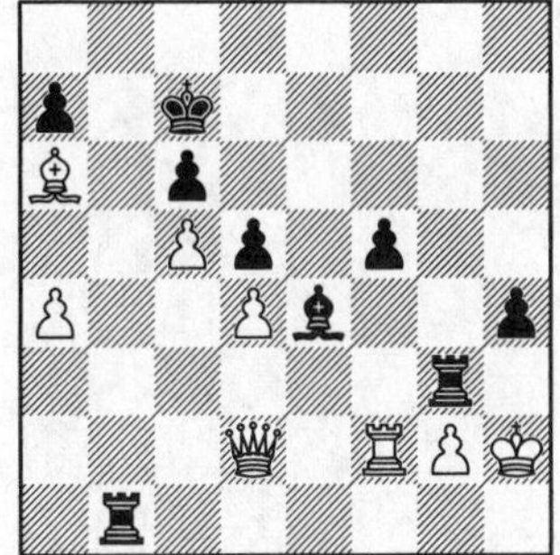

1316.

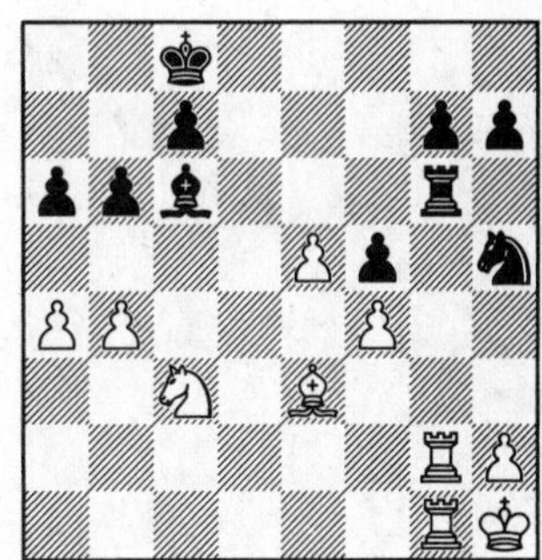

1317.

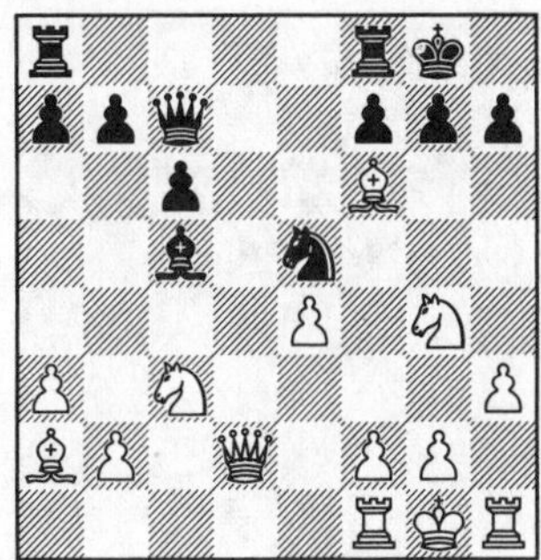

1318.

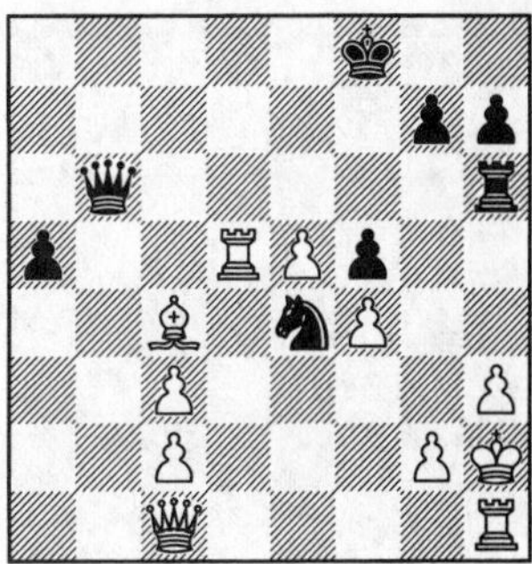

1319.

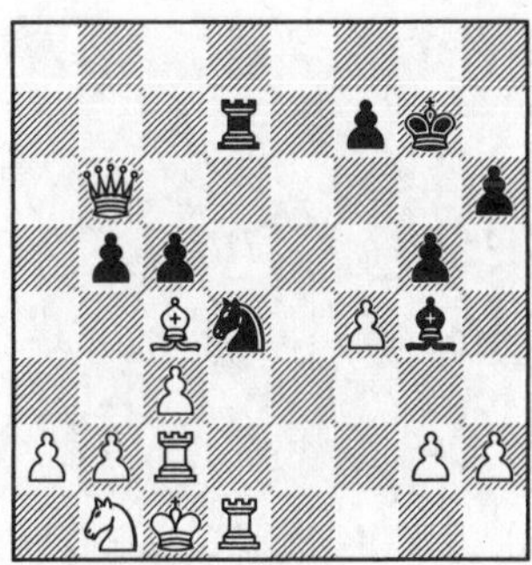

1320.

1321.

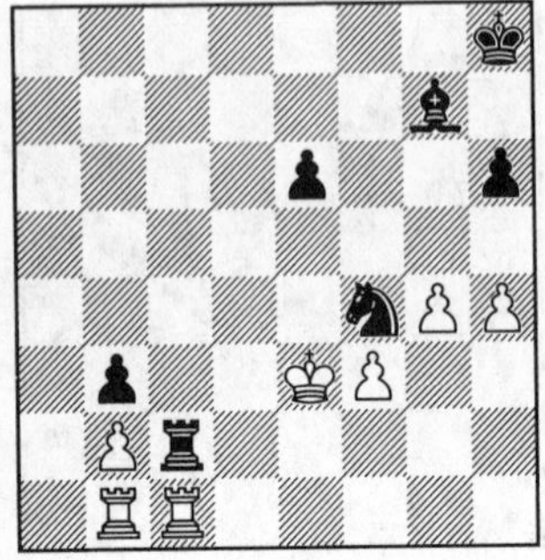

1322.

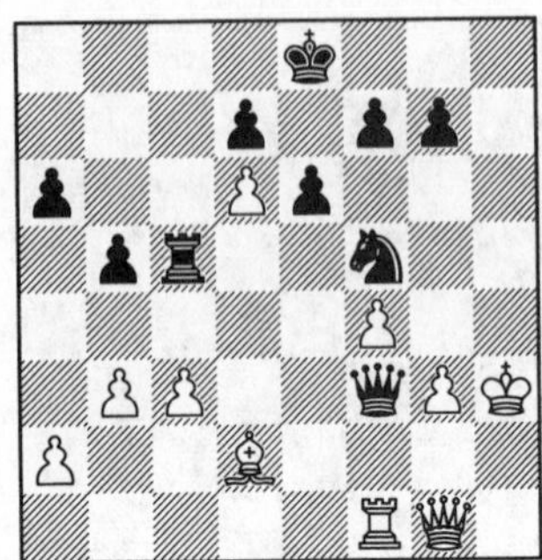

1323.

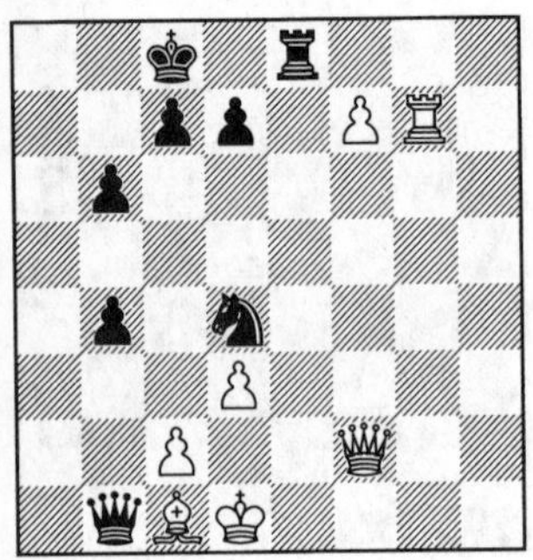

1324.

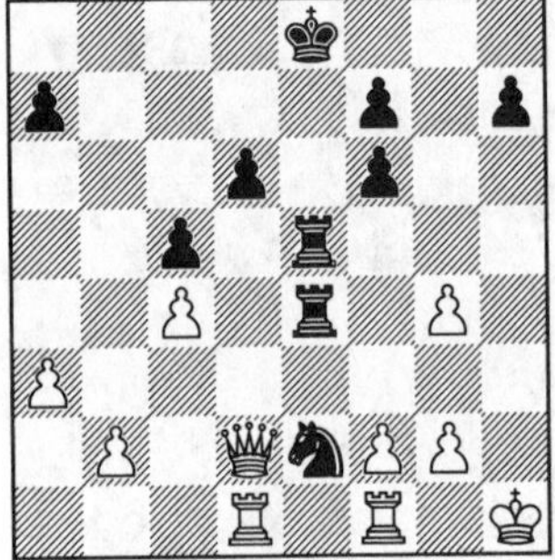

1325.

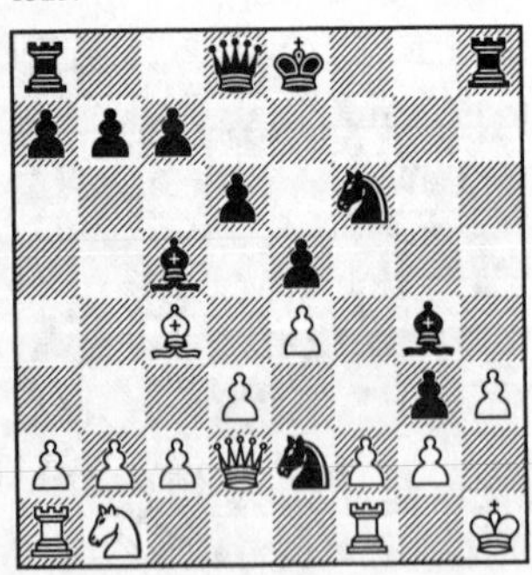

1326.

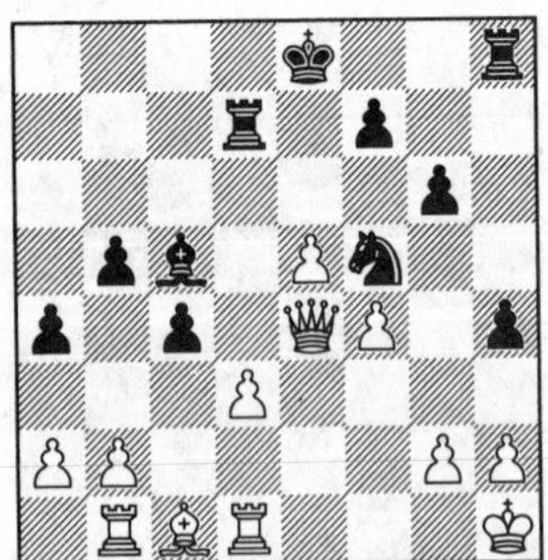

1327.

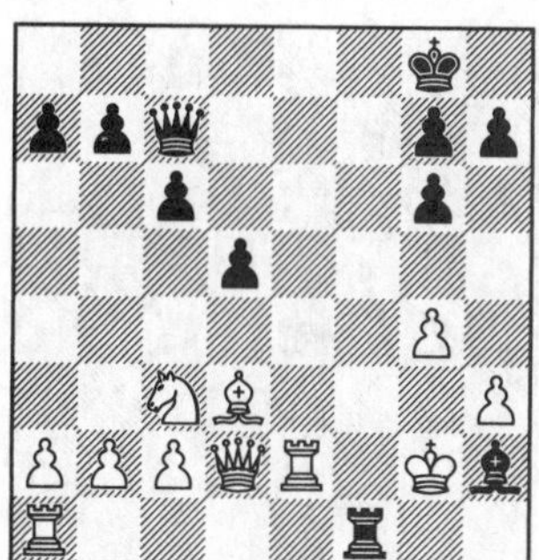

1328.

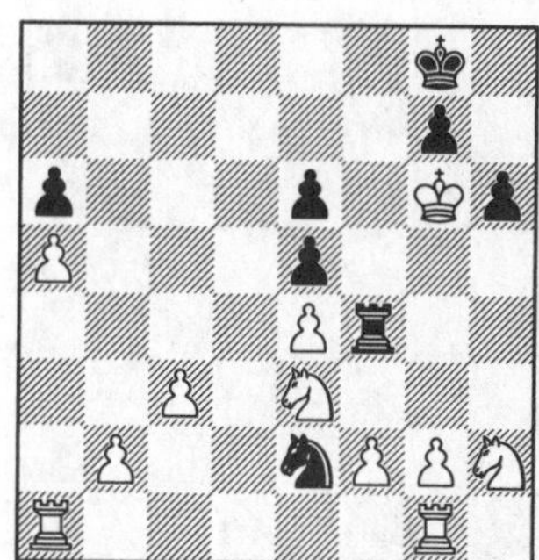

1329.

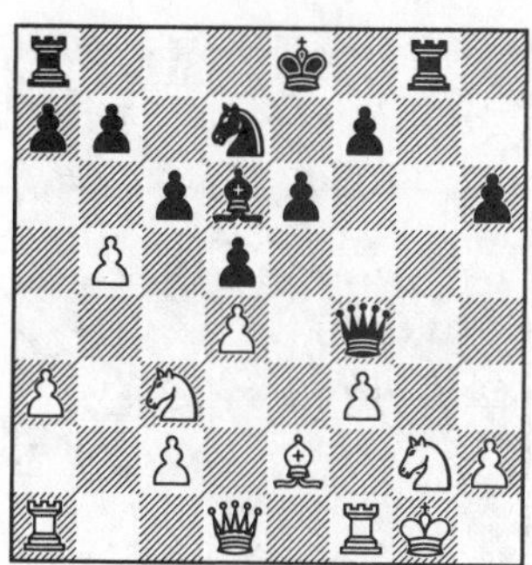

1330.

1331.

1332.

1333.

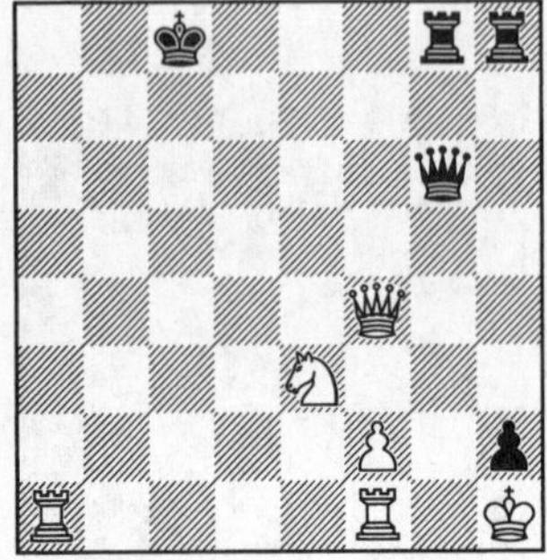

1334.

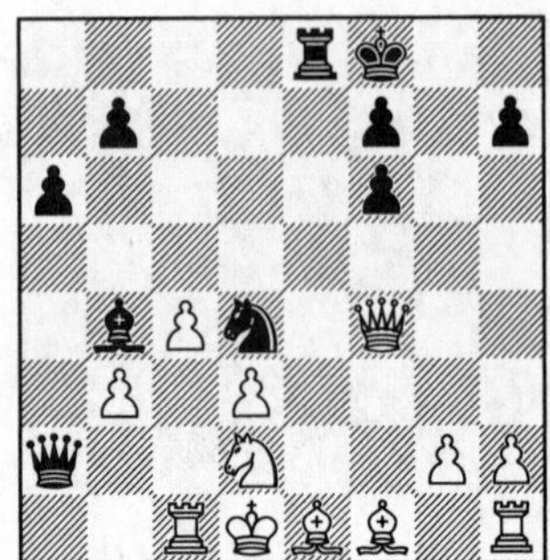

1335.

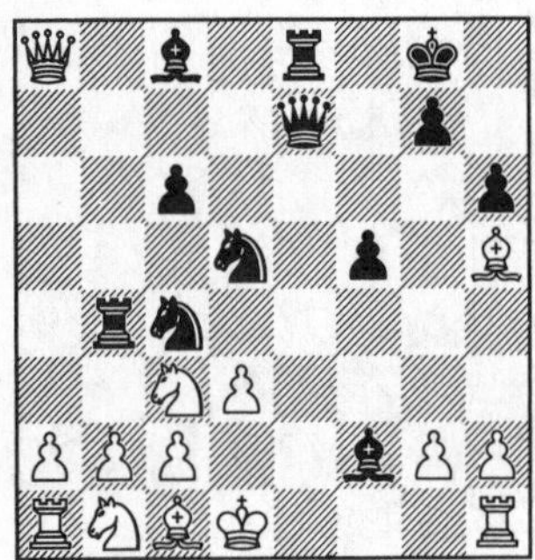

1336.

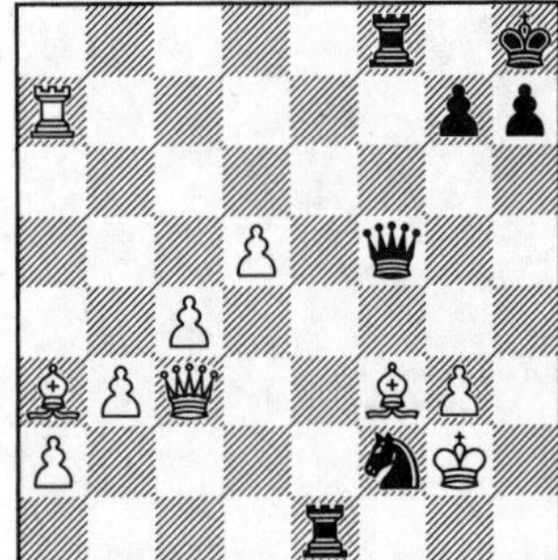

1337.

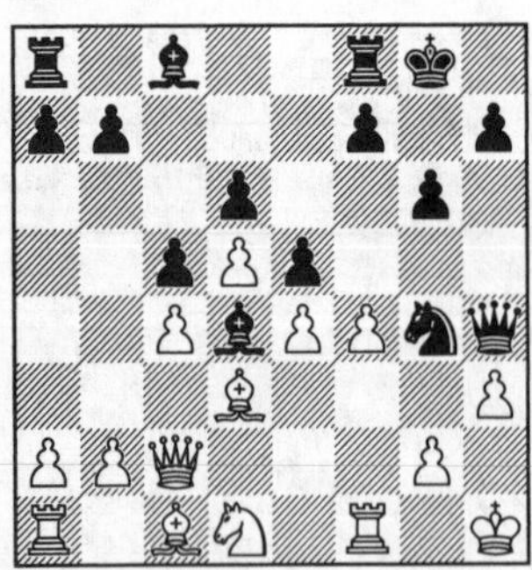

1338.

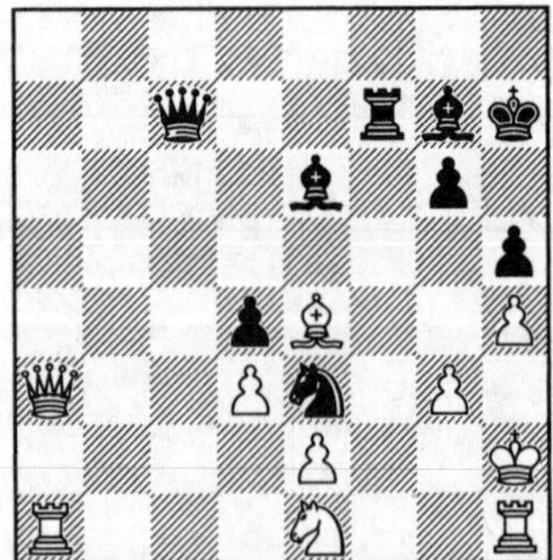

1339.

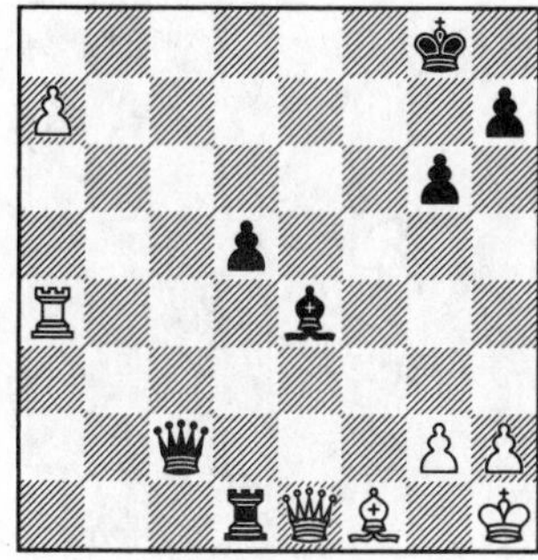

1340.

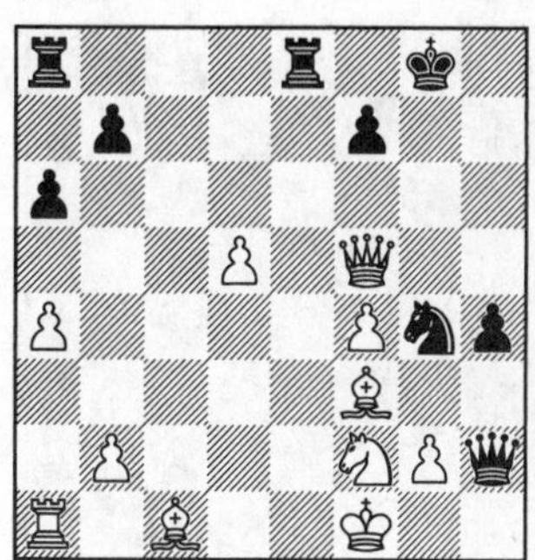

1341.

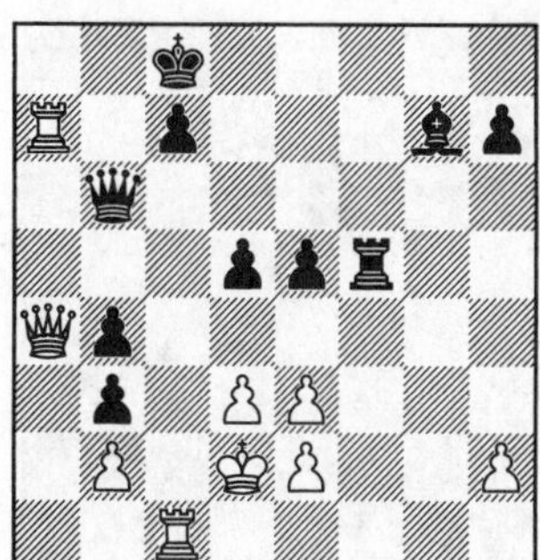

1342.

1343.

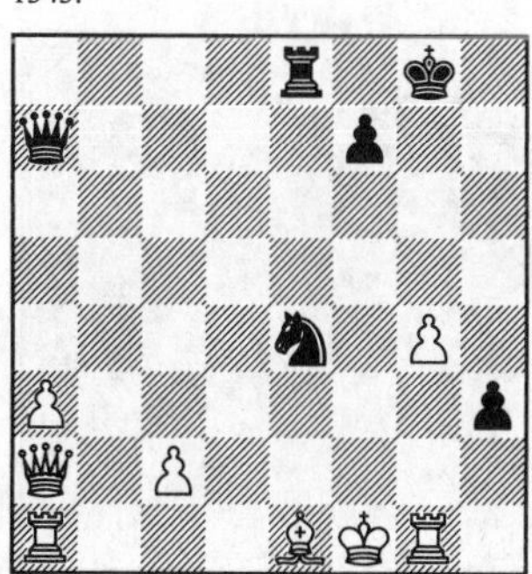

1344.

1345.

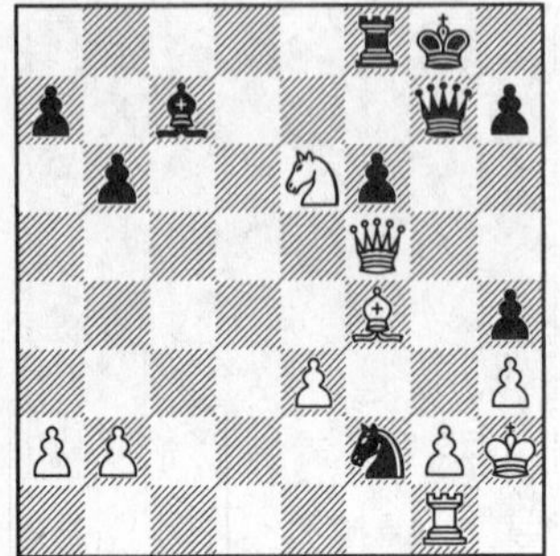

1346.

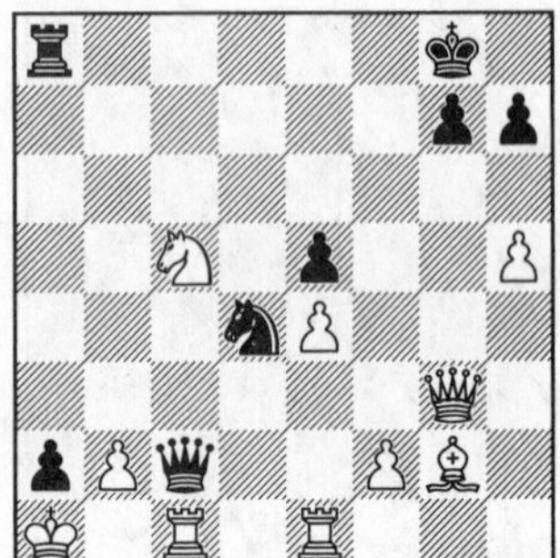

1347.

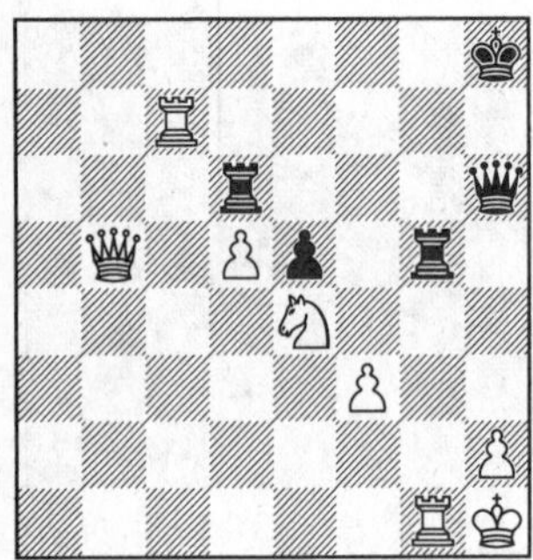

1348.

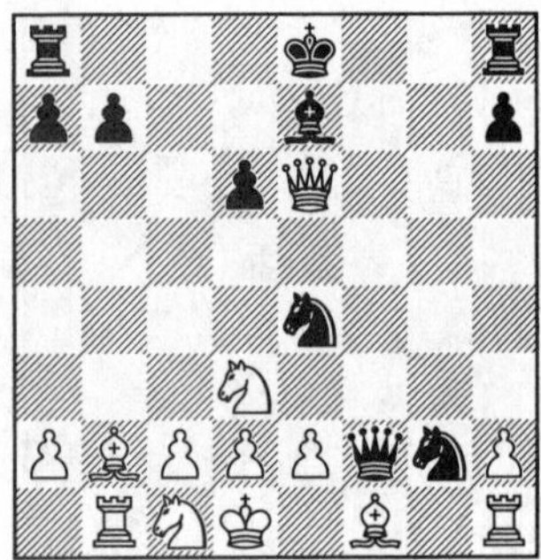

1349.

1350.

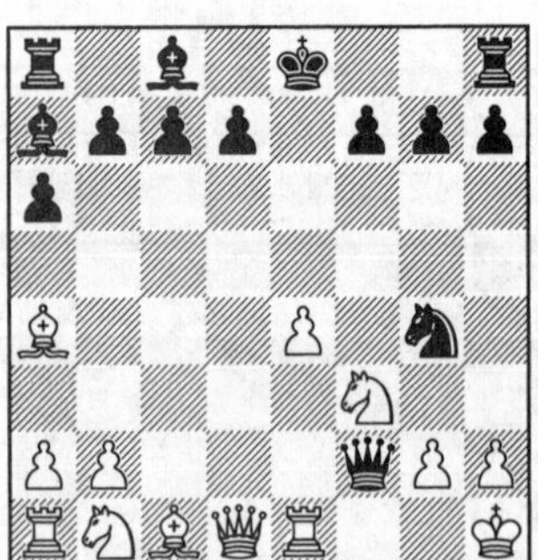

1351.

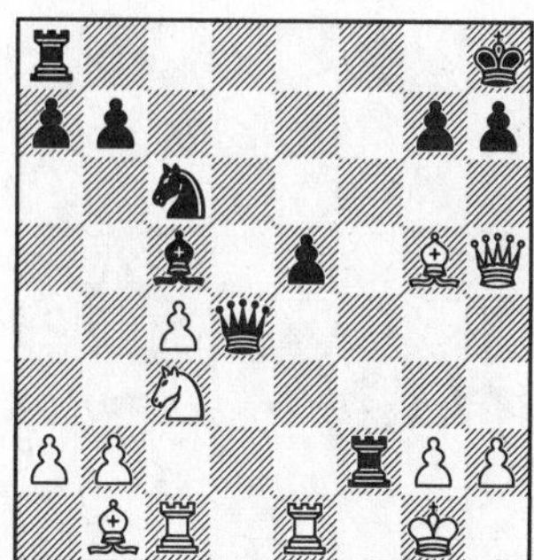

1352.

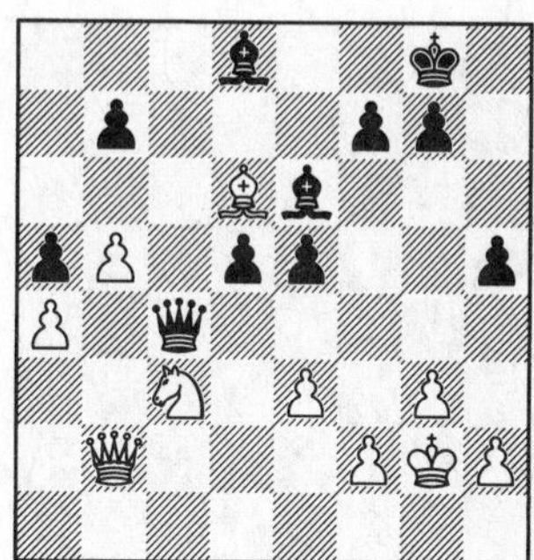

1353.

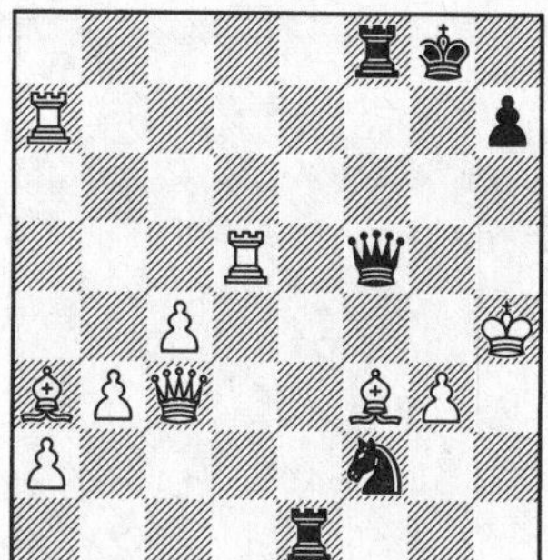

1354.

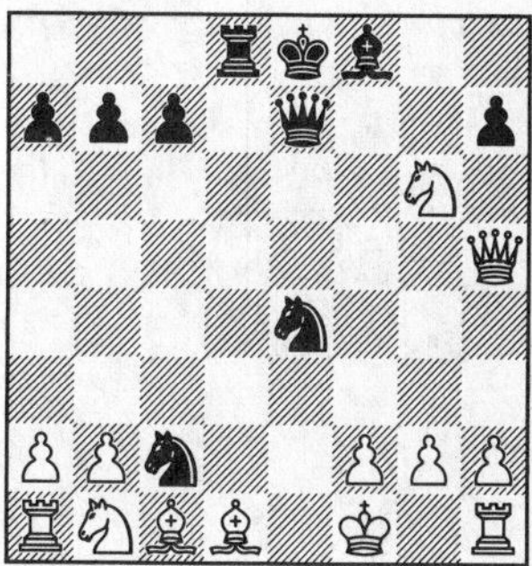

1355.

1356.

1357.

1358.

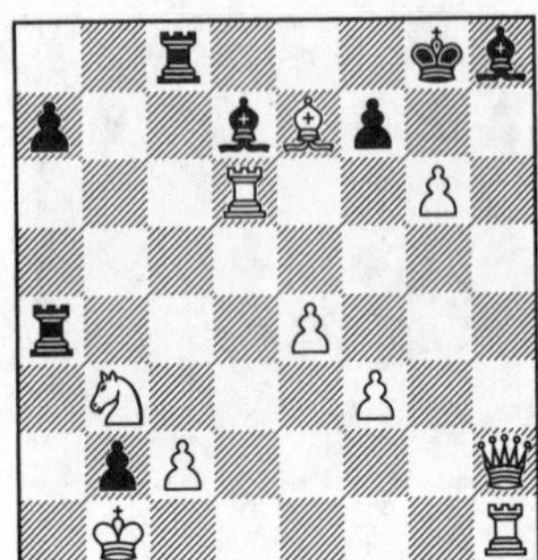

1359.

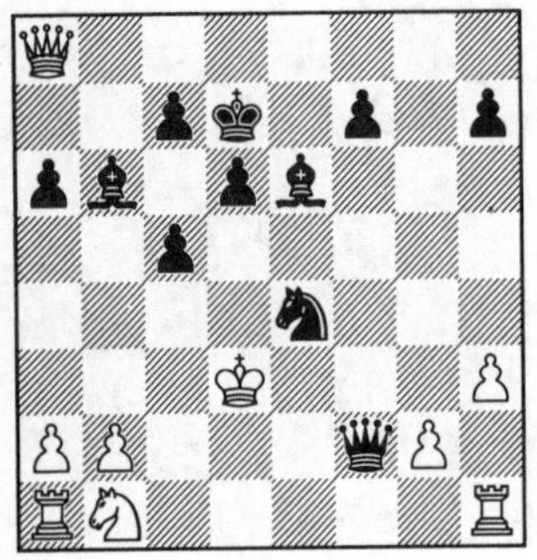

1360.

1361.

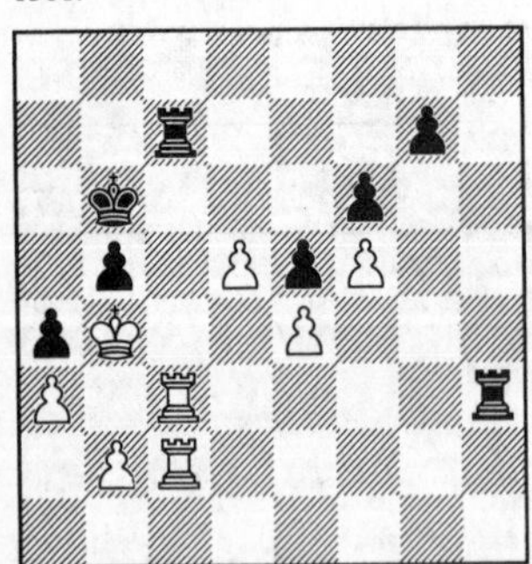

1362.

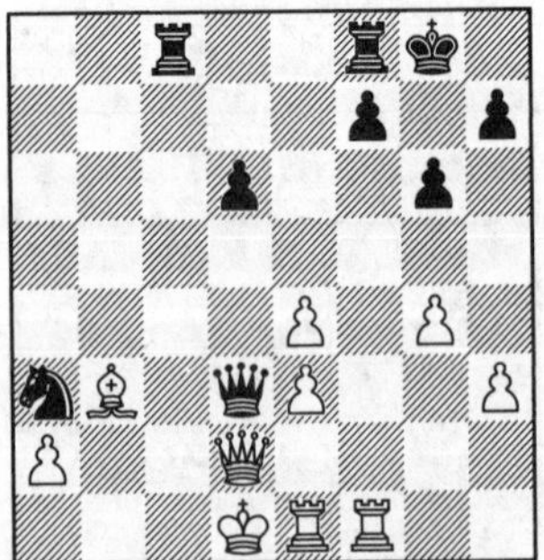

1363.

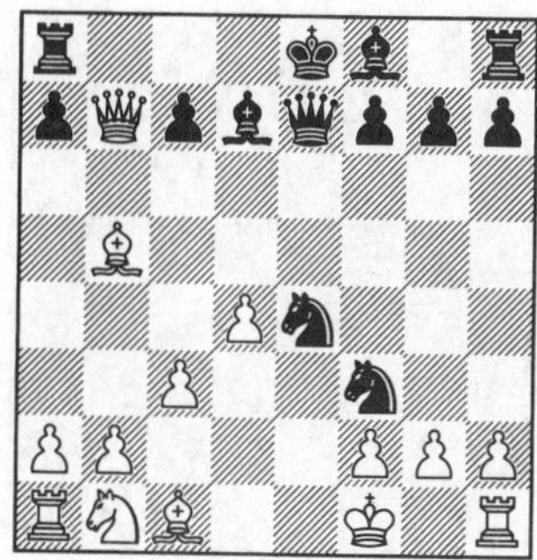

1364.

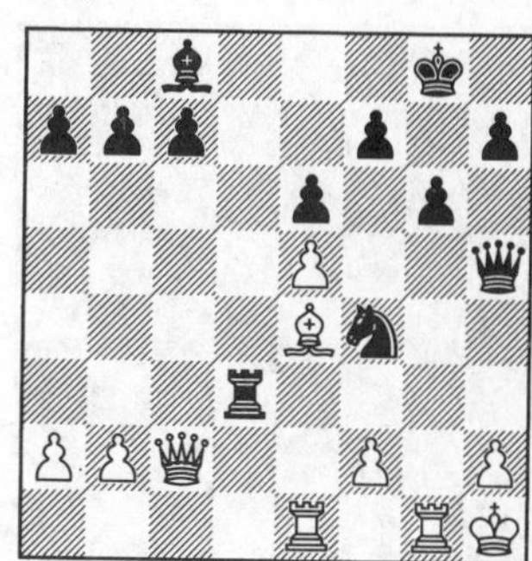

1365.

1366.

1367.

1368.

1369.

1370.

1371.

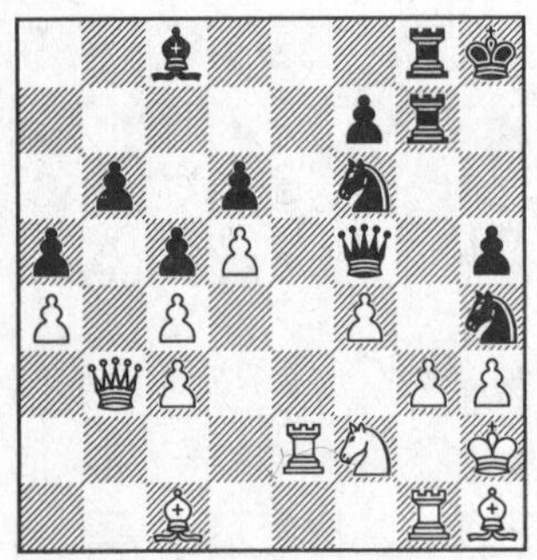

1372.

1373.

1374.

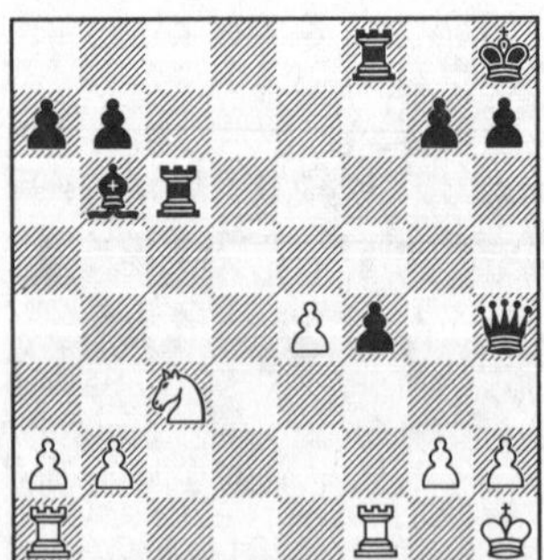

1375.

1376.

1377.

1378.

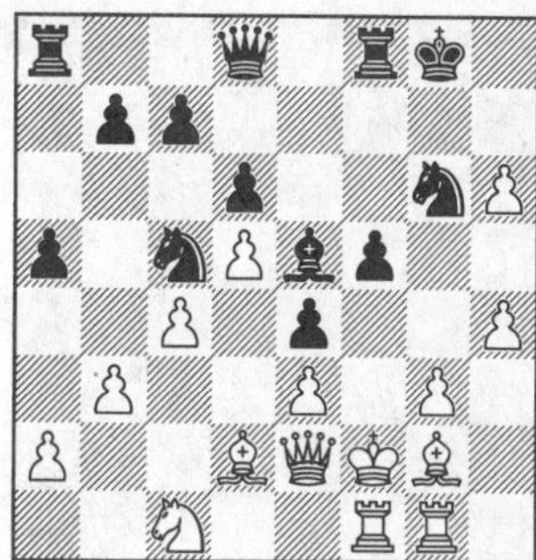

1379.

1380.

1381.

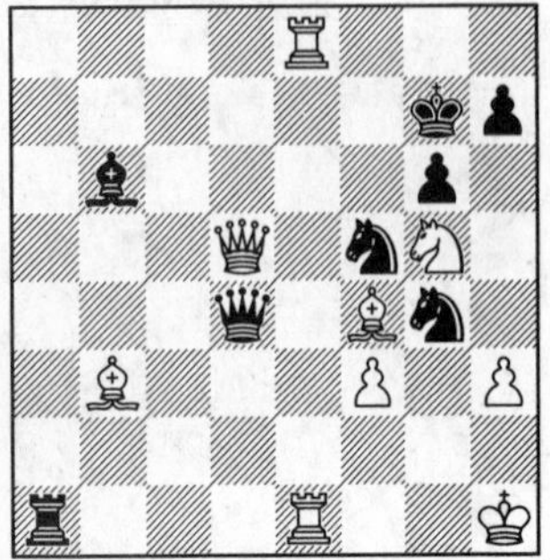

1382.

1383.

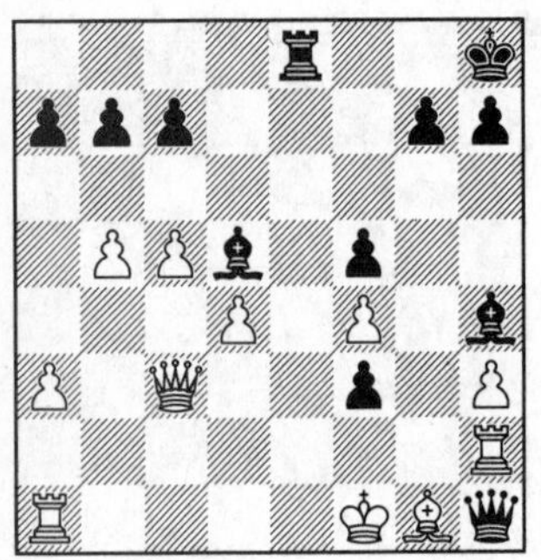

1384.

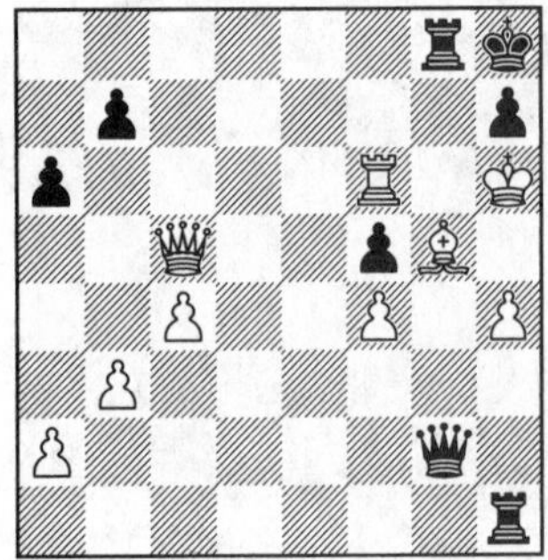

1385.

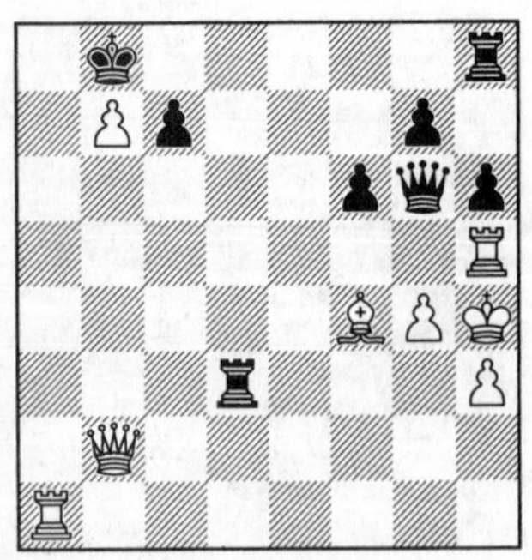

1386.

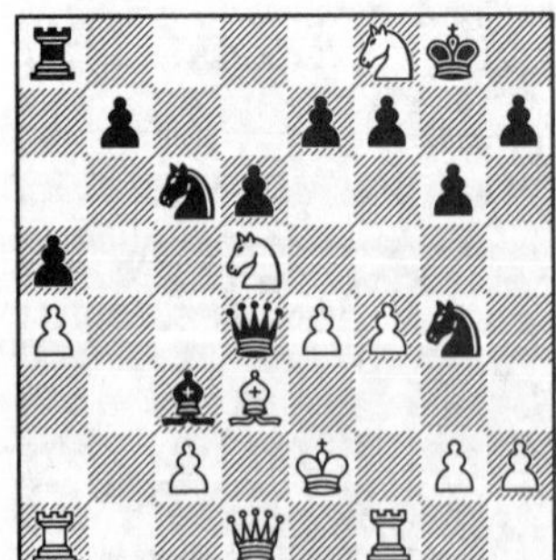

1387.

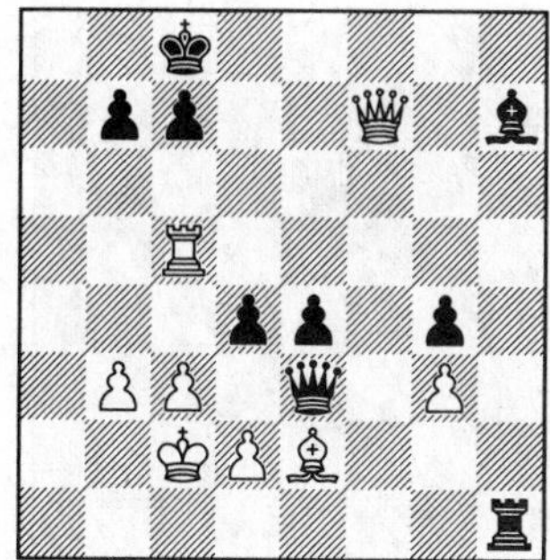

1388.

1389.

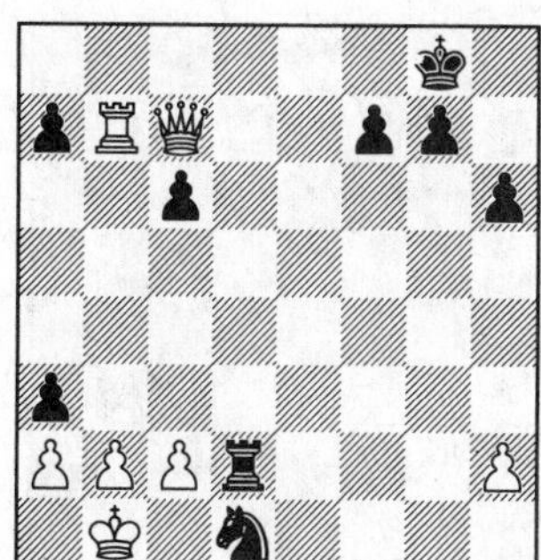

1390.

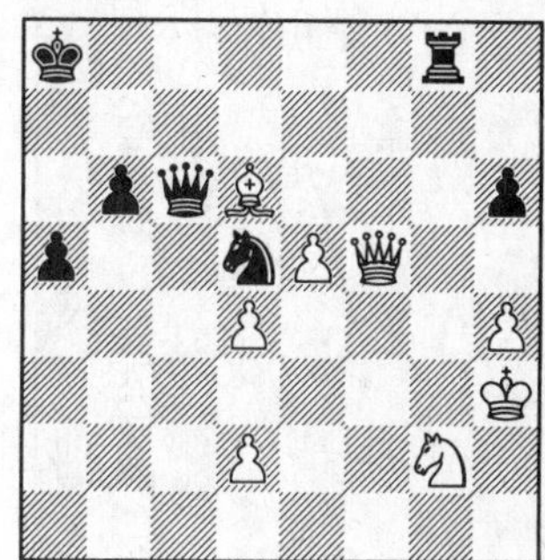

1391.

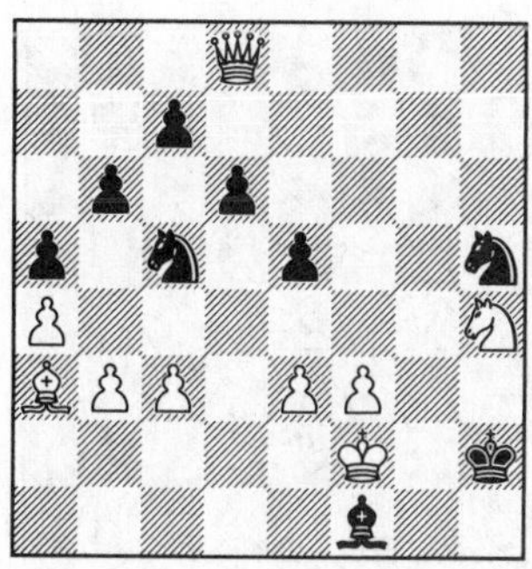

1392.

1393.

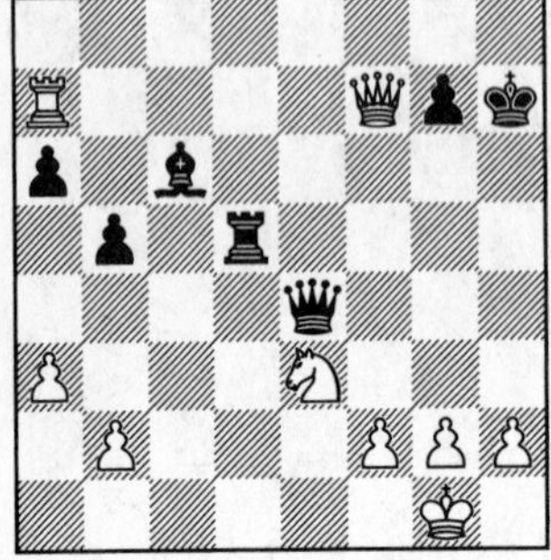

1394.

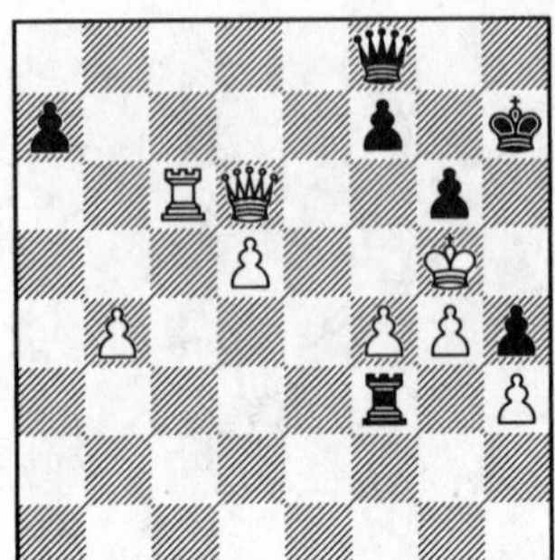

1395.

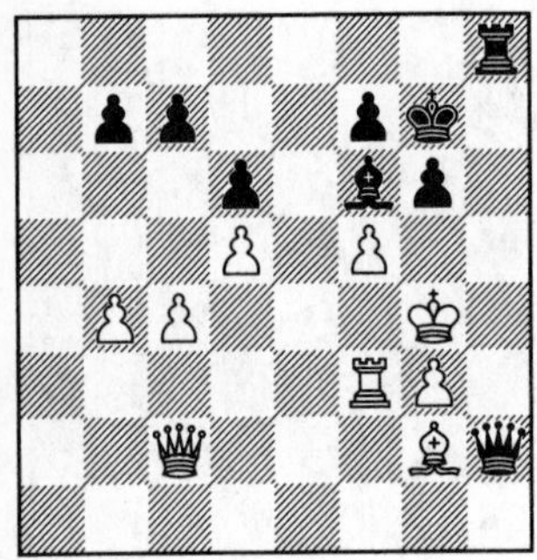

1396.

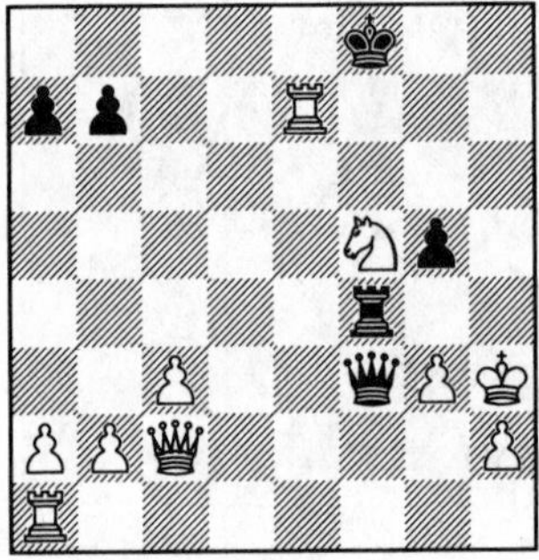

1397.

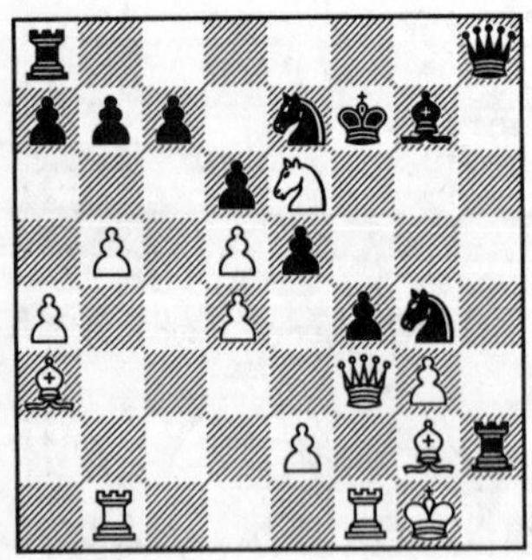

1398.

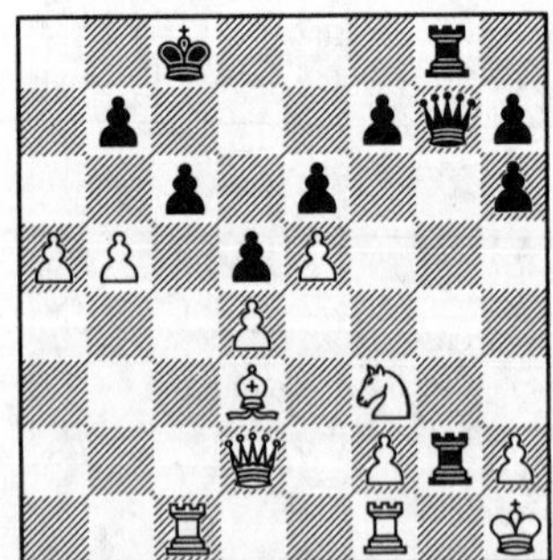

1399.

1400.

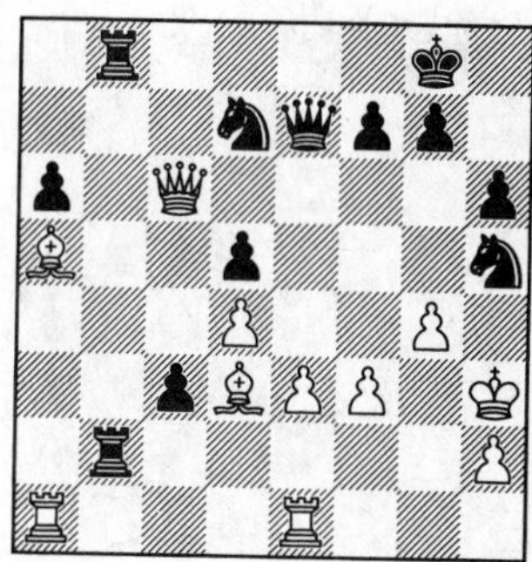

1401.

1402.

1403.

1404.

1405.

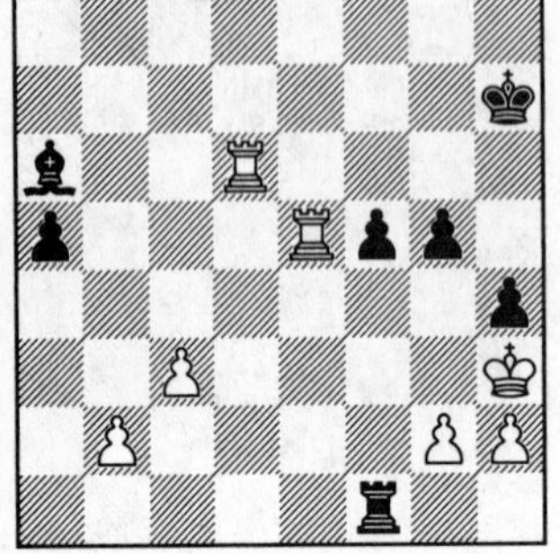

1406.

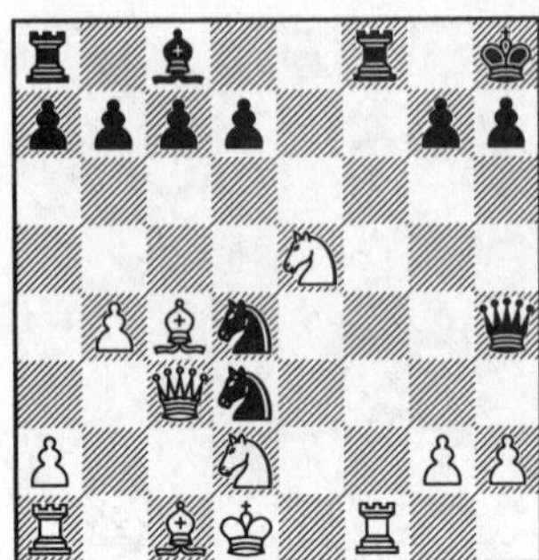

1407.

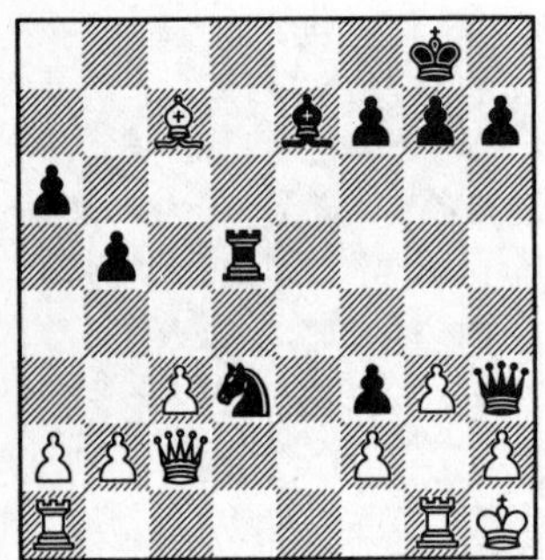

1408.

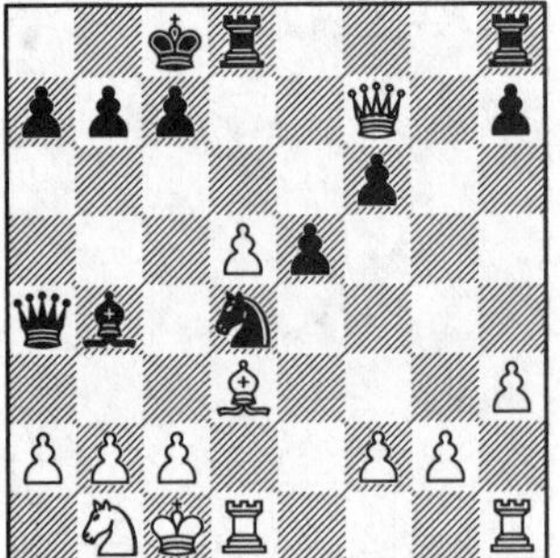

1409.

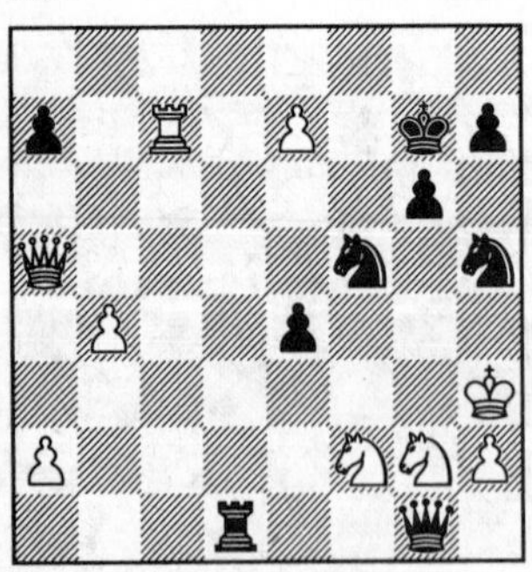

1410.

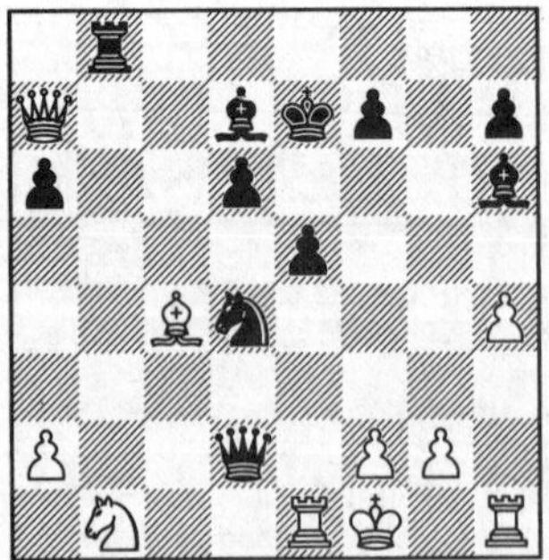

1411.

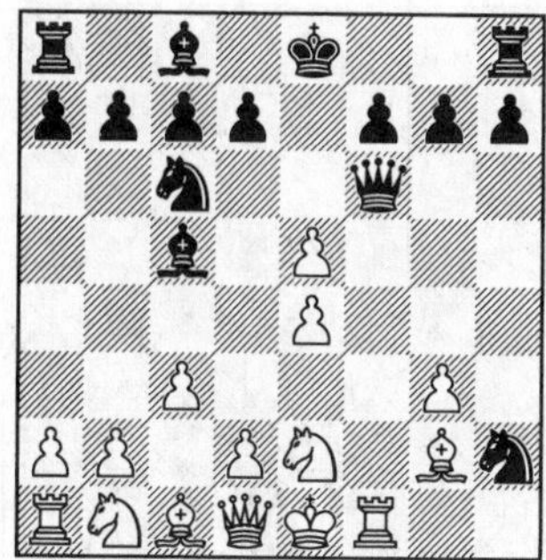

1412.

1413.

1414.

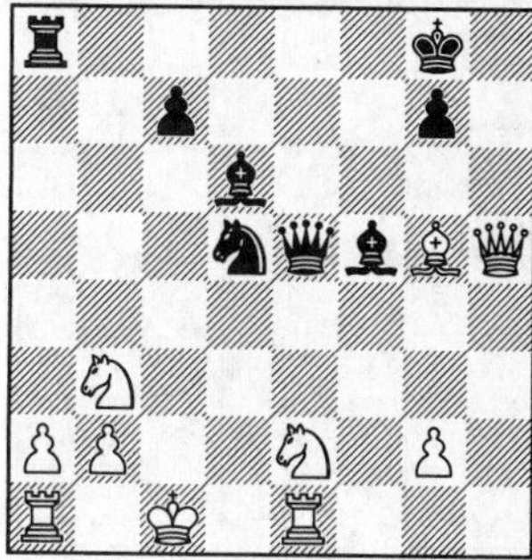

1415.

1416.

1417.

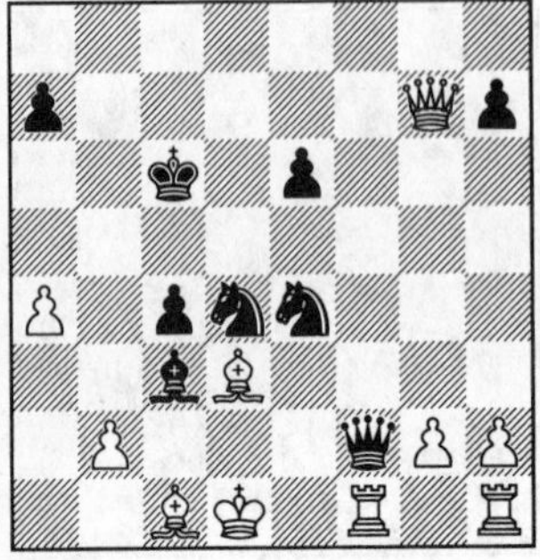

1418.

1419.

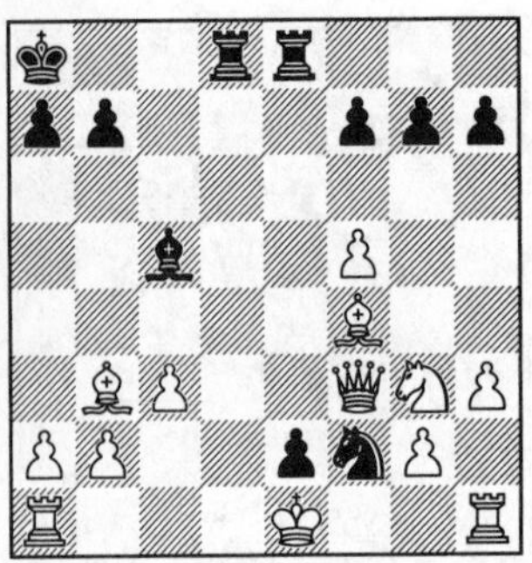

1420.

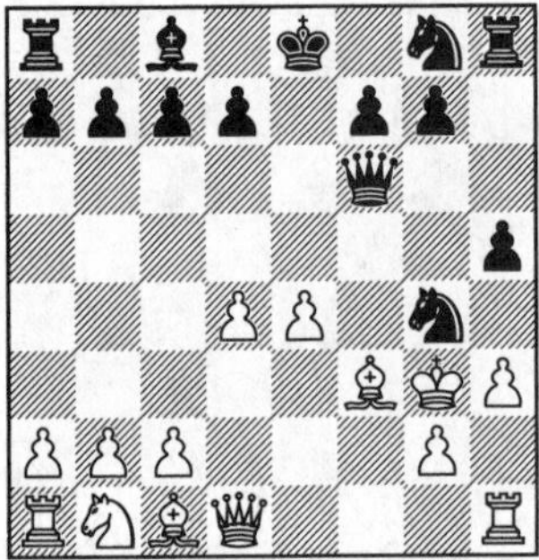

1421.

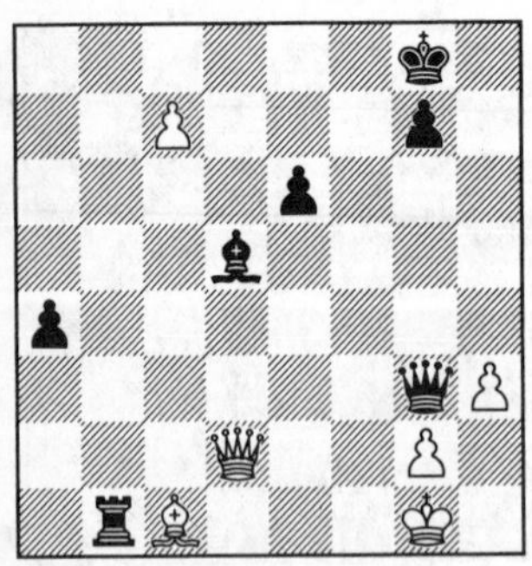

1422.

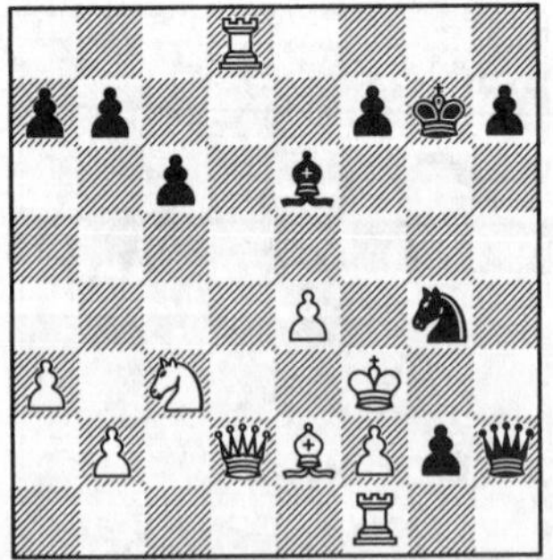

1423.

1424.

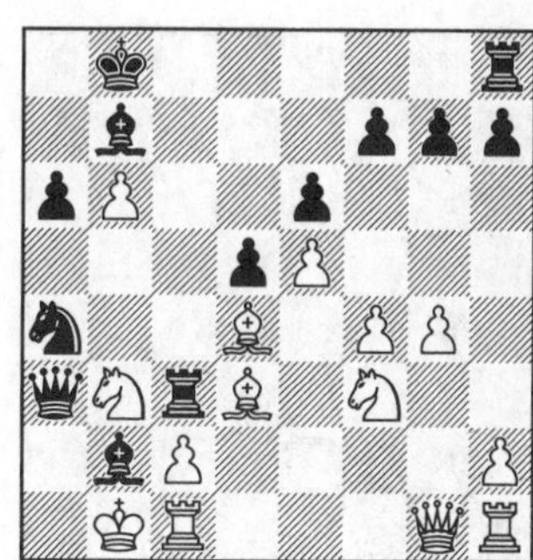

1425.

1426.

1427.

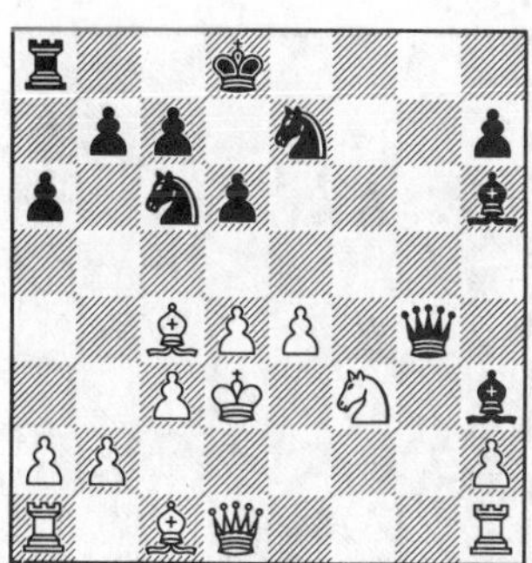

1428.

1429.

1430.

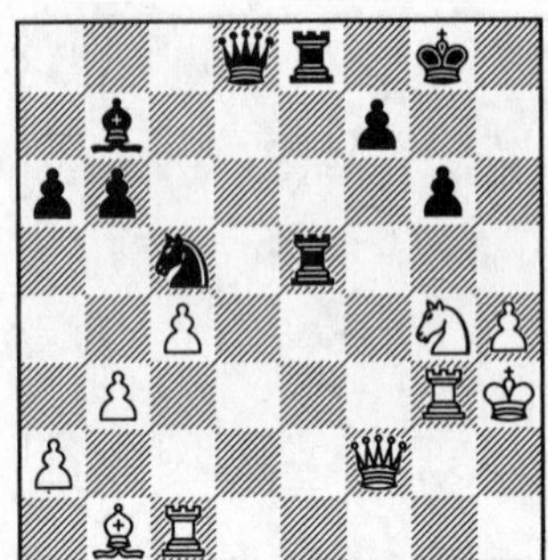

1431.

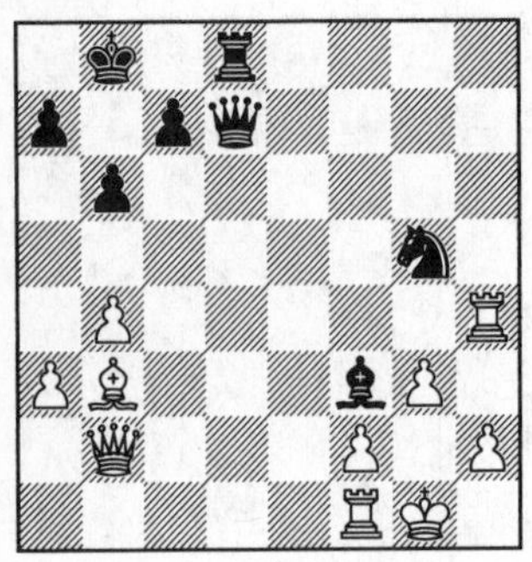

1432.

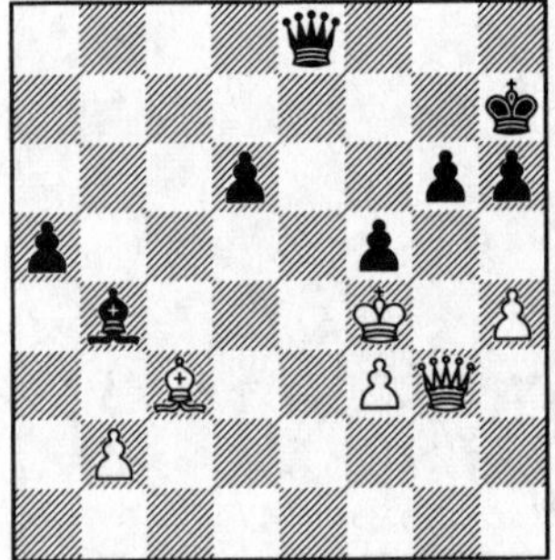

1433.

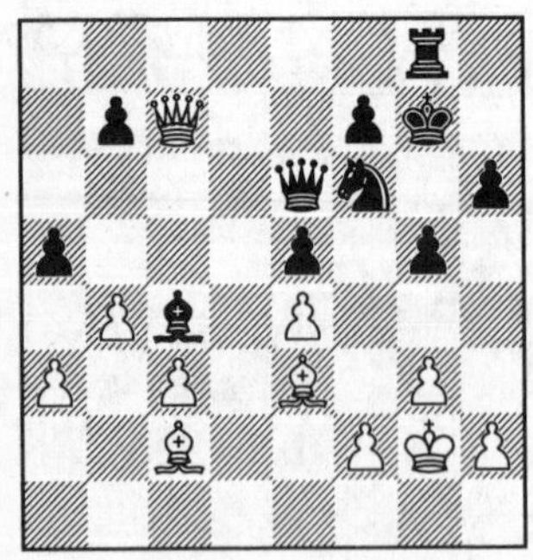

1434.

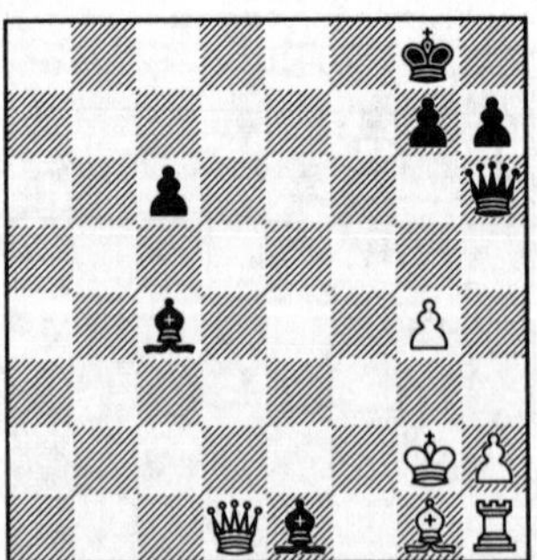

1435.

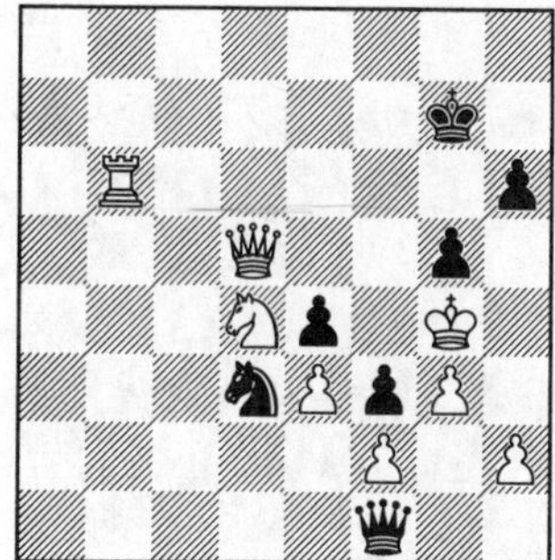

1436.

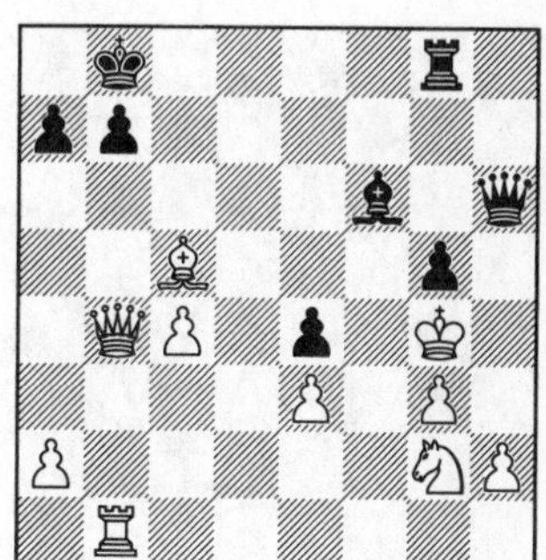

1437.

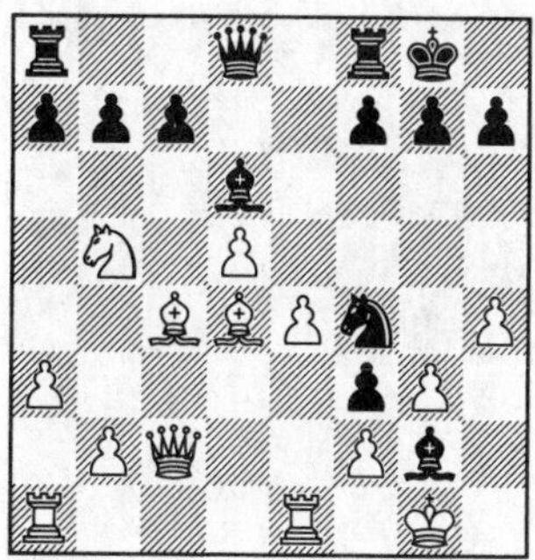

1438.

1439.

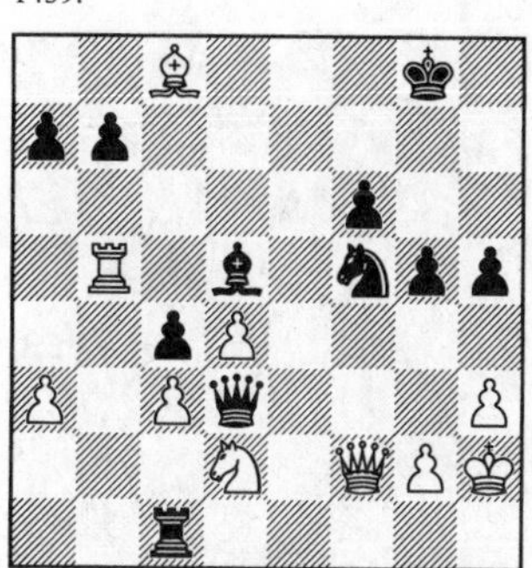

1440.

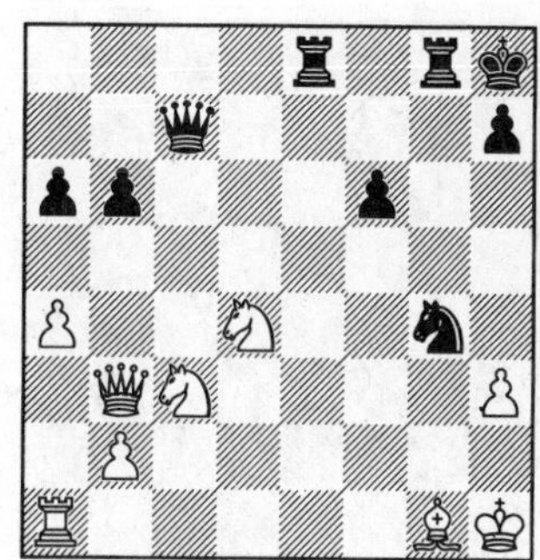

1441.

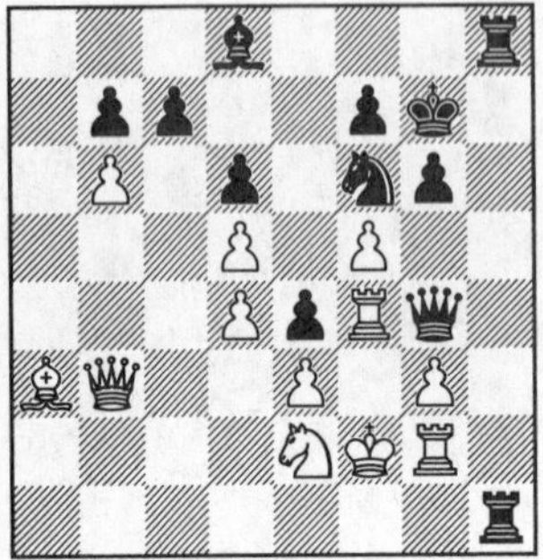

1442.

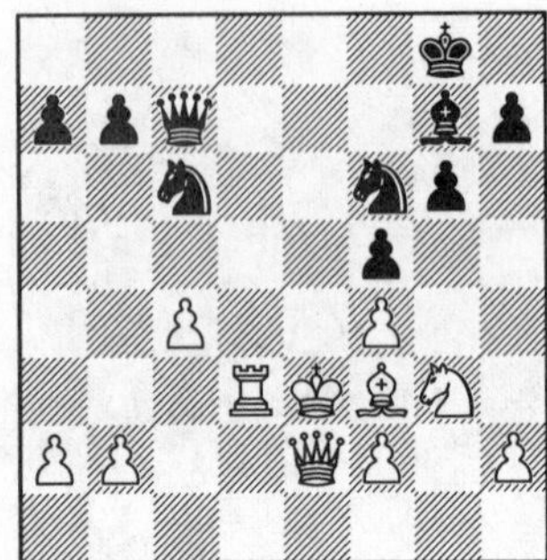

1443.

1444.

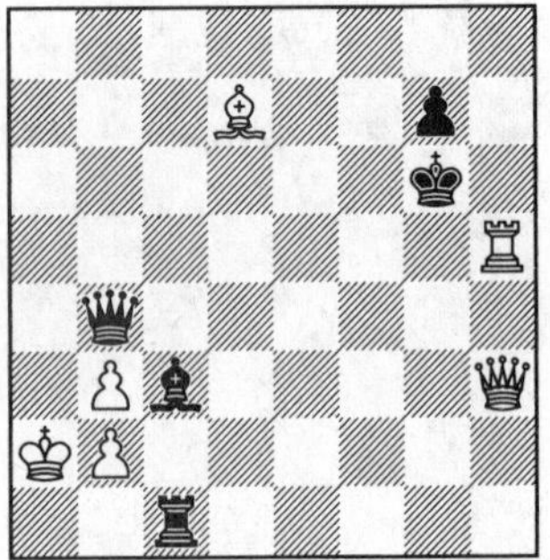

1445.

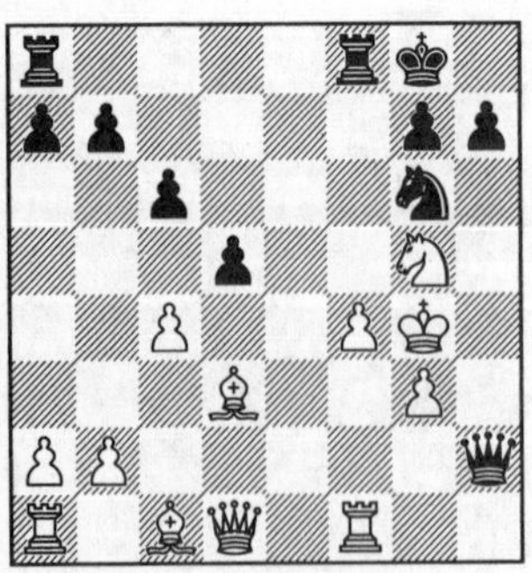

1446.

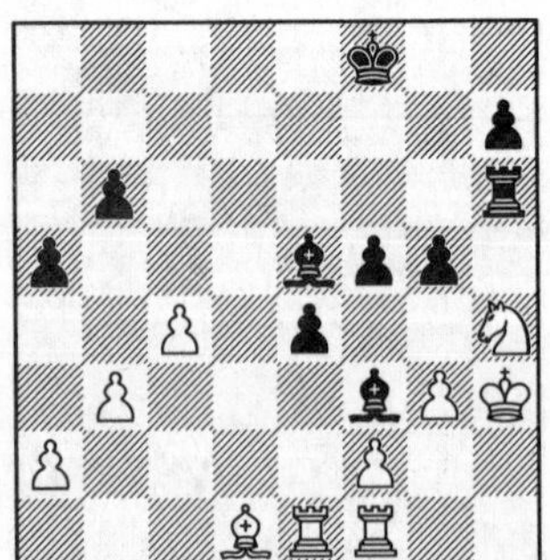

1447.

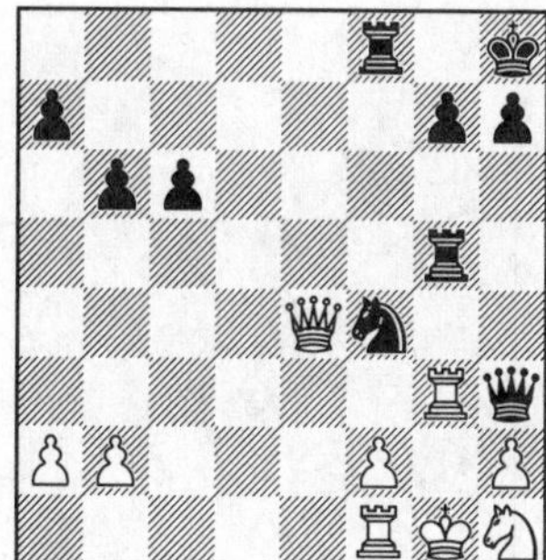

1448.

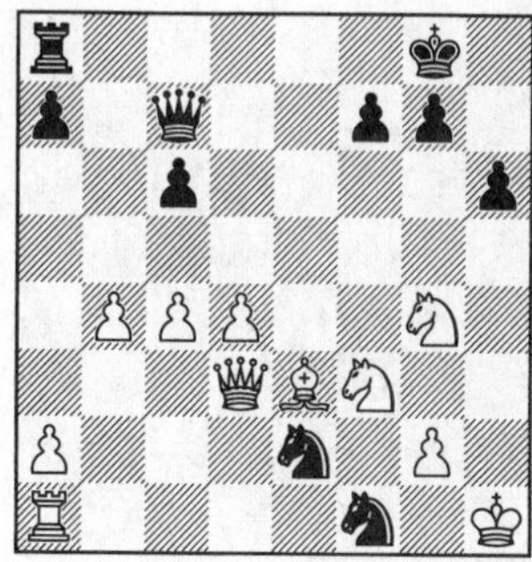

1449.

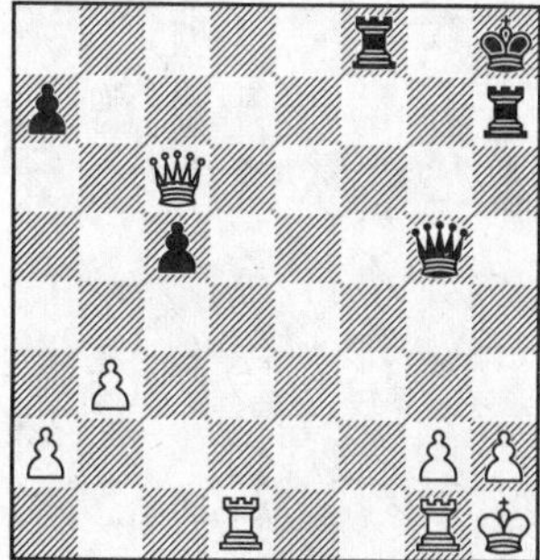

1450.

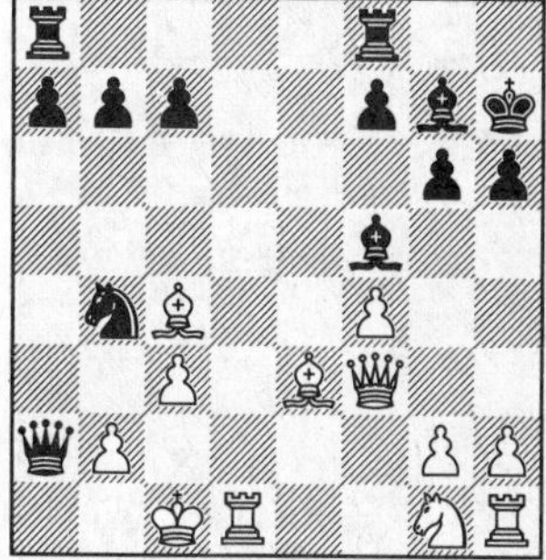

1451.

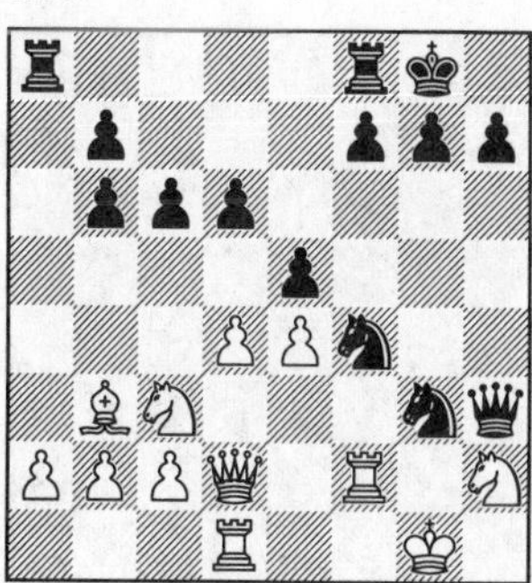

1452.

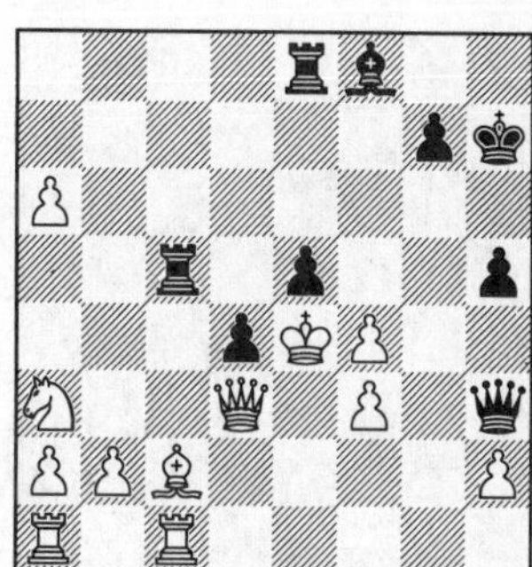

1453.

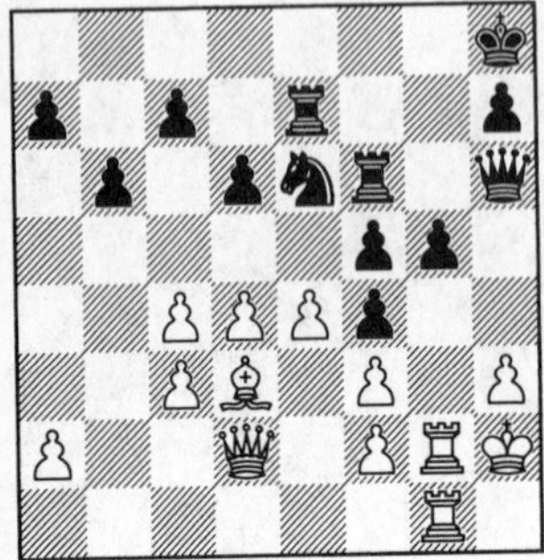

1454.

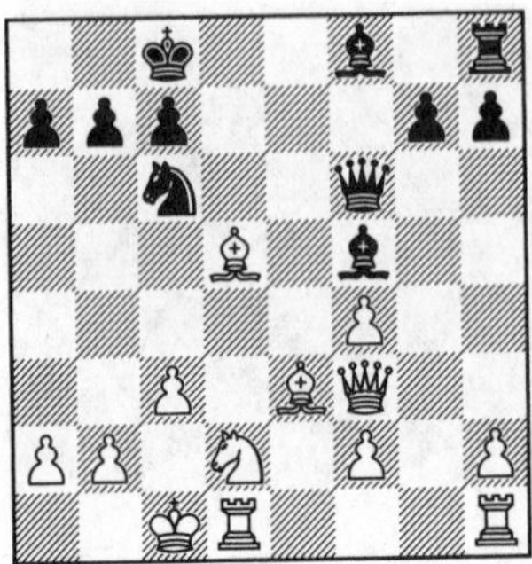

1455.

1456.

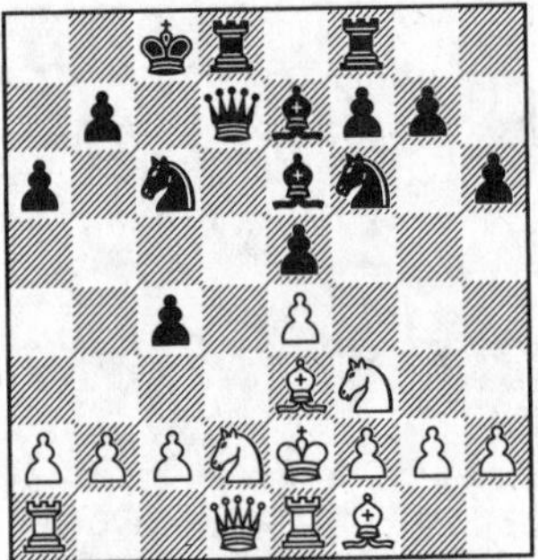

1457.

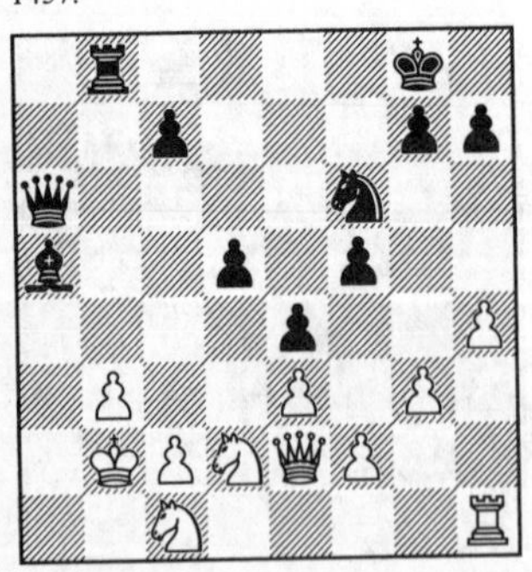

1458.

1459.

1460.

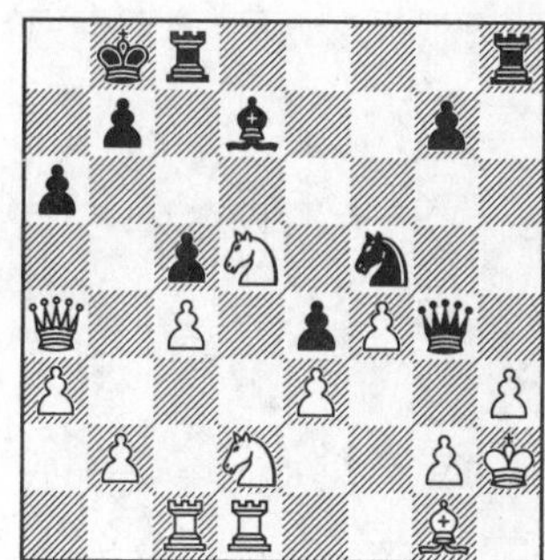

1461.

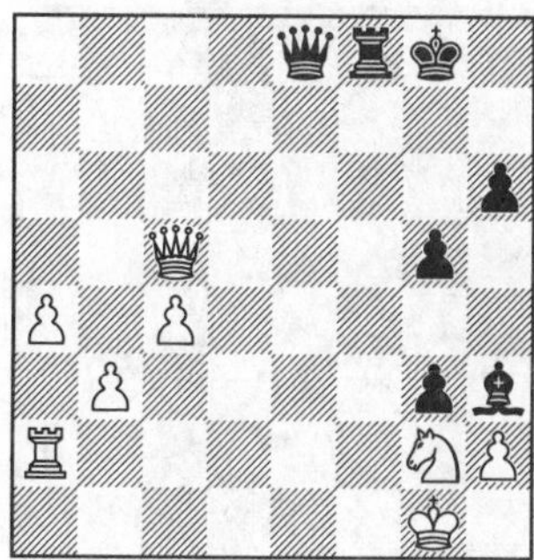

1462.

1463.

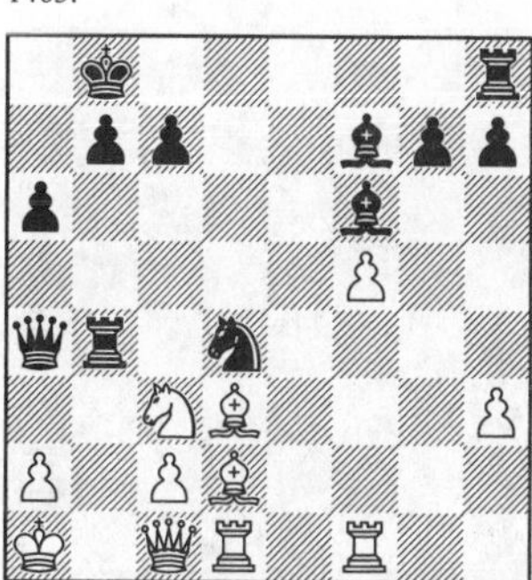

1464.

1465.

1466.

1467.

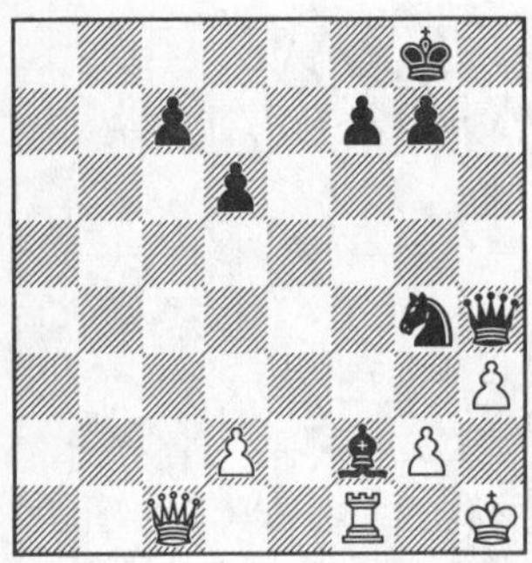

1468.

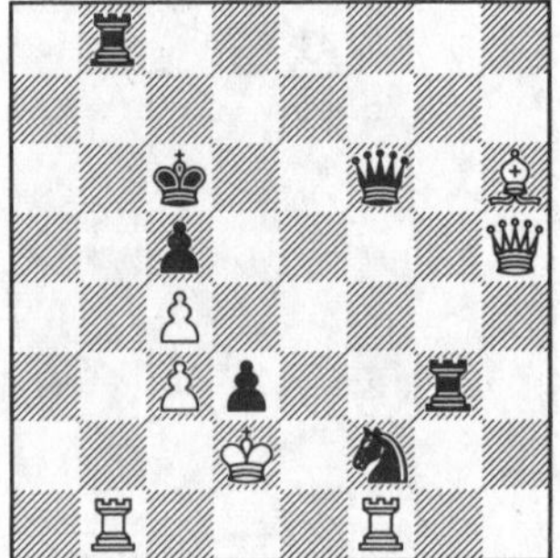

1469.

1470.

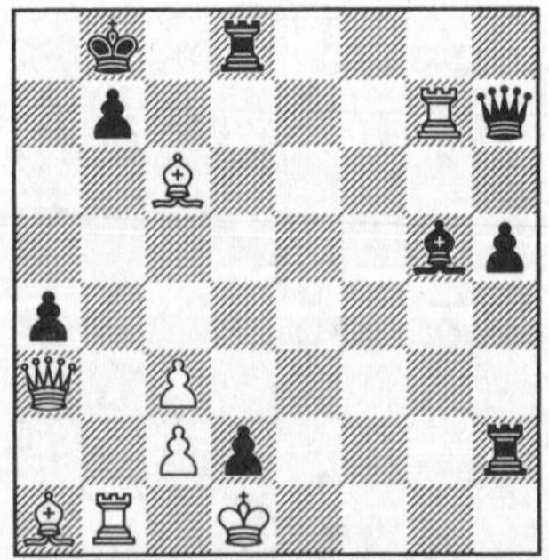

1471.

1472.

1473.

1474.

1475.

1476.

1477.

1478.

1479.

1480.

1481.

1482.

1483.

1484.

1485.

1486.

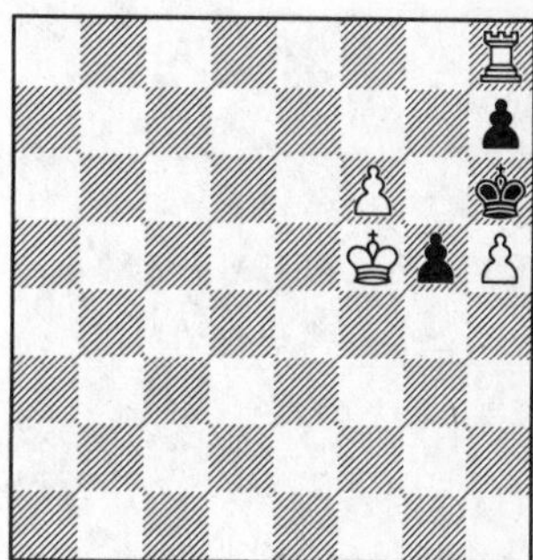

1487.

1488.

1489.

1490.

1491.

1492.

1493.

1494.

1495.

1496.

1497.

1498.

1499.

1500.

1501.

1502.

1503.

1504.

1505.

1506.

1507.

1508.

1509.

1510.

1511.

1512.

1513.

1514.

1515.

1516.

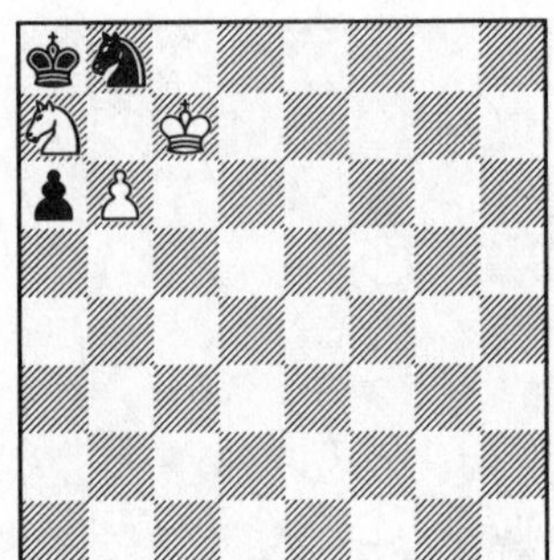

1517.

1518.

1519.

1520.

1521.

1522.

1523.

1524.

1525.

1526.

1527.

1528.

1529.

1530.

1531.

1532.

1533.

1534.

1535.

1536.

1537.

1538.

1539.

1540.

1541.

1542.

1543.

1544.

1545.

1546.

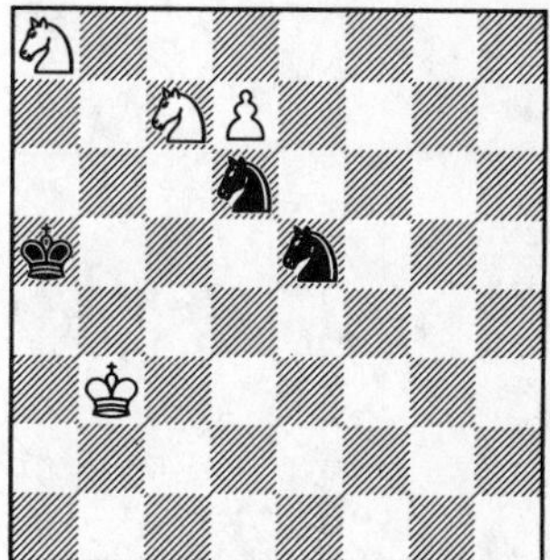

1547.

1548.

1549.

1550.

1551.

1552.

1553.

1554.

1555.

1556.

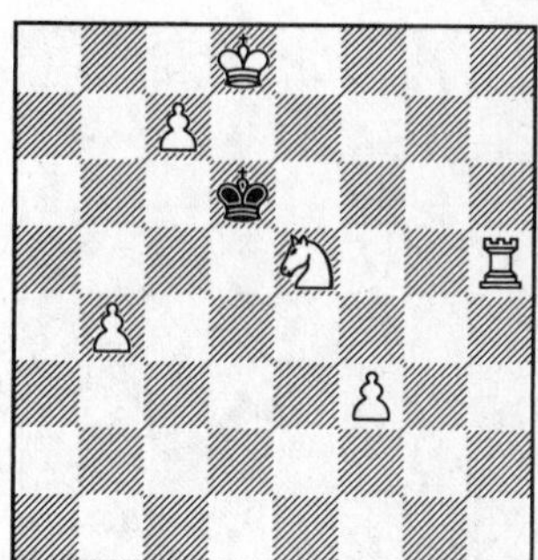

1557.

1558.

1559.

1560.

1561.

1562.

1563.

1564.

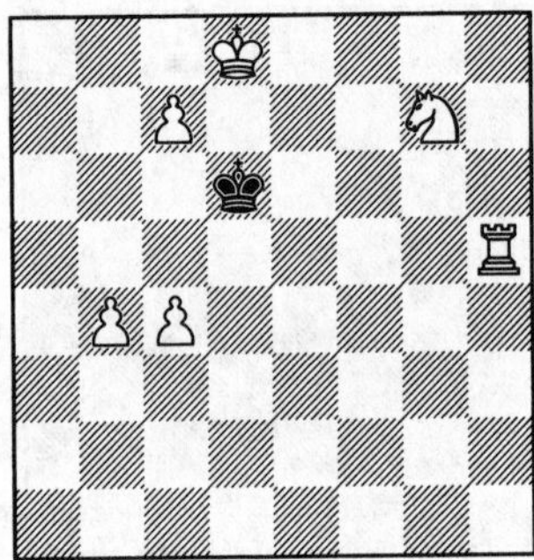

1565.

1566.

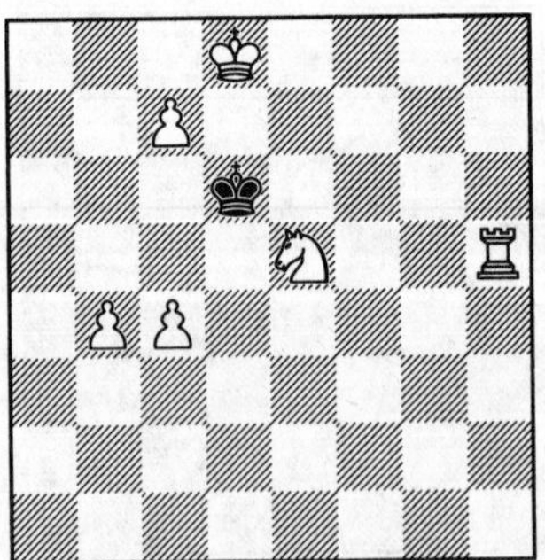

1567.

1568.

1569.

1570.

1571.

1572.

1573.

1574.

1575.

1576.

1577.

1578.

1579.

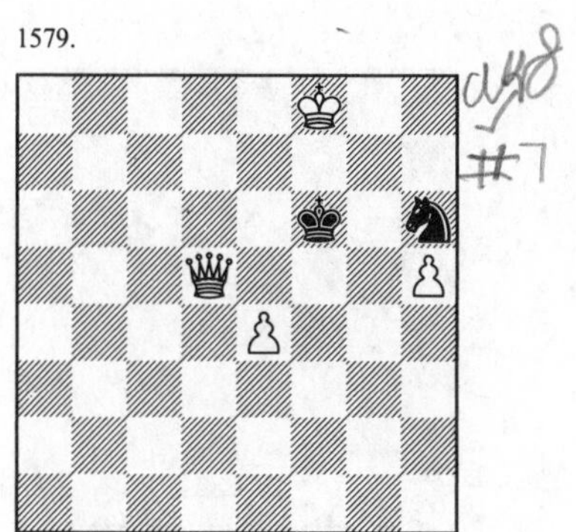

1580.

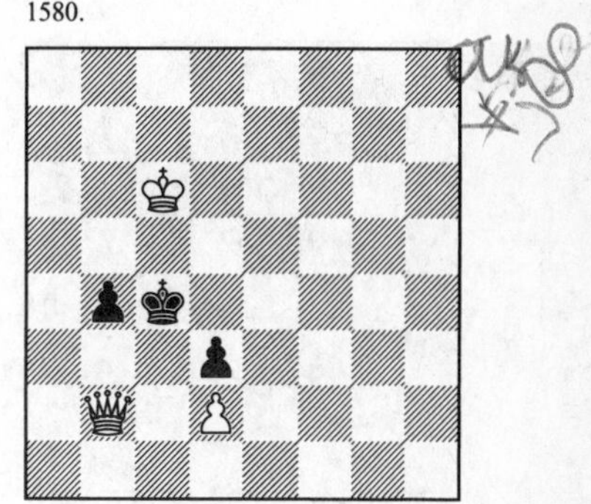

1581.

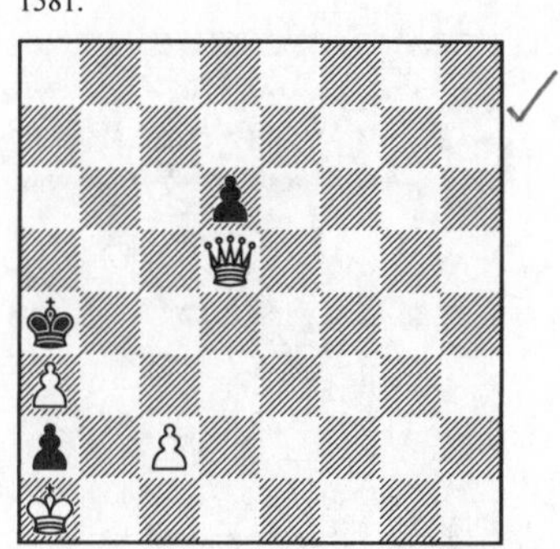

1582.

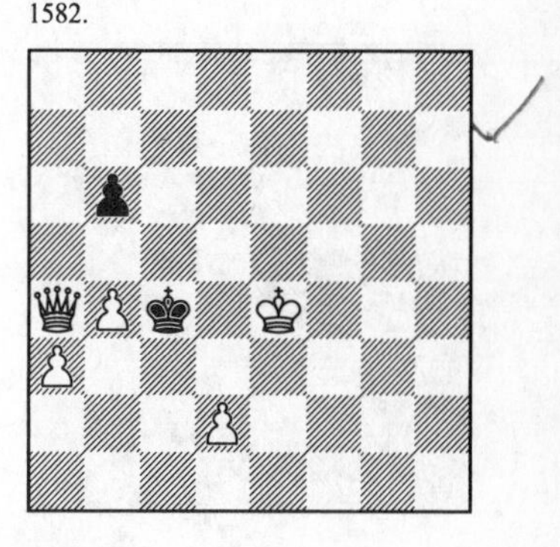

1583.

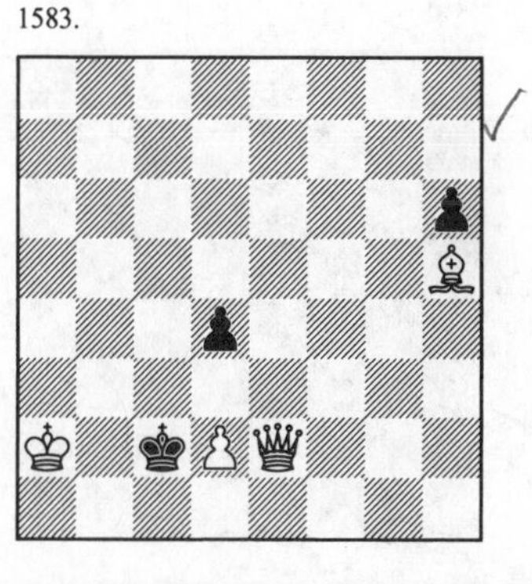

1584.

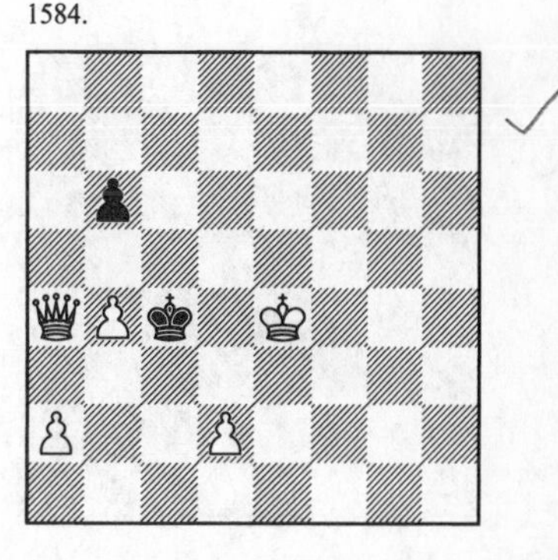

1585.

1586.

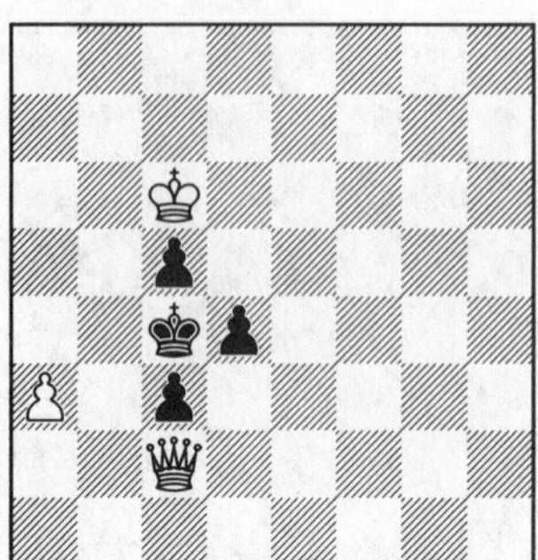

1587.

1588.

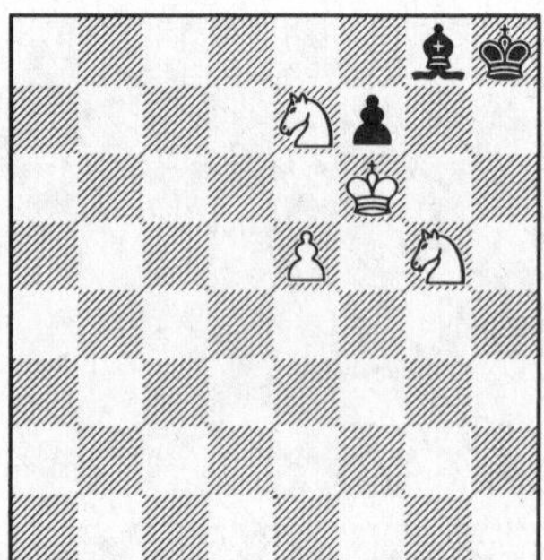

1589.

1590.

1591.

1592.

1593.

1594.

1595.

1596.

1597.

1598.

1599.

1600.

1601.

1602.

1603.

1604.

1605.

1606.

1607.

1608.

1609.

1610.

1611.

1612.

1613.

1614.

1615.

1616.

1617.

1618.

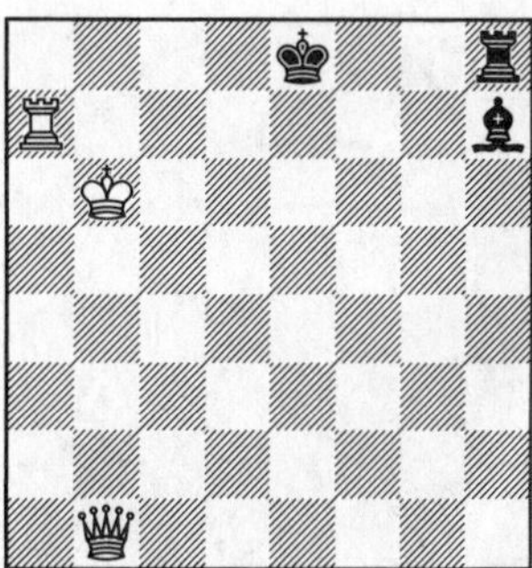

1619.

1620.

1621.

1622.

1623.

1624.

1625.

1626.

1627.

1628.

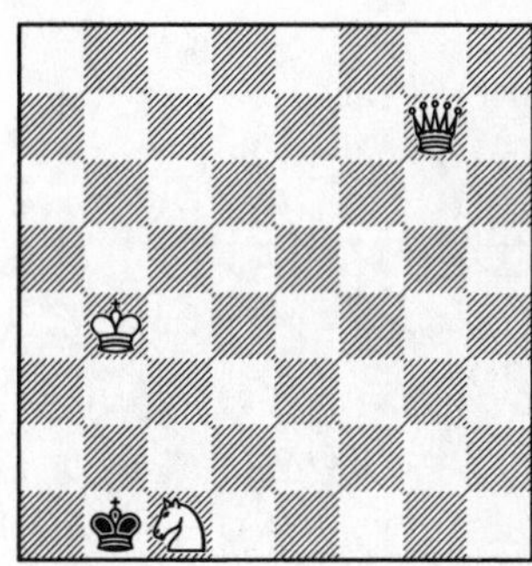

1629.

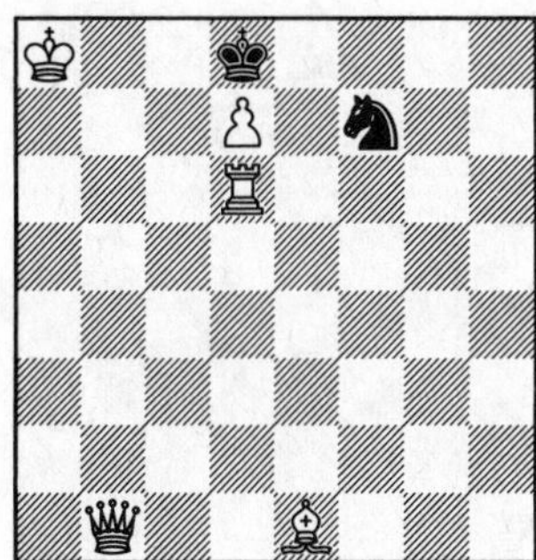

1630.

1631.

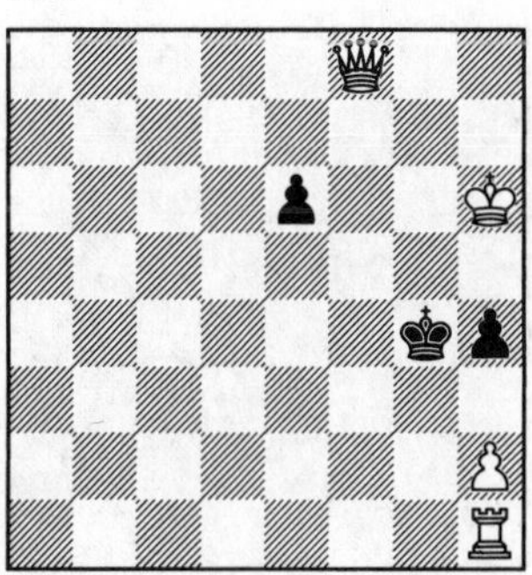

1632.

1633.

1634.

1635.

1636.

1637.

1638.

1639.

1640.

1641.

1642.

1643.

1644.

1645.

1646.

1647.

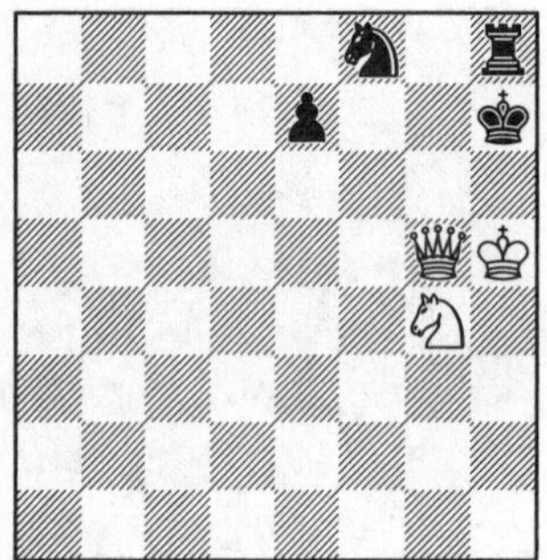

1648.

1649.

1650.

1651.

1652.

1653.

1654.

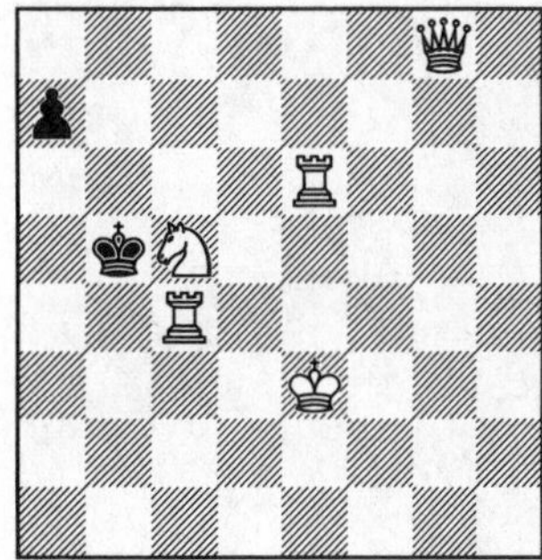

1655.

1656.

1657.

1658.

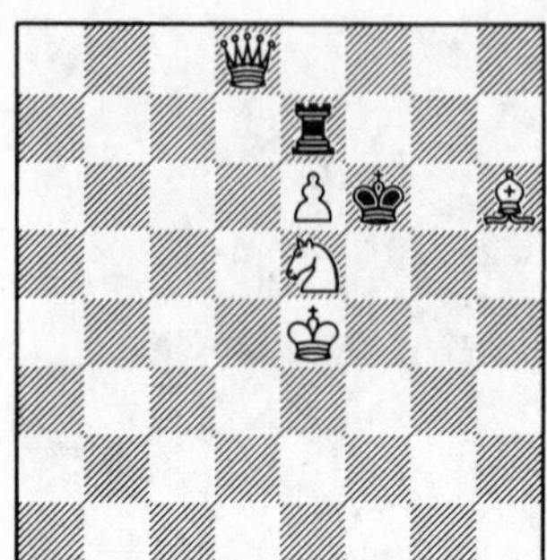

1659.

1660.

1661.

1662.

1663.

1664.

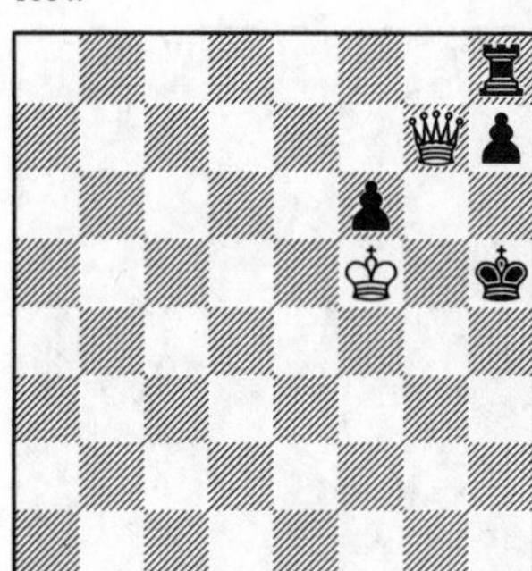

1665.

1666.

1667.

1668.

1669.

1670.

1671.

1672.

1673.

1674.

1675.

1676.

1677.

1678.

1679.

1680.

1681.

1682.

1683.

1684.

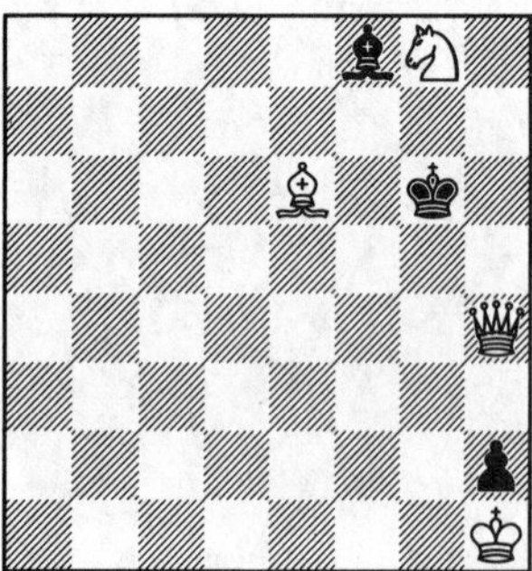

1685.

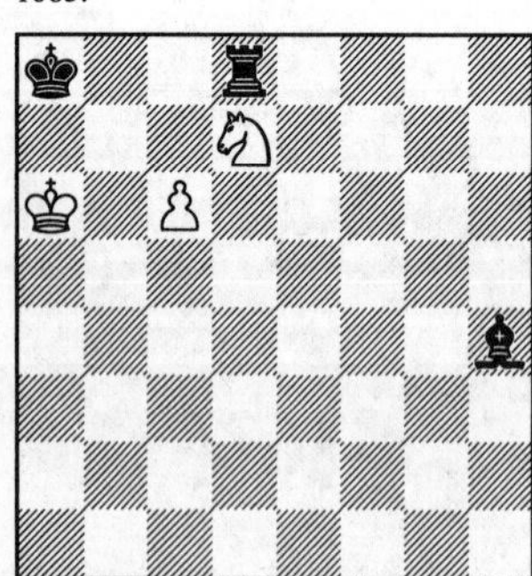

1686.

1687.

1688.

1689.

1690.

1691.

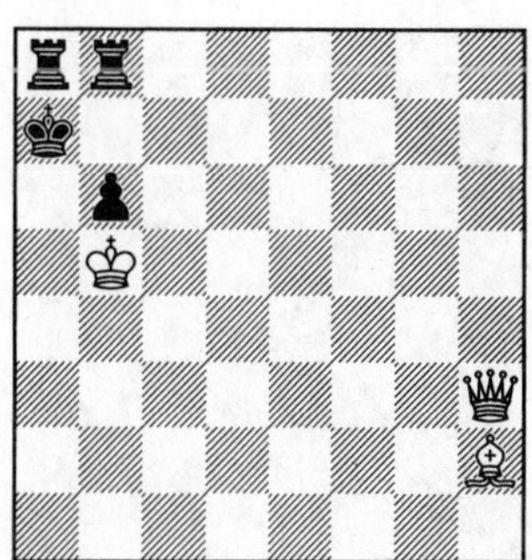

1692.

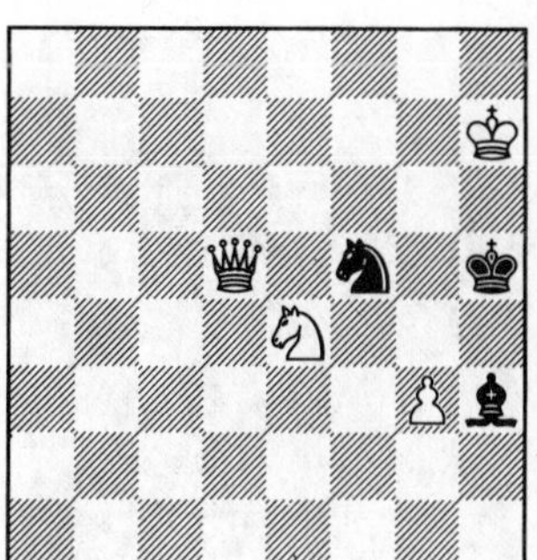

1693.

1694.

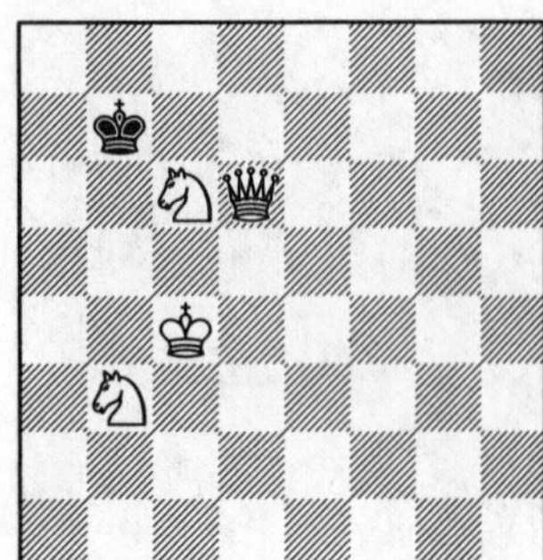

1695.

1696.

1697.

1698.

1699.

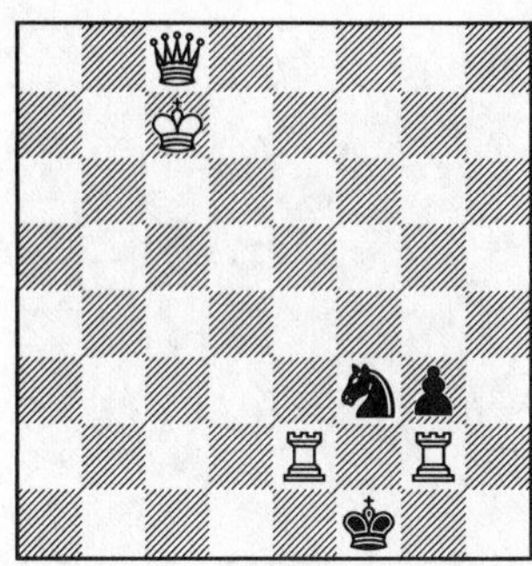

1700.

1701.

1702.

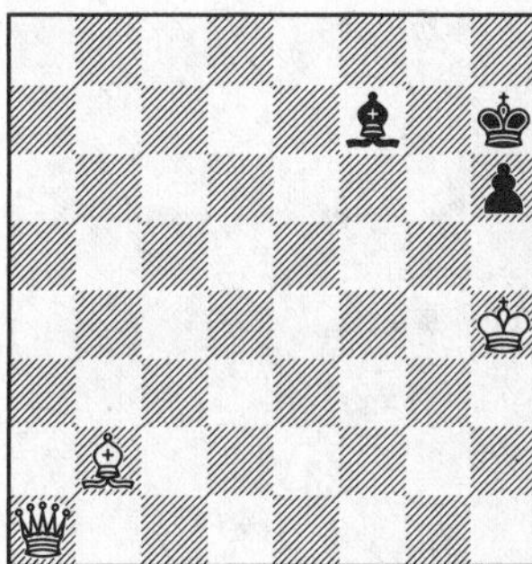

1703.

1704.

1705.

1706.

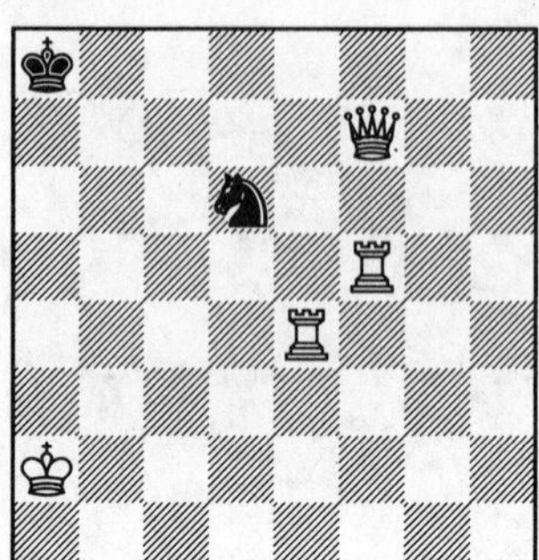

1707.

1708.

1709.

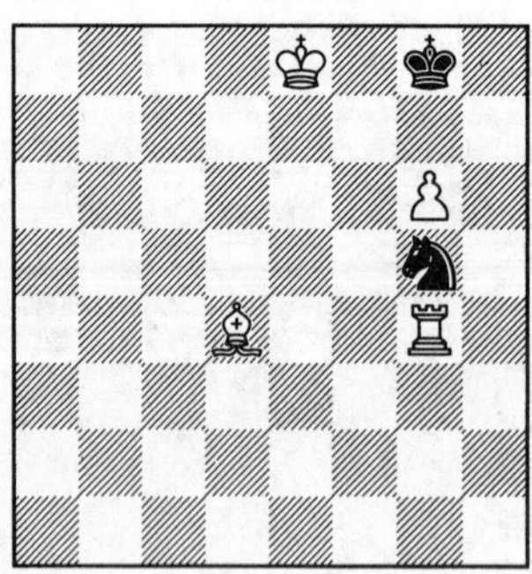

1710.

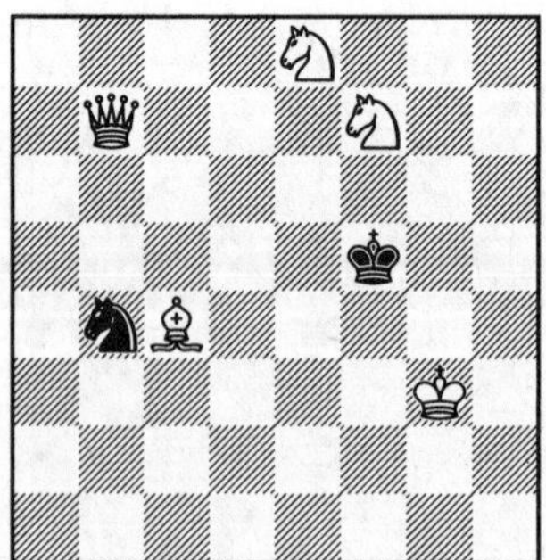

1711.

1712.

1713.

1714.

1715.

1716.

1717.

1718.

1719.

1720.

1721.

1722.

1723.

1724.

1725.

1726.

1727.

1728.

1729.

1730.

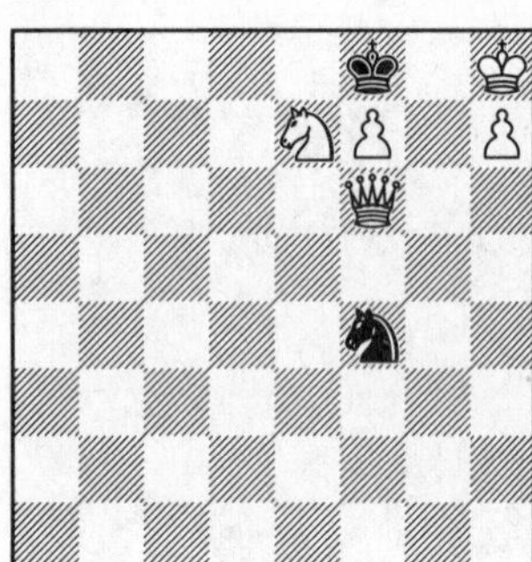

1731.

1732.

1733.

1734.

1735.

1736.

1737.

1738.

1739.

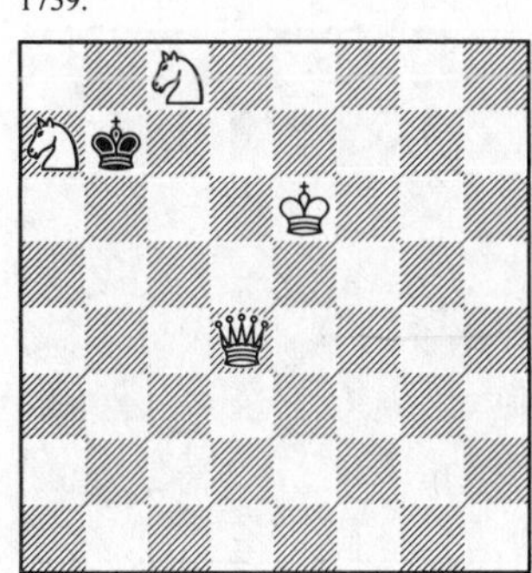

1740.

1741.

1742.

1743.

1744.

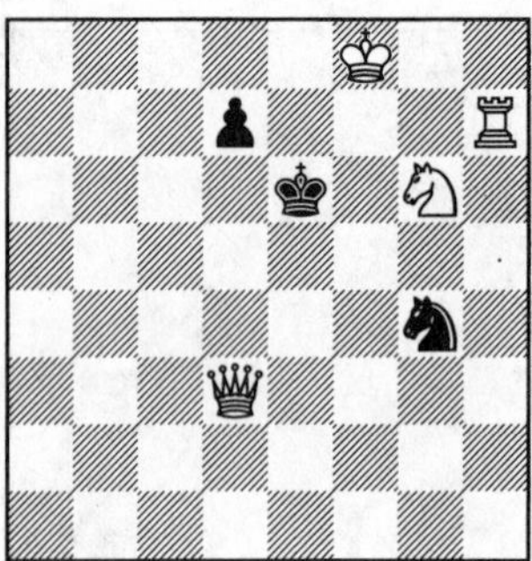

1745.

1746.

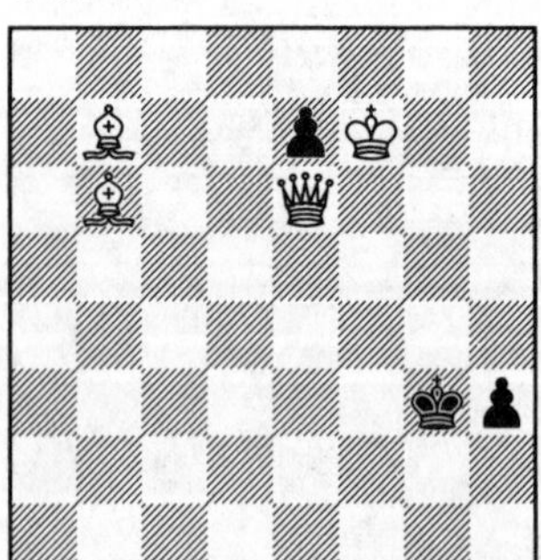

1747.

1748.

1749.

1750.

1751.

1752.

1753.

1754.

1755.

1756.

1757.

1758.

1759.

1760.

1761.

1762.

1763.

1764.

1765.

1766.

1767.

1768.

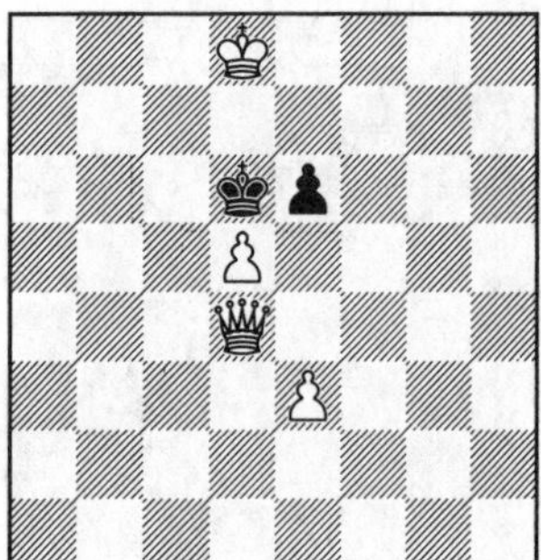

1769.

1770.

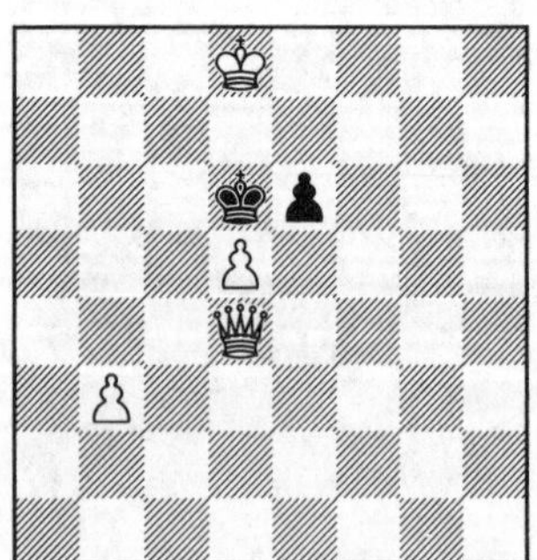

1771.

1772.

1773.

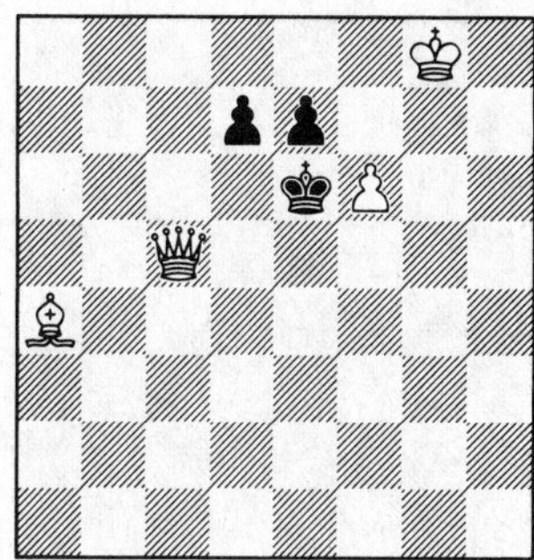

1774.

1775.

1776.

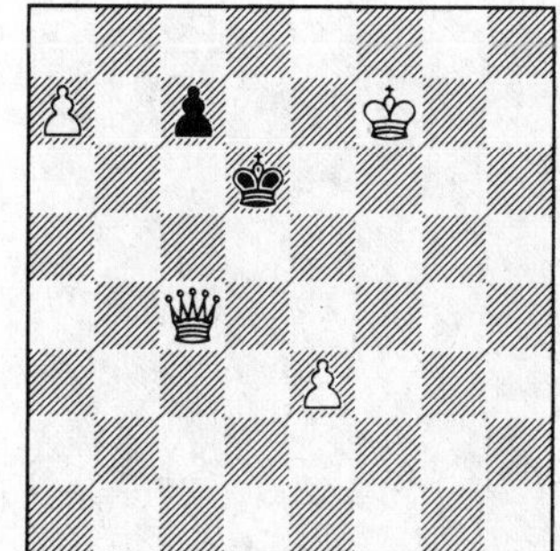

1777.

1778.

1779.

1780.

1781.

1782.

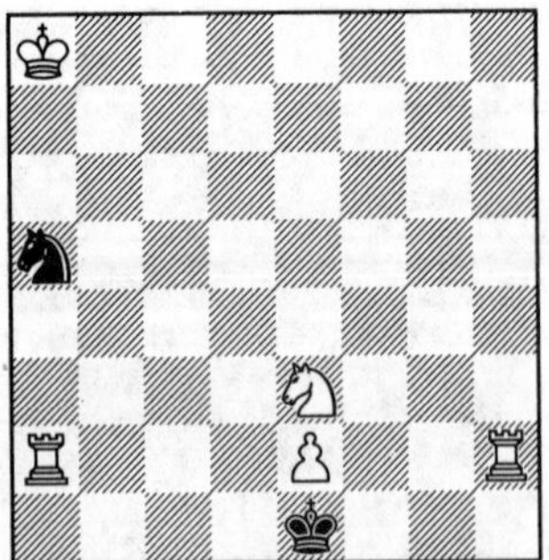

1783.

1784.

1785.

1786.

1787.

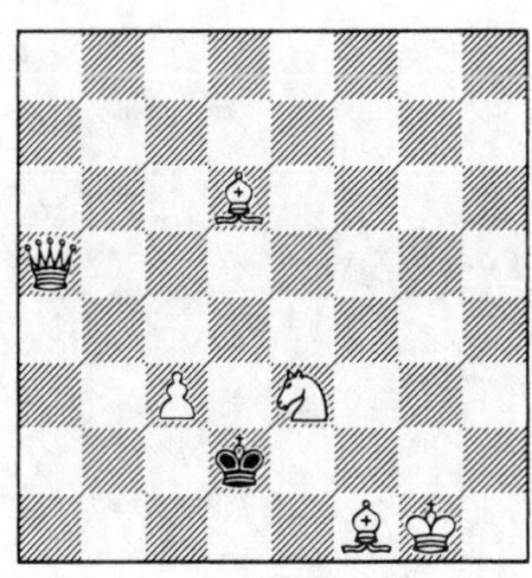

1788.

1789.

1790.

1791.

1792.

1793.

1794.

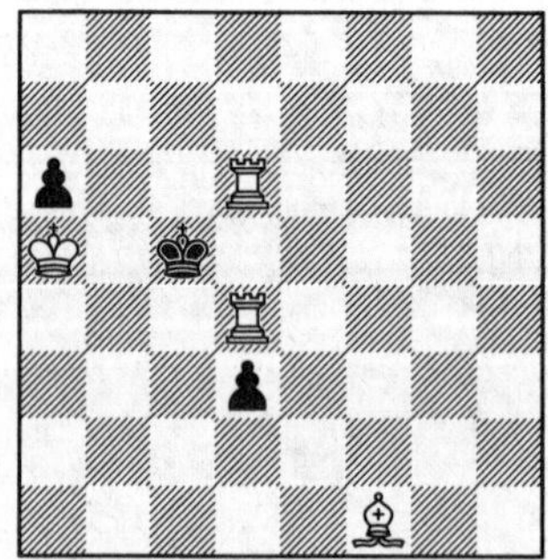

1795.

1796.

1797.

1798.

1799.

1800.

1801.

1802.

1803.

1804.

1805.

1806.

1807.

1808.

1809.

1810.

1811.

1812.

1813.

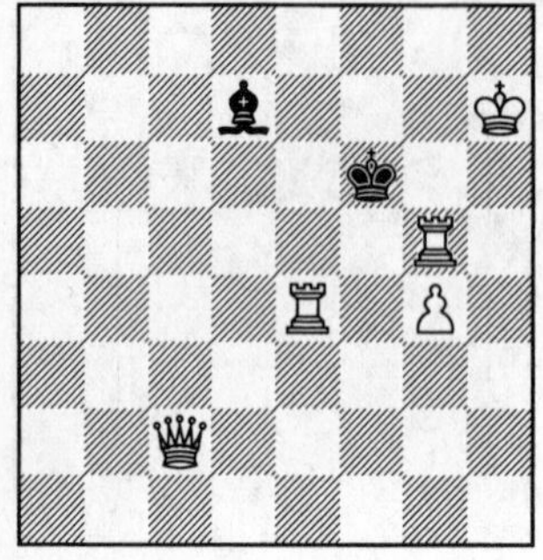

1814.

1815.

1816.

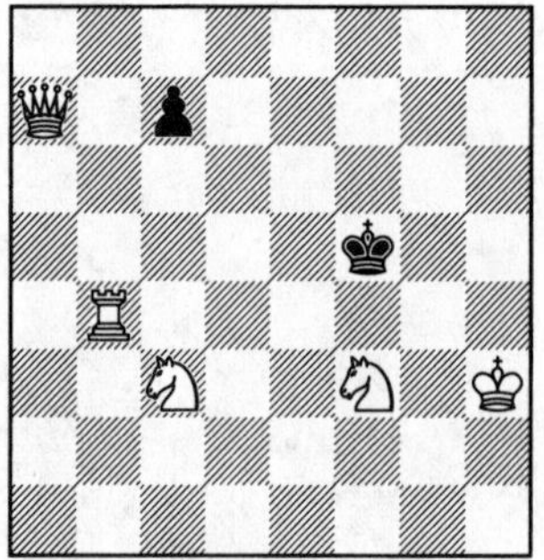

1817.

1818.

1819.

1820.

1821.

1822.

1823.

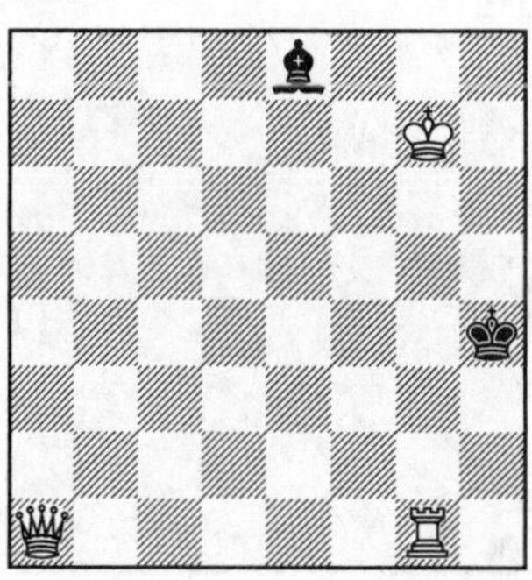

1824.

1825.

1826.

1827.

1828.

1829.

1830.

1831.

1832.

1833.

1834.

1835.

1836.

1837.

1838.

1839.

1840.

1841.

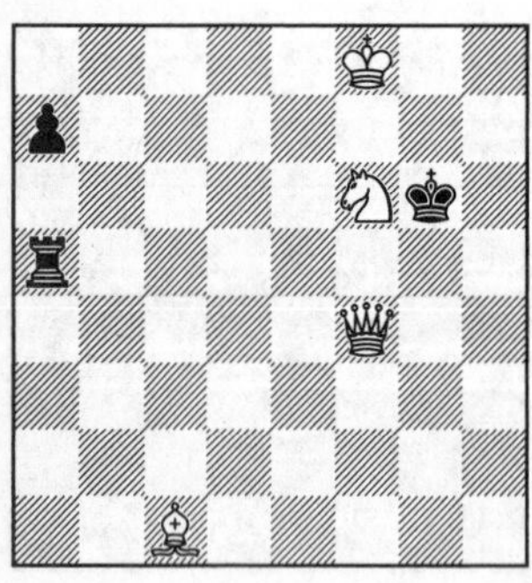

1842.

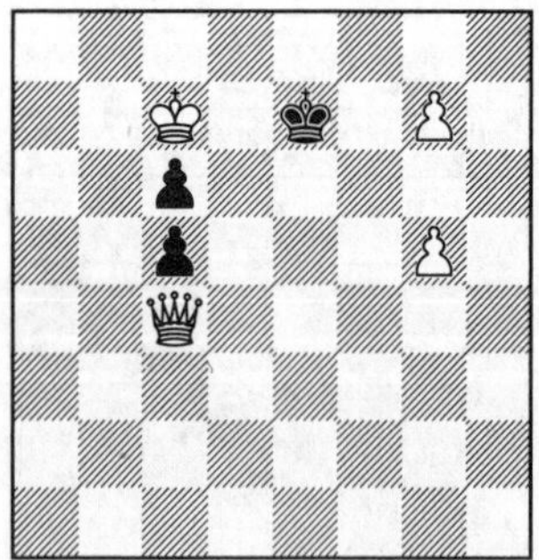

1843.

1844.

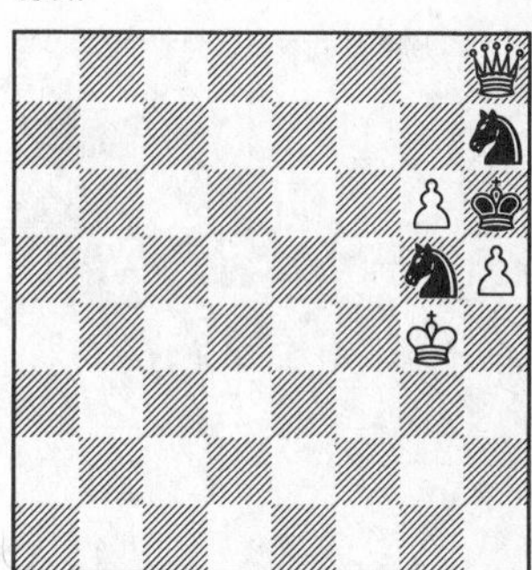

1845.

1846.

1847.

1848.

1849.

1850.

1851.

1852.

1853.

1854.

1855.

1856.

1857.

1858.

1859.

1860.

1861.

1862.

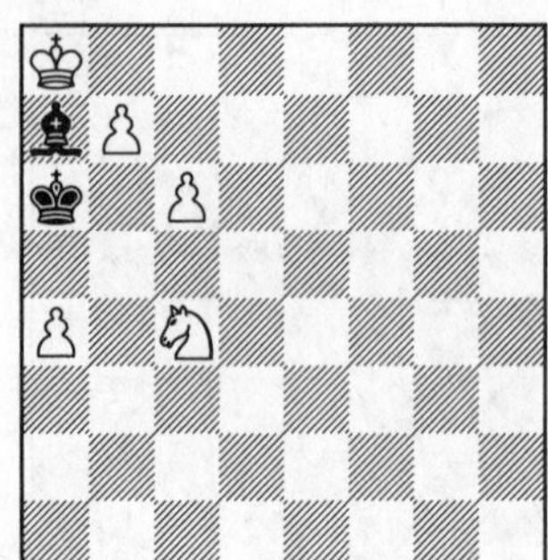

1863.

1864.

1865.

1866.

1867.

1868.

1869.

1870.

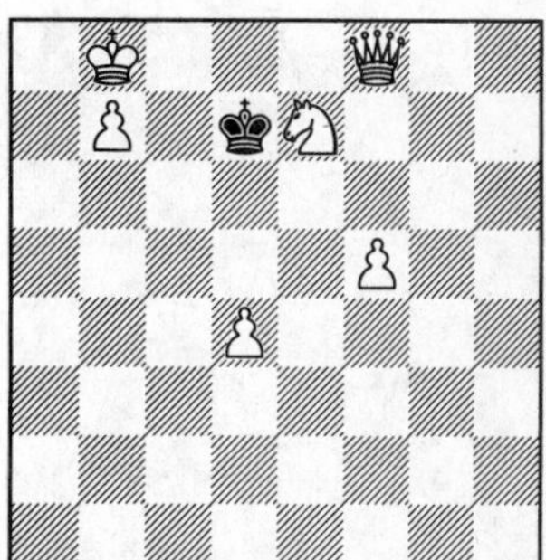

1871.

1872.

1873.

1874.

1875.

1876.

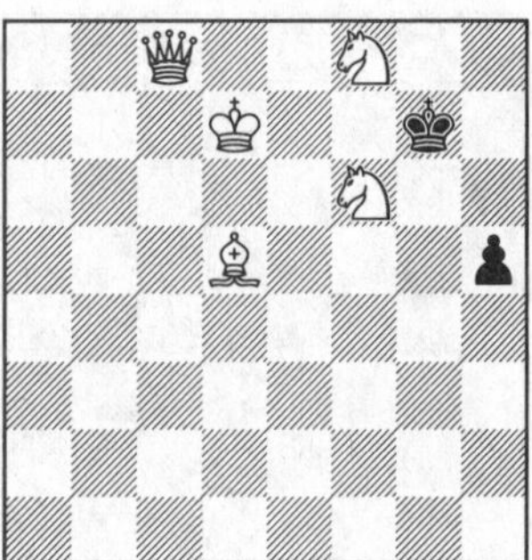

1877.

1878.

1879.

1880.

1881.

1882.

1883.

1884.

1885.

1886.

1887.

1888.

1889.

1890.

1891.

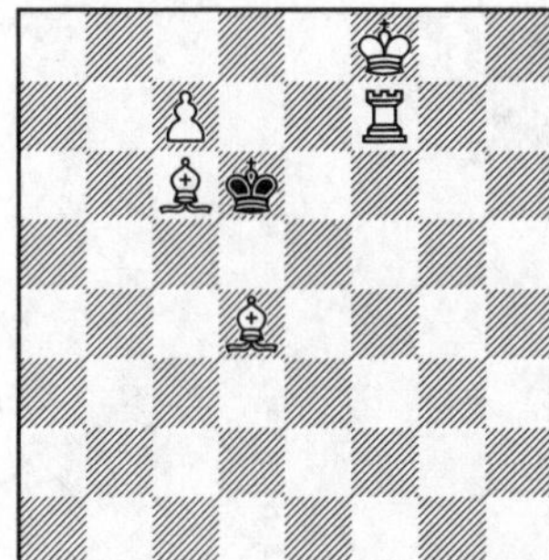

1892.

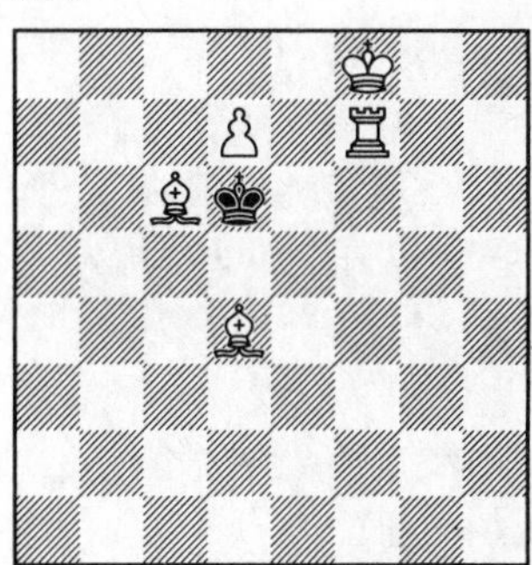

1893.

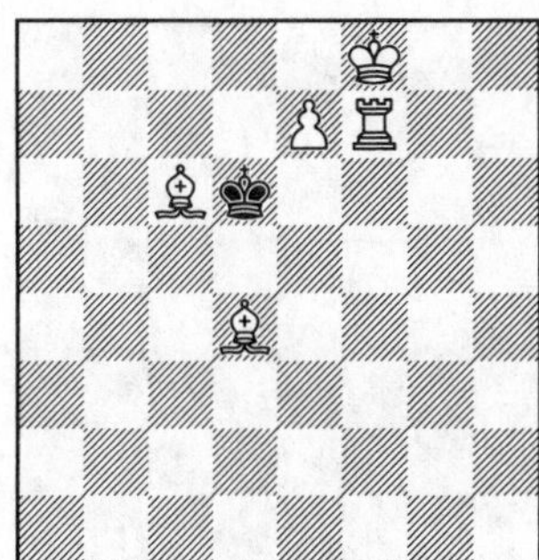

1894.

1895.

1896.

1897.

1898.

1899.

1900.

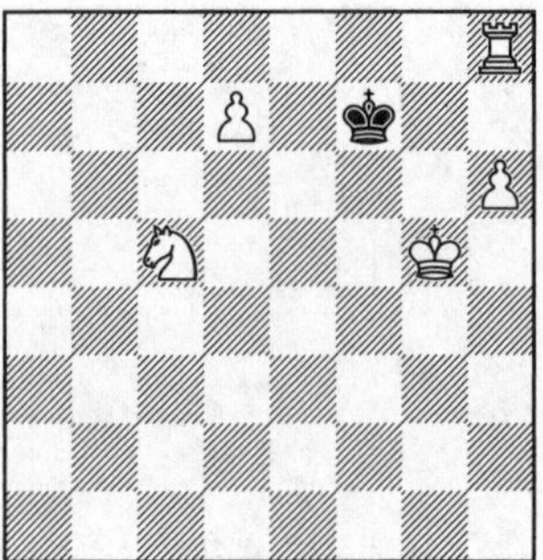

1901.

1902.

1903.

1904.

1905.

1906.

1907.

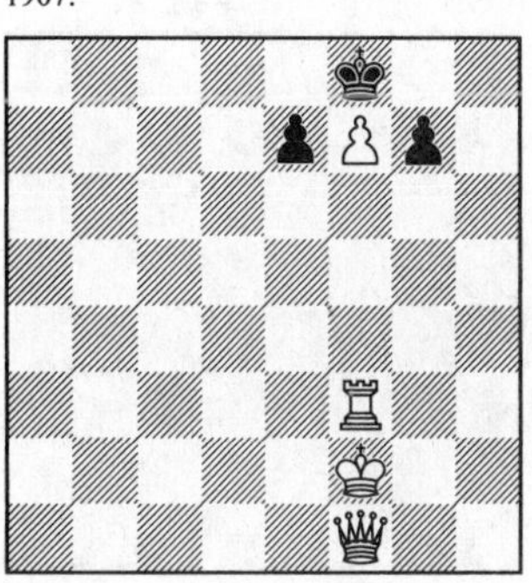

1908.

1909.

1910.

1911.

1912.

1913.

1914.

1915.

1916.

1917.

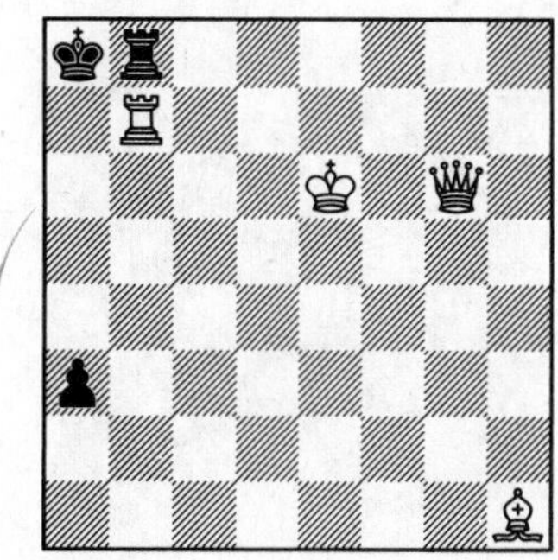

1918.

1919.

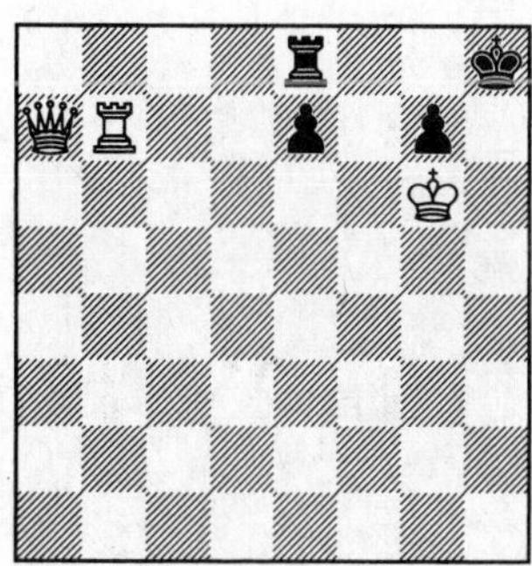

1920.

1921.

1922.

1923.

1924.

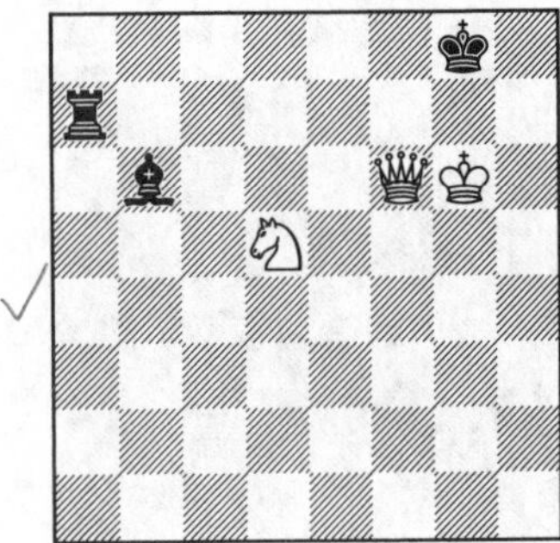

1925.

1926.

1927.

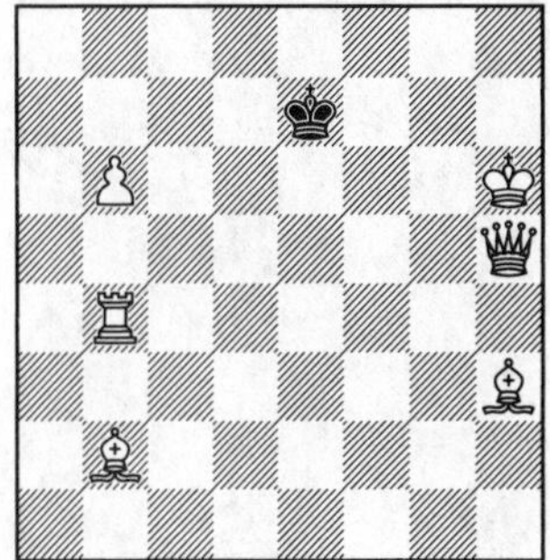

1928.

1929.

1930.

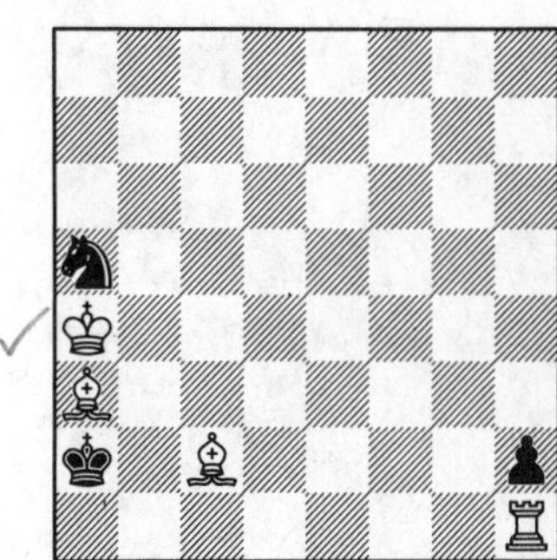

1931.

1932.

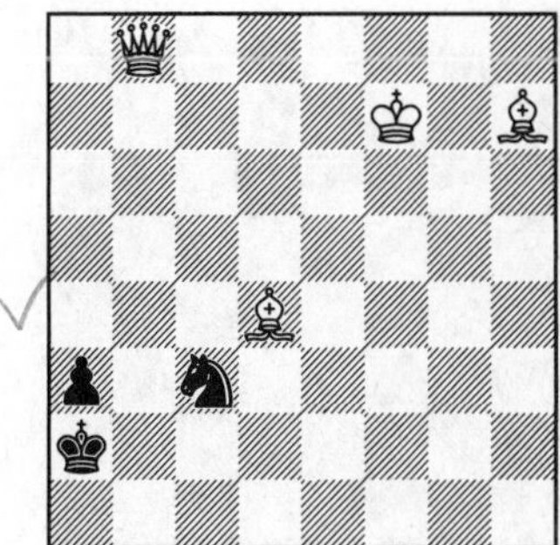

1933.

1934.

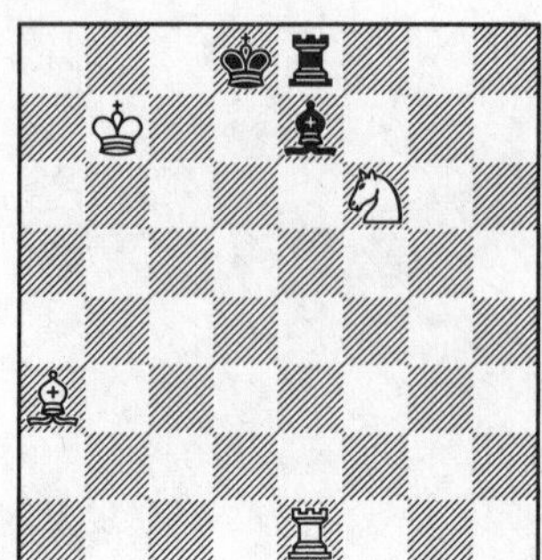

1935.

1936.

1937.

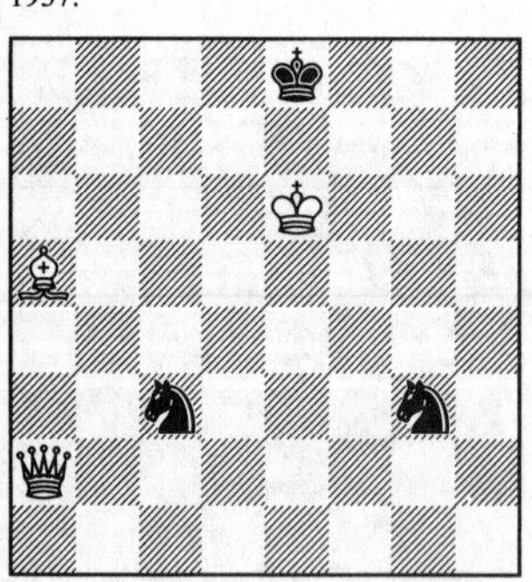

1938.

1939.

1940.

1941.

1942.

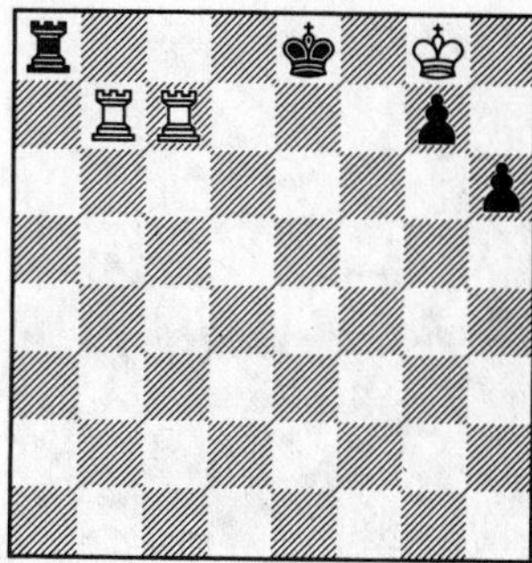

1943.

1944.

1945.

1946.

1947.

1948.

1949.

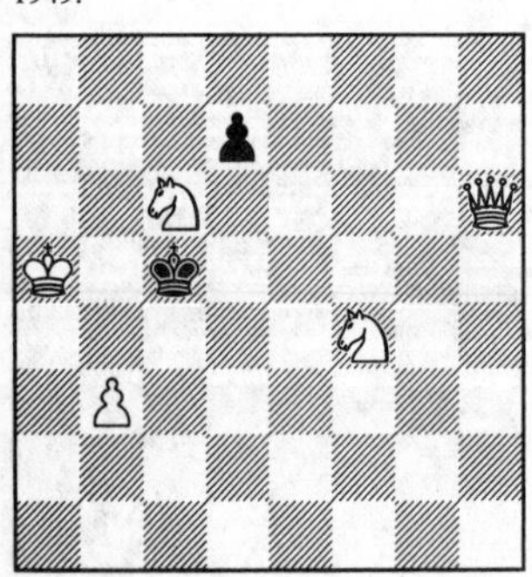

1950.

1951.

1952.

1953.

1954.

1955.

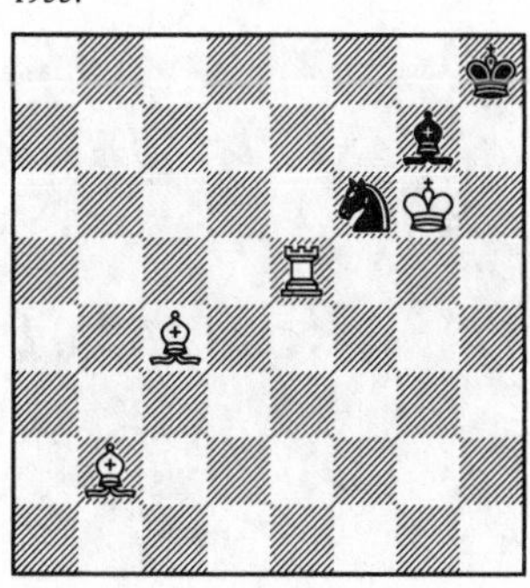

1956.

1957.

1958.

1959.

1960.

1961.

1962.

1963.

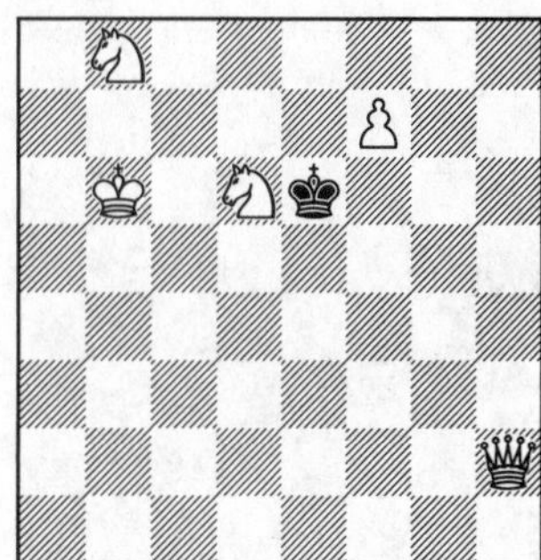

1964.

1965.

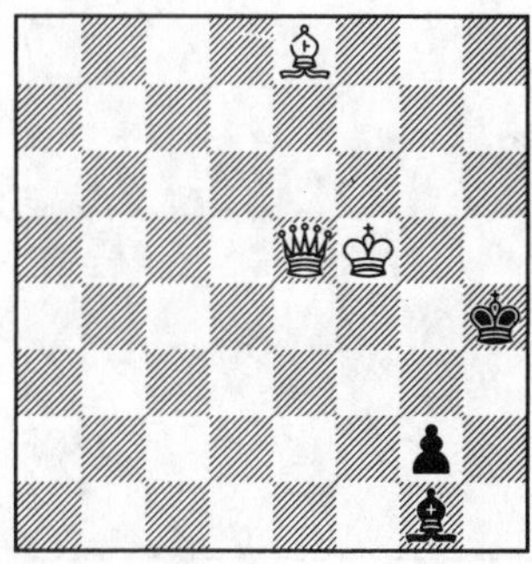

1966.

1967.

1968.

1969.

1970.

1971.

1972.

1973.

1974.

1975.

1976.

1977.

1978.

1979.

1980.

1981.

1982.

1983.

1984.

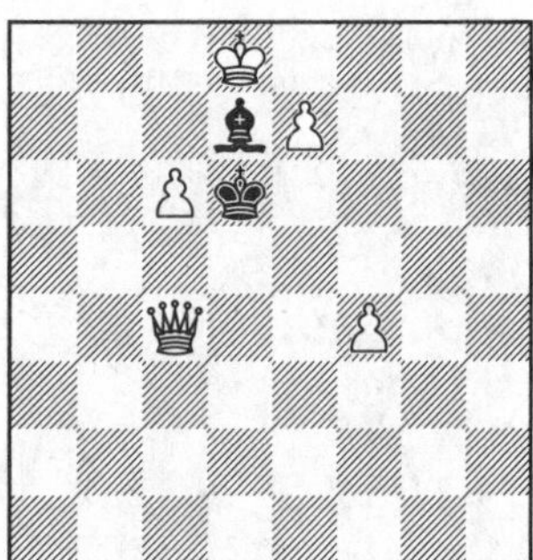

1985.

1986.

1987.

1988.

1989.

1990.

1991.

1992.

1993.

1994.

1995.

1996.

1997.

1998.

1999.

2000.

2001.

2002.

2003.

2004.

2005.

2006.

2007.

2008.

2009.

2010.

2011.

2012.

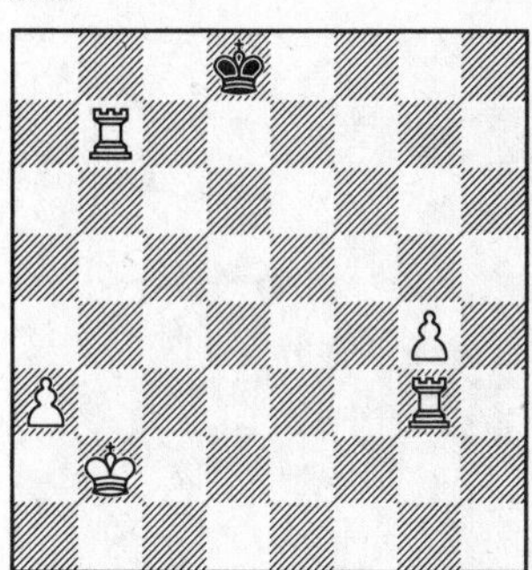

2013.

2014.

2015.

2016.

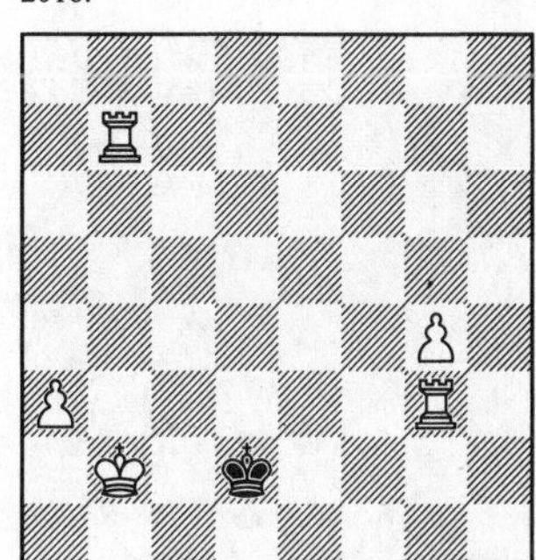

2017.

2018.

2019.

2020.

2021.

2022.

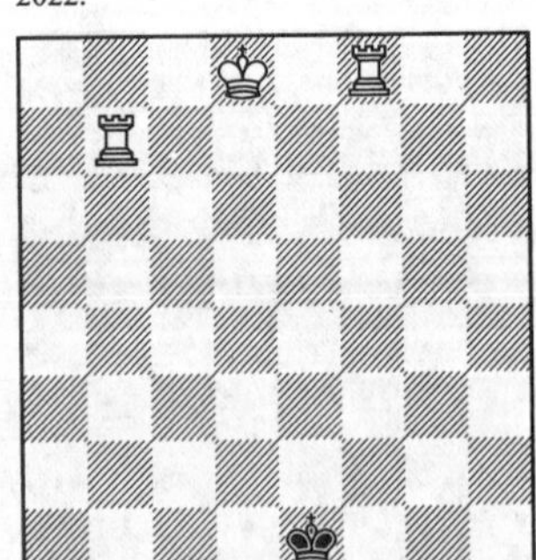

2023.

2024.

2025.

2026.

2027.

2028.

2029.

2030.

2031.

2032.

2033.

2034.

2035.

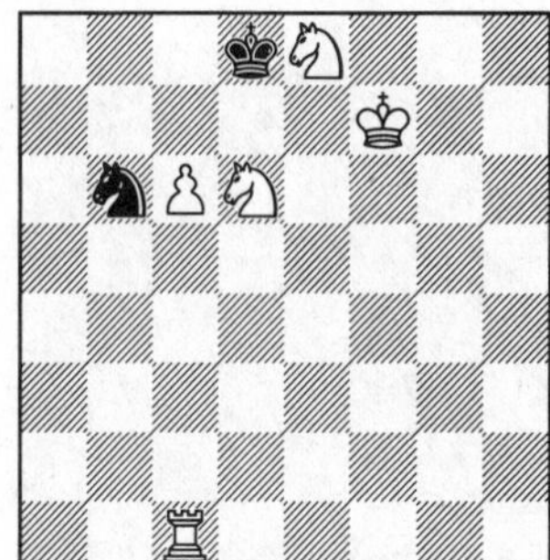

2036.

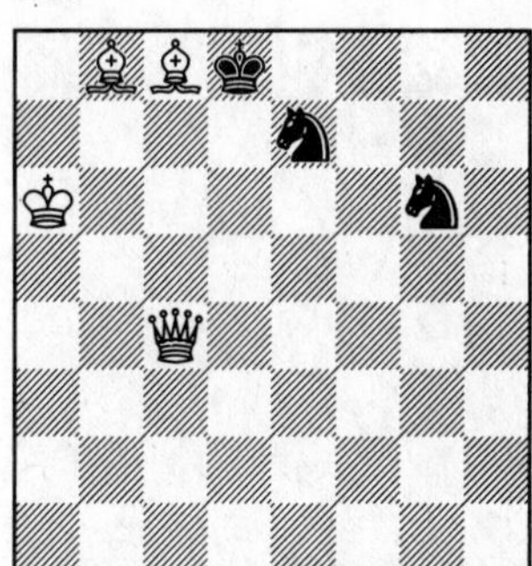

2037.

2038.

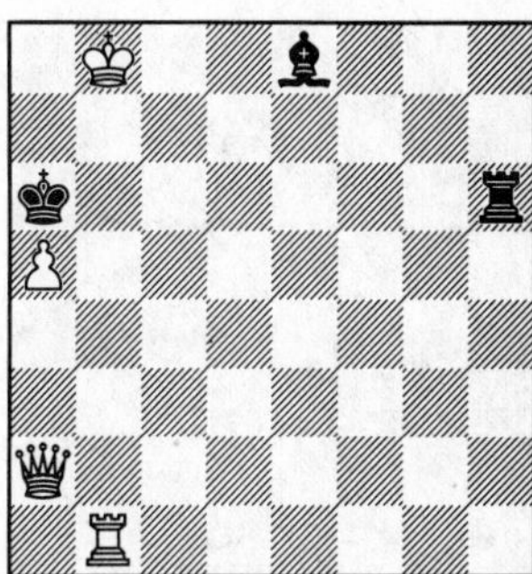

2039.

2040.

2041.

2042.

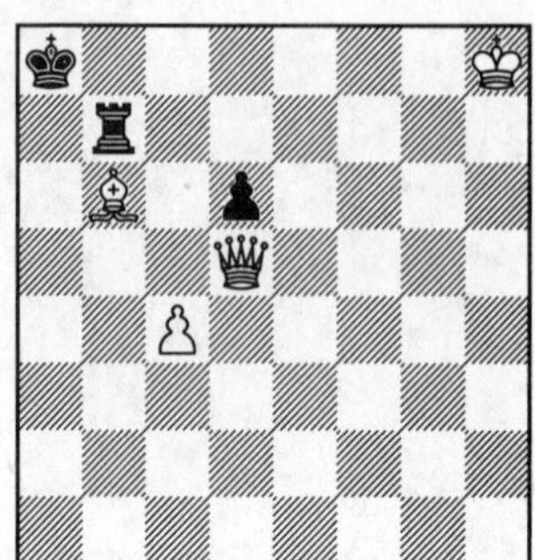

2043.

2044.

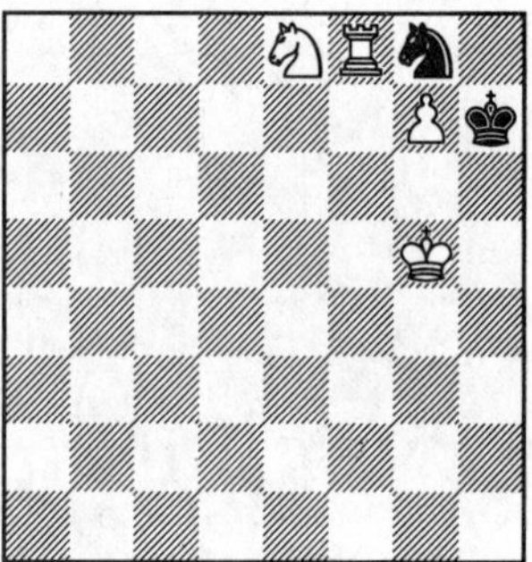

2045.

2046.

2047.

2048.

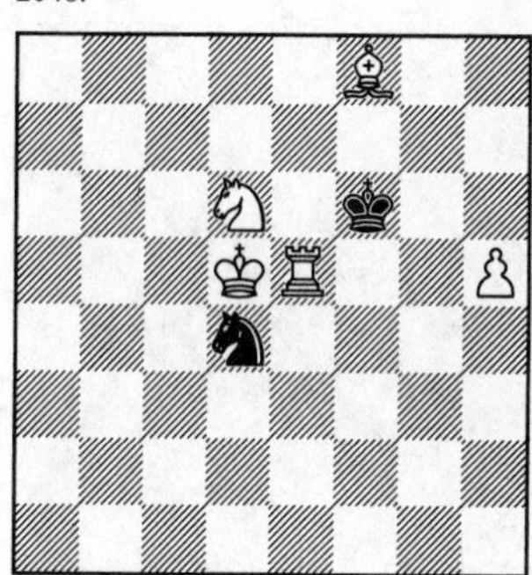

2049.

2050.

2051.

2052.

2053.

2054.

2055.

2056.

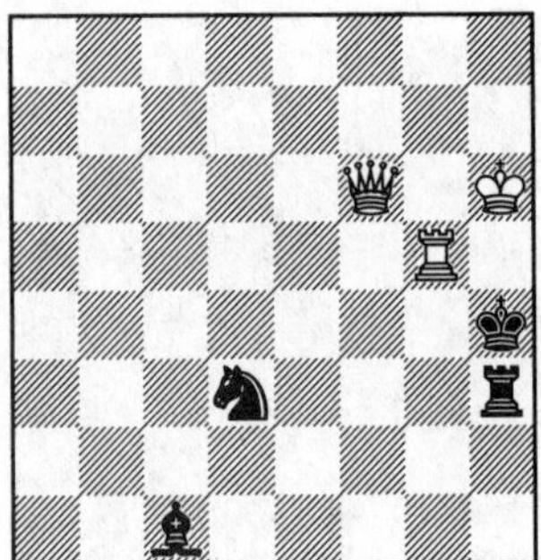

2057.

2058.

2059.

2060.

2061.

2062.

2063.

2064.

2065.

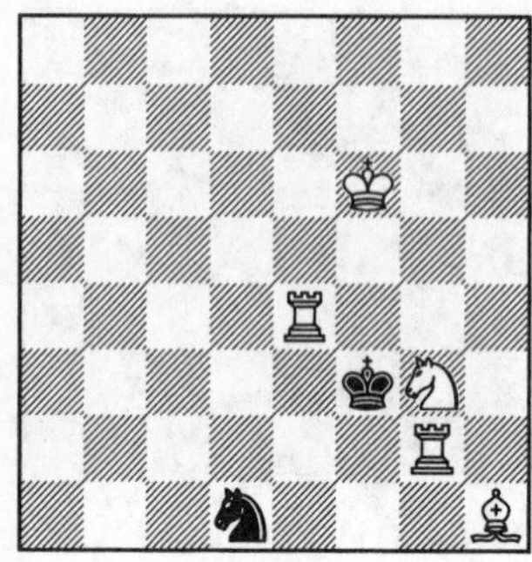

2066.

2067.

2068.

2069.

2070.

2071.

2072.

2073.

2074.

2075.

2076.

2077.

2078.

2079.

2080.

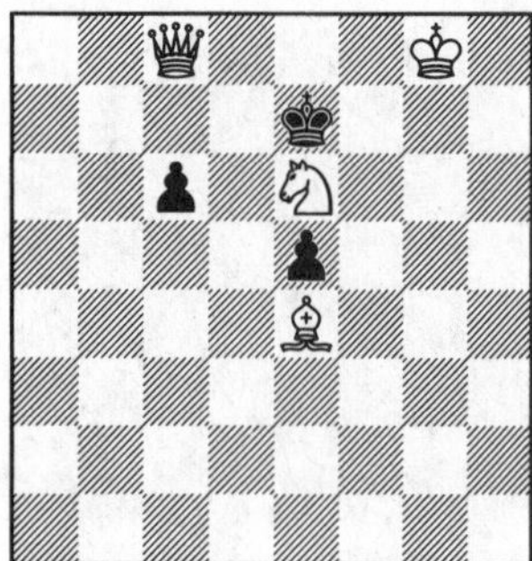

2081.

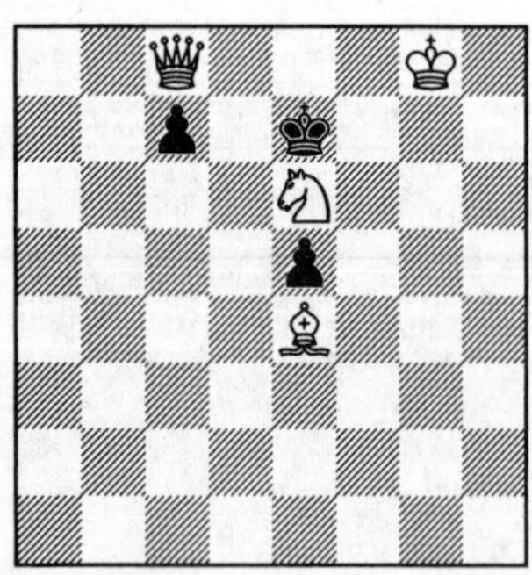

2082.

2083.

2084.

2085.

2086.

2087.

2088.

2089.

2090.

2091.

2092.

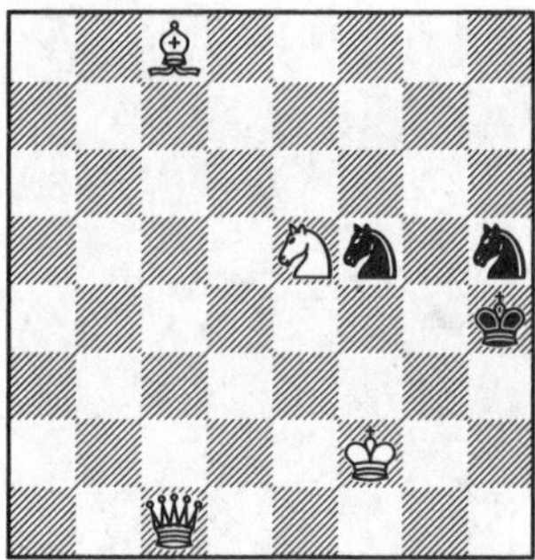

2093.

2094.

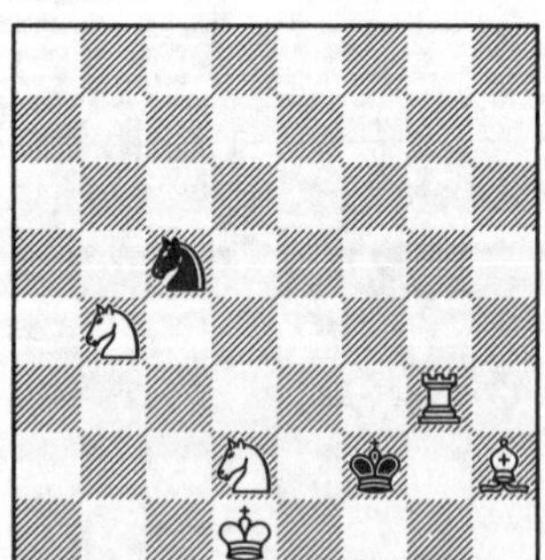

2095.

2096.

2097.

2098.

2099.

2100.

2101.

2102.

2103.

2104.

2105.

2106.

2107.

2108.

2109.

2110.

2111.

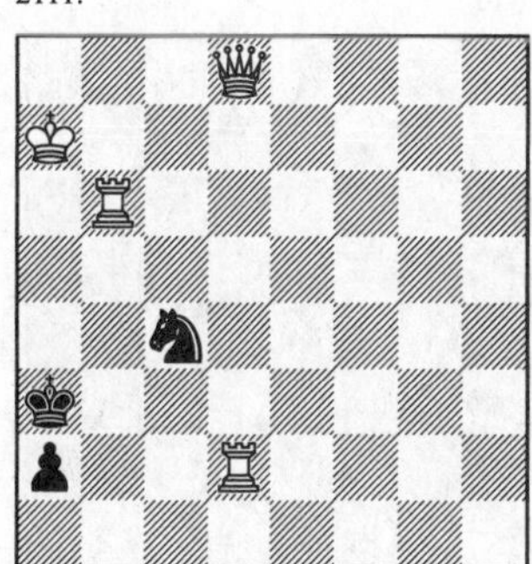

2112.

2113.

2114.

2115.

2116.

2117.

2118.

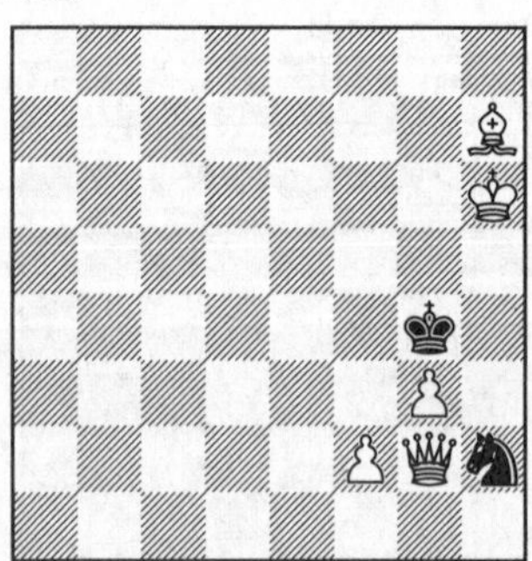

2119.

2120.

2121.

2122.

2123.

2124.

2125.

2126.

2127.

2128.

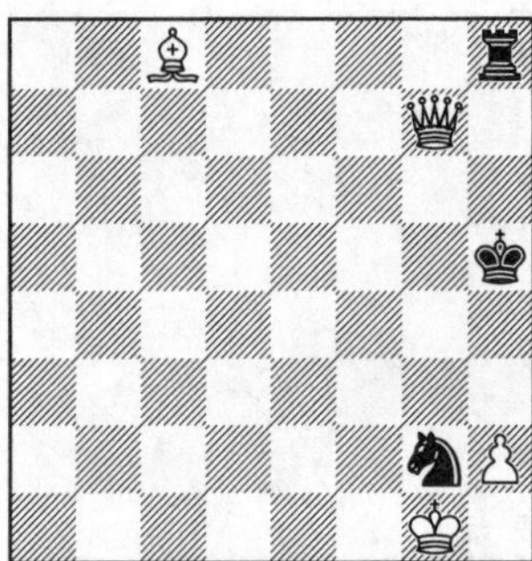

2129.

2130.

2131.

2132.

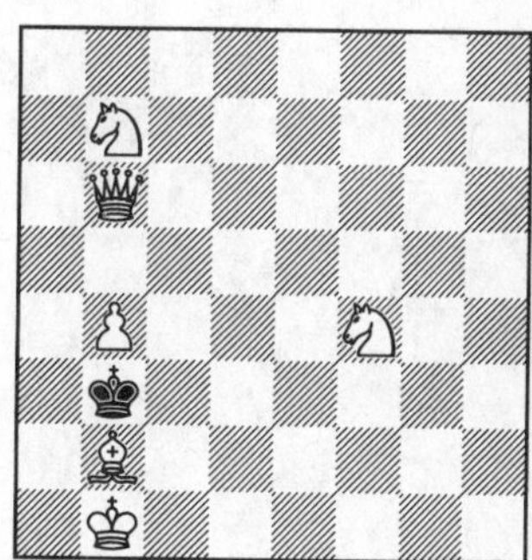

2133.

2134.

2135.

2136.

2137.

2138.

2139.

2140.

2141.

2142.

2143.

2144.

2145.

2146.

2147.

2148.

2149.

2150.

2151.

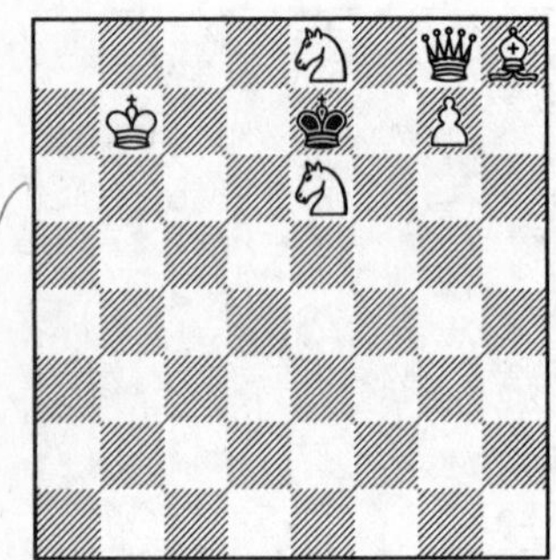

2152.

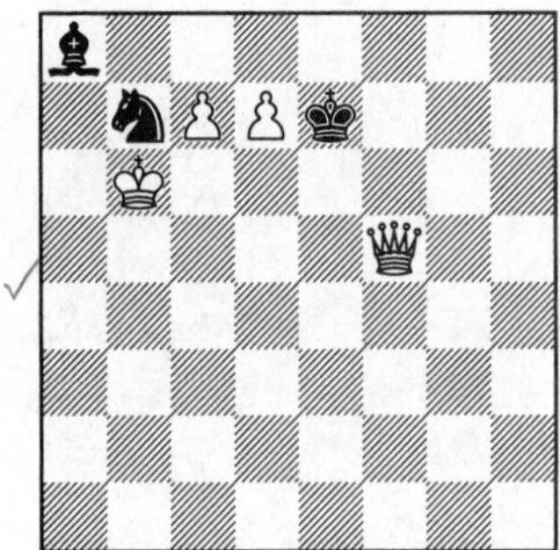

2153.

2154.

2155.

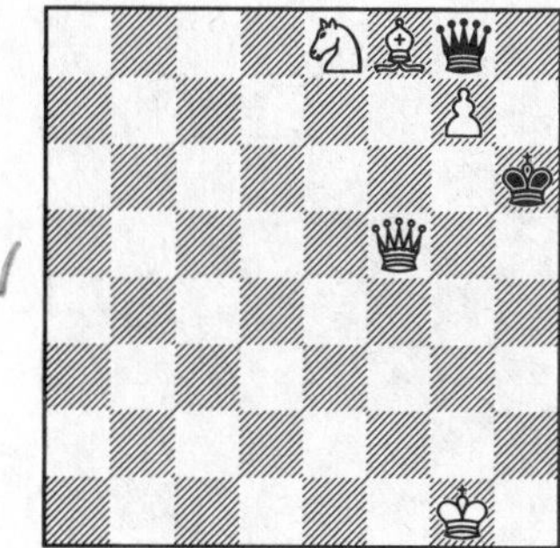

2156.

2157.

2158.

2159.

2160.

2161.

2162.

2163.

2164.

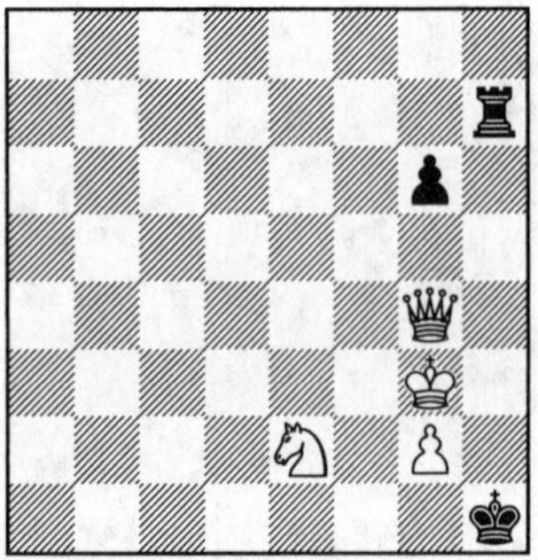

2165.

2166.

2167.

2168.

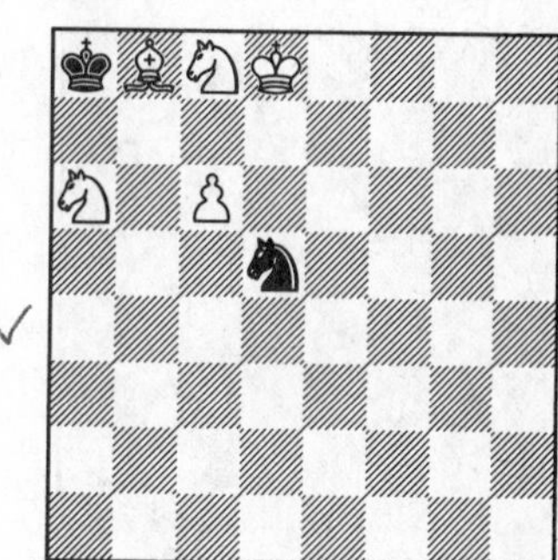

2169.

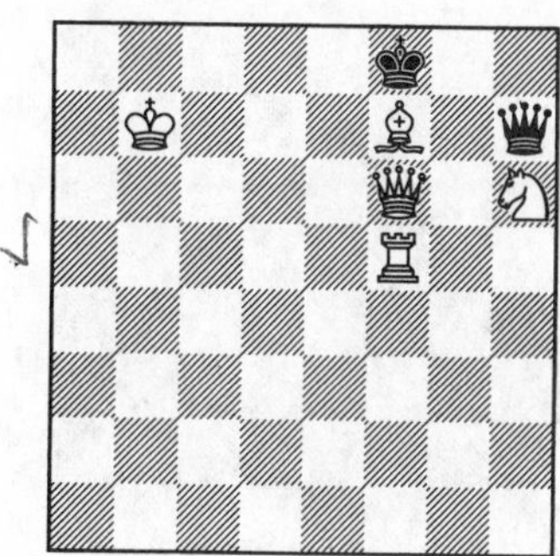

2170.

2171.

2172.

2173.

2174.

2175.

2176.

2177.

2178.

2179.

2180.

2181.

2182.

2183.

2184.

2185.

2186.

2187.

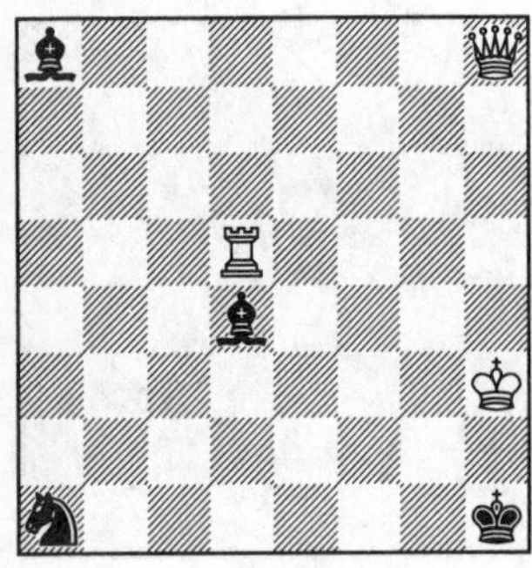

2188.

2189.

2190.

2191.

2192.

2193.

2194.

2195.

2196.

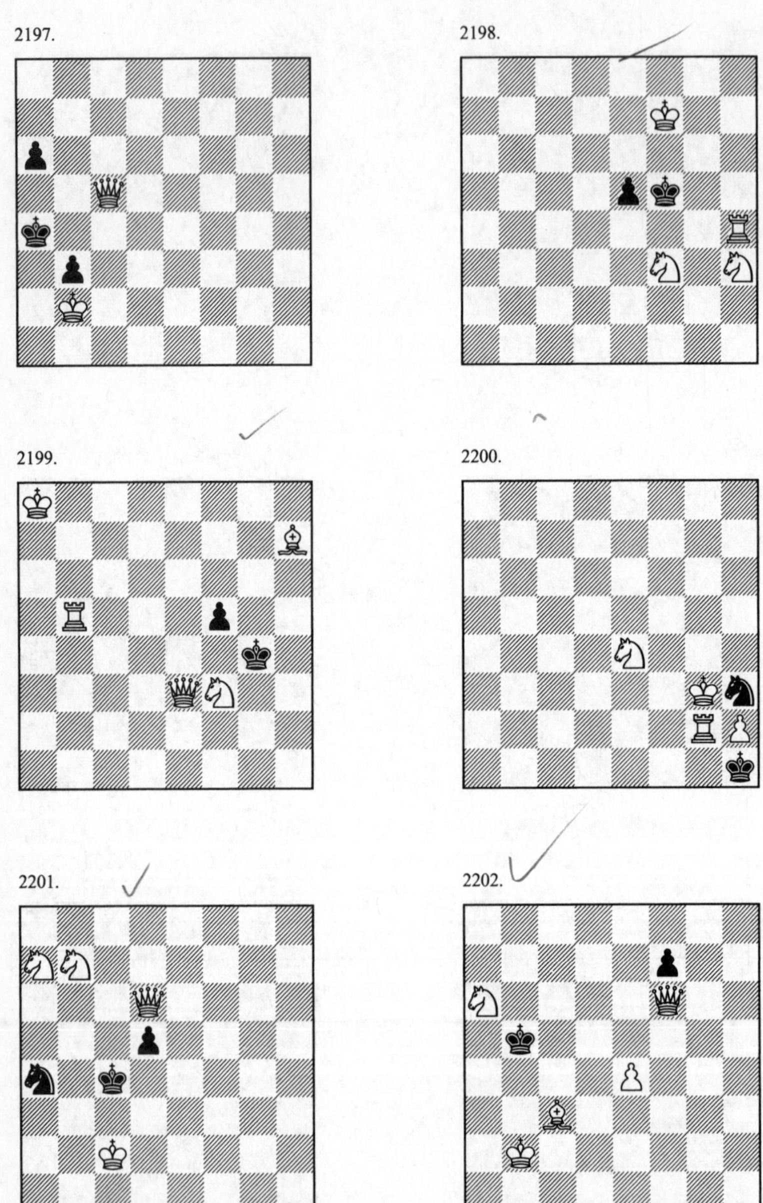

2197.

2198.

2199.

2200.

2201.

2202.

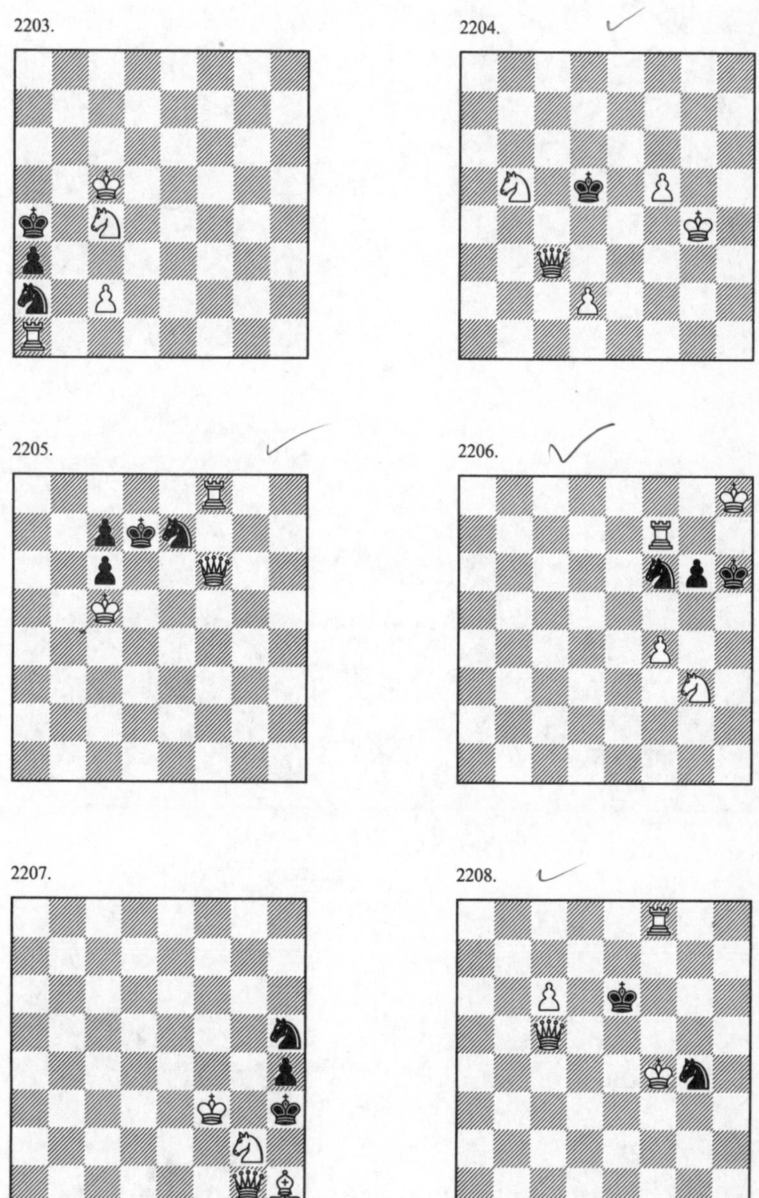
2203.
2204.
2205.
2206.
2207.
2208.

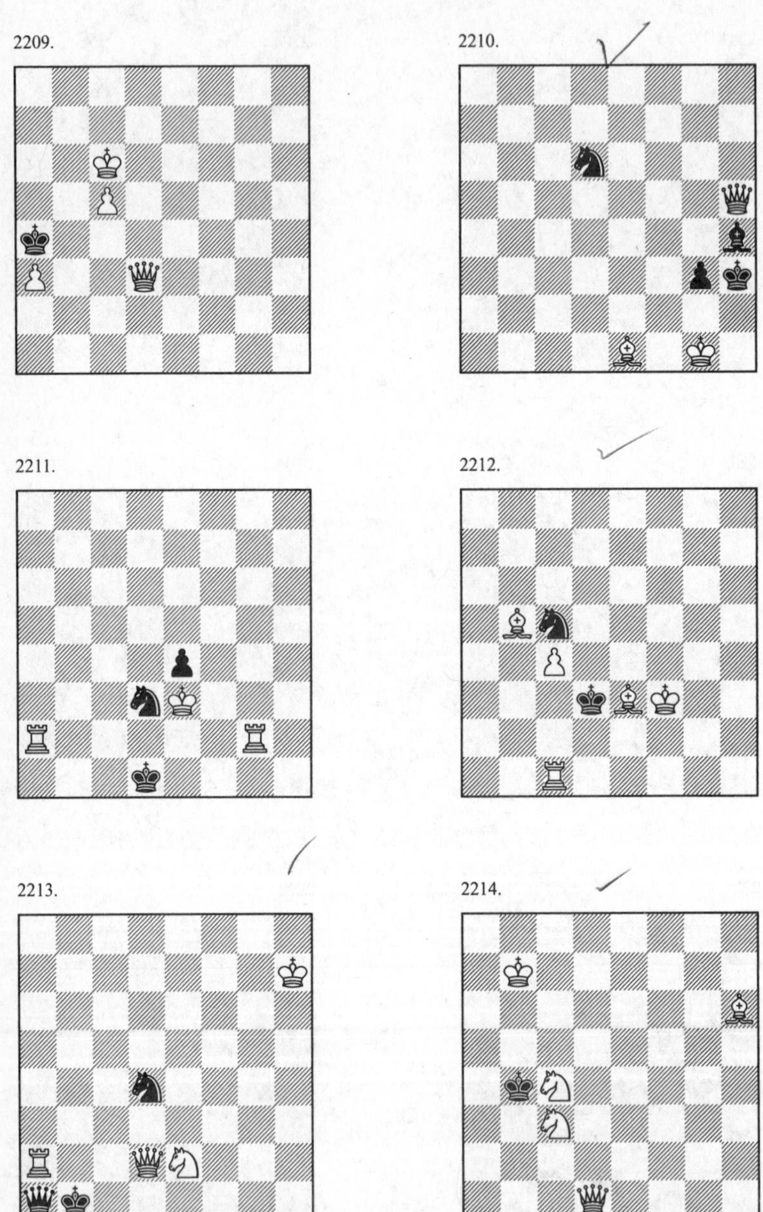

2209.

2210.

2211.

2212.

2213.

2214.

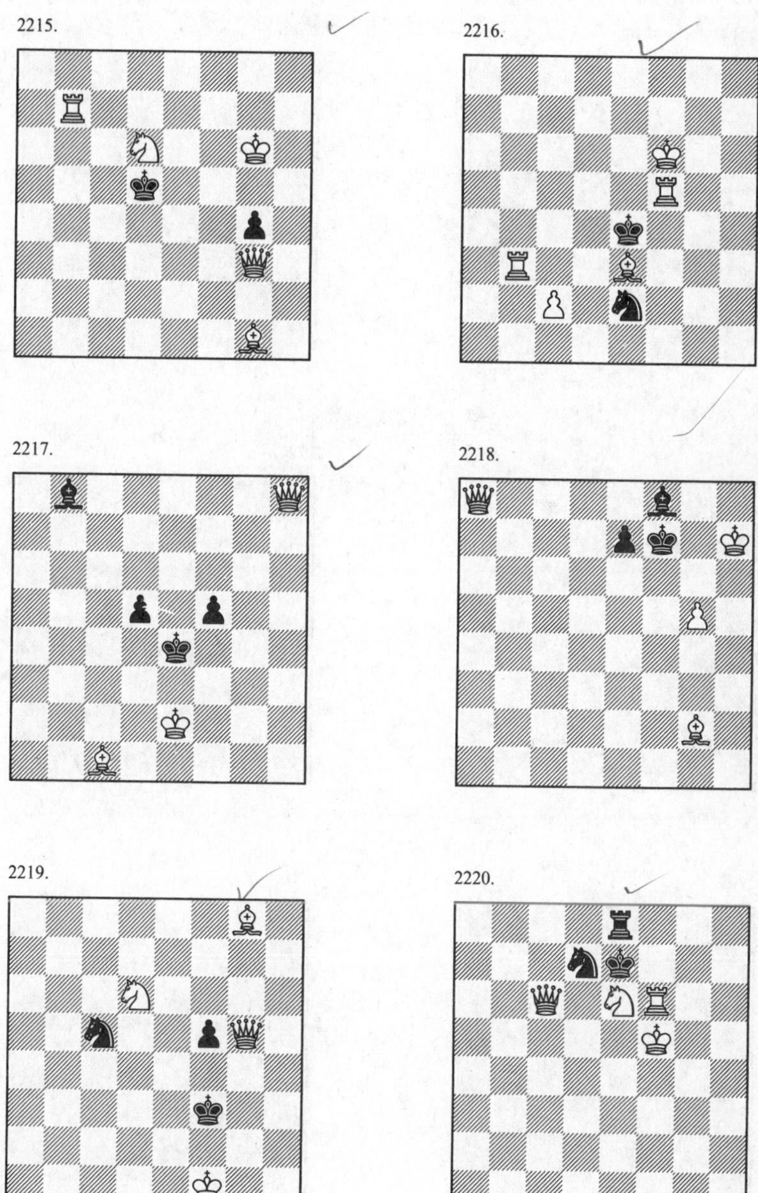
2215.
2216.
2217.
2218.
2219.
2220.

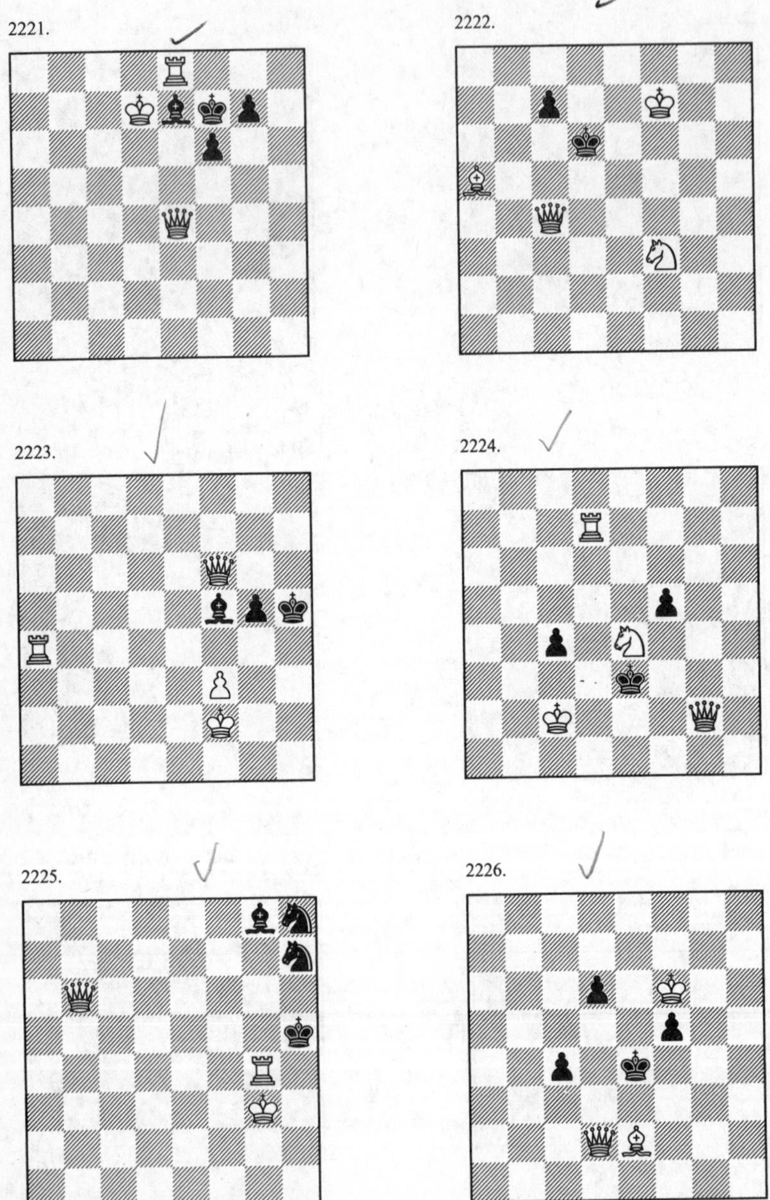

2221.

2222.

2223.

2224.

2225.

2226.

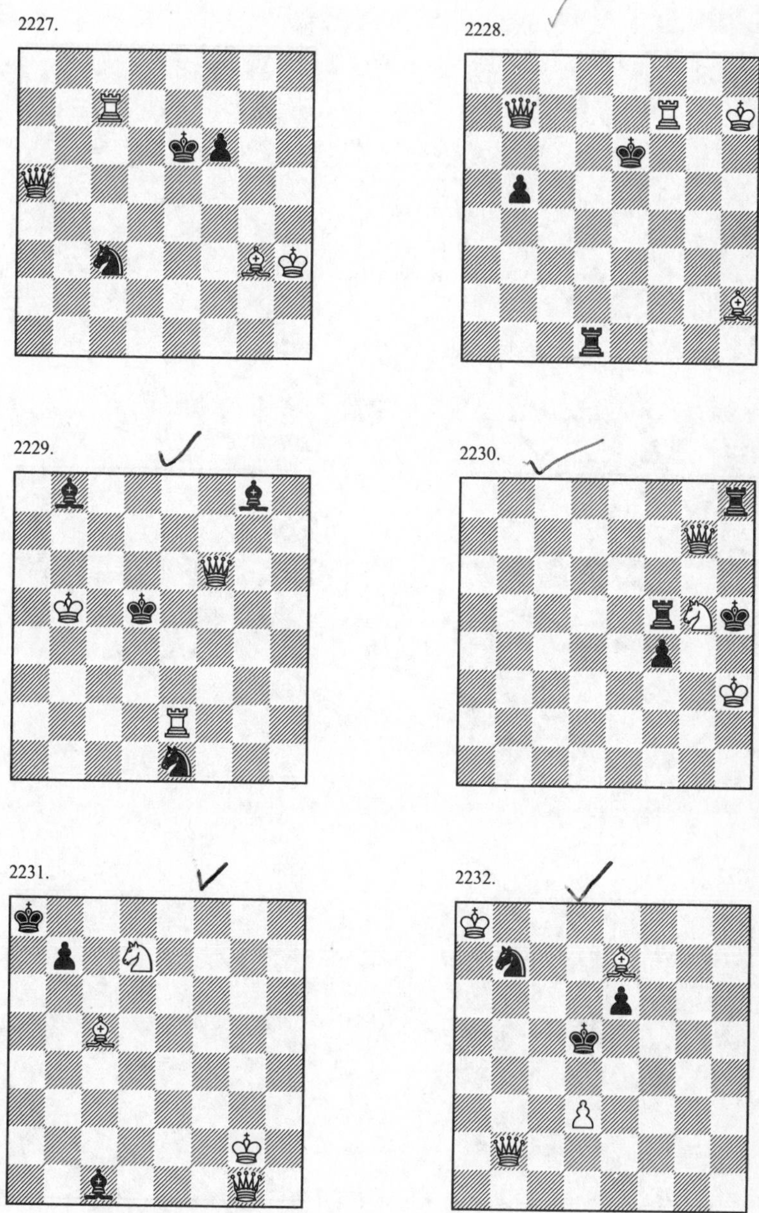
2227.
2228.
2229.
2230.
2231.
2232.

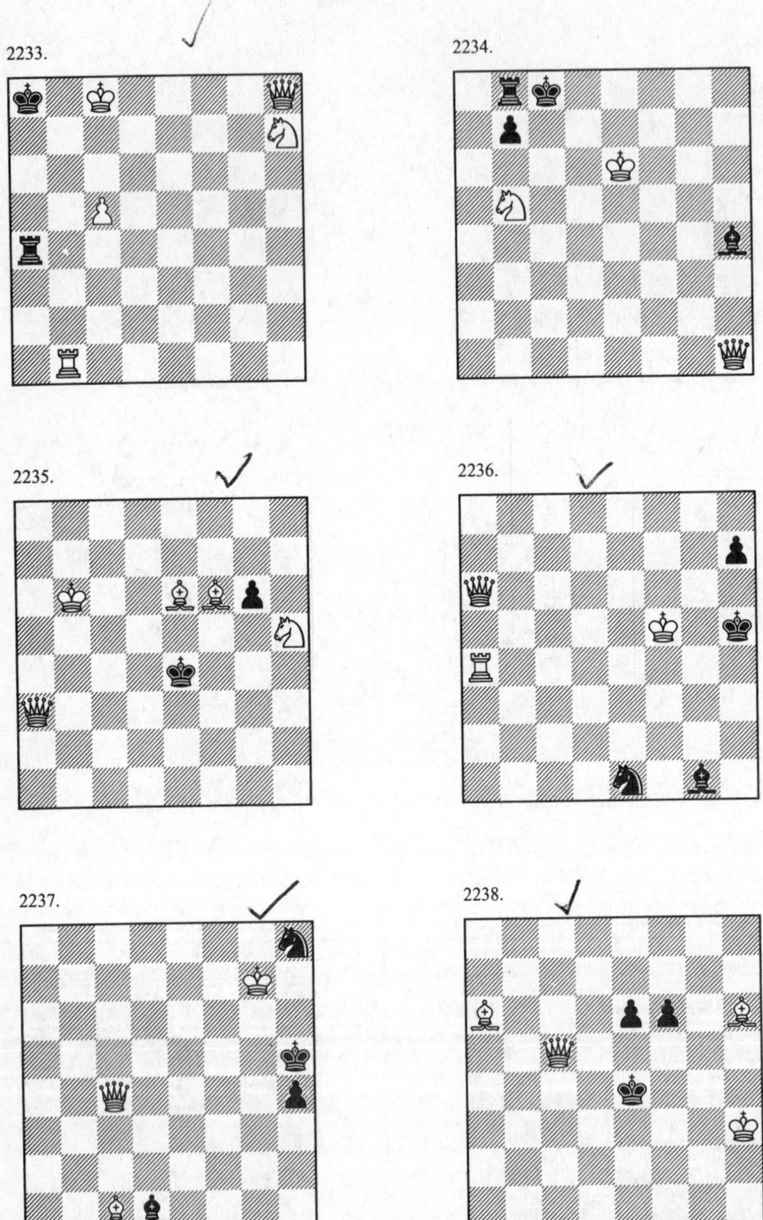
2233.
2234.
2235.
2236.
2237.
2238.

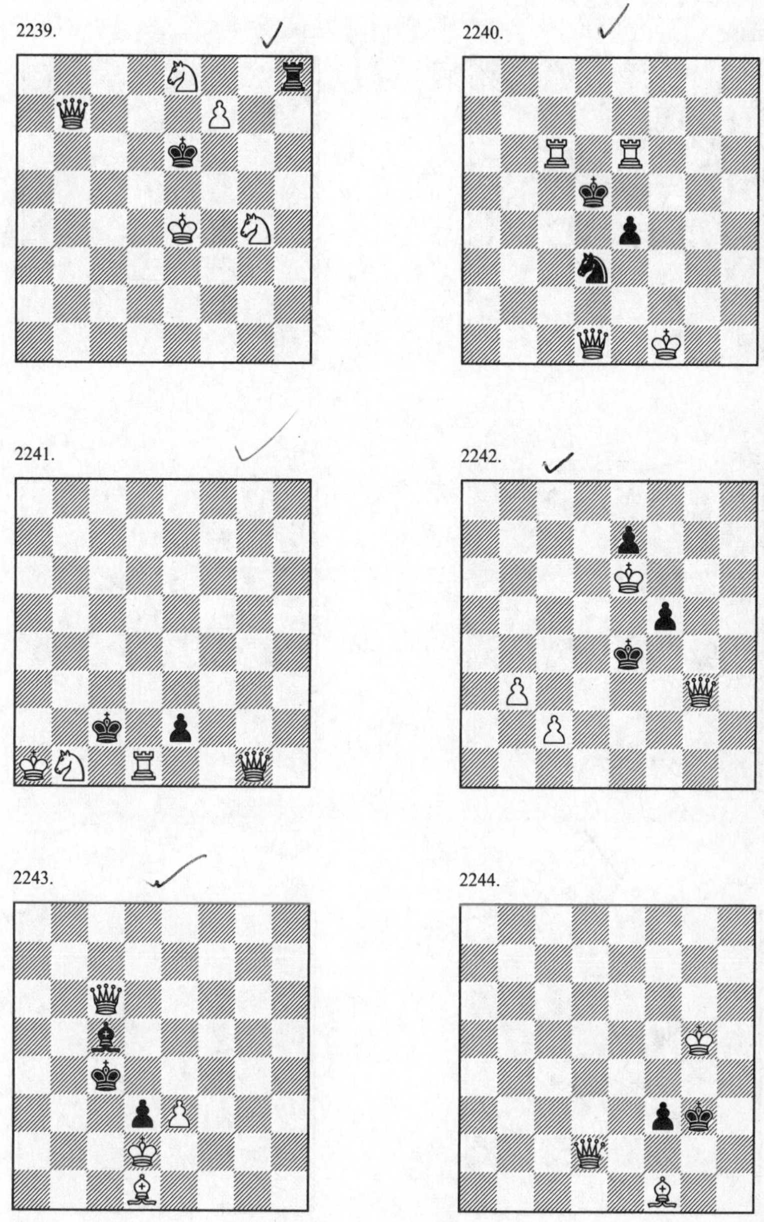
2239.
2240.
2241.
2242.
2243.
2244.

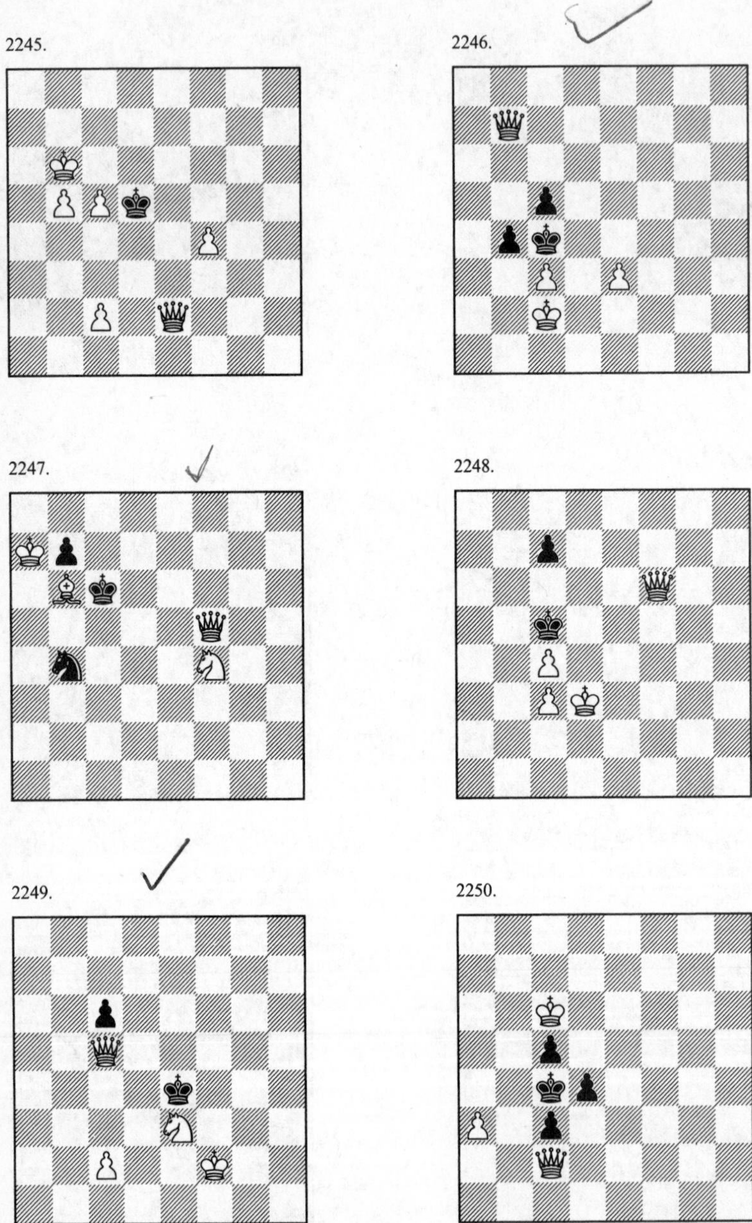
2245.

2246.

2247.

2248.

2249.

2250.

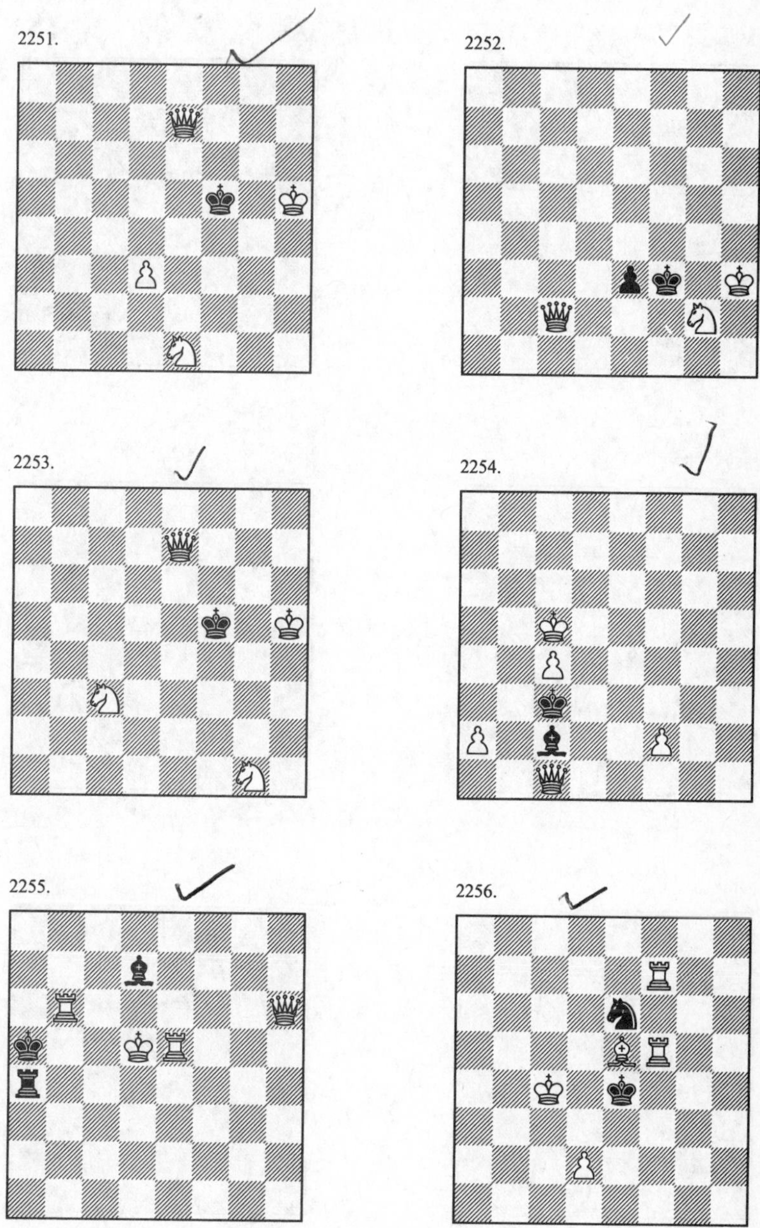
2251.
2252.
2253.
2254.
2255.
2256.

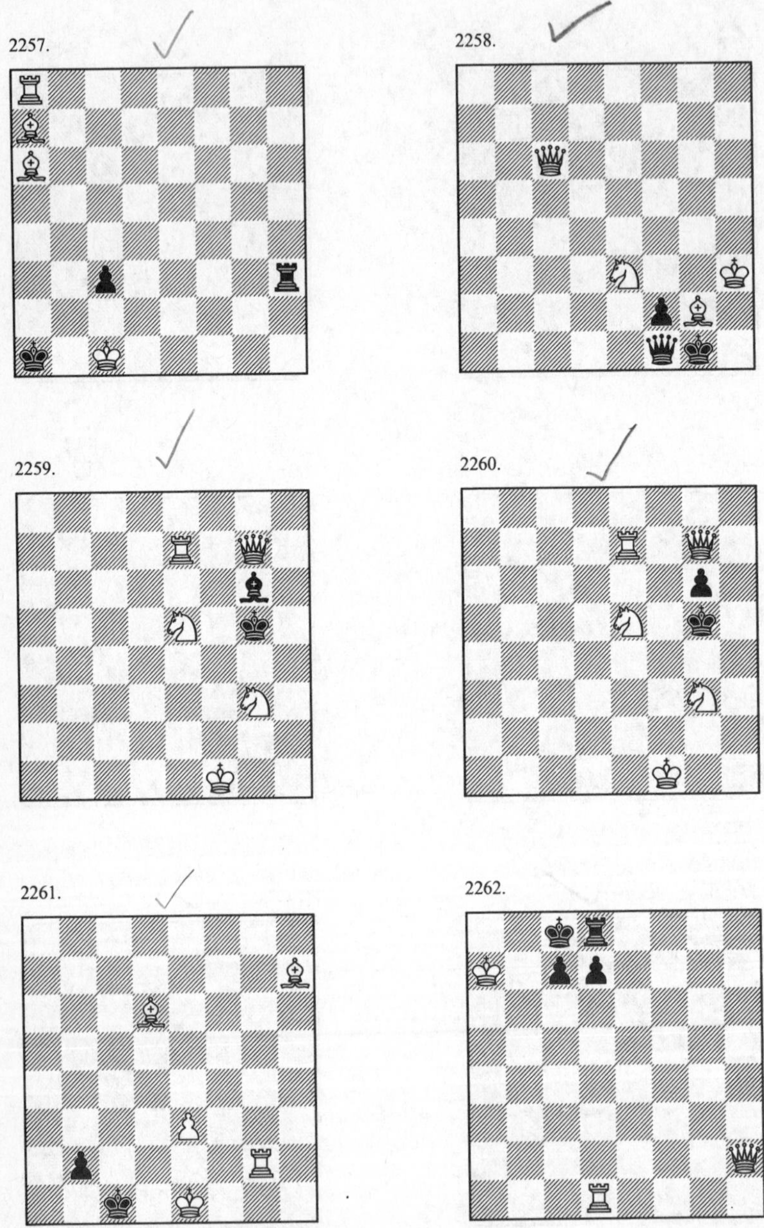
2257.
2258.
2259.
2260.
2261.
2262.

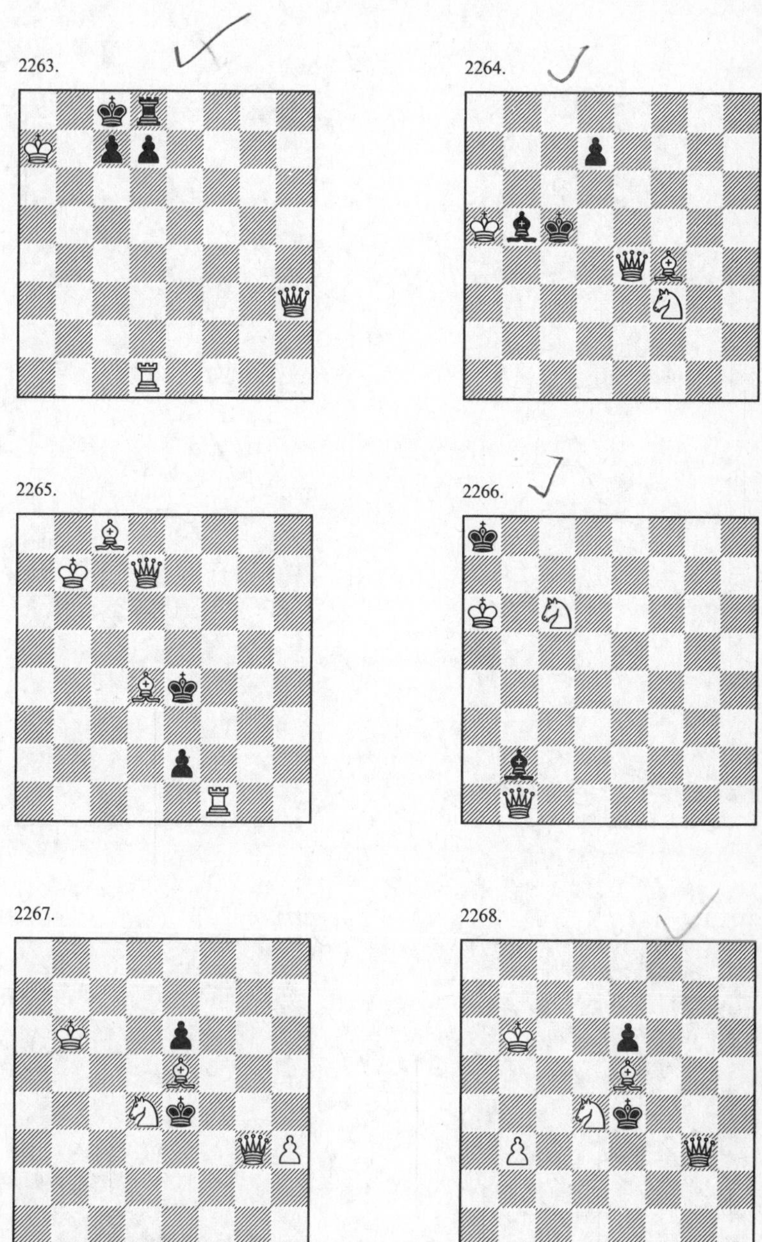
2263.
2264.
2265.
2266.
2267.
2268.

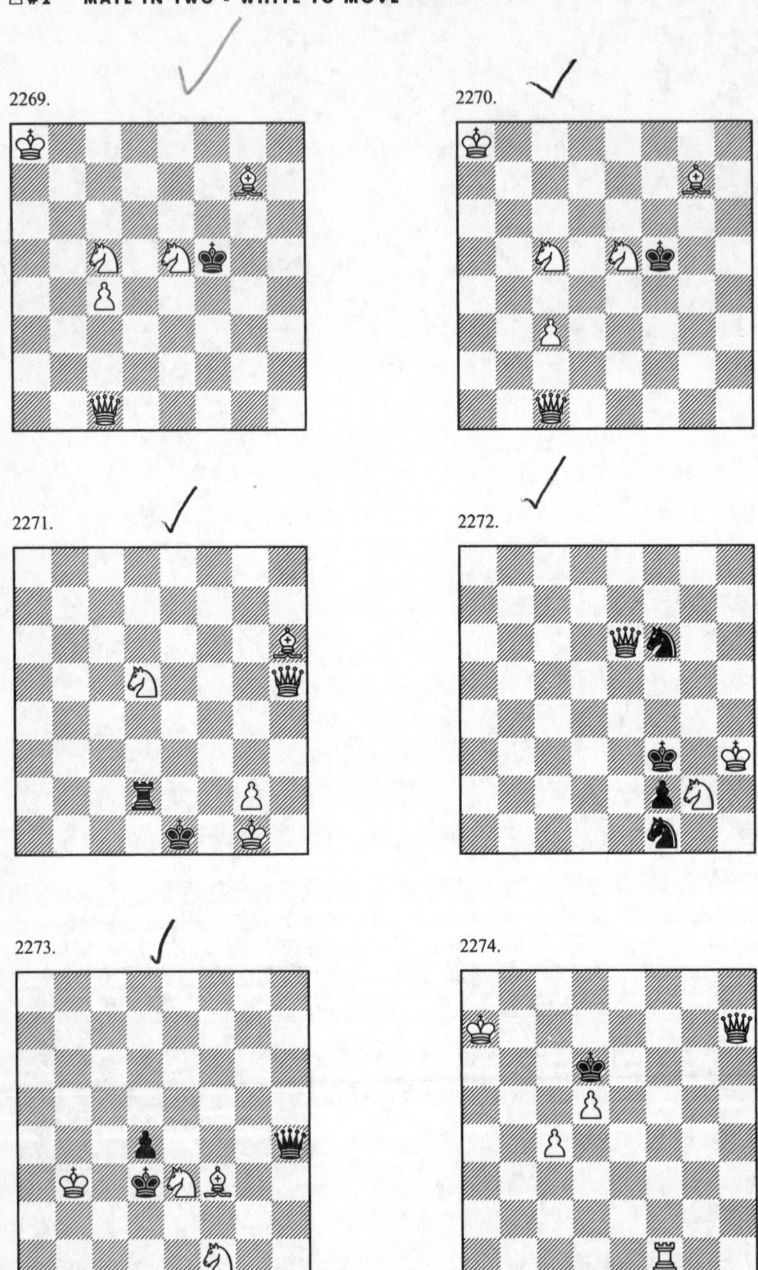
2269.
2270.
2271.
2272.
2273.
2274.

2275.
2276.
2277.
2278.
2279.
2280.

2281.

2282.

2283.

2284.

2285.

2286.

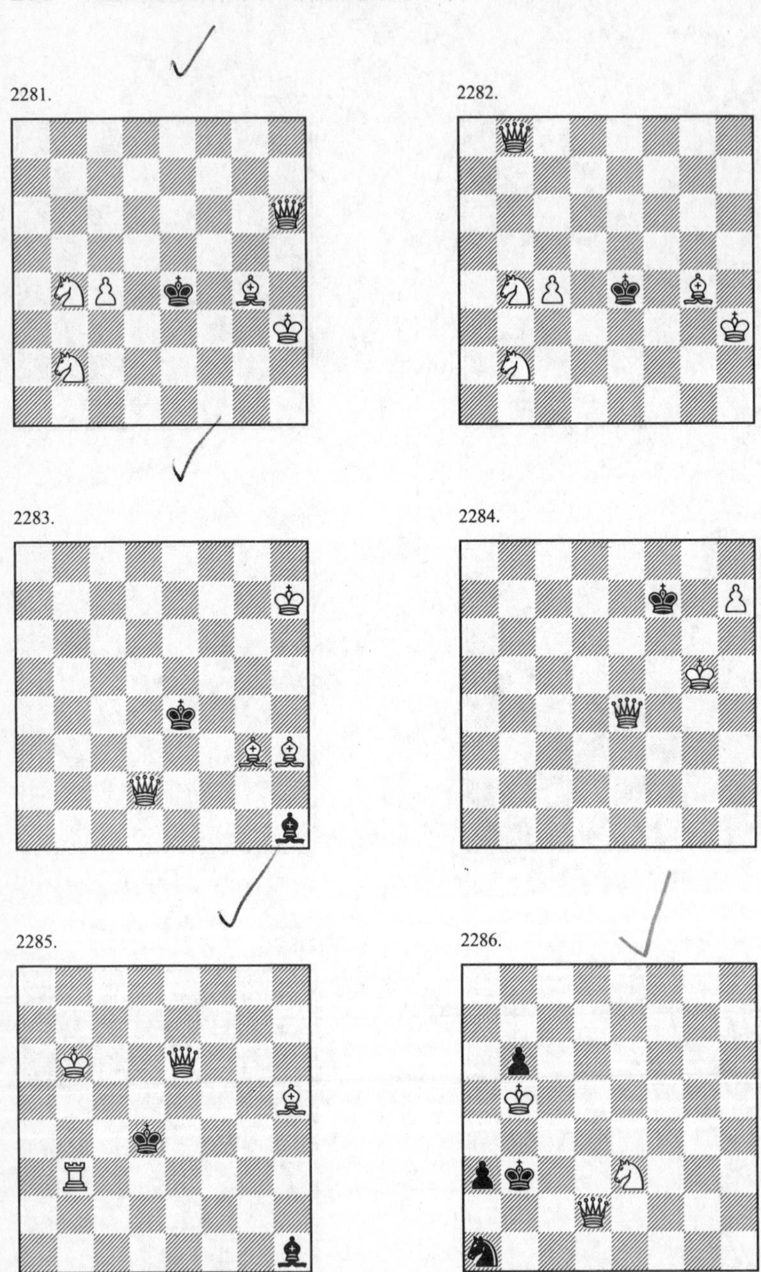

2287.

2288.

2289.

2290.

2291.

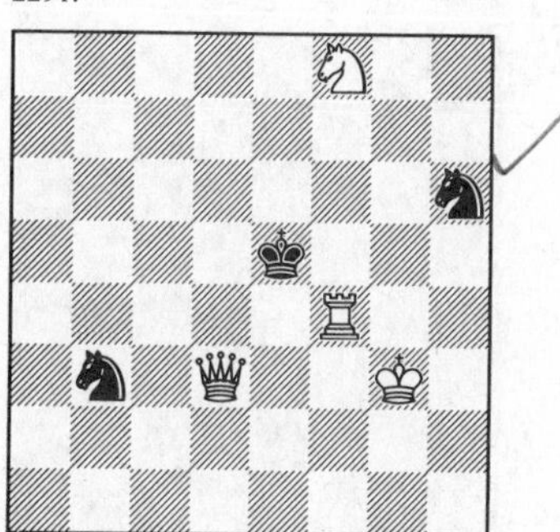

2292.

2293.

2294.

2295.

2296.

2297.

2298.

2299.

2300.

2301.

2302.

2303.

2304.

2305.

2306.

2307.

2308.

2309.

2310.

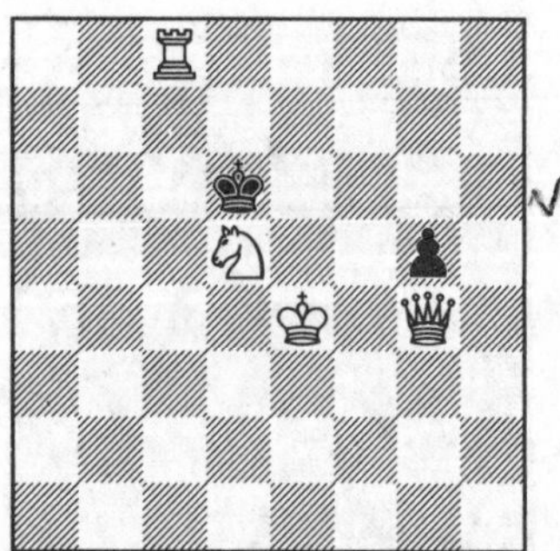

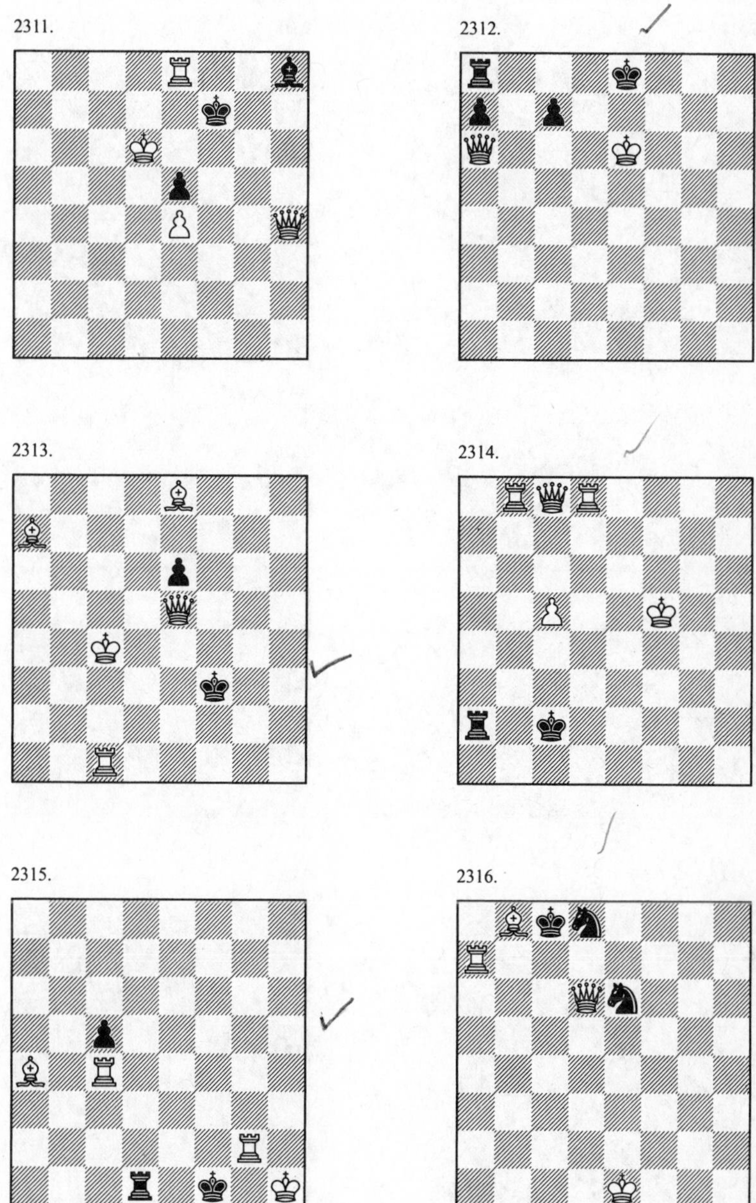

2311.

2312.

2313.

2314.

2315.

2316.

2317.

2318.

2319.

2320.

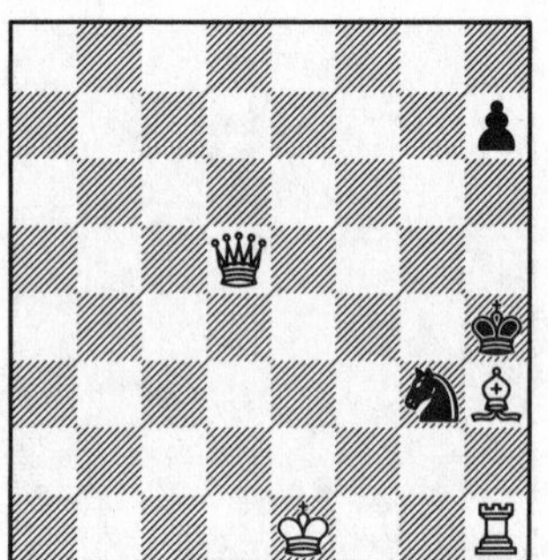

2321.

2322.

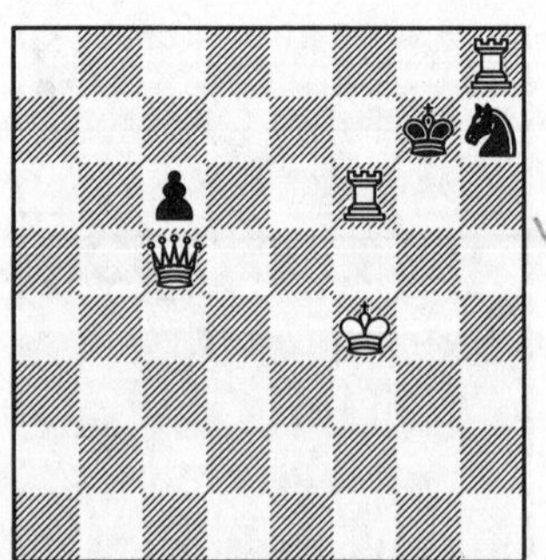

2323.

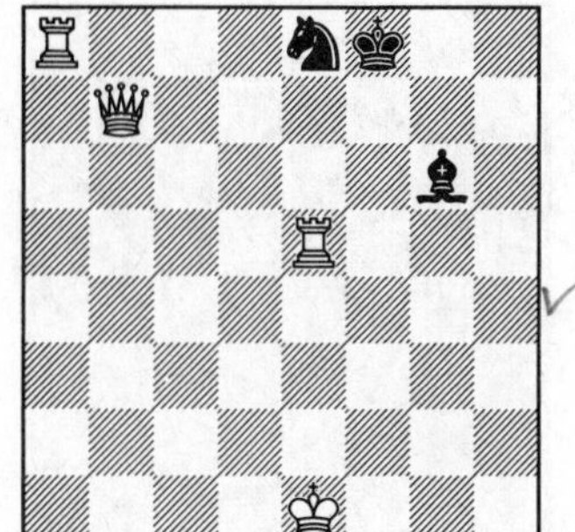

2324.

2325.

2326.

2327.

2328.

2329.

2330.

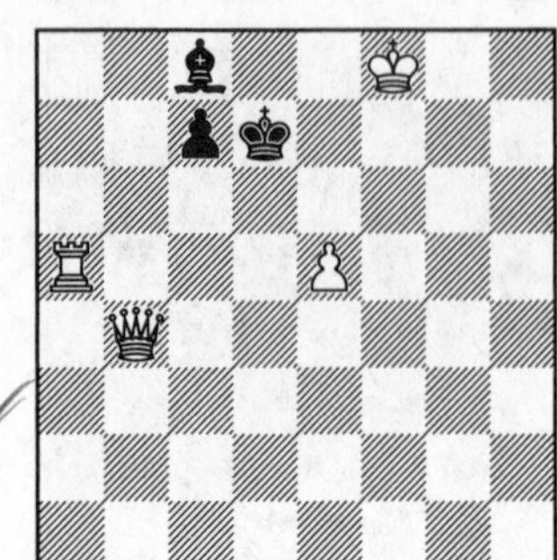

2331.

2332.

2333.

2334.

2335.

2336.

2337.

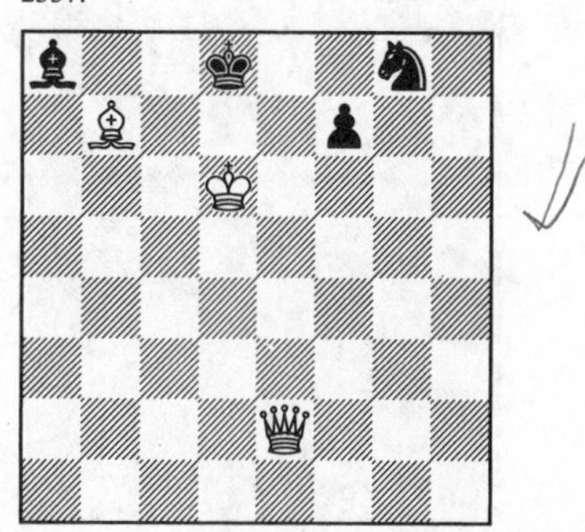

2338.

2339.

2340.

♙ #2 MATE IN TWO - WHITE TO MOVE

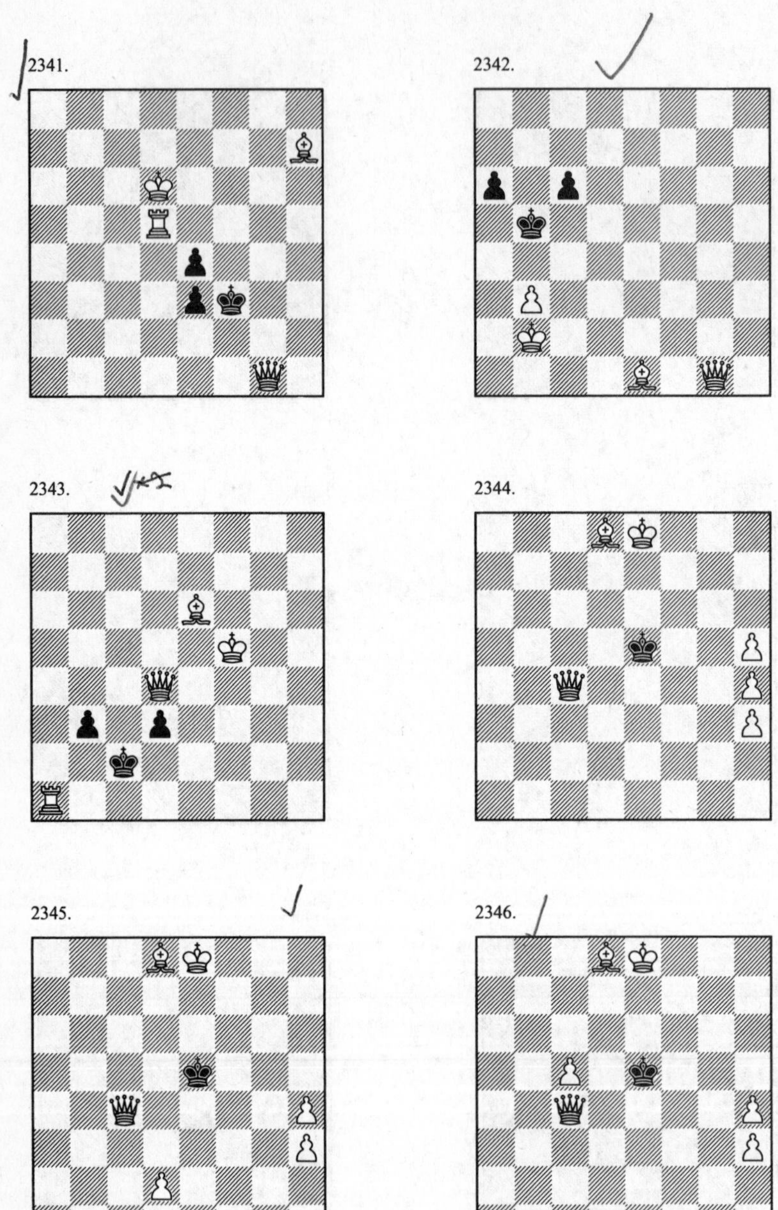

2347.

2348.

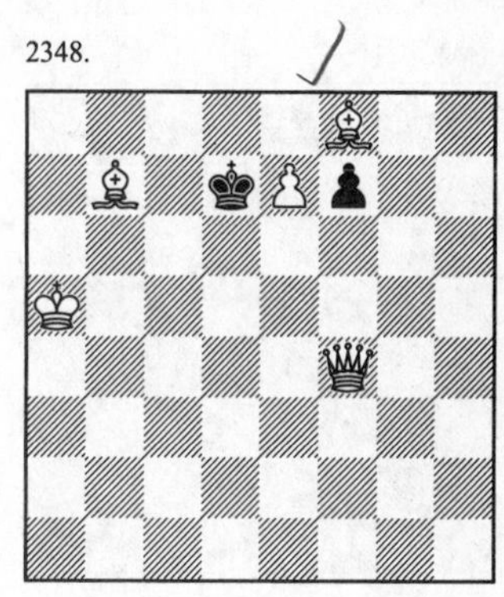

2349.

2350.

2351.

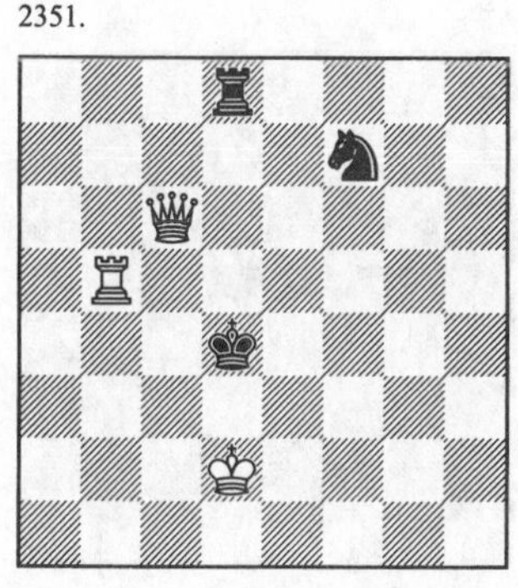

2352.

2353.

2354.

2355.

2356.

2357.

2358.

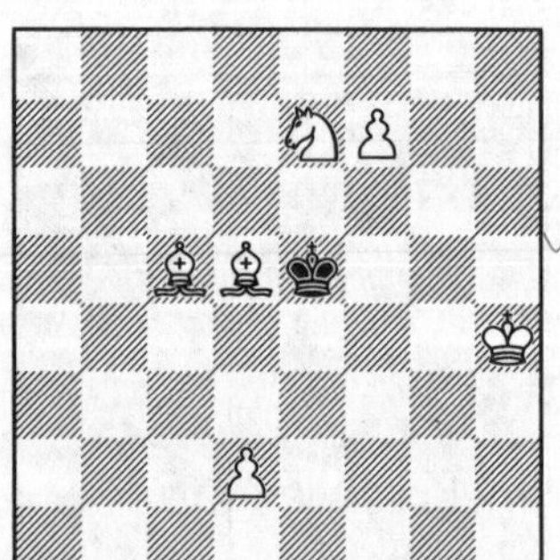

2359.

2360.

2361.

2362.

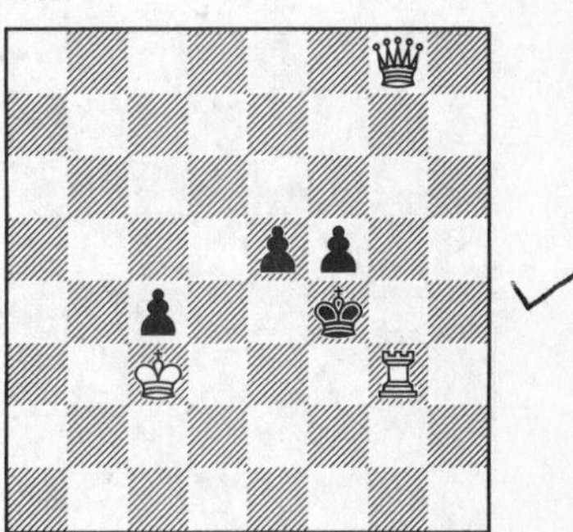

2363.

2364.

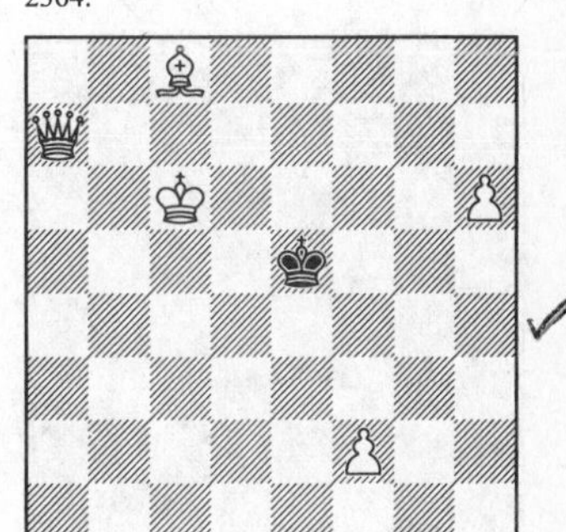

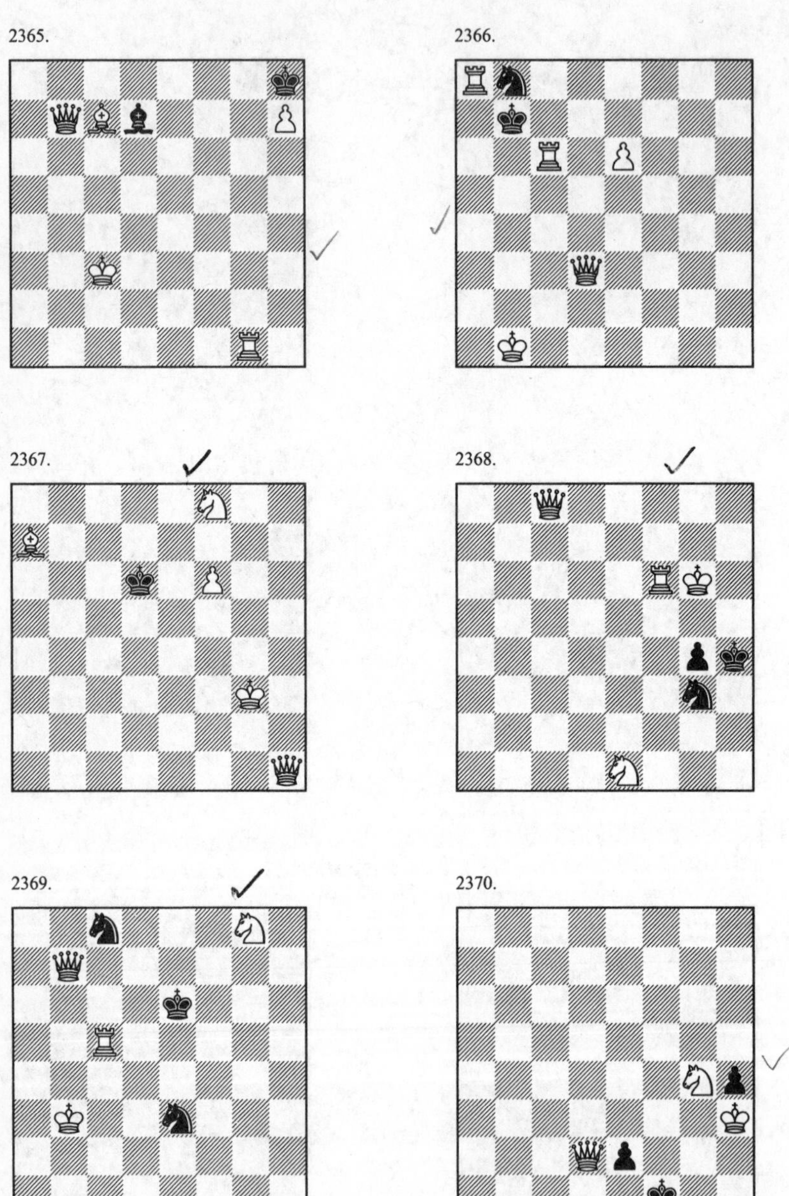
2365.
2366.
2367.
2368.
2369.
2370.

2371.

2372.

2373.

2374.

2375.

2376.

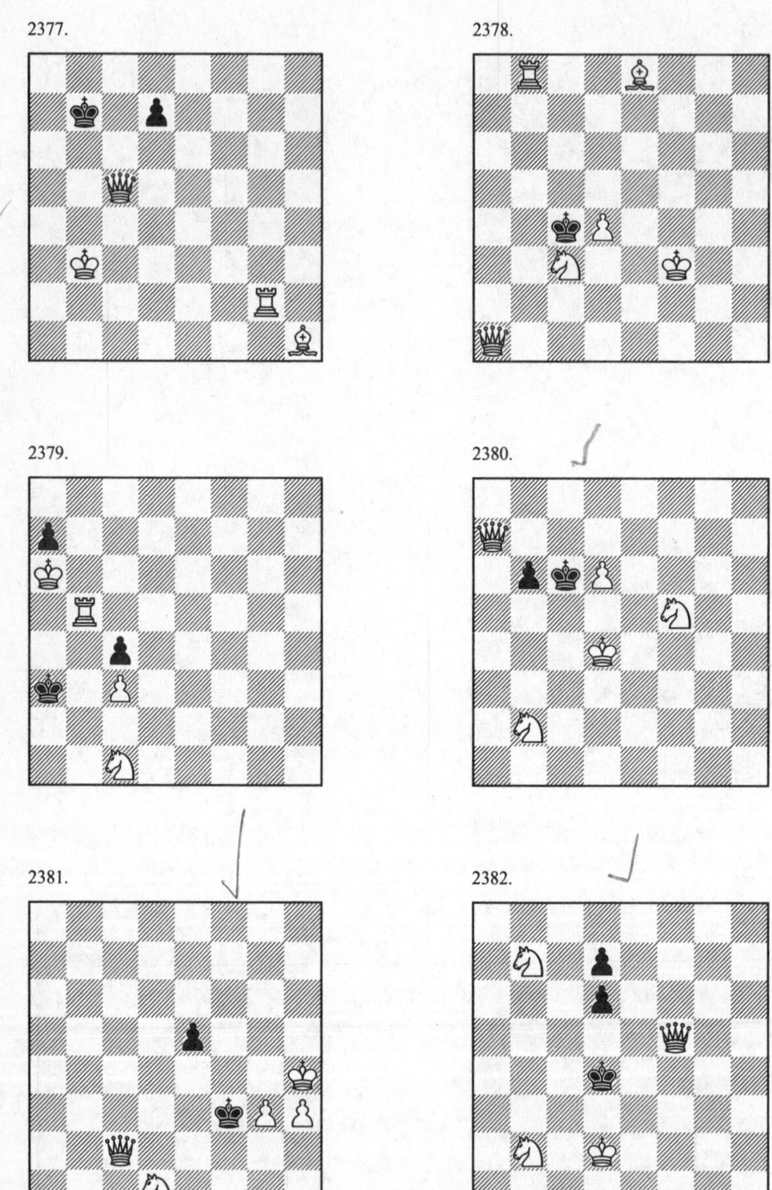
2377.
2378.
2379.
2380.
2381.
2382.

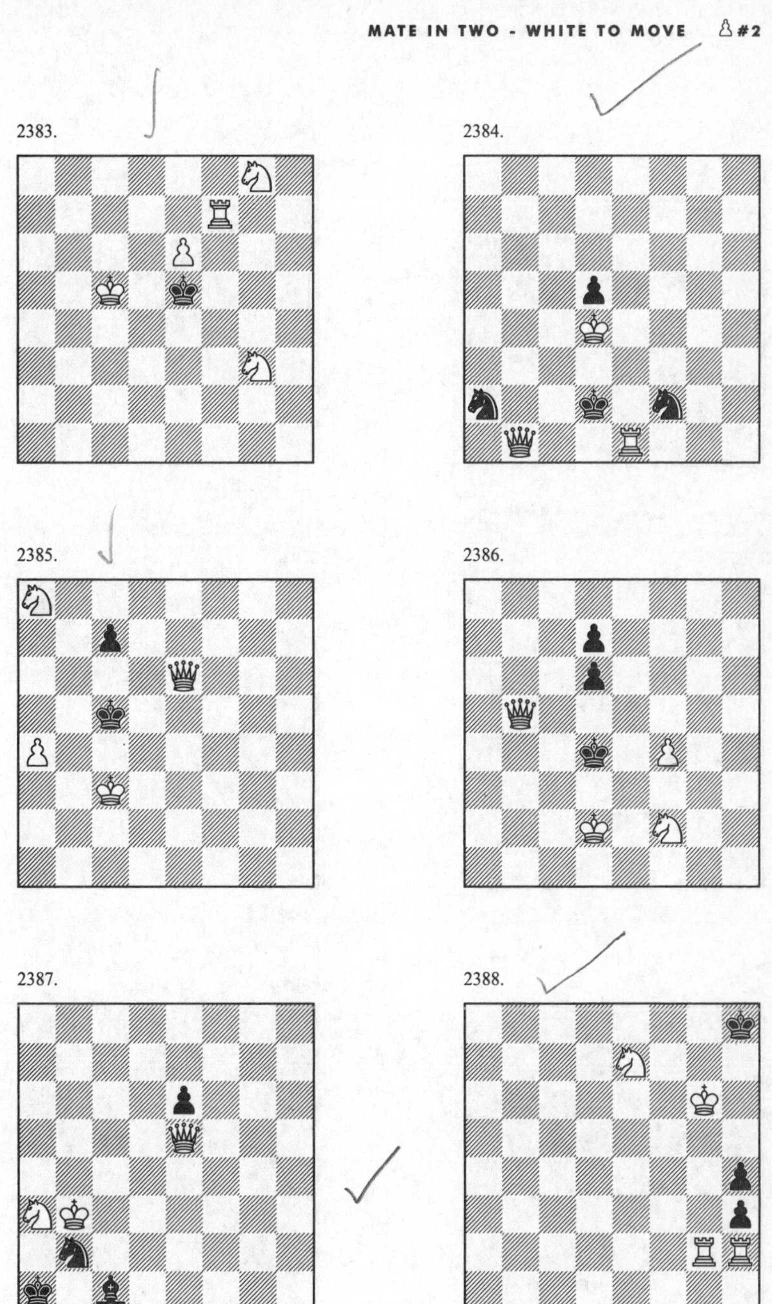
2383.
2384.
2385.
2386.
2387.
2388.

2389.

2390.

2391.

2392.

2393.

2394.

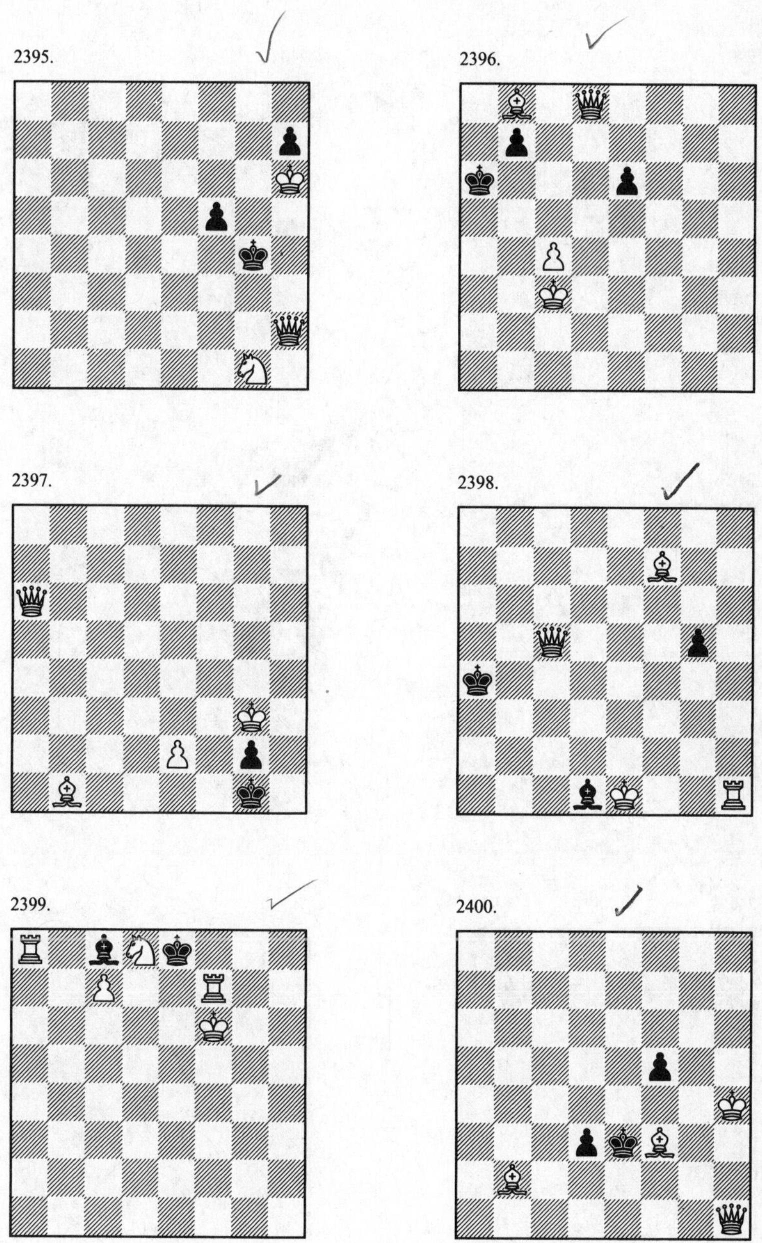
2395.
2396.
2397.
2398.
2399.
2400.

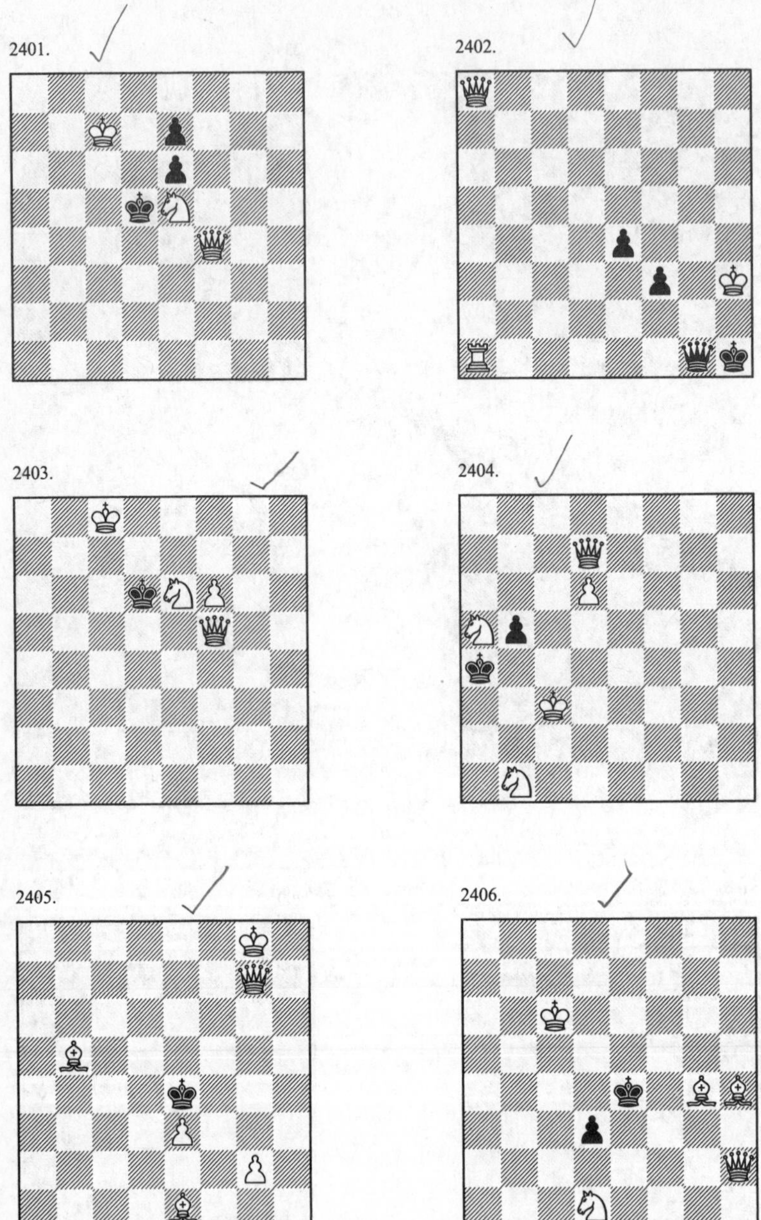
2401.
2402.
2403.
2404.
2405.
2406.

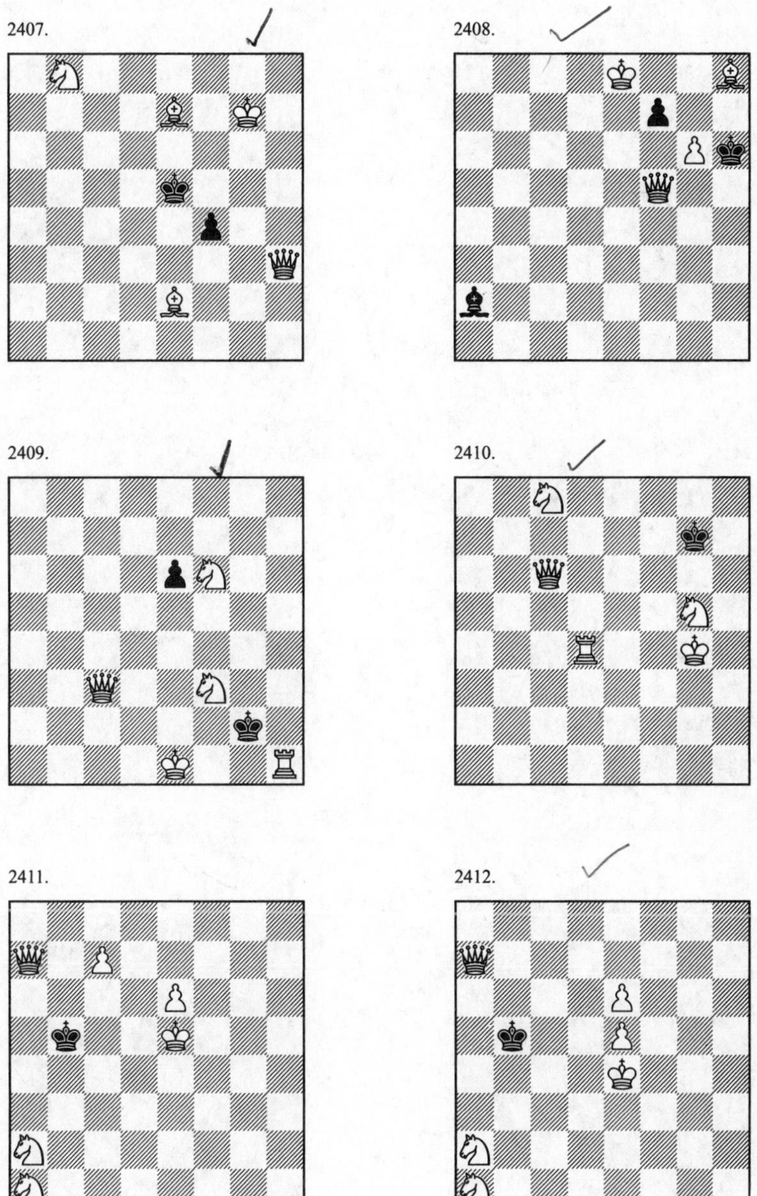
2407.
2408.
2409.
2410.
2411.
2412.

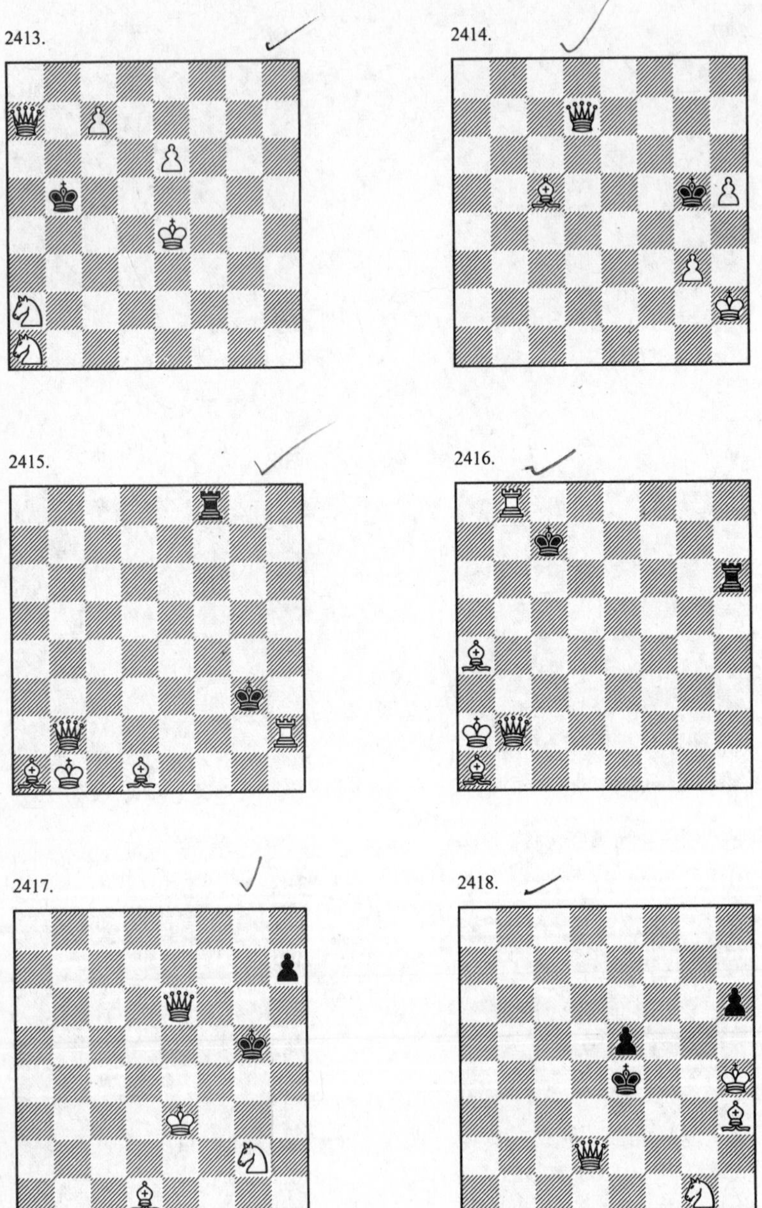
2413.

2414.

2415.

2416.

2417.

2418.

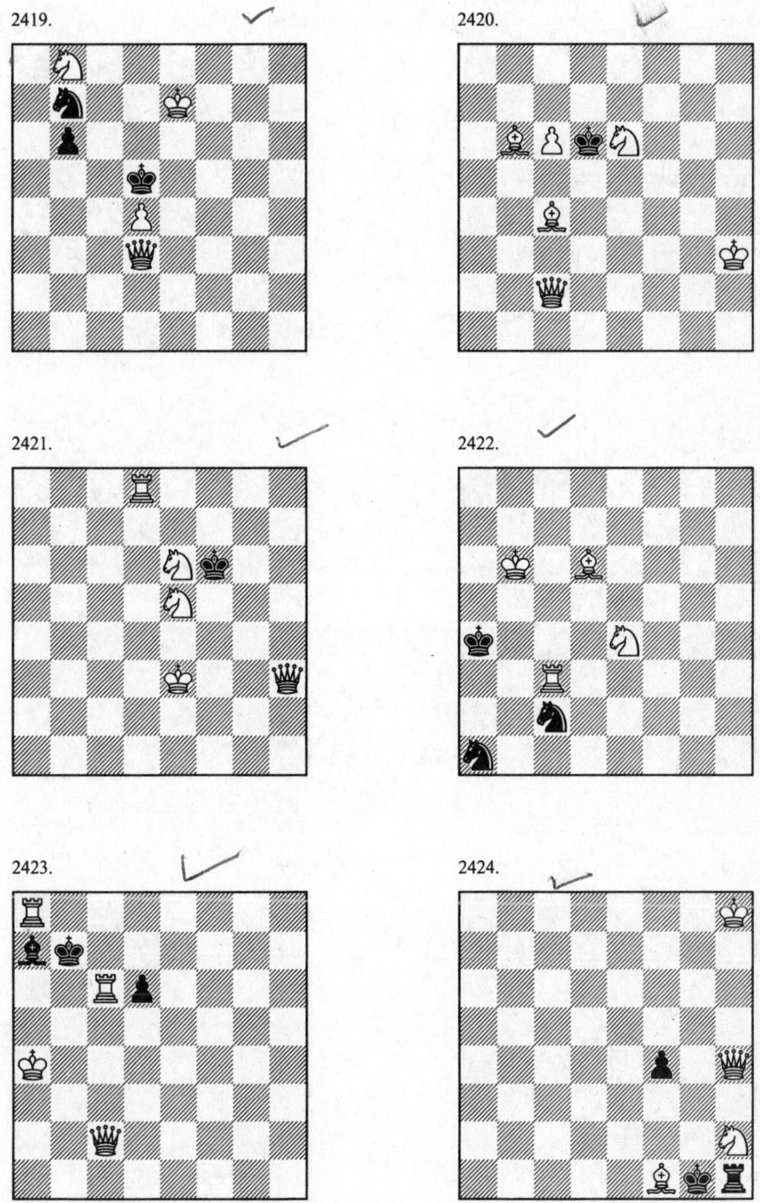
2419.
2420.
2421.
2422.
2423.
2424.

2425.

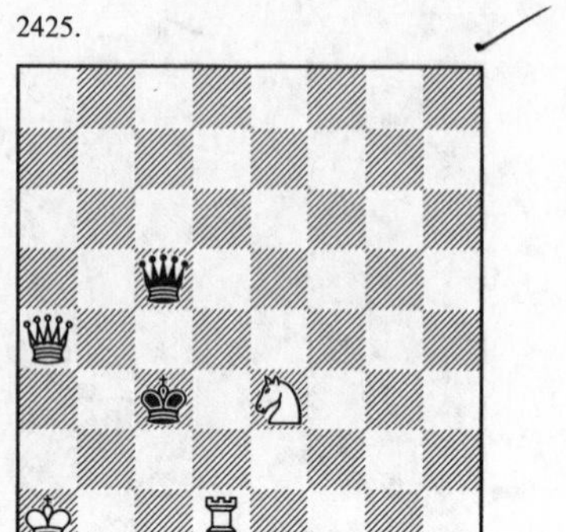

2426.

2427.

2428.

2429.

2430.

2431.

2432.

2433.

2434.

2435.

2436.

2437.

2438.

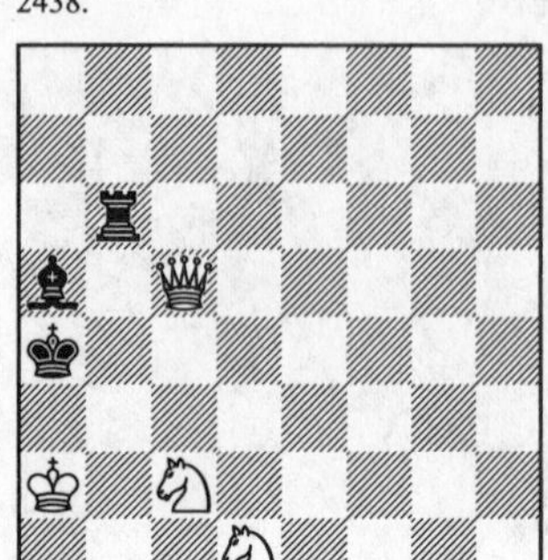

2439.

2440.

2441.

2442.

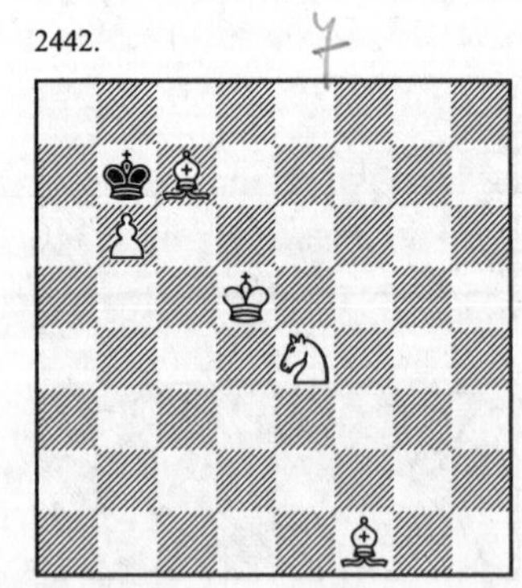

2443.

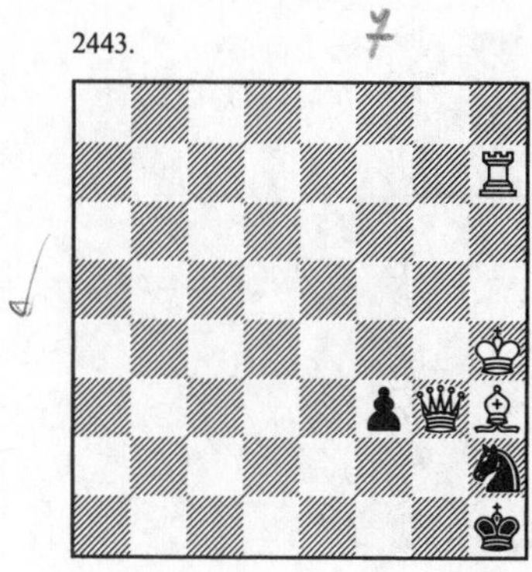

2444.

2445.

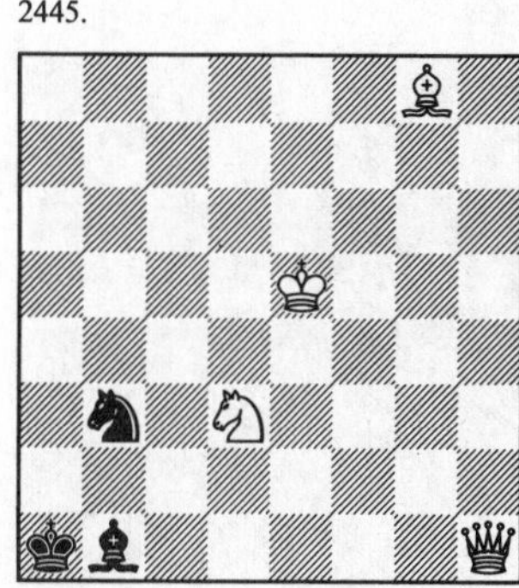

2446.

2447.

2448.

2449.

2450.

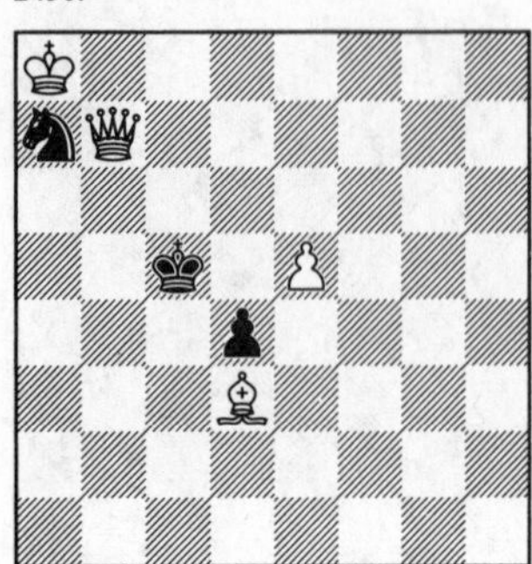

2451.

2452.

2453.

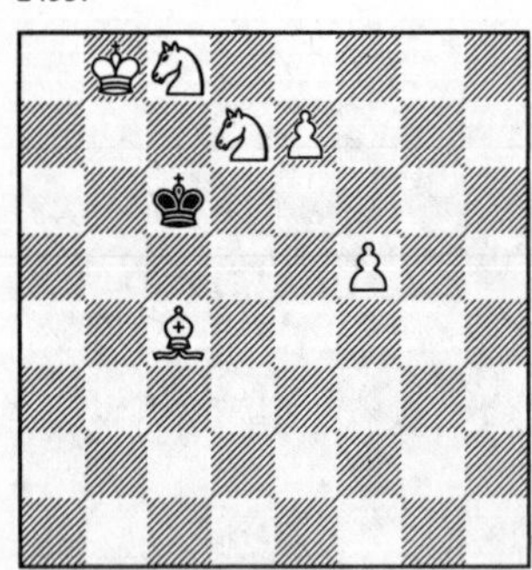

2454.

2455.

2456.

2457.

2458.

2459.

2460.

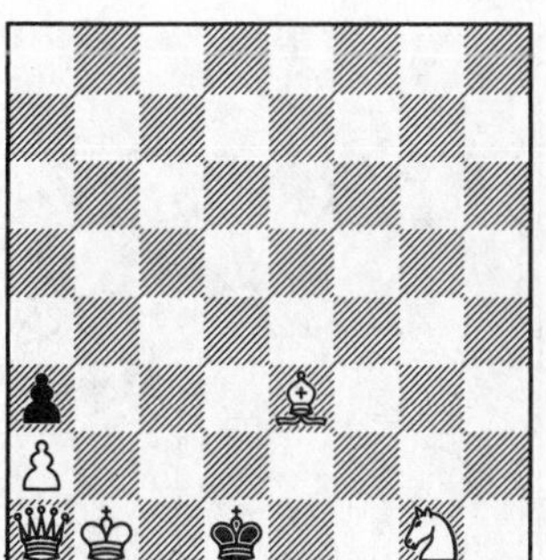

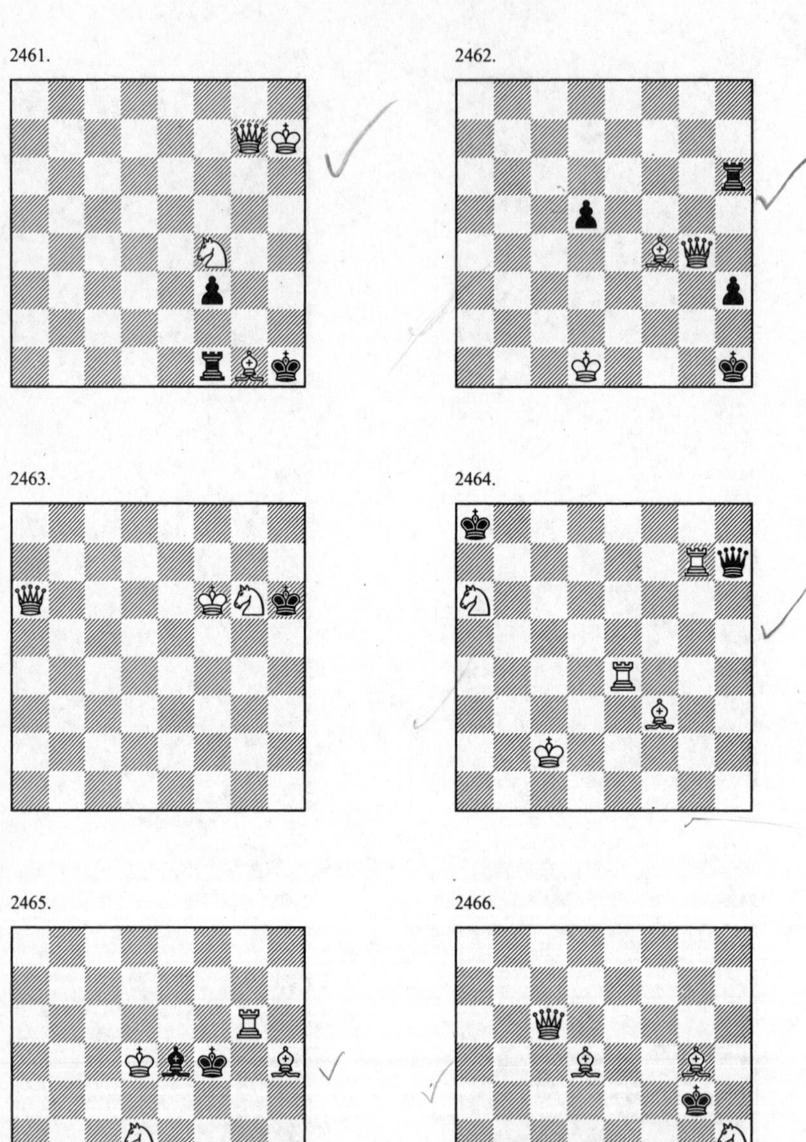

2461.

2462.

2463.

2464.

2465.

2466.

2467.

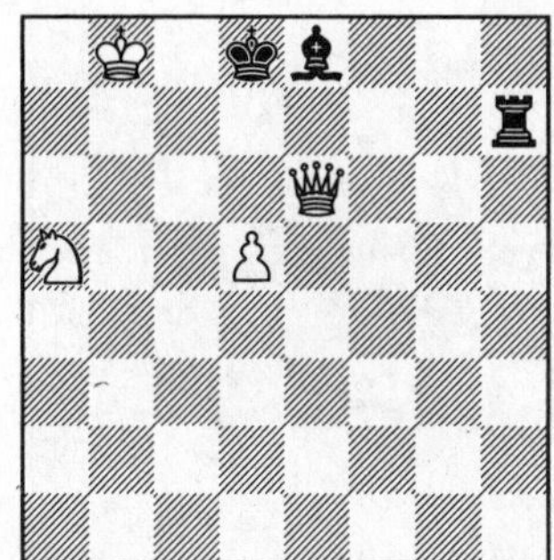

2468.

2469.

2470.

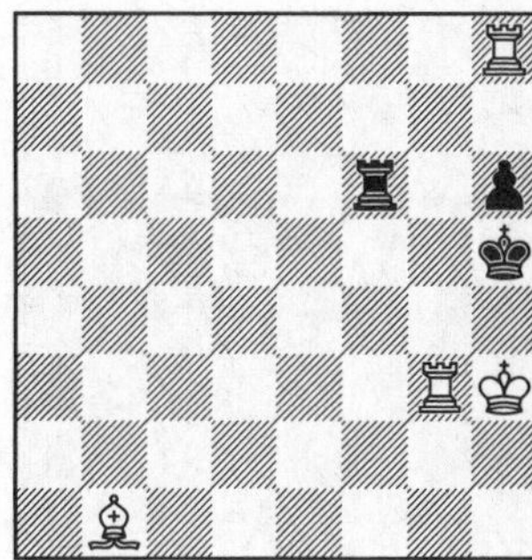

2471.

2472.

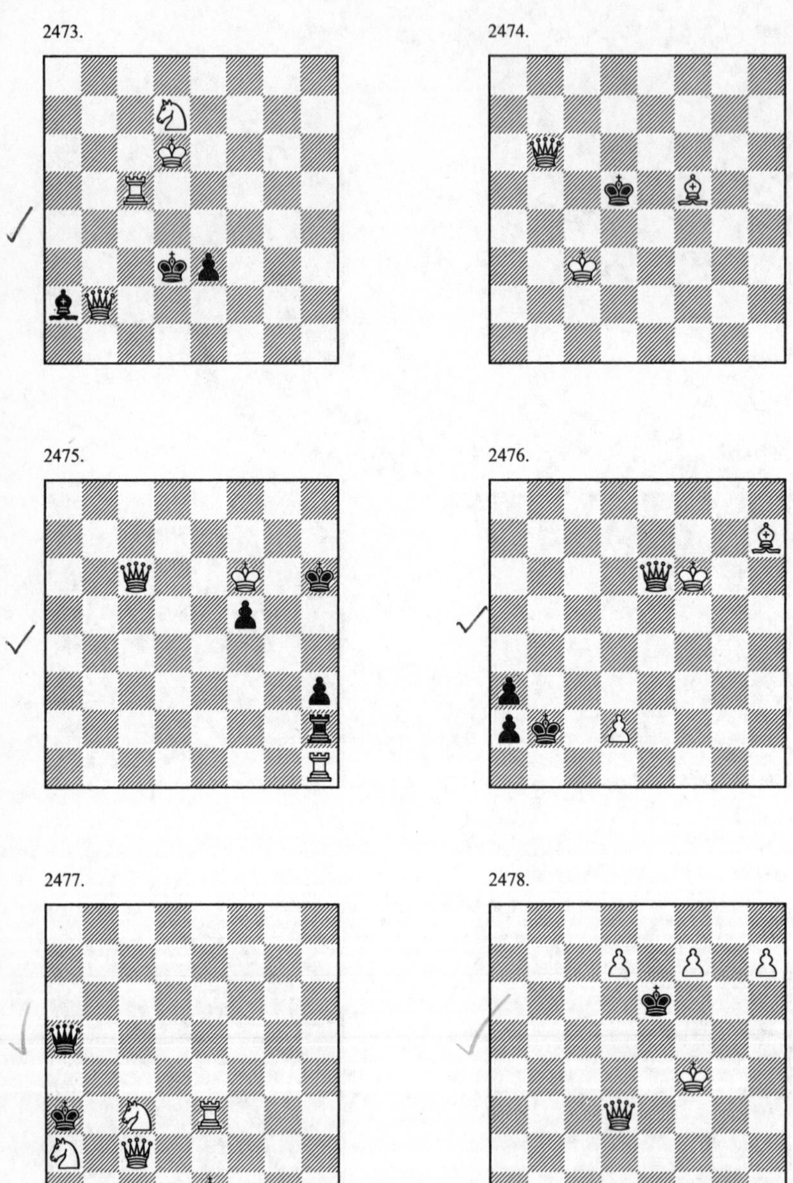
2473.
2474.
2475.
2476.
2477.
2478.

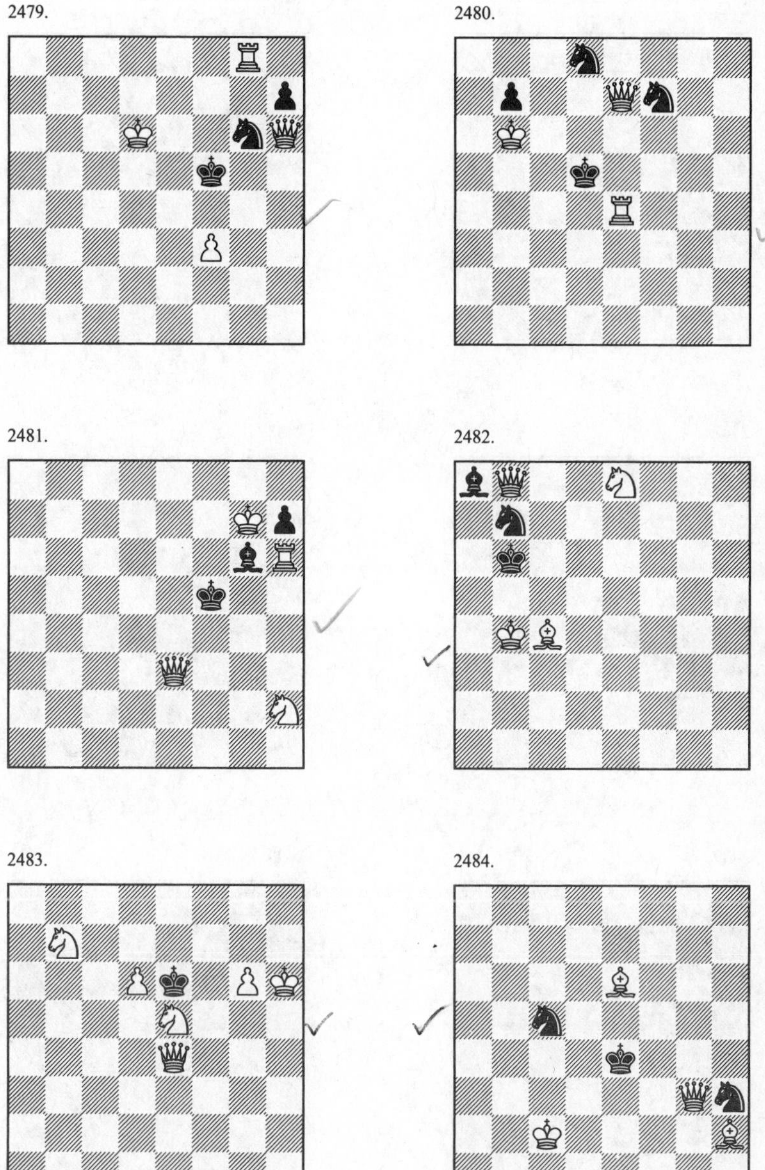

2479.

2480.

2481.

2482.

2483.

2484.

2485.

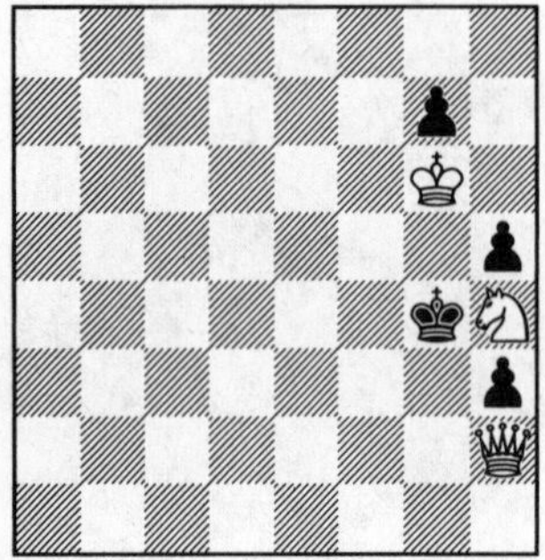

2486.

2487.

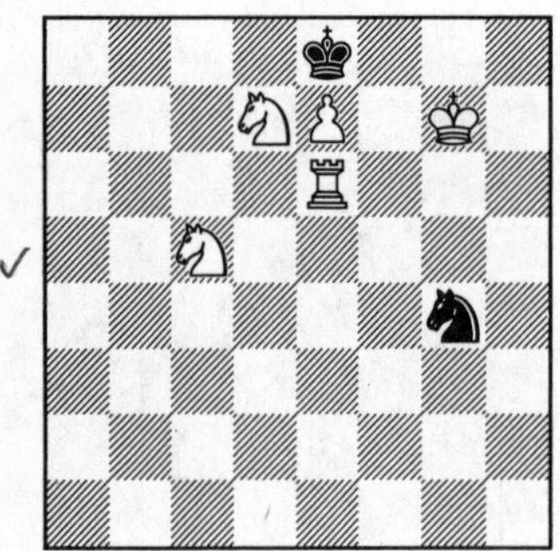

2488.

2489.

2490.

2491.

2492.

2493.

2494.

2495.

2496.

2497.

2498.

2499.

2500.

2501.

2502.

2503.

2504.

2505.

2506.

2507.

2508.

2509.

2510.

2511.

2512.

2513.

2514.

2515.

2516.

2517.

2518.

2519.

2520.

2521.

2522.

2523.

2524.

2525.

2526.

2527.

2528.

2529.

2530.

2531.

2532.

2533.

2534.

2535.

2536.

2537.

2538.

2539.

2540.

2541.

2542.

2543.

2544.

2545.

2546.

2547.

2548.

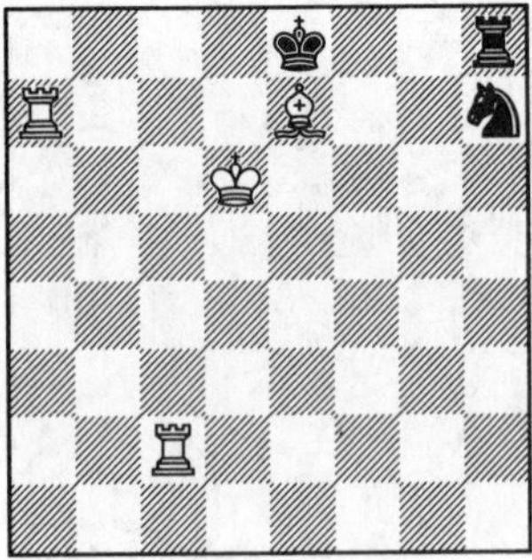

2549.

2550.

2551.

2552.

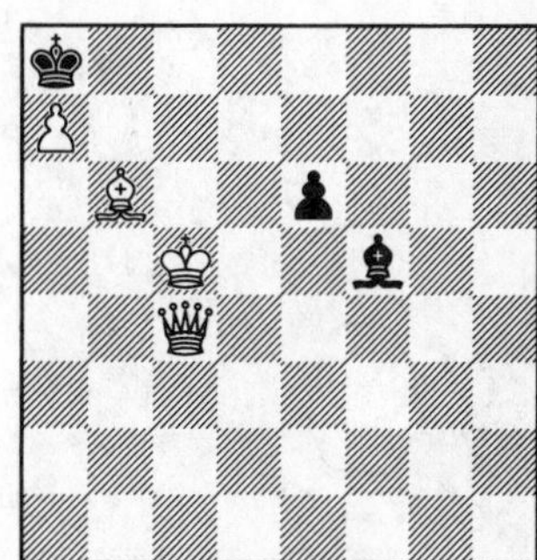

2553.

2554.

2555.

2556.

2557.

2558.

2559.

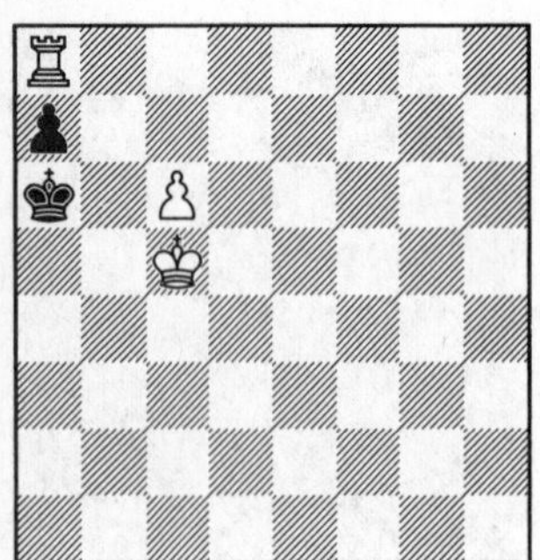

2560.

2561.

2562.

2563.

2564.

2565.

2566.

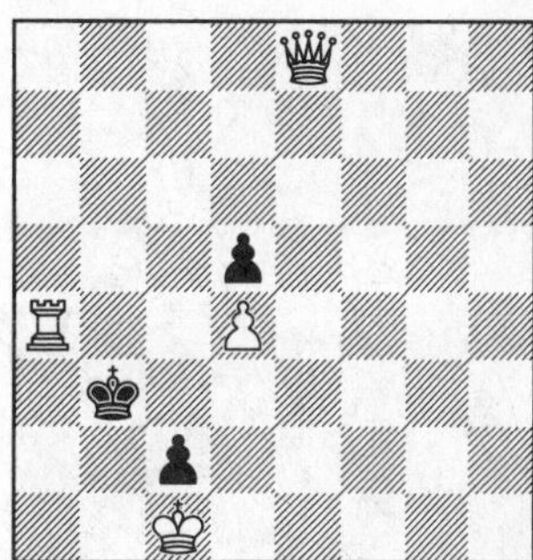

2567.

2568.

2569.

2570.

2571.

2572.

2573.

2574.

2575.

2576.

2577.

2578.

2579.

2580.

2581.

2582.

2583.

2584.

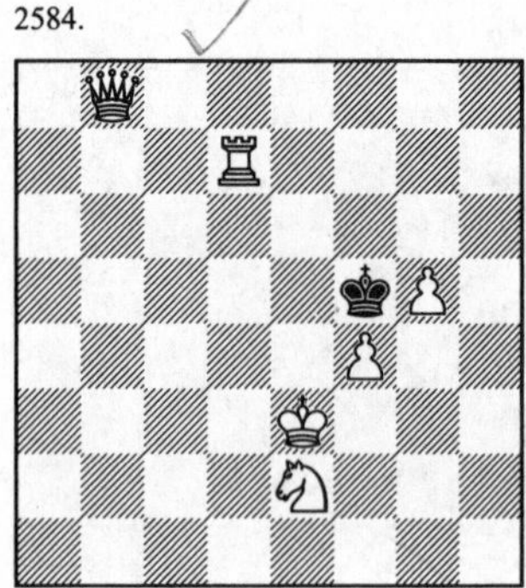

2585.

2586.

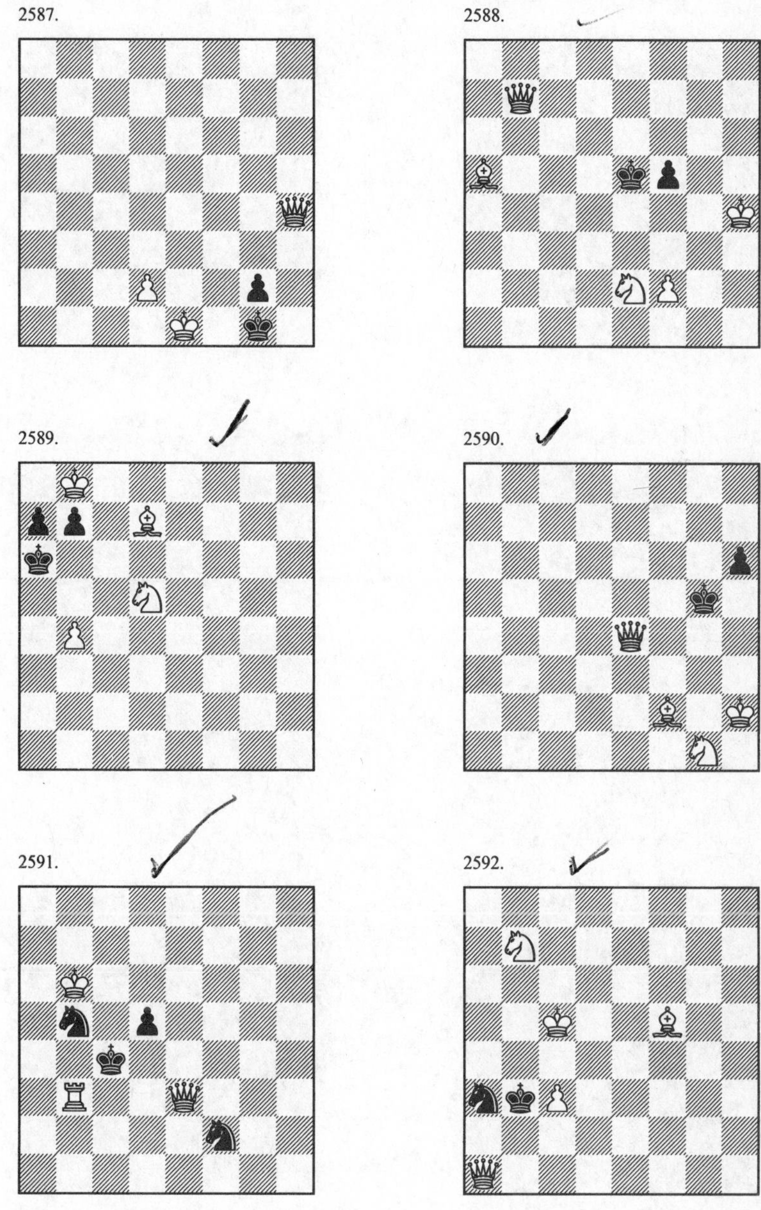
2587.
2588.
2589.
2590.
2591.
2592.

2593.

2594.

2595.

2596.

2597.

2598.

2599.

2600.

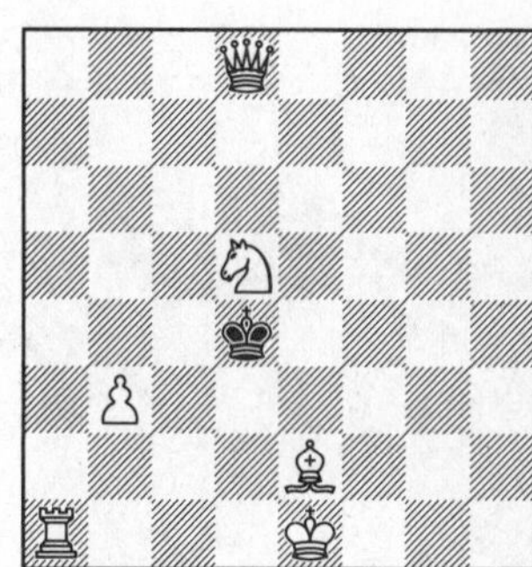

2601.

2602.

2603.

2604.

2605.

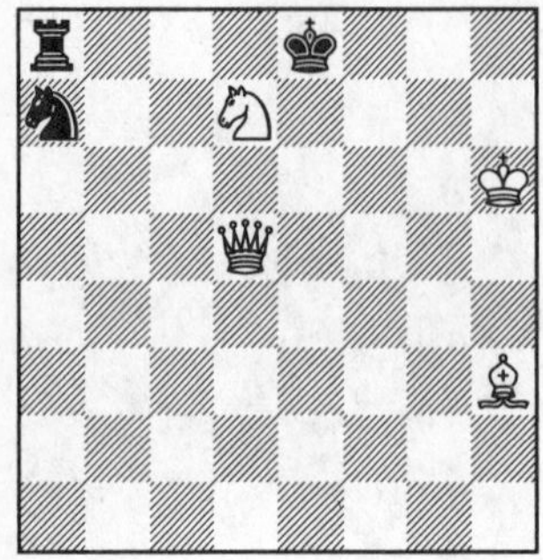

2606.

2607.

2608.

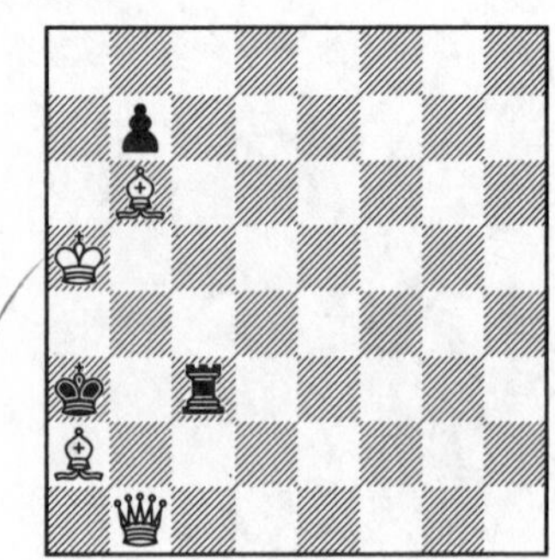

2609.

2610.

2611.

2612.

2613.

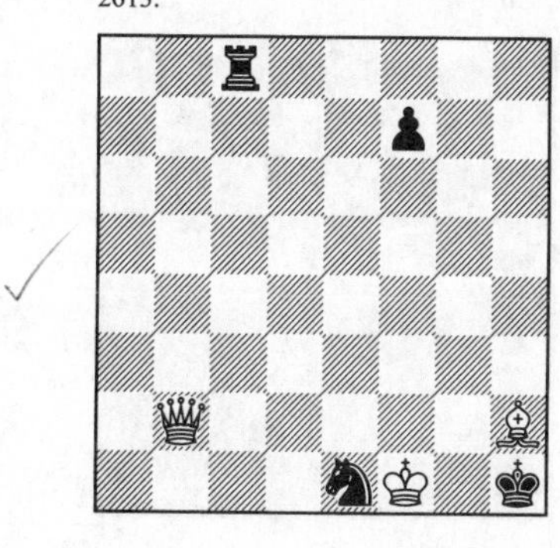

2614.

2615.

2616.

2617.

2618.

2619.

2620.

2621.

2622.

2623.

2624.

2625.

2626.

2627.

2628.

2629.

2630.

2631.

2632.

2633.

2634.

2635.

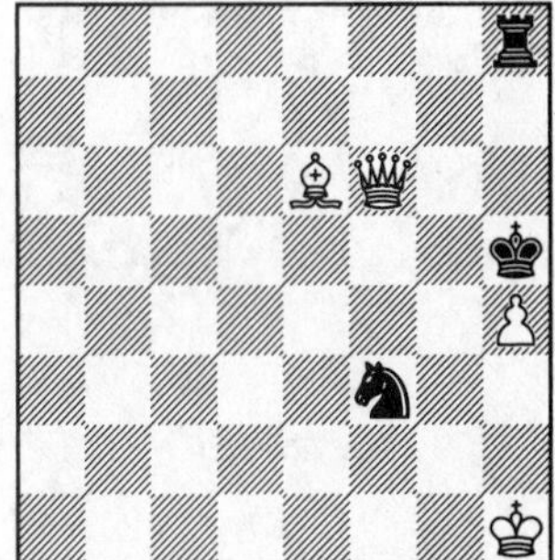

2636.

2637.

2638.

2639.

2640.

2641.

2642.

2643.

2644.

2645.

2646.

2647.

2648.

2649.

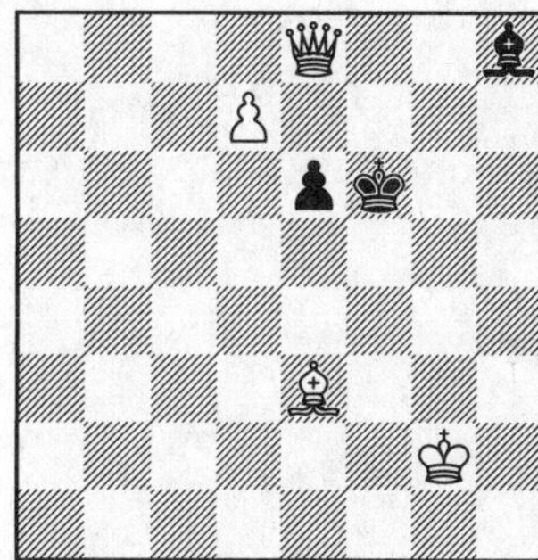

2650.

2651.

2652.

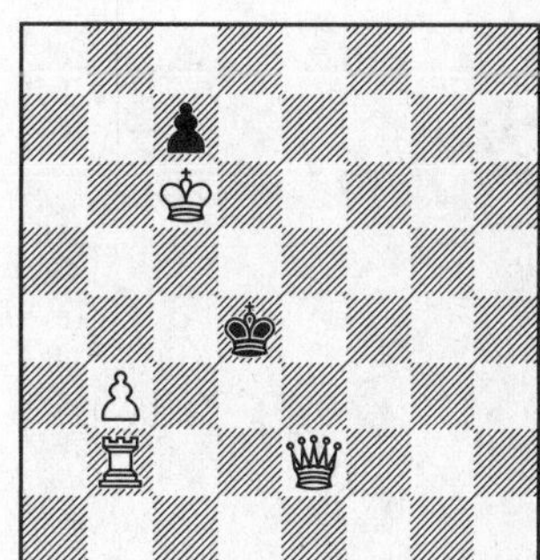

2653.

2654.

2655.

2656.

2657.

2658.

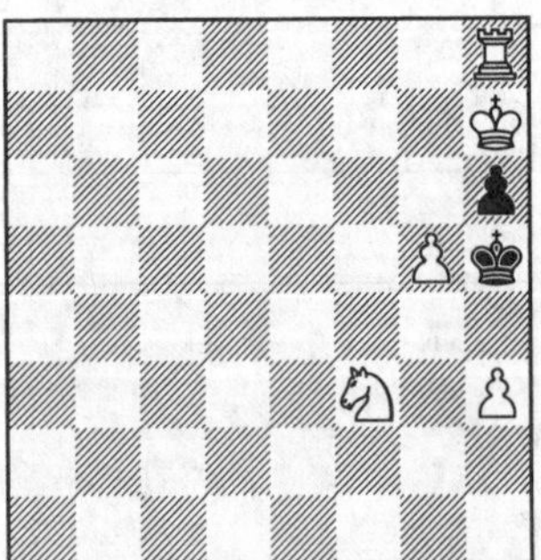

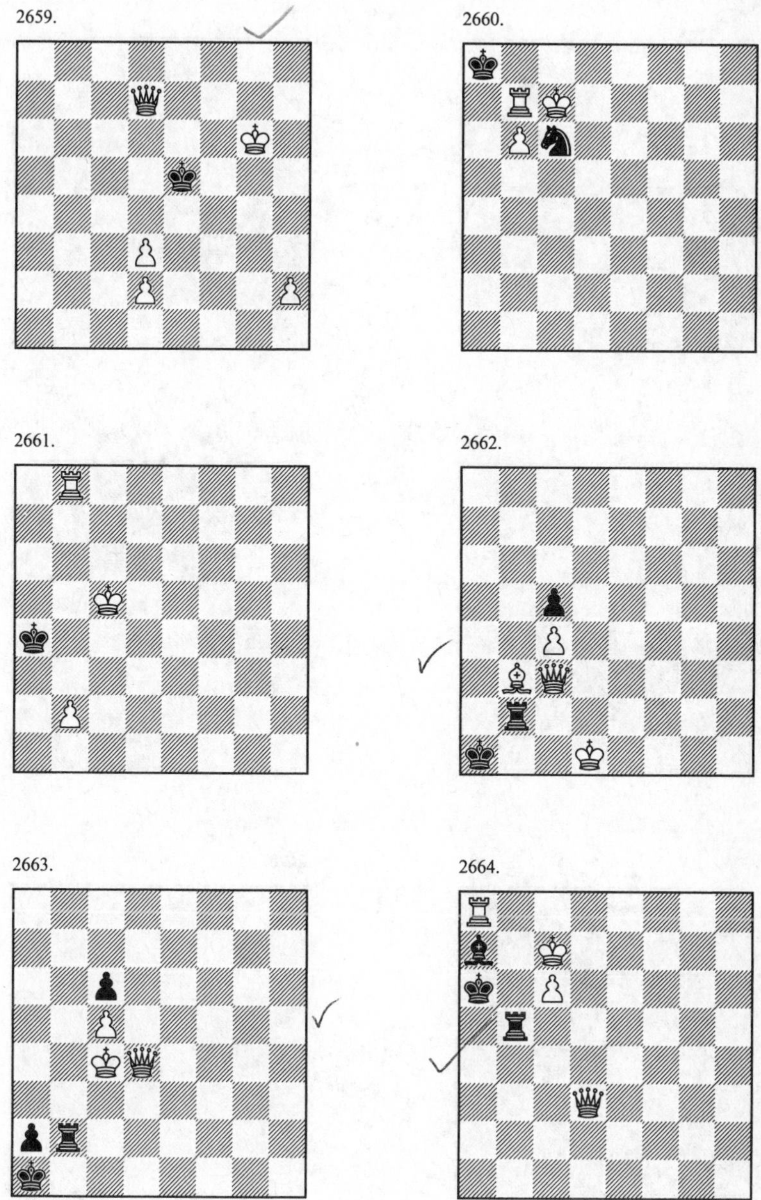

2659.

2660.

2661.

2662.

2663.

2664.

2665.

2666.

2667.

2668.

2669.

2670.

2671.

2672.

2673.

2674.

2675.

2676.

2677.

2678.

2679.

2680.

2681.

2682.

2683.

2684.

2685.

2686.

2687.

2688.

2689.

2690.

2691.

2692.

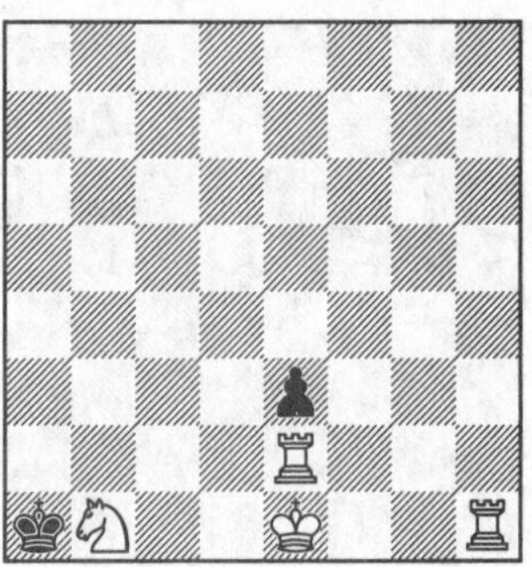

2693.

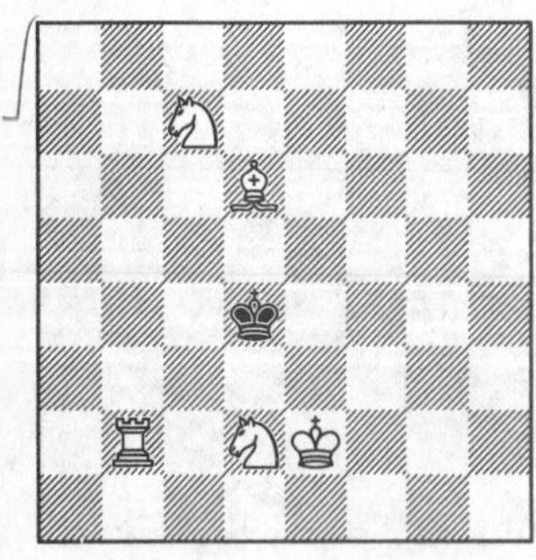

2694.

2695.

2696.

2697.

2698.

2699.

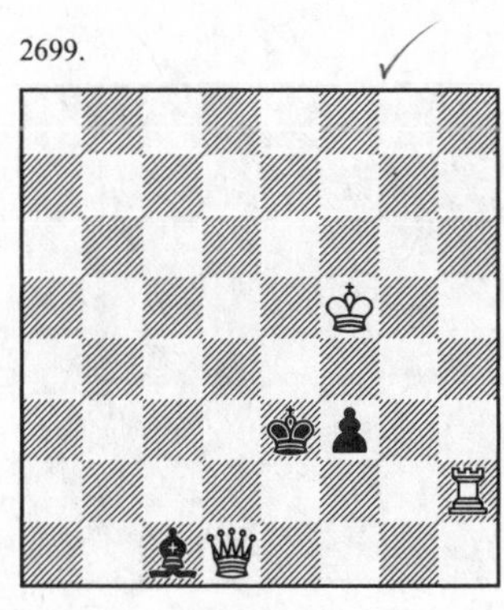

2700.

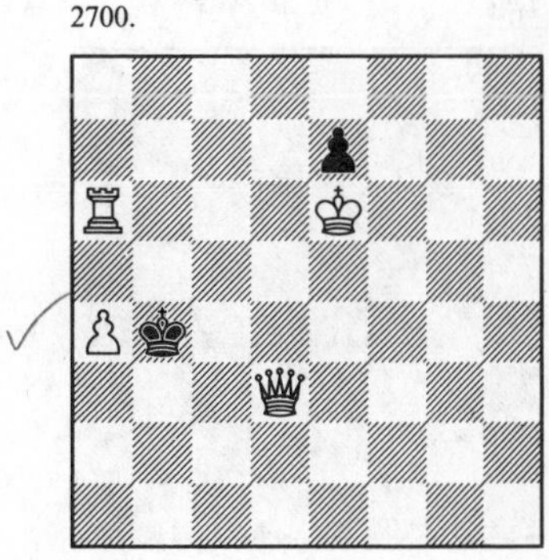

2701.

2702.

2703.

2704.

2705.

2706.

2707.

2708.

2709.

2710.

2711.

2712.

2713.

2714.

2715.

2716.

2717.

2718.

2719.

2720.

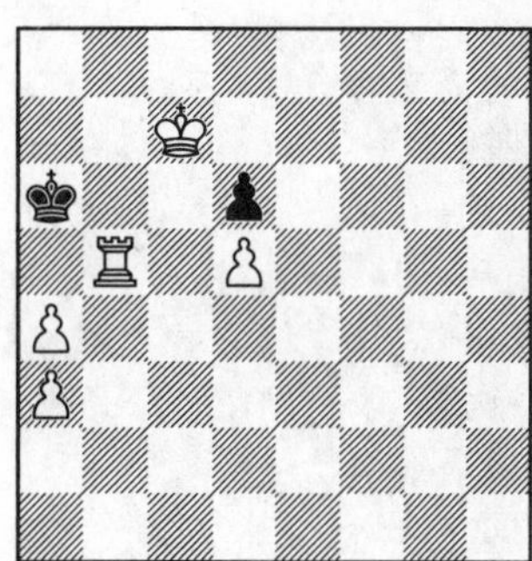

2721.

2722.

2723.

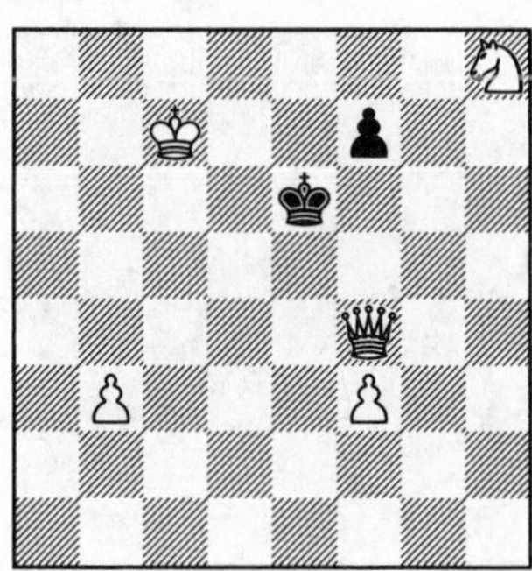

2724.

2725.

2726.

2727.

2728.

2729.

2730.

2731.

2732.

2733.

2734.

2735.

2736.

2737.

2738.

2739.

2740.

2741.

2742.

2743.

2744.

2745.

2746.

2747.

2748.

2749.

2750.

2751.

2752.

2753.

2754.

2755.

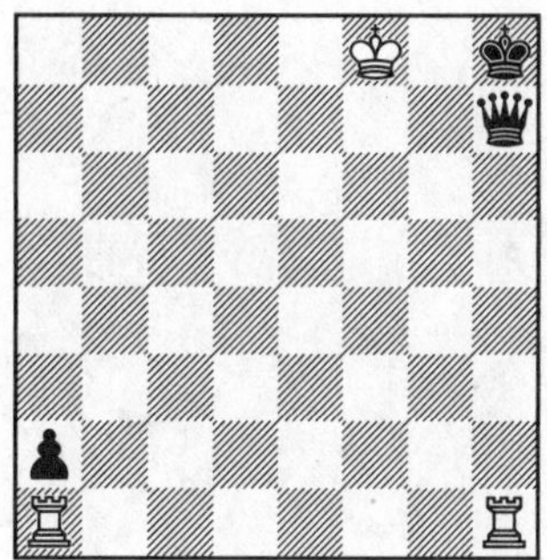

2756.

2757.

2758.

2759.

2760.

2761.

2762.

2763.

2764.

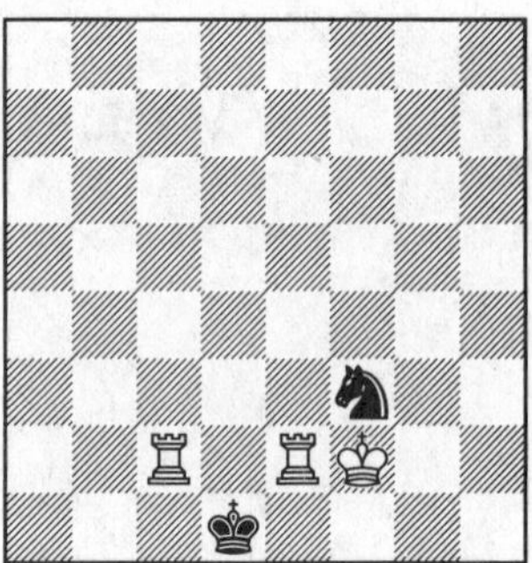

2765.

2766.

2767.

2768.

2769.

2770.

2771.

2772.

2773.

2774.

2775.

2776.

2777.

2778.

2779.

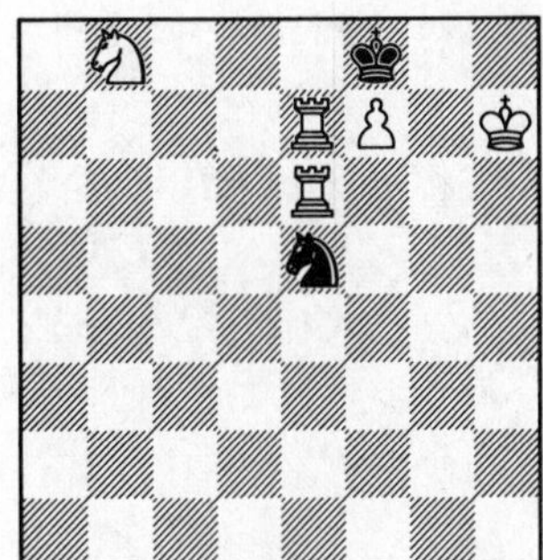

2780.

2781.

2782.

2783.

2784.

2785.

2786.

2787.

2788.

2789.

2790.

2791.

2792.

2793.

2794.

2795.

2796.

2797.

2798.

2799.

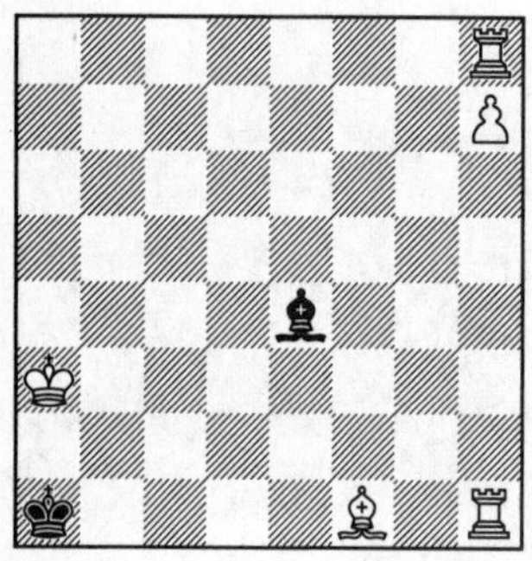

2800.

2801.

2802.

2803.

2804.

2805.

2806.

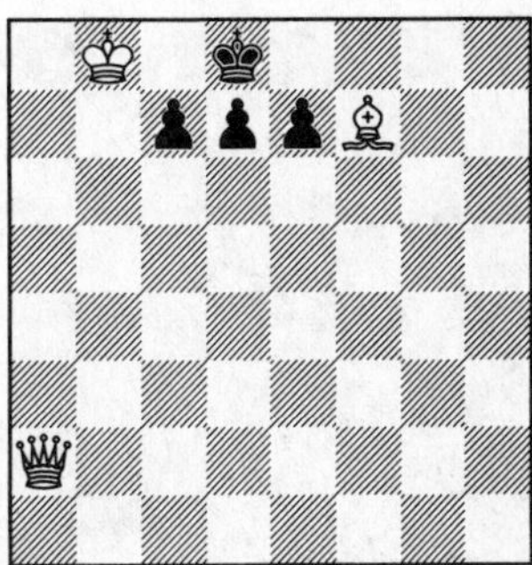

2807.

2808.

2809.

2810.

2811.

2812.

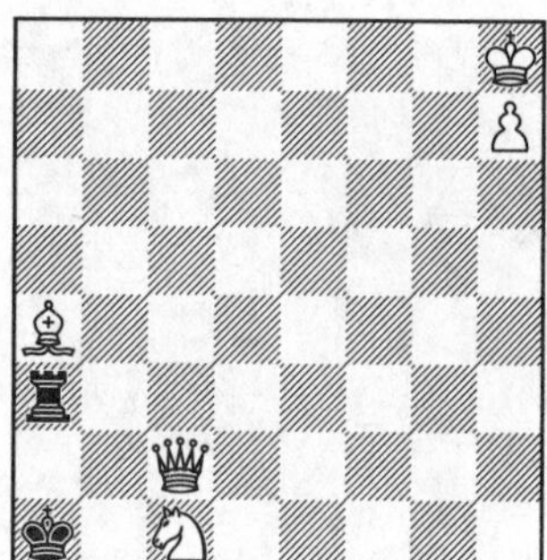

2813.

2814.

2815.

2816.

2817.

2818.

2819.

2820.

2821.

2822.

2823.

2824.

2825.

2826.

2827.

2828.

2829.

2830.

2831.

2832.

2833.

2834.

2835.

2836.

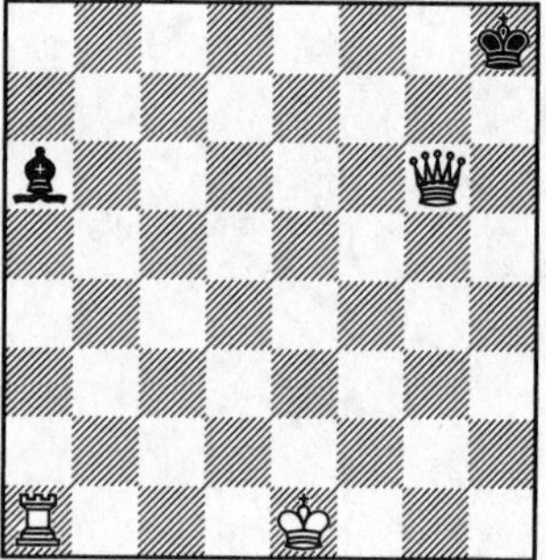

2837.

2838.

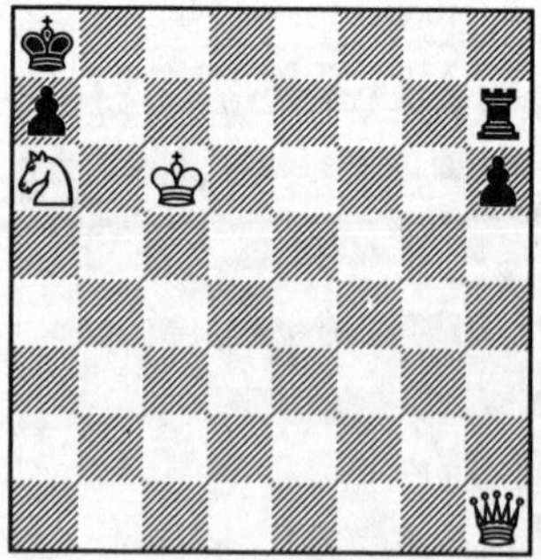

2839.

2840.

2841.

2842.

2843.

2844.

2845.

2846.

2847.

2848.

2849.

2850.

2851.

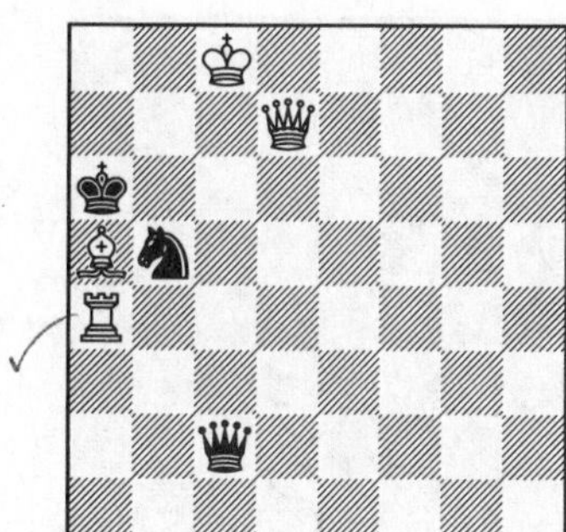

2852.

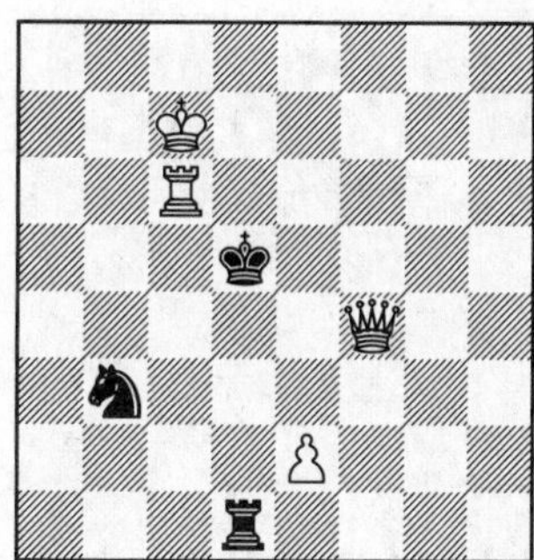

2853.

2854.

2855.

2856.

2857.

2858.

2859.

2860.

2861.

2862.

2863.

2864.

2865.

2866.

2867.

2868.

2869.

2870.

2871.

2872.

2873.

2874.

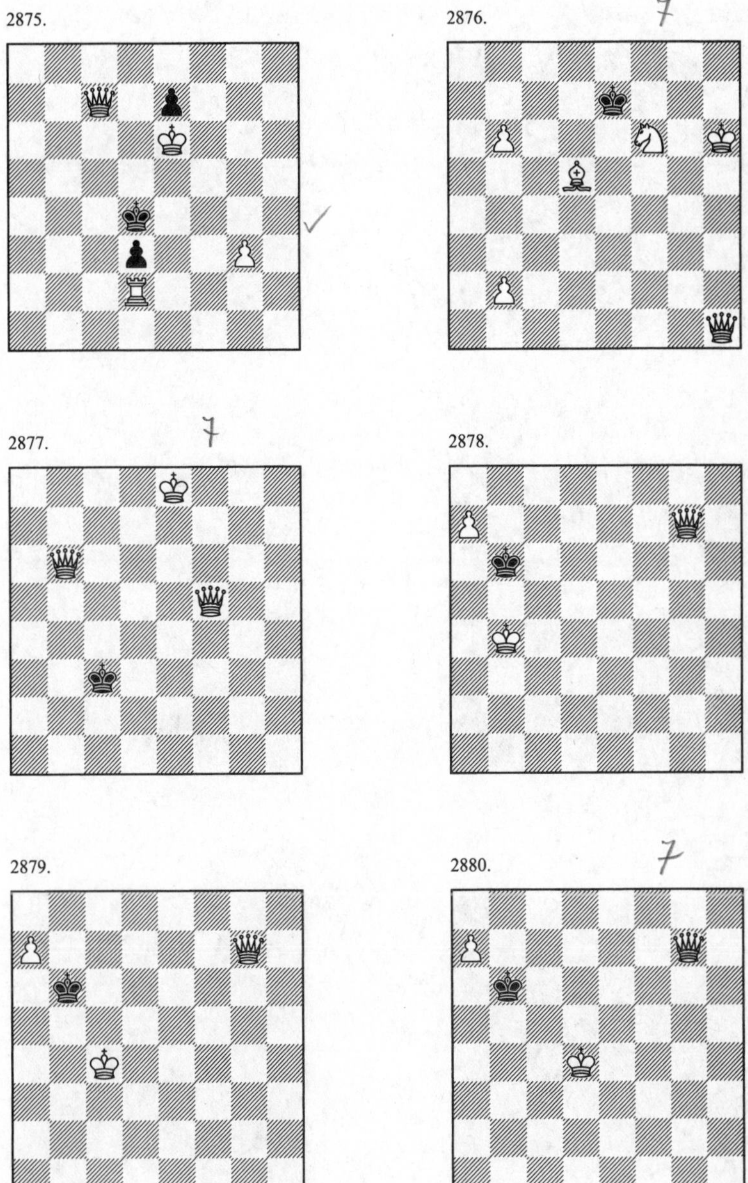
2875.
2876.
2877.
2878.
2879.
2880.

2881.

2882.

2883.

2884.

2885.

2886.

3007.

3008.

3009.

3010.

3011.

3012.

2893.

2894.

2895.

2896.

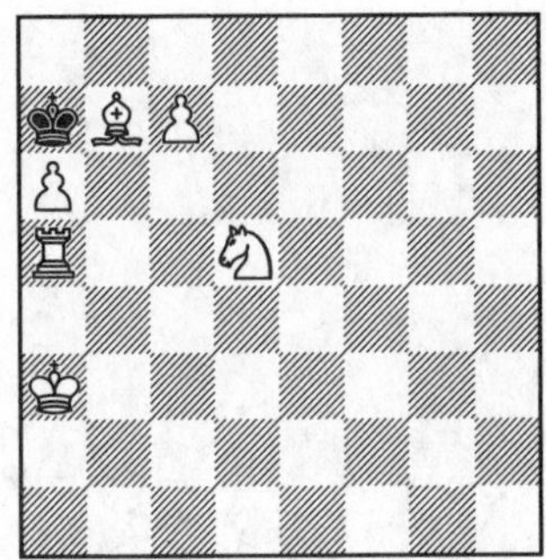

2897.

2898.

2899.

2900.

2901.

2902.

2903.

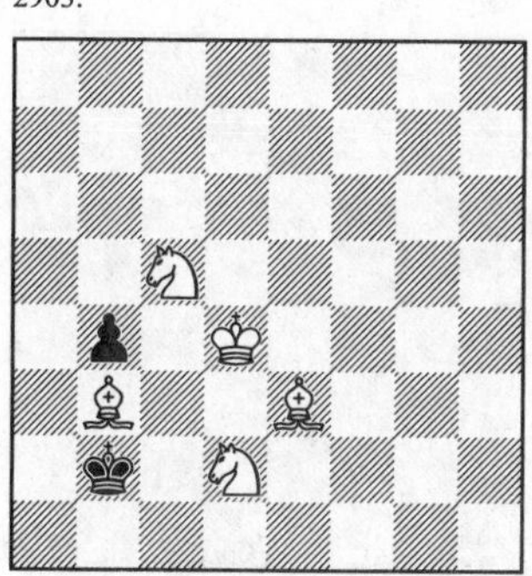

2904.

2905.

2906.

2907.

2908.

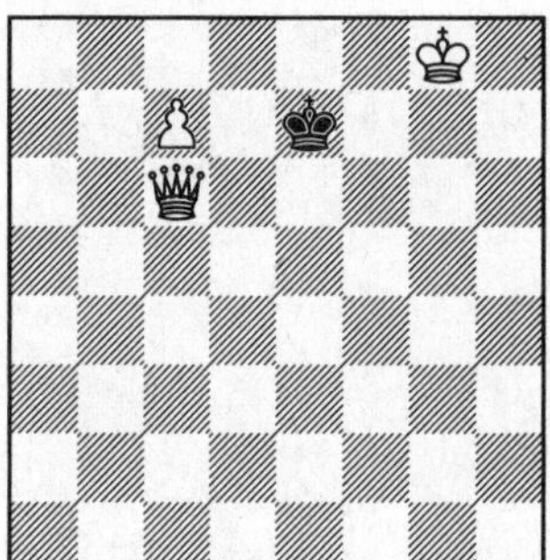

2909.

2910.

2911.

2912.

2913.

2914.

2915.

2916.

2917.

2918.

2919.

2920.

2921.

2922.

2923.

2924.

2925.

2926.

2927.

2928.

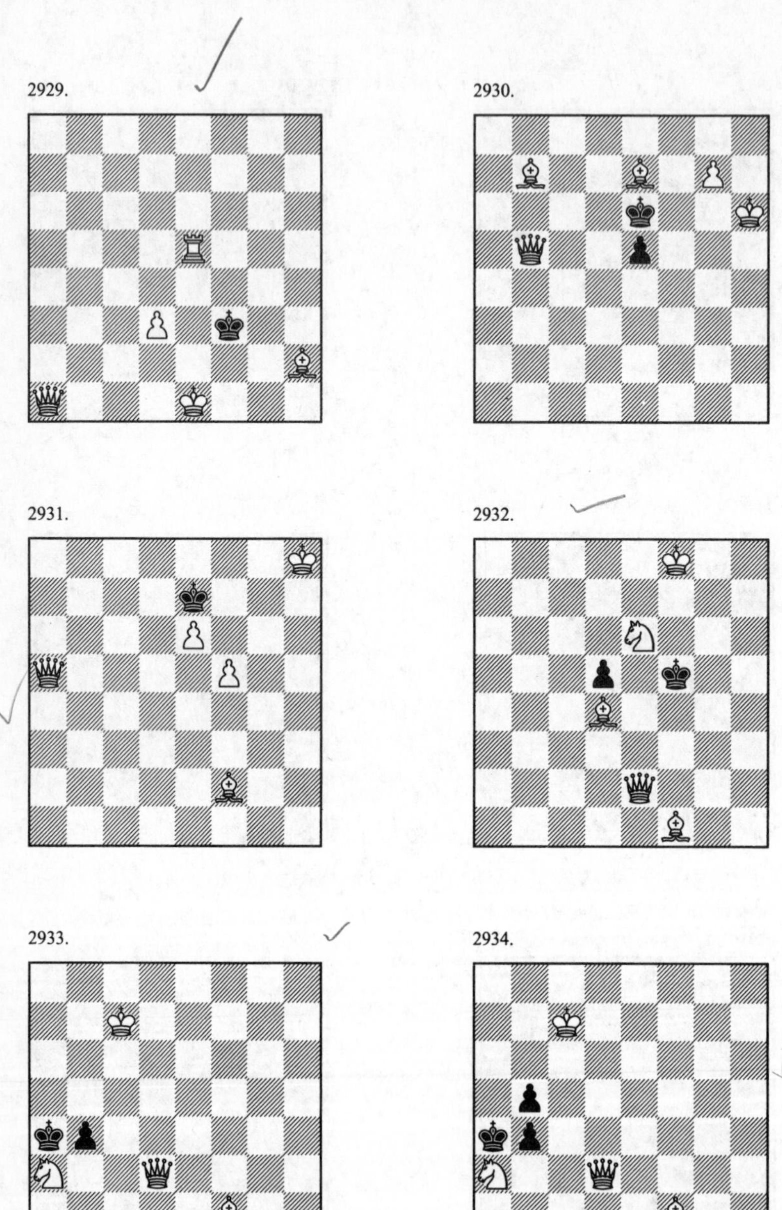
2929.
2930.
2931.
2932.
2933.
2934.

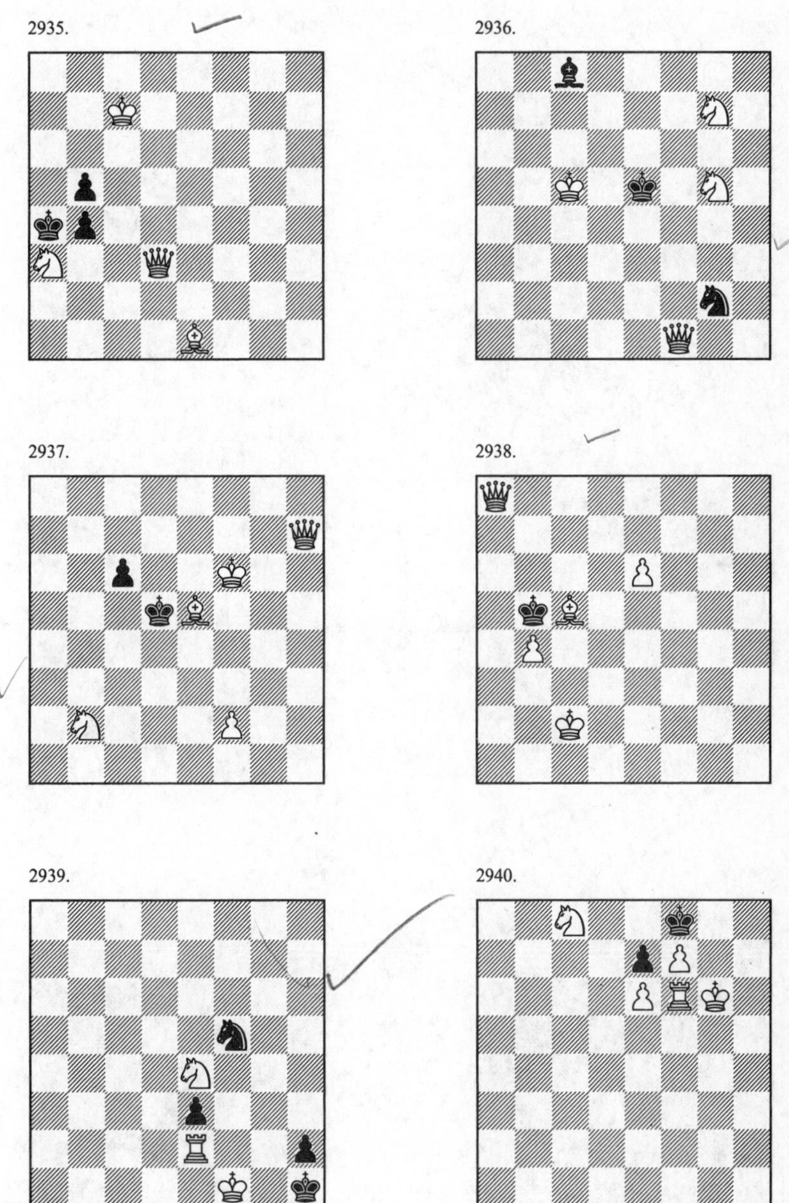
2935.
2936.
2937.
2938.
2939.
2940.

2941.

2942.

2943.

2944.

2945.

2946.

2947.

2948.

2949.

2950.

2951.

2952.

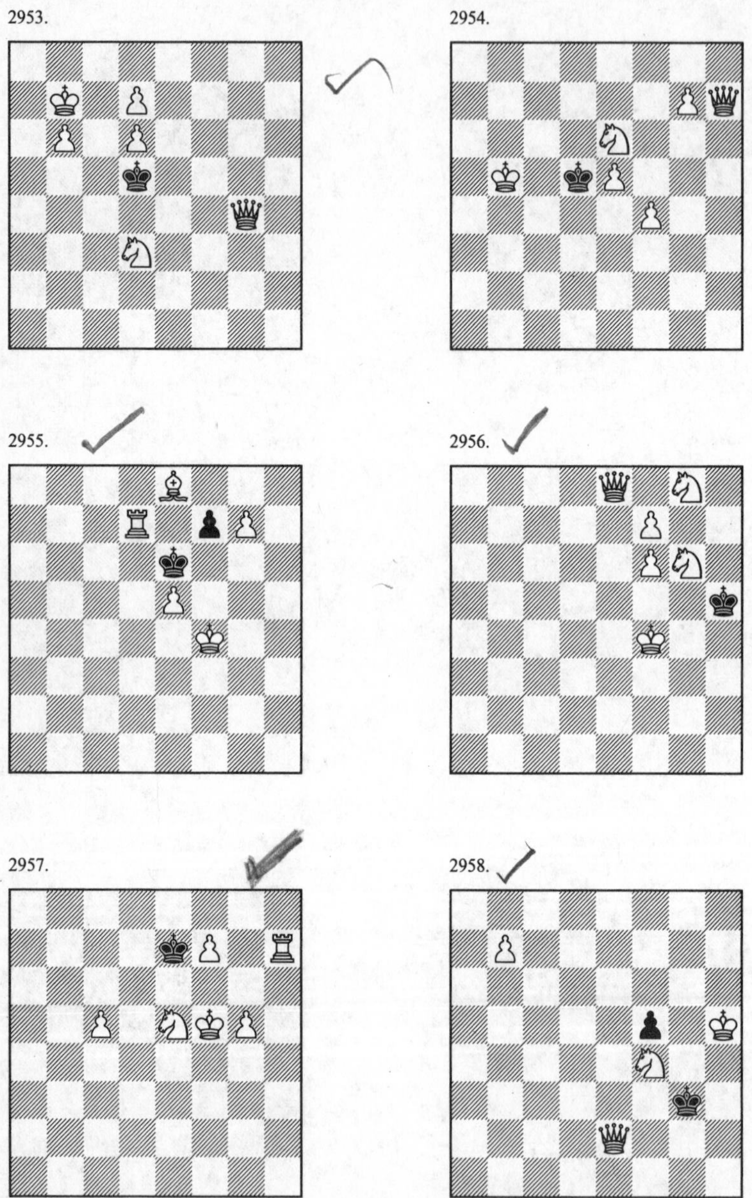

2953.
2954.
2955.
2956.
2957.
2958.

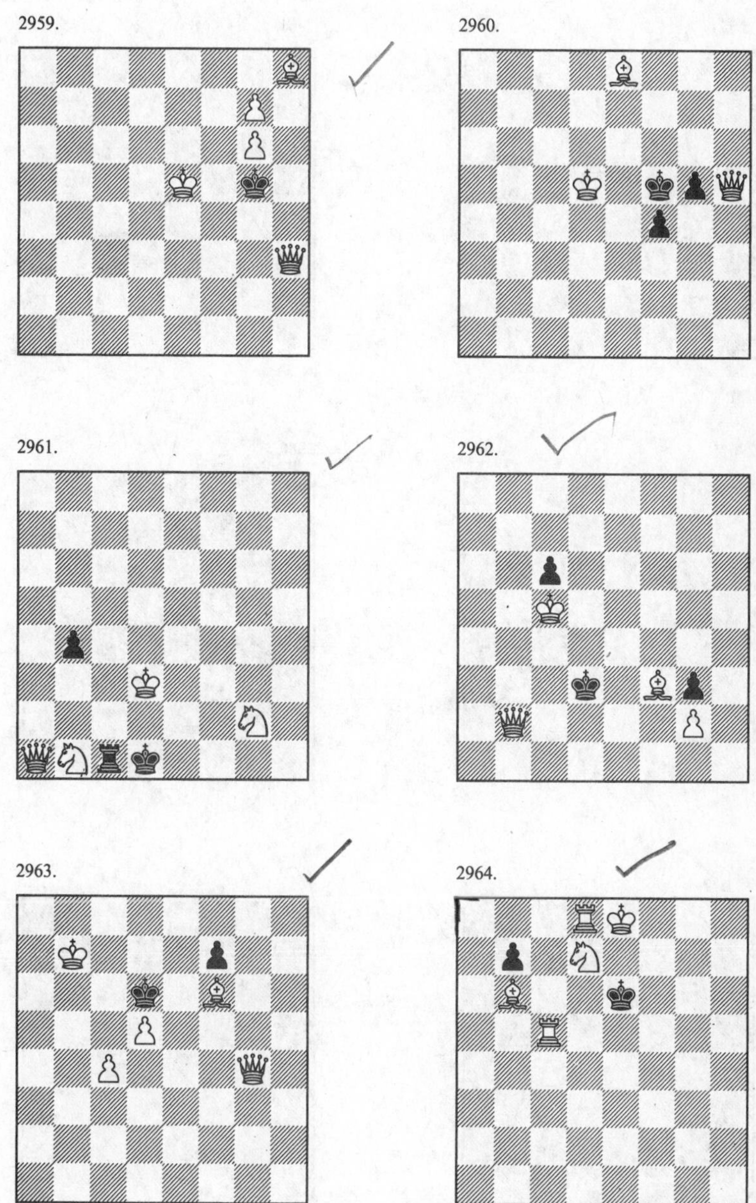

2959.

2960.

2961.

2962.

2963.

2964.

2965.

2966.

2967.

2968.

2969.

2970.

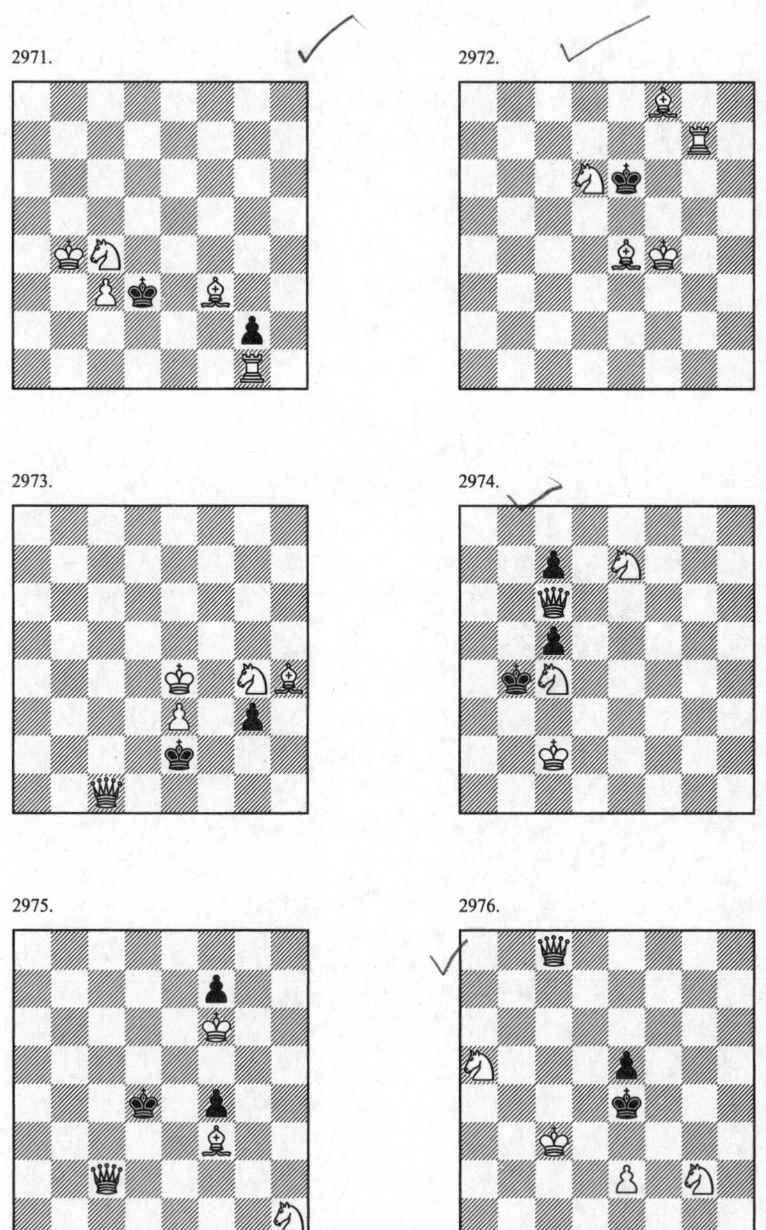
2971.

2972.

2973.

2974.

2975.

2976.

2977.

2978.

2979.

2980.

2981.

2982.

2983.

2984.

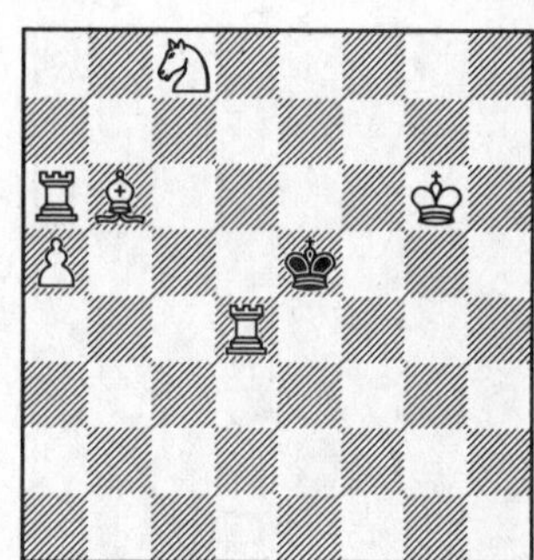

2985.

2986.

2987.

2988.

2989.

2990.

2991.

2992.

2993.

2994.

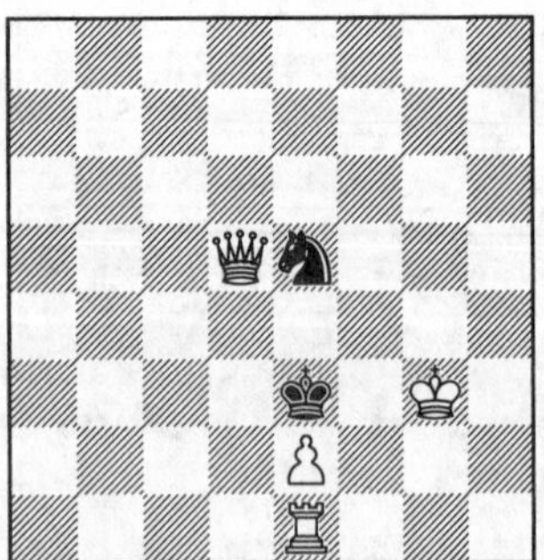

2995.

2996.

2997.

2998.

2999.

3000.

3001.

3002.

3003.

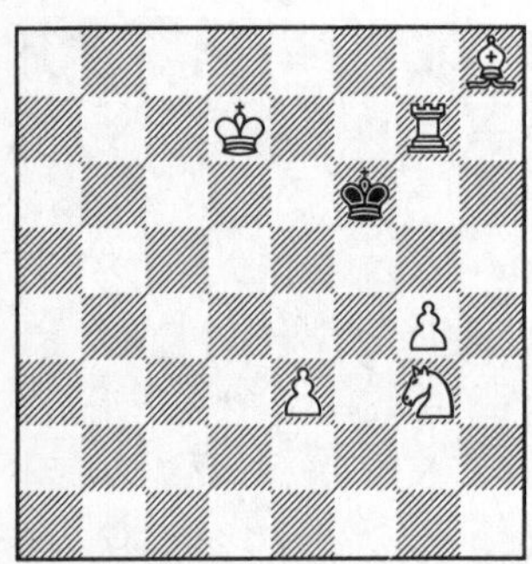

3004.

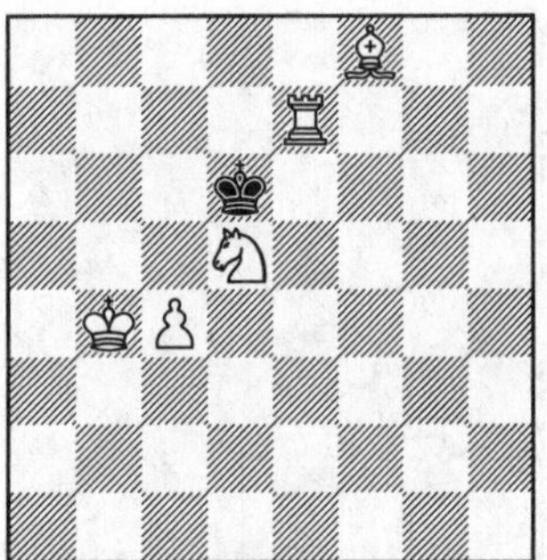

3005.

3006.

3007.

3008.

3009.

3010.

3011.

3012.

3013.

3014.

3015.

3016.

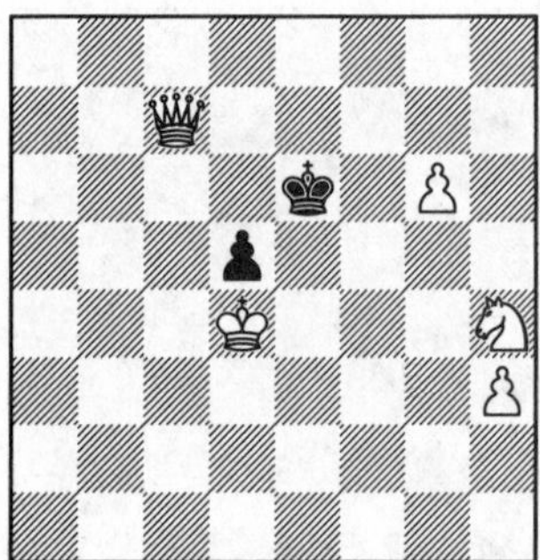

3017.

3018.

3019.

3020.

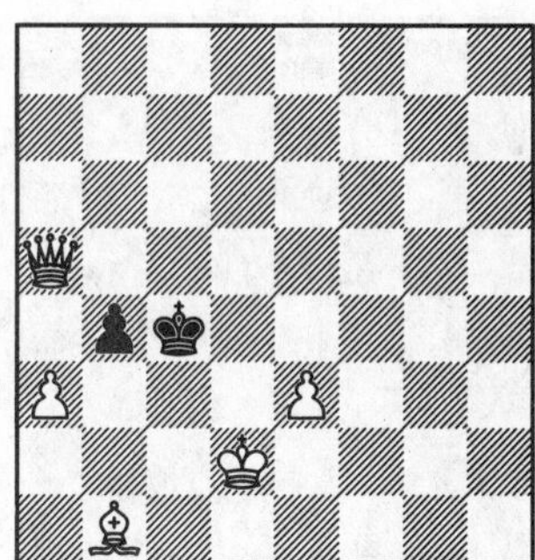

3021.

3022.

3023.

3024.

3025.

3026.

3027.

3028.

3029.

3030.

3031.

3032.

3033.

3034.

3035.

3036.

3037.

3038.

3039.

3040.

3041.

3042.

3043.

3044.

3045.

3046.

3047.

3048.

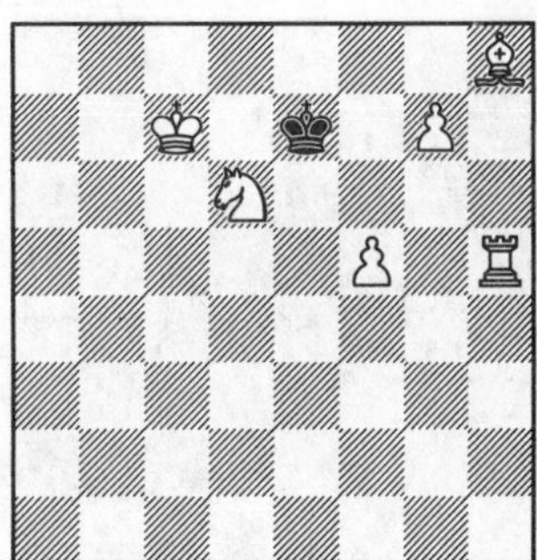

3049.

3050.

3051.

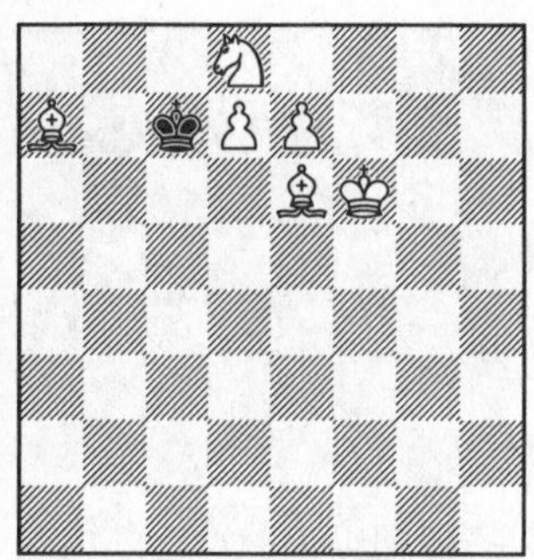

3052.

3053.

3054.

3055.

3056.

3057.

3058.

3059.

3060.

3061.

3062.

3063.

3064.

3065.

3066.

3067.

3068.

3069.

3070.

3071.

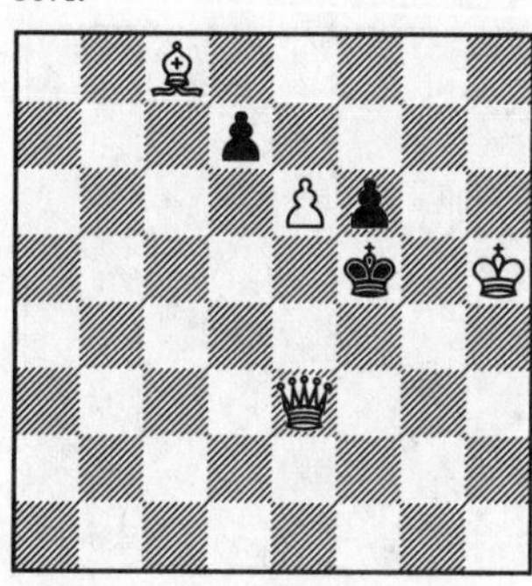

3072.

3073.

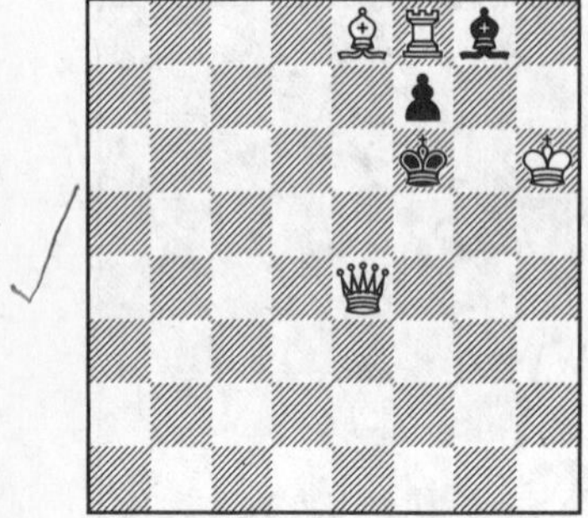

3074.

3075.

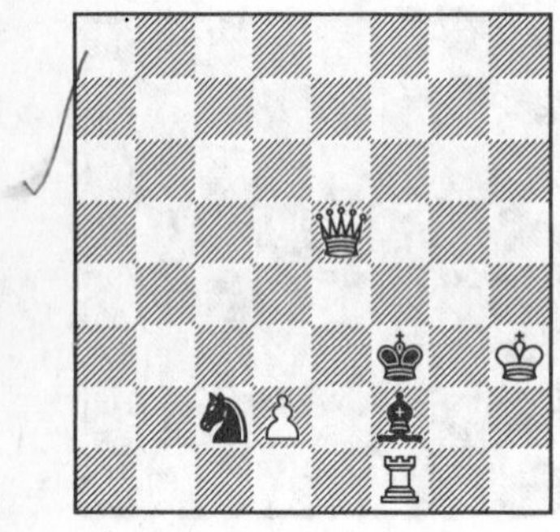

3076.

3077.

3078.

3079.

3080.

3081.

3082.

3083.

3084.

3085.

3086.

3087.

3088.

3089.

3090.

3091.

3092.

3093.

3094.

3095.

3096.

3097.

3098.

3099.

3100.

.3101.

3102.

3103.

3104.

3105.

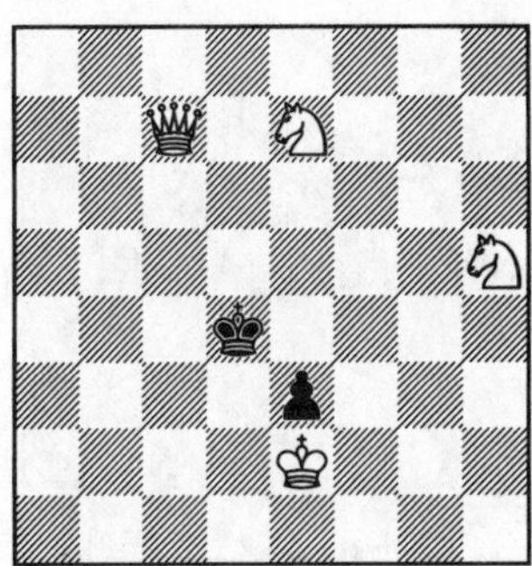

3106.

3107.

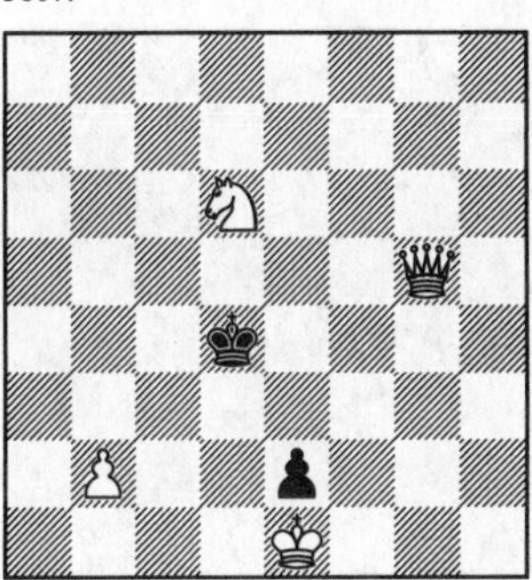

3108.

3109.

3110.

3111.

3112.

3113.

3114.

3115.

3116.

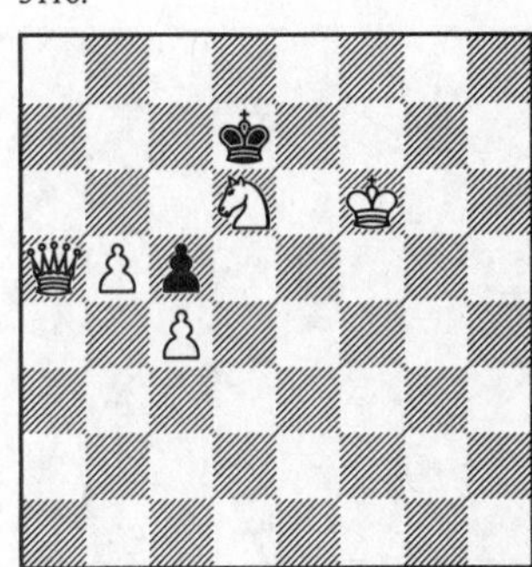

3117.

3118.

3119.

3120.

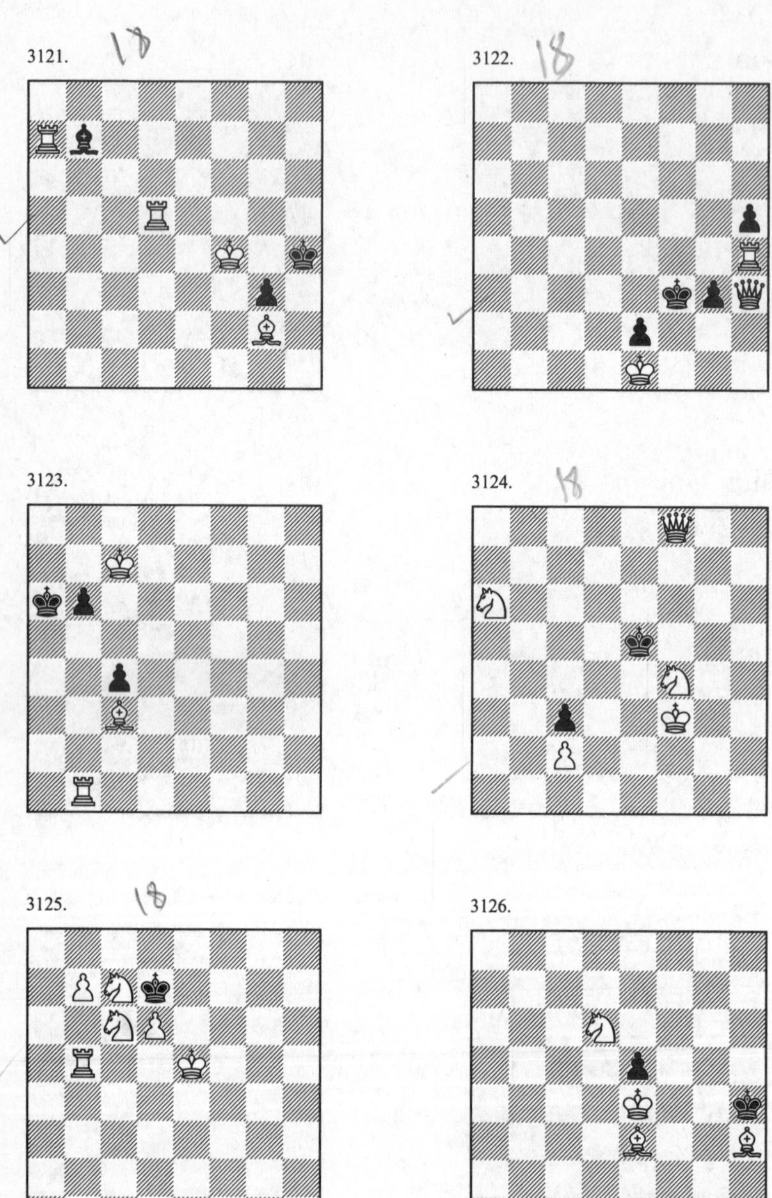
3121.
3122.
3123.
3124.
3125.
3126.

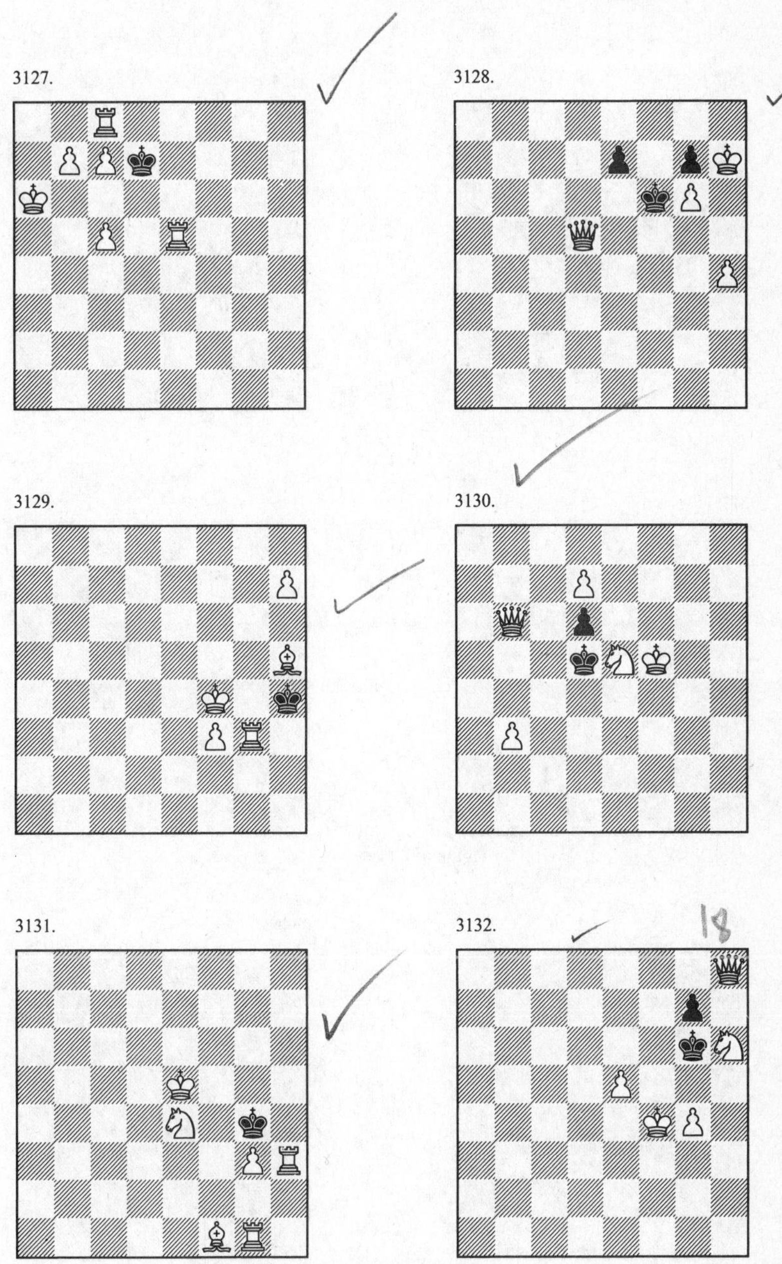
3127.
3128.
3129.
3130.
3131.
3132.

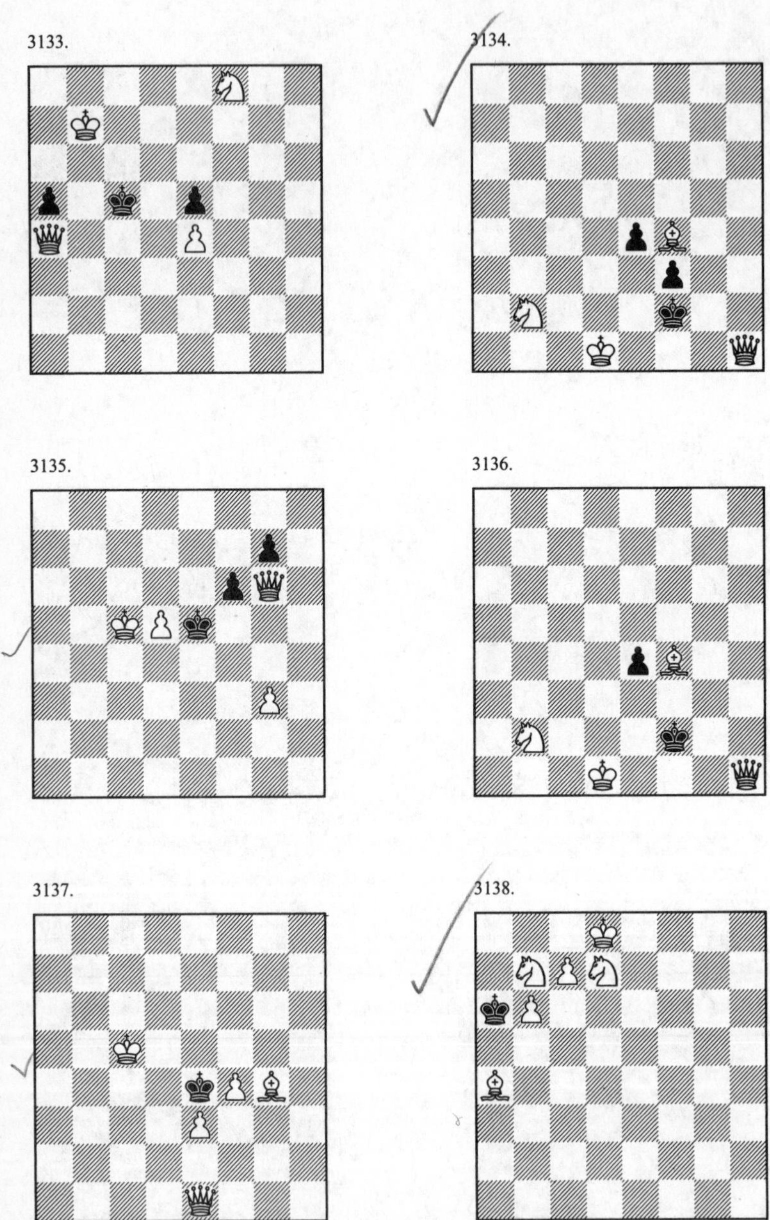
3133.
3134.
3135.
3136.
3137.
3138.

3139.

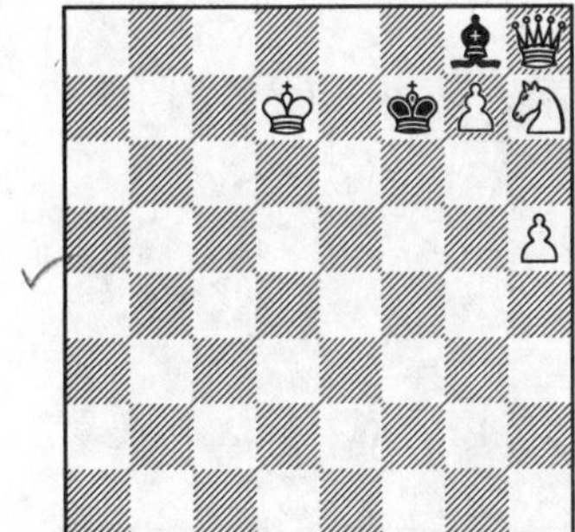

3140.

3141.

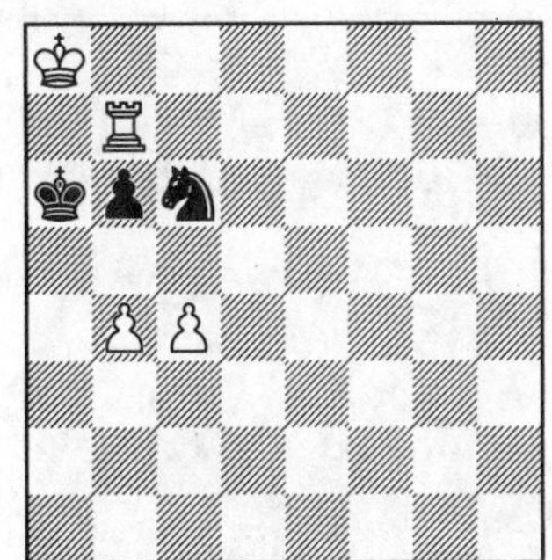

3142.

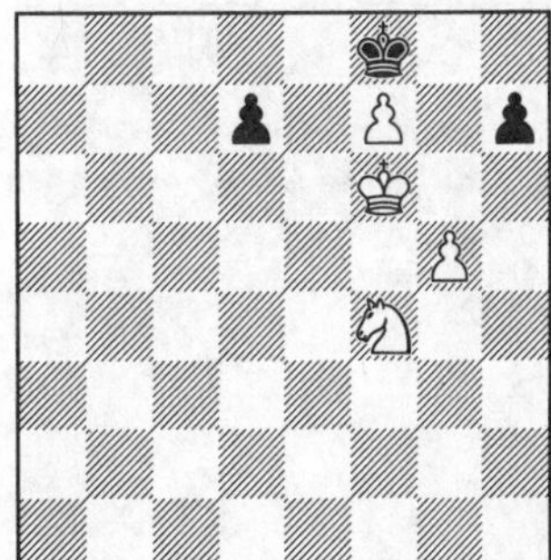

3143.

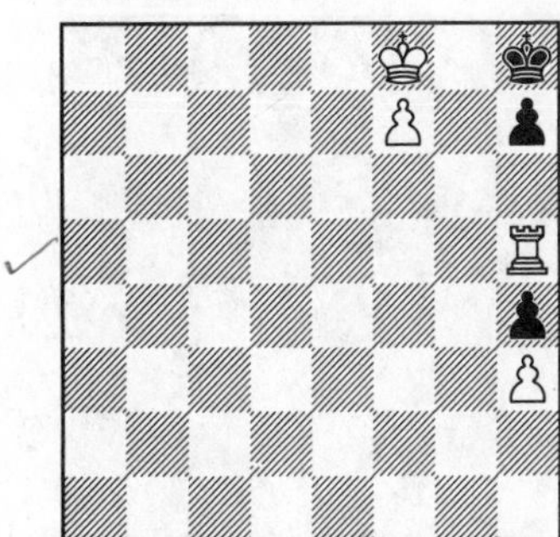

3144.

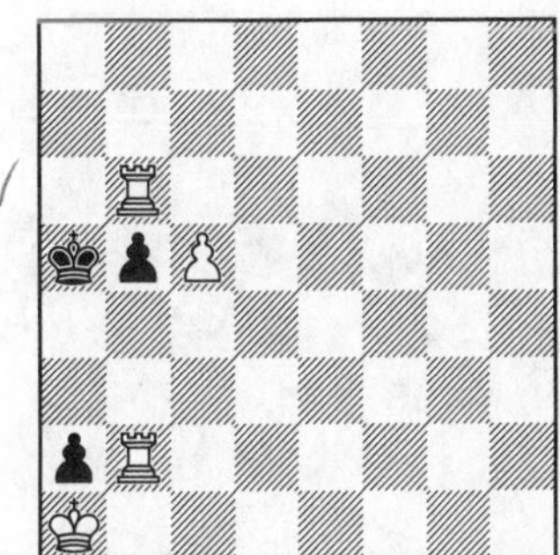

3145.

3146.

3147.

3148.

3149.

3150.

3151.

3152.

3153.

3154.

3155.

3156.

3157.

3158.

3159.

3160.

3161.

3162.

3163.

3164.

3165.

3166.

3167.

3168.

3169.

3170.

3171.

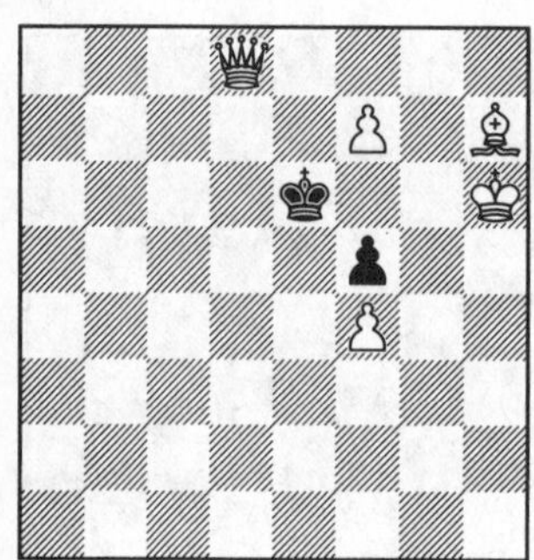

3172.

3173.

3174.

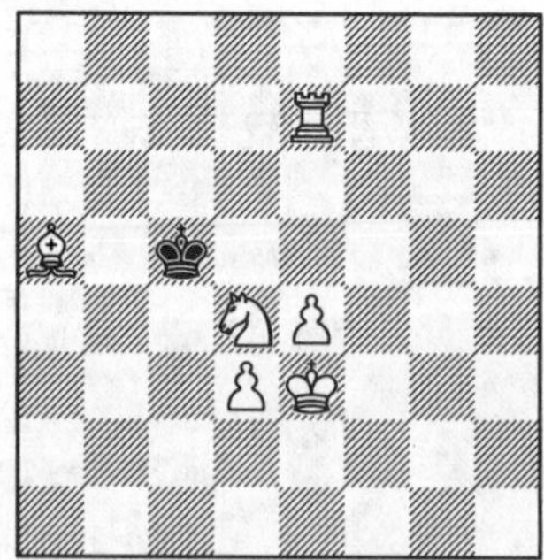

3175.

3176.

3177.

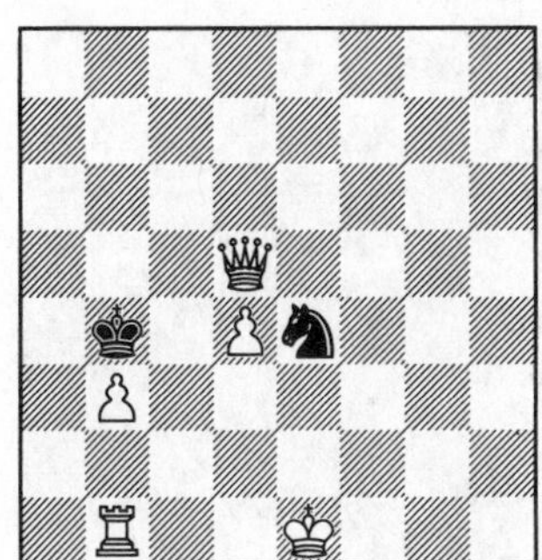

3178.

3179.

3180.

3181.

3182.

3183.

3184.

3185.

3186.

3187.

3188.

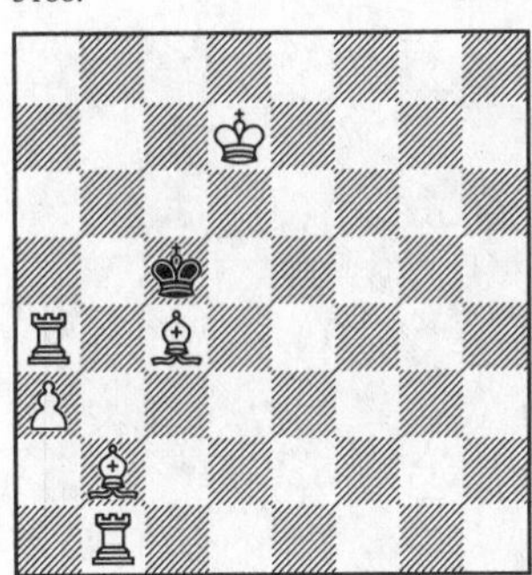

3189.

3190.

3191.

3192.

3193.

3194.

3195.

3196.

3197.

3198.

3199.

3200.

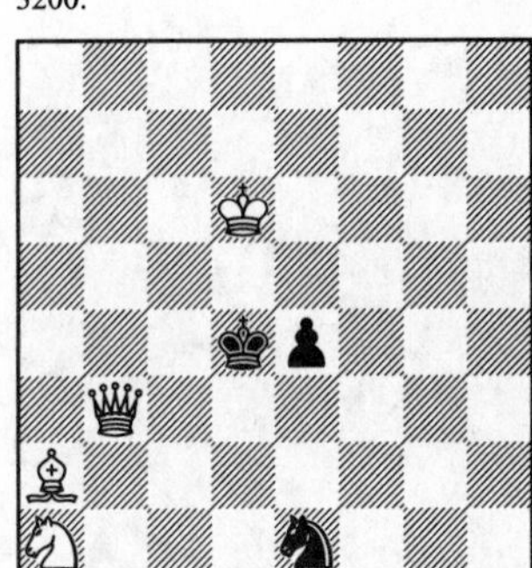

3201.

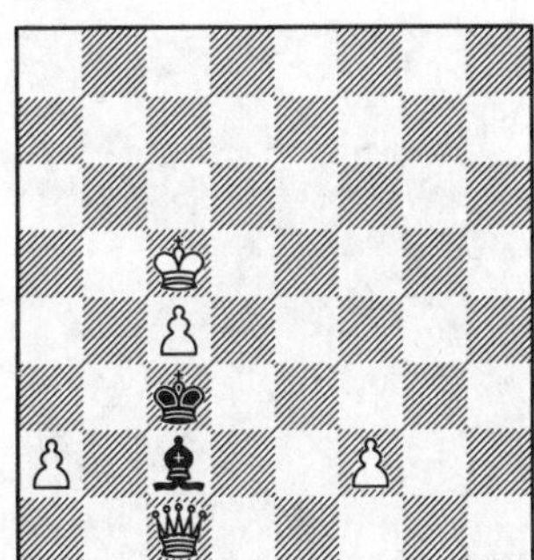

3202.

3203.

3204.

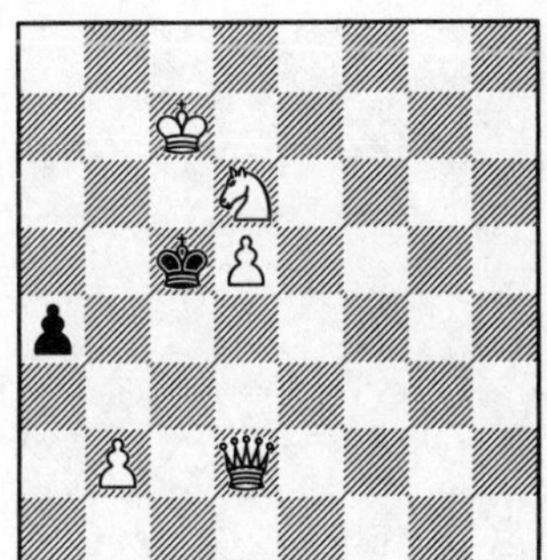

3205.

3206.

3207.

3208.

3209.

3210.

3211.

3212.

3213.

3214.

3215.

3216.

3217.

3218.

3219.

3220.

3221.

3222.

3223.

3224.

3225.

3226.

3227.

3228.

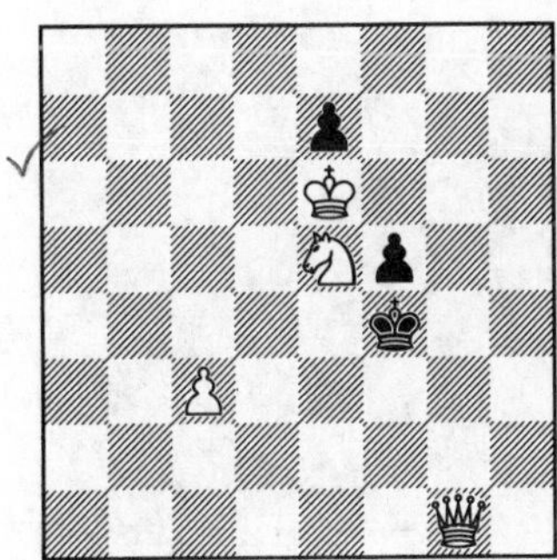

3229.

3230.

3231.

3232.

3233.

3234.

3235.

3236.

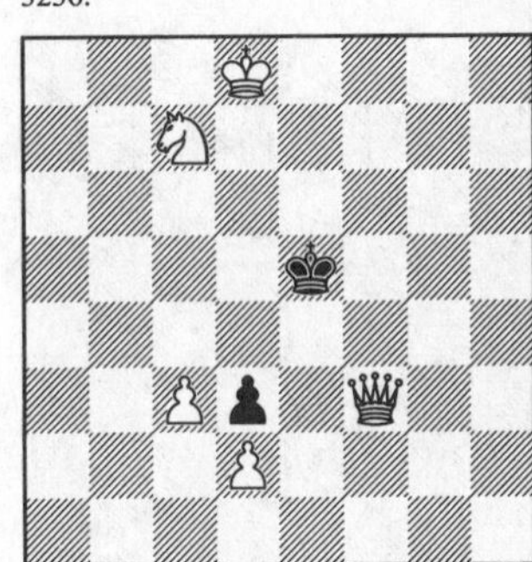

3237.

3238.

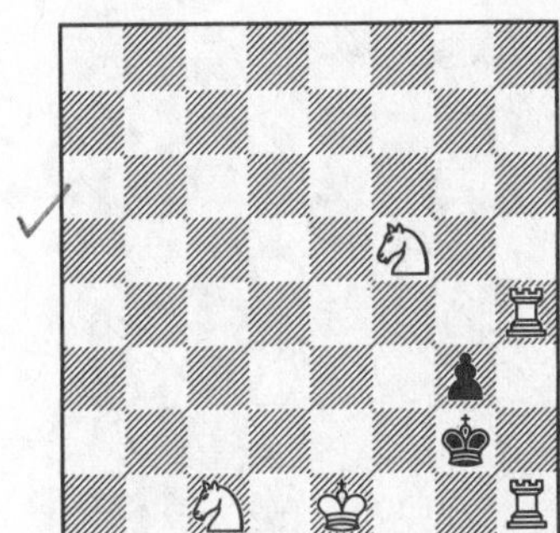

3239.

3240.

3241.

3242.

3243.

3244.

3245.

3246.

3247.

3248.

3249.

3250.

3251.

3252.

3253.

3254.

3255.

3256.

3257.

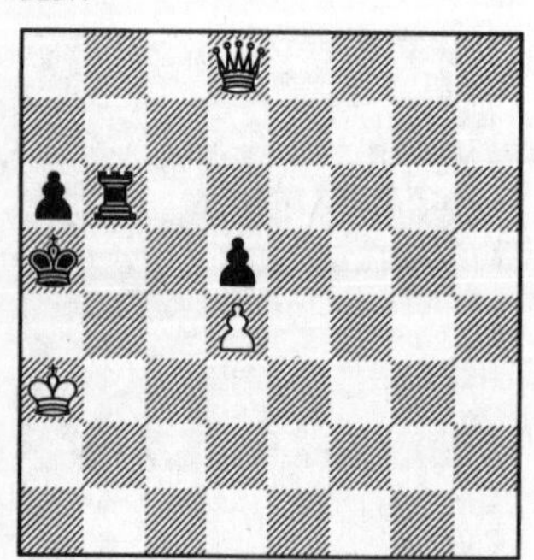

3258.

3259.

3260.

3261.

3262.

3263.

3264.

3265.

3266.

3267.

3268.

3269.

3270.

3271.

3272.

3273.

3274.

3275.

3276.

3277.

3278.

3279.

3280.

3281.

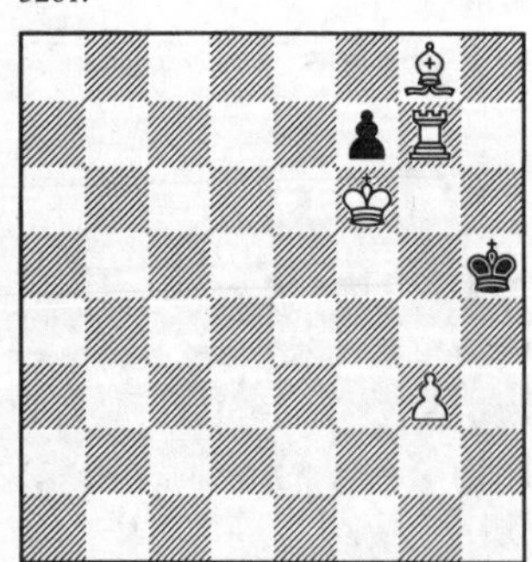

3282.

3283.

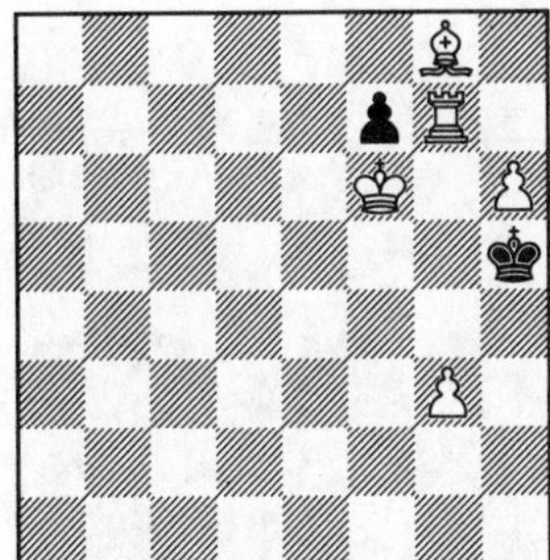

3284.

3285.

3286.

3287.

3288.

3289.

3290.

3291.

3292.

3293.

3294.

3295.

3296.

3297.

3298.

3299.

3300.

3301.

3302.

3303.

3304.

3305.

3306.

3307.

3308.

3309.

3310.

3311.

3312.

3313.

3314.

3315.

3316.

3317.

3318.

3319.

3320.

3321.

3322.

3323.

3324.

3325.

3326.

3327.

3328.

3329.

3330.

3331.

3332.

3333.

3334.

3335.

3336.

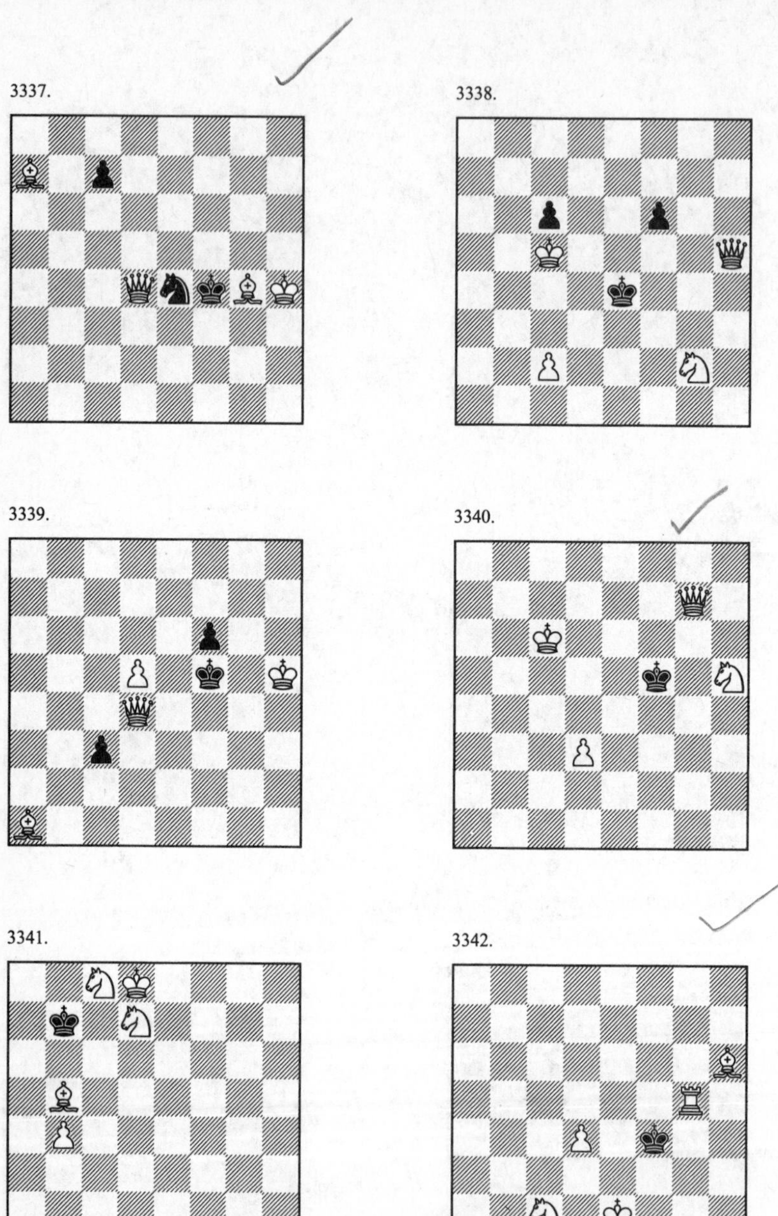
3337.
3338.
3339.
3340.
3341.
3342.

3343.

3344.

3345.

3346.

3347.

3348.

3349.

3350.

3351.

3352.

3353.

3354.

3355.

3356.

3357.

3358.

3359.

3360.

3361.

3362.

3363.

3364.

3365.

3366.

3367.

3368.

3369.

3370.

3371.

3372.

3373.

3374.

3375.

3376.

3377.

3378.

3379.

3380.

3381.

3382.

3383.

3384.

3385.

3386.

3387.

3388.

3389.

3390.

3391.

3392.

3393.

3394.

3395.

3396.

3397.

3398.

3399.

3400.

3401.

3402.

3403.

3404.

3405.

3406.

3407.

3408.

3409.

3410.

3411.

3412.

3413.

3414.

3415.

3416.

3417.

3418.

3419.

3420.

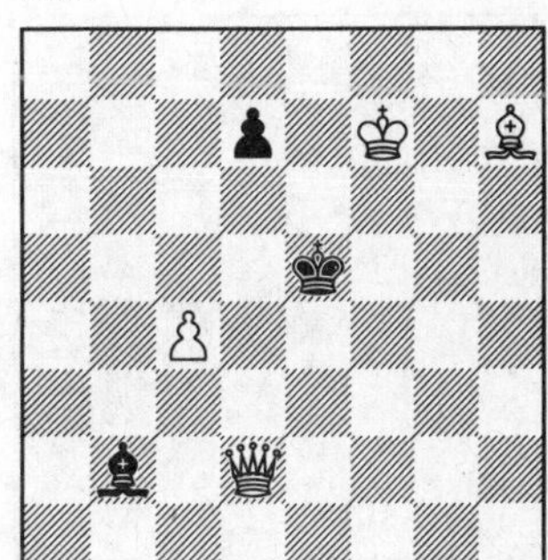

3421.

3422.

3423.

3424.

3425.

3426.

3427.

3428.

3429.

3430.

3431.

3432.

3433.

3434.

3435.

3436.

3437.

3438.

3439.

3440.

3441.

3442.

3443.

3444.

3445.

3446.

3447.

3448.

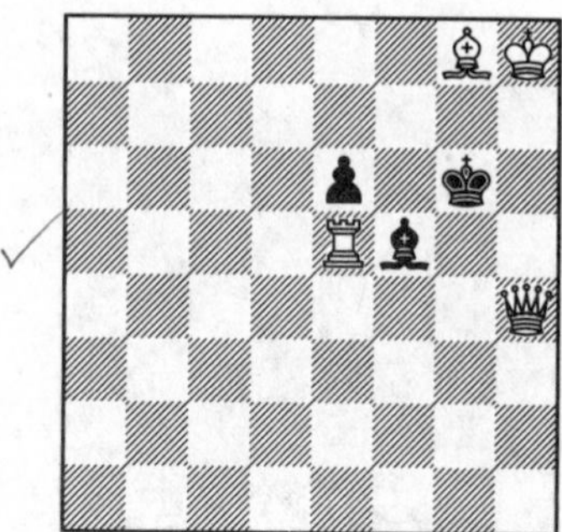

3449.

3450.

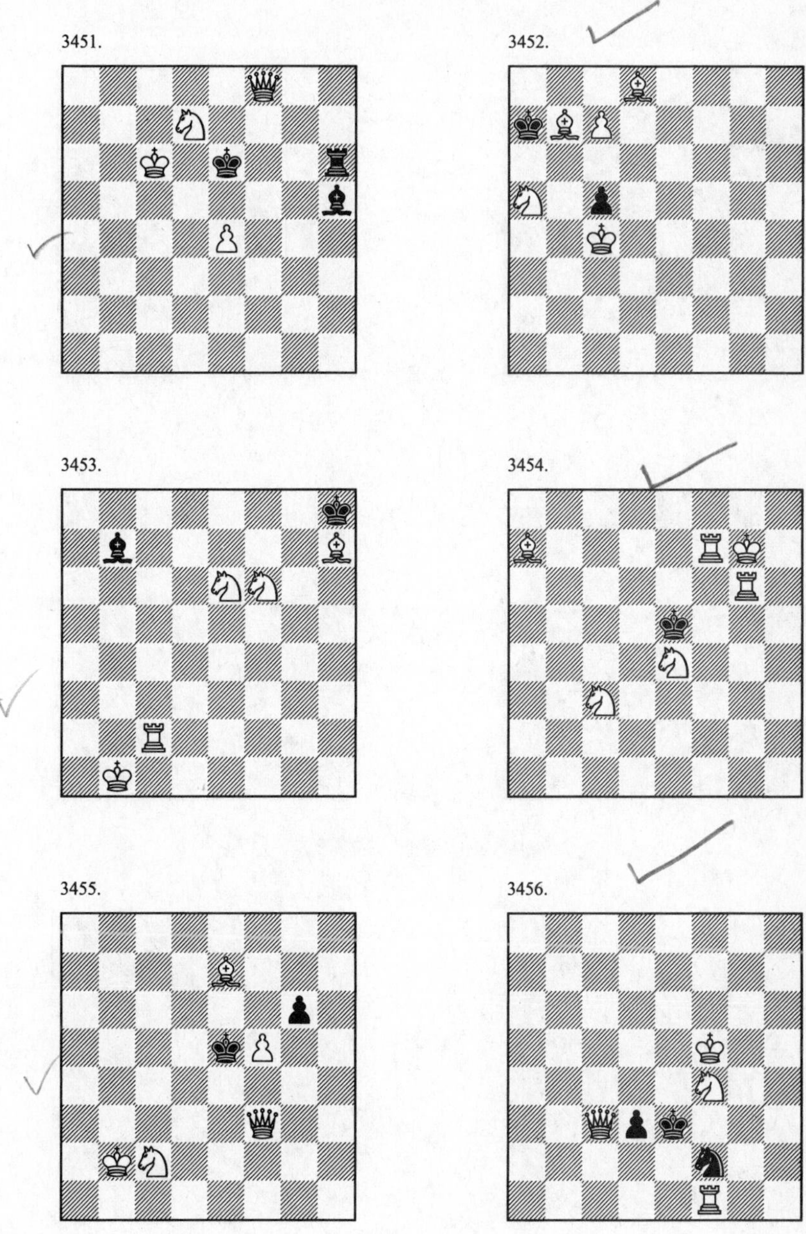

3451.

3452.

3453.

3454.

3455.

3456.

3457.

3458.

3459.

3460.

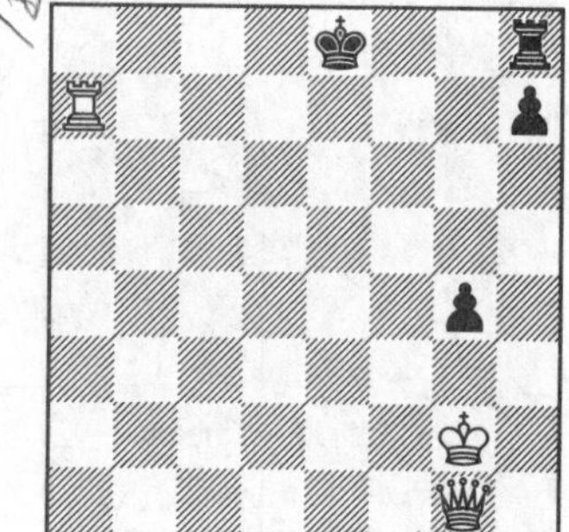

3461.

3462.

3463.

3464.

3465.

3466.

3467.

3468.

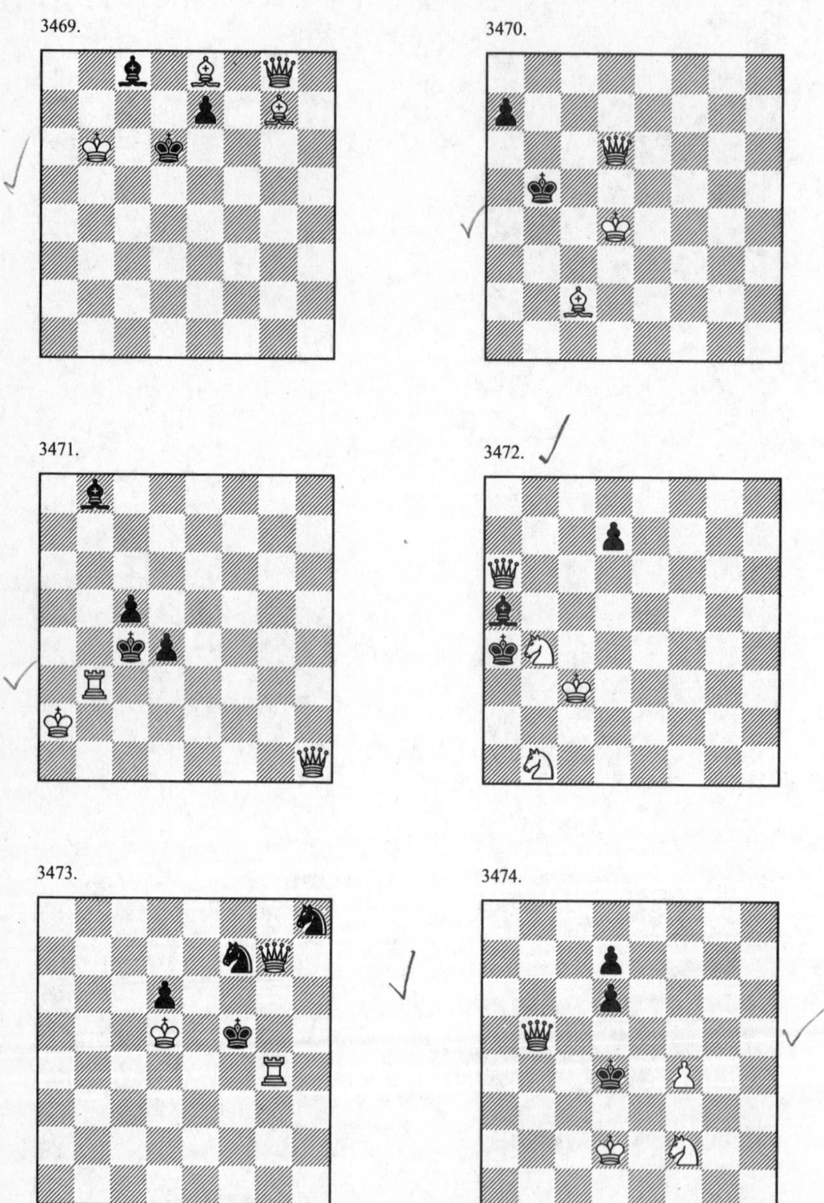
3469.
3470.
3471.
3472.
3473.
3474.

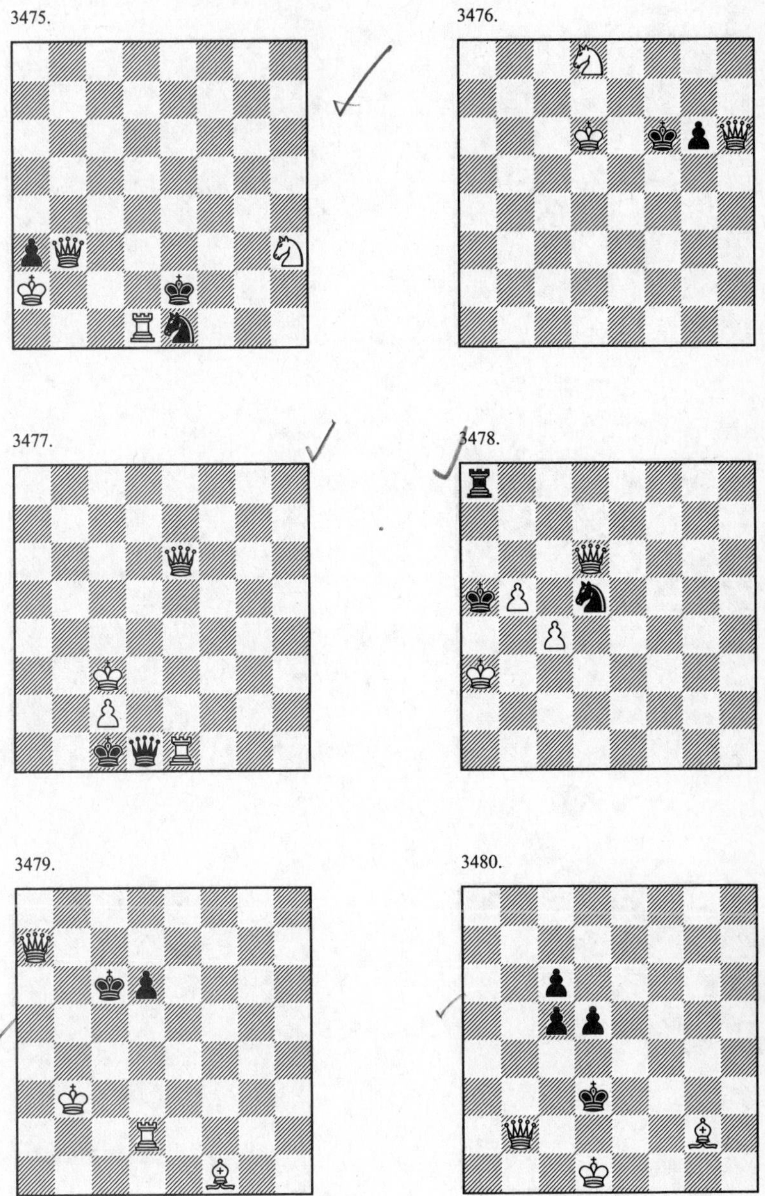
3475.
3476.
3477.
3478.
3479.
3480.

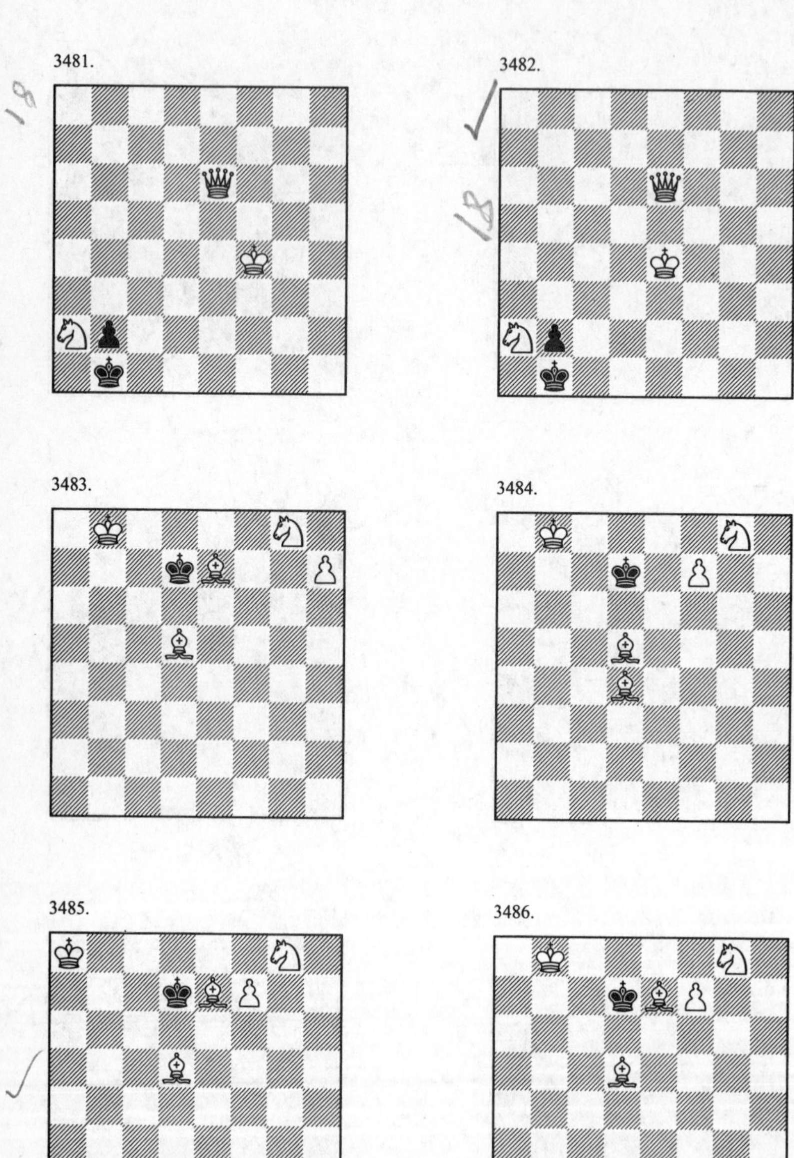

3481.

3482.

3483.

3484.

3485.

3486.

3487.

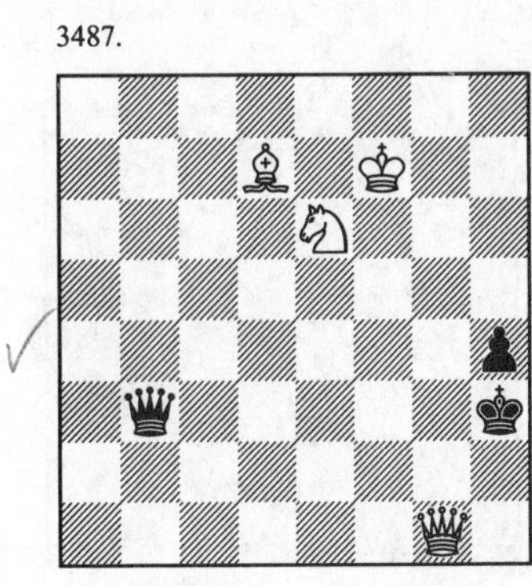

3488.

3489.

3490.

3491.

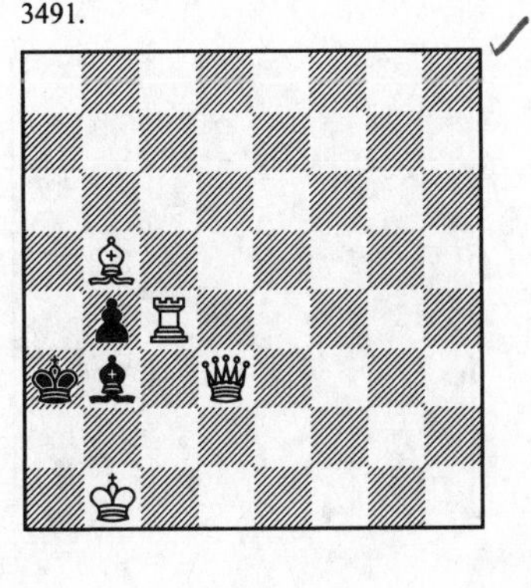

3492.

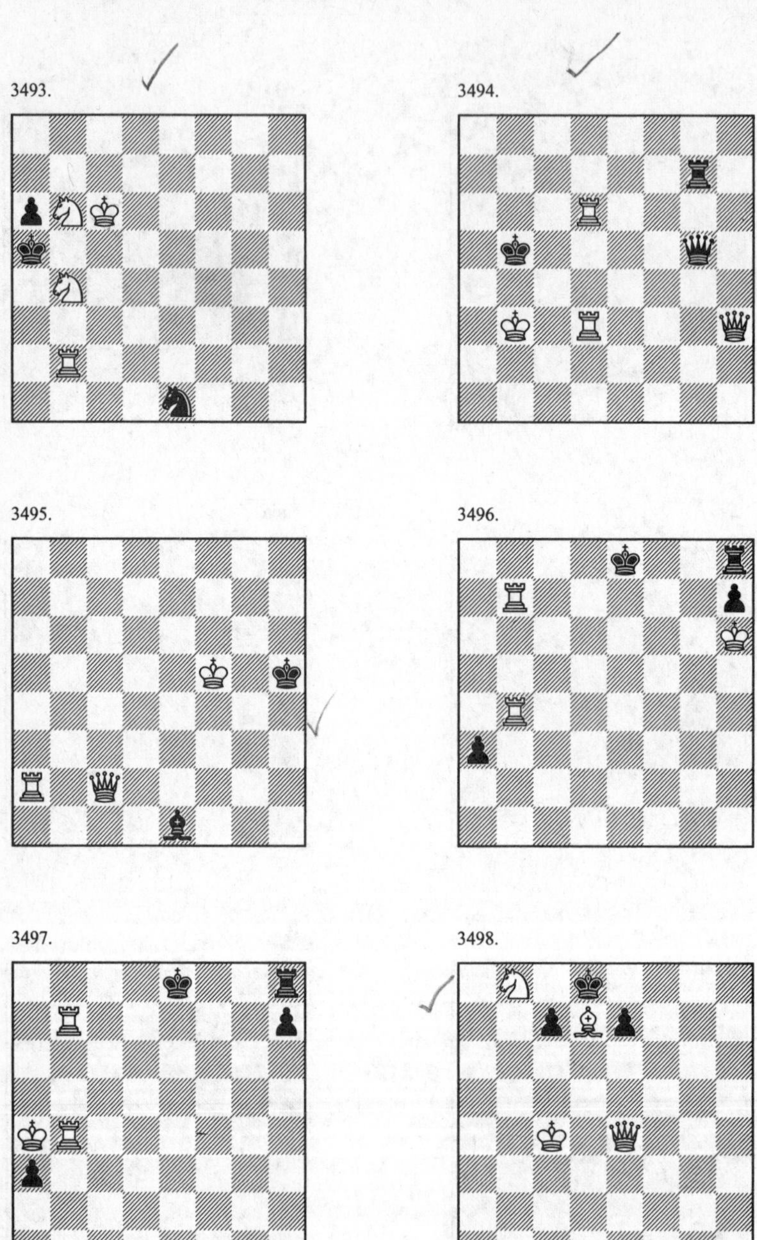
3493.
3494.
3495.
3496.
3497.
3498.

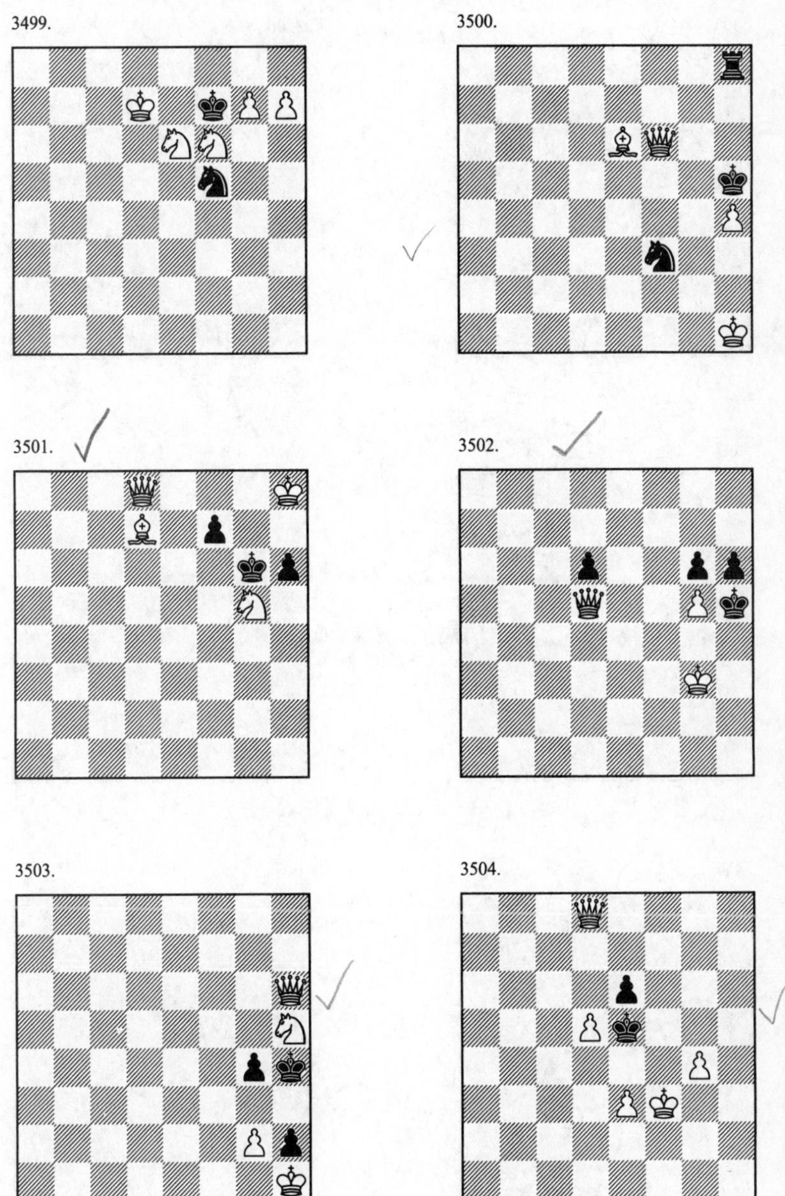
3499.
3500.
3501.
3502.
3503.
3504.

3505.

3506.

3507.

3508.

3509.

3510.

3511.

3512.

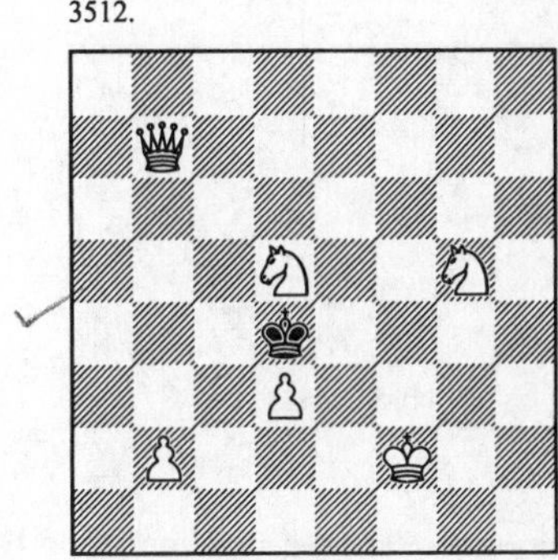

3513.

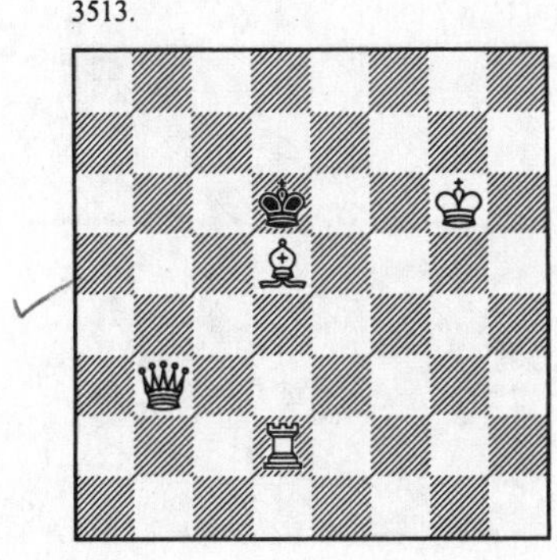

3514.

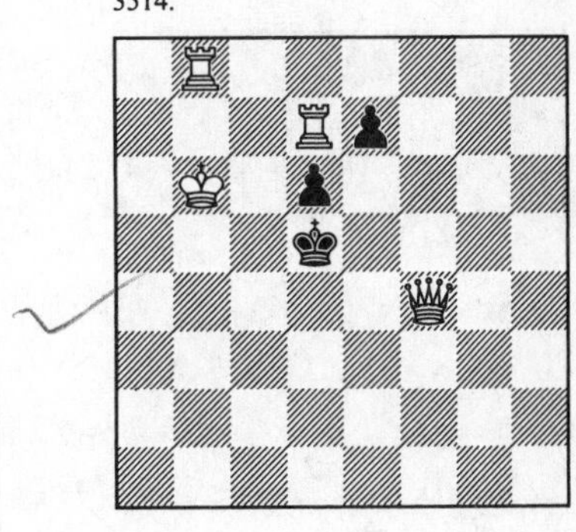

3515.

3516.

3517.

3518.

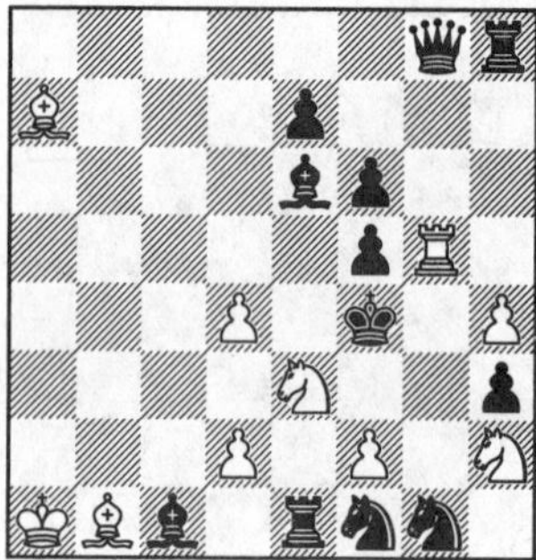

3519.

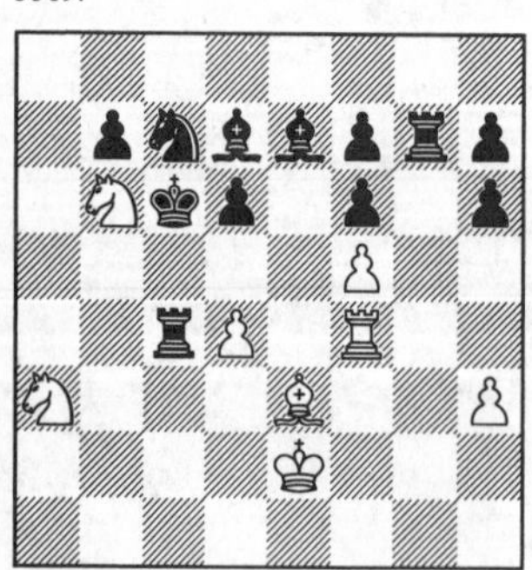

3520.

3521.

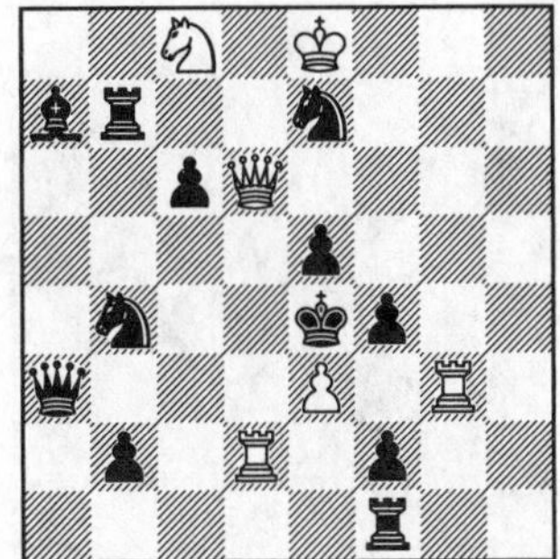

3522.

3523.

3524.

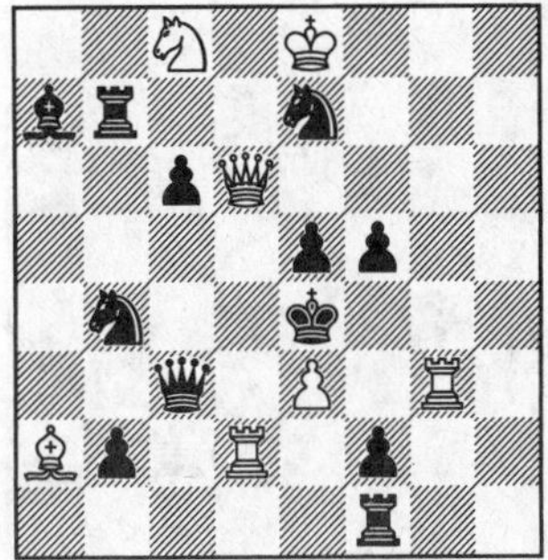

3525.

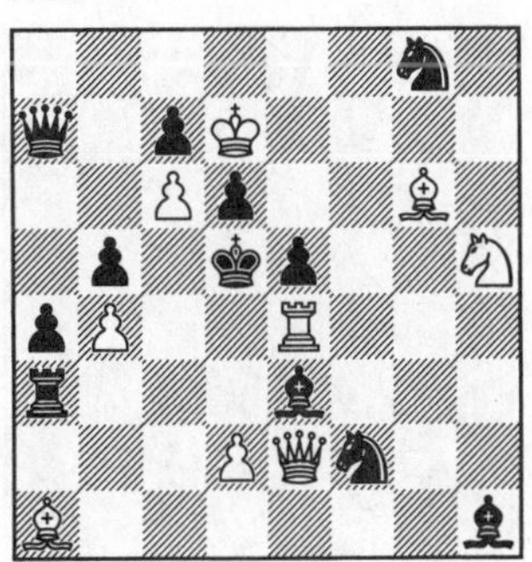

3526.

3527.

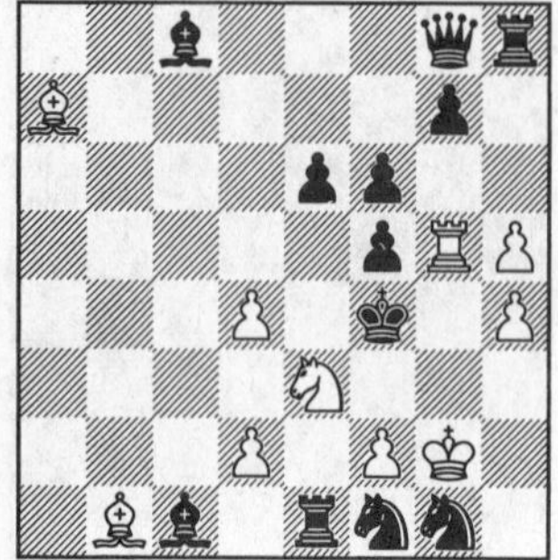

3528.

3529.

3530.

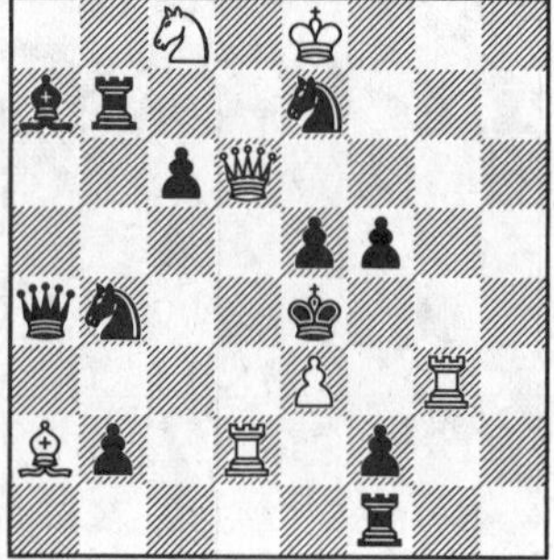

3531.

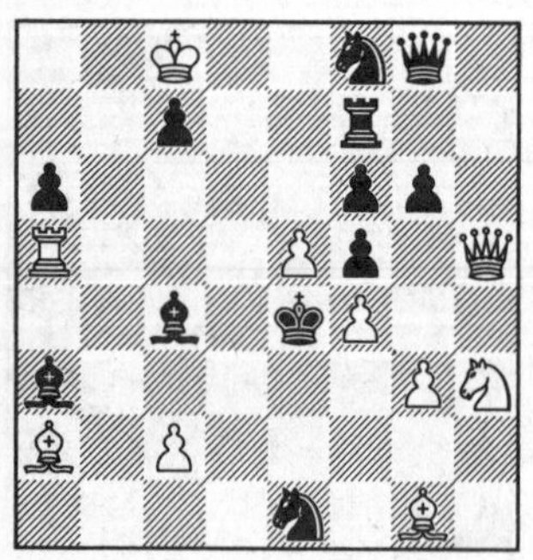

3532.

3533.

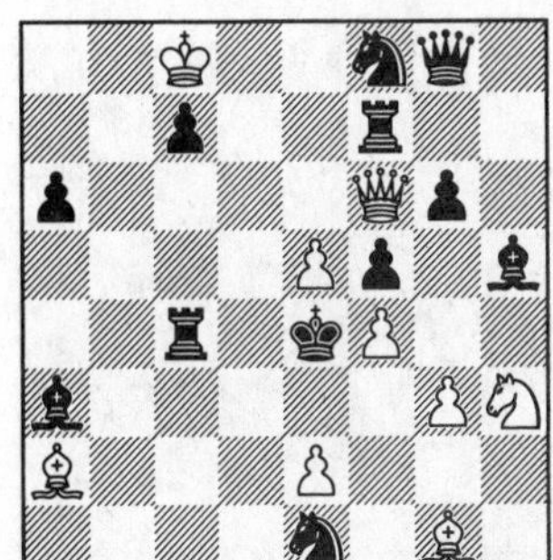

3534.

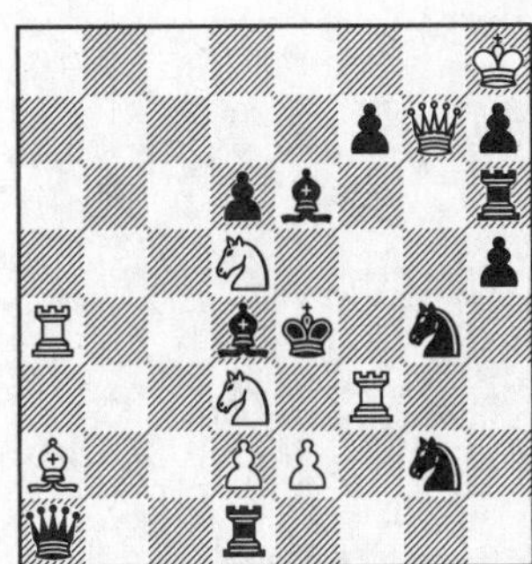

3535.

3536.

3537.

3538.

3539.

3540.

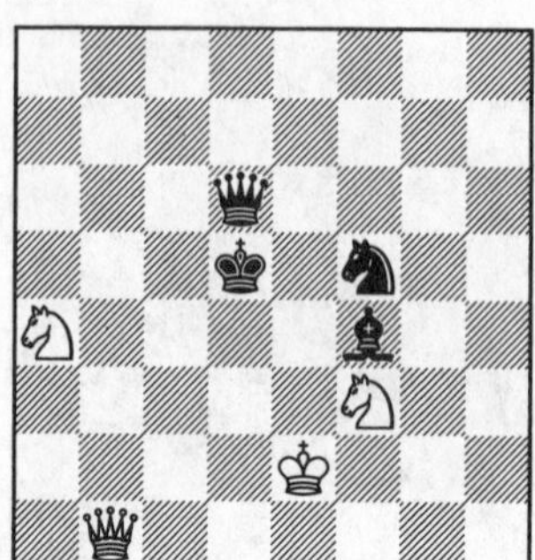

3541.

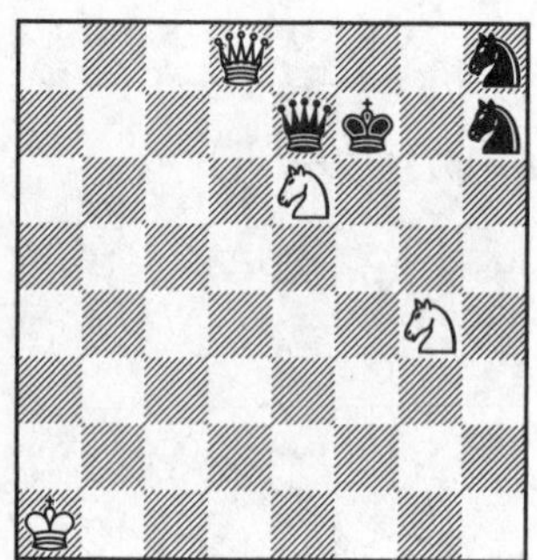

3542.

3543.

3544.

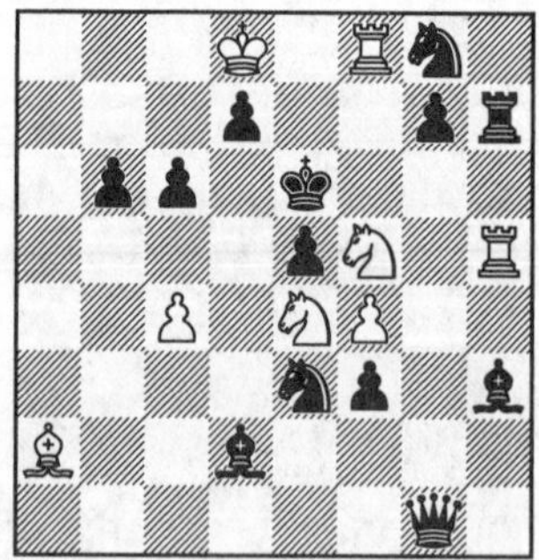

3545.

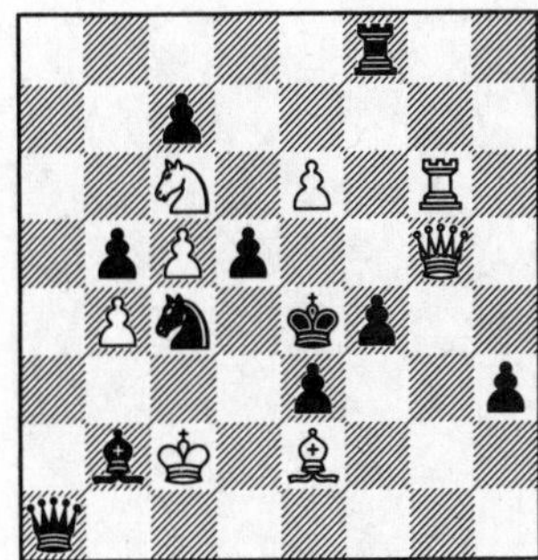

3546.

3547.

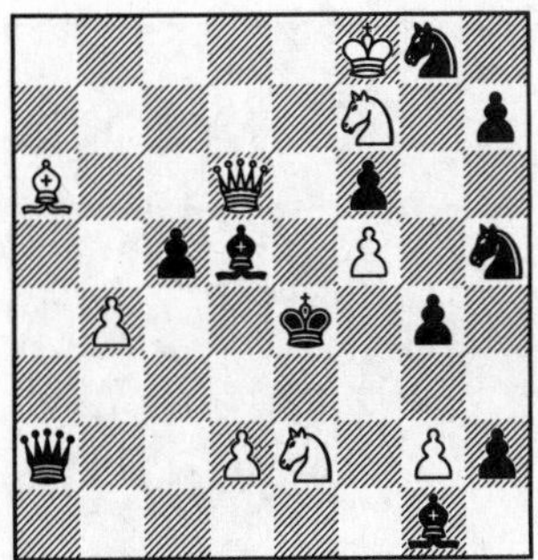

3548.

3549.

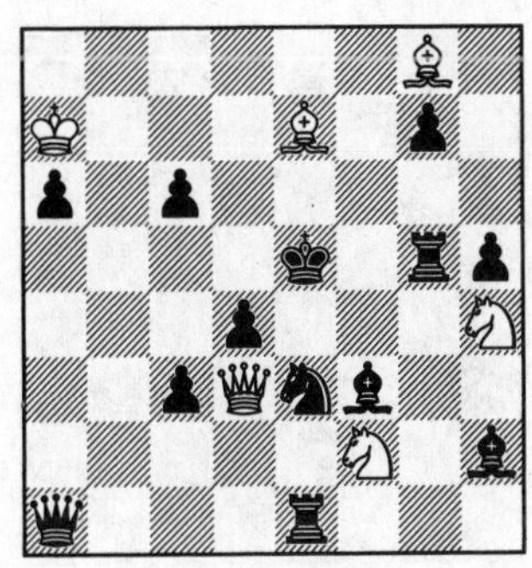

3550.

3551.

3552.

3553.

3554.

3555.

3556.

3557.

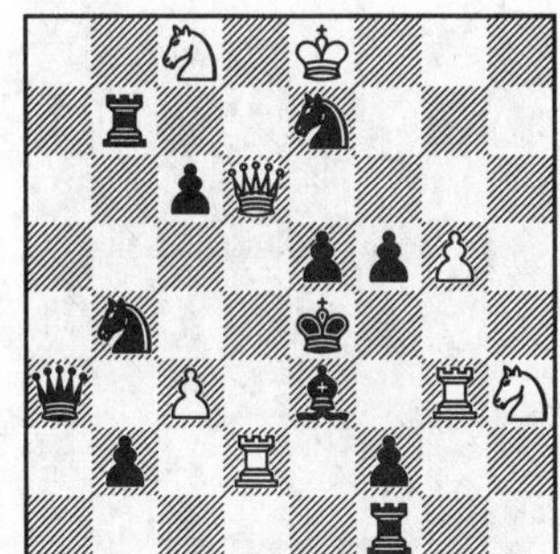

3558.

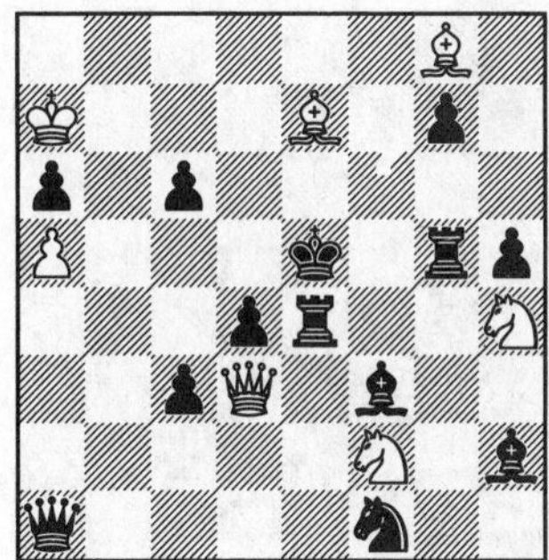

3559.

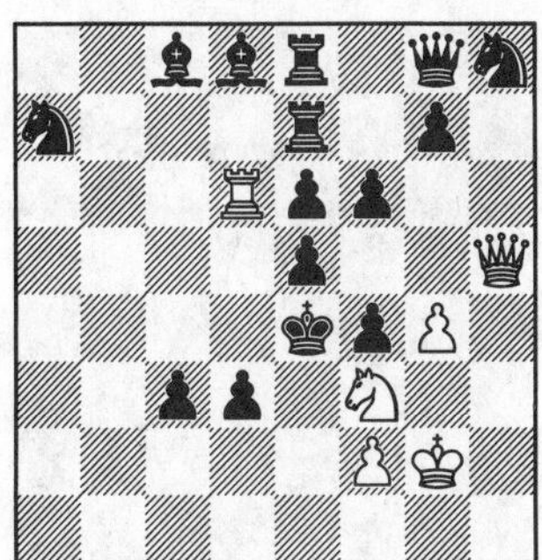

3560.

3561.

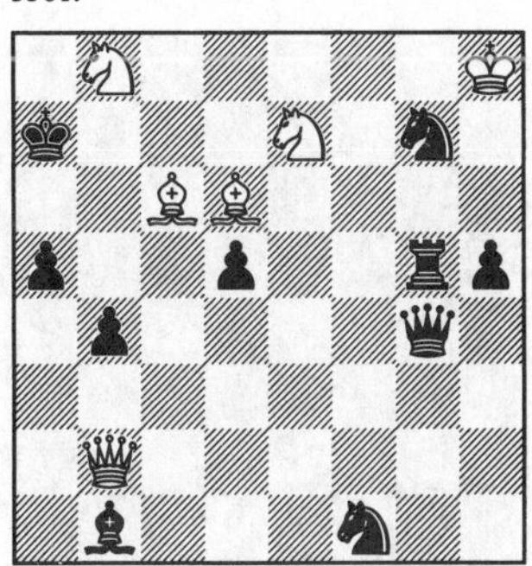

3562.

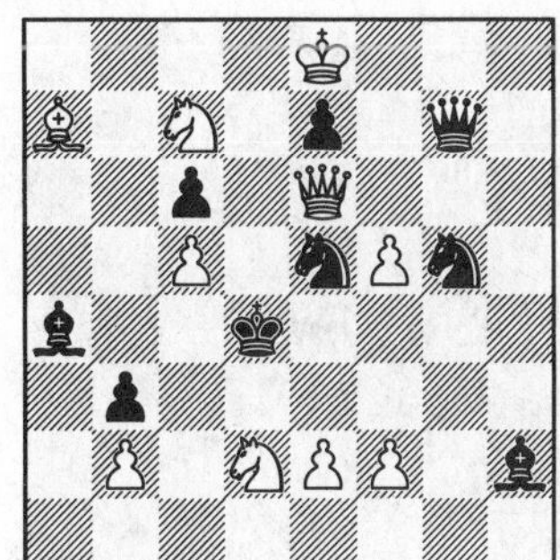

3563.

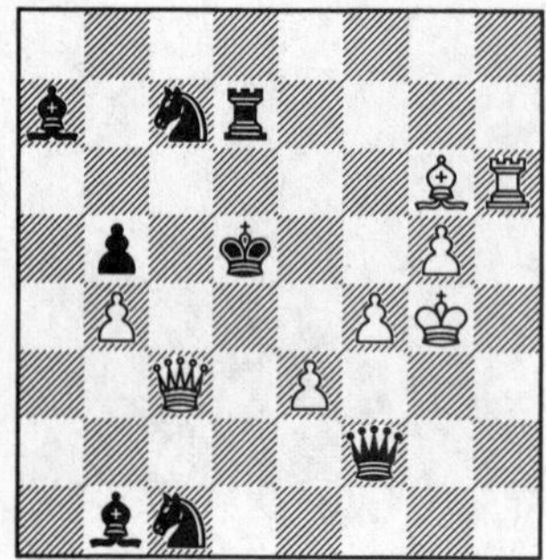

3564.

3565.

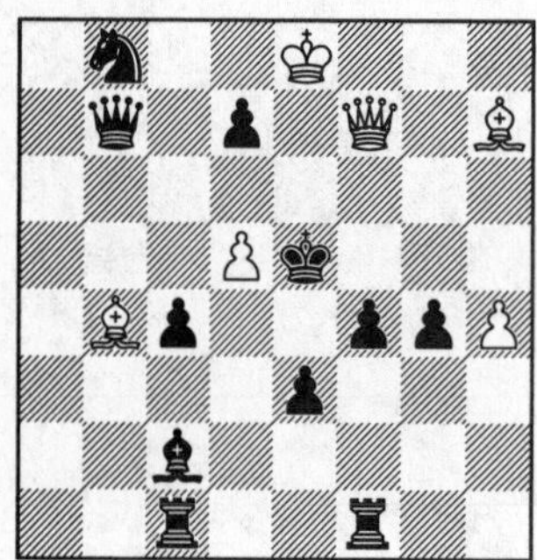

3566.

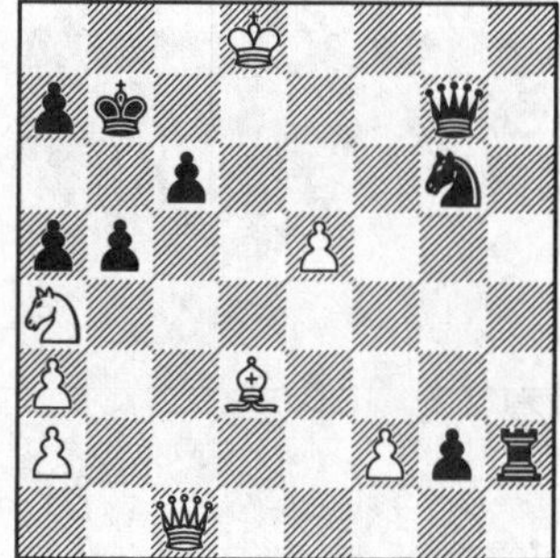

3567.

3568.

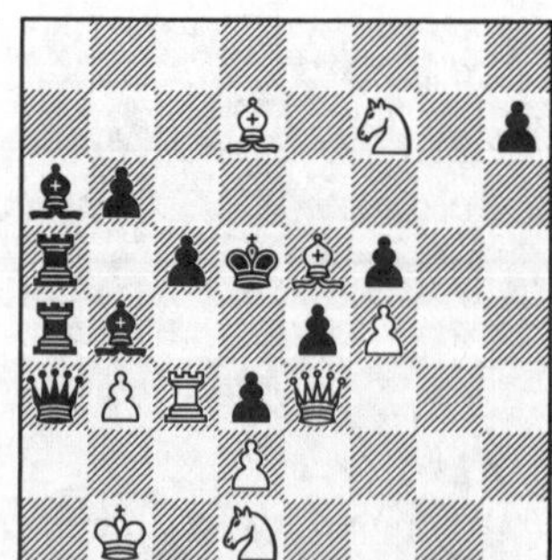

3569.

3570.

3571.

3572.

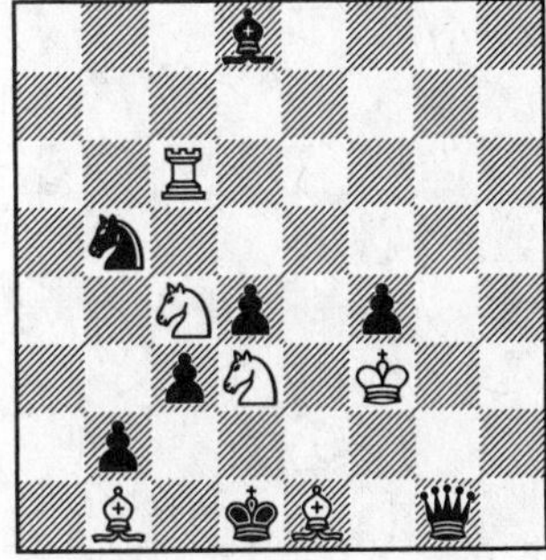

3573.

3574.

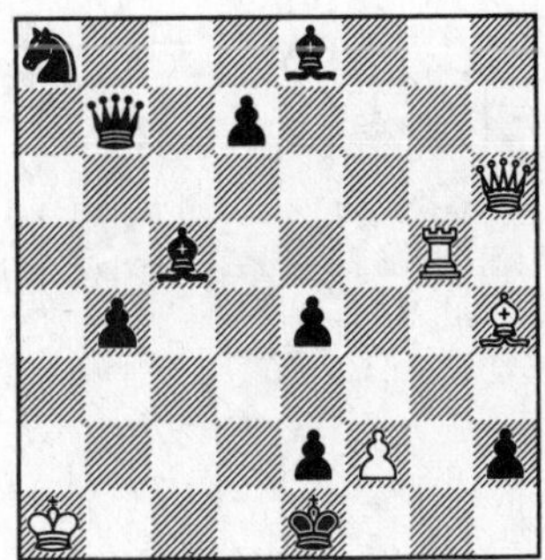

3575.

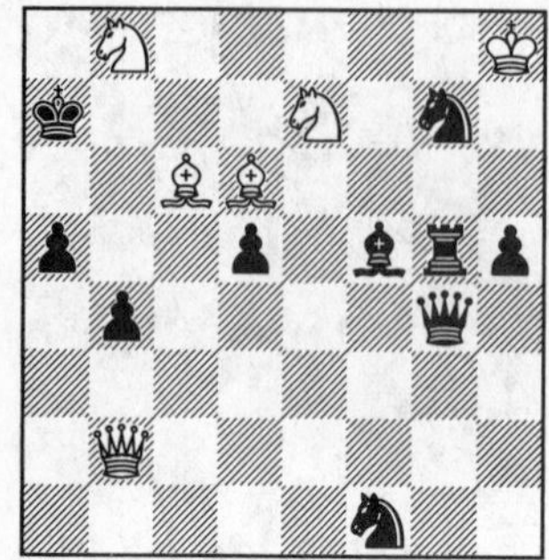

3576.

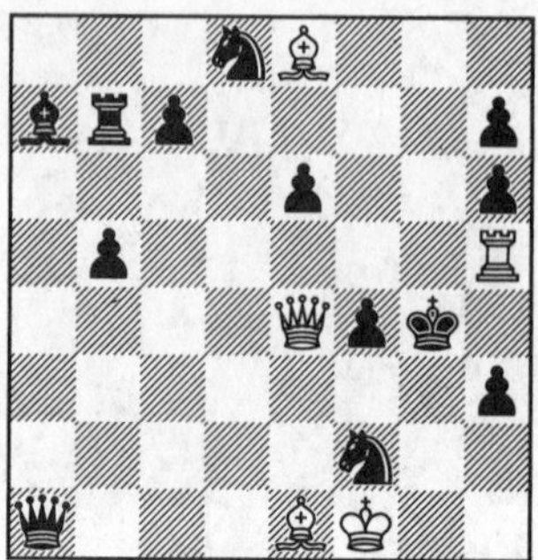

3577.

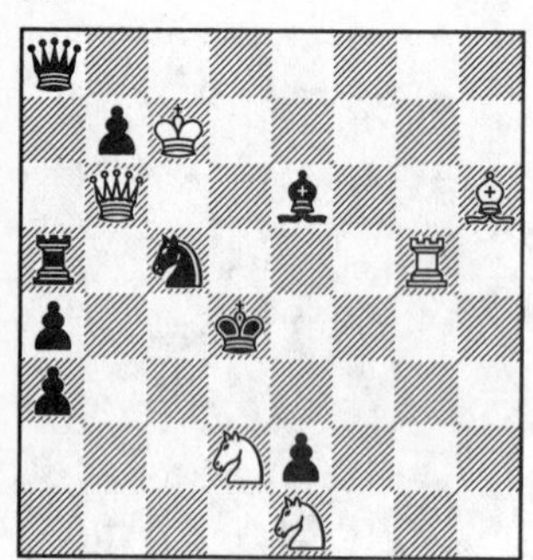

3578.

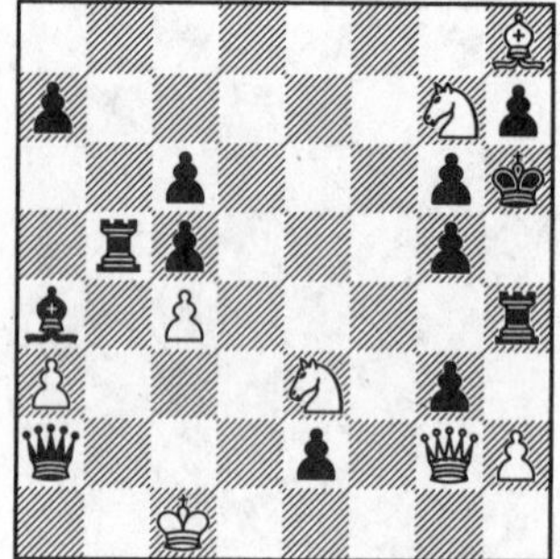

3579.

3580.

3581.

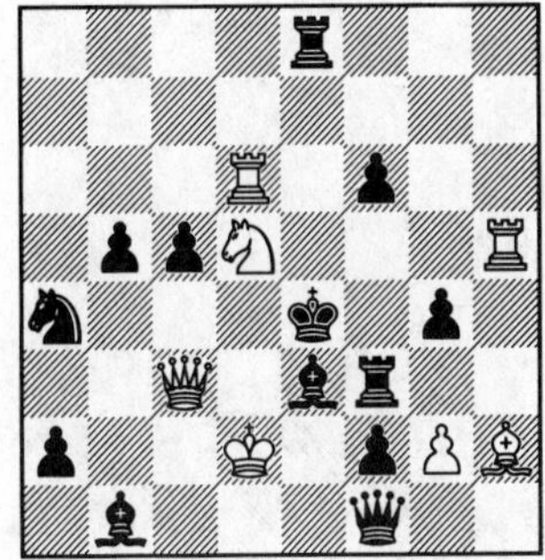

3582.

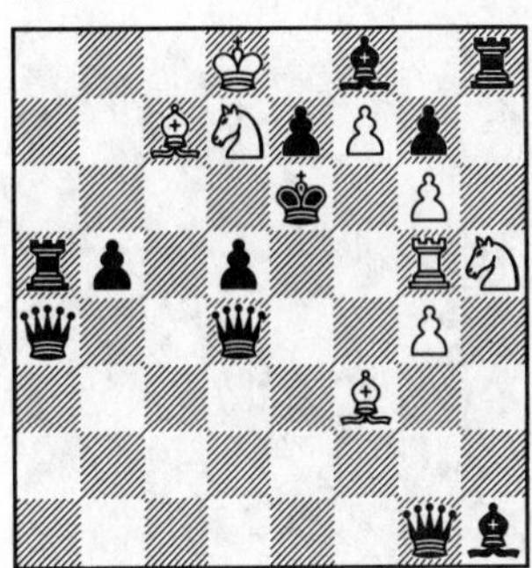

3583.

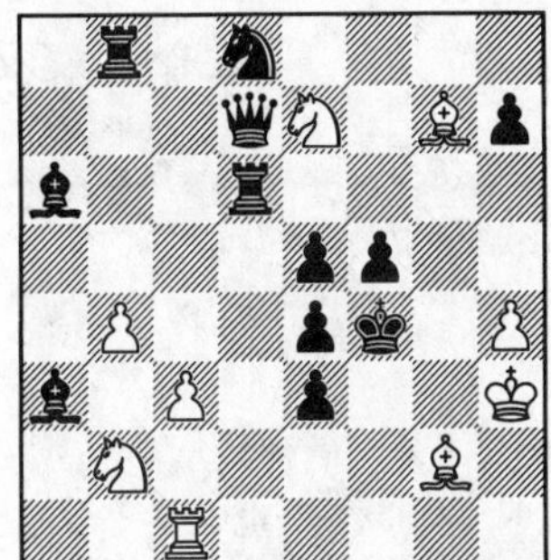

3584.

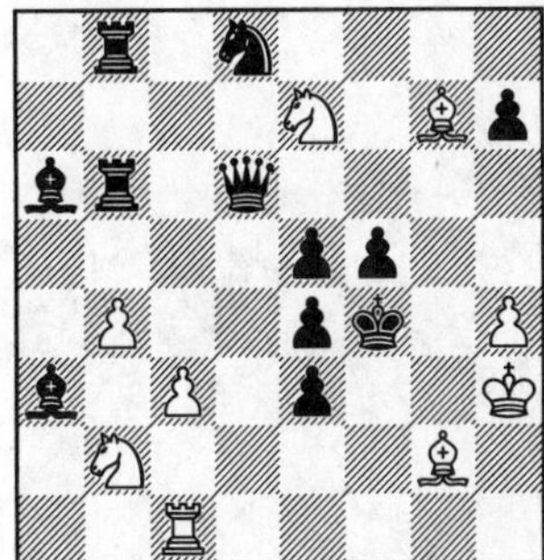

3585.

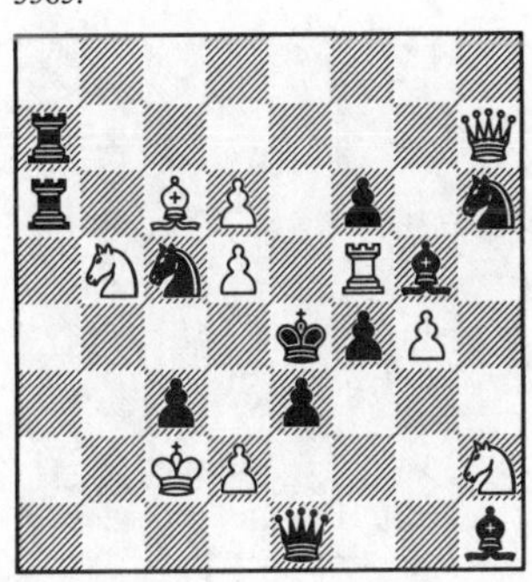

3586.

3587.

3588.

3589.

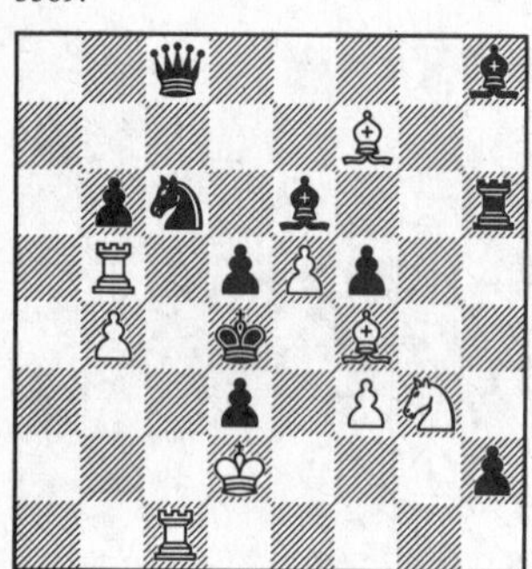

3590.

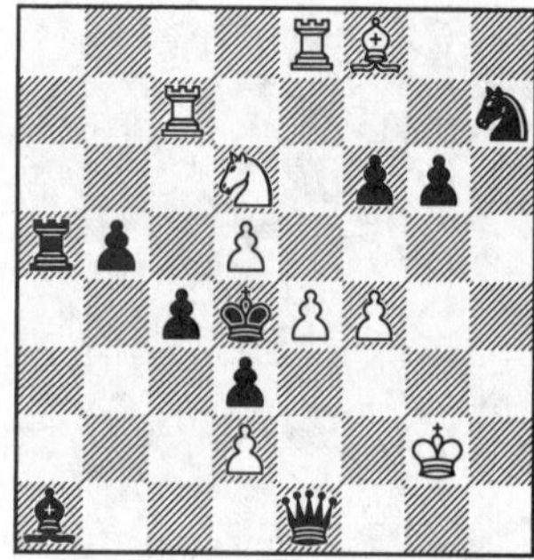

3591.

3592.

3593.

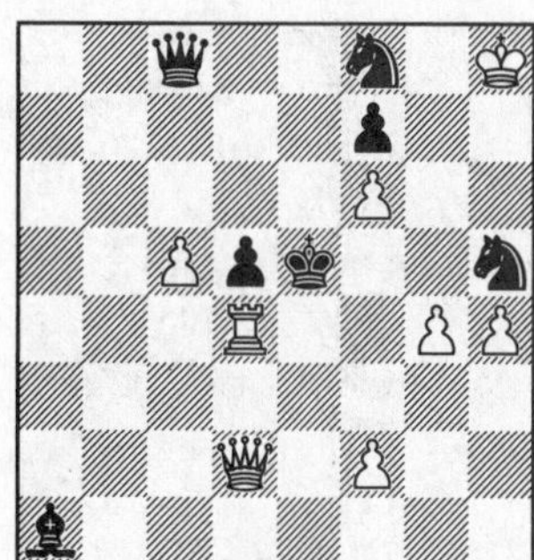

3594.

3595.

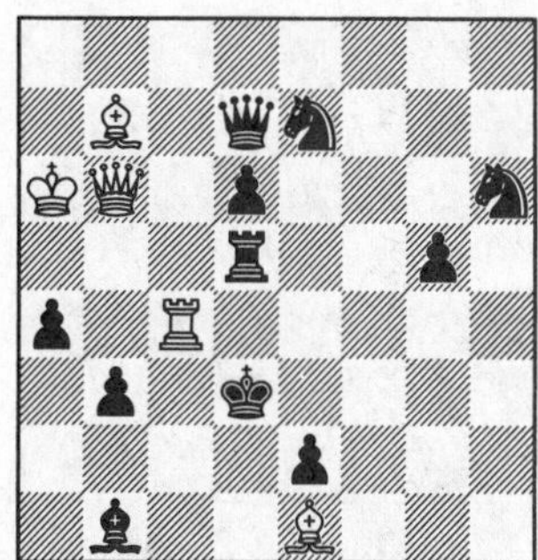

3596.

3597.

3598.

3599.

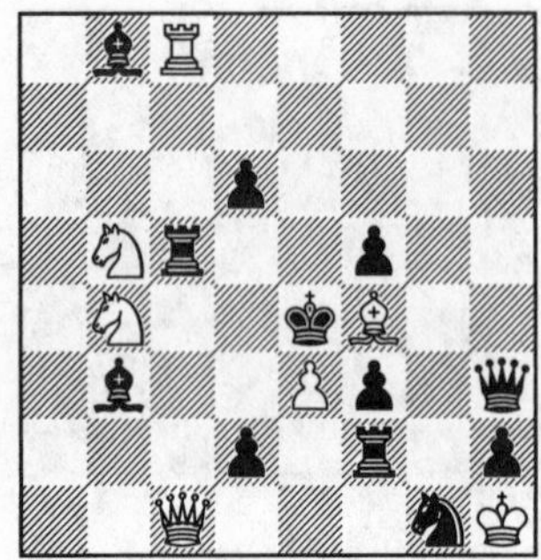

3600.

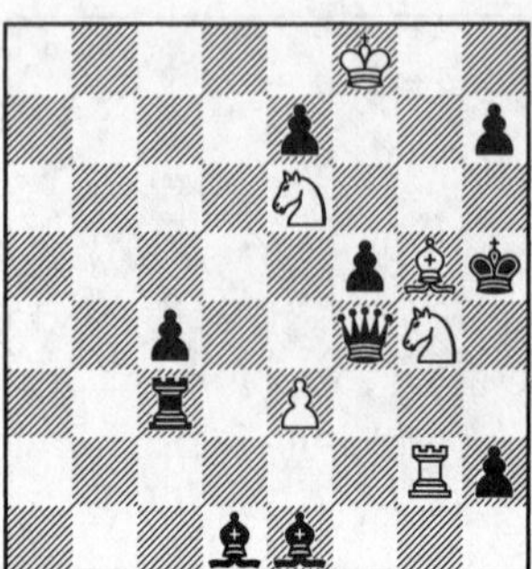

3601.

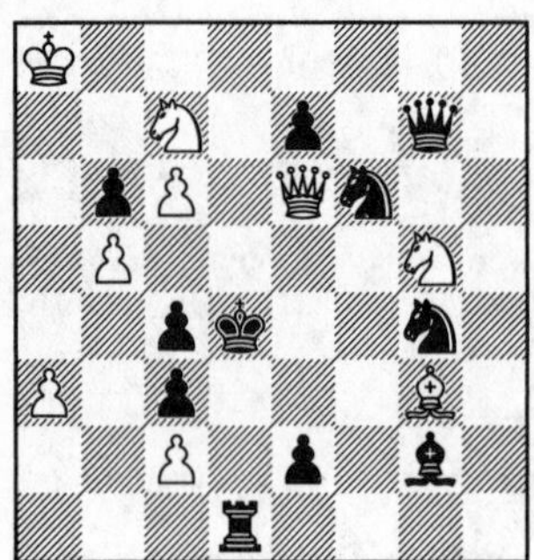

3602.

3603.

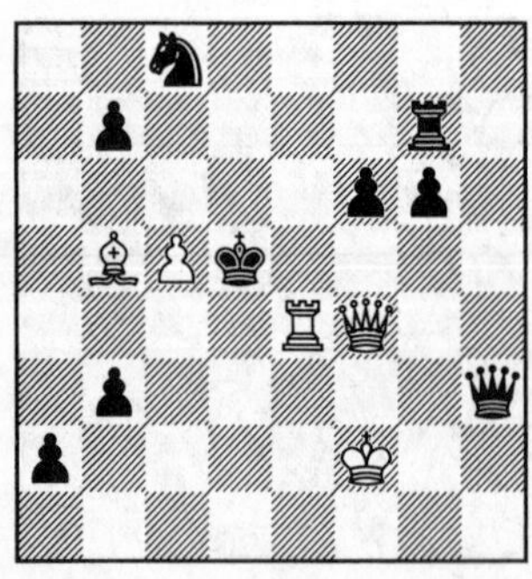

3604.

3605.

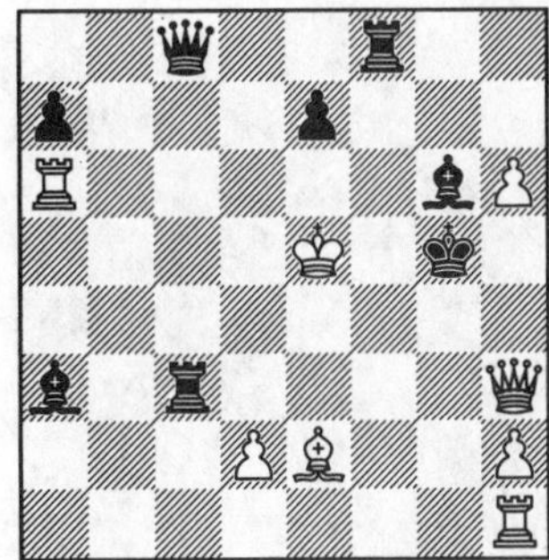

3606.

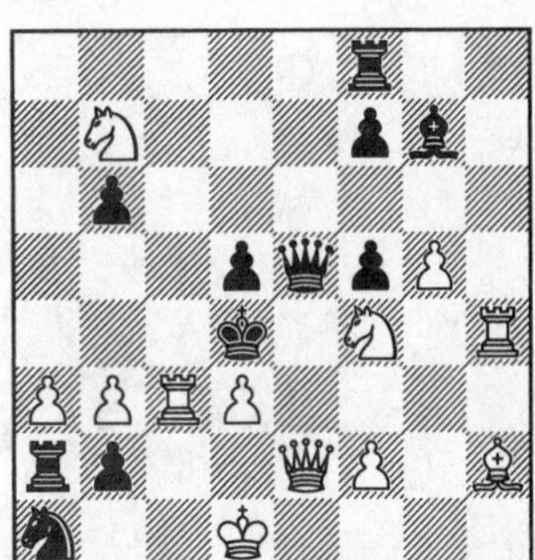

3607.

3608.

3609.

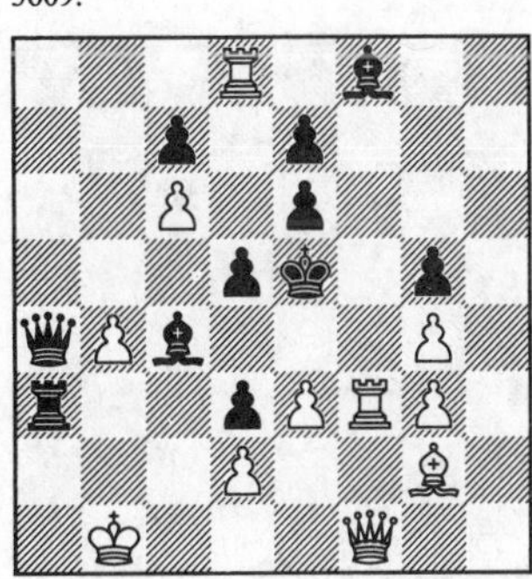

3610.

3611.

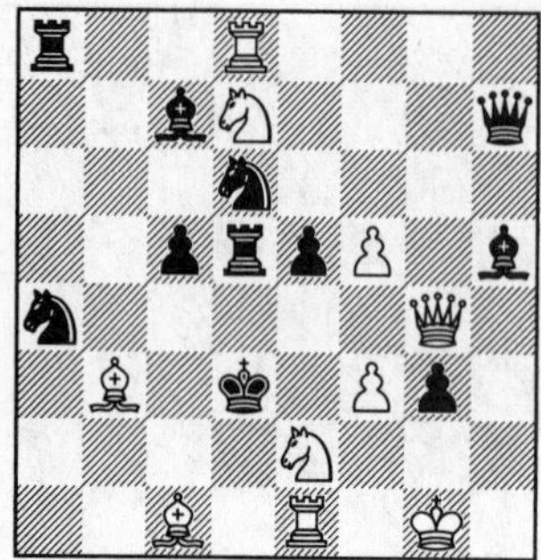

3612.

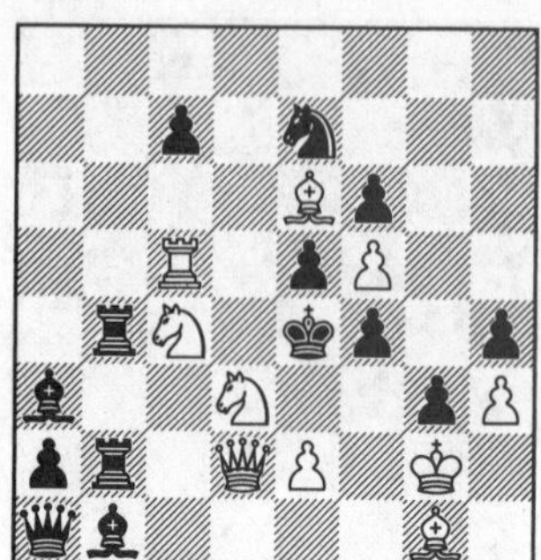

3613.

3614.

3615.

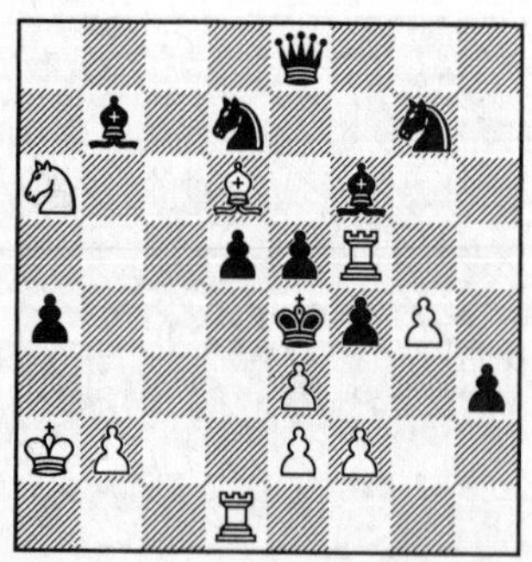

3616.

3617.

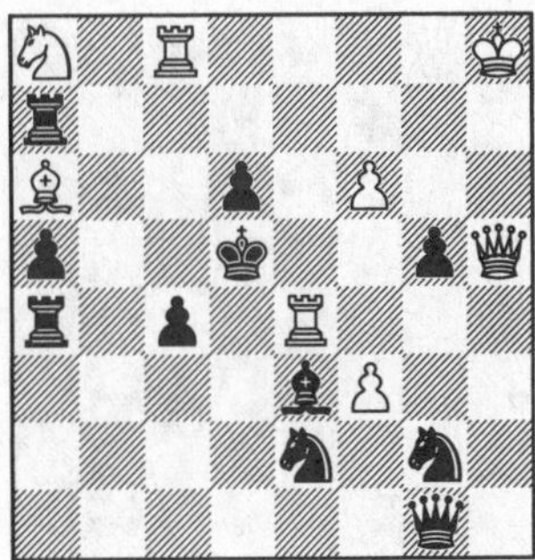

3618.

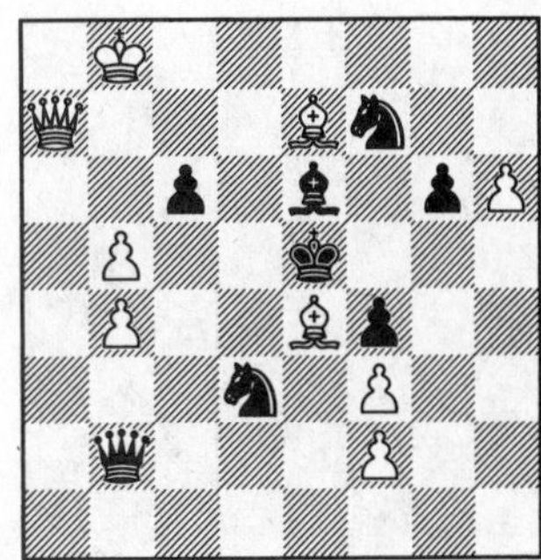

3619.

3620.

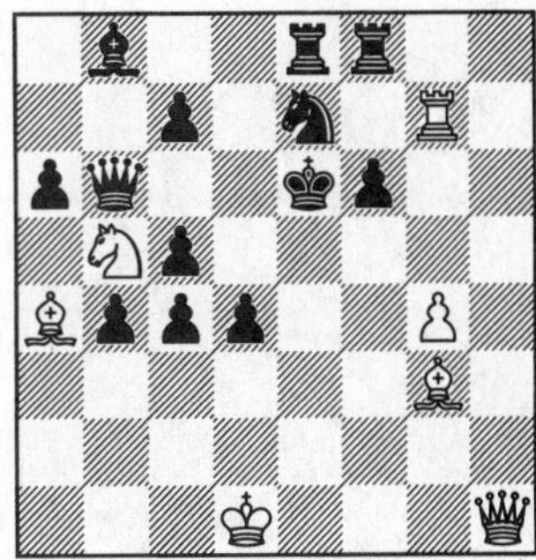

3621.

3622.

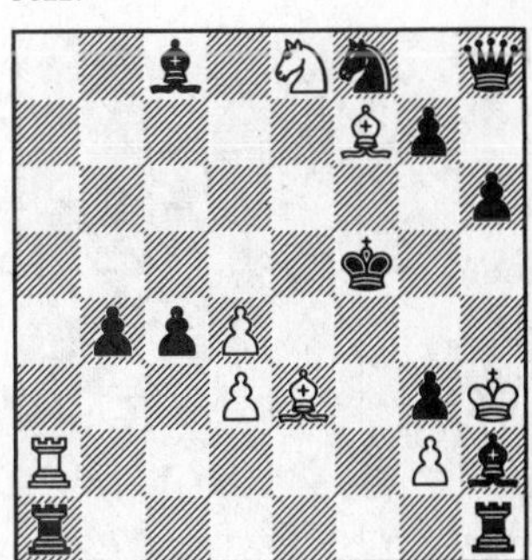

3623.

3624.

3625.

3626.

3627.

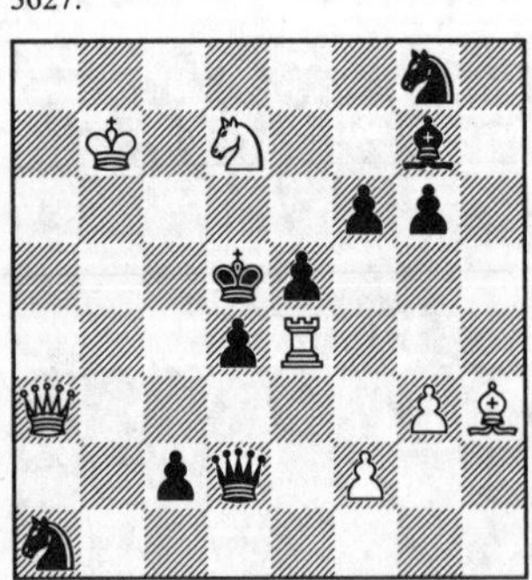

3628.

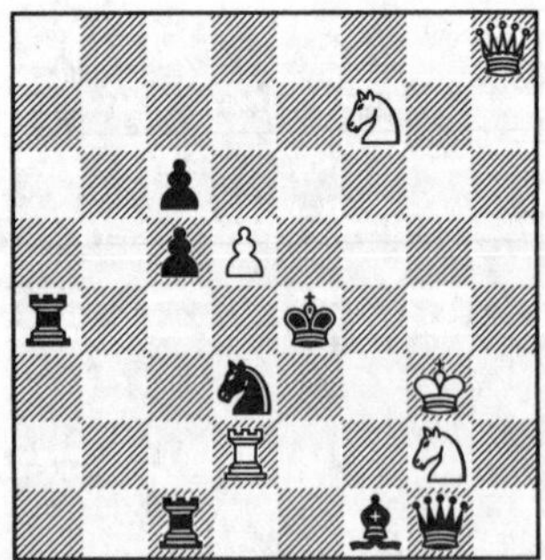

3629.

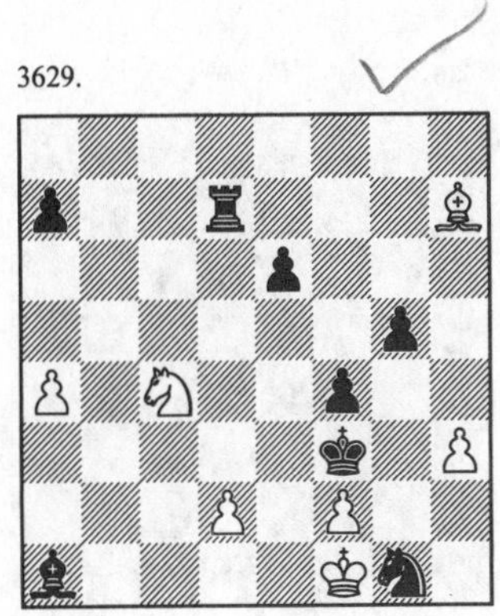

3630.

3631.

3632.

3633.

3634.

3635.

3636.

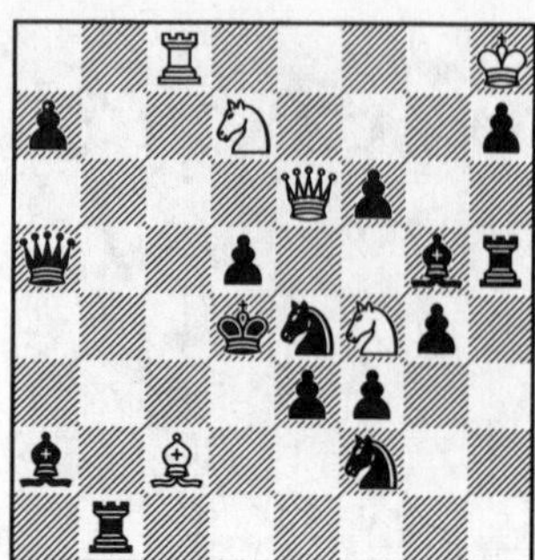

3637.

3638.

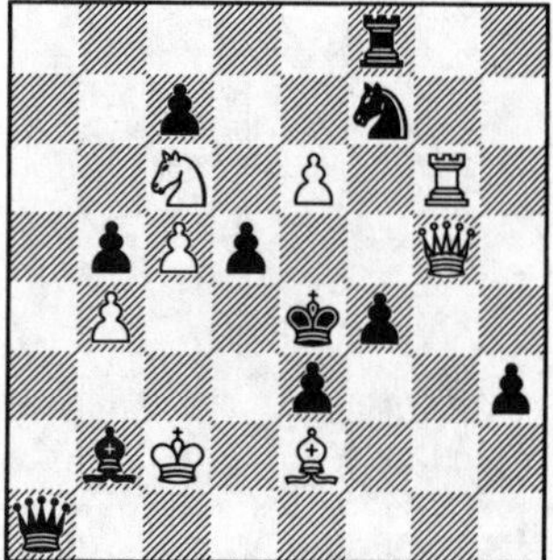

3639.

3640.

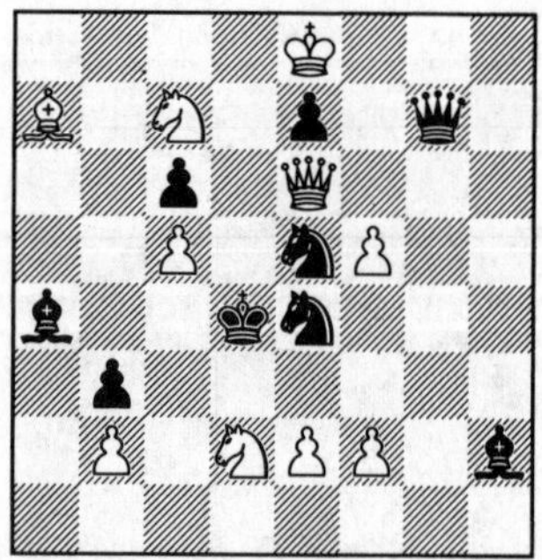

3641.

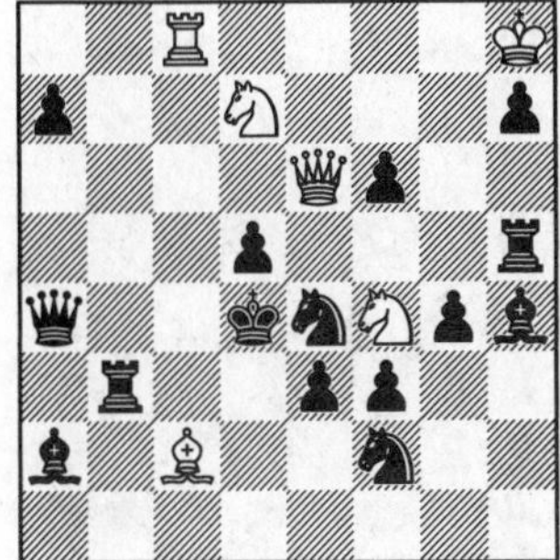

3642.

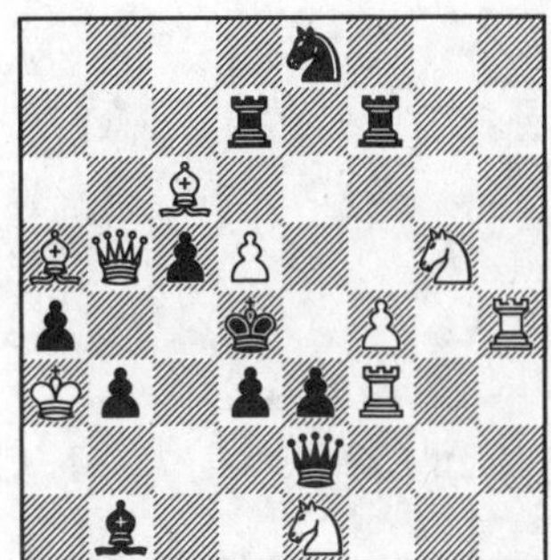

3643.

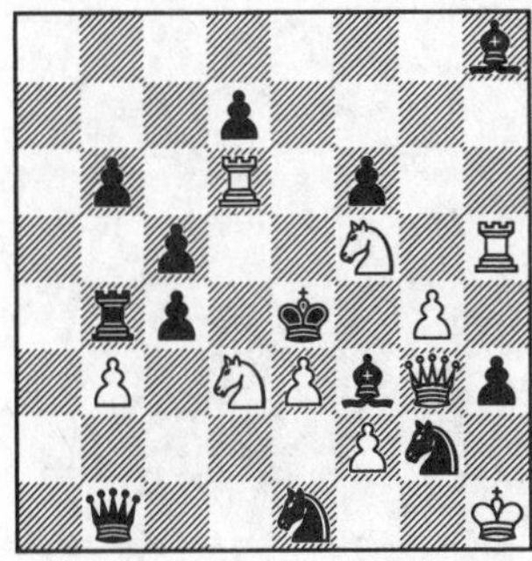

3644.

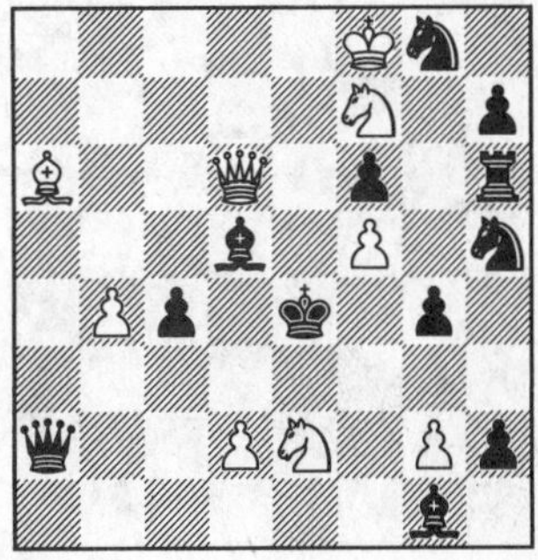

3645.

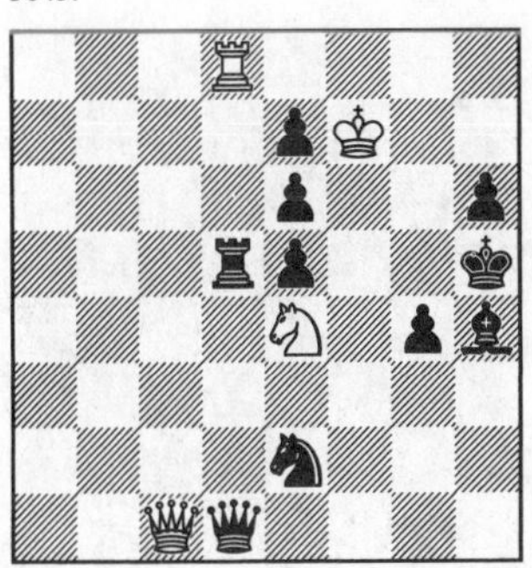

3646.

3647.

3648.

3649.

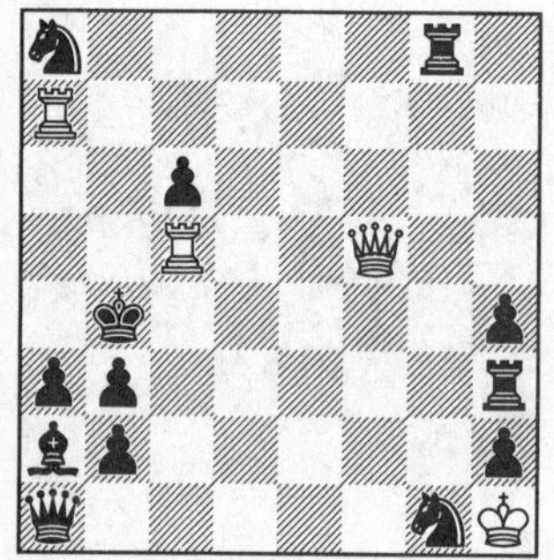

3650.

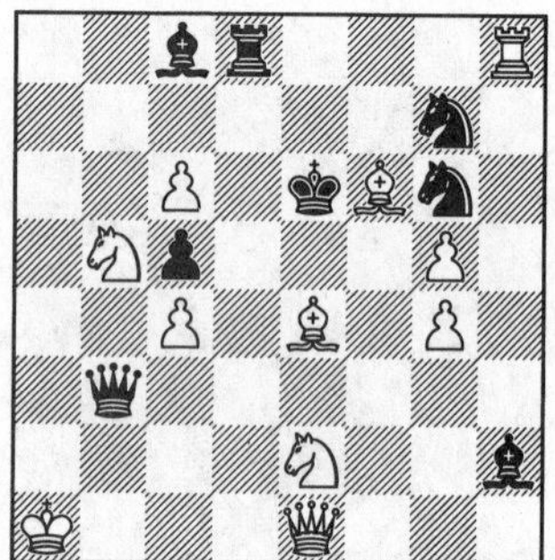

3651.

3652.

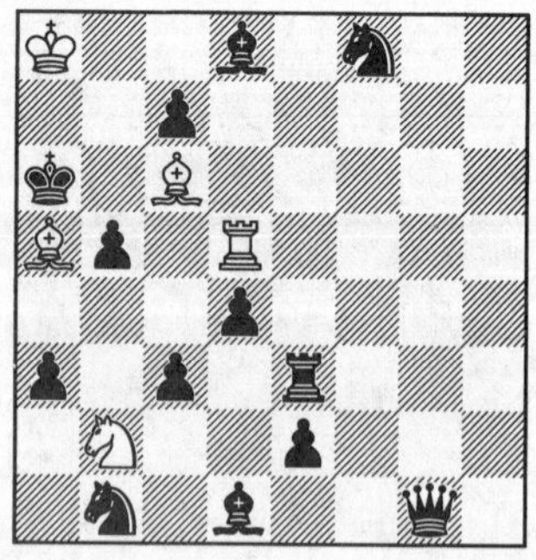

3653.

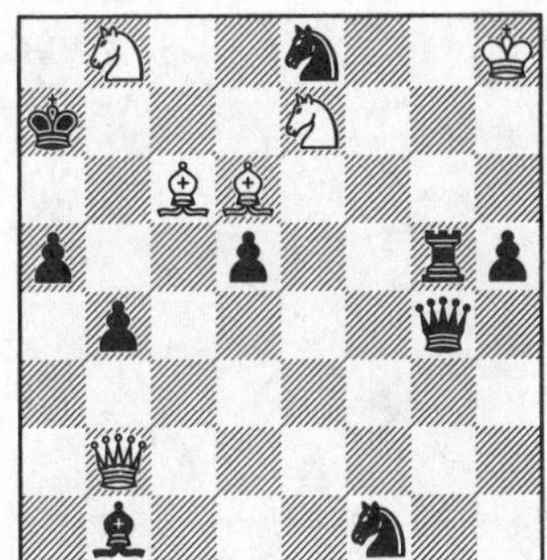

3654.

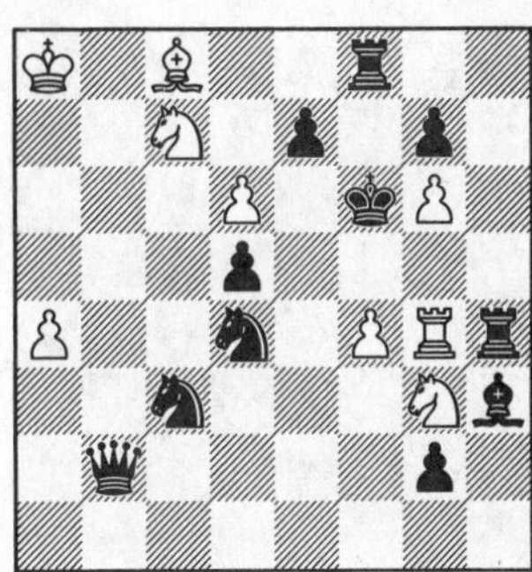

3655.

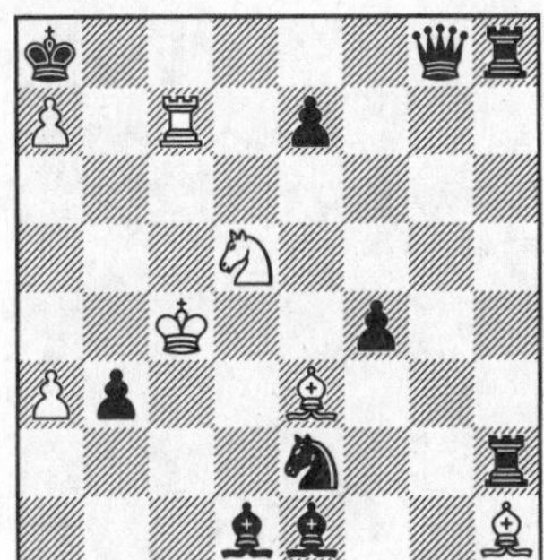

3656.

3657.

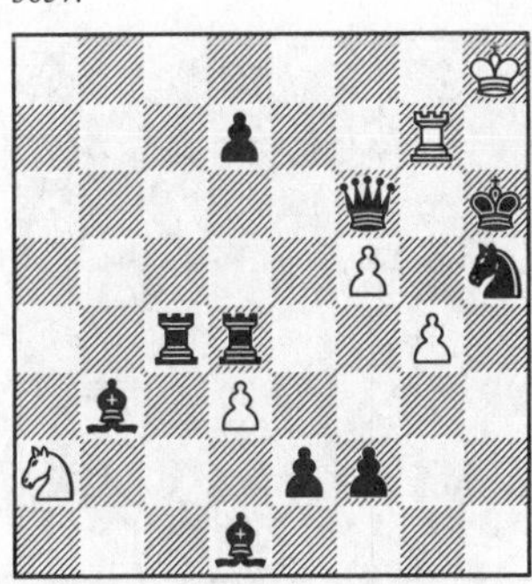

3658.

3659.

3660.

3661.

3662.

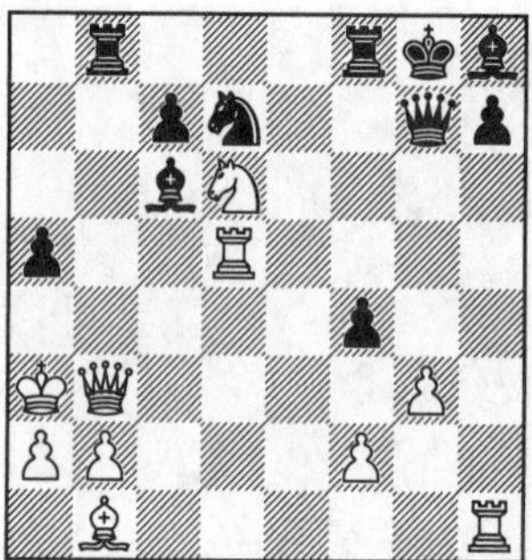

3663.

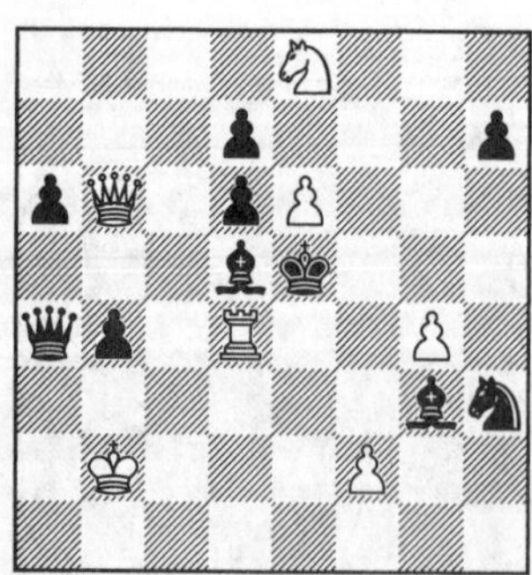

3664.

3665.

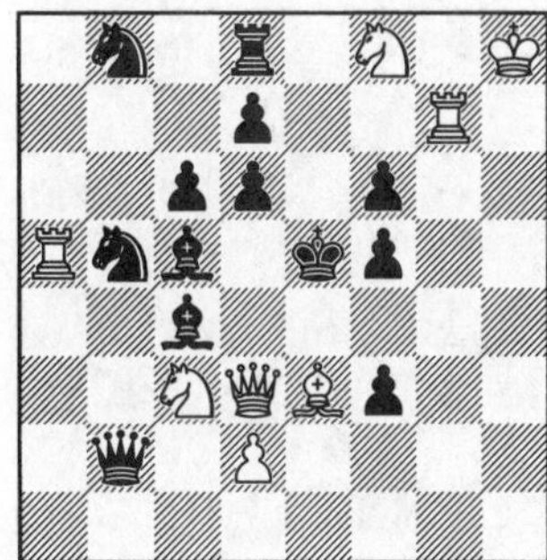

3666.

3667.

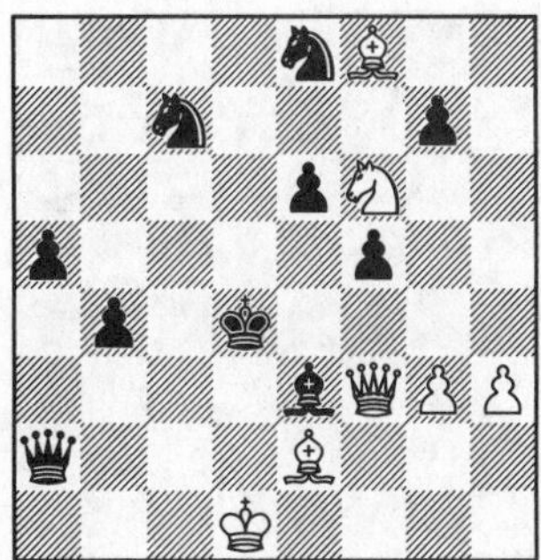

3668.

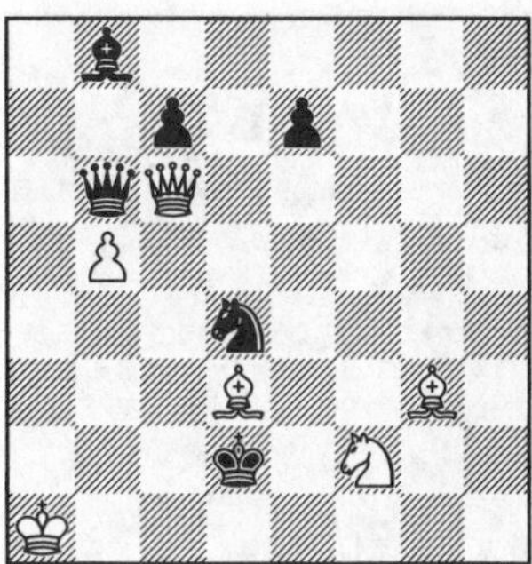

3669.

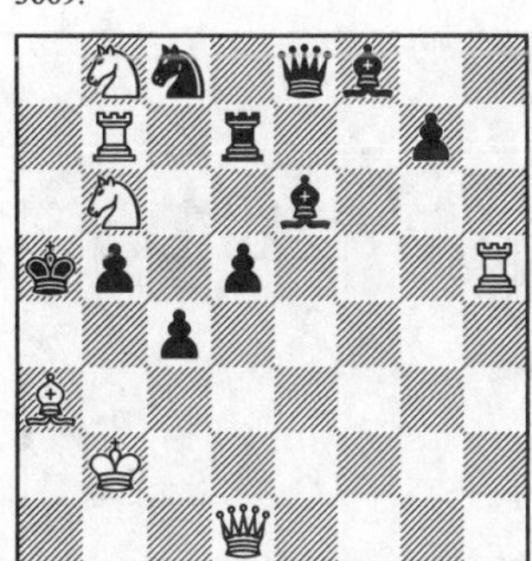

3670.

3671.

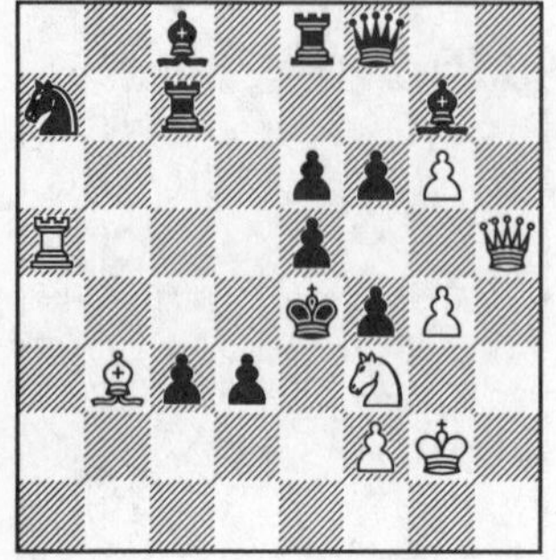

3672.

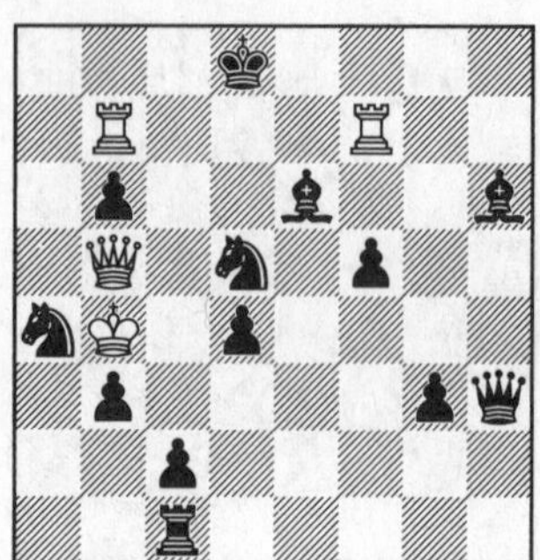

3673.

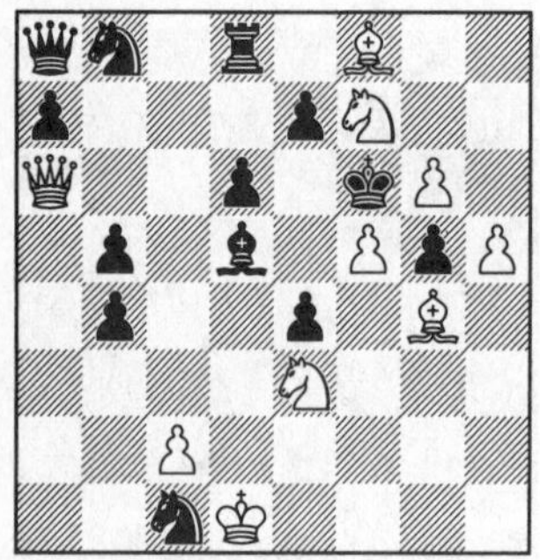

3674.

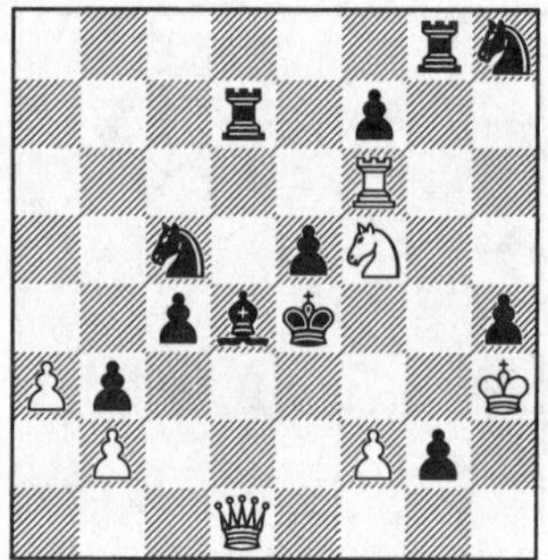

3675.

3676.

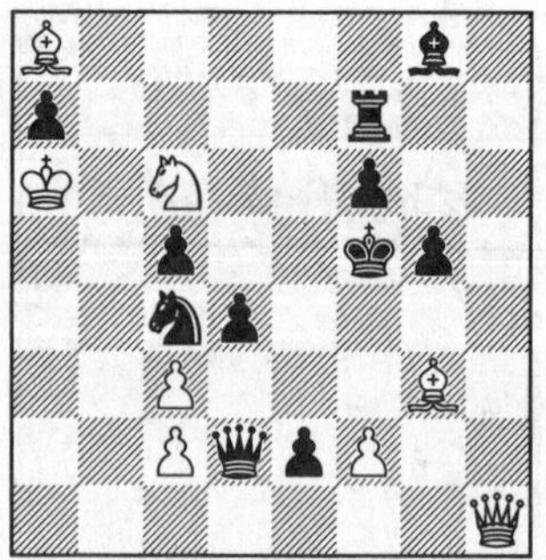

3677.

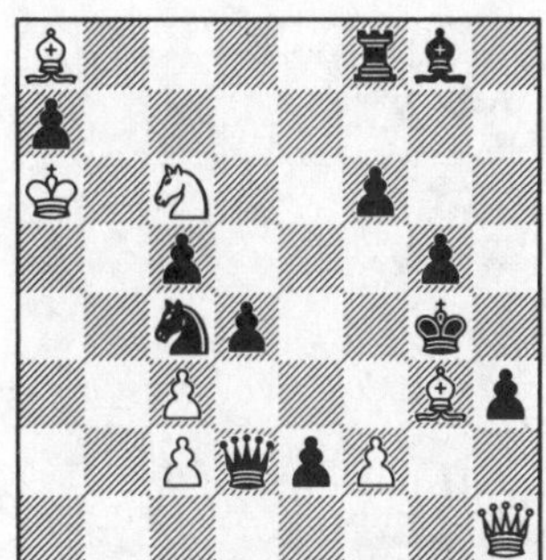

3678.

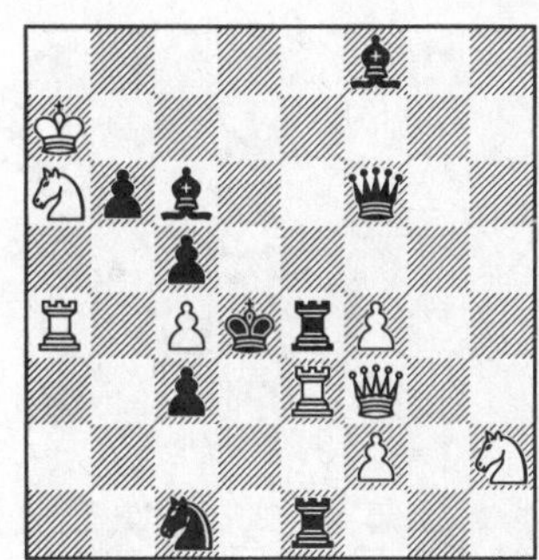

3679.

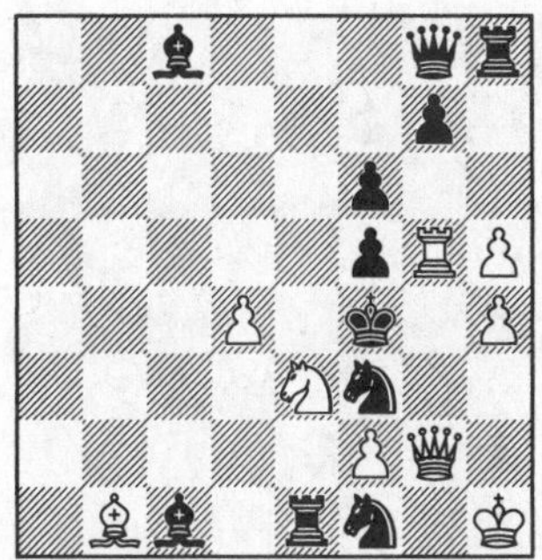

3680.

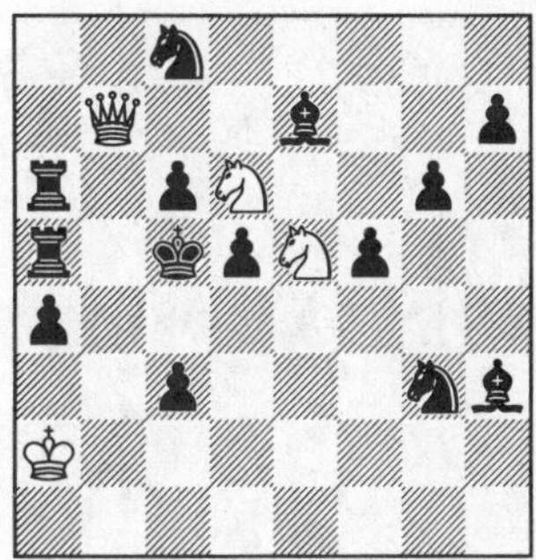

3681.

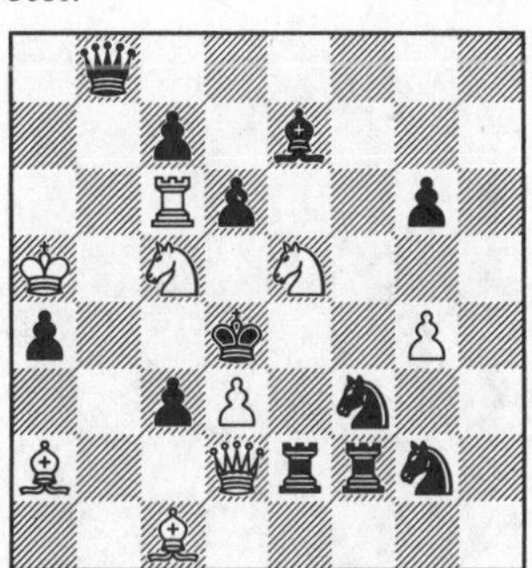

3682.

3683.

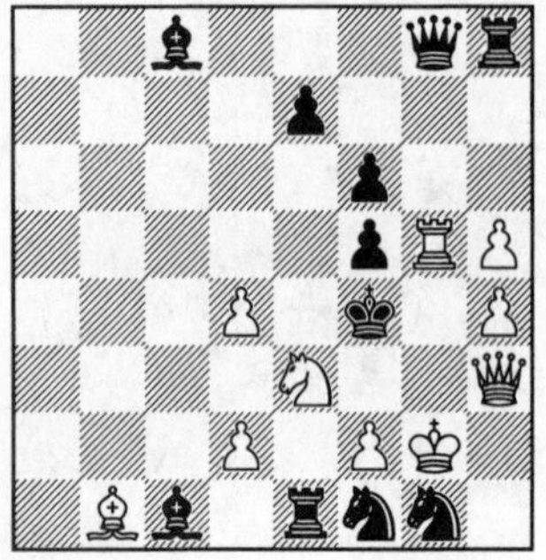

3684.

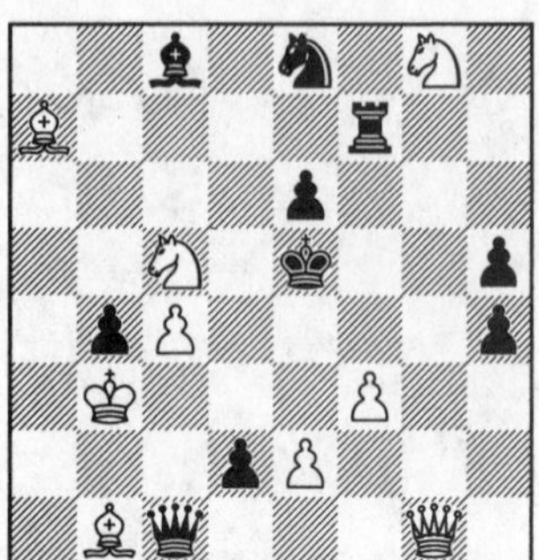

3685.

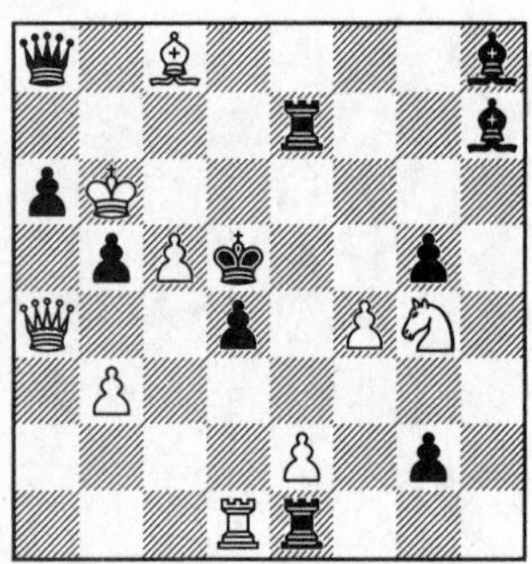

3686.

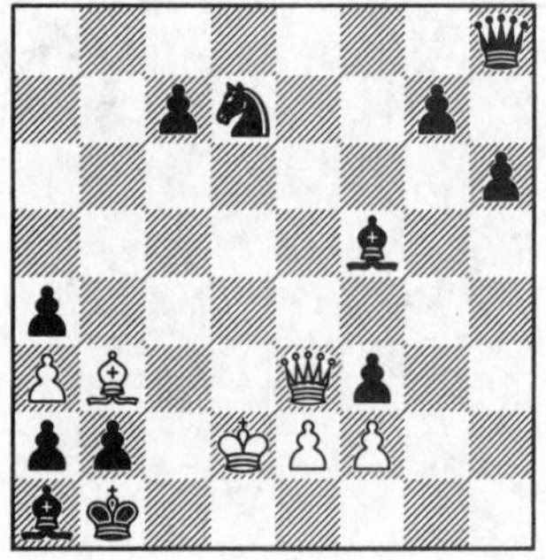

3687.

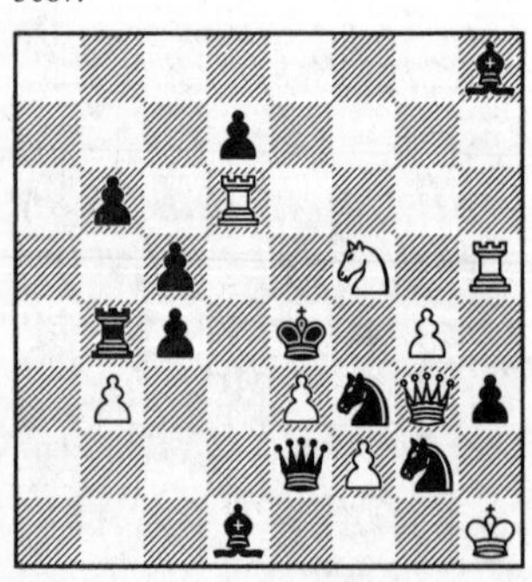

3688.

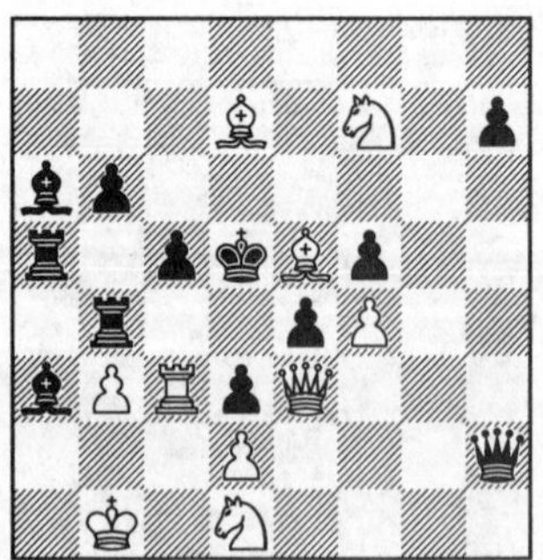

3689.

3690.

3691.

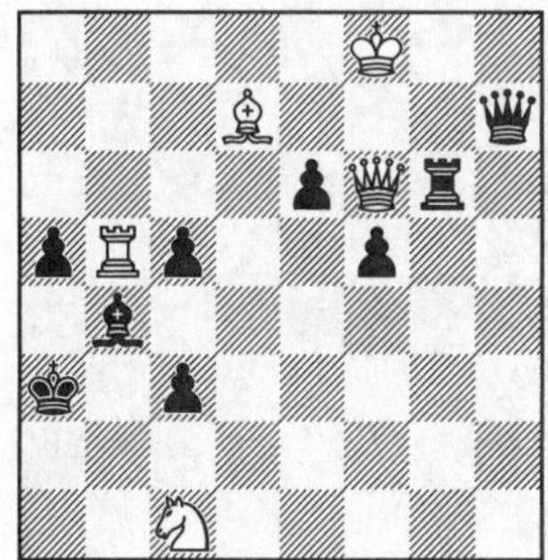

3692.

3693.

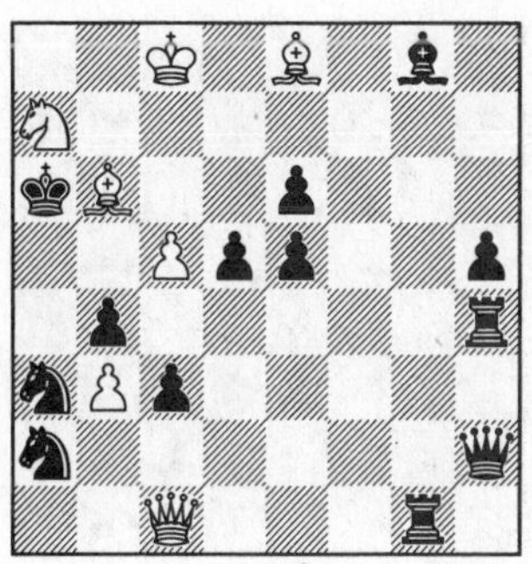

3694.

3695.

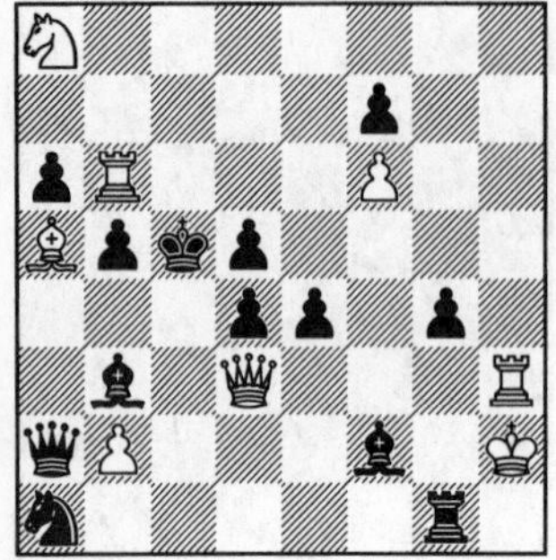

3696.

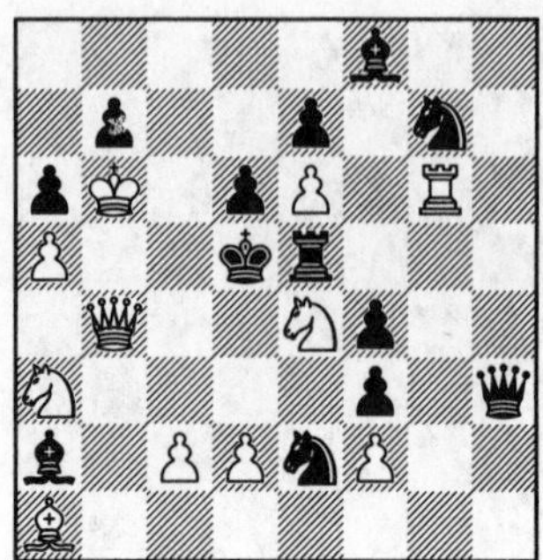

3697.

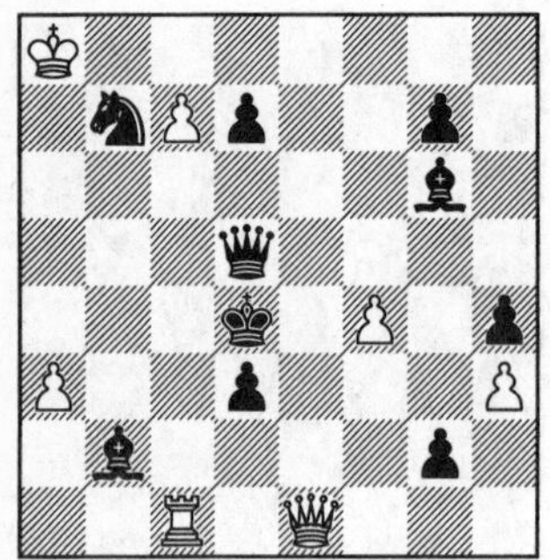

3698.

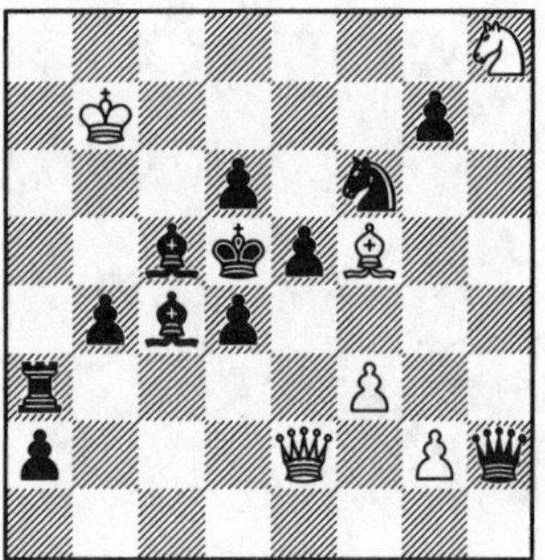

3699.

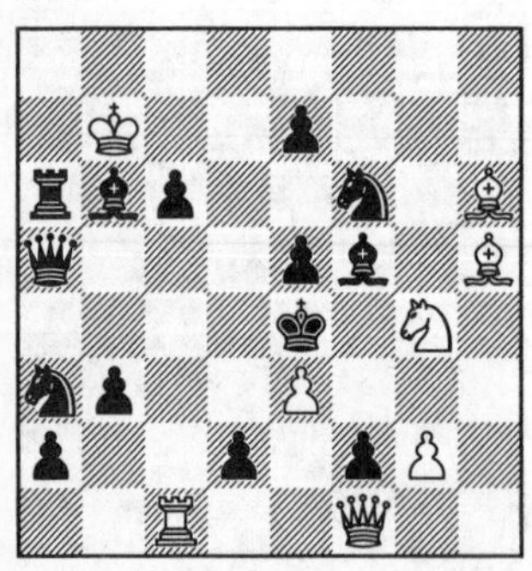

3700.

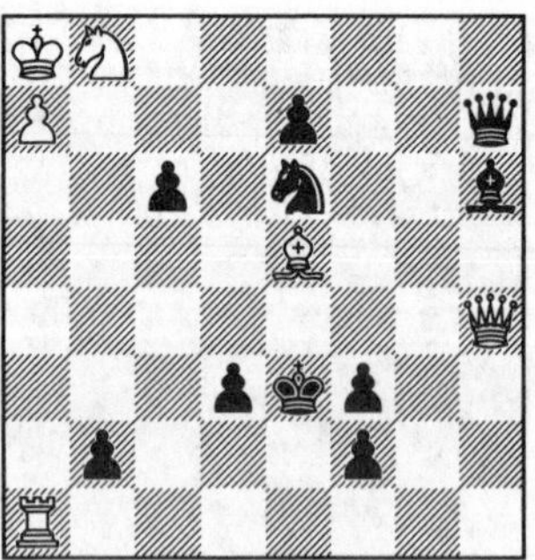

3701.

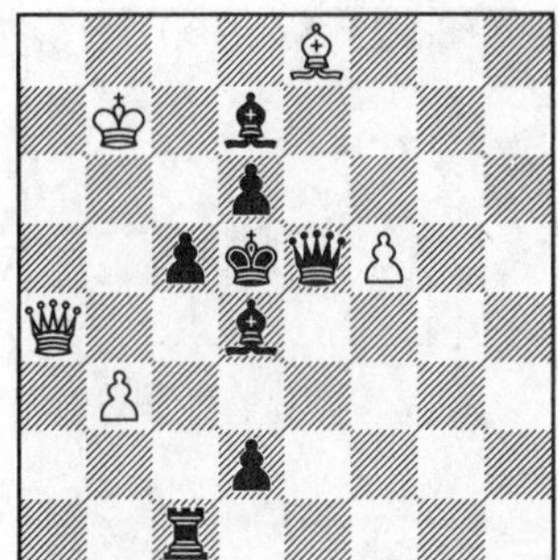

3702.

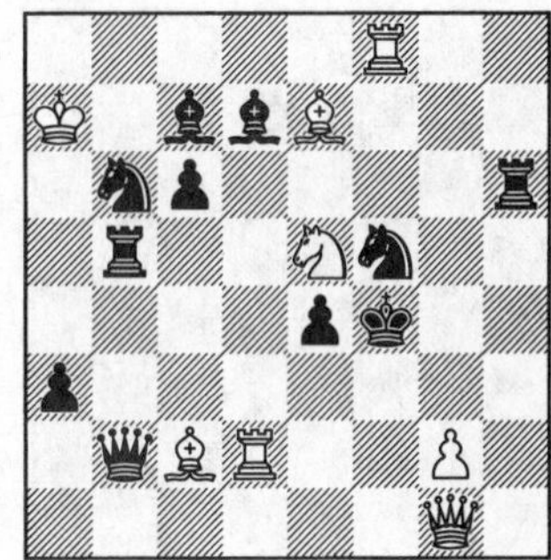

3703.

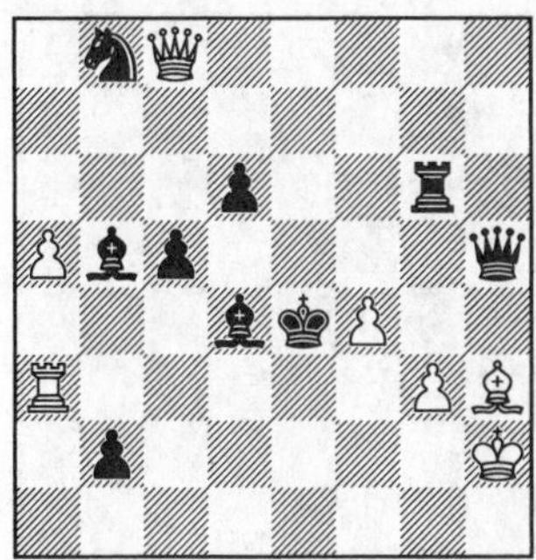

3704.

3705.

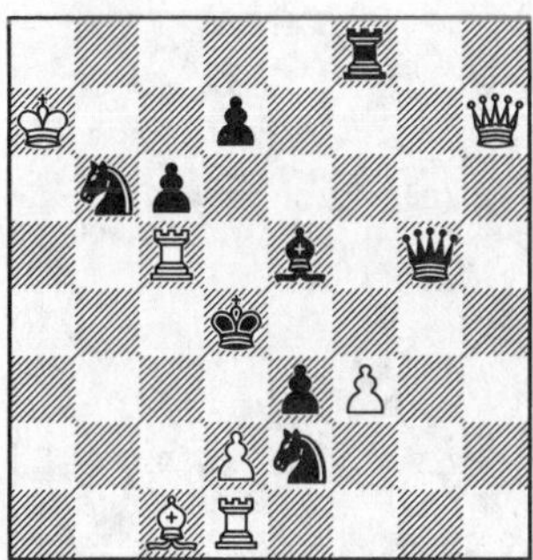

3706.

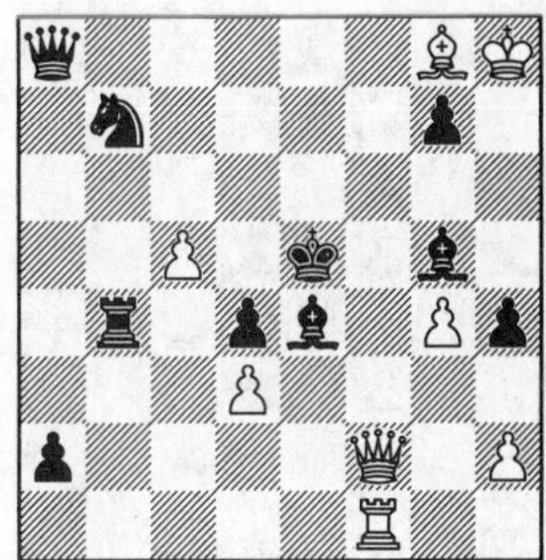

3707.

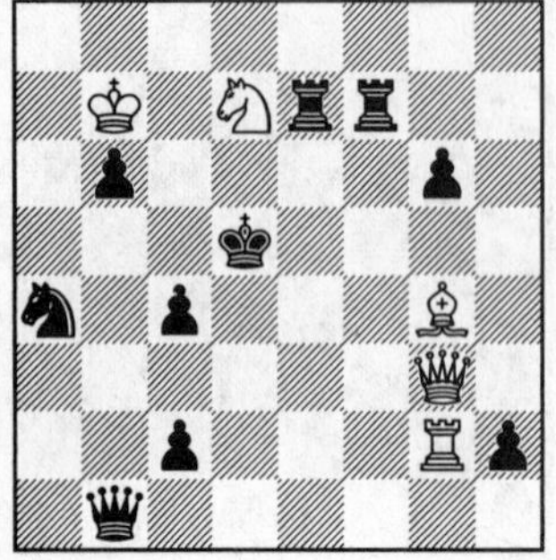

3708.

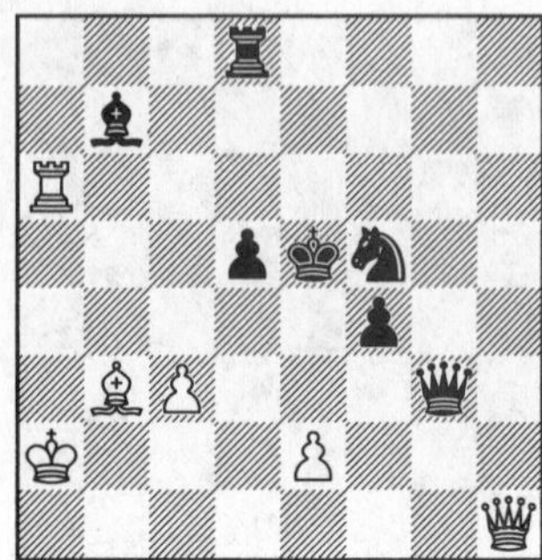

3709.

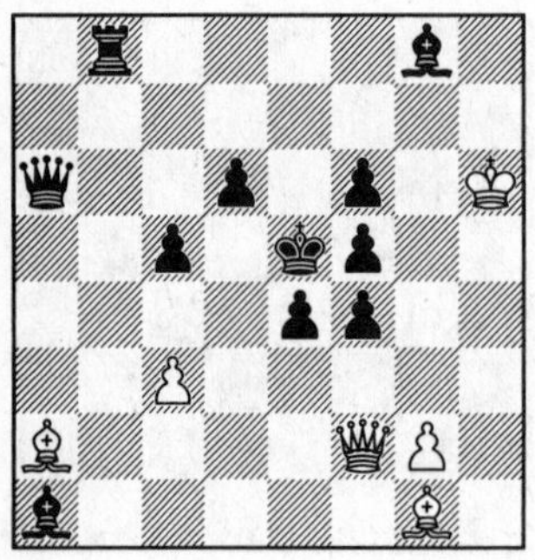

3710.

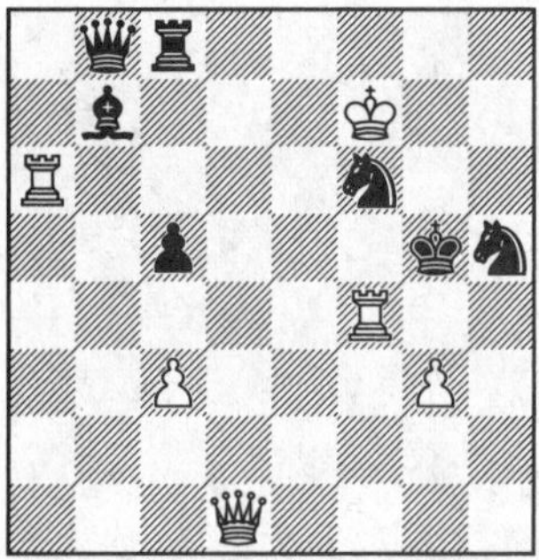

3711.

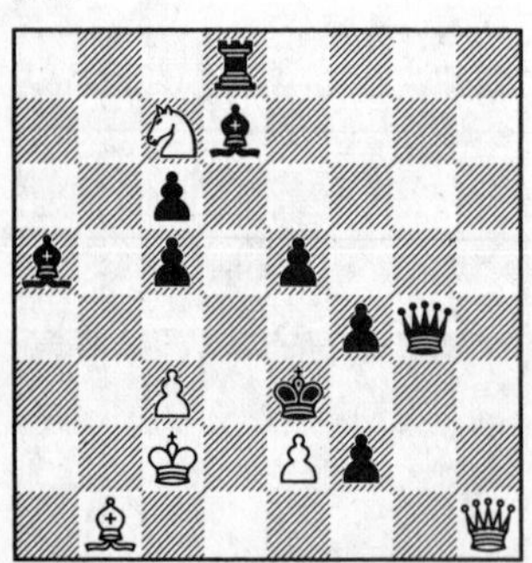

3712.

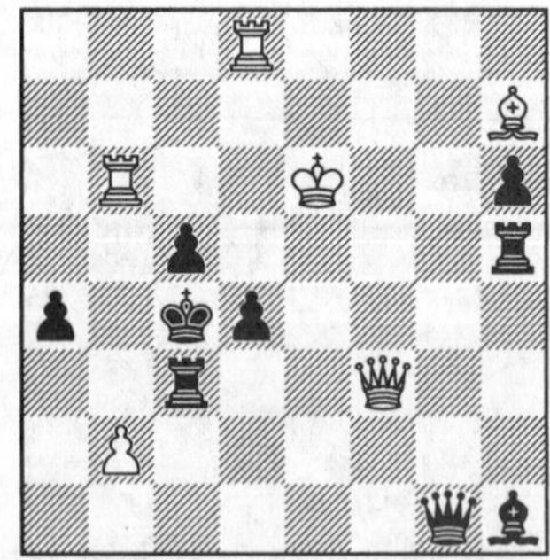

3713.

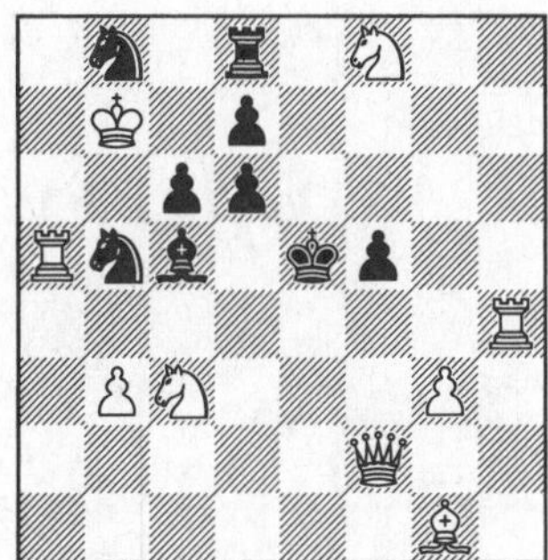

3714.

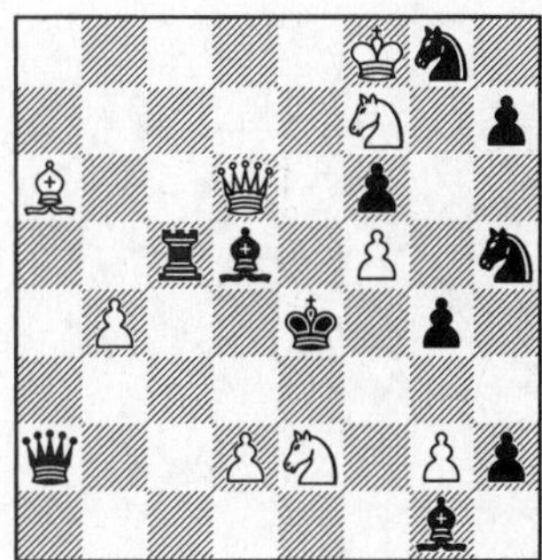

3715.

3716.

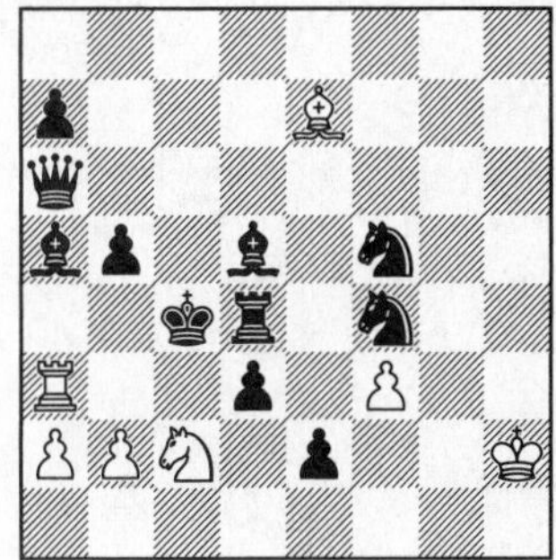

3717.

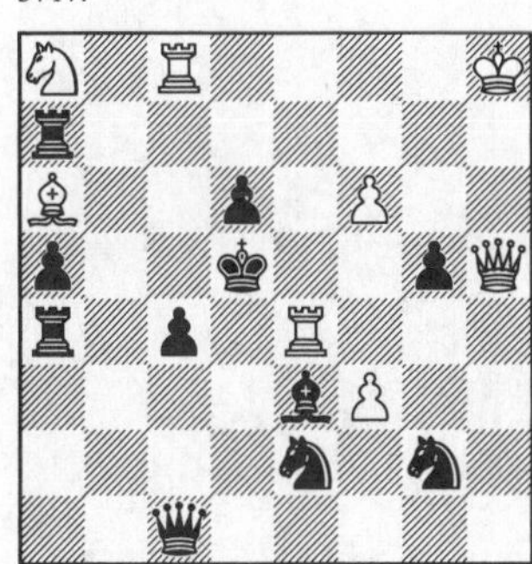

3718.

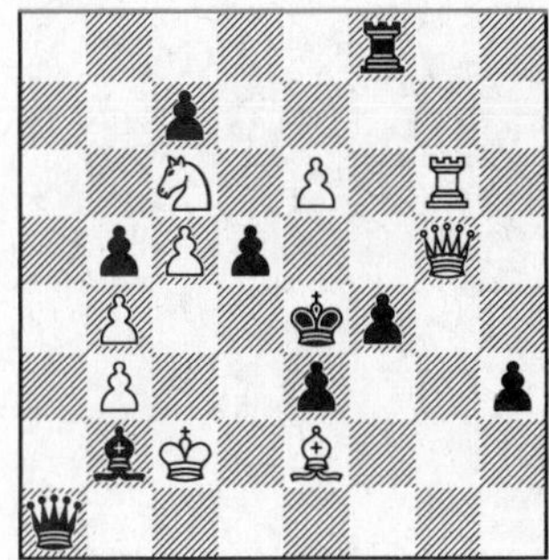

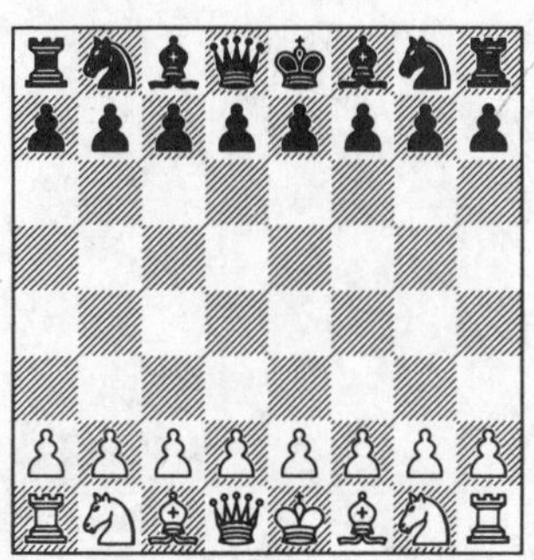

Mate in three

(3719–4462)

3719.

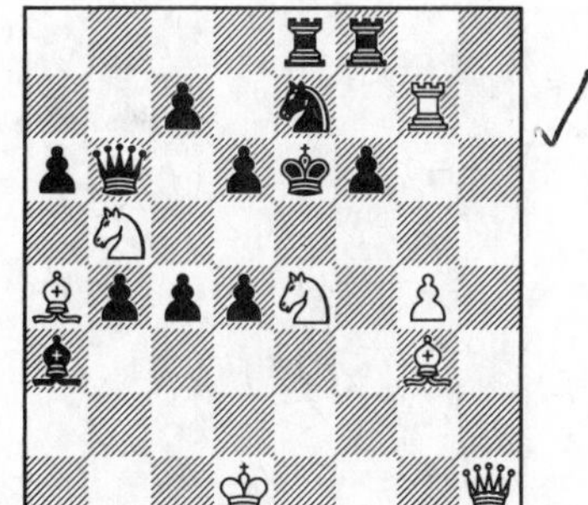

3720.

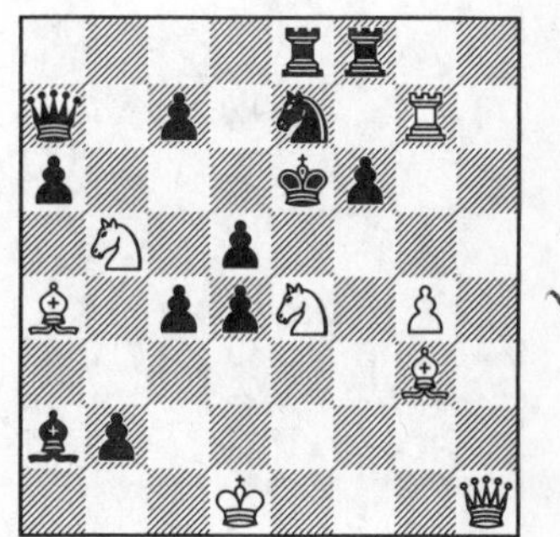

3721.

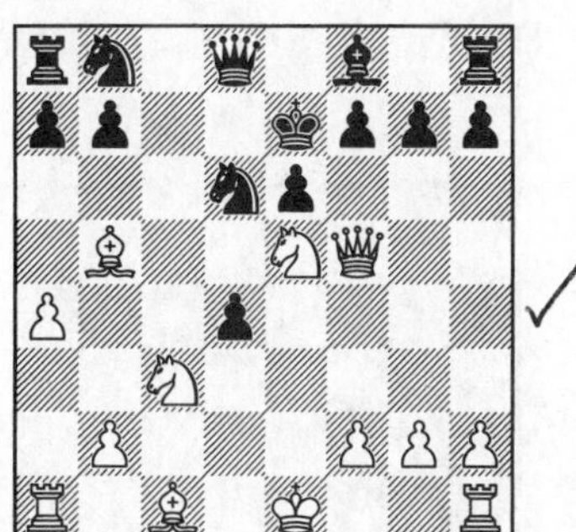

3722.

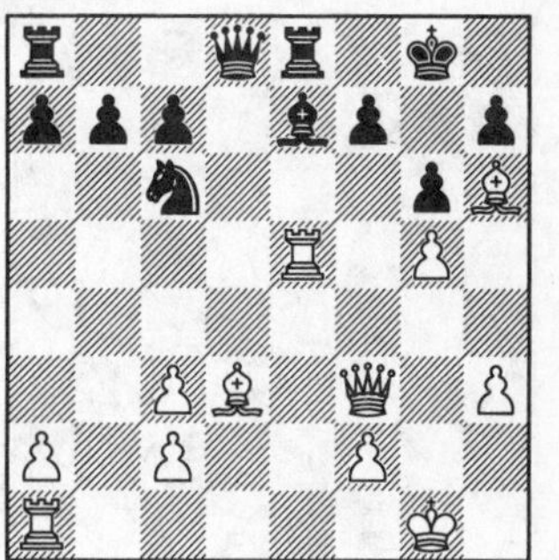

3723.

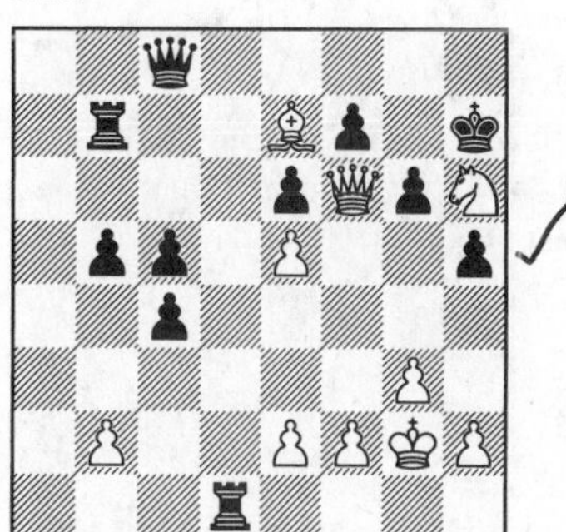

3724.

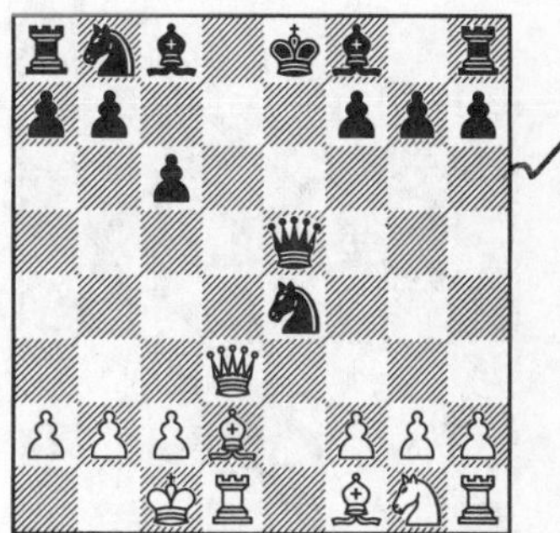

3725. 3726.

3727. 3728.

3729. 3730.

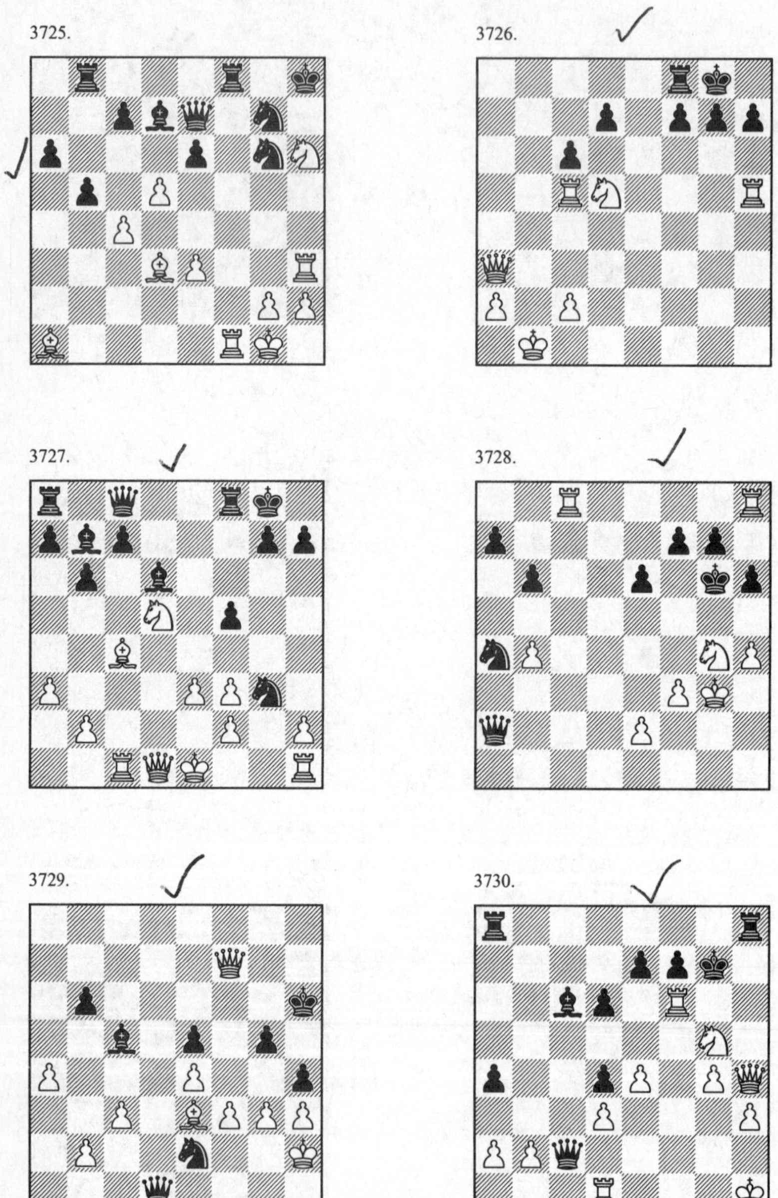

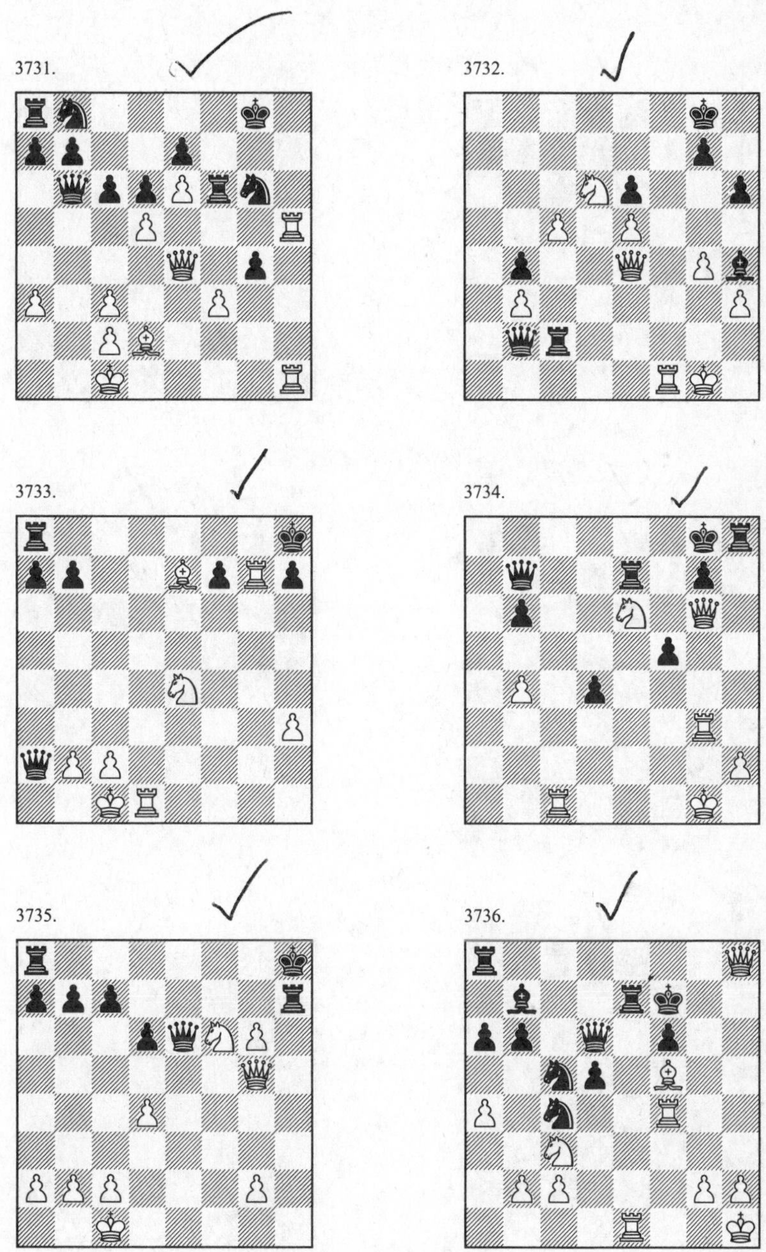
3731.
3732.
3733.
3734.
3735.
3736.

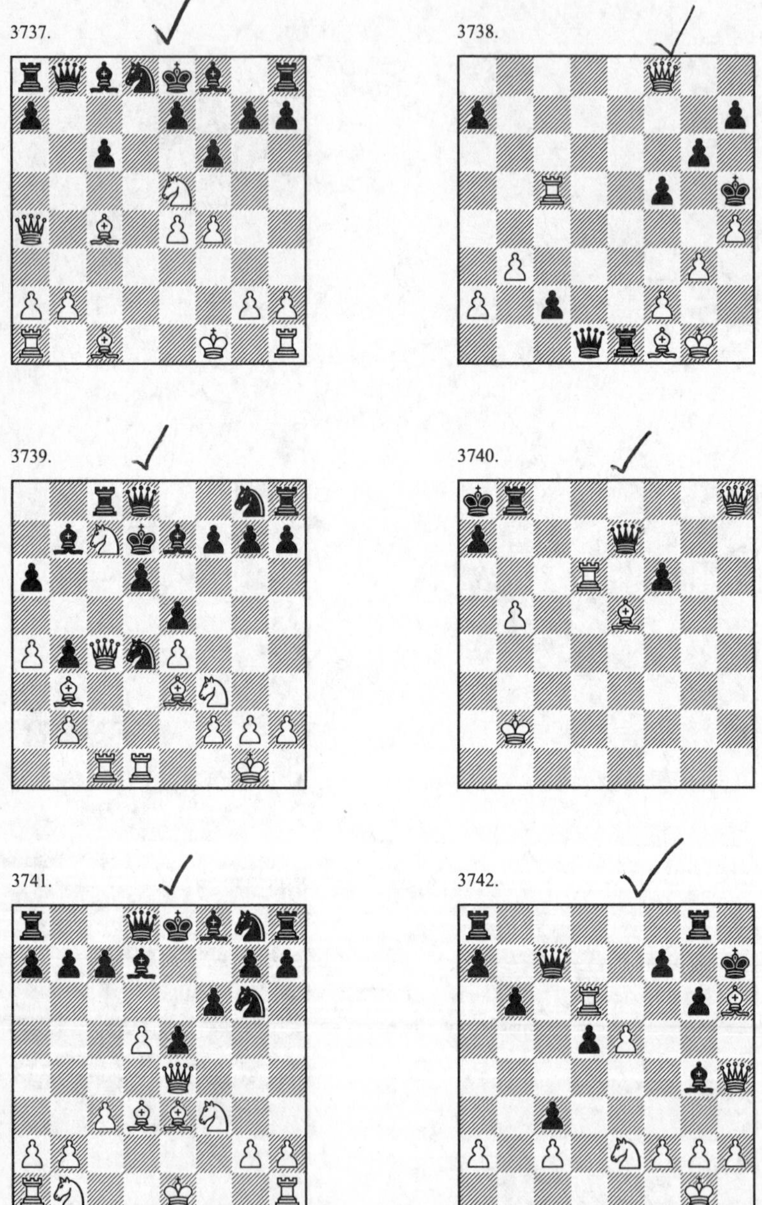
3737.
3738.
3739.
3740.
3741.
3742.

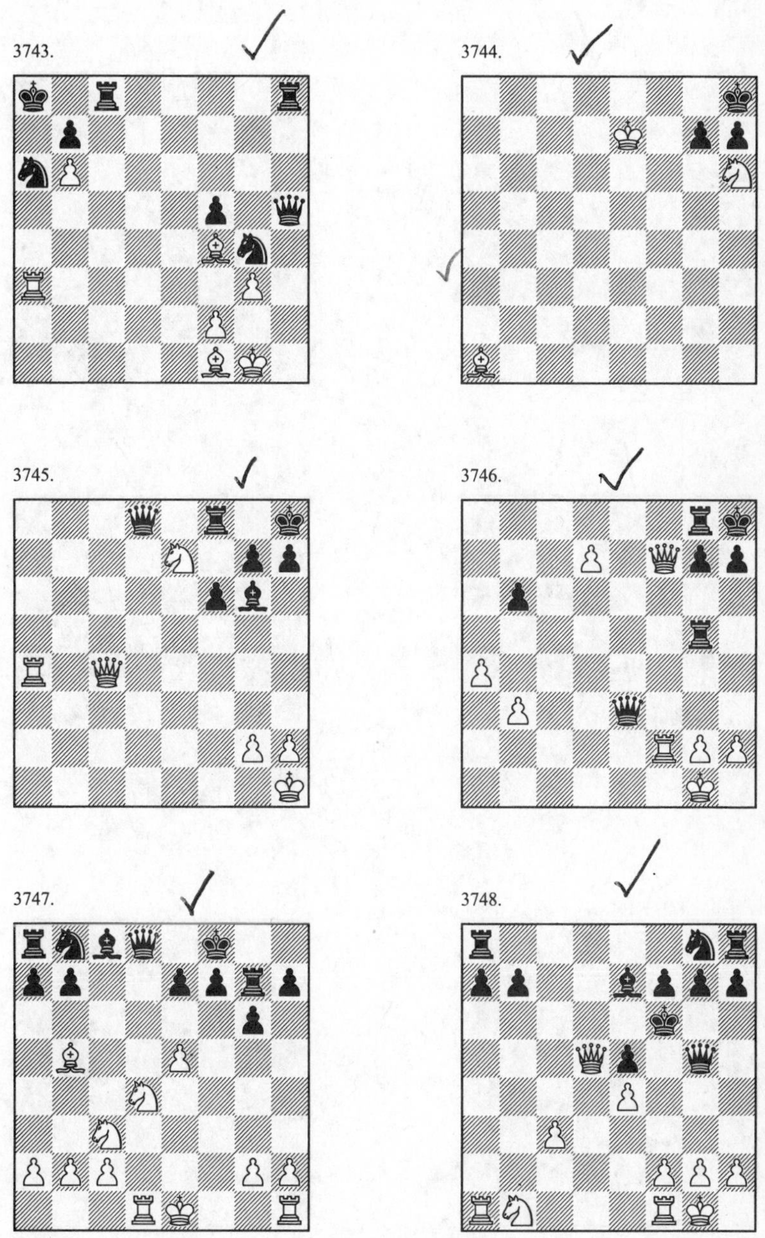

3743.

3744.

3745.

3746.

3747.

3748.

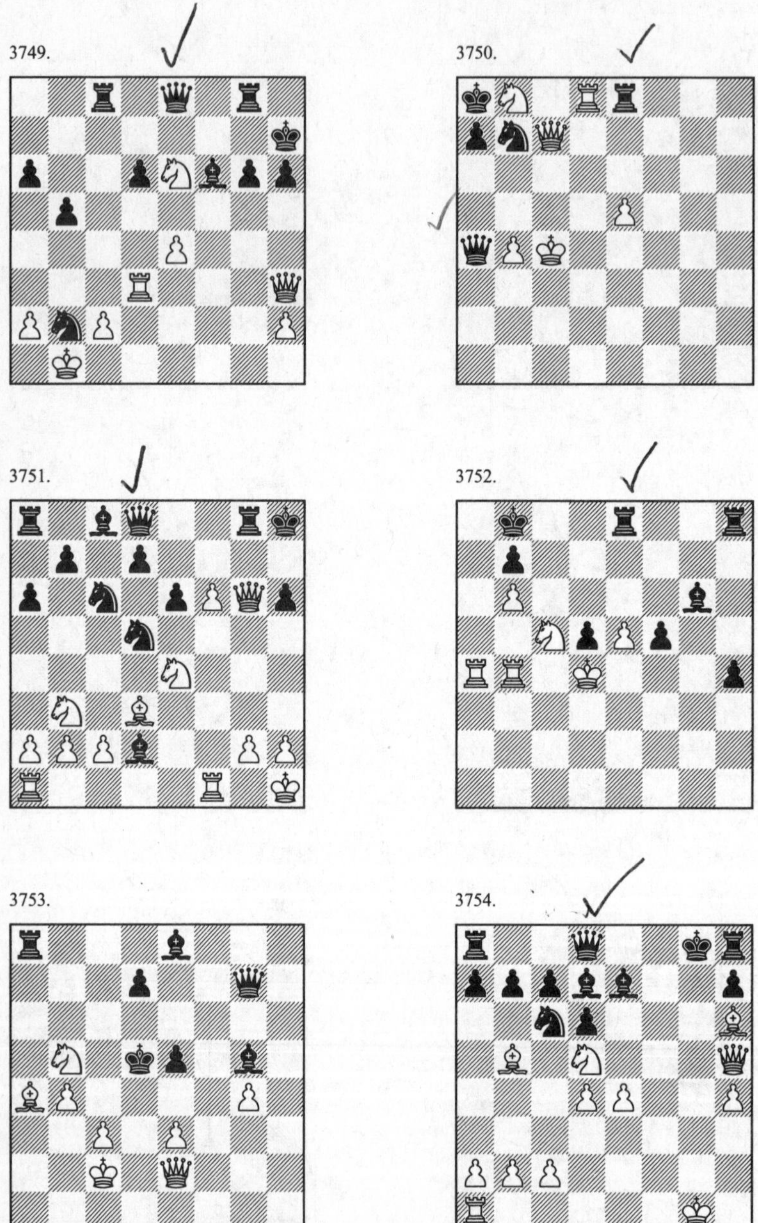
3749.
3750.
3751.
3752.
3753.
3754.

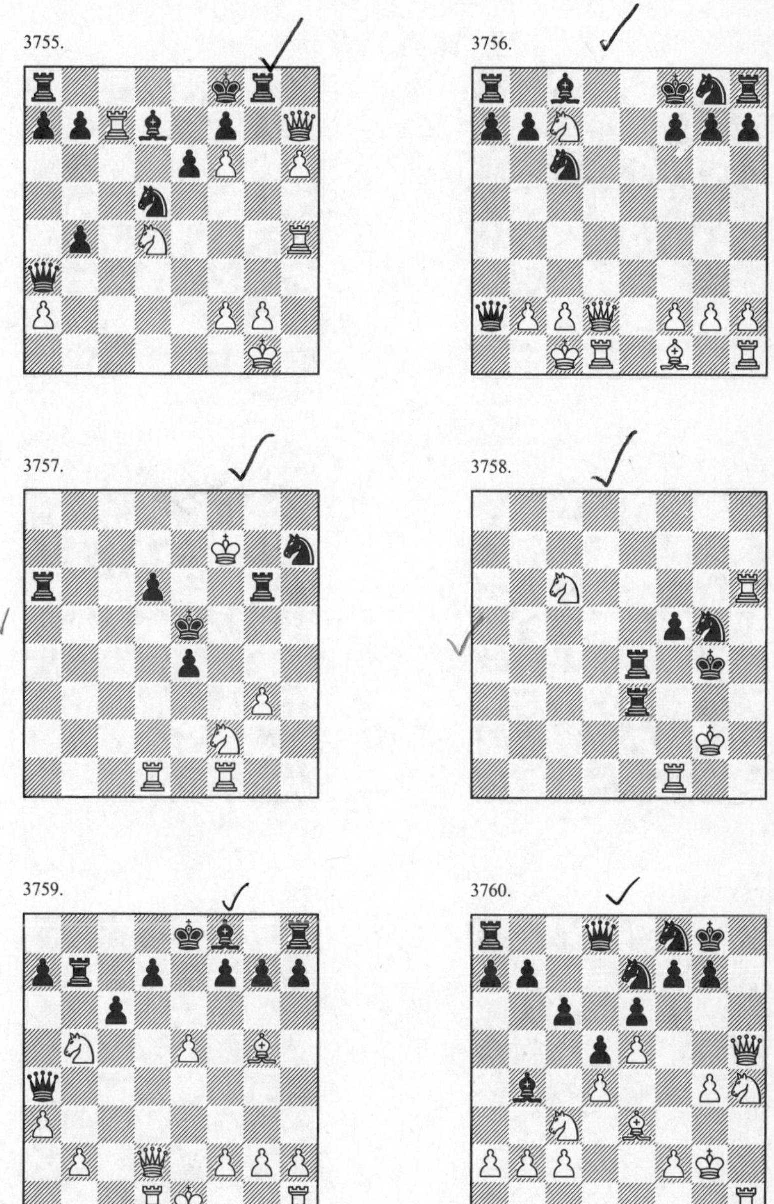
3755.
3756.
3757.
3758.
3759.
3760.

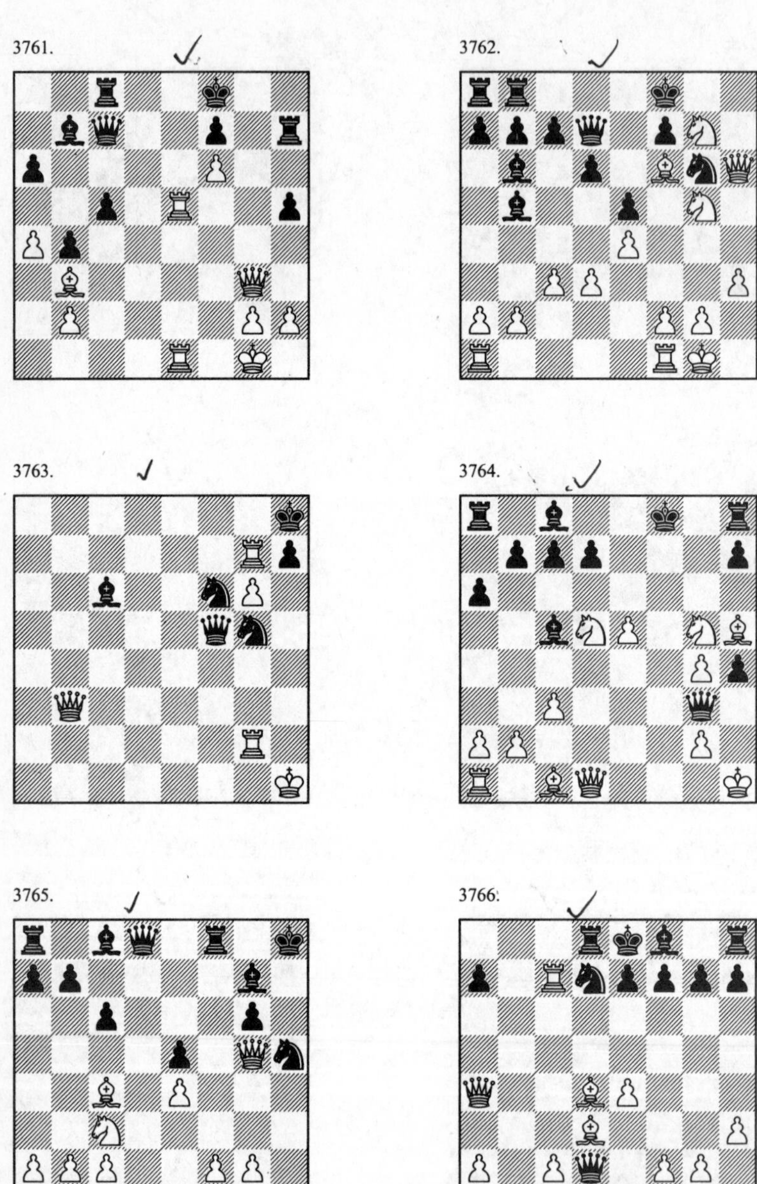

3761.

3762.

3763.

3764.

3765.

3766.

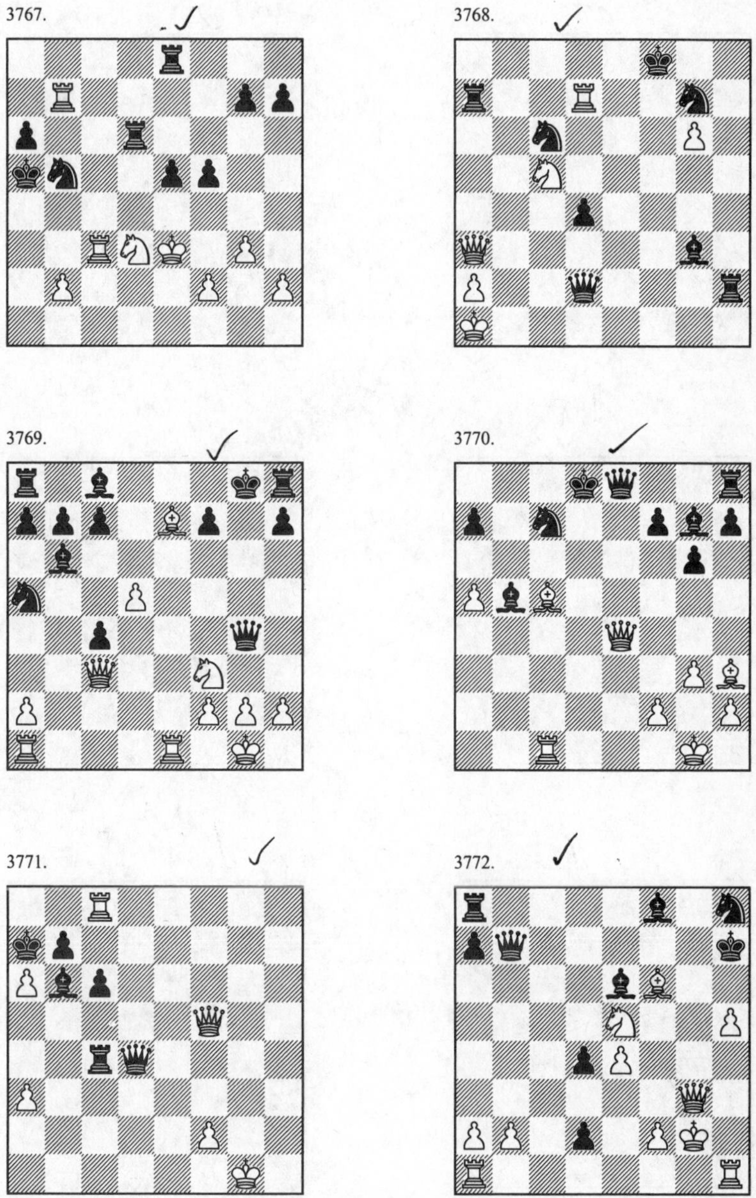
3767.
3768.
3769.
3770.
3771.
3772.

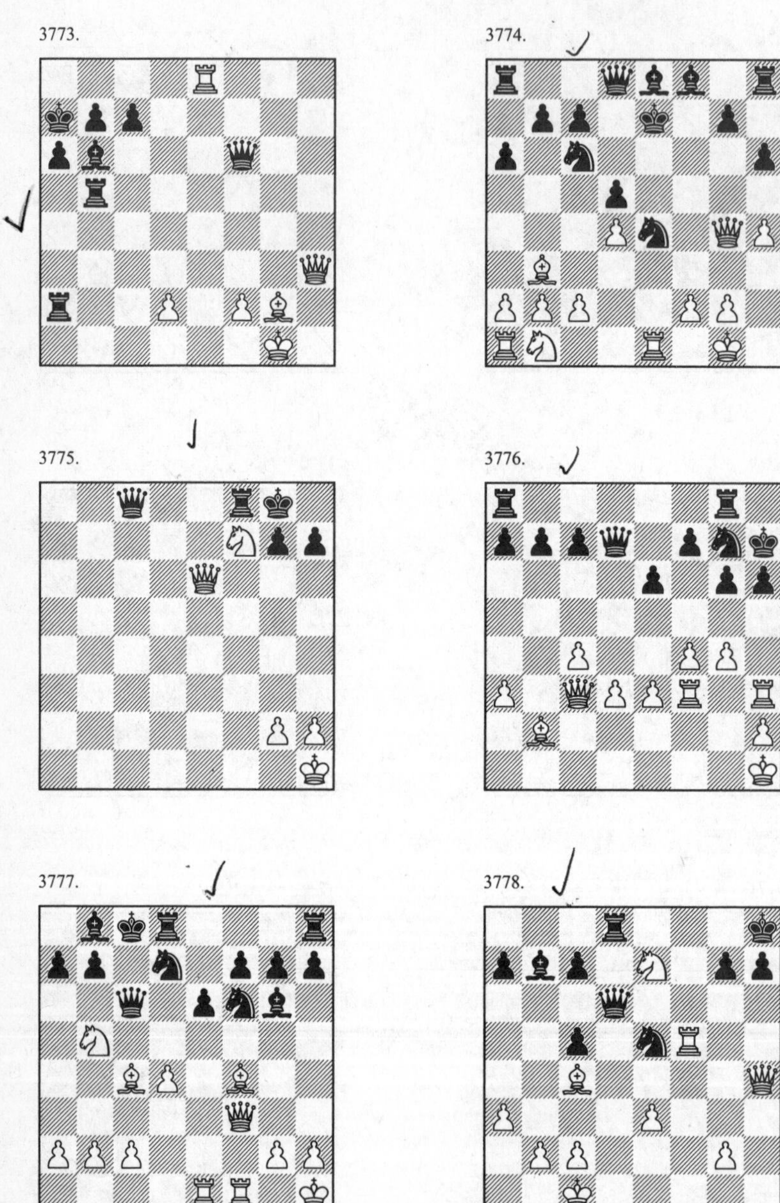
3773.
3774.
3775.
3776.
3777.
3778.

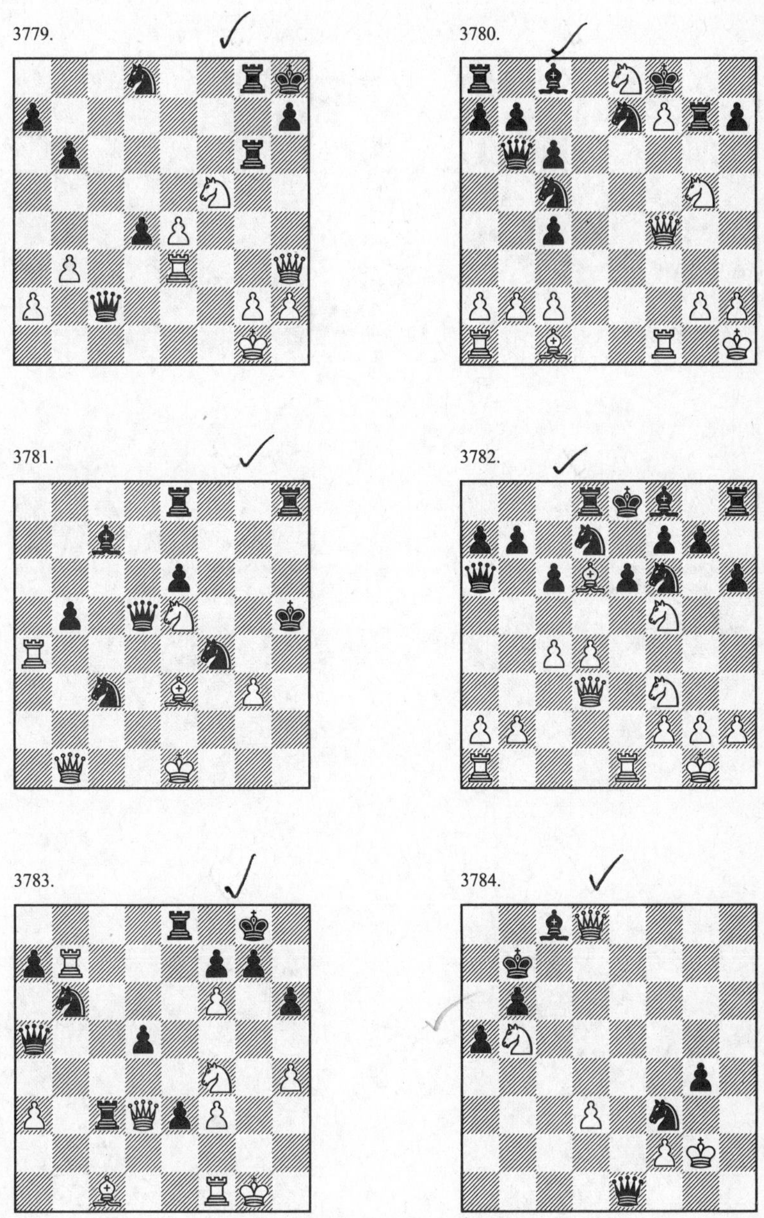
3779.
3780.
3781.
3782.
3783.
3784.

3785. 3786.

3787. 3788.

3789. 3790.

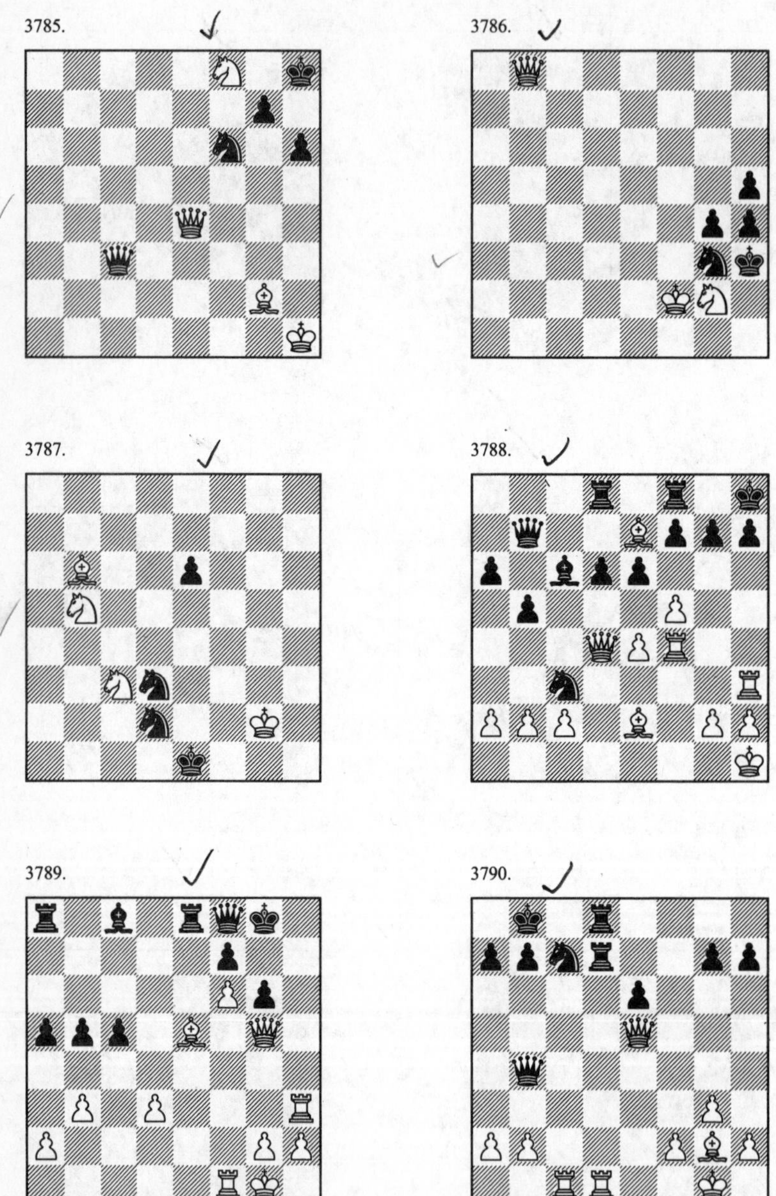

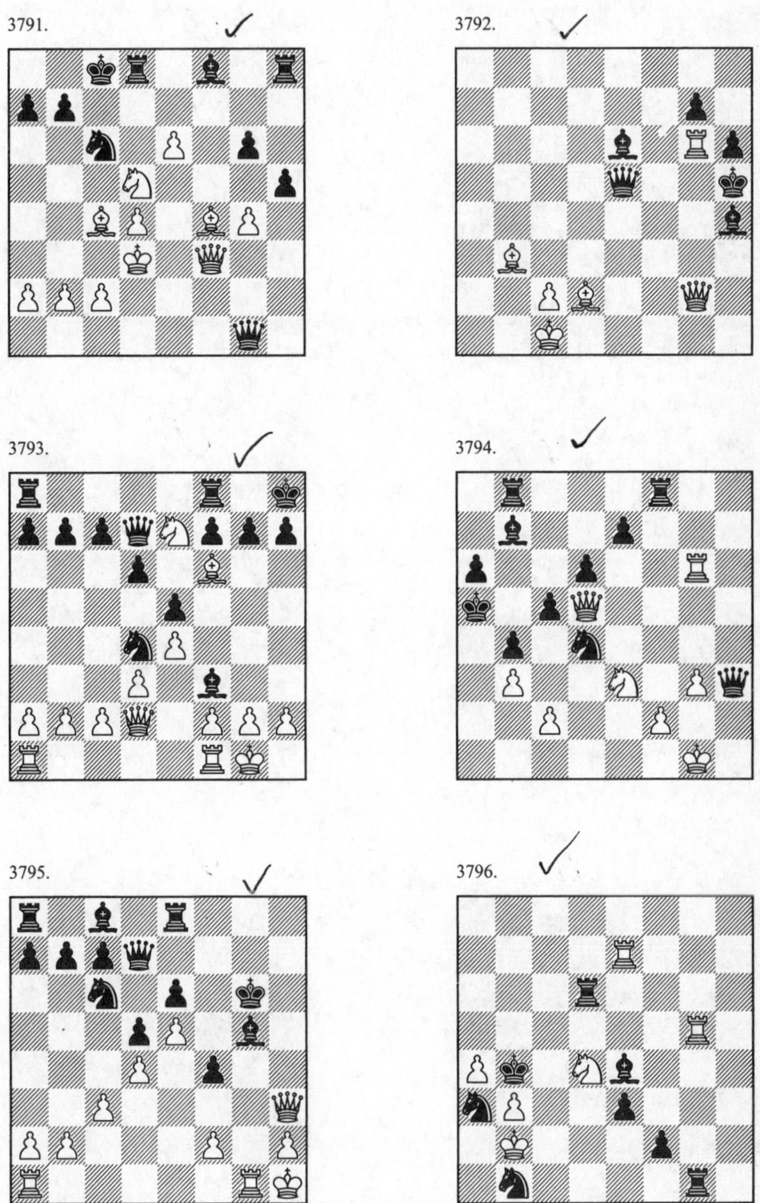
3791.
3792.
3793.
3794.
3795.
3796.

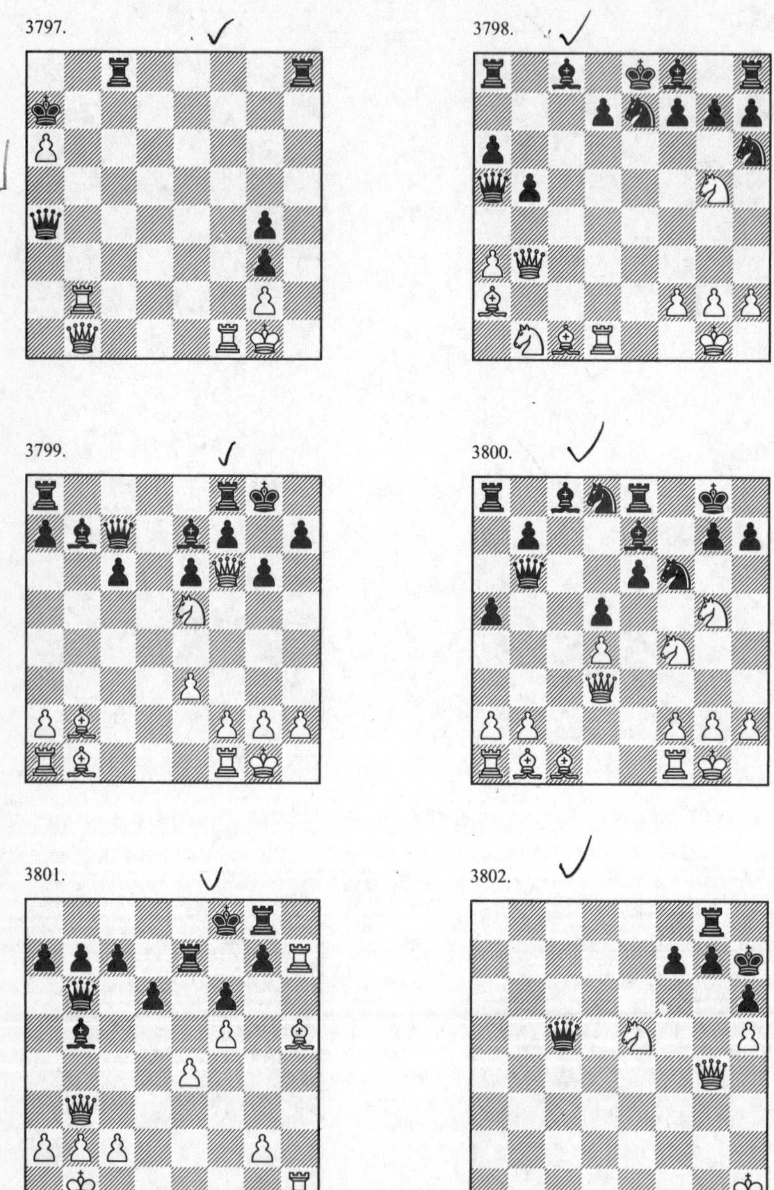
3797.
3798.
3799.
3800.
3801.
3802.

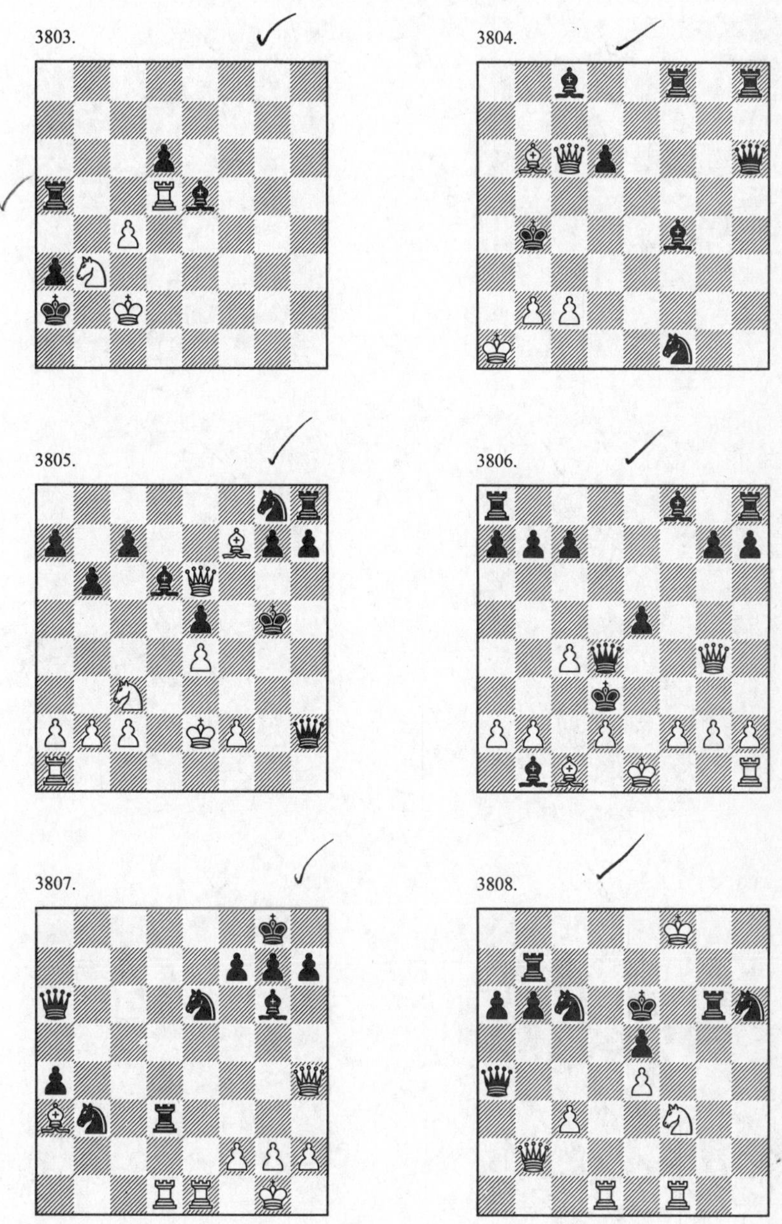
3803.
3804.
3805.
3806.
3807.
3808.

3809.

3810.

3811.

3812.

3813.

3814.

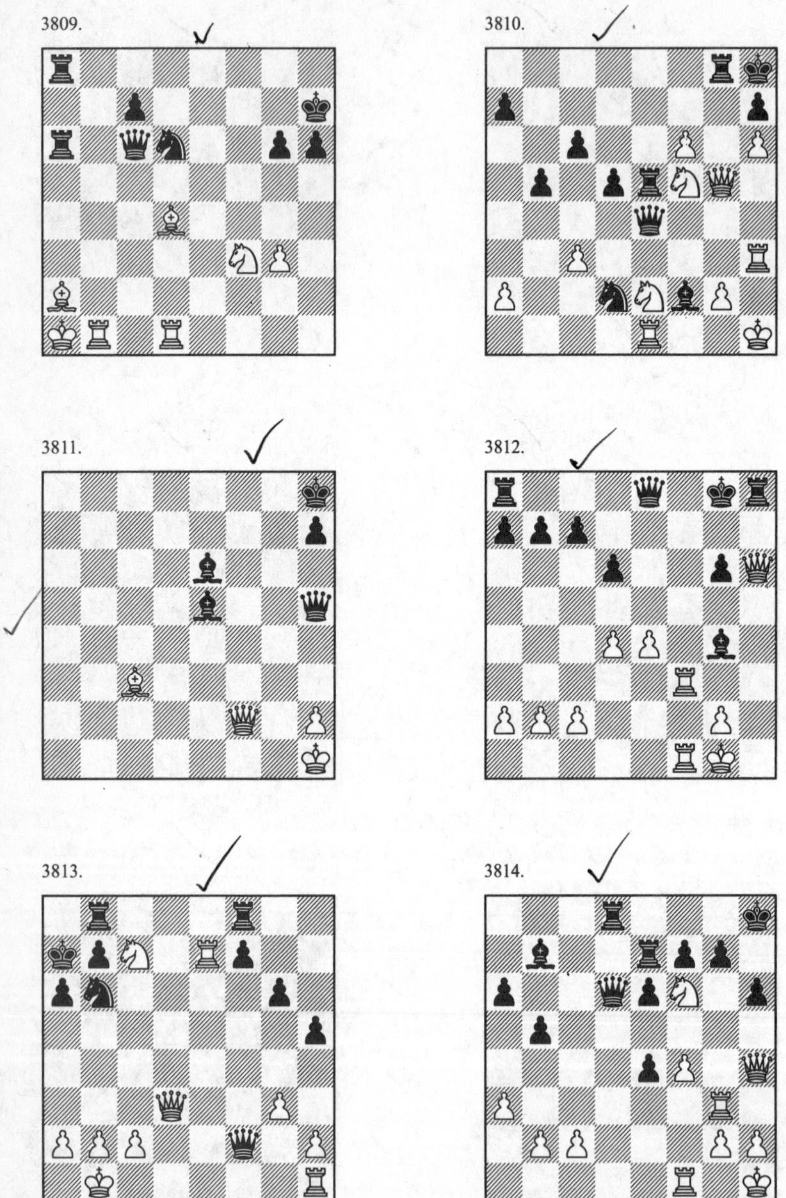

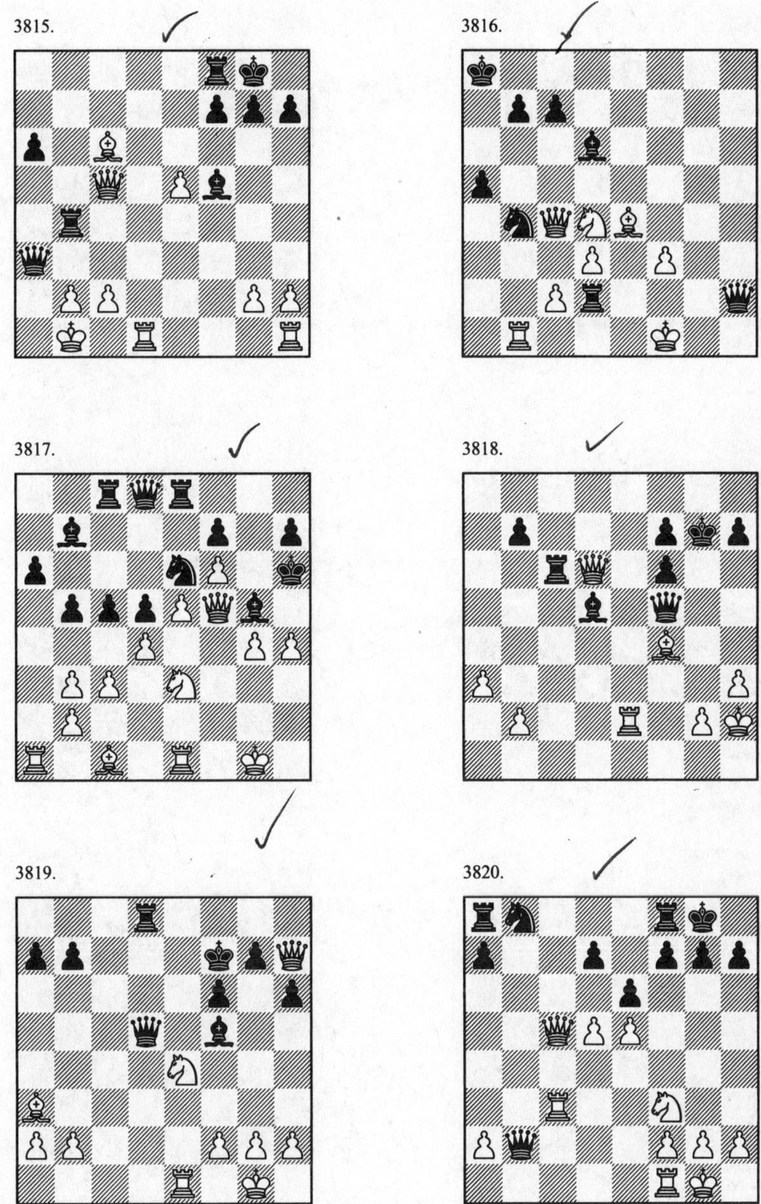
3815.
3816.
3817.
3818.
3819.
3820.

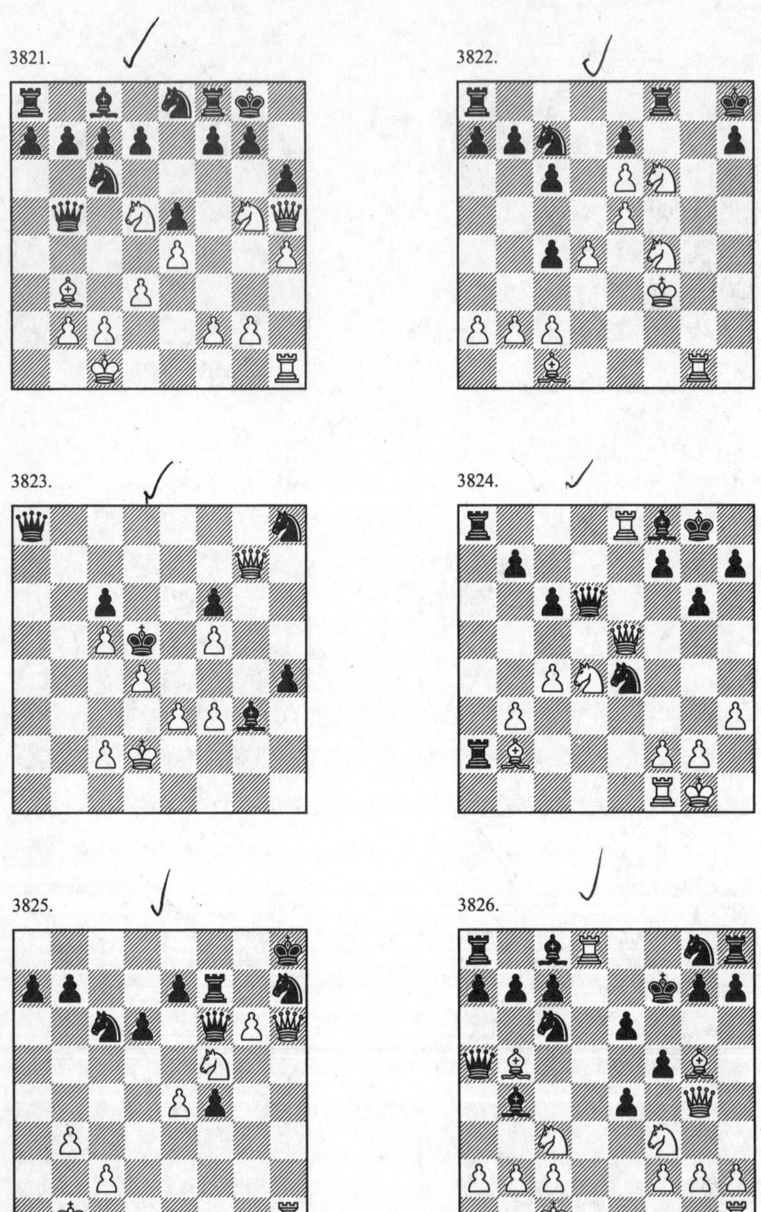
3821.
3822.
3823.
3824.
3825.
3826.

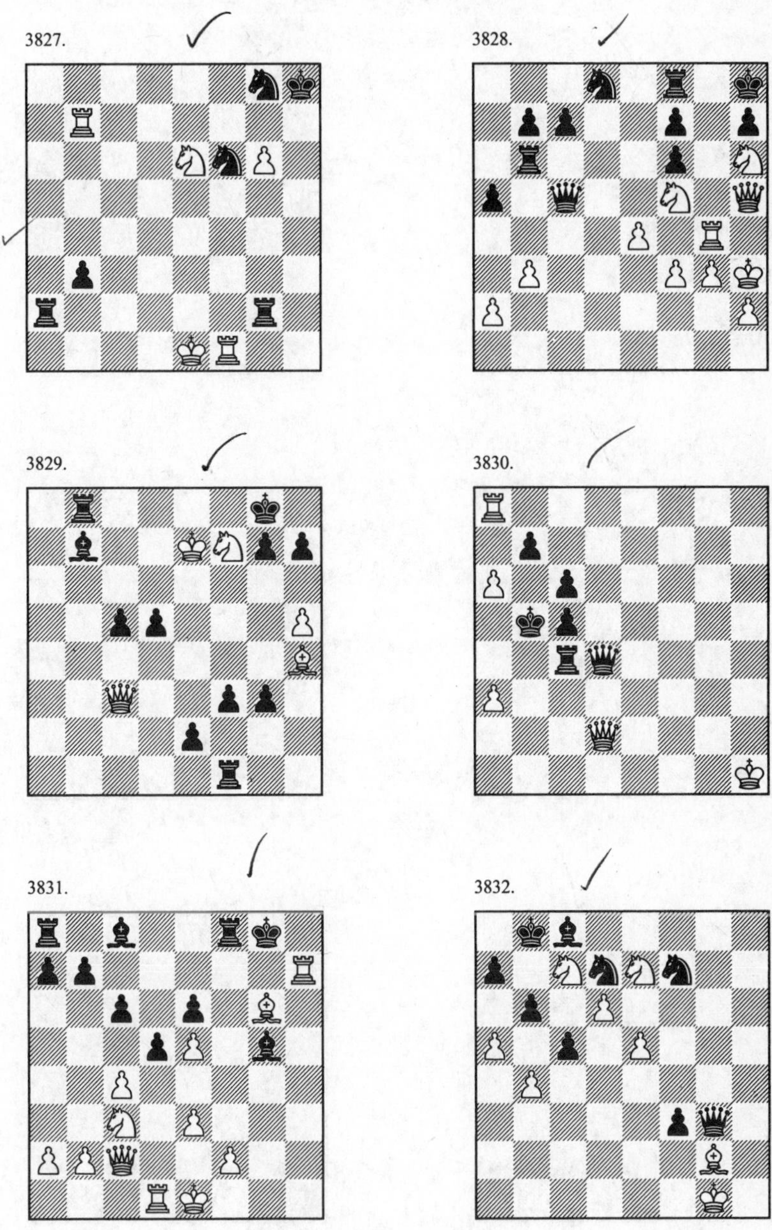

3827.

3828.

3829.

3830.

3831.

3832.

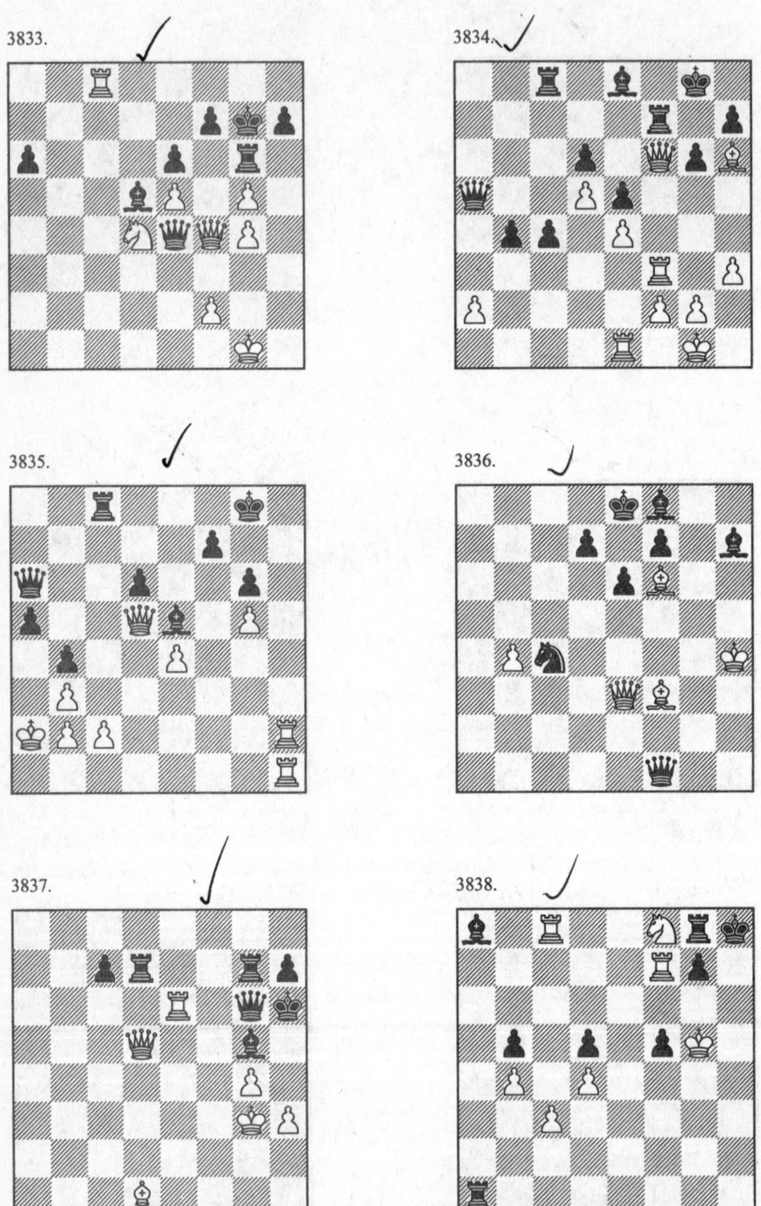
3833.
3834.
3835.
3836.
3837.
3838.

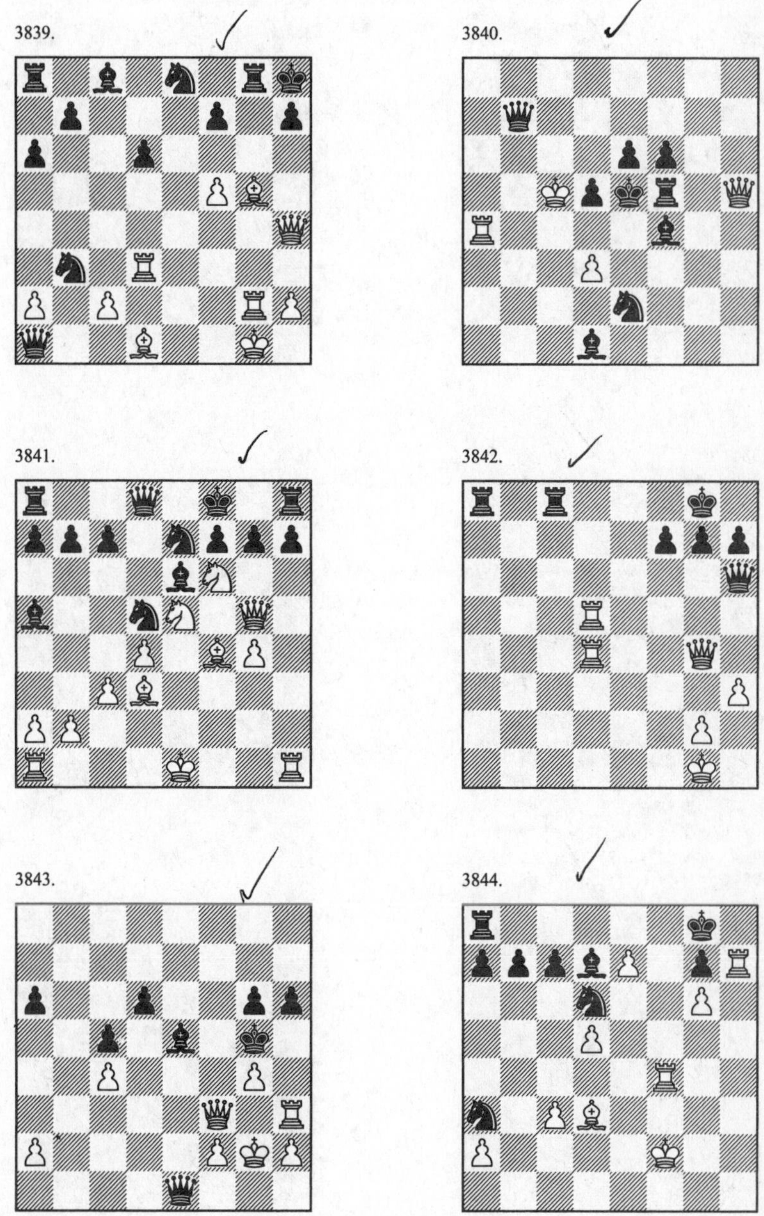
3839.
3840.
3841.
3842.
3843.
3844.

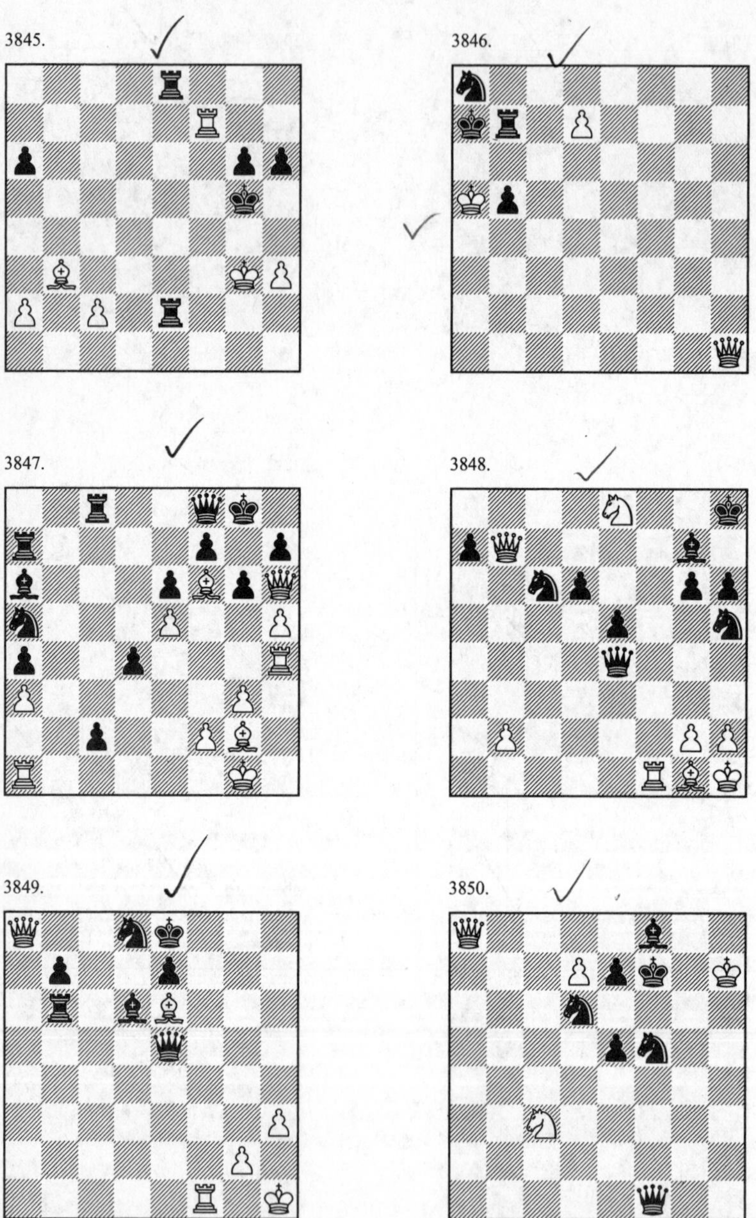
3845.
3846.
3847.
3848.
3849.
3850.

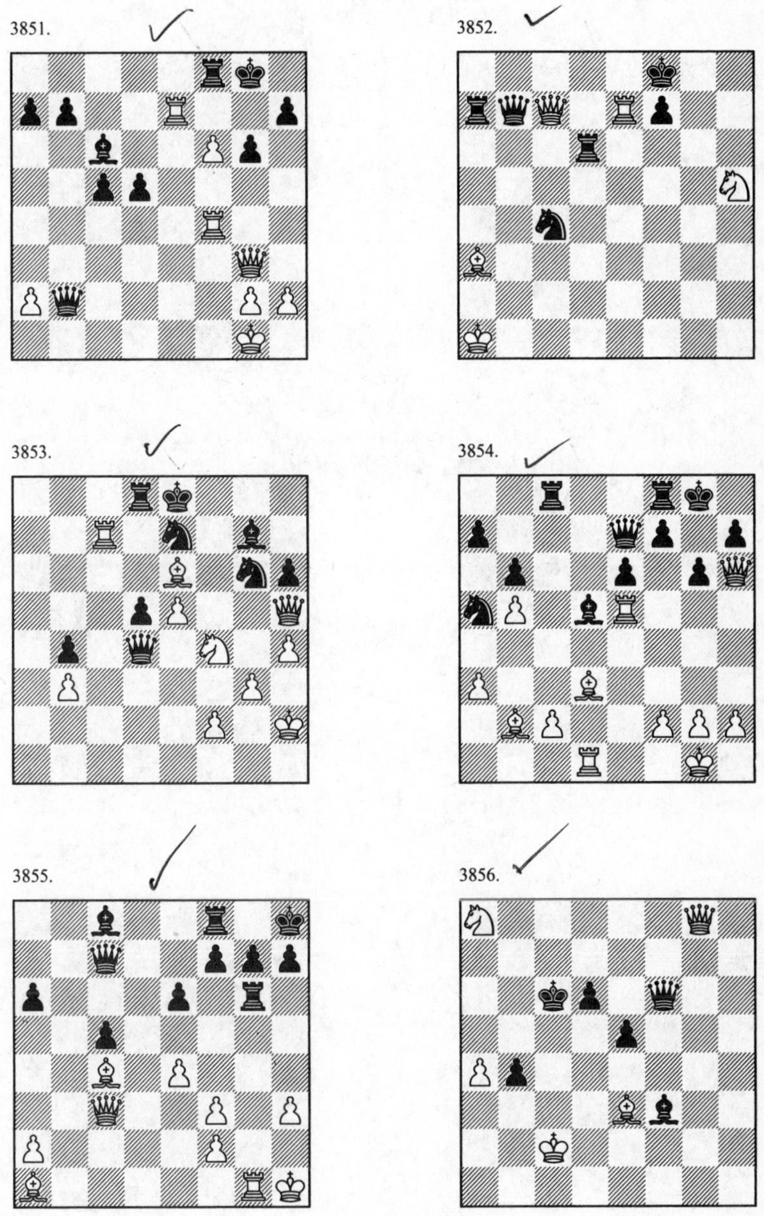
3851.
3852.
3853.
3854.
3855.
3856.

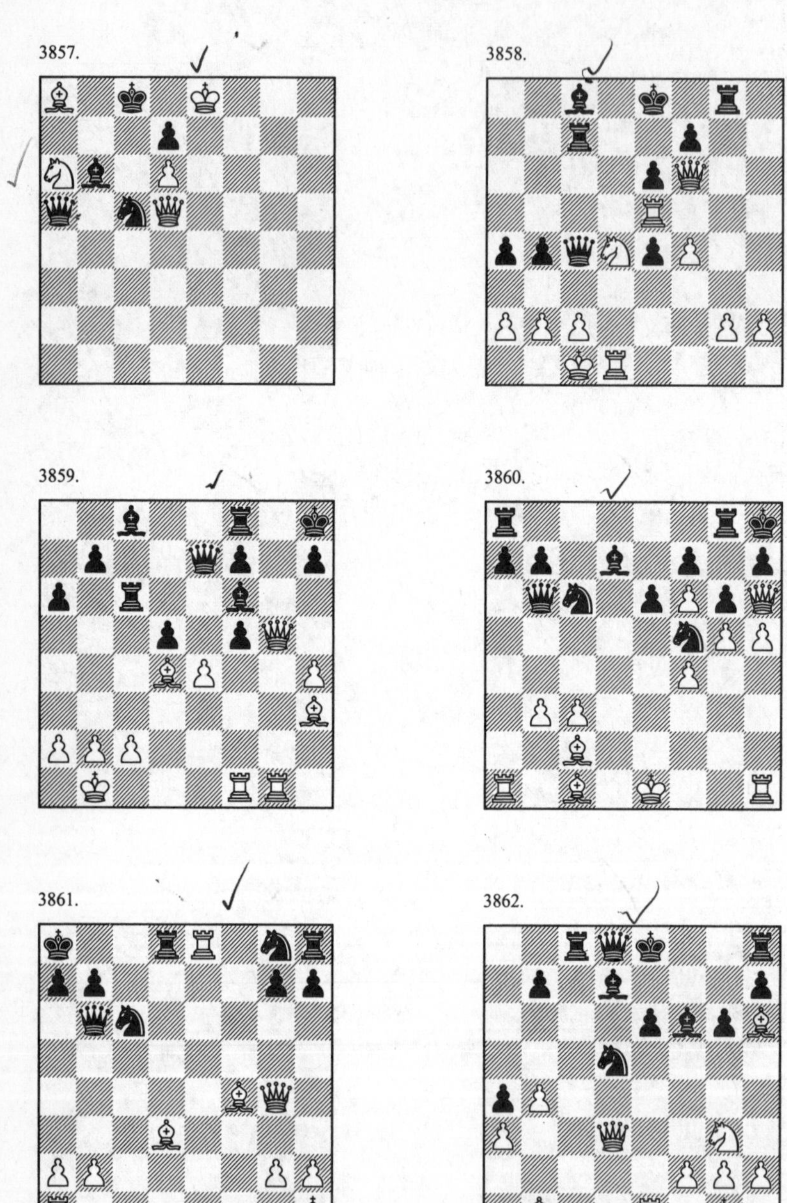
3857.
3858.
3859.
3860.
3861.
3862.

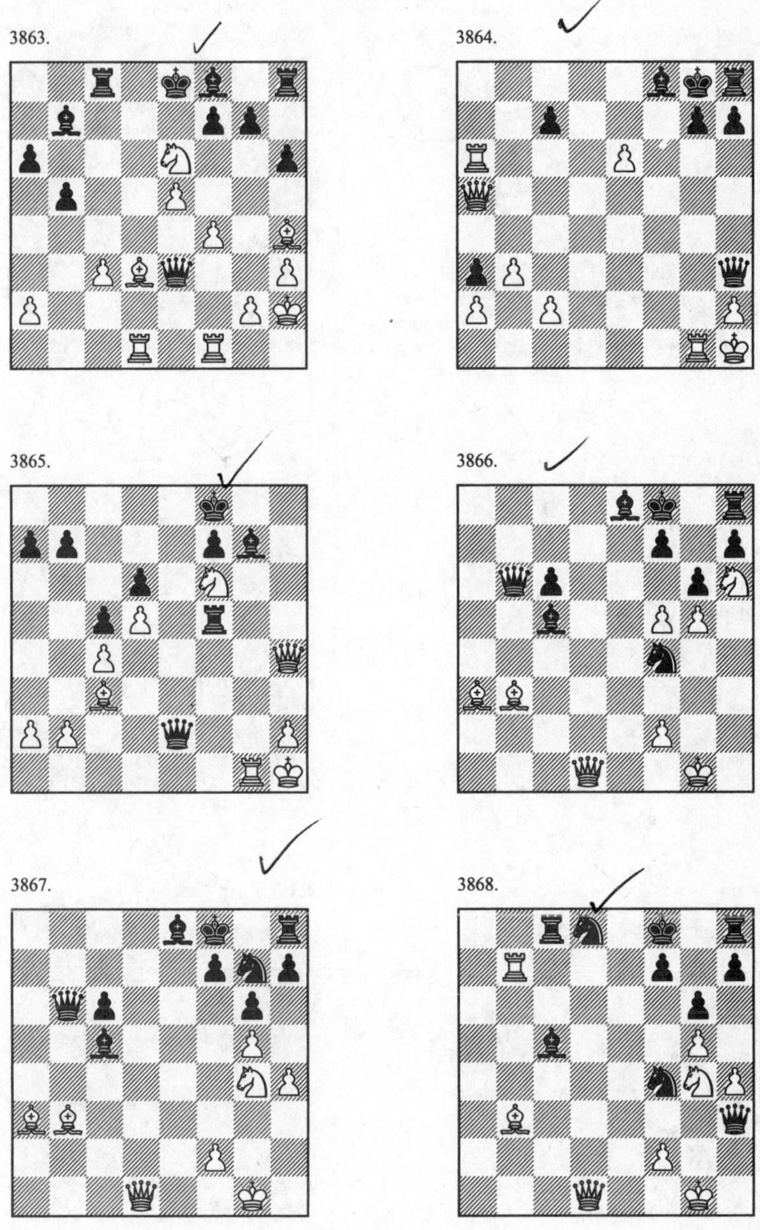

3863.

3864.

3865.

3866.

3867.

3868.

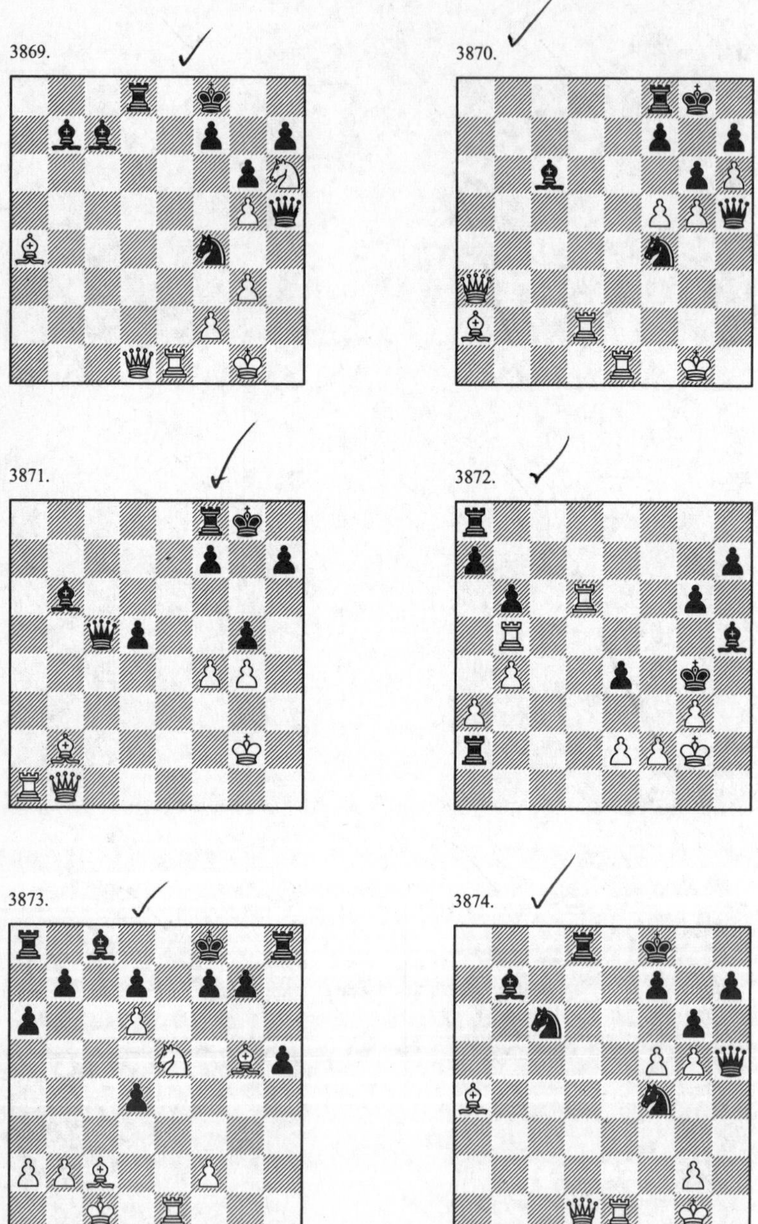
3869.
3870.
3871.
3872.
3873.
3874.

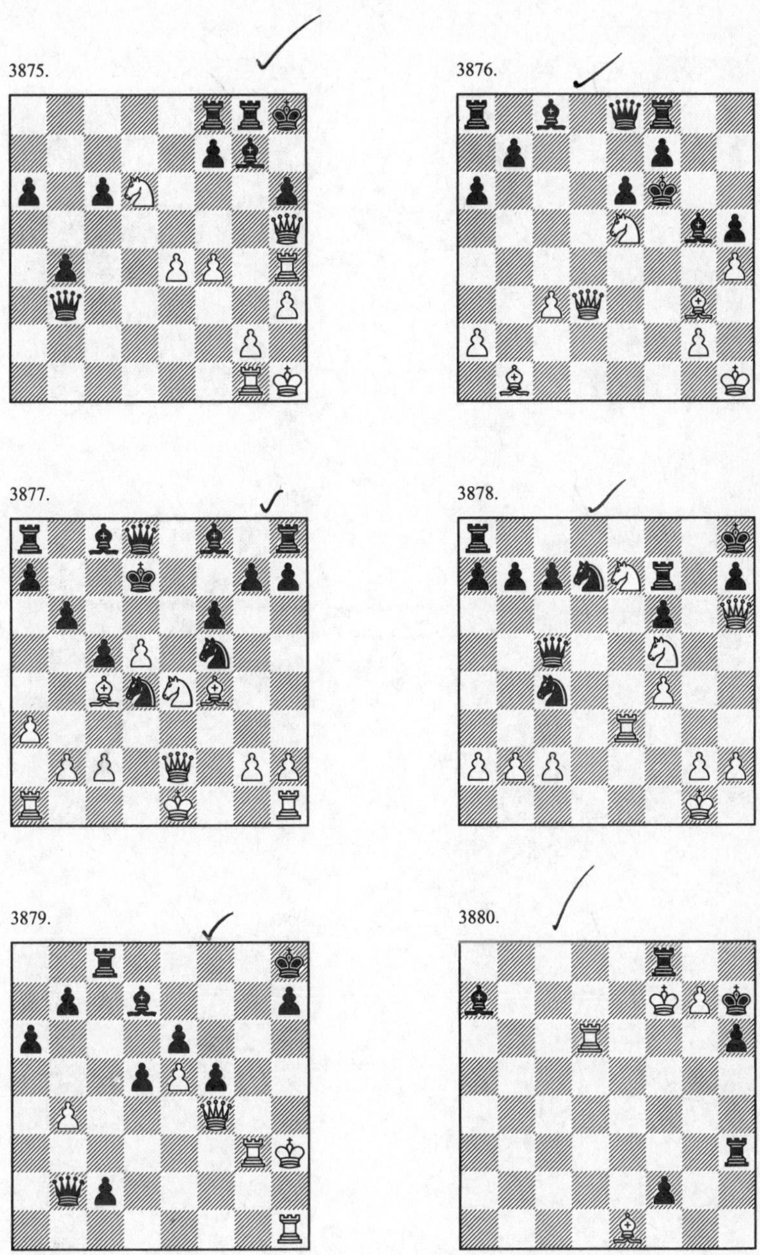
3875.
3876.
3877.
3878.
3879.
3880.

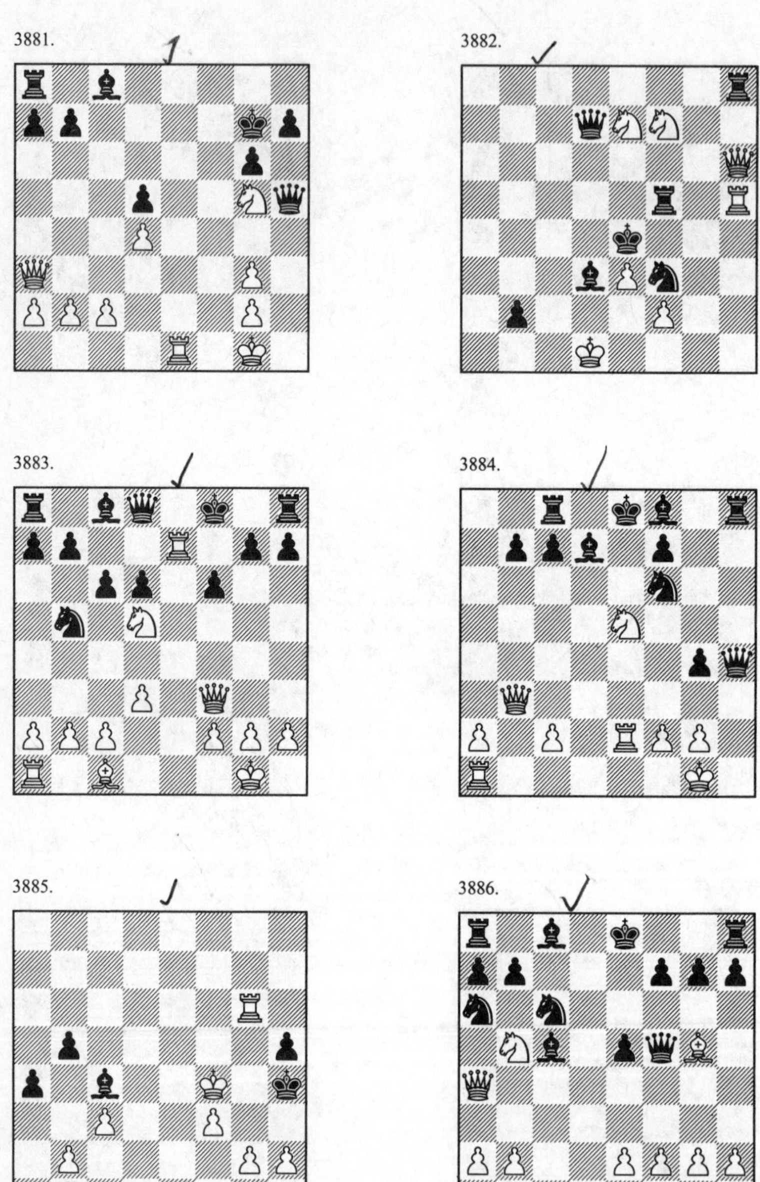

3881.

3882.

3883.

3884.

3885.

3886.

3887.

3888.

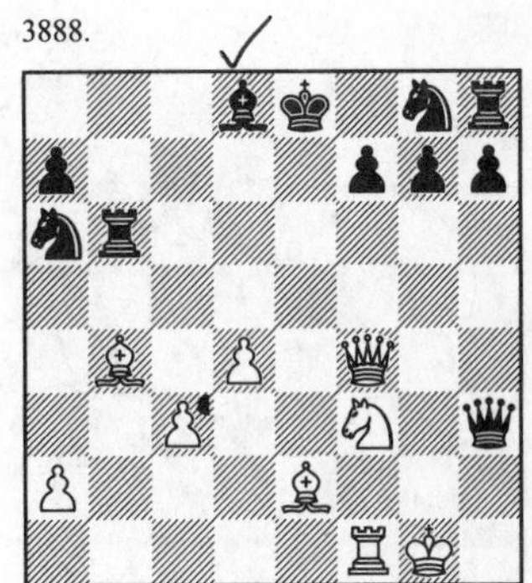

3889.

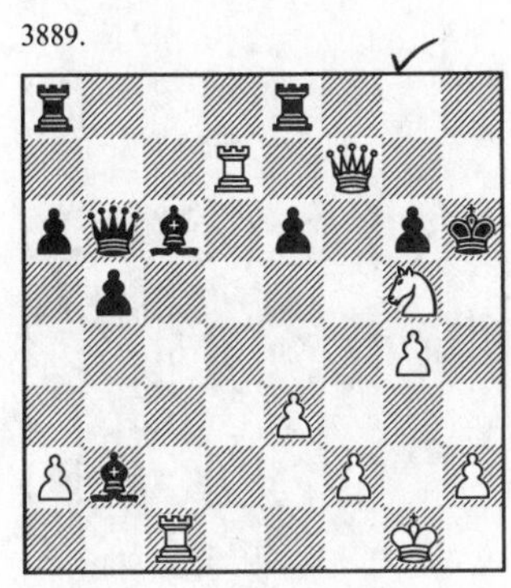

3890.

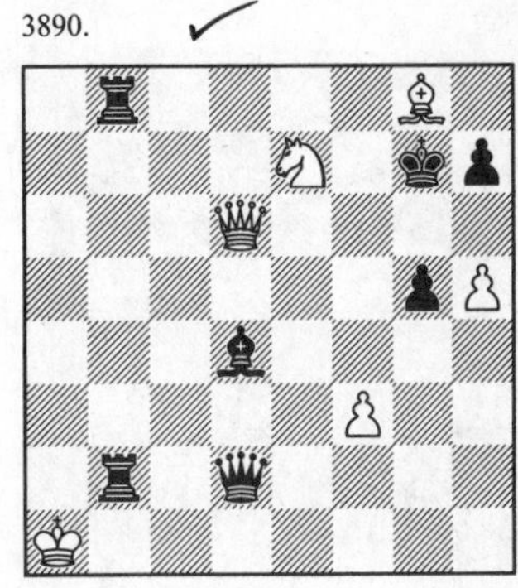

3891.

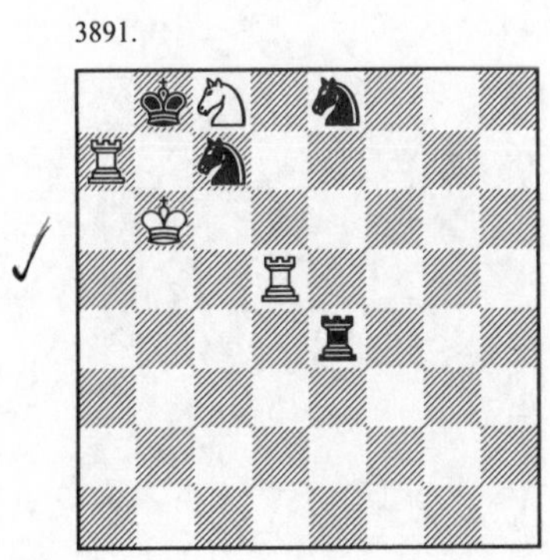

3892.

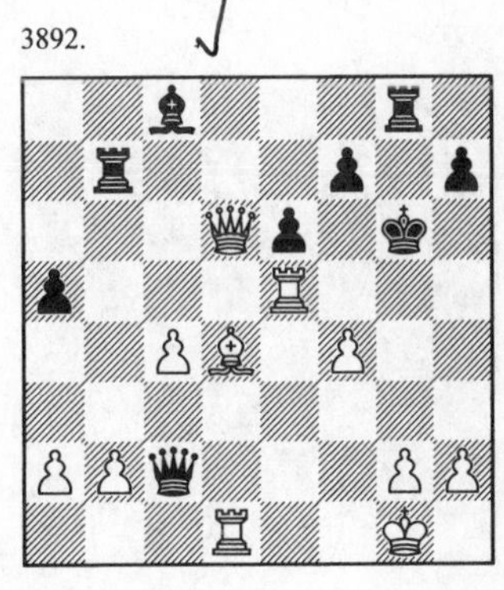

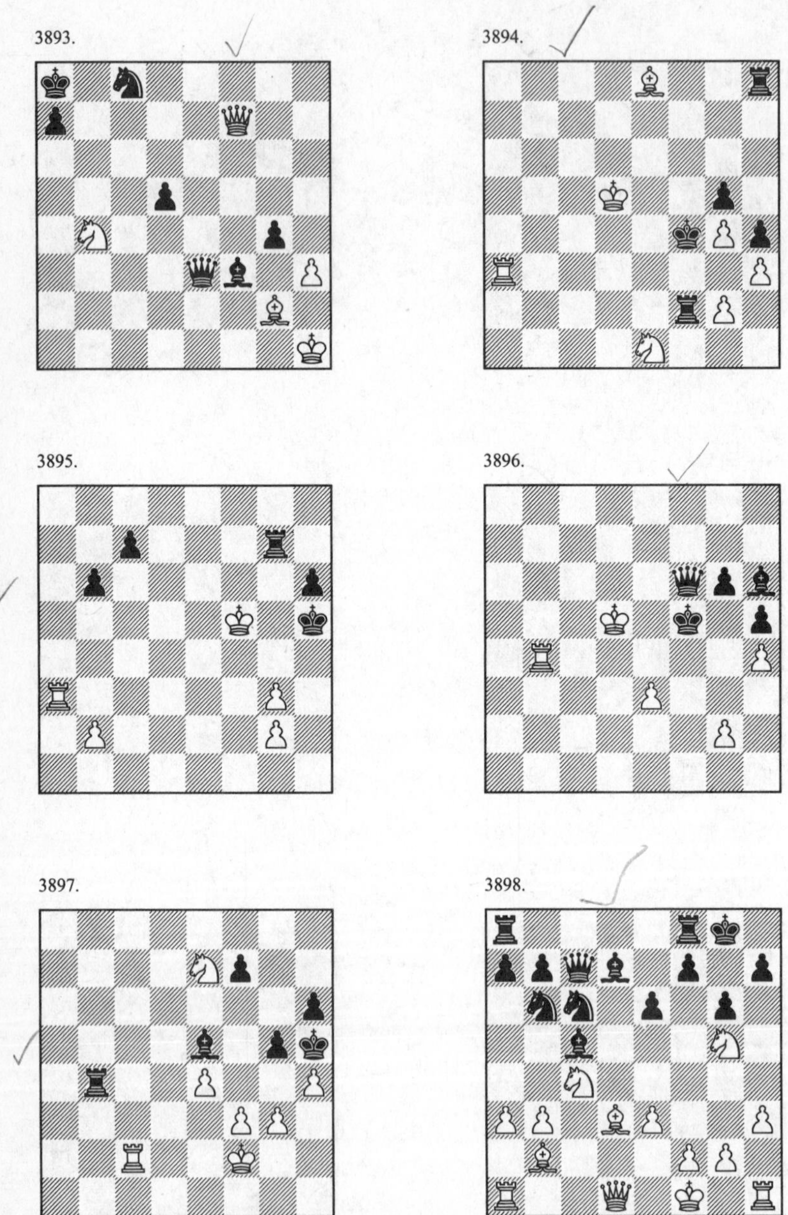
3893.
3894.
3895.
3896.
3897.
3898.

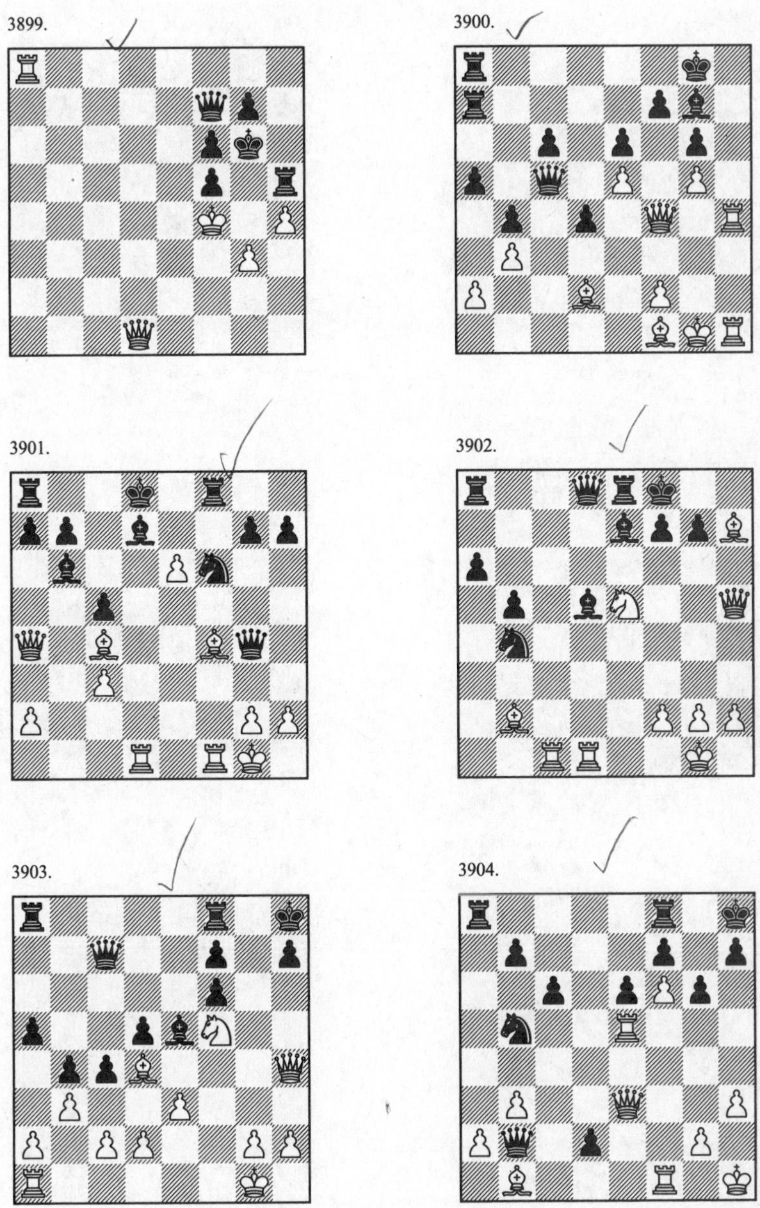
3899.
3900.
3901.
3902.
3903.
3904.

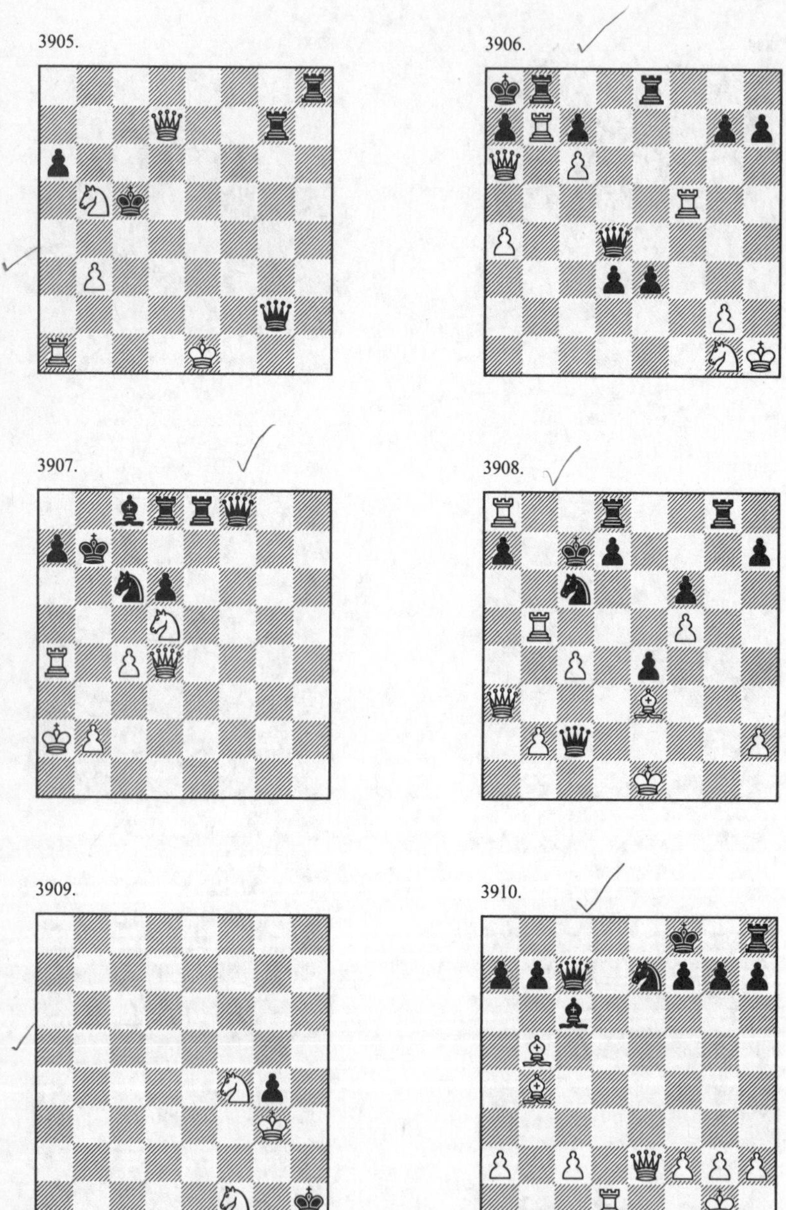

3905.

3906.

3907.

3908.

3909.

3910.

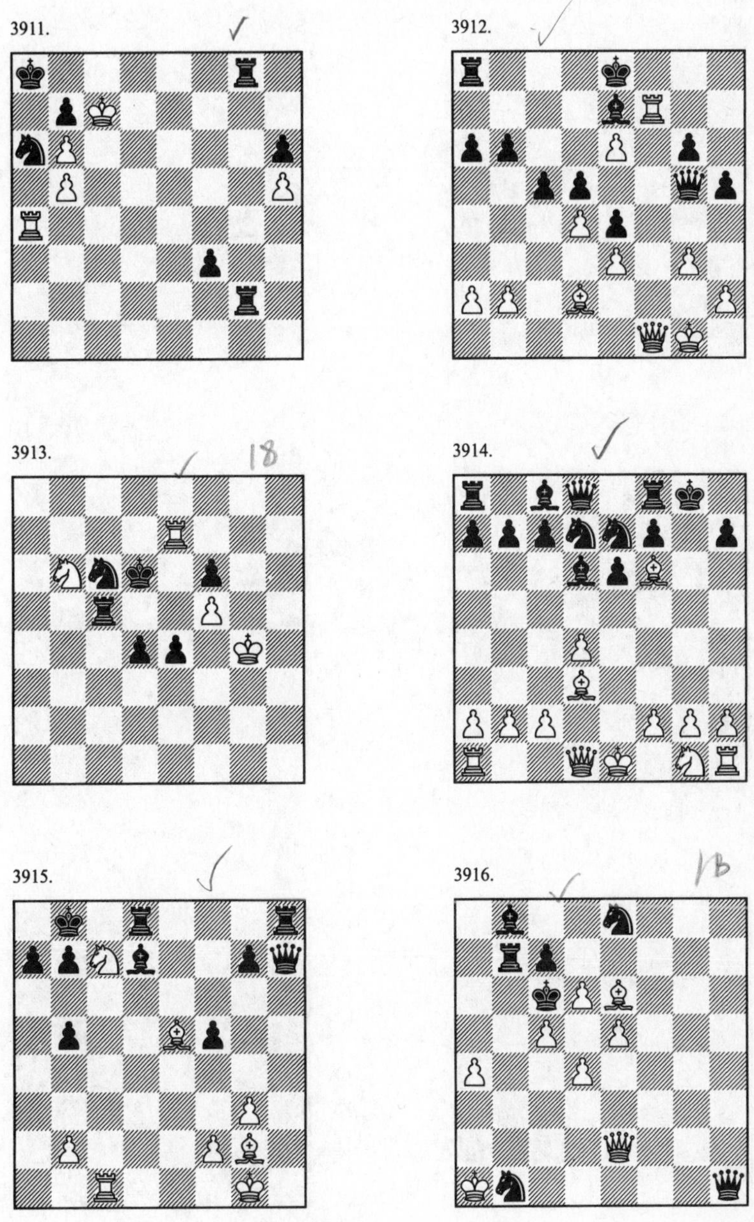
3911.
3912.
3913.
3914.
3915.
3916.

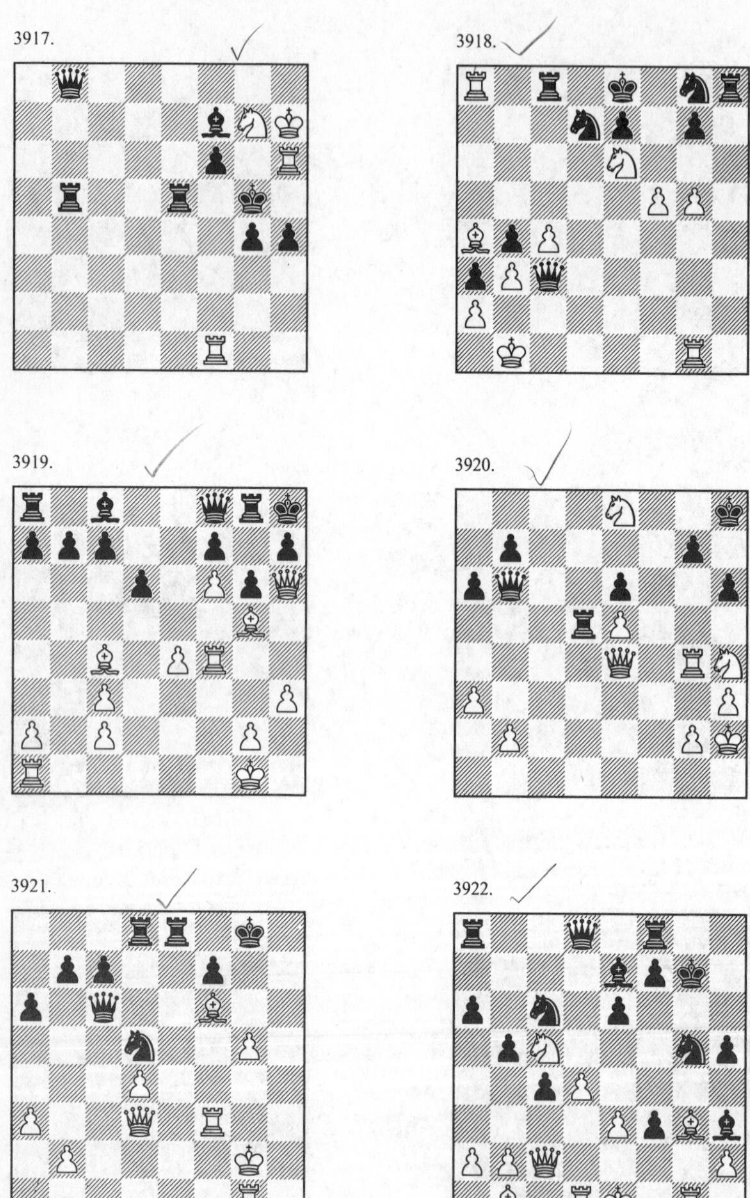
3917.
3918.
3919.
3920.
3921.
3922.

3923.

3924.

3925.

3926.

3927.

3928.

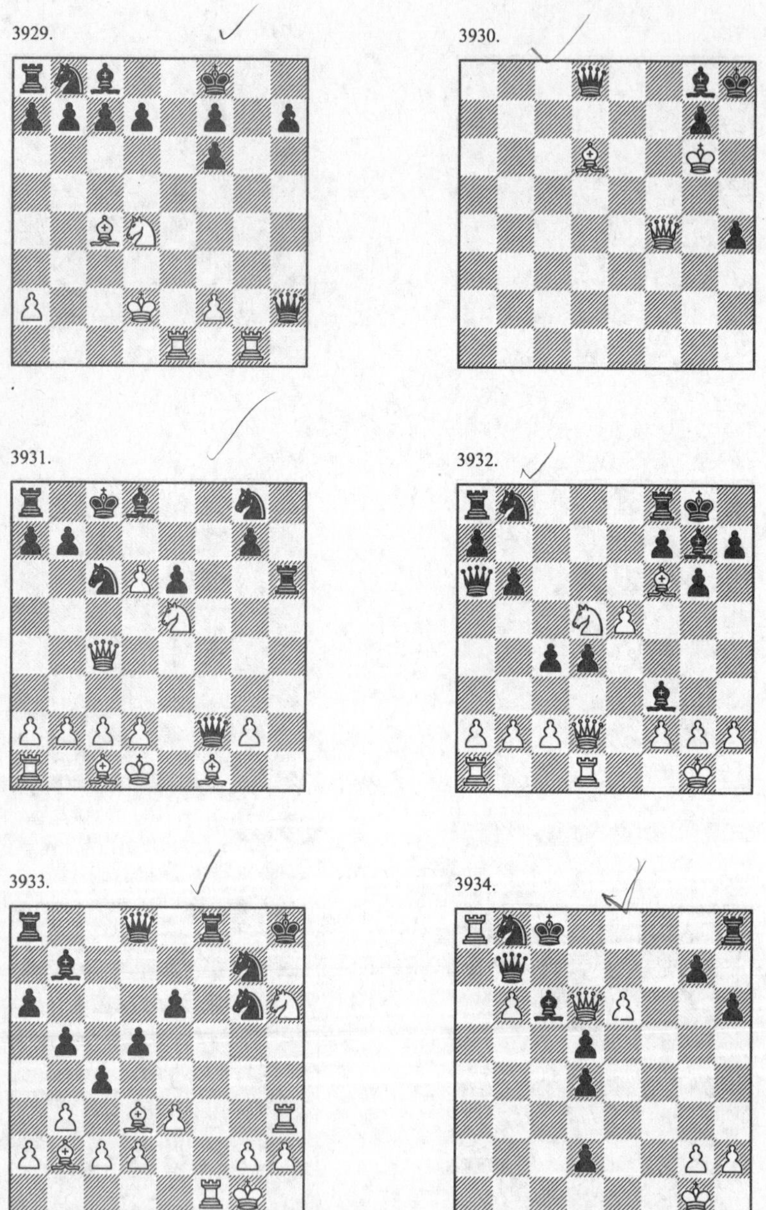
3929.
3930.
3931.
3932.
3933.
3934.

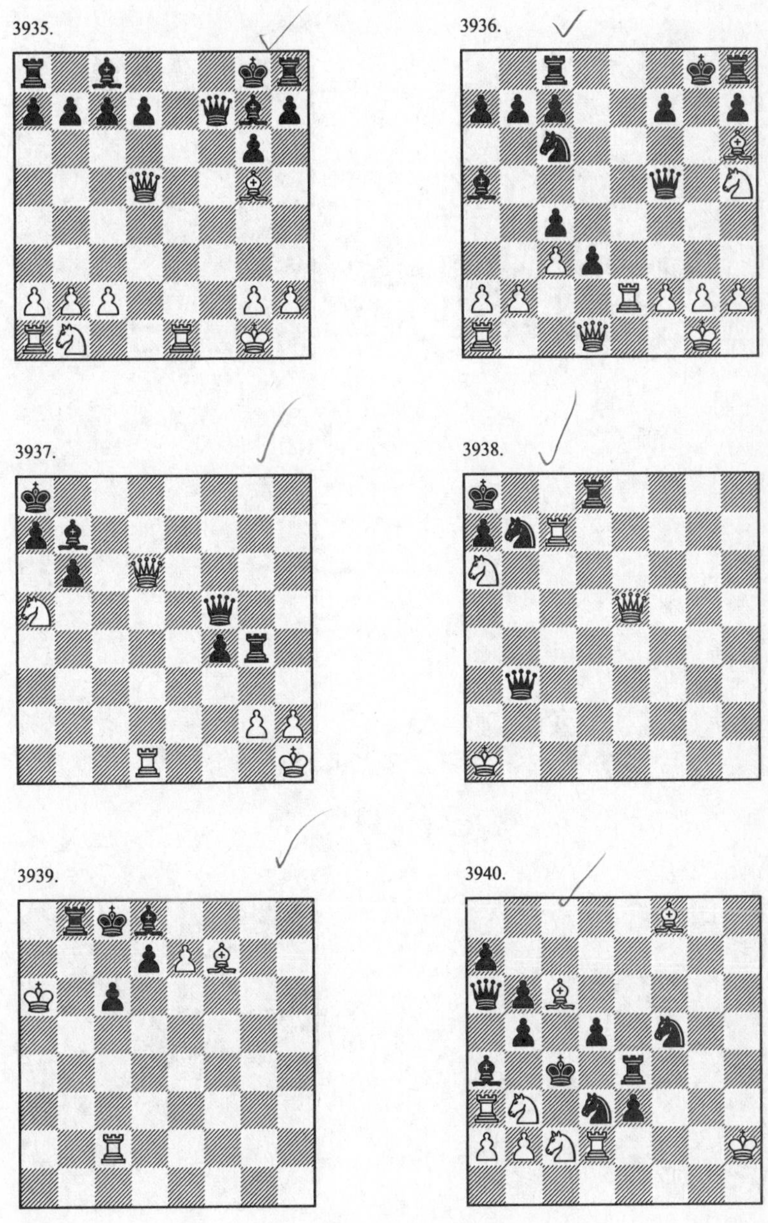
3935.
3936.
3937.
3938.
3939.
3940.

3941.

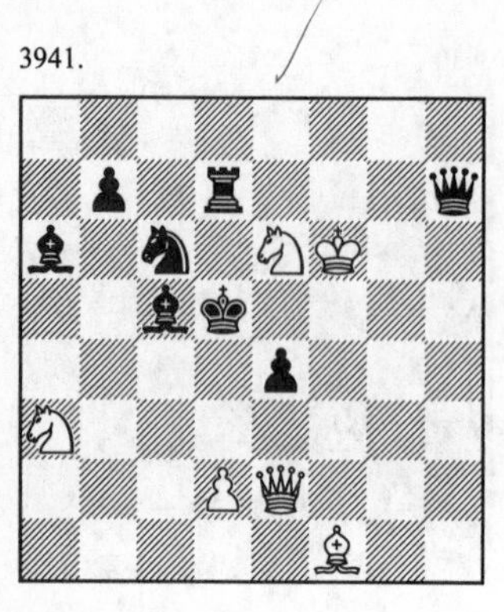

3942.

3943.

3944.

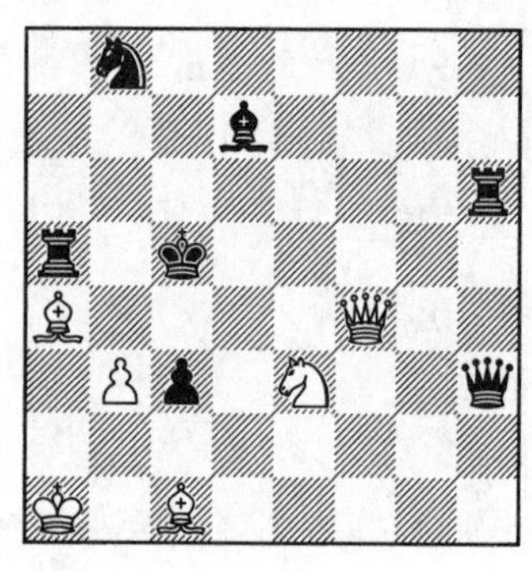

3945.

3946.

3947.

3948.

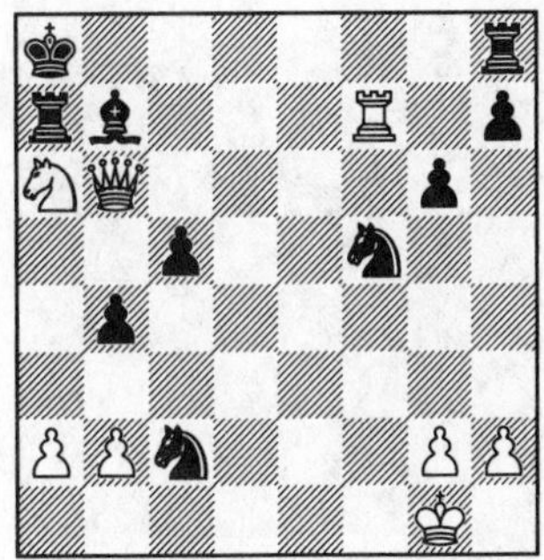

3949.

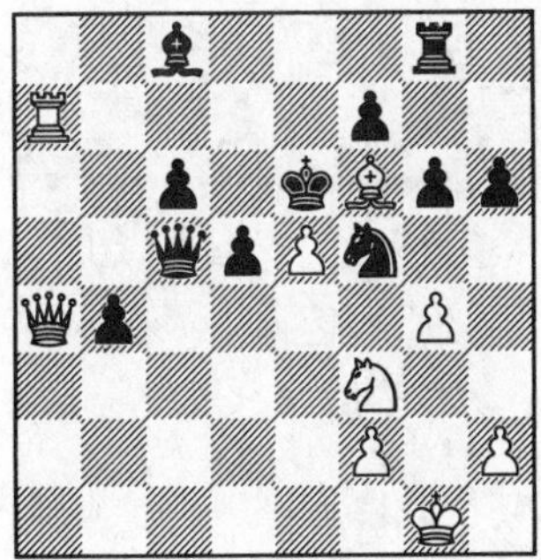

3950.

3951.

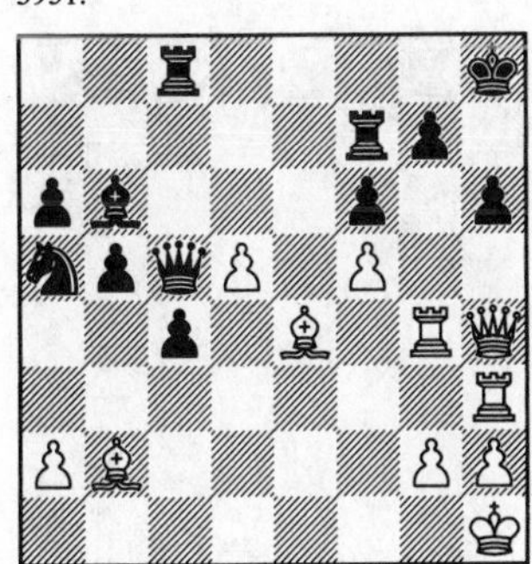

3952.

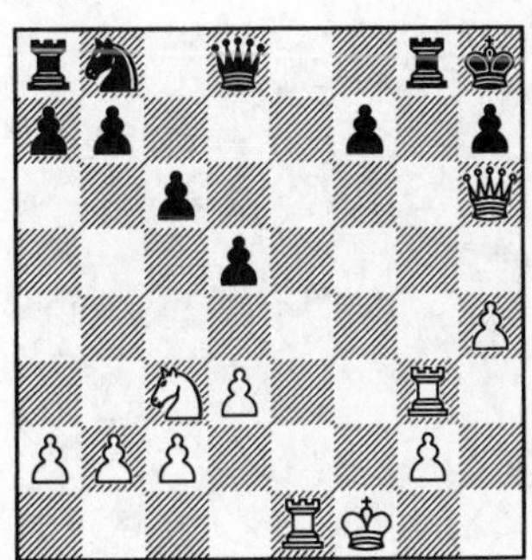

3953.

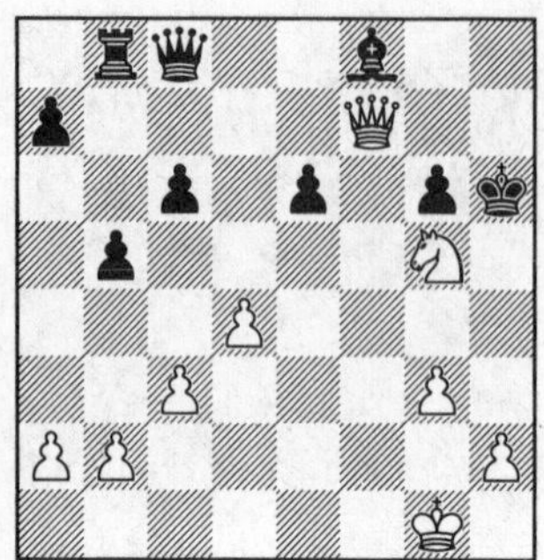

3954.

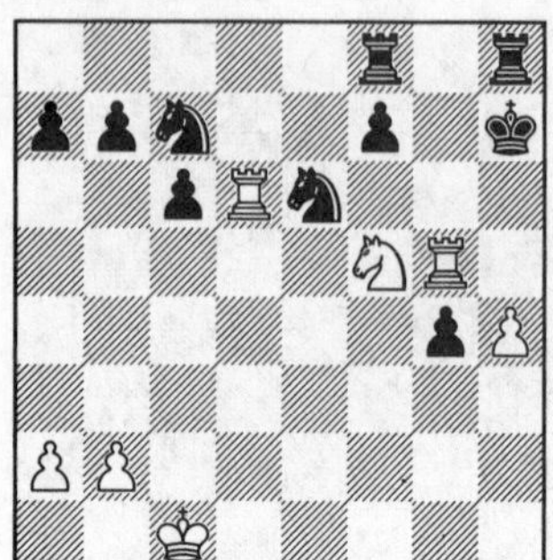

3955.

3956.

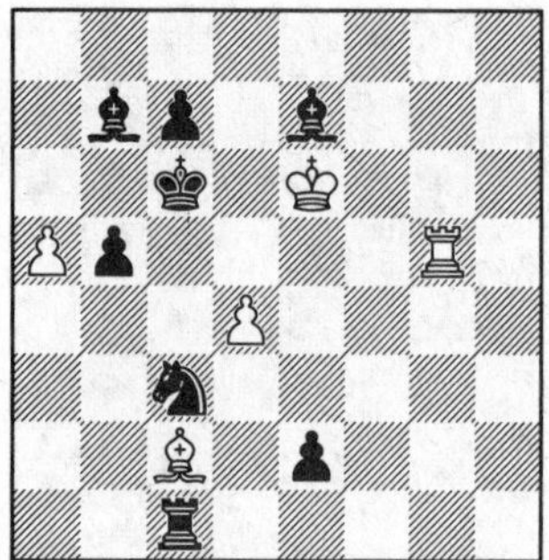

3957.

3958.

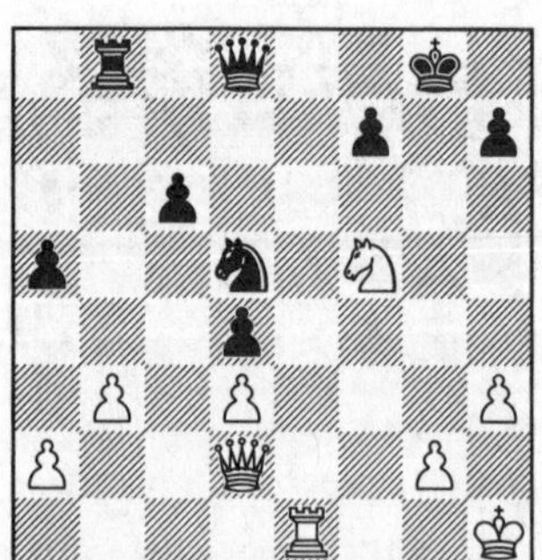

3959.

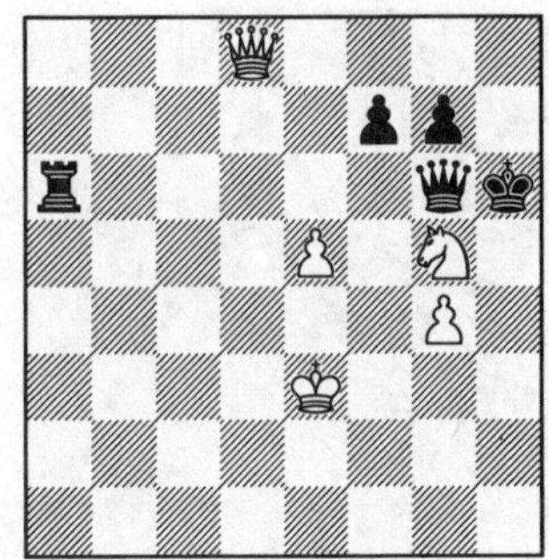

3960.

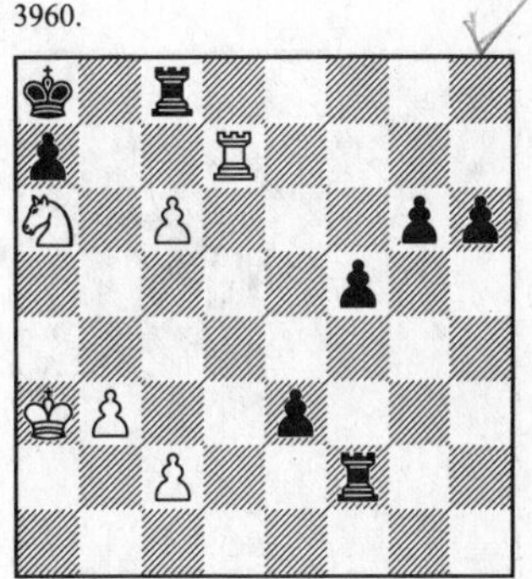

3961.

3962.

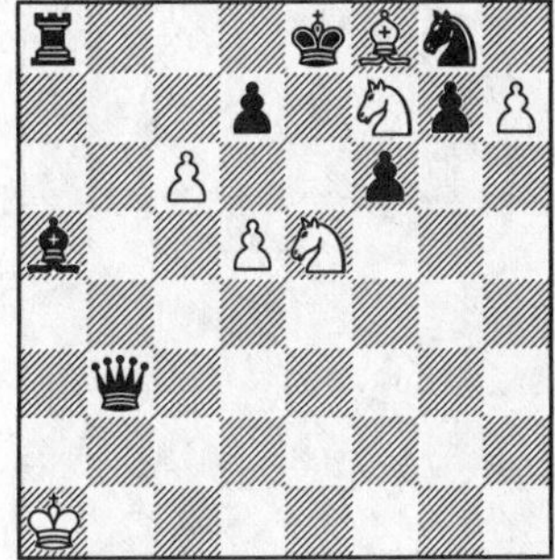

3963.

3964.

3965.

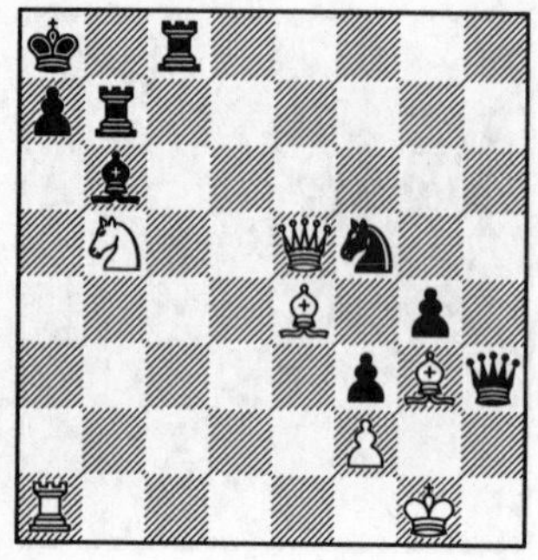

3966.

3967.

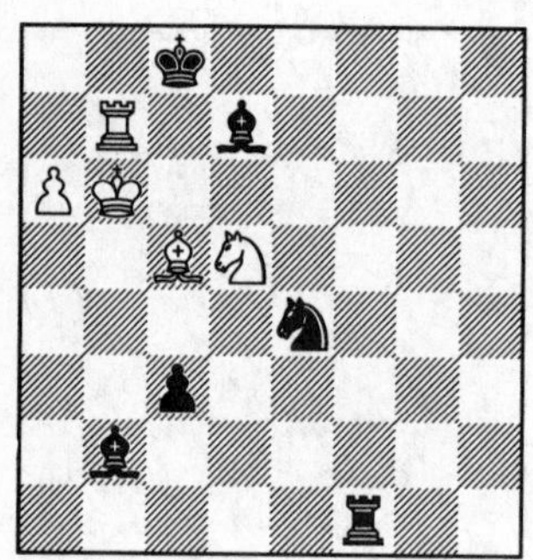

3968.

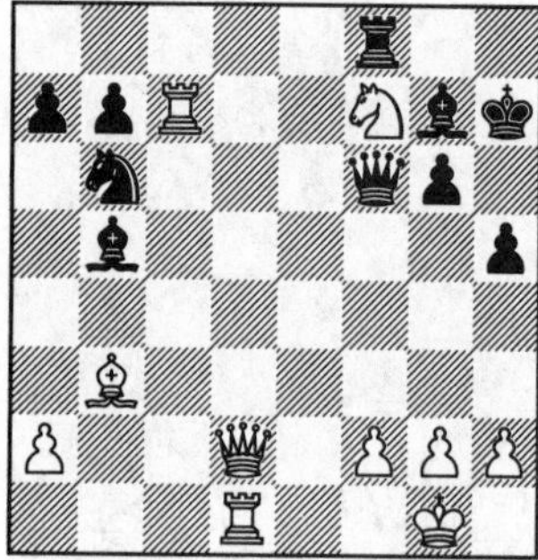

3969.

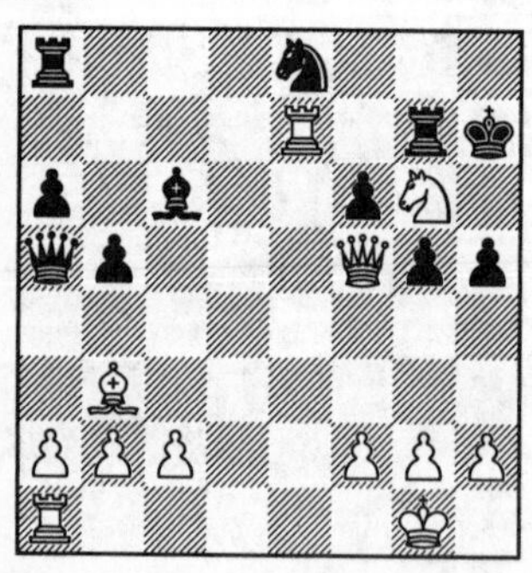

3970.

3971.

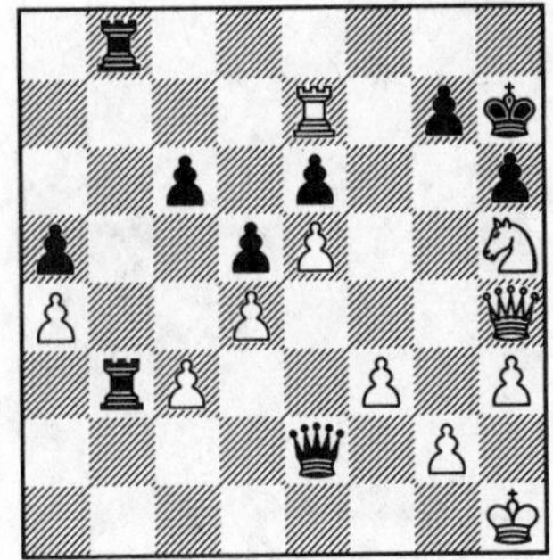

3972.

3973.

3974.

3975.

3976.

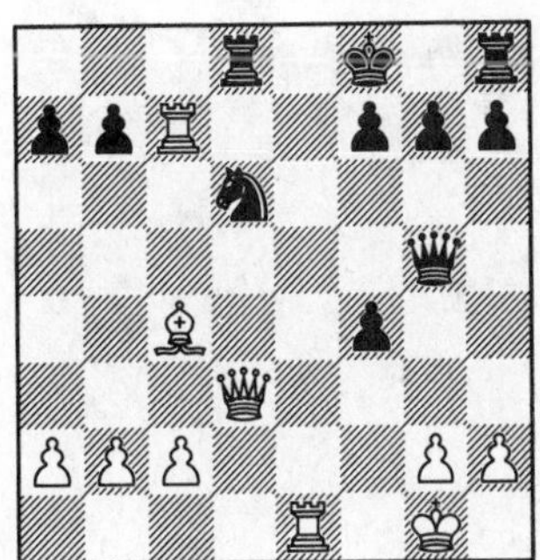

3977.

3978.

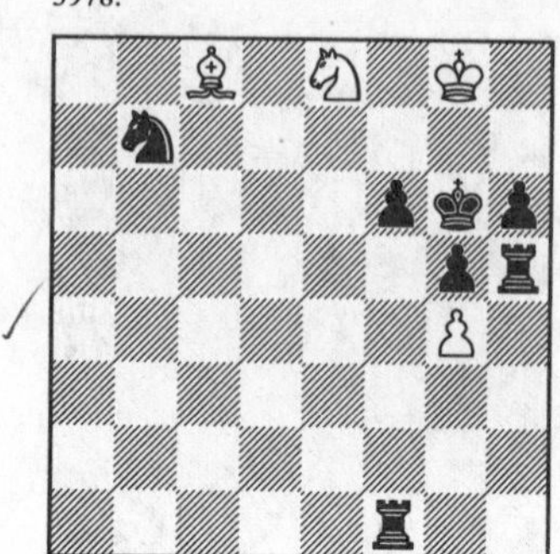

3979.

3980.

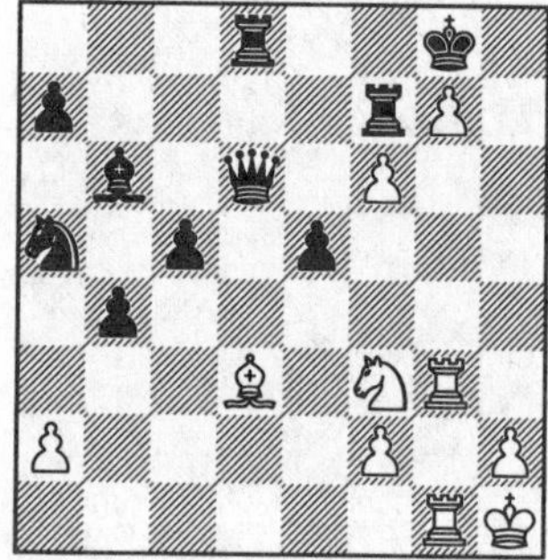

3981.

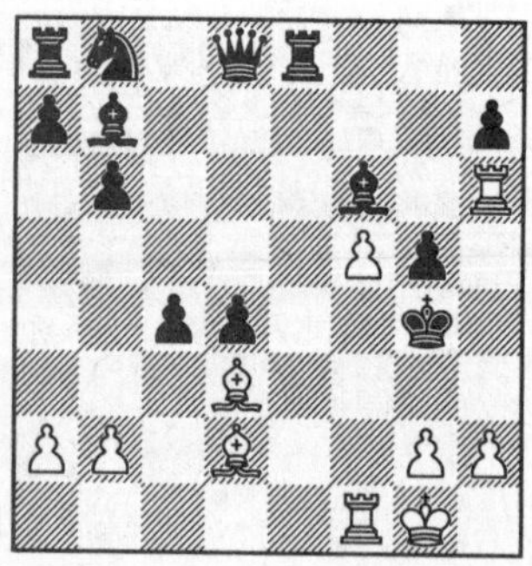

3982.

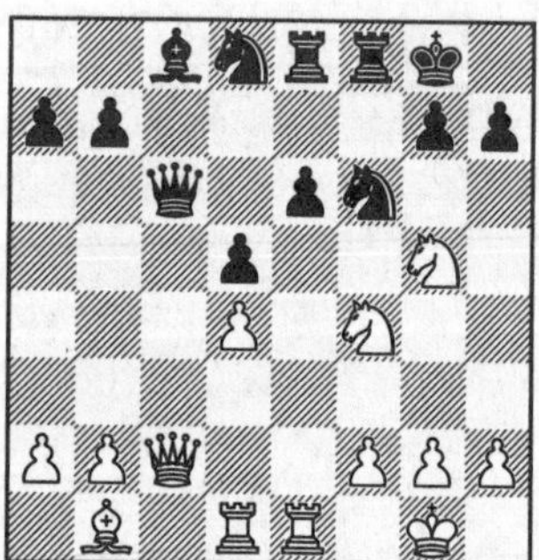

3983.

3984.

3985.

3986.

3987.

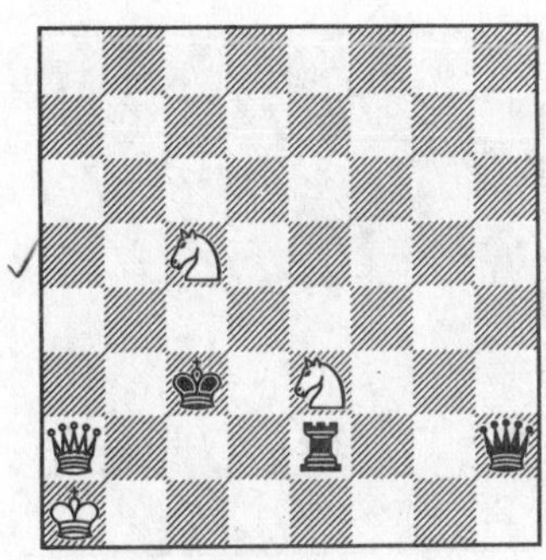

3988.

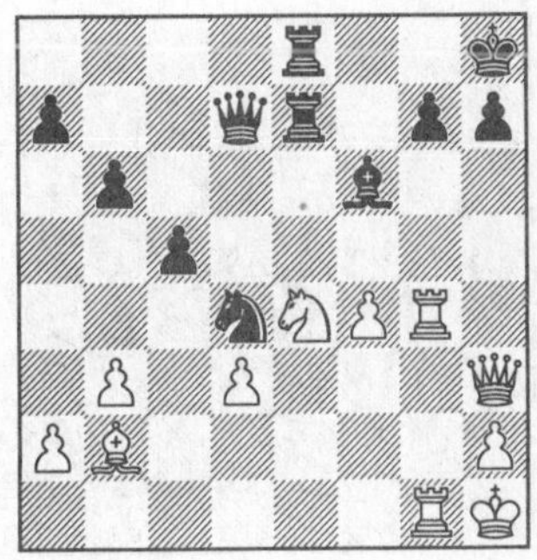

3989.

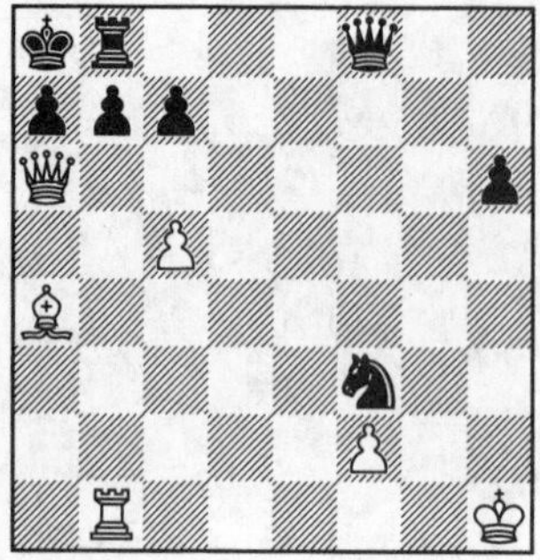

3990.

3991.

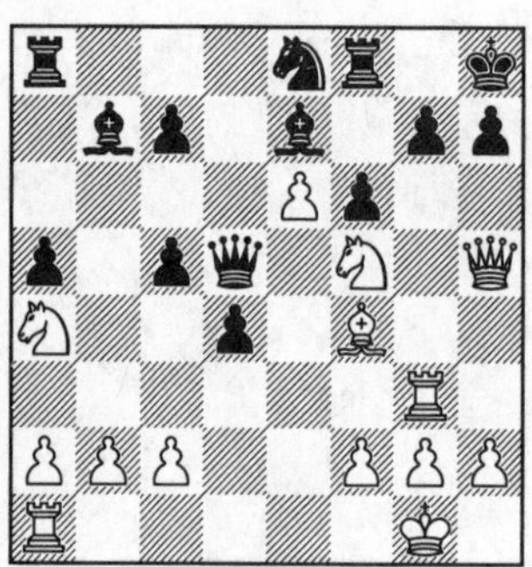

3992.

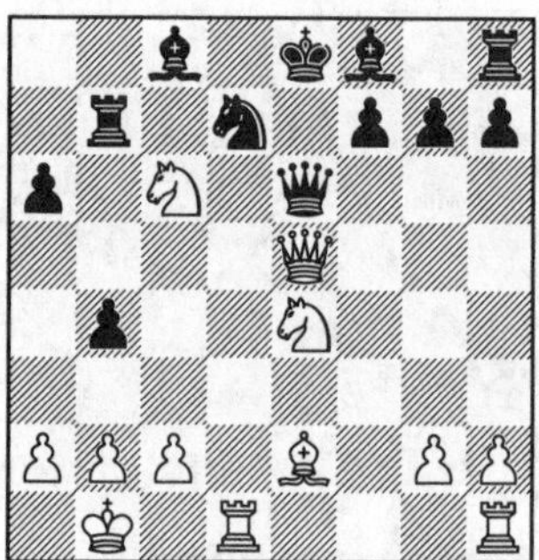

3993.

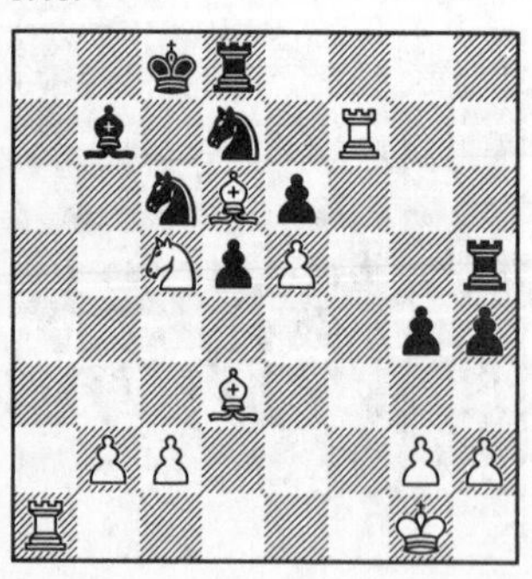

3994.

3995.

3996.

3997.

3998.

3999.

4000.

4001.

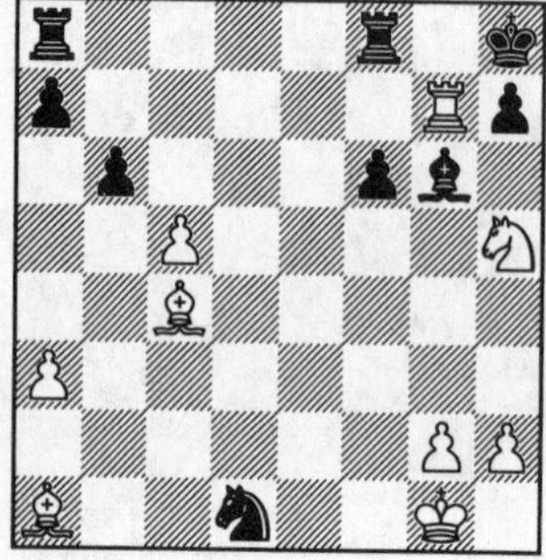

4002.

4003.

4004.

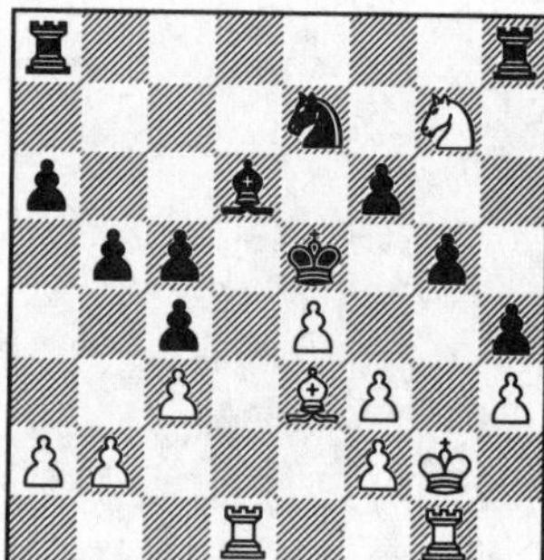

4005.

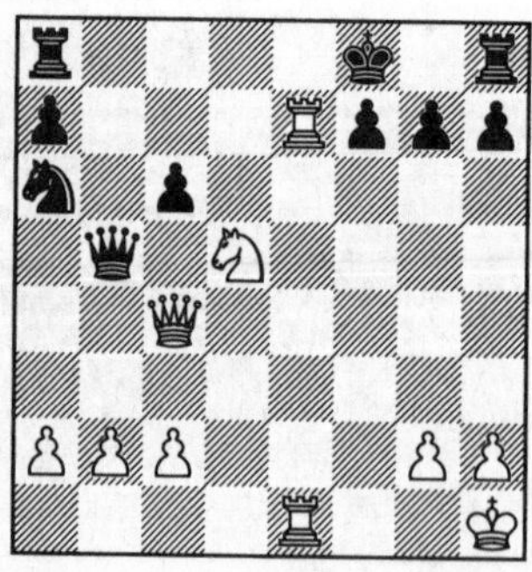

4006.

4007.

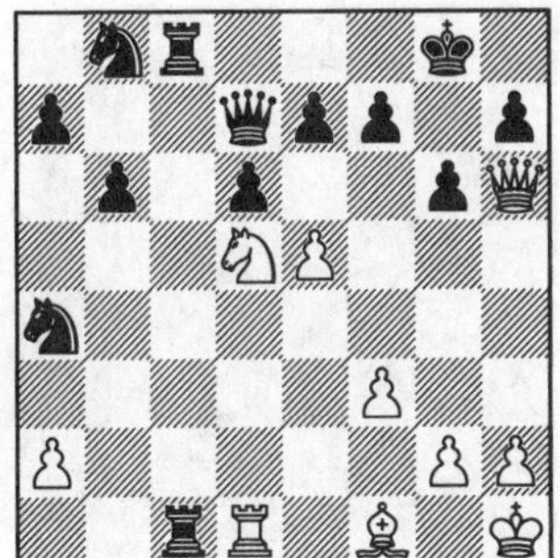

4008.

4009.

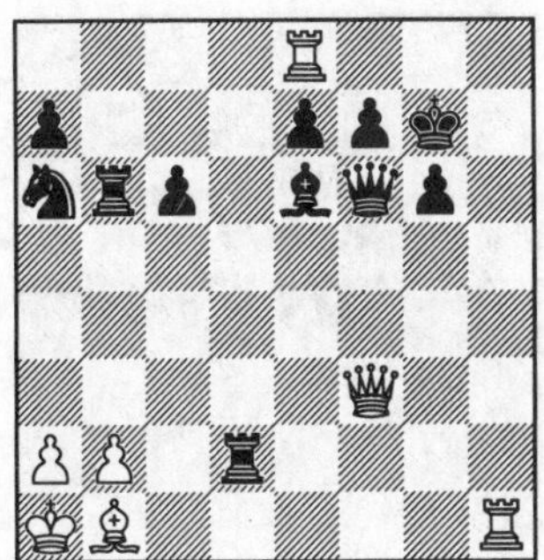

4010.

4011.

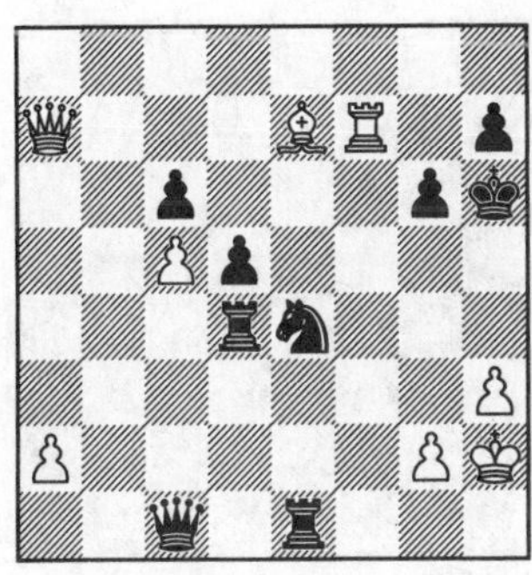

4012.

4013.

4014.

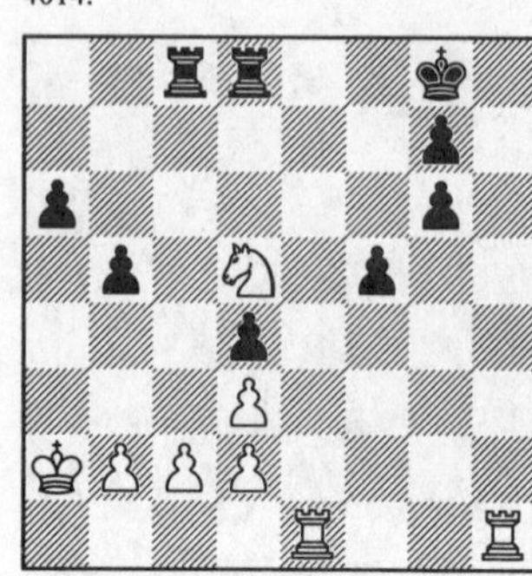

4015.

4016.

4017.

4018.

4019.

4020.

4021.

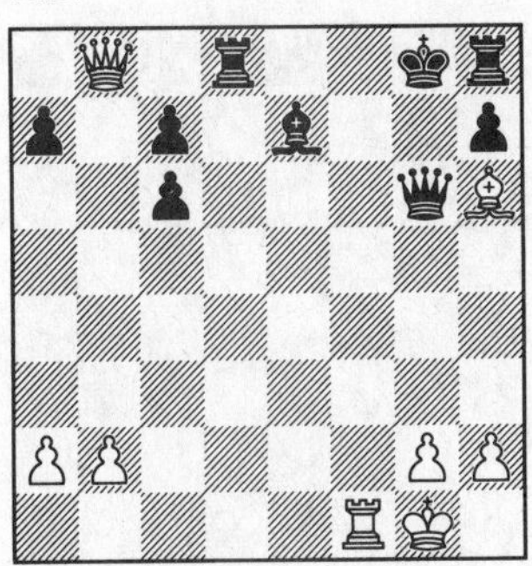

4022.

4023.

4024.

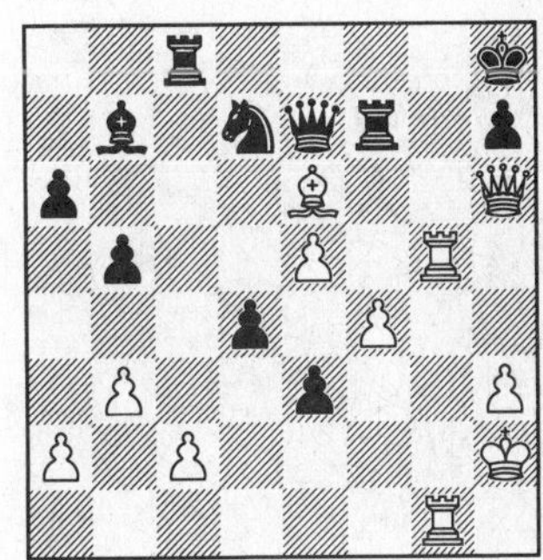

4025.

4026.

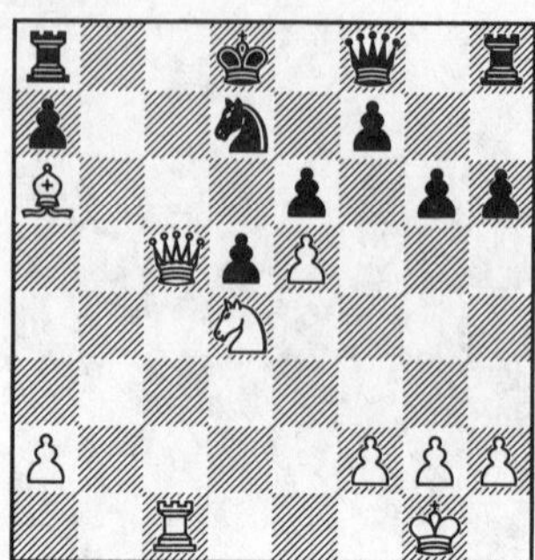

4027.

4028.

4029.

4030.

4031.

4032.

4033.

4034.

4035.

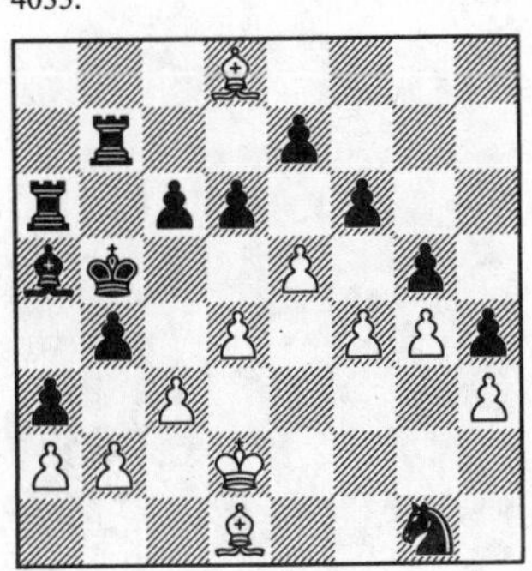

4036.

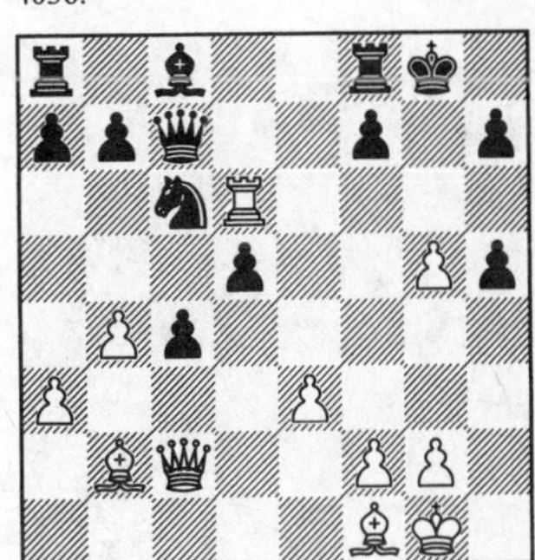

4037.

4038.

4039.

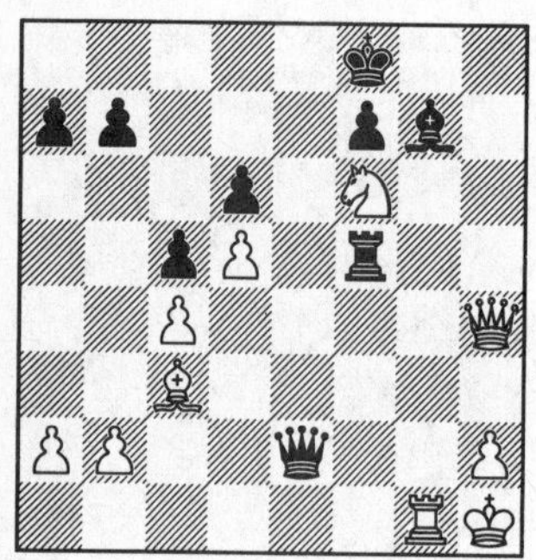

4040.

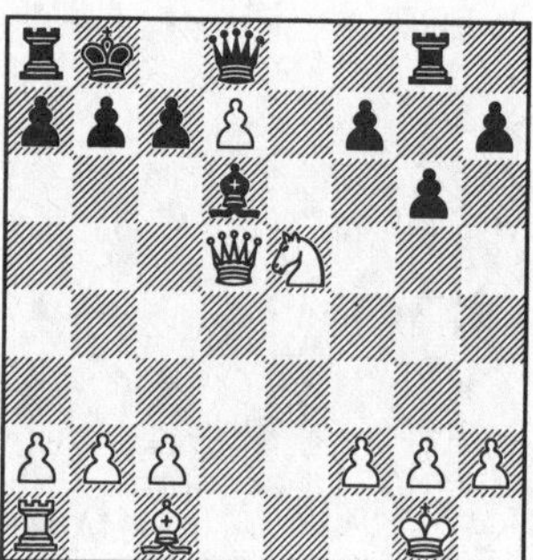

4041.

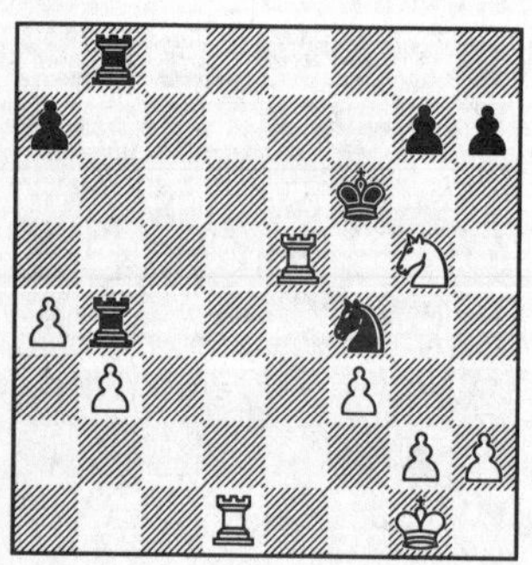

4042.

4043.

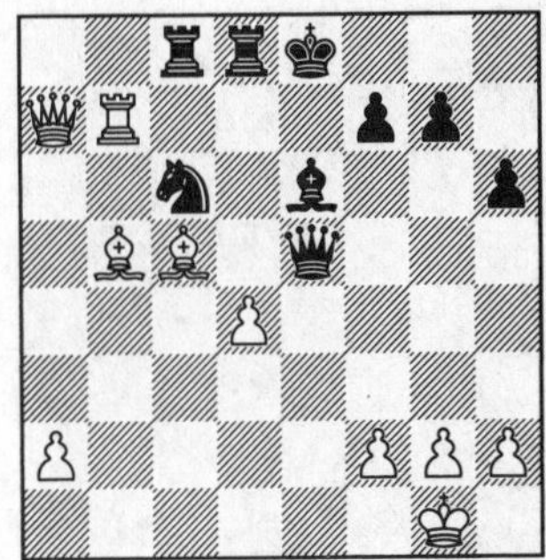

4044.

4045.

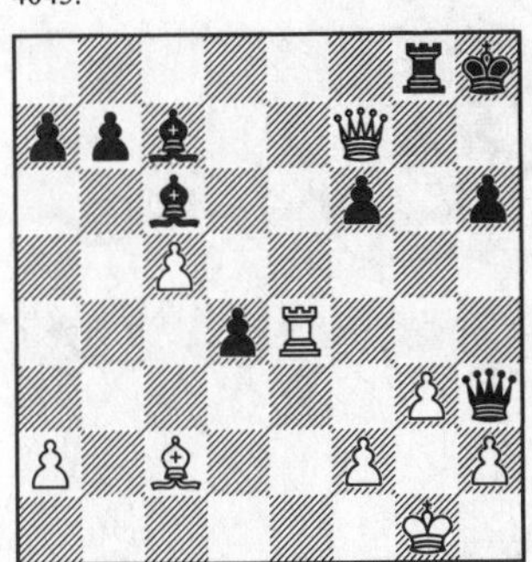

4046.

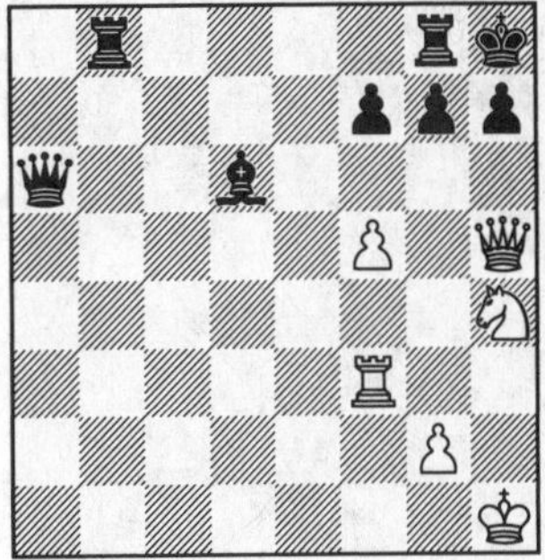

4047.

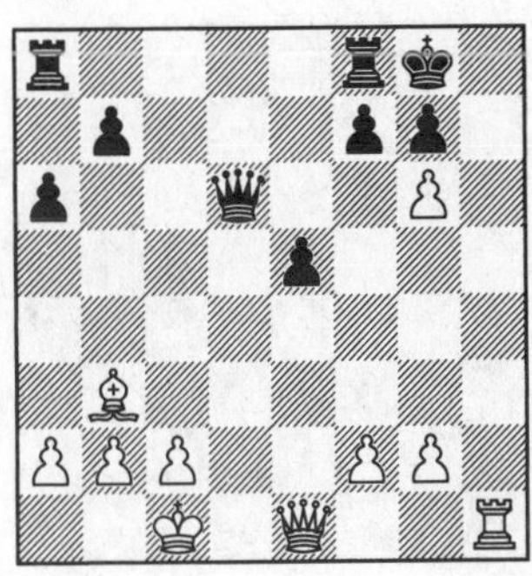

4048.

4049.

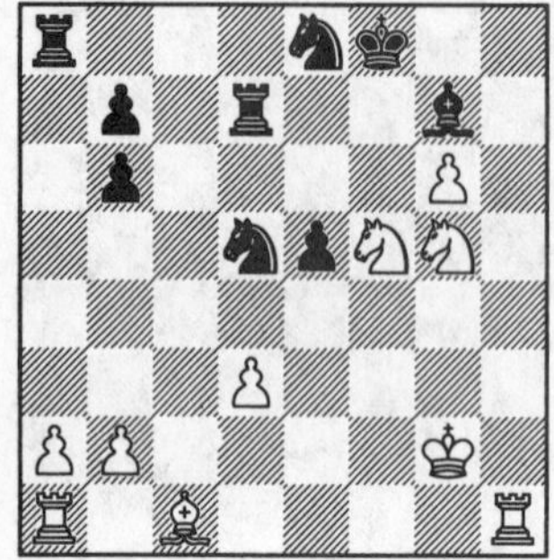

4050.

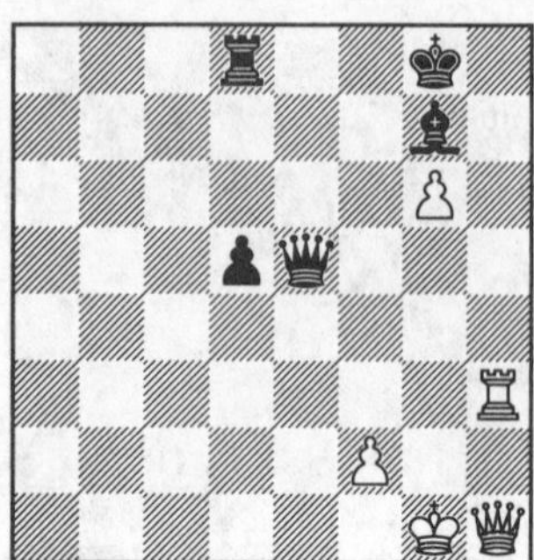

4051.

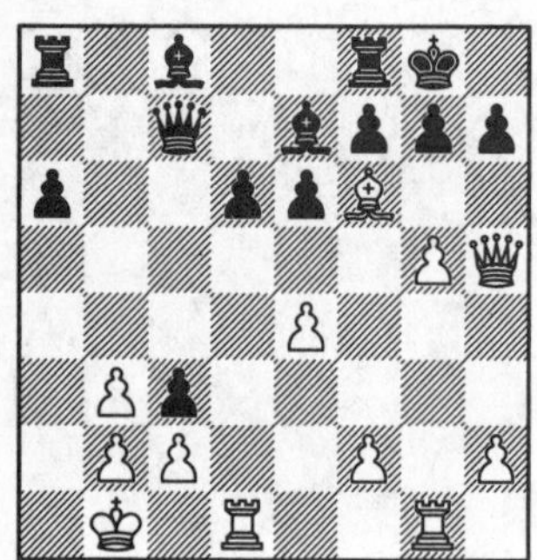

4052.

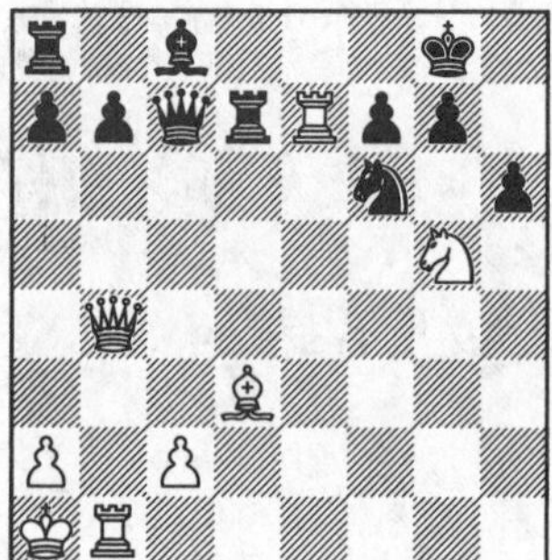

4053.

4054.

4055.

4056.

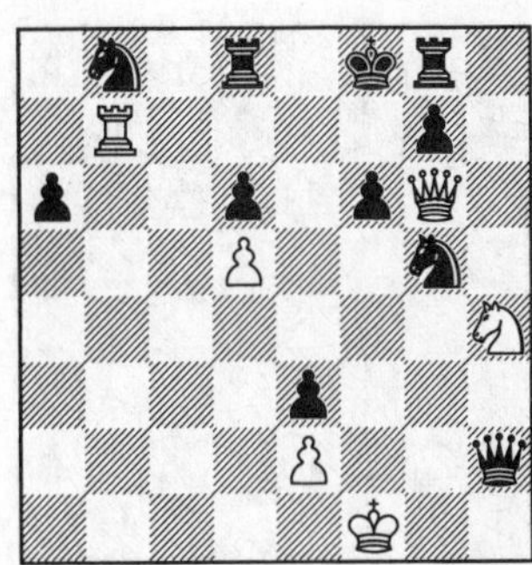

4057.

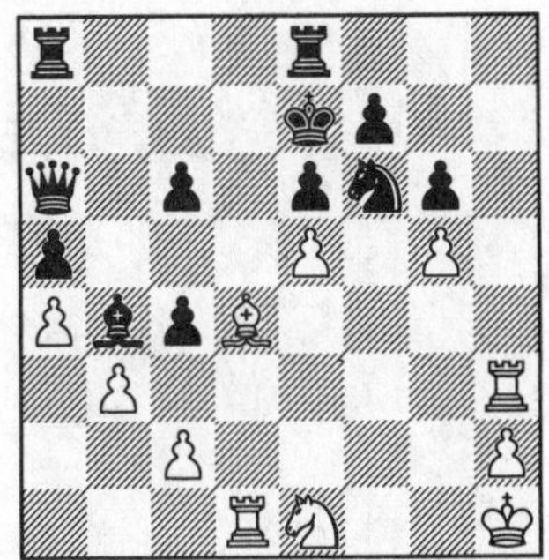

4058.

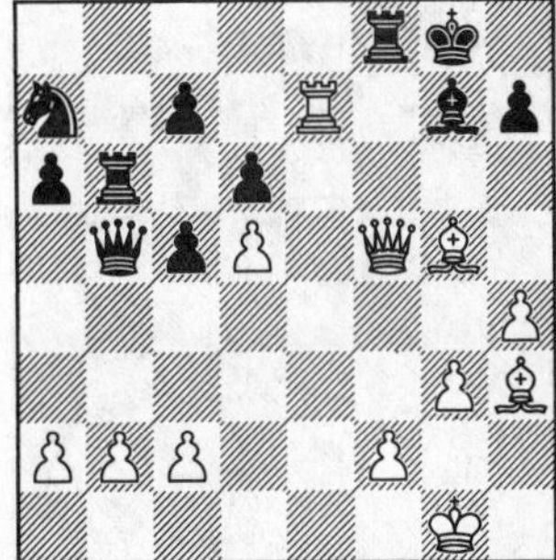

4059.

4060.

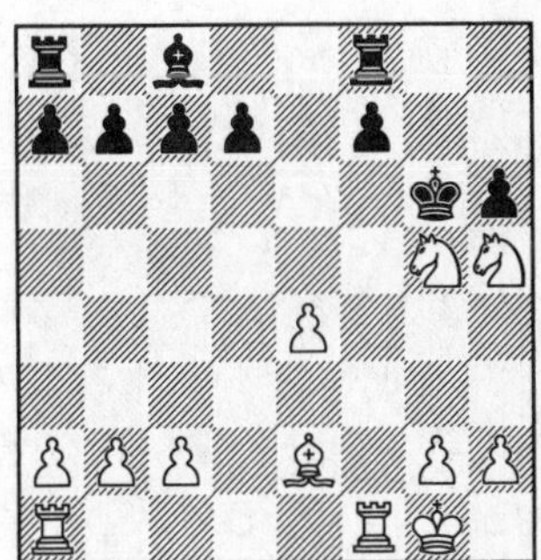

4061.

4062.

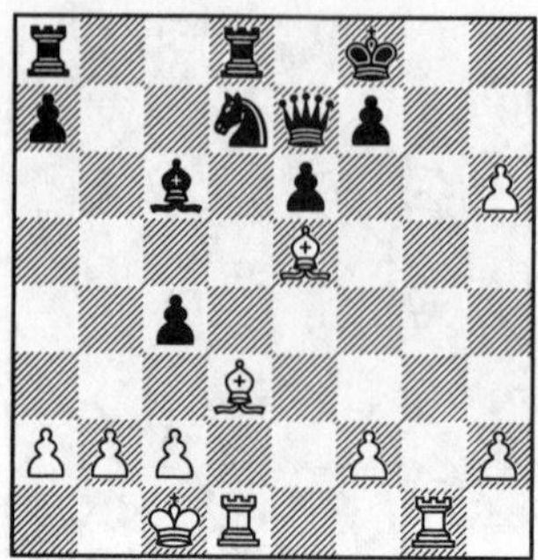

4063.

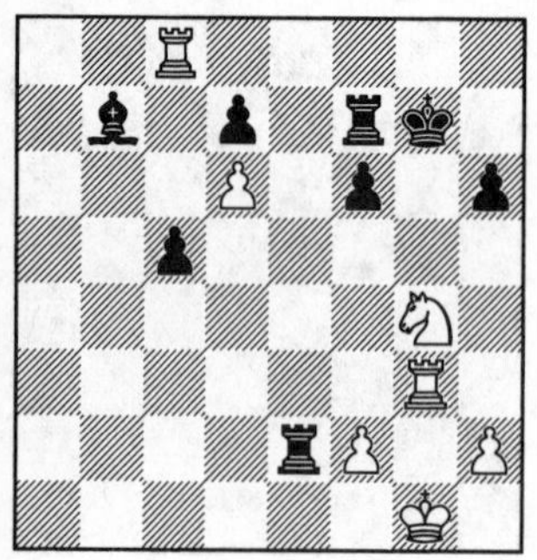

4064.

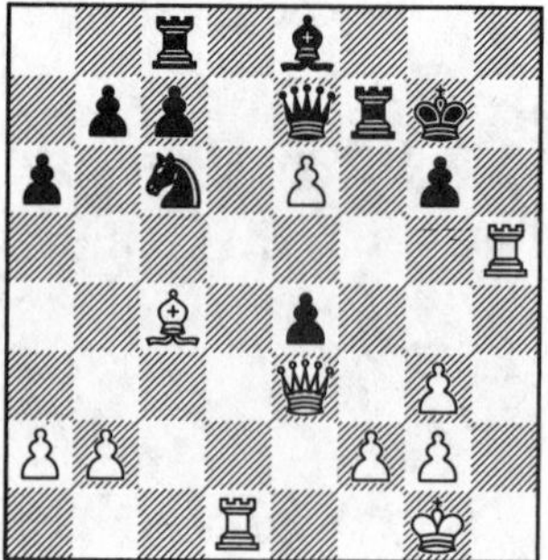

4065.

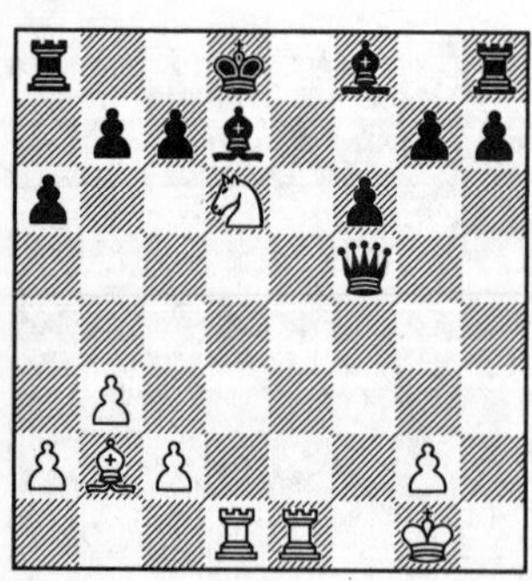

4066.

4067.

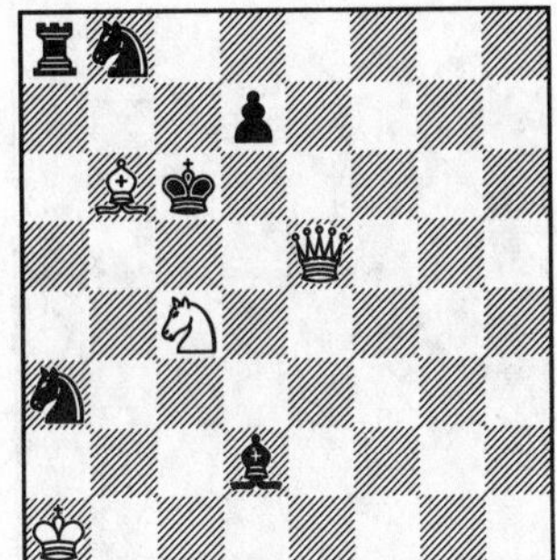

4068.

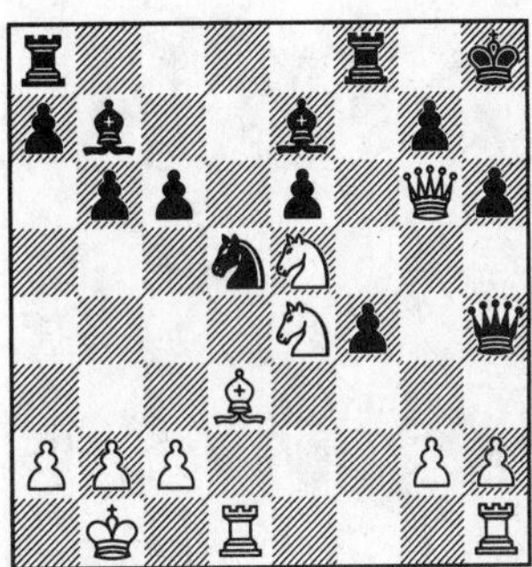

4069.

4070.

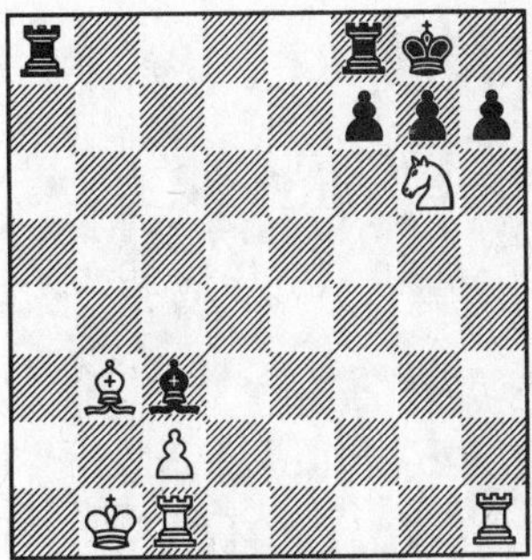

4071.

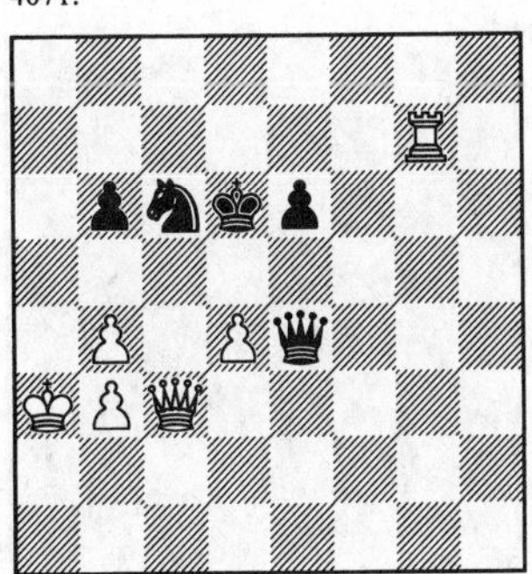

4072.

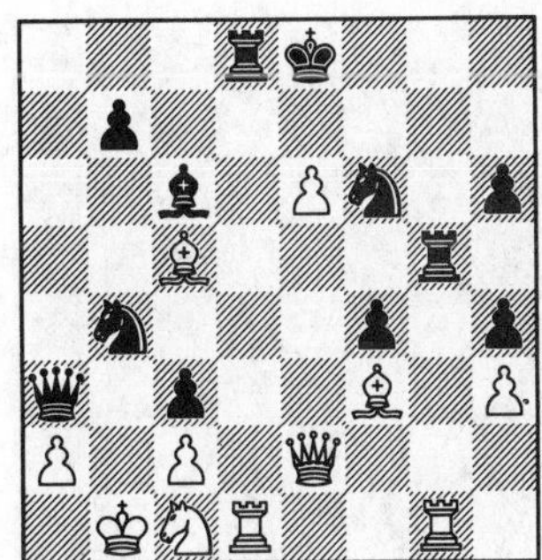

4073.

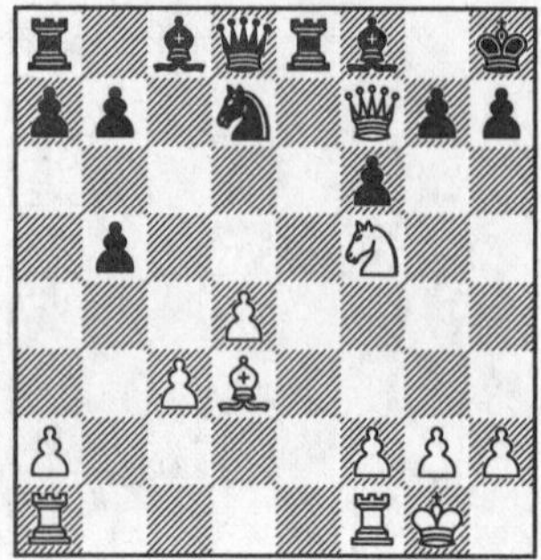

4074.

4075.

4076.

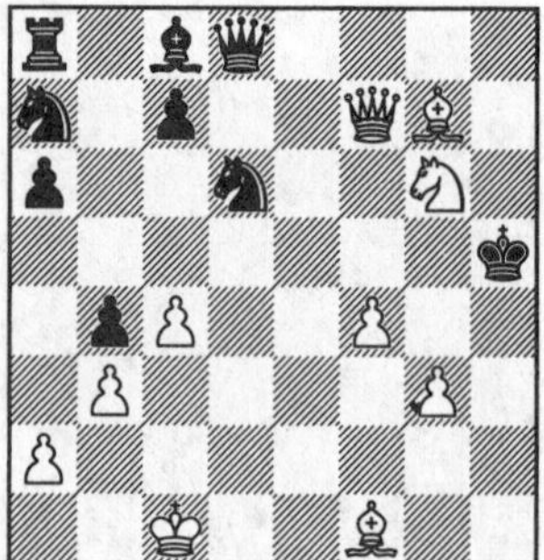

4077.

4078.

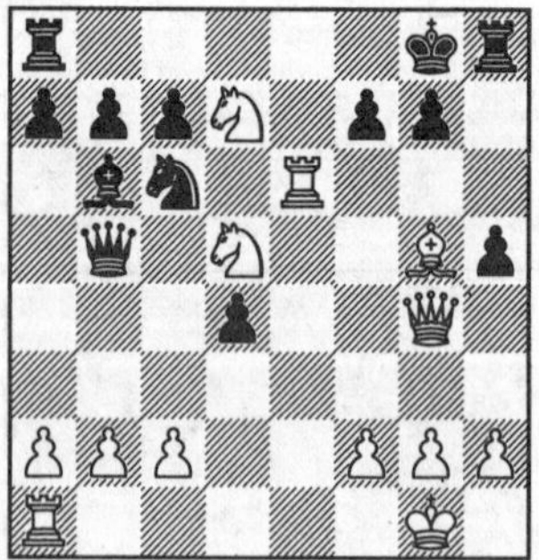

4079.

4080.

4081.

4082.

4083.

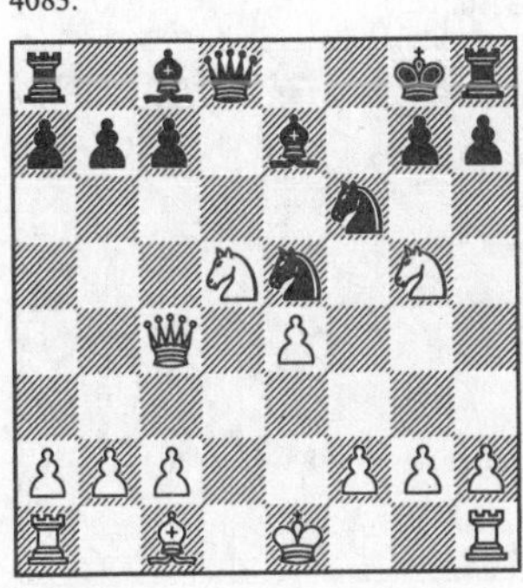

4084.

4085.

4086.

4087.

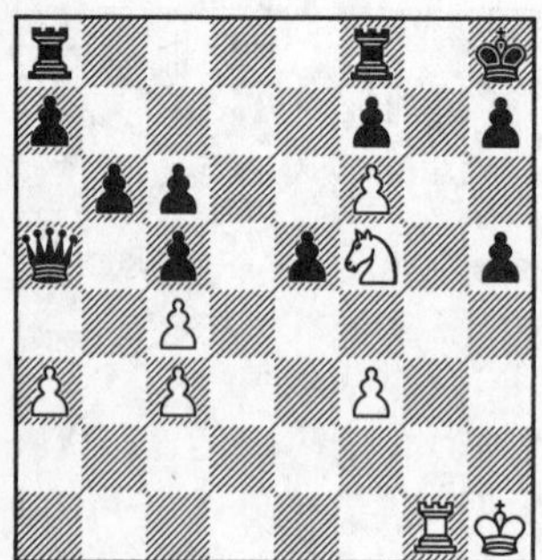

4088.

4089.

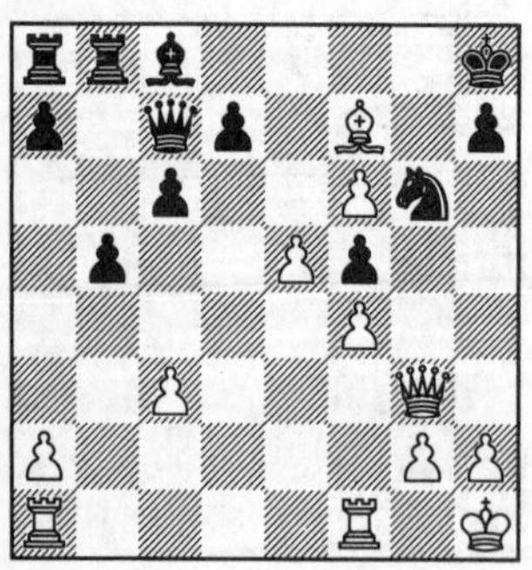

4090.

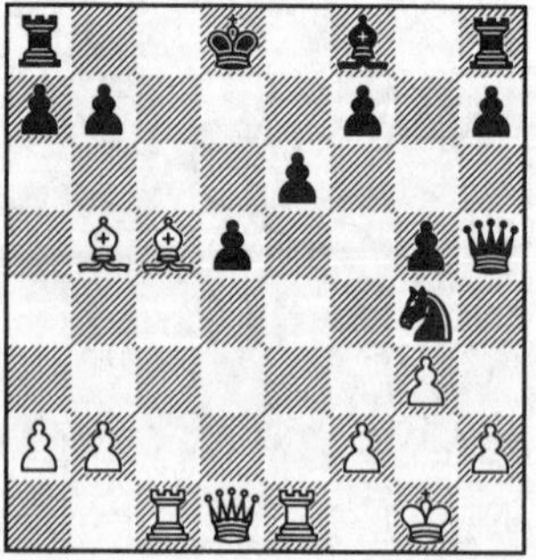

4091.

4092.

4093.

4094.

4095.

4096.

4097.

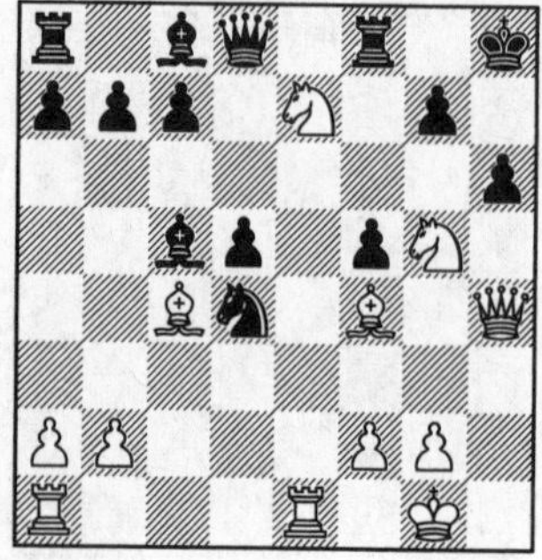

4098.

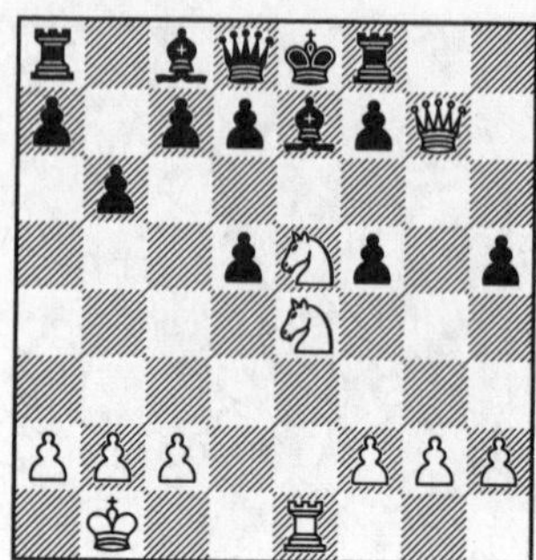

4099.

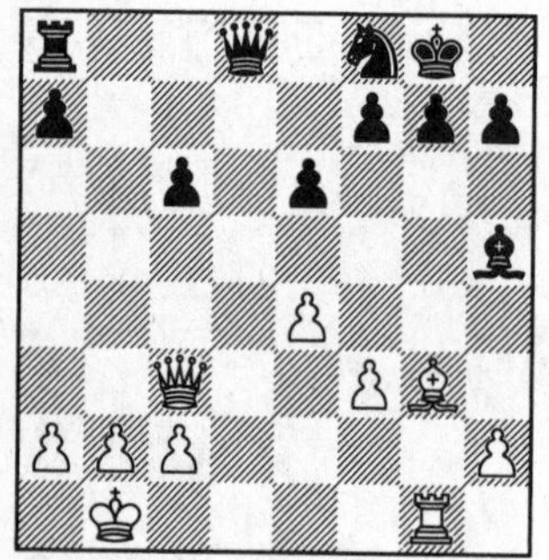

4100.

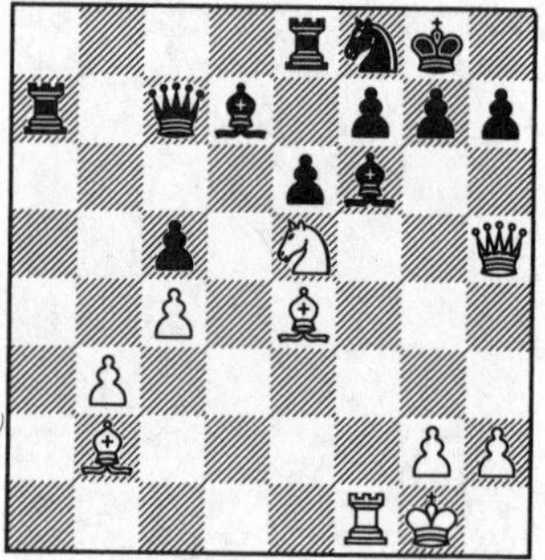

4101.

4102.

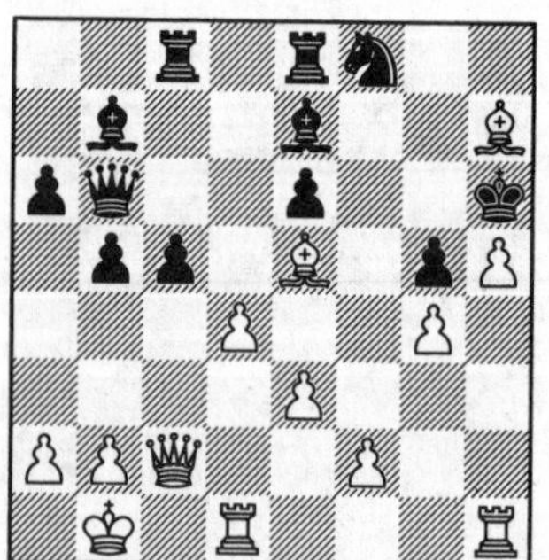

4103.

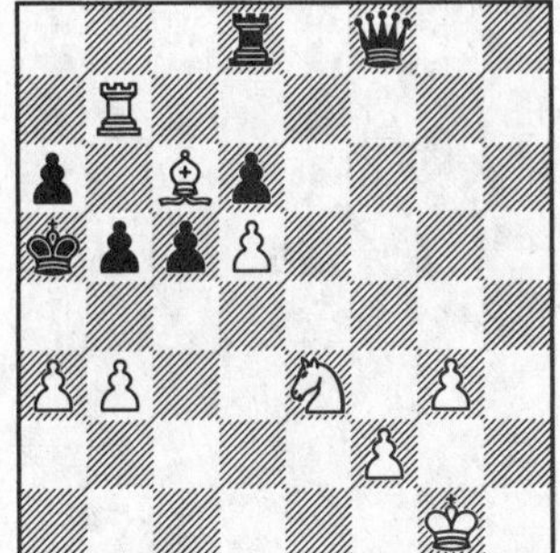

4104.

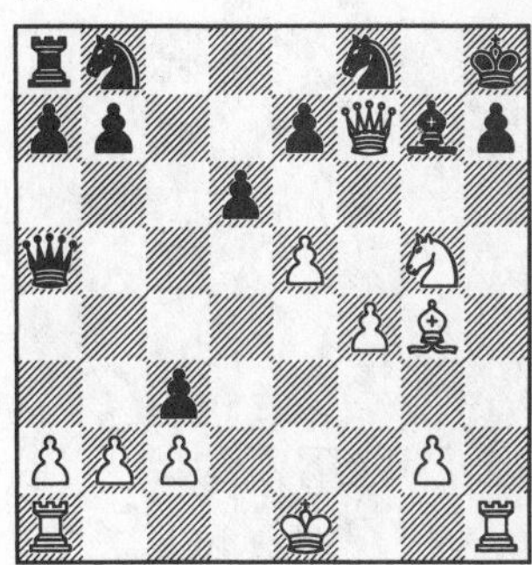

4105.

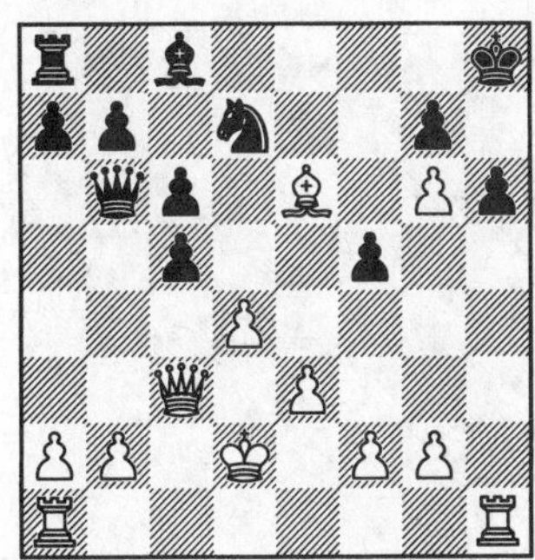

4106.

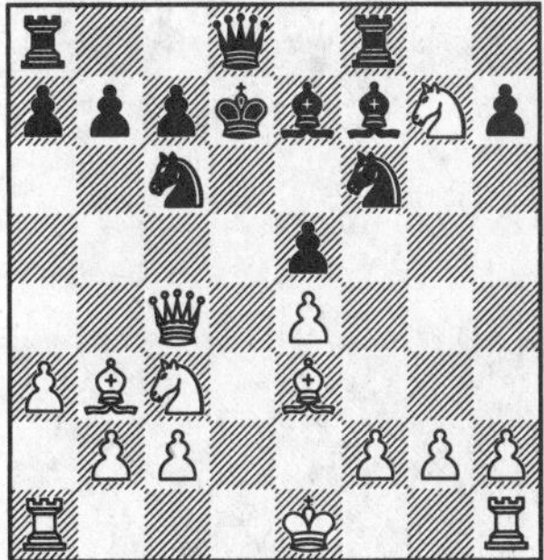

4107.

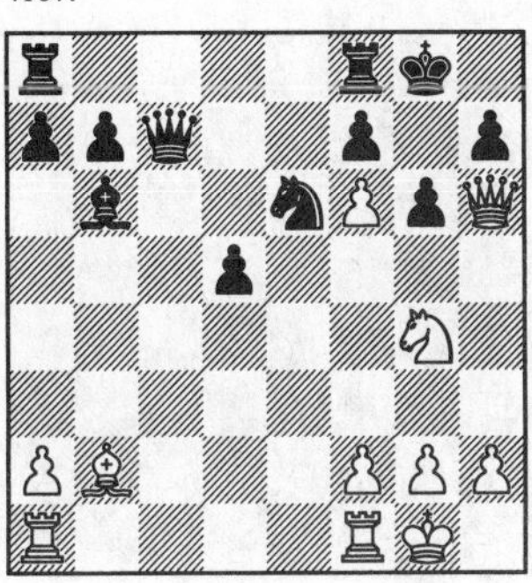

4108.

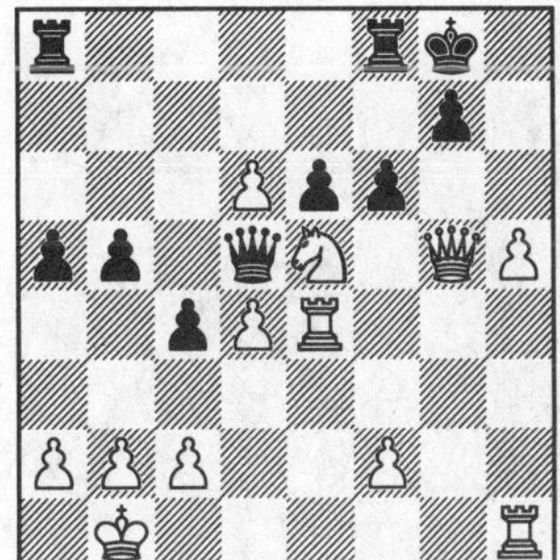

4109.

4110.

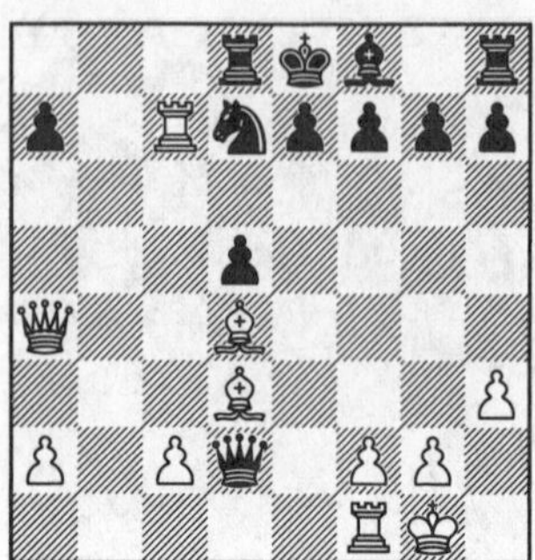

4111.

4112.

4113.

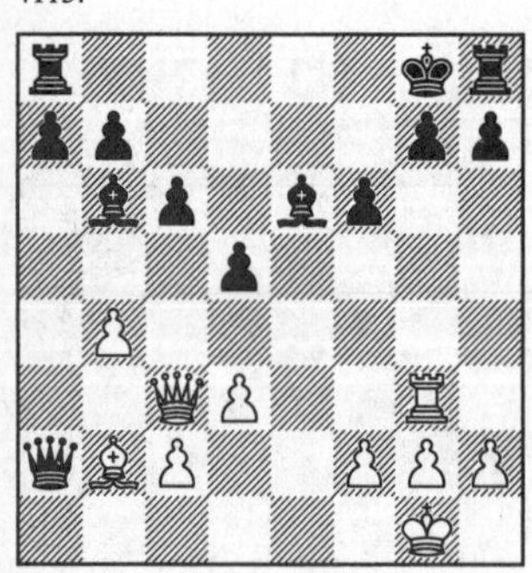

4114.

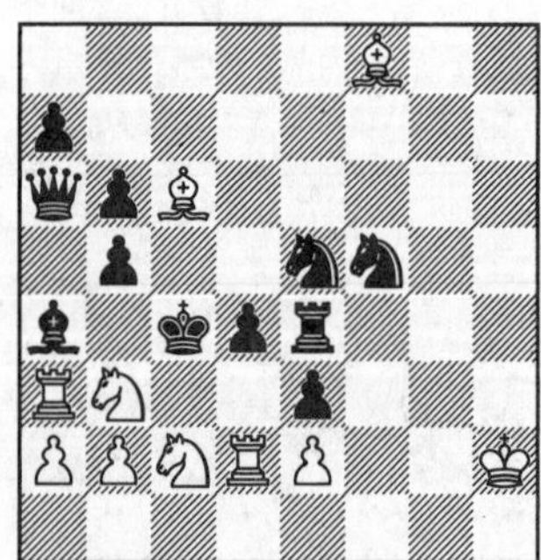

4115.

4116.

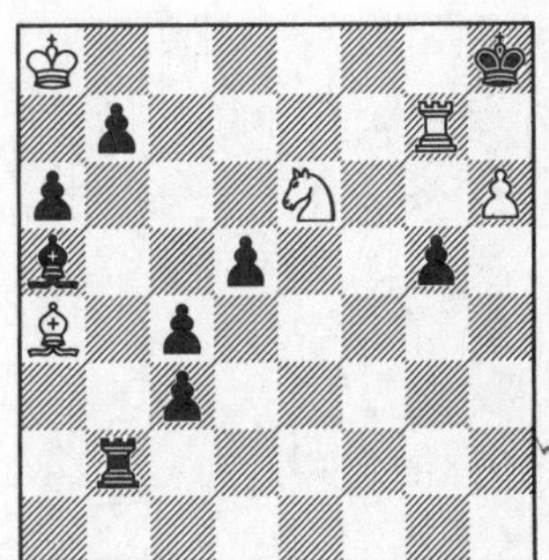

4117.

4118.

4119.

4120.

4121.

4122.

4123.

4124.

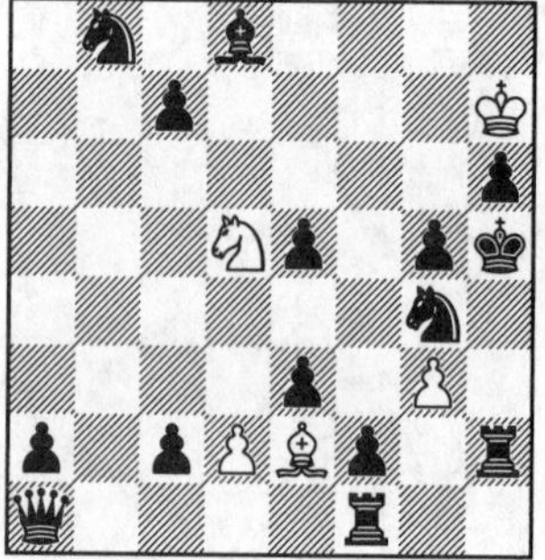

4125.

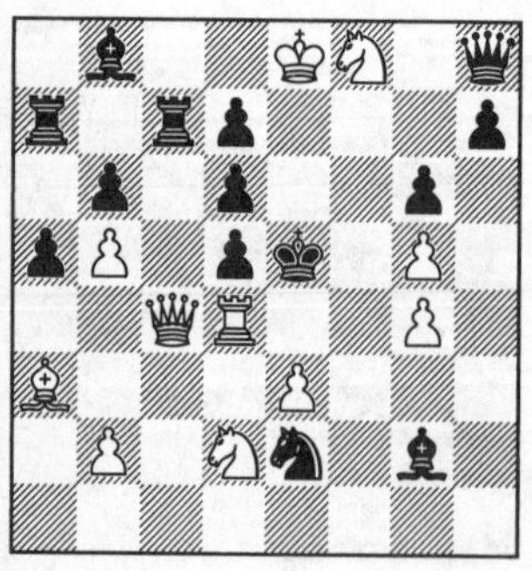

4126.

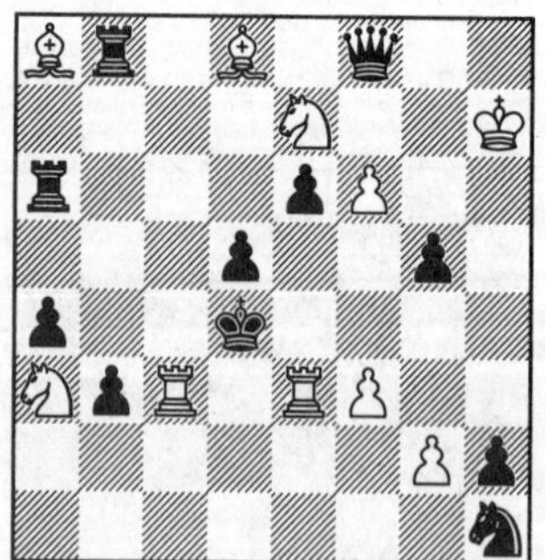

4127.

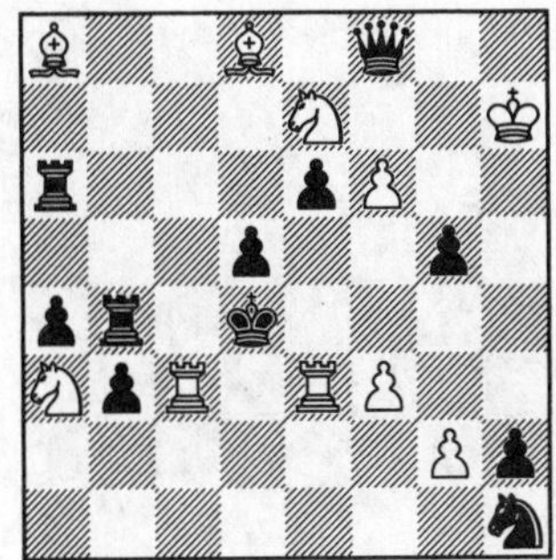

4128.

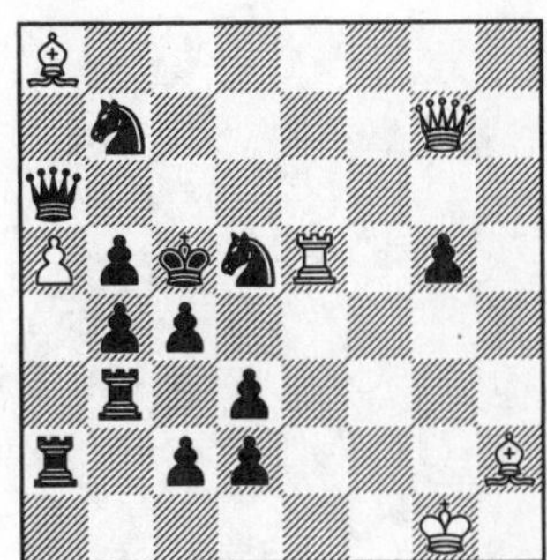

4129.

4130.

4131.

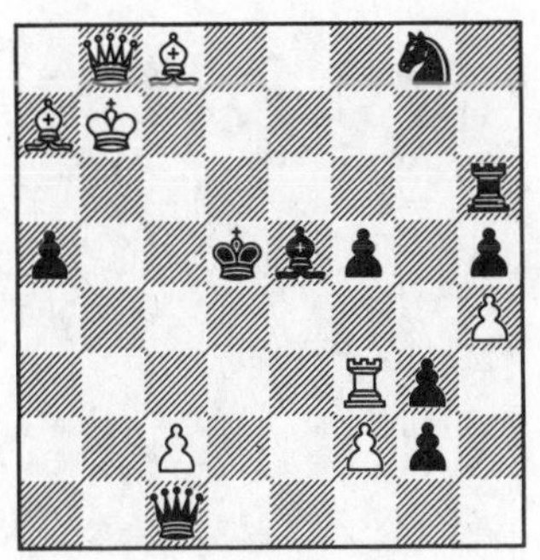

4132.

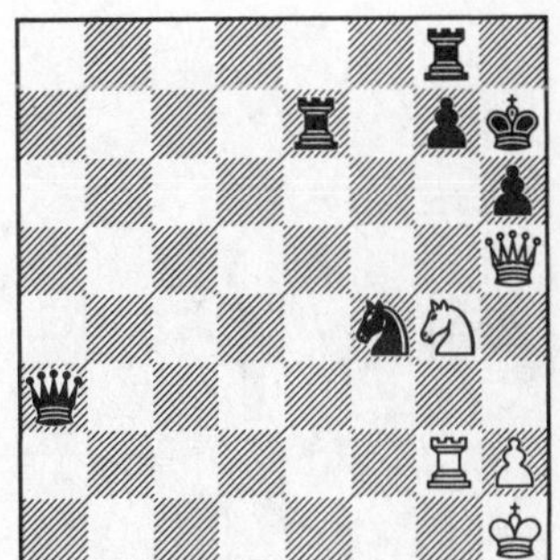

4133.

4134.

4135.

4136.

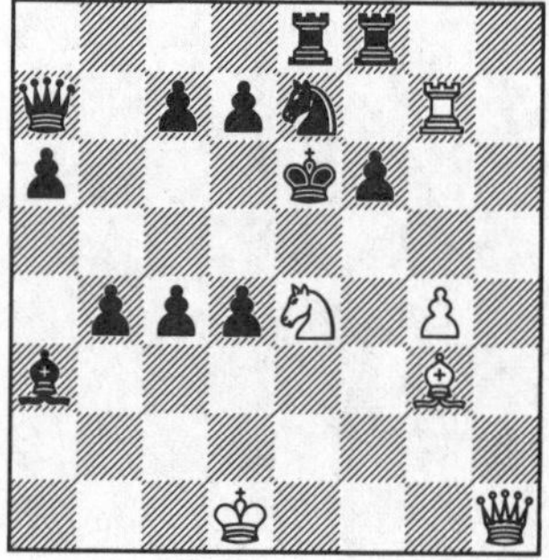

4137.

4138.

4139.

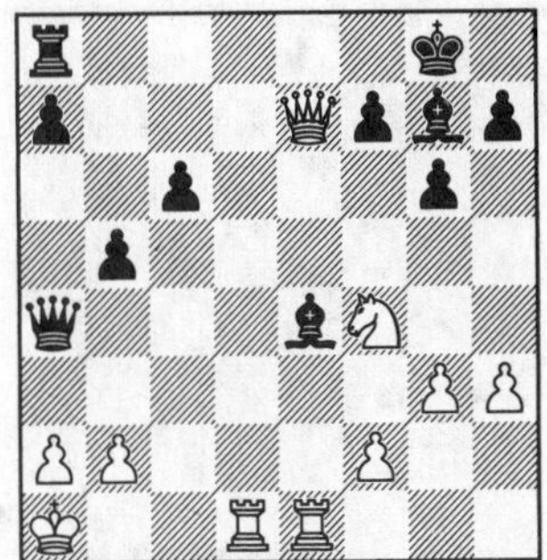

4140.

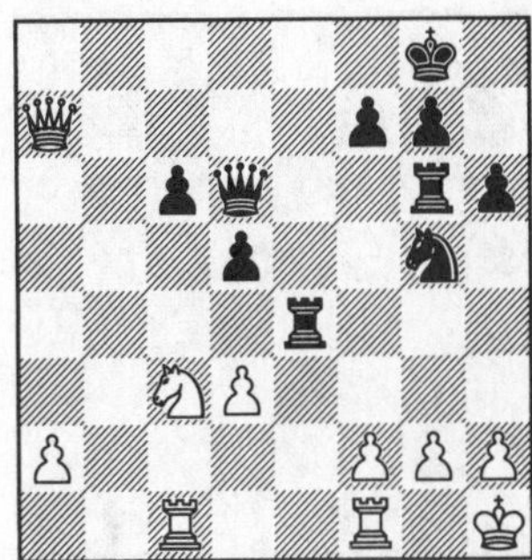

4141.

4142.

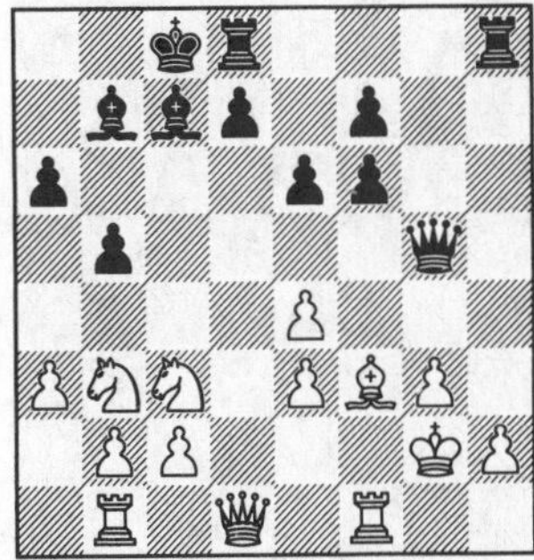

4143.

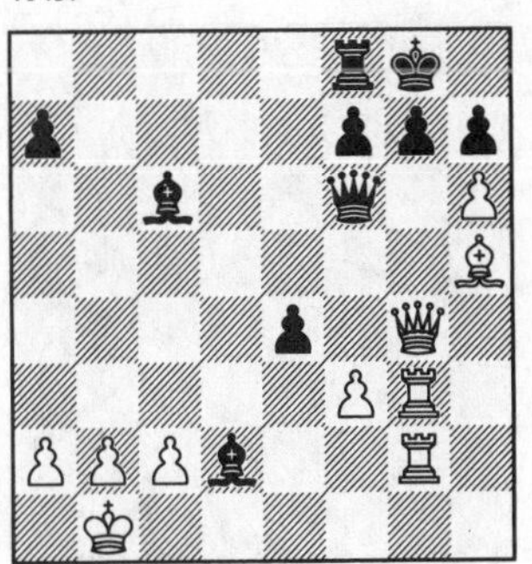

4144.

4145.

4146.

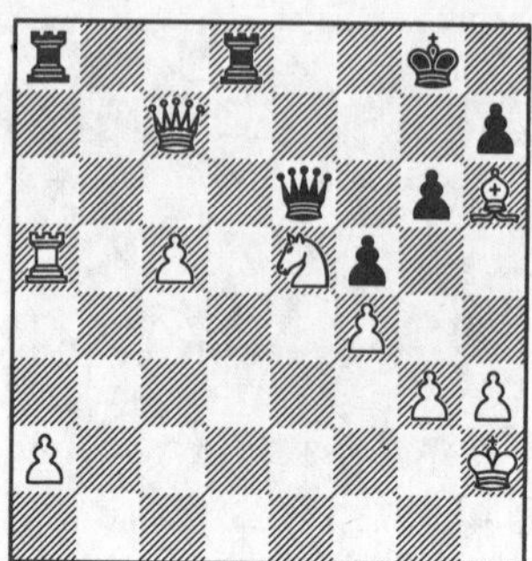

4147.

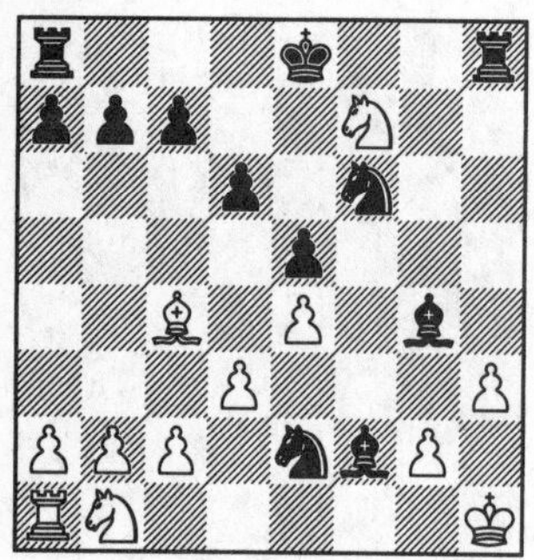

4148.

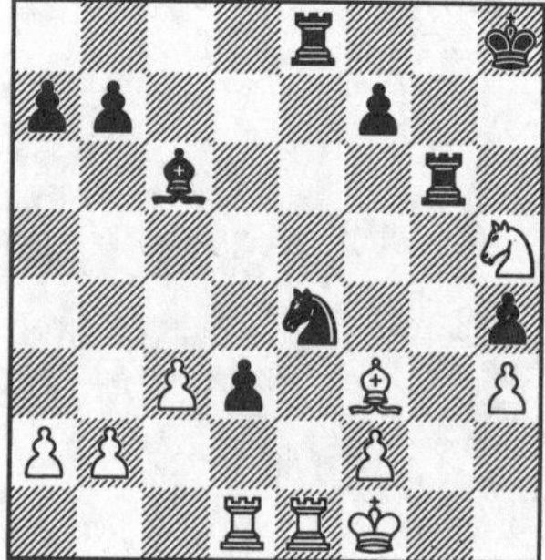

4149.

4150.

4151.

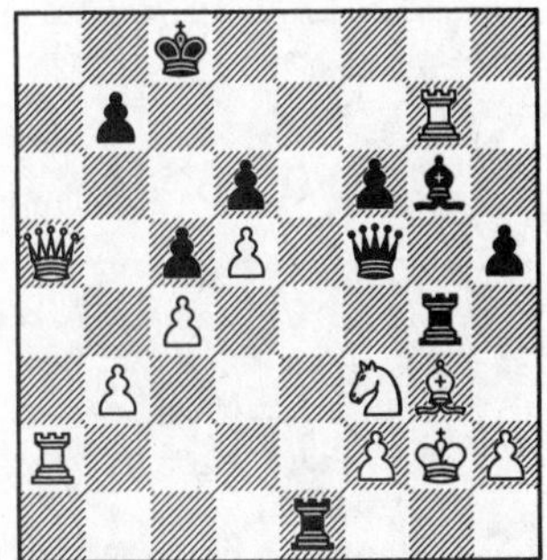

4152.

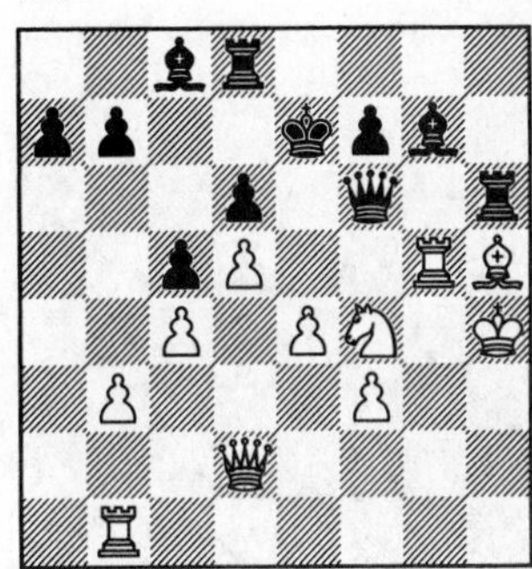

4153.

4154.

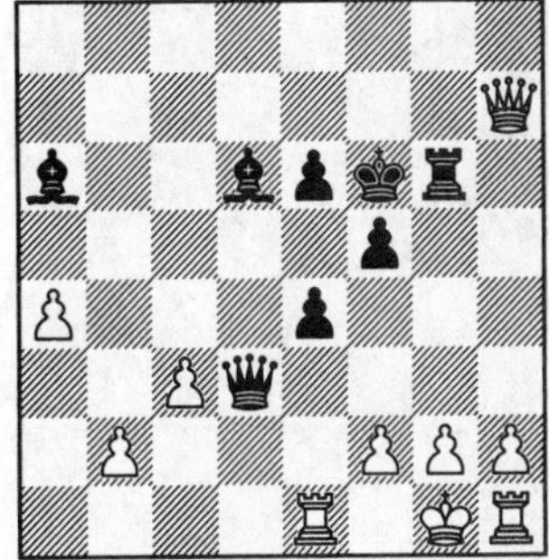

4155.

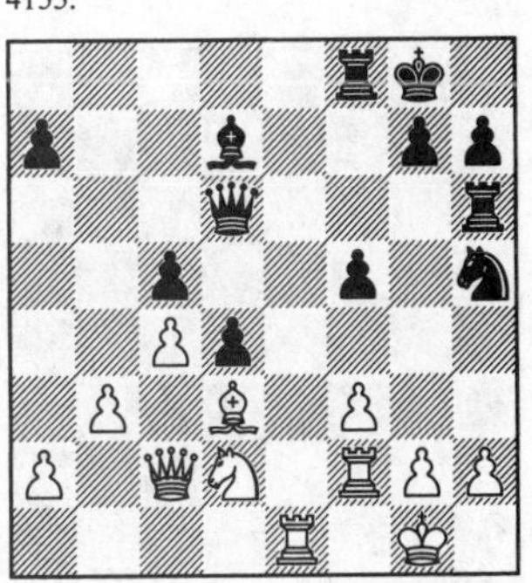

4156.

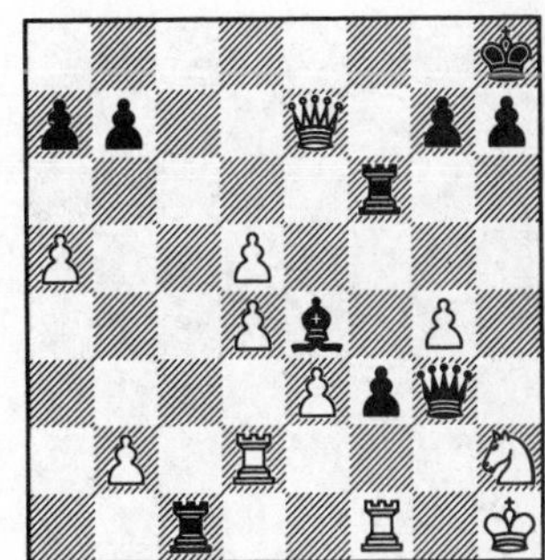

4157.

4158.

4159.

4160.

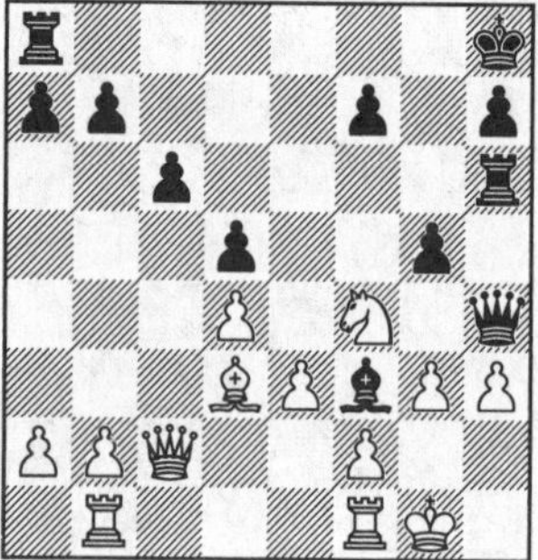

4161.

4162.

4163.

4164.

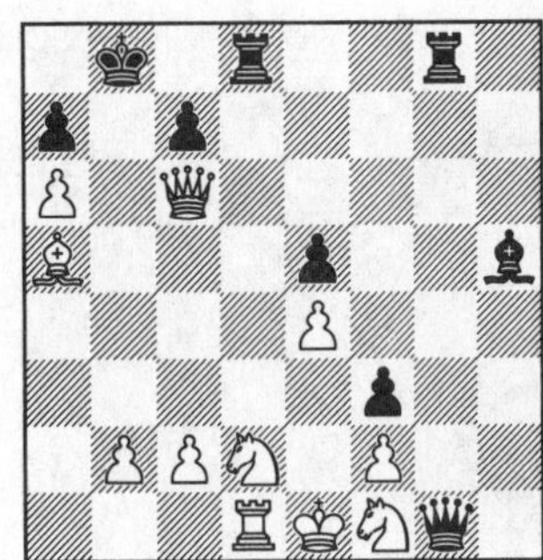

4165.

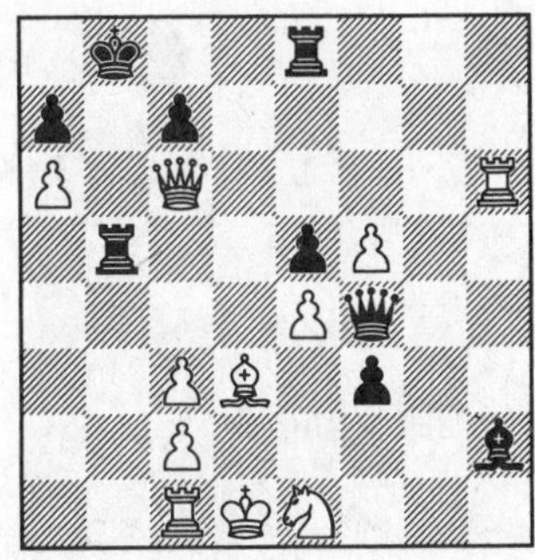

4166.

4167.

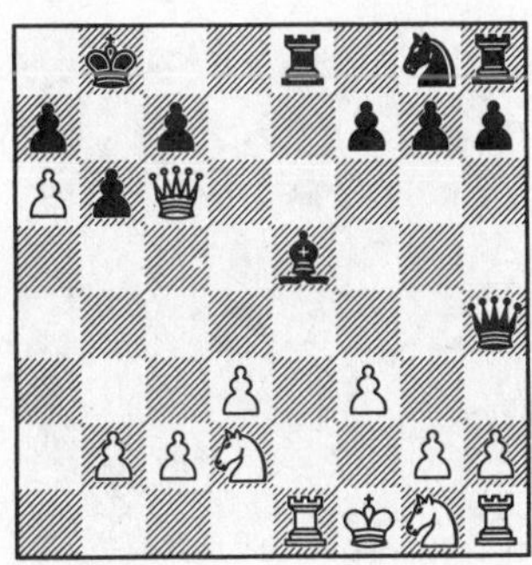

4168.

4169.

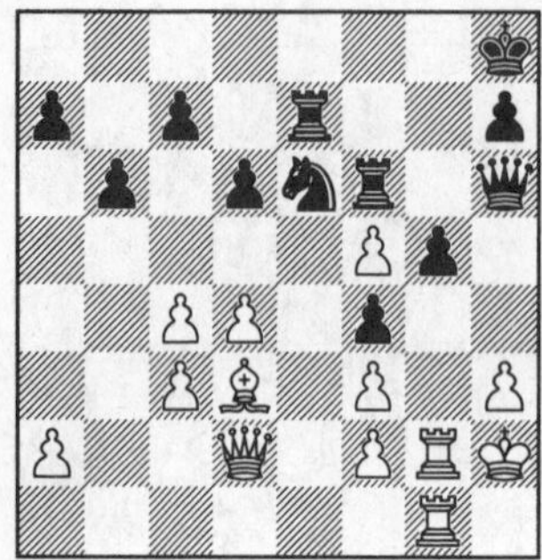

4170.

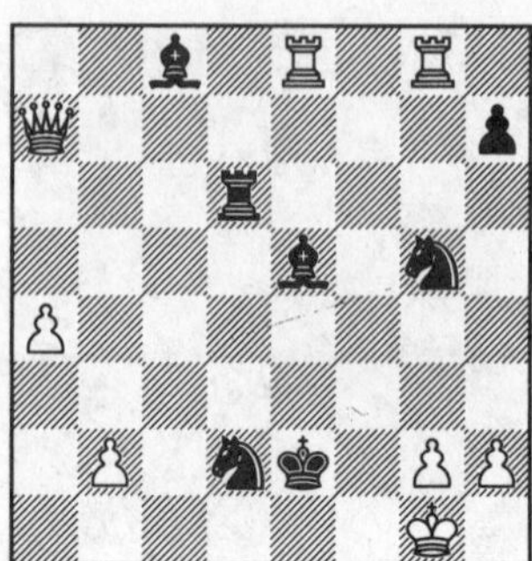

4171.

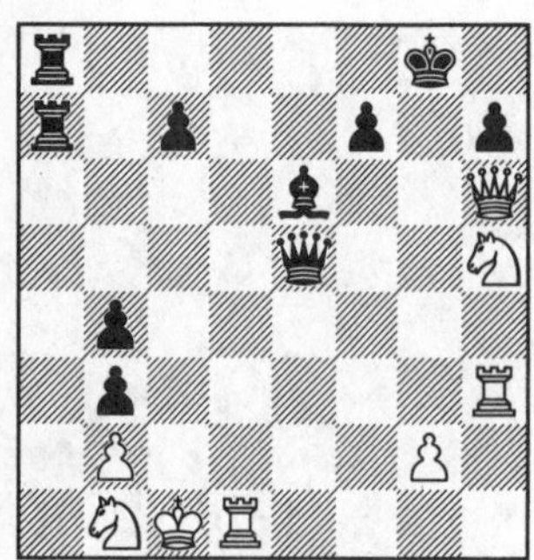

4172.

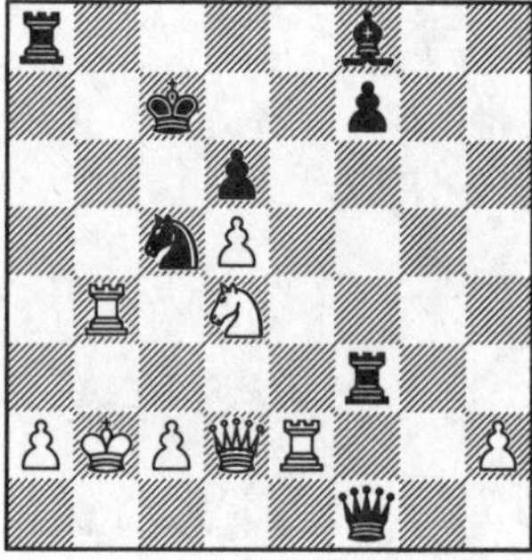

4173.

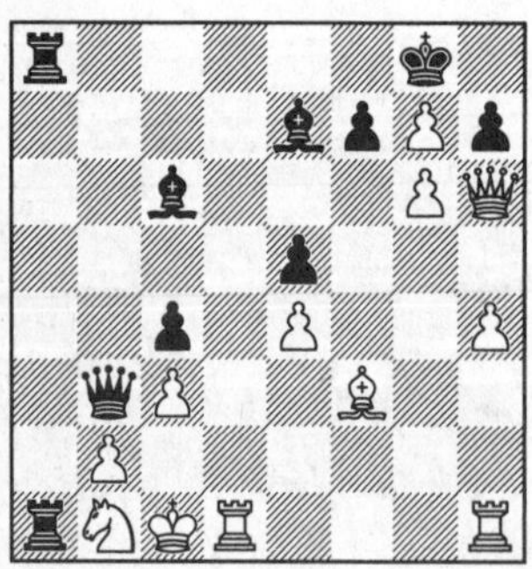

4174.

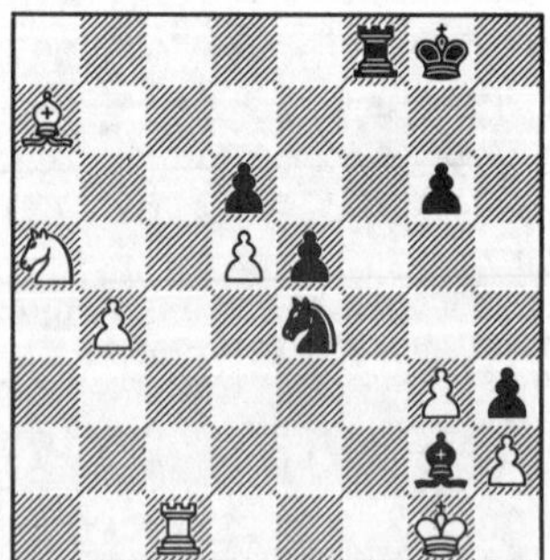

4175.

4176.

4177.

4178.

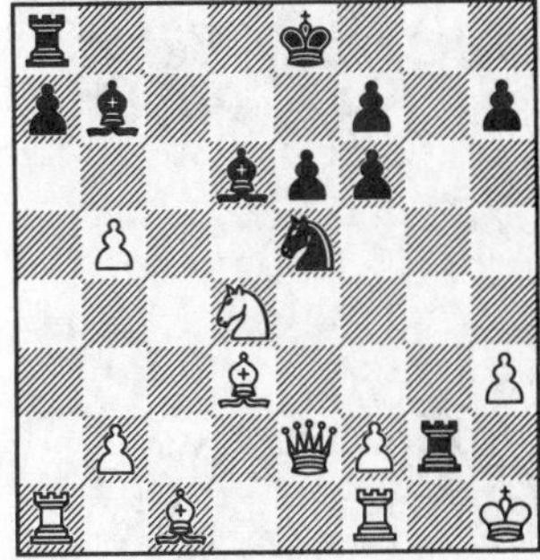

4179.

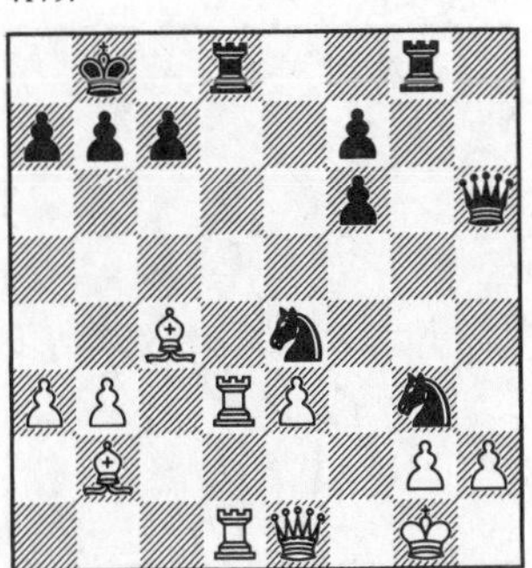

4180.

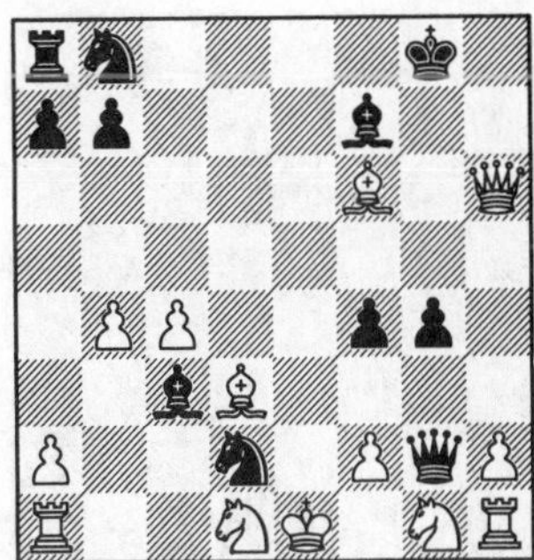

4181.

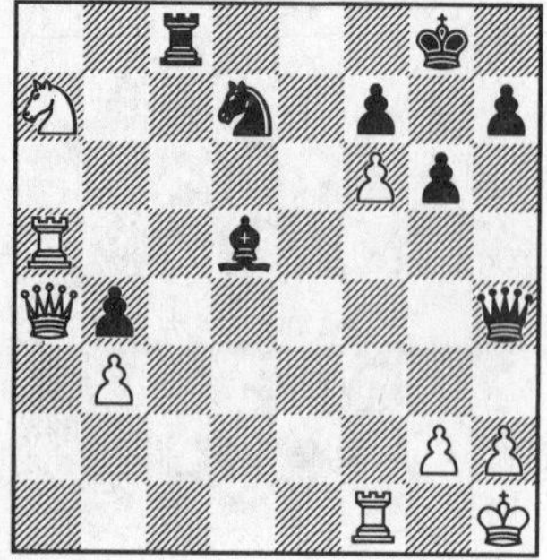

4182.

4183.

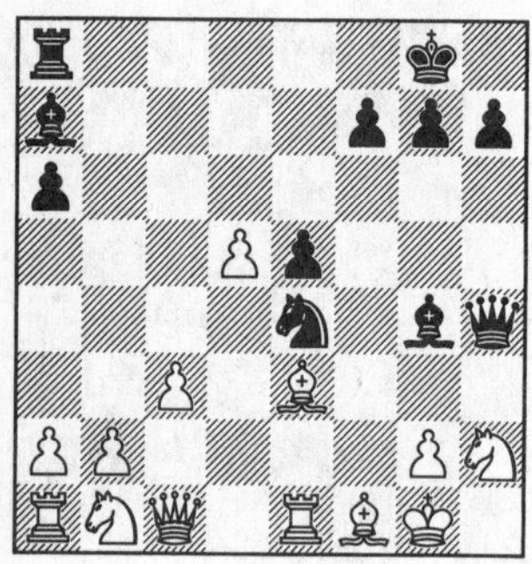

4184.

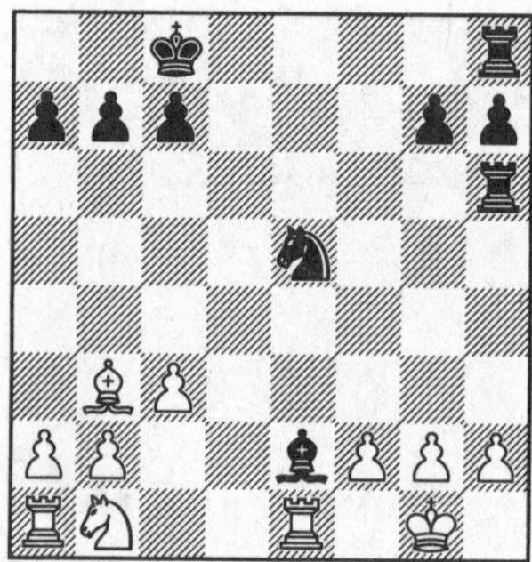

4185.

4186.

4187.

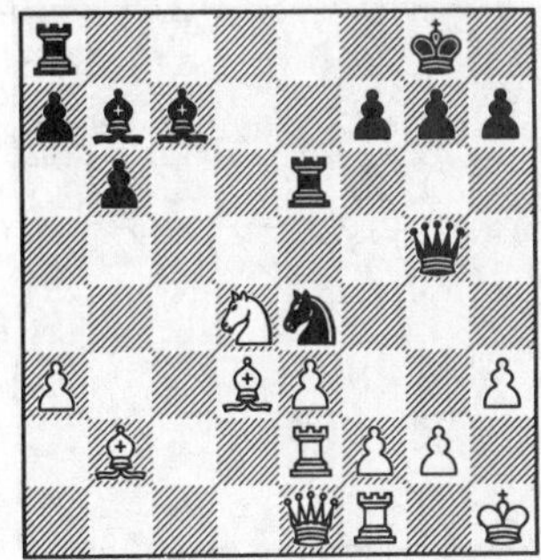

4188.

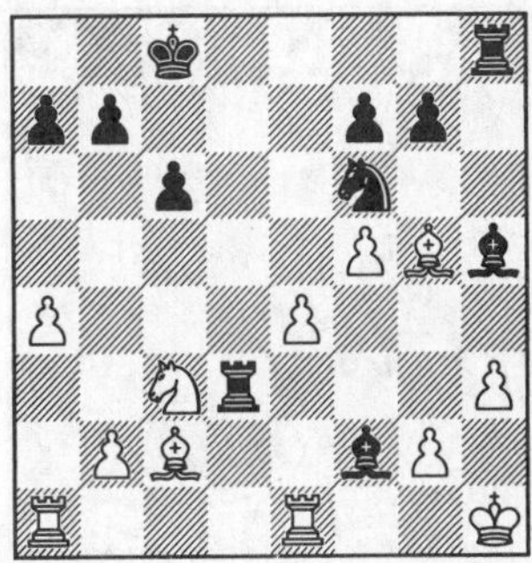

4189.

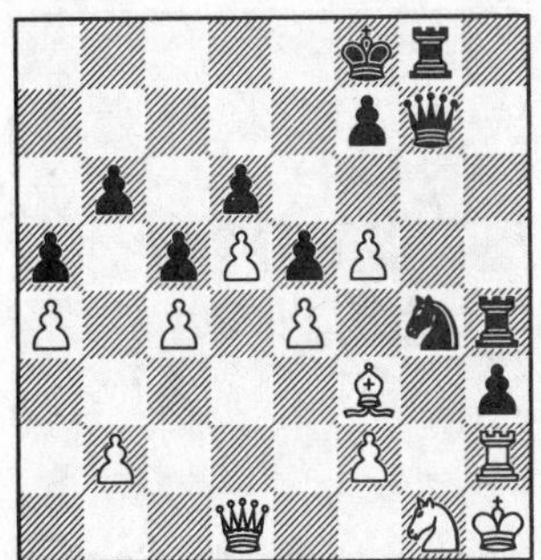

4190.

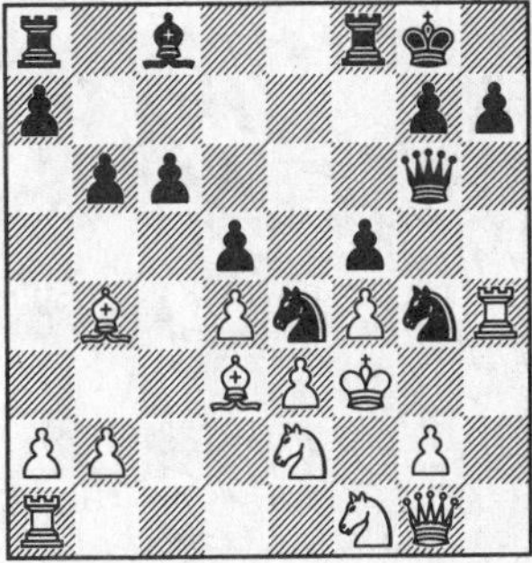

4191.

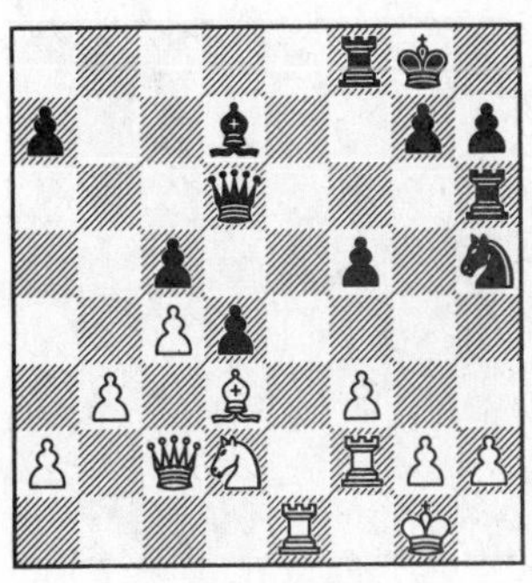

4192.

4193.

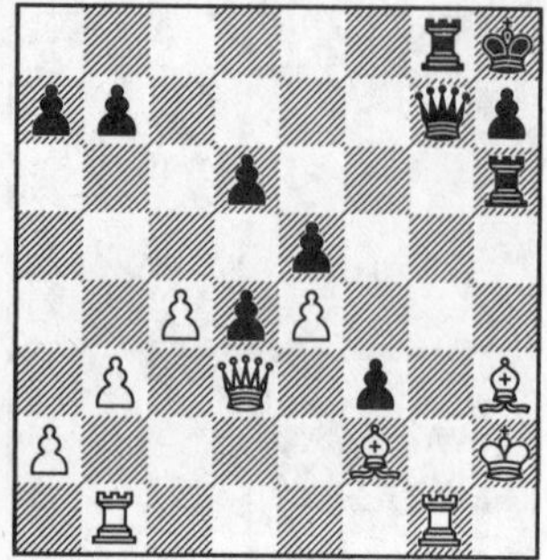

4194.

4195.

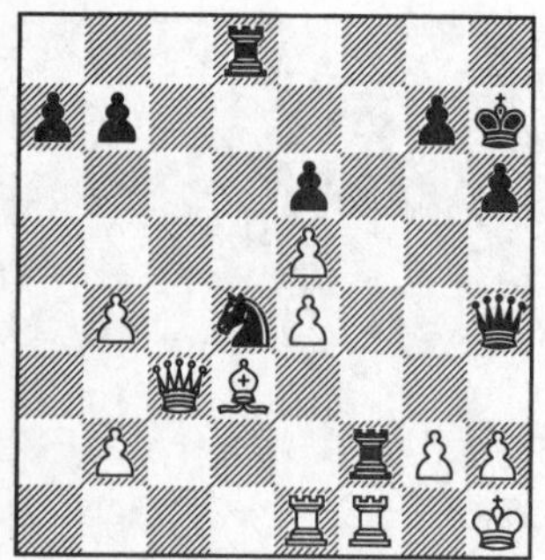

4196.

4197.

4198.

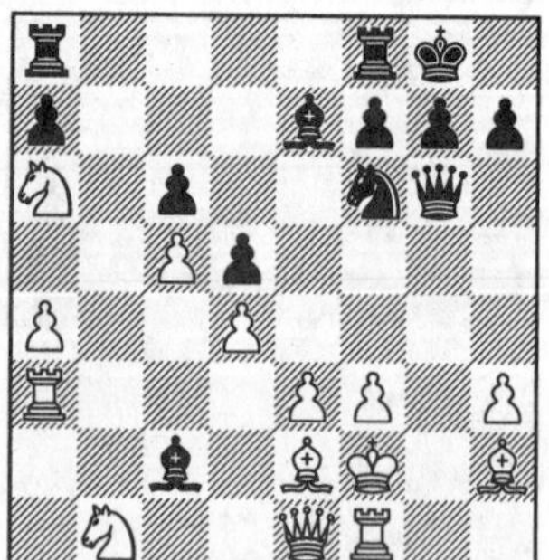

4199.

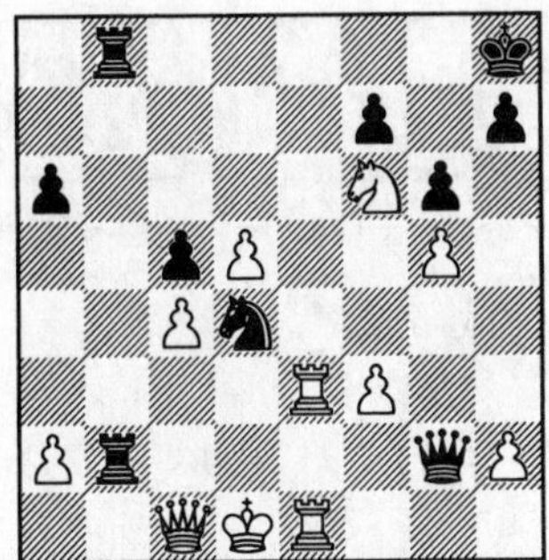

4200.

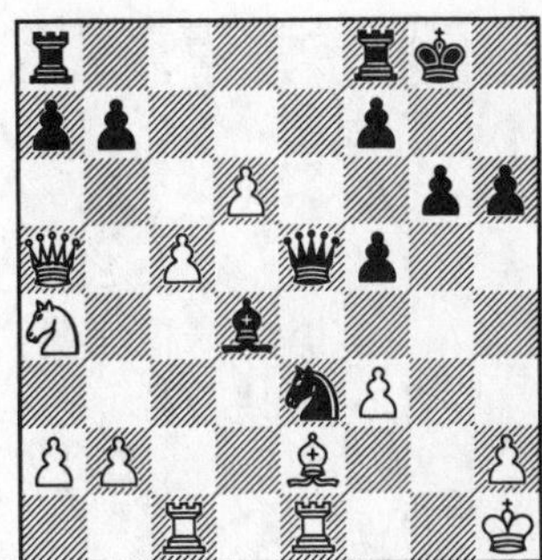

4201.

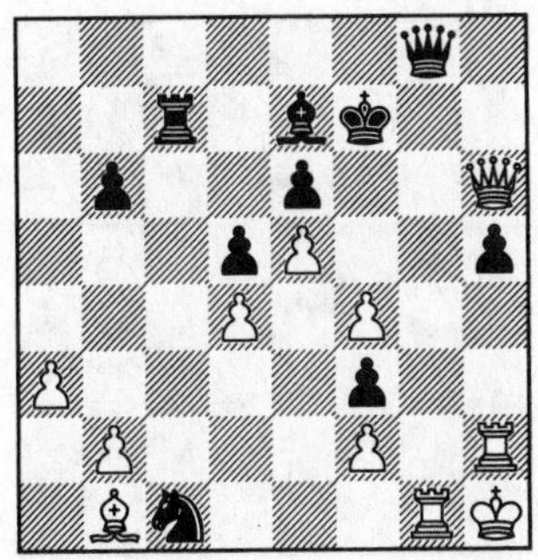

4202.

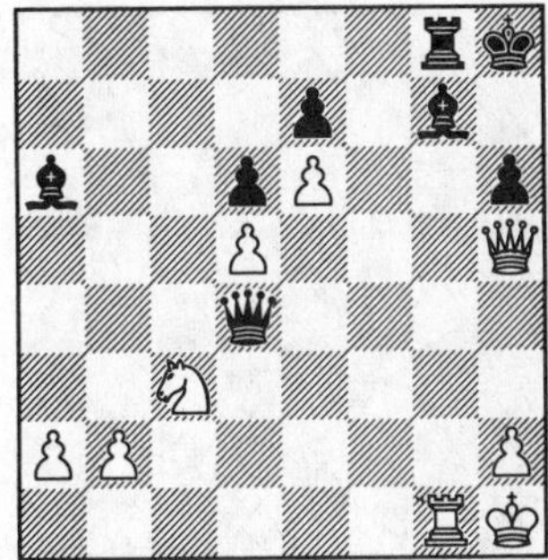

4203.

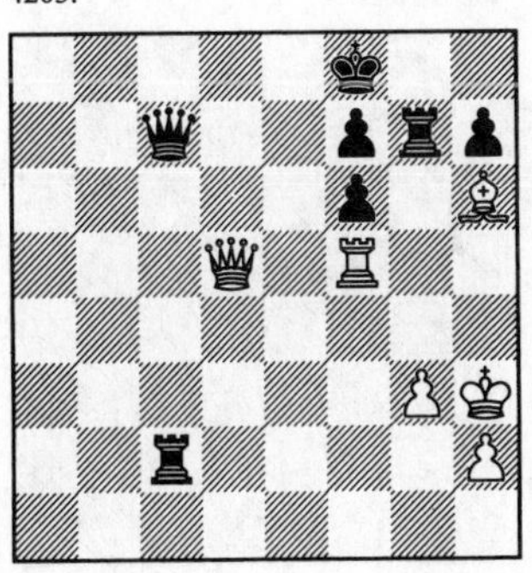

4204.

4205.

4206.

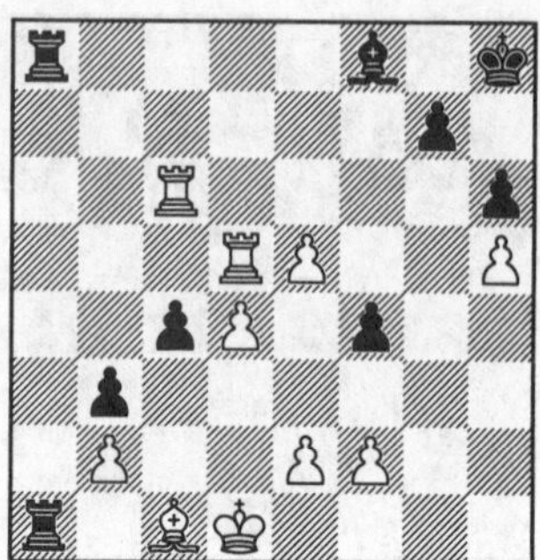

4207.

4208.

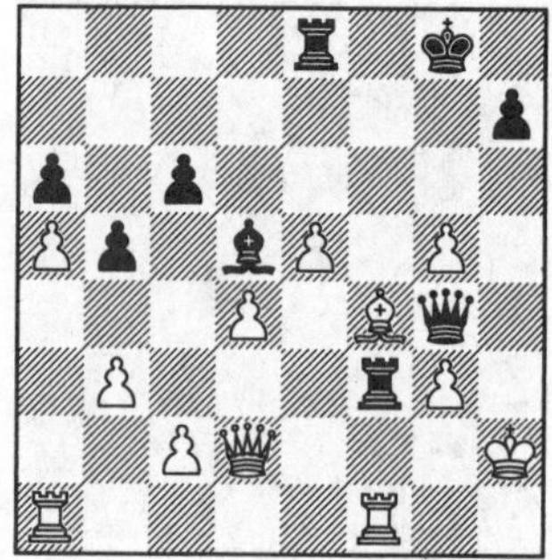

4209.

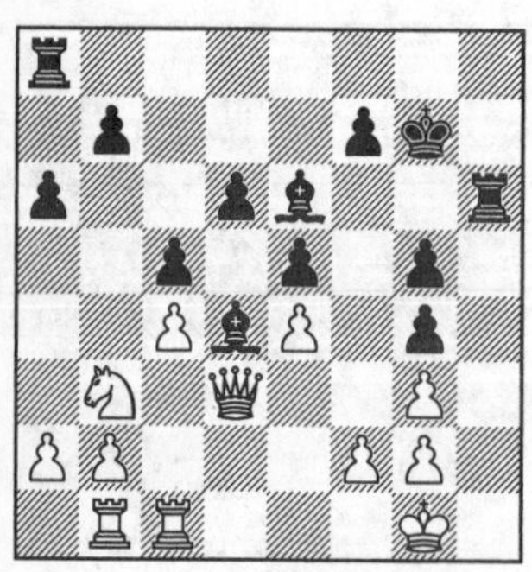

4210.

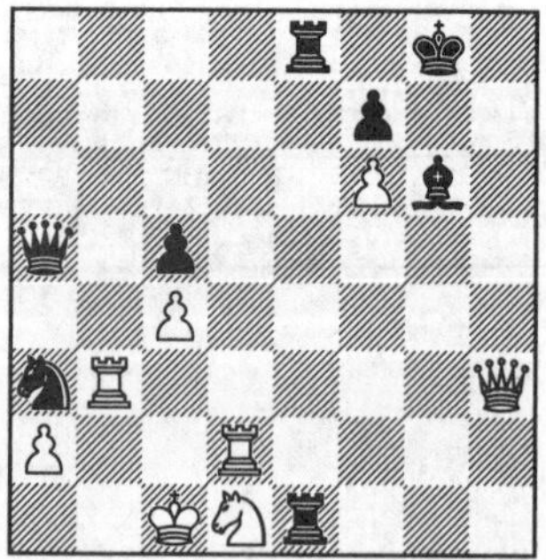

4211.

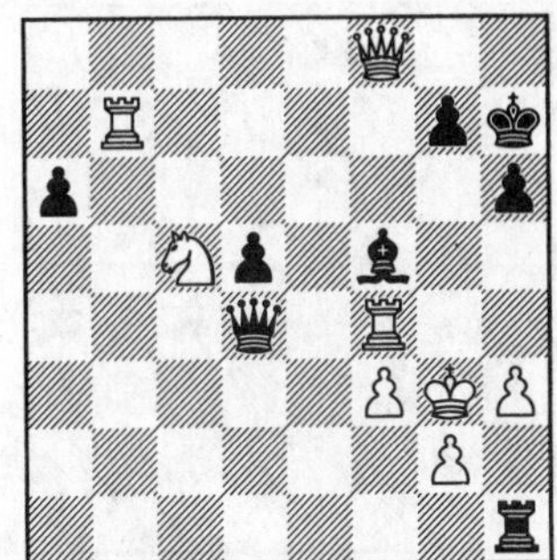

4212.

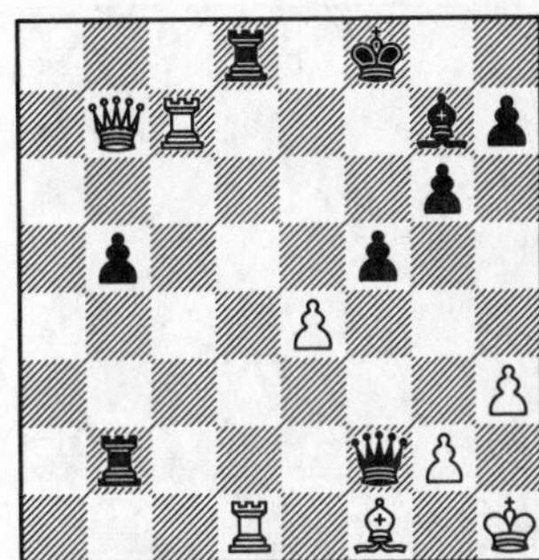

4213.

4214.

4215.

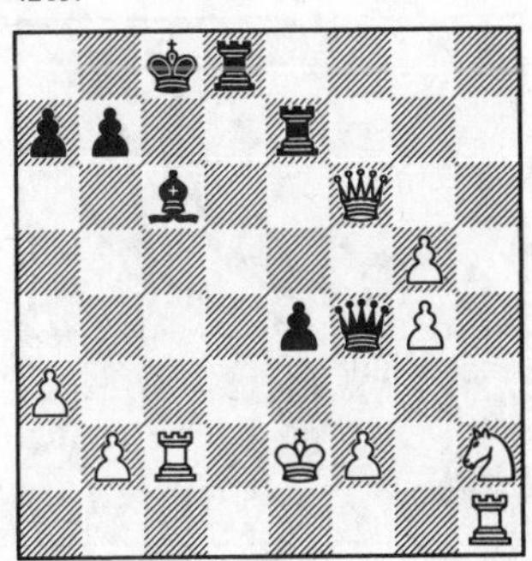

4216.

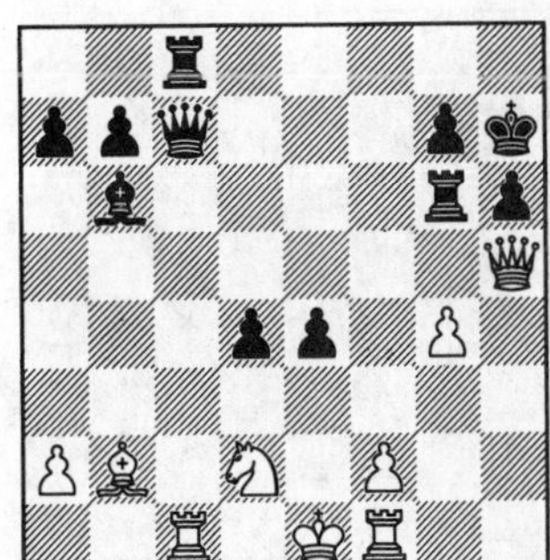

4217.

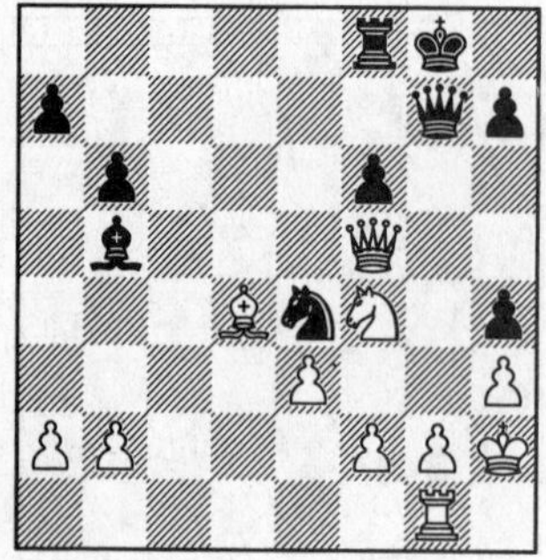

4218.

4219.

4220.

4221.

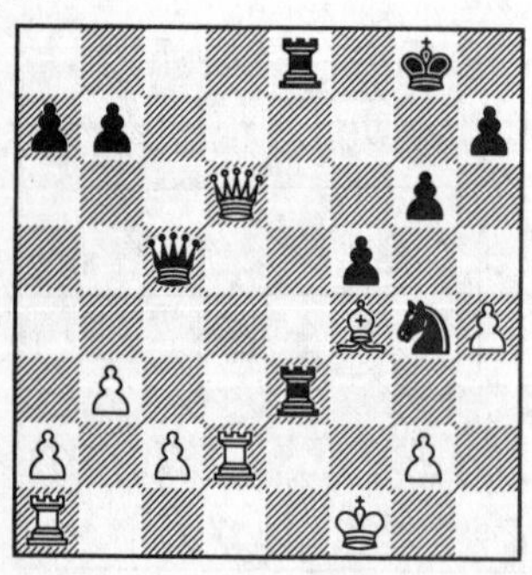

4222.

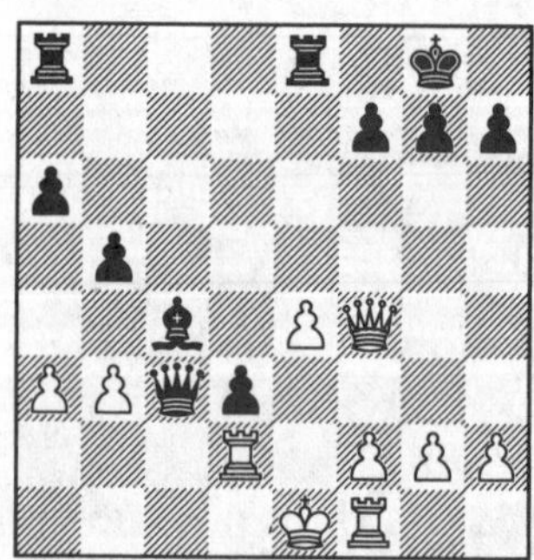

4223.

4224.

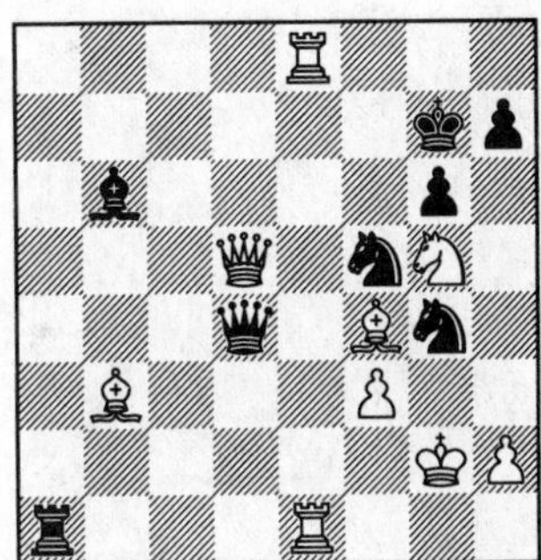

4225.

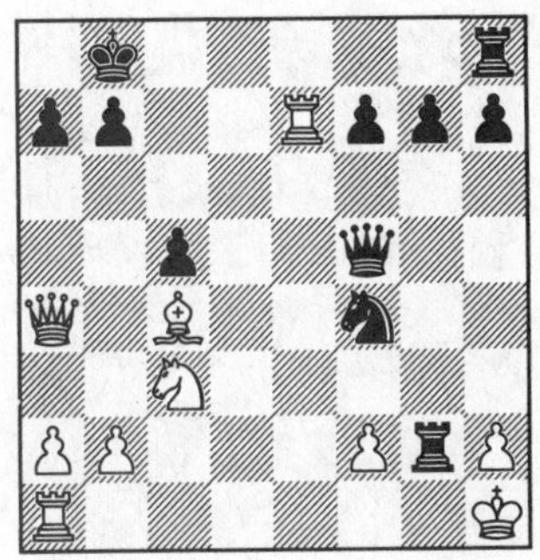

4226.

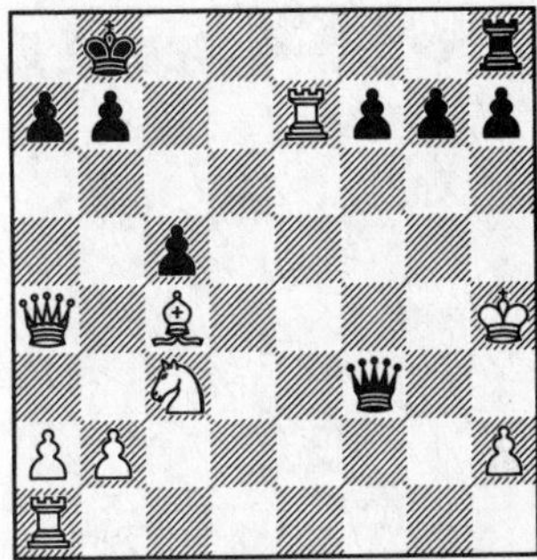

4227.

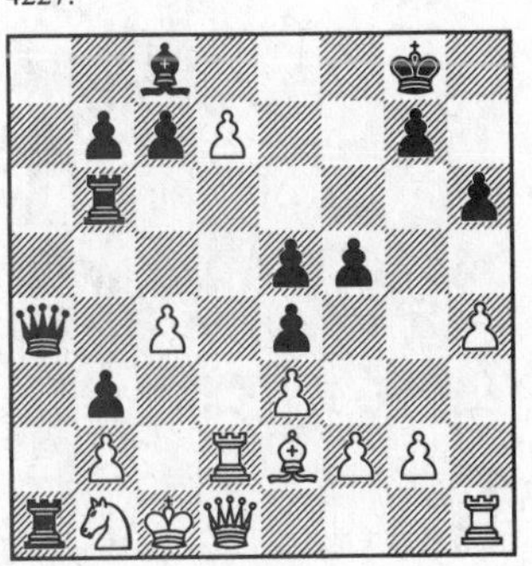

4228.

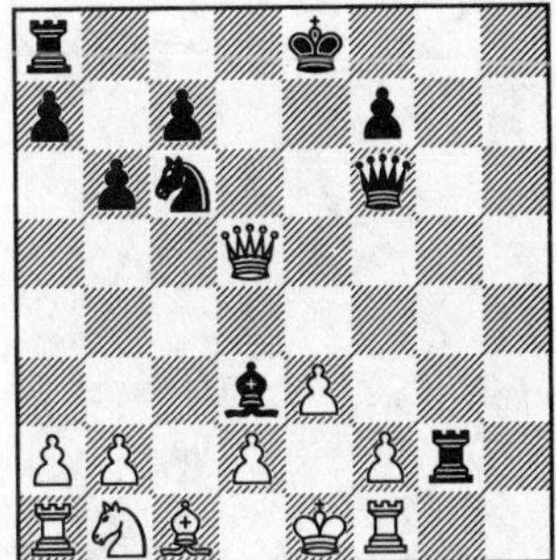

4229.

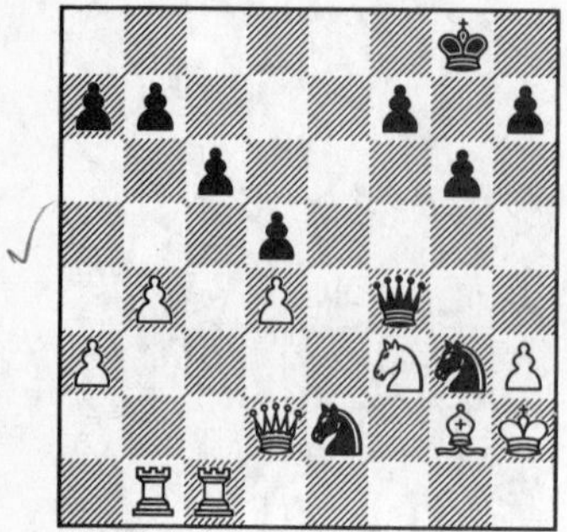

4230.

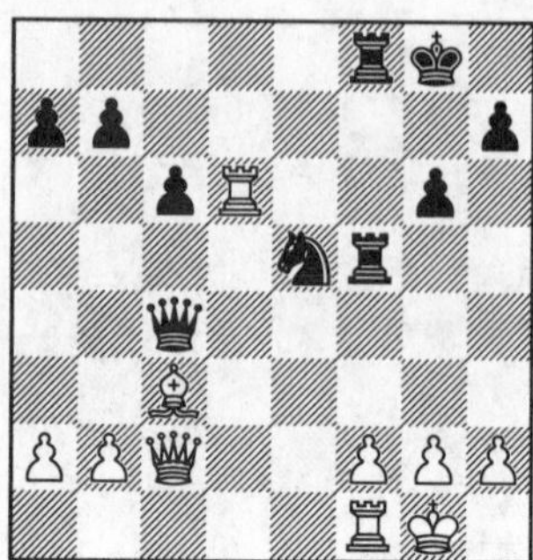

4231.

4232.

4233.

4234.

4235.

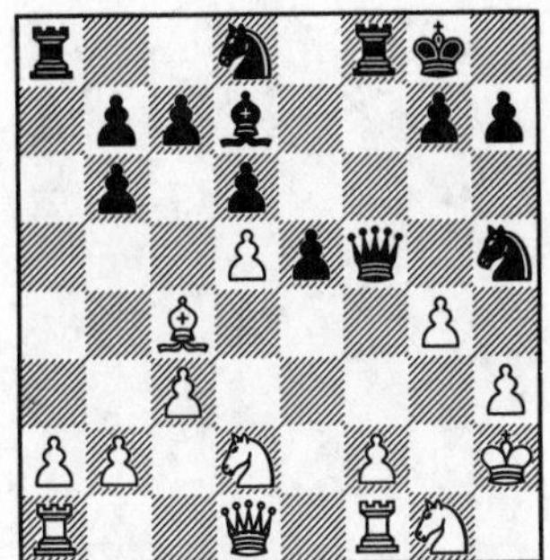

4236.

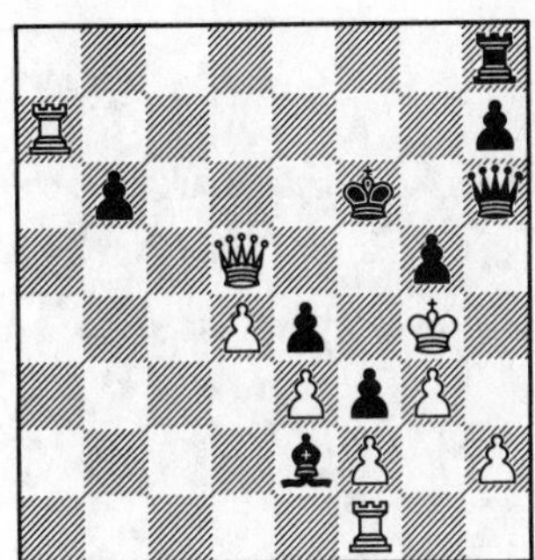

4237.

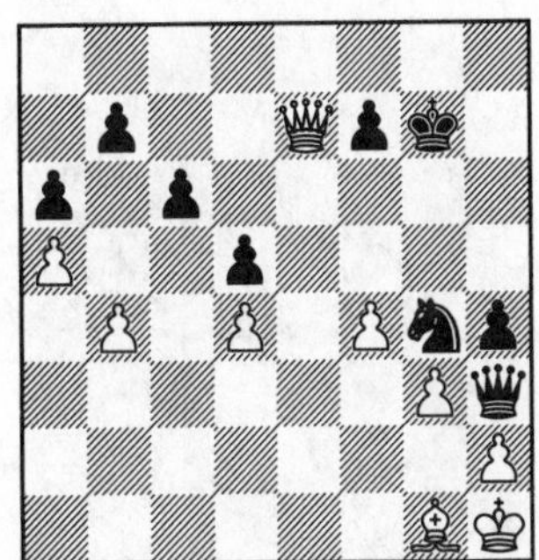

4238.

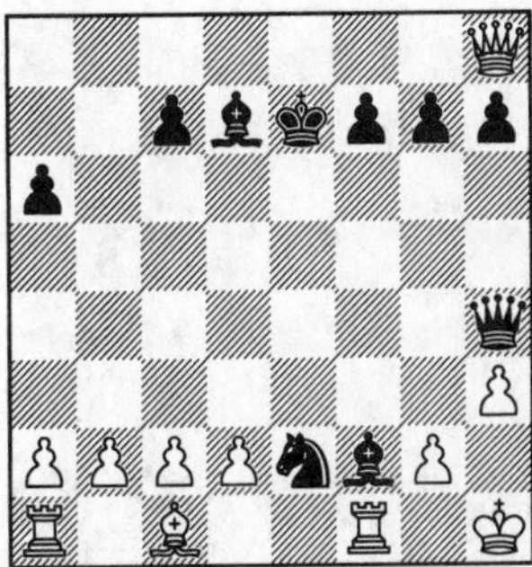

4239.

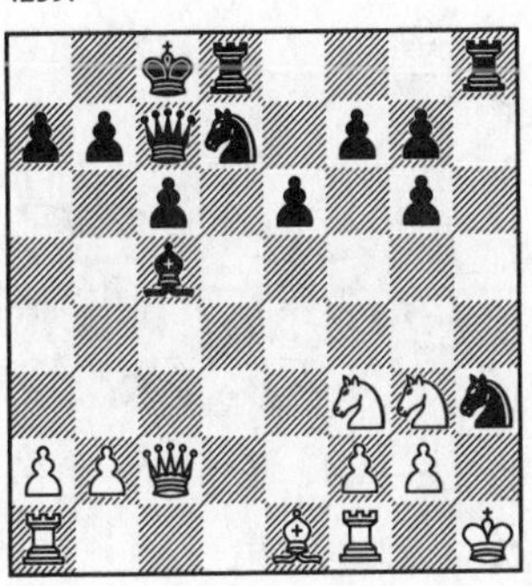

4240.

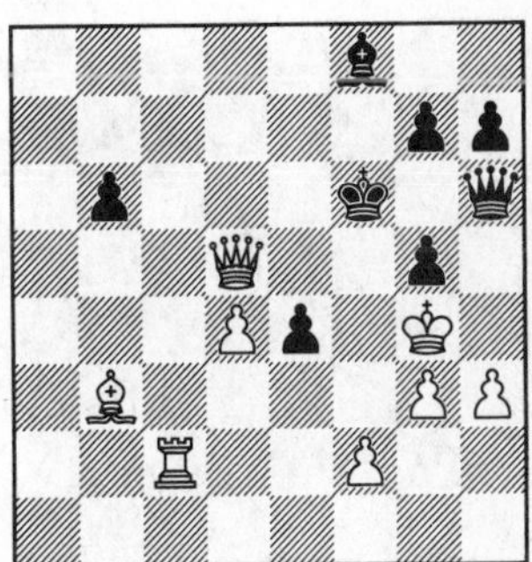

4241.

4242.

4243.

4244.

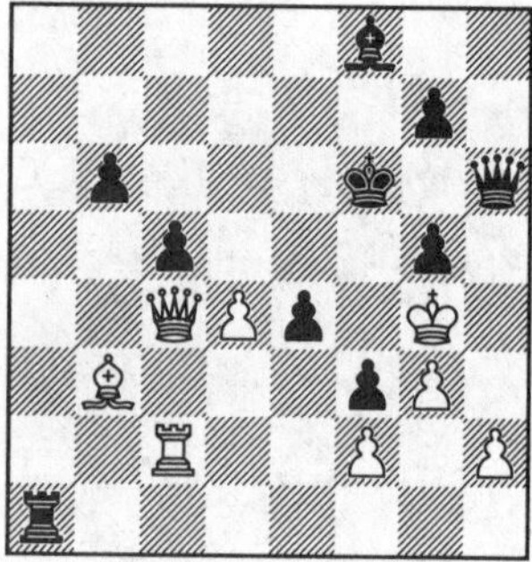

4245.

4246.

4247.

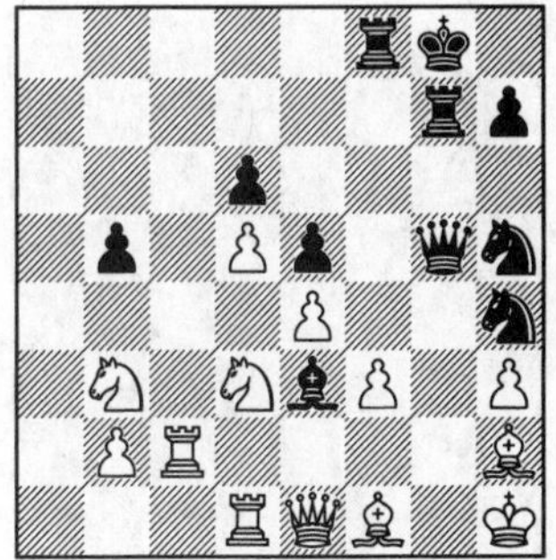

4248.

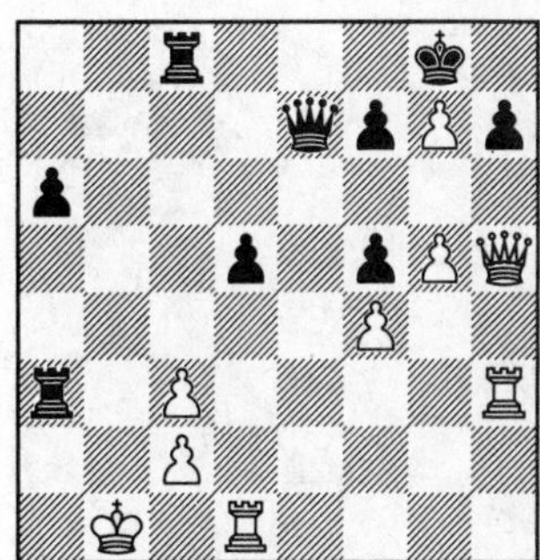

4249.

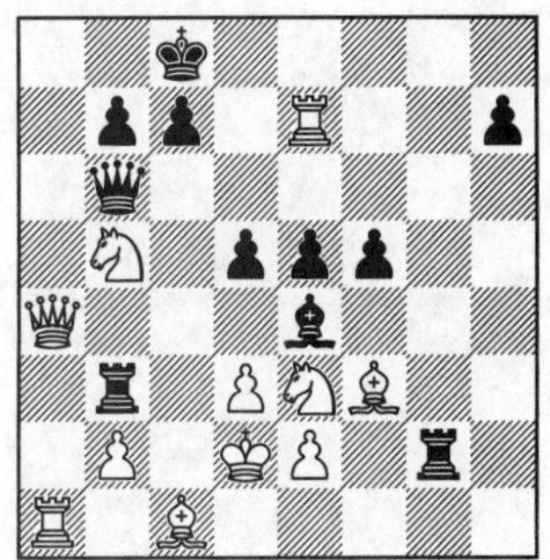

4250.

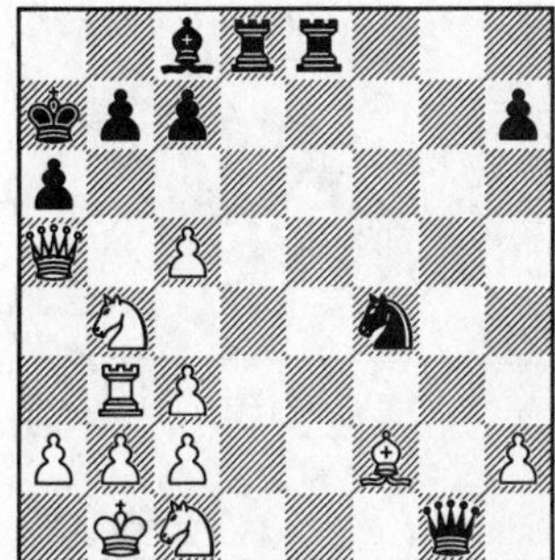

4251.

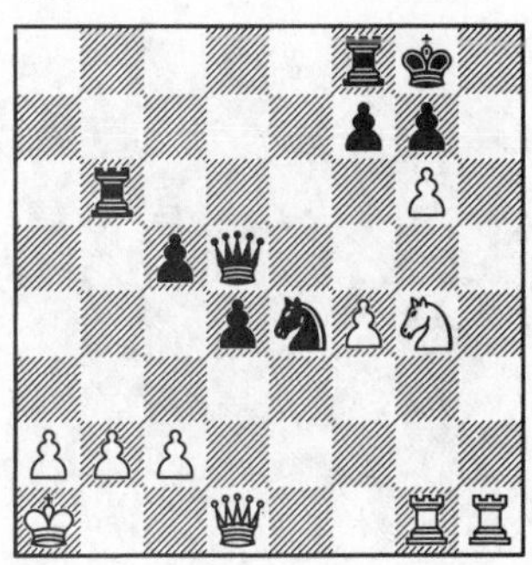

4252.

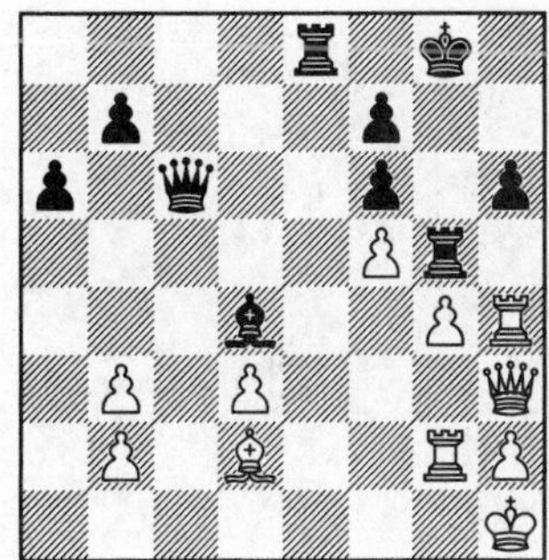

4253.

4254.

4255.

4256.

4257.

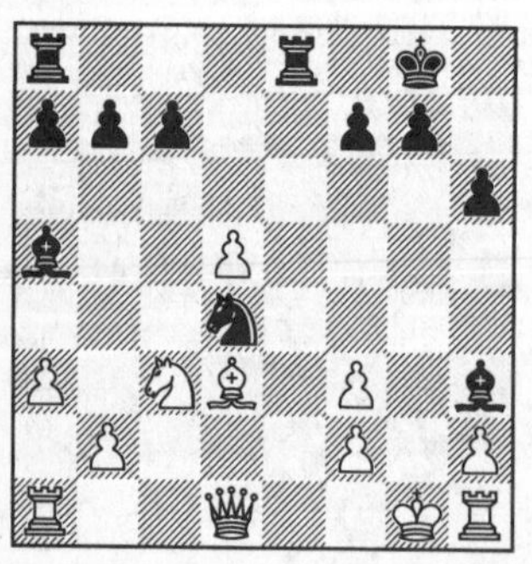

4258.

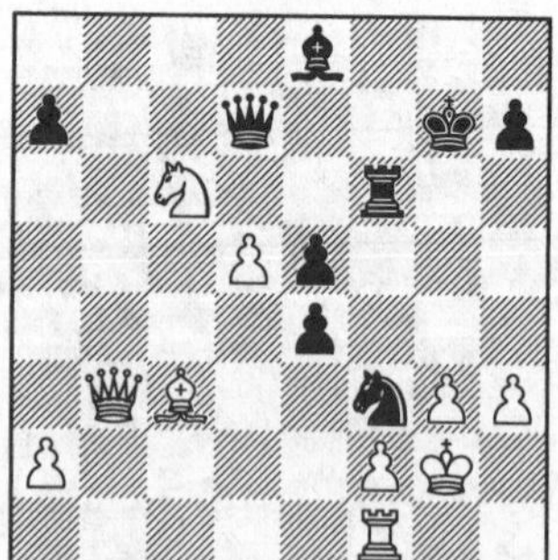

4259.

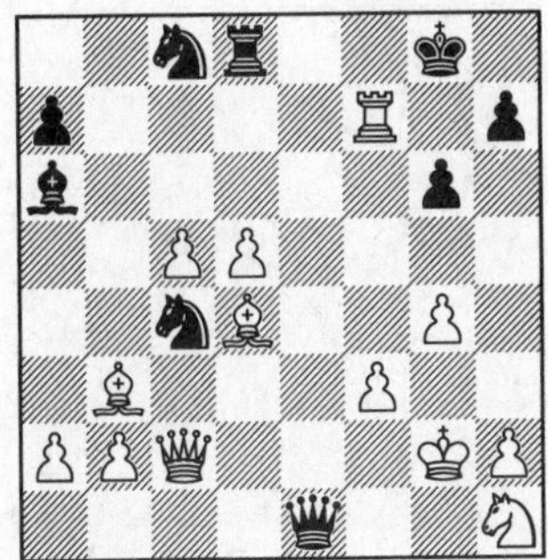

4260.

4261.

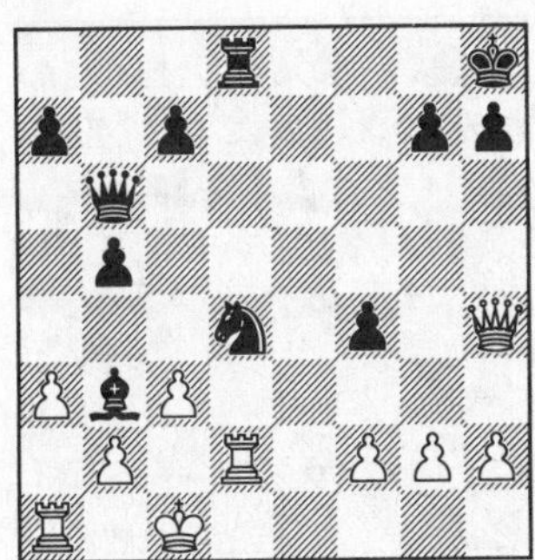

4262.

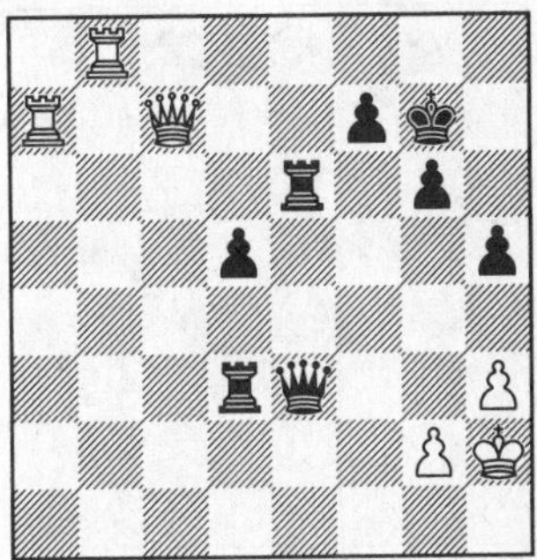

4263.

4264.

4265.

4266.

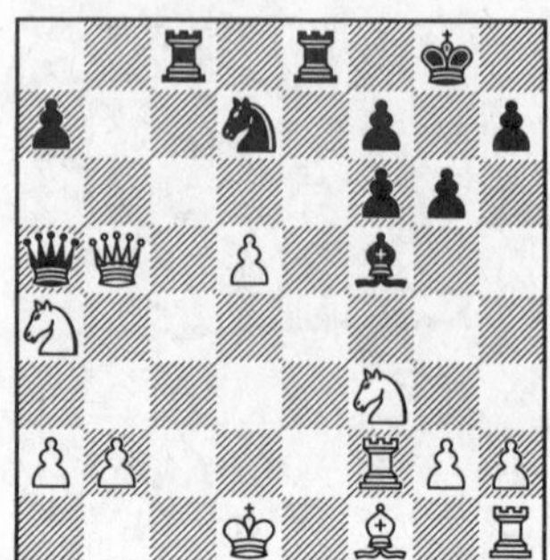

4267.

4268.

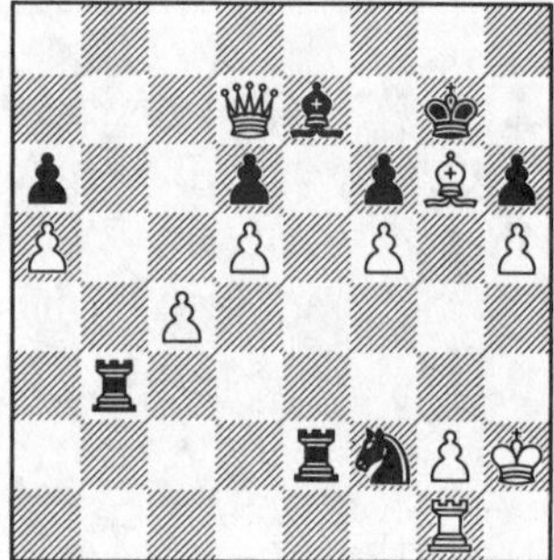

4269.

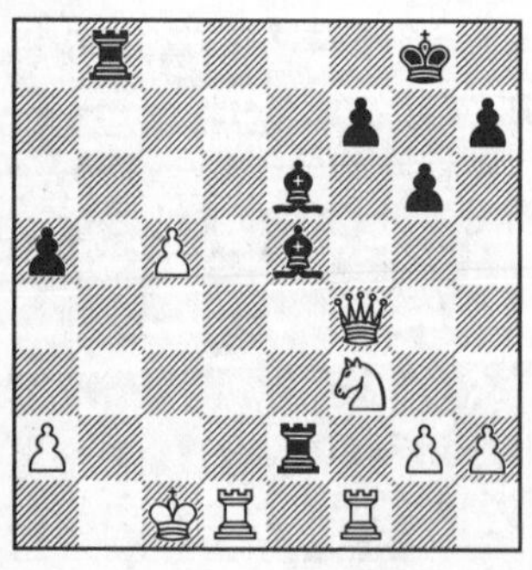

4270.

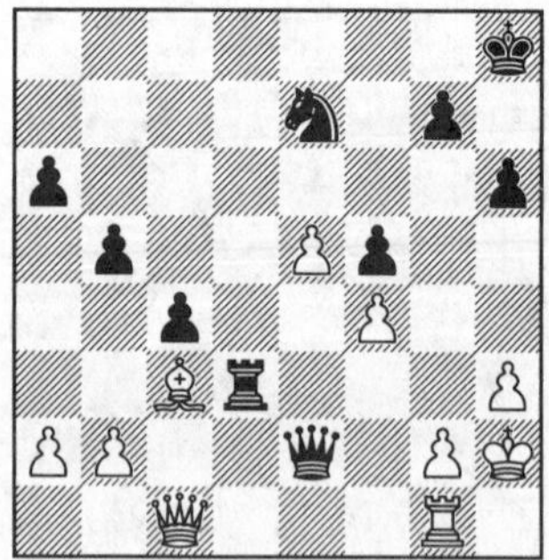

4271.

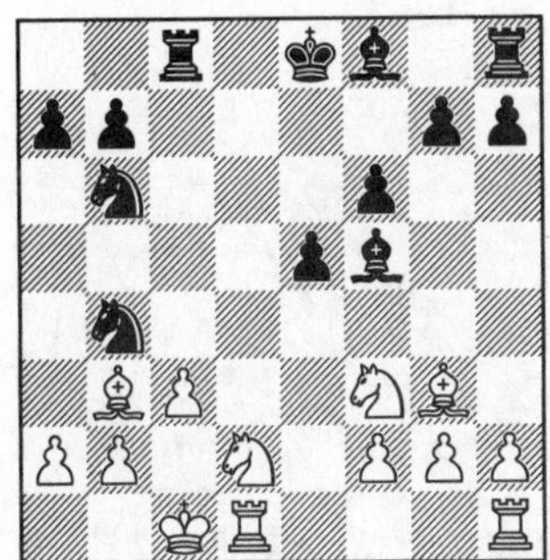

4272.

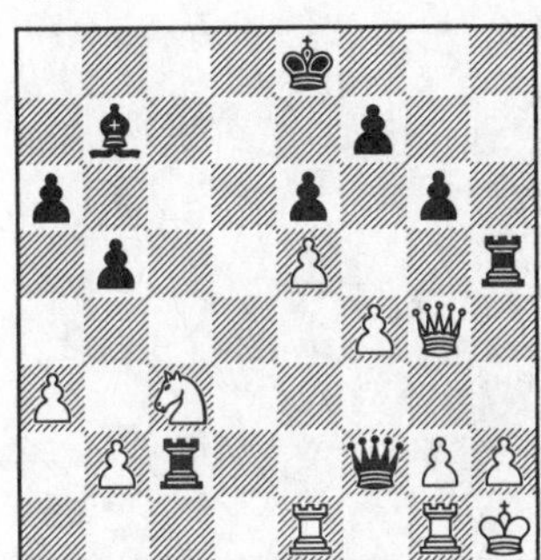

4273.

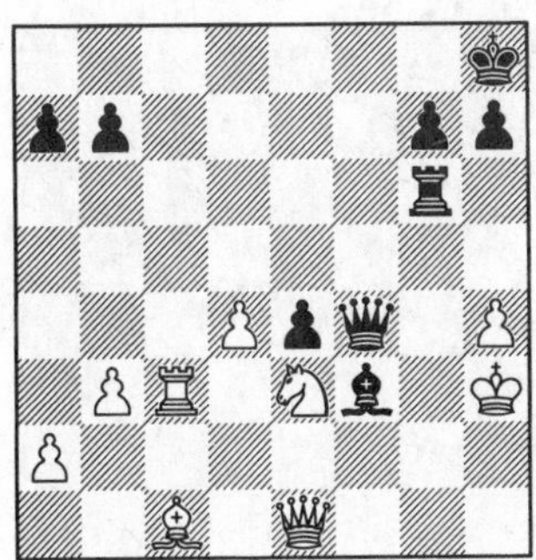

4274.

4275.

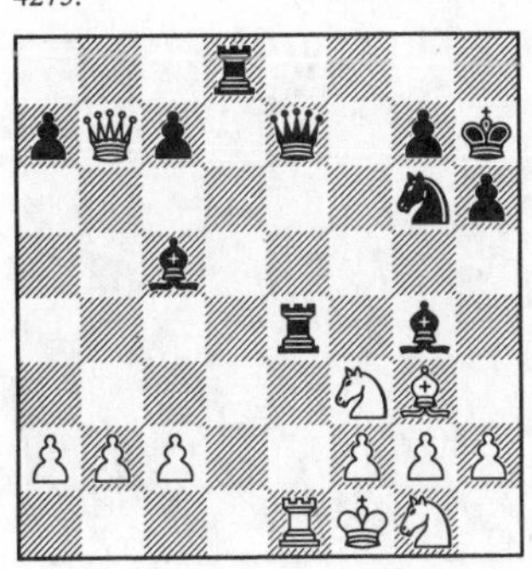

4276.

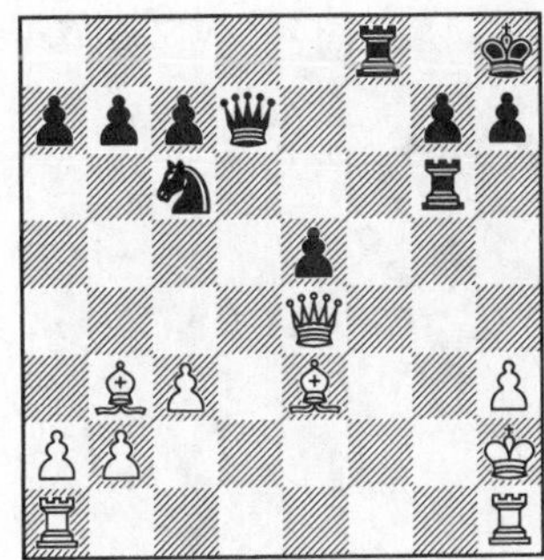

4277.

4278.

4279.

4280.

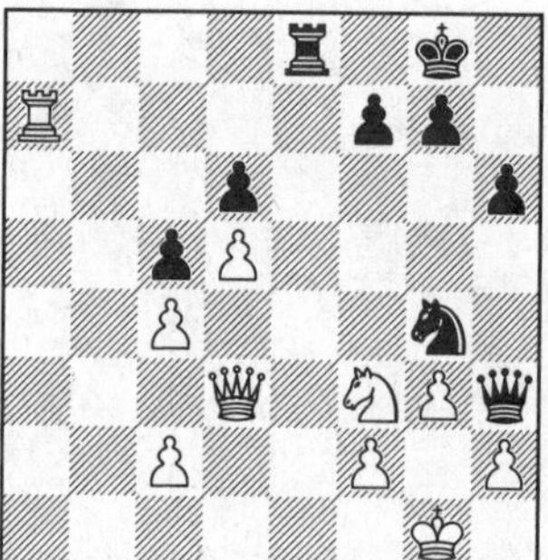

4281.

4282.

4283.

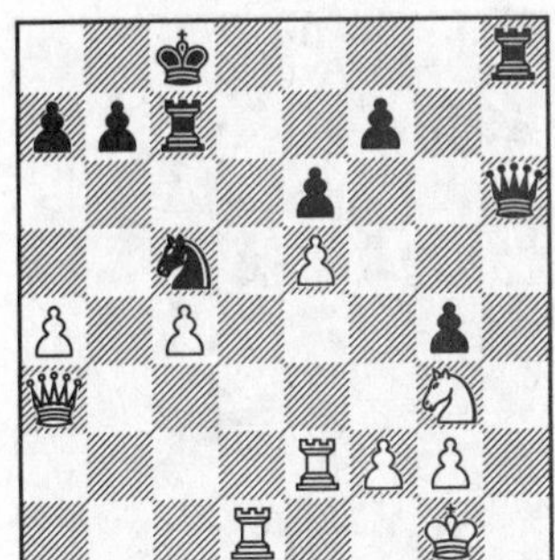

4284.

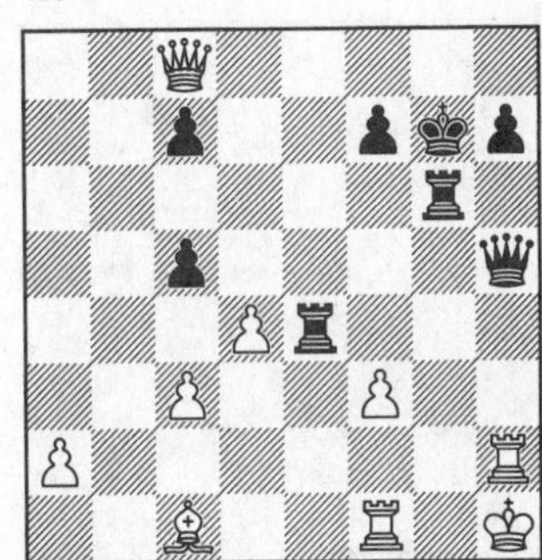

4285.

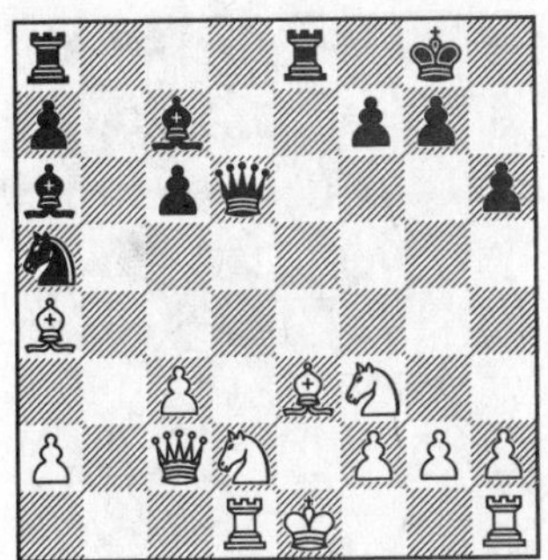

4286.

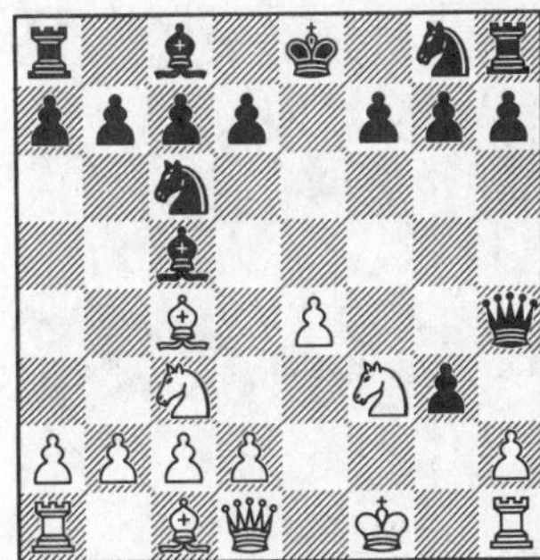

4287.

4288.

4289.

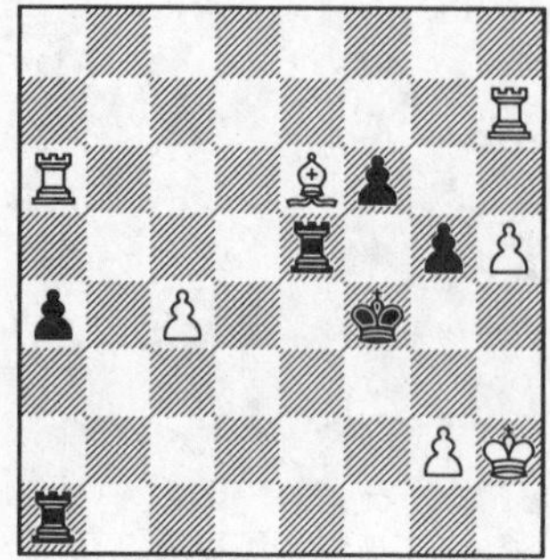

4290.

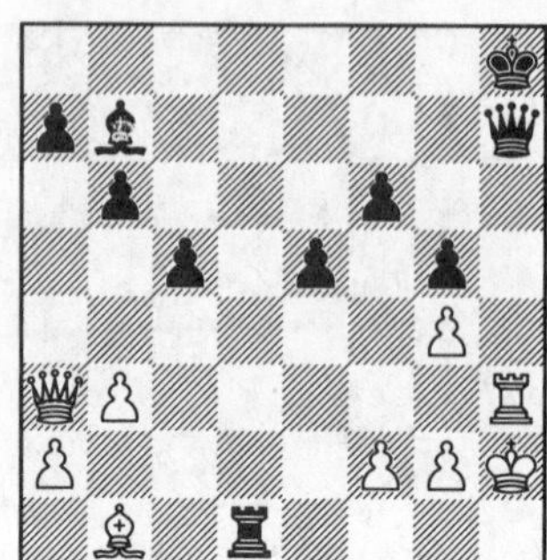

4291.

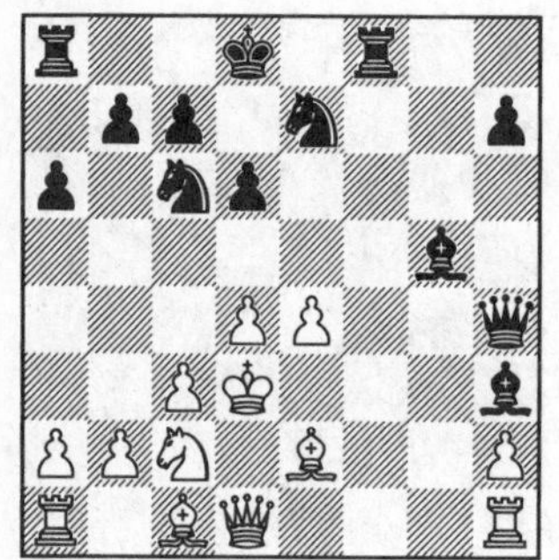

4292.

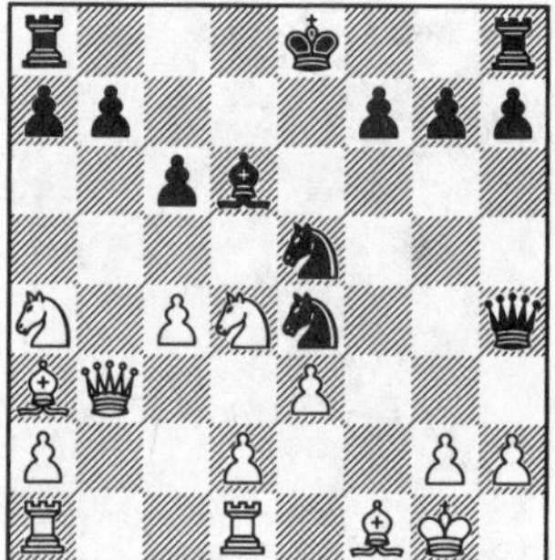

4293.

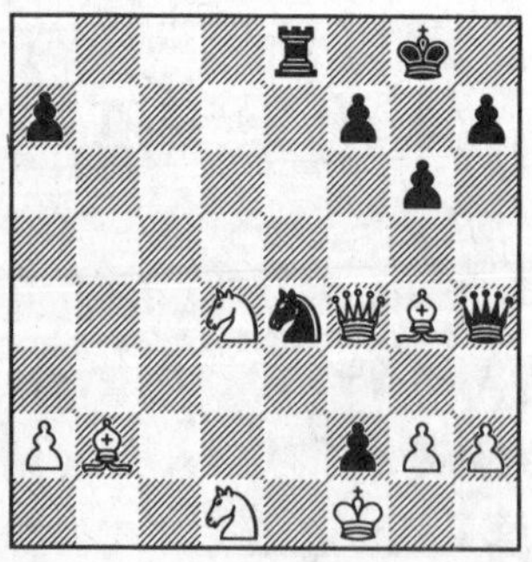

4294.

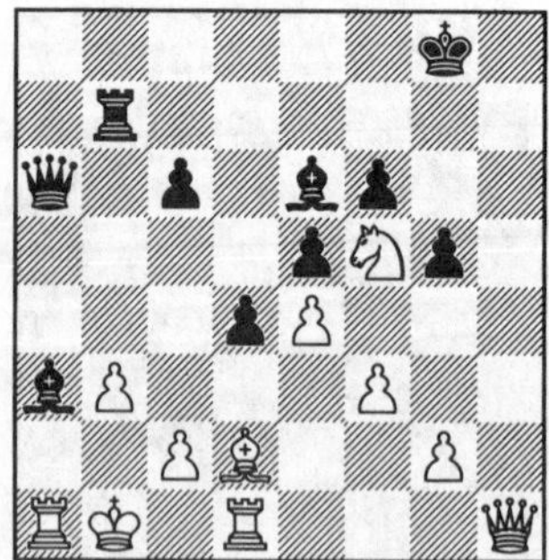

4295.

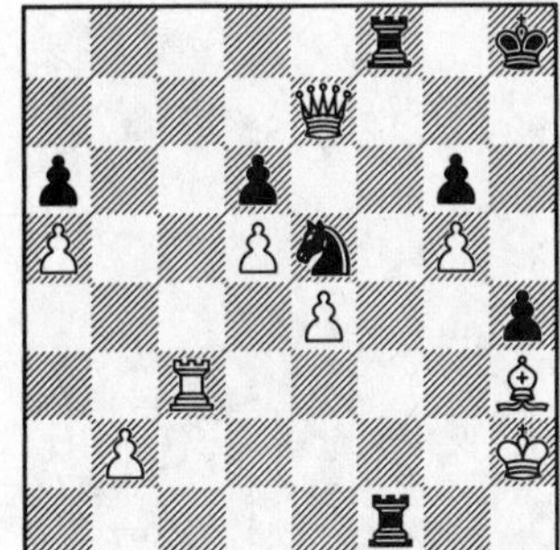

4296.

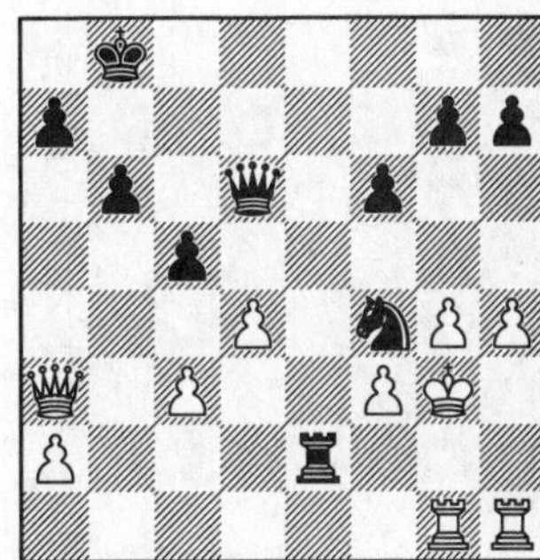

4297.

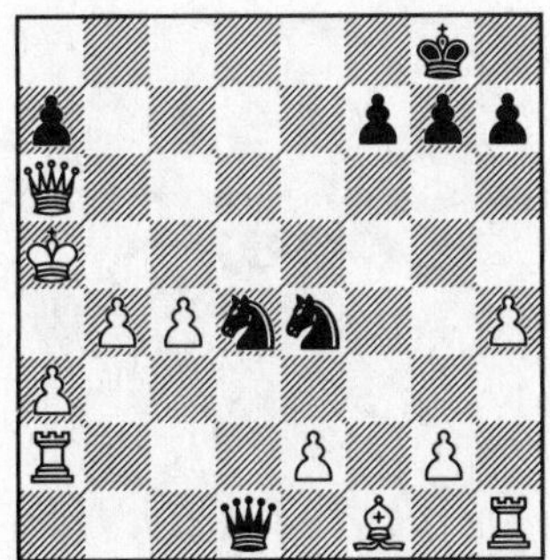

4298.

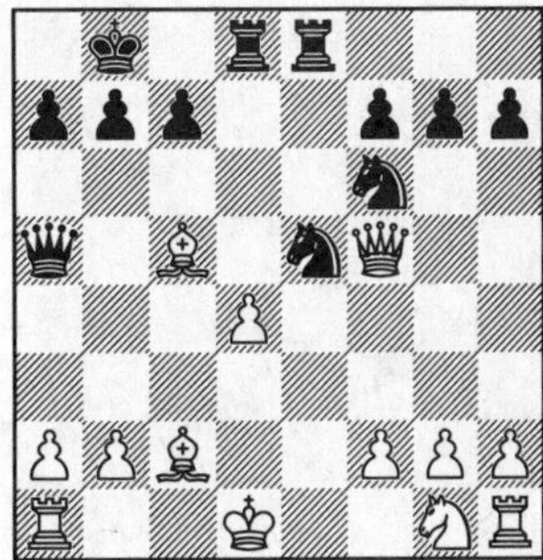

4299.

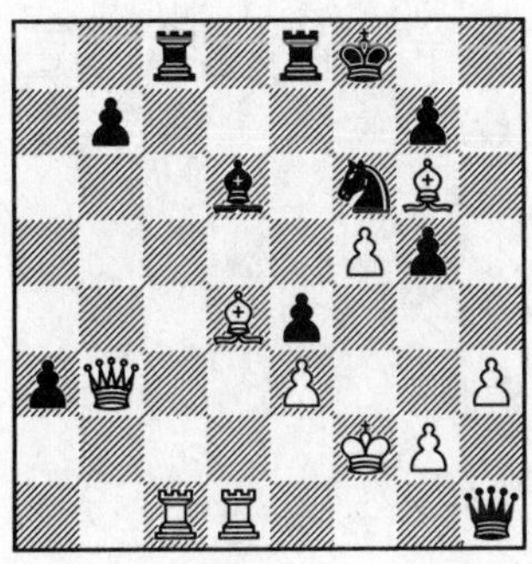

4300.

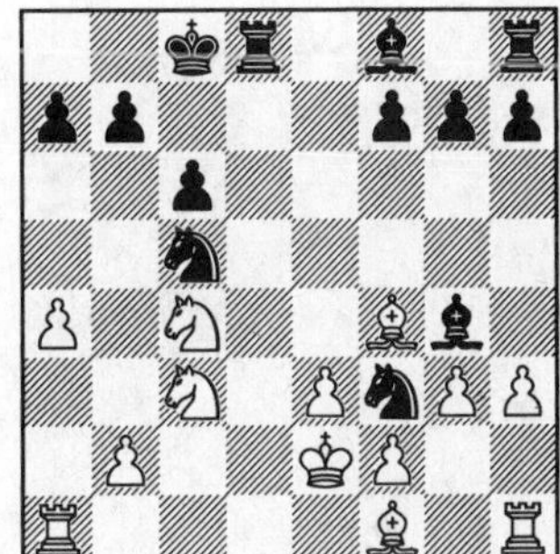

4301.

4302.

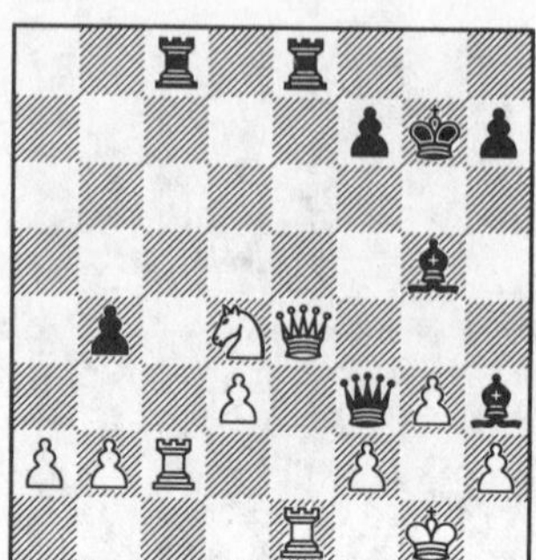

4303.

4304.

4305.

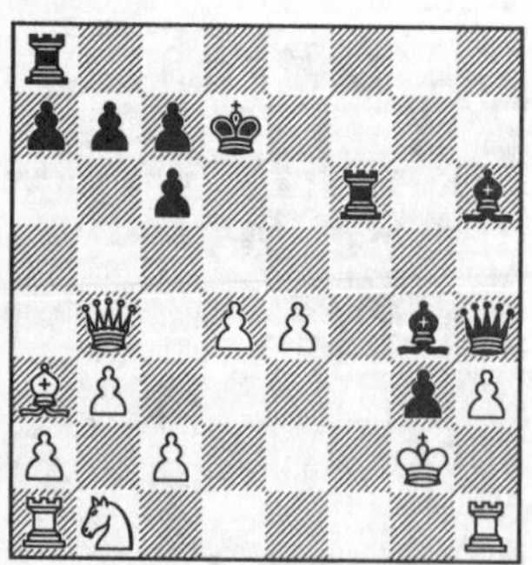

4306.

4307.

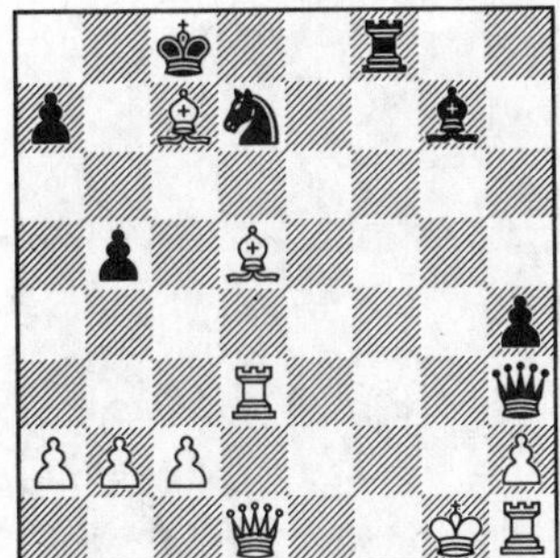

4308.

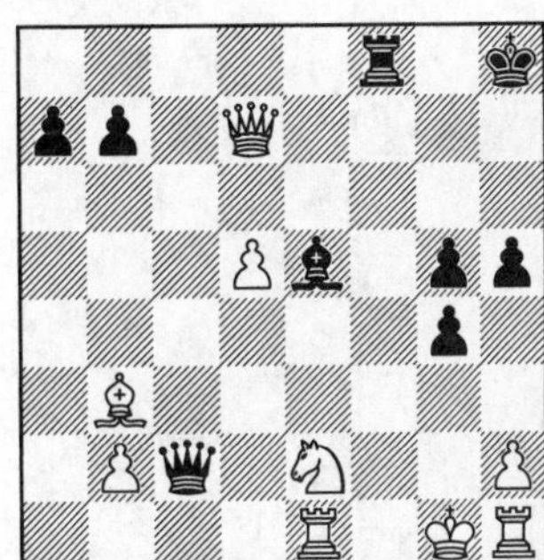

4309.

4310.

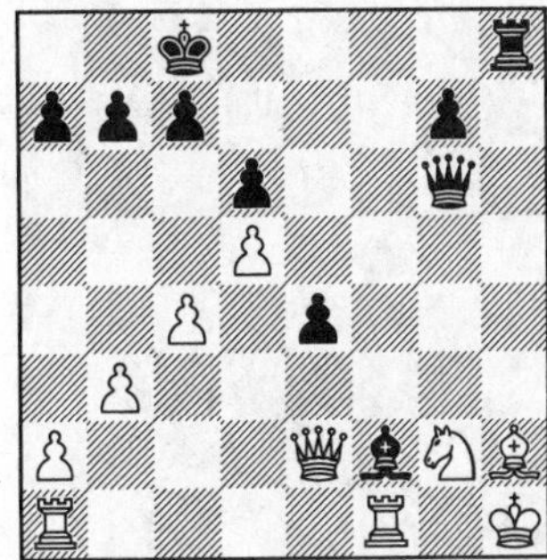

4311.

4312.

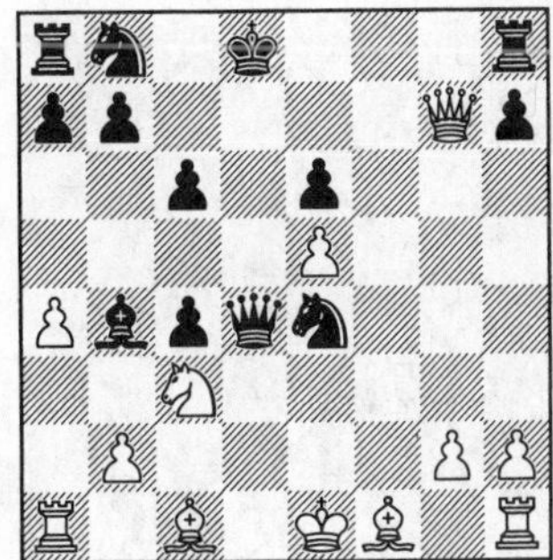

4313.

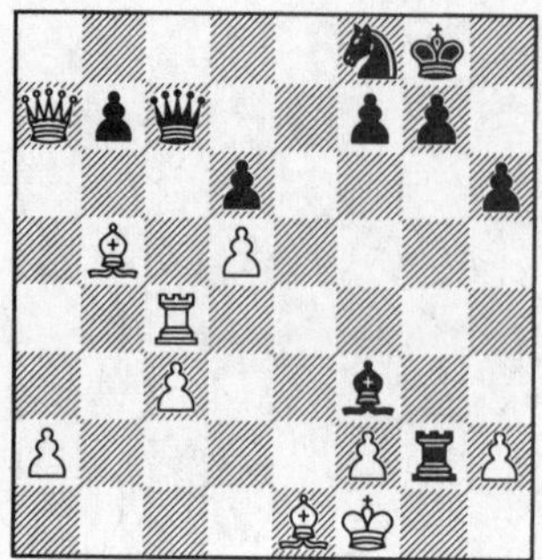

4314.

4315.

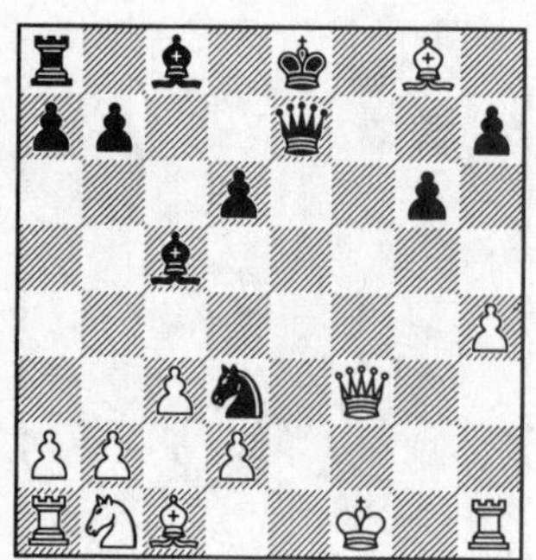

4316.

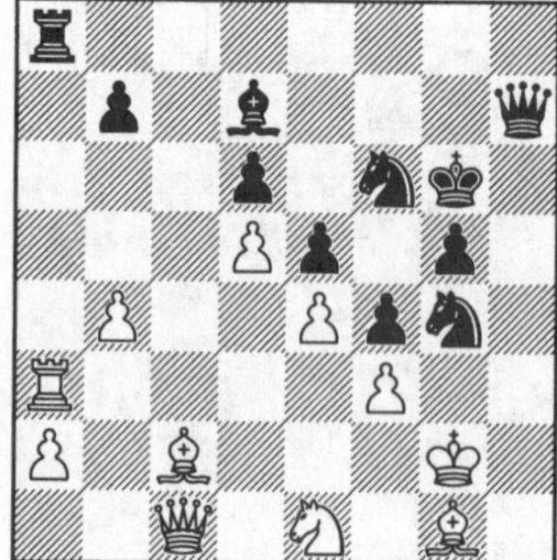

4317.

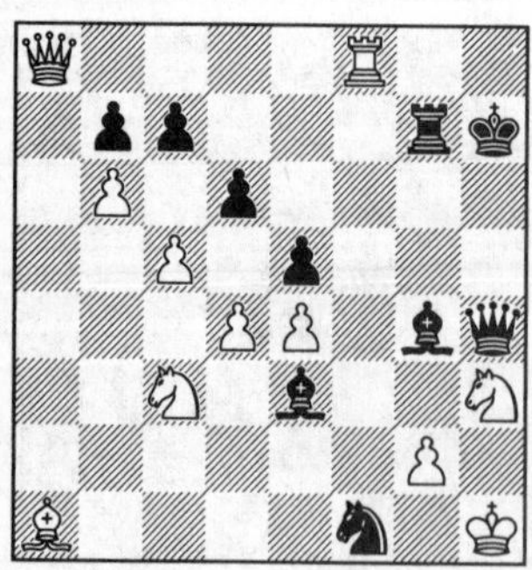

4318.

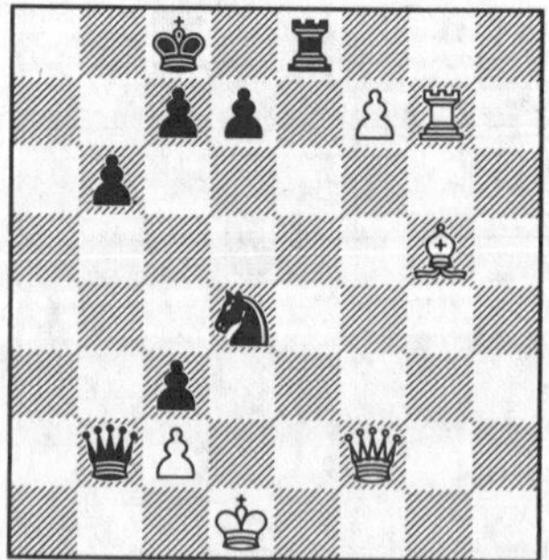

4319.

4320.

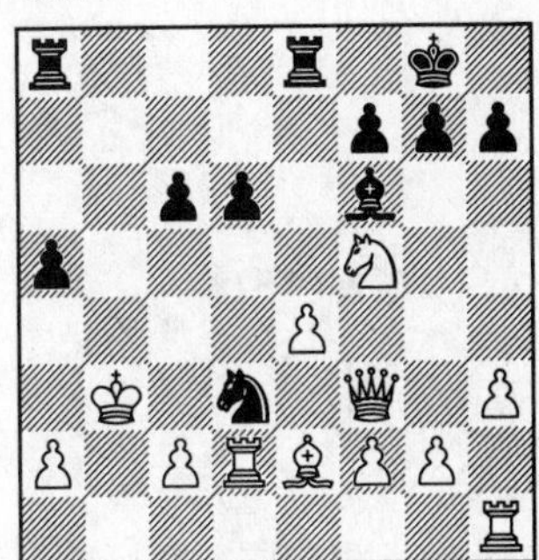

4321.

4322.

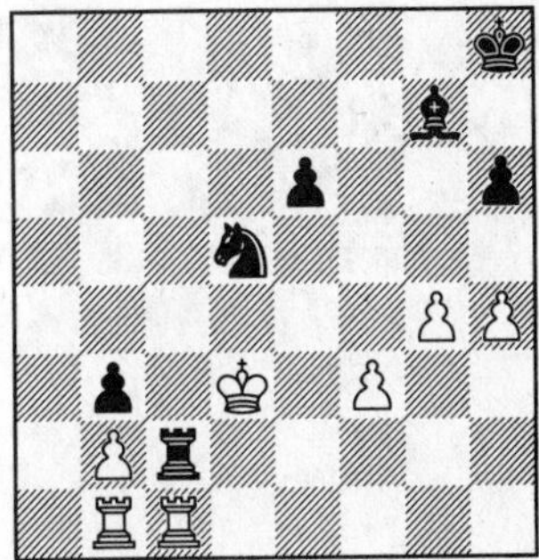

4323.

4324.

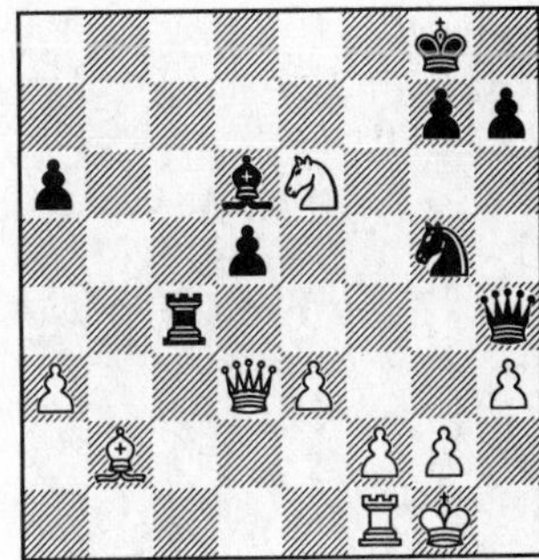

4325.

4326.

4327.

4328.

4329.

4330.

4331.

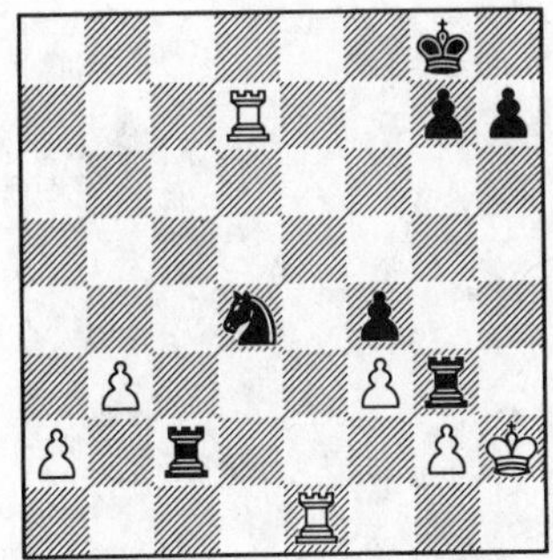

4332.

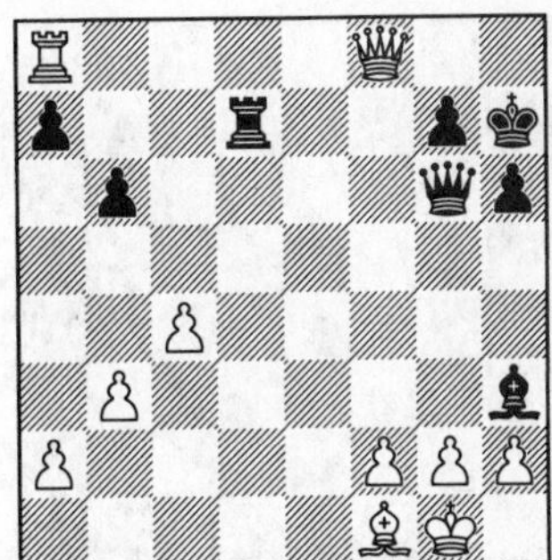

4333.

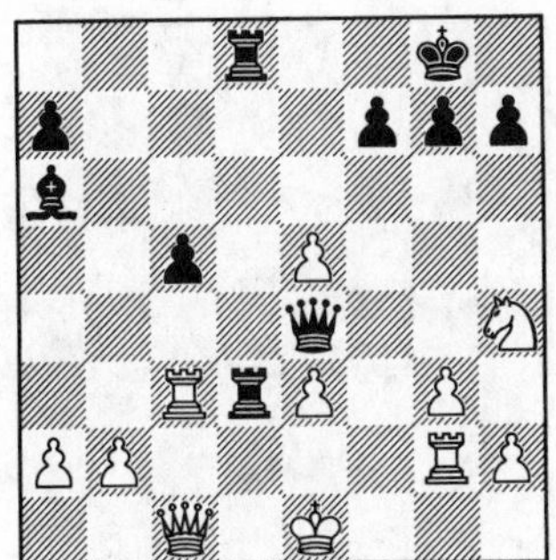

4334.

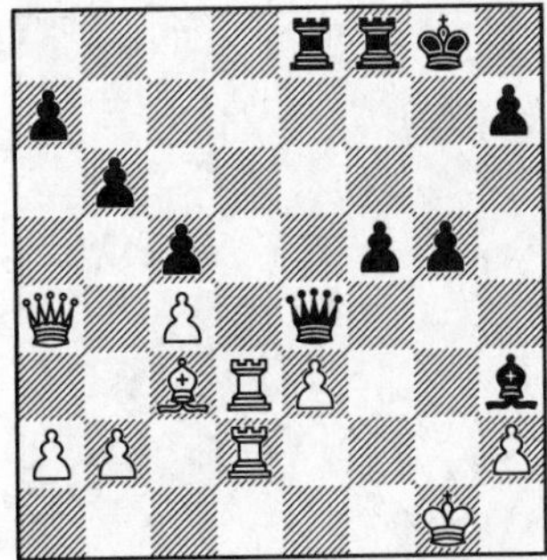

4335.

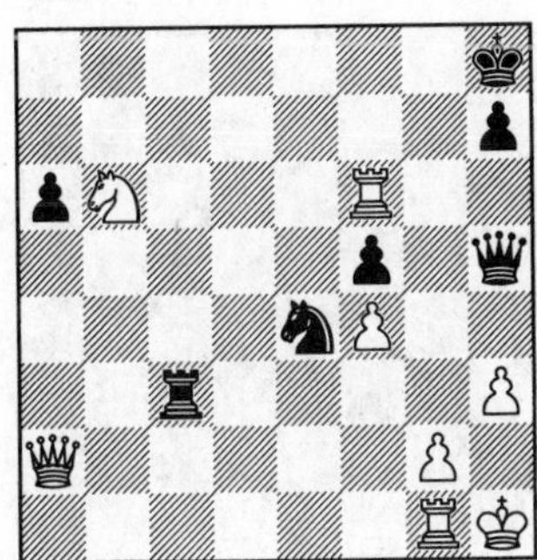

4336.

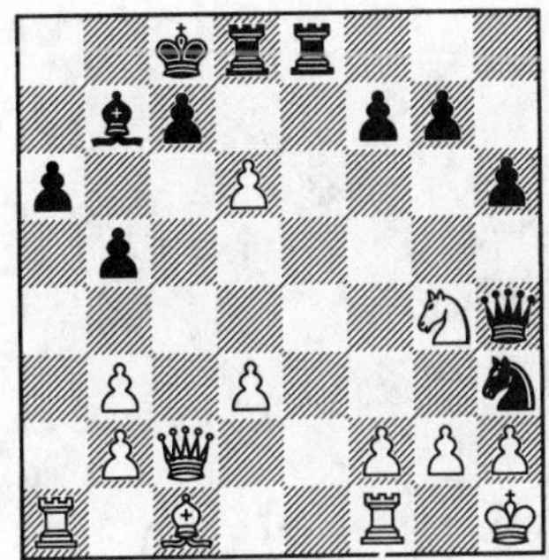

4337.

4338.

4339.

4340.

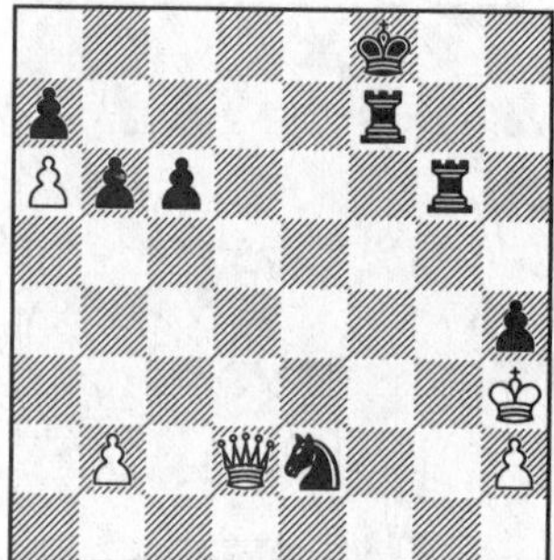

4341.

4342.

4343.

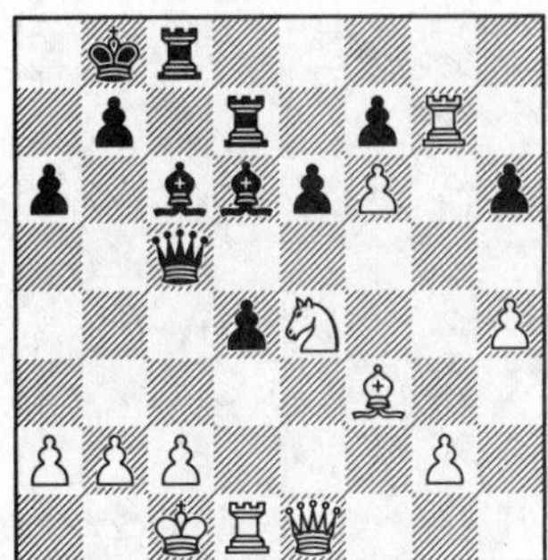

4344.

4345.

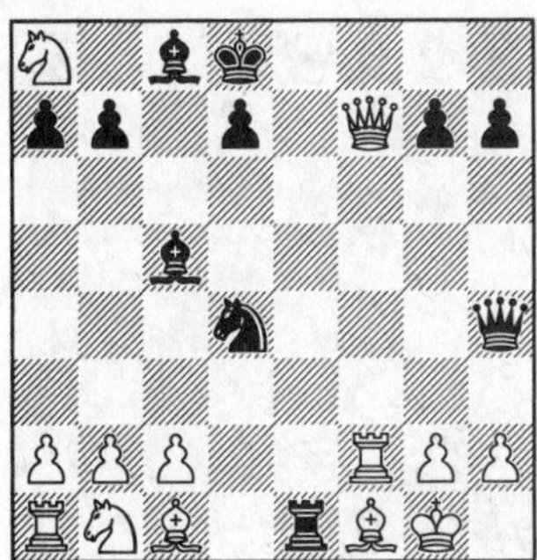

4346.

4347.

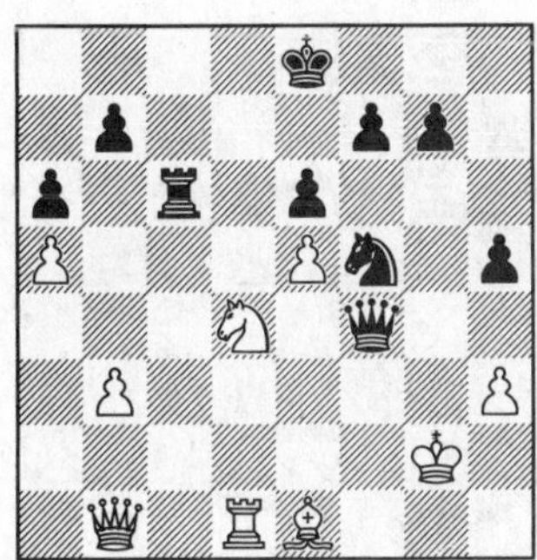

4348.

4349.

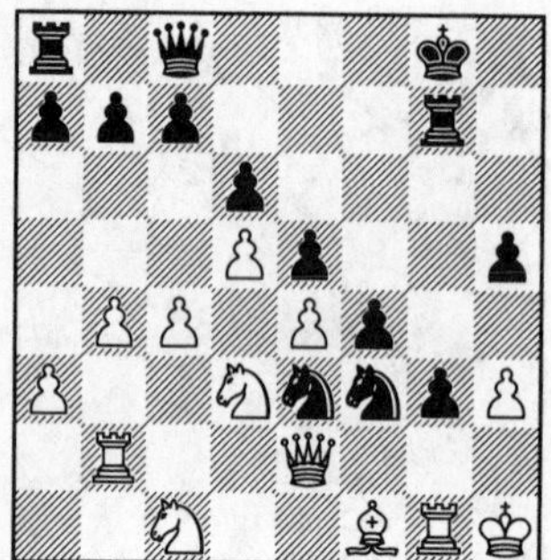

4350.

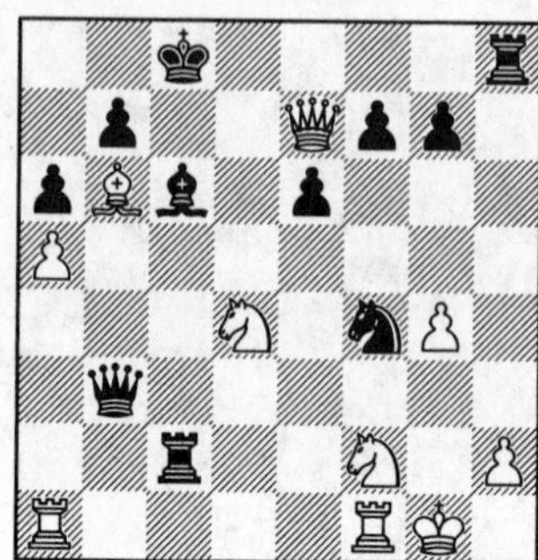

4351.

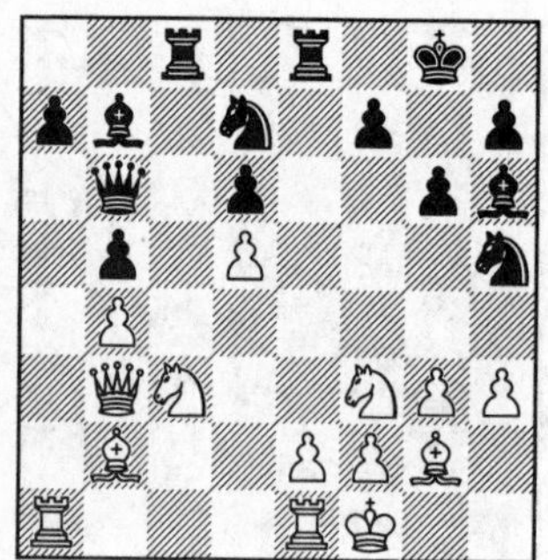

4352.

4353.

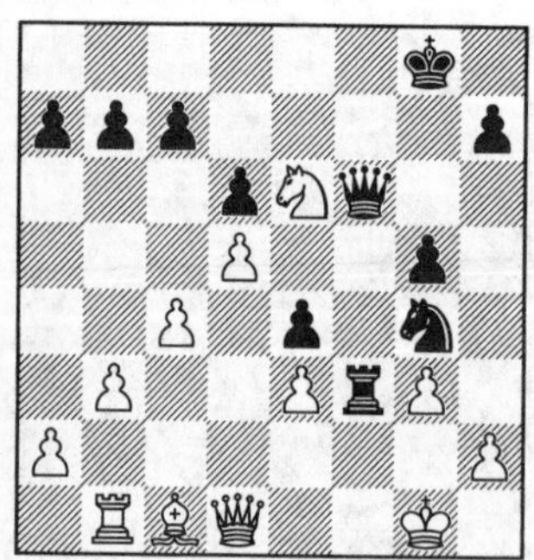

4354.

4355.

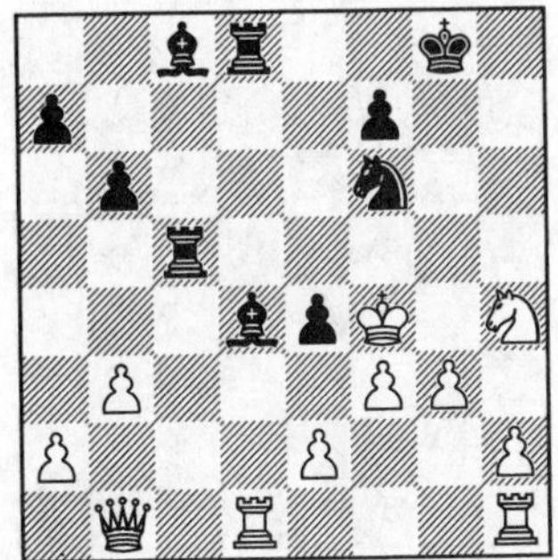

4356.

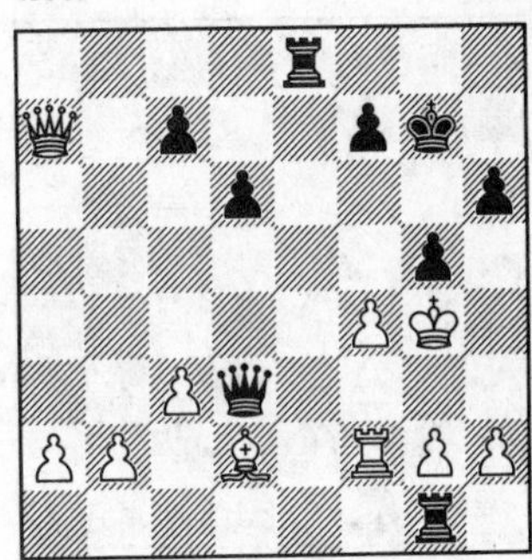

4357.

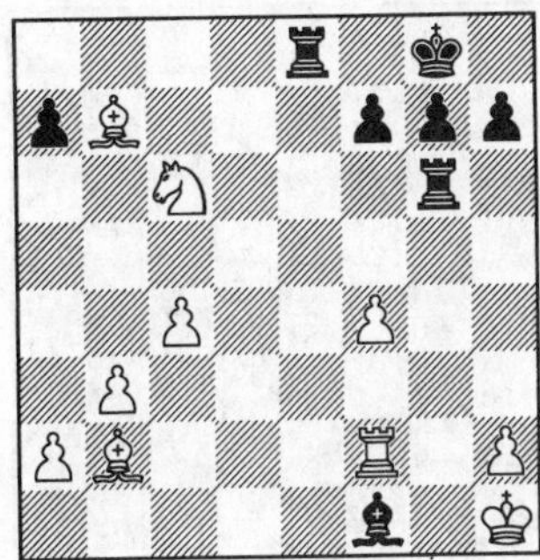

4358.

4359.

4360.

4361.

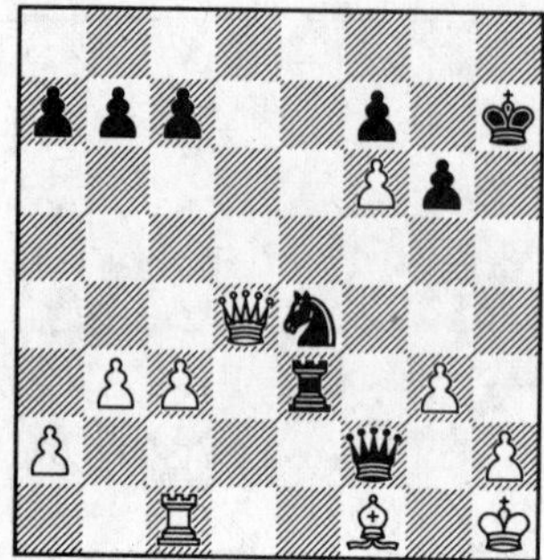

4362.

4363.

4364.

4365.

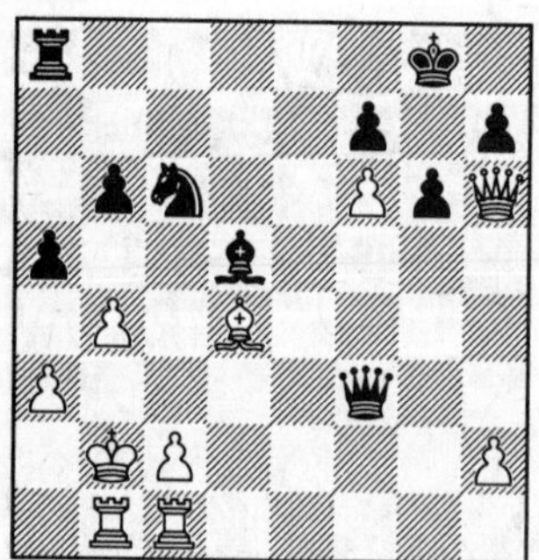

4366.

4367.

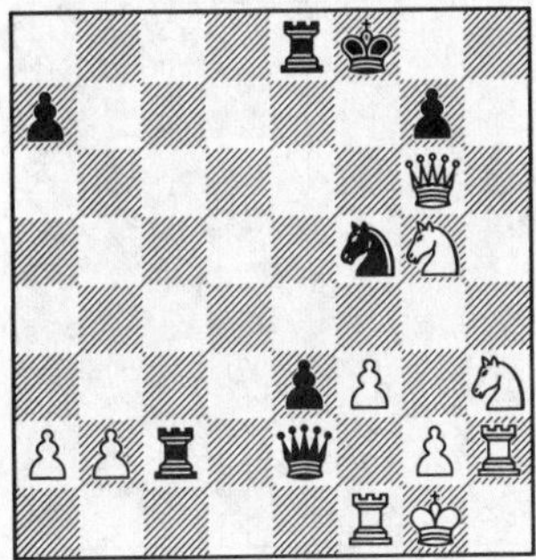

4368.

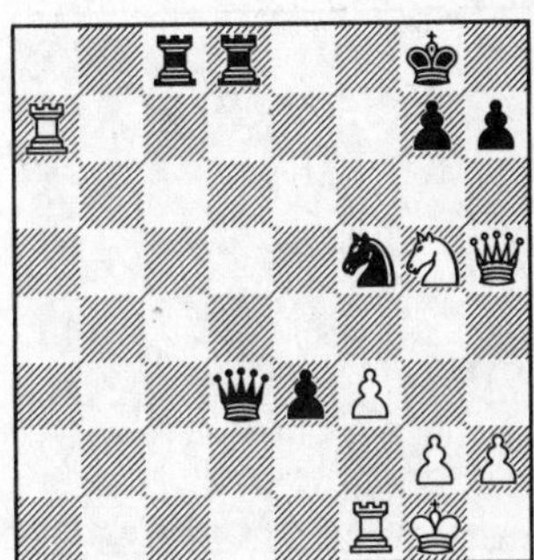

4369.

4370.

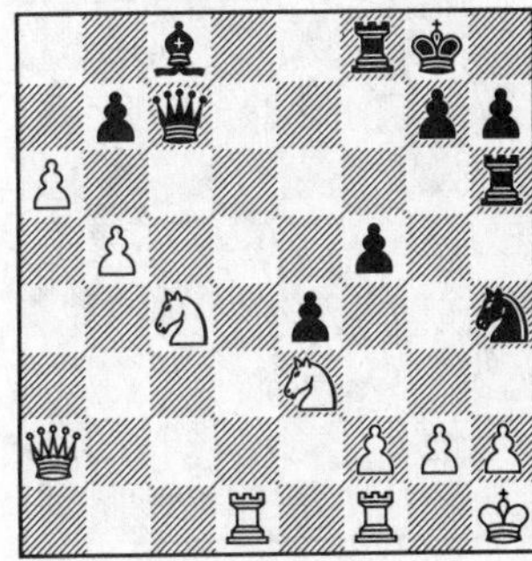

4371.

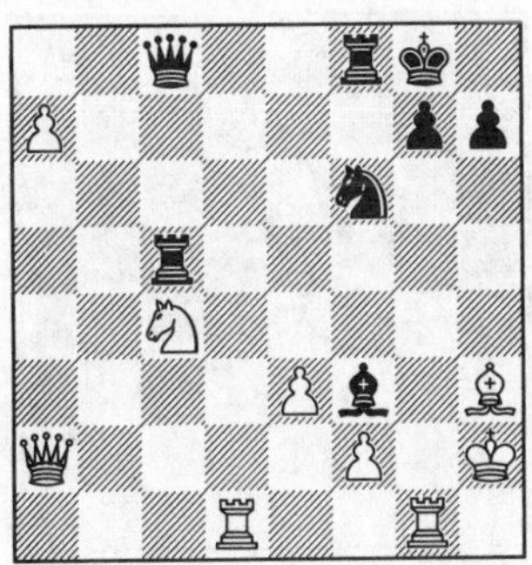

4372.

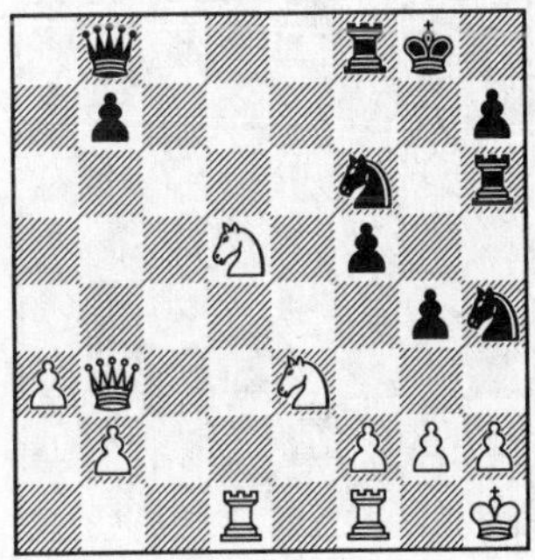

4373.

4374.

4375.

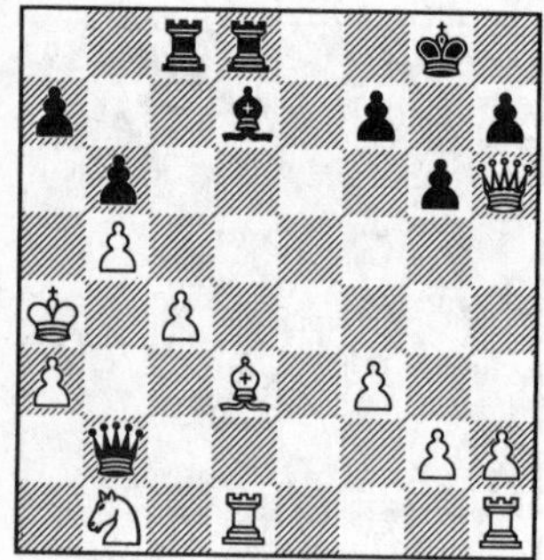

4376.

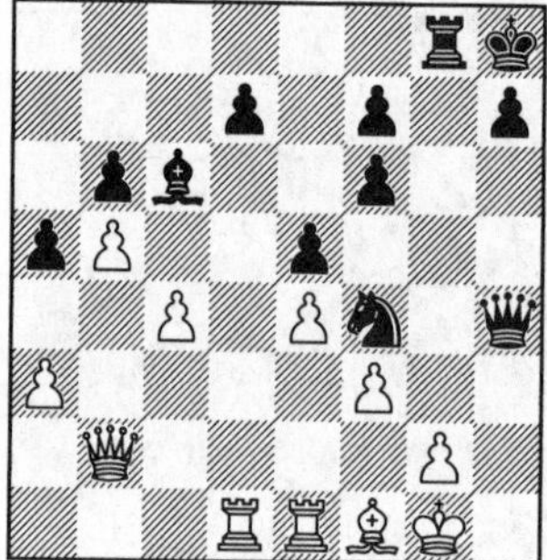

4377.

4378.

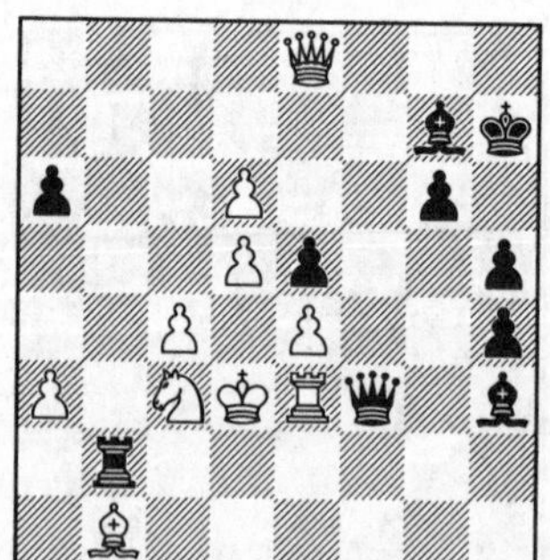

4379.

4380.

4381.

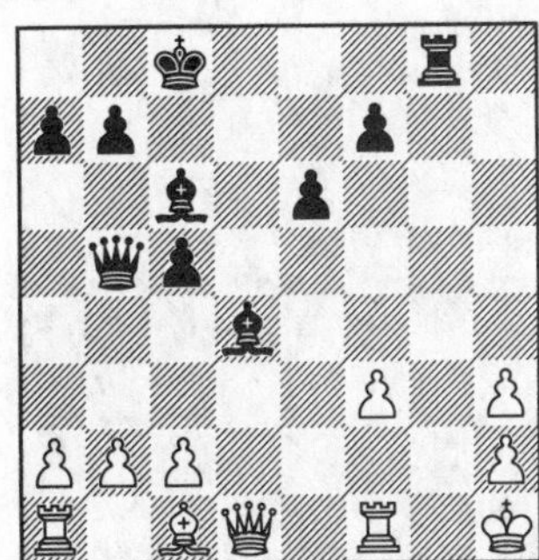

4382.

4383.

4384.

4385.

4386.

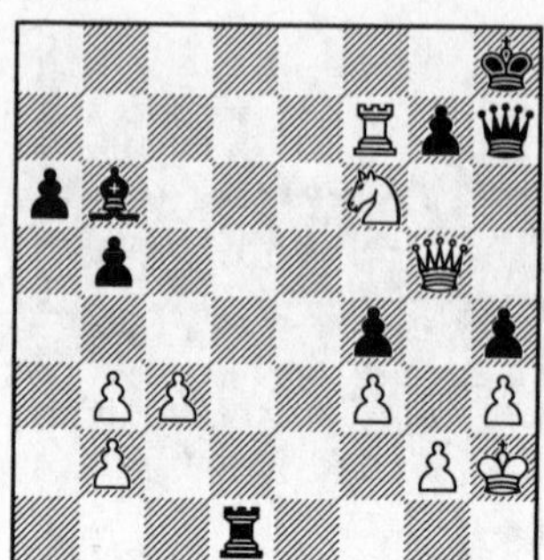

4387.

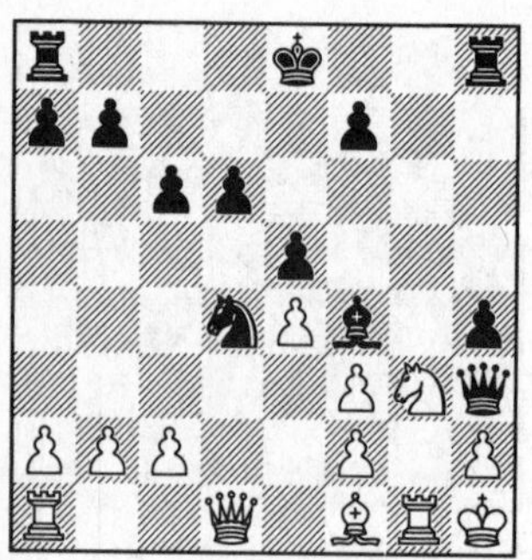

4388.

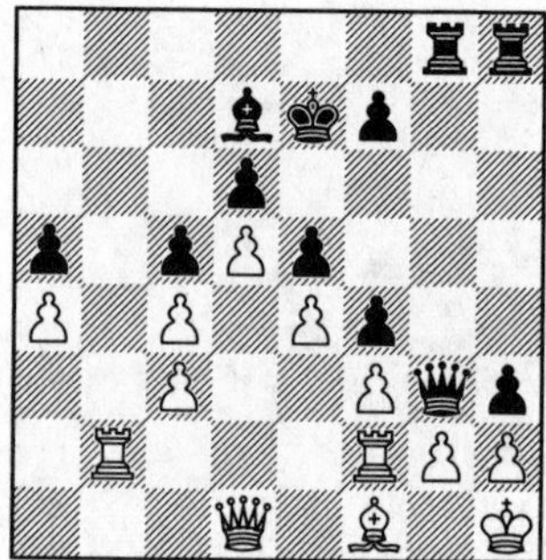

4389.

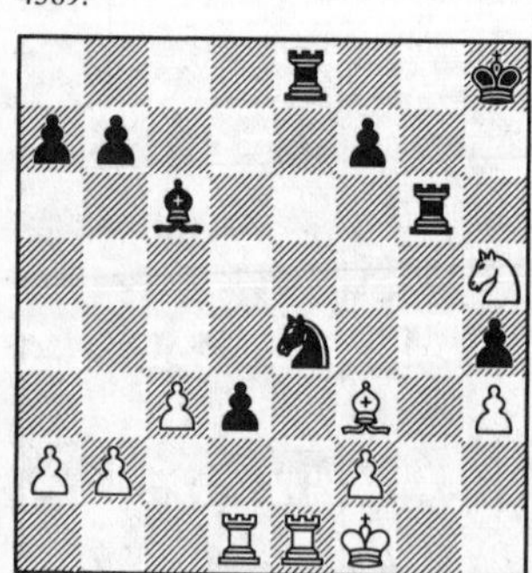

4390.

4391.

4392.

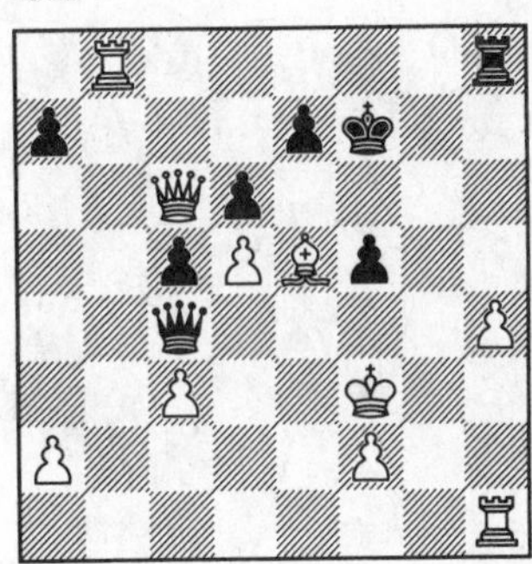

4393.

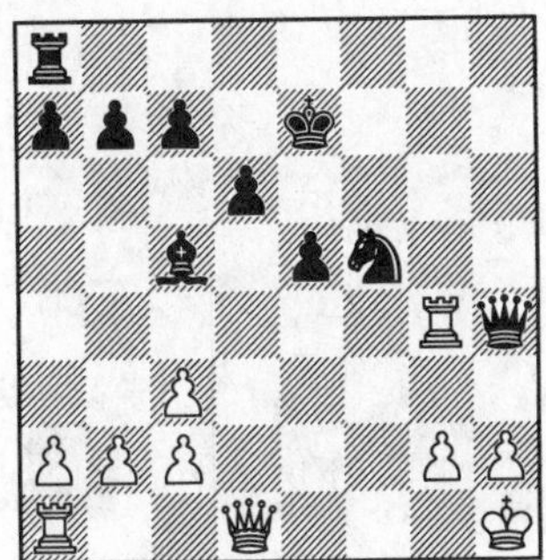

4394.

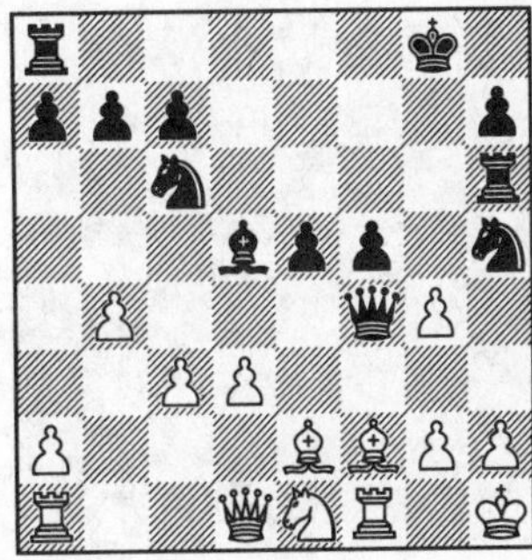

4395.

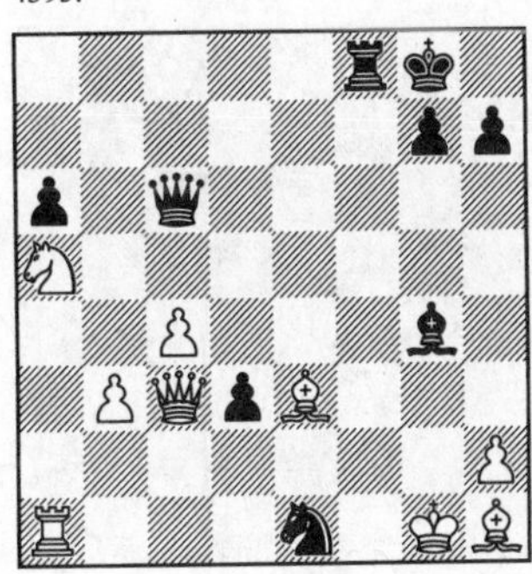

4396.

4397.

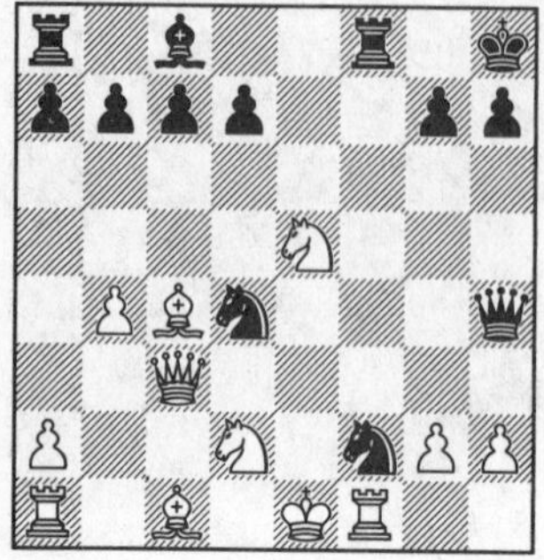

4398.

4399.

4400.

4401.

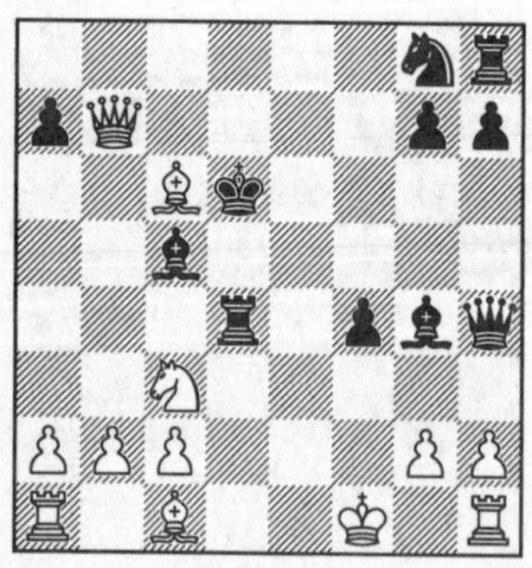

4402.

4403.

4404.

4405.

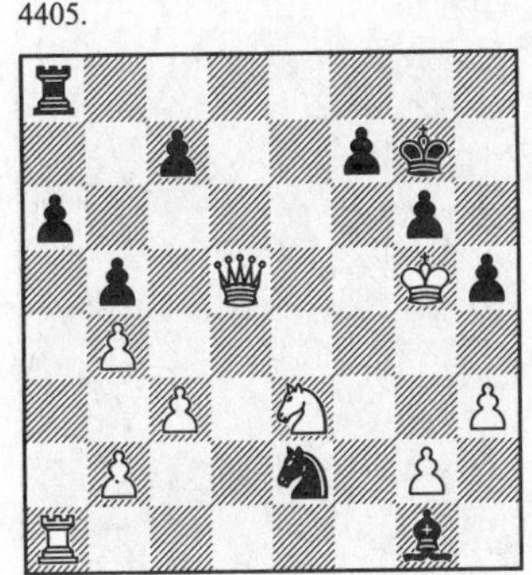

4406.

4407.

4408.

4409.

4410.

4411.

4412.

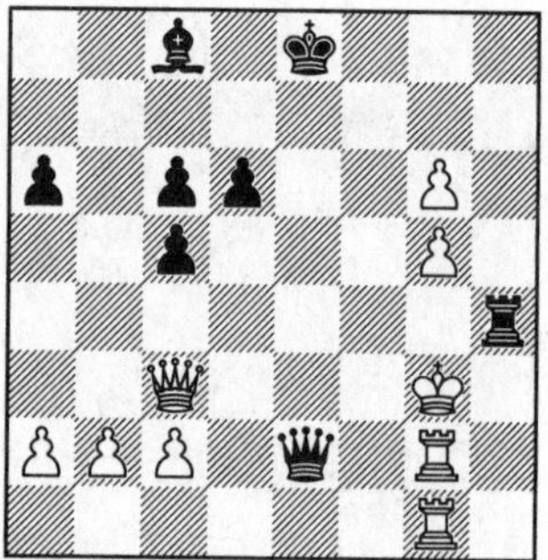

4413.

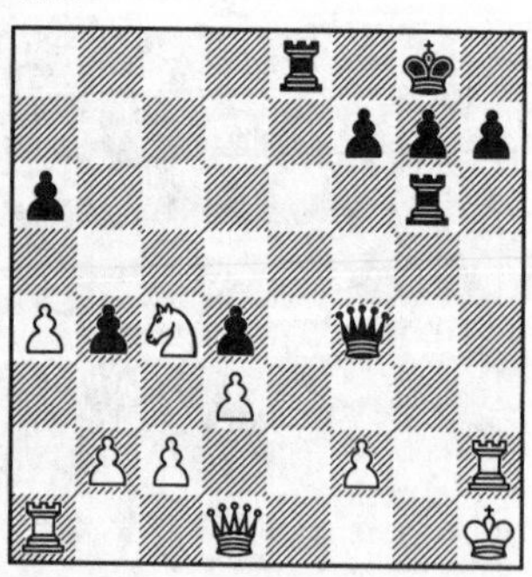

4414.

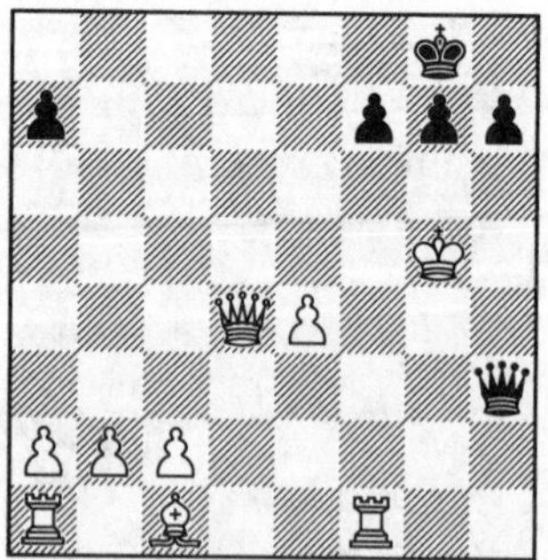

4415.

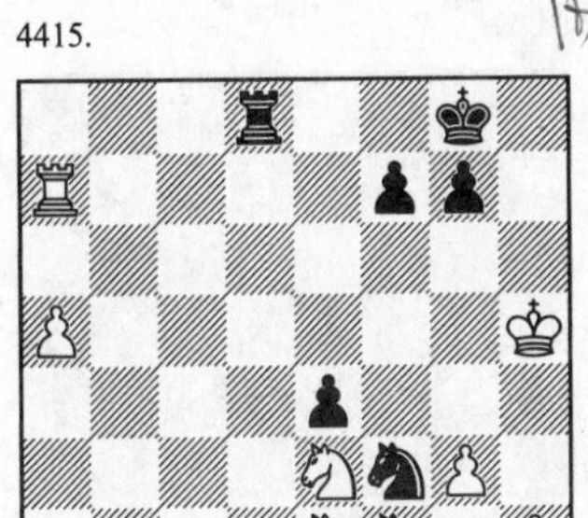

4416.

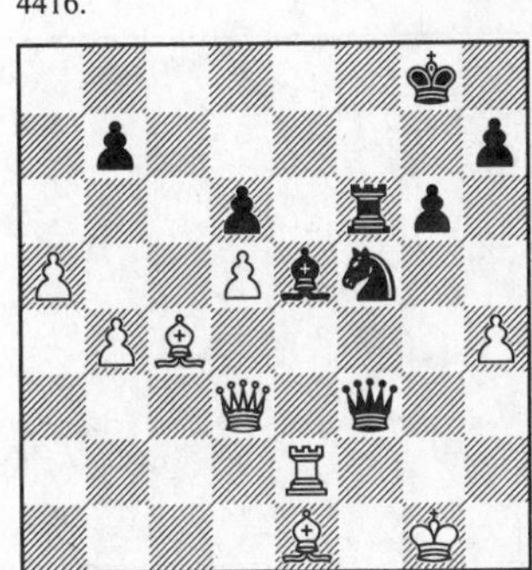

4417.

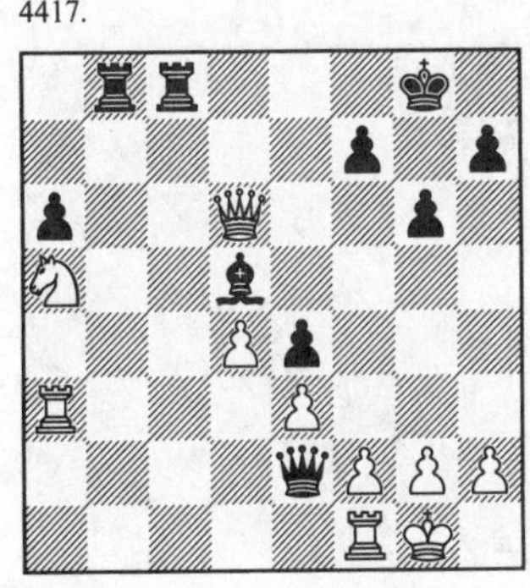

4418.

4419.

4420.

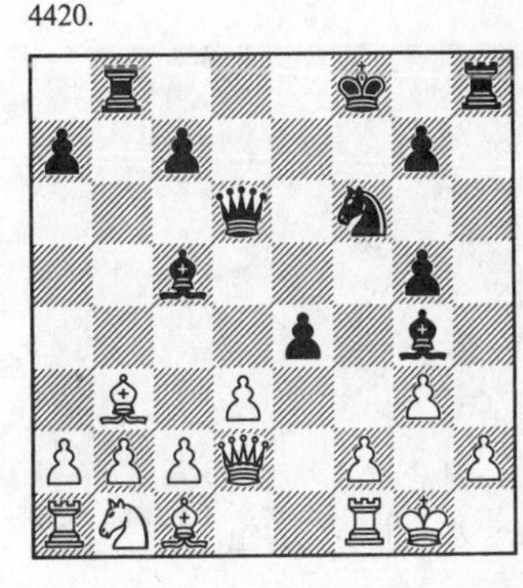

4421.

4422.

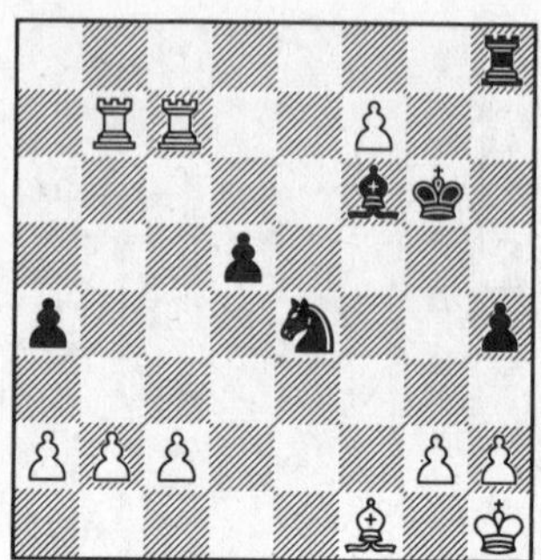

4423.

4424.

4425.

4426.

4427.

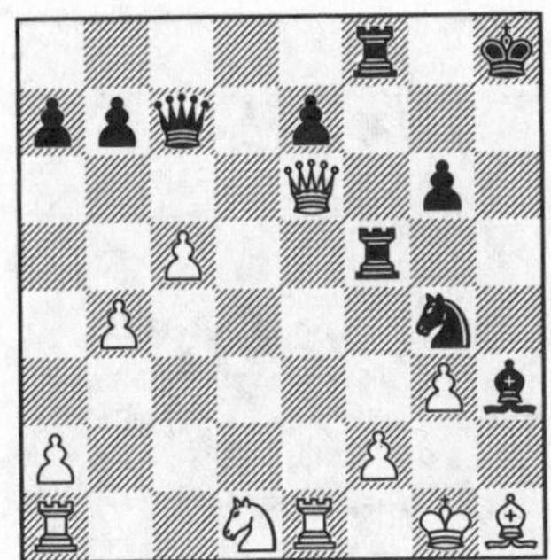

4428.

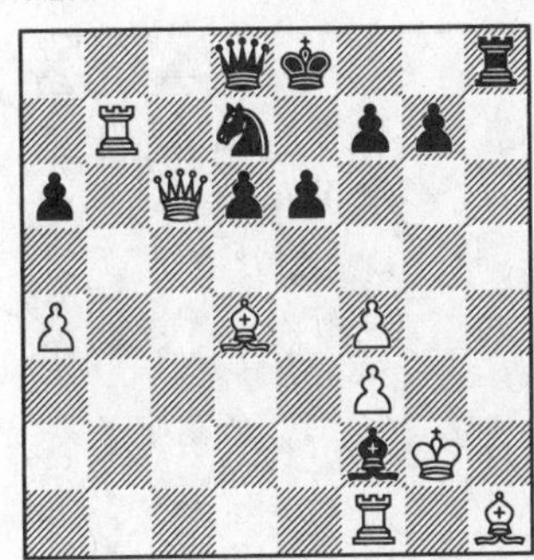

4429.

4430.

4431.

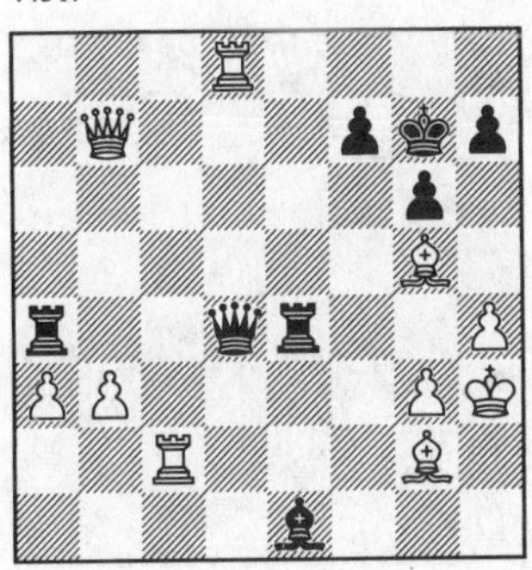

4432.

4433.

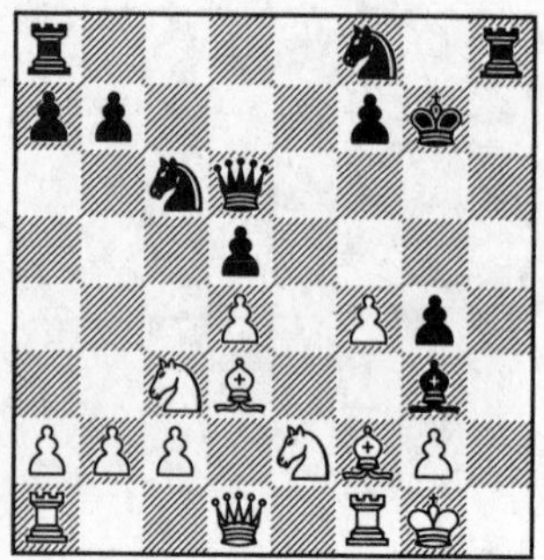

4434.

4435.

4436.

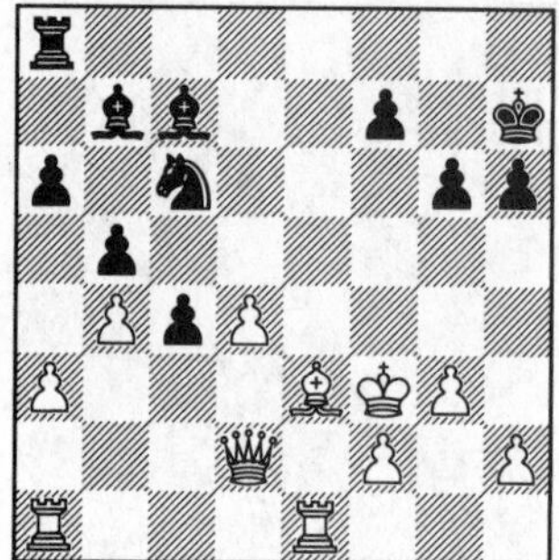

4437.

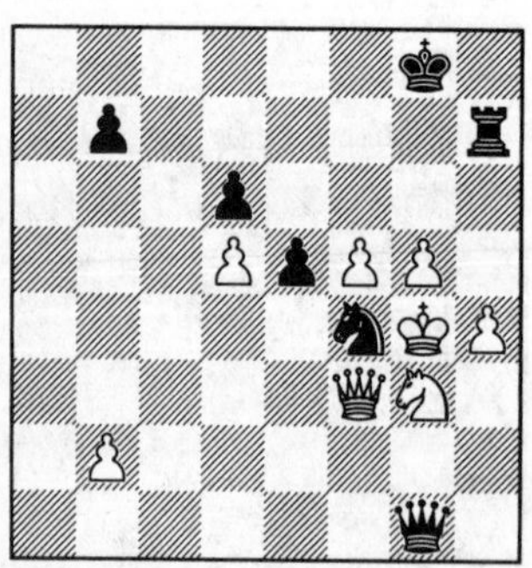

4438.

4439.

4440.

4441.

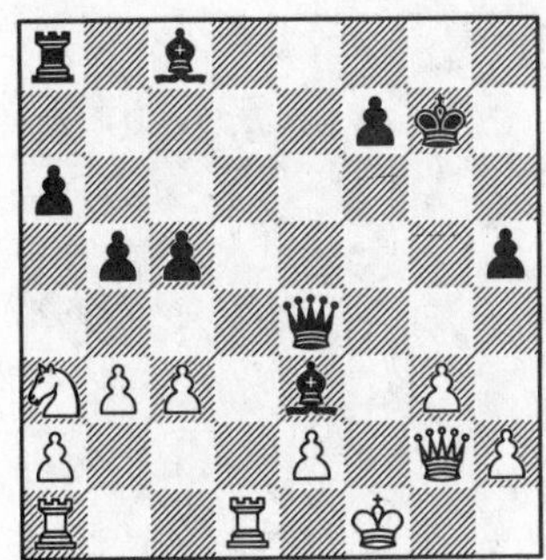

4442.

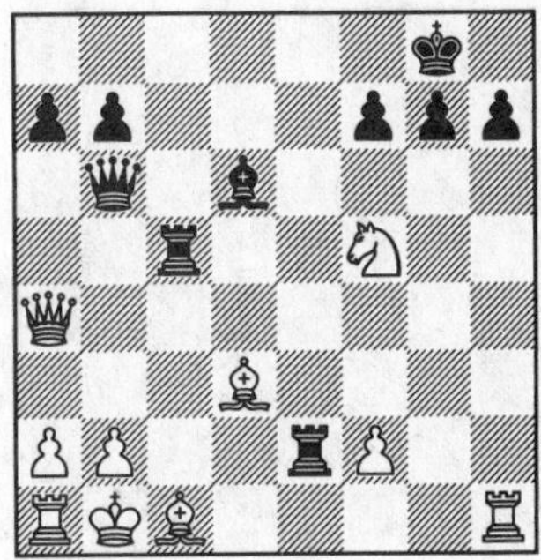

4443.

4444.

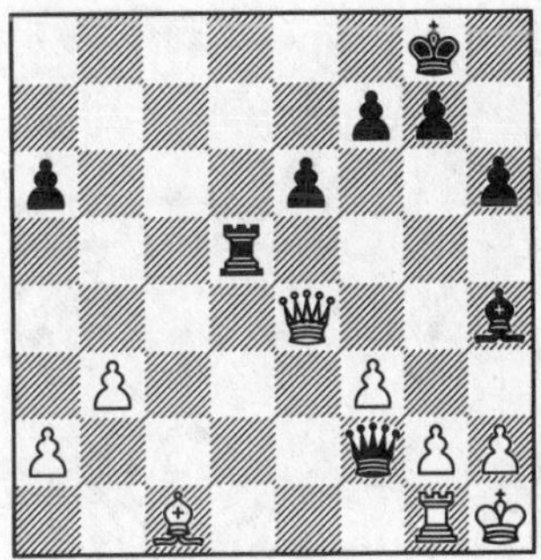

4445.

4446.

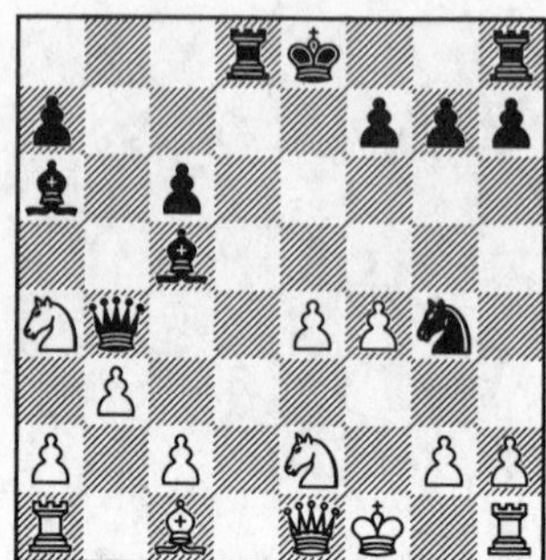

4447.

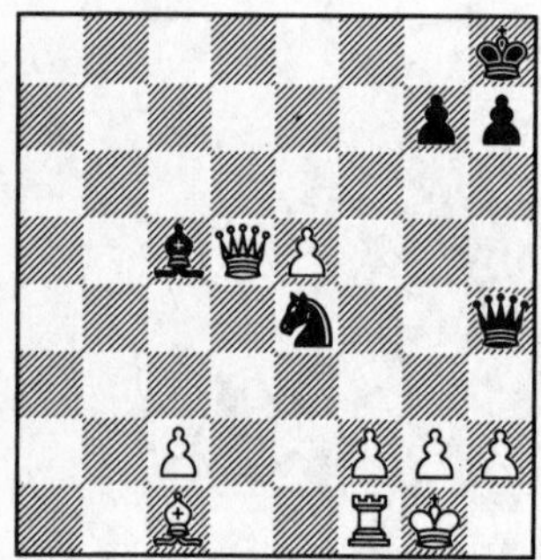

4448.

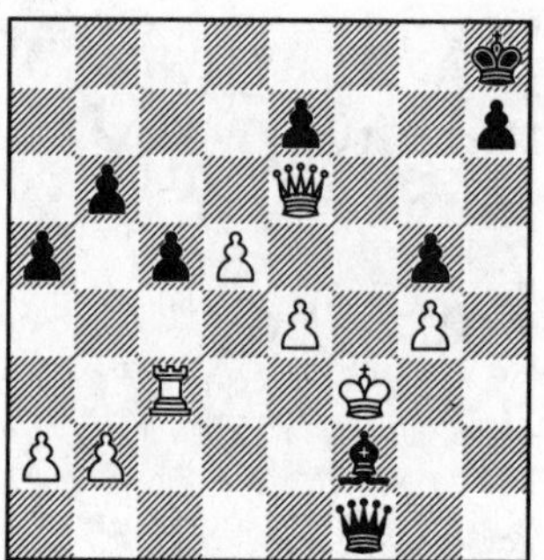

4449.

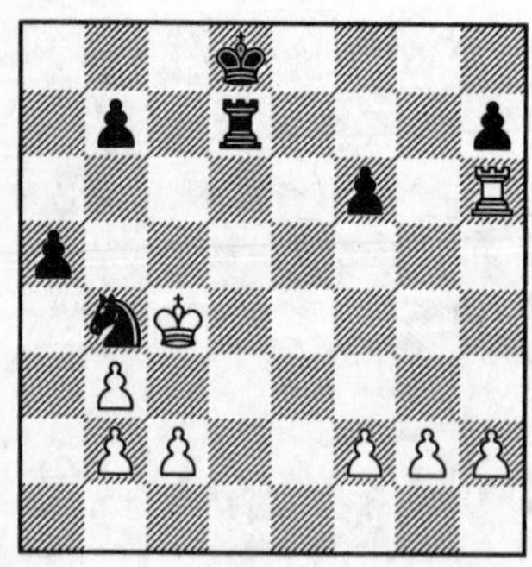

4450.

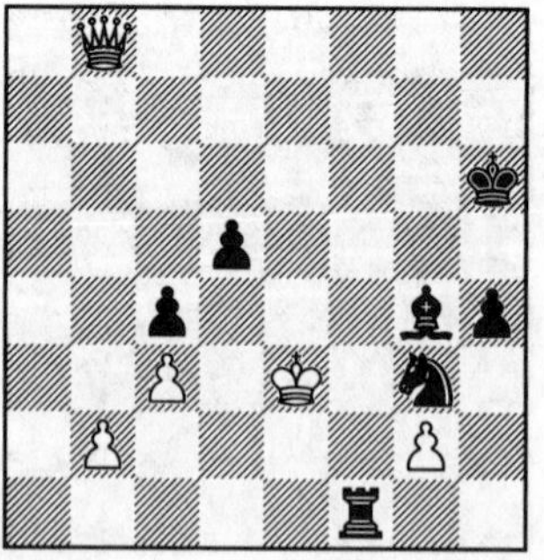

4451.

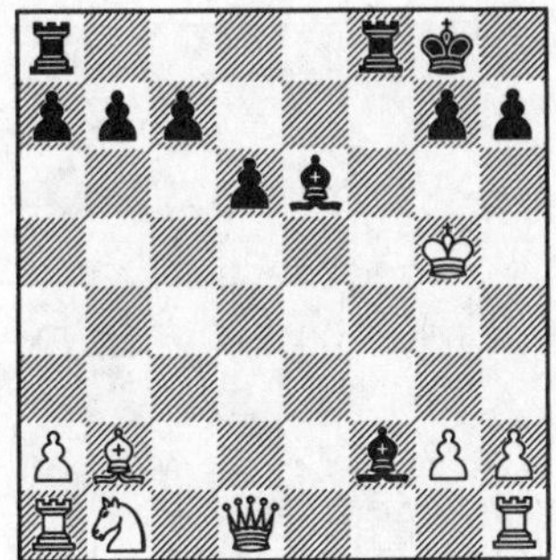

4452.

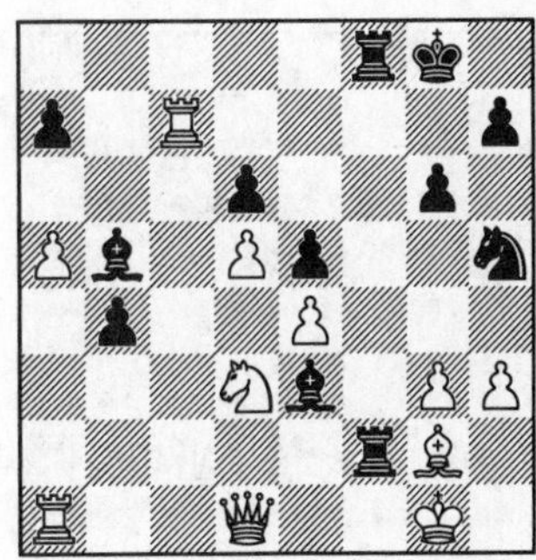

4453.

4454.

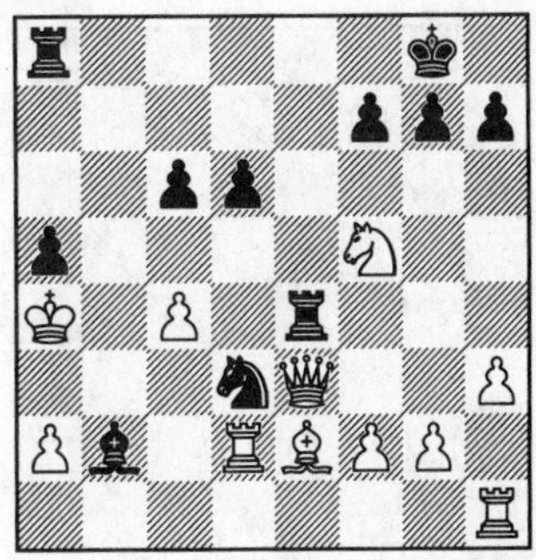

4455.

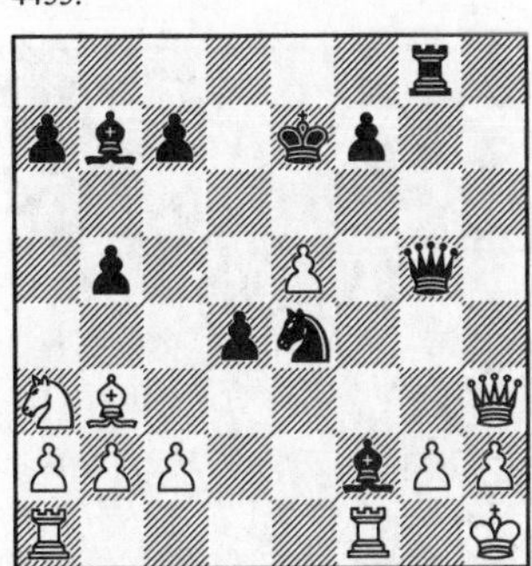

4456.

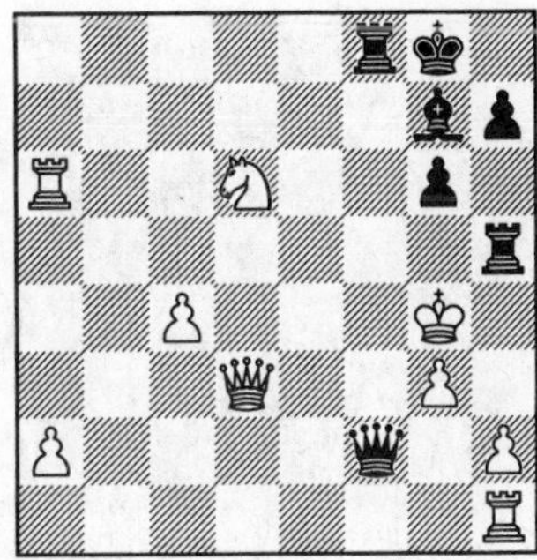

4457.

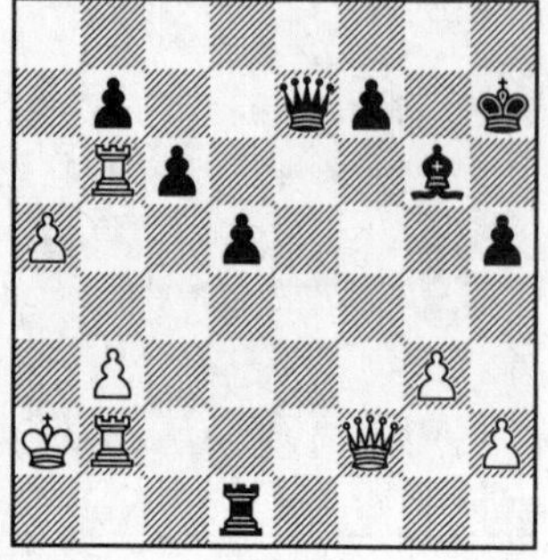

4458.

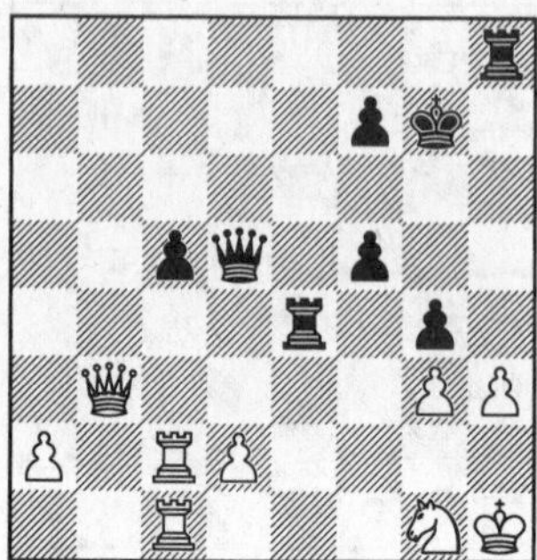

4459.

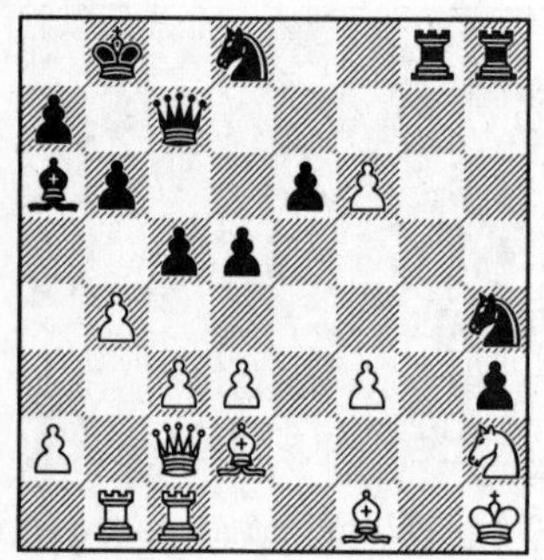

4460.

4461.

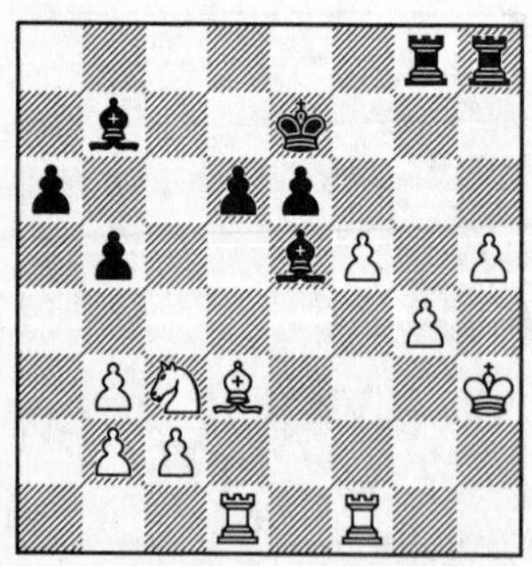

4462.

600 miniature games

(4463–5062)

4463. A 00

Sokolsky – Strugatsch

Belorus 1958

1. b4 e5 2. ♗b2 f6 3. e4 ♗b4 4. ♗c4 ♘c6 5. f4 ef4 6. ♘h3 ♘ge7 7. ♘f4 ♘a5

8. ♗f6! ♖f8[1] 9. ♘h5 ♘c4 10. ♘g7 ♔f7 11. 0-0 ♔g8 12. ♕h5 ♖f6 13. ♖f6 ♘g6[2] 14. ♖g6! hg6 15. ♕g6 ♔h8[3] 16. ♘e8! ♕e7 17. ♘f6[4] +–

[1] 8. – gf6 9. ♕h5 +–; 8. – ♘c4 9. ♗g7 ♖g8 10. ♕h5 ♘g6 11. ♘g6 hg6 12. ♕g6 ♔e7 13. ♗f6 ♔d6 14. ♕f5 +–; 11. – ♖g7 12. ♘e5 +–

[2] 13. – ♔g7 14. ♖f7 ♔g8 15. ♕h7#

[3] 15. – ♘e5 16. ♕g3 ♕f6 17. ♘h5 ♕g6 18. ♕e5 +–

[4] 17. – ♕g7 18. ♕h5 +–

4464. A 09

Soultanbeieff – Courtens

Visé 1931

1. ♘f3 d5 2. c4 e6 3. g3 b6 4. ♗g2 ♗b7 5. b3 dc4 6. bc4 c5 7. 0-0 ♘f6 8. ♗b2 ♘bd7 9. d3 ♕c7 10. ♘c3 a6 11. a4 ♖b8 12. e4 ♗d6 13. ♕e2 0-0 14. ♖ae1 ♘e5 15. ♘d2 ♖fd8 16. f4 ♘d3 17. ♕d3 ♗f4 18. ♘d5 ♗d2 19. ♕d2 ed5

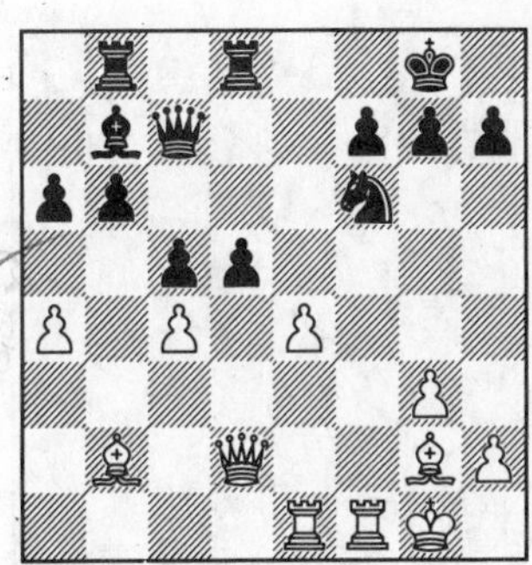

20. ♖f6! gf6 21. ♗f6 +–

4465. A 45

Panke – Kranenberg

Schmollenhagen 1941

1. d4 ♘f6 2. ♘c3 e6 3. e4 b6 4. f4 ♗b7 5. ♗d3 d6 6. ♘f3 ♗e7 7. e5 de5 8. fe5 ♘fd7 9. 0-0 c5 10. ♘e4 cd4 11. ♘d6 ♗d6 12. ed6 e5 13. ♘d4 ed4 14. ♕e2 ♔f8 15. ♖f7 ♔f7 16. ♗c4 ♔f8 17. ♗g5 ♘f6 18. ♖f1 ♘d7 19. ♕e6 ♕e8

20. ♖f6! ♘f6 21. ♕f6! gf6 22. ♗h6#

4466. A 46

Frink – Lecount

Brooklyn 1923

1. d4 ♘f6 2. ♘c3 d6 3. ♘f3 ♗f5 4. ♘h4 ♗g6 5. ♘g6 hg6 6. e4 ♘bd7 7. ♗c4 e5 8. 0-0 c6 9. a4 ed4 10. ♕d4 ♘g4 11. h3 ♘de5 12. hg4

12. – ♘f3!¹ 13. gf3 ♕h4 –+

¹ 12. – ♕h4? 13. f3!

4467. A 48
Karsens – Ullrich
Swinemünde 1932

1. d4 ♘f6 2. ♘f3 g6 3. e3 ♗g7 4. ♗d3 d6 5. 0-0 0-0 6. h3 ♘bd7 7. e4 c6 8. b3 ♕c7 9. ♗a3 ♖e8 10. ♘bd2 c5 11. ♗b2 a6 12. ♖c1 b5 13. ♖e1 ♖b8 14. c4 b4 15. d5 ♘e4 16. ♗g7 ♘d2 17. ♗h6 ♘f3 18. ♕f3 ♘e5 19. ♕f4 ♘d3

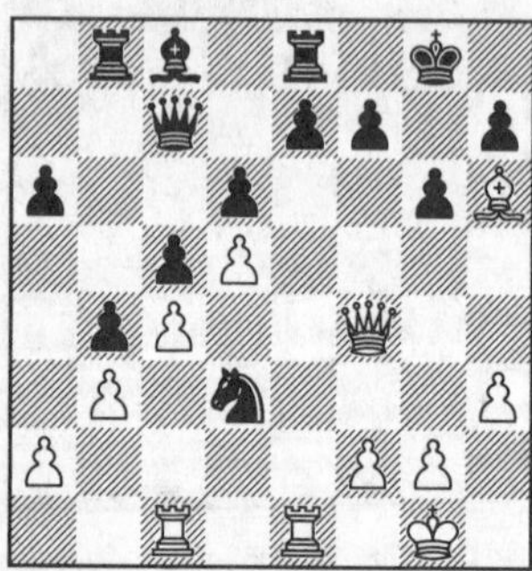

20. ♕f6! +–

4468. B 01
Labone – N. N.
Birmingham 1901

1. e4 d5 2. ed5 ♕d5 3. d4 e5 4. ♘f3 ed4 5. ♘d4 c5 6. ♕e2 ♗e7 7. ♘b5 ♘a6 8. ♘1c3 ♕e6 9. ♗e3 ♘f6 10. ♕f3 0-0 11. ♗d3 c4 12. ♘d4 ♕b6 13. ♘f5 ♕b2 14. ♘e7 ♔h8 15. ♗d4 cd3

16. ♕f6! ♖g8 17. ♕f7 ♗e6 18. ♕e6 ♕a1 19. ♔d2 ♕h1 20. ♘g6! +–

4469. B 05
Angel – Buschke
New York 1940

1. e4 ♘f6 2. e5 ♘d5 3. ♘f3 d6 4. d4 ♗g4 5. ♗e2 c5 6. 0-0 ♕b6 7. ed6 ed6 8. ♖e1 ♗e7 9. ♗c4 ♗f3 10. ♕f3 ♘f6 11. ♘c3 ♘c6 12. ♘d5-♕d8

13. ♕f6! gf6 14. ♘f6 ♔f8 15. ♗h6#

4470. B 06
Dorfman – Romanisin
Cienfuegos 1977

1. ♘f3 g6 2. e4 ♗g7 3. d4 d6 4. ♘c3 ♗g4 5. ♗e3 ♘c6 6. d5 ♘e5 7. ♗e2 ♘f3

8. gf3 ♗h5 9. ♗b5 ♔f8 10. 0-0 e6 11. de6 fe6 12. ♗e2 ♕h4 13. ♔h1

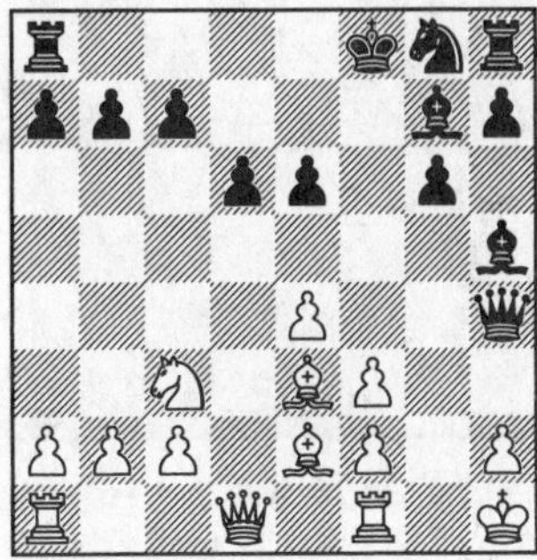

13. – ♗f3![1] –+

[1] 14. ♗f3 ♗e5 –+

4471. B 08
Zsuzsa Polgár – Endrődy
Budapest 1977

1. e4 d6 2. d4 ♘f6 3. ♘c3 g6 4. ♘f3 ♗g7 5. ♗e2 0-0 6. 0-0 ♗g4 7. ♗e3 ♘c6 8. ♕d2 e5 9. d5 ♘e7 10. h3 ♗f3 11. ♗f3 ♕d7 12. ♘e2 ♕b5 13. c4 ♕a6 14. ♘g3 b6 15. ♗h6 ♖fd8 16. ♗g7 ♔g7 17. ♗e2 c6 18. f4 cd5 19. cd5 ♕a4 20. fe5 de5

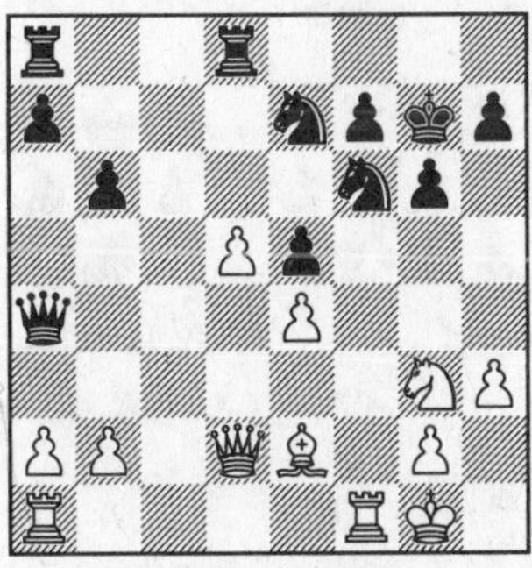

21. ♖f6! ♘d5[1] 22. ♖f7! ♔f7 23. ed5 ♕d4 24. ♕d4 ed4 25. ♖d1 +–

[1] 21. – ♔f6 22. ♕h6! ♘g8 23. ♖f1 ♔e7 24. ♕g7 ♖f8 25. ♕e5 ♔d8 26. ♕d6 ♔e8 27. ♘f5! gf5 28. ♖f5 ♘e7 29. ♖e5 ♕d7 30. ♗b5! +–; 24. – ♕e8 25. ♗b5! ♕b5 26. ♕f7 ♔d6 27. ♖c1 ♕d7 28. ♖c6! ♕c6 29. ♕e6 +–; 24. – ♕d4 25. ♔h2 ♖f8 26. ♖f5! +–

4472. B 17
Nimzovics – Nielson
Copenhagen 1930

1. e4 c6 2. d4 d5 3. ♘c3 de4 4. ♘e4 ♘d7 5. ♘f3 ♘gf6 6. ♘g3 e6 7. ♗d3 c5 8. 0-0 ♗e7 9. c3 0-0 10. ♖e1 b6 11. h3 ♗b7 12. ♗f4 ♗f3 13. ♕f3 cd4 14. cd4 ♘d5 15. ♗e4 ♘7f6 16. ♗e5 ♘e4 17. ♘e4 ♘f6 18. ♖ac1 ♘e4 19. ♖e4 ♕d5 20. ♖c7 ♗d6 21. ♖d7 ♖ad8 22. ♖d6 ♖d6

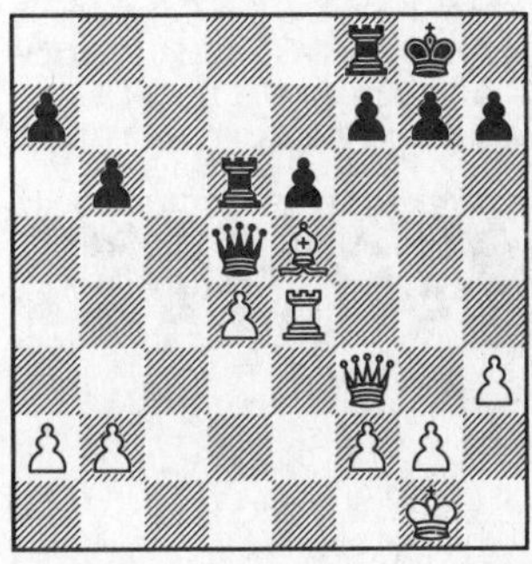

23. ♕f6![1] +–

[1] 23. – gf6 24. ♖g4 ♔h8 25. ♗f6#

4473. B 24
Ternplemeier – Müller
Germany 1950

1. e4 c5 2. ♘c3 ♘c6 3. g3 e6 4. ♗g2 g6 5. ♘ge2 ♗g7 6. 0-0 ♘ge7 7. d3 a6 8. ♗e3 b6 9. ♕d2 d5 10. ed5 ed5 11. ♗h6 0-0 12. ♗g7 ♔g7 13. ♘f4 d4 14. ♘cd5 ♖a7 15. ♖ae1 ♖e8 16. h4 ♖d7

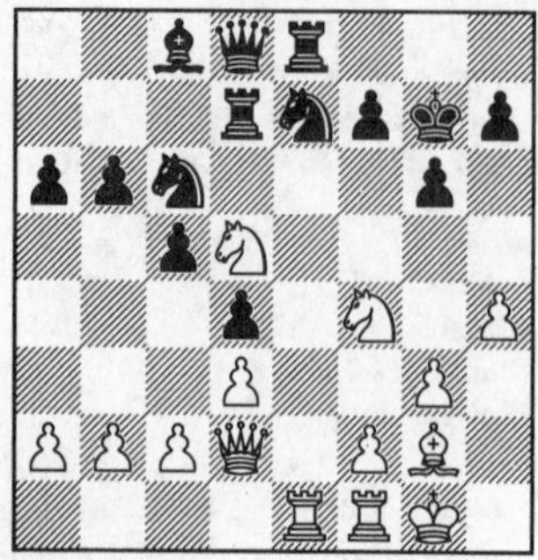

17. ♘f6! ♔f6 18. ♘h5! gh5 19. ♕g5#

4474. B 29

Usachy – Golyak

Cheljabinsk 1959

1. e4 c5 2. ♘f3 ♘f6 3. e5 ♘d5 4. d4 e6 5. c4 ♘b6 6. d5 d6 7. ♘c3 de5 8. ♘e5 ♗d6 9. ♗f4 0-0 10. de6 ♕e7 11. ♘e4 ♗e5 12. ♗e5 ♕e6 13. ♕d6 ♘c4 14. ♗c4 ♕c4

15. ♘f6![1] +–

[1] 15. – gf6 16. ♕f6 +–

4475. B 33

Galia – Grünfeld

Vienna 1946

1. e4 c5 2. ♘f3 ♘c6 3. d4 cd4 4. ♘d4 ♘f6 5. ♘c3 e5 6. ♘db5 d6 7. a4 a6 8. ♘a3 ♗e6 9. ♗c4 ♗c4 10. ♘c4 ♘e4 11. ♘e4 d5 12. ♗g5 f6

13. ♗f6! gf6 14. ♕d5 ♗e7 15. ♘cd6 ♗d6 16. ♘d6 ♔e7 17. 0-0-0 ♘d4[1] 18. ♖d4[2] +–

[1] 17. – ♖f8 18. ♘f5 ♔e8 19. ♘g7 ♔e7 20. ♕e6#

[2] 18. – ed4 19. ♖e1 ♔d7 20. ♘f7 +–

4476. B 45

Kinmark – Strom

Gothenburg 1927

1. e4 b6 2. d4 ♗b7 3. ♗d3 e6 4. ♘f3 c5 5. ♘c3 cd4 6. ♘d4 ♘c6 7. ♗e3 ♗b4 8. 0-0 ♗c3 9. bc3 ♘ge7 10. ♕h5 ♘g6 11. ♗g5 ♕c7 12. f4 a6 13. f5 ef5 14. ♘f5 ♕e5 15. ♗c4 0-0

16. ♗f6! ♕c5[1] 17. ♔h1 ♕c4 18. ♕h6! +–

[1] 16. – gf6 17. ♕h6 +–; 16. – ♕f6 17. ♘h6 +–

4477. B 45

Powers – N. N.

Jacksonville 1934

1. e4 c5 2. ♘c3 ♘c6 3. ♘f3 e6 4. d4 cd4 5. ♘d4 ♗c5 6. ♗e3 ♘d4 7. ♗d4 ♗d4 8. ♕d4 ♘f6 9. e5 ♘g8 10. ♗c4 ♘e7 11. ♘e4 0-0 12. 0-0 ♕c7 13. ♗d3 ♘c6

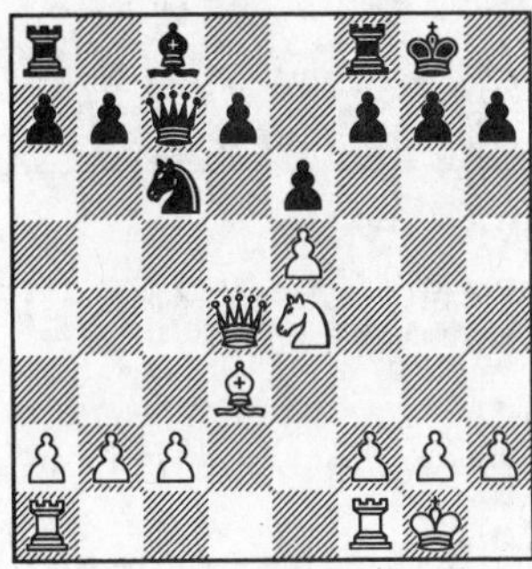

14. ♘f6! gf6 15. ♕g4 ♔h8 16. ef6 ♖g8 17. ♕h5 ♖g6 18. ♗g6 fg6 19. ♕h6 +−

4478. B 50

Buckley – N. N.

London 1840

1. e4 c5 2. ♘f3 d6 3. ♘c3 e5 4. ♗c4 ♘c6 5. d3 ♘ge7 6. ♗g5 ♗g4 7. ♘d5 ♘d4 8. ♘e5! ♗d1

9. ♘f6! gf6 10. ♗f7#

4479. B 71

Watts – Pierce

Corr. 1946

1. e4 c5 2. ♘e2 d6 3. d4 cd4 4. ♘d4 ♘f6 5. ♘c3 g6 6. f4 ♗g7 7. e5 de5 8. fe5 ♘d5 9. ♗b5 ♔f8 10. ♖f1 ♗e5 11. ♕f3 ♘f6 12. ♗h6 ♔g8 13. ♘de2 ♘c6 14. ♖d1 ♗c3 15. ♕c3 ♗d7

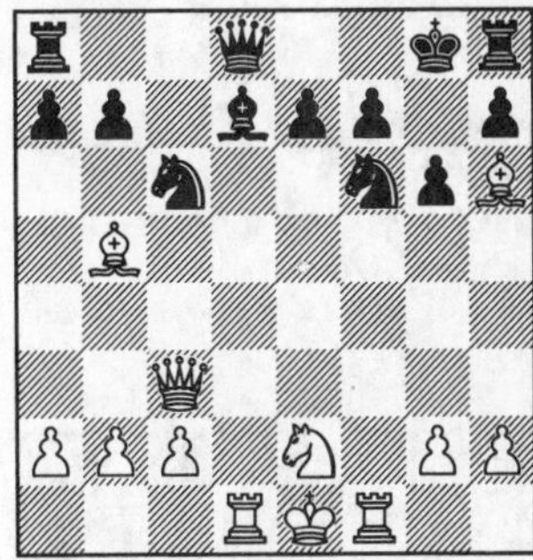

16. ♖f6! ef6 17. ♖d7! ♕a5[1] 18. ♗c6 ♕c3 19. ♘c3 bc6 20. ♘e4 +−

[1] 17. – ♕d7 18. ♕f6 +−

4480. B 85

Lasker – Pirc

Moscow 1935

1. e4 c5 2. ♘f3 ♘c6 3. d4 cd4 4. ♘d4 ♘f6 5. ♘c3 d6 6. ♗e2 e6 7. 0-0 a6 8. ♗e3 ♕c7 9. f4 ♘a5 10. f5 ♘c4 11. ♗c4 ♕c4 12. fe6 fe6

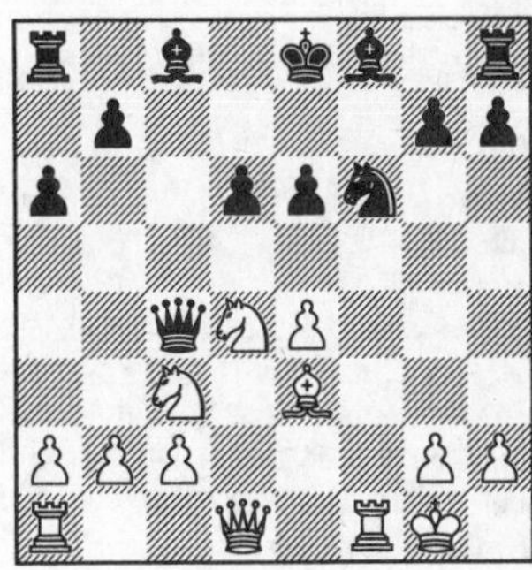

13. ♖f6! gf6 14. ♕h5 ♔d8[1] 15. ♕f7 ♗d7 16. ♕f6 ♔c7 17. ♕h8 ♗h6 18. ♘e6 ♕e6 19. ♕a8 ♗e3 20. ♔h1 +−

[1] 14. – ♔e7 15. ♘f5 ef5 16. ♘d5 ♔d8 17. ♗b6 ♔d7 18. ♕f7 +−

4481. B 97

Van der Wiel – Quinteros

Biel 1985

1. e4 c5 2. ♘f3 d6 3. d4 cd4 4. ♘d4 ♘f6 5. ♘c3 a6 6. ♗g5 e6 7. f4 ♕b6 8. ♕d3 ♘bd7 9. 0-0-0 ♗e7 10. ♕h3 ♘c5 11. e5 de5 12. fe5 ♘d5 13. ♗e7 ♘e7 14. ♗d3 ♗d7 15. ♕g3 ♘d3 16. ♖d3 0-0 17. ♘e4 ♔h8

18. ♘f6! gf6 19. ♕h4 ♘g8[1] 20. ♖g3 +−

[1] 19. – ♘g6 20. ♕h6 ♖g8 21. ♖h3 ♖g7 22. ef6 +−

4482. C 00

Freitag – Lilly

Corr. 1964

1. d4 e6 2. e4 d5 3. ♗e3 de4 4. f3 ef3 5. ♘f3 ♘f6 6. ♗c4 ♗d6 7. ♘c3 ♘bd7 8. ♕d2 ♘b6 9. ♗d3 ♗d7 10. 0-0 h6 11. ♘e5 ♕e7 12. ♘e4 ♗e5 13. de5 ♘e4 14. ♗e4 ♘c4 15. ♕c3 ♘e3 16. ♕e3 c6 17. ♖f3 0-0 18. ♖af1 ♔h8

19. ♖f6![1] +−

[1] 19. – ♔g8 20. ♖h6 gh6 21. ♕h6 f5 22. ef6 ♕c5 23. ♔h1 ♖f7 24. ♖f3 +−

4483. C 00

Harnach – Plaster

1967

1. e4 e6 2. d4 d5 3. ♗e3 de4 4. f3 ♘f6 5. ♘c3 ♗b4 6. a3 ♗c3 7. bc3 ♘d5 8. ♕d2 ef3 9. ♘f3 ♘e3 10. ♕e3 ♘d7 11. ♗d3 ♘f6 12. 0-0 ♗d7 13. ♘e5 0-0 14. ♖f3 b6 15. ♖af1 ♗e8

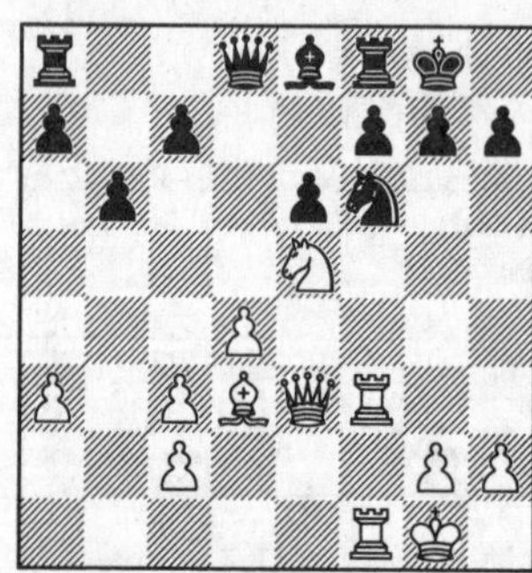

16. ♖f6! gf6 17. ♕h6 f5 18. ♖f5! f6 19. ♖g5! +−

4484. C 00

Sneiders – Zembachs

Corr. 1961

1. d4 e6 2. e4 d5 3. ♗e3 de4 4. ♘d2 ♘f6 5. f3 ef3 6. ♕f3 ♘c6 7. ♗b5 ♕d5 8. ♕e2

♗b4 9. c3 ♗e7 10. ♘h3 0-0 11. ♘f4 ♕d8 12. ♖d1 ♘d5 13. 0-0 ♗d6 14. ♘e4 ♘e3 15. ♕e3 ♘e7

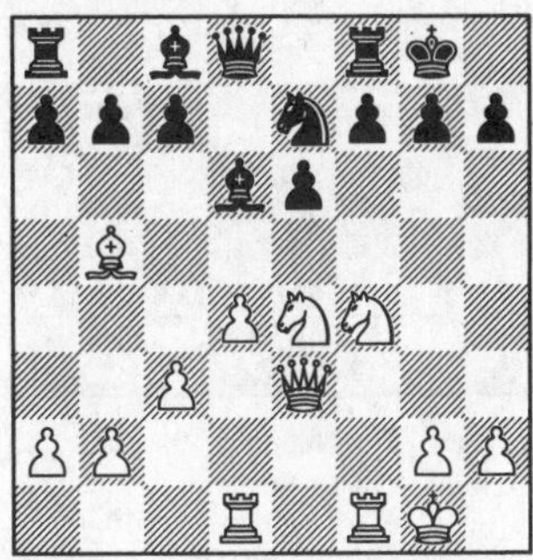

16. ♘f6! gf6 17. ♘h5 ♔h8[1] 18. ♕h6[2] +–

[1] 17. – ♘f5 18. ♖f5! ef5 19. ♕h6 +–
[2] 18. – ♘f5 19. ♖f5! +–; 18. – ♖g8 19. ♕f6 +–

4485. C 00

Webster – Ellis

Wisconsin 1971

1. d4 e6 2. e4 d5 3. ♗e3 de4 4. f3 ♘f6 5. ♘d2 ef3 6. ♘gf3 ♗d6 7. ♗d3 ♘bd7 8. 0-0 c6 9. ♕e2 0-0 10. ♘c4 ♗c7 11. ♗g5 ♘b6 12. ♘ce5 ♘bd5 13. c4 h6 14. cd5 hg5 15. ♘g5 ed5

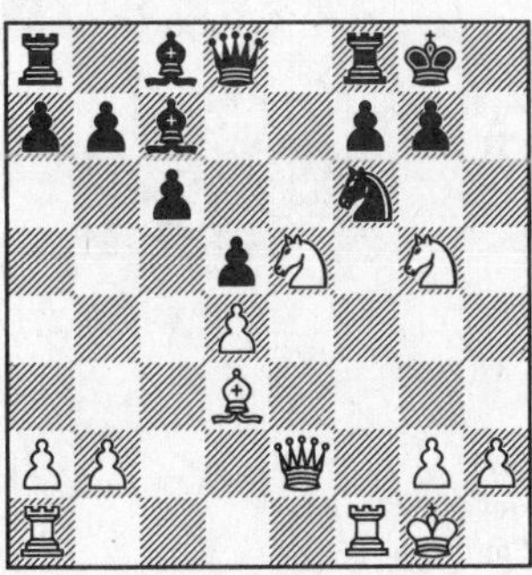

16. ♖f6! ♕f6[1] 17. ♗h7 ♔h8 18. ♕h5 +–

[1] 16. – gf6 17. ♕h5 +–

4486. C 00

Wirthensohn – Ott

Winterthur 1974

1. e4 e6 2. d3 c5 3. g3 ♘c6 4. ♗g2 d5 5. ♘d2 ♘f6 6. ♘gf3 de4 7. de4 ♗e7 8. 0-0 0-0 9. ♕e2 b6 10. e5 ♘d7 11. ♘e4 ♕c7 12. ♗f4 h6

13. ♘f6! gf6 14. ef6 ♗d6 15. ♗d6 ♕d6 16. ♖ad1 ♕c7 17. ♖d7 ♕d7 18. ♕e3 ♔h7 19. ♕e4 ♔h8 20. ♘g5! hg5 21. ♕g4 +–

4487. C 02

Tunbrige – Rithers

St. Bride 1938

1. e4 e6 2. d4 d5 3. e5 c5 4. ♘f3 ♘c6 5. ♘c3 cd4 6. ♘b5 f6 7. ef6 ♘f6 8. ♘bd4 ♗d6 9. ♗e2 0-0 10. c3 ♘e4 11. ♗e3 ♗d7 12. 0-0 ♘d4 13. ♗d4 ♕c7 14. ♕d3 ♗c6 15. ♗e3 ♖f6 16. g3 ♖af8 17. ♔g2 d4 18. cd4

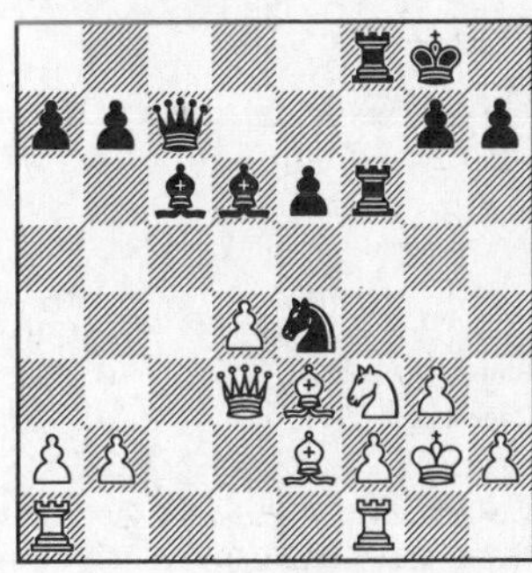

18. – ♖f3! 19. ♗f3 ♖f3 20. ♖ac1[1] ♘c5 21. ♕d2 ♖g3#

[1] 20. ♔f3 ♘c5 –+

4488. C 04

Frutsaert – Collins

Corr. 1948

1. e4 e6 2. d4 d5 3. ♘d2 ♘c6 4. ♘gf3 ♘f6 5. e5 ♘d7 6. c4 f6 7. cd5 ed5 8. ♗b5 fe5 9. ♘e5 ♕f6 10. ♘df3 ♗b4 11. ♗d2 ♗d2 12. ♕d2 0-0 13. ♗c6 bc6 14. ♖c1 c5 15. ♘d7 ♗d7 16. ♖c5 ♕g6 17. ♔f1

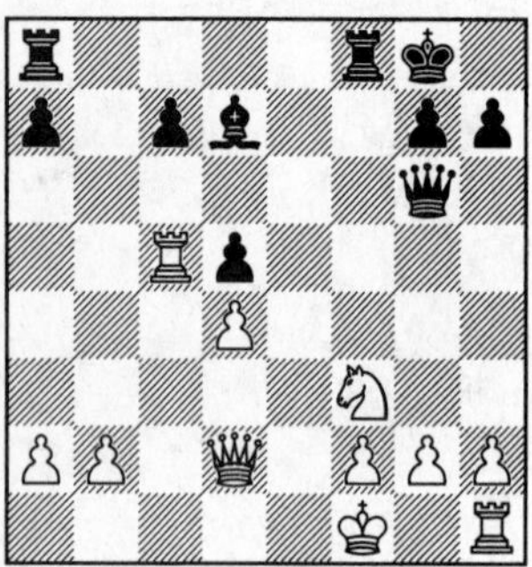

17. – ♖f3! 18. gf3 ♗h3 19. ♔e2 ♖e8 20. ♔d1 ♗g2[1] –+

[1] 21. ♖e1 ♕b1 22. ♕c1 ♕d3 23. ♕d2 ♖e1 24. ♔e1 ♕f1#

4489. C 10

Blom – Nielsen

Copenhagen 1934

1. e4 e6 2. d4 d5 3. ♘c3 de4 4. ♘e4 ♗d6 5. ♗d3 ♘e7 6. ♗g5 0-0

7. ♘f6! gf6 8. ♗f6 ♕d7 9. ♗h7! ♔h7 10. ♕h5 ♔g8 11. ♕h8#

4490. C 10

Engels – Zwetkoff

Munich 1936

1. e4 e6 2. d4 d5 3. ♘c3 de4 4. ♘e4 ♘d7 5. ♘f3 ♘gf6 6. ♗g5 ♗e7 7. ♘c3 c6 8. ♗c4 b5 9. ♗d3 ♗b7 10. ♕e2 a6 11. ♖d1 0-0 12. 0-0 ♔h8 13. ♘e4 ♕b6 14. ♘e5 ♖ad8 15. ♘d7 ♖d7 16. ♘f6 gf6

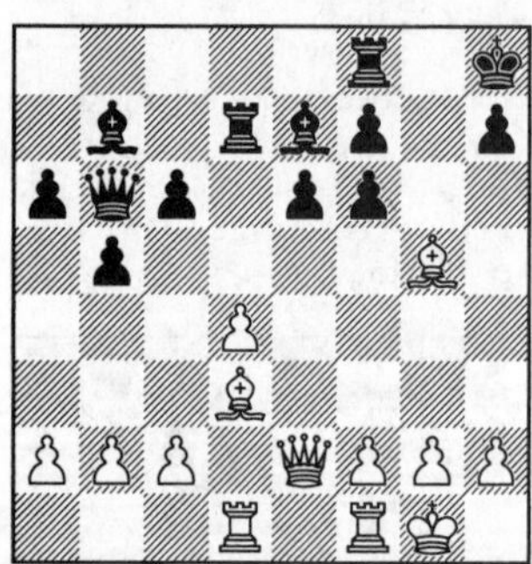

17. ♗f6![1] +–

[1] 17. – ♗f6 18. ♕h5 +–

4491. C 11

Hobson – Jameson

Corr. 1966

1. d4 ♘f6 2. ♘c3 e6 3. e4 ♗b4 4. f3 d5 5. ♗e3 de4 6. a3 ♗c3 7. bc3 ef3 8. ♘f3 0-0 9. ♗d3 ♘c6 10. 0-0 ♕d6 11. ♘g5 e5

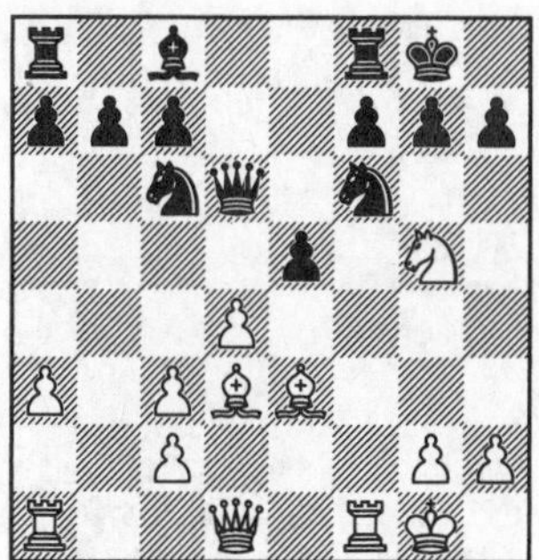

12. ♖f6![1] +–

[1] 12. – gf6 13. ♕h5 ♖e8 14. ♕f7 ♔h8 15. ♕h7#

4492. C 11

Whitehead – Maróczy

London 1923

1. e4 e6 2. d4 d5 3. ♘c3 ♘f6 4. e5 ♘fd7 5. ♘ce2 c5 6. c3 ♘c6 7. f4 ♕b6 8. ♘f3 f6 9. g3 cd4 10. cd4 fe5 11. fe5 ♗b4 12. ♔f2 0-0 13. ♗e3 ♘de5 14. de5

14. – ♖f3! 15. ♔f3 ♘e5 16. ♔f4[1] ♕d6 17. ♗h3 ♗d7 18. ♗d4 ♖f8 –+

[1] 16. ♔f2 ♘g4 –+

4493. C 11

Zilberstein – Tskitishvili

USSR 1977

1. e4 e6 2. ♘f3 d5 3. ♘c3 ♘f6 4. e5 ♘fd7 5. d4 c5 6. dc5 ♗c5 7. ♗d3 ♘c6 8. ♗f4 a6 9. 0-0 b5 10. ♖e1 ♕b6 11. ♗g3 ♘d4 12. ♘g5 h6 13. ♕h5 0-0 14. ♘h7 ♖e8

15. ♘f6! gf6 16. ef6 ♗f8[1] 17. ♗h7! ♔h7 18. ♕f7 ♔h8 19. ♘d5! ♕c5 20. ♘f4 ♕f5 21. ♘g6 +–

[1] 16. – ♘f6 17. ♕h6 ♗e7 18. ♗e5 +–

4494. C 13

Bogoljubov – Patigler

Corr. 1939

1. e4 e6 2. d4 d5 3. ♘c3 ♘f6 4. ♗g5 de4 5. ♘e4 ♗e7 6. ♗f6 ♗f6 7. c3 ♘d7 8. f4 ♗e7 9. ♘f3 b6 10. ♗b5 ♗b7 11. ♘e5 ♗e4 12. ♗d7 ♔f8 13. 0-0 c5 14. f5 ♗f6 15. fe6 fe6 16. ♗e6 ♔e7 17. ♕g4 ♗b7

18. ♖f6! gf6 19. ♗c4 ♗d5[1] 20. ♘c6! +–

[1] 19. fe5 20. ♕e6 ♔f8 21. ♕f7#

4495. C 15
Hasler – Robert
Swiss 1958

1. d4 d5 2. e4 e6 3. ♘c3 ♗b4 4. ♗e3 de4 5. f3 ♘f6 6. ♗c4 b6 7. a3 ♗c3 8. bc3 ♗b7 9. ♗g5 ♘bd7 10. ♕e2 0-0 11. fe4 ♗e4 12. ♘f3 ♕c8 13. 0-0 c5 14. ♘e5 ♗d5 15. ♗d3 h6 16. ♗f6 ♘f6

17. ♖f6! gf6 18. ♕g4 ♔h8 19. ♕h5 ♔g7 20. ♖f1 fe5 21. ♕g4 ♔h8 22. ♖f6 +–

4496. C 15
Perez – Fernandez
Aviles 1947

1. e4 e6 2. d4 d5 3. ♘c3 ♗b4 4. ♘e2 de4 5. a3 ♗d6 6. ♘e4 ♘e7 7. ♘2g3 ♘bc6 8. ♘h5 0-0

9. ♘hf6! gf6[1] 10. ♘f6 ♔g7 11. ♕h5! ♘f5[2] 12. ♗g5 h6 13. ♗h6! ♘h6 14. ♕g5 +–

[1] 9. – ♔h8 10. ♕h5 h6 11. ♗h6 +–
[2] 11. – ♔f6 12. ♕g5#

4497. C 17
Kaplan – Timman
Jerusalem 1967

1. e4 e6 2. d4 d5 3. ♘c3 ♗b4 4. e5 c5 5. ♕g4 ♘e7 6. dc5 ♘bc6 7. ♗d2 ♘f5 8. ♘f3 ♗c5 9. ♗d3 0-0 10. ♗g5 ♕b6 11. 0-0 ♕b4 12. ♕h3 h6 13. a3 ♕b6 14. ♗f5 ef5 15. ♘d5 ♕b2

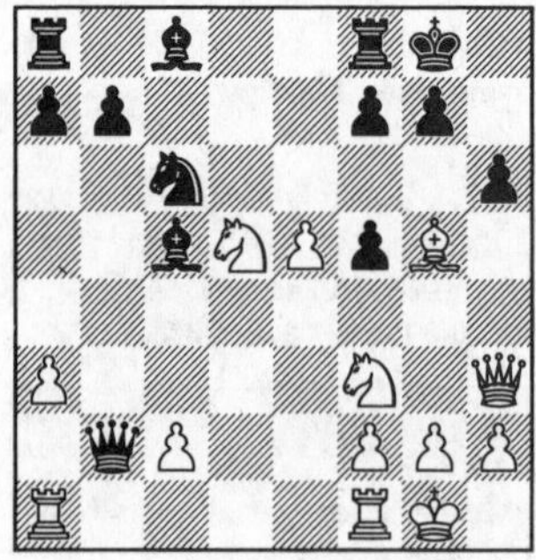

16. ♘f6! gf6[1] 17. ♗f6 ♔h7 18. ♕h5[2] +–

[1] 16. – ♔h8 17. ♗h6! +–
[2] (19. ♘g5) +–; 18. – ♖g8 19. ♕f7 +–

4498. C 21
Blackburne – N. N.
Kidderminsten 1863

1. e4 e5 2. d4 ed4 3. c3 dc3 4. ♗c4 d6 5. ♘c3 ♘c6 6. ♘f3 ♘e5 7. ♘e5 de5 8. ♗f7 ♔e7 9. ♗g5 ♘f6 10. ♕h5 c6 11. ♖d1 ♕a5 12. f4 ♕c5 13. fe5 ♕e5 14. 0-0 h6 15. ♗e8 ♗e6

16. ♖f6! gf6 17. ♖d7! ♗d7 18. ♕f7 ♔d6 19. ♕d7 ♔c5 20. ♗e3 ♔b4 21. ♕b7 ♔a5 22. b4! ♗b4 23. ♗b6! ab6 24. ♕a8#

4499. C 21

Denker – Gonzales

Detroit 1945

1. e4 e5 2. d4 ed4 3. c3 dc3 4. ♗c4 cb2 5. ♗b2 ♗b4 6. ♔f1 ♘f6 7. e5 ♘g8 8. ♕g4 ♗f8 9. ♕f3 ♘h6 10. ♘c3 ♗e7 11. ♘d5 0-0

12. ♘f6! ♔h8[1] 13. ♘h3 ♗f6 14. ef6 g6 15. ♕f4 ♘f5 16. ♘g5 ♘d6 17. ♘f7! ♘f7 18. ♕h6![2] +–

[1] 12. – gf6 13. ef6 ♗d6 14. ♕h5 +–

[2] 18. – ♘h6 19. f7 +–; 18. – ♖g8 19. ♗f7 +–

4500. C 21

Fidlow – Busse

USA 1952

1. e4 e5 2. d4 ed4 3. c3 dc3 4. ♗c4 cb2 5. ♗b2 ♘f6 6. ♘c3 ♗b4 7. ♘ge2 ♘e4 8. 0-0 ♘c3 9. ♘c3 0-0 10. ♘d5 ♗c5

11. ♘f6! gf6 12. ♕g4 ♔h8 13. ♕h4 ♗e7 14. ♗d3! +–

4501. C 21

Rasovsky – Mikyska

Corr. 1908

1. e4 e5 2. d4 ed4 3. c3 dc3 4. ♗c4 cb2 5. ♗b2 ♘f6 6. e5 ♗b4 7. ♘c3 ♕e7 8. ♘e2 ♘e4 9. 0-0 ♘c3 10. ♗c3 ♗c3 11. ♘c3 0-0 12. ♘d5 ♕d8 13. ♕h5 c6

14. ♘f6! gf6 15. ♗d3 ♖e8 16. ♕h7 ♔f8 17. ♕h8 ♔e7 18. ♕f6 ♔f8 19. ♕h6 ♔g8[1] 20. ♗h7 ♔h8 21. ♗g6 ♔g8 22. ♕h7 ♔f8 23. ♕f7#

[1] 19. – ♔e7 20. ♕d6#

4502. C 22

Kirdetzov – Kahn

Copenhagen 1918

1. e4 e5 2. d4 ed4 3. ♕d4 ♘c6 4. ♕d1 ♘f6 5. ♘c3 ♗b4 6. ♗d3 0-0 7. ♗g5 ♖e8 8. ♘f3 ♘e4 9. ♗e4 ♗c3 10. ♔f1 ♗f6 11. ♗f6 ♕f6 12. ♕d3 ♘b4 13. ♕c4 d5 14. ♗h7 ♔h7 15. ♕b4 c5 16. ♕c5

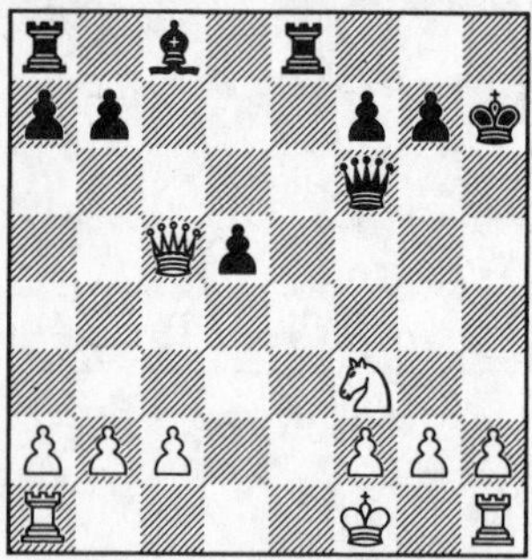

16. – ♕f3! 17. gf3 ♗h3 18. ♔g1 ♖e6 19. ♕c7 ♖ae8 20. ♖f1 ♖e1 –+

4503. C 23

Denker – Shayne

Rochester 1945

1. e4 e5 2. ♗c4 ♗c5 3. b4 ♗b4 4. c3 ♗c5 5. d4 ed4 6. ♘f3 ♘f6 7. e5 ♘e4 8. 0-0 ♘c3 9. ♘c3 dc3 10. ♗g5 ♗e7 11. ♕d5 ♖f8

12. ♗f6! gf6 13. ef6 ♗f6 14. ♖fe1 ♗e7 15. ♘g5 c6[1] 16. ♘f7 cd5[2] 17. ♘d6#

[1] 15. – d6 16. ♘h7 +–

[2] 16. – ♕c7 17. ♖e7 ♔e7 18. ♖e1 ♔f6 19. ♕g5#

4504. C 23

MacDonnel – Boden

London 1865

1. e4 e5 2. ♗c4 ♗c5 3. b4 ♗b4 4. c3 ♗c5 5. d4 ed4 6. cd4 ♗b4 7. ♔f1 ♗a5 8. ♕h5 d5 9. ♗d5 ♕e7 10. ♗a3 ♘f6 11. ♗f7 ♕f7 12. ♕a5 ♘c6 13. ♕a4 ♘e4 14. ♘f3 ♗d7 15. ♘bd2 ♘d2 16. ♘d2 0-0-0 17. ♖b1 ♕d5 18. ♘f3 ♗f5 19. ♖d1 ♖he8 20. ♗c5

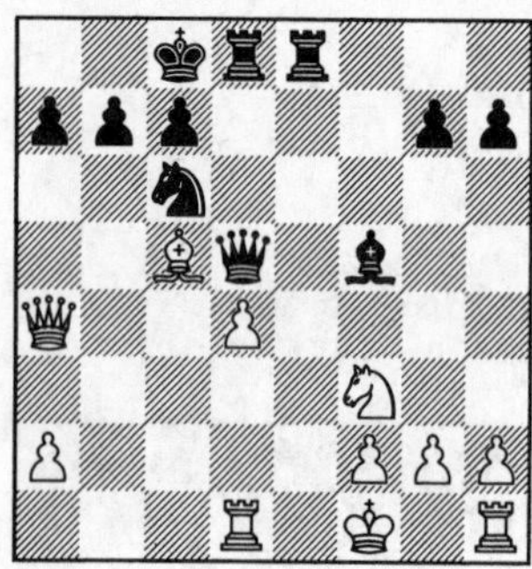

20. – ♕f3! 21. gf3 ♗h3 22. ♔g1 ♖e6 23. ♕c2[1] ♖d4! 24. ♗d4[2] ♘d4[3] –+

[1] 23. d5 ♖d5! –+

[2] 24. ♖b1 ♘e5! –+

[3] 25. ♕d3 ♖g6 26. ♕g6 ♘f3#

4505. C 24

Horowitz – N. N.

New York 1939

1. e4 e5 2. ♗c4 ♘f6 3. d4 ed4 4. ♘f3 ♗b4 5. c3 dc3 6. 0-0 0-0 7. e5 ♘e4 8. ♗d5 ♘c5 9. bc3 ♗a5 10. ♘g5 ♘e6 11. ♕h5 ♘g5 12. ♗g5 ♕e8

13. ♗f6! h6[1] 14. ♕g6 +–

[1] 13. – gf6 14. ♗e4! f5 15. ♕f5 +–

4506. C 24

Richardson – Delmar

New York 1887

1. e4 e5 2. ♗c4 ♘f6 3. ♘f3 ♘e4 4. ♘c3 ♘f2 5. ♔f2 ♗c5 6. d4 ed4 7. ♖e1 ♔f8 8. ♘e4 ♗b6 9. ♕d3 d5 10. ♕a3 ♔g8 11. ♗d5! ♕d5

12. ♘f6! gf6 13. ♕f8! ♔f8 14. ♗h6 ♔g8 15. ♖e8#

4507. C 29

N. N. – Löwy

Vienna 1905

1. e4 e5 2. ♘c3 ♘f6 3. f4 d5 4. fe5 ♘e4 5. ♘f3 ♘c6 6. ♗d3 f5 7. ef6 ♘f6 8. 0-0 ♗c5 9. ♔h1 0-0 10. ♗b5 ♘g4 11. ♗c6 bc6 12. d4 ♗d6 13. h3 ♗a6 14. hg4 ♗f1 15. ♕f1

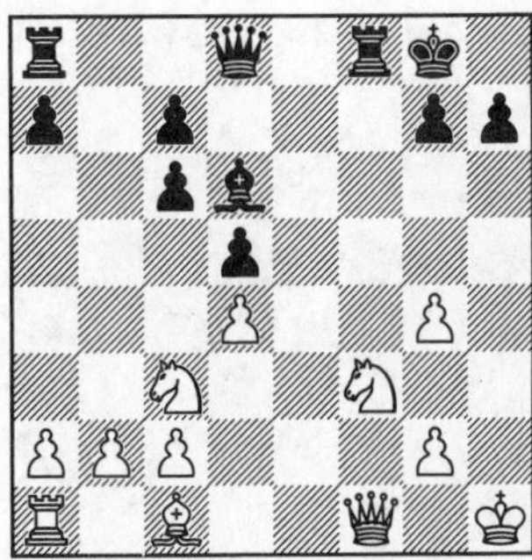

15. – ♖f3! 16. ♕e1[1] ♕h4! 17. ♕h4 ♖f1#

[1] 16. gf3 ♕h4 17. ♔g2 ♕h2#; 16. ♕f3 ♕h4 17. ♕h3 ♕e1#; 17. ♔g1 ♕e1 18. ♕f1 ♗h2 –+

4508. C 30

Grynfeld – Tarnowski

Poland 1948

1. e4 e5 2. f4 ♗c5 3. ♘f3 d6 4. d4 ed4 5. ♗c4 ♘f6 6. e5 de5 7. fe5 ♘d5 8. ♗g5 ♘e7 9. ♘bd2 ♘bc6 10. 0-0 0-0 11. ♘e4 ♗b6 12. ♗d3 ♗g4

13. ♘f6! gf6 14. ♗f6 ♕d7 15. ♕d2[1] +–

[1] 15. – ♖fe8 16. ♕h6 ♘f5 17. ♗f5 +–

4509. C 30

Janny – Steiner

1922

1. e4 e5 2. f4 ♗c5 3. ♘f3 d6 4. ♗c4 ♘c6 5. d3 ♗g4 6. ♘c3 ♘ge7 7. ♗f7 ♔f8 8.

♗b3 ♘d4 9. ♘d4 ed4 10. ♕g4 dc3 11. ♕e6 ♕e8 12. f5 ♗d4 13. f6 ♗f6

14. ♕f6! gf6 15. ♗h6#

4510. C 30

Zukertort – N. N.

Leipzig 1877

1. e4 e5 2. f4 d6 3. ♘c3 ♘f6 4. ♘f3 ♘c6 5. ♗c4 ♗g4 6. 0-0 ♗e7 7. d3 ♘h5 8. fe5 ♘e5 9. ♘e5! ♗d1 10. ♗f7 ♔f8 11. ♗h5 ♗f6

12. ♖f6! gf6[1] 13. ♗h6 ♔e7[2] 14. ♘d5 ♔e6 15. ♗f7 ♔e5 16. c3![3] +–

[1] 12. – ♕f6 13. ♘d7 +–

[2] 13. – ♔g8 14. ♗f7#

[3] 16. – c5 17. ♗f4#

4511. C 31

Cardiff – Bristol

Corr.

1. e4 e5 2. f4 d5 3. ed5 e4 4. ♗b5 c6 5. dc6 bc6 6. ♗a4 ♕d4 7. c3 ♕d6 8. ♘e2 ♗g4 9. 0-0 ♕d3 10. ♖e1 ♗c5 11. ♔f1

11. – ♕f3! 12. gf3 ♗h3#

4512. C 31

Ettlinger – Janovsky

New York 1898

1. e4 e5 2. f4 d5 3. ♘f3 de4 4. ♘e5 ♗c5 5. ♘c3 ♘f6 6. ♕e2 ♘c6 7. ♘f7 ♕e7 8. ♘h8 ♘d4 9. ♕d1

9. – ♘f3! 10. gf3 ef3 11. ♗e2 f2 12. ♔f1 ♗h3#

4513. C 32
Schulten – Morphy
New York 1857

1. e4 e5 2. f4 d5 3. ed5 e4 4. ♘c3 ♘f6 5. d3 ♗b4 6. ♗d2 e3 7. ♗e3 0-0 8. ♗d2 ♗c3 9. bc3 ♖e8 10. ♗e2 ♗g4 11. c4 c6 12. dc6 ♘c6 13. ♔f1 ♖e2! 14. ♘e2 ♘d4 15. ♕b1 ♗e2 16. ♔f2 ♘g4 17. ♔g1

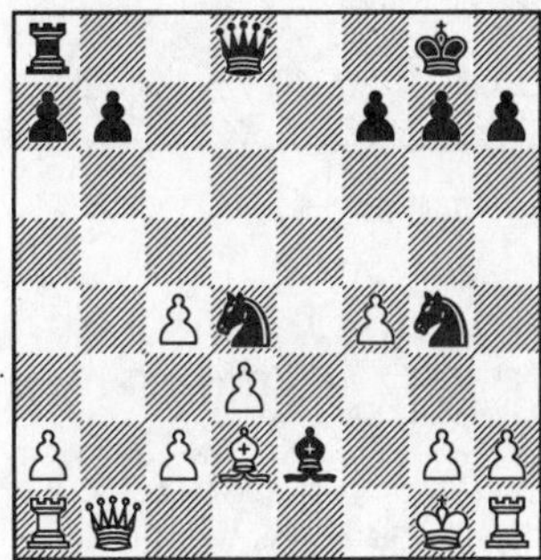

17. – ♘f3! 18. gf3 ♕d4 19. ♔g2 ♕f2 20. ♔h3 ♕f3 21. ♔h4 ♘h6 22. ♕g1 ♘f5 23. ♔g5 ♕h5#

4514. C 33
Courel – Blake
1904

1. e4 e5 2. f4 ef4 3. ♗c4 d5 4. ♗d5 ♕h4 5. ♔f1 g5 6. g3 fg3 7. ♔g2 ♗d6 8. h3 ♘e7 9. ♘f3 ♕h5 10. ♘c3 ♘g6 11. d4 ♗f4 12. ♘e2

12. – ♕f3![1] −+

[1] 13. ♔f3 ♘h4#

4515. C 33
Lokasto – Marcinkowski
Poland 1971

1. e4 e5 2. f4 ef4 3. ♗c4 d5 4. ♗d5 ♕h4 5. ♔f1 ♗d6 6. *Sf3 ♕h5 7. d4 ♘e7 8. ♘c3 f6 9. ♕e1 ♘bc6 10. ♘e2 g5 11. ♗c6 ♘c6 12. e5 fe5 13. de5 ♗b4 14. c3 ♗c5 15. b4

15. – ♕f3![1]

[1] 16. gf3 ♗h3#

4516. C 33
Reimann – Anderssen
Breslau 1876

1. e4 e5 2. f4 ef4 3. ♗c4 ♕h4 4. ♔f1 d5 5. ♗d5 ♘f6 6. ♘c3 ♗b4 7. e5 ♗c3 8. ef6 ♗f6 9. ♘f3 ♕h5 10. ♕e2 ♔d8 11. ♕c4 ♖e8 12. ♗f7

12. – ♕f3! 13. gf3 ♗h3 14. ♔f2[1] ♗h4 15. ♔g1 ♖e1 16. ♕f1 ♖f1#

[1] 14. ♔g1 ♖e1 15. ♔f2 ♗h4#

4517. C 33

Stean – Corden

England 1975

1. e4 e5 2. f4 ef4 3. ♗c4 d5 4. ♗d5 ♘f6 5. ♘c3 ♗b4 6.♘f3 0-0 7. 0-0 ♗c3 8. dc3 c6 9. ♗b3 ♕b6 10. ♔h1 ♘e4 11. ♕e1 ♗f5 12. ♘h4 ♖e8 13. ♗f4 ♘f6 14. ♕g3 ♗g6 15. ♘g6 hg6 16. ♗b8 ♖ab8

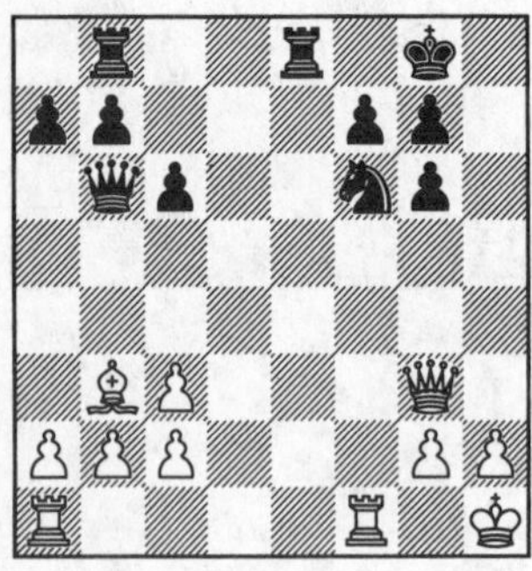

17. ♖f6! gf6 18. ♕g6 ♔h8 19. ♗f7 ♕e3 20. ♕f6 ♔h7 21. ♗g6[1] +–

[1] 21. – ♔g8 22. ♕f7 ♔h8 23. ♕h7#; 21. – ♔h6 22. ♗f5 ♔h5 23. g4#

4518. C 34

Ward – Brown

Nottingham 1874

(without ♘b1)

1. e4 e5 2. f4 ef4 3. ♘f3 f5 4. ♗c4 fe4 5. 0-0 ef3 6. ♕f3 ♗c5 7. d4 ♗d4 8. ♔h1 d6 9. ♗f4 ♘f6 10. ♖ae1 ♔f8 11. ♕d5 ♕d7 12. ♕d4 ♘c6

13. ♕f6! gf6 14. ♗h6 ♕g7 15. ♖f6#

4519. C 35

Spreckley – Mongredien

Liverpool 1846

1. e4 e5 2. f4 ef4 3. ♘f3 ♗e7 4. ♗c4 ♗h4 5. ♘h4 ♕h4 6. ♔f1 ♘f6 7. ♕f3 ♘c6 8. d3 ♘d4 9. ♕d1 ♘g4 10. ♕d2 ♘e3 11. ♔g1

11. – ♘f3! 12. gf3 ♕g5 13. ♔f2 ♕g2 14. ♔e1 ♕h1 15. ♔e2 ♕f1#

4520. C 41

Blake – Hooke

London 1891

1. e4 e5 2. ♘f3 d6 3. ♗c4 f5 4. d4 ♘f6 5. ♘c3 ed4 6. ♕d4 ♗d7 7. ♘g5 ♘c6 8. ♗f7 ♔e7

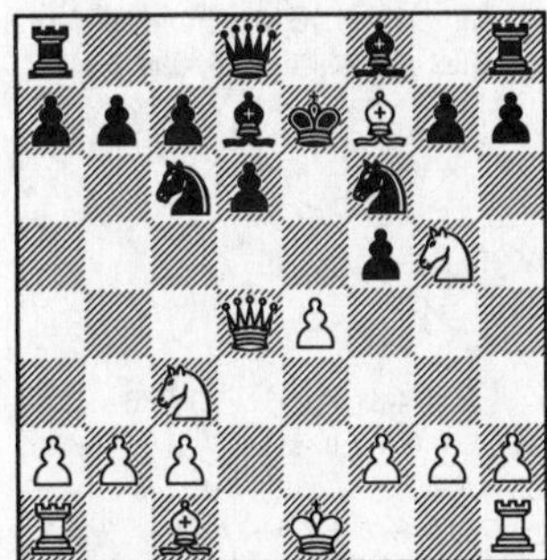

9. ♕f6! ♔f6 10. ♘d5 ♔e5 11. ♘f3 ♔e4 12. ♘c3#

4521. C 42

Bernstein – N. N.
Paris 1931

(without ♖a1)

1. e4 e5 2. ♘f3 ♘f6 3. ♘e5 ♘e4 4. ♕e2 ♕e7 5. ♕e4 d6 6. d4 f6 7. f4 ♘d7 8. ♗c4 fe5 9. fe5 de5 10. 0-0 ed4 11. ♗f7 ♔d8 12. ♗g5 ♘f6

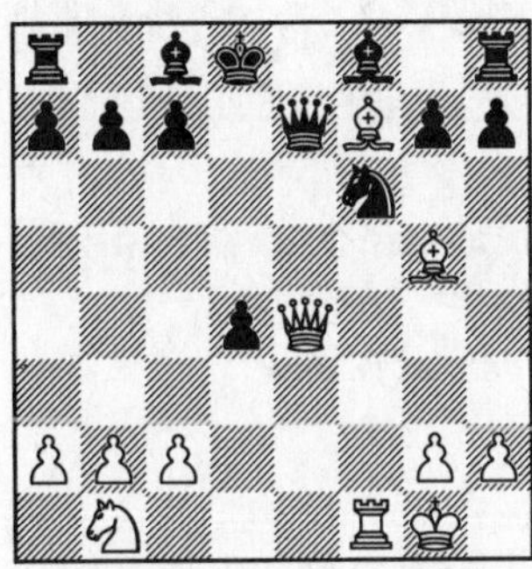

13. ♖f6! ♕e4[1] 14. ♖d6#

[1] 13. – gf6 14. ♗f6 +–

4522. C 42

Janovsky – Marshall
Biarritz 1912

1. e4 e5 2. ♘f3 ♘f6 3. ♘e5 d6 4. ♘f3 ♘e4 5. d4 d5 6. ♗d3 ♗d6 7. c4 ♗b4 8. ♔f1 0-0 9. cd5 ♕d5 10. ♕c2 ♖e8 11. ♘c3 ♘c3 12. bc3

12. – ♕f3! 13. cb4[1] ♘c6 14. ♗b2 ♘b4 15. ♗h7[2] ♔h8 16. gf3 ♗h3 17. ♔g1 ♘c2 18. ♗c2 ♖e2 19. ♖c1 ♖ae8 20. ♗c3 ♖8e3[3] 21. ♗b4[4] ♖f3 22. ♗d1 ♖f6 –+

[1] 13. gf3 ♗h3 14. ♔g1 ♖e1 –+
[2] 15. gf3 ♗h3 16. ♔g1 ♘c2 17. ♗c2 ♖e2 –+
[3] 20. – ♖c2! 21. ♖c2 ♖e6 –+
[4] 21. ♗b3 ♖c3 –+; 21. fe3 ♖g2 22. ♔f1 ♖c2 –+

4523. C 42

Kopetzky – Engert
Leipzig 1942

1. e4 e5 2. ♘f3 ♘f6 3. ♘c3 ♗b4 4. ♗c4 0-0 5. d3 d5 6. ed5 ♘d5 7. 0-0 ♘c3 8. bc3 ♗c3 9. ♖b1 ♘c6 10. ♘g5 h6 11. ♘e4 ♗d4 12. ♕h5 ♘a5 13. ♗g5 ♕e8

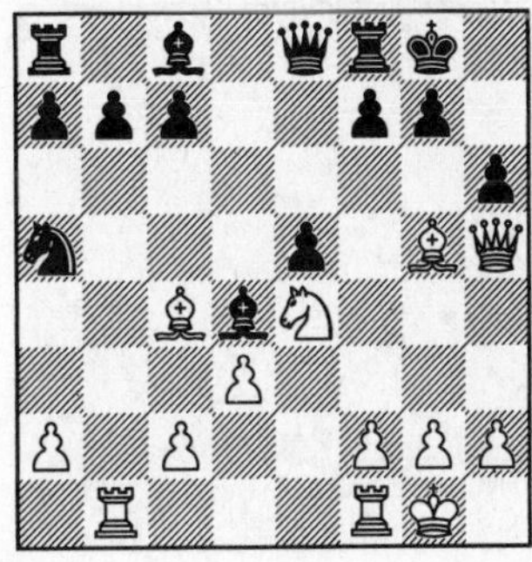

14. ♘f6![1] +−

[1] 14. – gf6 15. ♕g6 ♔h8 16. ♗f6#

4524. C 44

Heilpern – Pick

Vienna 1910

1. e4 e5 2. ♘f3 ♘c6 3. d4 ed4 4. c3 dc3 5. ♗c4 cb2 6. ♗b2 ♗b4 7. ♘c3 ♘f6 8. e5 ♘g4 9. 0-0 0-0 10. ♘d5 ♗c5 11. h3 ♘h6 12. ♕d2 ♘f5 13. ♕f4 d6 14. ♖ad1 ♗d7 15. ♕g4 ♘e3 16. ♕h5 ♘c4

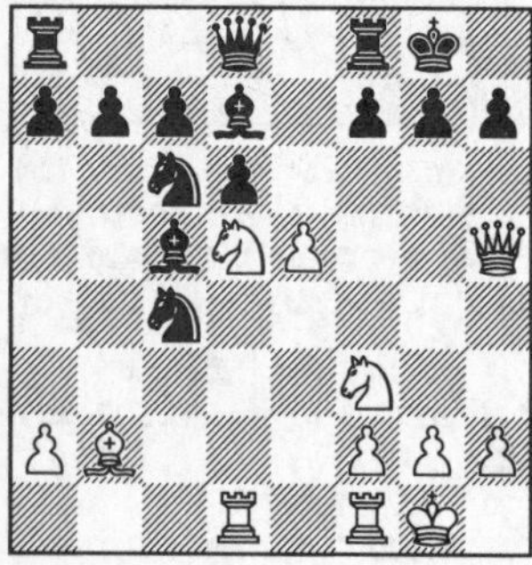

17. ♘f6! gf6 18. ef6 ♔h8 19. ♘g5 ♗f5 20. ♕h6 +−

4525. C 48

White – Serfőző

Corr. USA 1949

1. e4 e5 2. ♘f3 ♘c6 3. ♘c3 ♘f6 4. ♗b5 ♘d4 5. ♗a4 ♗c5 6. ♘e5 0-0 7. ♘f3 d5 8. ♘d4 ♗d4 9. ed5 ♗g4 10. f3 ♘h5 11. ♘e2

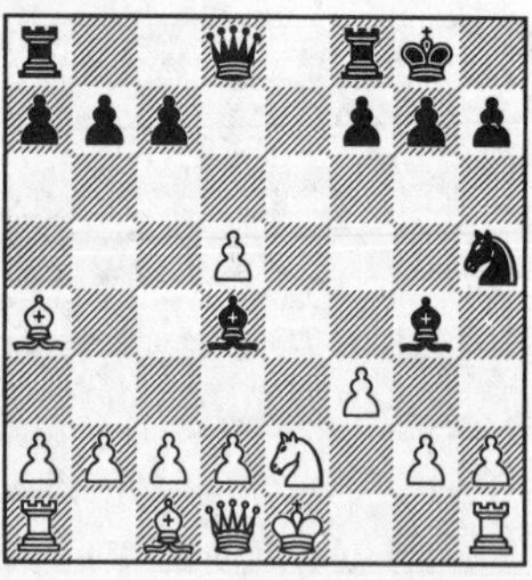

11. – ♗f3! 12. gf3 ♕h4 13. ♘g3 ♘g3 14. hg3 ♖fe8 15. ♗e8 ♖e8 16. ♔f1 ♕h1#

4526. C 49

Treybal – Moll

Berlin 1907

1. e4 e5 2. ♘f3 ♘c6 3. ♘c3 ♘f6 4. ♗b5 ♗b4 5. 0-0 0-0 6. d3 d6 7. ♗g5 ♘e7 8. ♘h4 c6 9. ♗a4 ♘g6 10. ♘g6 fg6 11. ♗b3 ♔h8 12. f4 ♕b6 13. ♔h1 ♘g4 14. ♕e1 h6 15. ♕h4 ♔h7 16. h3 ♘f6 17. fe5 de5

18. ♖f6! ♖f6[1] 19. ♗f6 gf6 20. ♕f6 ♕c7 21. ♗f7 +−

[1] 18. – gf6 19. ♕h6#

4527. C 50

Albin – Bernstein

Vienna 1904

1. e4 e5 2. ♘f3 ♘c6 3. ♗c4 ♗c5 4. ♘c3 d6 5. d3 ♘f6 6. ♗g5 ♗e6 7. ♘d5 ♗d5 8. ♗d5 h6 9. ♗f6 ♕f6 10. ♗c6 bc6 11. c3 ♖b8 12. b4 ♗b6 13. ♕a4 d5 14. ed5 e4 15. de4 ♕c3 16. ♔e2 ♕c4 17. ♔e1 ♕e4 18. ♔f1 0-0 19. ♕c6 ♖fe8 20. ♔g1 ♖e6 21. ♕d7 ♖d6 22. ♕a4 ♕e2 23. ♖f1

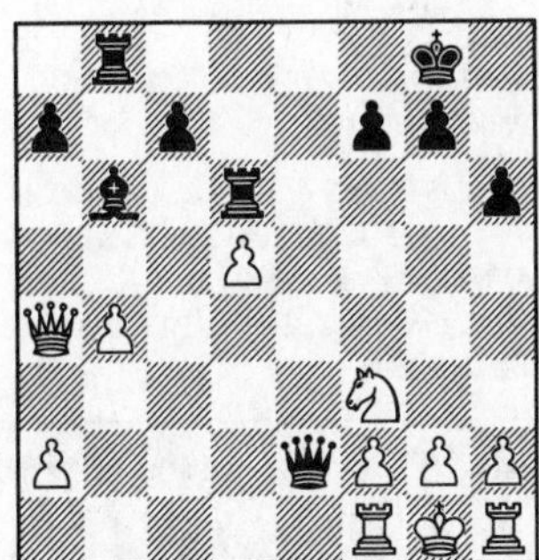

23. – ♕f3! 24. gf3 ♖g6#

4528. C 50

Grabiel – Mugridge

Los Angeles 1932

1. e4 e5 2. ♘f3 ♘c6 3. ♗c4 ♘f6 4. d3 ♗c5 5. 0-0 d6 6. ♗g5 h6 7. ♗h4 g5 8. ♗g3 h5 9. ♘g5 h4 10. ♘f7 hg3 11. ♘d8 ♗g4 12. ♕e1 ♘d4 13. ♘c3

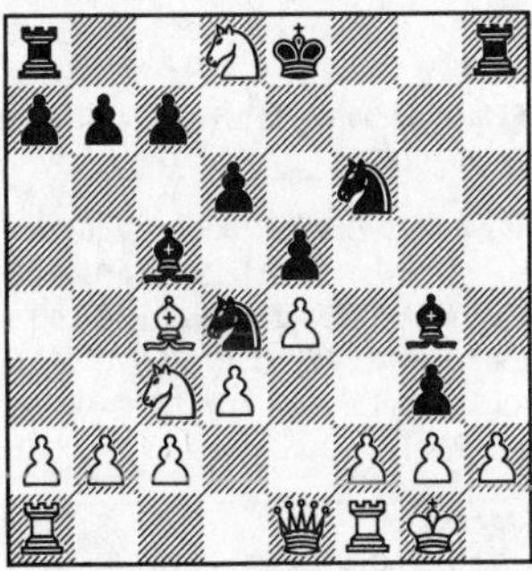

13. – ♘f3! 14. gf3 ♗f3 –+

4529. C 50

Müller – N. N.

Vienna 1927

1. ♘f3 ♘c6 2. e4 e5 3. ♗c4 ♗c5 4. d4 ♗d4 5. ♘d4 ♘d4 6. 0-0 ♘f6 7. f4 d6 8. fe5 de5 9. ♗g5 ♕e7 10. ♘c3 ♕c5 11. ♗f7 ♔f7 12. ♗f6 gf6 13. ♕h5 ♔e6

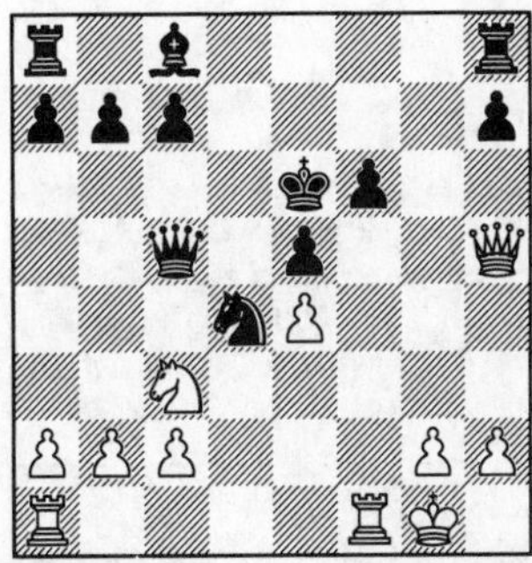

14. ♖f6! ♔f6 15. ♘d5 ♔e6[1] 16. ♕h6 ♔d7 17. ♕g7 ♔d6 18. ♕e7 ♔c6 19. ♕c7 ♔b5 20. a4 ♔c4 21. ♘e3 ♔b4 22. c3 ♔b3 23. ♕c5 ♔b2 24. ♕a3#

[1] 15. – ♔g7 16. ♕e5 ♔f7 17. ♕f6 ♔e8 18. ♕h8 ♔d7 19. b4! +–

4530. C 54

Lange – Lampert

Berlin 1903

1. e4 e5 2. ♘f3 ♘c6 3. ♗c4 ♗c5 4. c3 ♘f6 5. d4 ed4 6. cd4 ♗b4 7. ♘c3 ♘e4 8. 0-0 ♘c3 9. bc3 ♗e7 10. d5 ♘a5 11. d6 ♗d6 12. ♖e1 ♗e7 13. ♗g5 f6

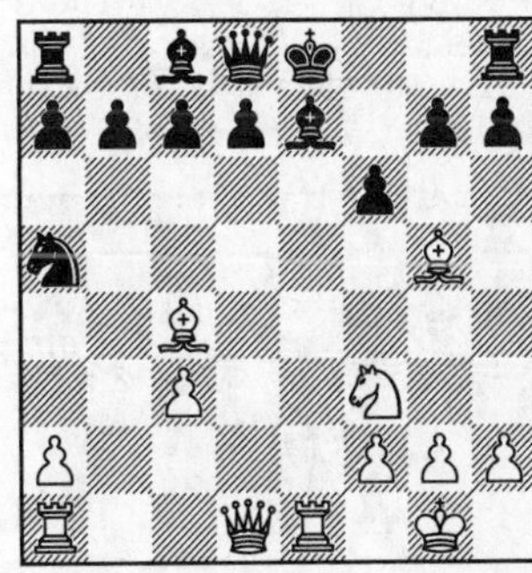

14. ♗f6! gf6 15. ♘e5 h5[1] 16. ♕d3 ♖h6 17. ♕d5 ♖h7 18. ♕g8 ♗f8 19. ♘g6 ♖e7 20. ♗f7#

[1] 15. – fe5 16. ♕h5 ♔f8 17. ♕f7#

4531. C 55

Boucek – Duras

Prague 1902

1. e4 e5 2. ♘f3 ♘c6 3. ♗c4 ♘f6 4. 0-0 ♗c5 5. d3 d6 6. ♗g5 ♗e6 7. ♗b3 ♕d7 8. ♗f6 gf6 9. ♘h4 ♗g4 10. ♕d2 0-0-0 11. ♔h1 ♗h5 12. f4 ♕g4 13. g3 ♖hg8 14. ♕g2 ♕h4 15. gh4 ♖g2 16. ♔g2 ♖g8 17. ♔h3 ♗g4 18. ♔g2 ♗e2 19. ♔h1

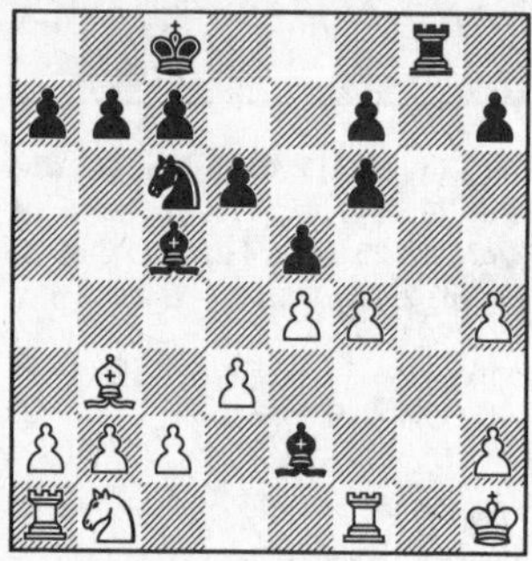

19. – ♗f3! 20. ♖f3 ♖g1#

4532. C 55

Peper – Pearce

Corr. 1944

1. e4 e5 2. ♘f3 ♘c6 3. ♗c4 ♘f6 4. d3 ♗c5 5. ♗g5 0-0 6. ♘c3 d6 7. ♘d5 ♔h8 8. ♗f6 gf6 9. ♕d2 f5 10. ♕h6 f6 11. ♘h4 ♖f7 12. ♘g6 ♔g8

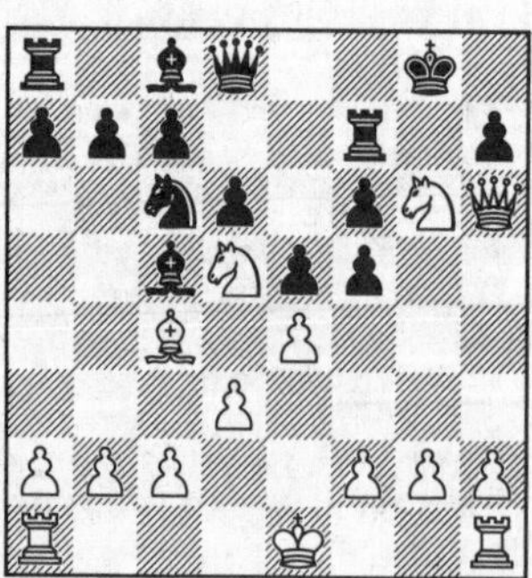

13. ♘f6! ♕f6 14. ♕f8#

4533. C 55

Reichhelm – N. N.

Philadelphia 1881

1. e4 e5 2. ♘f3 ♘c6 3. ♗c4 ♘f6 4. d4 ♘e4 5. de5 ♗c5 6. 0-0 ♗f2 7. ♔h1 0-0 8. ♕d5 ♘g3 9. hg3 ♗g3 10. ♘c3 ♘e5 11. ♗g5 ♕e8 12. ♘e4 ♘f3 13. ♖f3 ♕e5 14. ♖g3 ♕d5

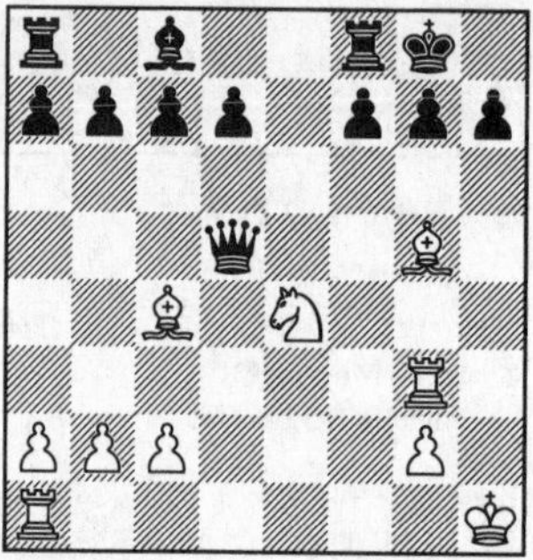

15. ♘f6! gf6 15. ♗f6 ♕g5 17. ♖g5#

4534. C 56

Prince Dadian of Mingrelia – Bitcham

Zugdidi 1892

1. e4 e5 2. ♘f3 ♘c6 3. ♗c4 ♘f6 4. d4 ed4 5. 0-0 ♘e4 6. ♖e1 d5 7. ♗d5 ♕d5 8. ♘c3 ♕c4 9. ♖e4 ♗e6 10. ♗g5 ♗c5 11. ♘d2 ♕a6 12. ♘b3 ♗b6 13. ♘d5 h6 14. ♘c5 ♕b5 15. ♖e6! ♔f8 16. ♘d7 ♔g8 17. ♕g4 h5

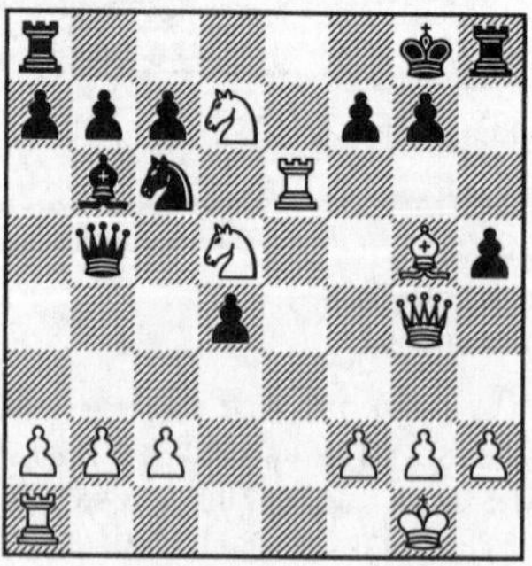

18. ♘5f6! gf6 19. ♗h6 ♕g5[1] 20. ♘f6#

[1] 19. – hg4 20. ♘f6#

4535. C 58
Durao – Prins
Malaga 1954

1. e4 e5 2. ♘f3 ♘c6 3. ♗c4 ♘f6 4. ♘g5 d5 5. ed5 ♘a5 6. ♗b5 c6 7. dc6 bc6 8. ♕f3 cb5 9. ♕a8 ♕d7 10. b4 ♘c6 11. a4 ♘b4 12. 0-0 ♘c2 13. ♗b2 b4 14. ♖c1 b3 15. ♕b8 ♕b7 16. ♕e5 ♗e7 17. ♘a3 0-0 18. ♘c2 bc2 19. ♖c2 ♗d8 20. ♘e4 ♕a8 21. ♖a3 ♗b7 22. ♘f6 ♗f6

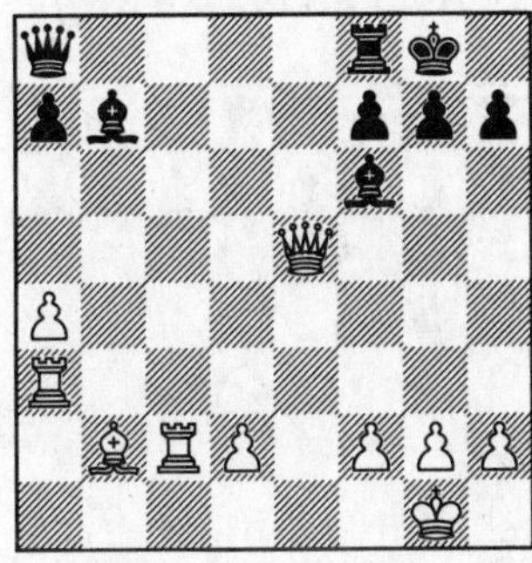

23. ♕f6[1] +–

[1] 23. – gf6 24. ♖g3 ♔h8 25. ♗f6#

4536. C 58
Field – Tenner
New York 1923

1. e4 e5 2. ♘f3 ♘c6 3. ♗c4 ♘f6 4. ♘g5 d5 5. ed5 ♘a5 6. d3 h6 7. ♘f3 e4 8. ♕e2 ♘c4 9. dc4 ♗c5 10. 0-0 0-0 11. ♘fd2 ♗g4 12. ♕e1 ♕d7 13. ♘b3

13. – ♗f3! 14. ♗f4[1] ♕g4 15. ♗g3 ♘h5 16. ♘c5 ♘f4 17. ♘e4 ♕h3! –+

[1] 14. gf3 ef3 15. ♔h1 ♕h3 16. ♖g1 ♗d6 –+

4537. C 58
de Rivière – Morphy
Paris 1863

1. e4 e5 2. ♘f3 ♘c6 3. ♗c4 ♘f6 4. ♘g5 d5 5. ed5 ♘a5 6. d3 h6 7. ♘f3 e4 8. ♕e2 ♘c4 9. dc4 ♗c5 10. h3 0-0 11. ♘h2 ♘h7 12. ♘d2 f5 13. ♘b3 ♗d6 14. 0-0 ♗h2 15. Kh2 f4 16. ♕e4 ♘g5 17. ♕d4

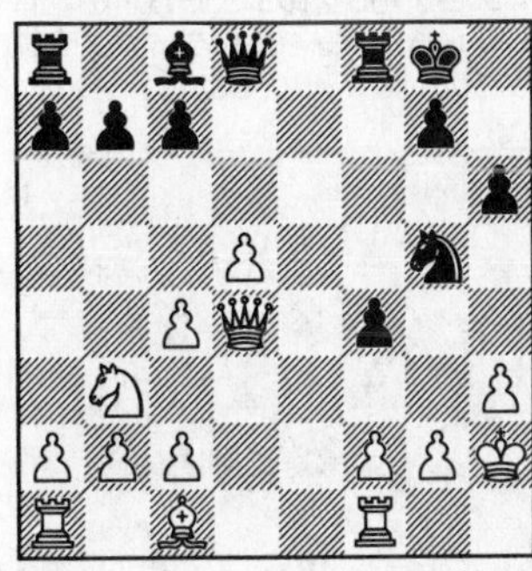

17. – ♘f3! 18. gf3 ♕h4 19. ♖h1 ♗h3 20. ♗d2 ♖f6 –+

4538. C 62
Réti – Capablanca
Berlin 1928

1. e4 e5 2. ♘f3 ♘c6 3. ♗b5 d6 4. c3 a6 5. ♗a4 f5 6. d4 fe4 7. ♘g5 ed4 8. ♘e4 ♘f6 9. ♗g5 ♗e7 10. ♕d4 b5 11. ♘f6 gf6 12. ♕d5 ba4 13. ♗h6 ♕d7 14. 0-0 ♗b7 15. ♗g7 0-0-0 16. ♗h8 ♘e5 17. ♕d1

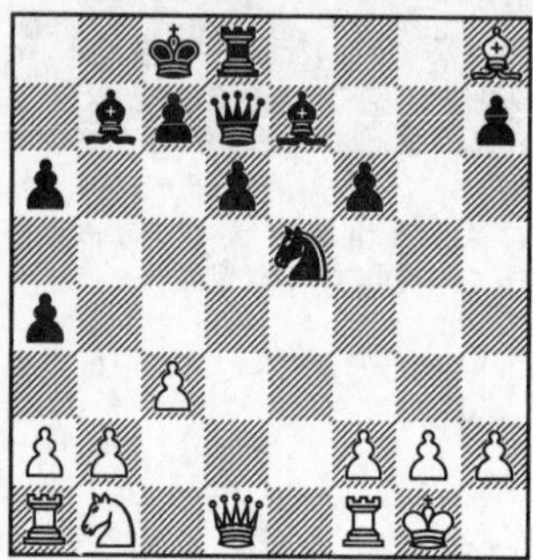

17. – ♗f3! 18. gf3[1] ♕h3 –+

[1] 18. ♕d4 ♕h3! 19. gh3 ♖g8 20. ♕g4 ♘g4 –+

4539. C 63
Balla – Réti
Budapest 1918

1. e4 e5 2. ♘f3 ♘c6 3. ♗b5 f5 4. d3 ♘f6 5. 0-0 ♗c5 6. ♗c4 d6 7. ♘g5 f4 8. ♘f7 ♕e7 9. ♘h8 ♗g4 10. ♕d2 ♘d4 11. ♔h1

11. – ♘f3! 12. ♕a5[1] ♘e4 13. g3[2] ♘f2 14. ♖f2 ♗f2 15. ♔g2 fg3 16. hg3 ♗g3 17. ♕b5[3] c6 18. ♕b4 ♕h4 19. ♗f7 ♔e7 20. ♕b7 ♔f6 –+

[1] 12. gf3 ♗f3 13. ♔g1 ♘e4 –+
[2] 13. de4 ♕h4 14. gf3 ♗f3 –+
[3] 17. ♔g3 ♕h4 18. ♔g2 ♕h2 19. ♔f1 ♗h3#

4540. C 64
Grossbach – Emerich
Vienna 1899

1. e4 e5 2. ♘f3 ♘c6 3. ♗b5 ♗c5 4. c3 ♘ge7 5. d4 ed4 6. cd4 ♗b6 7. d5 ♘b8 8. 0-0 0-0 9. d6 ♘g6 10. ♗g5 ♕e8 11. ♘c3 ♕e6 12. e5 ♘e5 13. ♘e5 ♕e5 14. ♗e7 ♖e8 15. ♖e1 ♕c5 16. ♘e4 ♕b5

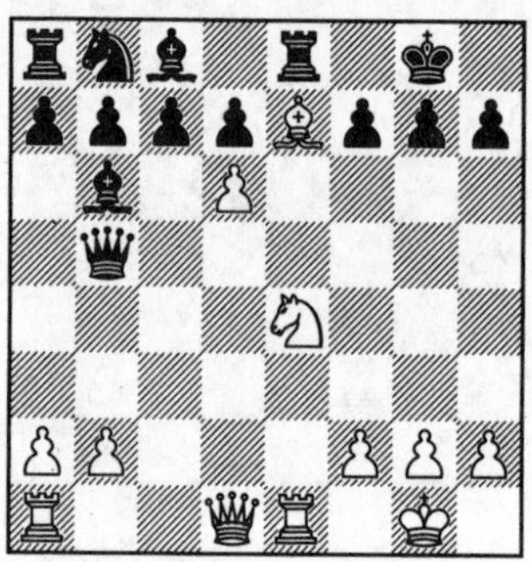

17. ♘f6! gf6 18. ♕g4 ♕g5 19. ♗f6! +–

4541. C 64
Williams – Helmke
Dayton 1976

1. e4 e5 2. ♘f3 ♘c6 3. ♗b5 ♗c5 4. c3 ♘f6 5. 0-0 ♘e4 6. ♗c6 bc6 7. ♘e5 0-0 8. d4 ♗d6 9. ♘f3 c5 10. ♕c2 ♘f6 11. ♗e3 cd4 12. cd4 ♗b7 13. ♘bd2 ♘g4 14. ♘g5 ♗h2 15. ♔h1 g6 16. ♘df3 ♗g3 17. ♘h3 ♕f6 18. ♗g5

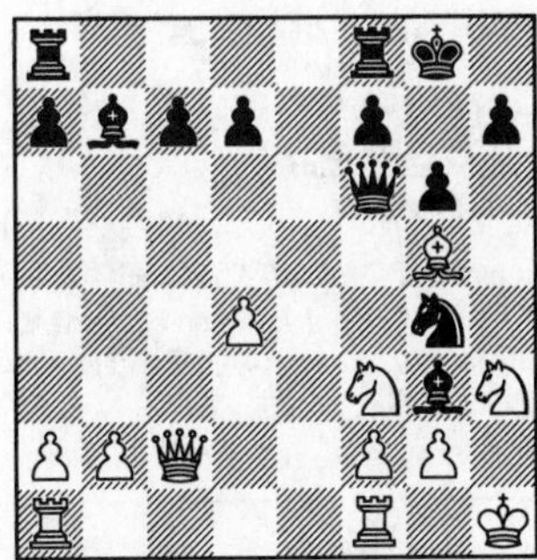

18. – ♕f3! 19. gf3 ♗f3 20. ♔g1 ♗h2#

4542. C 65

Joelson – Multhopp

Columbus Ohio 1983

1. e4 e5 2. ♘f3 ♘c6 3. ♗b5 ♗c5 4. 0-0 ♘f6 5. ♘c3 0-0 6. ♘e5 ♘d4 7. ♗a4 d6 8. ♘d3 ♗g4 9. ♕e1

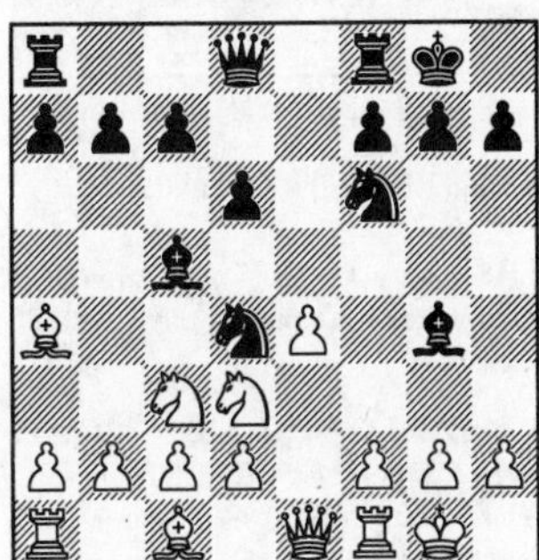

9. – ♘f3! 10. gf3 ♗f3 11. ♘c5[1] ♘g4 12. ♕e3[2] ♘e3 13. fe3 ♕g5 14. ♔f2 ♕g2 15. ♔e1 dc5 16. d3 ♗g4 17. ♖f2 ♕g1 –+

[1] 11. ♘e2 ♘g4 12. ♘g3 ♕h4 –+

[2] 12. d3 ♕h4 13. ♗f4 ♕h3 –+

4543. C 65

Lasker – N. N.

Moscow 1896

1. e4 e5 2. ♘f3 ♘c6 3. ♗b5 ♘f6 4. d4 ♘e4 5. ♕e2 ♘d6 6. ♗c6 bc6 7. de5 ♘b7 8. ♘c3 ♗e7 9. b3 0-0 10. ♗b2 ♘c5 11. 0-0-0 a5 12. ♘d4 ♗a6 13. ♕g4 ♕e8 14. ♘f5 ♘e6 15. ♘e4 ♔h8

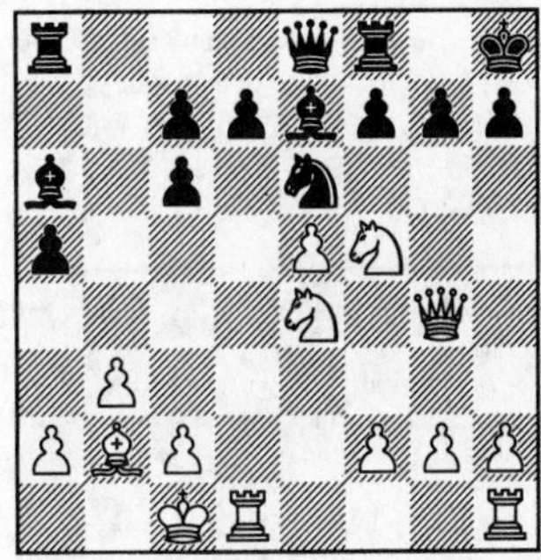

16. ♘f6 ♗f6[1] 17. ef6 g6 18. ♘e7 ♖g8 19. h4 ♖b8 20. h5 g5 21. ♘g6![2] +–

[1] 16. – gf6 17. ef6 ♗d8 18. ♕g7 ♘g7 19. fg7 ♔g8 20. ♘h6#

[2] 21. – hg6 22. hg6#; 21. – fg6 22. f7 +–

4544. C 67

Bachmann – Fiechtl

Regensburg 1887

1. e4 e5 2. ♘f3 ♘c6 3. ♗b5 ♘f6 4. 0-0 ♘e4 5. ♖e1 ♘d6 6. ♘e5 ♘e5 7. ♖e5 ♗e7 8. ♘c3 ♘b5 9. ♘d5 d6 10. ♖e7 ♔f8 11. ♕f3 f6 12. d3 c6

13. ♕f6! gf6 14. ♗h6 ♔g8 15. ♘f6#

4545. C 67
Tennant – Diesen
Corr. 1975

1. e4 e5 2. ♘f3 ♘c6 3. ♗b5 ♘f6 4. 0-0 ♘e4 5. d4 a6 6. ♗c6 dc6 7. ♕e2 ♗f5 8. g4 ♗g4 9. ♕e4 ♗f3 10. ♕f3 ♕d4 11. ♘c3 ♕d7 12. ♗e3 ♗d6 13. ♖ad1 ♖d8 14. ♘e4 0-0

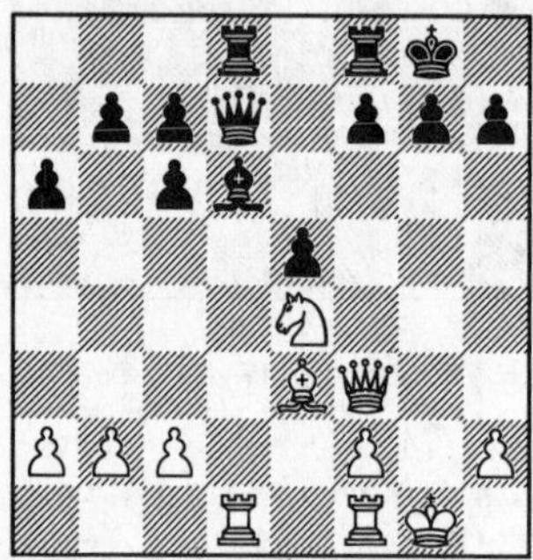

15. ♘f6! gf6 16. ♗h6 e4 17. ♕g2 +–

4546. C 67
Wolf – Haas
Vienna 1911

1. e4 e5 2. ♘f3 ♘c6 3. ♗b5 ♘f6 4. 0-0 ♘e4 5. d4 ♘d6 6. ♗c6 bc6 7. de5 ♘b7 8. ♘c3 ♗e7 9. ♘d4 0-0 10. ♗e3 c5 11. ♘f5 d6 12. ♘e7 ♕e7 13. ♘d5 ♕d8 14. ♕h5 ♖e8 15. ♗g5 ♖e5 16. ♖fe1 f6 17. f4 g6 18. ♕h6 ♖d5

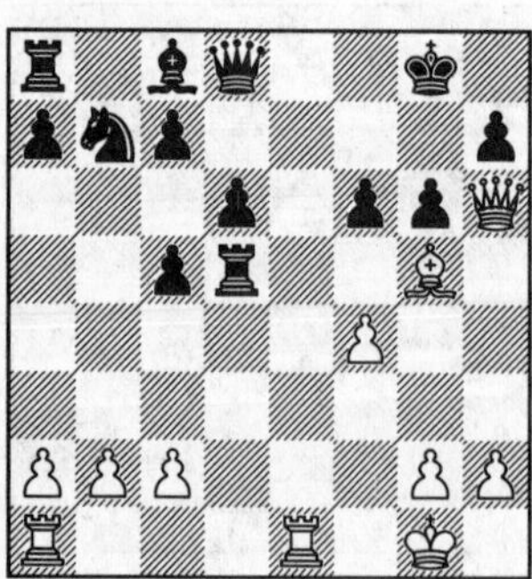

19. ♗f6! ♕f8[1] 20. ♖e8! +–

[1] 19. – ♕f6 20. ♖e8 ♔f7 21. ♕f8#

4547. C 68
Jacobs – Karklins
Lone Pine 1974

1. e4 e5 2. ♘f3 ♘c6 3. ♗b5 a6 4. ♗c6 dc6 5. 0-0 ♕d6 6. c3 ♗g4 7. d4 ed4 8. cd4 0-0-0 9. ♗e3 f5 10. e5 ♕g6 11. ♘c3 ♘e7 12. ♕e2 h6 13. ♘a4 f4 14. ♗c1 ♖d4 15. b3 Bd3 16. ♔h1 ♘f5 17. ♗f4

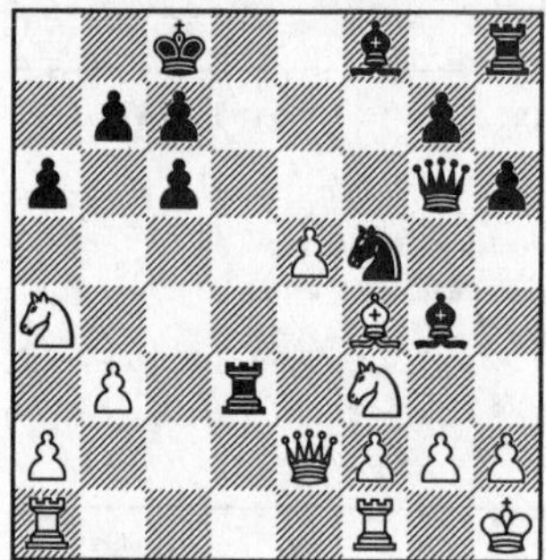

17. – ♖f3![1] –+

[1] 18. gf3 ♘d4 –+

4548. C 70
Kuzmichev – Kruminsh
Riga 1961

1. e4 e5 2. ♘f3 ♘c6 3. ♗b5 a6 4. ♗a4 f5 5. ♘c3 b5 6. ♗b3 b4 7. ♘d5 fe4 8. ♘h4 ♕h4 9. ♘c7 ♔d8 10. ♘a8 ♗c5 11. 0-0 ♘f6 12. h3 ♘d4 13. c3

13. – ♘f3! 14. ♔h1[1] d6 15. d4 ♗h3 16. gf3 ♗g4 –+

[1] 14. gf3 ♕g3 15. ♔h1 ♕h3 16. ♔g1 ef3 –+

4549. C 74
Bringman – Andrzejewski
Toledo 1976

1. e4 e5 2. ♘f3 ♘c6 3. ♗b5 a6 4. ♗a4 d6 5. c3 f5 6. d3 ♘f6 7. ♗g5 ♗e7 8. ♘bd2 0-0 9. 0-0 ♘g4 10. ♗e7 ♕e7 11. h3 fe4 12. de4 ♘f6 13. ♕e2 ♗d7 14. ♖ad1 ♘h5 15. ♗b3 ♔h8 16. ♘c4 ♘f4 17. ♕d2 ♘h3 18. ♔h2

18. – ♖f3! 19. gf3 ♕h4[1] –+

[1] 20. ♘e3 ♘g5 –+; 20. ♔g2 ♘f4 21. ♔g1 ♕g5 –+

4550. C 75
Svensson – Berg
Stockholm 1966

1. e4 e5 2. ♘f3 ♘c6 3. ♗b5 a6 4. ♗a4 d6 5. c3 ♗d7 6. 0-0 ♘ge7 7. ♖e1 ♘g6 8. d4 ♗e7 9. ♘bd2 0-0 10. ♗b3 ♕c8 11. d5 ♘d8 12. ♘f1 f5 13. ef5 ♗f5 14. ♘g3 ♗g4 15. ♗c2 ♘h4 16. ♗e4 ♗f3 17. ♗f3

17. – ♖f3! 18. gf3 ♕h3 –+

4551. C 77
Rösch – Schlage
Hamburg 1910

1. e4 e5 2. ♘f3 ♘c6 3. ♗b5 a6 4. ♗a4 ♘f6 5. ♕e2 b5 6. ♗b3 ♗e7 7. c3 0-0 8. 0-0 d5 9. ed5 ♘d5 10. ♘e5 ♘f4 11. ♕e4 ♘e5 12. ♕a8 ♕d3 13. ♗d1 ♗h3 14. ♕a6 ♗g2 15. ♖e1

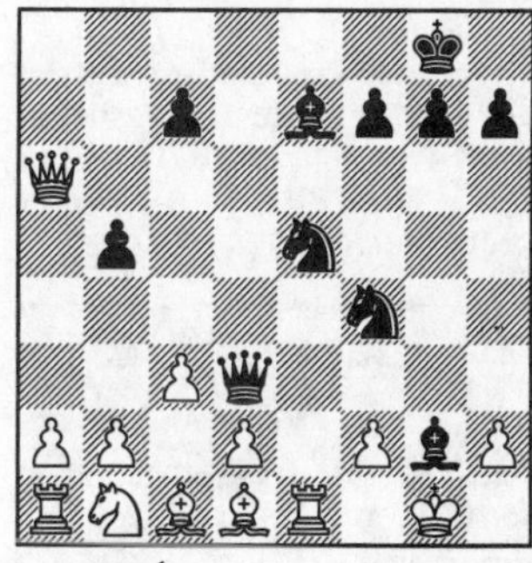

15. – ♕f3![1] –+

[1] 16. h4 ♕d1! –+; 16. – ♕h3 17. f3 ♕h1 18. ♔f2 ♕e1! 19. ♔e1 ♘ed3#

4552. C 80
Hanning – Langner
Berlin 1926

1. e4 e5 2. ♘f3 ♘c6 3. ♗b5 a6 4. ♗a4 ♘f6 5. 0-0 ♘e4 6. d4 ed4 7. ♖e1 d5 8. c4 dc3 9. ♘c3 ♗b4 10. ♗g5 f6 11. ♘e5 0-0 12. ♗c6 ♗c3 13. ♗d5 ♔h8 14. ♕h5 ♗e5 15.

♗e4 g6 16. ♗g6 ♕d7 17. ♖e5 fe5 18. ♖d1 ♕g7

19. ♗f6![1] +–

[1] 19. – ♖f6 20. ♖d8 ♖f8 21. ♖f8 ♕f8 22. ♕h7#

4553. C 88

Kaszowski – Daijot

Corr. 1948

1. e4 e5 2. ♘f3 ♘c6 3. ♗b5 a6 4. ♗a4 ♘f6 5. 0-0 ♗e7 6. ♖e1 b5 7. ♗b3 0-0 8. d4 ♘d4 9. ♘d4 ed4 10. e5 ♘e8 11. c3 dc3 12. ♘c3 c6 13. ♕f3 ♕c7 14. ♗f4 ♕b6 15. ♖ad1 ♘c7 16. ♘e4 ♘e6 17. ♗e3 c5

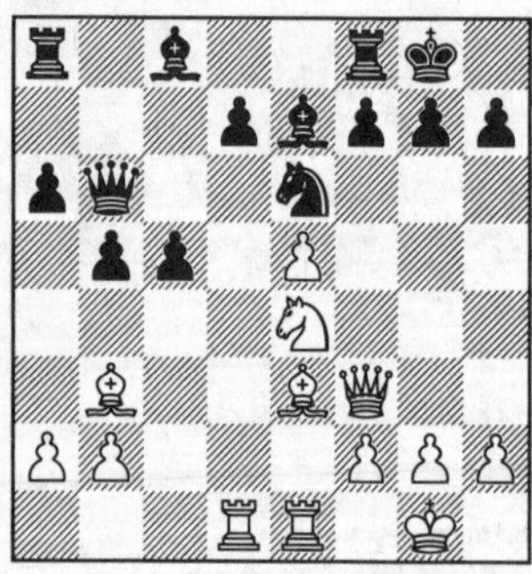

18. ♘f6! ♗f6[1] 19. ef6 ♘d4 20. ♗f7![2] +–

[1] 18. – gf6 19. ef6 (♗e6, ♕g3) +–
[2] 20. – ♔f7 21. fg7 ♔g7 22. ♕a8 ♗b7 23. ♗h6! +–

4554. C 89

Levchenkov – Sakhovich

USSR 1972

1. e4 e5 2. ♘f3 ♘c6 3. ♗b5 a6 4. ♗a4 ♘f6 5. 0-0 ♗e7 6. ♖e1 b5 7. ♗b3 0-0 8. c3 d5 9. ed5 e4 10. dc6 ef3 11. ♕f3 ♗g4 12. ♕g3 ♖e8 13. ♖e3 ♗d6 14. ♕h4 ♗f4 15. ♖e8 ♕e8 16. f3 ♕e2 17. c4 b4 18. g3 ♕e1 19. ♔g2

19. – ♗f3![1] –+

[1] 20. ♔f3 ♕f1#

4555. D 00

Erdős – Lichtner

Vienna 1922

1. d4 d5 2. e4 de4 3. f3 ef3 4. ♘f3 ♘f6 5. ♗c4 e6 6. ♗g5 ♗e7 7. ♘bd2 ♘c6 8. c3 0-0 9. 0-0 ♘g4 10. ♘e4 ♗g5 11. ♘fg5 ♘e3 12. ♕h5 h6 13. ♖f7 ♖f7 14. ♕f7 ♔h8

15. ♘f6! ♘e7[1] 16. ♕g6! ♕g8 17. ♕h7! ♕h7 18. ♘f7#

[1] 15. – ♕f6 16. ♕e8 +–

4556. D 05

Owen – Burn

London 1887

1. ♘f3 d5 2. d4 ♗f5 3. e3 e6 4. ♘c3 ♘f6 5. a3 c5 6. ♗b5 ♘bd7 7. ♘e5 ♗d6 8. g4 ♗e5 9. gf5 ♗d6 10. dc5 ♗c5 11. b4 ♗d6 12. ♗b2 ♖c8 13. ♕d4 0-0 14. ♗d7 ♕d7 15. ♘d5! ♘e8

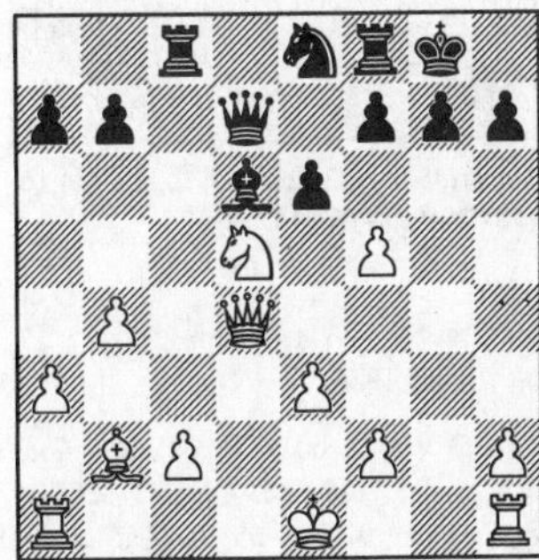

16. ♘f6! gf6[1] 17. ♖g1 ♔h8 18. ♕f6! ♘f6 19. ♗f6#

[1] 16. – ♘f6 17. ♕f6! e5 18. ♖g1 g6 19. ♖d1 +–

4557. D 05

Peters – Nightingale

Corr. 1949

1. d4 d5 2. ♘f3 ♘f6 3. e3 c5 4. c3 e6 5. ♗d3 ♘c6 6. ♘bd2 cd4 7. ed4 ♗d6 8. ♕e2 ♘d7 9. 0-0 0-0 10. ♖e1 e5 11. de5 ♘de5 12. ♘e5 ♘e5 13. ♗c2 ♕h4 14. ♘f1 ♗g4 15. ♕d2 ♖fe8 16. ♕g5

16. – ♘f3! 17. gf3 ♗h2! 18. ♘h2 ♖e1 19. ♘f1 ♖f1! 20. ♔f1 ♕h1 21. ♔e2 ♕f3 22. ♔f1 ♕h1#

4558. D 17

Kahn – N. N.

Paris 1932

1. d4 d5 2. c4 c6 3. ♘f3 ♘f6 4. ♘c3 dc4 5. a4 ♗f5 6. ♘h4 ♗d7 7. e3 b5 8. ♘f3 e6 9. ♘e5 a6 10. ♕f3 ♖a7 11. g4 ♗d6 12. g5 ♗e5 13. de5 ♘d5 14. ♘e4 0-0 15. ♖g1 ♘b4

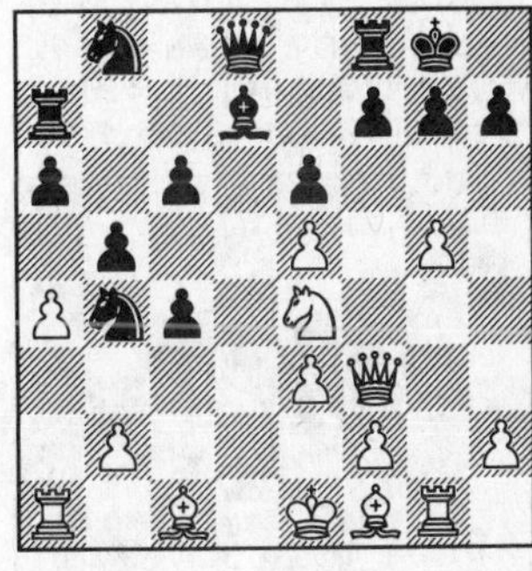

16. ♘f6! ♔h8[1] 17. ♕h5 h6[2] 18. g6 ♘c2 19. ♔e2 ♗e8 20. ♕h6! gh6 21. g7#

[1] 16. – gf6 17. gf6 ♔h8 18. ♖g7 ♖g8 19. ♖h7! ♔h7 20. ♕h5#

[2] 17. – gf6 18. ♖g3 +–

4559. D 31
Burn – Perlis
Ostende 1905

1. d4 d5 2. c4 e6 3. ♘c3 a6 4. cd5 ed5 5. ♘f3 b6 6. ♗f4 ♗b7 7. e3 ♗d6 8. ♘e5 ♘d7 9. ♕g4 ♗e5 10. de5 ♔f8 11. 0-0-0 ♘c5 12. e4 d4 13. ♗e3 ♘e6 14. ♗c4 ♗c8 15. f4 ♕e7 16. f5 dc3 17. fe6 ♕b4 18. ♖d8 ♔e7 19. ♗g5 f6

20. ♗f6[1] +–

[1] 20. – ♘f6 21. ♕g7 ♔d8 22. ♕f6 +–

4560. D 45
Heinicke – Reinhardt

1. ♘f3 ♘f6 2. c4 e6 3. ♘c3 d5 4. d4 c6 5. e3 ♘e4 6. ♗d3 f5 7. ♘e5 ♕h4 8. 0-0 ♘d7 9. f4 ♗d6 10. ♗e4 fe4 11. ♕g4 ♕h6 12. ♗d2 ♘f6 13. ♕e2 g5 14. fg5 ♕g5

15. ♖f6! ♕f6 16. ♕h5 ♔e7 17. ♖f1 ♗e5 18. ♖f6 ♗f6 19. ♘e4! de4 20. ♗b4 ♔d8 21. ♕h6 ♗d4 22. ed4 ♗d7 23. ♕f6 +–

4561. D 50
N. N. – Abrahams
Liverpool 1929

1. d4 ♘f6 2. c4 e6 3. ♘c3 d5 4. ♗g5 h6 5. ♗f6 ♕f6 6. ♘f3 ♗b4 7. ♕b3 ♘c6 8. a3 ♗a5 9. e3 0-0 10. ♗d3 e5 11. cd5 ed4 12. ed4 ♖e8 13. ♔f1

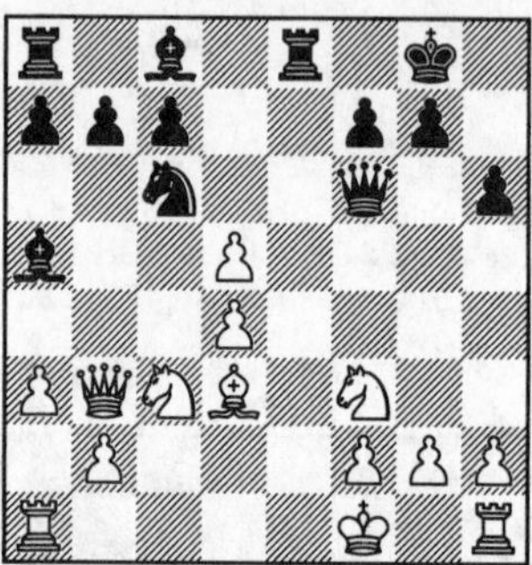

13. ♕f3! 14. gf3 ♗h3 15. ♔g1 ♘d4 16. ♕d1 ♖e1! 17. ♕e1 ♘f3#

4562. D 60
Ed. Lasker – Winkelman
New York 1926

1. ♘f3 ♘f6 2. d4 d5 3. c4 e6 4. ♘c3 ♗e7 5. ♗g5 0-0 6. e3 ♘bd7 7. ♗d3 a6 8. ♘e5 dc4 9. ♘c4 b5 10. ♘a5 c5 11. ♘c6 ♕e8 12. ♕f3 ♘b6 13. ♘e4 ♘fd5 14. ♘e7 ♘e7

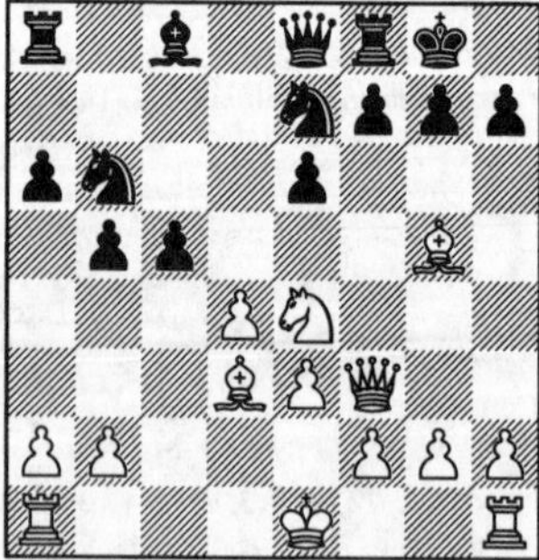

15. ♘f6![1] +–

[1] 15. – gf6 16. ♗h7! ♔h7 17. ♕h5 +–

4563. A 02

N. N. – du Mont

London 1913

1. f4 e5 2. fe5 d6 3. ed6 ♗d6 4. g3 ♕g5 5. ♘f3?

5. – ♕g3! 6. hg3 ♗g3#

4564. A 20

Martinec – Benes

Brno 1952

1. c4 e5 2. ♘f3 e4 3. ♘d4 ♘c6 4. ♘c6 bc6 5. ♕c2 ♘f6 6. g3 ♗e7 7. ♗g2 d5 8. 0-0 0-0 9. ♘c3 ♗f5 10. f3 ef3 11. ♕f5 fg2 12. ♔g2 d4 13. ♘e4 ♘e4 14. ♕e4 ♕d7 15. b3 ♖ae8 16. ♕d3 c5 17. ♗b2 ♗d6 18. ♖ae1 ♖e6 19. e4 ♖h6 20. ♔g1

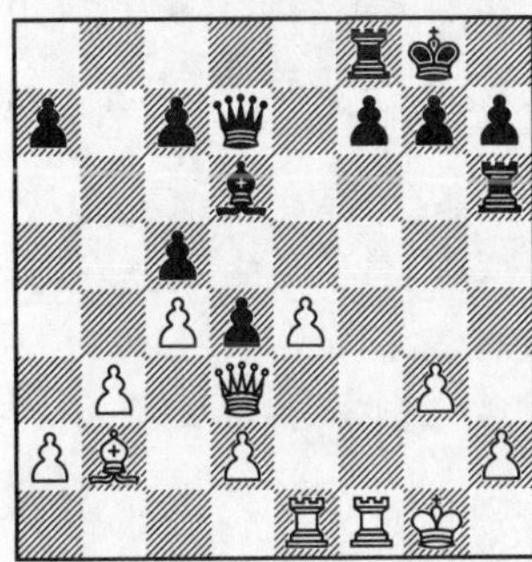

20. – ♗g3! 21. hg3[1] ♕h3 22. ♔f2[2] ♕h2 23. ♔f3 ♖f6 24. ♔g4 h5[3] –+

[1] 21. ♕g3 ♖g6 –+

[2] 22. ♖e2 ♕h1 23. ♔f2 ♖f6 –+

[3] 25. ♔g5 ♕h3 –+

4565. A 31

Dunkelblum – Henneberke

Beverwijk 1963

1. c4 c5 2. ♘f3 ♘f6 3. d4 e6 4. g3 b6 5. ♗g2 ♗b7 6. d5 ed5 7. ♘h4 g6 8. ♘c3 ♗g7 9. 0-0 0-0 10. ♗g5 h6 11. ♗f6 ♗f6

12. ♘g6! ♖e8[1] 13. ♘f4 ♘c6 14. ♗d5 ♖b8 15. ♕c2 ♘d4 16. ♕g6 +–

[1] 12. – fg6 13. ♗d5 ♗d5 14. ♕d5 +–

4566. A 40

N. N. – Charlik

1903

1. d4 e5 2. de5 d6 3. ♗f4 ♘c6 4. ed6 ♕f6 5. ♗c1 ♗d6 6. c3 ♗f5 7. e3 0-0-0 8. ♘d2 ♕g6 9. h3 ♘f6 10. ♘gf3 ♖he8 11. ♕a4 ♗c2 12. ♘b3 ♘e4 13. ♘h4

13. – ♕g3! 14. fg3[1] ♗g3 15. ♔e2 ♗d1#

[1] 14. ♕e4 ♖e4 15. fg3 ♗g3 16. ♔e2 ♗d1#

4567. A 48

Polau – Te Kolste

London 1927

1. ♘f3 ♘f6 2. d4 g6 3. ♘c3 d5 4. ♗f4 ♘h5 5. ♗e5 f6 6. ♗g3 ♘g3 7. hg3 ♗g7 8. e3 c6 9. ♗d3 e5 10. ♖h7 ♔f7

11. ♗g6! ♔g6 12. ♘e5! fe5[1] 13. ♕h5 ♔f6 14. ♕e5 ♔f7 15. ♕g7 ♔e6 16. ♕e5#

[1] 12. – ♔h7 13. ♕h5 ♔g8 14. ♕f7 ♔h7 15. 0-0-0 +–

4568. A 52

Rieol – Tóth

Tirgu Mures 1948

1. d4 ♘f6 2. c4 e5 3. de5 ♘g4 4. ♕d4 h5 5. f4 d6 6. ed6 ♘c6 7. ♕e4 ♗e6 8. f5 ♗d6 9. fe6 ♕h4 10. g3

10. – ♗g3! 11. hg3 ♕g3 12. ♔d1 0-0-0[1] 13. ♗d2 ♘e3 14. ♔c1 ♕e1! 15. ♗e1 ♖d1#

[1] 12. – ♘f2 –+

4569. A 81

Grünfeld – Torre

Baden–Baden 1975

1. d4 e6 2. ♘f3 f5 3. g3 ♘f6 4. ♗g2 d5 5. 0-0 ♗d6 6. c4 c6 7. ♕c2 0-0 8. b3 ♘e4 9. ♗b2 ♘d7 10. ♘e5 ♕f6 11. f3 ♘e5 12. de5 ♗c5 13. ♔h1

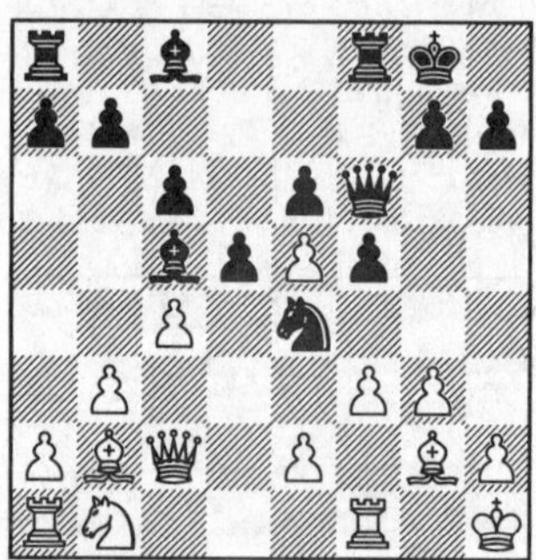

13. – ♘g3![1] –+

[1] 14. hg3 ♕h6 15. ♗h3 ♕h3#

4570. A 83

Caku – Birzoj

1970

1. d4 f5 2. e4 fe4 3. ♘c3 ♘f6 4. ♗g5 g6 5. f3 ef3 6. ♘f3 ♗g7 7. ♗d3 0-0 8. ♕d2 d6

9. 0-0-0 ♗e6 10. ♖de1 ♗f7 11. h4 ♖e8 12. ♗h6 ♕d7 13. h5 g5 14. ♗g7 ♔g7 15. ♕g5 ♔h8 16. ♗f5 ♕d8 17. ♘h4 ♖g8

18. ♘g6! ♗g6 19. hg6 ♖g7 20. ♘d5 ♘d7[1] 21. ♗d7 ♕d7 22. ♖e7! ♖e7 23. ♘f6 ♕e6 24. ♖h7[2] +–

[1] 20. – ♘d5 21. ♖h7 ♖h7 22. g7! ♖g7 23. ♖h1 ♔g8 24. ♗e6 +–

[2] 24. – ♖h7 25. g7! ♖g7 26. ♕h6 +–

4571. A 83

Palau – Nolbman

1. d4 f5 2. e4 fe4 3. ♘c3 ♘f6 4. ♗g5 d6 5. f3 ef3 6. ♘f3 ♗g4 7. ♗d3 ♘bd7 8. ♕e2 c6 9. ♘e4 ♘e4 10. ♕e4 ♗f3?

11. ♕g6! hg6 12. ♗g6#

4572. B 00

Chalupetzky – Rényi

Raab 1905

1. e4 b6 2. d4 ♗b7 3. ♗d3 e6 4. ♘c3 ♘f6 5. ♘ge2 d6 6. e5 ♘fd7 7. ♘f4 ♗e7 8. ♕g4 g6 9. ♘ce2 c5 10. ♘e6! fe6

11. ♗g6! hg6 12. ♕g6 ♔f8 13. ♘f4 +–

4573. B 00

Geller – Chiburdanidzhe

Aruba 1992

1. e4 b6 2. d4 ♗b7 3. ♗d3 e6 4. ♘f3 g6 5. ♗g5 ♕c8 6. ♘c3 ♗g7 7. 0-0 d6 8. ♖e1 ♘d7 9. e5 d5 10. a4 a6 11. ♘e2 ♘e7 12. ♘f4 ♘c6 13. c3 a5 14. h4 h6 15. ♗f6 ♗f6 16. ef6 ♘f6

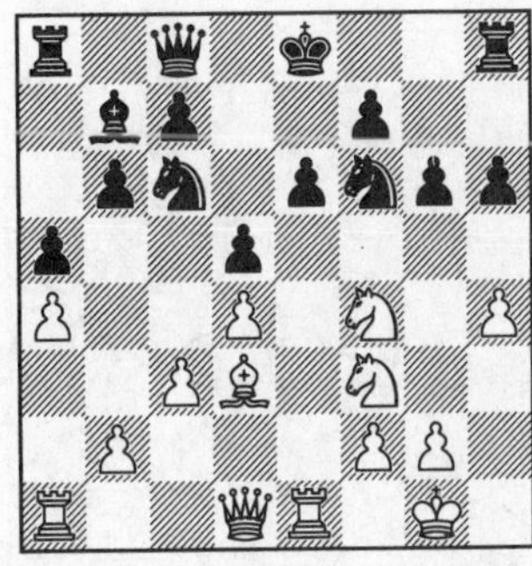

17. ♗g6! fg6 18. ♖e6 ♔f7 19. ♕d3 ♖g8 20. ♖ae1 ♘e4 21. B1e4 de4 22. ♕c4 +–

4574. B 01

Soultanbeieff – Marx

1. e4 d5 2. ed5 ♘f6 3. c4 c6 4. dc6 ♘c6 5. ♘f3 e5 6. d3 ♗f5 7. ♗g5 ♗b4 8. ♘c3 ♕d7 9. ♗f6 gf6 10. ♘h4 0-0-0 11. ♘f5 ♕f5 12. ♗e2 ♖hg8 13. 0-0 ♔b8 14. ♘d5 ♖g7 15. a3 ♖dg8 16. g3 ♗c5 17. ♗f3

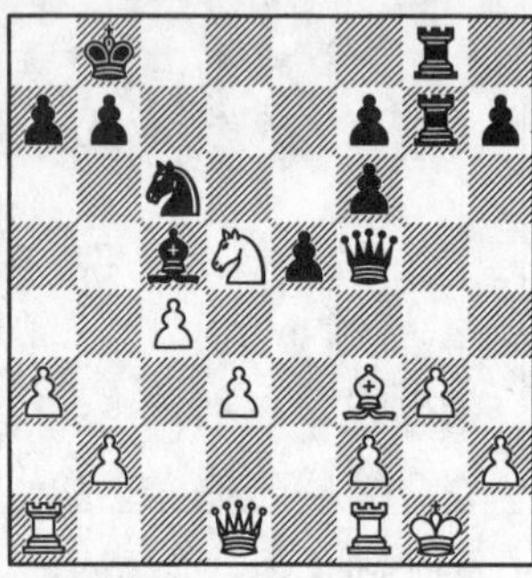

17. – ♖g3! 18. hg3 ♖g3 19. ♗g2 ♕h3 20. ♕f3 ♘d4![1] –+

[1] 21. fg3 ♘e2#

4575. B 02

Hill – Janeway

New York 1946

1. e4 ♘f6 2. e5 ♘d5 3. c4 ♘f4 4. d4 ♘g6 5. h4 h5 6. ♗e2 e6 7. ♗h5 ♘h4 8. ♕g4 ♘f5

9. ♕g6! ♕e7[1] 10. ♗g5 fg6 11. ♗g6 ♔d8 12. ♖h8 +–

[1] 9. – fg6 10. ♗g6 ♔e7 11. ♗g5#

4576. B 06

Bebciuk – Tomson

Moscow 1963

1. e4 g6 2. d4 ♗g7 3. ♘c3 d6 4. ♗e2 c6 5. h4 ♕b6 6. ♘f3 ♗g4 7. d5 ♘f6 8. a3 ♕a5 9. ♗d2 0-0 10. h5 ♗h5 11. ♘g5 ♗e2 12. ♕e2 ♕b6 13. 0-0-0 h6 14. ♘f3 h5 15. e5 ♘g4 16. e6 f5 17. ♘h4 ♔h7 18. f3 ♗c3 19. bc3 ♘e5 20. g4 fg4

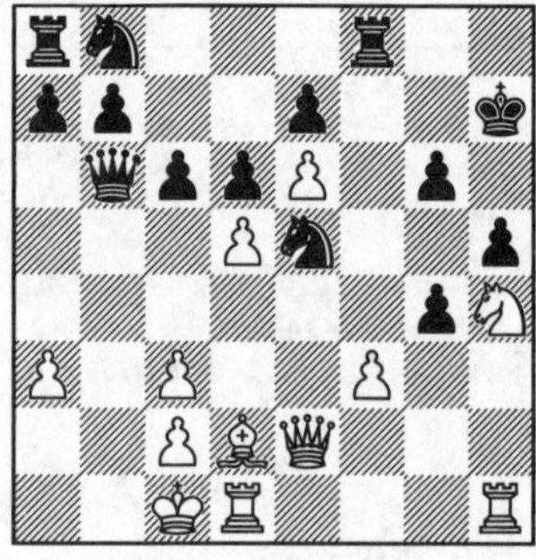

21. ♘g6! ♘g6[1] 22. ♖h5 ♔g8 23. ♕e4 ♖f6 24. ♖dh1[2]

[1] 21. – ♔g6 22. ♕e4 ♖f5 23. ♖h5! ♔h5 24. ♕f5 +–; 22. – ♔g7 23. ♖h5

[2] 24. – gf3 25. ♖h8! ♘h8 26. ♕h7 ♔f8 27. ♕h8#

4577. B 08

Gustone – Bluma

USSR 1979

1. e4 d6 2. d4 g6 3. ♘f3 ♗g7 4. ♘c3 ♘f6 5. ♗e2 0-0 6. h3 ♘bd7 7. ♗e3 c6 8. ♕d2 e5 9. de5 de5 10. 0-0-0 ♕e7 11. h4 ♘g4 12. h5 ♘e3 13. ♕e3 ♘f6 14. hg6 fg6 15. ♗c4 ♔h8 16. ♘h4 ♕e8 17. ♕g5 ♘h5

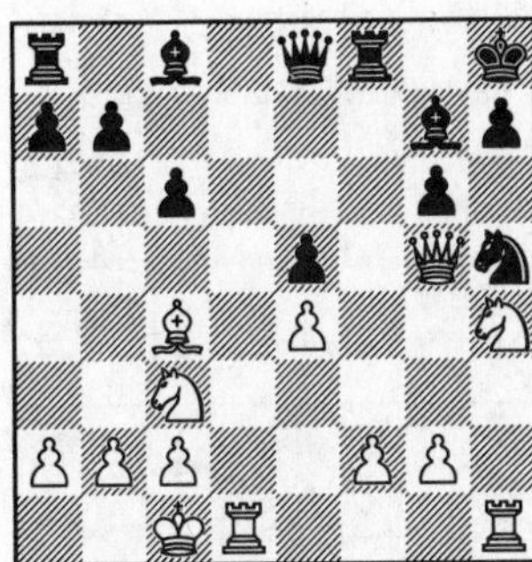

18. ♖d8! ♕d8 19. ♘g6! hg6 20. ♖h5 +–

4578. B 09

Marshall – Pillsbury

Cambridge–Springs 1904

1. d4 d6 2. e4 ♘f6 3. ♘c3 g6 4. f4 ♗g7 5. e5 de5 6. fe5 ♘d5 7. ♘f3 ♘c6 8. ♗c4 e6 9. ♗g5 ♘c3 10. bc3 ♘e7 11. 0-0 h6 12. ♗f6 ♗f6 13. ef6 ♘f5 14. ♕e2 ♕f6 15. g4 ♘d6 16. ♘e5 ♕e7 17. ♗d3 0-0 18. ♖f2 ♔g7 19. ♖af1 ♗d7 20. ♖f6 ♖g8

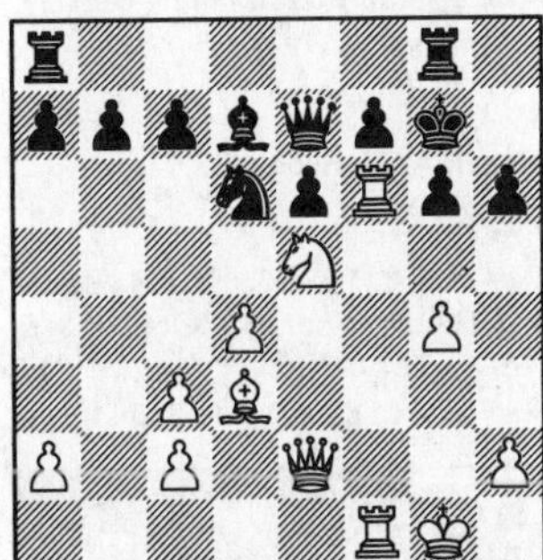

21. ♘g6! ♕f6[1] 22. ♖f6 ♔f6 23. ♕e5#

[1] 21. – fg6 22. ♖g6 ♔h8 23. ♖h6 ♔g7 24. ♖h7#

4579. B 14

Sprecher – Lutz

Bavaria 1937

1. e4 c6 2. d4 d5 3. ed5 cd5 4. c4 ♘f6 5. ♘c3 e6 6. ♘f3 ♗e7 7. ♗f4 0-0 8. c5 ♘e4 9. ♗d3 ♘c3 10. bc3 b6 11. cb6 ab6 12. h4 ♘d7 13. ♕c2 ♘f6 14. ♘g5 h6 15. ♗e5 ♖e8 16. ♗h7 ♔f8

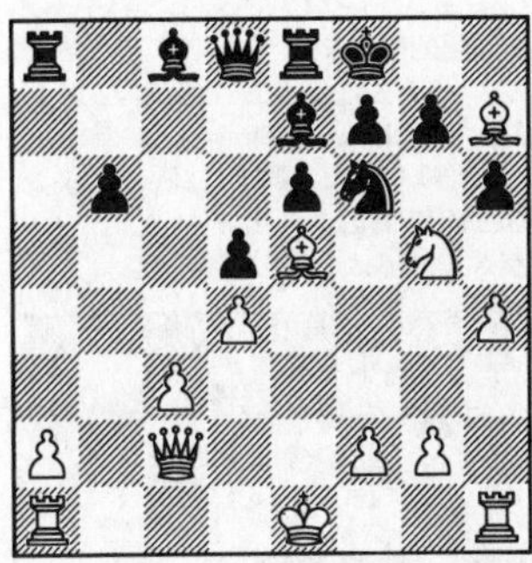

17. ♗g6! hg5[1] 18. hg5 ♘g8 19. ♗g7![2] +–

[1] 17. – fg6 18. ♕g6 ♔g8 19. ♘f7 ♕d7 20. ♘h6 ♔h8 21. ♘f7 ♔g8 22. h5! +–

[2] 19. – ♔g7 20. ♖h7 ♔f8 21. ♖f7#

4580. B 21

Konivicz – Levin

Belgrade 1924

1. e4 c5 2. f4 ♘c6 3. ♘f3 e6 4. c3 d5 5. e5 ♘h6 6. ♗b5 ♘f5 7. ♗c6 bc6 8. 0-0 h5 9. ♕a4 ♕b6 10. ♔h1 ♗a6 11. ♖e1 ♗d3 12. ♘a3 h4 13. ♘c2 ♗b5 14. ♕a3 c4 15. b4

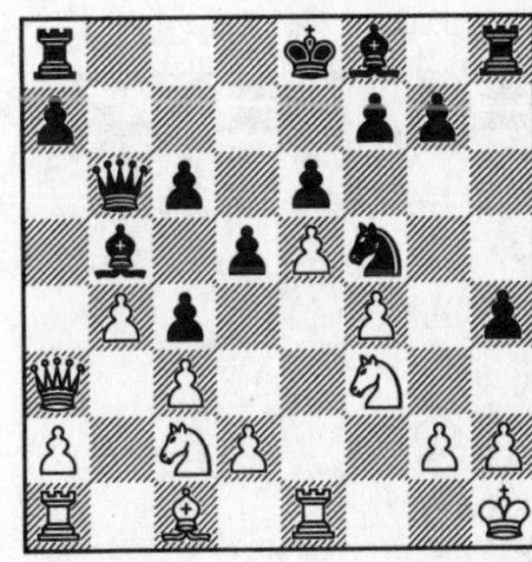

15. – ♘g3! 16. hg3 hg3 17. ♘h2 ♖h2#

4581. B 21

Vaisman – Vasile

Constanta 1958

1. e4 c5 2. d4 cd4 3. c3 dc3 4. ♘c3 d6 5. ♗c4 ♘f6 6. e5 ♘g8 7. ♘f3 e6 8. ed6 ♗d6 9. 0-0 ♘f6 10. ♗g5 ♗e7 11. ♕e2 a6 12. ♖fd1 ♕c7 13. ♗f6 ♗f6 14. ♘d5 ♕d8 15. ♖ac1 0-0 16. ♘f4 ♕b6 17. ♘h5 ♗e7 18. ♕e5 f6 19. ♕g3 ♖f7 20. ♘d4 ♗f8 21. ♘f4 ♔h8 22. ♘de6 ♘c6 23. ♘f8 ♖f8

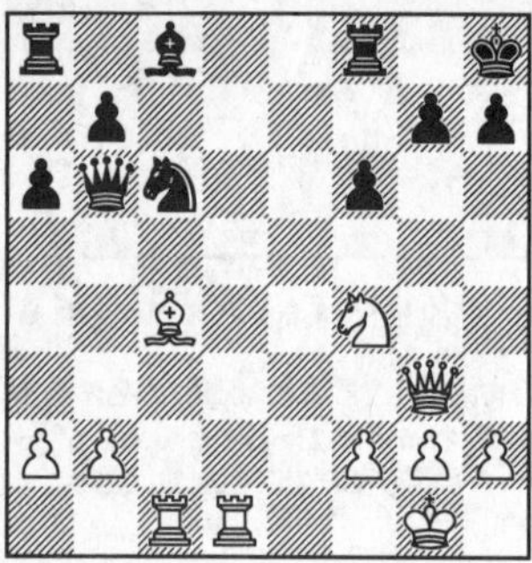

24. ♘g6! +–

4582. B 35

Parma – Konci

Bucharest 1960

1. e4 c5 2. ♘f3 ♘c6 3. d4 cd4 4. ♘d4 g6 5. ♘c3 ♗g7 6. ♗e3 ♘f6 7. ♗c4 0-0 8. ♗b3 d6 9. f3 a6 10. ♕d2 ♕c7 11. 0-0-0 ♗d7 12. h4 b5 13. h5 ♘h5 14. g4 ♘d4 15. ♗d4 ♗d4 16. ♘d5 ♕d8 17. gh5 ♗g7 18. hg6 hg6 19. ♕g5 ♖e8

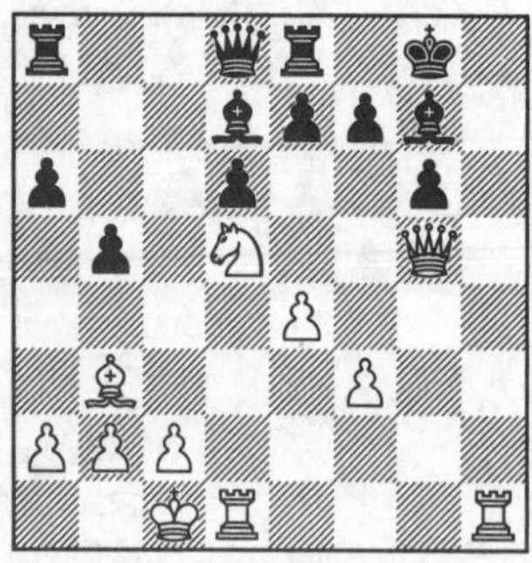

20. ♕g6! ♗f5[1] 21. ♘f6! +–

[1] 20. – fg6 21. ♘e7 ♔f8 22. ♘g6#

4583. B 40

Weenink – Schelfhout

1. e4 c5 2. ♘f3 e6 3. d4 cd4 4. ♘d4 ♘f6 5. ♘c3 ♗b4 6. ♗d3 ♘c6 7. ♘c6 bc6 8. e5 ♘d5 9. ♕g4 g6 10. 0-0 f5 11. ef6 ♗c3

12. ♗g6![1] +–

[1] 12. – hg6 13. ♕g6 ♔f8 14. ♕g7 ♔e8 15. f7 ♔e7 16. f8♕#; 12. – ♔f8 13. ♗h6 ♔g8 14. f7#

4584. B 44

Mardle – Gaprindashvili

Hastings 1965

1. e4 c5 2. ♘f3 ♘c6 3. d4 cd4 4. ♘d4 e6 5. ♗e3 ♘f6 6. ♘d2 e5 7. ♘c6 dc6 8. f3 ♗e7 9. ♗c4 0-0 10. 0-0 ♘h5 11. ♘b3 ♗g5 12. ♗c5 ♕f6 13. ♗f8 ♗e3 14. ♔h1

14. – ♘g3! –+

4585. B 93
Hearst – Franklin
Marshall Chess Club 1955

1. e4 c5 2. ♘f3 d6 3. d4 cd4 4. ♘d4 ♘f6 5. ♘c3 a6 6. f4 ♕c7 7. ♗d3 g6 8. 0-0 ♗g7 9. ♘f3 ♘c6 10. ♔h1 0-0 11. ♕e1 ♗e6 12. f5 ♗d7 13. ♕h4 ♘e5 14. ♘e5 de5 15. g4 ♗c6 16. g5 ♘d7 17. f6 ef6 18. gf6 ♗h8 19. ♖f3 b5 20. ♗g5 ♖fd8 21. ♖h3 ♘f8 22. ♖f1 ♖d6 23. ♘d5 ♗d5 24. ed5 ♕d8

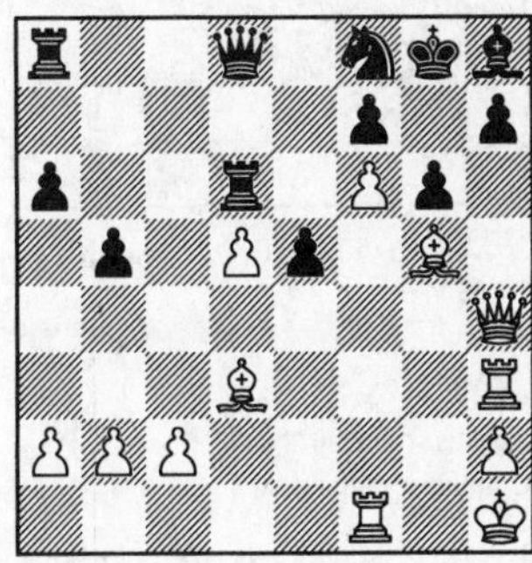

25. ♗g6![1] +–

[1] 25. – fg6 26. f7 ♔g7 27. ♗h6#; 25. – ♘g6 26. ♕h7 ♔f8 27. ♕h8! ♘h8 28. ♖h8#

4586. B 94
McDonald – Hodgson
London 1992

1. e4 c5 2. ♘f3 d6 3. d4 cd4 4. ♘d4 ♘f6 5. ♘c3 a6 6. ♗g5 ♘bd7 7. f4 g6 8. ♕e2 ♗g7 9. e5 de5 10. fe5 ♘h5 11. 0-0-0 ♗e5 12. ♘d5 ♘hf6 13. ♘f3 ♘d5 14. ♘e5 h6 15. ♗h4 ♘5b6

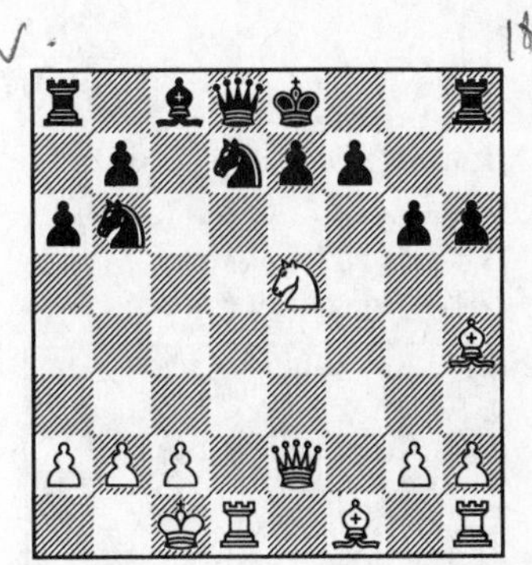

16. ♘g6! fg6 17. ♕e6 g5 18. ♕g6 +–

4587. C 00
Diemer – Illig
Corr. 1954

1. d4 e6 2. e4 d5 3. ♗e3 de4 4. f3 ef3 5. ♘f3 ♘f6 6. ♗d3 ♘bd7 7. 0-0 c5 8. c3 ♗e7 9. ♘e5 cd4 10 cd4 0-0 11. ♘c3 ♘b6 12. ♖f3 ♘bd5 13. ♖h3 g6 14. ♗h6 ♖e8 15. ♕d2 a6 16. ♖f1 ♘b4

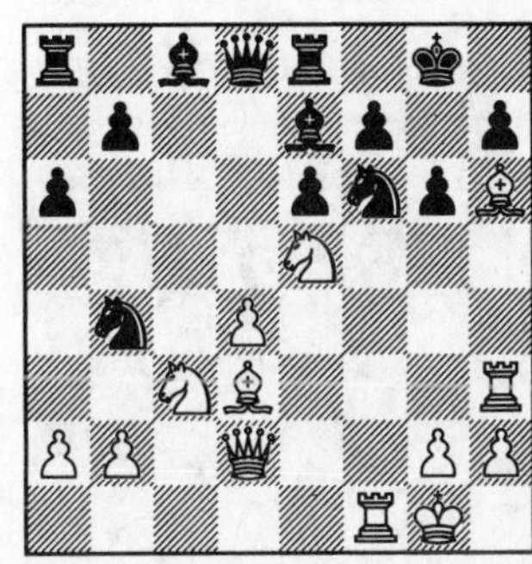

17. ♗g6! fg6[1] 18. ♗g7! ♘c6[2] 19. ♘c6 bc6 20. ♕h6 ♘h5 21. ♖h5[3] +–

[1] 17. – hg6 18. ♗g7 ♘h7 19. ♖h7! +–
[2] 18. – ♔g7 19. ♕h6 ♔g8 20. ♖f6 +–
[3] 21. – gh5 22. ♗e5 ♗f8 23. ♖f8! ♖f8 24. ♕g7#

4588. C 00

Kaiser – Strom

Karlstadt 1926

1. e4 b6 2. d4 ♗b7 3. ♘c3 e6 4. ♗d3 h6 5. ♘ge2 ♘f6 6. 0-0 c5 7. d5 ed5 8. ed5 ♘d5 9. ♘d5 ♗d5 10. ♘f4 ♗b7

11. ♘g6! fg6 12. ♗g6 ♔e7 13. ♖e1 ♔f6 14. ♕h5 +–

4589. C 00

Redlich – Treumann

1962

1. e4 e6 2. b3 d5 3. ♗b2 ♘f6 4. e5 ♘g8 5. f4 c5 6. ♘f3 ♘c6 7. ♗e2 ♘ge7 8. 0-0 h5 9. ♘c3 ♘f5 10. ♗d3 g6 11. a3 h4 12. ♔h1

12. – ♘g3! 13. hg3 hg3 14. ♔g1 c4! 15. ♘e2[1] ♗c5 16. ♘ed4 ♘d4 17. ♗d4 ♗d4 18. ♘d4 ♕h4 –+

[1] 15. ♗e2 ♗c5 16. d4 cd3 –+; 15. ♗c4 ♗c5 16. d4 ♗d4! 17. ♘d4 ♕h4 –+

4590. C 00

Webster – Sherman

Oshkosh 1963

1. e4 e6 2. ♘f3 d5 3. d3 de4 4. ♘g5 ed3 5. ♗d3 ♗b4 6. ♘c3 ♗c3 7. bc3 ♘f6 8. ♗a3 ♘c6 9. f4 ♘d5 10. 0-0 ♘e3 11. ♕h5 g6 12. ♕h6 ♘f1 13. ♕g7 ♔d7 14. ♕f7 ♘e7 15. ♕e6 ♔e8

16. ♗g6! hg6 17. ♕f7 ♔d7 18. ♖d1 ♔c6 19. ♕c4 +–

4591. C 01

Combs – Stampfli

Columbus Ohio 1980

1. e4 e6 2. d4 d5 3. ed5 ed5 4. ♗d3 ♘c6 5. c3 ♘f6 6. ♗g5 ♗e7 7. ♘f3 0-0 8. 0-0 ♘e4 9. ♗f4 ♗d6 10. ♘e5 ♕f6 11. ♘c6 ♗f4 12. ♘b4 c6 13. f3 ♗e3 14. ♔h1

14. – ♘g3! 15. hg3 ♕h6#

4592. C 01

Harrwitz – Champpe

England 1860

1. e4 e6 2. d4 d5 3. ed5 ed5 4. ♗d3 ♘f6 5. ♘f3 ♗d6 6. 0-0 0-0 7. c4 c6 8. ♘c3 ♗c7 9. ♗g5 ♗e6 10. cd5 ♗d5 11. ♘e5 ♕d6 12. ♗f6 ♕f6 13. f4 ♕d8 14. ♕h5 h6 15. ♕f5 g6

16. ♘g6! ♗e6[1] 17. ♘e7 ♔g7 18. ♕h7 ♔f6 19. ♘f5 ♗f5 20. ♕f5 ♔e7 21. ♖ae1 ♔d6 22. ♕e5 ♔d7 23. ♗f5#

[1] 16. – fg6 17. ♕g6 ♔h8 18. ♕h7#

4593. C 01

Sackmann – Ricard

New York 1950

1. e4 e6 2. d4 d5 3. ed5 ed5 4. ♘c3 ♘f6 5. ♘f3 ♗f5 6. ♗d3 ♗d3 7. ♕d3 ♗d6 8. 0-0 0-0 9. ♖e1 c6 10. ♘e5 ♕c7 11. ♕g3 ♘e4 12. ♘e4 de4 13. ♖e4 f6 14. ♕b3 ♔h8

15. ♘g6![1] +–

[1] 15. – hg6 16. ♖h3#

4594. C 02

Kashdan – Horneman

New York 1930

(without ♖a1)

1. e4 e6 2. d4 d5 3. e5 c5 4. ♕g4 cd4 5. ♘f3 ♘h6 6. ♕h3 ♗e7 7. ♗d3 b6 8. ♕g3 ♘f5 9. ♗f5 ef5 10. ♕g7 ♖f8 11. ♘d4 ♗a6 12. ♘f5 ♘d7 13. ♗g5 f6 14. e6 fg5

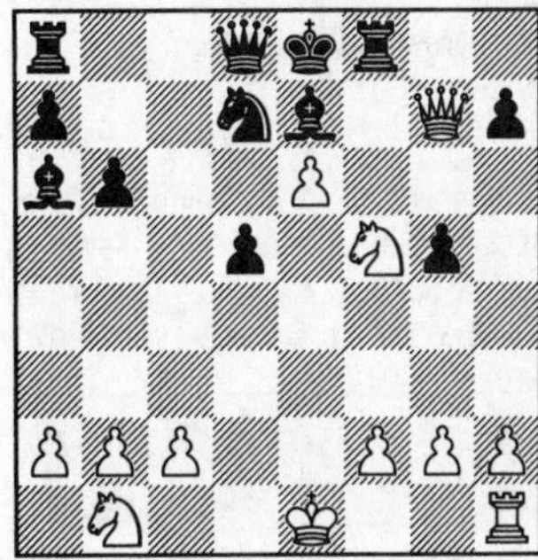

15. ♕g6! hg6 16. ♘g7#

4595. C 02

Koch – Nowarra

Bad Ceynhausen 1938

1. e4 e6 2. d4 d5 3. e5 c5 4. c3 ♘c6 5. ♘f3 ♕b6 6. ♗d3 ♗d7 7. dc5 ♗c5 8. 0-0 f6 9. b4 ♗e7 10. ♗f4 fe5 11. ♘e5 ♘e5 12. ♗e5 ♘f6 13. ♘d2 0-0 14. ♘f3 a5 15. ba5 ♖a5 16. ♖b1 ♕a7 17. ♘g5 g6 18. ♖b2 ♖a2 19. ♗d4 b6

20. ♗g6! ♖b2[1] 21. ♕h5! e5 22. ♗e5 ♖f2 23. ♕h7! ♘h7 24. ♗h7#

[1] 20. – hg6 21. ♕c2! +–

4596. C 02

Tarrasch – Kürschner
Nuremberg 1893

1. e4 e6 2. d4 d5 3. ♗d3 ♘f6 4. e5 ♘fd7 5. ♘f3 c5 6. c3 ♘c6 7. 0-0 f6 8. ♖e1 f5 9. ♗e3 c4 10. ♗c2 ♗e7 11. b3 b5 12. a4 ba4 13. bc4 dc4 14. d5 ♘ce5 15. de6 ♘f3 16. ♕f3 ♘b6 17. ♕f5 ♗f6 18. ♗c5 ♗b7

19. ♕g6! hg6 20. ♗g6#

4597. C 05

Helmuth – Seward
Cleveland 1982

1. e4 e6 2. d4 d5 3. ♘d2 ♘f6 4. e5 ♘fd7 5. ♗d3 c5 6. c3 ♘c6 7. ♘e2 ♗e7 8. 0-0 ♕b6 9. ♘f3 0-0 10. ♖e1 f5 11. ef6 ♗f6 12. ♘f4 ♘d8 13. ♕c2 g6

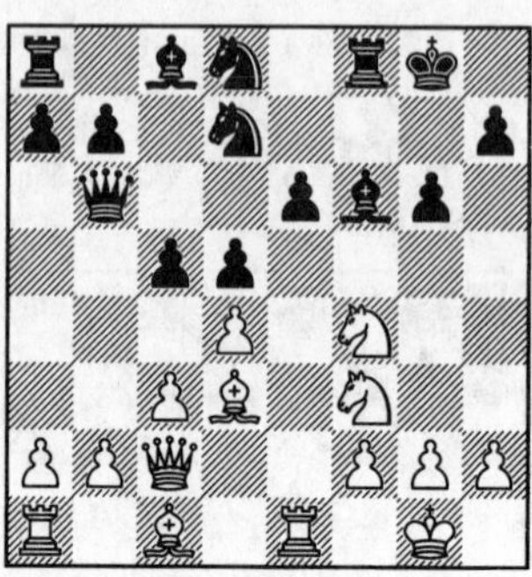

14. ♗g6! hg6 15. ♕g6 ♗g7[1] 16. ♘e6 ♘e6[2] 17. ♖e6 ♕c7 18. ♖e7 +–

[1] 15. – ♔h8 16. ♘g5! +–
[2] 16. – ♖f7 17. ♘fg5 +–

4598. C 05

Tal – Salnikov
1970

1. e4 e6 2. d4 d5 3. ♘d2 ♘f6 4. e5 ♘fd7 5. ♗d3 c5 6. c3 ♘c6 7. ♘gf3 cd4 8. cd4 f6 9. ♘g5 fg5 10. ♕h5 g6

11. ♗g6! hg6 12. ♕g6 ♔e7 13. ♘c4 ♗h6 14. ♗g5 ♗g5 15. ♕g7! ♔e8 16. ♘d6#

4599. C 07
Bolch – Pennel
N. Carolina 1974

1. e4 e6 2. d4 d5 3. ♘d2 c5 4. ed5 ♕d5 5. ♘gf3 cd4 6. ♘b3 e5 7. c3 ♕e4 8. ♗e2 dc3 9. bc3 ♘c6 10. 0-0 ♗e7 11. ♖e1 f6 12. ♘e5 ♕a4 13. ♗h5 g6

14. ♘g6! hg6 15. ♗g6 ♔f8 16. ♕d5 ♘d8 17. ♖e7 ♔e7?[1] 18. ♕c5 ♔d7 19. ♗a3 ♕a6 20. ♖d1 ♔e6 21. ♗f5 +–

[1] 17. – ♘e7! 18. ♕d8 ♔g7 19. ♕e7 ♔g6!

4600. C 10
Lewitzky – Marshall
Breslau 1912

1. d4 e6 2. e4 d5 3. ♘c3 c5 4. ♘f3 ♘c6 5. ed5 ed5 6. ♗e2 ♘f6 7. 0-0 ♗e7 8. ♗g5 0-0 9. dc5 ♗e6 10. ♘d4 ♗c5 11. ♘e6 fe6 12. ♗g4 ♕d6 13. ♗h3 ♖ae8 14. ♕d2 ♗b4 15. ♗f6 ♖f6 16. ♖ad1 ♕c5 17. ♕e2 ♗c3 18. bc3 ♕c3 19. ♖d5 ♘d4 20. ♕h5 ♖ef8 21. ♖e5 ♖h6 22. ♕g5 ♖h3 23. ♖c5

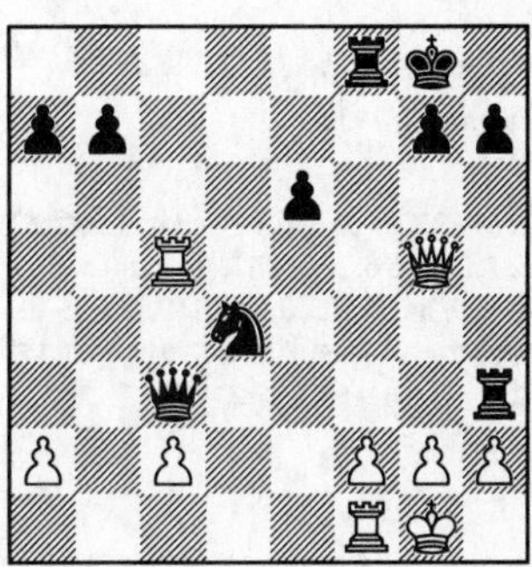

23. – ♕g3!![1] –+

[1] 24. hg3 ♘e2#; 24. fg3 ♘e2 25. ♔h1 ♖f1#; 24. ♕g3 ♘e2 25. ♔h1 ♘g3 26. ♔g1 ♘e2 –+

4601. C 12
Euwe – Bogoljubov
1921

1. e4 e6 2. d4 d5 3. ♘c3 ♘f6 4. ♗g5 ♗b4 5. e5 h6 6. ♗d2 ♗c3 7. bc3 ♘e4 8. ♕g4 g6 9. h4 c5 10. ♗d3 ♘d2 11. ♔d2 ♘c6 12. ♖h3 ♕a5

13. ♗g6![1] +–

[1] 13. – fg6 14. ♕g6 ♔f8 15. ♕f6 ♔g8 16. ♖g3 +–; 14. – ♔d7 15. ♕g7 +–

4602. C 13
Letelier – Niemi
Dubrovnik 1950

1. e4 e6 2. d4 d5 3. ♘c3 ♘f6 4. ♗g5 ♗e7 5. ♗f6 ♗f6 6. e5 ♗e7 7. ♕g4 0-0 8. ♘f3 ♘d7 9. h4 h6 10. ♖h3 g6 11. ♗d3 h5 12. ♕f4 ♔g7 13. g4 ♖h8 14. 0-0-0 c5 15. ♖g1 cd4 16. gh5 ♖h5

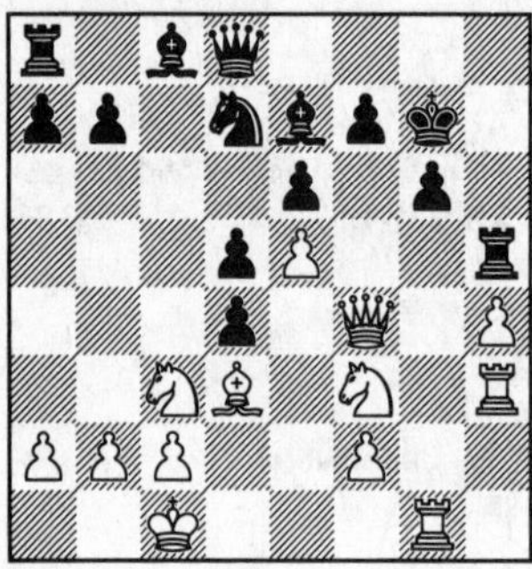

17. ♗g6! fg6 18. ♖g6![1] +–

[1] 18. – ♔g6 19. ♖g3 ♔h7 20. ♕f7 +–; 19. – ♗g5 20. ♘g5 +–

4603. C 15
Misteck – Wilkens
Kassel 1939

1. e4 e6 2. d4 d5 3. ♘c3 ♗b4 4. ♘ge2 de4 5. a3 ♗c3 6. ♘c3 f5 7. ♗f4 c6 8. ♕h5 g6 9. ♕h3 ♘d7 10. ♗c4 ♘b6 11. ♗b3 ♘f6 12. ♗e5 0-0 13. ♕h6 ♖f7 14. h3 ♗d7 15. 0-0-0 ♘bd5 16. ♘e2 b5 17. g4 fg4 18. hg4 ♘g4

19. ♕g6! +–

4604. C 21
From – Neumann
Paris 1867

1. e4 e5 2. d4 ed4 3. ♗c4 ♘c6 4. ♘f3 ♗c5 5. c3 ♘f6 6. 0-0 ♘e4 7. cd4 d5 8. ♖e1 ♗e7 9. ♗d3 f5 10. ♘e5 0-0 11. ♘c6 bc6 12. ♕e2 ♗d6 13. f3 ♕h4 14. g3

14. – ♘g3! 15. hg3 ♕g3 16. ♔h1[1] ♗d7 17. ♗e3 ♖f6 18. ♖d1 ♖g6 –+

[1] 16. ♔f1 f4!

4605. D 45
Ward – Flear
Mont St–Michel 1994

1. c4 c6 2. d4 d5 3. ♘c3 e6 4. e3 ♘f6 5. ♘f3 ♘bd7 6. ♕c2 ♗d6 7. g4 ♘g4 8. ♖g1 h5 9. h3 ♘h6 10. ♗d2 ♔f8 11. e4 e5 12.

0-0-0 ed4 13. ♘d4 ♘e5 14. cd5 cd5 15. ♘d5 ♗d7 16. f4 ♖c8 17. ♗c3 ♘g6

18. ♖g6! fg6 19. e5 ♗e7 20. ♕g6 ♕e8 21. ♕e4 ♖c5 22. ♘e7 ♕e7 23. ♕b7 ♘f5 24. ♘b3 +–

4606. C 22

L'Hermet – Hagemann

Magdeburg 1888

1. e4 e5 2. d4 ed4 3. ♕d4 ♘c6 4. ♕e3 ♗b4 5. c3 ♗a5 6. ♗c4 ♘ge7 7. ♕g3 0-0 8. h4 ♘g6 9. h5 ♘ge5 10. ♗g5 ♕e8 11. ♗f6 g6 12. hg6 ♘g6

13. ♕g6! hg6 14. ♖h8#

4607. C 24

Mandolpho – Kolisch

Paris 1859

1. e4 e5 2. ♗c4 ♘f6 3. ♘c3 c6 4. d3 b5 5. ♗b3 a5 6. a4 b4 7. ♘a2 d5 8. ed5 cd5 9. ♘f3 ♘c6 10. ♕e2 ♗g4 11. 0-0 ♗c5 12. ♗g5 h6 13. h3 h5 14. hg4 hg4 15. ♘e5 ♘d4 16. ♕e1

16. – ♘e4! 17. ♗d8[1] ♘g3! 18. ♘c6[2] ♘de2 –+

[1] 17. de4 ♕g5

[2] 18. fg3 ♘e2#

4608. C 25

Krejcik – N. N.

Wiedniu 1913

1. e4 e5 2. ♘c3 c6 3. d4 ed4 4. ♕d4 ♘f6 5. ♗g5 ♗e7 6. 0-0-0 0-0 7. e5 ♘e8 8. h4 ♗g5 9. hg5 ♕g5 10. f4 ♕g6 11. ♗d3 f5 12. ef6 ♕f6 13. ♗h7 ♔f7 14. ♕c4 d5 15. ♘d5 cd5 16. ♖d5 ♗e6 17. ♘f3 g6

18. ♗g6! ♔g6[1] 19. ♕e4 ♔f7[2] 20. ♘g5 ♔e7 21. ♕b4 +–

[1] 18. – ♕g6 19. ♘e5 +–

[2] 19. – ♗f5 20. ♖f5! +–

4609. C 27
Dunkan – Siegheim
London 1920

1. e4 e5 2. ♘c3 ♘f6 3. ♗c4 ♘e4 4. ♕h5 ♘d6 5. ♗b3 ♗e7 6. d3 0-0 7. ♘f3 ♘c6 8. ♗g5 ♘e8 9. 0-0-0 d6 10. ♗e7 ♕e7 11. ♘d5 ♕d8 12. ♘g5 h6 13. h4 ♘f6

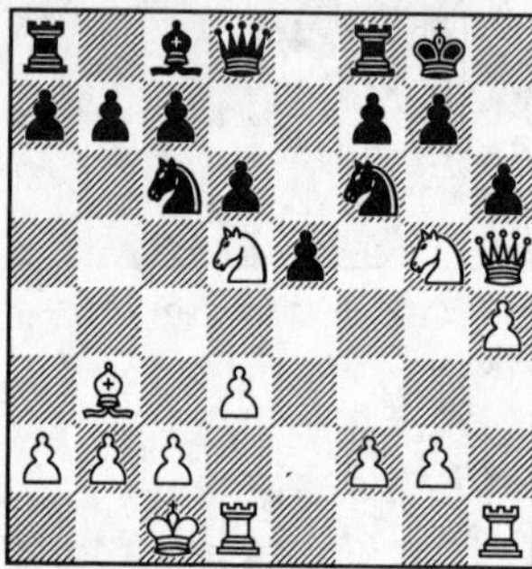

14. ♕g6! ♖e8[1] 15. ♘f7 +–

[1] 14. – fg6 15. ♘e7 ♔h8 16. ♘g6#; 14. – hg5 15. hg5 fg6 16. ♘e7#

4610. C 27
Mieses – N. N.
Liverpool 1900

1. e4 e5 2. ♘c3 ♘f6 3. ♗c4 ♘e4 4. ♕h5 ♘d6 5. ♗b3 ♗e7 6. d3 0-0 7. ♘f3 ♘c6 8. ♘g5 h6 9. h4 ♘e8 10. ♘d5 ♘f6

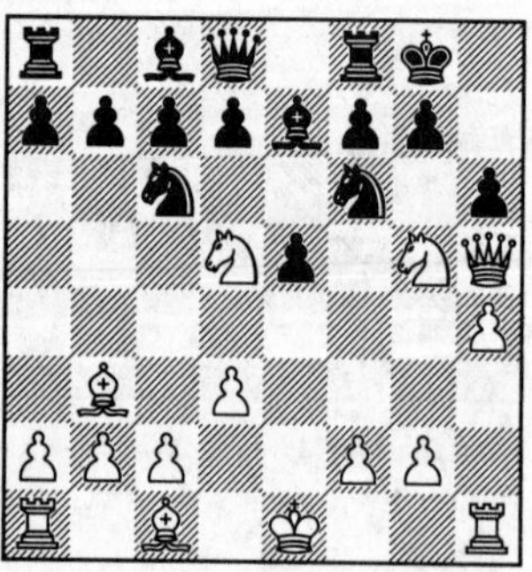

11. ♕g6! fg6[1] 12. ♘e7 ♔h8 13. ♘g6#

[1] 11. – hg5 12. hg5 +–

4611. C 30
Neumann – Dufresne
Berlin 1963

1. e4 e5 2. f4 ♗c5 3. ♘f3 d6 4. ♗c4 ♘f6 5. ♘c3 0-0 6. d3 ♘g4 7. ♖f1 ♘h2 8. ♖h1 ♘g4 9. ♕e2 ♗f2 10. ♔f1 ♘c6 11. f5 ♗c5 12. ♘g5 ♘h6 13. ♕h5 ♕e8 14. ♘h7! ♔h7 15. ♗h6 g6

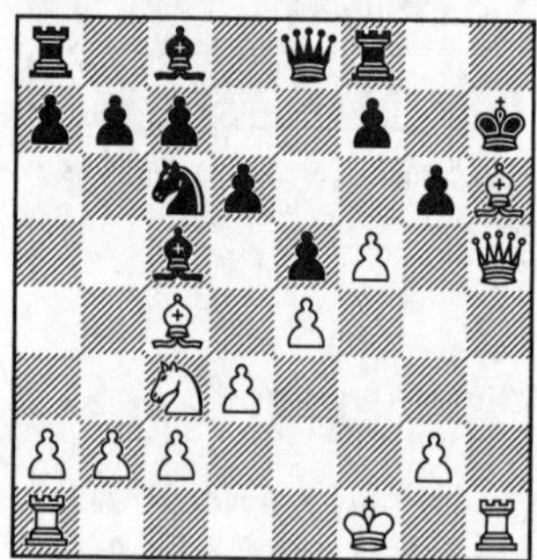

16. ♕g6! fg6 17. ♗f8#

4612. C 31
Murey – Nikitin
1971

1. e4 e5 2. f4 d5 3. ed5 e4 4. d3 ed3 5. ♗d3 ♕d5 6. ♘c3 ♕e6 7. ♘ge2 ♘h6 8. f5 ♘f5 9. 0-0 ♘e3 10. ♗e3 ♕e3 11. ♔h1 ♗d6 12. ♘f4 0-0 13. ♕h5 g6

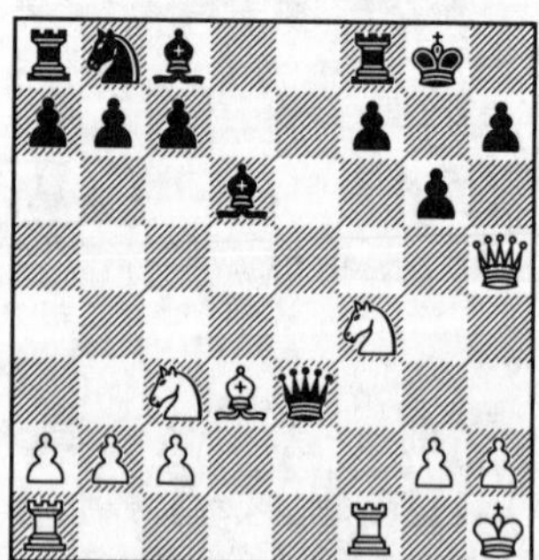

14. ♘g6! fg6 15. ♗g6! hg6 16. ♕g6 ♔h8 17. ♘d5 ♖f1 18. ♖f1 ♕e2 19. ♕h6 ♔g8 20. ♘f6 ♔f7 21. ♕h7[1] +–

[1] 21. – ♔e6 22. ♕g8 ♔e7 23. ♕e8#

4613. C 35

Multhopp – Bullock

Columbus Ohio 1982

1. e4 e5 2. f4 ef4 3. ♘f3 ♗e7 4. ♗c4 ♗h4 5. g3 fg3 6. 0-0 gh2 7. ♔h1 d5 8. ♗d5 ♘f6 9. ♗b3 0-0 10. ♘h4 ♘e4 11. ♕e1 ♘d6 12. ♘c3 ♖e8 13. ♕g3 ♗e6 14. d3 ♘c6 15. ♗h6 g6 16. ♘d5 ♗d5 17. ♗d5 ♖e5

18. ♘g6! hg6 19. ♕g6 +–

4614. C 39

Napier – N. N.

Hastings 1904

1. e4 e5 2. f4 ef4 3. ♘f3 g5 4. h4 g4 5. ♘e5 h6 6. ♘f7 ♔f7 7. d4 d5 8. ♗f4 ♗g7 9. ♘c3 de4 10. ♗c4 ♔g6 11. h5 ♔h7 12. ♘e4 ♕d4 13. ♗d3 ♗f5 14. ♕g4! ♕d7[1]

15. ♕g6! ♗g6 16. ♘g5! hg5 17. hg6#

[1] 14. – ♗g4 15. ♘f6#

4615. C 44

Tartakover – Schiffers

1. e4 e5 2. ♘f3 ♘c6 3. c3 d5 4. ♕a4 de4 5. ♘e5 ♕d5 6. ♗b5 ♘e7 7. f4 ♗d7 8. ♘d7 ♔d7 9. 0-0 ♘f5 10. b4 a5 11. ♔h1 ab4! 12. ♕a8 ♗c5! 13. ♕h8

13. – ♘g3! 14. hg3 ♕h5#

4616. C 45

Schwarzmüller – Kunsztowicz

Hamburg 1960

1. e4 e5 2. ♘f3 ♘c6 3. d4 ed4 4. ♘d4 ♘f6 5. ♘c6 bc6 6. e5 ♘d5 7. c4 ♘b4 8. a3 ♘a6 9. ♗d3 d6 10. ♗f4 ♗e7 11. 0-0 0-0 12. ♘c3 f6 13. ed6 cd6 14. ♖e1 ♖e8 15. ♕h5 g6

16. ♗g6! hg6 17. ♕g6 ♔h8 18. ♖e3 ♗f8[1] 19. ♖e8 +−

[1] 18. – ♘c5 19. ♖g3 +−

4617. C 45
Vogel – Stoppe
Bad Schandau 1932

1. e4 e5 2. ♘f3 ♘c6 3. d4 ed4 4. ♘d4 ♘f6 5. ♘c6 bc6 6. ♗d3 ♗e7 7. 0-0 0-0 8. b3 d5 9. ♘d2 ♖e8 10. ♗b2 ♗d6 11. f3 ♘h5 12. e5 ♗c5 13. ♔h1

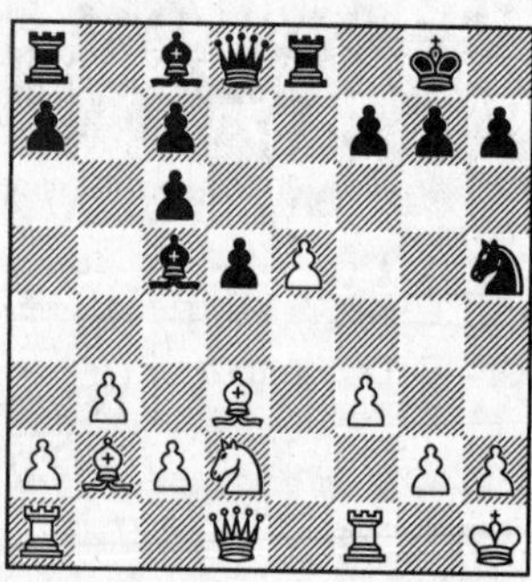

13. – ♘g3! 14. hg3 ♕g5 15. ♖f2 ♗f2 16. ♘f1 ♖e6[1] −+

[1] 17. ♗c1 ♖h6 18. ♘h2 ♖h2! 19. ♔h2 ♕h5#

4618. C 55
Lewis – Dayton
USA 1942

1. e4 e5 2. ♘f3 ♘c6 3. ♗c4 ♘f6 4. 0-0 ♘e4 5. ♖e1 d5 6. ♗b3 ♗c5 7. d4 ♘d4 8. ♘e5 ♕f6 9. ♗d5 ♕f2 10. ♔h1

10. – ♕g1! 11. ♔g1 ♘e2 12. ♔f1 ♘1g3! 13. hg3 ♘g3#

4619. C 55
Watkinson – N. N.
London 1863

1. e4 e5 2. ♘f3 ♘c6 3. ♗c4 h6 4. c3 ♘f6 5. d4 ed4 6. e5 ♘h7 7. 0-0 dc3 8. ♘c3 ♗e7 9. ♕d3 0-0 10. ♕g6 d5

11. ♘d5! fg6 12. ♘e7 ♔h8 13. ♘g6#

4620. C 57

MacMurray – Kussman

New York 1937

1. e4 e5 2. ♘f3 ♘c6 3. ♗c4 ♘f6 4. ♘g5 ♗c5 5. ♗f7 ♔f8 6. ♗b3 d5 7. 0-0 h6 8. ed5 hg5 9. dc6 e4 10. d3 ♕d6 11. g3 ♗g4 12. cb7 ♖b8 13. ♕d2

13. – ♕g3![1] –+

[1] 14. hg3 ♗f3 –+

4621. C 58

Balk – Barnes

New Zealand

1. e4 e5 2. ♘f3 ♘c6 3. ♗c4 ♘f6 4. ♘g5 d5 5. ed5 ♘a5 6. ♗b5 c6 7. dc6 bc6 8. ♗a4 h6 9. ♘f3 e4 10. ♘g1 ♗d6 11. d3 0-0 12. de4 ♘e4 13. ♗e3 ♗a6 14. ♘f3 ♕c7 15. ♘bd2 ♖fe8 16. c3 ♘c3 17. bc3

18. – ♖e3! 18. fe3 ♗g3! 19. hg3 ♕g3#

4622. C 61

Anderssen – Lange

Breslau 1859

1. e4 e5 2. ♘f3 ♘c6 3. ♗b5 ♘d4 4. ♘d4 ed4 5. ♗c4 ♘f6 6. e5 d5 7. ♗b3 ♗g4 8. f3 ♘e4 9. 0-0 d3 10. fg4 ♗c5 11. ♔h1

11. – ♘g3! 12. hg3 ♕g5 13. ♖f5 h5! 14. gh5[1] ♕f5 15. g4 ♖h5! 16. gh5 ♕e4 17. ♕f3 ♕h4 18. ♕h3 ♕e1 19. ♔h2 ♗g1 20. ♔h1 ♗f2 21. ♔h2 ♕g1#

[1] 14. ♖g5 hg4 –+

4623. C 63

Jonsson – Koskinen

Corr. 1981

1. e4 e5 2. ♘f3 ♘c6 3. ♗b5 f5 4. ♘c3 ♘d4 5. ♗a4 ♘f6 6. 0-0 ♗c5 7. ♘e5 0-0 8. ef5 d5 9. ♘e2 ♕d6 10. ♘d4 ♗d4 11. ♘f3 ♘g4 12. c3 ♖f5 13. cd4 ♖f3 14. g3 ♘h2 15. d3

15. – ♖g3! 16. fg3 ♕g3 17. ♔h1 ♕h3 18. ♔g1 ♗g4 19. ♕e1 ♗f3[1] −+

[1] 20. ♕f2 ♘g4! −+

4624. C 63

Larsen – Feeney

Corr. 1982

1. e4 e5 2. ♘f3 ♘c6 3. ♗b5 f5 4. ♘c3 ♘d4 5. ♘e5 ♕g5 6. 0-0 fe4 7. f4 ♕h4 8. d3 ♗c5 9. ♔h1 c6 10. ♗a4 ed3 11. ♕d3 d6 12. ♘e4 de5 13. ♘c5 ♘h6 14. c3 ♘df5 15. fe5 ♘g4 16. ♗f4 ♘f2 17. ♖f2 ♕f2 18. ♗g5 h6 19. ♖f1

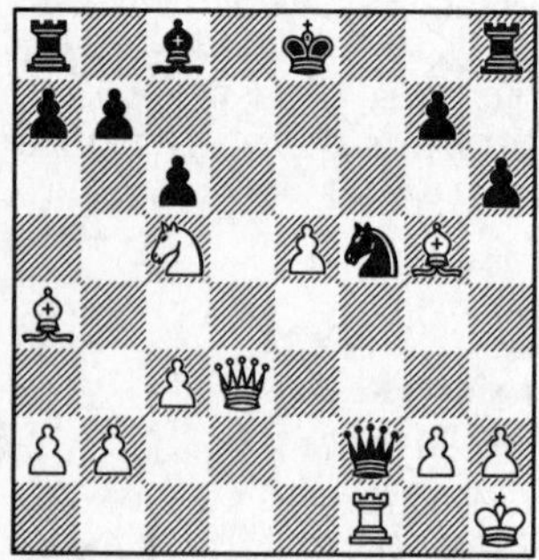

19. – ♘g3[1] −+

[1] 20. hg3 hg5# −+

4625. C 64

Bardeleben – Lebedev

St. Petersburg 1902

1. e4 e5 2. ♘f3 ♘c6 3. ♗b5 ♘f6 4. 0-0 ♗c5 5. ♘e5 ♘e5 6. d4 ♗e7 7. de5 ♘e4 8. ♕g4 ♘g5 9. f4 h5 10. ♕g3 h4 11. ♕g4 ♗c5 12. ♔h1 ♘e4 13. ♕g7 ♖h5 14. ♗e2

14. – ♘g3! 15. hg3 hg3 16. ♗h5 ♕h4#

4626. C 67

Fox – Bauer

Antwerpen 1901

1. e4 e5 2. ♘f3 ♘c6 3. ♗b5 ♘f6 4. 0-0 ♘e4 5. ♖e1 ♘d6 6. ♘e5 ♗e7 7. ♗f1 0-0 8. d4 ♘f5 9. c3 d5 10. ♕d3 ♖e8 11. f4 ♘d6 12. ♖e3 ♘a5 13. ♘d2 ♘f5 14. ♖h3 ♘h4 15. g4 ♘g6 16. ♖h5 ♘c6 17. ♘dc4 dc4

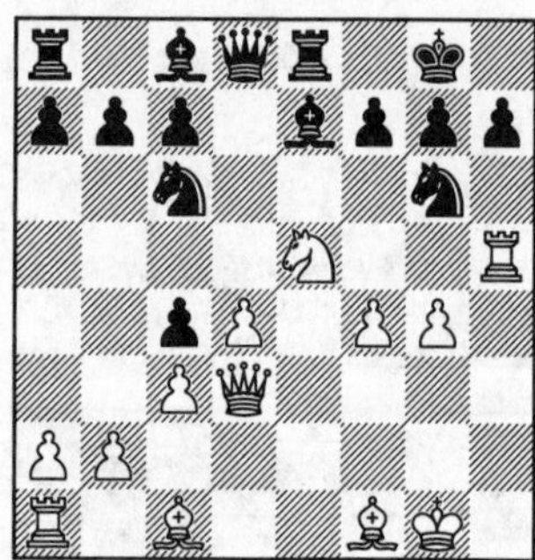

18. ♕g6! hg6 19. ♘g6! fg6 20. ♗c4 ♔f8 21. ♖h8#

4627. C 67

Leonhardt – Esser

Scheveningen 1905

1. e4 e5 2. ♘f3 ♘c6 3. ♗b5 ♘f6 4. 0-0 ♘e4 5. d4 ♘d6 6. ♗a4 e4 7. ♘e5 ♗e7 8. ♘c3 0-0 9. f4 f6 10. ♗b3 ♔h8

11. ♘g6! hg6 12. f5! ♘f5[1] 13. ♖f5 d5 14. ♖h5! ♔g8 15. ♖d5 ♕e8 16. ♖d8 +–

[1] 12. – ♘f7 13. ♗f7 ♖f7 14. fg6 ♖f8 15. ♕h5 ♔g8 16. ♕h7#

4628. C 78

Dutreeuv – Malanyuk

Citta di Forli 1991

1. e4 e5 2. ♘f3 ♘c6 3. ♗b5 a6 4. ♗a4 ♘f6 5. 0-0 b5 6. ♗b3 ♗b7 7. ♖e1 ♗c5 8. c3 d6 9. d4 ♗b6 10. ♗g5 h6 11. ♗h4 ♕d7 12. a4 0-0-0 13. ab5 ab5 14. ♗f6 gf6 15. ♗d5 ♖hg8 16. ♗c6 ♗c6 17. d5 ♗b7 18. c4 ♕h3 19. g3

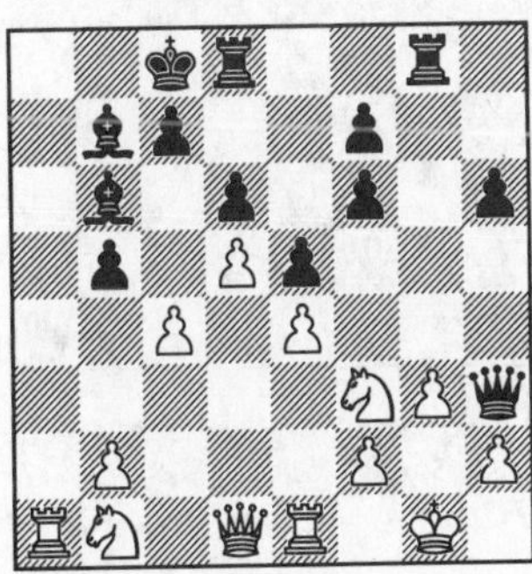

19. – ♖g3! 20. hg3 ♖g8! –+

4629. C 80

Honan–Zaichev – Mardle–Rochlin

Yaroslavl 1954

1. e4 e5 2. ♘f3 ♘c6 3. ♗b5 a6 4. ♗a4 ♘f6 5. 0-0 ♘e4 6. d4 b5 7. ♗b3 d5 8. ♘e5 ♘e5 9. de5 ♗b7 10. ♗e3 ♗c5 11. ♕g4 ♗e3 12. ♕g7 ♕g5 13. ♕h8 ♔e7 14. ♕h7 ♗f2 15. ♔h1 ♖g8 16. ♕h3 d4 17. ♘a3

17. – ♕g2! 18. ♕g2 ♘g3! 19. hg3 ♖h8#

4630. C 83

Fine – Helms

New York 1945

1. e4 e5 2. ♘f3 ♘c6 3. ♗b5 a6 4. ♗a4 ♘f6 5. 0-0 ♘e4 6. d4 b5 7. ♗b3 d5 8. de5 ♗e6 9. c3 ♗e7 10. ♘bd2 ♘d2 11. ♕d2 ♘a5 12. ♗c2 ♘c4 13. ♕f4 c5 14. a4 0-0 15. ♕g3 ♖e8 16. b3 ♘b6 17. ♗h6 g6 18. ♘g5 ♘d7 19. ♘e6 fe6

20. ♗g6! ♘f8[1] 21. ♗f7! ♔f7 22. ♕g7#

[1] 20. – hg6 21. ♕g6 ♔h8 22. ♕g7#

4631. C 89

Beckham – Siegal

Columbus Ohio 1976

1. e4 e5 2. ♘f3 ♘c6 3. ♗b5 a6 4. ♗a4 ♘f6 5. 0-0 b5 6. ♗b3 ♗e7 7. ♖e1 0-0 8. c3 d5 9. ed5 ♘d5 10. ♘e5 ♘e5 11. ♖e5 ♘f6 12. d4 ♗d6 13. ♖e2 ♘h5 14. ♖e1 ♕h4 15. g3

15. – ♘g3! 16. hg3 ♗g3 17. ♗f7[1] ♔h8 18. ♕h5 ♗f2 19. ♔g2 ♕g3 20. ♔h1 ♗g4 –+

[1] 17. fg3 ♕g3 18. ♔h1 ♗g4 –+

4632. C 89

Foltys – Thelen

Prague 1943

1. e4 e5 2. ♘f3 ♘c6 3. ♗b5 a6 4. ♗a4 ♘f6 5. 0-0 ♗e7 6. ♖e1 b5 7. ♗b3 0-0 8. c3 d5 9. ed5 ♘d5 10. ♘e5 ♘e5 11. ♖e5 c6 12. ♗d5 cd5 13. d4 ♗d6 14. ♖e1 ♕h4 15. g3 ♕h3 16. ♕f3 ♗f5 17. ♕d5 ♖ae8 18. ♖e3 ♕h5 19. f3 ♕g6 20. ♖e8 ♖e8 21. ♘d2

21. – ♗g3! 22. ♘e4[1] ♗h2[2] –+

[1] 22. hg3 ♕g3 23. ♔h1 Be1 –+

[2] 23. ♔h1 ♗h3 –+

4633. D 00

Fox – Hodges

New York 1937

(without ♘b1)

1. e4 d5 2. d4 ♘f6 3. e5 ♘fd7 4. e6 fe6 5. ♗d3 ♘f6 6. ♘f3 ♕d6 7. ♘e5 ♘bd7 8. ♗f4 ♕b4 9. c3 ♕b2 10. ♕c2 ♕a1 11. ♔e2 ♕h1

12. ♗g6! hg6[1] 13. ♕g6 ♔d8 14. ♘f7 ♔e8 15. ♘d6 ♔d8 16. ♕e8! ♘e8 17. ♘f7#

[1] 12. – ♔d8 13. ♘f7 ♔e8 14. ♘g5 ♔d8 15. ♘e6#

4634. D 00
Künitz – Salomon
Tanger 1907

1. d4 d5 2. e4 de4 3. f3 ef3 4. ♘f3 ♗g4 5. ♗e3 ♘c6 6. c3 e5 7. d5 ♘ce7 8. ♕a4 ♗d7 9. ♕e4 f6 10. ♗d3 ♘g6

11. ♕g6! hg6 12. ♗g6 ♔e7 13. ♗c5#

4635. D 00
Lovas – Asztalos
Budapest 1915

1. d4 d5 2. ♗f4 c5 3. e3 e6 4. c3 ♘c6 5. ♘d2 ♕b6 6. ♕b1 ♗d7 7. ♗d3 ♘f6 8. ♘gf3 ♗e7 9. 0-0 0-0 10. ♘e5 ♖fd8 11. ♘df3 ♗e8 12. ♘g5 h6 13. ♗h7 ♔f8

14. ♘g6! fg6 15. ♘e6 ♔f7 16. ♕g6! ♔e6 17. ♗g8 ♗f7 18. ♗f7 ♔d7 19. ♕f5#

4636. D 03
Saljowa – Dworakowska
Svitavy 1993

1. d4 ♘f6 2. ♘f3 g6 3. ♗g5 ♗g7 4. ♘bd2 d5 5. e3 b6 6. ♘e5 0-0 7. ♗d3 ♗b7 8. 0-0 c5 9. c3 ♘bd7 10. f4 ♖c8 11. ♕f3 ♖e8 12. ♕h3 a6 13. ♖ad1 b5 14. f5 ♘e5 15. de5 ♘d7 16. fg6 fg6

17. ♗g6! hg6 18. ♕e6 ♔h8 19. ♖f3 ♘f8 20. ♖h3 ♗h6 21. ♕f7 +–

4637. D 05
Bogdan – Kantor
Ploesti 1975

1. d4 d5 2. e3 ♘f6 3. ♗d3 e6 4. ♘f3 ♗e7 5. 0-0 0-0 6. ♘bd2 c5 7. c3 ♘c6 8. ♘e5 ♕c7 9. f4 ♖b8 10. ♕e2 a6 11. g4 b5 12. g5 ♘e8 13. ♗h7! ♔h7 14. ♕h5 ♔g8 15. ♖f3 g6 16. ♕h6 ♘g7 17. ♖h3 ♘h5

18. ♘g6! fg6 19. ♕g6 ♘g7 20. ♖h8! ♔h8 21. ♕h6 ♔g8 22. g6 +–

4638. D 05
Pillsbury – N. N.
Toronto 1899

1. d4 d5 2. ♘f3 e6 3. e3 ♘f6 4. ♗d3 ♘bd7 5. 0-0 b6 6. ♘bd2 ♗d6 7. e4 de4 8. ♘e4 ♗b7 9. ♘d6 cd6 10. ♗f4 ♗f3 11. ♕f3 d5 12. ♗d6 ♖c8 13. ♖fe1 ♖c6 14. ♗a3 a5 15. c4 ♘e4 16. cd5 ♘g5 17. ♕g3 ♖c8 18. de6 ♘e6

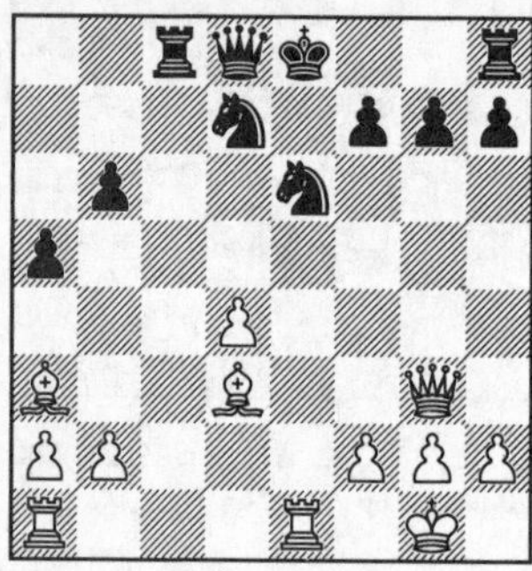

19. ♖e6! fe6 20. ♕g6! hg6 21. ♗g6#

4639. D 17
Nadel – Margulies
Berlin 1932

1. d4 d5 2. c4 c6 3. ♘f3 ♘f6 4. ♘c3 dc4 5. a4 ♗f5 6. ♘e5 c5 7. e4 ♘e4 8. ♕f3 cd4 9. ♕f5 ♘d6 10. ♗c4 e6 11. ♗b5 ♔e7

12. ♘g6! hg6[1] 13. ♘d5! ed5 14. ♕e5#

[1] 12. – fg6 13. ♗g5#

4640. D 24
Soultanbeieff – Defosse
Corr. 1941

1. d4 d5 2. c4 dc4 3. ♘f3 ♘f6 4. ♘c3 e6 5. ♕a4 ♘bd7 6. ♗g5 c5 7. e4 cd4 8. ♘d4 ♗e7 9. ♖d1 ♘e4 10. ♘e4 ♗g5 11. ♗c4 ♕e7 12. ♘g5 ♕g5 13. 0-0 0-0 14. ♘e6 ♘b6 15. ♘g5 ♘a4 16. ♖d4 ♘b2 17. ♗f7 ♔h8 18. ♗b3 a5 19. ♘f7 ♔g8 20. ♘e5 ♔h8

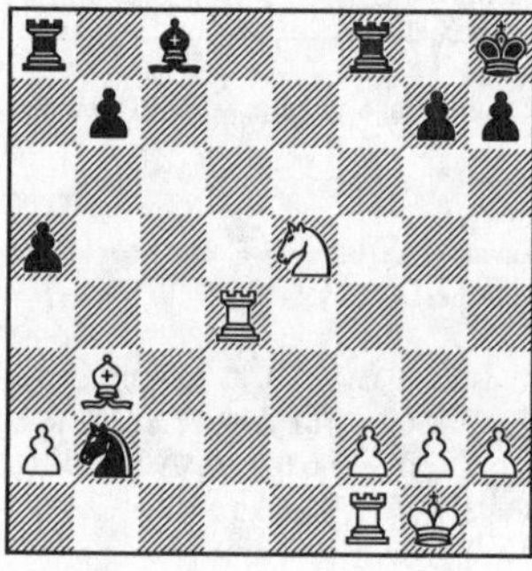

21. ♘g6! hg6 22. ♖h4#

4641. D 25
Sazonov – Krupsky
Minsk 1971

1. d4 d5 2. ♘f3 ♘f6 3. c4 dc4 4. e3 a6 5. ♗c4 e6 6. 0-0 c5 7. ♕e2 cd4 8. ed4 ♗e7 9. ♘c3 b5 10. ♗b3 ♗b7 11. ♗g5 0-0 12. ♖fe1 ♘bd7 13. ♖ad1 b4 14. d5 bc3 15. de6 fe6 16. ♕e6 ♔h8 17. ♘e5 ♕c8 18. ♖d7 ♗c5

19. ♘g6! hg6 20. ♗f6 ♕d7 21. ♕d7 gf6 22. bc3 ♖ab8 23. ♕c7 +–

4642. D 30

Aljechin – Apscheneek

Folkestone 1933

1. d4 ♘f6 2. c4 e6 3. ♘f3 d5 4. e3 c6 5. ♗d3 a6 6. ♘bd2 dc4 7. ♘c4 b5 8. ♘ce5 h6 9. a4 ♗b7 10. ♗d2 ♘bd7 11. ♕c2 ♖c8

12. ♗g6! ♘e5[1] 13. ♘e5 ♖c7 14. ♗a5 fg6 15. ♕g6 ♔e7 16. ♘f7 ♕e8 17. ♘h8 ♖c8 18. ♗b4 c5 19. ♗c5 ♖c5 20. dc5 ♗e4 21. ♕e8 ♔e8 22. f3 ♗d3 23. ab5 +–

[1] 12. – fg6 13. ♕g6 ♔e7 14. ♗b4 +–

4643. D 36

Byrne – Fieldsted

Reykjavik 1968

1. d4 d5 2. c4 e6 3. ♘c3 ♘f6 4. cd5 ed5 5. ♗g5 ♘bd7 6. e3 c6 7. ♗d3 ♗d6 8. ♘ge2 b5 9. ♕c2 ♗b7 10. ♘g3 ♗g3 11. hg3 h6 12. ♗f4 a6 13. ♗f5 ♕a5 14. ♗d6 g6 15. b4 ♕b6

16. ♗g6! fg6 17. ♕g6 ♔d8 18. ♖h6 ♖e8 19. ♕f7 ♘g4 20. ♖e6 ♖e6 21. ♕e6 +–

4644. D 36

Jones – Shapiro

New York 1947

1. d4 ♘f6 2. c4 e6 3. ♘c3 d5 4. ♗g5 ♗e7 5. e3 ♘bd7 6. ♘f3 0-0 7. cd5 ed5 8. ♖c1 c6 9. ♕c2 ♖e8 10. ♗d3 ♘f8 11. h3 ♘g6 12. 0-0 ♘h5 13. ♗e7 ♕e7 14. ♖b1 ♕f6 15. ♘d2 ♘h4 16. ♘e2 ♗h3 17. ♘g3 ♕g5 18. ♔h2 ♖e3 19. gh3 ♖g3 20. ♖g1 ♖g2 21. ♔h1

21. – ♘g3! 22. fg3 ♕g3 –+

4645. D 37

Johnston – Marshall

Chicago 1899

1. d4 d5 2. c4 e6 3. Nc3 Nc6 4. Nf3 Nf6 5. Bf4 Bd6 6. Bg3 Ne4 7. e3 0-0 8. Bd3 f5 9. a3 b6 10. Rc1 Bb7 11. cd5 ed5 12. Nd5 Nd4 13. Bc4 Nf3 14. gf3 Ng3 15. Ne7 Kh8

16. Ng6! hg6 17. hg3 +–

4646. D 37

Papantoniou – Koutilin

Athens 1935

1. d4 d5 2. Nf3 Nf6 3. c4 e6 4. Nc3 Nbd7 5. e3 Be7 6. a3 0-0 7. Qc2 a6 8. Bd3 Re8 9. 0-0 c6 10. c5 Nf8 11. Ne5 N6d7 12. f4 f6 13. Nf3 Qc7 14. f5 e5 15. Nh4 e4 16. Ne4 de4 17. Bc4 Kh8 18. Qe4 b6 19. Rf3 Bb7

20. Qe6! Ne6 21. Ng6! hg6 22. Rh3 Kg8 23. Be6 Kf8 24. Rh8#

4647. D 42

Kasparov – Begun

USSR 1978

1. d4 d5 2. c4 e6 3. Nc3 Nf6 4. Nf3 c5 5. cd5 Nd5 6. e3 Nc6 7. Bd3 Be7 8. 0-0 0-0 9. Nd5 Qd5 10. e4 Qd8 11. dc5 Bc5 12. e5 Be7 13. Qe2 Nb4 14. Bb1 Bd7 15. a3 Nd5 16. Qe4 g6 17. Bh6 Re8 18. h4 Qb6 19. h5 f5 20. ef6 Nf6 21. Qe1 Nh5 22. Ne5 Bb5

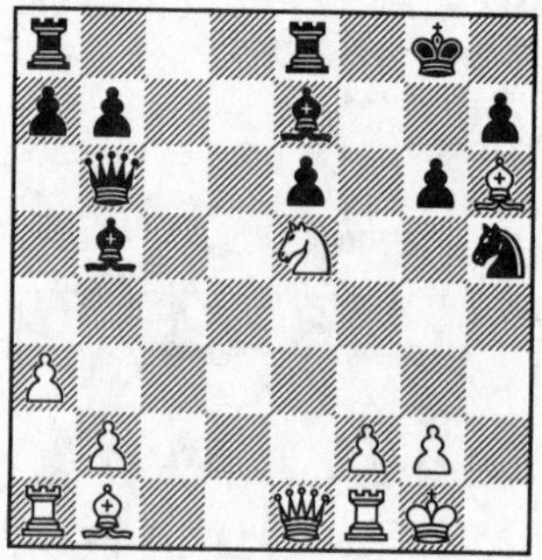

23. Bg6! Nf6[1] 24. Bh7[2] +–

[1] 23. – hg6 24. Qe4 +–

[2] 24. – Kh7 25. Qb1 Kh8 26. Qg6 Rg8 27. Nf7#; 24. – Nh7 25. Qe4 +–

4648. D 46

Sanguineti – Donoso

Fortaleza 1975

1. Nf3 Nf6 2. c4 c6 3. Nc3 d5 4. e3 e6 5. b3 Nbd7 6. Bb2 Be7 7. d4 0-0 8. Bd3 b6 9. 0-0 Bb7 10. Qe2 Rc8 11. Rad1 Qc7 12. Ne5 dc4 13. bc4 Ne5 14. de5 Ne8 15. Qh5 g6 16. Qg4 Qe5 17. Nd5 Qd6

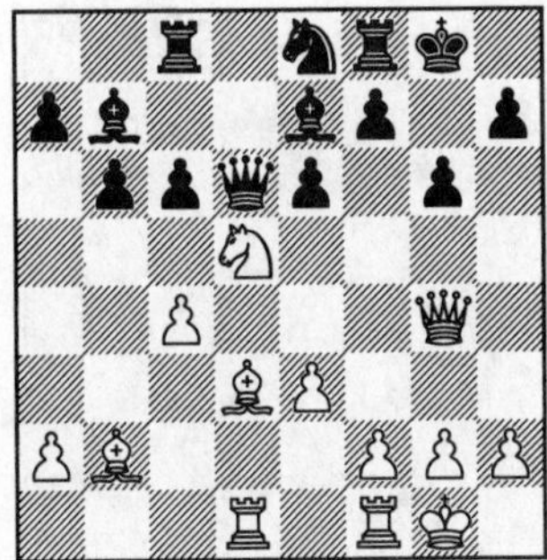

18. ♗g6! hg6 19. ♘f4 ♕b4 20. ♘g6! f5[1] 21. ♘e7 ♔f7 22. ♕g6 ♔e7 23. ♕g5[2] +−

[1] 20. – ♕b2 21. ♘e7 ♔h7 22. ♕h5 ♔g7 23. ♕g5 ♔h8 24. ♖d4 +−

[2] 23. – ♔f7 24. ♖d7; 23. – ♘f6 24. ♕g7 +−

4649. D 52

Sal Matera – Spital

New York 1971

1. d4 d5 2. c4 e6 3. ♘c3 ♘f6 4. ♗g5 ♘bd7 5. e3 c6 6. ♘f3 ♗d6 7. ♗d3 0-0 8. 0-0 b6 9. e4 de4 10. ♘e4 ♗e7 11. ♕e2 ♗b7 12. ♖ad1 ♖e8 13. ♗f6 ♘f6 14. ♘eg5 h6

15. ♗g6! hg5[1] 16. ♗f7 ♔f7 17. ♘g5 ♔g6 18. f4 c5 19. ♕e6 ♗f8[2] 20. ♕f7 ♔f5 21. g4 ♔g4 22. ♕g6 +−

[1] 15. – fg6 16. ♕e6 ♔h8 17. ♘f7 +−

[2] 19. – ♖f8 20. ♖d3 +−

4650. D 55

McInerney – Blas

Cincinnati 1974

1. d4 d5 2. c4 e6 3. ♘c3 ♘f6 4. ♗g5 ♗e7 5. ♘f3 0-0 6. e3 ♘c6 7. ♗d3 ♖e8 8. ♗f6 ♗f6 9. ♕c2 g6 10. h4 h5 11. 0-0-0 ♕d6 12. ♖dg1 dc4

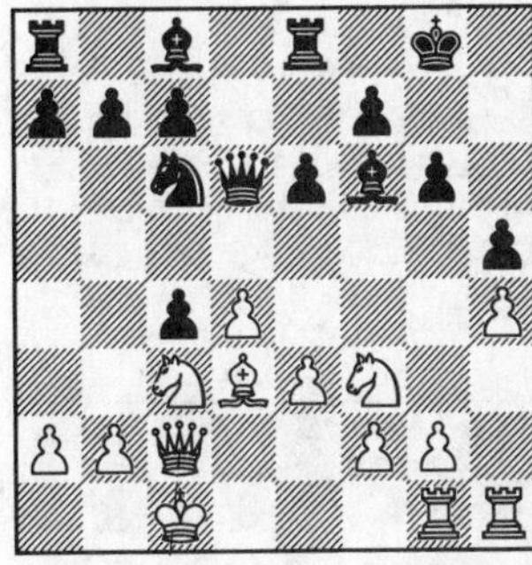

13. ♗g6! fg6 14. ♕g6 ♗g7 15. ♕e8 ♗f8 16. ♘g5 ♘b4 17. ♕f7 +−

4651. D 55

Schlechter – Przepiórka

Nuremberg 1906

1. d4 d5 2. c4 e6 3. ♘c3 ♘f6 4. ♘f3 ♘bd7 5. ♗g5 ♗e7 6. e3 b6 7. cd5 ♘d5 8. ♘d5 ed5 9. ♗f4 0-0 10. ♗d3 c5 11. 0-0 ♗b7 12. ♖c1 ♖e8 13. ♘e5 ♘e5 14. ♗e5 cd4 15. ♖c7 ♗c8 16. ♕h5 g6

17. ♗g6[1] +–

[1] 17. – fg6 18. ♕h6 +–

4652. D 61
Hernandez – Schumacher
Luzern 1982

1. d4 e6 2. ♘f3 d5 3. c4 ♘f6 4. ♘c3 ♗e7 5. ♗g5 ♘bd7 6. e3 0-0 7. ♕c2 c5 8. 0-0-0 dc4 9. ♗c4 ♕a5 10. ♔b1 cd4 11. ♖d4 e5 12. ♖h4 g6 13. ♗h6 ♘e8 14. ♖g4 ♘g7

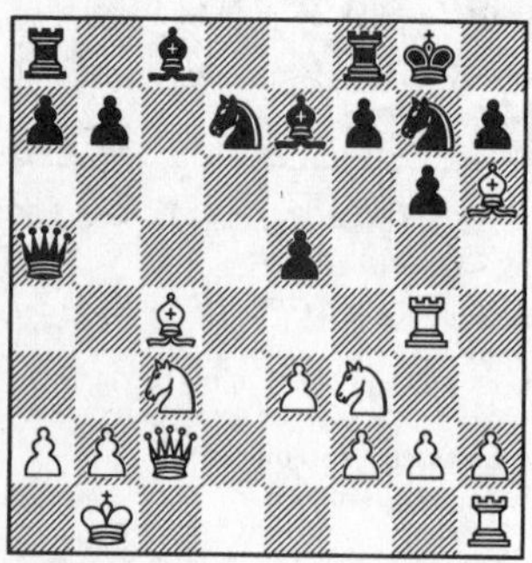

15. ♖g6! hg6 16. ♕g6 ♗f6 17. ♘g5[1] +–

[1] 17. – ♖d8 18. ♕f7 ♔h8 19. ♗g7 ♗g7 20. ♕h5 +–

4653. D 64
Barlage – Backman
Columbus Ohio 1965

1. d4 d5 2. c4 e6 3. ♘c3 ♘f6 4. ♘f3 ♘bd7 5. ♗g5 c6 6. e3 ♗e7 7. ♕c2 0-0 8. ♗d3 dc4 9. ♗c4 ♘d5 10. ♗e7 ♕e7 11. e4 ♘f4 12. ♕d2 ♘g2 13. ♔e2 ♘h4 14. ♖ag1 ♘f3 15. ♔f3 e5 16. ♕h6 g6

17. ♖g6! hg6 18. ♕g6 ♔h8 19. ♕h6 +–

4654. D 64
Engels – Büstgens
Düsseldorf 1935

1. d4 ♘f6 2. c4 e6 3. ♘c3 d5 4. ♗g5 ♘bd7 5. e3 ♗e7 6. ♘f3 c6 7. a3 0-0 8. ♖c1 ♖e8 9. ♕c2 a6 10. h3 dc4 11. ♗c4 ♘d5 12. ♗e7 ♕e7 13. ♘e4 ♘5f6 14. ♘g3 h6 15. 0-0 e5 16. ♘f5 ♕f8 17. ♘3h4 ♔h8

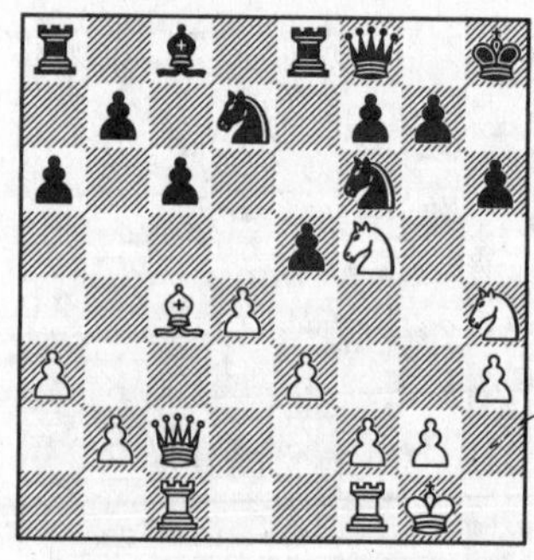

18. ♘g6! fg6 19. ♘h4 ♕d6 20. ♘g6 ♔h7 21. de5 ♕c7 22. ♘f8 ♔h8 23. ♕h7[1] +–

[1] 23. – ♘h7 24. ♘g6#

4655. D 87
Carrou – Knasoff
Corr. 1948

1. d4 ♘f6 2. c4 g6 3. ♘c3 d5 4. cd5 ♘d5 5. e4 ♘c3 6. bc3 c5 7. ♗c4 ♗g7 8. ♘e2 0-0 9. ♗e3 ♘d7 10. 0-0 ♕c7 11. ♗b3 b6 12. ♕d2 ♗a6 13. ♖fe1 ♖ad8 14. ♘g3 ♘b8 15. ♖ac1 ♘c6 16. ♗h6 ♗h6 17. ♕h6 cd4 18. cd4 ♖d4 19. ♖c6 ♕d7[1]

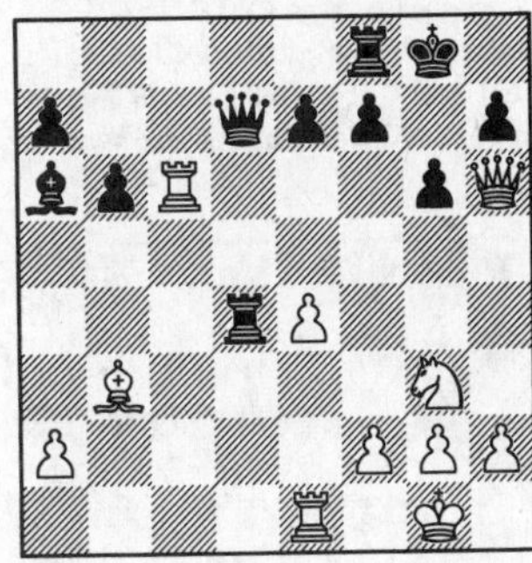

20. ♖g6! [2] +–

[1] 19. – ♕c6 20. ♘h5 +–
[2] 20. – hg6 21. ♕g6 ♔h8 22. ♕h6 ♔g8 23. ♘f5 +–

4656. D 88
Pedersen – Poulsen
Copenhagen 1949

1. d4 ♘f6 2. c4 g6 3. ♘c3 d5 4. cd5 ♘d5 5. e4 ♘c3 6. bc3 c5 7. ♗c4 ♗g7 8. ♘e2 cd4 9. cd4 ♘c6 10. ♗e3 0-0 11. 0-0 ♘a5 12. ♗d3 ♗e6 13. d5 ♗a1 14. ♕a1 f6 15. ♗h6 ♖e8 16. ♘f4 ♗d7 17. e5 ♖c8 18. e6 ♗a4 19. ♕b1 ♔h8

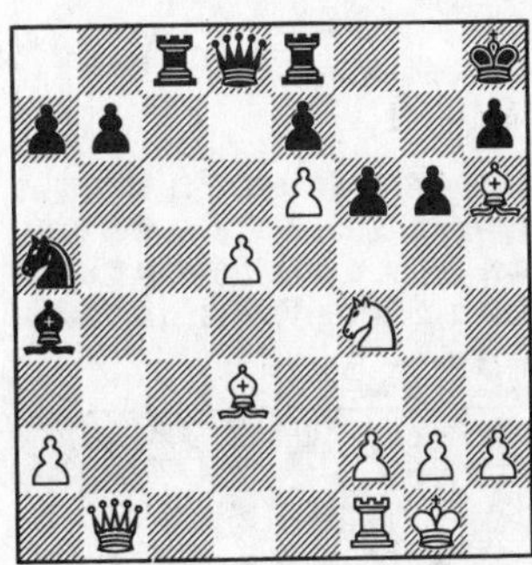

20. ♗g6! hg6 21. ♗g7![1] +–

[1] 21. – ♔g7 22. ♕g6 ♔h8 23. ♕h5 ♔g7 24. ♕f7 ♔h8 15. ♘g6#

4657. D 93
Bohosiewicz – Rödl
Munich 1936

1. d4 ♘f6 2. c4 g6 3. ♘f3 ♗g7 4. ♘c3 d5 5. ♗f4 0-0 6. e3 c5 7. ♕b3 cd4 8. ed4 dc4 9. ♗c4 ♘c6 10. ♘e5 ♘a5 11. ♗f7 ♔h8 12. ♕d1 ♘h5 13. ♗g3 ♘g3

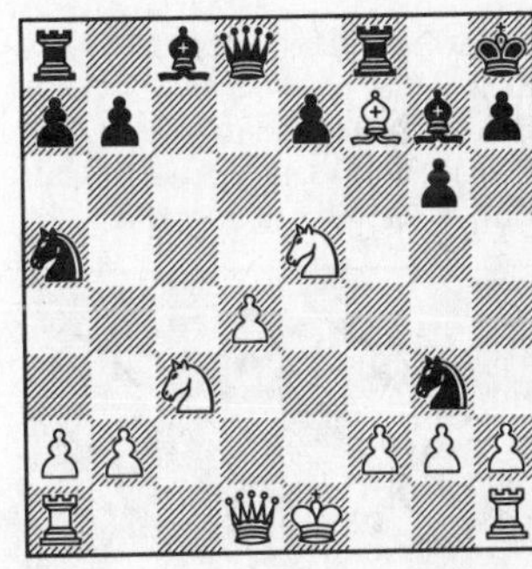

14. ♘g6! hg6 15. hg3 ♗h6 16. ♖h6 ♔g7 17. ♖g6 ♔f7 18. ♕h5 ♗f5 19. ♖d6 ♔g7 20. ♕g5 ♔h8 21. ♖h6 ♗h7 22. ♖h7![1] +–

[1] 22. – ♔h7 23. 0-0-0 +–

4658. E 12

Bobocov – Kolarov

Varna 1971

1. d4 ♘f6 2. c4 e6 3. ♘f3 b6 4. ♘c3 ♗b7 5. ♗g5 ♗e7 6. e3 c5 7. ♕c2 0-0 8. ♖d1 ♘c6 9. ♗e2 cd4 10. ♘d4 ♘d4 11. ♖d4 ♗g2 12. ♖g1 ♗c6 13. ♗d3 ♖e8 14. ♖h4 g6

15. ♗g6! hg6[1] 16. ♖h6 ♔g7 17. ♖g6! fg6 18. ♗h6 +–

[1] 15. – fg6 16. ♗f6 ♗f6 17. ♖h7! +–

4659. E 18

Porreca – Vecchio

Ferrara 1952

1. ♘f3 ♘f6 2. g3 b6 3. ♗g2 ♗b7 4. 0-0 e6 5. c4 ♗e7 6. d4 0-0 7. ♘c3 ♘e4 8. ♕d3 ♘c3 9. ♕c3 ♕c8 10. b3 d6 11. ♗b2 ♘d7 12. ♖fd1 e5 13. de5 de5 14. ♖d7 ♕d7 15. ♘e5 ♕c8 16. ♗b7 ♕b7

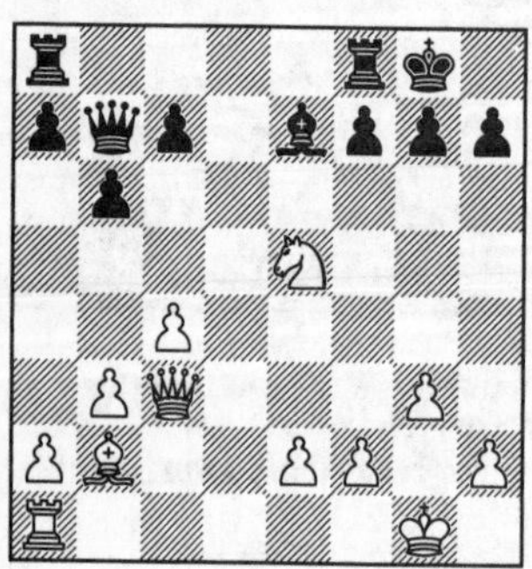

17. ♘g6![1] +–

[1] 17. – ♗f6 18. ♘e7 ♔h8 19. ♕f6 gf6 20. ♗f6#

4660. E 20

Argunov – Gossberg

Orenburg 1928

1. d4 ♘f6 2. c4 e6 3. ♘c3 ♗b4 4. e4 b6 5. e5 ♘e4 6. ♕g4 ♘c3 7. bc3 ♗c3 8. ♔d1 ♔f8 9. ♖b1 ♗b7 10. ♖b3 ♗a5 11. ♖g3 g6 12. ♗g5 ♕e8 13. ♗f6 ♖g8 14. ♕h4 h5 15. ♕g5 ♕c8 16. ♕h6 ♔e8

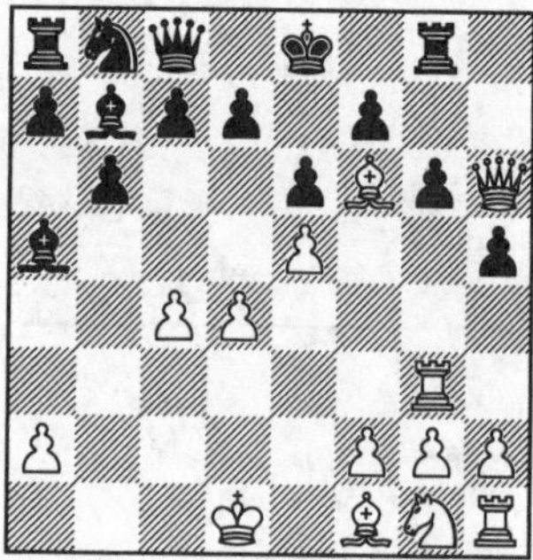

17. ♖g6! ♖f8[1] 18. ♕f8! ♔f8 19. ♖h6 +–

[1] 17. – fg6 18. ♕h7 +–

4661. E 51

Lanal – N. N.

Belgium 1929

1. d4 e6 2. ♘f3 ♘f6 3. c4 ♗b4 4. ♘c3 ♘e4 5. ♕c2 d5 6. e3 0-0 7. ♘d2 f5 8. cd5 ed5 9. ♘d5 ♗d2 10. ♗d2 ♘d2 11. ♔d2 c6 12. ♘f4 ♕a5 13. ♕c3 ♕b6 14. ♗c4 ♔h8 15. h4 ♘d7 16. h5 ♖f6

17. ♘g6! ♖g6 18. hg6 h6 19. ♖h6! +−

4662. E 94

Titz – Stanec

Germany 1991

1. c4 ♘f6 2. ♘c3 g6 3. e4 d6 4. d4 ♗g7 5. ♗e2 0-0 6. ♘f3 ♘bd7 7. 0-0 e5 8. ♗e3 ed4 9. ♘d4 ♘c5 10. f3 a5 11. ♕d2 a4 12. b4 ab3 13. ab3 ♖a1 14. ♖a1 ♘fd7 15. ♕c2 ♘e5 16. ♖d1 ♕h4 17. ♕d2 ♘c6 18. ♘c6 bc6 19. b4 ♘e6 20. b5 ♗e5 21. g3

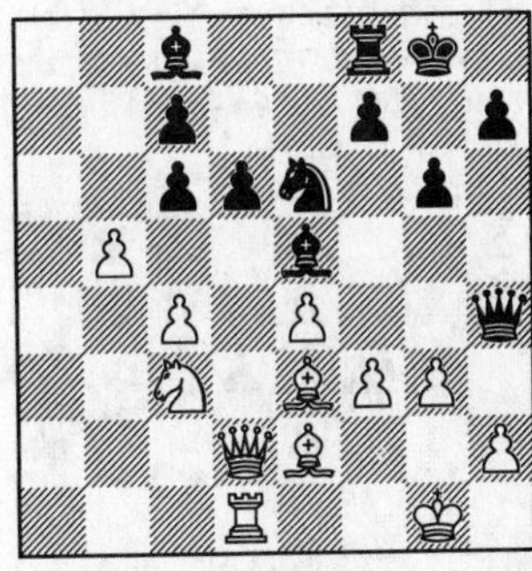

21. – ♗g3! 22. hg3 ♕g3 23. ♔f1 ♘f4 24. ♗f2 ♕h3 −+

4663. A 00
Hartlaub – N. N.
Bremen 1921

(without ♘b1)

1. b4 d5 2. ♗b2 ♘f6 3. e3 e6 4. a3 c5 5. c3 cb4 6. cb4 ♗d7 7. f4 ♘c6 8. ♘f3 ♗d6 9. ♖c1 0-0 10. ♘g5 ♘e4 11. h4 ♘g3 12. ♗d3 ♘h1 13. ♕h5 h6

14. ♕h6! gh6 15. ♗h7#

4664. A 02
Dear – Bauer
1895

1. f4 e5 2. fe5 d6 3. ed6 ♗d6 4. ♘f3 ♗g4 5. d3 ♘c6 6. c3 ♘ge7 7. e4 ♘g6 8. ♗e2 ♘f4 9. ♗f4 ♗f4 10. 0-0 ♕d6 11. h3

11. – ♗h3! 12. gh3 ♕g6 13. ♔h1[1] ♕g3 14. ♕e1[2] ♕h3 15. ♔g1 ♘e5[3] –+

[1] 13. ♔f2 ♕g3#

[2] 14. ♘g1 ♕h2#

[3] 16. ♖f2 ♘g4 17. ♕f1 ♕h5 –+

4665. A 16
Kramer – Rüster
Altheide 1926

1. c4 ♘f6 2. ♘c3 d5 3. cd5 ♘d5 4. e4 ♘c3 5. bc3 e5 6. ♘f3 ♗g4 7. ♕b3 ♗f3 8. gf3 ♗d6 9. ♖g1 0-0 10. d3 ♔h8 11. ♗h3 ♘d7 12. ♖g3 ♘c5 13. ♕c2 ♕e7 14. ♗e3 c6 15. ♔e2 f6 16. ♖ag1 ♖f7 17. ♕d2 ♗c7 18. ♗c5 ♕c5 19. ♗e6 ♖e7

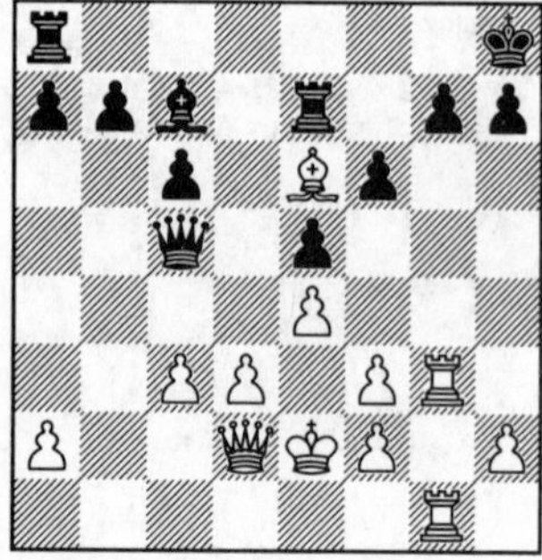

20. ♕h6![1] +–

[1] 20. – ♕c3 21. ♕h7 ♔h7 22. ♖h3#

4666. A 46
Janowski – Sämisch
Marienbad 1925

1. d4 ♘f6 2. ♘f3 e6 3. ♗g5 c5 4. e3 ♘c6 5. ♘bd2 b6 6. c3 ♗b7 7. ♗d3 cd4 8. ed4 ♗e7 9. ♘c4 0-0 10. ♕c2 ♕c7 11. h4 h6 12. ♕d2 ♘g4 13. ♗f4 d6 14. ♘e3 ♘e3 15. ♕e3 h5 16. ♖h3 e5 17. de5 ♘e5 18. ♘e5 de5 19. ♗e5 ♗d6

20. ♕h6![1] +–

[1] 20. – gh6 21. ♖g3#; 20. – f6 21. ♖g3 +–; 20. – f5 21. ♗c4 +–

4667. A 47

Vancsura – Bíró

Budapest 1985

1. d4 ♘f6 2. ♘f3 e6 3. ♗f4 b6 4. e3 ♗b7 5. ♗d3 ♗e7 6. ♘bd2 ♘h5 7. ♗g3 ♘g3 8. hg3 h6 9. ♕e2 d6 10. 0-0-0 ♘d7 11. e4 c5 12. ♘c4 a6 13. e5 d5 14. ♘d6 ♗d6 15. ed6 ♘f6 16. ♕e5 c4 17. ♗f5 ♕d7 18. ♘g5 ♗c8 19. ♖de1 0-0

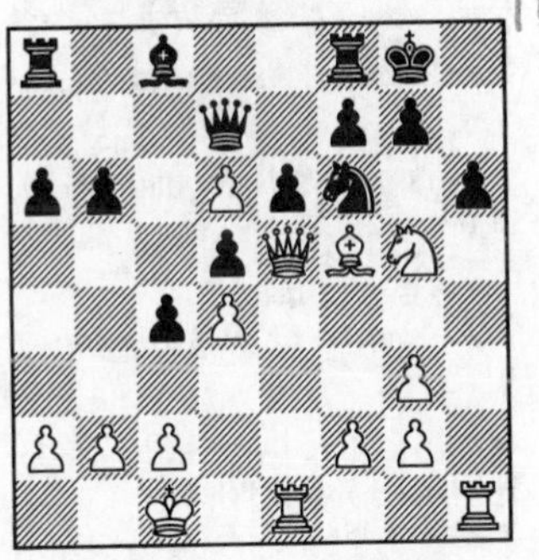

20. ♖h6! ef5 21. ♖eh1 gh6 22. ♕f6 +–

4668. A 50

Aljechin – König

Wiedniu 1922

1. d4 ♘f6 2. c4 b6 3. ♘c3 ♗b7 4. ♕c2 d5 5. cd5 ♘d5 6. ♘f3 e6 7. e4 ♘c3 8. bc3 ♗e7 9. ♗b5 c6 10. ♗d3 0-0 11. e5 h6 12. h4 c5 13. ♖h3 ♔h8

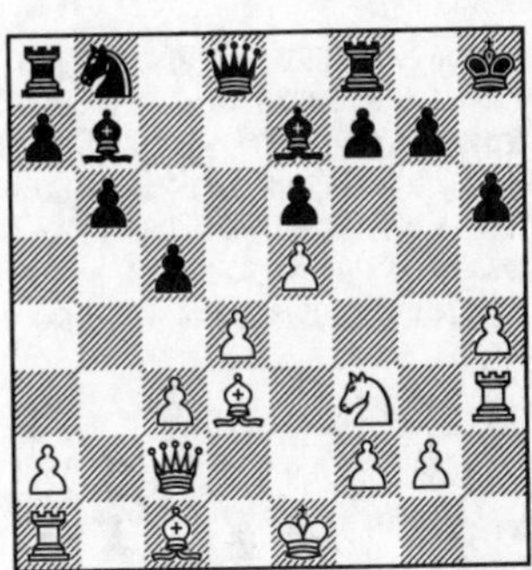

14. ♗h6! f5[1] 15. ef6 ♗f6 16. ♗g5 cd4 17. ♘e5 ♘c6 18. ♕e2 g6 19. ♗g6 ♔g7 20. ♗h6! ♔g8[2] 21. ♘c6 ♗c6 22. ♕e6 ♔h8 23. ♗f8 ♕f8 24. ♕c6 +–

[1] 14. – gh6 15. ♕d2 ♔g7 16. ♖g3 +–
[2] 20. – ♔h6 21. ♕h5 ♔g7 22. ♕h7#

4669. A 84

Kahn – Blanco

Buenos Aires 1939

1. d4 f5 2. c4 e6 3. ♘f3 ♘f6 4. e3 b6 5. ♗d3 ♗b7 6. 0-0 ♗d6 7. ♘e5 ♗e5 8. de5 ♘g4 9. e4 ♕h4 10. h3 ♘e5 11. ef5 0-0 12. fe6 ♖f3 13. e7 ♘a6 14. ♗e2

14. – ♖h3! 15. gh3 ♕h3 16. f3 ♕g3 17. ♔h1 ♘g4 –+

4670. B 09

Stein – Liberson

Erevan 1965

1. e4 d6 2. d4 ♘f6 3. ♘c3 g6 4. f4 ♗g7 5. ♘f3 0-0 6. e5 ♘fd7 7. h4 c5 8. h5 cd4 9. ♕d4 de5 10. ♕f2 e6 11. hg6 fg6 12. ♕g3 ef4 13. ♗f4 ♕a5 14. ♗d2 ♘f6 15. ♗c4 ♘c6 16. 0-0-0 ♕c5 17. ♕h4 ♘h5 18. ♘e4 ♕b6 19. c3 ♘a5 20. ♗e2 h6 21. g4 ♘f4 22. ♗f4 ♖f4 23. ♖d8 ♖f8 24. ♘f6 ♔h8

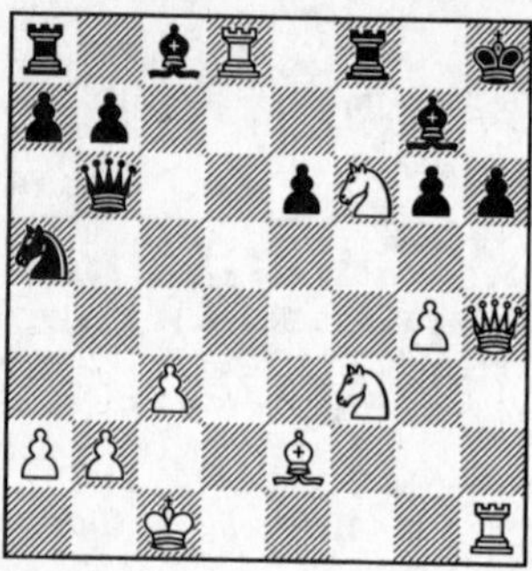

25. ♕h6![1] +–

[1] 25. – ♗h6 26. ♖h6 ♔g7 27. ♖h7 ♔f6 28. ♖f8#

4671. B 16

Guldin – Flohr

Leningrad 1951

1. e4 c6 2. ♘c3 d5 3. ♘f3 de4 4. ♘e4 ♘f6 5. ♘f6 gf6 6. d4 ♗g4 7. ♗e2 e6 8. c4 ♘d7 9. 0-0 ♕c7 10. d5 0-0-0 11. dc6 bc6 12. ♕c2 ♗c5 13. h3

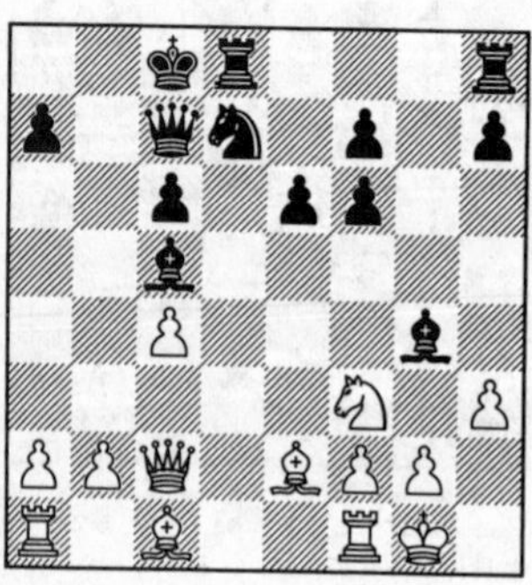

13. – ♗h3! 14. gh3 ♕g3 15. ♔h1 ♕h3 16. ♘h2 ♖hg8 17. ♕e4 ♘e5 18. ♗f3[1] ♖d4 19. ♗g2[2] ♖e4 20. ♗h3 ♖h4 –+

[1] 18. f4 f5 –+

[2] 19. ♕e2 ♕h2! 20. ♔h2 ♖h4#

4672. B 17

Bermanis – Meiers

Latvia 1939

1. e4 c6 2. d4 d5 3. ♘c3 de4 4. ♘e4 ♘d7 5. ♘f3 ♘gf6 6. ♘g3 ♕c7 7. ♗d3 e6 8. 0-0 ♗d6 9. ♖e1 b6 10. ♕e2 ♗b7 11. ♘e5 0-0 12. ♗g5 c5 13. c3 h6 14. ♘d7 ♘d7

15. ♗h6! gh6 16. ♕h5 ♗f4[1] 17. ♖e6! ♖fe8 18. ♖h6! ♗h6 19. ♕h6 ♕c6 20. ♗h7 ♔h8 21. ♗g6[2] +–

[1] 16. – ♔g7 17. ♖e6! +–

[2] 21. – ♔g8 22. ♕h7 ♔f8 23. ♕f7#

4673. B 33

Needham – Zsuzsa Polgár

Westergate 1981

1. e4 c5 2. ♘f3 ♘c6 3. d4 cd4 4. ♘d4 ♘f6 5. ♘c3 e5 6. ♘db5 d6 7. ♗g5 a6 8. ♗f6 gf6 9. ♘a3 b5 10. ♘d5 f5 11. ♗d3 ♗e6 12. c3 ♗g7 13. ♕e2 0-0 14. 0-0 ♖b8 15. ♖ad1 ♕d7 16. ♖fe1 f4 17. ♘c2 ♔h8 18. h3 ♖g8 19. ♔h2 ♘e7 20. ♘e7? ♕e7 21. ♘b4 ♕h4 22. ♘a6 ♗h6! 23. ♖h1

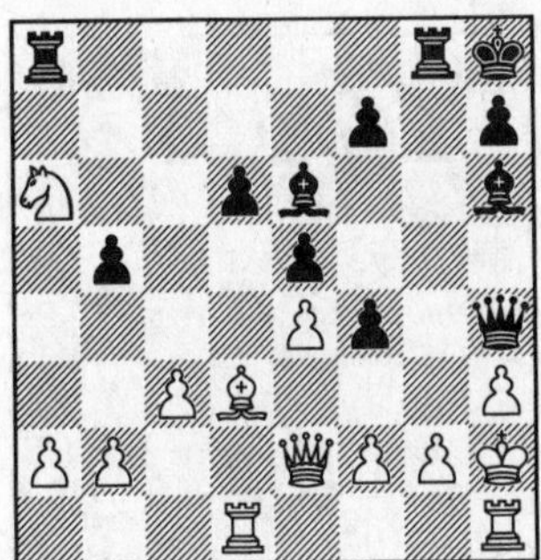

23. – ♗h3![1] –+

[1] 24. gh3 f3! –+

4674. B 44

Letelier – Najdorf
Buenos Aires 1964

1. c4 ♘f6 2. ♘c3 e6 3. e4 c5 4. ♘f3 ♘c6 5. d4 cd4 6. ♘d4 ♗b4 7. ♘c6 bc6 8. ♗d3 e5 9. 0-0 0-0 10. f4 ♗c5 11. ♔h1 d6 12. f5 h6 13. g4 d5 14. g5 dc4 15. gf6 ♕d3 16. fg7 ♔g7 17. ♕g4 ♔h7

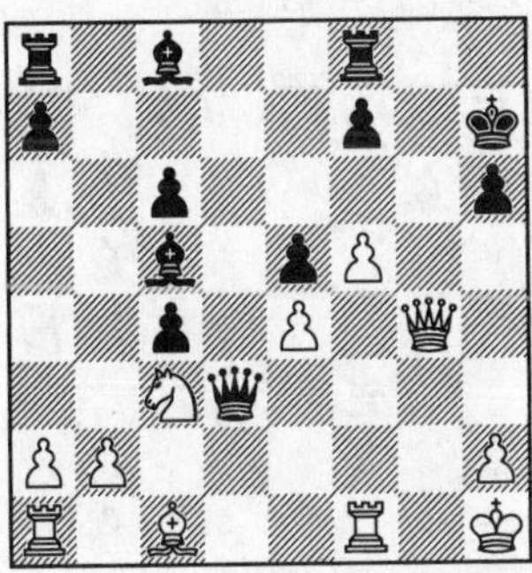

18. ♗h6![1] +–

[1] 18. – ♔h6 19. ♕h4 ♔g7 20. ♖g1! ♗g1 21. ♖g1 +–

4675. B 48

Spassky – Marsalek
Leningrad 1960

1. e4 c5 2. ♘f3 e6 3. ♘c3 a6 4. d4 cd4 5. ♘d4 ♕c7 6. ♗d3 ♘c6 7. ♗e3 b5 8. 0-0 ♗b7 9. ♘b3 ♘f6 10. f4 d6 11. ♕f3 ♗e7 12. a4 b4 13. ♘e2 e5 14. ♘g3 0-0 15. ♘f5 ef4 16. ♕f4 ♘e5 17. ♕g3 ♘g6 18. ♗d4 ♘e4 19. ♕g4 ♘f6 20. ♕g5 ♘e4

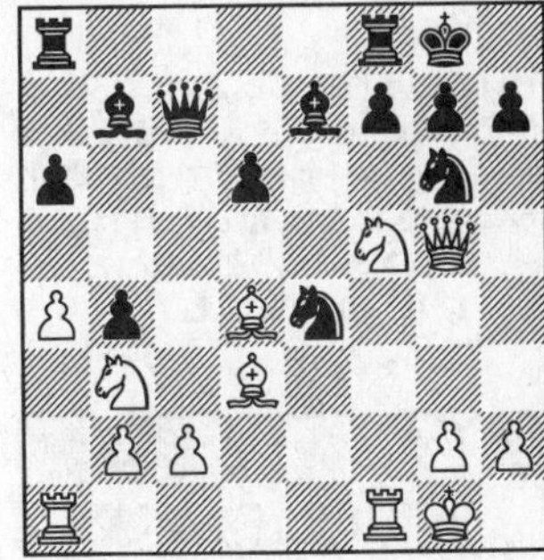

21. ♕h6! +–

4676. B 53

Kamsky – Lautier
Dortmund 1993

1. e4 c5 2. ♘f3 d6 3. d4 cd4 4. ♕d4 a6 5. ♗g5 ♘c6 6. ♕d2 ♘f6 7. ♗d3 e6 8. c4 h6 9. ♗f4 d5 10. ed5 ed5 11. 0-0 ♗e7 12. ♘c3 ♗g4 13. cd5 ♗f3 14. dc6 ♗c6 15. ♖ad1 0-0

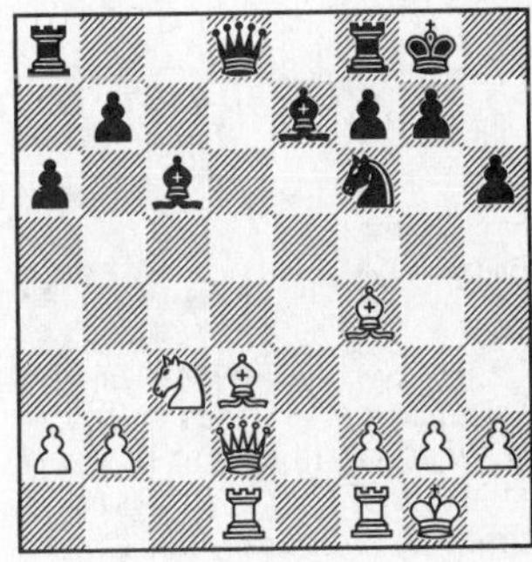

16. ♗h6! gh6 17. ♕h6 ♖e8 18. ♗c4 ♗d7 19. ♖d4 ♗f8 20. ♕g6 ♗g7 21. ♕f7 ♔h8 22. ♖h4 ♘h7 23. ♖h7 ♔h7 24. ♕h5 ♗h6 25. ♗d3 +–

4677. B 66

Fogarasi – Csom

Budapest, 1993

1. e4 c5 2. ♘f3 d6 3. d4 cd4 4. ♘d4 ♘f6 5. ♘c3 ♘c6 6. ♗g5 e6 7. ♕d2 a6 8. 0-0-0 h6 9. ♗e3 ♗d7 10. f4 b5 11. ♗d3 ♗e7 12. h3 ♕c7 13. ♖he1 ♘d4 14. ♗d4 b4 15. ♘e2 e5 16. ♗e3 ♗e6 17. ♔b1 ♕b7 18. ♘g3 0-0 19. f5 ♗d7

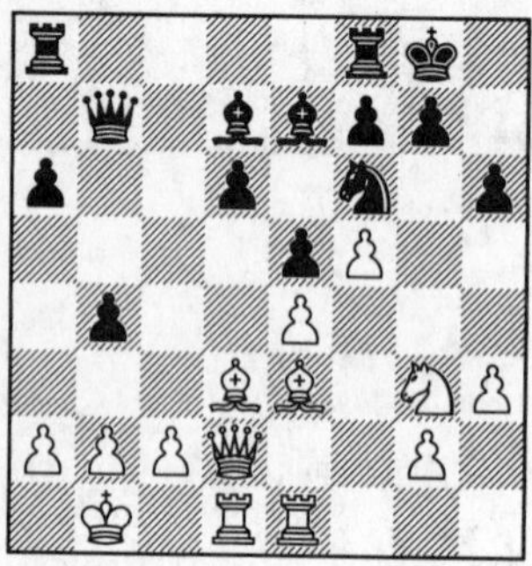

20. ♗h6! gh6 21. ♕h6 ♖fc8 22. ♖e3 ♕b6 23. ♘f1 +–

4678. B 93

Sax – Károlyi

Budapest 1981

1. e4 c5 2. ♘f3 d6 3. d4 cd4 4. ♘d4 ♘f6 5. ♘c3 a6 6. f4 e5 7. ♘f3 ♘bd7 8. a4 ♕c7 9. ♗d3 b6 10. 0-0 ♗b7 11. ♕e2 g6 12. ♗c4 ♗g7 13. f5 gf5 14. ♘g5 0-0 15. ♖f5 h6 16. ♘h3 ♘e4 17. ♘e4 d5

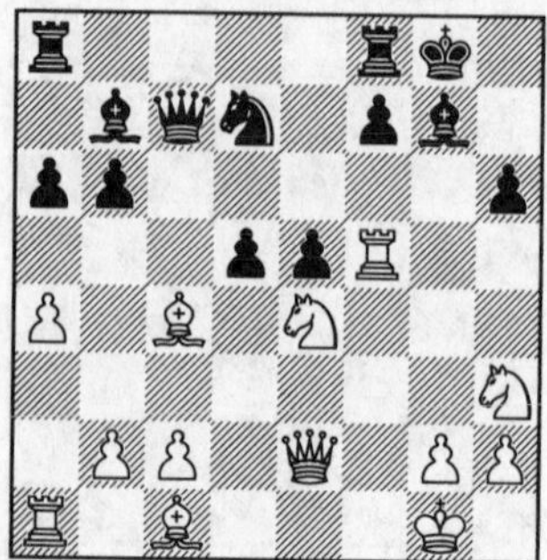

18. ♗h6! ♗h6 19. ♕g4 ♗g7 20. ♖g5 dc4 21. ♖g7 ♔h8 22. ♖h7! ♔h7 23. ♘hg5[1] +–

[1] 23. – ♔g6 24. ♘e6 ♔h6 25. ♕g5 ♔h7 26. ♕g7#

4679. C 00

Keres – Verbac

Corr.

1. e4 e6 2. d4 d5 3. ♗e3 de4 4. ♘d2 f5 5. f3 ef3 6. ♘gf3 ♘f6 7. ♗d3 c5 8. 0-0 cd4 9. ♘d4 f4 10. ♖f4 e5 11. ♗b5 ♔f7 12. ♕h5 g6 13. ♗c4 ♔g7

14. ♕h6![1] +–

[1] 14. – ♔h6 15. ♖h4 ♔g7 16. ♗h6#

4680. C 00
Mazuchowski – Drumm
Dayton 1965

1. e4 e6 2. d3 d5 3. Nd2 Nf6 4. Ngf3 c5 5. g3 Nc6 6. Bg2 de4 7. de4 Bd6 8. 0-0 Bb8 9. Re1 0-0 10. c3 b6 11. e5 Nd5 12. Nc4 b5 13. Ne3 Nde7 14. Qe2 Qb6 15. Ng4 b4 16. Ng5 h6 17. Ne4 bc3

18. Nh6! gh6 19. Nf6 Kh8[1] 20. Qh5 +–

[1] 19. – Kg7 20. Bh6! +–

4681. C 00
Wilke – Priwonitz
1933

1. e4 e6 2. c4 d5 3. cd5 ed5 4. ed5 Nf6 5. Bb5 Bd7 6. Bc4 Be7 7. Nc3 0-0 8. Nge2 Bd6 9. 0-0 Bh2 10. Kh2 Ng4 11. Kg1 Qh4 12. Re1 Qf2 13. Kh1 Re8 14. d4 Re3 15. Bd2

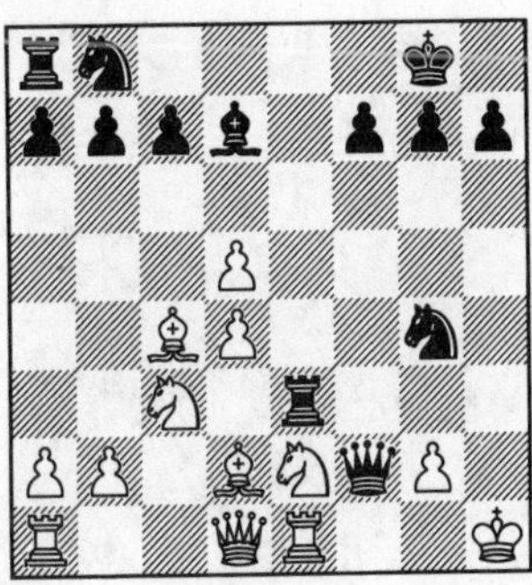

15. – Rh3! 16. gh3 Qh2#

4682. C 01
Rodatz – Krüger

1. e4 e6 2. d4 d5 3. ed5 ed5 4. Nf3 Nf6 5. Bd3 Bd6 6. 0-0 0-0 7. Bg5 Bg4 8. Nbd2 Nc6 9. c3 Qd7 10. h3

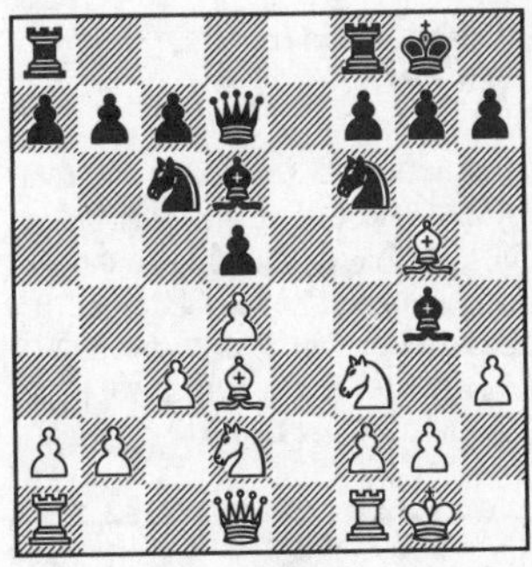

10. – Bh3! 11. gh3 Qh3 12. Re1[1] Ng4 13. Bf1 Bh2 14. Kh1 Nf2#

[1] 12. Bf6 gf6 13. Re1 Kh8! –+

4683. C 05
Chernov – Roots
USSR 1974

1. e4 e6 2. d4 d5 3. Nd2 Nf6 4. e5 Nfd7 5. f4 c5 6. c3 Nc6 7. Ndf3 Be7 8. Bd3 0-0 9. Ne2 f6 10. 0-0 b6 11. a3 a5 12. c4 Nd4 13. Ned4 cd4 14. cd5 fe5 15. fe5 ed5 16. Qc2 h6

17. ♗h6! ♘c5[1] 18. ♗h7 ♔h8 19. ♗g7! ♔g7 20. ♕g6 ♔h8 21. ♕h6 ♖f7 22. ♗g6 ♔g8 23. ♘g5 +–

[1] 17. – gh6 18. ♕c6! ♖b8 19. ♕g6 +–

4684. C 06

Gollasch – Fischer

GFR 1987

1. e4 e6 2. d4 d5 3. ♘d2 ♘f6 4. e5 ♘fd7 5. ♗d3 c5 6. c3 ♘c6 7. ♘e2 cd4 8. cd4 f6 9. ef6 ♘f6 10. ♘f3 ♗d6 11. 0-0 ♕c7 12. ♗d2 0-0 13. ♘c3 a6 14. ♖c1 ♗d7 15. ♘a4 ♘e4 16. ♘g5 ♗h2 17. ♔h1 ♖f2 18. ♗e4 ♖f1 19. ♕f1 de4 20. ♕f7 ♔h8 21. ♘c5 h6 22. ♘ge4 ♖d8

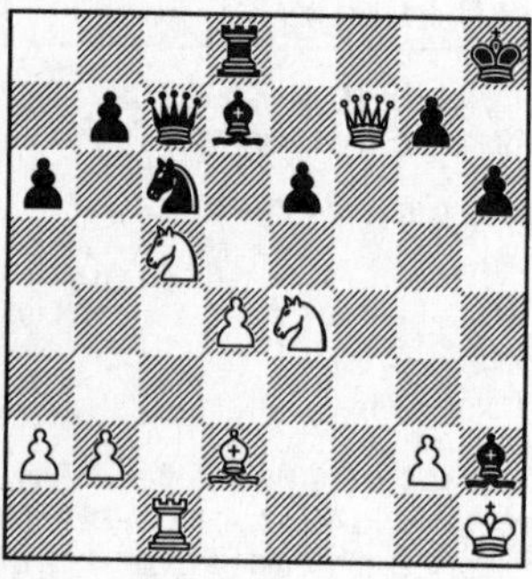

23. ♗h6! gh6 24. ♘f6 ♗c8 25. ♘cd7![1] +–

[1] 25. – ♖d7 26. ♕g8#

4685. C 06

Perfors – Seters

Wageningen 1955

1. e4 e6 2. d4 d5 3. ♘d2 ♘f6 4. e5 ♘fd7 5. ♗d3 c5 6. c3 ♘c6 7. ♘e2 ♕b6 8. ♘f3 cd4 9. cd4 f6 10. ef6 ♘f6 11. 0-0 ♗d6 12. ♘f4 0-0 13. ♖e1 ♗d7 14. ♘e6 ♖fe8 15. ♗f5 h6 16. ♘g7 ♔g7

17. ♗h6! ♔h8[1] 18. ♗d7 ♘d7 19. ♖e8 ♖e8 20. ♘g5 ♖e7 21. ♕h5 ♕b2 22. ♘f7[2] +–

[1] 17. – ♔h6 18. ♕d2 ♔g7 19. ♕g5 ♔f7 20. ♕g6 ♔f8 21. ♕f6 ♔g8 22. ♗h7! ♔h7 23. ♘g5 ♔g8 24. ♕f7 ♔h8 25. ♕f7#

[2] 22. – ♖f7 23. ♗c1 ♖h7 24. ♕e8 +–

4686. C 10

Georgiev – Kurtenkov

Elenite 1992

1. e4 e6 2. d4 d5 3. ♘d2 de4 4. ♘e4 ♗d7 5. ♘f3 ♗c6 6. ♗d3 ♘d7 7. 0-0 ♘gf6 8. ♘eg5 ♗f3 9. ♕f3 c6 10. ♖e1 ♗e7 11. ♕h3 ♘b6 12. ♗d2 ♘bd5 13. ♖e2 ♘c7 14. c3 h6 15. ♖ae1 0-0 16. ♘f3 ♘h7

17. ♗h6! gh6 18. ♕h6 f5 19. ♖e6 ♖f7 20. ♖g6 ♔h8 21. ♘e5 ♕f8 22. ♘f7 ♕f7 23. ♖g7 +–

4687. C 10

Spielmann – Grossman

New York 1922

1. e4 e6 2. d4 d5 3. ♘d2 de4 4. ♘e4 ♘d7 5. ♘f3 ♘gf6 6. ♘f6 ♘f6 7. ♗d3 h6 8. ♕e2 ♗d6 9. ♗d2 0-0 10. 0-0-0 ♗d7 11. ♘e5 c5 12. dc5 ♗e5 13. ♕e5 ♗c6 14. ♗f4 ♕e7 15. ♕d4 ♖fd8 16. ♗d6 ♕e8 17. ♖hg1 b6 18. ♕h4 bc5 19. ♗e5 ♕e7 20. g4 c4 21. g5 ♘d7

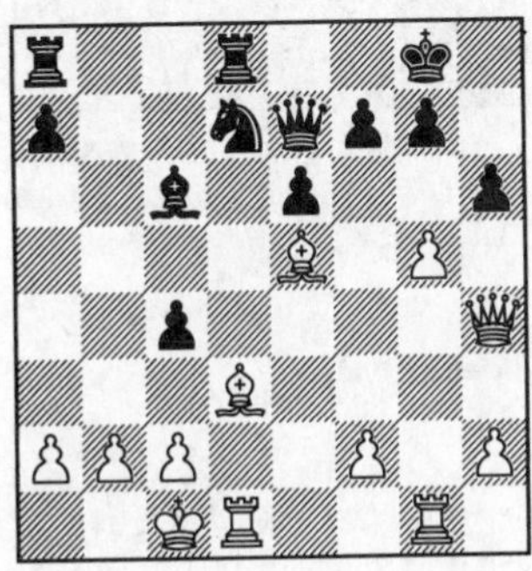

22. ♕h6! gh6[1] 23. gh6 ♔f8 24. ♖g8![2] +–

[1] 22. – ♘e5 23. ♕h7 ♔f8 24. ♕h8#

[2] 24. – ♔g8 25. h7 ♔f8 26. h8♕#

4688. C 10

Yates – Censer

London 1927

1. e4 e6 2. d4 d5 3. ♘c3 de4 4. ♘e4 ♘d7 5. ♘f3 ♘gf6 6. ♘g3 b6 7. ♗b5 ♗b7 8. ♘e5 ♗e7 9. ♗c6 ♕c8 10. ♘h5 0-0 11. ♗d7 ♘d7

12. ♗h6! ♘e5[1] 13. de5 ♖d8[2] 14. ♕g4 g6 15. ♕f4 c5 16. ♘f6 ♔h8 17. ♕h4[3] +–

[1] 12. – gh6 13. ♘d7 ♕d7 14. ♕g4 ♗g5 15. ♘f6 +–

[2] 13. – gh6 14. ♕g4 ♗g5 15. h4 +–

[3] 17. – ♖d4 18. ♗g7! +–

4689. C 11

Mackenzie – Mason

Paris 1878

1. e4 e6 2. d4 d5 3. ♘c3 ♘f6 4. ed5 ed5 5. ♘f3 ♗d6 6. ♗d3 0-0 7. 0-0 ♘c6 8. ♗g5 ♘e7 9. ♗f6 gf6 10. ♘h4 ♔g7 11. ♕h5 ♖h8 12. f4 c6 13. ♖f3 ♘g6 14. ♖af1 ♕c7 15. ♘e2 ♗d7 16. ♘g3 ♖ag8

17. ♕h6! ♔h6 18. ♘hf5 ♗f5 19. ♘f5 ♔h5 20. g4! ♔g4 21. ♖g3 ♔h5 22. ♗e2#

4690. C 15
Perez – Fernandez
Aviles 1947

1. e4 e6 2. d4 d5 3. ♘c3 ♗b4 4. ♘e2 de4 5. a3 ♗d6 6. ♘e4 ♘e7 7. ♘2g3 ♘bc6 8. ♘h5 0-0 9. ♘hf6 gf6 10. ♘f6 ♔g7 11. ♕h5 ♘f5 12. ♗g5 h6

♗h6! ♘h6[1] 14. ♕g5[2] +–

[1] 13. – ♔f6 14. ♕g5#
[2] 14. – ♔h8 15. ♕h6#

4691. C 21
Denker – Gonzalez
Detroit 1945

1. e4 e5 2. d4 ed4 3. c3 dc3 4. ♗c4 cb2 5. ♗b2 ♗b4 6. ♔f1 ♘f6 7. e5 ♘g8 8. ♕g4 ♗f8 9. ♕f3 ♘h6 10. ♘c3 ♗e7 11. ♘d5 0-0 12. ♘f6 ♔h8 13. ♘h3 ♗f6 14. ef6 g6 15. ♕f4 ♘f5 16. ♘g5 ♘d6 17. ♘f7 ♘f7

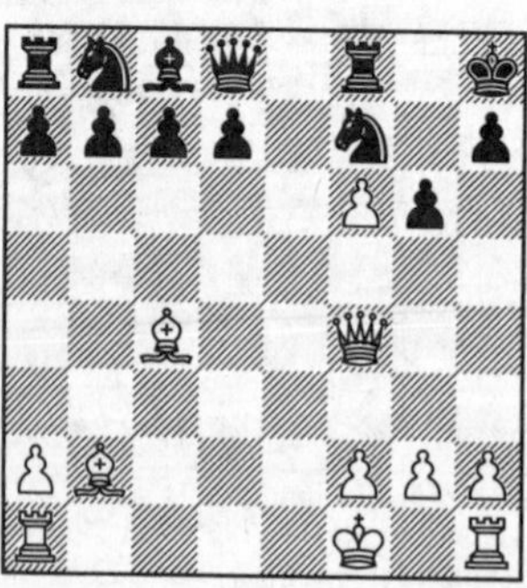

18. ♕h6! +–

4692. C 24
Hartlaub – Eisele
Freiburg 1889

1. e4 e5 2. ♗c4 ♘f6 3. ♘f3 ♘e4 4. ♘c3 ♘c3 5. dc3 f6 6. 0-0 ♗e7

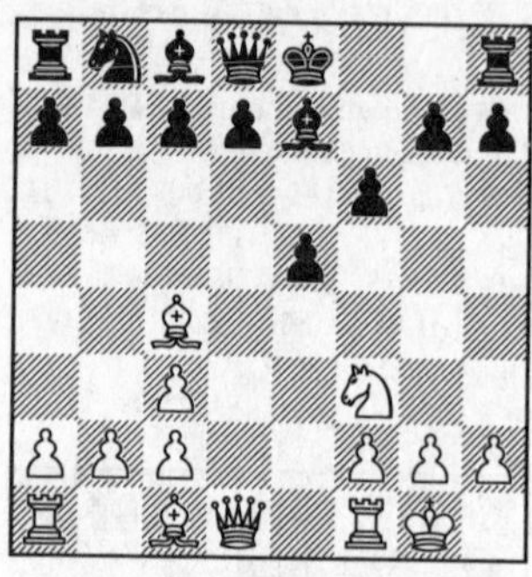

7. ♗h6! gh6 8. ♘e5! ♗c5 9. ♕h5 ♔e7 10. ♕f7 ♔d6 11. ♕d5 ♔e7 12. ♘g6! hg6 13. ♖ae1 ♔f8 14. ♕f7#

4693. C 25
Steinitz – Hodges
New York 1891

1. e4 e5 2. ♘c3 ♗c5 3. f4 ♗g1 4. ♖g1 ♕h4 5. g3 ♕e7 6. fe5 ♘c6 7. d4 f6 8. ♘d5 ♕d8 9. ♗f4 g5 10. ♕h5 ♔f8 11. ♗c4 gf4 12. gf4 ♘d4 13. ♘f6 ♕e7 14. ♗g8 ♘c2 15. ♔d1 h6 16. ♗b3 ♘a1

17. ♕h6! +–

4694. C 26

Mieses – Love

St. Louis 1908

(without ♘g1)

1. e4 e5 2. ♘c3 ♘f6 3. ♗c4 ♗c5 4. d3 d6 5. f4 ♘c6 6. f5 0-0 7. ♗g5 h6 8. ♗h4 ♘d4 9. ♘d5 ♔h7 10. ♘f6 gf6 11. c3 ♘c6 12. ♕h5 ♕e7 13. ♖f1 ♘d8 14. ♖f3 ♖g8 15. ♖h3 ♖g2

16. ♕h6! ♔g8[1] 17. ♕h8! ♔h8 18. ♗f6 ♔g8 19. ♖h8#

[1] 16. – ♔h6 17. ♗f6#

4695. C 28

Brandstätter – Geiger

1950

1. e4 e5 2. ♘c3 ♘f6 3. ♗c4 ♘c6 4. d3 ♗b4 5. ♗d2 ♗c3 6. bc3 d5 7. ed5 ♘d5 8. ♘e2 0-0 9. 0-0 ♔h8 10. ♔h1 b6 11. a3 ♗b7 12. f4 f5 13. fe5 ♘e5 14. ♗b3 ♘g4 15. h3 ♕h4 16. ♕e1

16. – ♕h3! 17. gh3 ♘f4 18. ♖f3 ♗f3 19. ♔g1 ♘h3 20. ♔f1 ♘h2#

4696. C 29

Dobias – Chalupetzky

Corr. 1904

1. e4 e5 2. ♘c3 ♘f6 3. f4 d5 4. fe5 ♘e4 5. ♘f3 ♗b4 6. ♕e2 ♗c3 7. bc3 0-0 8. ♕e3 ♕e7 9. d4 b6 10. ♗d3 ♗f5 11. c4 ♕b4 12. ♘d2 ♕c3 13. 0-0 ♕a1 14. ♘b3 ♕a2 15. ♖f5 c6 16. cd5 cd5 17. ♘d2 ♘d7 18. ♘e4 de4 19. ♗e4 ♕e6 20. ♖h5 h6 21. ♗f5 ♕c6 22. ♕g3 ♔h8

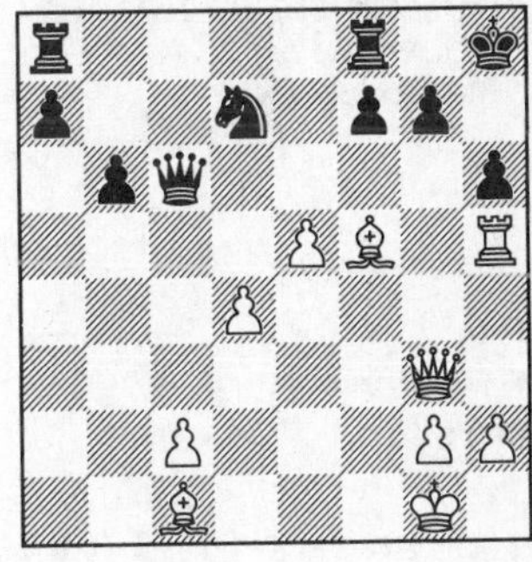

23. ♗h6![1] +–

[1] 23. – gh6 24. ♕g5! +–

4697. C 30
Ivanovic – Osterman
Yugoslavia 1979

1. e4 e5 2. f4 ♘f6 3. fe5 ♘e4 4. ♘f3 d5 5. d3 ♘c5 6. d4 ♘e4 7. ♗d3 ♗e7 8. 0-0 c5 9. c4 cd4 10. cd5 ♕d5 11. ♕c2 ♘c5 12. ♗c4 d3 13. ♕c3 ♕d8 14. ♘g5 ♘e6 15. ♘f7 ♕b6 16. ♔h1 0-0 17. ♗d3 ♕d4 18. ♕c2 ♕h4

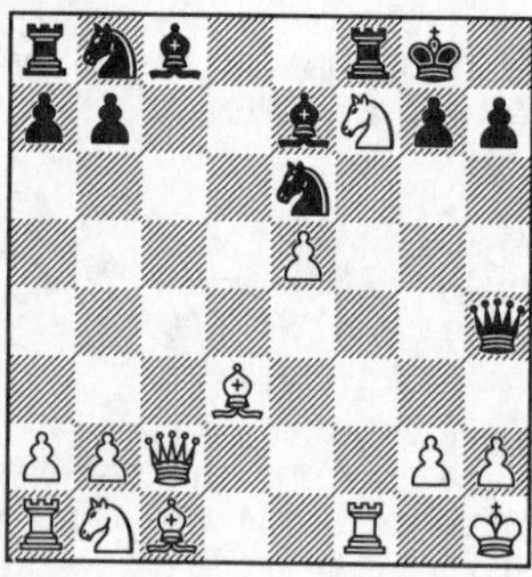

19. ♗h6![1] +–

[1] 19. – gh6 20. ♗h7 ♔g7 21. ♕g6#; 19. – ♘g5 20. ♗g5 ♗g5 21. ♗h7 ♕h7 22. ♕h7 ♔h7 23. ♘g5 +–

4698. C 30
Keller – Friedrich
Corr. 1976

1. e4 e5 2. f4 ♗c5 3. ♘f3 d6 4. c3 ♕e7 5. d4 ed4 6. cd4 ♗b6 7. ♘c3 ♘f6 8. e5 0-0 9. ♗e2 ♘fd7 10. ♘d5 ♕e8 11. 0-0 c5 12. ♗d3 h6 13. f5 cd4 14. f6 de5 15. fg7 ♔g7

♗h6![1] +–

[1] 16. – ♔h6 17. ♕d2 ♔g7 18. ♕g5 ♔h8 19. ♕h6 ♔g8 20. ♕h7#

4699. C 31
Farkas – Széll
1972

1. e4 e5 2. f4 d5 3. ♘f3 de4 4. ♘e5 ♘d7 5. d4 ed3 6. ♘d3 ♘gf6 7. ♕f3 ♗e7 8. ♘c3 0-0 9. ♗d2 c6 10. 0-0-0 ♕c7 11. g4 b5 12. g5 ♘d5 13. ♖g1 ♘7b6 14. f5 f6 15. g6 h6 16. ♘e4 ♖d8

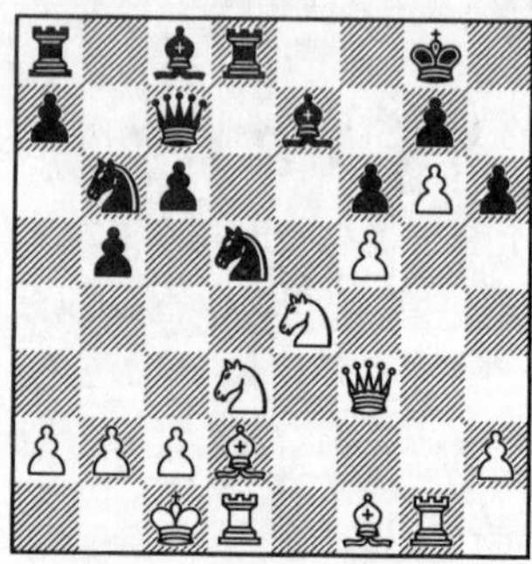

17. ♗h6! gh6 18. g7 +–

4700. C 31
Vértes – Steiner
1897

1. e4 e5 2. f4 d5 3. ♘f3 de4 4. ♘e5 ♗d6 5. ♗c4 ♗e5 6. ♕h5 ♕e7 7. fe5 g6 8. ♕e2 ♕e5 9. 0-0 ♗e6 10. ♗e6 ♕e6 11.

d3 ed3 12. ♕d3 ♘e7 13. ♗g5 0-0 14. ♘c3 ♘f5 15. ♖ae1 ♕d6 16. ♘d5 ♘c6 17. ♖f5 gf5 18. ♘f6 ♔g7 19. ♕f5 ♖h8

20. ♗h6![1] +–

[1] 20. – ♔h6 21. ♕h5 ♔g7 22. ♕g5 ♔f8 23. ♕h6#

4701. C 32

Loera – Jaffray

1959

1. e4 e5 2. f4 d5 3. ed5 e4 4. d3 ♘f6 5. ♘d2 ♗c5 6. ♘e4 ♗b6 7. ♘f3 0-0 8. ♗e2 ♖e8 9. ♘f6 ♕f6 10. d4 c6 11. 0-0 cd5 12. ♘e5 ♘c6 13. c3 ♘e5 14. fe5 ♕e5 15. ♗b5 ♖d8 16. ♗f4 ♕e7 17. ♕h5 h6 18. ♖ae1 ♗e6 19. ♗d3 ♕d7

20. ♗h6! gh6 21, ♕h6[1] +–

[1] 21. – ♗c7 22. ♗h7 ♔h8 23. ♗f5 ♔g8 24. ♖e6! +–

4702. C 33

Rudolph – Klein

1912

1. e4 e5 2. f4 ef4 3. ♗c4 ♗c5 4. d4 ♕h4 5. ♔f1 ♗b6 6. ♘f3 ♕d8 7. ♗f4 ♘e7 8. ♘g5 0-0 9. ♕h5 h6 10. ♗f7 ♔h8

11. ♕h6! gh6 12. ♗e5#

4703. C 33

Schulten – Kieseritzky

Paris 1844

1. e4 e5 2. f4 ef4 3. ♗c4 ♕h4 4. ♔f1 b5 5. ♗b5 ♘f6 6. ♘c3 ♘g4 7. ♘h3 ♘c6 8. ♘d5 ♘d4 9. ♘c7 ♔d8 10. ♘a8 f3 11. d3 f6 12. ♗c4 d5 13. ♗d5 ♗d6 14. ♕e1 fg2 15. ♔g2

15. – ♕h3! 16. ♔h3 ♘e3 17. ♔h4 ♘f3 18. ♔h5 ♗g4#

4704. C 34

Blackburne – Blanchard

London 1891

1. e4 e5 2. f4 ef4 3. ♘f3 ♘c6 4. ♘c3 ♗c5 5. d4 ♗b4 6. ♗f4 d5 7. e5 ♗c3 8. bc3 ♗e6 9. ♗d3 h6 10. 0-0 ♘ge7 11. ♖b1 b6 12. ♕d2 0-0

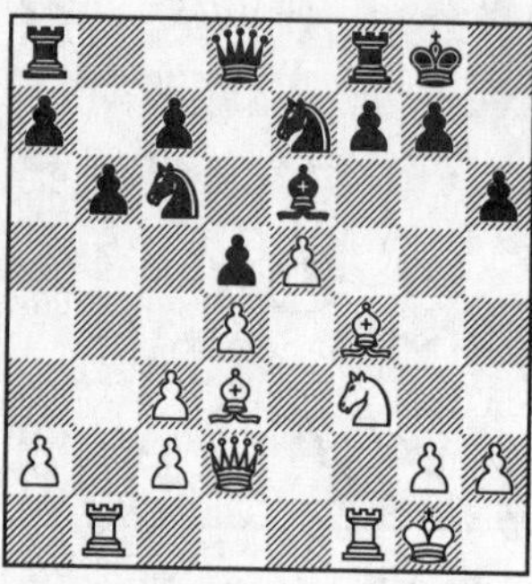

13. ♗h6! gh6 14. ♕h6 ♘g6 15. ♘g5 ♖e8 16. ♖f7! ♗f7 17. ♕h7 ♔f8 18. ♕f7#

4705. C 37

Itze – Reinte

1925

1. e4 e5 2. f4 ef4 3. ♘f3 g5 4. ♗c4 g4 5. ♘e5 ♕h4 6. ♔f1 ♗g7 7. ♘f7 d5 8. ♗d5 ♗d4 9. ♕e1 g3 10. h3 f3 11. ♘h8

11. – ♗h3! 12. ♖h3 ♕h3! 13. gh3 g2#

4706. C 43

Kurajica – Binder

Corr. 1966

1. e4 e5 2. ♘f3 ♘f6 3. d4 ed4 4. e5 ♘e4 5. ♕d4 d5 6. ed6 ♘d6 7. ♘c3 ♘c6 8. ♕f4 g6 9. ♗d2 ♗g7 10. 0-0-0 0-0 11. h4 ♖e8 12. h5 ♗f5 13. g4 ♗e4 14. hg6 fg6 15. ♘g5 ♗h1 16. ♗c4 ♔h8 17. ♖h1 h6 18. ♘b5 ♖f8

19. ♖h6! ♗h6 20. ♗c3 +–

4707. C 44

Abraham – Janny

Arad 1923

1. e4 e5 2. ♘f3 ♘c6 3. ♗e2 ♗c5 4. ♘e5 ♘e5 5. d4 ♗d6 6. de5 ♗e5 7. f4 ♗d6 8. 0-0 ♗c5 9. ♔h1 d6 10. ♗c4 ♕h4 11. ♕d5 ♗e6 12. ♕b7 ♗c4 13. ♕a8 ♔d7 14. ♖d1 ♘f6 15. ♕h8 ♘g4 16. h3 ♗e2 17. ♕g7

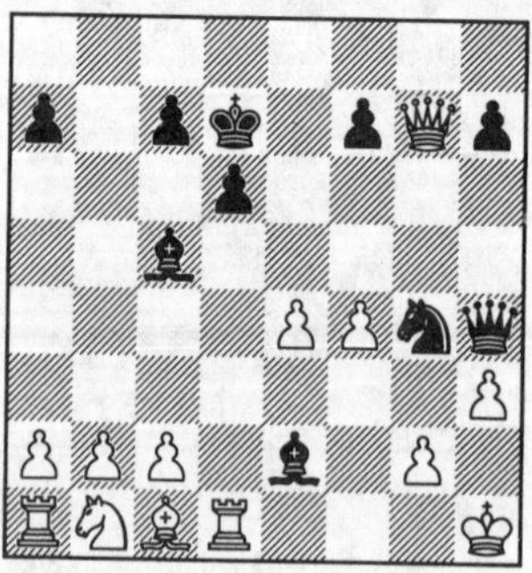

♕h3! 18. gh3 ♗f3#

4708. C 44

Blackburne – Giford

Hagen 1874

1. e4 e5 2. ♘f3 ♘c6 3. d4 ed4 4. ♗c4 ♗c5 5. ♘g5 ♘h6 6. ♕h5 ♕e7 7. f4 0-0 8. 0-0 d6 9. f5 d3 10. ♔h1 dc2 11. ♘c3 ♘e5 12. ♘d5 ♕d8 13. f6 ♘g6 14. fg7 ♔g7

15. ♕h6! ♔h6 16. ♘e6 ♔h5 17. ♗e2 ♔h4 18. ♖f4 ♘f4 19. g3 ♔h3 20. ♘df4#

4709. C 45

Astapovich – Golosov

Novosibirsk 1967

1. e4 e5 2. ♘f3 ♘c6 3. d4 ed4 4. ♘d4 ♘f6 5. ♘c3 ♗e7 6. ♘f5 0-0 7. ♗g5 ♖e8 8. ♗c4 ♘e4 9. ♗f7! ♔f7 10. ♕d5 ♔f8

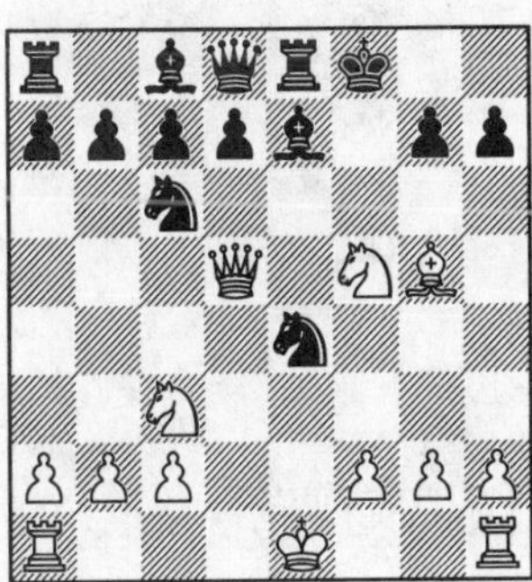

11. ♘h6![1] +–

[1] 11. – gh6 12. ♗h6#

4710. C 45

Dehen – Mallé

Dortmund 1964

1. e4 e5 2. ♘f3 ♘c6 3. d4 ed4 4. ♘d4 ♘f6 5. ♘c6 bc6 6. ♗d3 ♗c5 7. 0-0 d6 8. ♗g5 h6 9. ♗d2 ♘g4 10. h3 ♘e5 11. ♗e2 ♕h4 12. ♘c3

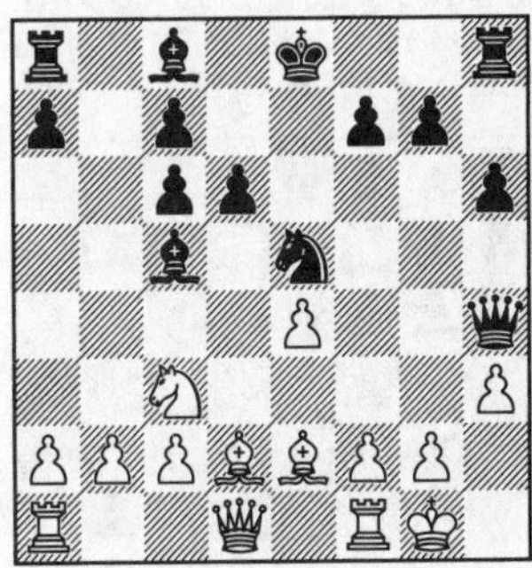

12. – ♗h3! 13. gh3 ♕g3 14. ♔h1 ♕h3 15. ♔g1 g5 16. ♗e3[1] g4 –+

[1] 16. ♘a4 g4 17. ♘c5 ♘f3 18. ♗f3 gf3 –+

4711. C 45

Delmar – Lipschuetz

New York 1888

1. e4 e5 2. ♘f3 ♘c6 3. d4 ed4 4. ♘d4 ♘f6 5. ♘c6 bc6 6. ♗d3 d5 7. e5 ♘g4 8. 0-0 ♗c5 9. h3 ♘e5 10. ♖e1 ♕f6 11. ♕e2 0-0 12. ♕e5 ♕f2 13. ♔h1

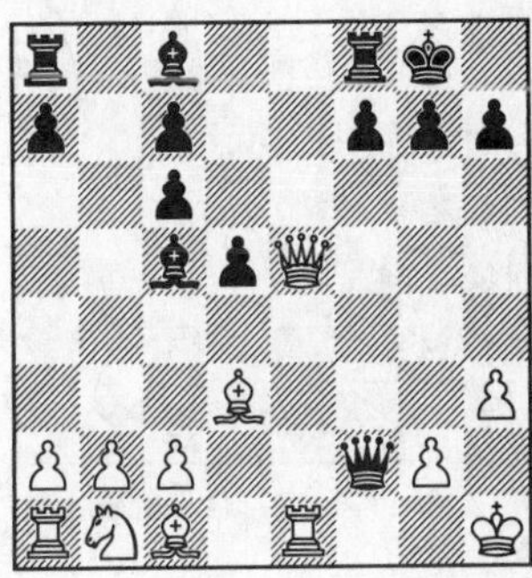

13. – ♗h3! 14. gh3 ♕f3 15. ♔h2 ♗d6 16. ♕d6 ♕f2 −+

4712. C 48

N. N. – Mau

1. e4 e5 2. ♘f3 ♘f6 3. ♘c3 ♘c6 4. ♗b5 ♘d4 5. ♘e5 ♘b5 6. ♘b5 c6 7. ♘c3 ♗b4 8. f3 d6 9. ♘d3 ♗a5 10. 0-0 ♗b6 11. ♔h1 h5 12. ♖e1 ♔f8 13. b3 h4 14. h3 ♘h5 15. g4 hg3 16. ♔g2

16. – ♗h3! 17. ♔h3 ♘f4 18. ♔g3 ♕h4#

4713. C 48

Pillsbury – Dickson

New Orleans 1900

1. e4 e5 2. ♘f3 ♘c6 3. ♘c3 ♘f6 4. ♗b5 ♗c5 5. 0-0 0-0 6. ♘e5 ♕e7 7. ♘c6 dc6 8. ♗d3 ♕e5 9. h3

9. – ♗h3! 10. gh3 ♕g3 11. ♔h1 ♕h3 12. ♔g1 ♘g4 −+

4714. C 50

Deutz – Scheidt

Aachen 1934

1. e4 e5 2. ♘f3 ♘c6 3. ♗c4 ♗c5 4. 0-0 ♘f6 5. d4 ♗d4 6. ♘d4 ♘d4 7. ♗g5 d6 8. f4 ♘e6 9. ♗e6 ♗e6 10. fe5 de5 11. ♘c3 ♕d4 12. ♔h1 ♘d7 13. ♕f3 0-0 14. ♖ad1 ♕b4 15. ♖d7 ♗d7 16. ♘d5 ♕b2 17. ♘f6 ♔h8 18. ♕h5 h6

19. ♗h6! gf6 20. ♗g5 +−

4715. C 50

Greville – Harrwitz

1. e4 e5 2. ♘f3 ♘c6 3. ♗c4 ♗c5 4. d3 d6 5. ♘g5 ♕f6 6. ♗f7 ♔f8 7. 0-0 h6 8. ♗g8 hg5 9. ♗d5 ♘d4 10. c3 ♕h6 11. h3

11. – ♗h3! 12. cd4[1] ♗g2! 13. ♔g2[2] ♕h3 14. ♔g1 ♕h1#

[1] 12. gh3 ♕h3 13. cd4 ♕h2#

[2] 13. f3 ♗d4 −+

4716. C 50

Michigan – Curran

Philadelphia 1876

1. e4 e5 2. ♘f3 ♘c6 3. ♗c4 ♗c5 4. 0-0 ♘f6 5. d3 d6 6. ♗g5 h6 7. ♗h4 g5 8. ♗g3 h5 9. ♘g5 h4 10. ♘f7 hg3 11. ♘d8 ♗g4 12. ♕d2 ♘d4 13. h3 gf2 14. ♖f2 ♘e2 15. ♔h1 ♗f2 16. ♘f7

16. – ♖h3! 17. gh3 ♗f3 18. ♔h2 ♗g1#

4717. C 50

N. N. – Blackburne

London 1880

1. e4 e5 2. ♘f3 ♘c6 3. ♗c4 ♗c5 4. ♗f7 ♔f7 5. ♘e5 ♘e5 6. ♕h5 g6 7. ♕e5 d6 8. ♕h8 ♕h4 9. 0-0 ♘f6 10. c3 ♘g4 11. h3 ♗f2 12. ♔h1 ♗f5 13. ♕a8

13. – ♕h3! 14. gh3 ♗e4#

4718. C 53

Dingeldein – Rosch

Munich 1939

1. e4 e5 2. ♘f3 ♘c6 3. ♗c4 ♗c5 4. c3 ♕e7 5. d4 ed4 6. 0-0 dc3 7. ♘c3 ♘f6 8. e5 ♘g4 9. ♘g5 ♘ge5 10. ♘d5 ♕d8 11. ♖e1 0-0 12. ♕c2 ♘g6 13. ♗f4 d6 14. ♕e4 ♔h8 15. h4 f5 16. ♕e2 ♘h4 17. ♕h5 h6 18. ♕h4 ♘d4 19. ♘e7 d5[1]

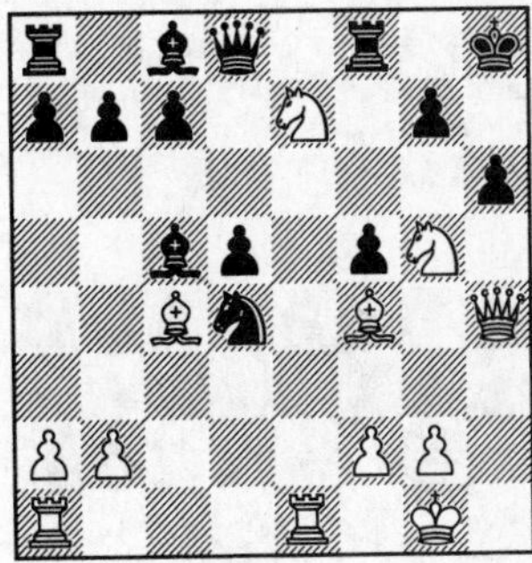

20. ♕h6! gh6 21. ♗e5 ♖f6 22. ♗f6#

4719. C 53

Scheve – Teichmann

Berlin 1907

1. e4 e5 2. ♘f3 ♘c6 3. ♗c4 ♗c5 4. c3 ♕e7 5. 0-0 d6 6. d4 ♗b6 7. a4 a6 8. a5 ♗a7 9. h3 ♘f6 10. de5 ♘e5 11. ♘e5 ♕e5 12. ♘d2

12. – ♗h3! 13. gh3 ♕g3 14. ♔h1 ♕h3 15. ♔g1 ♘g4 16. ♘f3 ♕g3 17. ♔h1 ♗f2 −+

4720. C 54

Marshall – Barnhart

Indianapolis 1907

1. e4 e5 2. ♘f3 ♘c6 3. ♗c4 ♗c5 4. 0-0 d6 5. c3 ♘f6 6. d4 ed4 7. cd4 ♗b6 8. h3 0-0 9. ♘c3 h6 10. ♗e3 ♘e7 11. ♕d2 ♗e6 12. ♗d3 ♘g6

13. ♗h6! gh6 14. ♕h6 ♖e8 15. e5 ♘h7 16. ♘e4 ♗f5 17. ♘eg5 ♘g5 18. ♘g5 ♕d7 19. ♘h7 +−

4721. C 54

N. N.– Hartlaub

Bremen 1889

1. e4 e5 2. ♘f3 ♘c6 3. ♗c4 ♗c5 4. c3 ♘f6 5. d4 ed4 6. cd4 ♗b4 7. ♗d2 ♘e4 8. ♗b4 ♘b4 9. ♗f7 ♔f7 10. ♕b3 d5 11. ♕b4 ♖e8 12. ♘e5 ♔g8 13. 0-0 c5! 14. ♕a4 b5 15. ♕b5 a5 16. ♕a4 ♖b8 17. f4 ♖b2 18. ♘c6

♗h3! 19. gh3[1] ♘g5 20. ♘c3[2] ♘h3 21. ♔h1 ♕h4 22. He5 ♖h2![3] −+

[1] 19. ♘d8 ♖g2 20. ♔h1 ♘f2 21. ♖f2 ♖e1 −+; 19. ♘d2 ♖d2 20. ♘d8 ♖g2 21. ♔h1 ♘g3! 22. hg3 ♖ee2 −+

[2] 20. ♘d8 ♘h3 21. ♔h1 ♖ee2 −+

[3] 23. ♔h2 ♘f4 24. ♔g1 ♕g3 25. ♔h1 ♕g2#

4722. C 57

Lvov – Radchenko

Krasnodar 1957

1. e4 e5 2. ♘f3 ♘c6 3. ♗c4 ♘f6 4. ♘g5 d5 5. ed5 b5 6. ♗f1 ♘d4 7. c3 ♘d5 8. ♘e4 ♕h4 9. ♘g3 ♗b7 10. cd4 0-0-0 11. ♗b5 ♘f4 12. 0-0

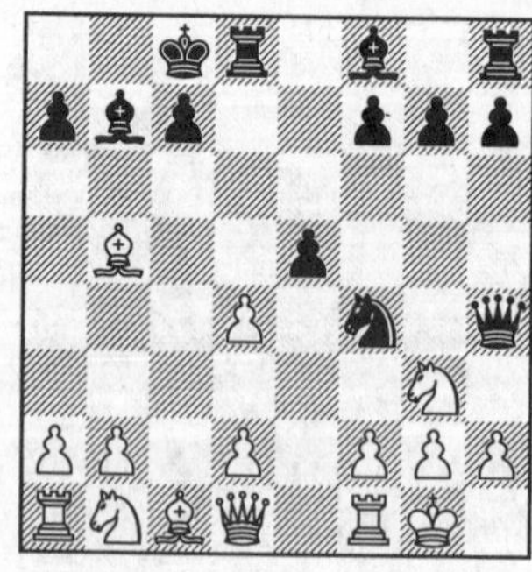

12. – ♕h3! −+

4723. C 58

Field – Techner

1933

1. e4 e5 2. ♘f3 ♘c6 3. ♗c4 ♘f6 4. ♘g5 d5 5. ed5 ♘a5 6. d3 h6 7. ♘f3 e4 8. ♕e2 ♘c4 9. dc4 ♗c5 10. 0-0 0-0 11. ♘fd2 ♗g4 12. ♕e1 ♕d7 13. ♘b3 ♗f3 14. ♗f4 ♕g4 15. ♗g3 ♘h5 16. ♘c5 ♘f4 17. ♘e4

17. – ♕h3! –+

4724. C 60

Stützkowski – Harmonist

Berlin 1898

1. e4 e5 2. ♘f3 ♘c6 3. ♗b5 ♗b4 4. c3 ♗a5 5. 0-0 ♘ge7 6. ♘a3 0-0 7. ♕a4 d5 8. ♗c6 ♘c6 9. ♘e5 ♘e5 10. ♕a5 ♘f3 11. ♔h1 ♕d6 12. gf3 ♕f4 13. ♔g2

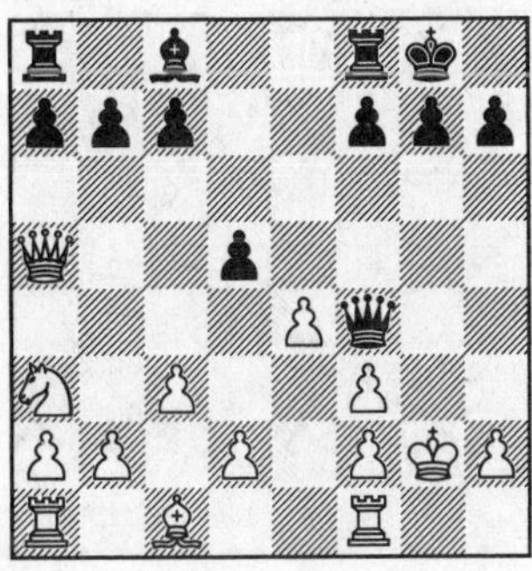

13. – ♗h3! 14. ♔h3 ♕f3 15. ♔h4 g5! 16. ♔g5 ♔h8 17. h3 ♖g8 18. ♔h4 ♕f6 19. ♔h5 ♕g5#

4725. C 61

Spassky – Barua

New York 1987

1. e4 e5 2. ♘f3 ♘c6 3. ♗b5 ♘d4 4. ♘d4 ed4 5. 0-0 ♗c5 6. d3 c6 7. ♗a4 ♘e7 8. f4 d5 9. f5 de4 10. de4 0-0 11. ♗b3 ♗d6 12. ♕h5 d3 13. cd3 ♗e5 14. ♖f3 ♕d4 15. ♗e3 ♕b2 16. ♖h3 h6

17. ♗h6! ♗d4[1] 18. ♔h1 ♕f2 19. ♘c3 g6 20. fg6 ♗h3 21. gh3 ♕f6[2] 22. ♗e3[3] +–

[1] 17. – gh6 18. ♕h6 ♖d8 19. ♕h7 ♔f8 20. ♕f7#; 17. – ♕a1 18. ♗c1! +–

[2] 21. – ♗c3 22. ♗e3 +–

[3] 22. – ♘g6 23. ♗d4 ♕d4 24. ♕g6 ♕g7 25. ♕f5 +–

4726. C 63

Canderay – Bonzanigo

Corr. 1980

1. e4 e5 2. ♘f3 ♘c6 3. ♗b5 f5 4. ♘c3 fe4 5. ♘e4 d5 6. ♘e5 de4 7. ♘c6 ♕d5 8. c4 ♕d6 9. ♘a7 ♗d7 10. ♗d7 ♕d7 11. ♘b5 ♘f6 12. 0-0 ♗c5 13. ♘c3 0-0 14. h3 c6 15. ♕e2 ♕d4 16. b3 ♖ad8 17. a4 ♖f7 18. ♖a2 ♕e5 19. ♖c2 ♖d3 20. ♗b2

20. – ♖h3![1] –+

[1] 21. gh3 ♕g3 22. ♔h1 ♕h3 23. ♔g1 ♘g4 –+

4727. C 65

Nedvidek – Kucera

Plzen 1914

1. e4 e5 2. ♘f3 ♘c6 3. ♗b5 ♘f6 4. d3 d6 5. c3 ♗e7 6. ♘bd2 0-0 7. h3 ♘d7 8. g4 ♘b6 9. ♘f1 a6 10. ♗c6 bc6 11. ♘e3 ♖e8 12. h4 ♗f8 13. ♘f5 d5 14. ♕e2 g6 15. ♖g1 gf5 16. gf5 ♔h8 17. ♘g5 ♕f6 18. ♕h5 h6 19. ♘f7 ♔h7

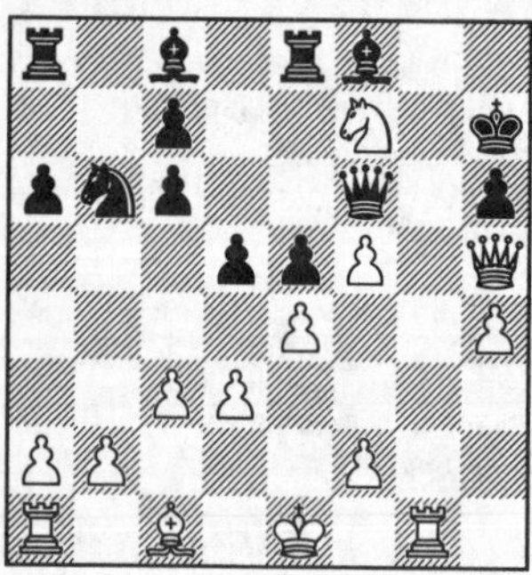

20. ♘h6![1] +–

[1] 20. – ♗h6 21. ♗h6 ♕h6 22. ♕f7 ♔h8 23. ♕e8 ♔h7 24. ♕g8#

4728. C 66

Varain – Salminger

Munich 1895

1. e4 e5 2. ♘f3 ♘c6 3. ♗b5 ♘f6 4. 0-0 d6 5. d4 ed4 6. ♘d4 ♗d7 7. ♘c3 ♗e7 8. f4 0-0 9. ♗e3 ♕e8 10. ♘c6 ♗c6 11. ♗d3 ♕d7 12. ♗d4 ♖fe8 13. ♔h1 h6 14. e5 ♕h3 15. ♘e4

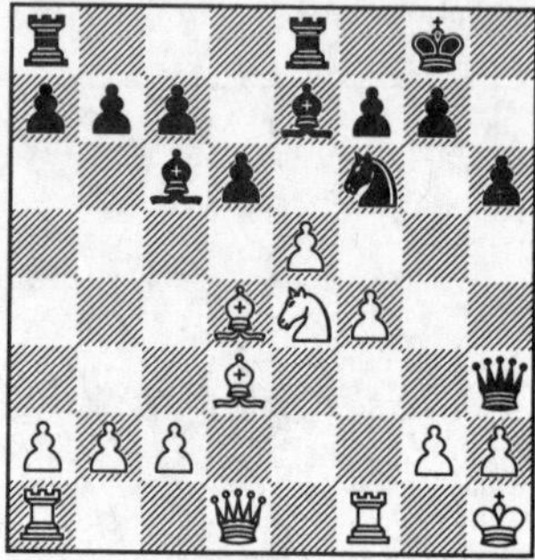

15. – ♘e4! 16. gh3 ♘f2 17. ♔g1 ♘h3#

4729. C 67

Hesse – N. N.

Bethlehem

1. e4 e5 2. ♘f3 ♘c6 3. ♗b5 ♘f6 4. 0-0 ♘e4 5. d4 ♗e7 6. d5 ♘d6 7. ♗a4 e4 8. ♘fd2 ♘b8 9. f3 ef3 10. ♘f3 0-0 11. ♕d3 ♘e8 12. c4 ♘f6 13. ♗c2 d6 14. ♘g5 g6 15. ♘c3 ♗f5 16. ♖f5 gf5 17. ♕h3 ♔g7 18. ♗f5 ♖h8

19. ♕h6![1] +–

[1] 19. – ♔h6 20. ♘e6 ♔h5 21. ♘g7 ♔h4 22. g3#; 19. – ♔g8 20. ♘e6! +–

4730. C 68

Jay – Finkelstein

USA 1949

1. e4 e5 2. ♘f3 ♘c6 3. ♗b5 a6 4. ♗c6 dc6 5. 0-0 ♗d6 6. h3 ♘e7 7. c3 ♗e6 8. d4 ♘g6 9. ♗g5 ♕d7 10. ♘bd2 ed4 11. ♘d4

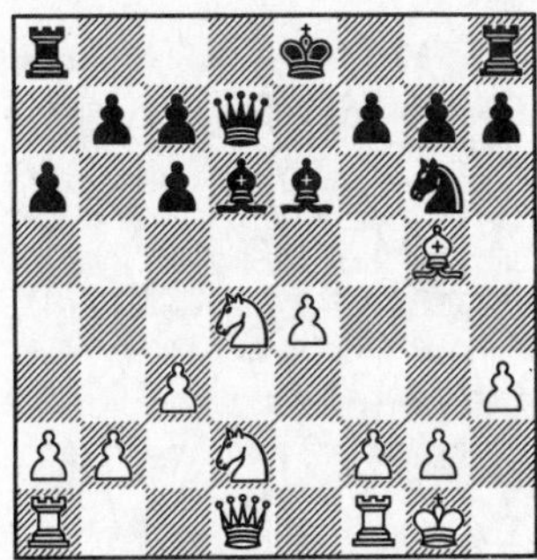

11. – ♗h3! 12. gh3 ♕h3 13. ♘2f3[1] h6[2] –+

[1] 14. f4 h6 –+

[2] 14. ♗e3 ♘h4 –+

4731. C 68

Rogers – Buchanan

Staunton 1978

1. e4 e5 2. ♘f3 ♘c6 3. ♗b5 a6 4. ♗c6 dc6 5. 0-0 ♗g4 6. h3 h5 7. d3 ♕f6 8. ♘bd2 ♘e7 9. b3 ♘g6 10. ♗b2 ♗d6 11. ♕e1 ♘f4 12. ♕e3

12. – ♗h3! 13. ♘e5[1] ♗e5 14. ♗e5 ♕e5 15. g3[2] ♘e6 16. ♘f3 ♕f6 17. e5 ♕h6 –+

[1] 13. gh3 ♕g6 –+

[2] 15. gh3 ♕g5 16. ♕g3 ♘e2 –+

4732. C 72

Shobert – Buchanan

Lima Ohio 1975

1. e4 e5 2. ♘f3 ♘c6 3. ♗b5 a6 4. ♗a4 d6 5. 0-0 ♗g4 6. h3 h5 7. d3 ♕f6 8. ♘bd2 ♘e7 9. a3 ♘g6 10. c3 ♘f4 11. d4 b5 12. ♗b3

12. – ♗h3! 13. gh3 ♕g6 –+

4733. C 77

Mosionzik – Nikolaevsky

Moscow 1972

1. e4 e5 2. ♘f3 ♘c6 3. ♗b5 a6 4. ♗a4 ♘f6 5. ♕e2 b5 6. ♗b3 ♘a5 7. d4 ♘b3 8. ab3 ed4 9. e5 ♘d5 10. ♘d4 ♗b7 11. 0-0 ♕h4 12. ♘f5 ♘f4 13. ♕d1

13. – ♛h3! 14. ♘e3[1] ♝g2 –+

[1] 14. gh3 ♞h3#

4734. C 78

Exner – Englund

Hannover 1920

1. e4 e5 2. ♘f3 ♘c6 3. ♗b5 a6 4. ♗a4 ♘f6 5. 0-0 ♗c5 6. c3 ♗a7 7. d4 ♘e4 8. de5 0-0 9. ♗f4 ♘e7 10. ♘g5 ♘f2 11. ♖f2 ♗f2 12. ♔f2 f6 13. ♗b3 ♔h8 14. ♘h7 ♕e8 15. ♘f8 fe5 16. ♕f3 d6 17. ♗g5 ♘f5 18. ♕h3 ♘h6

19. ♕h6! gh6 20. ♗f6#

4735. C 78

Neshesany – Vashichek

Prague 1947

1. e4 e5 2. ♘f3 ♘c6 3. ♗b5 a6 4. ♗a4 ♘f6 5. 0-0 b5 6. ♗b3 ♗c5 7. c3 0-0 8. d4 ♗b6 9. ♘e5 ♘e7 10. ♗g5 ♘e8 11. ♕h5 ♘d6 12. ♕h4 ♖e8 13. ♕f4 ♖f8 14. ♘g4 ♕e8 15. ♗f6 ♘g6

16. ♕h6! +–

4736. C 78

N. N. – Tarrasch

Munich 1932

1. e4 e5 2. ♘f3 ♘c6 3. ♗b5 a6 4. ♗a4 ♘f6 5. 0-0 ♗c5 6. ♘e5 ♘e4 7. ♘c6 dc6 8. ♕f3 ♕h4 9. ♘c3 ♘c3 10. ♗c6 bc6 11. ♕c6 ♗d7 12. ♕a8 ♔e7 13. ♕h8 ♘e2 14. ♔h1 ♗f2 15. h3

15. – ♛h3! 16. gh3 ♝c6 17. ♔h2 ♝g3#

4737. C 82

Brach – Horak

Corr. 1911–12

1. e4 e5 2. ♘f3 ♘c6 3. ♗b5 a6 4. ♗a4 ♘f6 5. 0-0 ♘e4 6. d4 b5 7. ♗b3 d5 8. de5 ♗e6 9. c3 ♗c5 10. ♘bd2 0-0 11.

♗c2 f5 12. ef6 ♘f6 13. ♘b3 ♗d6 14. ♘bd4 ♘d4 15. cd4 ♗g4 16. ♗d3 ♕e8 17. ♗e2 ♕h5 18. h3 ♖ae8 19. ♗e3 g5 20. ♖e1 ♖e3! 21. fe3[1]

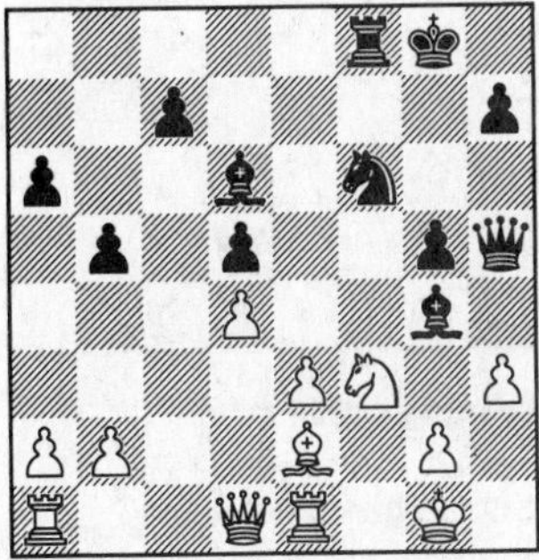

21. – ♗h3! 22. ♘e5 ♕h4 23. ♖f1 ♕g3 24. ♖f2 ♘g4 –+

[1] 21. hg4 ♘g4 22. fe3 ♗h2 23. ♔f1 ♘e3 –+

4738. C 83

Ansat – Radisch

Corr. 1936–37

1. e4 e5 2. ♘f3 ♘c6 3. ♗b5 a6 4. ♗a4 ♘f6 5. 0-0 ♘e4 6. d4 b5 7. ♗b3 d5 8. de5 ♗e6 9. c3 ♗e7 10. ♗f4 g5 11. ♗g3 h5 12. h3 g4 13. ♘fd2 ♘g3 14. fg3 ♗c5 15. ♔h2 h4 16. gh4 g3 17. ♔h1 ♕h4 18. ♘f3

18. – ♕h3! 19. gh3 ♗h3[1] –+

[1] 20. ♘h2 g2#

4739. C 86

Sax – Kalka

Dortmund 1993

1. e4 e5 2. ♘f3 ♘c6 3. ♗b5 a6 4. ♗a4 ♘f6 5. 0-0 ♗e7 6. ♕e2 b5 7. ♗b3 d6 8. c3 ♗g4 9. h3 ♗h5 10. d3 0-0 11. ♘bd2 ♘a5 12. ♗c2 c5 13. ♖d1 ♘c6 14. ♘f1 ♕c7 15. ♘g3 ♗g6 16. ♘h4 ♖fe8 17. ♘hf5 ♗f8 18. ♕f3 ♘d7 19. h4 ♘b6 20. h5 ♗f5 21. ♘f5 ♖ad8 22. ♗b3 ♘a5 23. ♗g5 ♖d7

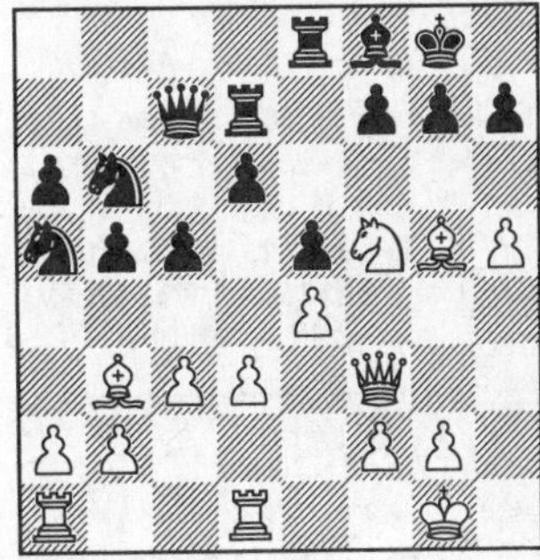

24. ♘h6! +–

4740. C 88

Thompson – Nunez

California 1976

1. e4 e5 2. ♘f3 ♘c6 3. ♗b5 a6 4. ♗a4 ♘f6 5. 0-0 ♗e7 6. ♖e1 b5 7. ♗b3 0-0 8. h3 d5 9. ed5 ♘d5 10. ♘e5 ♘e5 11. ♖e5 c6 12. ♗d5 cd5 13. d4 ♗d6 14. ♖e1 ♕h4 15. ♘c3

15. – ♗h3! 16. gh3 ♕h3 17. ♕d3[1] ♗h2 18. ♔h1 ♗g3[2] –+

[1] 17. ♖e2 ♕h2 18. ♔f1 ♕h1#; 17. ♖e5! ♗e5 18. de5 ♖ae8 –+

[2] 19. ♔g1 ♕h2 20. ♔f1 ♕f2#

4741. C 89

Gifford – Jordan

Dayton 1976

1. e4 e5 2. ♘f3 ♘c6 3. ♗b5 a6 4. ♗a4 ♘f6 5. 0-0 b5 6. ♗b3 ♗e7 7. ♖e1 0-0 8. c3 d5 9. ed5 ♘d5 10. ♘e5 ♘e5 11. ♖e5 ♘f6 12. ♖e1 ♗d6 13. f3 ♘h5 14. d4 ♕h4 15. h3 ♕g3 16. ♔f1 ♕h2 17. ♗e3

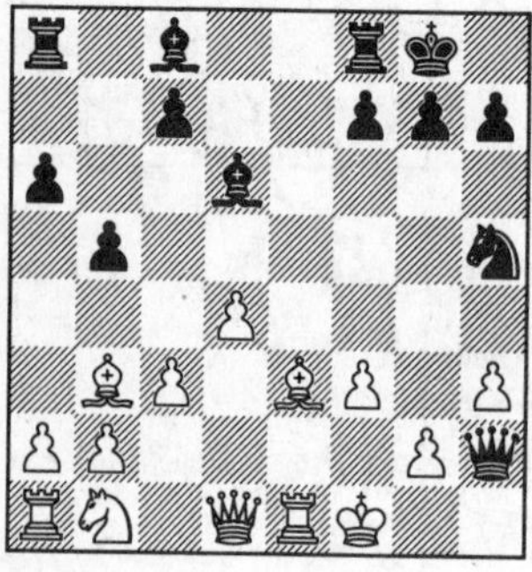

17. – ♗h3! 18. gh3 ♘g3#

4742. C 89

Koskinen – Grawe

Corr. 1965

1. e4 e5 2. ♘f3 ♘c6 3. ♗b5 a6 4. ♗a4 ♘f6 5. 0-0 ♗e7 6. ♖e1 b5 7. ♗b3 0-0 8. c3 d5 9. ed5 ♘d5 10. ♘e5 ♘e5 11. ♖e5 ♘f6 12. ♖e1 ♗d6 13. h3 ♘g4 14. ♕f3 ♕h4 15. d3 ♘f2 16. ♕f2 ♗h2 17. ♔f1 ♗g3 18. ♕e2

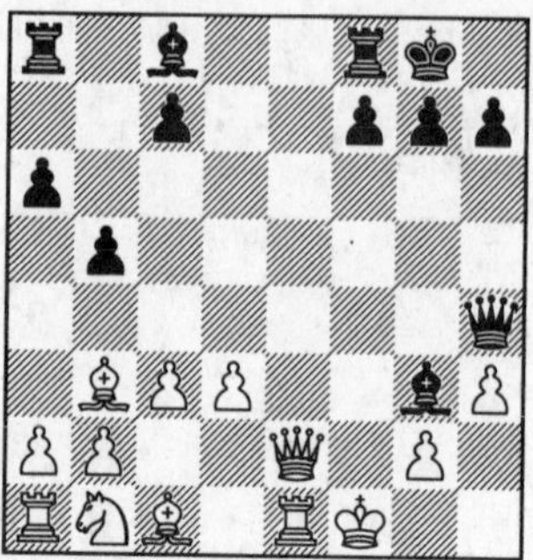

18. – ♗h3![1] –+

[1] 19. gh3 ♖ae8! 20. ♕e8 ♕h3 –+

4743. C 89

Zywicki – Carpenter

Michigan 1975

1. e4 e5 2. ♘f3 ♘c6 3. ♗b5 a6 4. ♗a4 ♘f6 5. 0-0 ♗e7 6. ♖e1 b5 7. ♗b3 0-0 8. c3 d5 9. ed5 ♘d5 10. ♘e5 ♘e5 11. ♖e5 ♘f6 12. d4 ♗d6 13. ♖e1 ♘g4 14. h3 ♕h4 15. ♕f3 ♘f2 16. ♗f7 ♖f7 17. ♖e8 ♗f8 18. ♕a8

18. – ♘h3! 19. gh3 ♕f2 20. ♔h1 ♕f1 21. ♔h2 ♕h3[1] –+

[1] 22. ♔g1 ♖f1#

4744. D 03
Richter – Moritz
Swinemünde 1931

1. d4 ♘f6 2. ♘f3 e6 3. ♗g5 c5 4. e3 ♘c6 5. ♗d3 d5 6. c3 ♗d6 7. ♘bd2 ♗d7 8. 0-0 ♕b6 9. dc5 ♗c5 10. b4 ♗e7 11. b5 ♘a5 12. ♘e5 ♕c7 13. ♘df3 0-0 14. ♕a4 b6 15. ♕h4 h6

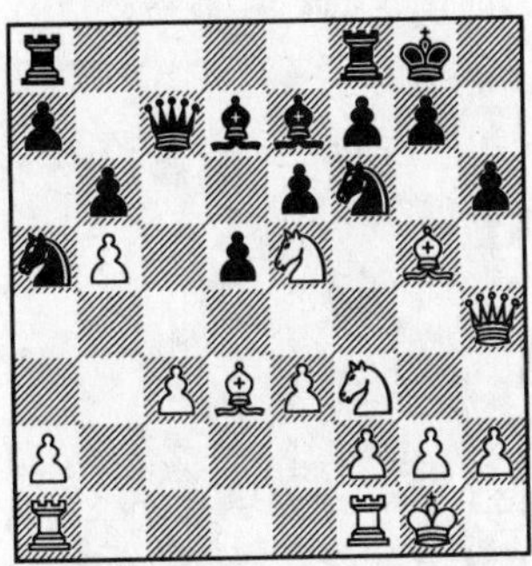

16. ♗h6! gh6 17. ♕g3 ♘g4[1] 18. ♕g4 ♔h8 19. ♕f4 +–

[1] 17. – ♔h8 18. ♘g6 +–

4745. D 05
Chernev – Jackson
New York 1940

1. d4 d5 2. ♘f3 ♘f6 3. e3 e6 4. ♗d3 ♗d6 5. ♘bd2 0-0 6. 0-0 ♘bd7 7. ♕e2 c5 8. c3 e5 9. e4 ed4 10. cd4 de4 11. ♘e4 ♘e4 12. ♕e4 ♘f6 13. ♕h4 cd4 14. ♗g5 h6

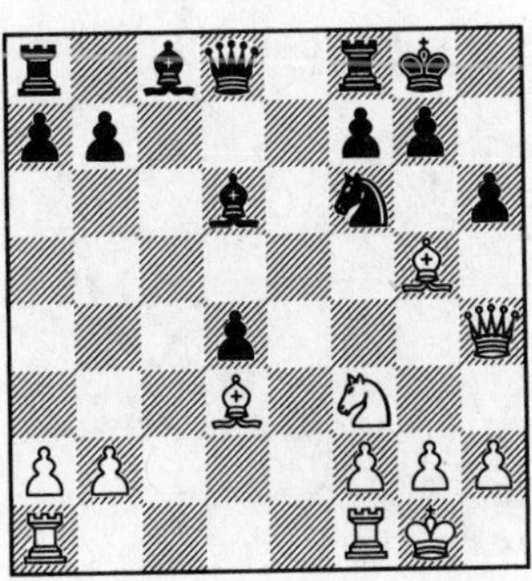

15. ♗h6! gh6 16. ♕h6 ♖e8 17. ♖ae1 ♗e6 18. ♘g5 ♗f4 19. ♖e6[1] +–

[1] 19. – fe6 20. ♕g6 ♔h8 21. ♘f7#; 20. – ♔f8 21. ♕f7#

4746. D 15
Lebedev – Goncek
Tulsk 1938

1. ♘f3 d5 2. d4 ♘f6 3. c4 c6 4. ♘c3 dc4 5. e3 e6 6. ♘e5 ♘bd7 7. f4 ♗b4 8. ♗c4 ♘e4 9. 0-0 ♘c3 10. bc3 ♗c3 11. ♖b1 ♘f6 12. f5 ♘d5 13. fe6 ♗e6 14. ♖b7 0-0 15. ♕b3 ♕c8 16. ♗a3 ♖e8 17. ♘f7 ♘e3 18. ♗e6 ♕e6

19. ♘h6! ♔h8[1] 20. ♖f8 ♖f8 21. ♗f8! ♖f8[2] 22. ♘f7[3] +–

[1] 19. – gh6 20. ♖f8! +–
[2] 21. – ♕b3 22. ♗g7#
[3] 22. – ♖f7 23. ♖b8 +–

4747. D 35
Jones – Shapiro
New York 1947

1. d4 ♘f6 2. c4 e6 3. ♘c3 d5 4. ♗g5 ♗e7 5. e3 ♘bd7 6. ♘f3 0-0 7. cd5 ed5 8. ♖c1 c6 9. ♕c2 ♖e8 10. ♗d3 ♘f8 11. h3 ♘g6 12. 0-0 ♘h5 13. ♗e7 ♕e7 14. ♖b1 ♕f6 15. ♘d2 ♘h4 16. ♘e2

16. – ♗h3! 17. ♘g3[1] ♕g5 18. ♔h2 ♖e3! 19. gh3 ♖g3! 20. ♖g1 ♖g2 21. ♔h1 ♘g3! 22. fg3 ♕g3 –+

[1] 17. gh3 ♕g5 18. ♘g3 ♘g3 –+

4748. D 36

Gereben – Komarov

Leningrad 1949

1. d4 d5 2. ♘f3 ♘f6 3. c4 e6 4. ♘c3 ♘bd7 5. ♗g5 ♗e7 6. e3 0-0 7. cd5 ed5 8. ♗d3 ♖e8 9. 0-0 c6 10. ♕c2 ♘f8 11. ♘e5 ♘g4 12. ♗e7 ♕e7 13. ♘g4 ♗g4 14. ♖ab1 ♕g5 15. ♔h1 ♖e6 16. ♘e2 ♖h6 17. ♘f4 ♕h4 18. h3 g5 19. g3 ♗f3 20. ♔h2

20. – ♕h3! 21. ♘h3 g4 –+

4749. D 37

Foltys – Book

Munich 1936

1. d4 d5 2. ♘f3 ♘f6 3. c4 e6 4. ♘c3 dc4 5. e4 ♗b4 6. e5 ♘d5 7. ♗d2 c5 8. dc5 ♘c3 9. bc3 ♗c5 10. ♗c4 ♘d7 11. 0-0 0-0 12. ♕e2 ♗e7 13. ♖ad1 ♕c7 14. ♗d3 ♘c5 15. ♗b1 b6 16. ♘g5 h6 17. ♘h7 ♗a6 18. c4 ♖fd8 19. ♘f6 ♔h8

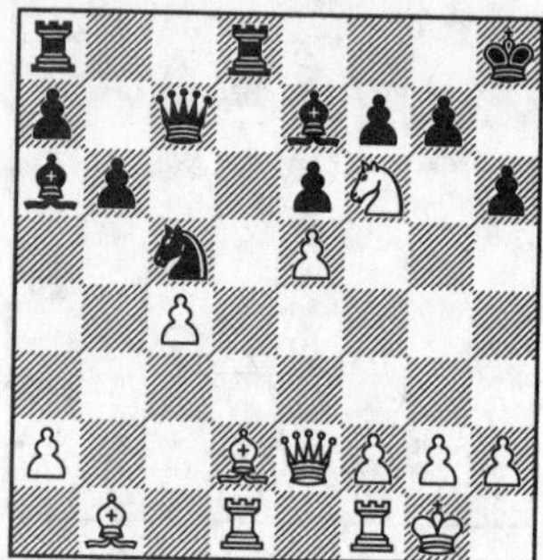

20. ♗h6! gh6 21. ♕c2 +–

4750. D 40

Gerasimov – Smyslov

Moscow 1935

1. d4 d5 2. ♘f3 ♘f6 3. e3 e6 4. ♗d3 c5 5. b3 ♘c6 6. ♗b2 ♗d6 7. 0-0 ♕c7 8. a3 b6 9. c4 ♗b7 10. ♘c3 a6 11. ♖e1 cd4 12. ed4 0-0 13. ♘a4 ♗f4 14. ♘e5 dc4 15. bc4 ♘e5 16. de5 ♕c6 17. ♗f1 ♖fd8 18. ♕b3 ♘g4 19. h3 ♖d3! 20. ♕b6

20. – ♖h3! 21. ♗d4[1] ♗h2 22. ♔h1 ♗e5[2] –+

[1] 21. ♕c6 ♗h2 22. ♔h1 ♘f2#
[2] 23. ♔g1 ♗h2 24. ♔h1 ♗c7 –+

4751. D 46
Capablanca – Jaffe
New York 1910

1. d4 d5 2. ♘f3 ♘f6 3. e3 c6 4. c4 e6 5. ♘c3 ♘bd7 6. ♗d3 ♗d6 7. 0-0 0-0 8. e4 de4 9. ♘e4 ♘e4 10. ♗e4 ♘f6 11. ♗c2 h6 12. b3 b6 13. ♗b2 ♗b7 14. ♕d3 g6 15. ♖ae1 ♘h5 16. ♗c1 ♔g7 17. ♖e6! ♘f6 18. ♘e5 c5

19. ♗h6! ♔h6 20. ♘f7! +–

4752. D 47
Denker – Willman
New York 1934

1. c4 ♘f6 2. ♘c3 e6 3. d4 d5 4. ♘f3 ♘bd7 5. e3 c6 6. ♗d3 dc4 7. ♗c4 b5 8. ♗d3 ♗e7 9. 0-0 0-0 10. e4 b4 11. ♘a4 h6 12. e5 ♘d5 13. ♕e2 ♖e8 14. ♕e4 ♘f8 15. ♘c5 ♕b6 16. ♕g4 f5 17. ef6 ♘f6 18. ♕g3 ♔h8

19. ♗h6! gh6 20. ♘e5 +–

4753. D 55
Charvat – Hakl
Prague 1908

1. d4 d5 2. c4 e6 3. ♘c3 ♘f6 4. ♗g5 ♗e7 5. e3 0-0 6. ♘f3 b6 7. ♕c2 ♗b7 8. cd5 ♘d5 9. h4 f6 10. ♗d3 ♘b4 11. ♗h7 ♔h8 12. ♕g6 ♗f3 13. gf3 fg5 14. hg5 ♗g5 15. ♕e6 ♗h6

16. ♖h6! gh6 17. ♕h6 ♕e7[1] 18. ♗c2 ♔g8 19. ♗b3 ♖f7 20. ♔e2 +–

[1] 17. – ♖f6 18. ♗g6 ♔g8 19. 0-0-0 +–

4754. D 58
Bruckner – Zude
GFR 1987

1. d4 ♘f6 2. ♘f3 d5 3. c4 e6 4. ♘c3 ♗e7 5. ♗g5 0-0 6. e3 h6 7. ♗h4 b6 8. ♗d3 ♗b7 9. 0-0 ♘bd7 10. ♕e2 c5 11.

♖fd1 ♘e4 12. ♗g3 cd4 13. ♘d4 ♘c3 14. bc3 ♕c8 15. cd5 ed5 16. ♘f5 ♗f6 17. ♕g4 ♔h8

18. ♘h6! g6[1] 19. ♗d6 ♔g7 20. ♘f7! +–

[1] 18. – gh6 19. ♕f5 +–

4755. D 61
Feierherst – Konradi
Dublin 1957

1. d4 ♘f6 2. c4 e6 3. ♘c3 d5 4. ♗g5 ♗e7 5. e3 0-0 6. ♕c2 h6 7. ♗f4 c5 8. ♘f3 a6 9. ♖d1 cd4 10. ♘d4 ♘d7 11. ♗g3 dc4 12. ♗c4 ♗b4 13. 0-0 ♗c3 14. bc3 e5 15. ♘f5 ♕c7 16. ♖d4 ♕c5

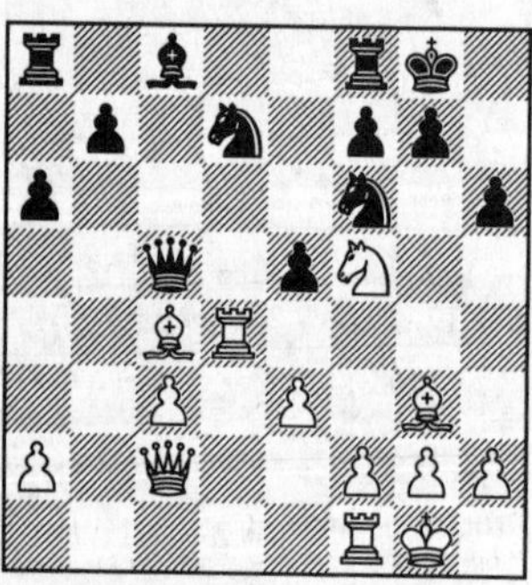

17. ♘h6! gh6 18. ♕g6 ♔h8 19. ♕h6[1] +–

[1] 19. – ♔g8 20. ♕g6 ♔h8 21. ♖h4 +–

4756. D 90
Abrahams – Spencer
England 1930

1. d4 d5 2. c4 c6 3. ♘f3 ♘f6 4. ♘c3 g6 5. ♗f4 ♗g7 6. e3 0-0 7. ♗d3 b6 8. ♘e5 ♗b7 9. cd5 cd5 10. h4 h6 11. h5 g5 12. ♗g5! hg5 13. h6 ♗h6 14. ♖h6 ♔g7

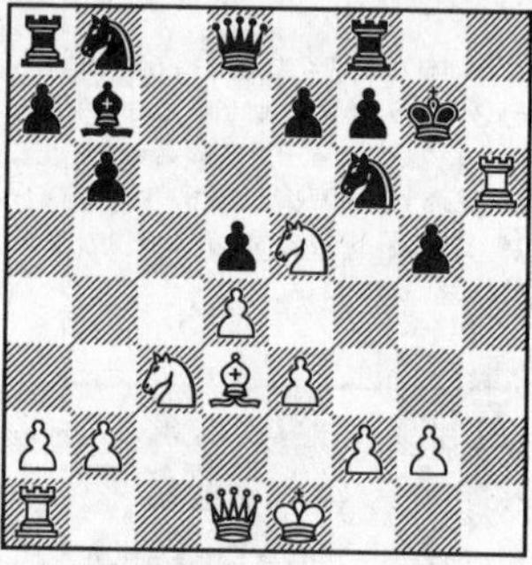

15. f4! ♔h6 16. fg5 ♔g5[1] 17. ♕f3 ♔h6[2] 18. ♕f4 ♔g7 19. ♕g5 ♔h8 20. 0-0-0 +–

[1] 16. – ♔g7 17. gf6 ♔f6 18. ♕f3 ♔e6 19. ♕f5 ♔d6 20. ♘b5#
[2] 17. – ♘h5 18. ♕f5 ♔h6 19. ♘f7 ♔g7 20. ♕g6#

4757. E 11
Costa – Schauwecker
Swiss 1993

1. d4 e6 2. c4 ♗b4 3. ♘d2 d5 4. a3 ♗d2 5. ♗d2 ♘f6 6. e3 0-0 7. ♘f3 ♘bd7 8. ♖c1 ♘e4 9. ♗d3 f5 10. 0-0 ♖f6 11. ♕c2 c6 12. ♗e1 ♖h6 13. ♘e5 ♘e5 14. de5 ♕h4 15. h3 ♘g5 16. f3

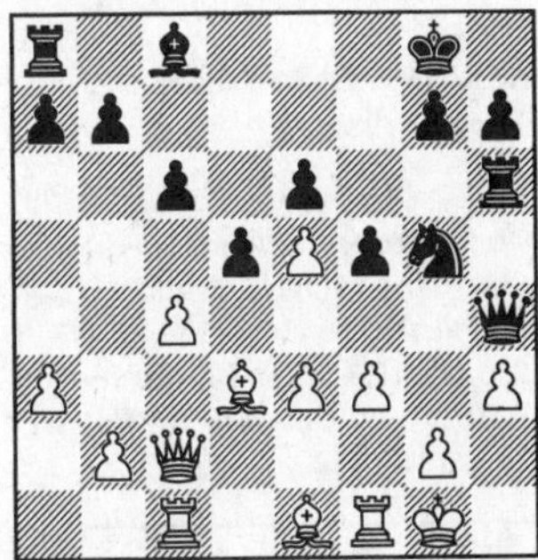

16. – ♘h3! 17. gh3 ♕h3 18. ♕g2 ♖g6 19. ♕g6 hg6 20. ♖c2 ♕h6 21. ♖e2 ♕g5 22. ♔h1 ♔f7 23. e4 ♗d7 24. ♗d2 –+

4758. E 14

Höi – Nielsen

Gausdal 1989

1. d4 ♘f6 2. ♘f3 e6 3. e3 c5 4. ♗d3 b6 5. 0-0 ♗e7 6. c4 ♗b7 7. ♘c3 cd4 8. ed4 d6 9. d5 e5 10. ♘g5 0-0 11. f4 ef4 12. ♗f4 ♘bd7 13. ♕f3 ♖e8 14. ♕h3 ♘f8 15. ♖ae1 a6 16. ♗d2 ♕c8 17. ♕h4 h6

18. ♘f7!! ♔f7 19. ♗h6 gh6 20. ♖e7! +–

4759. E 32

Johner – L. Steiner

Berlin 1928

1. d4 ♘f6 2. c4 e6 3. ♘c3 ♗b4 4. ♕c2 d6 5. e4 ♗c3 6. bc3 0-0 7. ♗d3 e5 8. ♘e2 ♕e7 9. 0-0 c5 10. f4 ♘fd7 11. ♘g3 ♖e8 12. ♘f5 ♕f8 13. fe5 de5 14. ♕f2 ♘b6

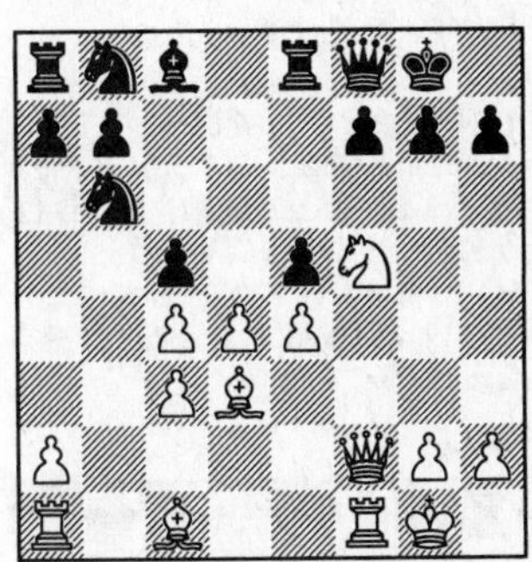

15. ♘h6![1] +–

[1] 15. – gh6 16. ♗h6! ♕h6 17. ♕f7 ♔h8 18. ♕e8 +–; 16. – ♕e7 17. ♕g3 +–

4760. E 34

Pártos – Gonzalez

New York 1946

1. d4 ♘f6 2. c4 e6 3. ♘c3 ♗b4 4. ♕c2 d5 5. e3 0-0 6. ♘f3 ♘bd7 7. a3 ♗e7 8. ♗d3 dc4 9. ♗c4 c5 10. 0-0 cd4 11. ♘d4 a6 12. ♗e6 fe6 13. ♘e6 ♕e8 14. ♘c7 ♕h5 15. ♘a8 ♘e5 16. ♘b6 ♗d6 17. h3

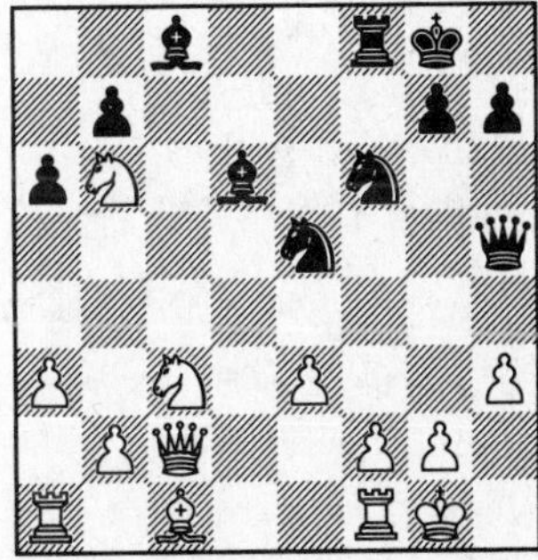

17. – ♗h3! 18. f4[1] ♗f5 19. e4 ♗c5 20. ♖f2 ♘eg4 21. ♘d1 ♘e4 22. ♘c4 ♘gf2 +–

[1] 18. gh3 ♘f3 19. ♔g2 ♘h4 20. ♔h1 ♕f3 –+

4761. E 48

Rabinovich – Goglidze

Leningrad – Moscow 1939

1. d4 ♘f6 2. c4 e6 3. ♘c3 ♗b4 4. e3 0-0 5. ♗d3 d5 6. ♘ge2 dc4 7. ♗c4 a6 8. 0-0 c5 9. a3 ♗c3 10. bc3 ♕c7 11. ♗d3 e5 12. de5 ♕e5 13. ♖b1 ♘c6 14. ♘g3 b5 15. c4 bc4 16. ♗c4 ♕e7 17. ♕f3 ♗e6 18. ♘f5 ♕d7 19. ♗e6 ♕e6 20. ♗b2 ♘e5 21. ♕f4 ♘g6 22. ♕g5 ♘e4

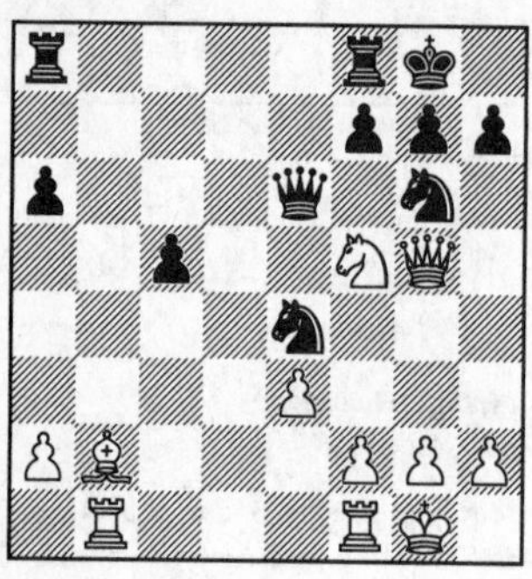

23. ♕h6! +–

4762. E 61

Nedelkov – Zsuzsa Polgár

Trgoviste 1982

1. d4 ♘f6 2. c4 g6 3. ♘c3 ♗g7 4. ♘f3 0-0 5. e3 d6 6. ♗d3 ♘bd7 7. e4 e5 8. d5 a5 9. 0-0 ♘c5 10. ♗c2 ♘h5 11. ♗g5 ♕e8 12. ♗e3 b6 13. h3 f5 14. ef5 gf5 15. ♗c5 bc5 16. ♘b5 ♕f7 17. ♖ab1 e4 18. ♘e1 ♘f4 19. ♕d2 ♗h6 20. ♕d1 ♕g7 21. ♔h2 ♔h8 22. g3

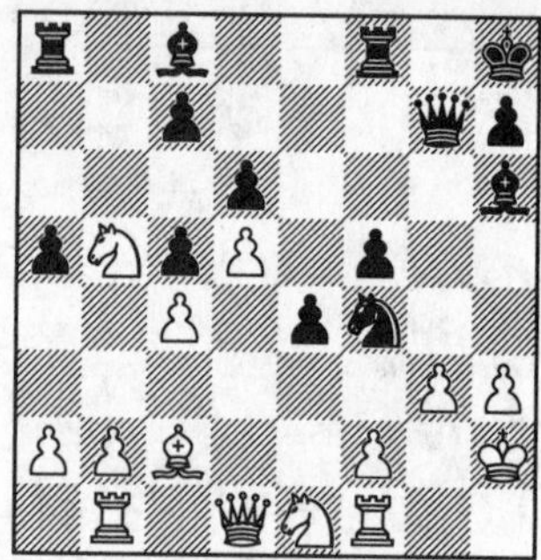

22. – ♘h3! 23. ♕h5 f4 24. ♘g2 ♘f2! –+

[1] 23. ♔h3 f4 24. ♔g2 f3 25. ♔g1 ♗e3! –+

4763. A 00

Hartlaub – Jost

Bad Kissingen 1921

(without ♘b1)

1. b4 e5 2. ♗b2 d6 3. e3 ♘f6 4. f4 ef4 5. ♘f3 fe3 6. ♗d3 ♗e7 7. 0-0 0-0 8. c4 ♗g4 9. ♖c1 c5 10. ♗b1 cb4 11. ♕c2 g6 12. ♘g5 ♗f5 13. ♖f5 ♘bd7 14. ♖f3 ♘g4 15. ♘h7! ♔h7

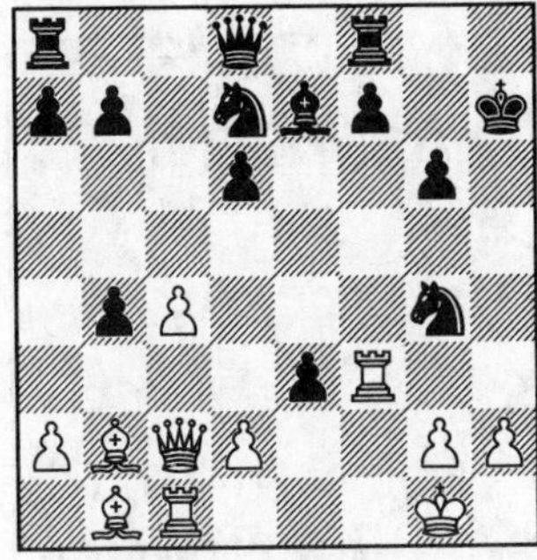

16. ♖f7! ♖f7 17. ♕g6#

4764. A 04

Vaisman – Stefanov

Roumania 1980

1. ♘f3 c5 2. b3 ♘f6 3. ♗b2 e6 4. e3 ♗e7 5. c4 0-0 6. ♘c3 b6 7. d4 cd4 8. ed4 d5 9. ♗d3 dc4 10. bc4 ♗b7 11. 0-0 ♘c6 12. ♖c1 ♖c8 13. ♕e2 ♕d6 14. ♖fd1 ♖fd8 15. ♗b1 ♘a5 16. ♘e5 ♗f8 17. ♘b5 ♕b8 18. d5! ed5

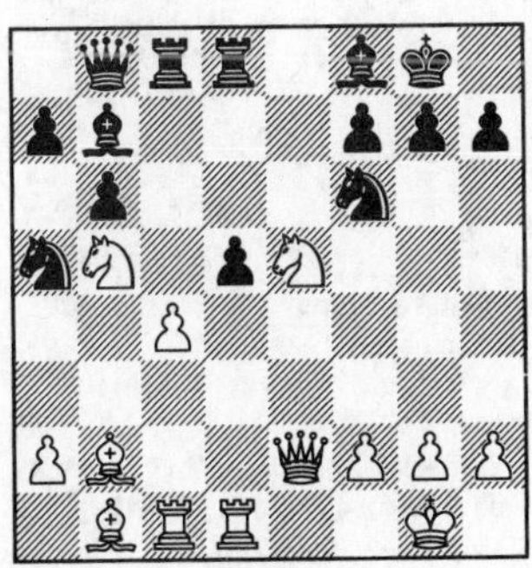

19. ♘f7! ♖e8[1] 20. ♕f3! dc4[2] 21. ♘h6 ♔h8 22. ♕f6! gf6 23. ♗f6 ♗g7 24. ♖d7! ♕e5 25. ♘f7[3] +–

[1] 19. – ♔f7 20. ♗f6 ♔f6 21. ♕f3 ♔e7 22. ♖e1 ♔d7 23. cd5 +–

[2] 20. – ♔f7 21. ♗f6 gf6 22. ♕h5 +–; 20. – d4 21. ♘h6 ♔h8 22. ♕h3! gh6 23. ♗d4 ♗g7 24. ♗f6 ♗f6 25. ♕h6 +–

[3] 25. – ♔g8 26. ♘e5 ♗f6 27. ♗h7 ♔f8 28. ♘g6#

4765. A 05

Ivkov – Torres

1984

1. ♘f3 ♘f6 2. c4 e6 3. g3 b6 4. ♗g2 ♗b7 5. 0-0 ♗e7 6. b3 0-0 7. ♗b2 d5 8. e3 c5 9. ♘c3 ♘c6 10. ♕e2 ♖c8 11. ♖fd1 ♕c7 12. ♖ac1 ♖fd8 13. cd5 ed5 14. d4 ♘a5 15. ♗h3 c4 16. ♘e5 a6 17. e4 b5 18. ed5 ♘d5

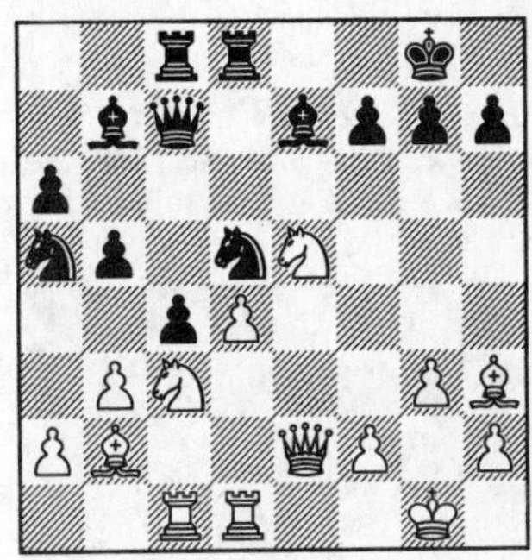

19. ♘f7! ♔f7 20. ♗e6 ♔f6[1] 21. ♕e4 g6 22. ♘d5 ♗d5 23. ♗d5 ♔g7 24. ♗e6 +–

[1] 20. – ♔f8 21. ♕h5 g6 22. ♕h7 +–

4766. A 09

D. Byrne – Pinkus

New York 1948

1. ♘f3 d5 2. c4 dc4 3. e3 c5 4. ♗c4 ♘c6 5. d4 e6 6. 0-0 ♘f6 7. ♕e2 a6 8. ♘c3 b5 9. ♗b3 ♗b7 10. ♖d1 ♕c7 11. d5 ed5 12. e4 de4 13. ♘e4 ♘e4 14. ♕e4 ♘e7

15. ♗f7![1] +–

[1] 15. – ♔f7 16. ♘g5 ♔e8 17. ♕e6 ♗d5 18. ♖d5 +–

4767. A 29

Wolz – Benzinger

Munich 1940

1. c4 e5 2. ♘c3 ♘c6 3. ♘f3 ♘f6 4. d4 ed4 5. ♘d4 ♗b4 6. g3 ♘e5 7. ♕b3 ♗c3 8. bc3 ♕e7 9. ♗e3 ♘fg4 10. ♘c2

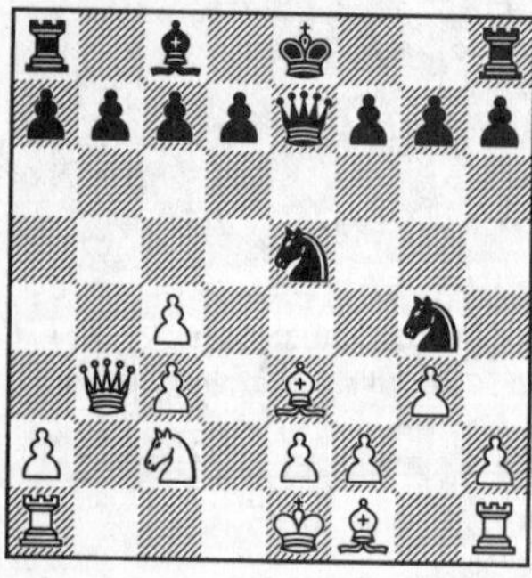

10. – ♘f2! 11. ♔f2[1] ♘g4 12. ♔f3 f5 13. ♔g2[2] ♕e4 14. ♔g1 ♘e3 15. ♘e3 ♕e3 16. ♔g2 b6 –+

[1] 11. ♗f2 ♘d3 –+

[2] 13. ♔f4 g5 14. ♔f5 d6#

4768. A 31

Petrosian – Korchnoy

Curacao 1962

1. c4 c5 2. ♘f3 ♘f6 3. d4 cd4 4. ♘d4 g6 5. ♘c3 d5 6. ♗g5 dc4 7. e3 ♕a5 8. ♗f6 ef6 9. ♗c4 ♗b4 10. ♖c1 a6 11. 0-0 ♘d7 12. a3 ♗e7 13. b4 ♕e5 14. f4 ♕b8

15. ♗f7! ♔f7 16. ♕b3 ♔e8[1] 17. ♘d5 ♗d6 18. ♘e6 b5 19. ♘dc7 ♔e7 20. ♘d4 ♔f8[2] 21. ♘a8 +–

[1] 16. – ♔g7 17. ♘e6 ♔h6 18. ♖f3 g5 19. f5 g4 20. ♕c4 ♘e5 21. ♖h3 gh3 22. ♕h4#

[2] 20. – ♗c7 21. ♘c6 ♔f8 22. ♘b8 ♗b8 23. ♕e6 +–; 20. – ♕c7 21. ♖c7 ♗c7 22. ♕e6 ♔f8 23. ♕c6 ♖a7 24. ♘e6 +–

4769. A 34

Pintér – Arkhipov

Balatonberény 1983

1. ♘f3 ♘f6 2. c4 c5 3. ♘c3 d5 4. cd5 ♘d5 5. e4 ♘b4 6. ♗c4 ♘d3 7. ♔e2 ♘f4 8. ♔f1 ♘e6 9. ♘e5 ♘d7

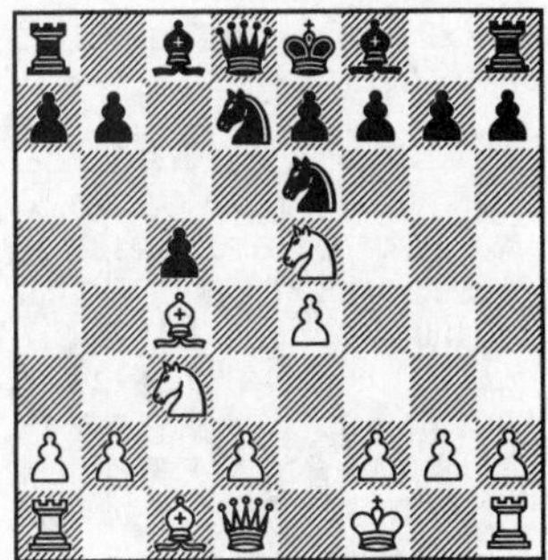

10. ♘f7! ♔f7 11. ♗e6 ♔e6 12. ♕b3 ♔f6[1] 13. ♘d5 ♔f7[2] 14. ♘c7 ♔g6 15. ♘e6! ♕e8 16. ♘f4 ♔g5[3] 17. h4 ♔h6[4] 18. ♕g3 ♕g6[5] 19. ♕g5! ♕g5 20. hg5 ♔g5 21. ♖h5 ♔f4[6] 22. d3 ♔g4 23. ♖g5 ♔h4 24. g3[7] +–

[1] 12. – ♔d6 13. ♘b5 ♔c6 14. ♕e6 ♔b5 15. a4 ♔a5 16. b4 cb4 17. ♕d5 ♔b6 18. ♗b2 e5 19. a5 ♔a6 20. ♕c4 b5 21. ♕c6 +–

[2] 13. – ♔g6 14. ♘f4 ♔h6 15. ♕h3 ♔g5 16. ♕h5 ♔f4 17. ♕f5#

[3] 16. – ♔f6 17. ♕e6 ♔g5 18. h4 ♔f4 19. d4#; 16. – ♔h6 17. ♕h3 ♔g5 18. ♘e6 ♔g6 19. ♕f5 ♔h6 20. ♕g5#

[4] 17. – ♔f4 18. d3 ♔g4 19. ♕d1#; 18. – ♔e5 19. ♕d5 ♔f6 20. ♕f5#

[5] 18. – g6 19. ♘e6 (♕g5#)

[6] 21. – ♔g4 22. f3 ♔f4 23. d4 ♔g3 24. ♖h3#

[7] 24. – ♔h3 25. ♖h5 ♔g4 26. ♖h4 ♔f3 27. ♖f4#

4770. A 51

Biegler – Peperle

Corr. 1952

1. d4 ♘f6 2. c4 e5 3. d5 ♗c5 4. h3

4. – ♗f2! 5. ♔f2 ♘e4 6. ♔f3 ♕h4 7. g3 ♕g3 8. ♔e4 f5! 9. ♔f5 d6 10. ♔e4 ♗f5! 11. ♔f5 ♕g6#

4771. A 51

Formanek – Oshana

Chicago 1970

1. d4 d5 2. c4 e5 3. de5 d4 4. g3 ♘c6 5. ♘f3 ♗g4 6. ♗g2 ♕d7 7. 0-0 0-0-0 8. ♘bd2 h5 9. ♕b3 h4 10. ♘h4 ♗e2 11. ♖e1 d3 12. ♘hf3 ♕f5 13. h4 g5 14. hg5 ♘ge7 15. ♘h4 ♖h4 16. gh4 ♕g4 17. ♘f1 ♘f5 18. ♘h2 ♕h4 19. ♗d2

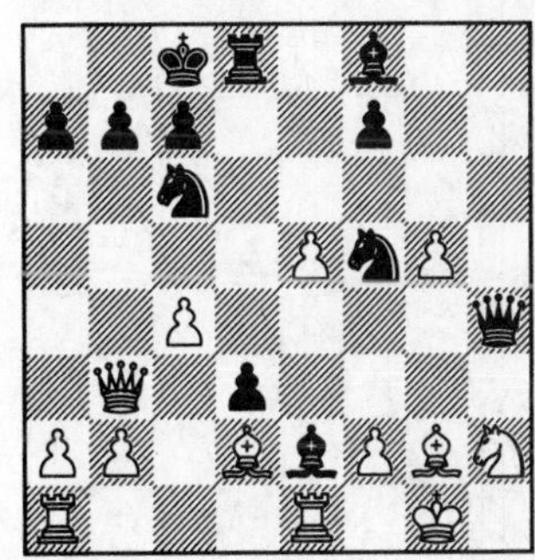

19. – ♕f2![1] –+

[1] 20. ♔f2 ♗c5 21. ♗e3 ♗e3#; 20. ♔h1 ♘g3#

4772. A 82
Koltanowsky – de Young
California

1. e4 e6 2. d4 f5 3. ♘c3 fe4 4. ♘e4 a6 5. ♘g5 ♗e7 6. ♘1f3 h6

7. ♘f7! ♔f7 8. ♘e5 ♔e8[1] 9. ♕h5 ♔f8 10. ♕f7#

[1] 8. – ♔f8 9. ♕h5 ♕e8 10. ♘g6 +–

4773. A 85
Carls – Hartlaub
1920–21

1. c4 b6 2. d4 ♗b7 3. ♘c3 e6 4. e4 f5 5. ef5 ♘f6 6. fe6 ♗e7 7. ♘f3 0-0 8. ♗d3 de6 9. 0-0 ♘c6 10. ♖e1 ♘b4 11. ♗e2 ♘g4 12. a3

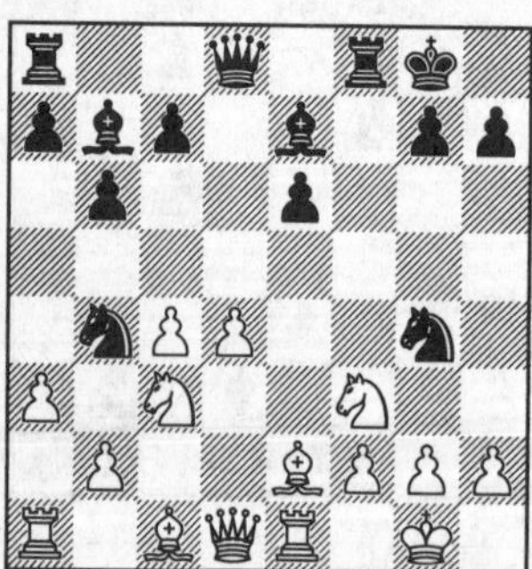

12. – ♘f2! 13. ♔f2 ♘c2! 14. ♕c2 ♕d4 15. ♗e3[1] ♗h4 16. g3 ♕g4 17. ♘d5[2] ed5 18. ♔g1 ♗g3 19. h3 ♕h3 20. ♗f1 ♕g4 21. ♗g2 d4 22. ♘h2 ♗f2 23. ♕f2 ♖f2 24. ♘g4 ♖g2 25. ♔f1 de3 –+

[1] 15. ♔f1 ♗c5 16. ♗d1 ♕g1 17. ♔e2 ♕g2 18. ♔d3 ♖f3 19. ♗f3 ♕f3 20. ♗e3 ♖d8 21. ♘d5 ed5 22. ♕e2 dc4 23. ♔c2 ♗e4 24. ♔c1 ♗e3 25. ♕e3 ♖d1 26. ♖d1 ♕e3 –+; 16. ♗d3 ♕g1 17. ♔e2 ♕g2 18. ♔d1 ♗f3 19. ♘e2 ♖ad8 20. ♗d2 ♗f2 –+; 15. ♔g3 ♗d6 16. ♔h3 ♕f2 17. ♕d1 e5 (♗c8) –+

[2] 17. gh4 ♖f3 18. ♗f3 ♕f3 19. ♔g1 ♕h1 20. ♔f2 ♕g2#

4774. B 01
Junek – Stickel
Prague 1941

1. e4 d5 2. ed5 ♕d5 3. ♘c3 ♕a5 4. ♘f3 ♗f5 5. d4 c6 6. ♗d3 ♘f6 7. 0-0 e6 8. ♖e1 ♘bd7 9. ♘e5 ♗g6 10. ♕e2 ♕d8 11. ♗g6 hg6

12. ♘f7! +–

4775. B 07
Addiks – Gudju
Prague 1931

1. d4 ♘f6 2. ♘f3 g6 3. ♘bd2 ♗g7 4. e4 d6 5. ♗c4 0-0 6. ♕e2 d5 7. ed5 ♘d5 8. ♘e4 ♘b6 9. ♗b3 ♗d4 10. ♘d4 ♕d4 11. ♗h6 ♖e8 12. c3 ♕h8 13. h4 ♘8d7

14. ♗f7![1] +–

[1] 14. – ♔f7 15. ♘g5 ♔g8 16. ♕e6#

4776. B 07

Cheron – Polikier

Chamonix 1927

1. d4 g6 2. e4 ♗g7 3. ♘f3 d6 4. ♘c3 ♘d7 5. ♗c4 ♘gf6 6. e5 de5 7. de5 ♘h5

8. ♗f7! ♔f7 9. ♘g5[1] +–

[1] 9. – ♔g8 10. ♕d5 +–

4777. B 07

Santasiere – R. Byrne

New York 1946

1. ♘f3 g6 2. e4 ♗g7 3. d4 d6 4. ♗c4 ♘d7

5. ♗f7! ♔f8[1] 6. ♘g5 ♘h6 7. ♕f3 ♘f6 8. e5 de5 9. de5 ♗g4 10. ef6 ♗f3 11. fg7 ♔g7 12. ♘e6 +–

[1] 5. – ♔f7 6. ♘g5 ♔e8 7. ♘e6 +–; 6. – ♔f6 7. ♕f3#

4778. B 10

Aljechin – Pirc

1938

1. e4 c6 2. ♘c3 d5 3. ♘f3 de4 4. ♘e4 ♗f5 5. ♘g3 ♗g6 6. h4 h6 7. ♘e5 ♗h7 8. ♕h5 g6

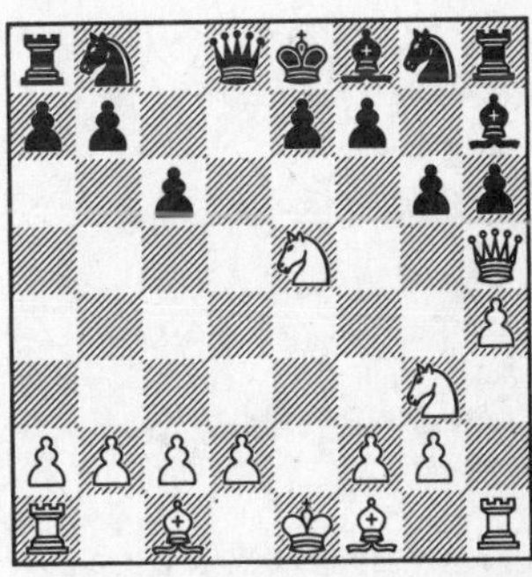

9. ♗c4! e6 10. ♕e2 ♘f6 11. ♘f7! ♔f7 12. ♕e6 ♔g7 13. ♕f7#

4779. B 12

Baturinsky – Toleshko

Corr. 1933

1. e4 c6 2. d4 d5 3. f3 e6 4. ♗e3 de4 5. ♘d2 ef3 6. ♘gf3 ♘f6 7. ♗d3 ♗d6 8. 0-0 ♕c7 9. ♘c4 ♗e7 10. ♘g5 h6

11. ♘f7! ♔f7 12. ♕h5 ♔g8[1] 13. ♖f6! gf6[2] 14. ♘e5! ♗d6[3] 15. ♖f1 ♗e5 16. de5 f5 17. ♖f5! ef5[4] 18. ♕g6[5] +–

[1] 12. – ♔f8 13. ♘e5 ♗d8 14. d5! cd5 15. ♗b5 ♘c6 16. ♗c5 ♔g8 17. ♖f6! gf6 18. ♕e8 +–; 14. – ed5 15. ♗c5 ♔g8 16. ♕e8! ♘e8 17. ♖f8#

[2] 13. – ♗f6 14. ♕e8#

[3] 14. – fe5 15. ♕g6 ♔f8 16. ♖f1 +–; 14. – ♗f8 15. ♖f1 +–

[4] 17. – ♕e7 18. ♕g6 ♕g7 19. ♕e8 ♔h7 20. ♖f7#

[5] 18. – ♕g7 19. ♗c4 ♔f8 20. ♗c5 +–

4780. B 13

Gerschenkron – Fischer

Vienna 1935

1. e4 d5 2. ed5 c6 3. d4 cd5 4. ♗d3 ♘f6 5. c3 ♘bd7 6. ♘f3 e6 7. 0-0 ♗e7 8. ♕e2 0-0 9. ♗f4 a6 10. ♘bd2 b5 11. ♘e5 ♖e8 12. ♘df3 ♗b7

13. ♘f7! ♔f7 14. ♕e6! ♔e6[1] 15. ♘g5#

[1] 14. – ♔f8 15. ♘g5 +–

4781. B 13

Kaila – Kiwi

Helsinki 1949

1. e4 c6 2. d4 d5 3. ed5 cd5 4. c4 dc4 5. ♗c4 e6 6. ♘c3 ♘d7 7. d5 e5 8. ♘f3 a6 9. 0-0 ♘gf6 10. d6 ♘b6

11. ♗f7! ♔f7 12. ♕b3 ♔g6[1] 13. ♘e5 ♔h5 14. ♘e2 ♕e8 15. ♕f3 ♗g4 16. ♕g4 ♘g4 17. ♘g3 ♔h4 18. ♘f3#

[1] 12. – ♔e8 13. ♘e5 ♕d6 14. ♕f7 ♔d8 15. ♖d1 +–

4782. B 14

Kiuru – Salo

Helsinki

1. e4 c6 2. d4 d5 3. ed5 cd5 4. c4 ♘f6 5. ♘c3 e6 6. ♘f3 ♗e7 7. ♗d3 dc4 8. ♗c4 0-0 9. 0-0 a6 10. ♕e2 b5 11. ♗b3 ♗b7 12. ♗g5 ♕d6 13. ♖ad1 ♘bd7 14. ♘e5 ♖ac8 15. a3 ♘b6 16. ♖d3 ♘bd5 17. ♗d2 b4 18. ab4 ♘b4 19 ♖g3 ♖fe8

20. ♘f7! ♔f7 21. ♗e6 ♔f8[1] 22. ♗h6! gh6 23. ♕h5![2] +–

[1] 21. – ♕e6 22. ♖g7 ♔g7 23. ♕e6 +–

[2] 23. – ♘h5 24. ♖g8#; 23. – ♕e6 24. ♕h6 ♔f7 25. ♕g7#

4783. B 14

Scaravella – Kirschstein

USA 1949

1. e4 c6 2. d4 d5 3. ed5 cd5 4. c4 ♘f6 5. ♘c3 e6 6. ♗g5 dc4 7. ♗c4 ♗e7 8. ♘f3 a6 9. 0-0 0-0 10. ♖c1 ♘bd7 11. ♕e2 b5 12. ♗b3 ♗b7 13. ♖fd1 ♖e8 14. ♘e5 b4

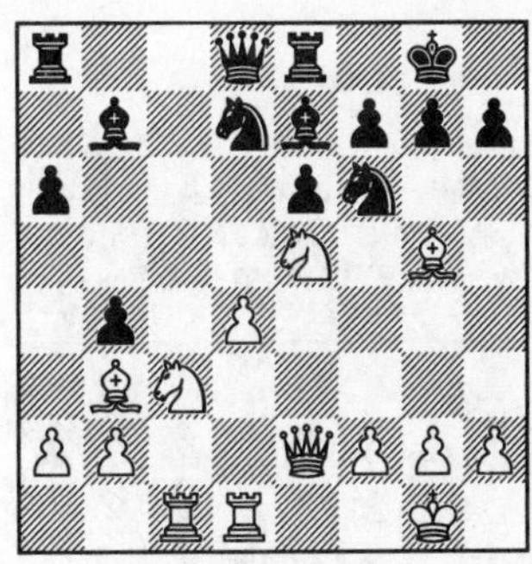

15. ♘f7! ♔f7 16. ♕e6 ♔g6 17. ♕f7 ♔g5 18. ♕g7 ♔f5 19. ♗c2 ♔e6 20. ♖e1 ♔d6 21. ♕g3 ♔c6 22. ♖e6 +–

4784. B 14

Verlinsky – Budo

USSR 1931

1. e4 c6 2. d4 d5 3. ed5 cd5 4. c4 ♘f6 5. ♘c3 e6 6. ♘f3 ♗e7 7. ♗d3 0-0 8. 0-0 dc4 9. ♗c4 a6 10. a4 ♘c6 11. ♗e3 ♘b4 12. ♕e2 ♗d7 13. ♘e5 ♖c8 14. ♖fd1 ♗c6 15. ♖ac1 ♗d5 16. ♘d5 ♘fd5 17. ♗b3 ♖c1 18. ♖c1 ♗g5

19. ♘f7! ♗e3[1] 20. ♘d8 ♖f2 21. ♕f2 ♗f2 22. ♔f1 ♗d4 23. ♘e6 ♗b2 24. ♖c8 ♔f7 25. ♘f4 +–

[1] 19. – ♔f7 20. ♕h5 +–; 19. – ♖f7 20. ♗g5 ♕g5 21. ♖c8 ♖f8 22. ♕e6 +–

4785. B 15
Zhulkov – Gavemann
Moscow 1947

1. e4 c6 2. ♘c3 d5 3. d4 de4 4. ♘e4 ♘f6 5. ♘f6 ef6 6. ♗c4 ♗e7 7. ♕h5 0-0 8. ♘e2 g6 9. ♕f3 ♘d7 10. ♗h6 ♖e8

11. ♗f7! ♔f7 12. ♕b3#

4786. B 21
Horváth – Ketsits
Hungary 1987

1. e4 c5 2. d4 cd4 3. c3 ♘f6 4. e5 ♘d5 5. ♘f3 e6 6. cd4 ♗b4 7. ♗d2 ♗e7 8. ♘c3 ♘c3 9. bc3 d6 10. ♗d3 ♘d7 11. ♕e2 ♕c7 12. ed6 ♕d6 13. 0-0 ♘f6 14. ♘e5 0-0 15. ♕f3 ♕d5 16. ♕h3 ♖d8 17. ♖fe1 ♗d7 18. ♖e3 ♗c6 19. ♗c4 ♕d6

20. ♘f7! +–

4787. B 29
Seidman – Santasiere
New York 1939

1. e4 c5 2. ♘f3 ♘f6 3. e5 ♘d5 4. ♘c3 ♘c3 5. dc3 b6 6. ♗c4 e6 7. ♗f4 ♕c7 8. 0-0 ♗b7 9. ♕e2 a6 10. a4 ♘c6 11. ♖ad1 ♗e7 12. ♖d2 0-0 13. ♖fd1 ♖fd8 14. ♘g5 h6

15. ♘f7! ♔f7 16. ♖d7 ♕d7[1] 17. ♖d7 ♖d7 18. ♕h5 g6 19. ♗e6! ♔e6 20. ♕g6 ♗f6 21. ♕f6 ♔d5 22. ♕f5[2] +–

[1] 16. – ♖d7 17. ♗e6 ♔e6 18. ♕c4 ♔f5 19. ♕f7 ♔g4 20. h3 ♔h4 21. ♗g3 ♔g5 22. f4#

[2] 22. – ♖dd8 23. ♕f7 ♔e4 24. f3#

4788. B 40
Baird – Lipke
Vienna 1898

1. e4 c5 2. ♘f3 e6 3. d4 cd4 4. ♘d4 ♘f6 5. ♗d3 ♘c6 6. ♘c6 bc6 7. 0-0 d5 8. ♘c3 ♗e7 9. b3 0-0 10. e5 ♘d7 11. ♖e1 f5 12. ef6 ♘f6 13. h3 ♗d6 14. ♗b2 e5 15. ♘a4 e4 16. ♗f1 ♘g4 17. ♖e2

17. – ♖f2! 18. hg4[1] ♕h4 19. ♕d4 ♕h2 20. ♔f2 ♗g3 21. ♔e3 ♕h6#

[1] 18. ♖f2 ♗h2 –+

4789. B 57

Stefanov – Sofia Polgár

Teteven 1984

1. e4 c5 2. ♘f3 ♘c6 3. d4 cd4 4. ♘d4 ♘f6 5. ♘c3 d6 6. ♗c4 e6 7. 0-0 ♗e7 8. f4 0-0 9. ♗b3 ♕c7 10. a3 ♘d4 11. ♕d4 d5 12. ♕d3 de4 13. ♘e4 ♖d8 14. ♕f3 ♗d7 15. ♘g5 h6 16. ♘e4 ♗c6 17. ♘f6 ♗f6 18. ♕h5 g6! 19. ♕h6 ♕b6 20. ♔h1

20. – ♕f2! 21. ♖g1 ♖d1! –+

4790. B 91

Klovan – Pukudruva

Latvian Championship 1955

1. e4 c5 2. ♘f3 d6 3. d4 cd4 4. ♘d4 ♘f6 5. ♘c3 a6 6. g3 ♕c7 7. ♗g2 e6 8. 0-0 ♗e7 9. ♗e3 0-0 10. ♕e2 ♗d7 11. ♖ad1 ♘c6 12. ♘b3 b5 13. a3 ♖ab8 14. f4 ♖fd8 15. g4 b4 16. ab4 ♘b4 17. g5 ♘e8 18. f5 ♘c2 19. ♕c2 ♗a4 20. fe6 ♗b3 21. ef7 ♗f7

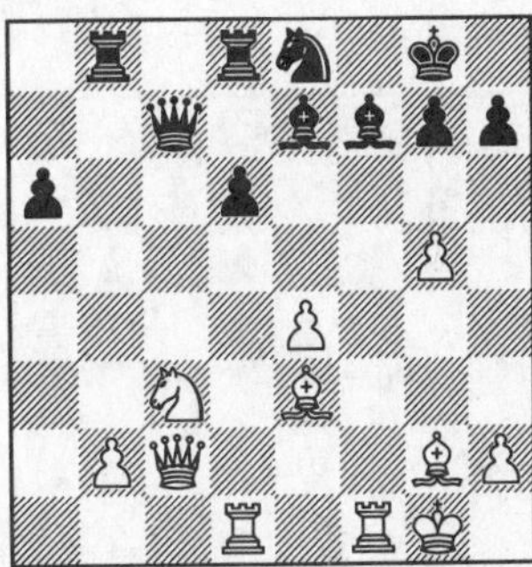

22. ♖f7! ♔f7 23. ♖f1 ♔g8[1] 24. ♘d5! ♕b7[2] 15. ♕f2! +–

[1] 23. – ♔e6 24. ♗h3; 23. – ♔g6 24. e5 +–

[2] 24. – ♕c2 25. ♘e7 ♔h8 26. ♖f8#

4791. C 00

Bahr – Zukaitis

Dayton, Ohio 1960

1. e4 e6 2. ♘f3 d5 3. ed5 ♕d5 4. d4 ♘f6 5. ♘d2 b6 6. ♗c4 ♕d8 7. 0-0 ♗b7 8. c3 h6 9. ♘e5 ♘bd7 10. f4 c5 11. ♕e2 cd4

12. ♘f7! ♔f7 13. ♕e6 ♔g6 14. ♗d3[1] +–

[1] 14. – ♔h5 15. ♕h3#

4792. C 02
Hartlaub – Terschek
Hamburg 1916

1. e4 e6 2. d4 d5 3. e5 c5 4. f4 ♘c6 5. ♘f3 ♕b6 6. a3 cd4 7. ♗d3 ♘ge7 8. g4 h5 9. gh5 ♖h5 10. ♘g5 g6

11. ♘f7! ♔f7 12. ♗g6 ♔g6[1] 13. ♖g1 ♔h7[2] 14. ♕h5 ♗h6 15. f5 +–

[1] 12. – ♘g6 13. ♕h5 +–
[2] 13. – ♔h6 14. f5 +–

4793. C 67
Vere – Minchin
London 1871

1. e4 e5 2. ♘f3 ♘c6 3. ♗b5 ♘f6 4. 0-0 ♘e4 5. ♖e1 ♘d6 6. ♘e5 ♗e7 7. ♘c3 0-0 8. d4 ♗f6 9. ♗d3 h6 10. ♘d5 ♘e8 11. ♕g4 d6 12. ♕e4 g6

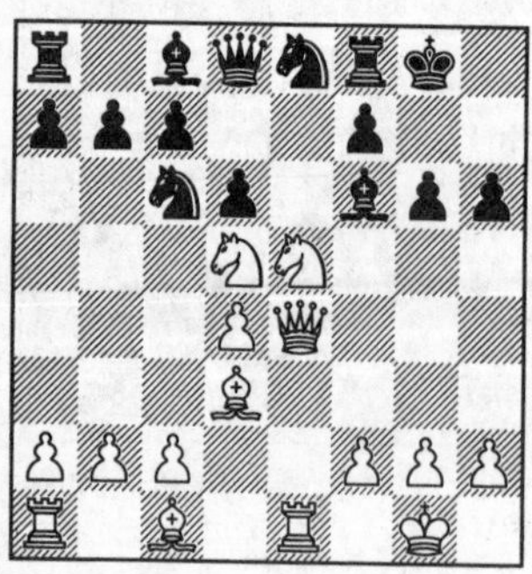

13. ♘f7! ♖f7 14. ♕g6 ♔f8[1] 15. ♗h6 ♗g7 16. ♕h7 ♘e7 17. ♕h8 ♘g8 18. ♗h7 +–

[1] 14. – ♘g7 15. ♕h7 ♔f8 16. ♕h8#; 14. – ♗g7 15. ♕h7 ♔f8 16. ♗h6 (♕h8#)

4794. C 10
Najdorf – Shapiro
Lodz 1929

1. e4 e6 2. d4 d5 3. ♘c3 de4 4. ♘e4 ♘d7 5. ♘f3 ♘gf6 6. ♗d3 ♗e7 7. 0-0 b6 8. ♘e5 ♗b7 9. ♘f6 gf6 10. ♘f7! ♔f7 11. ♕h5 ♔g8 12. ♖e1 ♘f8

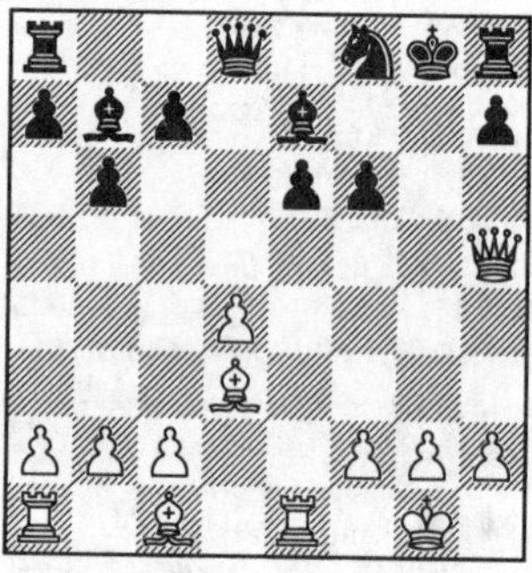

13. ♖e6[1] ♘e6 14. ♗c4 ♕d6 15. ♗h6 ♗f8 16. ♖e1 ♗c8 17. ♕e8 ♗d7 18. ♖e6 ♖e8 19. ♖e8 ♗e6 20. ♗e6 ♕e6 21. ♖f8#

[1] 13. ♗h6 – Strautmanis – Hasenfluss

4795. C 10
Strautmanis – Hasenfluss
Riga 1934

1. e4 e6 2. d4 d5 3. ♘c3 de4 4. ♘e4 ♘d7 5. ♘f3 ♘gf6 6. ♗d3 ♗e7 7. 0-0 b6 8. ♘e5 ♗b7 9. ♘f6 gf6

10. ♘f7! ♔f7 11. ♕h5 ♔g8 12. ♖e1 ♘f8[1] 13. ♗h6 f5[2] 14. ♖e3! ♕e8 15. ♖g3 ♘g6 16. ♗c4 ♗f8[3] 17. ♕f5 ♗h6 18. ♗e6 ♔g7 19. ♕e5 ♔f8 20. ♕f6 +–

[1] 12. – ♕e8 12. ♕g4 ♔f7 13. ♕e6 +–
[2] 13. – ♕d5 14. ♕g4 ♔f7 15. ♗e4 +–
[3] 16. – ♗d5 17. ♗d5 ed5 18. ♖e1 ♕f7 19. ♖e7! ♕e7 20. ♖g6! +–

4796. C 19

Dueball – Burnett

GFR 1970

1.e4 e6 2. d4 d5 3. ♘c3 ♗b4 4. e5 c5 5. a3 ♗c3 6. bc3 ♕c7 7. ♘f3 ♘e7 8. a4 b6 9. ♗b5 ♗d7 10. ♗d3 ♘bc6 11. 0-0 ♘a5 12. ♖e1 cd4 13. cd4 ♘c4 14. ♘g5 h6 15. ♕h5 g6 16. ♕h4 ♘f5 17. ♗f5 gf5

18. ♘f7! ♔f7 19. ♕f6 ♔g8 20. ♖a3 f4[1] 21. ♗f4 ♘a3 12. ♖e3 ♖e8 23. ♖g3 +–

[1] 20. – ♘a3 21. ♕g6 ♔f8 21. ♗a3 +–

4797. C 21

Barnes – N. N.

New York 1873

1. e4 e5 2. d4 ed4 3. c3 dc3 4. ♗c4 ♘c6 5. a3 ♘ce7 6. ♘f3 a6 7. 0-0 b5 8. ♗a2 c6 9. ♘g5 ♘h6 10. ♕b3 ♕a5 11. ♖e1 cb2 12. ♖d1 ba1♕

13. ♕f7! ♘f7 14. ♗f7 ♔d8 15. ♘e6#

4798. C 21

Charousek – Wollner

Kassa 1893

1. e4 e5 2. d4 ed4 3. c3 dc3 4. ♗c4 ♘f6 5. ♘f3 ♗c5 6. ♘c3 d6 7. 0-0 0-0 8. ♘g5 h6

9. ♘f7! ♖f7 10. e5 ♘g4 11. e6 ♕h4 12. ef7 ♔f8 13. ♗f4 ♘f2 14. ♕e2 ♘g4 15. ♔h1 ♗d7 16. ♖ae1 ♘c6 17. ♕e8 ♖e8 18. fe8♕ ♗e8 19. ♗d6#

4799. C 24
Hartlaub – Wielandt
Oldenburg 1920

1. e4 e5 2. ♗c4 ♘f6 3. ♘f3 ♘c6 4. d4 ed4 5. 0-0 ♗c5 6. e5 ♘g8 7. c3 dc3 8. ♘c3 ♘ge7 9. ♘e4 ♗b6 10. ♘f6 gf6 11. ef6 ♘g6 12. ♖e1 ♔f8 13. ♗h6 ♔g8

14. ♗f7! ♔f7 15. ♕d5[1] +–

[1] 15. – ♔f6 16. ♕g5 ♔f7 17. ♕f5 ♕f6 18. ♘g5 ♔g8 19. ♖e8 ♘f8 20. ♕d5 ♕e6 21. ♖f8#

4800. C 25
Kihn – Hartlaub
Germany 1920

1. e4 e5 2. ♘c3 ♗c5 3. ♘f3 d6 4. ♗c4 ♗e6 5. ♕e2 ♘e7 6. ♘g5 ♘g6 7. ♘e6 fe6 8. ♗e6 ♘f4 9. ♕g4 ♕f6 10. ♗c8

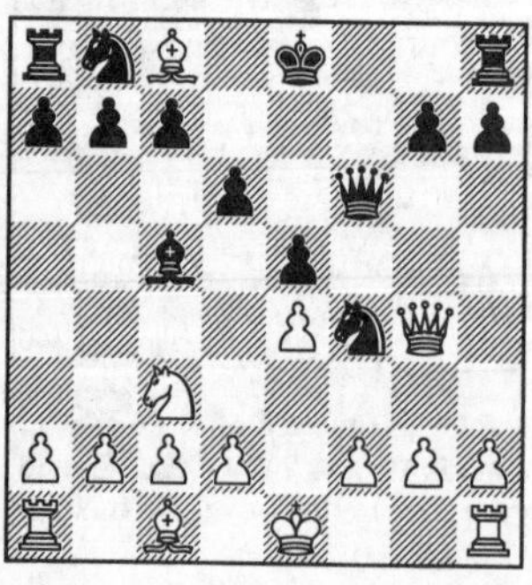

10. – ♗f2! 11. ♔f2 0-0 12. ♔e1[1] ♘c6 13. ♗b7 ♘d4 14. ♗a8?[2] ♘d3 15. cd3 ♕f2 16. ♔d1 ♕f1 17. ♖f1 ♖f1#

[1] 12. ♗f5 g6 13. g3 h5 14. ♕d1 gf5 15. gf4 ♕h4 –+

[2] 14. d3 ♖ab8 15. ♗d5 ♘d5 16. ♘d5 ♕f2 17. ♔d1 ♕c2 18. ♔e1 ♖b2! –+

4801. C 25
Weisz – Briefner
1896

1. e4 e5 2. ♘c3 ♘c6 3. f4 d6 4. ♗b5 ♗d7 5. ♘f3 ♗e7 6. 0-0 ef4 7. d4 g5 8. ♘d5 ♘h6 9. h4 g4 10. ♗f4 gf3 11. ♗h6 fg2

12. ♖f7! ♔f7 13. ♕h5 ♔g8 14. ♘f6! ♗f6 15. ♕d5 ♗e6 16. ♕e6#

4802. C 27
Kristensen – Andersen
Rajamäki 1977

1. e4 e5 2. ♗c4 ♘f6 3. ♘c3 ♘e4 4. ♕h5 ♘d6 5. ♗b3 ♗e7 6. ♘f3 ♘c6 7. ♘e5 0-0 8. 0-0 ♘e5 9. ♕e5 ♗f6 10. ♕f4 ♗g5 11. ♕f3 c6 12. d4 ♕f6 13. ♗g5 ♕g5 14. ♖fe1 ♘f5 15. d5 d6 16. dc6 ♘d4

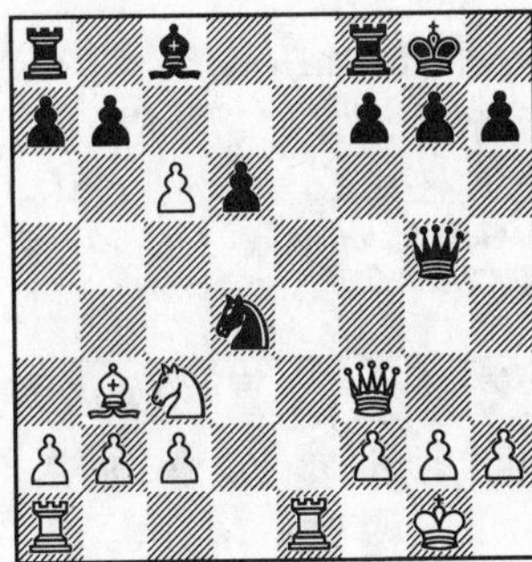

17. Qf7! Rf7 18. Re8#

4803. C 31

Kujoth – Élő

Milwaukee

1. e4 e5 2. f4 d5 3. Nf3 Nf6 4. fe5 Ne4 5. d3 Nc5 6. d4 Ne4 7. Bd3 Be7 8. 0-0 c5 9. c4 cd4 10. cd5 Qd5 11. Qc2 Nc5 12. Bc4 Qd7

13. Bf7! Kf7 14. e6 Ke6[1] 15. Qc4 Kd6[2] 16. Bf4 Kc6 17. Ne5 Kb6 18. Nd7 +–

[1] 14. – Ne6 15. Ne5 +–

[2] 15. – Qd5 16. Nd4 Kd6 17. Bf4 +–

4804. C 35

Hotimirsky – Robine

1. e4 e5 2. f4 ef4 3. Nf3 Be7 4. Bc4 Bh4 5. g3 fg3 6. 0-0 gh2 7. Kh1 d5 8. ed5 Bf6 9. d4 Nge7 10. Ng5 h6

11. Nf7! Kf7 12. d6 Kf8 13. Qh5 Qe8 14. Rf6! gf6 15. Qh6 Rh6 16. Bh6#

4805. C 37

Bródy – Bánya

Budapest 1901

1. e4 e5 2. f4 ef4 3. Nf3 g5 4. Bc4 g4 5. 0-0 gf3 6. Qf3 Qf6 7. e5 Qe5 8. b3 Bh6 9. Nc3 Ne7 10. Bb2 0-0 11. Rae1 Qd4 12. Kh1 Ng6 13. Ba1 Qh8 14. Nd5 Bg7 15. Qf4! Nc6[1]

16. Qf7! Rf7 17. Re8 Nf8 18. Ne7! Ne7 19. Bf7#

[1] 15. – Ba1 16. Ne7 Ne7 17. Bf7 Kg7 18. Ra1 (Qf6) +–

4806. C 37
MacDonnel – Labourdonnais
London 1834

1. e4 e5 2. f4 ef4 3. ♘f3 g5 4. ♗c4 g4 5. ♘c3 gf3 6. 0-0 c6 7. ♕f3 ♕f6 8. e5 ♕e5

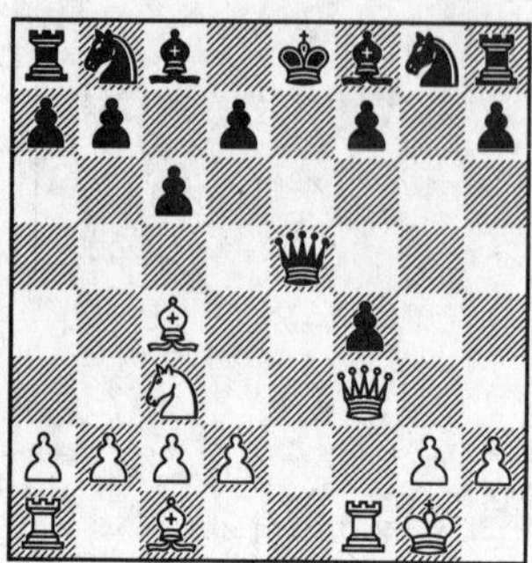

9. ♗f7! ♔f7 10. d4 ♕d4 11. ♗e3 ♕g7 12. ♗f4 ♘f6 13. ♘e4 ♗e7 14. ♗g5 ♖g8 15. ♕h5 ♕g6 16. ♘d6 ♔e6 17. ♖ae1 ♔d6 18. ♗f4#

4807. C 39
Maurian – N. N.
New Orleans 1866

1. e4 e5 2. f4 ef4 3. ♘f3 g5 4. h4 g4 5. ♘e5 h6

6. ♘f7! ♔f7 7. d4 d6 8. ♗f4 ♘c6 9. ♗c4 ♔g7 10. 0-0 ♕h4 11. ♕d3 ♘f6 12. e5 ♘h5 13. ♗g3 ♕e7 14. ♗h4 ♕e8 15. ♗f6 ♘f6 16. ef6#

4808. C 41
Holzhausen – Tarrasch
Hamburg 1910

1. e4 e5 2. ♘f3 d6 3. ♗c4 ♗e7 4. d4 ed4 5. ♘d4 ♘f6 6. ♘c3 ♘c6 7. 0-0 0-0 8. h3 ♖e8 9. ♖e1 ♘d7

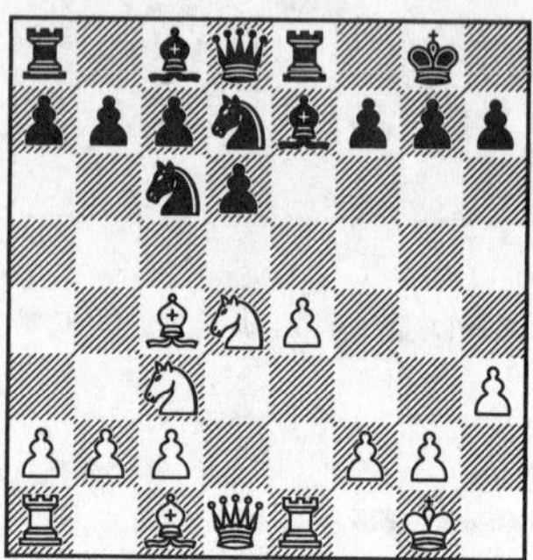

10. ♗f7! ♔f7 11. ♘e6! ♔e6 12. ♕d5 ♔f6 13. ♕f5#

4809. C 41
Leonhardt – N. N.
Hamburg 1912

1. e4 e5 2. ♘f3 d6 3. d4 ♘d7 4. ♗c4 c6 5. ♘g5 ♘h6 6. a4 ♗e7

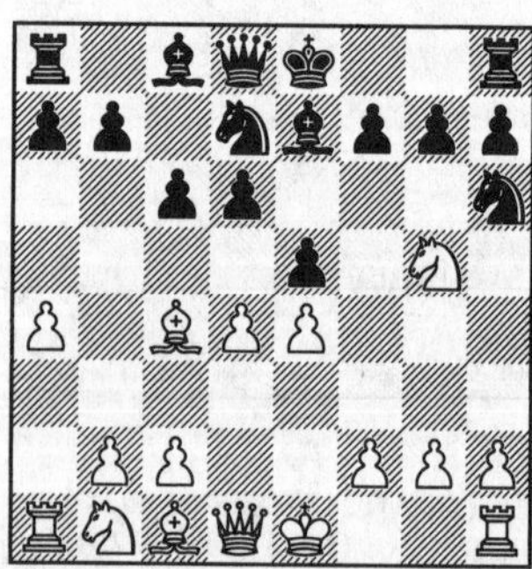

7. ♗f7! ♘f7 8. ♘e6 ♕b6 9. a5 ♕b4 10. c3 ♕c4 11. ♘c7 ♔d8 12. b3 +−

4810. C 42

Borochov – McCudden

New York 1918

1. e4 e5 2. ♘c3 ♘f6 3. ♘f3 d6 4. d4 ♘bd7 5. ♗c4 ♗e7 6. 0-0 0-0 7. ♗e3 c6 8. a4 ♕c7 9. de5 de5 10. ♕e2 h6 11. ♘h4 ♖e8

12. ♗f7! ♔f7 13. ♕c4 ♘d5[1] 14. ♘d5 ♕d6 15. ♘f4 +–

[1] 13. – ♔f8 14. ♘g6#

4811. C 42

Hamlisch – N. N.

Vienna 1899

(without ♖a1)

1. a3 e5 2. e4 ♘f6 3. ♘c3 ♗c5 4. ♘f3 a6 5. ♘e5 ♕e7 6. d4 ♗a7 7. ♗g5 h6 8. ♘d5 ♕d8 9. ♗f6 gf6 10. ♕h5 ♖f8 11. ♗c4 ♗d4

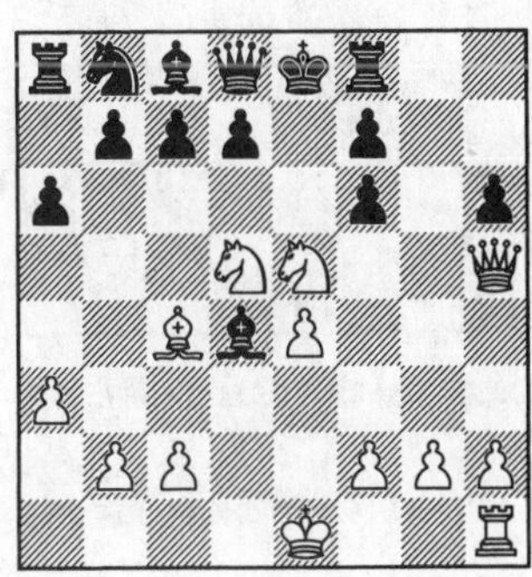

12. ♘f7! ♖f7 13. ♘c7 ♔e7[1] 14. ♕f7 ♔d6 15. ♘e8 ♔c6 16. ♕d5 ♔b6 17. ♕d4 ♔c6 18. ♗d5 ♔b5 19. ♕b4#

[1] 13. – ♕c7 14. ♕f7 ♔d8 15. ♕f8#

4812. C 45

Astapovich – Golosov

Novosibirsk 1967

1. e4 e5 2. ♘f3 ♘c6 3. d4 ed4 4. ♘d4 ♘f6 5. ♘c3 ♗e7 6. ♘f5 0-0 7. ♗g5 ♖e8 8. ♗c4 ♘e4

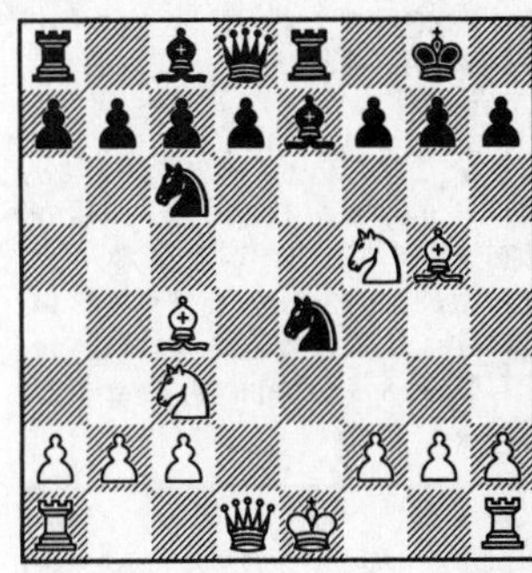

9. ♗f7! ♔f7 10. ♕d5 ♔f8[1] 11. ♘h6! +–

[1] 10. – ♔g6 11. ♘h4 ♔h5 12. ♗e7 +–

4813. C 49

Nimzovitsch – Tartakover

Karlsbad 1911

1. e4 e5 2. ♘f3 ♘c6 3. ♘c3 ♘f6 4. ♗b5 ♗b4 5. 0-0 ♘d4 6. ♘d4 ed4 7. e5 dc3 8. dc3 ♗e7 9. ef6 ♗f6 10. ♖e1 ♔f8 11. ♗c4 d6 12. ♕h5 g6 13. ♗h6 ♗g7 14. ♕f3 ♕d7 15. ♕f6 ♖g8 16. ♗g7 ♖g7

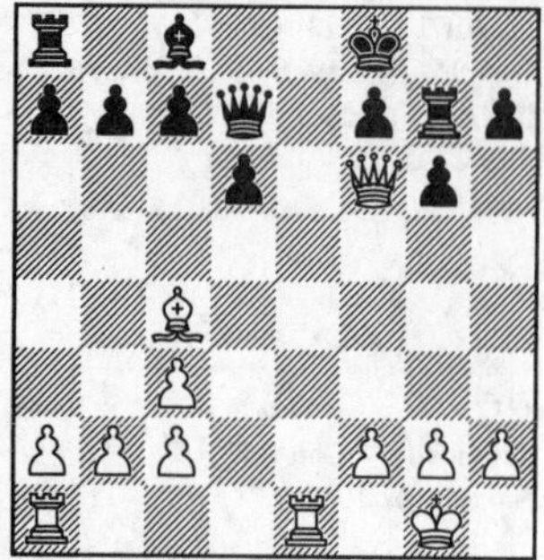

17. ♗f7![1] +–

[1] 17. – ♕f7 18. ♕d8 +–; 17. – ♖f7 18. ♕h8#

4814. C 49
Wittekopf – Wait
Libau 1911

1. e4 e5 2. ♘f3 ♘c6 3. ♘c3 ♘f6 4. ♗b5 ♗b4 5. 0-0 0-0 6. d3 d6 7. ♗g5 ♘e7 8. ♗f6 gf6 9. ♘e2 c6 10. ♗a4 ♘g6 11. ♘g3 ♔h8 12. c3 ♗a5 13. ♕d2 ♖g8 14. ♔h1 ♘f4 15. ♖g1 ♕f8 16. ♘h4 ♖g5 17. ♘f3 ♖g4 18. ♗d1 ♕h6 19. h3 ♗e6 20. ♘h2 ♘h3 21. ♕c2

21. – ♘f2! 22. ♕f2 ♕h2! 23. ♔h2 ♖h4#

4815. C 51
Karsher – Friedmann
Nuremberg 1897
(without ♖a1)

1. e4 e5 2. ♘f3 ♘c6 3. ♗c4 ♗c5 4. b4 ♗b4 5. c3 ♗c5 6. 0-0 d6 7. d4 ed4 8. cd4 ♗b6 9 ♗a3 ♗g4 10. e5 ♘d4 11. ♘bd2 de5 12. ♘e5 ♗e6 13. ♕g4 ♘f6 14. ♕g7 ♖g8

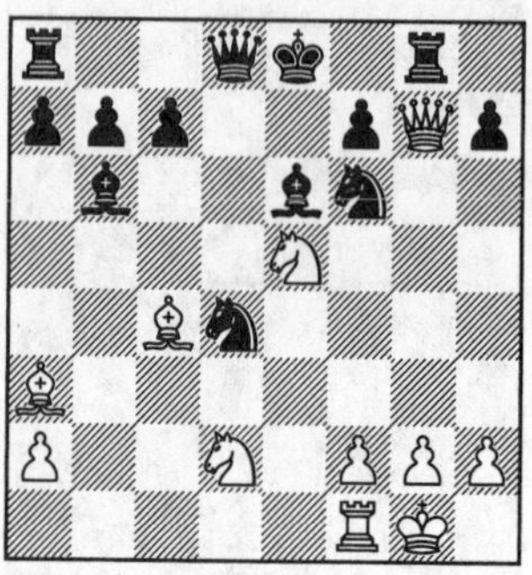

15. ♕f7! ♗f7 16. ♗f7#

4816. C 54
Bramley – Burgess
Surrey 1946

1. e4 e5 2. ♘f3 ♘c6 3. ♗c4 ♗c5 4. c3 ♘f6 5. d4 ed4 6. cd4 ♗b4 7. ♘c3 ♘e4 8. 0-0 ♘c3 9. bc3 ♗e7 10. ♖e1 0-0 11. d5 ♘b8 12. d6 ♗f6 13. dc7 ♕c7 14. ♗a3 ♗c3

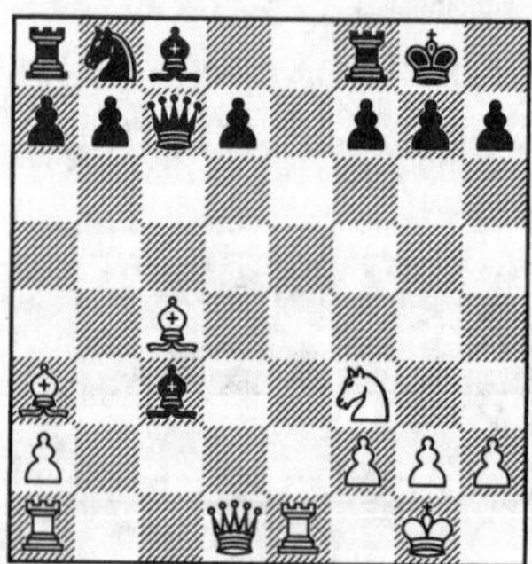

15. ♗f7! ♔h8[1] 16. ♗f8 ♗e1 17. ♕d4 +–

[1] 15. – ♔f7 16. ♕d5 ♔f6 17. ♗e7 ♔g6 18. ♘h4 +–

4817. C 54

Marco – N. N.

Vienna 1889

(without ♖a1, the WPa2 stands on a3)

1. e4 e5 2. ♘f3 ♘c6 3. ♗c4 ♗c5 4. c3 ♘f6 5. d4 ed4 6. cd4 ♗b6 7. e5 ♘e4 8. ♗d5 f5 9. 0-0 ♘e7 10. ♗a2 d5 11. ed6 ♘d6 12. ♘e5 h5 13. ♗g5 g6 14. ♘c3 c6 15. ♖e1 ♘e4 16. ♖e4 fe4 17. ♕b3 ♖f8

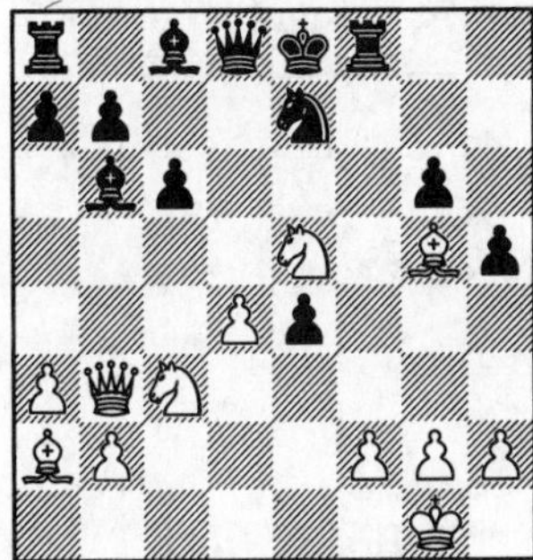

18. ♕f7! ♖f7 19. ♗f7 ♔f8 20. ♗h6#

4818. C 54

Mohr – Hartlaub

Wiesbaden 1913

1. e4 e5 2. ♘f3 ♘c6 3. ♗c4 ♗c5 4. c3 ♘f6 5. d4 ed4 6. e5 ♘e4 7. ♕e2 d5 8. ed6 0-0 9. dc7 ♕d7 10. 0-0 ♖e8 11. ♕d3 ♘e5 12. ♘e5 ♖e5 13. ♗f4

13. – ♘f2! 14. ♖f2[1] ♖e1 15. ♖f1 dc3 16. ♔h1 ♕d3 17. ♗d3 ♖f1 18. ♗f1 cb2 –+

[1] 14. ♔f2 dc3 15. ♔f3 ♕g4#

4819. C 54

Thompson – Bishop

Corr. 1950

1. e4 e5 2. ♘f3 ♘c6 3. ♗c4 ♗c5 4. c3 ♗b6 5. d4 ♕e7 6. 0-0 ♘f6 7. ♗g5 0-0 8. ♘bd2 ed4 9. e5 ♘e5 10. ♖e1 ♘f3 11. ♕f3 ♕d8 12. ♘e4 d6 13. ♘f6 gf6 14. ♗f6 ♕d7 15. ♕h5 ♕f5

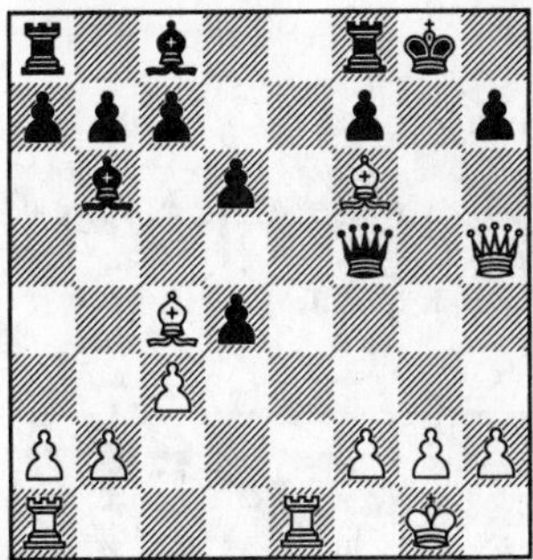

16. ♕f7! +–

4820. C 55

Denker – Avram

New York 1939

1. e4 e5 2. ♘f3 ♘c6 3. ♗c4 ♘f6 4. d4 ed4 5. 0-0 ♗c5 6. e5 d5 7. ef6 dc4 8. ♖e1 ♗e6 9. ♘g5 ♕d5 10. ♘c3 ♕f5 11. ♘ce4 ♗f8 12. g4 ♕d5

13. ♘f7! ♔f7[1] 14. ♘g5 ♔g8 15. ♘e6 ♘e5 16. f7 ♔f7[2] 17. ♘g5 ♔g8 18. ♖e5! ♕e5 19. ♕f3[3] +–

[1] 13. – ♗f7 14. ♘c3 +–
[2] 16. – ♘f7 17. ♘c7 +–
[3] 19. – ♕e7 20. ♕d5 +–

4821. C 55
Finn – Nugent
New York 1900

1. e4 e5 2. ♘f3 ♘c6 3. d4 ed4 4. ♗c4 ♗c5 5. 0-0 ♘f6 6. e5 d5 7. ef6 dc4 8. ♖e1 ♗e6 9. ♘g5 ♕d5 10. ♘c3 ♕f5 11. ♘ce4 ♗f8 12. ♘f7! ♔f7 13. ♘g5 ♔g8 14. g4 ♕f6 15. ♖e6 ♕d8 16. ♕f3 ♕d7

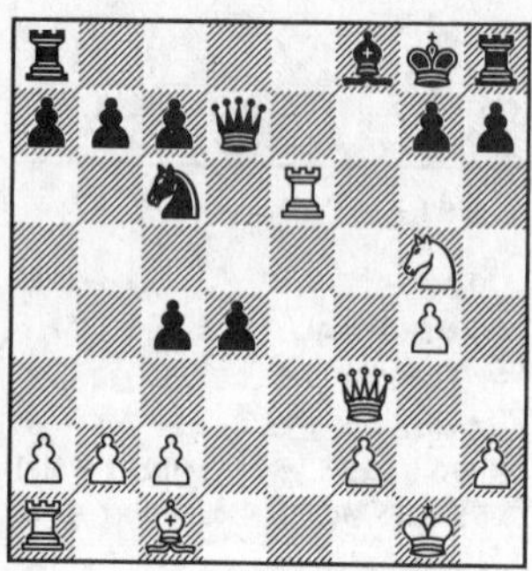

17. ♖e7!! ♕e7 18. ♕d5 ♕f7 19. ♕f7#

4822. C 57
Schröder – Illgen
Dresden 1926

1. e4 e5 2. ♗c4 ♘f6 3. ♘f3 ♘c6 4. d4 ed4 5. ♘g5 d5 6. ed5 ♘d5 7. 0-0 ♗e6 8. ♖e1 ♕d7

9. ♘f7! ♔f7[1] 10. ♕f3 ♔g6[2] 11. ♖e6! ♕e6 12. ♗d3 +–

[1] 9. – ♕f7 10. ♗d5 +–
[2] 10. – ♔g8 11. ♖e6 +–

4823. C 57
Soyka – Toth
Vienna 1948

1. e4 e5 2. ♘f3 ♘c6 3. ♗c4 ♘f6 4. ♘g5 ♗c5 5. ♘f7

5. – ♗f2! 6. ♔f2 ♘e4 7. ♔g1 ♕h4 8. ♕f1[1] ♖f8 9. d3 ♘d6 10. ♘d6 cd6 11. ♕e2 ♘d4 12. ♕d2 ♕g4 –+

[1] 8. g3! ♘g3 9. ♘h8 d5 10. hg3 ♕g3 11. ♔f11 ♗h3 12. ♖h3 ♕h3 13. ♔g1 ♕g3 14. ♔h1 =

4824. C 58
Allies – Blackburne
Hastings 1895

1. e4 e5 2. ♘f3 ♘c6 3. 3. ♗c4 ♘f6 4. ♘g5 d5 5. ed5 ♘a5 6. ♗b5 c6 7. dc6 bc6 8. ♕f3 cb5 9. ♕a8 ♗c5 10. b4 ♗b4 11. ♕a7 0-0 12. 0-0 ♘c6 13. ♕e3 ♘d4 14. ♘e4 ♘g4 15. ♕d3 f5 16. ♘g3 e4 17. ♘e4 fe4 18. ♕e4 ♗f5 19. ♕b7 ♕h4 20. ♕c7 ♘e2 21. ♔h1

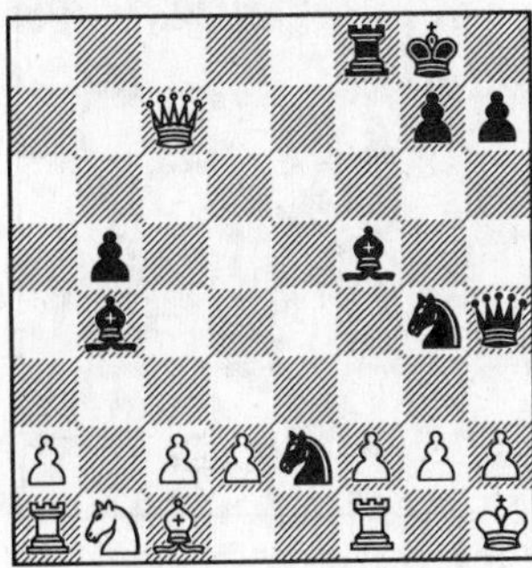

21. – ♕f2![1] –+

[1] 22. ♖d1 ♕g1! 23. ♖g1 ♘f2#

4825. C 60
Kupreichik – Planinc
Sombor 1970

1. e4 e5 2. ♘f3 ♘c6 3. ♗b5 ♗e7 4. 0-0 ♘f6 5. ♖e1 d6 6. c3 0-0 7. d4 ♗d7 8. h3 ♖e8 9. ♘bd2 ♗f8 10. ♗c4 ed4 11. cd4 d5 12. ♗b3 de4

13. ♗f7! ♔f7 14. ♕b3 ♔g6[1] 15. ♘h4 ♔h5 16. ♘e4 ♖e4 17. ♖e4 g5 18. ♕f7 ♔h6 19. ♘f5![2] +–

[1] 14. – ♗e6 15. ♘g5 +–
[2] 19. – ♗f5 20. ♖h4 +–

4826. C 61
Boden – Bird
London 1873

1. e4 e5 2. ♘f3 ♘c6 3. ♗b5 ♘d4 4. ♘d4 ed4 5. 0-0 ♗c5 6. c3 ♘e7 7. d3 c6 8. ♗c4 0-0 9. ♗g5 ♔h8 10. ♕h5 f6 11. ♗f6 d5 12. ♗e7 ♕e7 13. ed5

13. – ♖f2! 14. ♘d2[1] dc3 15. ♘b3 cb2 16. ♖ae1 ♖f1 17. ♔f1 ♕f6 18. ♕f3 ♕f3 19. gf3 ♗h3 20. ♔e2 ♖e8 –+

[1] 14. ♔f2 ♕e3#; 14. ♖f2 ♕e1 15. ♖f1 dc3 –+

4827. C 61
Zaitzev – Timchenko
Moscow 1956

1. e4 e5 2. ♘f3 ♘c6 3. ♗b5 ♘d4 4. ♗c4 b5 5. ♗f7 ♔f7 6. ♘d4 ♕h4 7. ♕f3 ♘f6 8. ♘b5 ♗c5 9. ♘5c3 ♖f8 10. d3 ♔g8 11. ♕g3 ♕h5 12. 0-0 ♘g4 13. ♘d1

13. – ♘f2! 14. ♗e3[1] ♘h3 15. gh3 ♖f1 16. ♔f1 ♕d1 17. ♔g2 ♗e3 18. ♕e3 d6 19. ♕d2 ♗h3! 20. ♔h3 ♕f3 21. ♔h4 h6 –+

[1] 14. ♘f2 ♖f2 15. ♖f2 ♕d1#

4828. C 63

Neukirch – Möhring

Corr.

1. e4 e5 2. ♘f3 ♘c6 3. ♗b5 f5 4. ♘c3 fe4 5. ♘e4 d5 6. ♘e5 de4 7. ♘c6 ♕d5 8. c4 ♕d6 9. ♘a7 ♗d7 10. ♗d7 ♕d7 11. ♕h5 g6 12. ♕e5 ♔f7 13. ♘b5 c6 14. ♘c3 ♖e8 15. ♕h8 ♘f6 16. 0-0 ♕d3 17. ♖e1 ♗c5 18. ♕e8 ♔e8 19. b3

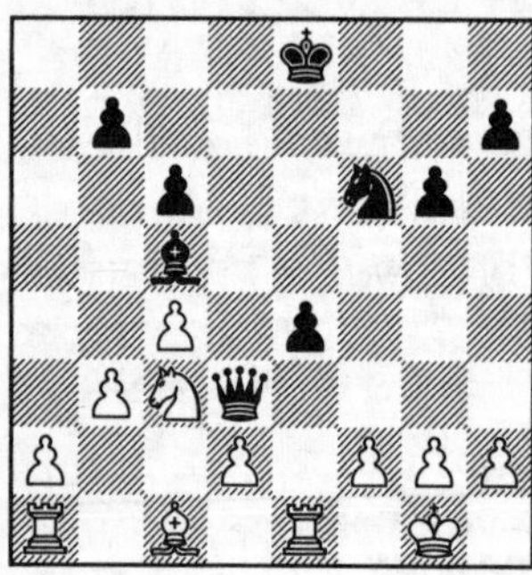

19. – ♗f2![1] –+

[1] 20. ♔f2 ♘g4 21. ♔g1 ♕d4 22. ♔h1 ♘f2 23. ♔g1 ♘h3 24. ♔h1 ♕g1! 25. ♖g1 ♘f2#

4829. C 65

Pond – Lumsden

Corr. 1952

1. e4 e5 2. ♘f3 ♘c6 3. ♗b5 ♘f6 4. 0-0 ♗e7 5. d4 ♘d4 6. ♘d4 ed4 7. e5 ♘d5 8. ♕d4 c6 9. ♗c4 ♘b4 10. ♗d3 0-0 11. a3 ♕b6 12. ♗h7 ♔h8 13. ♕e4 ♘d5 14. ♗f5 d6 15. ♗c8 ♖ac8 16. ♘d2 f5 17. ef6 ♗f6 18. ♘c4 ♕c5 19. c3 ♖ce8 20. ♕g4

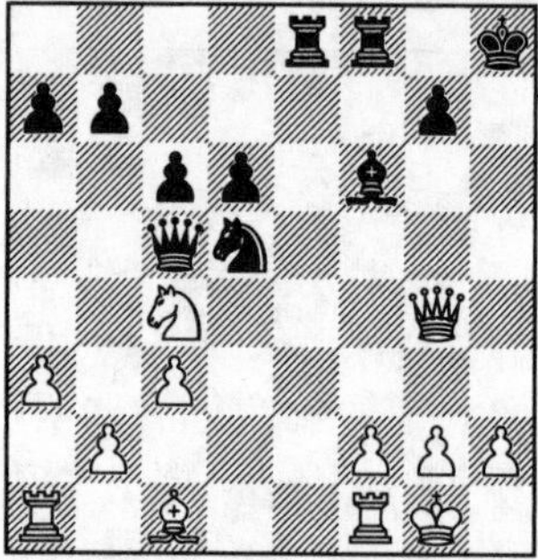

20. – ♕f2![1] –+

[1] 21. ♔f2 ♗h4 22. ♔g1 ♖f1 23. ♔f1 ♖e1#; 21. ♖f2 ♖e1 22. ♖f1 ♗d4 –+

4830. C 67

Varain – N. N.

Leipzig 1890

1. e4 e5 2. ♘f3 ♘c6 3. ♗b5 ♘f6 4. 0-0 ♘e4 5. d4 ♘d6 6. ♘e5 ♘b5 7. ♖e1 ♘e7 8. ♕h5 g6 9. ♕f3 ♘d6 10. ♘c3 c6 11. ♘e4 ♕c7

12. ♕f7! ♔d8[1] 13. ♘f6 ♕b6 14. ♗f4 ♘ef5 15. ♕e8! ♔c7 16. ♘f7 ♗g7 17. ♘d6 ♖e8 18. ♘b5 ♔d8 19. ♖e8#

[1] 12. – ♘f7 13. ♘f6 ♔d8 14. ♘f7#

4831. C 70

Karklins – Tautvaisas

Illinois 1970

1. e4 e5 2. ♘f3 ♘c6 3. ♗b5 a6 4. ♗a4 f5 5. ♘c3 ♘f6 6. ef5 e4 7. ♕e2 ♕e7 8. ♗c6 bc6 9. ♘g5 d5 10. ♘e6 ♗e6 11. fe6 ♕e6 12. d3 ♗d6 13. ♗g5 ♗e5 14. ♗f6 ♕f6 15. 0-0 ed3 16. ♕d3 0-0 17. ♖ae1 ♖ae8 18. ♖e3 ♗d4 19. ♖h3

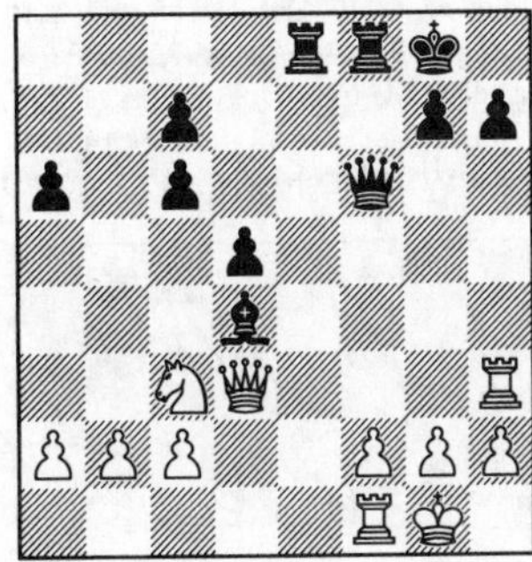

19. – ♕f2![1]

[1] 20. ♖f2 ♖e1 21. ♕f1 ♗f2 –+

4832. C 70

Oberbuchhagen – Unger

Corr. 1956

1. e4 e5 2. ♘f3 ♘c6 3. ♗b5 a6 4. ♗a4 ♘f6 5. 0-0 ♘e4 6. d4 d5 7. ♘e5 ♗d7

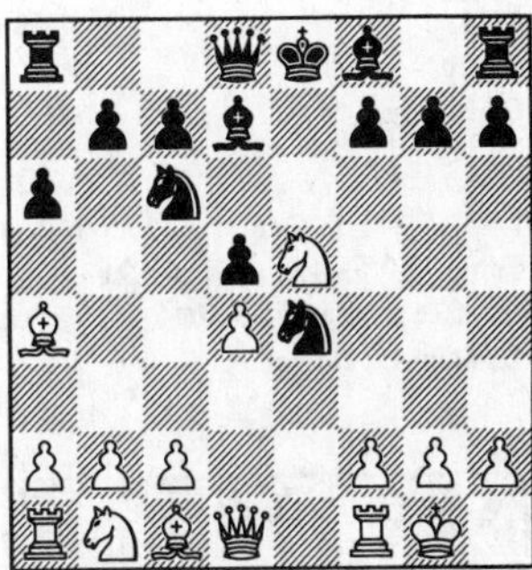

8. ♘f7! ♔f7 9. ♕h5 ♔e6[1] 10. ♘c3 ♘c3 11. ♖e1 ♘e4 12. ♖e4![2] +–

[1] 9. – g6 10. ♕d5 ♔g7 11. ♕e4 +–

[2] 12. – de4 13. d5 ♔d6 14. ♗f4 ♔c5 15. dc6 ♔b6 16. ♗e3 +–; 15. – ♔b4 16. c3 ♔a4 17. b3 ♔a3 18. ♗c1#

4833. C 77

Belson – Martin

Toronto 1933

1. e4 e5 2. ♘f3 ♘c6 3. ♗b5 a6 4. ♗a4 ♘f6 5. ♘c3 ♗c5 6. ♘e5 ♘e5 7. d4 ♗d6 8. 0-0 c5 9. de5 ♗e5 10. ♗b3 ♕c7 11. f4 ♗c3 12. bc3 c4 13. e5 ♘g8 14. ♗a3 cb3 15. ♗d6 ♕b6 16. ♔h1 ♘h6 17. f5 g6 18. ♕d2 ♘g8 19. fg6 hg6

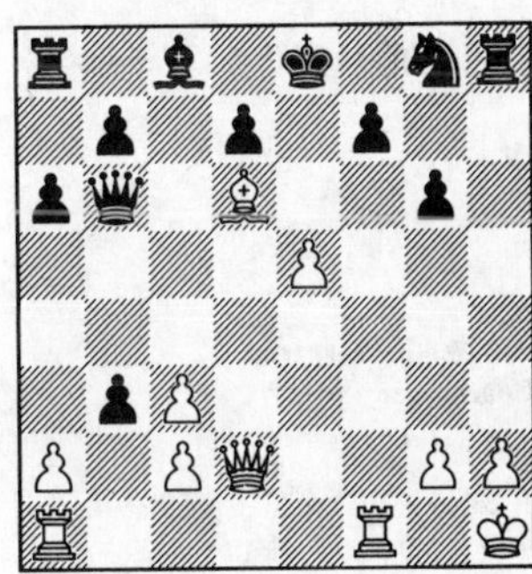

20. ♖f7![1] +–

[1] 20. – ♔f7 21. ♕f4 ♔e6 22. ♕c4 +–; 21. – ♔g7 22. ♗f8 ♔h7 23. ♕f7#

4834. C 81

Adam – Seibold

Corr. 1939

1. e4 e5 2. ♘f3 ♘c6 3. ♗b5 a6 4. ♗a4 ♘f6 5. 0-0 ♘e4 6. d4 b5 7. ♗b3 d5 8. de5 ♗e6 9. ♕e2 ♘a5 10. ♘bd2 c5 11. ♘e4 de4 12. ♗e6 ef3

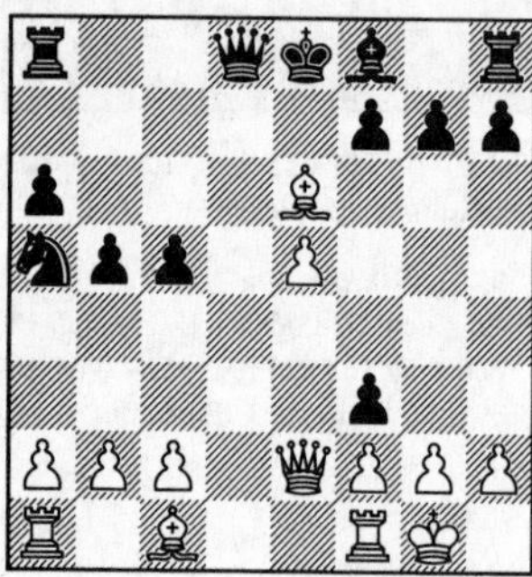

13. ♗f7! ♔f7 14. ♕f3 ♔e8[1] 15. ♖d1 ♕c8 16. e6! ♕b7 17. ♖d5! ♕e7[2] 18. ♔f1! +−

[1] 14. – ♔g8 15. ♖d1 ♕e8 16. ♕d5 +−; 14. – ♔g6 15. ♖d1 ♕e8 16. e6! (♕g4) +−

[2] 17. – ♖d8 18. ♕f7 +−; 17. – ♘c6 18. ♖d7 +−

4835. C 89

Janowski – Tornerup

Copenhagen 1947

1. e4 e5 2. ♘f3 ♘c6 3. ♗b5 a6 4. ♗a4 ♘f6 5. 0-0 ♗e7 6. ♖e1 b5 7. ♗b3 0-0 8. c3 d5 9. ed5 e4 10. ♘g5 ♗d6 11. ♘e4 ♘e4 12. ♖e4 ♗f5 13. ♖e3 ♘e5 14. h3 ♕h4 15. ♗c2 ♘g4 16. ♖e2 ♗c2 17. ♕c2 ♖ae8 18. d4

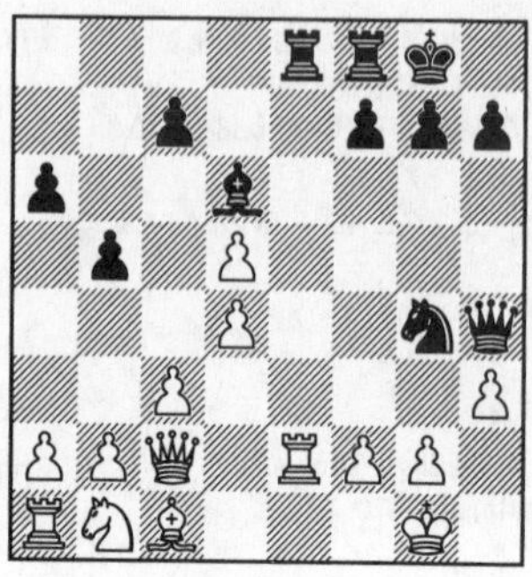

18. – ♕f2![1] −+

[1] 19. ♖f2 ♖e1 20. ♖f1 ♗h2 −+

4836. C 89

Sery – Vécsey

Brno 1921

1. e4 e5 2. ♘f3 ♘c6 3. ♗b5 a6 4. ♗a4 ♘f6 5. 0-0 ♗e7 6. ♖e1 b5 7. ♗b3 0-0 8. c3 d5 9. ed5 ♘d5 10. ♘e5 ♘e5 11. ♖e5 ♘f6 12. ♖e1 ♗d6 13. h3 ♘g4 14. ♕f3 ♕h4 15. d4 ♘f2 16. ♕f2 ♗g3

17. ♕f7! ♖f7 18. ♖e8#

4837. C 92

Muller – Riegler

Corr. 1947

1. e4 e5 2. ♘f3 ♘c6 3. ♗b5 a6 4. ♗a4 ♘f6 5. 0-0 ♗e7 6. ♖e1 d6 7. c3 b5 8. ♗b3 0-0 9. h3 ♗e6 10. ♗c2 d5 11. d4

♘e4 12. ♘e5 ♘e5 13. de5 f5 14. ef6 ♖f6 15. ♗e4 de4 16. ♘d2

16. – ♖f2! 17. ♔f2 ♗c5 18. ♔f1[1] ♕f6 19. ♘f3 ♗c4 20. ♖e2 ef3 21. gf3 ♕f3 22. ♔e1 ♕h1 −+

[1] 18. ♔g3 ♕d6 19. ♔h4 ♕f4 20. ♔h5 g6#

4838. D 00

Wirthensohn – Forster

Swiss 1993

1. d4 ♘f6 2. ♘f3 g6 3. ♘c3 d5 4. ♗f4 ♗g7 5. e3 c6 6. h3 ♘fd7 7. ♗d3 ♕b6 8. ♖b1 a5 9. 0-0 0-0 10. ♖e1 ♘f6 11. e4 de4 12. ♘e4 ♘bd7 13. ♘ed2 ♖e8 14. ♘c4 ♕d8 15. c3 c5 16. ♘ce5 ♘d5 17. ♗h2 cd4 18. ♗b5 ♘c7

19. ♘f7! ♔f7 20. ♘g5 ♔g8 21. ♗c7 ♕c7 22. ♕b3 e6 23. ♖e6 +−

4839. D 00

Benzinger – N. N.

Munich

1. d4 ♘f6 2. c3 e6 3. ♘f3 d5 4. ♗f4 ♘bd7 5. e3 a6 6. ♘bd2 c5 7. ♗d3 ♗e7 8. ♘e5 0-0 9. ♕f3 ♖e8 10. ♕h3 b5 11. ♘df3 ♗b7

12. ♘f7! ♔f7 13. ♕e6![1] +−

[1] 13. – ♔e6 14. ♘g5#; 13. – ♔f8 14. ♘g5 +−

4840. D 00

Davies – Berg

1981

1. d4 e6 2. e4 d5 3. ♘d2 de4 4. ♘e4 ♘d7 5. ♘f3 ♘gf6 6. ♗d3 b6 7. ♘f6 ♘f6 8. ♘e5 a6 9. ♗g5 ♗b7 10. ♕e2 ♗g2 11. ♖g1 ♗b7 12. 0-0-0 ♗e7 13. f4 g6 14. f5 ♘d5

15. ♘f7! ♔f7 16. ♕e6 ♔e8 17. fg6 ♕d6 18. g7 ♕e6 19. gh8♕ +–

4841. D 02
Helbig – Schröder
Hamburg 1933

1. d4 d5 2. ♘f3 ♘f6 3. ♘bd2 e6 4. a3 c5 5. dc5 ♗c5 6. b4

6. – ♗f2 7. ♔f2 ♘g4 8. ♔g3[1] h5 9. ♘h4[2] ♕c7 10. ♔f3 ♕c3 11. ♔f4 ♕e3#

[1] 8. ♔g1 ♕b6 –+
[2] 9. ♕e1 e5 –+

4842. D 15
Seiffert – Jakubowski
Berlin 1936

1. d4 ♘f6 2. c4 d5 3. ♘c3 c6 4. ♘f3 a6 5. e3 ♘bd7 6. ♗d3 dc4 7. ♗c4 b5

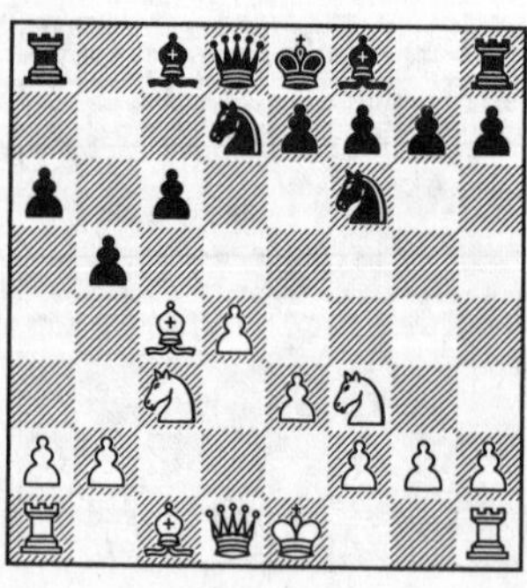

8. ♗f7!? ♔f7 9. ♘g5 ♔g6[1] 10. h4 ♘b6 11. h5 ♘h5 12. ♕h5 ♔f6 13. ♘ce4 ♔f5 14. g4#

[1] 9. – ♔e8 10 ♕b3 ♕b6? 11. ♕f7 ♔d8 12. ♘e6#; 10. – ♘b6 11. ♘f7; 10. – e6 11. ♕e6 ♕e7 12. ♕c6 =

4843. D 25
Zsuzsa Polgár – Shoengold
New York 1985 (simultan)

1. ♘f3 d5 2. d4 ♘f6 3. c4 dc4 4. e3 g6 5. ♗c4 e6? 6. 0-0 ♗e7 7. ♘c3 0-0 8. e4 c6 9. ♗h6 ♖e8 10. ♘e5 ♘bd7

11. ♘f7! ♔f7 12. ♗e6! ♔e6 13. ♕b3 ♔d6[1] 14. ♗f4 ♘e5 15. de5 ♔c7[2] 16. ef6 ♗d6 17. ♕f7 ♗d7 18. ♗d6 ♔c8 19. ♖fd1 b6 20. ♗e7 ♕c7 21. ♖d7![3] +–

[1] 13. – ♘d5 14. ♘d5 cd5 15. ed5 ♔d6 16. ♗f4 ♘e5 17. de5 ♔d7? 18. e6#; 15. – ♔f7 16. d6 ♔f6 17. ♕f3 ♔e6 18. ♖fe1 ♔d6 19. ♗f4 ♘e5 20. ♗e5 ♔d7 21. ♕d5 +–
[2] 15. – ♔d7?? 16. e6#
[3] 21. – ♕d7 22. ♖d1 +–

4844. D 25
Wall – Gantt
Hickory 1978

1. d4 d5 2. c4 dc4 3. ♘f3 ♘f6 4. e3 g6 5. ♗c4 ♗g7 6. ♕a4 ♘bd7

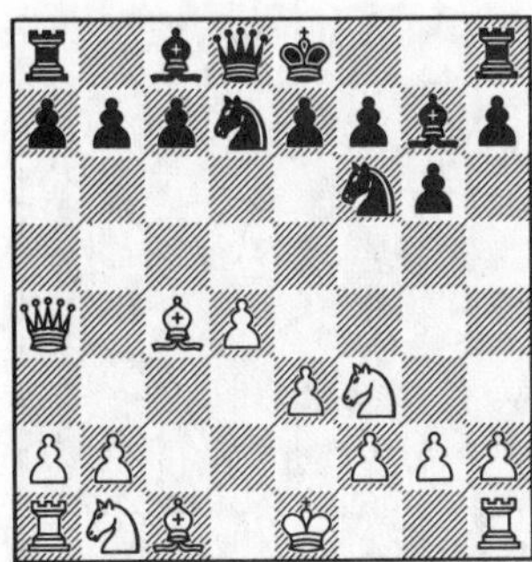

7. ♗f7! ♔f7 8. ♘g5 ♔g8[1] 9. ♕c4 +–

[1] 8. – ♔e8 9. ♘e6 +–

4845. D 26

Gesov – Manolyov

USSR 1935

1. d4 d5 2. ♘f3 ♘f6 3. c4 dc4 4. e3 b6 5. ♗c4 ♗b7 6. ♘c3 e6 7. 0-0 c5 8. ♘e5 ♘bd7 9. f4 ♕c7 10. ♘b5 ♕c8

11. ♘f7! ♔f7 12. f5 ♗d5[1] 13. fe6 ♔g8 14. ♗d5 ♘d5 15. ♕h5 h6 16. ♕d5 ♘f6 17. ♖f6! gf6 18. e7 ♔g7 19. ♘d6 +–

[1] 12. – cd4 13. fe6 ♔e7 14. b3 +–

4846. D 26

Wirthensohn – Niklasson

Reggio Emilia 1978

1. d4 d5 2. c4 dc4 3. e3 ♘f6 4. ♗c4 e6 5. ♘f3 c5 6. 0-0 ♘c6 7. ♘c3 ♗e7 8. a3 cd4 9. ed4 0-0 10. ♖e1 b6 11. ♕d3 ♗b7 12. ♗a2 ♖c8 13. ♗f4 ♖e8 14. ♖ad1 ♗f8 15. d5 ed5 16. ♖e8 ♕e8 17. ♖e1 ♕d8 18. ♘d5 h6 19. ♗g5 hg5 20. ♘g5 ♘e4

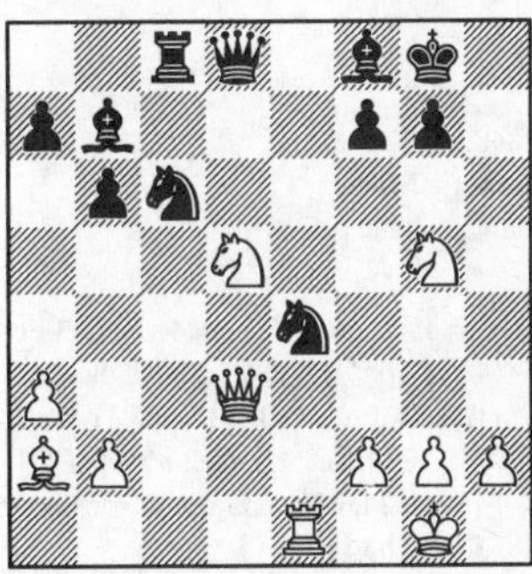

21. ♘f7! ♘c5[1] 22. ♘g5![2] +–

[1] 21. – ♔f7 22. ♘f4 ♔f6 23. ♕e4 +–

[2] 22. – ♘d3 23. ♘f6 ♔h8 24. ♘f7#

4847. D 30

Hanauer – Bartha

New York 1929

1. d4 ♘f6 2. ♘f3 e6 3. c4 d5 4. ♗g5 ♘bd7 5. e3 ♗e7 6. ♘bd2 0-0 7. ♗d3 a6 8. 0-0 c5 9. ♕e2 ♖e8 10. ♖fe1 cd4 11. ed4 dc4 12. ♘c4 b5 13. ♘ce5 ♗b7 14. ♗f6 gf6

15. ♘f7! ♔f7 16. ♘e5 fe5[1] 17. ♕h5 ♔g7 18. ♕h7 ♔f8 19. ♕h6 ♔g8 20. ♗h7[2] +–

[1] 16. – ♘e5 17. ♕h5 ♘g6 18. ♕h7 +–
[2] 20. – ♔f7 21. ♕g6 ♔f8 22. ♕g8#; 20. – ♔h8 21. ♗g6 ♔g8 22. ♕h7 ♔f8 23. ♕f7#

4848. D 31
James – Miles
New Zealand 1911

1. ♘f3 d5 2. d4 ♘f6 3. ♗f4 e6 4. e3 b6 5. c4 ♗b7 6. ♘c3 a6 7. cd5 ed5 8. ♗d3 ♘bd7 9. ♖c1 ♗e7 10. 0-0 ♘h5 11. ♘e2 ♘f4 12. ♘f4 ♖c8 13. ♕e2 a5 14. ♖fd1 0-0 15. ♕c2 h6 16. ♗f5 c5 17. ♘e5 cd4 18. ♗h7 ♔h8

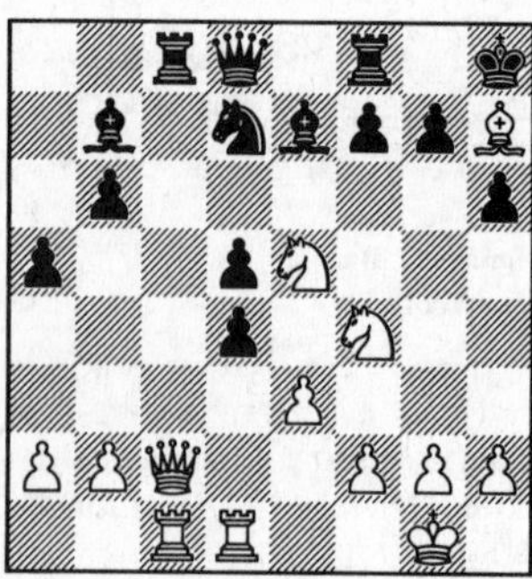

19. ♘f7! ♖f7 20. ♘g6 ♔h7 21. ♘f8 ♔g8 22. ♕h7! ♔f8 23. ♕h8#

4849. D 31
Palmer – Gunsberg
England 1908

1. d4 d5 2. c4 e6 3. ♘c3 ♗e7 4. ♘f3 c6 5. e4 de4 6. ♘e4 ♘f6 7. ♗d3 ♘bd7 8. 0-0 ♘e4 9. ♗e4 ♘f6 10. ♗c2 0-0 11. ♕d3 g6 12. ♗h6 ♖e8 13. ♘e5 ♘d7 14. f4 c5 15. d5 ♗f6 16. ♖ae1 ♗e5 17. fe5 ed5

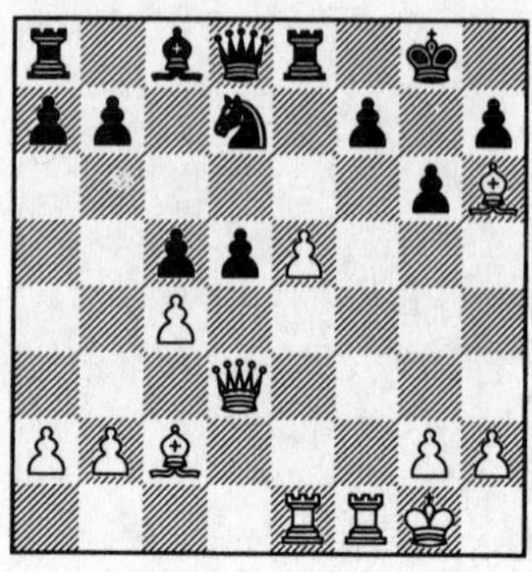

18. ♖f7! ♘e5[1] 19. ♖g7 ♔h8 20. ♖e5![2] +–

[1] 18. – ♔f7 19. ♕d5 ♖e6 20. ♖f1 ♔e7 21. ♗g5 +–
[2] 20. – ♖e5 21. ♖h7! ♔h7 22. ♕g6 ♔h8 23. ♕g7#

4850. D 36
Wikström – Wood
Corr. 1947

1. d4 d5 2. c4 e6 3. ♘c3 ♘f6 4. ♗g5 ♘bd7 5. ♘f3 c6 6. cd5 ed5 7. e3 ♗e7 8. ♗d3 0-0 9. 0-0 ♖e8 10. ♕c2 h6 11. ♗f6 ♘f6 12. ♘e5 ♘g4 13. f4 ♘e3 14. ♗h7 ♔f8 15. ♕e2 ♘f5 16. ♕h5 ♘d6

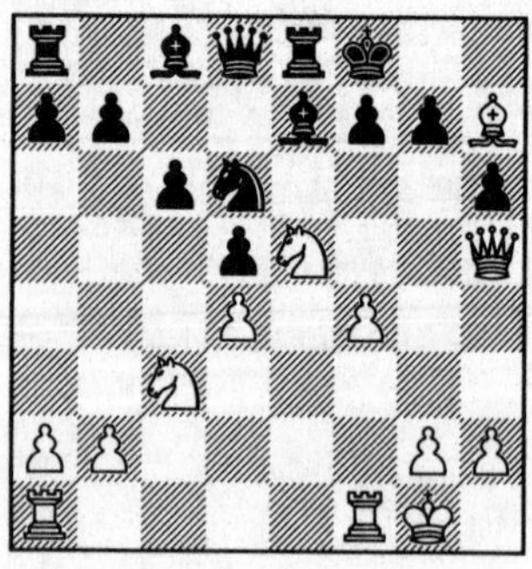

17. ♕f7! ♘f7 18. ♘g6#

4851. D 40

Ghitescu – Donner

Beverwijk 1967

1. d4 ♘f6 2. c4 e6 3. ♘f3 d5 4. ♘c3 c5 5. e3 ♘c6 6. a3 cd4 7. ed4 ♗e7 8. cd5 ♘d5 9. ♗d3 ♘f6 10. 0-0 0-0 11. ♗g5 b6 12. ♕e2 ♗b7 13. ♖ad1 g6 14. ♗c4 ♘d5 15. ♗h6 ♘c3 16. bc3 ♖e8 17. d5 ed5 18. ♗d5 ♕c7

19. ♗f7! ♔f7 20. ♕c4 ♔f6 21. ♘g5[1] +–

[1] 21. – ♗f8 22. ♘h7 ♕h7 23. ♕f4 ♔e6 24. ♖fe1 +–

4852. D 41

Bolbochan – Pachman

Moscow 1956

1. d4 ♘f6 2. c4 e6 3. ♘f3 d5 4. ♘c3 c5 5. cd5 ♘d5 6. e3 ♘c6 7. ♗c4 cd4 8. ed4 ♗e7 9. 0-0 0-0 10. ♖e1 ♘c3 11. bc3 b6 12. ♗d3 ♗b7 13. ♕c2 g6 14. ♗h6 ♖e8 15. ♕d2 ♖c8 16. ♖ac1 ♗f6 17. ♕f4 ♘a5 18. ♘e5 ♘c6 19. ♘g4 ♗h4 20. g3 ♗e7 21. ♗c4 ♖c7

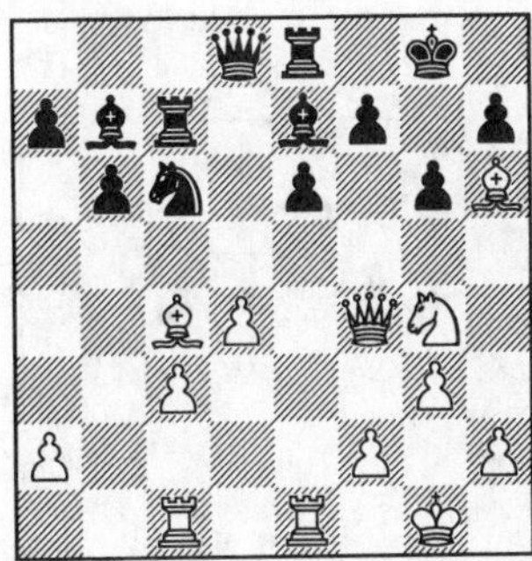

22. ♕f7! +–

4853. D 45

Conquest – Moser

Argentina 1981

1. d4 d5 2. c4 e6 3. ♘c3 ♘f6 4. ♘f3 c6 5. e3 ♘bd7 6. ♘e5 dc4 7. f4 ♗b4 8. ♕c2 c5 9. ♗c4 ♕b6 10. a3 cd4 11. ed4 ♗d6 12. ♕e2 a6

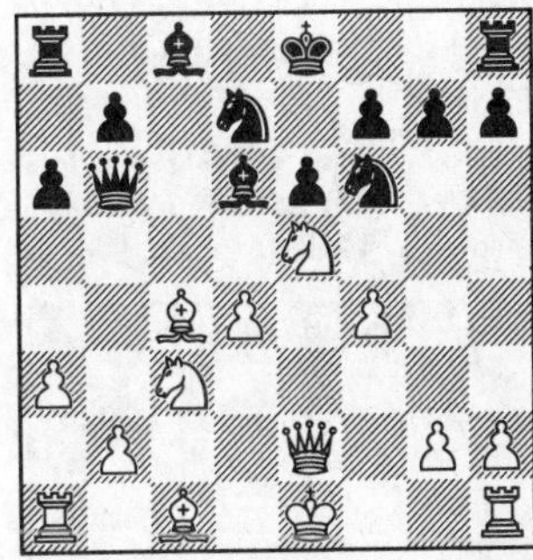

13. ♘f7! ♖f8[1] 14. ♘g5 ♗e7 15. ♘e6 ♖f7 16. f5 h6 17. ♗f4 ♖a7 18. ♘c7 ♔f8 19. ♕e6 +–

[1] 13. – ♔f7 14. ♕e6 ♔g6 15. g4 +–

4854. D 46

Benzinger – N. N.

Munich

1. d4 d5 2. ♘f3 ♘f6 3. c4 e6 4. ♘c3 c6 5. e3 ♘bd7 6. ♗d3 ♗e7 7. ♕e2 b6 8. e4

de4 9. ♘e4 ♗b7 10. ♘eg5 ♕c7 11. 0-0 ♖d8

12. ♘f7! ♔f7 13. ♘g5 ♔e8 14. ♘e6 ♕b8[1] 15. ♖e1 ♘g8 16. ♕h5 g6 17. ♕g6! hg6 18. ♗g6#

[1] 14. – ♕d6 15. ♗f4 ♕b4 16. a3 ♕b3 17. ♗c2 ♕b2 18. ♗g6 +–; 16.– ♕a4 17. ♗c2 ♕a5 18. b4 +–

4855. D 46

Sigurdsson – Thornwaldson

Iceland 1934

1. d4 d5 2. c4 e6 3. ♘c3 ♘f6 4. ♘f3 ♘bd7 5. e3 c6 6. ♗d3 ♗e7 7. 0-0 0-0 8. e4 dc4 9. ♗c4 ♖e8 10. ♘g5 h6

11. ♘f7! ♔f7 12. ♗e6 ♔e6[1] 13. ♕b3 ♘d5[2] 14. ♘d5 ♔f7 15. ♘f4 +–

[1] 12. – ♔f8 13. e5 +–

[2] 13. – ♔d6 14. ♗f4 +–

4856. D 85

Dake – Schmitt

Washington 1949

1. d4 ♘f6 2. c4 g6 3. f3 d5 4. cd5 ♘d5 5. e4 ♘b6 6. ♗e3 ♗g7 7. ♘c3 0-0 8. f4 ♘c6 9. d5 ♘b8 10. ♘f3 e6? 11. ♗c5 ♖e8 12. d6 ♘6d7 13. ♗a3 cd6 14. ♗d6 ♕b6 15. ♕d2 e5 16. ♗c4 ef4

17. ♗f7! ♔f7 18. ♘g5 ♔f6[1] 19. ♕f4#

[1] 18. – ♔g8 19. ♕d5 ♔h8 20. ♘f7 ♔g8 21. ♘h6 ♔h8 22. ♕g8 ♖g8 23. ♘f7#

4857. E 04

Veitch – Penrose

Burton 1950

1. d4 ♘f6 2. c4 e6 3. ♘f3 d5 4. g3 dc4 5. ♘bd2 c5 6. dc5 ♗c5 7. ♗g2

7. – ♗f2! 8. ♔f2 ♘g4 9. ♔e1[1] ♘e3[2] –+

[1] 9. ♔g1 ♕b6 –+
[2] 10. ♕a4 ♗d7 11. ♕b4 ♘c2 –+

4858. E 12
Bidan – Short
1981

1. d4 ♘f6 2. c4 e6 3. ♘f3 b6 4. a3 c5 5. d5 ♗a6 6. ♕c2 ed5 7. cd5 ♗b7 8. e4 ♕e7 9. ♗d3 ♘d5 10. 0-0 ♘c7 11. ♘c3 ♕d8 12. ♘d5 ♘e6 13. ♘e5 ♘c6 14. f4 ♘e5 15. fe5 ♗e7 16. ♕e2 h6 17. ♕h5 ♖f8 18. ♗h6! gh6

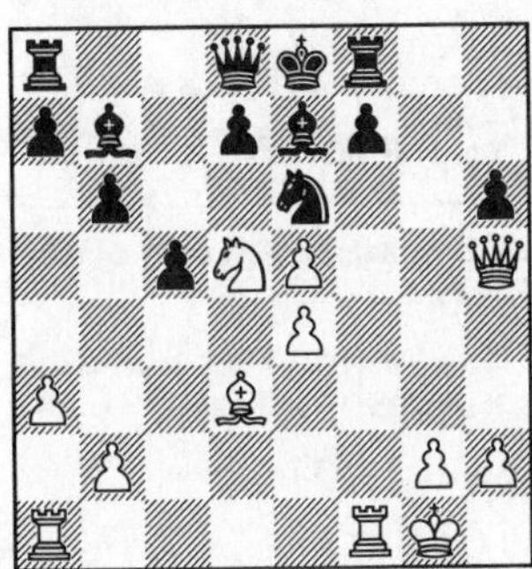

19. ♖f7! ♖f7 20. ♖f1 ♘g5 21. ♗c4 ♔f8 22. ♖f7 ♘f7 23. ♘f6 +–

4859. E 15
Gheorghiu – Prewoznik
Timisoara 1987

1. d4 ♘f6 2. c4 e6 3. ♘f3 b6 4. g3 ♗a6 5. ♘bd2 b5 6. cb5 ♗b5 7. ♗g2 d5 8. 0-0 c5 9. dc5 ♗c5 10. b4 ♗d6 11. a4 ♗d7 12. ♖b1 ♕e7 13. b5 a6 14. e4 de4 15. ♘g5 ♘d5 16. ♘de4 ♗b4

17. ♘f7! 0-0[1] 18. ♘eg5 ♖f7 19. ♗d5 +–

[1] 17. – ♕f7 18. ♖b4 +–; 17. – ♔f7 18. ♘g5 ♔f8 19. ♗d5 ed5 20. ♕d5 ♖a7 21. ♖b4 +–

4860. E 32
Tóth – Tatai
1984

1. d4 ♘f6 2. c4 e6 3. ♘c3 ♗b4 4. ♕c2 d6 5. ♘f3 ♘bd7 6. e3 b6 7. ♗e2 ♗b7 8. 0-0 ♗c3 9. ♕c3 0-0 10. b3 ♘e4 11. ♕c2 f5 12. ♗b2 ♘df6 13. ♘e1 ♘g4 14. d5 ed5 15. cd5 ♕e7 16. ♗g4 fg4 17. ♖d1 ♖ae8 18. ♖d4

18. – ♘f2! 19. ♘d3[1] ♘d3 20. ♖f8 ♖f8 21. ♖d3 ♗a6 22. ♖d1 ♕e3 23. ♔h1 ♕e2 24. ♕e2 ♗e2 25. ♖e1 ♖f1 –+

[1] 19. ♖f2 ♕e3 20. ♘d3 ♕e1! –+

4861. E 72

Riley – Parr

Felixstowe 1949

1. d4 ♘f6 2. c4 g6 3. ♘c3 ♗g7 4. g3 0-0 5. ♗g2 d6 6. e4 ♘c6 7. ♘ge2 e5 8. 0-0 ed4 9. ♘d4 ♘e4 10. ♘c6 ♘c3 11. bc3 bc6 12. ♗c6 ♗h3 13. ♖e1 ♗c3 14. ♗h6 ♕f6 15. ♗a8 ♗e1 16. ♗f3 ♖e8 17. ♖b1 ♗a5 18. ♗c6 ♖d8 19. ♖b5 ♗b6 20. ♕e1

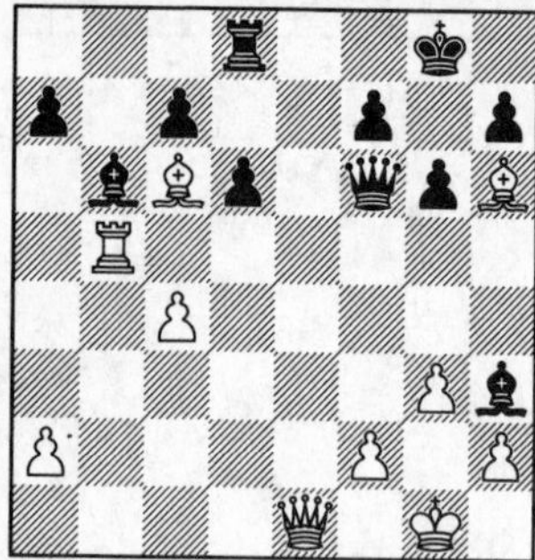

20. – ♗f2![1] –+

[1] 21. ♕f2 ♕a1 –+

4862. E 94

Ogaard – Rantanen

1981

1. d4 ♘f6 2. c4 g6 3. ♘c3 ♗g7 4. ♘f3 0-0 5. e4 d6 6. ♗e2 c6 7. 0-0 ♘bd7 8. ♗e3 e5 9. h3 ed4 10. ♗d4 ♖e8 11. ♕c2 ♕e7 12. ♖fe1 c5 13. ♘d5 ♕d8 14. ♗c3 ♘e4 15. ♗g7 ♔g7 16. ♗d3 ♘ef6 17. ♕c3 ♔g8 18. ♖e8 ♘e8 19. ♖e1 ♘df6 20. ♘e7 ♔g7 21. ♘g5 h6

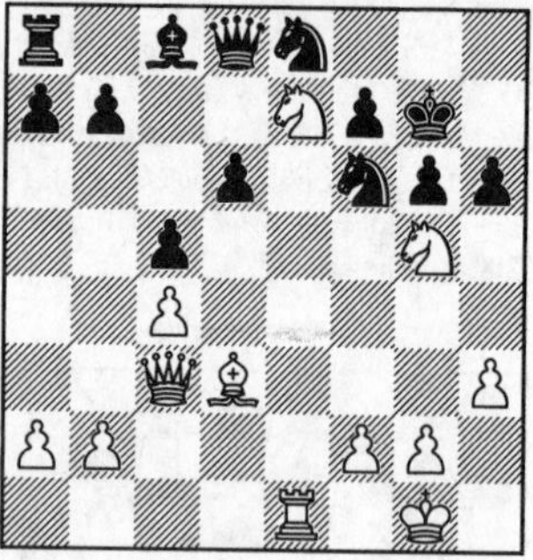

22. ♘f7! ♔f7 23. ♗g6 ♔g7[1] 24. ♕g3 ♘c7 25. ♗f5 +–

[1] 23. – ♔f8 24. ♗e8 ♘e8 25. ♕h8 ♔f7 26. ♕g8 ♔f6 27. ♕g6#

4863. A 03
Browne – H. Smith
London 1932

1. f4 d5 2. e3 ♘f6 3. ♘f3 e6 4. b3 ♗d6 5. ♗b2 0-0 6. ♗d3 ♘bd7 7. 0-0 ♖e8 8. ♘e5 c6 9. ♖f3 ♕c7 10. ♘d7 ♘d7 11. ♗h7! ♔h7 12. ♖h3 ♔g8 13. ♕h5 ♔f8

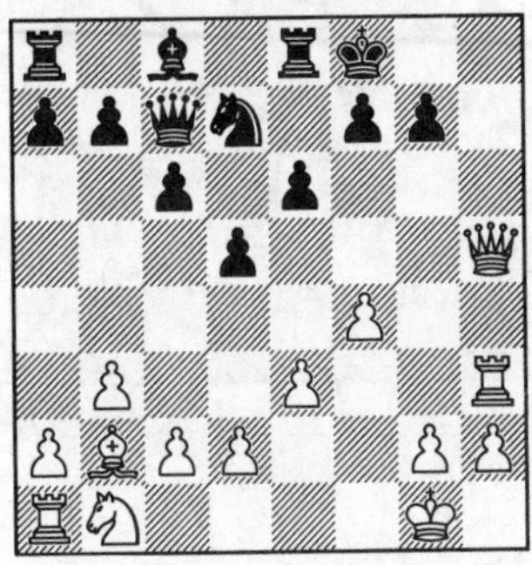

14. ♗g7! ♔e7[1] 15. ♕g5 f6 16. ♗f6! ♘f6 17. ♖h7 +–

[1] 14. – ♔g7 15. ♕h6 ♔g8 16. ♕h8#

4864. A 03
Lasker – Bauer
Amsterdam 1889

1. f4 d5 2. e3 ♘f6 3. b3 e6 4. ♗b2 ♗e7 5. ♗d3 b6 6. ♘c3 ♗b7 7. ♘f3 ♘bd7 8. 0-0 0-0 9. ♘e2 c5 10. ♘g3 ♕c7 11. ♘e5 ♘e5 12. ♗e5 ♕c6 13. ♕e2 a6 14. ♘h5 ♘h5 15. ♗h7! ♔h7 16. ♕h5 ♔g8

17. ♗g7! ♔g7 18. ♕g4 ♔h7 19. ♖f3 e5 20. ♖h3 ♕h6 21. ♖h6 ♔h6 22. ♕d7 ♗f6 23. ♕b7 +–

4865. A 09
Aljechin – Drewitt
Portsmouth 1923

1. ♘f3 d5 2. b4 e6 3. ♗b2 ♘f6 4. a3 c5 5. bc5 ♗c5 6. e3 0-0 7. c4 ♘c6 8. d4 ♗b6 9. ♘bd2 ♕e7 10. ♗d3 ♖d8 11. 0-0 ♗d7 12. ♘e5 ♗e8 13. f4 ♖ac8 14. ♖ac1 ♘d7 15. ♘c6 ♖c6 16. c5 ♘c5 17. dc5 ♗c5 18. ♖f3 ♗a3 19. ♖c6 ♗c6 20. ♗h7! ♔h7 21. ♖h3 ♔g8

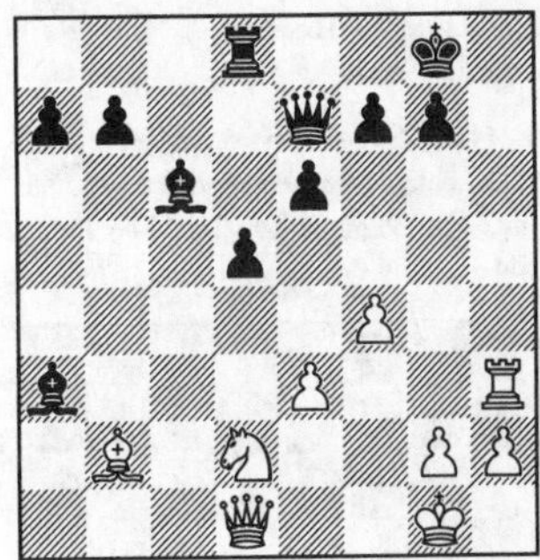

22. ♗g7![1] +–

[1] 22. – ♔g7 23. ♕g4

4866. A 28
Ragozin – Sozin
1937

1. c4 e5 2. ♘f3 ♘c6 3. ♘c3 ♘f6 4. d4 ed4 5. ♘d4 ♗b4 6. ♗g5 h6 7. ♗h4 ♗c3 8. bc3 ♘e5 9. ♘b5 a6 10. ♕d4 d6 11. ♗f6 gf6 12. ♘a3 c5 13. ♕d2 ♕a5 14. ♕b2 ♗d7 15. e4 ♗c6 16. f3 ♖g8 17. ♔f2 f5 18. ef5

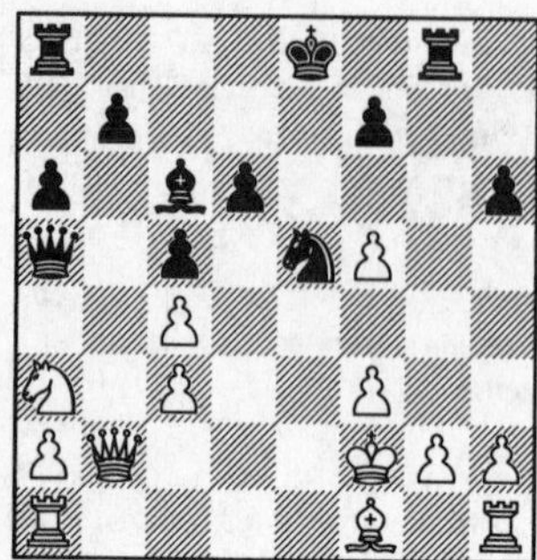

18. – ♖g2! 19. ♔g2[1] ♗f3 20. ♔g1 0-0-0 21. ♘c2 ♖g8 22. ♔f2 ♕d8 23. h4 ♗h1 –+

[1] 19. ♗g2 ♘d3 –+

4867. A 41

Vasilchenko – Hauser
Policka 1993

1. d4 d6 2. ♘f3 ♗g4 3. g3 ♕c8 4. h3 ♗f3 5. ef3 d5 6. ♗e3 e6 7. ♗d3 ♗d6 8. ♕e2 ♘e7 9. ♘d2 ♘d7 10. f4 c5 11. c4 cd4 12. ♗d4 0-0 13. ♖c1 ♘c6

14. ♗g7! ♔g7 15. cd5 ed5 16. ♕g4 ♔h8 17. ♕f5 ♔g7 18. ♕h7 ♔f6 19. ♕h4 +–

4868. A 47

Zsuzsa Polgár – N. N.
Budapest 1982

1. d4 ♘f6 2. ♘f3 b6 3. ♗f4 d5 4. ♘bd2 c5 5. e3 e6 6. ♘e5 cd4 7. ed4 ♗e7 8. ♘df3 0-0 9. ♗d3 ♗b7 10. 0-0 ♘c6 11. c3 ♖c8 12. h3 ♘e5 13. de5 ♘e4 14. ♘d4 ♘c5 15. ♗c2 ♗a6 16. ♖e1 ♕e8 17. ♖e3 ♗b7 18. ♖g3 a6 19. ♗h7! ♔h7 20. ♕h5 ♔g8

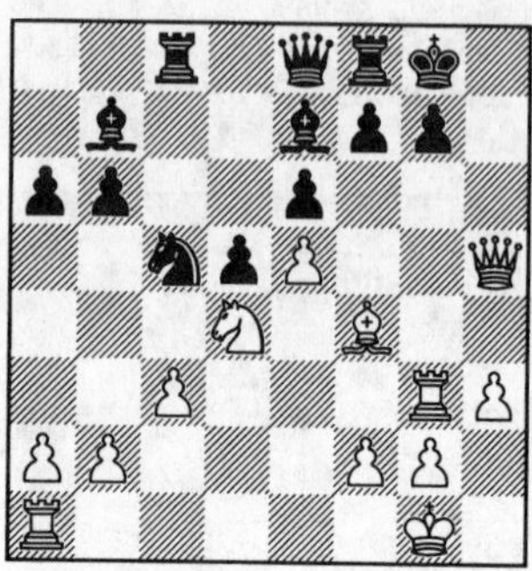

21. ♖g7! ♔g7 22. ♗h6 ♔h8 23. ♗g5 ♔g8 24. ♗f6! ♗f6 25. ef6 +–

4869. A 48

Torma – Sofia Polgár
Budapest 1983

1. d4 ♘f6 2. ♘f3 g6 3. ♗f4 ♗g7 4. e3 0-0 5. h3 d6 6. ♗c4 ♘bd7 7. 0-0 b6 8. a4 a6 9. ♘bd2 ♗b7 10. c3 ♖e8 11. ♕b3 d5 12. ♗e2 ♘h5 13. ♗h2 e5 14. de5 ♘e5 15. ♘e5 ♗e5 16. ♗h5 gh5 17. ♖ad1 ♔h8 18. f3 ♗h2 19. ♔h2 ♖e3 20. ♘e4 ♕h4 21. ♕b4 c5 22. ♕b6 de4 23. ♕b7 ♖g8 24. ♕c7

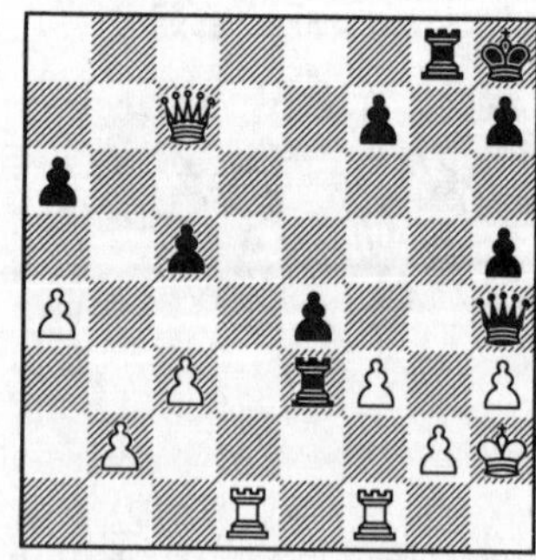

24. – ♖g2![1] –+

[1] 25. ♔g2 ♖e2 26. ♔g1 ♕g5 −+

4870. A 60

Sjoholm – Spielmann

Kalmaar 1941

1. d4 ♘f6 2. ♘f3 c5 3. d5 e6 4. de6 fe6 5. c4 b5 6. e3 ♗b7 7. cb5 d5 8. b3 ♗d6 9. ♗b2 0-0 10. ♗e2 ♘bd7 11. 0-0 e5 12. ♘g5 ♕e7 13. ♗g4 ♘g4 14. ♕g4 ♖f6 15. ♘h3 ♖g6 16. ♕e2 ♕h4 17. ♔h1 e4 18. ♘d2 d4 19. ed4 e3 20. ♘f3

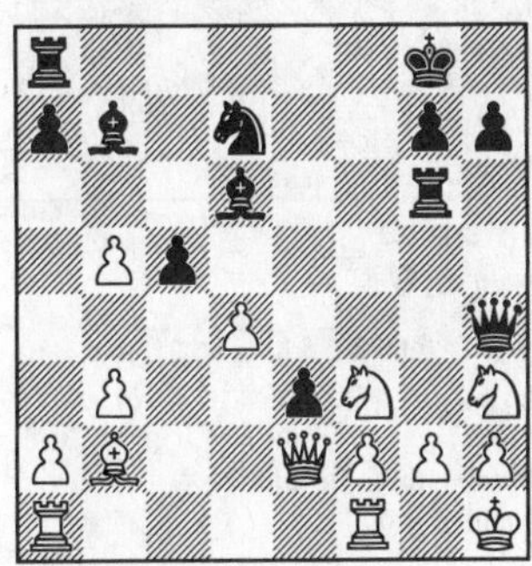

20. – ♖g2![1] −+

[1] 21. ♔g2 ♕g4 22. ♔h1 ♗f3 −+; 21. ♘h4 ♖h2 22. ♔g1 ♖h1#

4871. A 83

Gubanov – Uraev

St. Petersburg 1910

1. d4 f5 2. e4 fe4 3. ♘c3 ♘f6 4. f3 ef3 5. ♘f3 e6 6. ♗g5 ♗e7 7. ♗d3 b6 8. 0-0 0-0 9. ♘e5 ♗b7 10. ♗f6 ♗f6 11. ♗h7! ♔h7 12. ♕h5 ♔g8 13. ♘g6 ♖e8 14. ♕h8 ♔f7 15. ♘e5 ♔e7

16. ♕g7! ♗g7 17. ♖f7 ♔d6 18. ♘b5 ♔d5 19. c4 ♔e4 20. ♖e1#

4872. A 85

Smyth – Helms

New York 1915

1. d4 f5 2. ♘f3 ♘f6 3. c4 e6 4. ♘c3 b6 5. e3 ♗b7 6. ♗d3 ♗d6 7. a3 a5 8. 0-0 0-0 9. ♕c2 ♘c6 10. e4 fe4 11. ♘e4 ♘e4 12. ♗e4 ♘d4 13. ♗h7 ♔h8 14. ♘d4 ♕h4 15. g3 ♕d4 16. ♗d3 ♖f3 17. ♗e3 ♕e5 18. ♖ae1 ♖af8 19. ♗b6 ♕h5 20. ♗e3 ♕h3 21. ♗e4 ♖8f5 22. ♗f5

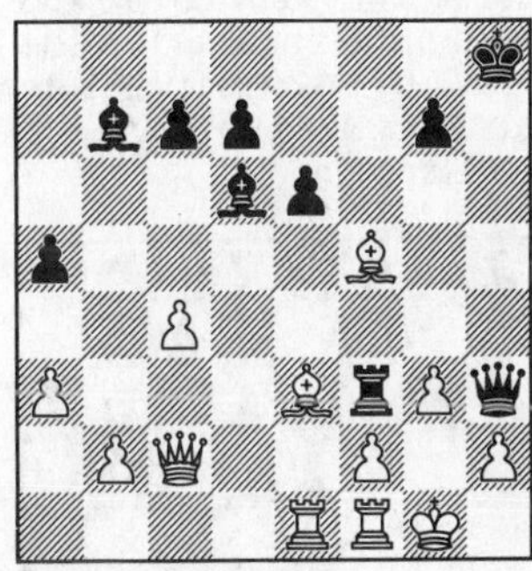

22. – ♕g2! 23. ♔g2 ♖g3#

4873. B 02

Mestrovic – Cvitan

Zurich 1983

1. ♘c3 d5 2. e4 ♘f6 3. ed5 ♘d5 4. ♕f3 e6 5. ♗c4 c6 6. ♗b3 ♗e7 7. ♘ge2 0-0 8.

d4 ♘c3 9. ♕c3 b6 10. ♕f3 ♗b7 11. 0-0 ♕c8 12. ♕g3 ♘d7 13. c3 c5 14. ♗h6 ♗f6 15. ♗c2 ♖d8 16. ♘f4 ♔h8 17. ♗g5 ♗g5 18. ♕g5 cd4 19. ♕h5 ♘f6 20. ♕f7 e5 21. ♘h3

21. – ♗g2! –+

4874. B 14

Sprecher – Lutz

Fürth 1937

1. e4 c6 2. d4 d5 3. ed5 cd5 4. c4 ♘f6 5. ♘c3 e6 6. ♘f3 ♗e7 7. ♗f4 0-0 8. c5 ♘e4 9. ♗d3 ♘c3 10. bc3 b6 11. cb6 ab6 12. h4 ♘d7 13. ♕c2 ♘f6 14. ♘g5 h6 15. ♗e5 ♖e8 16. ♗h7 ♔f8 17. ♗g6 hg5 18. hg5 ♘g8

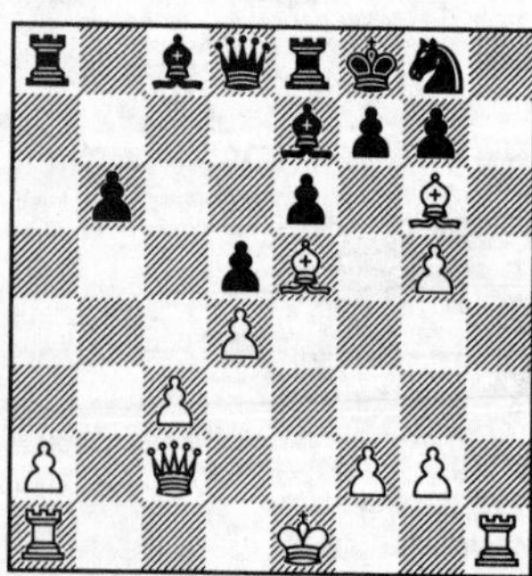

19. ♗g7![1] +–

[1] 19. – ♔g7 20. ♖h7 ♔f8 21. ♖f7#

4875. B 15

Spielmann – Nimzowitsch

Monachium 1905

1. e4 c6 2. d4 d5 3. ♘c3 de4 4. ♘e4 ♘f6 5. ♘g3 e5 6. ♘f3 ed4 7. ♘d4 ♗e7 8. ♗e2 0-0 9. 0-0 ♔h8 10. b3 ♘g8 11. ♗b2 ♘d7 12. ♘df5 ♗f6 13. ♗a3 ♗a1 14. ♕a1 ♘df6 15. ♖d1 ♕e8

16. ♘g7! ♔g7 17. ♖d6 ♗e6 18. ♘h5 ♔g6 19. ♕e5[1] +–

[1] 19. – ♘h5 20. ♗h5 ♔h6 21. ♗c1#

4876. B 15

Spielmann – Tartakover

Munich 1909

1. e4 c6 2. d4 d5 3. ♘c3 de4 4. ♘e4 ♘f6 5. ♘g3 e5 6. ♘f3 ed4 7. ♘d4 ♗c5 8. ♗e3 ♕b6 9. ♕e2 0-0 10. 0-0-0 ♘d5 11. ♕h5 ♘f6 12. ♕h4 ♗g4 13. ♗d3 ♗d1 14. ♖d1 ♘bd7 15. ♘gf5 ♘e5

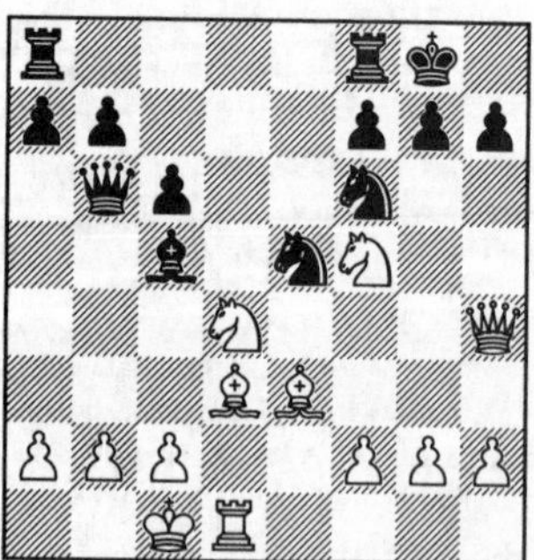

16. ♘g7! ♕d8[1] 17. ♘gf5 ♘g6 18. ♕h6 ♘e8 19. ♘f3 ♗e3 20. fe3 ♕f6 21. ♘g5 ♕h8 22. ♘e7! +–

[1] 16. – ♔g7 17. ♗h6 ♔g8 18. ♕f6 +–

4877. B 17
Shamkovich – Holmov
Baku 1961

1. e4 c6 2. d4 d5 3. ♘c3 de4 4. ♘e4 ♘d7 5. ♗c4 ♘gf6 6. ♘g5 e6 7. ♕e2 ♘b6 8. ♗b3 h6 9. ♘5f3 c5 10. ♗e3 ♕c7 11. ♘e5 ♗d6 12. ♘gf3 0-0 13. g4 c4 14. ♘c4 ♘c4 15. ♗c4 ♘g4 16. ♖g1 e5 17. 0-0-0 ♘e3 18. ♕e3 ♔h8

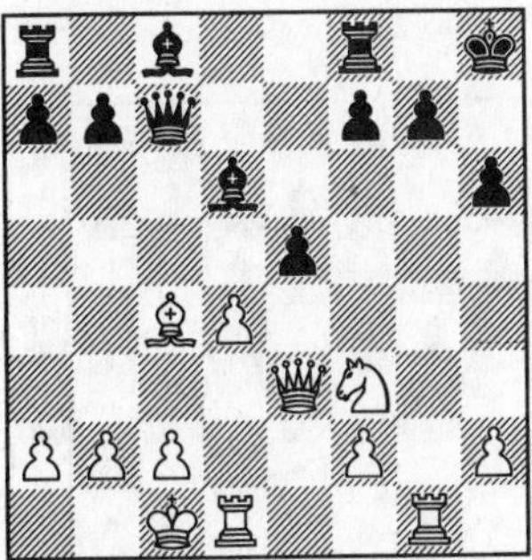

19. ♖g7! ♔g7 20. ♖g1 ♔f6[1] 21. de5 ♗e5 22. ♕h6 ♔e7 23. ♖e1 ♗e6 24. ♘e5 ♖e8 25. ♕g5[2] +–

[1] 20. – ♔h7 21. ♘g5! ♔h8 22. ♘f7 ♖f7 23. ♕h6 ♖h7 24. ♖g8#; 21. – hg5 22. ♕g5 +–

[2] 25. – ♔f8 26. ♕f6 ♖ec8 27. ♘g6 ♔e8 28. ♗b5 +–

4878. B 19
Gaprindashvili – Nikolac
Wijk aan Zee 1979

1. e4 c6 2. d4 d5 3. ♘d2 de4 4. ♘e4 ♗f5 5. ♘g3 ♗g6 6. h4 h6 7. h5 ♗h7 8. ♘f3 ♘d7 9. ♗d3 ♗d3 10. ♕d3 e6 11. ♗f4 ♕a5 12. c3 ♘gf6 13. a4 c5 14. 0-0 ♖c8 15. ♖fe1 c4 16. ♕c2 ♗e7 17. ♘e5 0-0 18. ♘f5 ♖fe8[1]

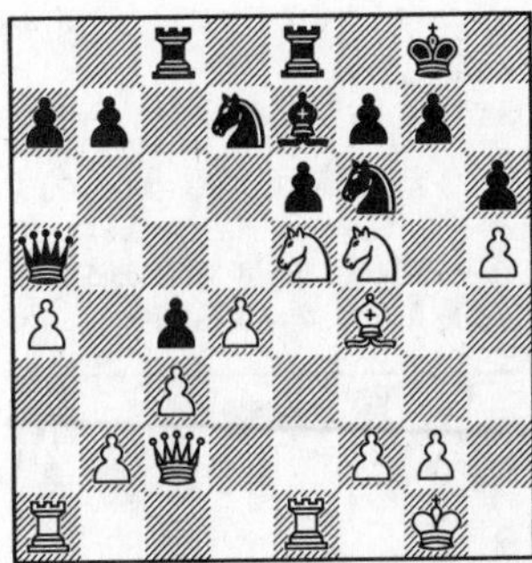

19. ♘g7! ♔g7 20. ♗h6! ♔h6 21. ♘f7 ♔h5[2] 22. g4 ♔h4[3] 23. f3 ♘g4 24. ♖e4 +–

[1] 18. – ef5 19. ♘d7 +–

[2] 21. – ♔g7 22. ♕g6 ♔f8 23. ♘g5 +–

[3] 22. – ♘g4 23. ♕h7 +–

4879. B 20
Keres – Foltys
Salzburg 1943

1. e4 c5 2. ♘e2 ♘f6 3. ♘bc3 ♘c6 4. g3 d5 5. ed5 ♘d5 6. ♗g2 ♘c3 7. bc3 e6 8. 0-0 ♗e7 9. ♖b1 0-0 10. c4 ♕d7 11. ♗b2 b6 12. d4 ♗b7 13. d5 ♘a5 14. ♘f4 ♘c4

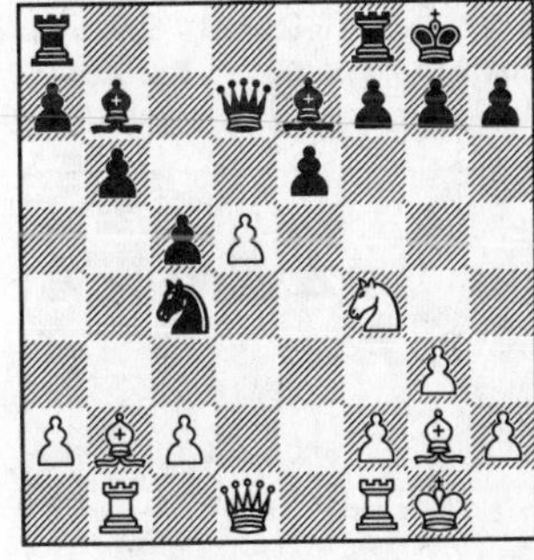

15. ♗g7! ♔g7 16. ♕g4 ♔h8 17. de6 ♖g8[1] 18. ♕g8! +–

[1] 17. – fe6 18. ♗b7 ♕b7 19. ♘e6 ♗f6 20. ♕c4 +–

4880. B 20
Araiza – Larea
USA 1948

1. e4 c5 2. ♘e2 e5 3. ♘bc3 ♘c6 4. ♘g3 ♘d4 5. ♗c4 a6 6. a4 b6 7. 0-0 ♗b7 8. d3 ♘f6 9. f4 d6 10. fe5 de5 11. ♗g5 ♗e7 12. ♗f6 ♗f6 13. ♘h5 ♖c8

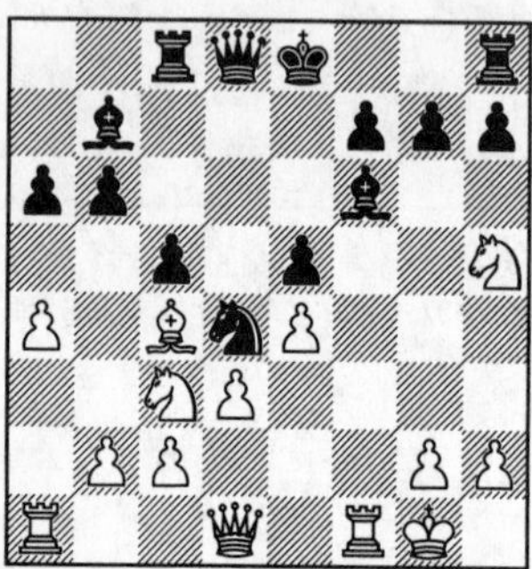

14. ♘g7! ♗g7 15. ♖f7 ♖c7 16. ♕h5 ♘f3 17. ♖f3 ♔d7 18. ♕g4 ♔c6 19. ♕e6[1] +–

[1] 19. – ♕d6 20. ♗d5#

4881. B 27
Schlosser – Katchev
Corr. 1968

1. e4 c5 2. ♘f3 b6 3. d4 cd4 4. ♘d4 ♗b7 5. ♘c3 a6 6. ♗c4 b5 7. ♗b3 b4 8. ♘a4 ♗e4 9. ♘c5 ♗g2 10. ♖g1 ♗c6 11. ♗f7! ♔f7

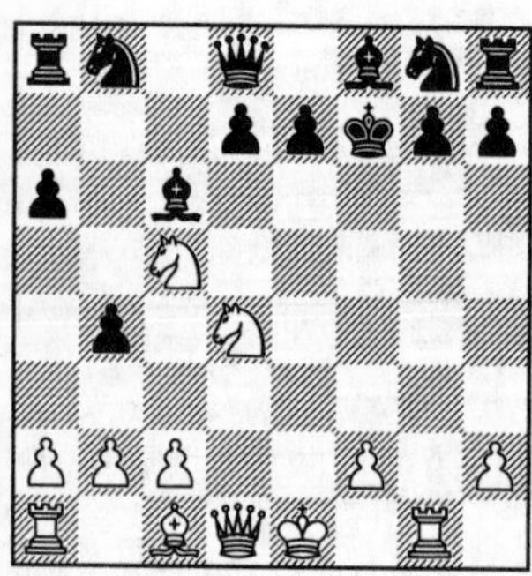

12. ♖g7![1] +–

[1] 12. – ♗g7 13. ♕h5 ♔f6 14. ♕f5#; 13. – ♔f8 14. ♘ce6 de6 15. ♘e6#; 12. – ♔g7 13. ♕g4 ♔f7 14. ♕h5 ♔g7 15. ♘ce6 de6 16. ♘e6 ♔f6 17. ♘d8 h6 18. ♕f7 ♔e5 19. ♕e6 ♔d4 20. ♗e3#

4882. B 29
de Rooi – Crabbendam
Beverwijk 1965

1. e4 c5 2. ♘f3 ♘f6 3. ♘c3 d5 4. ed5 ♘d5 5. ♗b5 ♗d7 6. ♕e2 ♘c3 7. dc3 ♘c6 8. ♗f4 a6 9. ♗c4 e6 10. 0-0-0 ♗e7 11. ♖d2 b5 12. ♗d3 ♖a7 13. ♗e4 ♕c8 14. ♘e5 ♘e5 15. ♗e5 0-0 16. ♗h7! ♔h7 17. ♕h5 ♔g8

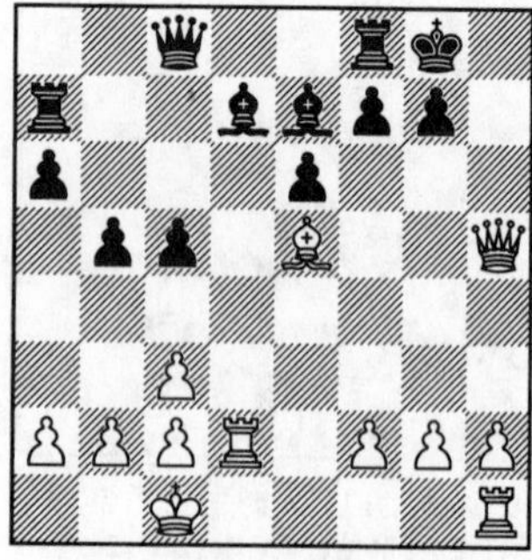

18. ♗g7! ♔g7[1] 19. ♖d3 f5 20. ♖g3 ♔f6 21. ♕g6 ♔e5 22. ♖d3 ♗c6 23. f4 ♔f4 24. ♕g3 ♔e4 25. ♕e3#

[1] 18. – f6 19. ♗h6 +–; 18. – f5 19. ♗e5 +–

4883. B 29
Pietzsch – Schröder
1961

1. e4 c5 2. ♘f3 ♘f6 3. e5 ♘d5 4. ♘c3 e6 5. ♘d5 ed5 6. d4 ♘c6 7. ♗d3 d6 8. ed6 cd4 9. 0-0 ♗d6 10. ♖e1 ♗e7 11. ♘g5 h6 12. ♕h5 0-0 13. ♘h7 ♖e8 14. ♗h6! ♗e6[1]

15. ♗g7![2] +–

[1] 14. – gh6 15. ♕h6 f5 16. ♕g6 ♔h8 17. ♘f6! +–

[2] 15. – ♔g7 16. ♖e6! fe6 17. ♕g6 ♔h8 18. ♘f6 +–

4884. B 33

Estino – Keeng

1968

1. e4 c5 2. ♘f3 ♘c6 3. d4 cd4 4. ♘d4 ♘f6 5. ♘c3 e5 6. ♘de2 ♗b4 7. a3 ♗a5 8. ♘g3 0-0 9. b4 ♗c7 10. ♘d5 ♘e8 11. ♗c4 b5 12. ♗a2 a5 13. h4 ab4 14. ♗g5 ♘f6 15. ♘h5 ♗b6 16. ♘hf6 ♔h8 17. ♕h5 h6 18. ♗h6 g6

19. ♗g7! ♔g7 20. ♕h7#

4885. B 41

Zimmermann – Schlott

1974

1. e4 c5 2. ♘f3 e6 3. d4 cd4 4. ♘d4 a6 5. c4 ♘f6 6. ♘c3 ♗b4 7. e5 ♘e4 8. ♕g4 ♘c3 9. a3 ♗f8 10. bc3 d6 11. ed6 ♕d6 12. ♗f4 ♕d8 13. ♖d1 ♘d7 14. ♗d3 ♘f6 15. ♕e2 ♗a3 16. ♘f5 0-0

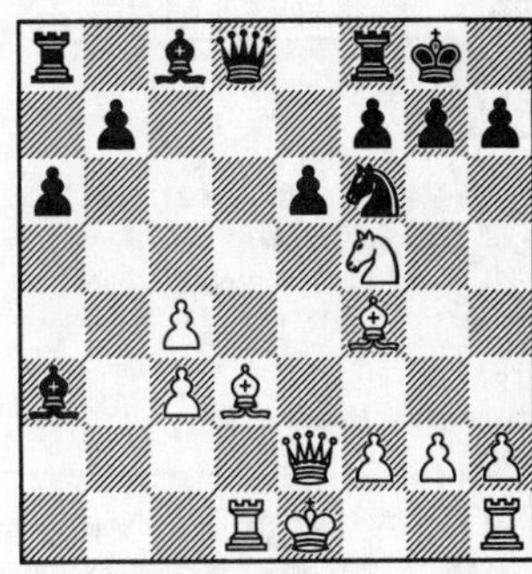

17. ♘g7! ♔g7 18. ♗e5 ♗e7[1] 19. ♕g4 ♔h8 20. ♕h4 +–

[1] 18. – ♔g8 19. ♕f3 +–; 18. – ♖e8 19. ♕g4 ♔f8 20. ♕h4 ♘d7 21. ♕h6 ♔e7 22. ♕g5 ♔f8 23. ♗g7 ♔g8 24. ♗f6 +–

4886. B 42

Stein – Portisch

Stockholm 1962

1. e4 c5 2. ♘f3 e6 3. d4 cd4 4. ♘d4 a6 5. ♗d3 ♘f6 6. 0-0 ♕c7 7. ♘d2 ♘c6 8. ♘c6 bc6 9. f4 ♗c5 10. ♔h1 d6 11. ♘f3 e5 12. fe5 de5 13. ♘h4 0-0 14. ♘f5 ♗e6 15. ♕e2 a5 16. ♗c4 ♔h8 17. ♗g5 ♘d7 18. ♖ad1 ♘b6

19. ♘g7! ♗c4[1] 20. ♗f6! ♗e7[2] 21. ♕f3 +–

[1] 19. – ♔g7 20. ♗f6 ♔g8 21. ♕h5 ♖e8 22. ♕h6 ♗f8 23. ♕g5 +–

[2] 20. – ♗e2 21. ♘f5 ♔g8 22. ♘h6#

4887. B 43

Vaslanjan – Hodos

1964

1. e4 c5 2. ♘f3 e6 3. d4 cd4 4. ♘d4 a6 5. ♘c3 ♕c7 6. ♗e2 ♘f6 7. ♗e3 ♗b4 8. 0-0 ♗c3 9. bc3 ♘e4 10. ♗d3 ♘f6 11. ♖e1 ♘c6 12. ♘f5 0-0

13. ♘g7! ♔g7 14. ♗h6! ♔h6[1] 15. ♕d2 ♔h5[2] 16. ♖e3 ♕h2 17. ♔h2 ♘g4 18. ♔h1 ♘e3 19. ♕e3 +–

[1] 14. – ♔g8 15. ♕f3! +–

[2] 15. – ♔g7 16. ♕g5 ♔h8 17. ♕f6 ♔g8 18. ♖e3 d5 19. ♗h7! 1–0 Heinsohn – Bauer, Wittenberg 1965

4888. B 45

Brunnemer – Failing

USA 1920

1. e4 c5 2. ♘f3 ♘c6 3. d4 cd4 4. ♘d4 ♘f6 5. ♘c3 e6 6. ♗e2 ♗b4 7. 0-0 ♗c3 8. bc3 ♘e4 9. ♗f3 ♘c3 10. ♕d3 ♘d5 11. ♗d5 ed5 12. ♖e1 ♔f8 13. ♘f5 d6

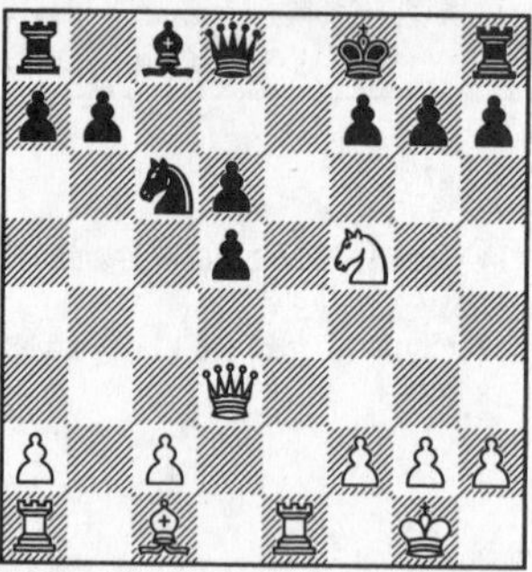

14. ♘g7! ♘e5[1] 15. ♘h5! ♗e6[2] 16. ♖e5! de5 17. ♗h6 ♔e8 18. ♕b5[3] +–

[1] 14. – ♔g7 15. ♕g3 ♔f8 16. ♗h6#; 15. – ♔f6 16. ♕g5#

[2] 15. – ♘d3 16. ♗h6 ♔g8 17. ♖e8 ♕e8 18. ♘f6#

[3] 18. – ♗d7 19. ♕d5 ♕e7 20. ♖d1!+–

4889. B 47

Hector – Plachetka

Gausdal 1989

1. e4 c5 2. ♘f3 e6 3. d4 cd4 4. ♘d4 ♘c6 5. ♘c3 ♕c7 6. f4 a6 7. ♘c6 ♕c6 8. ♗d3 b5 9. ♕e2 ♗b7 10. 0-0 ♘f6 11. a3 ♗c5 12. ♔h1 ♗d4 13. ♘a2 d6 14. a4 ba4 15. ♘b4 ♕b6 16. ♖a4 0-0 17. c3 ♗c5 18. e5 ♘d7 19. ♘a6 de5 20. fe5 ♘e5 21. ♕e5 ♗a6

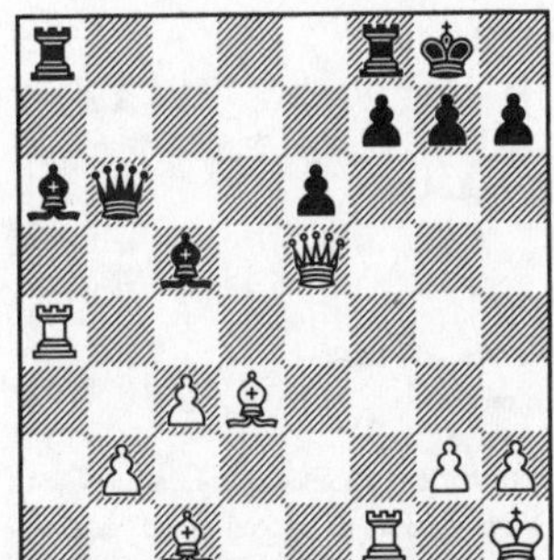

22. ♗h7! ♔h7 23. ♖g4! f6 24. ♕h5 ♔g8 25. ♖g7! +–

4890. B 49

Kuzmin – Sveshnikov

Moscow 1973

1. e4 c5 2. ♘f3 e6 3. d4 cd4 4. ♘d4 ♘c6 5. ♘c3 a6 6. ♗e2 ♕c7 7. 0-0 ♘f6 8. ♗e3 ♗b4 9. ♘c6 bc6 10. ♘a4 0-0 11. c4 ♗d6 12. f4 ♘e4 13. c5 ♗e7 14. ♗d3 ♘f6 15. ♗d4 ♘d5 16. ♘b6 ♘b6 17. ♗h7! ♔h7 18. ♕h5 ♔g8

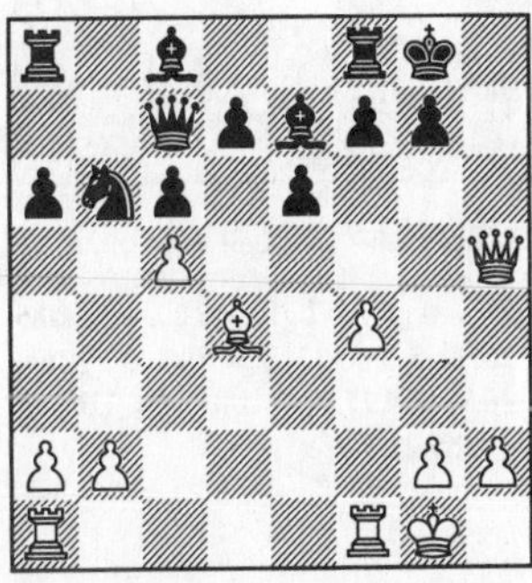

19. ♗g7! ♔g7 20. ♕g4 ♔h7 21. ♖f3 ♗c5 22. ♔h1 +–

4891. B 53

Martius – Darga

Hamburg–Berlin 1958

1. e4 c5 2. ♘f3 d6 3. d4 cd4 4. ♕d4 ♘c6 5. ♗b5 ♗d7 6. ♗c6 ♗c6 7. ♘c3 e6 8. ♗g5 ♘f6 9. 0-0-0 ♗e7 10. e5 de5 11. ♕e5 ♕b8 12. ♕e2 0-0 13. ♘e5 ♕c7 14. ♖d3 ♘d5 15. ♗d2 ♖ac8 16. ♖h3 ♘c3 17. ♗c3 ♗g2 18. ♖g3 ♗h1

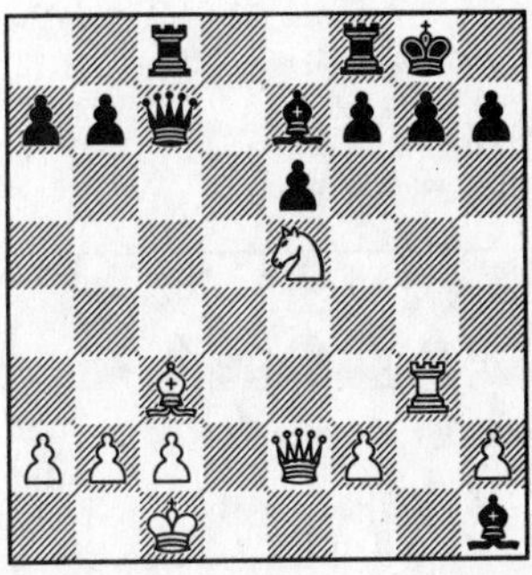

19. ♖g7![1] +–

[1] 19. – ♔g7 20. ♕g4 ♔h6 21. ♗d2 +–; 20. – ♔h8 21. ♘f7#

4892. B 56

Gording – Ginsburg

New York 1953

1. e4 c5 2. ♘f3 ♘c6 3. d4 cd4 4. ♘d4 ♘f6 5. ♘c3 d6 6. h3 g6 7. ♗e3 ♗g7 8. ♕d2 0-0 9. g4 a6 10. ♗e2 e5 11. ♘f5 gf5 12. gf5 ♘a5 13. ♖g1 ♔h8

14. ♖g7!? ♔g7 15. ♗g5 ♔h8?[1] 16. ♘d5! ♘e4 17. ♗d8 ♘d2 18. ♗f6 ♔g8 19. ♘e7#

[1] 15. – ♖g8!

4893. B 57

André – Knoll
Fos 1975

1. e4 c5 2. ♘f3 ♘c6 3. d4 cd4 4. ♘d4 ♘f6 5. ♘c3 d6 6. ♗c4 e6 7. ♗e3 ♗e7 8. ♕e2 0-0 9. 0-0-0 a6 10. ♖hg1 ♘d4 11. ♗d4 b5 12. e5 de5 13. ♗e5 ♕b6 14. ♗d3 ♗b7 15. g4 ♖fd8 16. g5 ♘d7 17. ♗h7! ♔h7 18. ♕h5 ♔g8

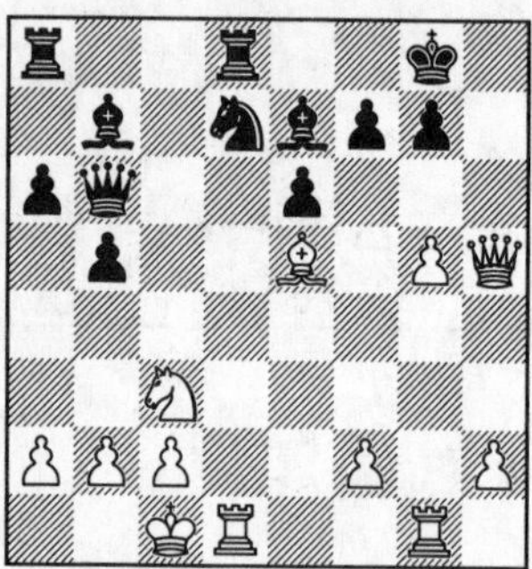

19. ♗g7! ♔g7 20. ♕h6 ♔g8 21. g6 ♘f8 22. gf7 ♔f7 23. ♖g7 ♔e8 24. ♕h5 +–

4894. B 57

Sarbaai – Weingold

1. e4 c5 2. ♘f3 ♘c6 3. d4 cd4 4. ♘d4 ♘f6 5. ♘c3 d6 6. ♗c4 e6 7. ♗e3 ♗e7 8. ♕e2 0-0 9. 0-0-0 a6 10. ♗b3 ♕c7 11. ♖hg1 b5 12. g4 b4 13. ♘c6 ♕c6 14. ♘d5 ed5 15. g5 de4 16. gf6 ♗f6 17. ♗d5 ♕a4 18. ♕h5 ♗e6

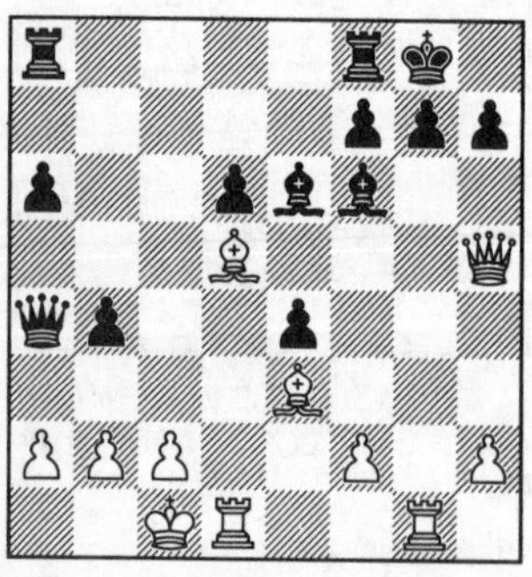

19. ♖g7![1] +–

[1] 19. – ♔g7 20. ♕h6 +–; 19. – ♗g7 20. ♖g1 ♖fc8 21. ♖g7! ♔g7 22. ♕h6 ♔g8 23. ♗e4 +–

4895. B 64

Moran – Franco
Gijon 1955

1. e4 c5 2. ♘f3 ♘c6 3. d4 cd4 4. ♘d4 ♘f6 5. ♘c3 d6 6. ♗g5 e6 7. ♕d2 ♗e7 8. 0-0-0 0-0 9. f4 e5 10. ♘f3 ♗g4 11. h3 ♗f3 12. gf3 ♘d4 13. fe5 de5 14. f4 ♕a5 15. fe5 ♘f3 16. ef6 ♘d2 17. fe7 ♘f1 18. ef8♕ ♔f8 19. ♗f4 ♘g3 20. ♖hg1 ♘e4 21. ♘e4 ♕a2 22. ♗d6 ♔g8

23. ♖g7! ♔h8[1] 24. ♖g8! +–

[1] 23. – ♔g7 24. ♖g1 ♔h6 25. ♗f4 ♔h5 26. ♖g5 ♔h6 27. ♖g4 ♔h5 28. ♘f6#; 26. – ♔h4 27. ♖g4! ♔h3 28. ♘f2#; 27. – ♔h5 28. ♘f6#

4896. B 64

Keres – Szabó
Budapest 1955

1. e4 c5 2. ♘f3 d6 3. d4 cd4 4. ♘d4 ♘f6 5. ♘c3 ♘c6 6. ♗g5 e6 7. ♕d2 ♗e7 8. 0-0-0 0-0 9. f4 a6 10. e5! de5 11. ♘c6 bc6 12. fe5 ♘d7 13. h4 ♖b8 14. ♕e3 ♖e8 15. ♖h3 ♕a5 16. ♗e7 ♖e7 17. ♖g3 ♖e8 18. ♖d7 ♗d7 19. ♗d3 h6 20. ♕f4 ♔f8

21. ♖g7! ♔g7 22. ♕f6 ♔f8[1] 23. ♗g6[2] +–

[1] 22. – ♔g8 23. ♕h6 ♕e5 24. ♗h7 ♔h8 25. ♗g6 ♔g8 26. ♕h7 ♔f8 27. ♕f7#
[2] 23. – ♖e7 24. ♕h8#

4897. B 83
Patrici – Torres
Corr. 1973

1. e4 c5 2. ♘f3 d6 3. d4 cd4 4. ♘d4 ♘f6 5. ♘c3 e6 6. ♗e2 ♘c6 7. ♗e3 ♗e7 8. 0-0 0-0 9. f4 ♘d4 10. ♗d4 a6 11. ♕e1 e5 12. fe5 de5 13. ♗e5 ♕b6 14. ♔h1 ♕b2 15. ♕g3 ♔h8 16. ♘d5 ♕c2 17. ♘e7 ♕e2 18. ♖f6 ♕e4

19. ♕g7! ♔g7 20. ♖g6#

4898. B 85
Yates – Naegeli
London 1927

1. e4 c5 2. ♘f3 e6 3. d4 cd 4. ♘d4 ♘f6 5. ♘c3 d6 6. ♗e2 ♘c6 7. 0-0 ♗e7 8. ♗e3 a6 9. ♔h1 0-0 10. f4 ♕c7 11. ♕e1 ♗d7 12. ♕g3 ♖ac8 13. ♖ad1 ♖fd8 14. e5 ♘d5 15. ♘d5 ed5 16. ♘c6 bc6 17. ♗d4 ♗f8 18. f5 c5 19. ed6 ♕d6 20. ♗e5 ♕c6

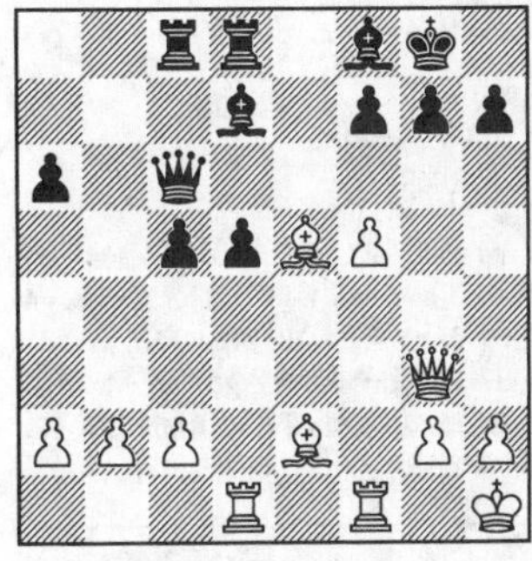

20. ♗g7![1] +–

[1] 21. – ♗g7 22. f6 +–

4899. B 86
Mészáros – Poloch
Fonyód 1982

1. e4 c5 2. ♘f3 d6 3. d4 cd4 4. ♘d4 ♘f6 5. ♘c3 a6 6. ♗c4 e6 7. ♗g5 ♗e7 8. ♗b3 0-0 9. f4 h6 10. h4 ♘c6 11. ♘c6 bc6 12. ♕f3 ♕c7 13. g4 c5 14. ♗f6 ♗f6 15. g5 ♗c3 16. ♕c3 h5 17. 0-0-0 ♗b7 18. f5 ♗e4 19. fe6 fe6 20. ♗e6 ♔h7 21. g6 ♗g6 22. ♖hg1 ♖ae8 23. ♗c4 ♗e4 24. ♖d6 ♖f3 25. ♕d2 ♖e7

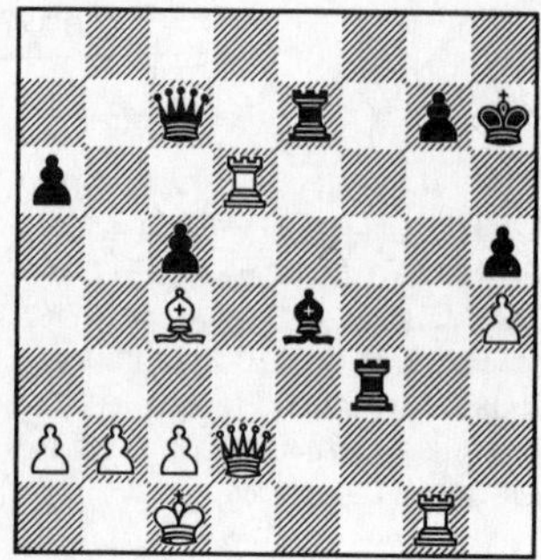

26. ♖g7! +–

4900. B 86

List – Lösche

Corr. 1968

1. e4 c5 2. ♘f3 d6 3. d4 cd4 4. ♘d4 ♘f6 5. ♘c3 a6 6. ♗c4 e6 7. ♗b3 ♗e7 8. g4 ♘c6 9. g5 ♘d4 10. ♕d4 e5 11. ♕d1 ♘d7 12. ♕h5 0-0 13. g6 hg6 14. ♕g6 ♘f6 15. ♖g1 ♘e8 16. ♗h6 ♗f6 17. 0-0-0 ♕a5

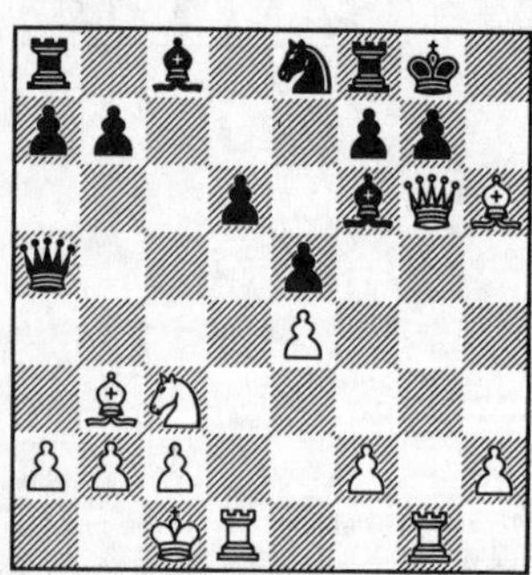

18. ♗g7! ♗g7 19. ♕h6[1] +–

[1] 19. – ♗e6 20. ♖g7! ♘g7 21. ♖g1 +–

4901. B 93

Kupper – Olafsson

1. e4 c5 2. ♘f3 d6 3. d4 cd4 4. ♘d4 ♘f6 5. ♘c3 a6 6. f4 e5 7. ♘f3 ♕c7 8. ♗d3 ♘bd7 9. a4 b6 10. 0-0 ♗b7 11. ♕e1 g6 12. ♕h4 ♗g7 13. fe5 de5 14. ♗h6 0-0 15. ♘g5 ♘h5 16. ♗g7 ♔g7 17. ♖f7 ♔g8

18. ♖g7! ♔h8 19. ♖h7 ♔g8 20. ♖g7 +–

4902. C 01

Mieses – Sergeant

Hastings 1947

1. e4 e6 2. d4 d5 3. ed5 ed5 4. ♗d3 ♗d6 5. ♘c3 c6 6. ♘ge2 ♘e7 7. ♘g3 ♘d7 8. ♗e3 ♘f6 9. ♗g5 ♘d7 10. ♘h5 0-0

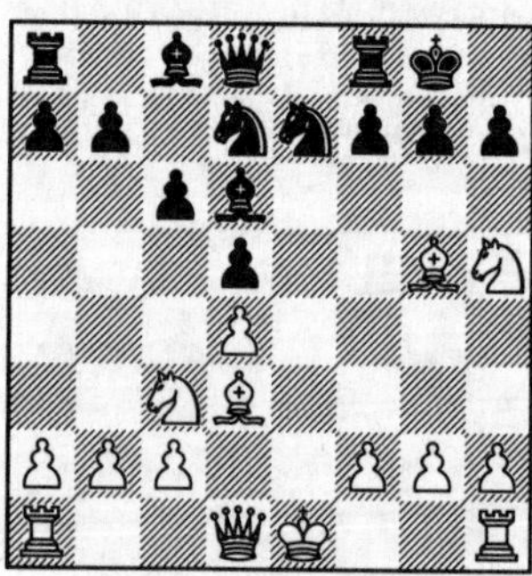

11. ♘g7! ♔g7 12. ♕h5 f5 13. ♕h6 ♔h8 14. ♕d6 ♕e8 15. 0-0[1] ♘g6 16. ♖fe1 ♕f7 17. ♗h6 ♖e8 18. ♖e8 ♕e8 19. ♗f5 +–

[1] 15. ♕e7 ♕e7 16. ♗e7 ♖e8!

4903. C 06

Perfors – Seters

Wageningen 1955

1. e4 e6 2. d4 d5 3. ♘d2 ♘f6 4. e5 ♘fd7 5. ♗d3 c5 6. c3 ♘c6 7. ♘e2 ♕b6 8. ♘f3 cd4 9. cd4 f6 10. ef6 ♘f6 11. 0-0 ♗d6 12. ♘f4 0-0 13. ♖e1 ♗d7 14. ♘e6 ♖fe8 15. ♗f5 h6

16. ♘g7! ♔g7 17. ♗h6! ♔h8[1] 18. ♗d7 ♘d7 19. ♖e8 ♖e8 20. ♘g5 ♖e7 21. ♕h5 ♕b2 22. ♘f7[2] +–

[1] 17. – ♔h6 18. ♕d2 ♔g7 19. ♕g5 ♔f7 20. ♕g6 ♔f8 21. ♕f6 ♔g8 22. ♗h7! ♔h7 23. ♘g5 ♔g8 24. ♕f7 ♔h8 25. ♕h7#

[2] 22. – ♖f7 23. ♗c1 ♖h7 24. ♕e8 +–

4904. C 07

Donaldson – Ornstein

Gausdal 1979

1. e4 e6 2. d4 d5 3. ♘d2 c5 4. ed5 ♕d5 5. ♘gf3 cd4 6. ♗c4 ♕d6 7. 0-0 ♘f6 8. ♘b3 ♘c6 9. ♘bd4 ♘d4 10. ♘d4 ♗d7 11. b3 a6 12. ♗b2 ♕c7 13. ♕e2 ♗d6 14. ♘f5 ♗h2 15. ♔h1 0-0

16. ♘g7! ♕f4[1] 17. ♘h5 +–

[1] 16. – ♔g7 17. ♕g4 ♔h6 18. ♗c1 +–

4905. C 10

Platz – Hegebarth

Magdeburg 1938

1. e4 e6 2. d4 d5 3. ♘c3 de4 4. ♘e4 ♘f6 5. ♘f6 ♕f6 6. ♘f3 h6 7. ♗d3 ♗d6 8. 0-0 0-0 9. ♕e2 ♘c6 10. c3 ♕e7 11. ♖e1 ♖e8 12. ♘e5 ♘e5 13. de5 ♗c5 14. ♕h5 ♗d7 15. ♖e4 ♕f8 16. ♖g4 ♔h8

17. ♖g7! ♗f2[1] 18. ♔f2 ♕c5 19. ♔f1 +–

[1] 17. – ♕g7 18. ♗h6

4906. C 10

Boleslavsky – Ufimcev

Omsk 1943

1. e4 e6 2. d4 d5 3. ♘c3 de4 4. ♘e4 ♘f6 5. ♘f6 gf6 6. ♘f3 b6 7. ♗b5 c6 8. ♗c4

♗a6 9. ♗b3 ♕c7 10. c4 ♘d7 11. 0-0 0-0-0 12. ♕e2 ♗d6 13. a4 ♖hg8 14. a5 c5 15. ab6 ♕b6 16. ♗e3 ♗b7 17. dc5 ♘c5 18. ♗d1 ♖g4 19. ♕d2 ♘e4 20. ♕a5 ♖g8 21. ♘e1[1]

21. – ♖g2! 22. ♘g2 ♘d2! 23. ♕d5[2] ♗d5 24. cd5 ♕b2 25. ♗d2 ♕a1 26. ♗f3 ♗h2 –+

[1] 21. ♗b6 ♖g2 22. ♔h1 ♘f2 23. ♗f2 ♖h2#

[2] 21. ♗b6 ♖g2 22. ♔h1 ♖h2 23. ♔g1 ♖h1#

4907. C 13

Ljuboshit – Shagalovich

Minsk 1956

1. e4 e6 2. d4 d5 3. ♘c3 ♘f6 4. ♗g5 ♗e7 5. e5 ♘fd7 6. h4 c5 7. ♘b5 f6 8. ♗d3 ♕a5 9. ♗d2 ♕b6 10. ♕h5 ♔f8 11. ♖h3 cd4 12. ♖g3 fe5

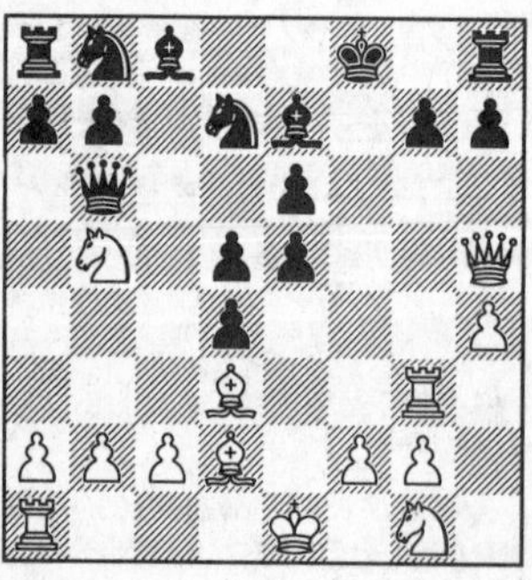

13. ♖g7! ♔g7 14. ♗h6 ♔g8[1] ♗g6! +–

[1] 14. – ♔f6 15. ♕f3#

4908. C 13

Rossetto – Stahlberg

Vinn de Mar 1947

1. e4 e6 2. d4 d5 3. ♘c3 ♘f6 4. ♗g5 ♗e7 5. e5 ♘fd7 6. h4 c5 7. ♘b5 f6 8. ♗d3 a6 9. ♕h5 ♔f8 10. ♖h3 ab5 11. ♗h6 ♕a5 12. ♗d2 ♕c7 13. ♖g3 cd4 14. ♘f3 ♘e5

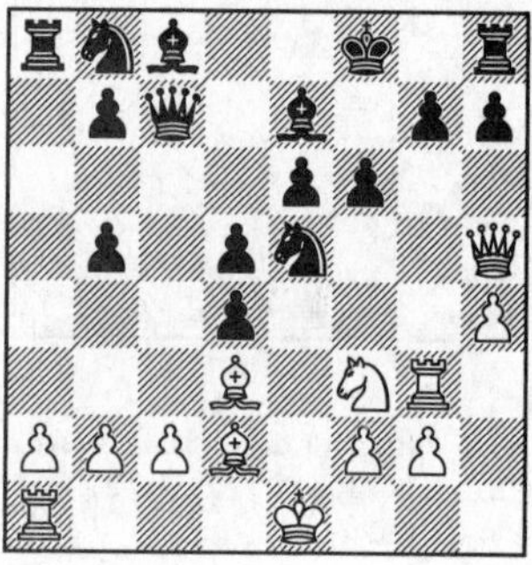

15. ♖g7! h6[1] 16. ♗h7 ♔g7[2] 17. ♕h6[3] =

[1] 15. – ♔g7 16. ♗h6 ♔g8 17. ♕e8 +–

[2] 16. – ♘f3 17. ♔d1! (17. gf3 ♕e5)

[3] 17. – ♔f7 18. ♕h5 ♔g7 (18. – ♔f8 19. ♗h6#) 19. ♕h6 =

4909. C 14

Harmonist – Burn

Frankfurt 1887

1. e4 e6 2. d4 d5 3. ♘c3 ♘f6 4. ♗g5 ♗e7 5. e5 ♘fd7 6. ♗e7 ♕e7 7. ♕d2 0-0 8. ♘d1 f5 9. ef6 ♕f6 10. ♗d3 e5 11. de5 ♘e5 12. ♘e2 c5 13. c3 ♗h3 14. ♔f1

14. – ♗g2! 15. ♔g2 ♕f3 16. ♔g1 ♕g4 17. ♔f1 ♕h3 –+

4910. C 15

Lane – Tisdall

Gausdal 1987

1. e4 e6 2. d4 d5 3. ♘c3 ♗b4 4. a3 ♗c3 5. bc3 de4 6. ♕g4 ♘f6 7. ♕g7 ♖g8 8. ♕h6 ♘bd7 9. ♘h3 b6 10. ♗e2 ♗b7 11. 0-0 ♕e7 12. a4 0-0-0 13. ♗a3 c5 14. ♕h4 ♕d6 15. ♖fd1 e3 16. f3

16. – ♖g2! 17. ♔g2 ♖g8 18. ♘g5[1] h6 19. ♖g1 hg5 20. ♕g3 ♕d5 21. ♔f1 ♘h5 22. ♕g2 ♘f4 23. ♕g3 ♘e2 24. ♔e2 ♕c4 –+

[1] 18. ♔h1 ♘g4

4911. C 18

Törber – Menke

Corr. 1950

1. e4 e6 2. d4 d5 3. ♘c3 ♗b4 4. a3 ♗c3 5. bc3 ♘f6 6. e5 ♘fd7 7. a4 c5 8. ♕g4 ♔f8 9. h4 ♕c7 10. ♖h3 cd4 11. ♗a3 ♔g8

12. ♕g7![1] +–

[1] 12. – ♔g7 13. ♖g3 ♔h6 14. ♗c1 ♔h5 15. ♗e2 ♔h4 16. ♖h3#

4912. C 21

Hartlaub – Fahrni

Norymberdze 1906

1. e4 e5 2. d4 ed4 3. c3 dc3 4. ♗c4 cb2 5. ♗b2 ♕e7 6. ♘c3 ♕b4 7. ♕e2 ♘c6 8. ♘f3 ♗c5 9. 0-0 ♗d4 10. ♘d4 ♘d4 11. ♕g4 ♕f8 12. ♘d5 ♘e6 13. f4 ♕c5 14. ♔h1 ♕c4 15. f5 h5 16. fe6! hg4 17. ef7 ♔f8

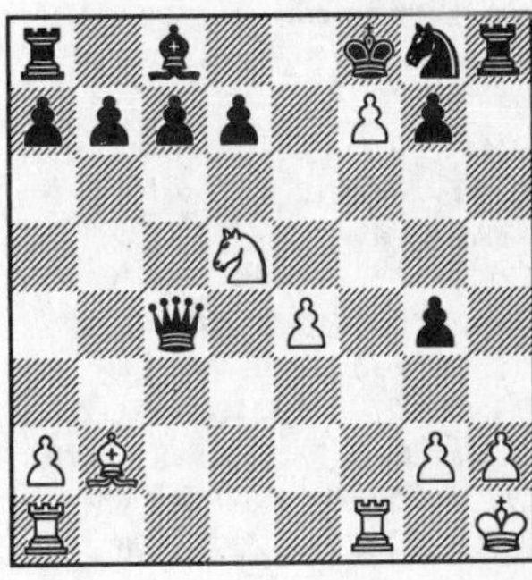

18. ♗g7![1] +–

[1] 18. – ♔g7 19. f8♕ ♔g6 20. ♕f7 ♔h6 21. ♖f5 +–

4913. C 21
Hartlaub – Testa
Bremen 1912

1. e4 e5 2. d4 ed4 3. c3 dc3 4. ♗c4 cb2 5. ♗b2 ♗b4 6. ♘c3 d6 7. ♘f3 ♘f6 8. 0-0 ♗c3 9. ♗c3 0-0 10. e5 ♘e4 11. ♗b2 ♗g4 12. ♕d4 ♗f3 13. gf3 ♘g5 14. ♔h1 ♘f3 15. ♕d3 ♘e5 16. ♖g1 ♕d7[1] 17. ♕d2 ♘g6 18. ♕d4 ♘e5

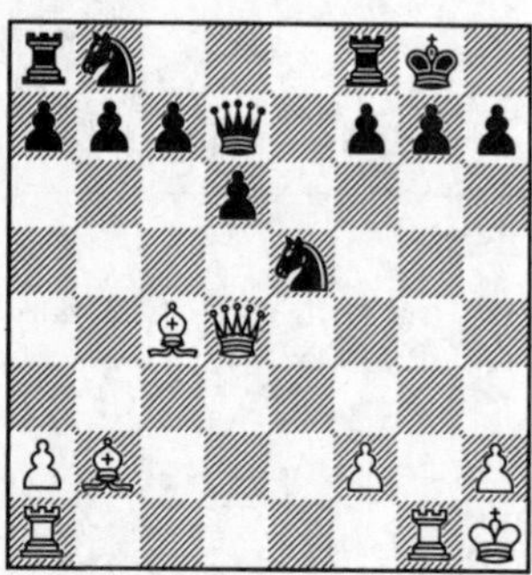

19. ♖g7! ♔g7 20. ♖g1 ♔h8[2] 21. ♕e5! de5 22. ♗e5 f6 23. ♗f6 ♖f6 24. ♖g8#

[1] 16. – ♘d3 17. ♖g7 ♔h8 18. ♖g8 ♔g8 19. ♖g1 ♕g5 20. ♖g5#
[2] 20. – ♔f6 21. ♕h4 +–

4914. C 23
Hewitt – Steinitz
London 1866

1. e4 e5 2. ♗c4 f5 3. d3 ♘f6 4. ♘e2 ♗c5 5. c3 ♘c6 6. d4 ed4 7. ♘d4 fe4 8. ♗f4 d5 9. ♗b5 ♗d4 10. ♕d4 0-0 11. ♗c6 bc6 12. ♕a4 ♗d7 13. ♗g5 ♕e8 14. ♕b3 ♘g4 15. ♗h4 e3 16. 0-0 ♕h5 17. ♗g3 e2 18. ♖e1 ♖f2 19. ♘d2 ♖af8 20. c4

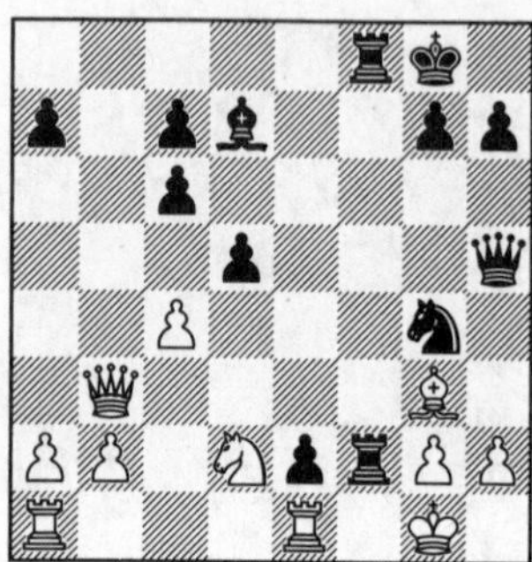

20. – ♖g2! 21. ♔g2 ♕h3 22. ♔h3[1] ♘e3 23. ♔h4 ♘g2 24. ♔g5 ♖f5 25. ♔g4 h5[2] –+

[1] 22. ♔g1 ♖f2! 23. ♕f3 ♖f3 24. ♘f3 ♘h2 –+
[2] 26. ♔h3 ♖f2#

4915. C 23
Staunton – Harrison
London 1844

(without ♘b1)

1. e4 e5 2. ♗c4 ♗c5 3. ♕e2 ♘c6 4. c3 ♘f6 5. f4 ♗g1 6. ♖g1 0-0 7. d3 d5 8. ♗b3 de4 9. de4 ef4 10. ♗c2 ♘g4 11. ♗f4 ♕h4 12. ♗g3 ♕h6 13. h3 ♘e3 14. ♗d3 ♗e6 15. ♗f2 ♘g2 16. ♖g2 ♕h3 17. ♖g3 ♕h1 18. ♖g1 ♕h2 19. 0-0-0 ♘e5 20. ♔b1 ♘d3

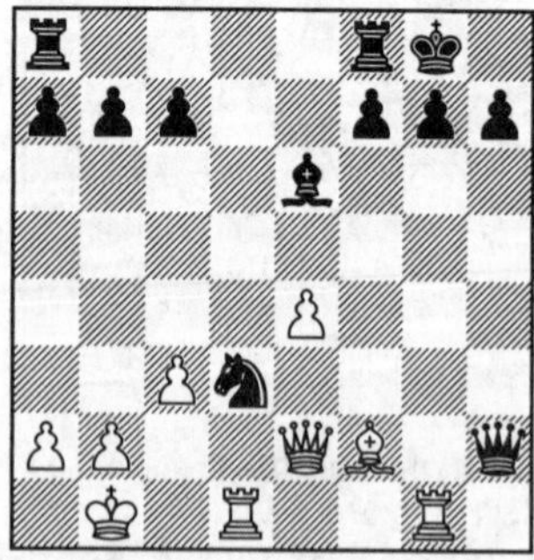

21. ♖g7! ♔h8[1] 22. ♗d4! ♕e2 23. ♖f7 ♔g8 24. ♖g7 ♔h8 25. ♖g6 +–

[1] 21. – ♔g7 22. ♗d4 +–

4916. C 25

Horowitz – N. N.

Los Angeles 1940

1. e4 e5 2. ♘c3 ♘c6 3. ♗c4 ♗c5 4. ♕g4 ♕f6 5. ♘d5 ♕f2 6. ♔d1 ♔f8 7. ♘h3 ♕d4 8. d3 ♗b6 9. ♖f1 ♘f6 10. ♖f6 d6

11. ♕g7! ♔g7 12. ♗h6 ♔g8 13. ♖g6! hg6 14. ♘f6#

4917. C 25

Kolbe – Rotenstein

Berlin 1921

1. e4 e5 2. ♘c3 ♘c6 3. f4 ef4 4. ♗c4 ♕h4 5. ♔f1 ♗c5 6. g3 fg3 7. ♘f3

7. – g2! 8. ♔g2[1] ♕f2 9. ♔h3 d5#

[1] 8. ♔e2 ♕f2 9. ♔d3 ♘b4#

4918. C 29

Milner-Barry – Hanninen

Moscow 1956

1. e4 e5 2. ♘c3 ♘f6 3. f4 d5 4. fe5 ♘e4 5. d3 ♘c3 6. bc3 d4 7. ♘f3 c5 8. ♗e2 ♗e7 9. 0-0 0-0 10. ♕e1 f6 11. ♕g3 fe5 12. ♗h6 ♗f6 13. ♘e5 ♗e5 14. ♕e5 ♖f6

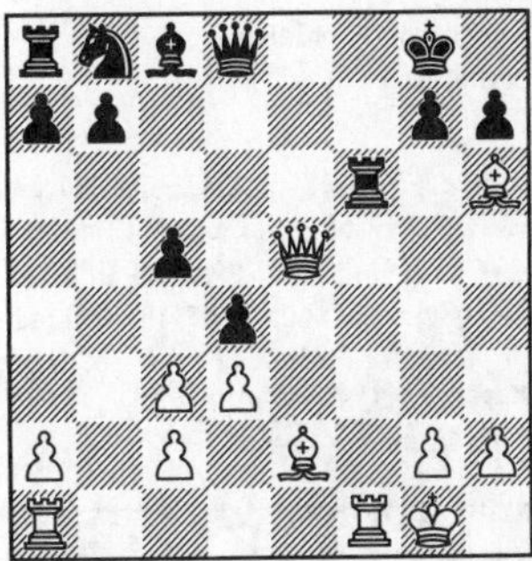

15. ♗g7! ♖e6[1] 16. ♕h5 ♕e7 17. ♗h6 ♘d7 18. ♗g4 ♖e5 19. ♕h3 ♘b6 20. ♖f8 ♕f8 21. ♗f8 ♔f8 22. ♕h7 +–

[1] 15. – ♔g7 16. ♕g5 ♔f7 17. ♗h5 +–

4919. C 32

Soldatenkov – Villenchik

1. e4 e5 2. f4 d5 3. ed5 e4 4. d3 ♘f6 5. de4 ♘e4 6. ♕e2 ♕d5 7. ♘d2 f5 8. g4 ♘c6 9. c3 ♗e7 10. ♗g2 g6 11. ♘e4 fe4 12. ♗e4 ♕e6 13. f5 gf5 14. gf5 ♗h4 15. ♔f1 0-0 16. ♘f3 ♕e7 17. ♖g1 ♔h8 18. ♗h6 ♖f7 19. ♘h4 ♕h4 20. ♕g2 ♗f5

21. ♕g7![1] +–

[1] 21. – ♖g7 22. ♗g7 ♔g8 23. ♗d5 ♗e6 24. ♗e6#

4920. C 39

Anderssen – Schallopp
Berlin 1865

1. e4 e5 2. f4 ef4 3. ♘f3 g5 4. h4 g4 5. ♘e5 ♗g7 6. ♘g4 d5 7. ♘f2 de4 8. ♘e4 ♘f6 9. ♘bc3 0-0 10. d3 ♖e8 11. ♗e2 ♘d5 12. ♘d5 ♕d5 13. ♗f4 f5 14. ♘c3 ♕g2 15. ♔d2 ♘c6 16. ♖g1 ♕f2 17. ♘d5 ♘d4

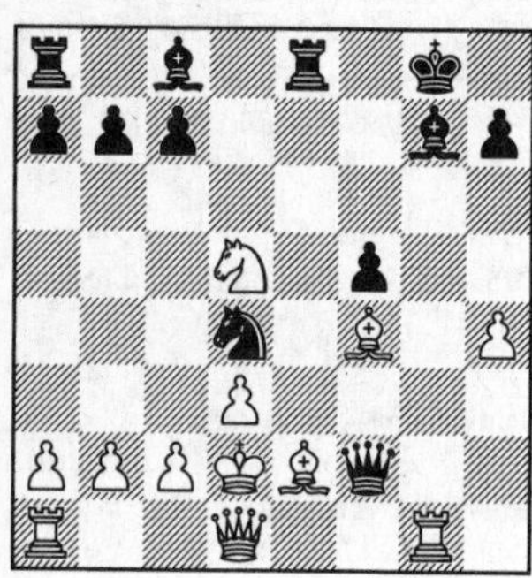

18. ♖g7! ♔g7 19. ♕g1 ♕g1 20. ♖g1 ♔h8 21. ♗h5 +–

4921. C 42

Hülsmann – Engert
Düsseldorf 1965

1. e4 e5 2. ♘f3 ♘f6 3. ♘c3 ♗b4 4. a3 ♗c3 5. dc3 d6 6. ♗g5 ♗e6 7. ♗d3 ♘bd7 8. 0-0 h6 9. ♗h4 ♘f8 10. ♕e2 ♘g6 11. ♗g3 ♘h5 12. ♖fd1 ♘hf4 13. ♕f1 0-0 14. ♗c4 ♕f6 15. ♖d2

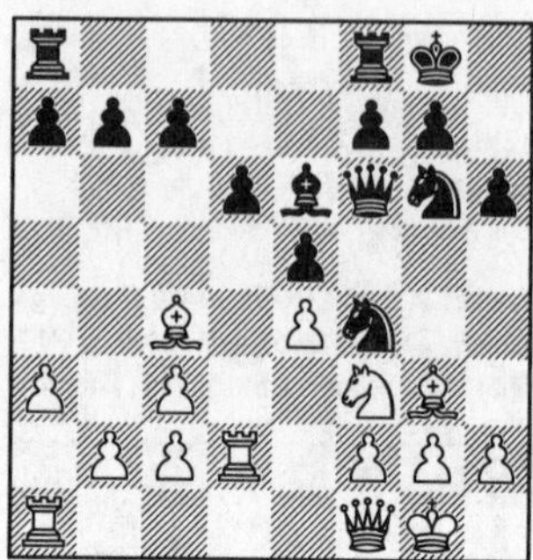

15. – ♘g2! 16. ♔g2 ♗h3! 17. ♔h3 ♕f3[1] –+

[1] 18. ♖d3 ♘f4 19. ♔h4 ♕h5#

4922. C 44

Marten – Winterburn
London 1930

1. e4 e5 2. ♘f3 ♘c6 3. c3 d5 4. ♕a4 ♗d7 5. ed5 ♘d4 6. ♕d1 ♘f3 7. ♕f3 f5 8. d4 e4 9. ♕e3 ♗d6 10. ♕g5 ♘f6 11. ♕g7 ♖g8 12. ♕h6 ♗f8 13. ♕d2 ♘d5 14. ♗c4 ♗c6 15. 0-0

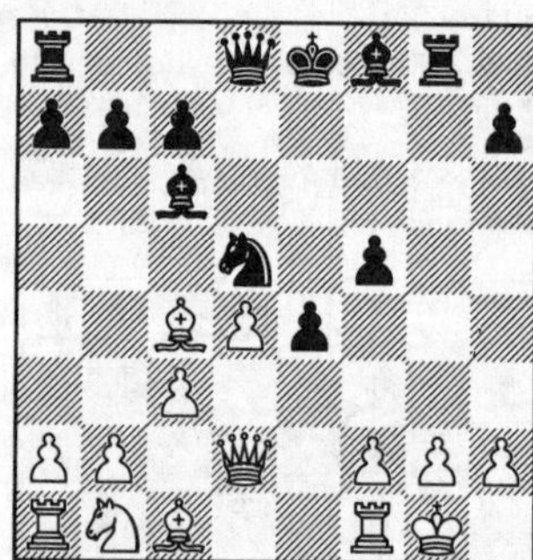

15. – ♖g2! 16. ♔g2 e3! 17. fe3 ♘f4 18. ♔g3 ♕g5 19. ♔f2 ♕h4 20. ♔g1 ♘h3#

4923. C 45

Vukovic – Mozetic
Banja Vrucica 1991

1. e4 e5 2. ♘f3 ♘c6 3. d4 ed4 4. ♘d4 ♕h4 5. ♘c3 ♗b4 6. ♗e2 ♘f6 7. 0-0 ♗c3

8. ♘f5 ♕e4 9. ♗d3 ♕g4 10. f3 ♕a4 11. bc3 0-0

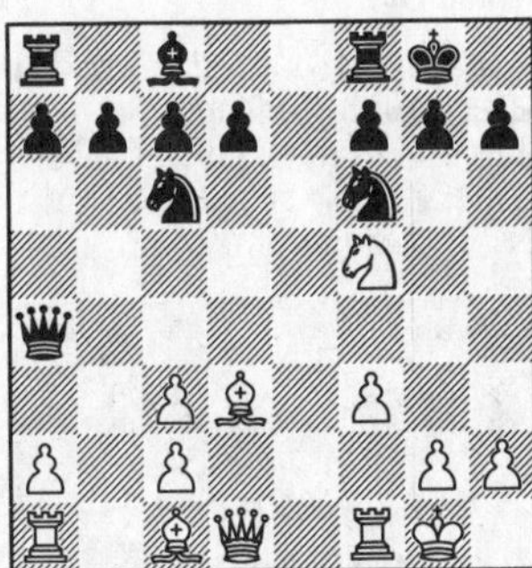

12. ♘g7! ♔g7 13. ♗h6 ♔h8 14. ♗f8 d6 15. ♕d2 ♕h4 16. ♗h6 ♘g8 17. ♗g5 ♕a4 18. c4 ♕a5 19. ♕f4 ♘e5 20. ♗f6 ♘f6 21. ♕f6 ♔g8 22. ♕g5 +–

4924. C 48

I. Horváth – Boguslavsky

Veszprém 1983

1. e4 e5 2. ♘f3 ♘c6 3. ♘c3 ♘f6 4. ♗b5 d6 5. d4 ed4 6. ♘d4 ♗d7 7. 0-0 ♘d4 8. ♗d7 ♕d7 9. ♕d4 ♗e7 10. ♗f4 0-0 11. ♖ad1 ♕g4 12. ♗g3 a6 13. ♕c4 ♖ac8 14. h3 ♕g6 15. a4 ♖fe8 16. ♖d3 ♗f8 17. ♖e1 ♘d7 18. ♘d5 ♘c5 19. ♖de3 ♕e6 20. e5 de5 21. ♗e5 ♖ed8

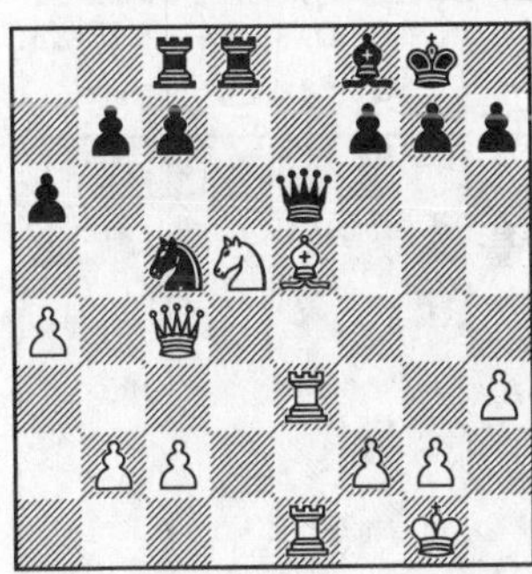

22. ♗g7! ♕d5 23. ♗f6! ♘e6[1] 24. ♖g3 ♘g7 25. ♖g7![2] +–

[1] 23. – ♕c4 24. ♖g3 ♗g7 25. ♖g7 ♔f8 26. ♖h7 +–

[2] 25. – ♗g7 26. ♕g4 +–

4925. C 48

Posch – Dorrer

Viden 1958

1. e4 e5 2. ♘f3 ♘c6 3. ♘c3 ♘f6 4. ♗b5 ♘d4 5. ♘d4 ed4 6. e5 dc3 7. ef6 cd2 8. ♗d2 ♕f6 9. 0-0 ♗e7 10. ♗c3 ♕g5 11. ♖e1 0-0 12. ♖e5 ♕f6 13. ♗d3 h6 14. ♕g4 ♕h4

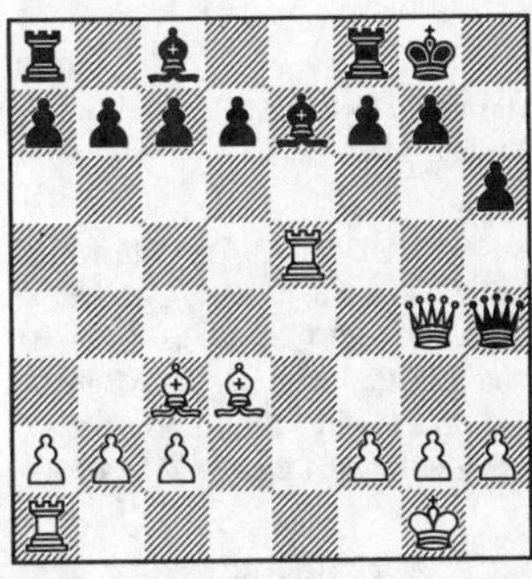

15. ♕g7! ♔g7 16. ♖g5#

4926. C 49

Coria – Capablanca

Buenos Aires 1914

1. e4 e5 2. ♘f3 ♘c6 3. ♘c3 ♘f6 4. ♗b5 ♗b4 5. 0-0 0-0 6. d3 d6 7. ♗g5 ♗c3 8. bc3 ♕e7 9. ♘d2 h6 10. ♗h4 ♘d8 11. d4 ♘e6 12. de5 de5 13. ♗d3 ♘f4 14. ♘c4 ♖d8 15. ♗f6 ♕f6 16. ♕d2 ♗h3 17. ♘e3

17. – ♗g2! 18. ♘f5[1] ♗e4 19. ♘g3 ♘h3#

[1] 18. ♘g2 ♕g5 19. f3 ♘h3 –+

4927. C 49

Maróczy – Swiderski

Wiedniu 1908

1. e4 e5 2. ♘f3 ♘c6 3. ♘c3 ♘f6 4. ♗b5 ♗b4 5. 0-0 0-0 6. d3 d6 7. ♘e2 ♕e7 8. c3 ♗c5 9. ♘g3 ♘e8 10. h3 h6 11. d4 ♗b6 12. ♖fe1 ♔h7 13. ♗e3 ♖g8 14. ♘g5 ♔h8 15. ♕h5 ♖f8 16. ♗c6 bc6 17. ♘f3 ♘f6 18. ♕h4 ♔g8 19. ♘h5 ♖fe8

20. ♘g7![1] +–

[1] 20. – ♔g7 21. ♗h6 ♔h7 22. ♗g5 +–

4928. C 50

Trofimov – Boleslavsky

Taskent 1965

1. e4 e5 2. ♘f3 ♘c6 3. ♗c4 ♗e7 4. d4 ed4 5. 0-0 d6 6. ♘d4 ♘f6 7. ♘c3 0-0 8. f4 d5 9. ♘c6 bc6 10. ed5 ♗c5 11. ♔h1 cd5 12. ♗d5 ♗a6 13. ♗a8 ♕a8 14. ♖e1 ♗b7 15. ♖e2 ♘g4 16. ♕e1

16. – ♗g2! 17. ♖g2 ♘f2 18. ♔g1 ♘d3 19. ♗e3 ♘e1 20. ♖e1 ♖e8 –+

4929. C 51

Anderssen – N. N.

Breslau 1860

1. e4 e5 2. ♘f3 ♘c6 3. ♗c4 ♗c5 4. b4 ♗b4 5. c3 ♗c5 6. 0-0 d6 7. d4 ed4 8. cd4 ♗b6 9. d5 ♘a5 10. ♗b2 ♘f6 11. ♗d3 ♗g4 12. ♘c3 c6 13. ♘e2 0-0 14. ♕d2 ♖c8 15. ♕g5 ♗f3 16. gf3 cd5 17. ♔h1 ♘c4 18. ♖g1 ♘e8

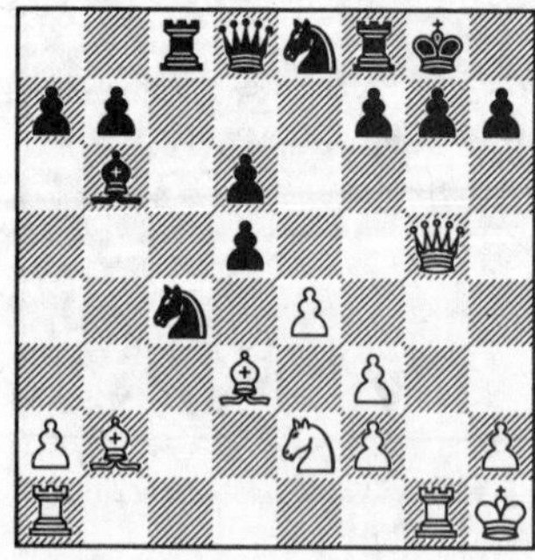

19. ♕g7! ♘g7 20. ♖g7 ♔h8 21. ♖g8! ♔g8 22. ♖g1 ♕g5 23. ♖g5#

4930. C 54

Guila – Cardinal Pezzi

Perouse 1875

1. e4 e5 2. ♘f3 ♘c6 3. ♗c4 ♗c5 4. c3 ♘f6 5. d4 ed4 6. e5 d5 7. ef6 dc4 8. ♕e2 ♗e6 9. fg7 ♖g8 10. cd4 ♘d4 11. ♘d4 ♗d4 12. ♕h5 ♕f6 13. 0-0 ♖g7 14. ♕b5 c6 15. ♕b7

15. – ♖g2! 16. ♔g2 ♕g6 17. ♔h1[1] ♗d5 18. f3 ♗f3 19. ♖f3 ♕g1#

[1] 17. ♔f3 ♕g4#

4931. C 54

B. Sörensen – S. A. Sörensen

1. e4 e5 2. ♘f3 ♘c6 3. ♗c4 ♗c5 4. c3 ♘f6 5. d4 ed4 6. e5 d5 7. ef6 dc4 8. fg7 ♖g8 9. 0-0 ♖g7 10. ♖e1 ♗e7 11. cd4 ♗g4 12. d5 ♕d7 13. ♘c3 0-0-0 14. dc6 ♕d1 15. cb7 ♔b7 16. ♘d1 ♗f3 17. ♘e3 ♗b4 18. ♖f1

18. – ♖g2! 19. ♘g2 ♖g8 20. ♖d1 ♖g2 21. ♔f1[1] −+

[1] 21. – ♖h2 −+

4932. C 55

Bird – Pinkerley

London 1850

(without ♖a1)

1. e4 e5 2. ♘f3 ♘c6 3. ♗c4 ♗c5 4. 0-0 ♘f6 5. c3 0-0 6. d4 ed4 7. cd4 ♗b6 8. e5 d5 9. ef6 dc4 10. ♗g5 g6 11. d5 ♘b8 12. b3 cb3 13. ♕b3 ♗g4 14. ♘bd2 ♗f3 15. ♘f3 ♕d7 16. ♗c1 c6 17. ♘e5 ♕c7 18. ♗b2 ♘d7 19. ♘g4 ♘c5 20. ♕e3 ♘e6 21. ♕h6 cd5

22. ♕g7! ♘g7 23. ♘h6 ♔h8 24. fg7#

4933. C 55

Capablanca – N. N.
New York 1918

1. e4 e5 2. ♘f3 ♘c6 3. ♘c3 ♘f6 4. ♗c4 ♗c5 5. 0-0 0-0 6. d3 d6 7. ♗g5 ♗g4 8. ♘d5 ♘d4 9. ♕d2 ♕d7 10. ♗f6 ♗f3 11. ♘e7 ♔h8

12. ♗g7! ♔g7 13. ♕g5 ♔h8 14. ♕f6#

4934. C 55

Haág – Varnusz
Budapest 1958

1. e4 e5 2. ♘f3 ♘c6 3. ♗c4 ♘f6 4. d4 ed4 5. e5 d5 6. ♗b5 ♘e4 7. ♘d4 ♗d7 8. ♗c6 bc6 9. 0-0 ♗c5 10. ♗e3 ♗b6 11. ♘d2 ♘d2 12. ♕d2 0-0 13. ♗g5 ♕e8 14. ♖fe1 c5 15. ♗f6! h6 16. ♖e3 ♔h7 17. ♕d3 ♔h8 18. ♖h3! ♗h3 19. ♕h3 ♔h7

20. ♗g7![1] +–

[1] 20. – ♔g7 21. ♘f5 ♔g6 22. ♕g4 ♔h7 23. ♕g7#

4935. C 56

Euwe – Réti
Amsterdam 1920

1. e4 e5 2. ♘f3 ♘c6 3. ♗c4 ♘f6 4. d4 ed4 5. 0-0 ♘e4 6. ♖e1 d5 7. ♗d5 ♕d5 8. ♘c3 ♕a5 9. ♘d4 ♘d4 10. ♕d4 f5 11. ♗g5 ♕c5 12. ♕d8 ♔f7 13. ♘e4 fe4 14. ♖ad1 ♗d6 15. ♕h8 ♕g5 16. f4 ♕h4 17. ♖e4 ♗h3 18. ♕a8 ♗c5 19. ♔h1

19. – ♗g2! 20. ♔g2 ♕g4[1] –+

[1] 21. ♔h1 ♕f3#

4936. C 58

Young – Barden
Corr. 1945

1. e4 e5 2. ♘f3 ♘c6 3. ♗c4 ♘f6 4. ♘g5 d5 5. ed5 ♘a5 6. ♗b5 c6 7. dc6 bc6 8. ♕f3 cb5 9. ♕a8 ♗c5 10. ♘e4 ♘e4 11. ♕e4 0-0 12. 0-0 ♖e8 13. ♕e2 ♗b7 14. ♕b5 ♗b6 15. d3 ♖e6 16. ♔h1

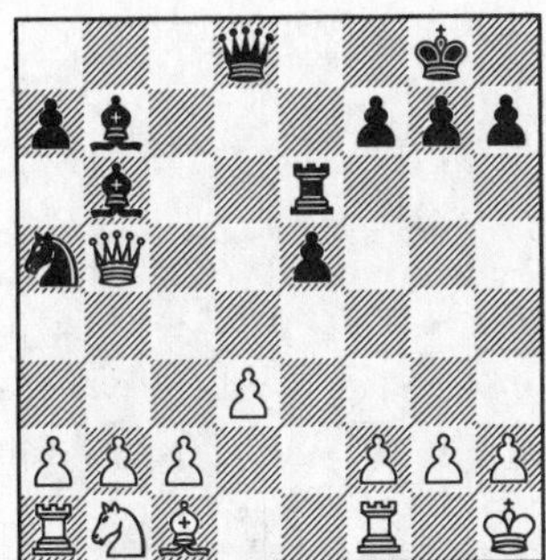

16. – ♗g2![1] –+

[1] 17. ♔g2 ♖g6? 18. ♔h1 ♕a8 19. f3 ♕f3! 20. ♖f3 ♖g1#; 18. ♔f3!; 17. – ♕a8! 18. f3 ♖g6 –+; 18. ♔g3 ♖g6 –+

4937. C 60

Anderssen – Suhle

Breslau 1859

1. e4 e5 2. ♘f3 ♘c6 3. ♗b5 ♘ge7 4. d4 ed4 5. 0-0 ♘g6 6. ♘d4 ♗e7 7. ♘f5 0-0 8. ♘c3 ♗c5 9. ♕h5 d6 10. ♗g5 ♕e8

11. ♘g7! ♔g7 12. ♕h6[1] +–

[1] 12. – ♔g8 13. ♗f6 and 14. ♕g7#

4938. C 64

Kieninger – Aljechin

Munich 1941

1. e4 e5 2. ♘f3 ♘c6 3. ♗b5 ♗c5 4. c3 ♕f6 5. 0-0 ♘ge7 6. d3 h6 7. ♘bd2 0-0 8. ♘c4 ♘g6 9. d4 ed4 10. ♗c6 dc6 11. ♘d4 ♖e8 12. ♘b3 ♗f8 13. ♕c2 ♕e6 14. ♘cd2 ♘h4 15. f3 c5 16. ♖d1

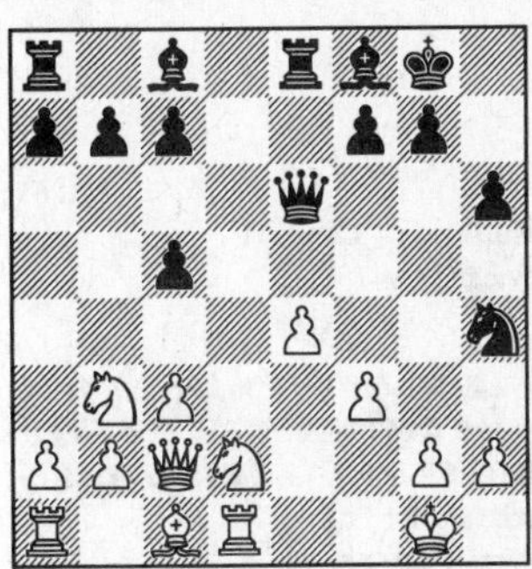

16. – ♘g2! 17. ♔g2 ♕h3 18. ♔g1[1] ♗d6 19. ♘f1 ♕f3 20. ♖d3 ♕e4 21. ♖d2 ♕h4 22. ♖g2 ♗h3 23. ♕f2 ♕e4 24. ♗d2 ♕g2 25. ♕g2 ♗g2 –+

[1] 18. ♔h1 ♗d6 19. f4 ♗f5 20. ef5 ♖e2 –+; 20. ♖f1 ♗e4 –+; 20. ♖e1 ♗f4 21. ♘f1 ♖e4 –+

4939. C 64

Maki – Rantanen

Helsinki 1983

1. e4 e5 2. ♘f3 ♘c6 3. ♗b5 ♗c5 4. c3 f5 5. d4 fe4 6. ♘e5 ♘e5 7. ♕h5 ♘g6 8. dc5 ♘f6 9. ♕e2 0-0 10. 0-0 d5 11. cd6 ♕d6 12. h3 ♔h8 13. ♗g5 ♕e5 14. ♗e3 ♘h5 15. ♗a4 ♘gf4 16. ♕b5 ♕f6 17. ♔h2

17. – ♘g2![1]

[1] 18. ♔g2 ♕f3 19. ♔g1 ♗h3 –+

4940. C 65
Hoffmann – Lambert
Wiedinu 1947

1. e4 e5 2. ♘f3 ♘c6 3. ♗b5 ♘f6 4. 0-0 ♗c5 5. ♘e5 ♘e5 6. d4 c6 7. ♗a4 ♘e4 8. ♖e1 d5 9. dc5 0-0 10. ♕d4 ♕f6 11. ♗e3 ♗h3 12. ♔h1

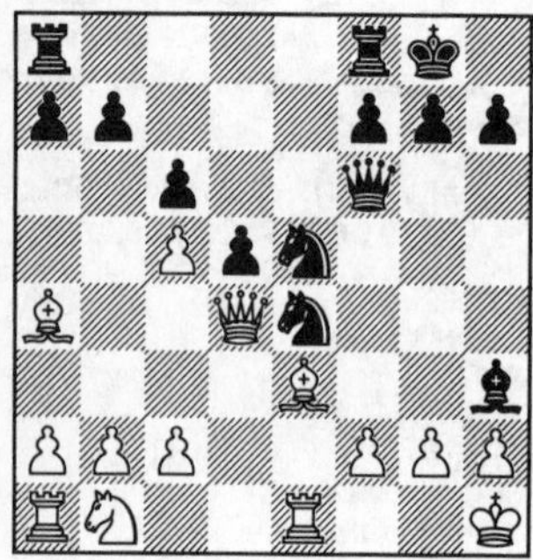

12. – ♗g2! 13. ♔g2 ♕f3 14. ♔g1 ♕g4 15. ♔h1[1] ♘f3 16. ♕d1 ♕h3 –+

[1] 15. ♔f1 ♘f3 16. ♕d3 ♘h2#

4941. C 67
Ljubojevic – Calwo
Arrecife 1973

1. e4 e5 2. ♘f3 ♘c6 3. ♗b5 ♘f6 4. 0-0 ♘e4 5. ♖e1 ♘d6 6. ♘e5 ♗e7 7. ♕h5 ♘e5 8. ♕e5 ♘b5 9. ♕g7 ♖f8 10. a4 ♘d6 11. ♘c3 ♘f5

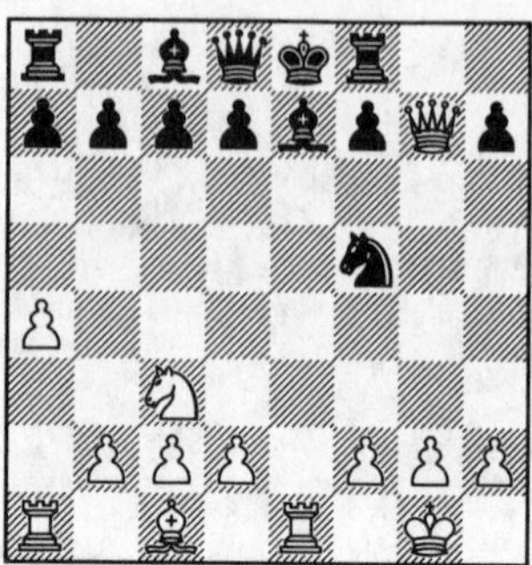

12. ♘d5! f6[1] 13. ♕h7 d6 14. ♕g6[2] +–

[1] 12. – ♘g7 13. ♘f6#
[2] 14. – ♖f7 15. ♘f6

4942. C 68
Bondum – Iskov
Denmark 1972

1. e4 e5 2. ♘f3 ♘c6 3. ♗b5 a6 4. ♗c6 dc6 5. 0-0 ♗g4 6. h3 h5 7. d3 ♕f6 8. ♘bd2 ♘e7 9. ♖e1 ♘g6 10. d4 ♘f4 11. hg4 hg4 12. ♘h2

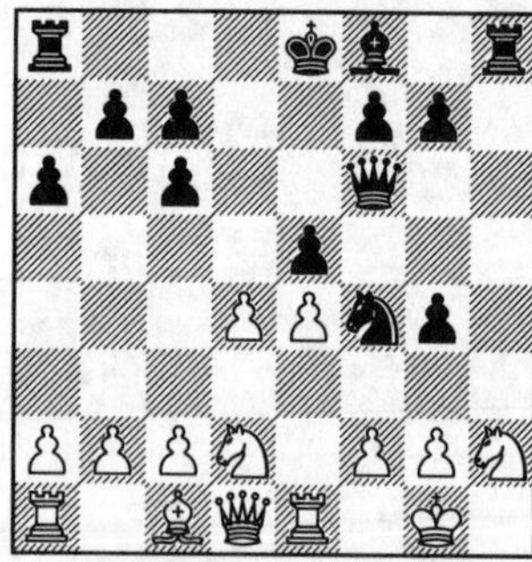

12. – ♘g2! 13. ♘df1[1] ♘e1 14. ♕e1 0-0-0 15. ♗d2 ♗d6 16. d5 cd5 17. ed5 e4 18. ♕e4 ♗h2 19. ♘h2 ♖h2![2] –+

[1] 13. ♔g2 ♖h2! 14. ♔h2 ♕f2 15. ♔h1 g3 16. ♕e2 g2 17. ♔h2 g1♕ –+
[2] 20. ♔h2 ♕f2 21. ♔h1 ♖h8 –+

4943. C 77
Kirilov – Furman
Vilnius 1949

1. e4 e5 2. ♘f3 ♘c6 3. ♗b5 a6 4. ♗a4 ♘f6 5. ♕e2 b5 6. ♗b3 ♗e7 7. a4 b4 8. ♗d5 ♘d5 9. ed5 ♘d4 10. ♘d4 ed4 11. 0-0 0-0 12. ♕c4 c5 13. dc6 dc6 14. ♕c6 ♖a7 15. ♕f3 ♖c7 16. d3 ♗b7 17. ♕d1 ♗d6 18. ♘bd2 ♖e8 19. ♘c4 ♗h2! 20. ♔h2 ♕h4 21. ♔g1

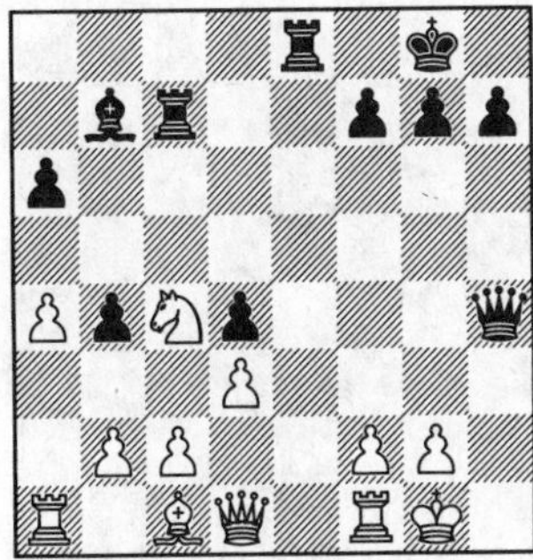

21. – ♗g2! 22. ♔g2 ♖c6 23. ♗f4[1] ♕f4 24. ♖h1 ♖f6 25. ♖h2 ♖g6[2] –+

[1] 23. ♖h1 ♖g6 24. ♔f3 ♕g4#
[2] 26. ♔h1 ♖e1! 27. ♕e1 ♕f3 –+

4944. C 77
Thomas – Keres
Margate 1937

1. e4 e5 2. ♘f3 ♘c6 3. ♗b5 a6 4. ♗a4 ♘f6 5. ♘c3 b5 6. ♗b3 d6 7. ♘g5 d5 8. ♘d5 ♘d4 9. ♘e3 ♘b3 10. ab3 h6 11. ♘f3 ♘e4 12. ♘e5 ♕f6 13. ♘f3 ♗b7 14. ♕e2 0-0-0 15. 0-0 ♗d6 16. ♘g4 ♕f5 17. d3 ♘g5 18. ♘h4 ♕d5 19. c4 ♘h3 20. ♔h1 ♕h5 21. c5 ♖he8 22. ♕c2 ♕h4 23. cd6

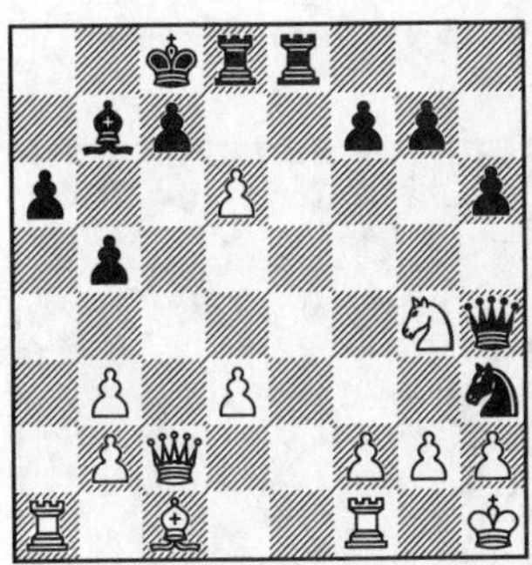

23. – ♗g2! 24. ♔g2 ♕g4 25. ♔h1 ♕f3#

4945. C 80
Zaitzev – Rokhlin
Yaroslavl 1954

1. e4 e5 2. ♘f3 ♘c6 3. ♗b5 a6 4. ♗a4 ♘f6 5. 0-0 ♘e4 6. d4 b5 7. ♗b3 d5 8. ♘e5 ♘e5 9. de5 ♗b7 10. ♗e3 ♗c5 11. ♕g4 ♗e3 12. ♕g7 ♕g5 13. ♕h8 ♔e7 14. ♕h7 ♗f2 15. ♔h1 ♖g8 16. ♕h3 d4 17. ♘a3

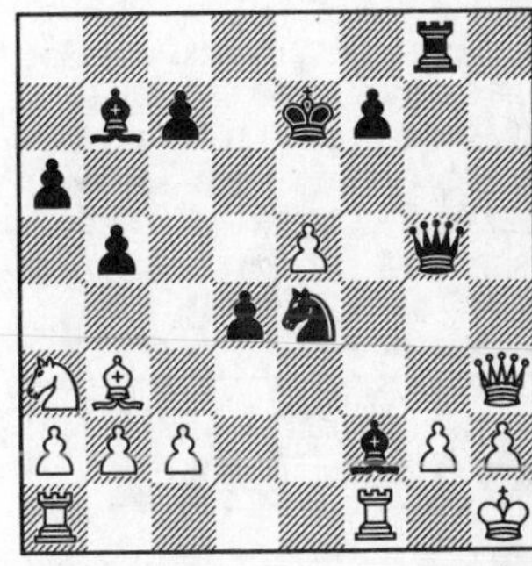

17. – ♕g2! 18. ♕g2 ♘g3! 19. hg3 ♖h8#

4946. C 84
Showalter – Pillsbury
New York 1898

1. e4 e5 2. ♘f3 ♘c6 3. ♗b5 ♘f6 4. d4 ed4 5. 0-0 a6 6. ♗a4 ♗e7 7. ♖e1 0-0 8. e5 ♘e8 9. ♘d4 ♘d4 10. ♕d4 d5 11. b4 c6 12. ♗b2 ♘c7 13. a3 a5 14. c3 ♗f5 15.

♕b6 ♕c8 16. ♘d2 ♘e6 17. ♘f3 ♘f4 18. ♗b3

18. – ♘g2! 19. ♖ed1[1] ♗e4 20. ♘d2 ♖a6 21. ♕d4 ♘f4 22. ♕e3 ♕g4 23. ♔f1 ♗d3 24. ♕d3 ♕g2 25. ♔e1 ♘d3 –+

[1] 19. ♔g2 ♗h3 20. ♔h1 ♕g4

4947. C 90

Short – Hebden

Hastings 1983/84

1. e4 e5 2. ♘f3 ♘c6 3. ♗b5 a6 4. ♗a4 ♘f6 5. 0-0 ♗e7 6. ♖e1 b5 7. ♗b3 0-0 8. d3 d6 9. c3 ♘a5 10. ♗c2 c5 11. ♘bd2 ♖e8 12. ♘f1 h6 13. ♘g3 ♗f8 14. h3 ♗b7 15. d4 cd4 16. cd4 ed4 17. ♘d4 ♖c8 18. b3 d5 19. e5 ♘e4 20. ♗b2 ♖e5 21. f3 ♖g5 22. ♘f1

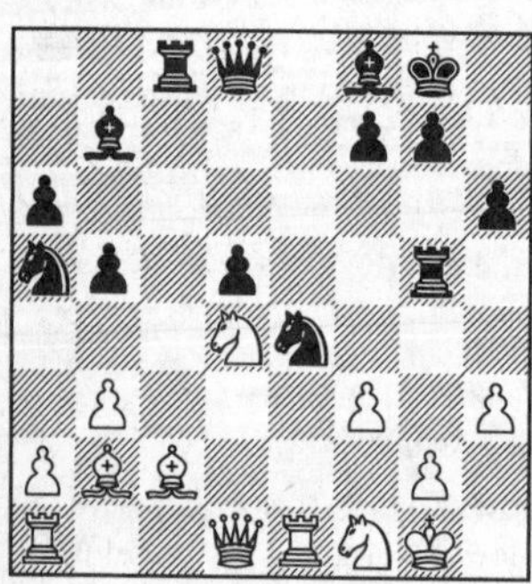

22. – ♖g2![1] –+

[1] 23. ♔g2 ♕g5 24. ♔h2 ♗d6 25. ♔h1 ♘f2#

4948. D 02

Lustenberger – Mooser

Bern 1992

1. d4 d5 2. ♘f3 e6 3. ♗f4 ♘f6 4. ♘bd2 c5 5. e3 ♘c6 6. c3 ♗e7 7. ♗d3 0-0 8. ♘e5 ♗d7 9. g4 ♘e8 10. g5 ♘e5 11. de5 ♗g5 12. ♕h5 h6 13. ♖g1 ♗f4 14. ef4 ♔h8 15. 0-0-0 f5 16. ef6 ♕f6 17. ♖g6 ♕f4 18. ♖dg1 e5

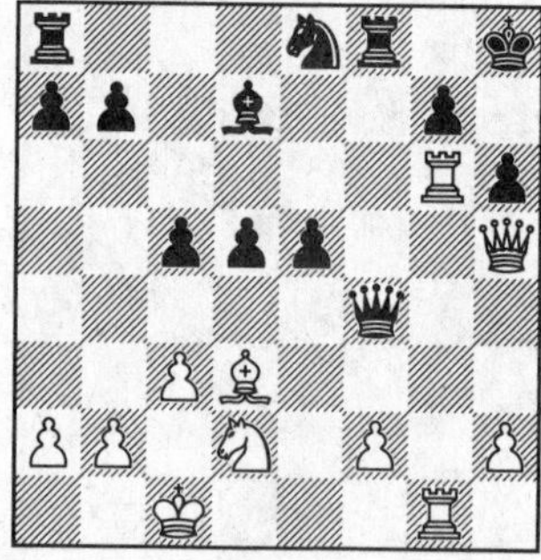

19. ♖g7! ♘g7 20. ♕g6 +–

4949. D 07

Kibbermann – Keres

Tartu 1935

1. d4 ♘c6 2. ♘f3 d5 3. c4 e5 4. ♘e5 ♘e5 5. de5 d4 6. e4 ♘e7 7. ♗d3 ♘g6 8. f4 ♗b4 9. ♗d2 ♘f4 10. ♗b4 ♘g2 11. ♔f2

11. – ♕h4! 12. ♔g1[1] ♕g5 13. ♕f3 ♘e1 14. ♕g3 ♕g3 15. hg3 ♘d3 16. ♗a3 ♘e5 –+

[1] 12. ♔g2 ♗h3 13. ♔g1 ♕g5 14. ♔f2 ♕e3#

4950. D 21

Spielmann – Grünfeld

Karlsbad 1929

1. d4 d5 2. c4 e6 3. ♘c3 dc4 4. e4 c5 5. ♘f3 cd4 6. ♘d4 a6 7. ♗c4 ♗d7 8. 0-0 ♘c6 9. ♘f3 ♕c7 10. ♕e2 ♗d6 11. ♖d1 ♘ge7 12. ♗e3 ♘e5 13. ♘e5 ♗e5 14. g3 ♗c3 15. bc3 ♘g6 16. ♗b3 0-0 17. ♗d4 b5 18. ♕e3 ♗c6 19. h4 ♕b7 20. h5 ♘e7

21. ♗g7! ♔g7 22. ♕g5 ♘g6 23. h6 +–

4951. D 36.

Marshall – Kupchik

Chicago 1926

1. d4 ♘f6 2. c4 c6 3. ♘c3 d5 4. ♘f3 e6 5. ♗g5 ♘bd7 6. cd5 ed5 7. e3 ♗e7 8. ♗d3 0-0 9. ♕c2 ♖e8 10. 0-0 ♘f8 11. ♖ae1 ♗e6 12. ♘e5 ♘6d7 13. ♗e7 ♖e7 14. f4 f6 15. ♘d7 ♕d7 16. f5 ♗f7 17. ♖f3 ♖ae8 18. ♕f2 h6 19. a3 ♘h7 20. h4 ♕d6 21. ♖g3 ♔h8 22. ♘e2 b6 23. ♘f4 c5?

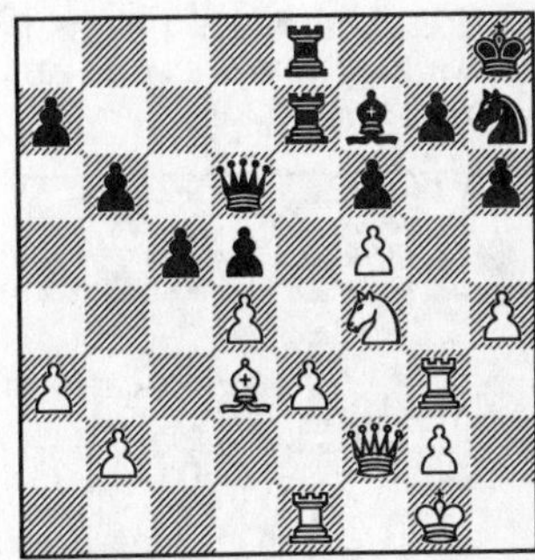

24. ♖g7! ♕d8[1] 25. ♖g3 +–

[1] 24. – ♔g7 25. ♕g3 ♔h8 26. ♘g6

4952. D 40

Miles – Browne

Luzern 1982

1. ♘f3 c5 2. c4 ♘f6 3. ♘c3 e6 4. e3 ♘c6 5. d4 d5 6. dc5 ♗c5 7. a3 a6 8. b4 ♗a7 9. ♗b2 0-0 10. ♖c1 d4 11. ed4 ♘d4 12. c5 ♘f3 13. ♕f3 ♗d7 14. ♗d3 ♗c6 15. ♘e4 ♘e4 16. ♗e4 ♕c7 17. 0-0 ♖ad8 18. ♗h7! ♔h7 19. ♕h5 ♔g8

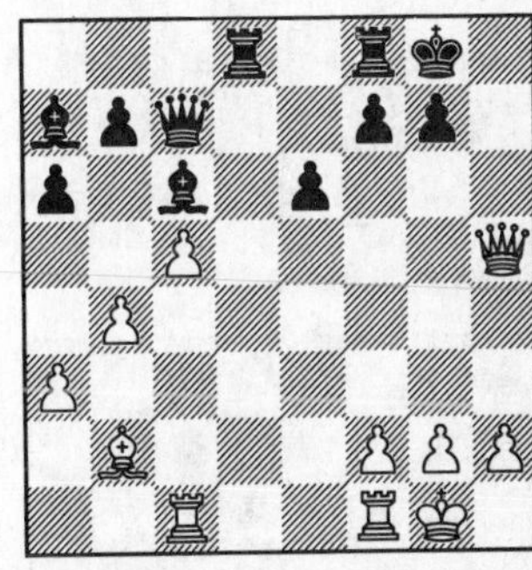

20. ♗g7! ♔g7 21. ♕g5 ♔h8 22. ♕f6 ♔g8 23. ♖c4 +–

4953. D 43

Honlinger – Palda

Vienna 1933

1. d4 d5 2. c4 c6 3. ♘f3 ♘f6 4. e3 e6 5. ♘bd2 ♘bd7 6. ♗d3 ♗e7 7. 0-0 0-0 8. b3

b6 9. ♗b2 ♗b7 10. ♕e2 c5 11. ♘e5 ♕c7 12. ♖ac1 ♖ac8 13. f4 ♗d6 14. cd5 ed5 15. ♘d7 ♘d7 16. dc5 bc5

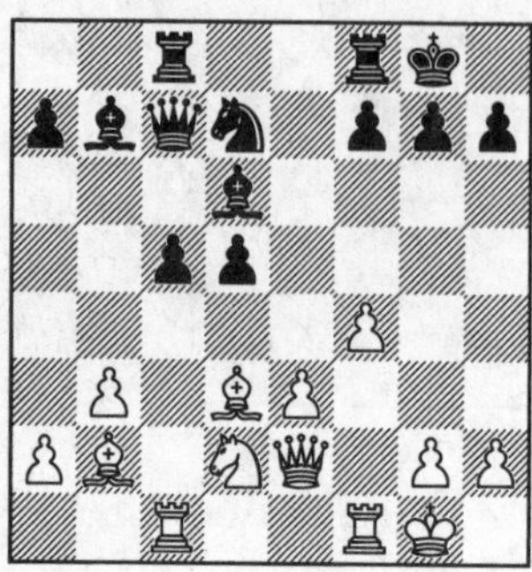

17. ♗g7! ♔g7 18. ♕g4 ♔h8[1] 19. ♕f5[2] +–

[1] 18. – ♔f6 19. ♕g5 ♔e6 20. ♗f5#
[2] 19. – ♔g7 20. ♕h7 ♔f6 21. ♕h4 ♔e6 22. f5 ♔e5 23. ♘f3#

4954. D 43
Lilienthal – Shamkovich
Baku 1951

1. d4 d5 2. c4 e6 3. ♘c3 c6 4. ♘f3 ♘f6 5. ♗g5 h6 6. ♗f6 ♕f6 7. ♕b3 ♘d7 8. e4 de4 9. ♘e4 ♕f4 10. ♗d3 ♗e7 11. 0-0 0-0 12. ♖fe1 b6 13. ♗c2 ♗b7 14. ♕d3 ♕c7 15. ♘g3 ♘f6 16. ♘e5 ♖fd8 17. ♘h5 ♔f8 18. ♕g3 ♘e8

19. ♘g7![1]

[1] 19. – ♘g7 20. ♘g6! +–

4955. D 54
Fuderer – Milic
Zagreb 1955

1. c4 e6 2. ♘c3 d5 3. d4 ♘f6 4. ♗g5 ♗e7 5. e3 0-0 6. ♖c1 h6 7. ♗h4 ♘e4 8. ♗e7 ♕e7 9. ♕c2 c6 10. ♗d3 ♘c3 11. ♕c3 ♕g5 12. ♘f3 ♕g2 13. ♔e2 ♕h3 14. ♖cg1 f5 15. ♖g3 ♕h5 16. ♖hg1 ♖f7 17. ♕a3 ♘d7 18. ♔e1 dc4 19. ♗c4 f4

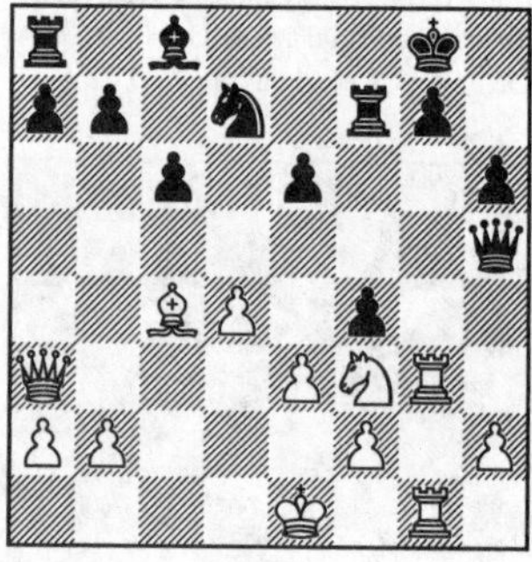

20. ♖g7! ♖g7 21. ♗e6 ♔h8 22. ♖g7 ♔g7 23. ♕e7 ♔h8 24. ♘e5 fe3[1] 25. f4 +–

[1] 24. – ♘e5 25. ♕f8 ♔h7 26. ♕g8#

4956. E 18
Aljechin – Stoltz
Nice 1931

1. d4 ♘f6 2. c4 e6 3. ♘f3 b6 4. g3 ♗b7 5. ♗g2 ♗e7 6. 0-0 0-0 7. b3 ♕c8 8. ♘c3 d5 9. cd5 ♘d5 10. ♗b2 c5 11. ♖c1 ♘c3 12. ♗c3 ♖d8 13. ♕d2 ♗d5 14. ♕f4 ♕b7 15. dc5 bc5 16. e4 ♗c6 17. ♘e5 ♗e8 18. ♘g4 ♘a6

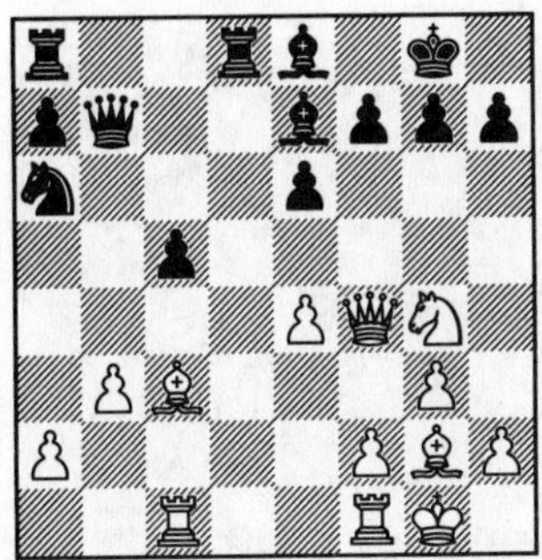

19. ♗g7![1] +–

[1] 19. – ♔g7 20. ♕h6 ♔h8 21. ♘f6 ♗f6 22. ♕f8#; 20, – ♔g8 21. e5 ♗c6 22. ♘f6 ♗f6 23. ef6 +–

4957. E 20

Lukomsky – Pogebin

Moscow 1929

1. d4 ♘f6 2. c4 e6 3. ♘c3 b6 4. e4 ♗b4 5. e5 ♘e4 6. ♕g4 ♘c3 7. bc3 ♗c3 8. ♔d1 ♔f8 9. ♖b1 ♘c6 10. ♗a3 ♔g8 11. ♖b3 ♗d4

12. ♕g7! ♔g7 13. ♖g3 ♔h6 14. ♗c1 ♔h5 15. ♗e2 ♔h4 16. ♖h3#

4958. E 21

Hüppin – Baumgartner

Biel 1993

1. d4 ♘f6 2. ♘f3 b6 3. c4 e6 4. ♘c3 ♗b4 5. ♗d2 ♗b7 6. e3 0-0 7. ♗e2 d6 8. 0-0 ♘bd7 9. a3 ♗c3 10. ♗c3 ♘e4 11. ♖c1 a5 12. b4 f5 13. ♗b2 ♕e7 14. h3 g5 15. ♘d2 ♘df6 16. ♔h2 ♕g7 17. f3 ♘d2 18. ♕d2 g4 19. ♗d3 ♕g5 20. f4 ♕h5 21. ♗e2

21. – ♗g2! 22. ♔g2 ♕h3 23. ♔g1 ♘e4 24. ♕e1 ♕e3 25. ♔g2 ♕h3 –+

4959. E 28

Botvinnik – Keres

1948

1. d4 ♘f6 2. c4 e6 3. ♘c3 ♗b4 4. e3 0-0 5. a3 ♗c3 6. bc3 ♖e8 7. ♘e2 e5 8. ♘g3 d6 9. ♗e2 ♘bd7 10. 0-0 c5 11. f3 cd4 12. cd4 ♘b6 13. ♗b2 ed4 14. e4 ♗e6 15. ♖c1 ♖e7 16. ♕d4 ♕c7 17. c5 dc5 18. ♖c5 ♕f4 19. ♗c1 ♕b8 20. ♖g5 ♘bd7

21. ♖g7! ♔g7 22. ♘h5 ♔g6[1] 23. ♕e3 +–

[1] 22. – ♔f8 23. ♘f6 ♘f6 24. ♕f6 ♖d7 25. ♗g5! ♔g8 26. ♗h6 +–

4960. E 32

Euwe – Colle

Amsterdam 1928

1. d4 ♘f6 2. c4 e6 3. ♘c3 ♗b4 4. ♕c2 b6 5. e4 ♗b7 6. ♗d3 ♗c3 7. bc3 d6 8. ♘e2 h6 9. 0-0 0-0 10. f4 ♘bd7 11. e5 ♘e8 12. ♘g3 c5 13. ♕e2 ♕h4 14. f5 cd4 15. ♖f4 ♕d8 16. cd4 de5 17. de5 ♘c7 18. ♖g4 ♕e7

19. ♖g7![1] +–

[1] 19. – ♔g7 20. f6 ♘f6 21. ef6 ♔f6 22. ♘h5#

4961. E 33

Sullivan – Cook

USA 1949

1. d4 ♘f6 2. c4 e6 3. ♘c3 ♗b4 4. ♕c2 ♘c6 5. ♘f3 d6 6. e3 0-0 7. ♗d3 e5 8. d5 ♗c3 9. ♕c3 e4 10. dc6 ef3 11. gf3 bc6 12. b3 ♗e6 13. ♗b2 a5 14. ♖g1 ♔h8 15. ♔e2 a4 16. ♕d4 ab3

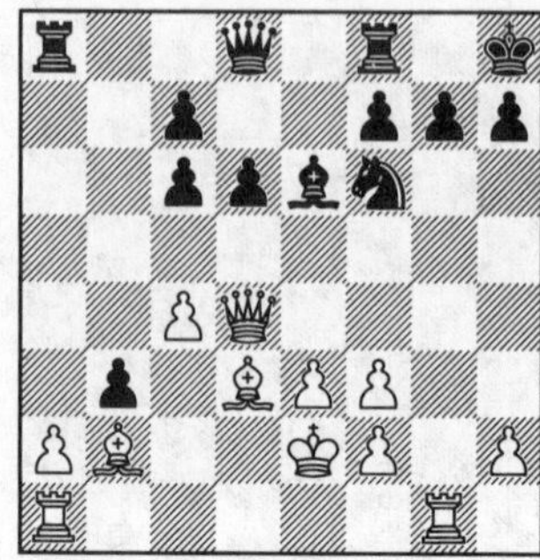

17. ♖g7! ♔g7 18. ♖g1 ♔h6 19. ♕f4 +–

4962. E 43

Soto–Larrea – Brunner

Mexico 1932

1. d4 ♘f6 2. c4 e6 3. ♘f3 b6 4. ♘c3 ♗b7 5. e3 ♗b4 6. ♗d2 0-0 7. ♗d3 d6 8. 0-0 ♘bd7 9. a3 ♗c3 10. ♗c3 ♕e7 11. ♕c2 e5 12. e4 ed4 13. ♘d4 ♘e5 14. ♘f5 ♕e6 15. f3 a5 16. ♖ad1 ♘d3 17. ♕d3 ♗a6 18. ♕e3 ♘e8

19. ♗g7! ♘g7 20. ♕g5 ♕e5 21. f4[1]

[1] 21. – ♕b2 22. e5 +–

4963. A 03
N. N. – Crepeaux
Nice 1923

1. f4 d5 2. ♘f3 ♘c6 3. e3 ♗g4 4. b3 e5 5. fe5 ♘e5 6. ♗e2 ♗f3 7. ♗f3 ♘f6 8. ♗b2 ♗d6 9. 0-0 ♘e4 10. ♘c3 ♕h4 11. ♘d5

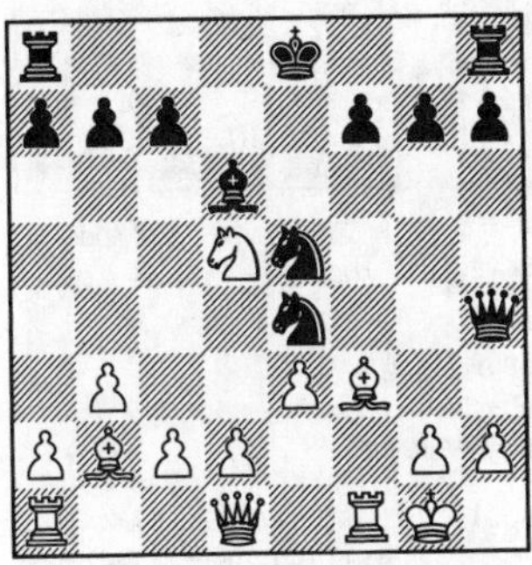

11. – ♕h2! 12. ♔h2 ♘f3 13. ♔h3 ♘eg5 14. ♔g4 h5 15. ♔f5 g6 16. ♔f6 ♔f8![1] –+

[1] (17. – ♘h7#)

4964. A 03
Cook – N. N.
Bristol 1906

(without ♘b1)

1. f4 d5 2. e3 c5 3. b3 e6 4. ♗b2 ♘f6 5. ♘f3 ♗e7 6. ♗d3 0-0 7. ♘g5 g6 8. h4 ♘h5 9. ♕h5! ♗g5

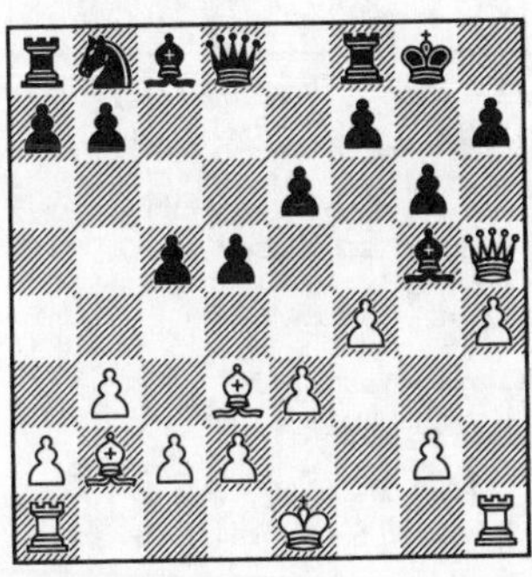

10. ♕h7! ♔h7 11. hg5 ♔g8 12. ♖h8#

4965. A 03
Lasker – Bauer
1889

1. f4 d5 2. e3 ♘f6 3. b3 e6 4. ♗b2 ♗e7 5. ♗d3 b6 6. ♘f3 ♗b7 7. ♘c3 ♘bd7 8. 0-0 0-0 9. ♘e2 c5 10. ♘g3 ♕c7 11. ♘e5 ♘e5 12. ♗e5 ♕c6 13. ♕e2 a6 14. ♘h5! ♘h5

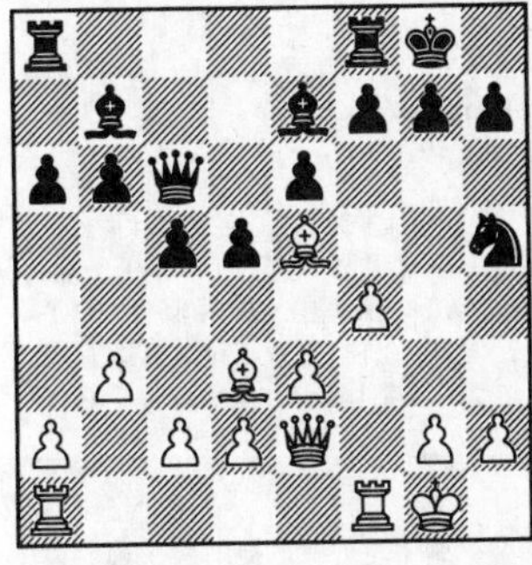

15. ♗h7! ♔h7 16. ♕h5 ♔g8 17. ♗g7! ♔g7 18. ♕g4 ♔h7 19. ♖f3 e5 20. ♖h3 ♕h6 21. ♖h6 ♔h6 22. ♕d7 +–

4966. A 03
Tarrasch – Satzinger
Munich 1915

1. f4 e6 2. ♘f3 d5 3. e3 c5 4. b3 ♗e7 5. ♗b2 ♗f6 6. ♘e5 ♗e5 7. fe5 ♘e7 8. ♗d3 ♘bc6 9. 0-0 0-0 10. ♕h5 ♘g6 11. ♖f3 ♘ce7 12. ♘c3 a6 13. ♖af1 b5 14. ♘d1 ♗b7 15. ♘f2 c4 16. ♘g4 f5 17. ef6 ♘f5 18. fg7 ♘g7

19. ♕h7! ♔h7 20. ♖h3 ♔g8 21. ♘h6 ♔h8 22. ♘f7 ♔g8 23. ♖h8! ♘h8 24. ♘h6#

4967. A 21

Nivergenit – Kapello

Milan 1970

1. c4 e5 2. g3 d6 3. ♗g2 ♗e7 4. ♘c3 c6 5. d3 f5 6. ♖b1 ♘f6 7. e4 0-0 8. ♘e2 ♘a6 9. b4 ♕e8 10. 0-0 ♕h5 11. f3 fe4 12. fe4 ♗h3 13. ♗h3 ♕h3 14. ♖f2 ♘g4 15. ♖g2 ♖f7 16. ♕e1 ♖af8 17. ♘d1

17. – ♘h2! –+

4968. A 22

Shaw – Whitney

USA 1949

1. c4 e5 2. ♘c3 ♘f6 3. e4 ♘c6 4. f4 d6 5. d3 ♗e7 6. ♗e3 ♘g4 7. ♗d2

7. – ♘h2! 8. ♖h2 ♗h4 9. ♔e2 ♘d4 10. ♔e3 ef4 11. ♔d4 ♕f6 12. ♔d5 ♗e6#

4969. A 47

Henschell – Karf

New York 1946

1. d4 ♘f6 2. ♘f3 b6 3. e3 ♗b7 4. ♘bd2 e6 5. ♗d3 d5 6. 0-0 ♘bd7 7. ♕e2 ♘e4 8. c4 ♗d6 9. ♘e1 0-0 10. ♕f3 f5 11. ♕e2 ♖f6 12. f3

12. – ♗h2! 13. ♔h2 ♖h6 14. ♔g1 ♘g3 15. ♕d1 ♕h4 16. ♕c2?[1] ♘h1! –+

[1] 16. f4! ♕h2 17. ♔f2 ♘f6 18. ♘ef3 ♘g4 19. ♔e1 ♕g2 –+

4970. A 47

Zsuzsa Polgár – N. N.

Budapest 1982

1. d4 ♘f6 2. ♘f3 b6 3. ♗f4 d5 4. ♘bd2 c5 5. e3 e6 6. ♘e5 cd4 7. ed4 ♗e7 8.

♘df3 0-0 9. ♗d3 ♗b7 10. 0-0 ♘c6 11. c3 ♖c8 12. h3 ♘e5 13. de5 ♘e4 14. ♘d4 ♘c5 15. ♗c2 ♗a6 16. ♖e1 ♕e8 17. ♖e3 ♗b7 18. ♖g3 a6

19. ♗h7! ♔h7 20. ♕h5 ♔g8 21. ♖g7! ♔g7 22. ♗h6 ♔h8 23. ♗g5 ♔g8 24. ♗f6! ♗f6 25. ef6 +−

4971. A 48

Marshall – Burn

Ostende 1907

1. d4 ♘f6 2. ♘f3 d6 3. ♗f4 ♘bd7 4. e3 g6 5. ♗d3 ♗g7 6. ♘bd2 0-0 7. h4 ♖e8 8. h5 ♘h5 9. ♖h5 gh5

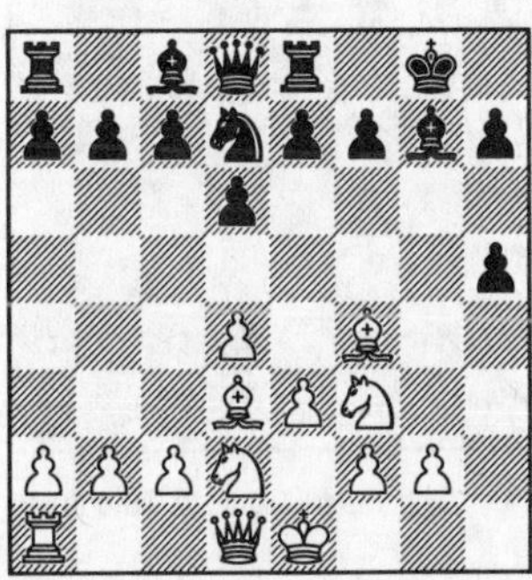

10. ♗h7! ♔h7 11. ♘g5 ♔g6[1] 12. ♘df3 e5 13. ♘h4 ♔f6 14. ♘h7 ♔e7 15. ♘f5 ♔e6 16. ♘g7 ♔e7 17. ♘f5 ♔e6 18. d5 ♔f5 19. ♕h5 ♔e4 20. 0-0-0 +−

[1] 11. – ♔g8 12. ♕h5 ♘f6 13. ♕f7 ♔h8 14. 0-0-0 +−

4972. A 82

Denker – Robbins

USA 1934

1. d4 f5 2. e4 fe4 3. ♘c3 ♘f6 4. f3 ef3 5. ♘f3 e6 6. ♗d3 ♗b4 7. 0-0 ♗c3 8. bc3 b6 9. ♗g5 ♗b7 10. ♘e5 0-0 11. ♘g4 d6 12. ♘f6 gf6

13. ♗h7! ♔g7[1] 14. ♕h5 fg5 15. ♕g6 ♔h8 16. ♗g8! +−

[1] 13. – ♔h7 14. ♕h5 ♔g7 15. ♕h6 ♔g8 16. ♕g6 ♔h8 17. ♖f6 +−

4973. A 83

Jaglom – Kahn

USA 1949

1. d4 f5 2. e4 fe4 3. ♘c3 ♘f6 4. ♗g5 b6 5. f3 ef3 6. ♘f3 ♗b7 7. ♗d3 e6 8. 0-0 ♗e7 9. ♕e2 0-0 10. ♘e5 ♘d5

11. ♗h7! ♔h7 12. ♕h5 ♔g8 13. ♖f8 ♕f8 14. ♘g6 ♕e8 15. ♕h8 ♔f7 16. ♘e5#

4974. A 83
Ed. Lasker – Thomas
London 1912

1. d4 f5 2. e4 fe4 3. ♘c3 ♘f6 4. ♗g5 e6 5. ♘e4 ♗e7 6. ♗f6 ♗f6 7. ♘f3 0-0 8. ♗d3 b6 9. ♘e5 ♗b7 10. ♕h5 ♕e7

11. ♕h7! ♔h7 12. ♘f6 ♔h6 13. ♘eg4 ♔g5 14. h4 ♔f4 15. g3 ♔f3 16. ♗e2 ♔g2 17. ♖h2 ♔g1 18. ♔d2#

4975. A 85
Abramovich – Botvinnik
Leningrad 1924

1. d4 f5 2. ♘f3 ♘f6 3. c4 e6 4. ♘c3 b6 5. ♗g5 ♗e7 6. e3 ♗b7 7. ♗d3 0-0 8. 0-0 ♘e4 9. ♗e7 ♕e7 10. ♖c1 ♘a6 11. a3 ♖f6 12. ♕a4 ♘c3 13. ♖c3 ♗f3 14. ♕a6 ♖g6 15. g3 ♕g5 16. ♖fe1 ♕h5 17. e4

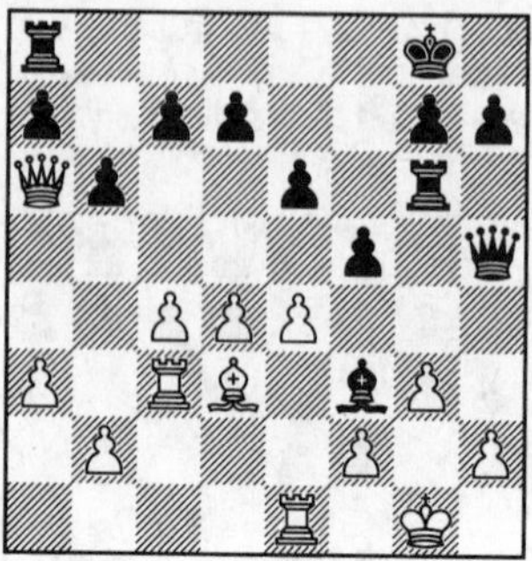

17. – ♕h2! –+

4976. A 85
Tarrasch – Richter
Halle 1883

1. d4 f5 2. c4 e6 3. ♘f3 ♘f6 4. e3 b6 5. ♗d3 ♗b7 6. 0-0 ♗d6 7. ♘c3 ♘c6 8. e4 fe4 9. ♘e4 ♗e7 10. ♘e5 ♘d4 11. ♘f6 ♗f6 12. ♕h5 g6 13. ♗g6 hg6 14. ♕g6 ♔e7 15. ♘g4 ♕f8 16. ♘f6 ♕f6 17. ♗g5 ♘e2 18. ♔h1

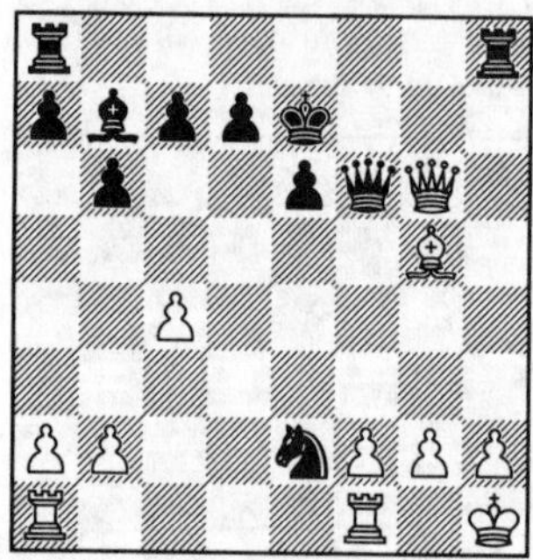

18. – ♖h2! 19. ♔h2 ♖h8 20. ♗h6 ♕h4#

4977. A 89
Zsuzsa Polgár – Chirakov
Targoviste 1981

1. d4 f5 2. c4 ♘f6 3. ♘f3 g6 4. g3 ♗g7 5. ♗g2 0-0 6. 0-0 d6 7. ♘c3 c6 8. d5 e5 9. de6 ♗e6 10. ♕d3 ♘a6 11. ♘g5 ♕e7

12. ♗f4 ♖ad8 13. ♖fd1 ♘e8 14. ♕e3 ♘c5 15. b4 ♘e4 16. ♘ce4! ♗a1

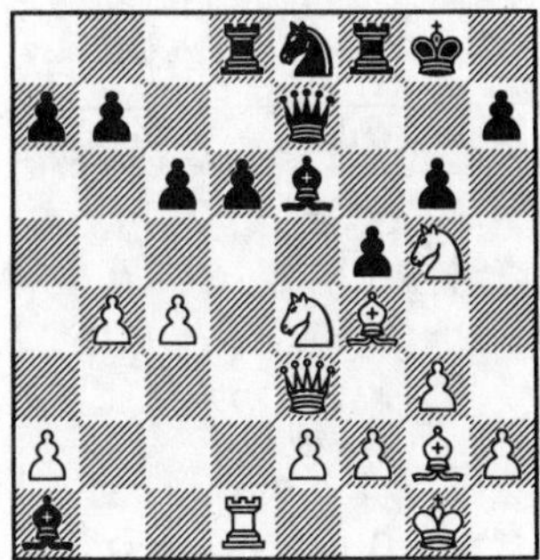

17. ♘h7!! fe4[1] 18. ♘f8 ♔f8 19. ♖a1 ♗c4? 20. ♗g5 ♘f6 21. ♕d4 +–

[1] 17. – ♔h7 18. ♘g5 +–

4978. B 09

Judit Polgár – N. N.

Budapest 1987

1. e4 g6 2. d4 ♗g7 3. f4 c6 4. ♘f3 d6 5. ♘c3 ♘f6 6. ♗d3 0-0 7. 0-0 ♘bd7 8. e5 ♘e8 9. ♗e3 ♘c7 10. ♘e4 d5 11. ♘eg5 f6

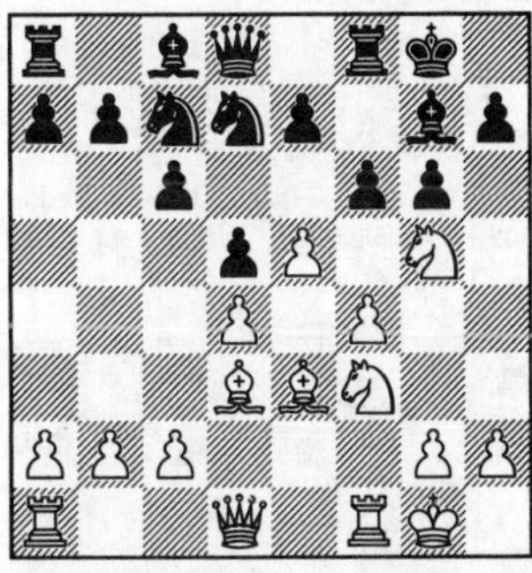

12. ♘h7! ♔h7 13. ♘g5! ♔g8[1] 14. ♗g6 fg5 15. ♕h5 ♘f6[2] 16. ef6 ♖f6 17. ♕h7 ♔f8 18. fg5 ♗g4 19. gf6 ef6 20. ♗h6![3] +–

[1] 13. – fg5 14. ♕h5 +–

[2] 15. – ♖f6 16. ♕h7 ♔f8 17. fg5 +–

[3] 20. – ♕e7 21. ♕h8#

4979. B 09

Korelov – Kremenetzky

Moscow 1977

1. e4 d6 2. d4 ♘f6 3. ♘c3 g6 4. f4 ♗g7 5. ♘f3 0-0 6. ♗d3 ♘a6 7. 0-0 c5 8. d5 ♘c7 9. a4 ♖b8 10. ♕e1 e6 11. de6 ♘e6 12. f5 ♘d4 13. ♕h4 gf5 14. ♗g5 ♘e6 15. ef5 ♘g5 16. ♘g5 d5

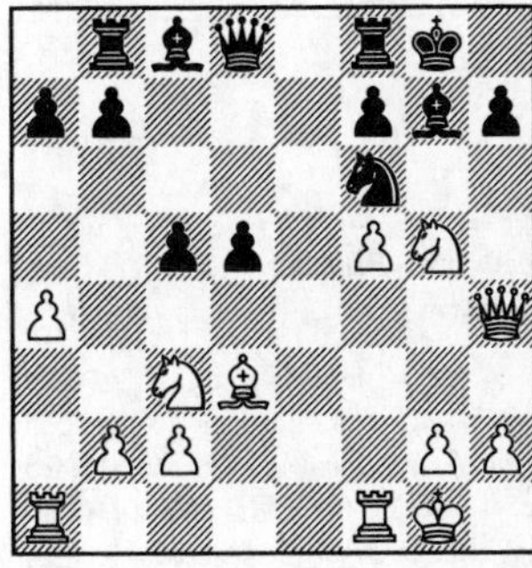

17. ♘h7! c4[1] 18. ♘d5! ♘e4 19. ♘e7 +–

[1] 17. – ♘h7 18. f6 ♘f6 19. ♖f6 +–

4980. B 21

Tarrasch – Schröder

Nuremberg 1892

(without ♘b1)

1. e4 d5 2. e5 c5 3. f4 e6 4. ♘f3 ♘c6 5. c3 ♕b6 6. ♗d3 ♗d7 7. ♗c2 c4 8. b3 cb3 9. ab3 d4 10. ♕e2 dc3 11. dc3 ♘h6 12. ♘d2 ♗e7 13. ♘e4 0-0 14. ♕d3 g6 15. ♕h3 ♘f5 16. g4 ♘g7 17. ♘f6 ♗f6 18. ef6 ♘e8 19. g5 ♔h8 20. ♕h6 ♖g8 21. h4 ♘d6 22. h5 ♘f5

23. ♕h7![1] +–

[1] 23. – ♔h7 24. hg6 ♔g6 25. ♖h6#

4981. B 33

Medvegy – Hilmer

Stockerau 1992

1. e4 c5 2. ♘f3 ♘c6 3. d4 cd4 4. ♘d4 ♘f6 5. ♘c3 e5 6. ♘db5 d6 7. ♘d5 ♘d5 8. ed5 ♘b8 9. f4 a6 10. ♘c3 ♗e7 11. fe5 de5 12. ♗e3 0-0 13. ♕f3 ♘d7 14. 0-0-0 ♖b8 15. g4 ♗g5 16. ♖g1 b5 17. ♘e4 ♗e3 18. ♕e3 a5 19. g5 b4 20. ♗d3 a4 21. ♖df1 ♕b6 22. ♕h3 a3 23. ♖g3 ♕d4

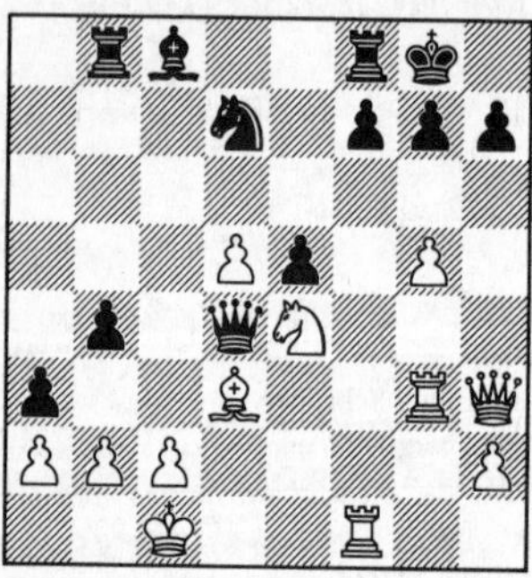

24. ♕h7! ♔h7 25. ♘f6 +–

4982. B 35

Judit Polgár – Metodiev

Albena 1986

1. e4 c5 2. ♘f3 ♘c6 3. d4 cd4 4. ♘d4 ♘f6 5. ♘c3 g6 6. ♗e3 a6 7. ♗c4 ♗g7 8. ♗b3 0-0 9. f3 ♕c7 10. ♕d2 e6 11. 0-0-0 b5 12. h4 ♘a5 13. h5 ♘c4 14. ♗c4 bc4 15. hg6 fg6 16. ♗h6 ♖f7 17. ♗g7 ♖g7 18. ♕g5 ♖f7 19. e5 ♘d5 20. ♘d5 ed5

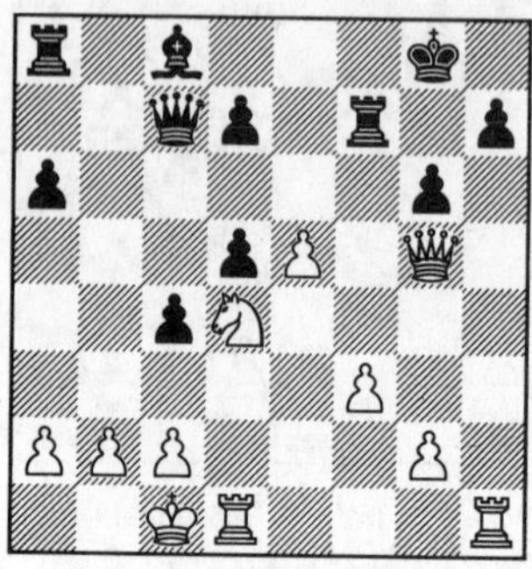

21. ♖h7! ♖h7[1] 22. ♕g6 ♔h8 23. ♕e8[2] +–

[1] 21. – ♔h7 22. ♖h1 ♔g7 23. ♕h6 ♔g8 24. ♕h8#; 22. – ♔g8 23. ♕g6 ♖g7 24. ♕e8#; 23. – ♔f8 24. ♖h8 ♔e7 25. ♕g5 +–

[2] 23. – ♔g7 24. ♘f5#

4983. B 40

Böök – Ingerslev

Gothenburg 1929

1. e4 e6 2. ♘c3 ♗b4 3. ♘f3 c5 4. ♗d3 d5 5. a3 ♗a5 6. b4 cb4 7. ab4 ♗b4 8. ♗b2 ♘e7 9. 0-0 0-0 10. ♖e1 b6 11. ed5 ♘d5 12. ♘d5 ed5 13. ♘d4 ♗b7 14. ♖e3 ♘d7 15. ♘c6! ♗c6

16. ♗h7! ♔h7 17. ♕h5 ♔g8 18. ♗g7! ♔g7 19. ♖g3 ♔f6 20. ♖e1[1] +–

[1] 20. – ♖g8 21. ♖f3 ♔g7 22. ♖f7#

4984. B 42

Tal – N. N.

1974

1. e4 c5 2. ♘f3 e6 3. d4 cd4 4. ♘d4 a6 5. ♗d3 ♘f6 6. 0-0 ♕c7 7. ♔h1 d6 8. f4 ♘bd7 9. ♘d2 ♗e7 10. ♘2f3 0-0 11. ♕e2 ♘c5 12. e5 de5 13. fe5 ♘fd7 14. ♗g5 ♘e5 15. ♗e7 ♘f3 16. ♖f3 ♕e7

17. ♗h7! ♔h7 18. ♖h3 ♔g8 19. ♘f5 ♕g5 20. ♕h5![1] +–

[1] 20. – ♕h5 21. ♘e7 ♔h7 22. ♖h5#

4985. B 45

Solymár – Navarovszky

Budapest 1953

1. e4 c5 2. ♘f3 e6 3. d4 cd4 4. ♘d4 ♘f6 5. ♗d3 ♘c6 6. ♘c6 bc6 7. c4 e5 8. ♘c3 ♗c5 9. 0-0 ♖b8 10. ♔h1 d6 11. ♕e2 ♘g4 12. f3

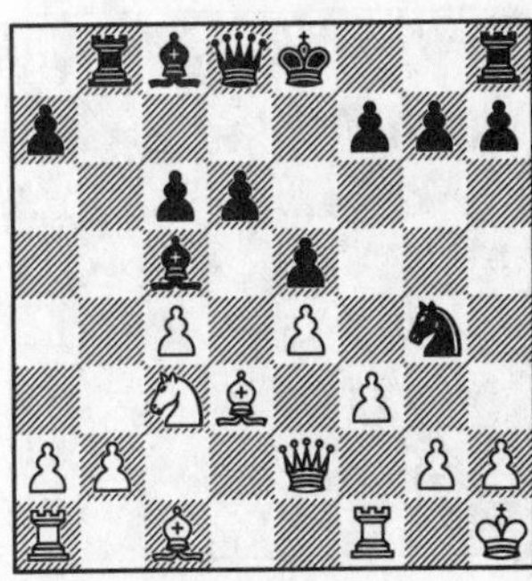

12. – ♘h2! 13. ♖e1[1] ♘f1![2] –+

[1] 13. ♔h2 ♕h4#

[2] 14. ♕f1 ♕h4#

4986. B 64

Hort – Rabulov

1974

1. e4 c5 2. ♘f3 ♘c6 3. d4 cd4 4. ♘d4 ♘f6 5. ♘c3 d6 6. ♗g5 e6 7. ♕d2 ♗e7 8. f4 d5 9. e5 ♘d7 10. ♗e7 ♕e7 11. 0-0-0 a6 12. ♘f3 0-0 13. ♗d3 ♖b8 14. ♕e3 b5

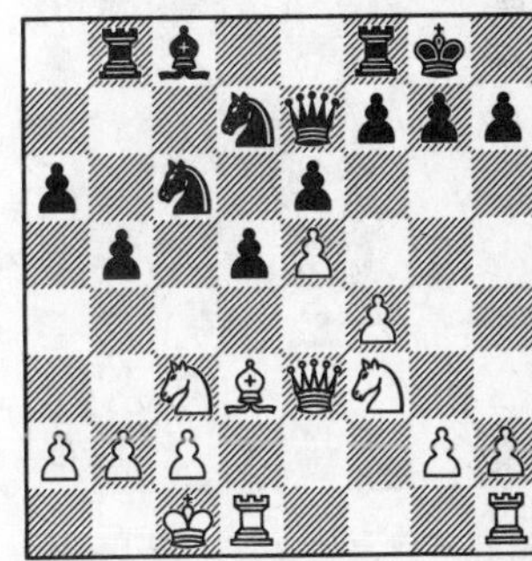

15. ♗h7! ♔h7 16. ♘g5 ♔g6 17. ♕h3 +–

4987. C 00

Bielfeldt – Daley

Corr. 1964

1. e4 e6 2. d4 d5 3. ♗e3 ♘f6 4. f3 ♘c6 5. e5 ♘d7 6. f4 ♗e7 7. c3 0-0 8. ♗d3 b6 9. ♘f3 ♗b7 10. h4 f6 11. ♘g5 fg5

12. ♗h7! ♔h7 13. ♕h5 ♔g8 14. hg5 ♗a6[1] 15. g6 +–

[1] 14. ♖f5 15. g4 +–

4988. C 00

Curdo – Richards

Springvale 1977

1. e4 e6 2. d4 c5 3. c3 cd4 4. cd4 d5 5. e5 ♘c6 6. ♘f3 ♗b4 7. ♘c3 ♘ge7 8. ♗d3 0-0

9. ♗h7! ♔h7 10. ♘g5 ♔g6 11. ♕g4 f5 12. ♕g3 f4 13. ♕g4 ♔h6 14. ♕h4 ♔g6 15. ♕h7 ♔g5 16. h4 ♔g4 17. ♕g7 ♘g6 18. ♕g6 +–

4989. C 00

Diemer – Zeller

1. d4 d5 2. e4 e6 3. ♗e3 de4 4. ♘d2 f5 5. f3 ♘f6 6. fe4 fe4 7. ♘h3 ♗d6 8. ♗c4 0-0 9. 0-0 ♔h8 10. ♘g5 ♕e7 11. ♘de4 e5 12. ♘f6 gf6 13. de5 ♗e5 14. ♕h5 fg5 15. ♖f8 ♕f8 16. ♖f1 ♗f6 17. ♗d3 ♕e7 18. ♖f6 ♕e3 19. ♔h1 ♕e7 20. ♖f7 ♕e1 21. ♖f1 ♕e7

22. ♗h7![1] +–

[1] 21. – ♕h7 22. ♖f8 ♔g7 23. ♖f7 ♔g8 24. ♕h7#

4990. C 00

Greco – N. N.

Roma 1619

1. e4 e6 2. d4 ♘f6 3. ♗d3 ♘c6 4. ♘f3 ♗e7 5. h4 0-0 6. e5 ♘d5

7. ♗h7! ♔h7 8. ♘g5 ♗g5 9. hg5 ♔g6 10. ♕h5 ♔f5 11. ♕h7 g6 12. ♕h3 ♔e4 13. ♕d3#

4991. C 00

Kaufmann – Rheti

Vienna 1914

1. e4 e6 2. d4 d5 3. ♗d3 c5 4. ed5 ♕d5 5. ♘c3 ♕d4 6. ♘f3 ♕d8 7. ♗f4 ♘f6 8. ♕e2 ♘c6 9. 0-0 ♗d6 10. ♗d6 ♕d6 11. ♖ad1 ♕f4 12. ♘e4 ♘e4 13. ♗e4 0-0 14. ♖fe1 ♕c7 15. ♕c4 b6 16. ♖d3 ♗b7

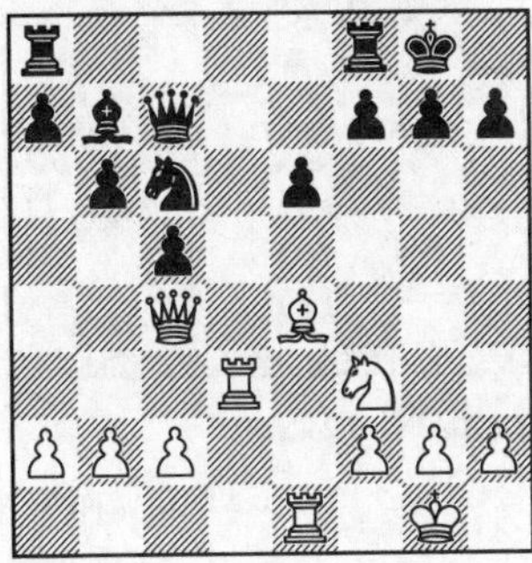

17. ♗h7! ♔h7 18. ♕h4 ♔g8 19. ♘g5[1] +–

[1] 19. – ♖fd8 20. ♕h7 ♔f8 21. ♕h8 ♔e7 22. ♖e6 fe6 23. ♕g7 ♔e8 24. ♕g8 ♔e7 25. ♕f7#

4992. C 00

Lasker – Traxler

Luzern 1935

1. e4 e6 2. ♘c3 ♘f6 3. ♘f3 d5 4. e5 ♘fd7 5. d4 ♗e7 6. ♗e3 0-0 7. ♗d3 b6 8. h4 f6

9. ♗h7! ♔h7 10. ♘g5 fg5[1] 11. hg5 ♔g8 12. ♕h5 ♖f5 13. f4 ♘f8 14. g4 g6 15. ♕h8 ♔f7 16. gf5 ef5 17. 0-0-0 a5 18. ♖h6 ♔e6 19. ♕g8 ♔d7 20. ♕d5 +–

[1] 11. – ♔g6 12. h5 ♔h6 13. ♘e6 +–

4993. C 00

Szasz – Sheppards

USA 1976

1. d4 d5 2. e4 e6 3. ♗e3 b6 4. e5 ♗b7 5.f4 ♘d7 6. ♘f3 c5 7. c3 cd4 8. cd4 ♖c8 9. a3 ♘e7 10. ♗d3 ♘c6 11. 0-0 ♘cb8 12. ♘c3 ♗a6 13. ♘b5 ♗b5 14. ♗b5 a6 15. ♗d3 ♗e7 16. ♗d2 0-0 17. ♗b4 ♗b4 18. ab4 a5

19. ♗h7! ♔h8[1] 20. ♘g5 g6 21. ♕g4 f5 22. ♕h4 ♔g7 23. ♘e6 +–

[1] 19. – ♔h7 20. ♘g5 ♔g8 21. ♕h5 ♖e8 22. ♕f7 ♔h8 23. ♕h5 ♔g8 24. ♕h7 ♔f8 25. ♕h8 ♔e7 16. ♕g7#; 20. – ♔g6 21. ♕g4 +–

4994. C 02

Balogh – Redeleit

1928

1. e4 e6 2. d4 d5 3. e5 c5 4. ♕g4 cd4 5. ♘f3 ♘c6 6. ♗d3 ♘ge7 7. 0-0 ♘g6 8. ♖e1 ♗e7 9. a3 0-0 10. h4 ♘h4

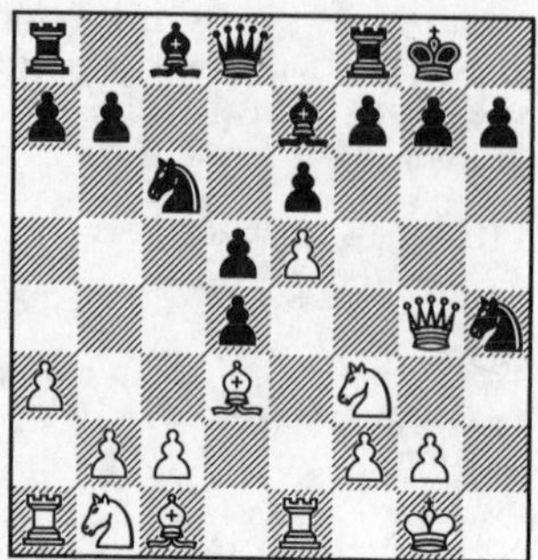

11. ♗h7! ♔h7 12. ♘g5 ♗g5 13. ♗g5 ♕c7 14. ♕h4 ♔g8 15. ♗f6! gf6 16. ef6 +–

4995. C 02

Gaudersen – Faul

Melbourne 1928

1. e4 e6 2. d4 d5 3. e5 c5 4. c3 cd4 5. cd4 ♗b4 6. ♘c3 ♘c6 7. ♘f3 ♘ge7 8. ♗d3 0-0

9. ♗h7! ♔h7 10. ♘g5 ♔g6 11. h4 ♘d4 12. ♕g4 f5 13. h5 ♔h6 14. ♘e6 g5 15. hg6#

4996. C 02

Gurgenidze – Simchak

Varna 1975

1. e4 e6 2. ♘f3 d5 3. ♘c3 ♘f6 4. e5 ♘fd7 5. d4 c5 6. ♗b5 ♘c6 7. 0-0 ♗e7 8. ♘e2 cd4 9. ♘ed4 ♘d4 10. ♘d4 0-0 11. ♖e1 ♕c7 12. ♗f4 ♘b8 13. ♕g4 ♖d8 14. ♖e3 ♘c6 15. ♘c6 bc6 16. ♗d3 g6 17. h4 ♔h8 18. ♗g5 ♖e8 19. ♖f3 ♖f8 20. h5 ♗b7 21. ♖h3 ♖g8 22. hg6 fg6

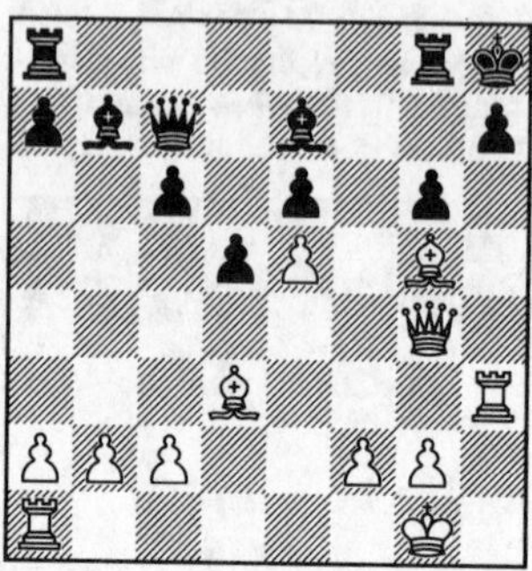

23. ♖h7! ♔h7 24. ♕h5[1] +–

[1] 24. – ♔g7 25. ♕g6 ♔f8 26. ♗h6 ♖g7 27. ♗g7 +–

4997. C 02

Petersen – Rochelle

Omaha 1976

1. e4 e6 2. d4 d5 3. e5 c5 4. c3 ♘c6 5. ♘f3 ♕b6 6. ♗e2 cd4 7. cd4 ♗b4 8. ♘c3 ♘ge7 9. 0-0 ♘f5 10. ♗e3 ♗d7 11. a3 ♘e3 12. fe3 ♗c3 13. bc3 ♘a5 14. ♘g5 0-0 15. ♗d3 g6 16. ♕f3 ♕d8 17. ♕g3 f6

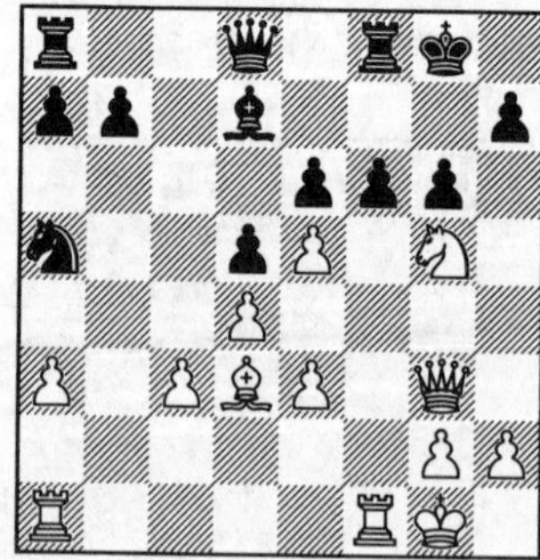

18. ♘h7![1] +–

[1] 18. – ♔h7 19. ♕g6 ♔h8 20. ♕h7#

4998. C 02

Wolk – Bukrejev

1936

1. e4 e6 2. d4 d5 3. e5 c5 4. ♘f3 ♘c6 5. dc5 ♗c5 6. ♗d3 ♘ge7 7. ♗f4 ♕b6 8. 0-0 ♕b2 9. ♘bd2 0-0 10. ♘b3 ♕a3

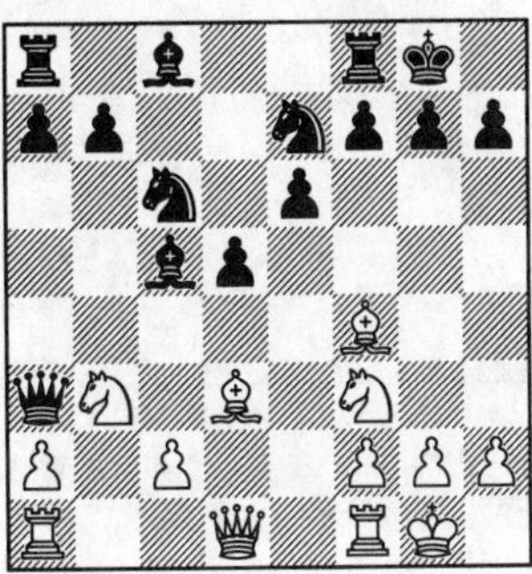

11. ♗h7! ♔h7 12. ♘g5 ♔g6 13. ♕d3 ♘f5 14. ♕h3[1] +–

[1] 14. – ♘h6 15. g4 +–

4999. C 03

Jansa – Marovic

Madonna di Campiglio 1974

1. e4 e6 2. d4 d5 3. ♘d2 b6 4. ♘gf3 ♘f6 5. ♗d3 de4 6. ♘e4 ♗b7 7. ♕e2 ♗e7 8. ♗f4 ♘bd7 9. 0-0-0 ♘d5 10. ♗d2 ♘b4 11. ♗b4 ♗b4 12. ♘e5 0-0 13. ♘d7 ♕d7 14. ♘f6! gf6

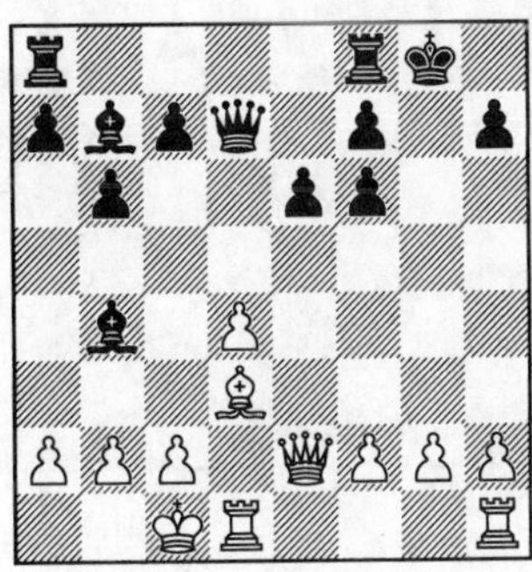

15. ♗h7! ♔h8[1] 16. ♕h5 ♔g7 17. ♕g4 ♔h8 18. d5 ♖fd8 19. ♖d3 ♗d5 20. ♗f5 ♗g2 21. ♕g2 +–

[1] 15. – ♔h7 16. ♕h5 ♔g7 17. ♕g4 ♔h7 18. ♖d3 +–

5000. C 04

Machatschek – Springhot

1962

1. e4 e6 2. d4 d5 3. ♘d2 ♘c6 4. ♘gf3 ♘f6 5. e5 ♘d7 6. ♘b3 ♗e7 7. ♗d3 0-0 8. h4 f6

9. ♘g5! fg5 10. ♗h7! ♔h7 11. hg5 ♔g8 12. ♖h8 ♔h8[1] 13. ♕h5 ♔g8 14. g6 ♗b4 15. c3 ♖f5 16. ♕h7 ♔f8 17. ♕h8 ♔e7 18. ♗g5! +–

[1] 12. – ♔f7 13. ♕h5 g6 14. ♕h7 ♔e8 15. ♕g6#

5001. C 04

Perez – Sanz

Madrid 1948

1. e4 e6 2. d4 d5 3. ♘d2 ♘c6 4. ♘gf3 ♘f6 5. e5 ♘d7 6. ♗e2 ♗e7 7. ♘f1 f6 8. ef6 ♗f6 9. ♘e3 0-0 10. c3 b6 11. ♕c2 ♖e8 12. 0-0 ♗b7 13. ♗d3 ♘f8 14. b3 ♖c8 15. ♘g4 a6 16. ♖e1 ♘b8 17. ♘fe5 c5

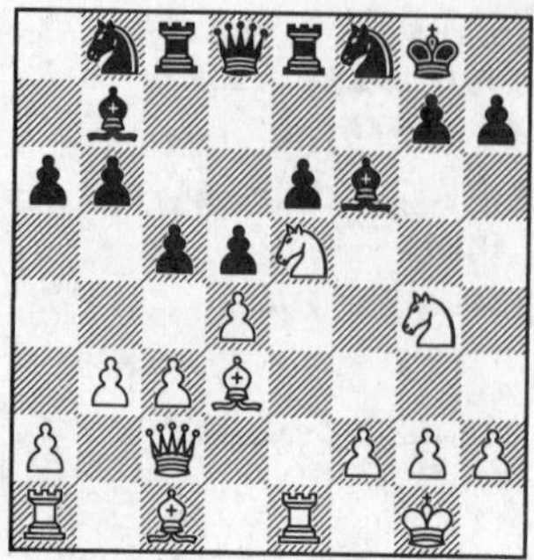

18. ♗h7! ♘h7 19. ♘h6![1] +–

[1] 19. – gh6 20. ♕g6 ♗g7 21. ♕f7 ♔h8 22. ♘g6#

5002. C 05

Foltys – Mohyla

Ostrava 1940

1. e4 e6 2. d4 d5 3. ♘d2 ♘f6 4. ♗d3 c5 5. e5 ♘fd7 6. c3 ♘c6 7. ♘e2 ♗e7 8. ♘f3 0-0 9. ♘f4 b6 10. h4 ♖e8

11. ♗h7! ♔h7 12. ♘g5 ♔g8 13. ♕h5 ♗g5 14. hg5 ♔f8 15. ♕h8 ♔e7 16. ♘g6! fg6 17. ♕g7#

5003. C 05

Miller – Blaine

Wabash 1981

1. e4 e6 2. d4 d5 3. ♘d2 ♘f6 4. e5 ♘fd7 5. ♗d3 c5 6. c3 ♘c6 7. ♘gf3 ♕b6 8. dc5 ♗c5 9. ♕e2 ♕c7 10. ♘b3 ♗b6 11. ♗f4 0-0 12. ♘bd4 ♖e8 13. ♘b5 ♕b8 14. ♘d6 ♖e7

15. ♗h7! ♔h7 16. ♘g5 ♔g8[1] 17. ♕h5 ♘f8 18. ♘df7 +–

[1] 16. – ♔g6 17. ♕d3 f5 18. ef6 ♔f6 19. ♘h7#

5004. C 10

Blom – Jensen

1934

1. e4 e6 2. d4 d5 3. ♘c3 de4 4. ♘e4 ♗d6 5. ♗d3 ♘e7 6. ♗g5 0-0 7. ♘f6! gf6 8. ♗f6 ♕d7

9. ♗h7! +–

5005. C 10

Hayden – N. N.

1952

1. e4 e6 2. d4 d5 3. ♘c3 c5 4. ♘f3 cd4 5. ♘d4 ♗b4 6. ♗e2 ♘f6 7. 0-0 ♗c3 8. bc3 de4 9. ♗a3 b6 10. f3 ♘d5 11. fe4 ♘e3 12. ♗b5 ♗d7 13. ♘e6! fe6 14. ♕h5 g6

15. ♕h7! +–

5006. C 10

Holcomb – Elmore

Dayton 1980

1. e4 e6 2. d4 d5 3. ♘d2 de4 4. ♘e4 ♘c6 5. ♘f3 ♗b4 6. c3 ♗a5 7. ♗d3 ♗b6 8. 0-0 a6 9. ♖e1 ♘ge7 10. ♘g3 0-0

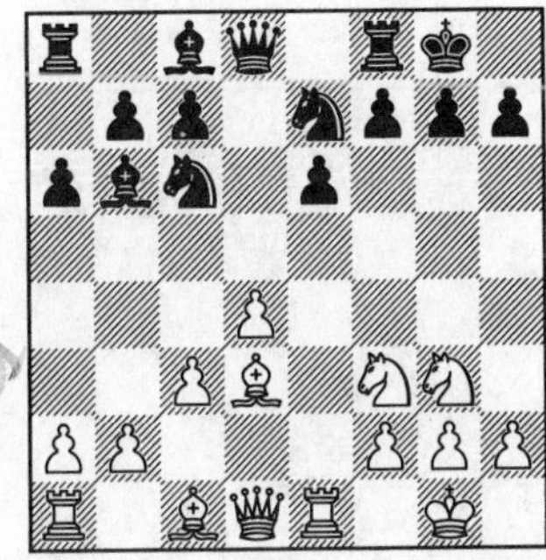

11. ♗h7! ♔h7 12. ♘g5 ♔g8 13. ♕h5 +–

5007. C 10

Ilievsky – Graul

Sandomierz 1976

1. e4 e6 2. d4 d5 3. ♘d2 de4 4. ♘e4 ♘f6 5. ♘f6 ♕f6 6. ♘f3 ♗d6 7. ♗g5 ♕f5 8. ♗d3 ♕a5 9. c3 ♘d7 10. 0-0 c6 11. ♖e1 ♕c7 12. ♕e2 0-0 13. ♕e4 g6 14. ♕h4 ♖e8 15. ♖e4 ♘b6 16. ♗f6 ♘d5

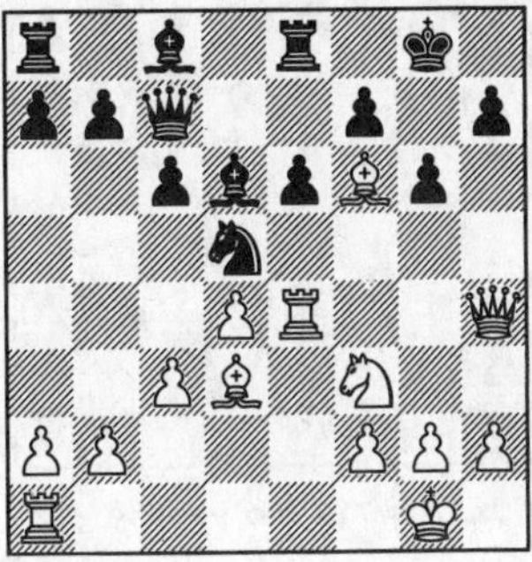

17.♕h7! +–

5008. C 11

Dake – Cranston

1935

1. e4 e6 2. d4 d5 3. ♘c3 ♘f6 4. ♗g5 de4 5. ♘e4 ♗e7 6. ♗f6 ♗f6 7. ♘f3 ♘d7 8. c3 0-0 9. ♕c2 ♗e7 10. 0-0-0 c6 11. h4 ♘f6 12. ♘f6 ♗f6 13. ♗d3 g6 14. h5 ♔g7 15. ♖h2 ♖g8 16. ♕d2 ♔h8 17. ♕h6 ♗g7

18. ♕h7 ♔h7 19. hg6#

5009. C 13
Bogoljubov – N. N.
1956

1. e4 e6 2. d4 d5 3. Nc3 Nf6 4. Bg5 Be7 5. e5 Nfd7 6. h4 0-0 7. Bd3 c5 8. Nh3 cd4 9. Be7 Qe7

10. Bh7! Kh7 11. Qh5 Kg8 12. Ng5 +–

5010. C 13
Fritz – Bozener
1899

1. e4 e6 2. d4 d5 3. Nc3 Nf6 4. Bg5 Be7 5. Bf6 Bf6 6. Nf3 0-0 7. Bd3 c5 8. e5 Be7 9. h4 f5 10. ef6 gf6 11. Ng5 fg5 12. hg5 Rf7

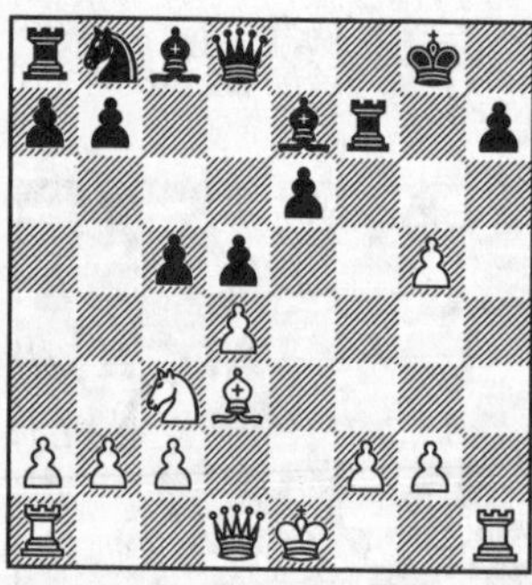

13. Bh7! Rh7 14. Rh7 Bg5[1] 15. Qh5 Qf6 16. Nb5 +–

[1] 14. – Kh7 15. Qh5 Kg7 16. Qh6 Kg8 17. g6 +–

5011. C 13
Euwe – Maróczy

1. e4 e6 2. d4 d5 3. Nc3 Nf6 4. Bg5 Be7 5. e5 Nfd7 6. h4 0-0 7. Bd3 c5 8. Qh5 g6 9. Qh6 Re8 10. Be7 Qe7 11. h5 Nf8 12. Nf3 cd4 13. Ng5 Nbd7

14. Nh7! Ne5[1] 15. hg6 Nfg6 16. Bg6 Ng6 17. g4 dc3 18. 0-0-0[2] +–

[1] 14. – Nh7 15. hg6 fg6 16. Qg6 +–
[2] (g5, Nf6 +–)

5012. C 13
Kasperski – Mazel
Minsk 1925

1. e4 e6 2. d4 d5 3. Nc3 Nf6 4. Bg5 Be7 5. e5 Nfd7 6. h4 Bg5 7. hg5 Qg5 8. Nh3 Qe7 9. Qg4 g6 10. Nf4 a6 11. Bd3 Qf8 12. 0-0-0 c5 13. Ncd5 ed5 14. Nd5 Ra7 15. e6 fe6 16. Nc7 Kf7 17. Qe6 Kg7

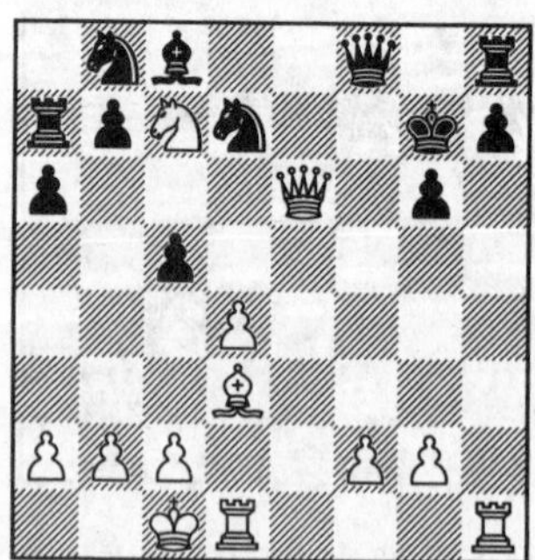

18. ♖h7! ♖h7 19. ♕g6 ♔h8 20. ♕h7#

5013. C 13

Keres – Wade

London 1954

1. e4 e6 2. d4 d5 3. ♘c3 ♘f6 4. ♗g5 ♗e7 5. e5 ♘fd7 6. h4 ♗g5 7. hg5 ♕g5 8. ♘h3 ♕e7 9. ♘f4 a6 10. ♕g4 ♔f8 11. ♕f3 ♔g8 12. ♗d3 c5

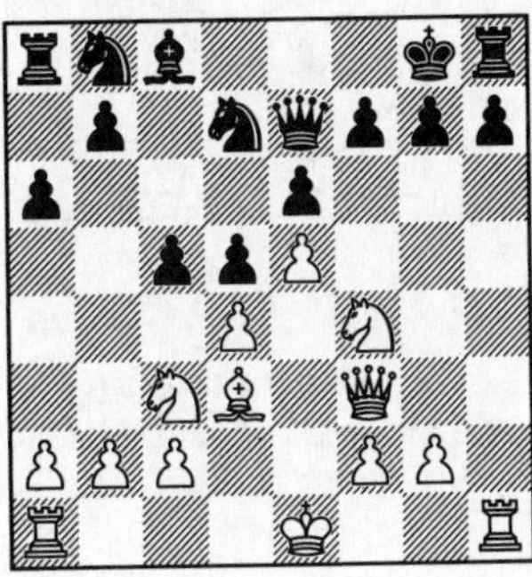

13. ♗h7! ♖h7 14. ♖h7 ♔h7 15. 0-0-0 f5 16. ♖h1 ♔g8 17. ♖h8 +–

5014. C 13

Pestalozzi – Duhm

1908

1. e4 e6 2. d4 d5 3. ♘c3 ♘f6 4. ♗g5 ♗e7 5. ♗f6 ♗f6 6. e5 ♗e7 7. ♗d3 c5 8. dc5 ♗c5 9. ♕g4 0-0 10. ♘f3 ♘c6

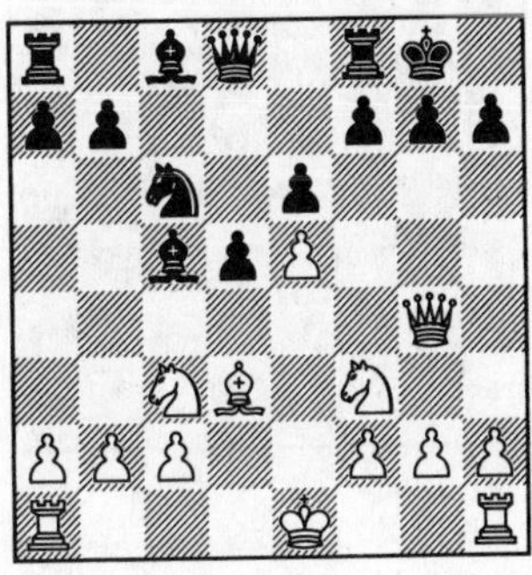

11. ♗h7! ♔h7 12. ♕h5 ♔g8 13. ♘g5 ♖e8 14. ♕f7 ♔h8 15. ♕h5 ♔g8 16. ♕h7 ♔f8 17. ♕h8 ♔e7 18. ♕g7#

5015. C 13

Richter – Darga

1. e4 e6 2. d4 d5 3. ♘c3 ♘f6 4. ♗g5 ♗e7 5. ♗f6 ♗f6 6. e5 ♗e7 7. ♕g4 0-0 8. ♗d3 c5 9. dc5 ♘d7 10. ♘f3 ♘c5 11. 0-0-0 ♕a5

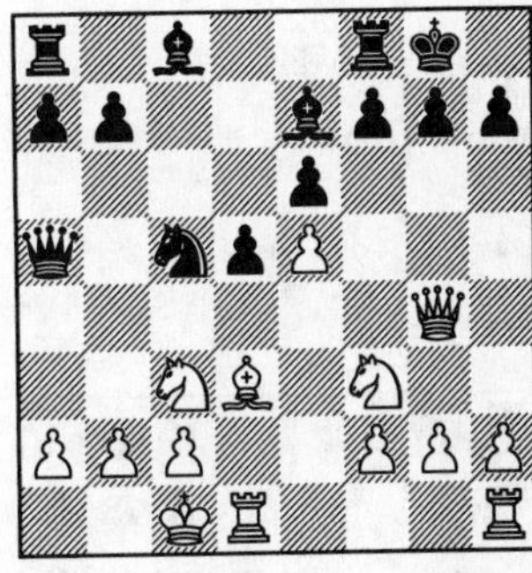

12. ♗h7! ♔h7 13. ♕h5 ♔g8 14. h4 ♖e8 15. ♘g5 ♗g5 16. hg5 ♔f8 17. g6 fg6 18. ♕g6 ♗d7 19. ♖h7[1] +–

[1] 19. – ♔e7 20. ♕g5 ♔f7 21. ♕g7#

5016. C 13

Schröder – Hendricks

Corr. 1959

1. e4 e6 2. d4 d5 3. ♘c3 ♘f6 4. ♗g5 ♗e7 5. e5 ♘fd7 6. ♗e3 ♗b4 7. ♘f3 c5 8. dc5 ♘c5 9. ♗c5 ♗c5 10. ♗d3 ♕b6 11. ♕d2 ♗d7 12. 0-0 ♗c6 13. ♕g5 0-0

14. ♗h7! ♔h7 15. ♕h5 ♔g8 16. ♘g5 ♖e8 17. ♕f7 ♔h8 18. ♖ad1 ♕d8 19. ♖d3 +–

5017. C 13

Yates – Marin

1930

1. e4 e6 2. d4 d5 3. ♘c3 ♘f6 4. ♗g5 ♗e7 5. e5 ♘e4 6. ♗e7 ♕e7 7. ♗d3 ♘c3 8. bc c5 9. ♕g4 0-0 10. ♘f3 c4

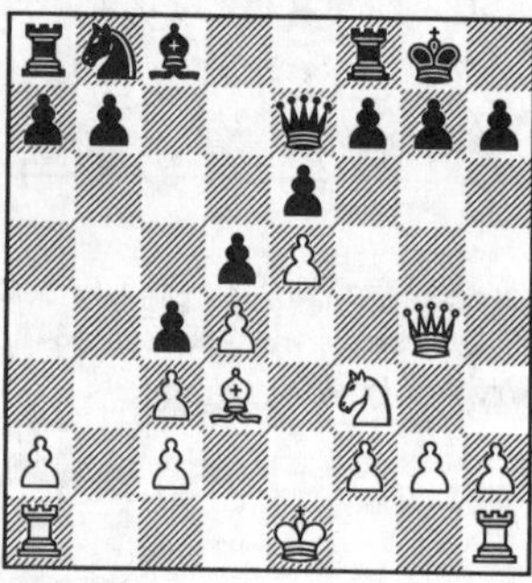

11. ♗h7![1] +–

[1] 11. – ♔h7 12. ♕h5 ♔g8 13. ♘g5 ♖d8 14. ♕h7 ♔f8 15. ♕h8#

5018. C 14

Schlechter – Stubenrauch

1901

1. e4 e6 2. d4 d5 3. ♘c3 ♘f6 4. ♗g5 ♗e7 5. e5 ♘fd7 6. ♗e7 ♕e7 7. &Fd3 c5 8. ♘b5 0-0 9. c3 ♘b6 10. f4 ♘c6 11. ♘f3 cd4 12. cd4 ♗d7 13. 0-0 a6 14. ♘d6 ♗e8

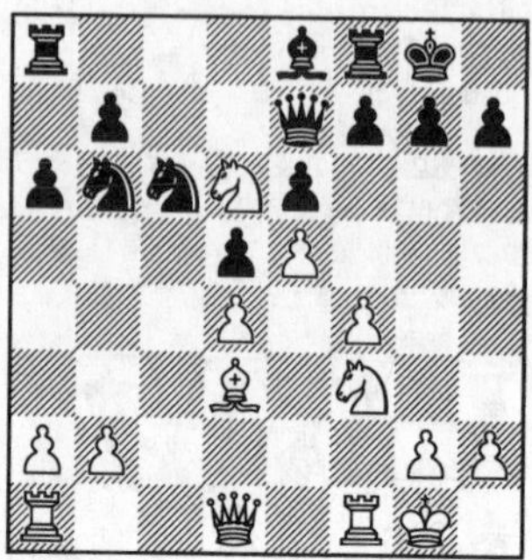

15. ♗h7! ♔h7 16. ♘g5[1] +–

[1] 16. – ♔g8 17. ♕h5 +–; 16. – ♔g6 17. ♕d3 ♔h6 18. ♕h3 ♔g6 19. ♕h7#; 17. – f5 18. ef6 ♔f6 19. ♘h7#

5019. C 14

Speyer – Couvee

1. e4 e6 2.d4 d5 3. ♘c3 ♘f6 4. ♗g5 ♗e7 5. e5 ♘fd7 6. ♗e7 ♕e7 7. ♘b5 ♕d8 8. ♗d3 c6 9. ♘d6 ♔e7 10. ♕h5 g6 11. ♕h4 f6 12. ♘h3 ♘a6 13. ♘f4 g5

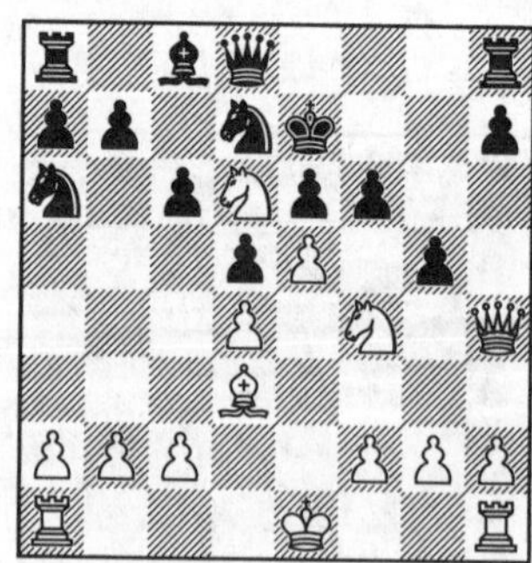

14. ♕h7! ♖h7 15. ♘g6#

5020. C 15
Blackburne – West
Hamilton 1885

1. e4 e6 2. d4 d5 3. ♘c3 ♗b4 4. ed5 ed5 5. ♗d3 ♘c6 6. ♘f3 ♘f6 7. 0-0 ♗c3 8. bc3 0-0 9. ♘e5 ♗e6 10. f4 ♘e4 11. f5 ♘e5 12. de5 ♗d7 13. f6 g6 14. ♗a3 ♖e8 15. ♗e4 de4 16. ♕d2 ♔h8 17. ♕g5 c6 18. ♖f4 ♕a5 19. ♕h6 ♖g8

20. ♕h7! ♔h7 21. ♖h4#

5021. C 15
Tzeitlin – Vladimirov
USSR 1981

1. e4 e6 2. d4 d5 3. ♘c3 ♗b4 4. ♘e2 de4 5. a3 ♗c3 6. ♘c3 ♘c6 7. ♗b5 ♘e7 8. ♘e4 0-0 9. c3 ♕d5 10. ♕e2 f5 11. ♘g5 ♕g2 12. ♖f1 e5 13. ♗c4 ♔h8

14. ♘h7! ♖e8 15. ♘f6![1] +–

[1] 15. – gf6 16. ♕h5 ♔g7 17. ♕h6#

5022. C 17
Giusti – Cipriani
Corr. 1954/55

1. e4 e6 2. d4 d5 3. ♘c3 ♗b4 4. e5 c5 5. ♗d2 ♘e7 6. ♘b5 ♗d2 7. ♕d2 0-0 8. c3 ♘bc6 9. ♘d6 ♕b6 10. ♘f3 ♗d7 11. ♗d3 ♘c8

12. ♗h7! ♔h7 13. ♘g5 ♔g6 14. ♕d3 ♔g5 15. f4[1] +–

[1] 15. ♔h6 16. ♕h3 ♔g6 17. g4 +–

5023. C 17
Knox – Pytel
1981

1. e4 e6 2. d4 d5 3. ♘c3 ♗b4 4. e5 c5 5. ♗d2 ♘e7 6. ♘b5 0-0 7. c3 ♗a5 8. dc5 ♗c7 9. f4 ♘d7 10. b4 b6 11. cb6 ♘b6 12. ♘f3 ♗b7 13. ♗d3 ♘c4

14. ♗h7! ♔h7 15. ♘g5 ♔g6[1] 16. ♕g4 f5 17. ♕g3 ♕d7[2] 18. ♘e6 ♔f7 19. ♕g7 ♔e6 20. ♘d4#

[1] 15. – ♔h6 16. ♕g4 +–; 15. – ♔g8 16. ♕h5 +–
[2] 17. – ♕c8 18. ♘c7 +–

5024. C 18
Steinshouer – Martinez
Colorado 1970

1. e4 e6 2. d4 d5 3. ♘c3 ♗b4 4. e5 c5 5. a3 ♗c3 6. bc3 ♘e7 7. ♕g4 0-0 8. ♗g5 ♕c7 9. ♗d3 cd4 10. ♘e2 dc3 11. h4 ♘f5 12. ♗f6 ♔h8 13. ♗f5 gf6

14. ♗h7! ♕e5[1] 15. f4 ♕e3 16. ♖h3 ♕h3[2] 17. gh3 ♔h7 18. ♔f2 ♖g8 19. ♕h5 ♔g7 20. ♖g1 ♔f8 21. ♕h6 +–

[1] 14. – ♔h7 15. ef6 ♖g8 16. ♕h5#
[2] 16. – ♕d2 17. ♔f2 ♔h7 18. ♕h5 ♔g7 19. ♖g3#

5025. C 19
Strucic – Janzek
Radovljca 1958

1. e4 e6 2. d4 d5 3. ♘c3 ♗b4 4. e5 c5 5. a3 ♗c3 6. bc3 ♕c7 7. ♘f3 ♘e7 8. ♗d3 ♘c6 9. 0-0 0-0

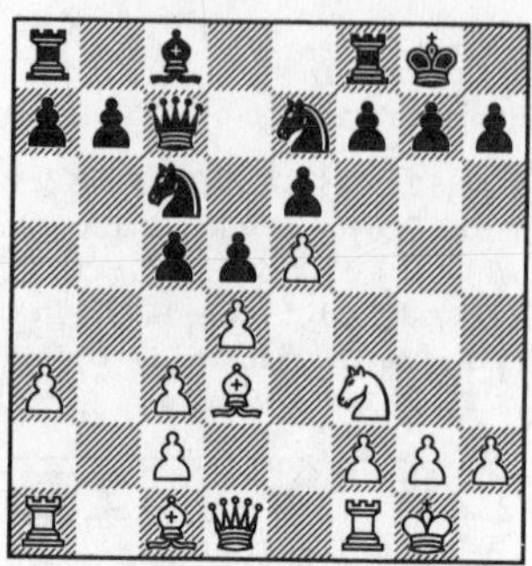

10. ♗h7! ♔h7 11. ♘g5 ♔g6[1] 12. ♕g4 f5 13. ♕h4 f4 14. ♕h7! ♔g5 15. h4 ♔g4 16. f3 ♔g3 17. ♗d2 +–

[1] 11. – ♔g8 12. ♕h5 ♖d8 13. ♕f7 ♔h8 14. f4 ♕d7 15. ♖f3 ♘g8 16. ♕g6 +–

5026. C 21
Hodurski – Rückert
Corr. 1951

1. e4 e5 2. d4 ed4 3. c3 dc3 4. ♘c3 d6 5. ♗c4 ♘c6 6. ♘f3 ♗e6 7. ♗e6 fe6 8. ♕b3 ♕c8 9. ♘g5 ♘d8 10. f4 c6 11. 0-0 d5 12. f5 ♗c5 13. ♔h1 ef5 14. ed5 cd5 15. ♘d5 ♕c6 16. ♗f4 ♖c8 17. ♖fe1 ♔f8 18. ♖ad1 h6

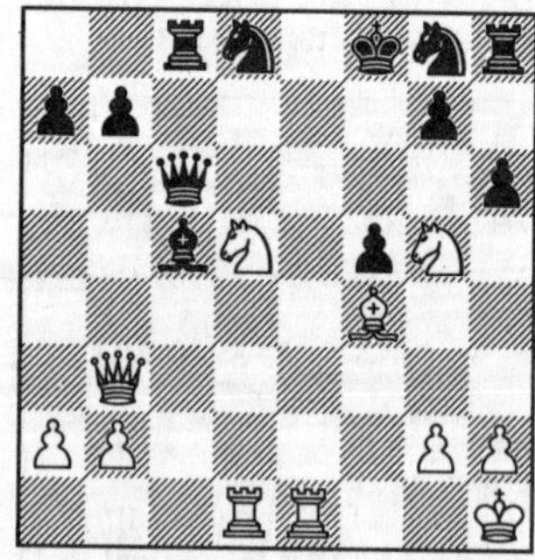

19. ♘e7! ♕f6[1] 20. ♘h7! +–

[1] 19. – ♗e7 20. ♖d8 +–

5027. C 21

Lasker – Breyer

Budapest 1911

1. e4 e5 2. d4 ed4 3. c3 d5 4. ed5 ♛d5 5. cd4 ♞f6 6. ♘f3 ♝b4 7. ♘c3 0-0 8. ♗e2 ♞e4 9. ♗d2 ♝c3 10. bc3 ♞c6 11. 0-0 ♛a5 12. ♕c2 ♝f5 13. ♗d3 ♜fe8 14. ♗e1 ♛d5 15. c4 ♛d6 16. d5 ♞e5 17. ♗e4 ♞f3 18. gf3 ♛g6 19. ♔h1 ♜e4 20. ♕c3 ♜h4 21. ♖g1

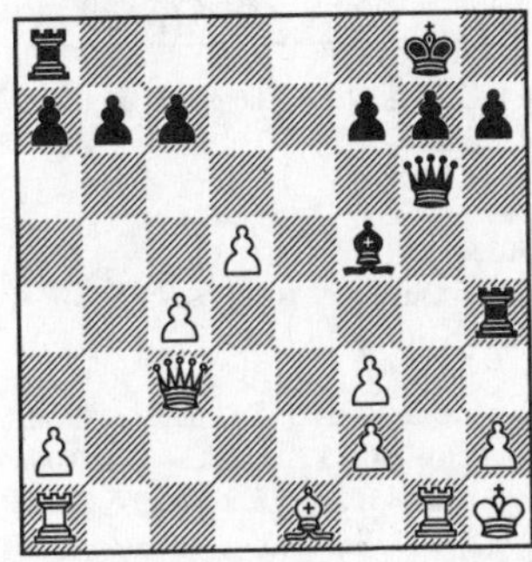

21. – ♜h2! 22. ♔h2 ♛h5 23. ♔g3 ♛g5 24. ♔h2 ♛h4 25. ♔g2 ♛h3#

5028. C 25

Alexandrov – N. N.

Moscow 1895

1. e4 e5 2. ♘c3 ♞c6 3. f4 ef4 4. ♘f3 ♝c5 5. d4 ♝b4 6. ♗f4 ♞f6 7. ♗d3 0-0 8. 0-0 d5 9. e5 ♞d7

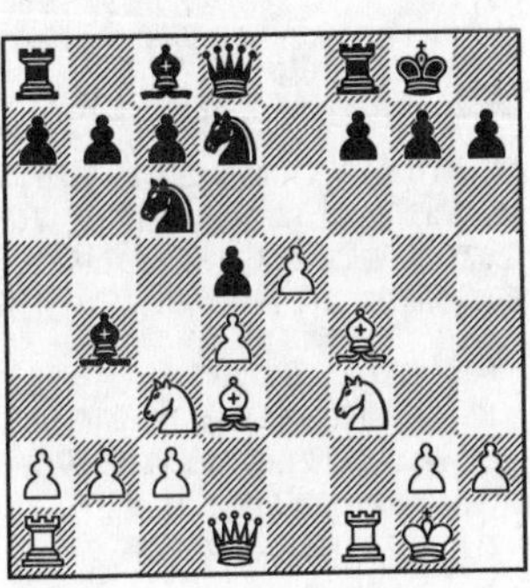

10. ♗h7! ♚h7 11. ♘g5 ♚g6 12. ♕d3 f5 13. ef6 ♚f6 14. ♘d5#

5029. C 29

Beckers – Denckens

Corr. 1959/60

1. e4 e5 2. ♘c3 ♞f6 3. f4 d5 4. fe5 ♞e4 5. d3 ♞c3 6. bc3 d4 7. ♘f3 c5 8. ♗e2 ♝e7 9. 0-0 0-0 10.♕e1 f6 11. ♕g3 fe5 12. ♗h6 ♝f6 13. ♘g5 ♛e7 14. ♗f3 ♚h8

15. ♘h7![1] gh6 16. ♘f8 ♛f8 17. ♗d5 ♞d7 18. ♕g6 ♛g7 19. ♕e8 ♞f8 20. ♖f6! ♛f6 21. ♖f1 +−

[1] 15. – ♚h7 16. ♗e4 ♚g8 17. ♕g6 ♞d7 18. ♗g5! ♝g5 19. ♕h7#; 18. – ♞b6 19. ♗f6 ♜f6 20. ♖f6 ♛f6 21. ♕e8 ♛f8 22. ♗h7 +−; 18. – ♛e6 19. c4 ♛g4 20. ♕h7 ♚f7 21. ♗g6 ♚e7 22. ♕g7 +−

5030. C 30

Magee – Aker

Corr. 1981

1. e4 e5 2. f4 ♝c5 3. ♘f3 d6 4. ♘c3 ♞f6 5. ♗c4 ♞c6 6. d3 ♞g4 7. ♘g5 0-0 8. f5 ♞f6 9. ♘d5 h6 10. ♘f6 ♛f6 11. h4 ♞a5

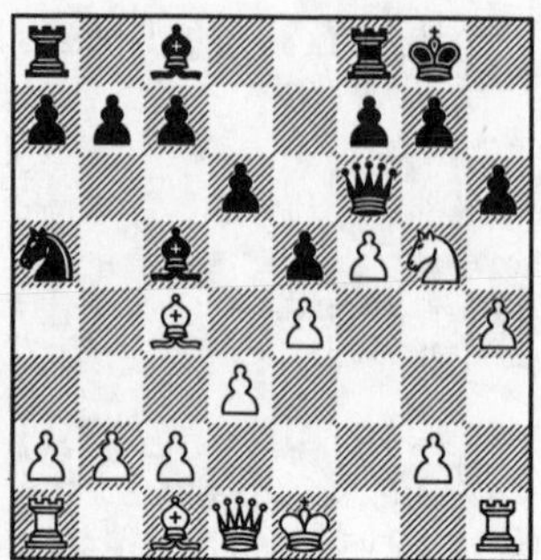

12. ♘h7! ♔h7 13. ♗g5! +–

5031. C 30

Mason – N. N.
London 1900

1. e4 e5 2. f4 ♗c5 3. ♘f3 d6 4. c3 ♗g4 5. ♗c4 ♘c6 6. d4 ed4 7. 0-0 dc3 8. ♔h1 ♘d4 9. ♘c3 ♗f3 10. gf3 ♘e7 11. ♗e3 ♘e6 12. f5 ♗e3 13. fe6 0-0 14. ef7 ♔h8 15. f4 ♘g6 16. ♘d5 ♗c5 17. b4 ♗b6 18. f5 ♘e5 19. ♕h5 ♕d7 20. ♖f4 ♕f7

21. ♕h7! ♔h7 22. ♖h4 ♔g8 23. ♘e7#

5032. C 30

Neumann – Dufresne
1863

1. e4 e5 2. f4 ♗c5 3. ♘f3 d6 4. ♗c4 ♘f6 5. ♘c3 0-0 6. d3 ♘g4 7. ♖f1 ♘h2 8. ♖h1 ♘g4 9. ♕e2 ♗f2 10. ♔f1 ♘c6 11. f5 ♗c5 12. ♘g5 ♘h6 13. ♕h5 ♕e8

14. ♘h7! ♔h7 15. ♗h6 g6 16. ♕g6! fg6 17. ♗f8#

5033. C 36

Réti – Duras
Abbazia 1912

1. e4 e5 2. f4 ef4 3. ♘f3 d5 4. ed5 ♘f6 5. c4 c6 6. d4 cd5 7. ♗f4 ♗b4 8. ♘c3 0-0 9. ♗d3 ♖e8 10. ♗e5 ♗e6 11. c5 ♘g4 12. 0-0 ♗c3 13. bc3 ♘e3

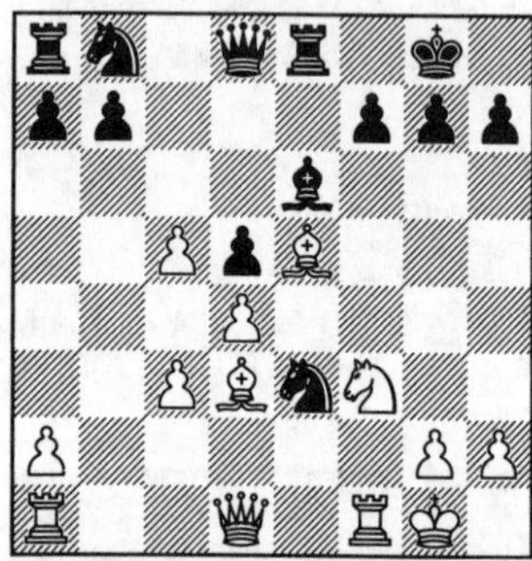

14. ♗h7! ♔h8[1] 15. ♕d2 ♘f1 16. ♕h6 f6 17. ♕h5 ♗g4[2] 18. ♕g4 ♔h7 19. ♖f1 ♘d7[3] 20. ♕h5 ♔g8 21. ♘g5 ♘f8[4] 22. ♕f7 ♔h8 23. ♖f4 +–

[1] 14. – ♔h7 15. ♕d3 ♘f5 16. g4 +–
[2] 17. – fe5 18. ♗g6 ♔g8 19. ♕h7 ♔f8 20. ♕h8 ♔e7 21. ♕g7 ♗f7 22. ♕f7#
[3] 19. – fe5 20. ♕h5 ♔g8 21. ♘g5 +–
[4] 21. – fg5 22. ♕f7 ♔h7 23. ♕g7#

5034. C 38

Cazenove – Tomalin

1818

1. e4 e5 2. f4 ef4 3. ♘f3 g5 4. ♗c4 ♗g7 5. d4 ♕e7 6. 0-0 h6 7. ♘c3 c6 8. e5 ♕b4 9. ♘e4 ♗f8 10. ♕e2 g4 11. ♘d6 ♗d6 12. ed6 ♔d8 13. ♘e5 ♖h7 14. c3 f3 15. ♕e4 ♘f6

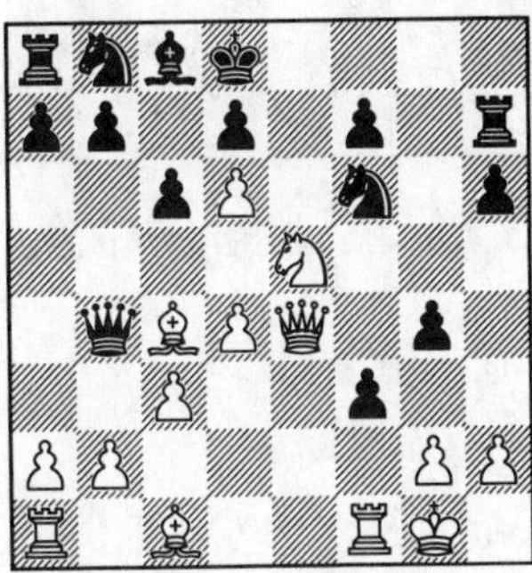

16. ♕h7! ♘h7 17. ♘f7[1] +–

[1] 17. – ♔e8 18. ♖e1 ♔f8 19. ♗h6 ♔g8 20. ♖e8 ♘f8 21. ♖f8 ♔h7 22. ♗d3#

5035. C 38

Mayet – Hirschfeld

Berlin 1861

1. e4 e5 2. f4 ef4 3. ♘f3 g5 4. ♗c4 ♗g7 5. d4 d6 6. c3 g4 7. ♕b3 gf3 8. ♗f7 ♔f8 9. ♗g8 ♖g8 10. 0-0 ♗d4 11. cd4 ♖g2 12. ♔h1

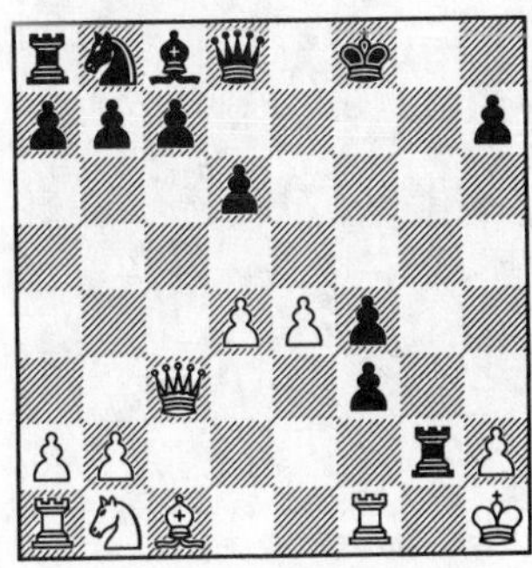

12. – ♖h2! 13. ♔h2 ♕h4 14. ♔g1 ♕g3 15. ♔h1 ♕g2#

5036. C 44

Rainer – Steinitz

Viden 1860

1. e4 e5 2. ♘f3 ♘c6 3. d4 ed4 4. ♗c4 ♗c5 5. 0-0 d6 6. c3 ♗g4 7. ♕b3 ♗f3 8. ♗f7 ♔f8 9. ♗g8 ♖g8 10. gf3 g5 11. ♕e6 ♘e5 12. ♕f5 ♔g7 13. ♔h1 ♔h8 14. ♖g1 g4 15. f4 ♘f3 16. ♖g4

16. – ♕h4! 17. ♖g2 ♕h2! 18. ♖h2 ♖g1#

5037. C 45

Martin – Kozma

Oslo 1954

1. e4 e5 2. ♘f3 ♘c6 3. d4 ed4 4. ♘d4 ♗c5 5. ♘b3 ♗b6 6. a4 a6 7. a5 ♗a7 8. ♘c3 ♘f6 9. ♗g5 h6 10. ♗h4 d6 11. ♗e2 ♗e6 12. 0-0 g5 13. ♗g3 h5 14. h4 ♘g4 15. hg? ♕g5 16. ♕d2 ♘e3 17. ♗f3 h4 18. ♗h2 h3 19. ♘d1 hg2 20. ♖e1

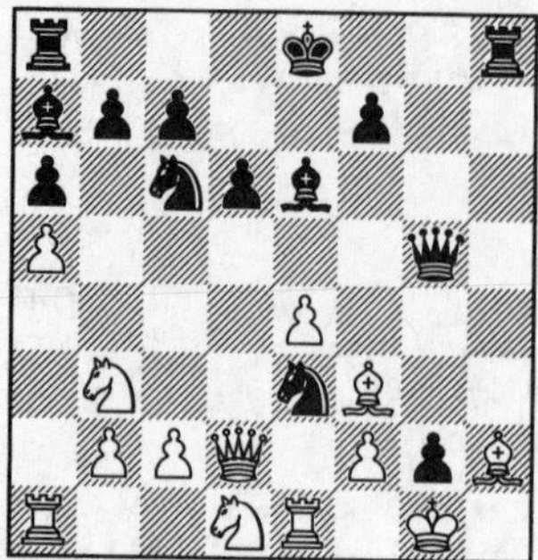

20. – ♜h2![1] –+

[1] 21. ♔h2 ♛h4 22. ♔g1 ♛h1#

5038. C 48

Belitzmann – Rubinstein

Warsaw 1917

1. e4 e5 2. ♘f3 ♘c6 3. ♘c3 ♘f6 4. ♗b5 ♘d4 5. ♗c4 ♗c5 6. ♘e5 ♛e7 7. ♘d3 d5 8. ♘d5 ♛e4 9. ♘e3 ♗d6 10. 0-0 b5 11. ♗b3 ♗b7 12. ♘e1 ♛h4 13. g3 ♛h3 14. c3 h5 15. cd4 h4 16. ♕e2

16. – ♛h2! 17. ♔h2 hg3 18. ♔g1 ♜h1#

5039. C 53

Kan – N. N.

Moscow 1914

1. e4 e5 2. ♘f3 ♘c6 3. ♗c4 ♗c5 4. c3 ♛e7 5. d4 ed4 6. 0-0 dc3 7. ♘c3 d6 8. ♘d5 ♛d8 9. b4 ♗b6 10. ♗b2 ♘f6 11. ♘f6 gf6 12. ♘g5 0-0

13. ♘h7! ♔h7 14. ♕h5 ♔g7[1] 15. ♕g5 +–

[1] 14. – ♔g8 15. ♕g6 +–

5040. C 53

Maczynsky – Pratten

Portsmouth 1948

1. e4 e5 2. ♘f3 ♘c6 3. ♗c4 ♗c5 4. c3 ♛e7 5. 0-0 d6 6. d4 ♗b6 7. b4 ♗g4 8. a4 a5 9. b5 ♘d8 10. ♗a3 f6 11. ♖a2 ♘e6 12. de5 fe5 13. ♕d5 ♗f3 14. ♕b7 ♛g5 15. ♕a8 ♔e7 16. g3 ♘f4 17. ♖e1 ♛h5 18. ♘d2 ♘f6 19. ♕h8

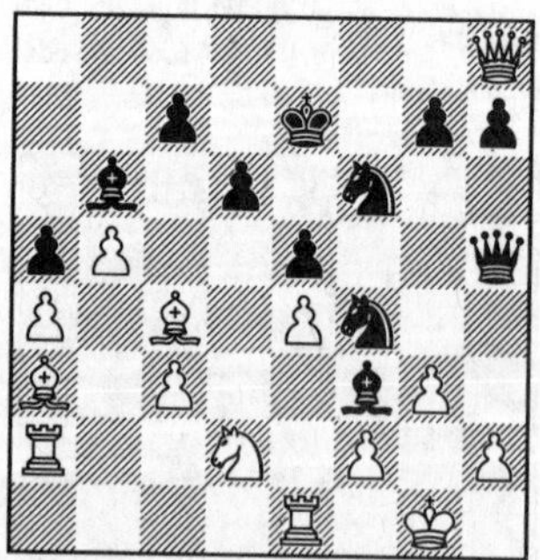

19. – ♛h2! 20. ♔h2 ♘g4 21. ♔g1 ♘h3 22. ♔f1 ♘h2#

5041. C 57

Silberstein – Grossman

USA 1949

1. e4 e5 2. ♘f3 ♘c6 3. ♗c4 ♘f6 4. ♘g5 d5 5. ed5 b5 6. ♗b5 ♕d5 7. ♗c6 ♕c6 8. 0-0 ♗b7 9. ♘f3 0-0-0 10. d3 ♗d6 11. ♗g5 e4 12. ♗f6 ef3 13. ♗h4

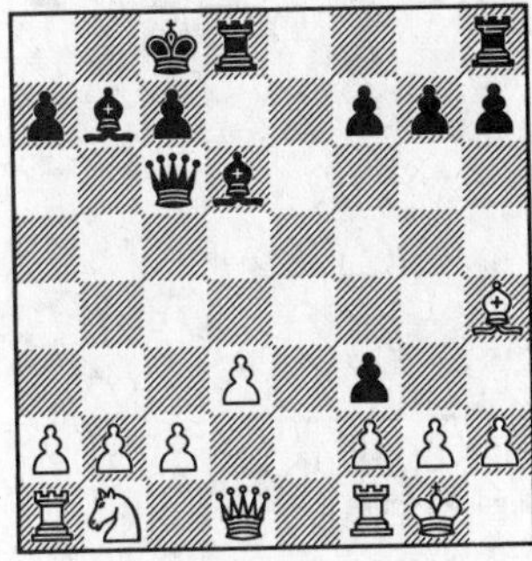

13. – ♗h2! 14. ♔h2 ♖d4 15. g3[1] ♕h6 16. ♖h1 ♖h4 17. gh4 ♕h4 18. ♔g1 ♕g4 19. ♔f1 ♕g2 20. ♔e1 ♕h1 21. ♔d2 ♕h6 22. ♔c3 ♖e8 23. ♘d2 ♖e2 –+

[1] 15. ♗g3 ♕h6 16. ♔g1 fg2 –+; 15. ♗g5 ♕g6 –+

5042. C 58

Fütterer–Blind – Zinkl

Berlin 1897

1. e4 e5 2. ♘f3 ♘c6 3. ♗c4 ♘f6 4. ♘g5 d5 5. ed5 ♘a5 6. d3 h6 7. ♘f3 ♗d6 8. c3 0-0 9. b4 ♘c4 10. dc4 c5 11. a3 e4 12. ♘fd2 ♖e8 13. 0-0

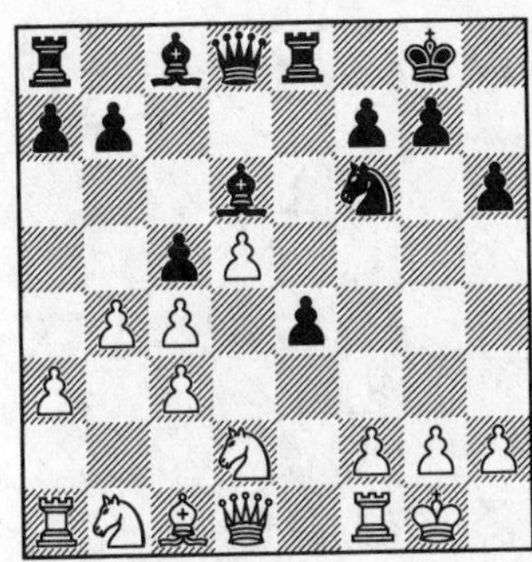

13. – ♗h2! 14. ♔h2 ♘g4 15. ♔g3 ♕d6 16. f4 ef3 17. ♔f3 ♖e3#

5043. C 61

Pencil – Goltsoff

Dayton 1976

1. e4 e5 2. ♘f3 ♘c6 3. ♗b5 ♘d4 4. ♘d4 ed4 5. 0-0 c6 6. ♗a4 ♗c5 7. d3 ♘e7 8. ♗b3 0-0 9. f4 ♔h8 10. f5 f6 11. ♕h5 d6 12. ♖f3 ♕e8

13. ♕h7! +–

5044. C 63

Stern – Roehll

Columbus Ohio 1979

1. e4 e5 2. ♘f3 ♘c6 3. ♗b5 f5 4. ef5 e4 5. ♗c6 dc6 6. ♘e5 ♗f5 7. 0-0 ♘f6 8. ♘c4 ♗d6 9. ♘c3

9. – ♗h2! 10. ♔h2 ♘g4 11. ♔g3 ♕g5 12. a3 ♘e3 –+

5045. C 63

Trifunovic – Kostic

Yugoslavia 1937

1. e4 e5 2. ♘f3 ♘c6 3. ♗b5 f5 4. ♘c3 fe4 5. ♘e4 ♗e7 6. d4 ed4 7. 0-0 ♘f6 8. ♘f6 ♗f6 9. ♖e1 ♘e7 10. ♘g5 0-0

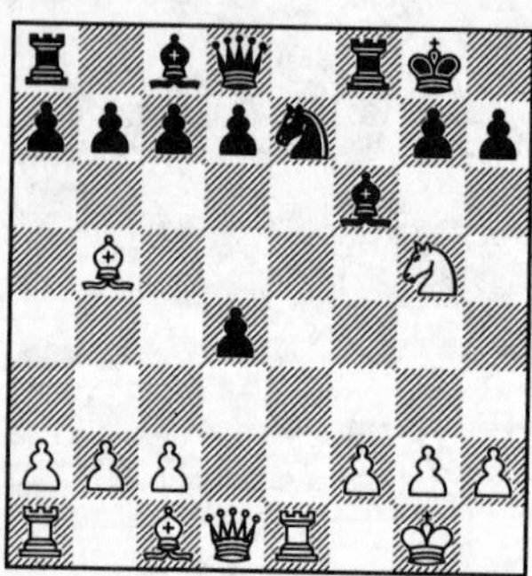

11. ♘h7! ♔h7 12. ♕h5 ♔g8 13. ♗d3 ♖e8[1] 14. g4 d6 15. g5 ♗e5 16. ♗h7 ♔f8 17. ♕f3 +–

[1] 13. – ♕e8 14. ♗h7 ♔h8 15. ♗g6 ♔g8 16. ♕h7#; 13. – g6 14. ♗g6 ♘g6 15. ♕g6 ♗g7 16. ♗g5 +–

5046. C 67

Cole – Ward

London 1898

1. e4 e5 2. ♘f3 ♘c6 3. ♗b5 ♘f6 4. 0-0 ♘e4 5. ♖e1 ♘d6 6. ♘e5 ♘e5 7. ♖e5 ♗e7 8. ♗d3 0-0 9. ♘c3 ♘e8 10. b3 c6 11. ♗b2 d5 12. ♕h5 g6

13. ♘d5! gh5[1] 14. ♗h7! ♔h7 15. ♖h5 ♔g6 16. ♘f4#

[1] 13. – cd5 14. ♕h7! ♔h7 15. ♖h5 ♔g8 16. ♖h8#

5047. C 78

Bender – Nalborczyk

Indiana 1977

1. e4 e5 2. ♘f3 ♘c6 3. ♗b5 a6 4. ♗a4 ♘f6 5. 0-0 b5 6. ♗b3 ♗e7 7. ♖e1 0-0 8. c3 d6 9. h3 ♗d7 10. d4 ed4 11. cd4 ♕c8 12. ♘c3 ♘a5 13. ♗c2 c5 14. e5 ♘e8 15. ♘d5 ♗d8 16. ♘g5 ♗g5 17. ♗g5 ♕b7

18. ♗h7![1] +–

[1] 18. – ♔h7 19. ♕h5 ♔g8 20. ♘e7#

5048. C 80

Bailey – Brown

Sydney 1951

1. e4 e5 2. ♘f3 ♘c6 3. ♗b5 a6 4. ♗a4 ♘f6 5. 0-0 ♘e4 6. ♖e1 ♘c5 7. ♘c3 ♘a4 8. ♘e5 ♘e5 9. ♖e5 ♗e7 10. ♘d5 0-0 11. ♘e7 ♔h8 12. ♕h5 d6

13. ♕h7! +–

5049. C 84

Fink – Dickman

Columbus Ohio 1973

1. e4 e5 2. ♘f3 ♘c6 3. ♗b5 a6 4. ♗a4 ♘f6 5. d4 ed4 6. 0-0 ♗e7 7. ♕e2 0-0 8. e5 ♘d5 9. ♖d1 ♗c5 10. c3 b5 11. ♗c2 ♖e8 12. cd4 ♗f8

13. ♗h7! ♔h7 14. ♘g5 ♔g6 15. ♕e4 f5 16. ♕d5 ♕e7 17. ♕g8 ♘e5 18. de5 ♕e5 19. ♕f7 +–

5050. C 89

Kaplan – Palmert

Dayton 1977

1. e4 e5 2. ♘f3 ♘c6 3. ♗b5 a6 4. ♗a4 b5 5. ♗b3 ♘f6 6. 0-0 ♗e7 7. c3 0-0 8. ♖e1 d5 9. ed5 ♘d5 10. ♘e5 ♘e5 11. ♖e5 ♘f6 12. d4 ♗d6 13. ♖e1 ♘g4 14. g3

14. – ♘h2! 15. ♔h2 ♕h4 16. ♔g1 ♗g3 17. fg3 ♕g3 18. ♔h1 ♗g4 –+

5051. C 99

Leininger – Massum

Corr. 1960

1. e4 e5 2. ♘f3 ♘c6 3. ♗b5 a6 4. ♗a4 ♘f6 5. 0-0 ♗e7 6. ♖e1 b5 7. ♗b3 d6 8. c3 0-0 9. h3 ♘a5 10. ♗c2 c5 11. d4 ♕c7 12. ♘bd2 cd4 13. cd4 ♗b7 14. ♘f1 ♖ac8 15. ♗b1 d5 16. ed5 ed4 17. ♗g5 ♘d5 18. ♗e7 ♘e7

19. ♗h7! ♔h8[1] 20. ♕d4 ♗f3 21. ♕h4 +–

[1] 19. – ♔h7 20. ♘g5 ♔g6 21. ♕g4 f5 22. ♕h4 ♗d5 23. ♕h7 ♔g5 24. ♕g7 +–; 23. – ♔f6 24. ♘g3 ♖h8 25. ♘h5 ♔g5 26. f4 ♔h4 27. ♕g7 ♖h5 28. ♔h2 +–; 22. – ♖f6 23. ♕h7 ♔g5 24. f4 ♔f4 25. ♕h4#

5052. D 00

Santasiere – Adams

New York 1926

1. d4 d5 2. e3 e6 3. ♗d3 ♘f6 4. ♘d2 ♗d6 5. f4 ♘c6 6. c3 ♘e7 7.♘h3 0-0 8. 0-0 ♗d7 9. e4 de4 10. ♘e4 ♘g6 11. ♘f6 gf6 12. f5 ef5 13. ♗f5 ♗f5 14. ♖f5 ♔h8 15. ♕h5 ♖g8

16. ♕h7! ♔h7 17. ♖h5 ♔g7 18. ♗h6 ♔h7 19. ♗f8#

5053. D 05

Colle – Burger

Hastings 1928

1. d4 ♘f6 2. ♘f3 d5 3. e3 e6 4. ♗d3 ♗e7 5. ♘bd2 0-0 6. 0-0 ♘bd7 7. e4 de4 8. ♘e4 ♘e4 9. ♗e4 ♘f6 10. ♗d3 c5 11. dc5 ♗c5 12. ♗g5 ♗e7 13. ♕e2 ♕c7 14. ♖ad1 ♖d8 15. ♘e5 ♗d7

16. ♗h7! ♔h7 17. ♗f6 ♗f6 18. ♕h5 ♔g8 19. ♕f7 ♔h7 20. ♖d3 +–

5054. D 29

Geller – Hermlin

1973

1. d4 d5 2. c4 dc4 3. ♘f3 ♘f6 4. e3 e6 5. ♗c4 c5 6. 0-0 a6 7. ♕e2 b5 8. ♗b3 ♗b7 9. ♖d1 ♘bd7 10. ♘c3 ♕b8 11. d5 ♘d5 12. ♘d5 ♗d5 13. ♗d5 ed5 14. ♖d5 ♗e7 15. e4 ♘b6 16. ♖h5 0-0 17. e5 ♖e8 18. e6 f6 19. ♘h4 ♕d6 20. ♘f5 ♕c7

21. ♖h7![1] +–

[1] 21. – ♔h7 22. ♕h5 ♔g8 23. ♕f7 ♔h7 24. ♕g7#

5055. D 37

Najdorf – N. N.

Rafaela 1942

1. d4 d5 2. c4 e6 3. ♘c3 ♘f6 4. ♘f3 a6 5. cd5 ♘d5 6. e4 ♘c3 7. bc3 ♗e7 8. ♗d3 0-0 9. 0-0 c5 10. ♕e2 cd4 11. cd4 ♖e8 12. ♖d1 ♘c6 13. ♗b2 b5 14. ♖ac1 ♗b7 15. d5 ed5 16. ed5 ♘b4 17. ♗e4 ♘a2 18. ♘e5 ♘b4

19. ♗h7! ♔f8[1] 20. ♕h5 ♗d5 21. ♖d5! ♕d5 22. ♕f7 ♕f7 23. ♘d7#

[1] 19. – ♔h7 20. ♕h5 ♔g8 21. ♕f7 ♔h7 22. ♘g4 ♗f8 23. ♘f6 +–

5056. D 38

Heller – Wechsler

Stuttgart

1. d4 d5 2. c4 e6 3. ♘c3 ♘f6 4. ♗f4 ♘c6 5. ♘f3 ♗b4 6. e3 0-0 7. ♖c1 ♘e4 8. ♗d3 ♘c3 9. bc3 dc4

10. ♗h7! ♔h7 11. ♘g5 ♔g6[1] 12. ♕g4 e5 13. ♘e6 ♔f6 14. ♗g5 ♔g6 15. ♗h4 ♔h6 16. ♕g7 ♔h5 17. ♕h7 ♔g4 18. h3#

[1] 11. – ♔g8 12. ♕h5 ♖e8 13. ♕f7 ♔h8 14. ♕h5 ♔g8 15. ♕h7 ♔f8 16. ♕h8 ♔e7 17. ♕g7#

5057. D 43

Collins – Hearst

New York 1949

1. d4 d5 2. ♘f3 e6 3. c4 f5 4. ♘c3 ♘f6 5. g3 c6 6. ♗g2 ♗d6 7. ♗f4 0-0 8. 0-0 ♘e4 9. ♕c2 ♗f4 10. gf4 ♘d7 11. ♔h1 ♖f6 12. e3 ♖h6 13. ♖g1 ♘df6 14. ♘e5 ♘g4 15. ♘g4 fg4 16. ♗e4

16. – ♖h2! 17. ♔h2 ♕h4 18. ♔g2 ♕h3#

5058. D 46
Rasmusson – Pulkkinen
Helsinki 1933

1. d4 d5 2. ♘f3 ♘f6 3. c4 c6 4. ♘c3 e6 5. e3 ♘bd7 6. ♗d3 ♗e7 7. 0-0 0-0 8. b3 b6 9. ♗b2 ♗b7 10. ♕e2 c5 11. ♘e5 ♕c7 12. f4 ♖ae8 13. ♘b5 ♕b8 14. ♘d7 ♘d7 15. dc5 bc5

16. ♗h7! ♔h7 17. ♕h5 ♔g8 18. ♗g7! ♔g7 19. ♕g4 ♔h8 20. ♖f3 ♘f6 21. ♖h3 ♘h7 22. ♕h5 +–

5059. D 47
Kóródy – Benkő
Budapest 1951

1. d4 d5 2. c4 c6 3. ♘f3 ♘f6 4. ♘c3 e6 5. e3 ♘bd7 6. ♗d3 dc4 7. ♗c4 b5 8. ♗d3 ♗b7 9. 0-0 b4 10. ♘e4 c5 11. ♘f6 gf6 12. ♕e2 ♕b6 13. a3 ♗d6 14. ab4 cd4 15. ed4 ♖g8 16. b5 ♕d4! 17. h3 ♘e5 18. ♘d4 ♖g2! 19. ♔h1

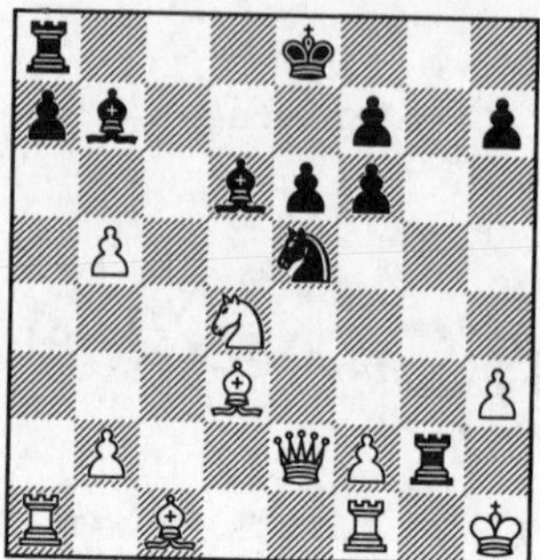

19. – ♖h2![1] –+

[1] 20. ♔h2 ♘g4 21. ♔g1 ♗h2#

5060. D 48
Kraizer – Dyment
Corr. 1952

1. d4 d5 2. c4 c6 3. ♘c3 ♘f6 4. e3 e6 5. ♘f3 ♘bd7 6. ♗d3 dc4 7. ♗c4 b5 8. ♗d3 a6 9. e4 ♗e7 10. e5 ♘d5 11. ♘d5 cd5 12. ♗d2 0-0 13. h4 f6

14. ♘g5! fg5 15. ♗h7! ♔h7 16. hg5 ♔g8[1] 17. ♖h8[2] +–

[1] 16. – ♔g6 17. ♕h5 ♔f5 18. ♕h7 g6 19. ♕h3 ♔e4 20. ♕e3 ♔f5 21. ♕f3#
[2] 17. – ♔h8 18. ♕h5 ♔g8 19. g6 ♖f5 20. ♕h7 ♔f8 21. ♕h8#; 17. – ♔f7 18. ♕h5 g6 19. ♕h7 ♔e8 20. ♕g6#

5061. D 66

Negrea – Cretzulescu

Bucharest 1960

1. d4 ♘f6 2. c4 e6 3. ♘c3 ♗b4 4. ♗g5 d5 5. e3 0-0 6. a3 ♗e7 7. ♘f3 ♘bd7 8. ♖c1 c6 9. ♗d3 dc4 10. ♗c4 b5 11. ♗d3 b4 12. ab4 ♗b4 13. 0-0 c5 14. ♘a2 ♕b6 15. ♘b4 ♕b4 16. dc5 ♘c5 17. ♖c4 ♕b6 18. ♗f6 gf6

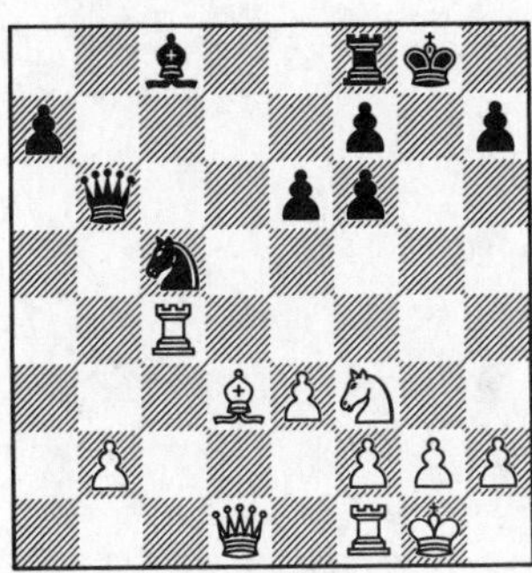

19. ♗h7! ♔h7 20. ♘e5[1] +–

[1] 20. – fe5 21. ♕h5 ♔g7 22. ♖g4 ♔f6 23. ♕g5#

5062. E 28

Lilienthal – Najdorf

Budapest 1950

1. d4 ♘f6 2. c4 e6 3. ♘c3 ♗b4 4. a3 ♗c3 5. bc3 c5 6. e3 ♘c6 7. ♗d3 b6 8. ♘e2 0-0 9. e4 ♘e8 10. 0-0 d6 11. e5 de5 12. de5 ♗b7 13. ♘f4 f5 14. ef6 e5 15. fg7 ♖f4 16. ♗f4 ef4

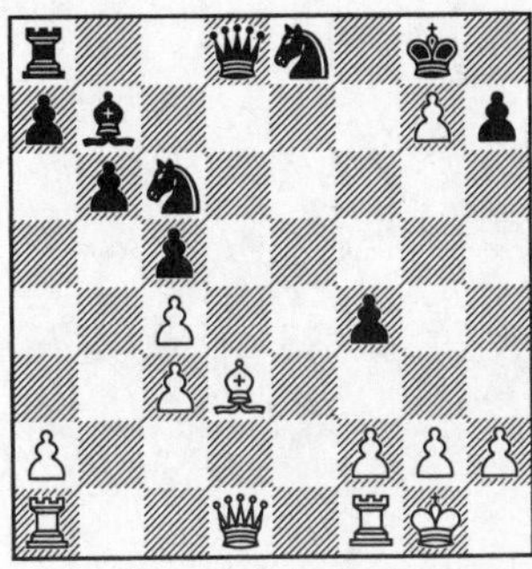

17. ♗h7! ♔h7 18. ♕h5 ♔g7 19. ♖ad1 ♕f6[1] 20. ♖d7 ♔f8 21. ♖b7 ♘d8 22. ♖d7 ♘f7 23. ♕d5 ♖b8 24. ♖e1 f3 25. ♖e3 +–

[1] 19. – ♕e7 20. ♕g4 ♔f8 21. ♖d7 +–

Simple endgames

(5063–5206)

5063.

5064.

5065.

5066.

5067.

5068.

5069.

5070.

5071.

5072.

5073.

5074.

5075.

5076.

5077.

5078.

5079.

5080.

5081.

5082.

5083.

5084.

5085.

5086.

5087.

5088.

5089.

5090.

5091.

5092.

5093.

5094.

5095.

5096.

5097.

5098.

5099.

5100.

5101.

5102.

5103.

5104.

5105.

5106.

5107.

5108.

5109.

5110.

5111.

5112.

5113.

5114.

5115.

5116.

5117.

5118.

5119.

5120.

5121.

5122.

5123.

5124.

5125.

5126.

5127.

5128.

5129.

5130.

5131.

5132.

5133.

5134.

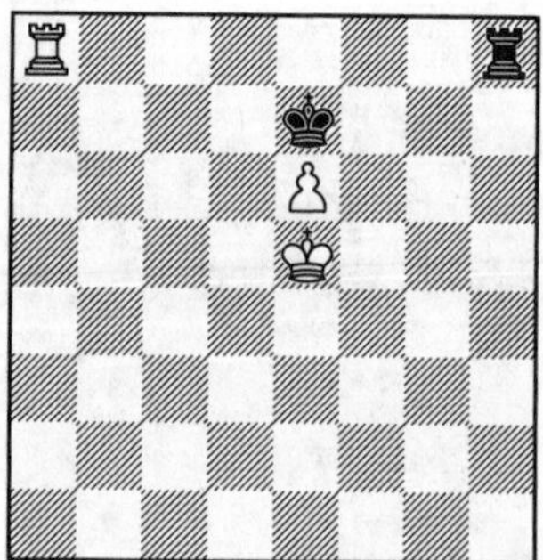

5135.

5136.

5137.

5138.

5139.

5140.

5141.

5142.

5143.

5144.

5145.

5146.

5147.

5148.

5149.

5150.

5151.

5152.

5153.

5154.

5155.

5156.

5157.

5158.

5159.

5160.

5161.

5162.

5163.

5164.

5165.

5166.

5167.

5168.

5169.

5170.

5171.

5172.

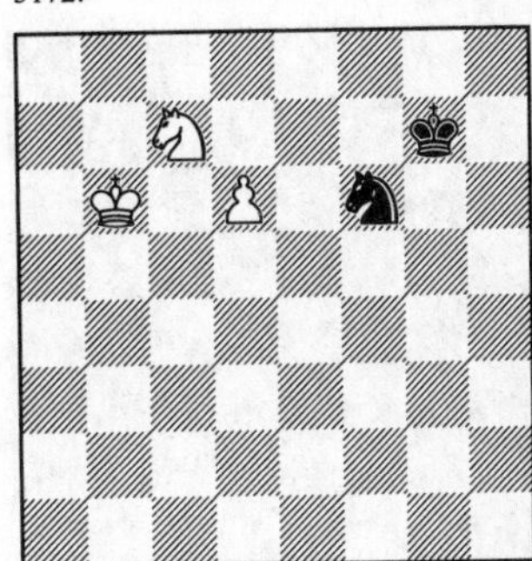

5173.

5174.

5175.

5176.

5177.

5178.

5179.

5180.

5181.

5182.

5183.

5184.

5185.

5186.

5187.

5188.

5189.

5190.

5191.

5192.

5193.

5194.

5195.

5196.

5197.

5198.

5199.

5200.

5201.

5202.

5203.

5204.

5205.

5206.

Polgár sisters' tournament-game combinations

(5207–5334)

5207.

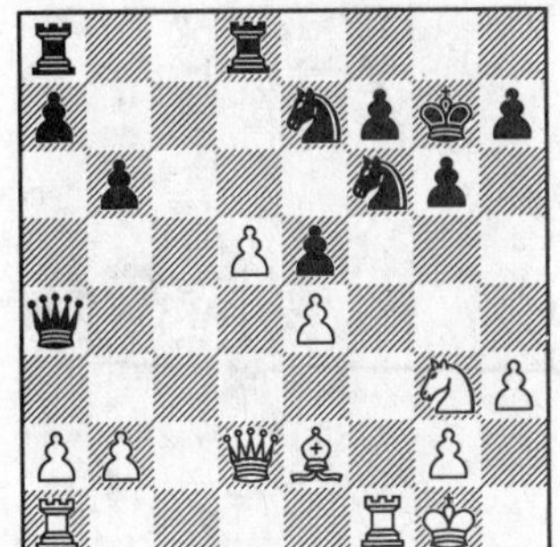

5208.

5209.

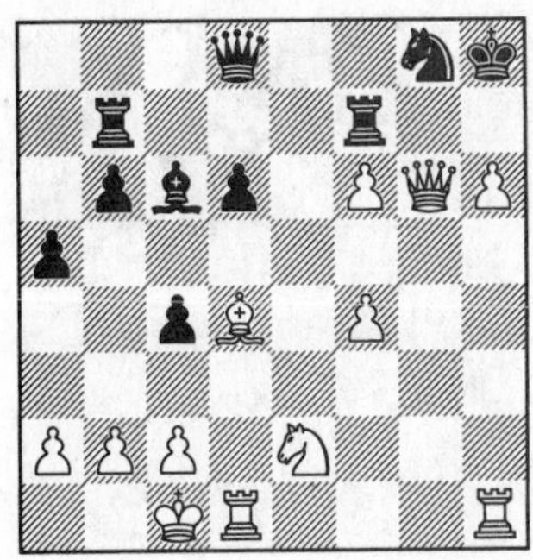

5210.

5211.

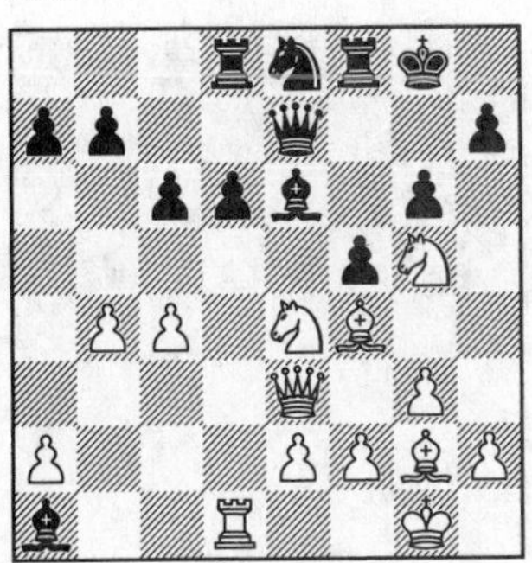

5212.

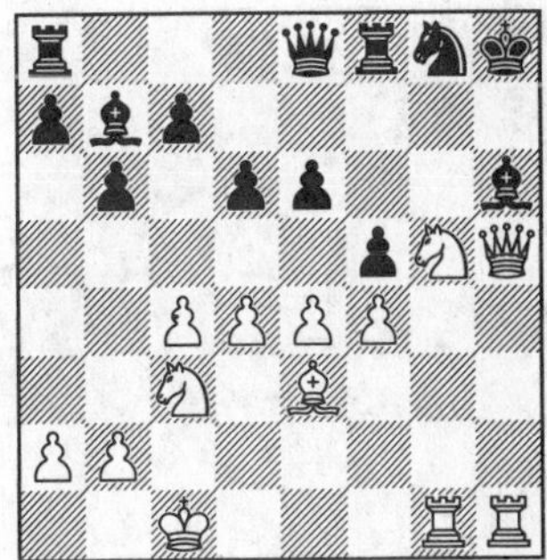

5213.

5214.

5215.

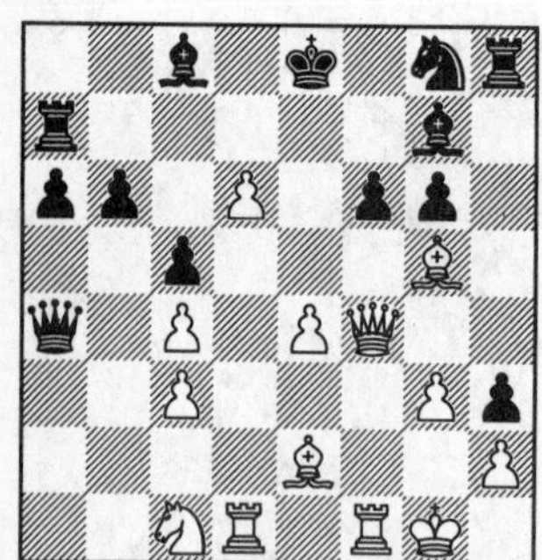

5216.

5217.

5218.

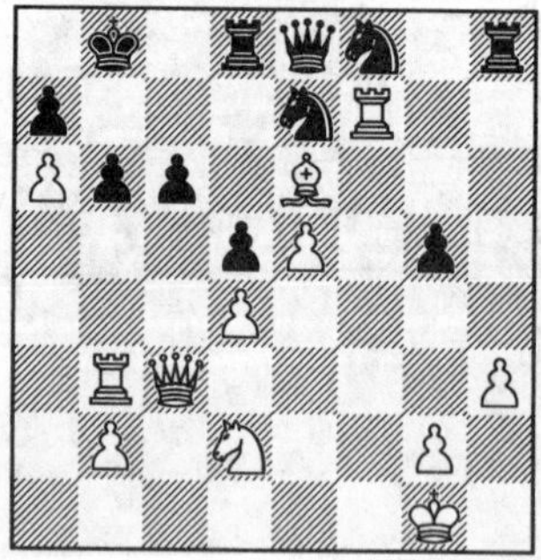

5219.

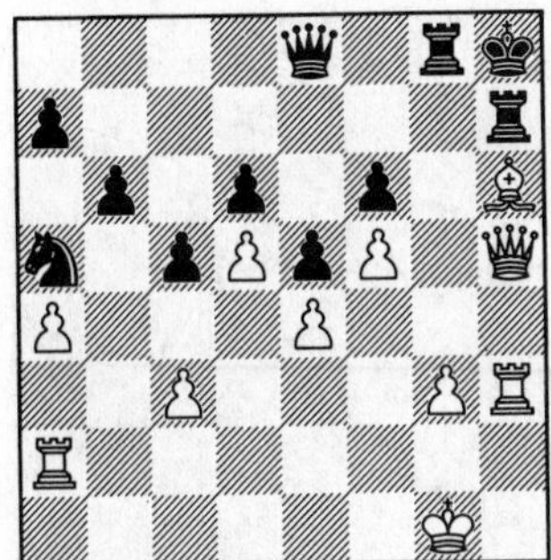

5220.

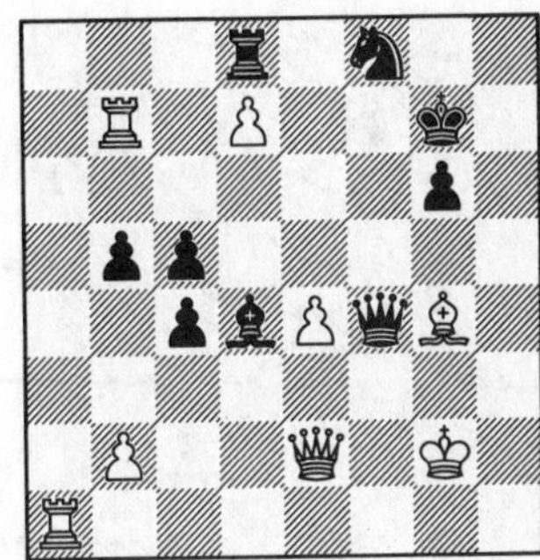

5221.

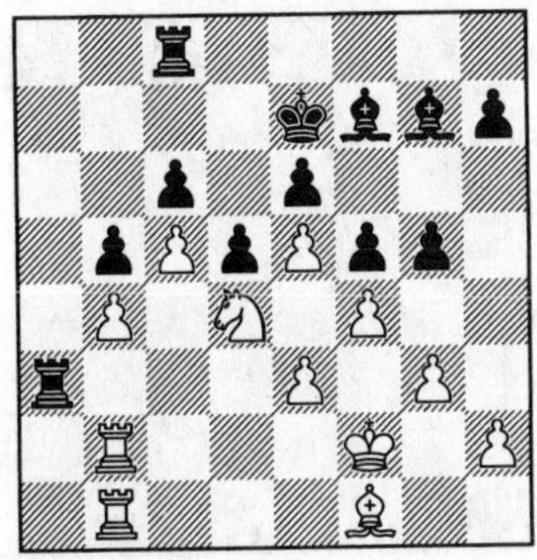

5222.

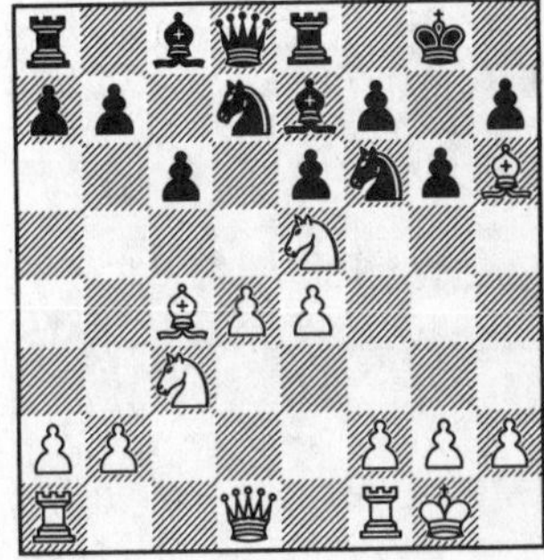

5223.

5224.

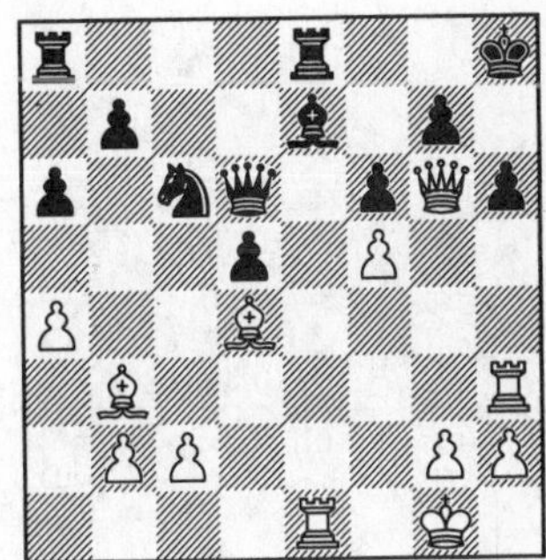

5225.

5226.

5227.

5228.

5229.

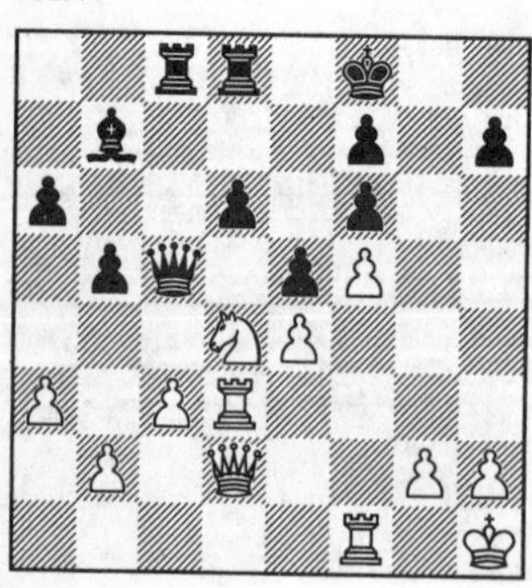

5230.

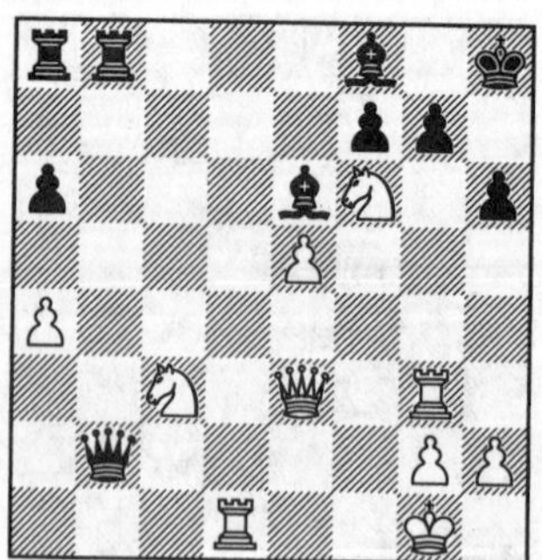

5231.

5232.

5233.

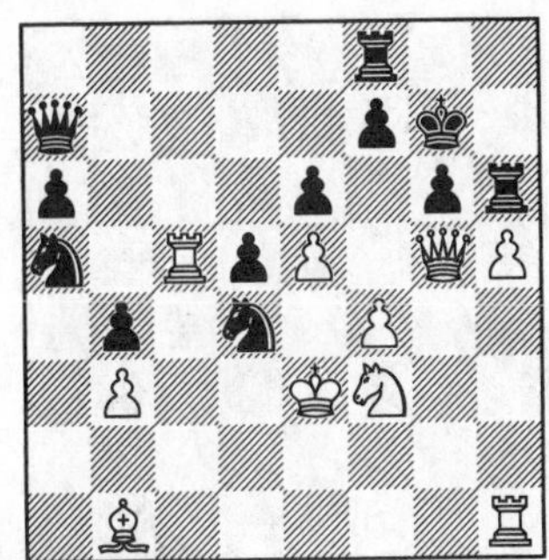

5234.

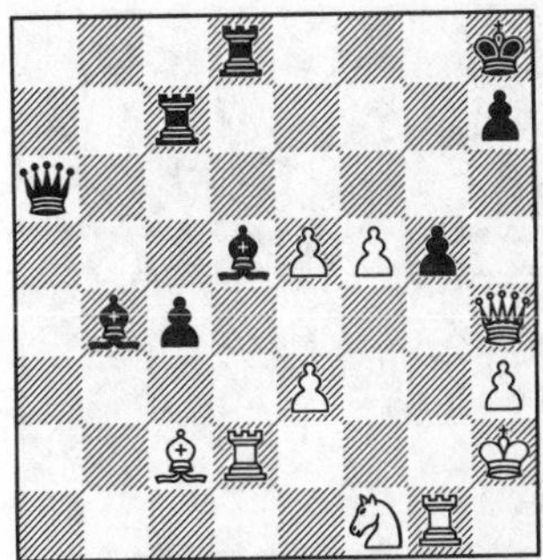

5235.

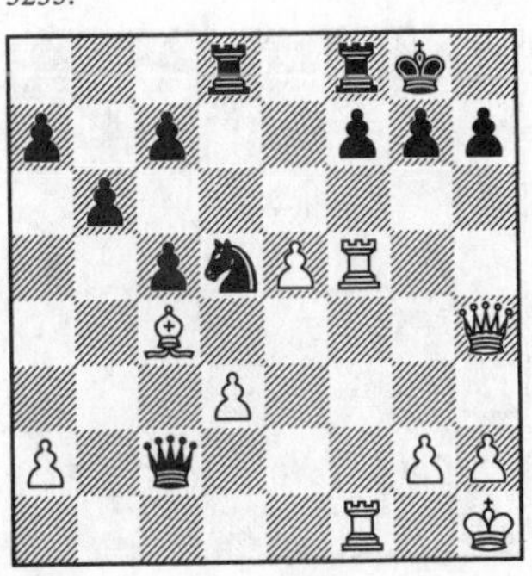

5236.

5237.

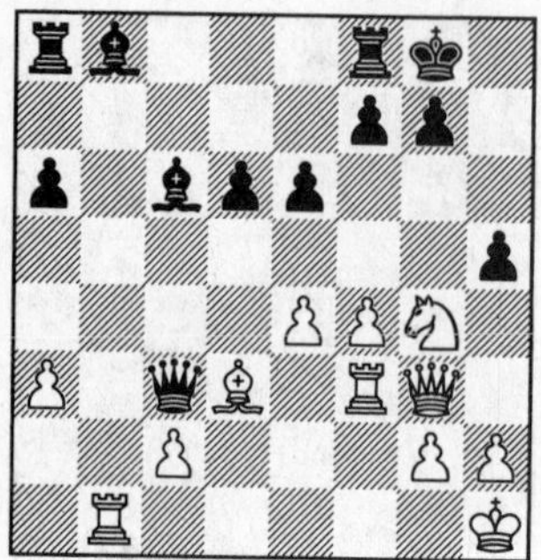

5238.

5239.

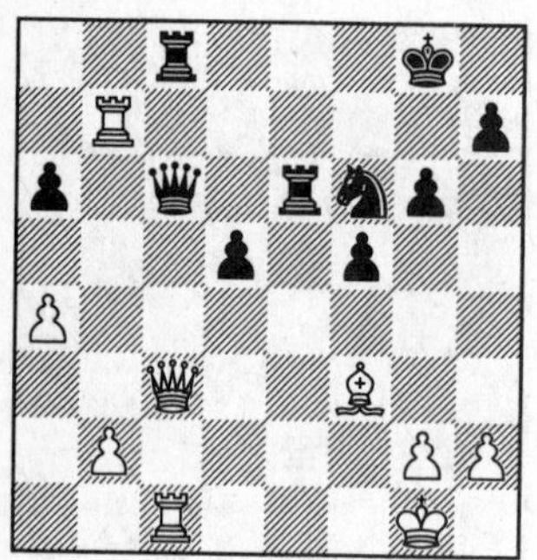

5240.

5241.

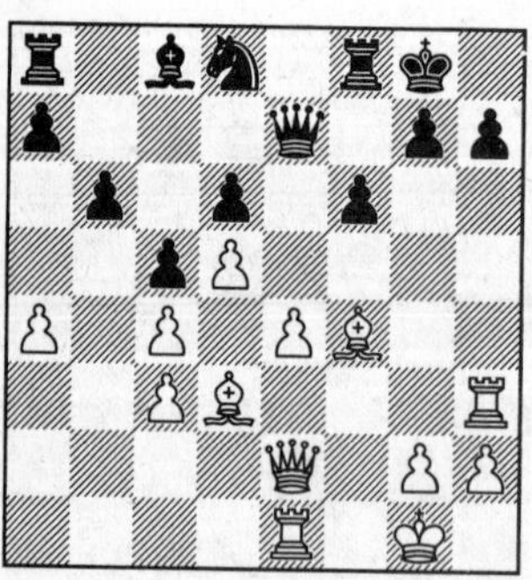

5242.

5243.

5244.

5245.

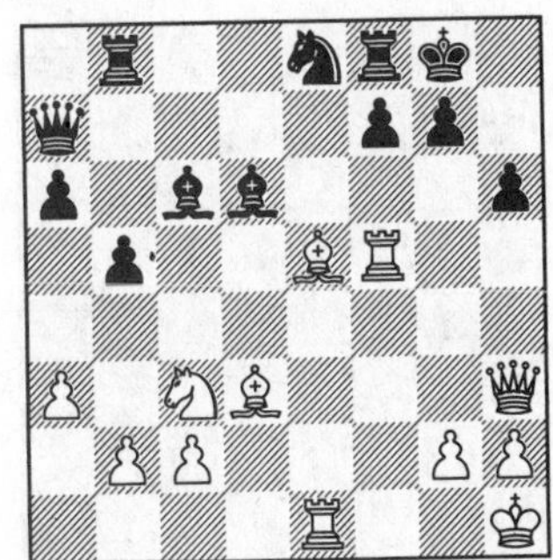

5246.

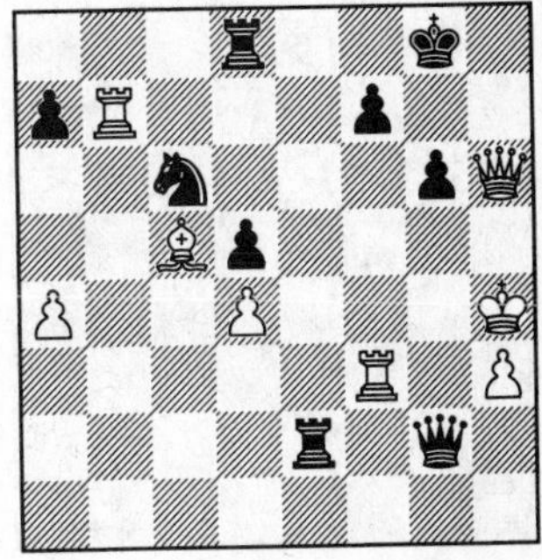

5247.

5248.

5249.

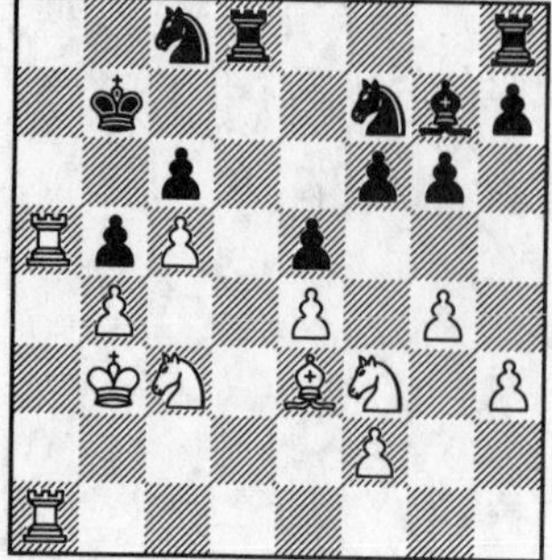

5250.

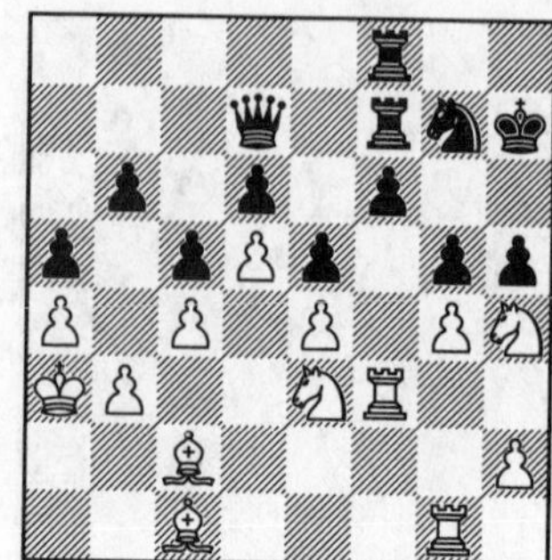

5251.

5252.

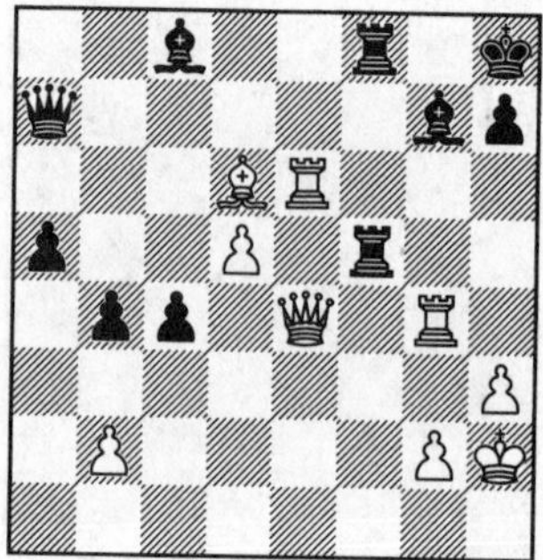

5253.

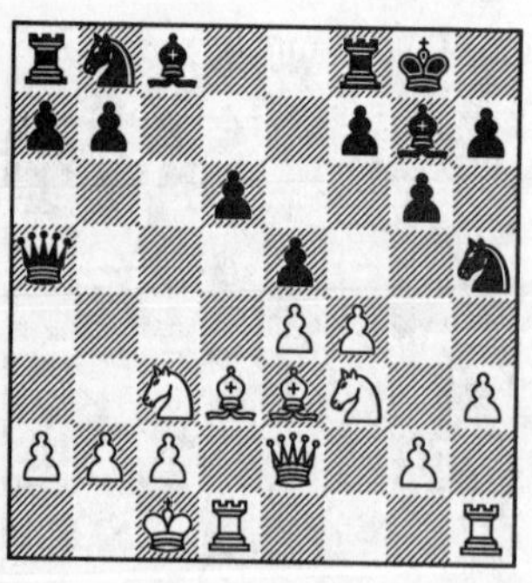

5254.

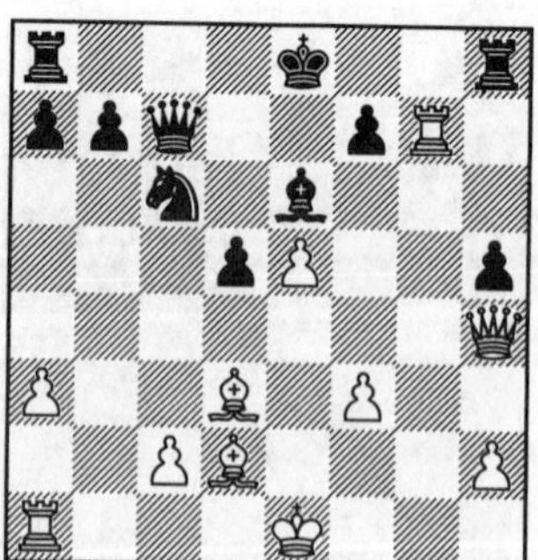

5255.

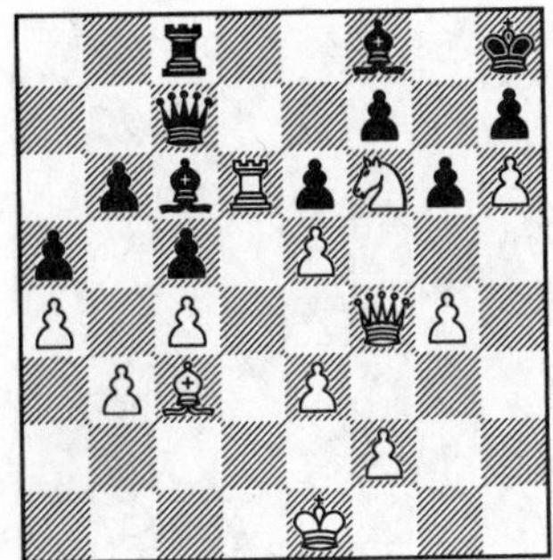

5256.

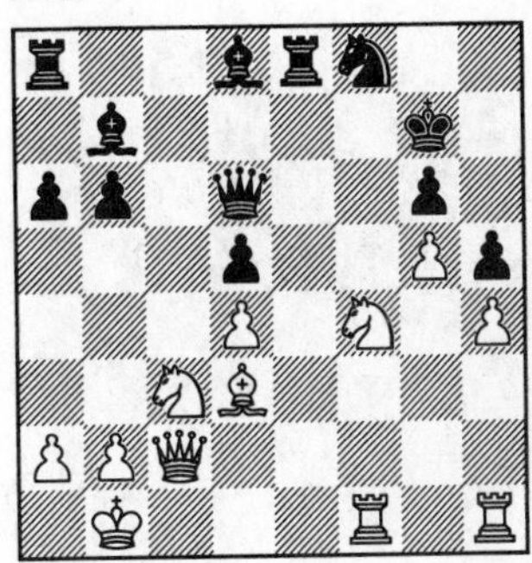

5257.

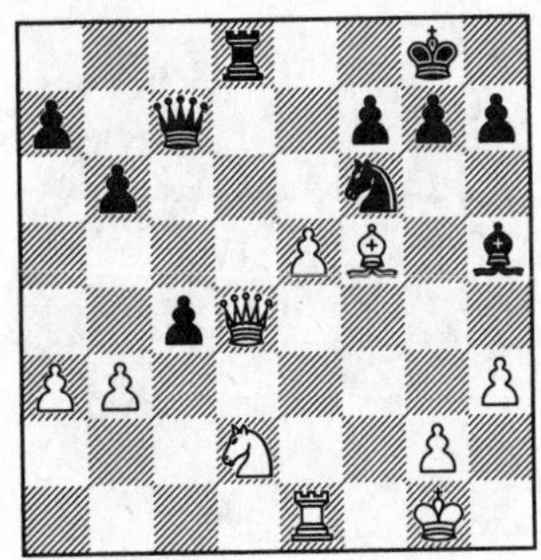

5258.

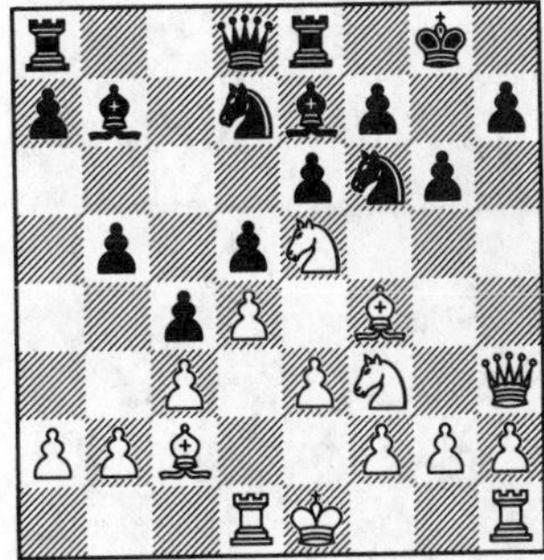

5259.

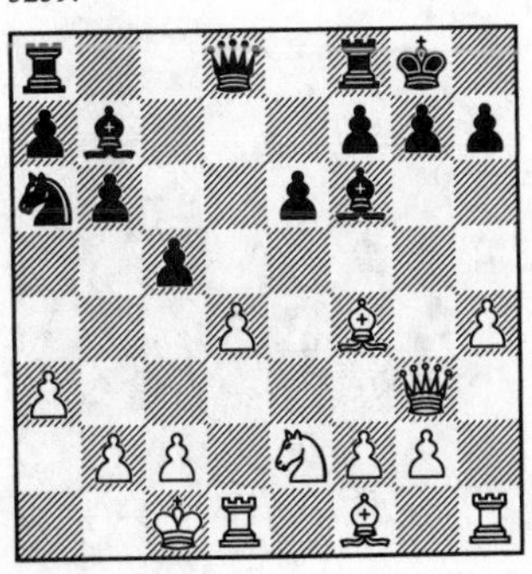

5260.

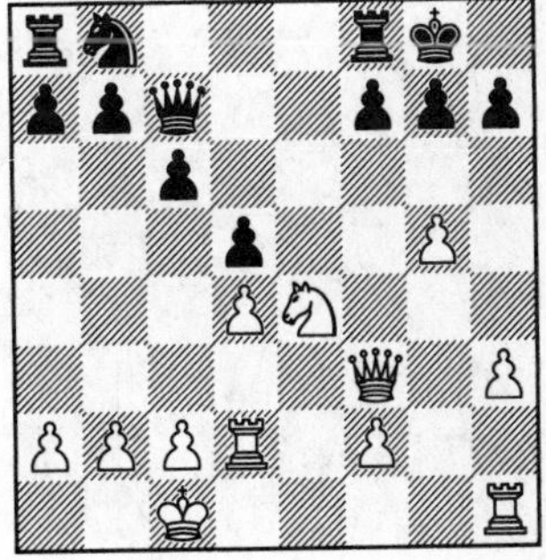

5261.

5262.

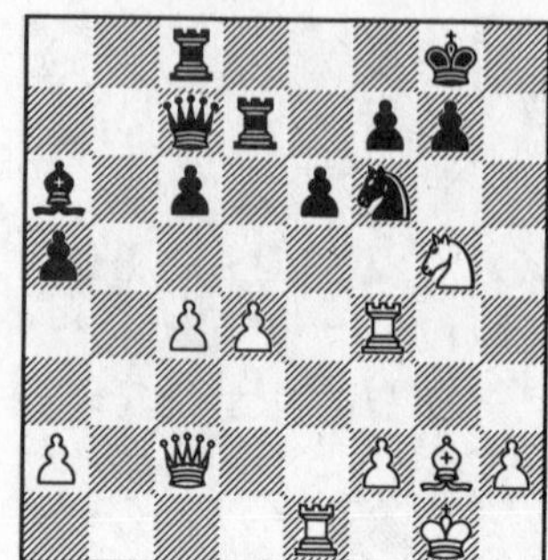

5263.

5264.

5265.

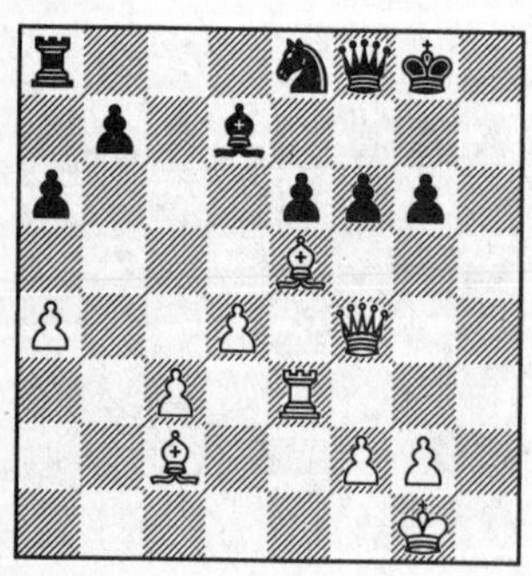

5266.

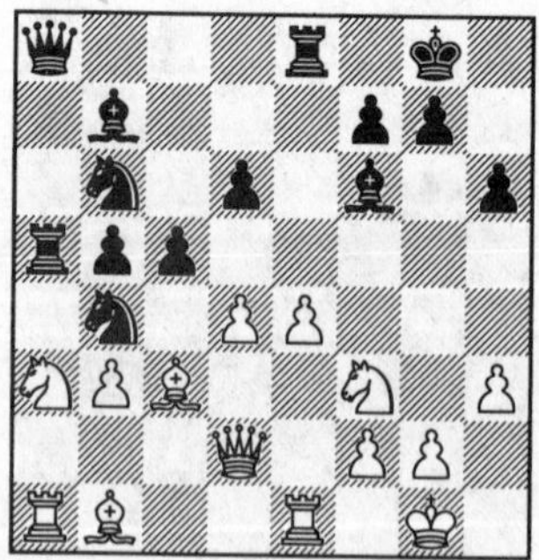

5267.

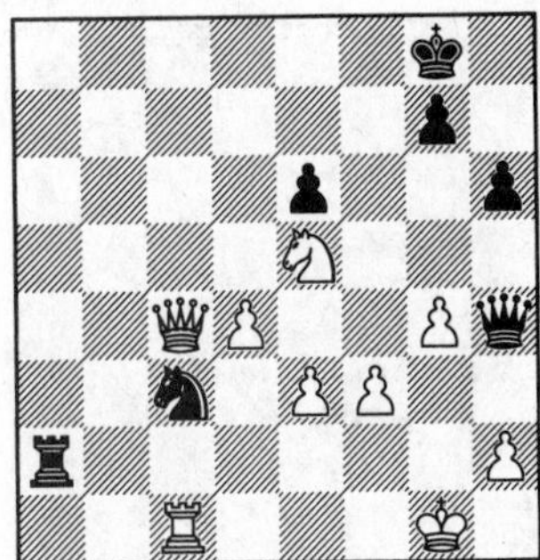

5268.

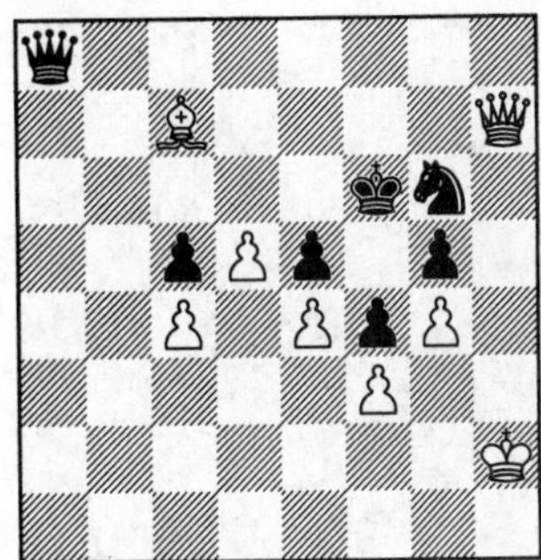

5269.

5270.

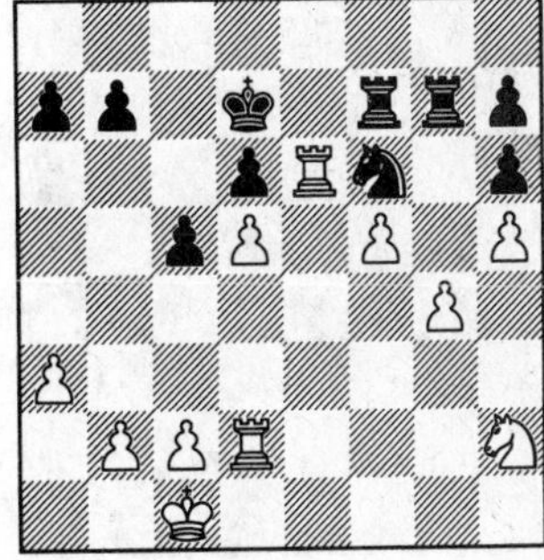

5271.

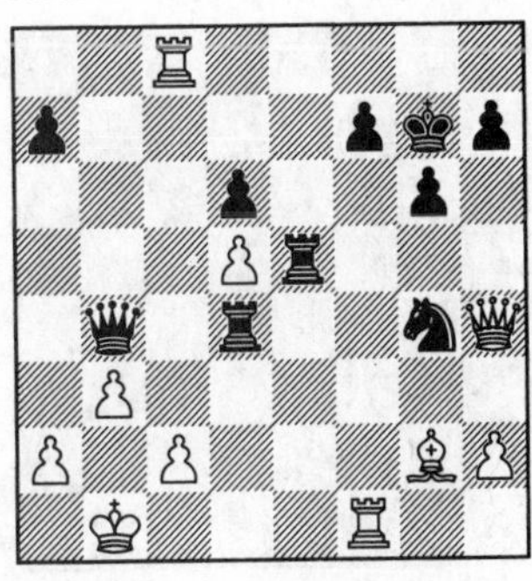

5272.

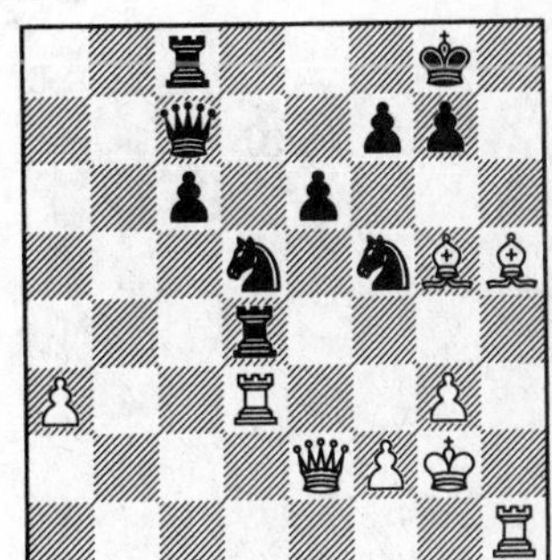

5273.

5274.

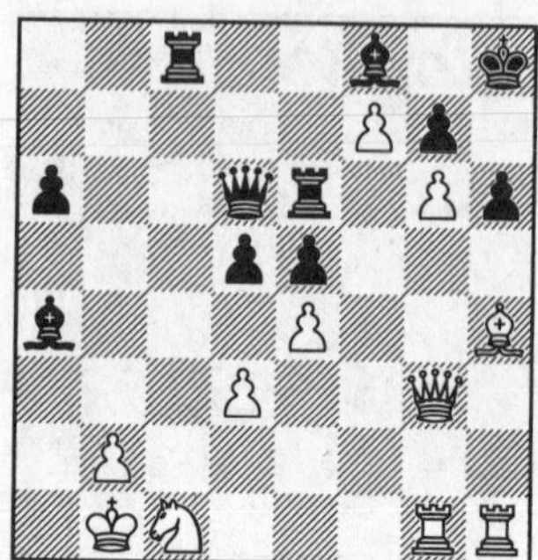

5275.

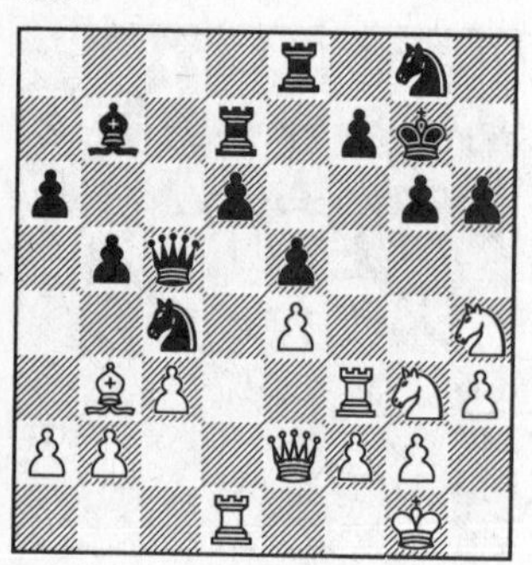

5276.

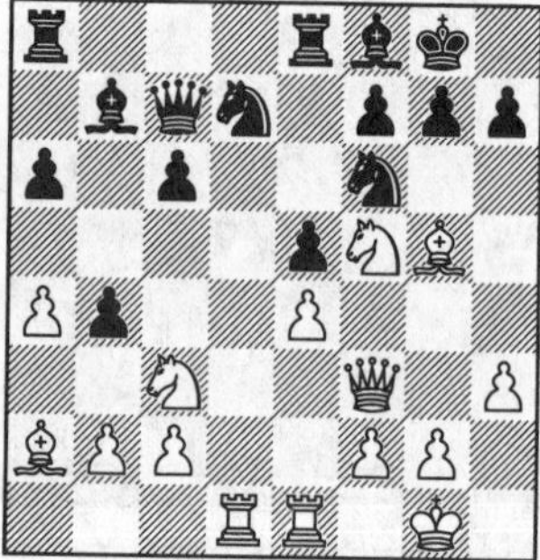

5277.

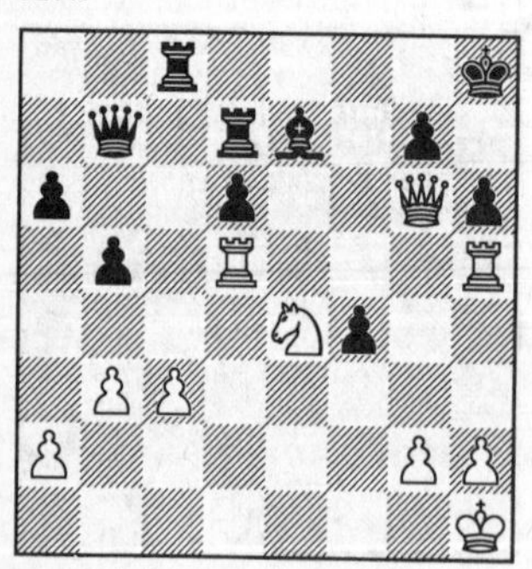

5278.

5279.

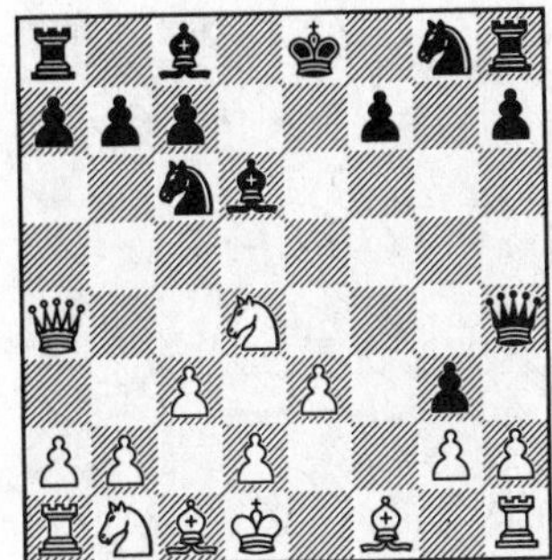

5280.

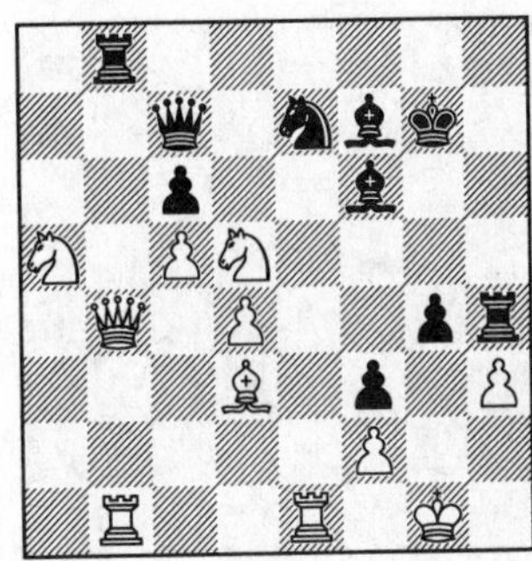

5281.

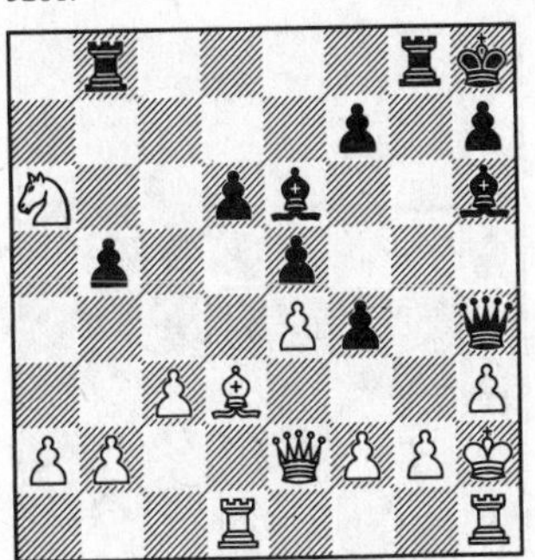

5282.

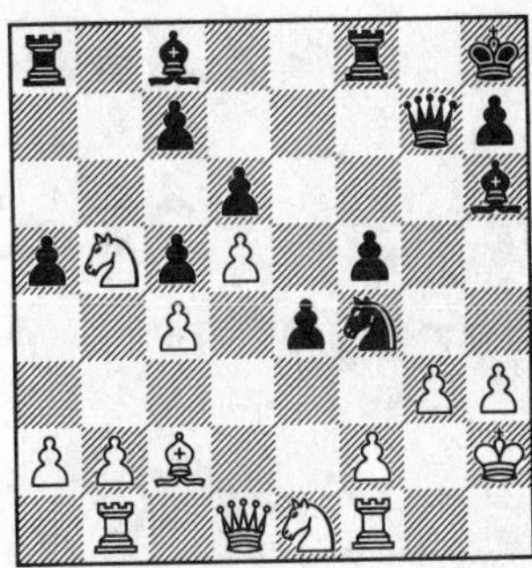

5283.

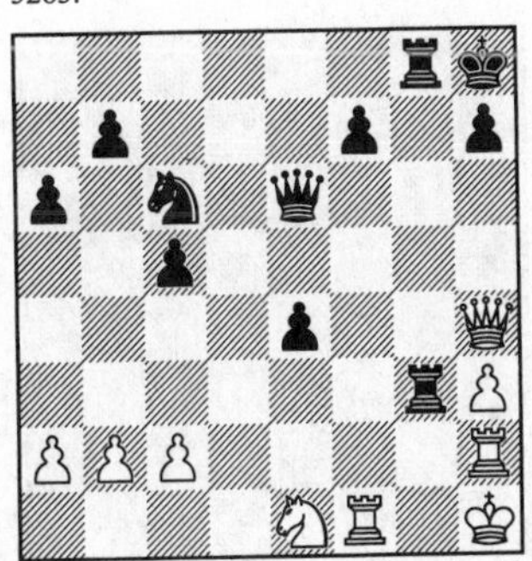

5284.

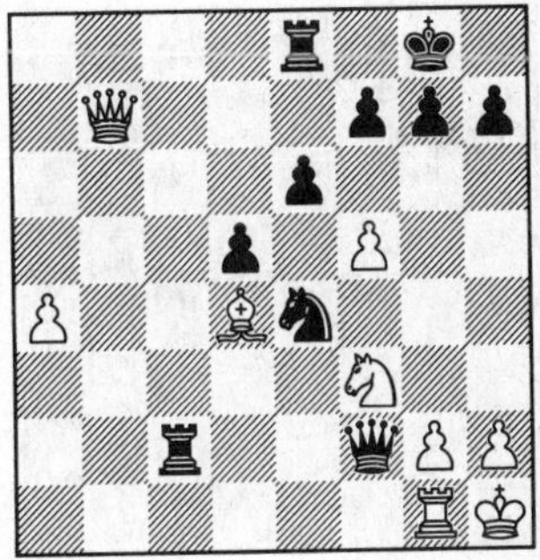

5285.

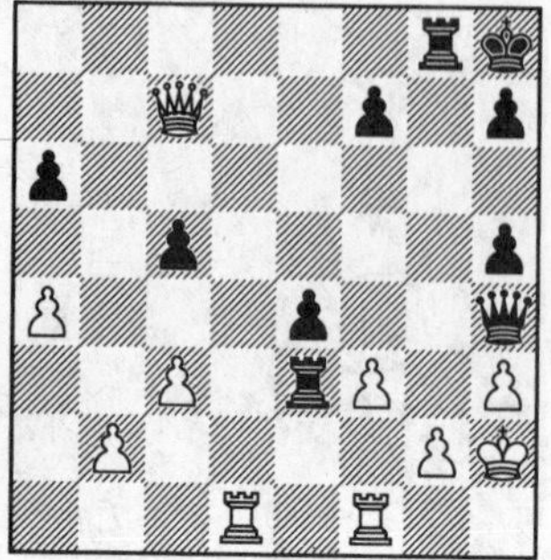

5286.

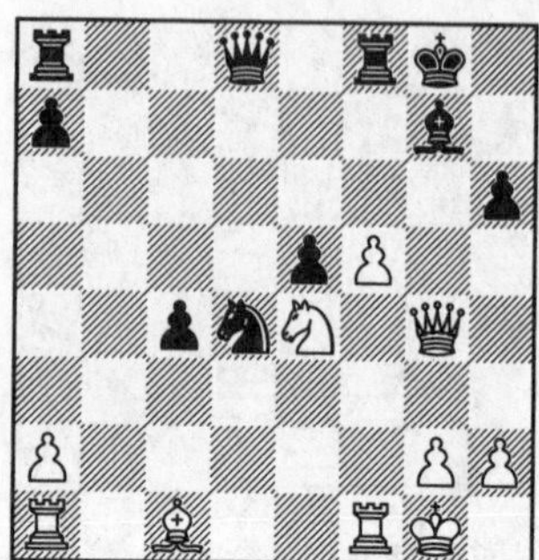

5287.

5288.

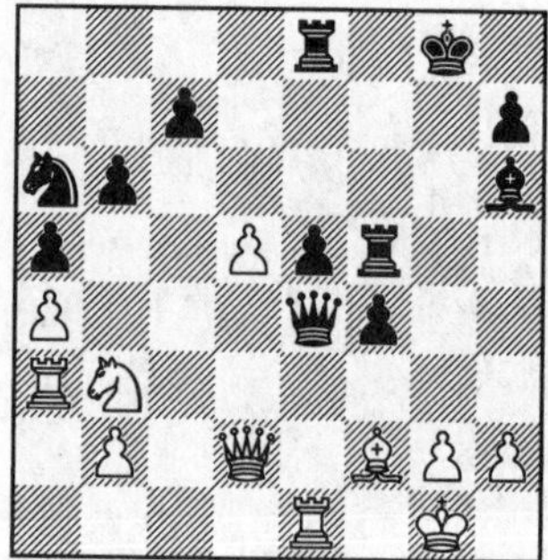

5289.

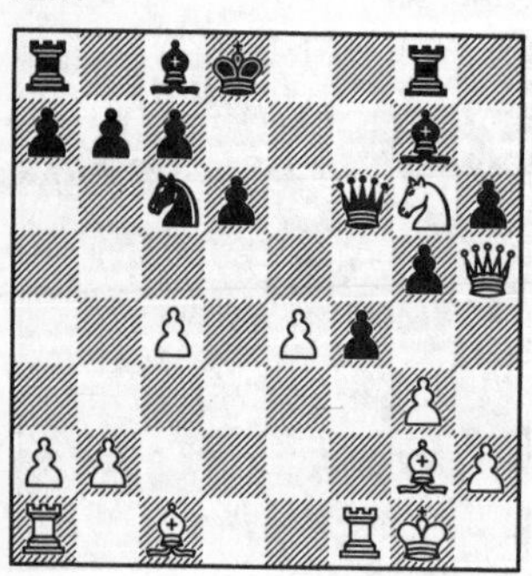

5290.

5291.

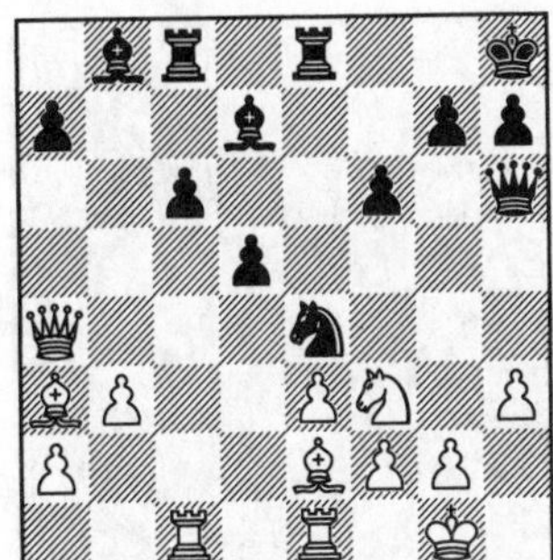

5292.

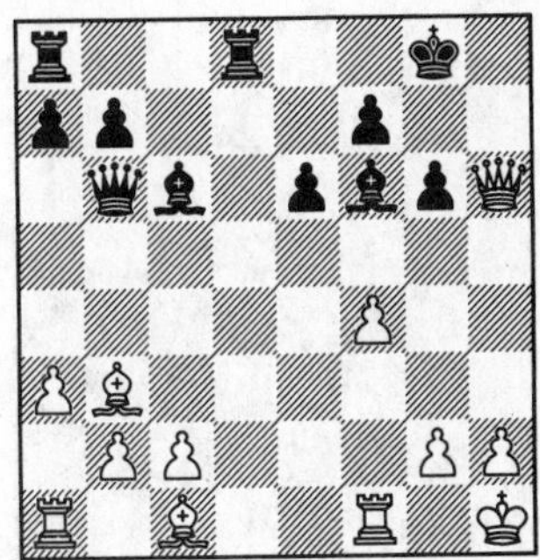

5293.

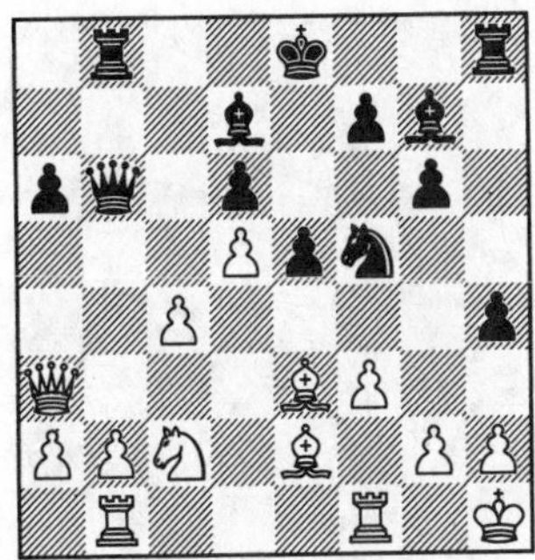

5294.

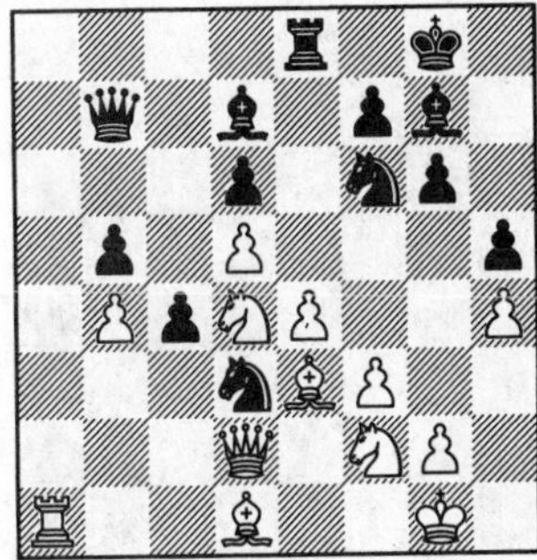

5295.

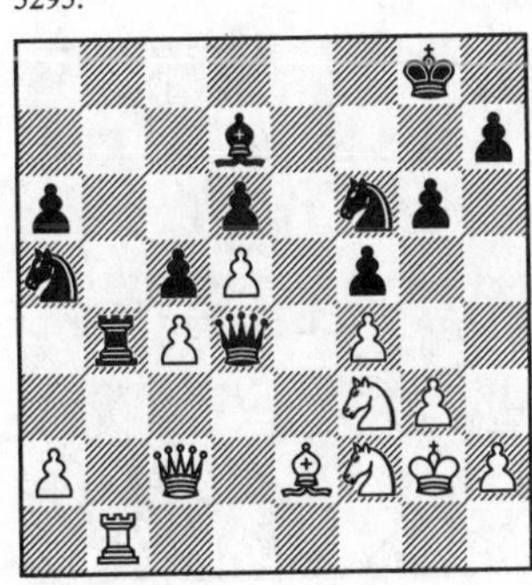

5296.

5297.

5298.

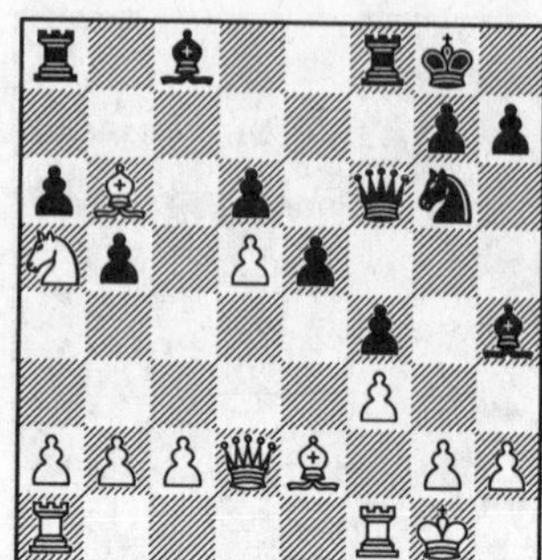

5299.

5300.

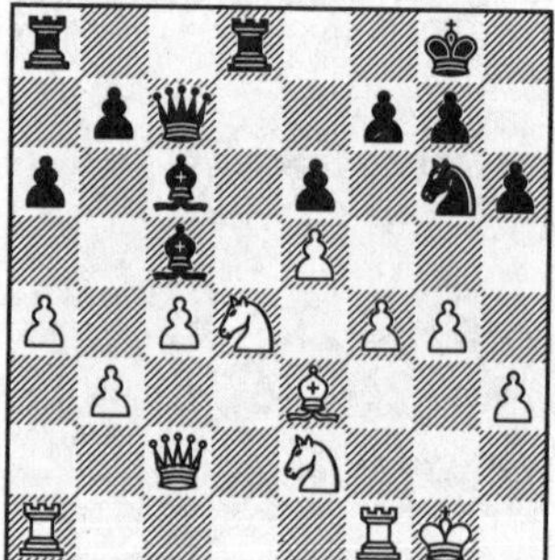

5301.

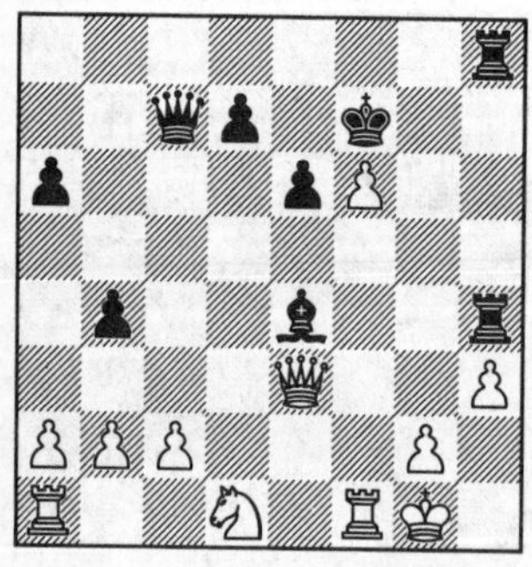

5302.

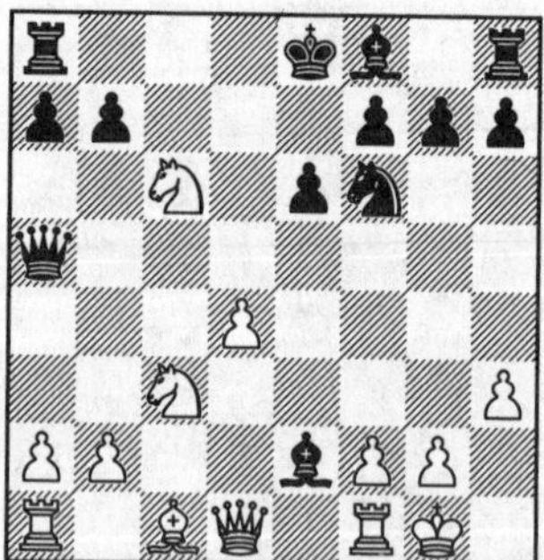

5303.

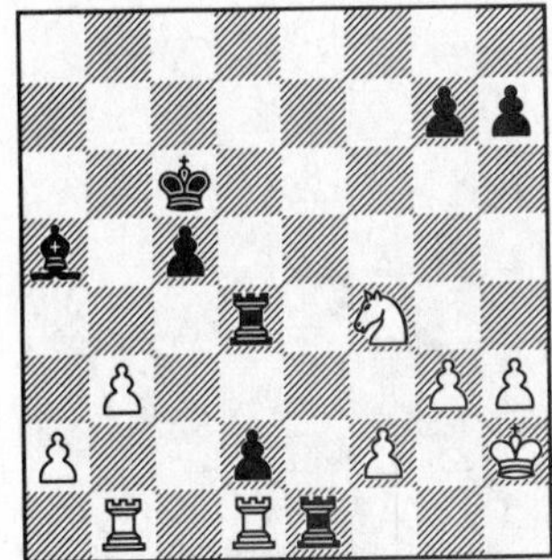

5304.

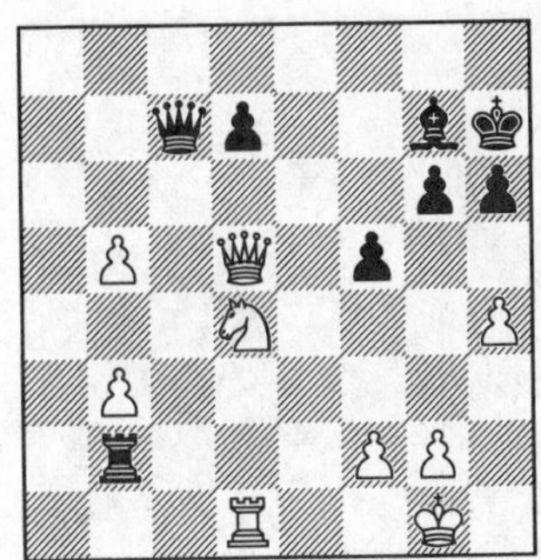

5305.

5306.

5307.

5308.

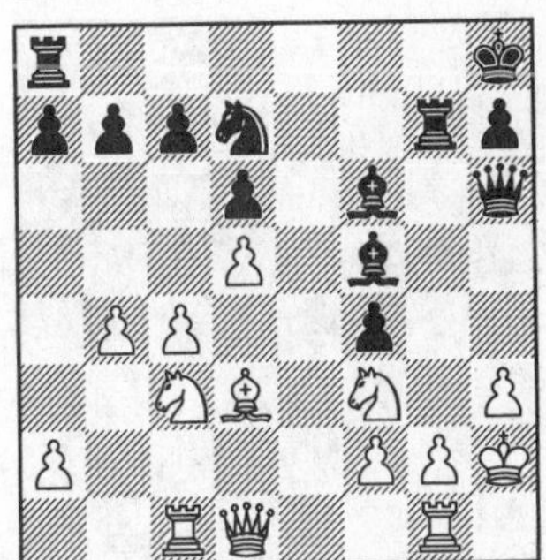

5309.

5310.

5311.

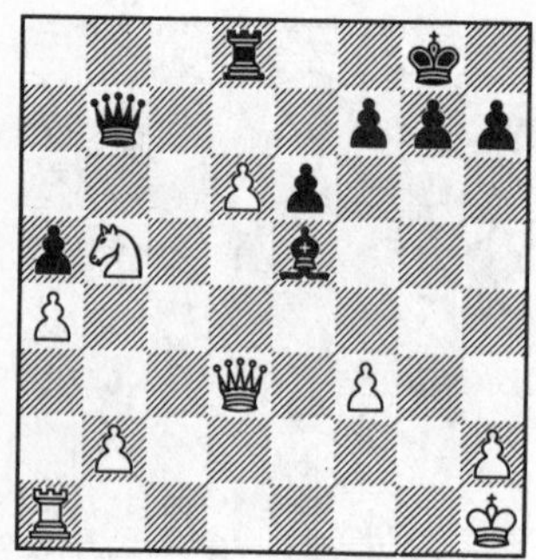

5312.

5313.

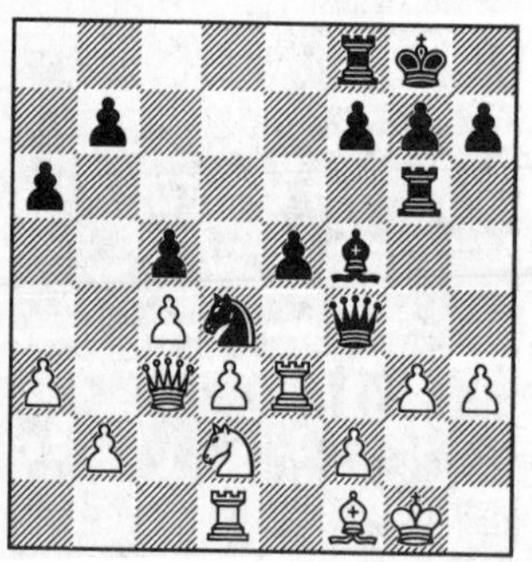

5314.

5315.

5316.

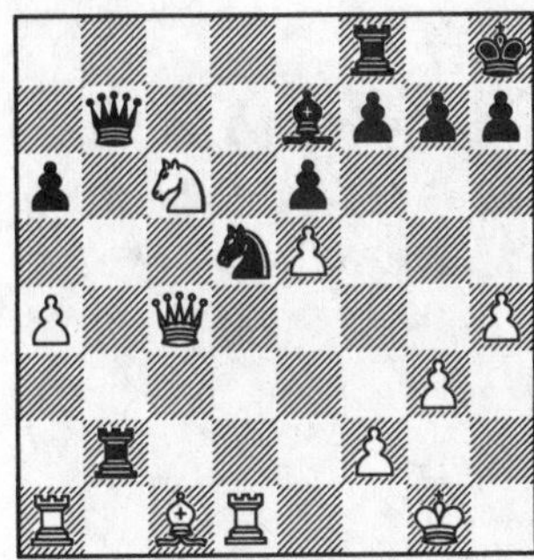

5317.

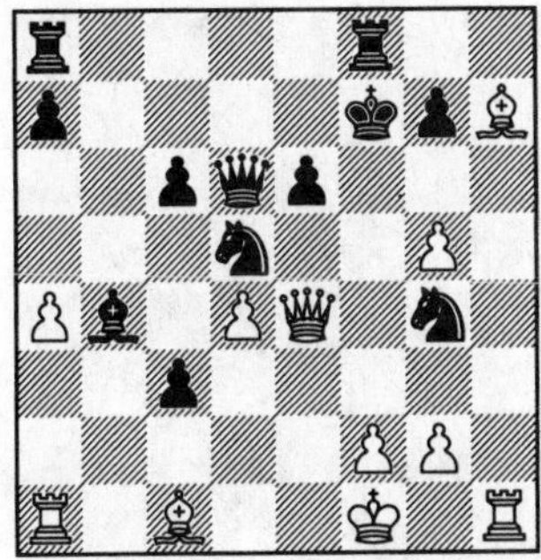

5318.

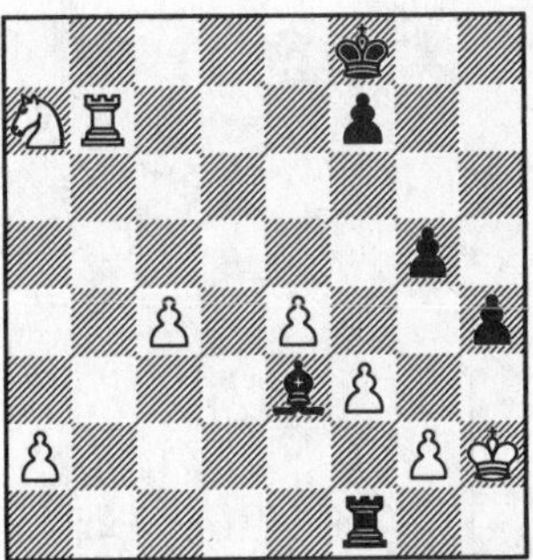

5319.

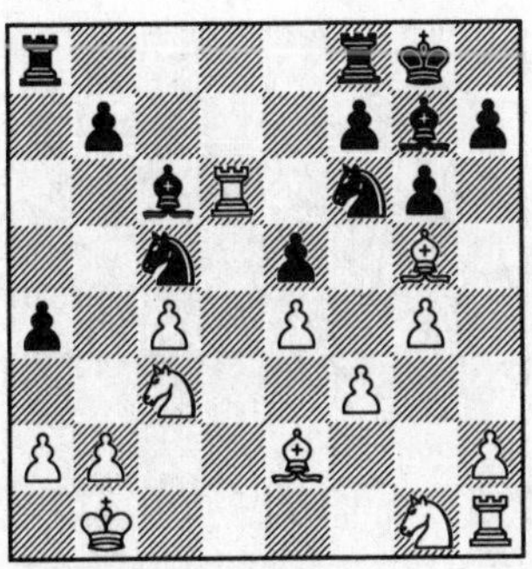

5320.

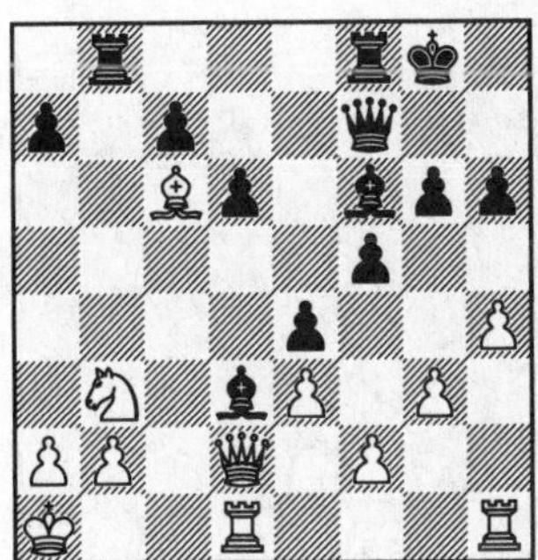

5321.

5322.

5323.

5324.

5325.

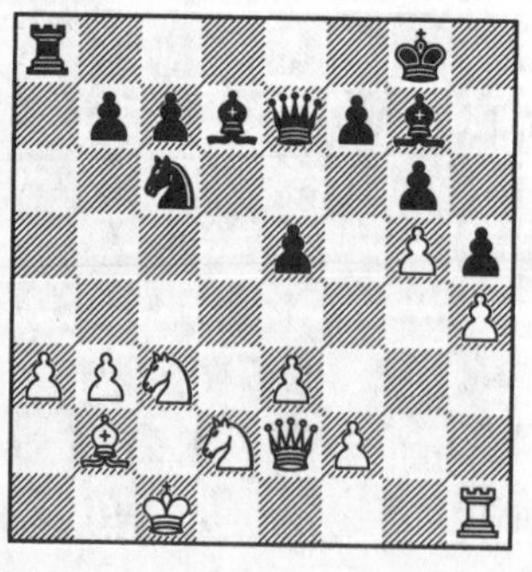

5326.

5327.

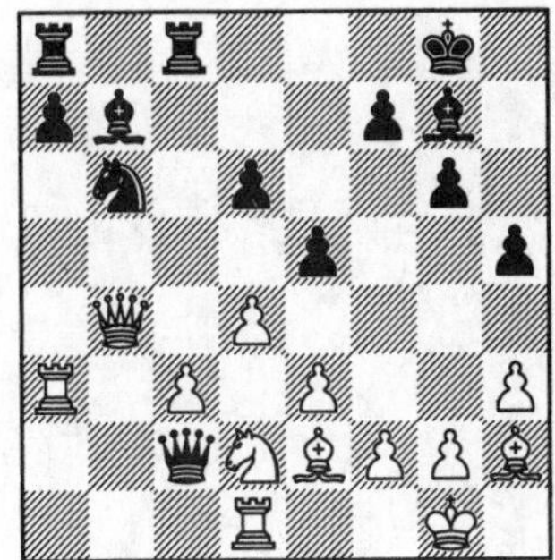

5328.

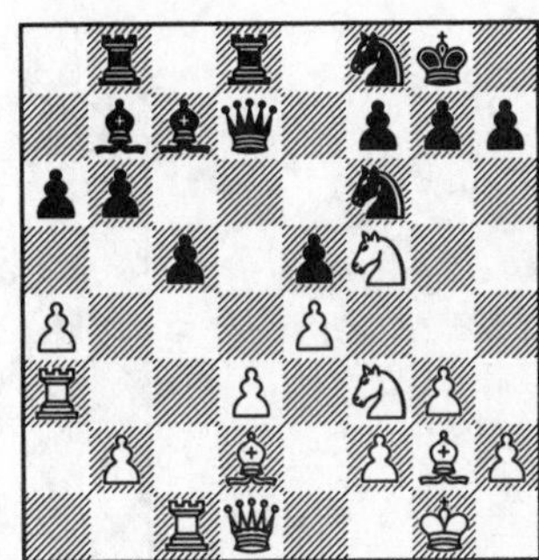

5329.

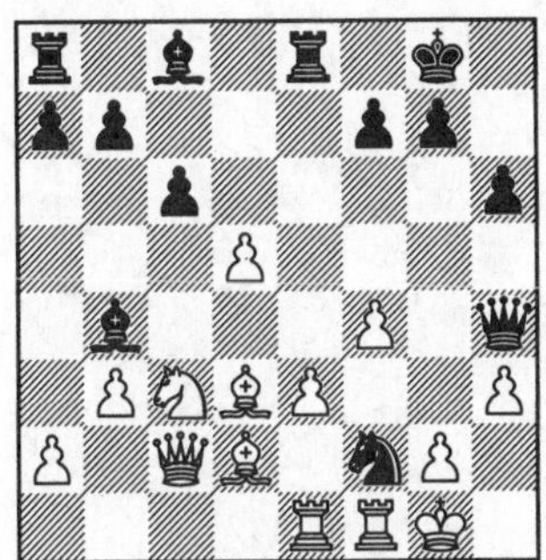

5330.

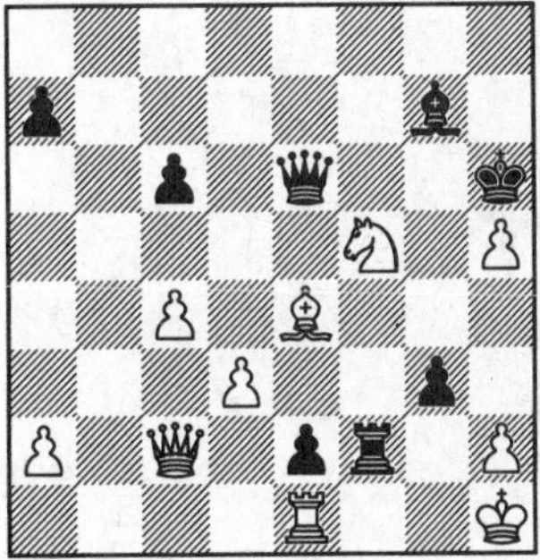

5331.

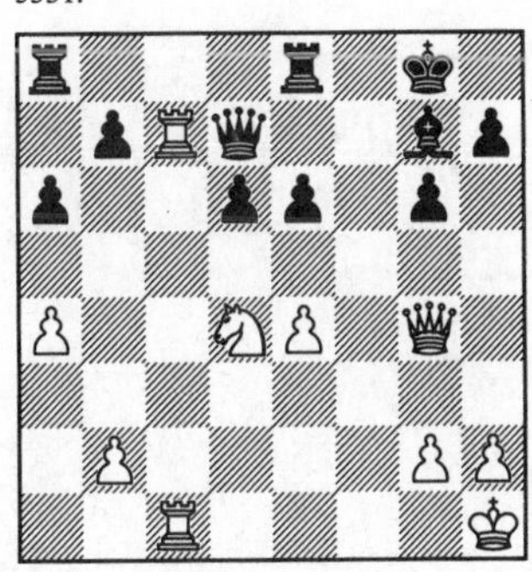

5332.

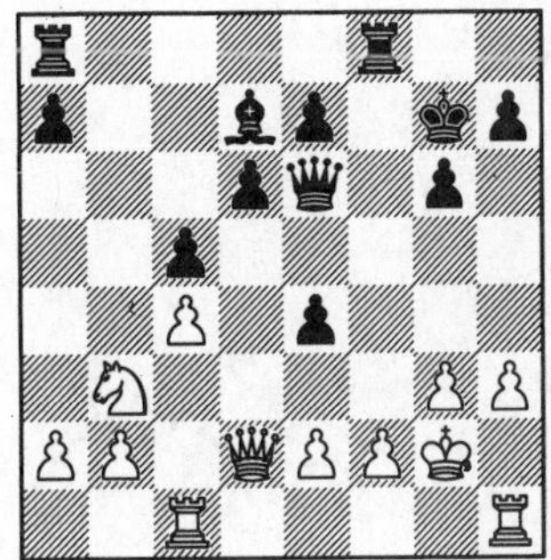

5333.

5334

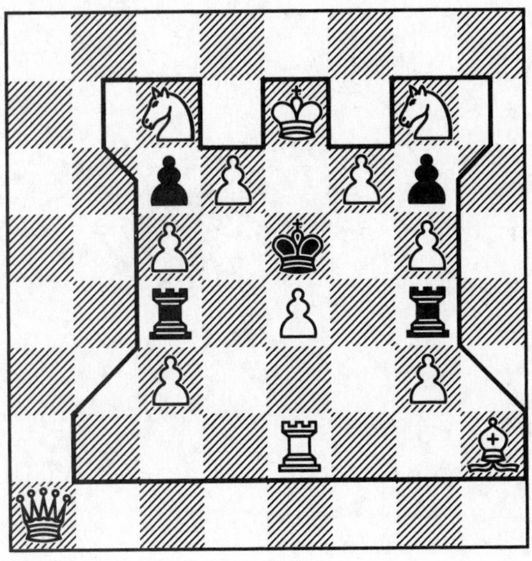

Solutions

1.
1. ♛g7

2.
1. ♛h7

3.
1. ♛g7

4.
1. ♜h3

5.
1. ♜h8

6.
1. ♜g1

7.
1. ♜h1

8.
1. ♜h1

9.
1. ♜g1

10.
1. ♗a6

11.
1. ♜d8

12.
1. ♗h7

13.
1. ♛f7

14.
1. ♛h5

15.
1. ♜e8

16.
1. ♛h7

17.
1. ♜d8

18.
1. ♘g6

19.
1. fg8♘

20.
1. f8♛

21.
1. de8♘

22.
1. ♗a6

23.
1. ♛c6

24.
1. ♘d7

25.
1. ♘f5

26.
1. ♛h7

27.
1. ♘e6

28.
1. ♘f7

29.
1. ♗f7

30.
1. h7

31.
1. ♛g7

32.
1. d7

33.
1. ♛b7

34.
1. ♜c7

35.
1. ♜h7

36.
1. ♜h6

37.
1. ♛d7

38.
1. ♘g7

39.
1. ♛c7

40.
1. ♜g8

41.
1. ♜h7

42.
1. ♘d6

43.
1. ♜e5

44.
1. f3

45.
1. ♛d6

46.
1. f4

47.
1. ♛e5

48.
1. f8♛

49.
1. ♛g6

50.
1. ♛b7

51.
1. ♛g2

52.
1. ♘f6

53.
1. g8♘

54.
1. ♛h7

55.
1. d7

56.
1. b7

57.
1. ♛e4

58.
1. ♛g8

59.
1. ♛e5

60.
1. ♛g6

61.
1. ♘c7

62.
1. ♘b6

63.
1. ♘c6

64.
1. b7

65.
1. ♘b6

66.
1. ♘c7

67.
1. ♜h2

68.
1. ♔f7

69.
1. ♜f5

70.
1. ♗f3

71.
1. ♘e5

72.
1. ♗d4

73.
1. ♜b7

74.
1. ♜h7

75.
1. ♜g8

76.
1. ♜e6

77.
1. ♜e7

78.
1. ♜d6

79.
1. ♗f1

80.
1. ♗f5

81.
1. ♗f1

82.
1. ♗d8

83.
1. ♗e8

84.
1. ♜a8

85.
1. ♛h6

86.
1. ♛f5

87.
1. ♛f7

88.
1. ♛e6

89.
1. ♜b8

90.
1. ♜a8

91.
1. ♛f3

92.
1. ♛f3

93.
1. ♛e8

94.
1. ♛e8

95.
1. ♛e5

96.
1. ♛g5

97.
1. ♛f1

98.
1. ♛b1

99.
1. ♛h1

100.
1. ♛e2

101.
1. ♛e2

102.
1. ♛f3

103.
1. ♛c8

104.
1. ♛d5

105.
1. ♛d6

106.
1. ♛f1

107.
1. ♛h6

108.
1. ♛f1

109.
1. ♛d4

110.
1. ♛h8

111.
1. ♛h8

112.
1. ♛h8

SOLUTIONS (113 - 224)

113.
1. ♕c4

114.
1. ♕f3

115.
1. ♕d8

116.
1. ♕e4

117.
1. ♕d1

118.
1. ♗f8

119.
1. ♕e5

120.
1. ♗d4

121.
1. ♕e6

122.
1. ♕e6

123.
1. ♕e5

124.
1. ♕g4

125.
1. ♕g4

126.
1. ♕e5

127.
1. ♕e7

128.
1. ♕f8

129.
1. ♕ae6

130.
1. ♕ae8

131.
1. ♕he5

132.
1. ♖a7

133.
1. ♗e5

134.
1. ♗f6

135.
1. ♗d4

136.
1. ♘f7

137.
1. ♘d6

138.
1. ♕e7

139.
1. ♖a8

140.
1. ♖h7

141.
1. ♖a3

142.
1. ♖h3

143.
1. ♖a3

144.
1. ♖a6

145.
1. ♕b3

146.
1. ♕a3

147.
1. ♕b3

148.
1. ♕b3

149.
1. ♕b3

150.
1. ♕e5

151.
1. ♕a8

152.
1. ♕a8

153.
1. ♕d6

154.
1. ♕g7

155.
1. ♕b3

156.
1. ♕b3

157.
1. ♕a4

158.
1. ♗a4

159.
1. ♕g8

160.
1. ♖g8

161.
1. ♘d7

162.
1. ♕b7

163.
1. h6

164.
1. ♕e6

165.
1. ♕e6

166.
1. ♗a5

167.
1. e7

168.
1. ♖g6

169.
1. ♘f6

170.
1. ♗d4

171.
1. ef7

172.
1. f7

173.
1. ♕g6

174.
1. ♘f6

175.
1. ♘f7

176.
1. ♕e8

177.
1. ♗a6

178.
1. ♘e7

179.
1. ♕h8

180.
1. ♘b6

181.
1. ♕h7

182.
1. ♘d6

183.
1. ♖e8

184.
1. ♗g7

185.
1. ♖g2

186.
1. ♕d5

187.
1. ♕g6

188.
1. ♕h7

189.
1. ♗e5

190.
1. ♘d6

191.
1. ♖h8

192.
1. ♘g6

193.
1. ♕g6

194.
1. ♘g7

195.
1. ♕g7

196.
1. ♗g5

197.
1. ♘d5

198.
1. ♘g6

199.
1. ♗d5

200.
1. ♘f6

201.
1. ♘e7

202.
1. ♘d6

203.
1. ♘d6

204.
1. ♕f5

205.
1. ♖d8

206.
1. ♗f6

207.
1. ♘h6

208.
1. ♗b5

209.
1. ♕h7

210.
1. ♘h6

211.
1. ♗f6

212.
1. ♖g1

213.
1. ♖e8

214.
1. ♖e8

215.
1. ♖h1

216.
1. ♗c5

217.
1. ♗d6

218.
1. ♗a7

219.
1. ♖h4

220.
1. ed8♕ (♖)

221.
1. ed8♘

222.
1. gf7

223.
1. ♕g7

224.
1. ♖c7

225.
1. ♕e3
226.
1. ♔b1
227.
1. ♕d6
228.
1. ♔d6
229.
1. ♖h8
230.
1. f7
231.
1. ♖hg7
232.
1. ♖d7
233.
1. ♘a7
234.
1. ♕h8
235.
1. ♕f7
236.
1. ♖g4
237.
1. g7
238.
1. ♘h6
239.
1. ♗d6
240.
1. cb4
241.
1. c3
242.
1. ♖a5
243.
1. c4
244.
1. f6
245.
1. ♘b5
246.
1. ♖g6
247.
1. e7
248.
1. ♖a7
249.
1. ♖e7
250.
1. ♗h5
251.
1. ♘d5
252.
1. a3

253.
1. ♘f5
254.
1. ♘f7
255.
1. ♕c6
256.
1. c5
257.
1. ♘d7
258.
1. gh7
259.
1. ♕b8
260.
1. ♕g8
261.
1. d8♕ (♖)
262.
1. ♗c7
263.
1. ♗f7
264.
1. ♘b4
265.
1. ♗f7
266.
1. ♘fe7
267.
1. ed6
268.
1. c6
269.
1. fg5
270.
1. ♕g7
271.
1. ♗b8
272.
1. ♘f6
273.
1. c5
274.
1. g5
275.
1. c6
276.
1. ♖d8
277.
1. ♕b4
278.
1. ♖h5
279.
1. ♗h3
280.
1. ♘f5

281.
1. c7
282.
1. ♖h8
283.
1. ♘f7
284.
1. ♕h1
285.
1. e8♘
286.
1. ♘g6
287.
1. ♕f8
288.
1. f8♘
289.
1. ♗g7
290.
1. ♕a7
291.
1. ♖h8
292.
1. g5
293.
1. ♘e6
294.
1. ♖a5
295.
1. ♗d8
296.
1. ♖c6
297.
1. ♗c4
298.
1. gf5
299.
1. ♕f7
300.
1. ♗h6
301.
1. ♘c7
302.
1. ♕c6
303.
1. f7
304.
1. ♘c5
305.
1. 0-0
306.
1. ab6
307.
T. Kardos 1971
1. ♔c3

308.
T. Kardos 1971
1. ♔c4
309.
W. A. Shinkman 1871
1. ♕b1
310.
T. Schönberger 1925
1. ♕h2
311.
T. Schönberger 1925
1. ♕f5
312.
M. Locker 1971
1. ♘h5
313.
E. Szentgyörgyi 1925
1. ♘f7
314.
T. Schönberger 1926
1. ♕h3
315.
F. Lazard 1925
1. ♗b3
316.
T. Kardos 1968
1. ♕a2
317.
Gy. Schiffert 1931
1. ♕d3
318.
T. Schönberger 1925
1. ♗g4
319.
B. Schwarz 1927
1. ♕e1
320.
L. Lindner 1933
1. ♘e5
321.
T. Schönberger 1935
1. ♔c3
322.
T. Schönberger – Gy. Schiffert 1926
1. ♔f2
323.
T. Schönberger 1935
1. ♘d5
324.
T. Schönberger 1935
1. ♘e6
325.
E. Szentgyörgyi 1935
1. ♘g6
326.
E. Szentgyörgyi 1933
1. ♕f2

SOLUTIONS (327 - 402)

327.
E. Szentgyörgyi 1927
1. ♕g2

328.
E. Szentgyörgyi 1927
1. ♖a2

329.
E. Szentgyörgyi 1927
1. ♕f7

330.
L. Szász 1927
1. ♕f6

331.
Gy. Schiffert 1927
1. ♖e7

332.
E. Szentgyörgyi 1928
1. ♗f8

333.
E. Szentgyörgyi 1928
1. ♘f7

334.
E. Szentgyörgyi 1928
1. ♘e6

335.
E. Szentgyörgyi 1933
1. ♖g1

336.
E. Szentgyörgyi 1933
1. ♕c3

337.
E. Szentgyörgyi 1933
1. ♘h4

338.
E. Szentgyörgyi 1929
1. ♕e6

339.
E. Szentgyörgyi 1931
1. ♘d4

340.
E. Szentgyörgyi 1931
1. ♕h7

341.
S. Boros 1929
1. ♘f5

342.
E. Szentgyörgyi 1929
1. ♖f6

343.
E. Szentgyörgyi 1929
1. ♘a5

344.
E. Szentgyörgyi 1933
1. ♖a1

345.
E. Szentgyörgyi 1933
1. ♘g4

346.
E. Szentgyörgyi 1933
1. ♕e3

347.
T. Schönberger 1926
1. ♕g3

348.
T. Schönberger 1926
1. ♕f3

349.
T. Schönberger 1925
1. ♕f5

350.
T. Schönberger 1935
1. ♕c2

351.
T. Schönberger 1935
1. ♘g4

352.
T. Schönberger 1935
1. ♕b3

353.
E. Szentgyörgyi 1935
1. ♗c5

354.
E. Szentgyörgyi 1933
1. ♕g4

355.
Z. Zilahi 1928
1. ♕c5

356.
Z. Zilahi 1928
1. ♕f4

357.
T. Schönberger 1926
1. ♖h8

358.
Gy. Balló 1926
1. ♕f3

359.
E. Szentgyörgyi 1926
1. b8♕

360.
E. Szentgyörgyi 1926
1. ♗d6

361.
Gy. Balló 1927
1. ♕d5

362.
E. Szentgyörgyi 1927
1. ♘e4

363.
E. Szentgyörgyi 1927
1. ♕d5

364.
E. Szentgyörgyi 1927
1. g5

365.
E. Szentgyörgyi 1927
1. ♖hg3

366.
Gy. Schiffert 1927
1. ♗h7

367.
Gy. Schiffert 1927
1. ♘c6

368.
E. Szentgyörgyi 1927
1. ♖c6

369.
E. Szentgyörgyi L. Szász 1927
1. ♘c8

370.
L. Szász 1927
1. ♕f3

371.
E. Szentgyörgyi 1927
1. ♕h1

372.
E. Szentgyörgyi 1927
1. ♘d5

373.
E. Szentgyörgyi 1928
1. ♕h2

374.
E. Szentgyörgyi 1925
1. ♘c3

375.
Gy. Schiffert 1926
1. ♖b7

376.
Gy. Schiffert 1926
1. ♕d3

377.
Gy. Schiffert 1931
1. ♘f3

378.
E. Szentgyörgyi 1930
1. ♘g4

379.
E. Szentgyörgyi 1930
1. ♖a1

380.
E. Szentgyörgyi 1929
1. ♘e2

381.
T. Schönberger 1926
1. ♕b5

382.
R. Pikler 1926
1. ♗f3

383.
R. Pikler 1926
1. c5

384.
Z. Zilahi 1926
1. ♕h1

385.
E. Goldschmiedt 1926
1. ♗c5

386.
E. Goldschmiedt 1926
1. ♘b5

387.
Z. Zilahi 1926
1.Hg4

388.
T. Schönberger 1926
1. ♕a3

389.
E. Goldschmiedt 1927
1. ♕e2

390.
Z. Zilahi 1927
1. ♕e3

391.
Z. Zilahi 1926
1. ♘f6

392.
T. Feldmann 1932
1. ♕f4

393.
T. Schönberger 1935
1. ♕g8

394.
T. Schönberger 1935
1. ♘c4

395.
T. Schönberger 1935
1. ♖h2

396.
T. Schönberger 1935
1. ♖a2

397.
S. Boros 1929
1. ♔e3

398.
E. Szentgyörgyi 1931
1. ♕f6

399.
E. Szentgyörgyi 1933
1. ♖c1

400.
E. Szentgyörgyi 1933
1. ♕e2

401.
E. Szentgyörgyi 1933
1. ♕e4

402.
E. Szentgyörgyi 1933
1. ♕h5

403.
E. Szentgyörgyi 1933
1. ♘b7
404.
E. Szentgyörgyi 1933
1. ♖d8
405.
E. Szentgyörgyi 1933
1. ♕e2
406.
E. Szentgyörgyi 1933
1. ♘b1
407.
E. Szentgyörgyi 1933
1. ♕e6
408.
B. Lincoln 1992
1. ♗g2
409.
L. Talabér 1933
1. ♖c7
410.
I. Kiss 1982
1. ♖f1
411.
E. Szentgyörgyi 1931
1. ♗g3
412.
Gy. Balló 1926
1. ♘e6
413.
Gy. Balló 1926
1. ♕g6
414.
E. Szentgyörgyi 1926
1. ♖f8
415.
E. Szentgyörgyi 1926
1. ♖g4
416.
E. Szentgyörgyi 1926
1. ♔d8
417.
E. Szentgyörgyi 1926
1. b7
418.
E. Szentgyörgyi 1927
1. ♘c8
419.
L. Szász 1927
1. ♘c6
420.
E. Szentgyörgyi 1927
1. ♖a1
421.
E. Szentgyörgyi 1927
1. ♖bc7
422.
E. Szentgyörgyi 1927
1. ♘d6
423.
E. Szentgyörgyi 1927
1. ♖g2
424.
E. Szentgyörgyi 1927
1. ♖c4
425.
M. Papp 1927
1. ♘e3
426.
Gy. Schiffert 1927
1. ♕b3
427.
I. Katkó 1927
1. ♕g8
428.
E. Szentgyörgyi 1927
1. ♕c7
429.
E. Szentgyörgyi – L. Szász 1927
1. ♖c7
430.
L. Szász 1927
1. ♕d5
431.
E. Szentgyörgyi 1927
1. ♕d3
432.
E. Szentgyörgyi 1927
1. ♕d8
433.
E. Szentgyörgyi 1928
1. ♔g5
434.
E. Szentgyörgyi 1928
1. ♗g8
435.
E. Szentgyörgyi 1928
1. ♘b3
436.
E. Szentgyörgyi 1928
1. ♗g8
437.
E. Szentgyörgyi 1928
1. ♖a6
438.
E. Szentgyörgyi 1928
1. ♖h3
439.
L. Szász 1927
1. ♗f1
440.
E. Szentgyörgyi 1928
1. ♘f5
441.
S. Boros 1926
1. ♕c6
442.
S. Boros 1926
1. ♘e3
443.
E. Szentgyörgyi 1931
1. ♕h2
444.
E. Szentgyörgyi 1925
1. ♖c2
445.
Gy. Balló – E. Szentgyörgyi 1928
1. ♖b5
446.
Gy. Balló 1928
1. ♘c4
447.
Gy. Balló – Gy. Schiffert 1926
1. ♕f6
448.
Gy. Balló 1925
1. ♕f3
449.
E. Szentgyörgyi 1931
1. d7
450.
E. Szentgyörgyi 1933
1. ♕f5
451.
1. ♘c5
452.
1. ♘f6
453.
1. ♘f6
454.
1. ♖h7
455.
1. ♘b5
456.
1. ♖e8
457.
1. ♗f5
458.
1. ♗d1
459.
1. ♖f7
460.
1. ♖h5
461.
1. ♖f5
462.
1. h4
463.
1. ♗g6
464.
1. ♘e7
465.
1. ♖g7
466.
1. ♖d7
467.
1. ♗f8
468.
1. ♕h8
469.
1. ♕f6
470.
1. ♕f7
471.
1. ♕f7
472.
1. ♕h5
473.
1. ♕c4
474.
1. ♕f8
475.
1. ♕a8
476.
1. ♕h6
477.
1. ♕a6
478.
1. ♕b8
479.
1. ♕f8
480.
1. ♕h8
481.
1. ♕g8
482.
1. ♕f6
483.
1. ♕c3
484.
1. b8♕ (♖)
485.
1. ♕f8
486.
1. ♕e8
487.
1. ♕h7
488.
1. ♕c5
489.
1. ♕h7
490.
1. ♕f8
491.
1. ♕h6

492.
1. ♕c8
493.
1. ♕f6
494.
1. ♕g7
495.
1. ♕h6
496.
1. ♕h7
497.
1. ♕h8
498.
1. ♕d5
499.
1. ♕d5
500.
1. ♕f6
501.
1. d4
502.
1. ♘c7
503.
1. ♖e8
504.
1. ♖f8
505.
1. ♕e7
506.
1. ♖d6
507.
1. ♘c7
508.
1. ♕d5
509.
1. ♕g7
510.
1. ♘d5
511.
1. ♕c5
512.
1. ♖f6
513.
1. ♖f3
514.
1. ♕a6
515.
1. ♕e8
516.
1. ♖f8
517.
1. ♖g8
518.
1. ♖f8
519.
1. ♖d4
520.
1. ♖e3
521.
1. ♖e8
522.
1. ♕d7
523.
1. ♖g8
524.
1. ♖d8
525.
1. ♖f5
526.
1. ♖h5
527.
1. ♖f6
528.
1. ♘e2
529.
1. g7
530.
1. ♖d4
531.
1. ♖h7
532.
1. ♖g4
533.
1. ♕b7
534.
1. ♕b6
535.
1. ♕b6
536.
1. ♕d8
537.
1. ♕d8
538.
1. ♕d8
539.
1. ♕a1
540.
1. ♖h7
541.
1. ♖b5
542.
1. h4
543.
1. ♖d6
544.
1. ♖g5
545.
1. ♖b5
546.
1. ♘g3
547.
1. ♖d7
548.
1. ♖f1
549.
1. ♖f5
550.
1. ♘6b5
551.
1. ♗d5
552.
1. ♗d8
553.
1. ♖b1
554.
1. ♘d6
555.
1. ♘c7
556.
1. ♘g6
557.
1. ♖h8
558.
1. ♖c8
559.
1. ♖d8
560.
1. ♖h7
561.
1. ♘f6
562.
1. ♖e8
563.
1. ♕g6
564.
1. ♕g6
565.
1. ♗b2
566.
1. ♖h6
567.
1. ♖h7
568.
1. ♖g6
569.
1. ♖c7
570.
1. ♘b5
571.
1. ♗f6
572.
1. ♖e8
573.
1. ♗h6
574.
1. ♖h8
575.
1. ♕h8
576.
1. ♕a4
577.
1. ♕a6
578.
1. ♕f6
579.
1. ♕f8
580.
1. ♕g7
581.
1. ♘d5
582.
1. ♖e5
583.
1. ♘b6
584.
1. ♘b6
585.
1. ♕b5
586.
1. ♖f5
587.
1. ♕d5
588.
1. ♕f5
589.
1. ♕d5
590.
1. ♕d6
591.
1. ♕f4
592.
1. ♕d6
593.
1. ♕e6
594.
1. ♕b6
595.
1. ♕c4
596.
1. ♘b2
597.
1. ♖c6
598.
1. ♖h6
599.
1. g4
600.
1. ♘c4
601.
1. ♕g5
602.
1. ♗b4
603.
1. ♘f3

604.
1. ♕f6
605.
1. ♗g6
606.
1. ♖e2
607.
1. ♖e5
608.
1. ♕f6
609.
1. ♕c6
610.
1. ♕g4
611.
1. ♖h6
612.
1. ♕e6
613.
1. ♗c5
614.
1. ♕f8
615.
1. ♖g7
616.
1. ♕h7
617.
1. ♕h7
618.
1. ♕c4
619.
1. ♕g6
620.
1. ♖c5
621.
1. ♕c4
622.
1. ♖d4
623.
1. ♖e4
624.
1. ♘f3
625.
1. ♖a6
626.
1. ♘c5
627.
1. ♖c8
628.
1. ♕f6
629.
1. ♘e5
630.
1. ♕g1
631.
1. ♕e4
632.
1. ♕c3
633.
1. ♕c5
634.
1. ♕e4
635.
1. ♕c6
636.
1. ♕e6
637.
1. ♕f3
638.
1. ♕e5
639.
1. ♕d6
640.
1. ♕g7
641.
1. 0-0-0
642.
1. ♖e6
643.
1. ♖h6
644.
1. ♖h5
645.
1. g7
646.
1. ♖a2
647.
1. ♗d3
648.
1. ♖h8
649.
1. ♘e8
650.
1. ♖b5
651.
1. ♖h5
652.
1. ♘g6
653.
1. ♖a5
654.
1. ♖g8
655.
1. ♖h7
656.
1. ♗c3
657.
1. ♖g6
658.
1. ♗h6
659.
1. ♘fg6
660.
1. ♖h4
661.
1. ♘g5
662.
1. ♖h7
663.
1. ♖h7
664.
1. ♘g6
665.
1. ♘a6
666.
1. ♘g6
667.
1. ♖c8
668.
1. ♘f6
669.
1. ♖f5
670.
1. ♗h6
671.
1. ♖f8
672.
1. f7
673.
1. ♕e7
674.
1. ♕c6
675.
1. ♕b7
676.
1. ♕d5
677.
1. ♕e3
678.
1. ♕e5
679.
1. ♕h6
680.
1. ♗f6
681.
1. ♖e8
682.
1. ♖g3
683.
1. ♖c2
684.
1. ♗d5
685.
1. ♖e8
686.
1. ♘d6
687.
1. ♖c8
688.
1. ♖h5
689.
1. ♖h7
690.
1. ♖d8
691.
1. ♘5f6
692.
1. ♖d6
693.
1. ♖h8
694.
1. ♖d8
695.
1. ♖f5
696.
1. ♗b6
697.
1. ♖a5
698.
1. ♖e4
699.
1. ♖f6
700.
1. ♖f5
701.
1. ♖d5
702.
1. ♖e8
703.
1. ♖g7
704.
1. ♖f4
705.
1. ♖f7
706.
1. ♖b8
707.
1. ♖f4
708.
1. ♖g8
709.
1. ♖h6
710.
1. ♖d4
711.
1. ♕f8
712.
1. ♕b4
713.
1. ♕h7
714.
1. ♕f4
715.
1. ♕f8

716.
1. ♕c5

717.
1. ♕c5

718.
1. f8♕

719.
1. ♕g6

720.
1. ♕e7

721.
1. ♕h6

722.
1. ♕c5

723.
1. ♕c6

724.
1. ♕d5

725.
1. ♕b8

726.
1. ♕d7

727.
1. ♕c8

728.
1. ♕b1

729.
1. ♕g8

730.
1. ♕d5

731.
1. ♕c5

732.
1. ♕d8

733.
1. ♕g8

734.
1. ♕f6

735.
1. e8♕

736.
1. f8♕

737.
1. ♕c2

738.
1. ♕b5

739.
1. ♕f6

740.
1. ♕b4

741.
1. ♕g1

742.
1. ♕h7

743.
1. ♕c5

744.
1. ♕b7

745.
1. ♗c8

746.
1. ♖e5

747.
1. ♖b6

748.
1. ♖b8

749.
1. g4

750.
1. ♗e5

751.
1. ♖e8

752.
1. ♖c8

753.
1. ♖h8

754.
1. c7

755.
1. ♗g7

756.
1. ♖f8

757.
1. ♖g8

758.
1. ♖h5

759.
1. ♖h4

760.
1. ♖h4

761.
1. ♖a4

762.
1. ♘f5

763.
1. ♘c5

764.
1. ♗b8

765.
1. ♕d6

766.
1. ♕b4

767.
1. ♕b6

768.
1. ♕e4

769.
1. ♕g6

770.
1. ♕g7

771.
1. ♕e7

772.
1. ♕h7

773.
1. ♗g8

774.
1. b5

775.
1. ♘c6

776.
1. ♖e6

777.
1. h6

778.
1. ♘a6

779.
1. d4

780.
1. f4

781.
1. ♖h4

782.
1. ♖g6

783.
1. ♘d6

784.
1. ♖f7

785.
1. ♖h6

786.
1. ♖a4

787.
1. ♖f4

788.
1. ♕f8

789.
1. ♕h6

790.
1. ♕f8

791.
1. ♕b6

792.
1. ♕f5

793.
1. ♕g6

794.
1. ♕c3

795.
1. ♕c5

796.
1. ♕e7

797.
1. ♕d8

798.
1. ♕c5

799.
1. ♕h7

800.
1. ♕h4

801.
1. ♕d5

802.
1. ♕h7

803.
1. ♕h7

804.
1. ♕f4

805.
1. ♕f5

806.
1. ♕f8

807.
1. ♕f8

808.
1. ♕d5

809.
1. ♕d6

810.
1. ♕g6

811.
1. ♕f7

812.
1. ♕f6

813.
1. ♕f6

814.
1. ♕f6

815.
1. ♕g5

816.
1. ♕g7

817.
1. ♕e7

818.
1. ♕e6

819.
1. ♕e6

820.
1. ♕e4

821.
1. ♕d4

822.
1. ♕c6

823.
1. ♕e5

824.
1. ♕b5

825.
1. ♕h6

826.
1. ♕a6

827.
1. ♕f7

828.
1. ♕c7
829.
1. ♕e3
830.
1. ♖e8
831.
1. ♕h6
832.
1. ♕d3
833.
1. ♕e4
834.
1. ♕g5
835.
1. ♕h7
836.
1. ♕h6
837.
1. ♕h7
838.
1. ♕h4
839.
1. ♕b5
840.
1. ♖h8
841.
1. ♖g8
842.
1. ♖h8
843.
1. ♘c5
844.
1. ♖d8
845.
1. ♗g4
846.
1. ♘f6
847.
1. ♖c4
848.
1. ♖d3
849.
1. ♗c6
850.
1. ♘c7
851.
1. b4
852.
1. ♕e5
853.
1. ♕f8
854.
1. ♕g6
855.
1. ♕c2
856.
1. ♕h7
857.
1. ♕h7
858.
1. ♕g8
859.
1. ♕a5
860.
1. ♕c4
861.
1. ♕d3
862.
1. ♕d7
863.
1. ♕f6
864.
1. ♕g7
865.
1. ♕g5
866.
1. ♕h6
867.
1. ♕h6
868.
1. ♕h7
869.
1. ♕h7
870.
1. ♖1d6
871.
1. ♕f6
872.
1. ♕f8
873.
1. ♕h4
874.
1. ♕f6
875.
1. ♕a5
876.
1. ♕f8
877.
1. ♕c6
878.
1. ♕h8
879.
1. ♘b6
880.
1. ♘h6
881.
1. ♖h6
882.
1. g5
883.
1. ♗g7
884.
1. ♖e8
885.
1. ♕h6
886.
1. ♕d5
887.
1. ♕c8
888.
1. ♕c7
889.
1. ♕h6
890.
1. ♖h1
891.
1. b3
892.
1. f4
893.
1. ♕g5
894.
1. ♕g5
895.
1. ♕g5
896.
1. ♕g5
897.
1. ♕g5
898.
1. ♕g7
899.
1. ♕g7
900.
1. ♕g7
901.
1. ♕c5
902.
1. ♕c1
903.
1. ♕c8
904.
1. ♕c8
905.
1. ♕c8
906.
1. ♕c8
907.
1. ♕e6
908.
1. ♕c6
909.
1. ♕c6
910.
1. ♕c6
911.
1. ♕c6
912.
1. ♕c7
913.
1. ♕h5
914.
1. ♕g8
915.
1. ♕b5
916.
1. ♕g8
917.
1. ♕g8
918.
1. ♕g7
919.
1. ♕g7
920.
1. ♕c6
921.
1. ♕d6
922.
1. ♕c6
923.
1. ♕c6
924.
1. ♕g7
925.
1. ♕g7
926.
1. ♕g7
927.
1. ♕h5
928.
1. ♕h5
929.
1. ♕h5
930.
1. ♕h5
931.
1. ♕h5
932.
1. ♕h5
933.
1. ♕h5
934.
1. ♕h5
935.
1. ♕f1
936.
1. ♕h7
937.
1. ♕d5
938.
1. ♕g3
939.
1. ♕e5

SOLUTIONS (940 – 1051)

940.
1. ♕f8

941.
1. ♕g6

942.
1. ♘f6

943.
1. ♘g6

944.
1. ♕f7

945.
1. ♕f7

946.
1. ♖g6

947.
1. ♗f7

948.
1. ♕f8

949.
1. ♕g4

950.
1. ♕f7

951.
1. ♕e6

952.
1. ♕g6

953.
1. ♘g5

954.
1. ♘e5

955.
1. f7

956.
1. ♘h7

957.
1. ♕g6

958.
1. ♕e5

959.
1. ♕e3

960.
1. ♕h4

961.
1. ♕f8

962.
1. ♕h6

963.
1. ♕h6

964.
1. ♕h6

965.
1. ♕b8

966.
1. ♕f7

967.
1. ♕d8

968.
1. ♕c8

969.
1. ♕g6

970.
1. ♕f8

971.
1. ♕f5

972.
1. ♘g6

973.
1. ♖g7

974.
1. ♖a7

975.
1. f4

976.
1. ♗f6

977.
1. ♖e7

978.
1. ♖e4

979.
1. ♘e7

980.
1. ♘b4

981.
1. ♖h5

982.
1. ♘d7

983.
1. ♕e5

984.
1. ♕f6

985.
1. ♖f5

986.
1. ♖h8

987.
1. ♖c3

988.
1. ♕e3

989.
1. ♕b5

990.
1. ♖e6

991.
1. ♕d5

992.
1. ♘d6

993.
1. ♕e6

994.
1. ♕d4

995.
1. ♕h7

996.
1. ♕h6

997.
1. ♕h6

998.
1. ♕h8

999.
1. ♕f5

1000.
1. ♖h8

1001.
1. ♗e5

1002.
1. ♕e6

1003.
1. ♕h7

1004.
1. ♕e6

1005.
1. ♕h7

1006.
1. ♗h5

1007.
1. ♘g5

1008.
1. ♘g6

1009.
1. ♕h3

1010.
1. ♘g5

1011.
1. ♖d8

1012.
1. ♖c8

1013.
1. ♖a6

1014.
1. ♖h6

1015.
1. ♖h5

1016.
1. ♘g3

1017.
1. ♖h5

1018.
1. ♘e6

1019.
1. ♖h5

1020.
1. ♖d8

1021.
1. ♖a4

1022.
1. ♖g8

1023.
1. ♖d5

1024.
1. ♘g6

1025.
1. ♖e5

1026.
1. ♗b5

1027.
1. ♘f6

1028.
1. ♘d5

1029.
1. ♖h3

1030.
1. ♖f7

1031.
1. ♗h7

1032.
1. ♕h7

1033.
1. ♕g6

1034.
1. ♕c8

1035.
1. ♗h6

1036.
1. ♘c7

1037.
1. ♘g6

1038.
1. ♕h7

1039.
1. ♕g5

1040.
1. ♕d7

1041.
1. ♕b7

1042.
1. ♕d7

1043.
1. ♕b7

1044.
1. ♕d7

1045.
1. ♖g7

1046.
1. ♗f6

1047.
1. ♖a7

1048.
1. ♖d5

1049.
1. ♖b5

1050.
1. ♖b7

1051.
1. ♕d6

1052.
1. ♕e4
1053.
1. ♖e8
1054.
1. ♗g5
1055.
1. ♕a5
1056.
1. ♕h7
1057.
1. ♕f5
1058.
1. ♖e8
1059.
1. ♕f5
1060.
1. ♖f4
1061.
1. ♖h5
1062.
1. ♖f6
1063.
1. ♕h7
1064.
1. ♘g6
1065.
1. ♕f6
1066.
1. ♕g8
1067.
1. ♘h6
1068.
1. ♘h5
1069.
1. ♖h8
1070.
1. ♕h6
1071.
1. ♘f6
1072.
1. ♕f8
1073.
1. ♕b8
1074.
1. ♕h7
1075.
1. ♖e8
1076.
1. ♖f6
1077.
1. ♗e8
1078.
1. ♗g7
1079.
1. ♖e8
1080.
1. ♗e4
1081.
1. ♖e5
1082.
1. ♖e4
1083.
1. ♖e7
1084.
1. ♗h6
1085.
1. ♗h5
1086.
1. ♕g6
1087.
1. ♕h7
1088.
1. ♗c6
1089.
1. ♗f6
1090.
1. ♗f6
1091.
1. ♗h6
1092.
1. ♗f6
1093.
1. ♗e6
1094.
1. ♗h5
1095.
1. ♖e8
1096.
1. ♗h7
1097.
1. ♗e4
1098.
1. ♗h6
1099.
1. ♖e6
1100.
1. ♖e6
1101.
1. ♖e6
1102.
1. ♕d6
1103.
1. ♖f5
1104.
1. ♘d6
1105.
1. ♖f5
1106.
1. ♗g7
1107.
1. ♖f5
1108.
1. ♗f8
1109.
1. ♖f5
1110.
1. ♗e5
1111.
1. ♗f4
1112.
1. d4
1113.
1. ♖g5
1114.
1. ♖g7
1115.
1. f6
1116.
1. ♖g5
1117.
1. ♕f8
1118.
1. ♖g3
1119.
1. ♕d6
1120.
1. c5
1121.
1. ♖a5
1122.
1. ♕d6
1123.
1. ♖g7
1124.
1. ♕g6
1125.
1. ♕f8
1126.
1. ♕d5
1127.
1. ♖a7
1128.
1. ♖a1
1129.
1. ♖d7
1130.
1. ♖a8
1131.
1. ♕d4
1132.
1. ♕g6
1133.
1. ♖ab8
1134.
1. ♖d6
1135.
1. ♖b7
1136.
1. ♖b6
1137.
1. ♕b1
1138.
1. ♘e6
1139.
1. ♘c6
1140.
1. f6
1141.
1. ♕f8
1142.
1. ♘c6
1143.
1. ♗d8
1144.
1. ♘c7
1145.
1. f4
1146.
1. ♕g6
1147.
1. ♕g6
1148.
1. ♖d6
1149.
1. ♕g6
1150.
1. ♖h6
1151.
1. ♕e8
1152.
1. ♖d8
1153.
1. ♕e7
1154.
1. ♖d8
1155.
1. ♕e7
1156.
1. ♖c7
1157.
1. ♕g6
1158.
1. ♖d4
1159.
1. ♕e6
1160.
1. ♕e8
1161.
1. ♘f7
1162.
1. ♖f7
1163.
1. ♘f6

1164.
1. ♖h6
1165.
1. ♘f6
1166.
1. ♕e6
1167.
1. ♖f7
1168.
1. ♘h6
1169.
1. ♕h6
1170.
1. ♖h6
1171.
1. ♕b8
1172.
1. ♘b3
1173.
1. ♖h5
1174.
1. ♖f6
1175.
1. ♕h8
1176.
1. ♖f6
1177.
1. ♖e4
1178.
1. ♕h8
1179.
1. ♖e4
1180.
1. ♕h6
1181.
1. ♖h8
1182.
1. ♕h8
1183.
1. ♘f6
1184.
1. ♗g5
1185.
1. ♖f7
1186.
1. ♕g8
1187.
1. ♘b6
1188.
1. ♕h6
1189.
1. ♘b6
1190.
1. ♕h6
1191.
1. ♖h5
1192.
1. ♘g6
1193.
1. ♕h6
1194.
1. ♕e1
1195.
1. ♕h6
1196.
1. ♖e8
1197.
1. ♖h5
1198.
1. ♕d8
1199.
1. ♕e6
1200.
1. ♕f6
1201.
1. ♕g6
1202.
1. ♕e5
1203.
1. ♖f7
1204.
1. ♕a6
1205.
1. ♘d3
1206.
1. ♗e5
1207.
1. d3
1208.
1. ♕e5
1209.
1. ♕a7
1210.
1. ♗f6
1211.
1. ♕a6
1212.
1. ♕g6
1213.
1. ♕e5
1214.
1. ♖h8
1215.
1. a6
1216.
1. ♘f6
1217.
1. ♕a7
1218.
1. ♖h3
1219.
1. ♕h4
1220.
1. ♕e6
1221.
1. ♕e6
1222.
1. ♖g8
1223.
1. ♖h2
1224.
1. ♖h6
1225.
1. ♕e6
1226.
1. ♘f6
1227.
1. ♕e6
1228.
1. ♕f7
1229.
1. ♕g6
1230.
1. ♘d2
1231.
1. ♕f7
1232.
1. ♕f7
1233.
1. ♕f4
1234.
1. ♕f7
1235.
1. ♘d5
1236.
1. ♘d6
1237.
1. ♕f7
1238.
1. ♘d1
1239.
1. ♘d7
1240.
1. ♕f7
1241.
1. ♕d8
1242.
1. ♘g5
1243.
1. ♕f8
1244.
1. ♕h6
1245.
1. ♕h7
1246.
1. ♕f6
1247.
1. ♕e6
1248.
1. ♖h5
1249.
1. ♕h3
1250.
1. ♖c6
1251.
1. ♘c5
1252.
1. ♕f8
1253.
1. ♖a8
1254.
1. ♕b6
1255.
1. – ♖d1
1256.
1. – ♖c2
1257.
1. – ♖h6
1258.
1. – g4
1259.
1. – ♖e1
1260.
1. – ♕f1
1261.
1. – ♘e3
1262.
1. – ♖h3
1263.
1. – ♕c1
1264.
1. – ♕e1
1265.
1. – ♗h3
1266.
1. – ♗d4
1267.
1. – ♕g1
1268.
1. – ♕d3
1269.
1. – ♕f1
1270.
1. – ♕f1
1271.
1. – ♕g1
1272.
1. – ♕h3
1273.
1. – ♕e1
1274.
1. – ♕f1
1275.
1. – ♕f1

1276.
1. – ♕g2
1277.
1. – ♕e1
1278.
1. – ♗e2
1279.
1. – ♘a3
1280.
1. – ♖a1
1281.
1. – ♖f1
1282.
1. – ♖f2
1283.
1. – ♗d3
1284.
1. – ♘d3
1285.
1. – ♕h3
1286.
1. – ♕g1
1287.
1. – ♕g2
1288.
1. – ♗g1
1289.
1. – ♖f1
1290.
1. – ♖h3
1291.
1. – ♘a3
1292.
1. – ♖e5
1293.
1. – ♗c4
1294.
1. – ♖g2
1295.
1. – ♘g5
1296.
1. – ♖f3
1297.
1. – ♘c5
1298.
1. – ♕b2
1299.
1. – ♕g4
1300.
1. – ♖g2
1301.
1. – ♘f3
1302.
1. – ♖h3
1303.
1. – ♖f1
1304.
1. – ♗e4
1305.
1. – ♖h3
1306.
1. – ♖b1
1307.
1. – ♘c5
1308.
1. – ♘f4
1309.
1. – ♗d4
1310.
1. – ♖c1
1311.
1. – ♖h4
1312.
1. – ♕f1
1313.
1. – ♕h3
1314.
1. – ♖g1
1315.
1. – ♖h3
1316.
1. – ♘g3
1317.
1. – ♘f3
1318.
1. – ♖h3
1319.
1. – ♘b3
1320.
1. – ♘g4
1321.
1. – ♖e2
1322.
1. – ♘e3
1323.
1. – ♖e1
1324.
1. – ♖h5
1325.
1. – ♖h3
1326.
1. – ♘g3
1327.
1. – ♕g3
1328.
1. – ♖f5
1329.
1. – ♖g2
1330.
1. – ♖c2
1331.
1. – ♘e3
1332.
1. – ♕c3
1333.
1. – ♕g1
1334.
1. – ♕b3
1335.
1. – ♕e1
1336.
1. – ♕h3
1337.
1. – ♕g3
1338.
1. – ♕g3
1339.
1. – ♕g2
1340.
1. – ♕g1
1341.
1. – ♕e3
1342.
1. – ♕g3
1343.
1. – ♕f2
1344.
1. – ♕f2
1345.
1. – ♕g3
1346.
1. – ♕b1
1347.
1. – ♕h2
1348.
1. – ♕e1
1349.
1. – ♕g6
1350.
1. – ♕g1
1351.
1. – ♖f1
1352.
1. – ♗h3
1353.
1. – ♖h1
1354.
1. – ♘g3
1355.
1. – ♘g3
1356.
1. – ♖a3
1357.
1. – ♘c4
1358.
1. – ♖a1
1359.
1. – c4
1360.
1. – ♘e2
1361.
1. – ♖c4
1362.
1. – ♖c1
1363.
1. – ♘g3
1364.
1. – ♕h2
1365.
1. – ♖h1
1366.
1. – ♗g4
1367.
1. – ♖h1
1368.
1. – ♕b4
1369.
1. – ♖h4
1370.
1. – ♗f4
1371.
1. – ♕h3
1372.
1. – ♖h3
1373.
1. – ♘f2
1374.
1. – ♕h2
1375.
1. – ♕h1
1376.
1. – ♕g4
1377.
1. – ♖h1
1378.
1. – ♗g3
1379.
1. – ♕g1
1380.
1. – ♕c1
1381.
1. – ♕g1
1382.
1. – ♕g2
1383.
1. – ♕g2
1384.
1. – ♕g5
1385.
1. – ♕g5
1386.
1. – ♕e3
1387.
1. – ♕d2

1388.
1. – ♖f5
1389.
1. – ♘c3
1390.
1. – ♘f4
1391.
1. – ♘d3
1392.
1. – ♘d3
1393.
1. – ♖d1
1394.
1. – f6
1395.
1. – ♖h4
1396.
1. – ♖h4
1397.
1. – ♖h1
1398.
1. – ♖h2
1399.
1. – ♖d1
1400.
1. – ♕h4
1401.
1. – ♖d1
1402.
1. – ♕e3
1403.
1. – ♕e1
1404.
1. – ♕g3
1405.
1. – ♖f3
1406.
1. – ♕e1
1407.
1. – ♕h2
1408.
1. – ♕c2
1409.
1. – ♕g2
1410.
1. – ♕e1
1411.
1. – ♕f1
1412.
1. – ♗f5
1413.
1. – ♕h3
1414.
1. – ♕b2
1415.
1. – ♖c1
1416.
1. – ♕d7
1417.
1. – ♕e1
1418.
1. – ♕f3
1419.
1. – ♖d1
1420.
1. – h4
1421.
1. – ♖c1
1422.
1. – g1♘
1423.
1. – ♕b1
1424.
1. – ♗a1
1425.
1. – ♘f6
1426.
1. – ♕e4
1427.
1. – ♕e4
1428.
1. – ♕e4
1429.
1. – ♕d3
1430.
1. – ♕h4
1431.
1. – ♕h3
1432.
1. – ♕e5
1433.
1. – ♕h3
1434.
1. – ♕h3
1435.
1. – ♕h3
1436.
1. – ♕h3
1437.
1. – ♕h4
1438.
1. – ♕h4
1439.
1. – ♕h3
1440.
1. – ♕h2
1441.
1. – ♕f3
1442.
1. – ♕f4
1443.
1. – ♕f4
1444.
1. – ♕a3
1445.
1. – ♘e5
1446.
1. – ♖h4
1447.
1. – ♕g2
1448.
1. – ♕h2
1449.
1. – ♖h2
1450.
1. – ♗c3
1451.
1. – ♕g2
1452.
1. – ♕f5
1453.
1. – ♕h3
1454.
1. – ♕c3
1455.
1. – h4
1456.
1. – ♕d3
1457.
1. – ♗c3
1458.
1. – ♘d2
1459.
1. – ♗f3
1460.
1. – ♘h4
1461.
1. – ♕e1
1462.
1. – ♖f5
1463.
1. – ♕a2
1464.
1. – ♕a2
1465.
1. – ♖h3
1466.
1. – ♖g2
1467.
1. – ♕g3
1468.
1. – ♕c3
1469.
1. – ♕e2
1470.
1. – ♕c2
1471.
J. W. Abbott 1875
1. ♕h8
1472.
F. Abdurahmanovic 1958
1. ♘e4
1473.
M. Adabashev 1936
1. ♕g8
1474.
M. Adabashev 1936
1. ♖g7
1475.
Y. Afek 1978
1. ♖d3
1476.
I. Agapov 1986
1. ♕e3
1477.
F. Aitov 1979
1. ♕d4
1478.
M. Akchurin 1982
1. ♕f3
1479.
V. Alaikov 1966
1. ♖a1
1480.
M. Albasi 1992
1.♕d7
1481.
H. Albrecht 1938
1. ♘c5
1482.
R. Aliovsadzade 1989
1. ♗a4
1483.
R. Aliovsadzade – M. Vagidov 1980
1. ♘d2
1484.
R. Aliovsadzade – M. Vagidov 1978
1. ♕a2
1485.
J. Almay 1926
1. ♖d2
1486.
F. Amelung 1897
1. hg6
1487.
T. Amirov 1981
1. ♘c3
1488.
J. Andersen 1941
1. ♘c3

1489.
J. de Andrade 1923
1. ♖e6

1490.
A. Andreyev 1983
1. ♕f5

1491.
H. Animitza 1978
1. ♘e6

1492.
Y. Arefyev 1992
1. ♕f4

1493.
H. Aschehoug 1902
1. ♔a6

1494.
R. Asplund 1957
1. ♖d8

1495.
A. Azushin 1970
1. ♘g5

1496.
A. Azushin 1972/73
1. f8♕

1497.
A. Azushin 1972/73
1. f8♘

1498.
A. Azushin 1972/73
1. f8♖

1499.
A. Azushin 1971
1. b8♗

1500.
A. Azushin 1971
1. b8♘

1501.
A. Azushin 1972/73
1. f8♗

1502.
A. Azushin 1971
1. b8♖

1503.
A. Azushin 1971
1. b8♕

1504.
V. Babichev 1972/73
1. ♕c1

1505.
V. Babichev 1972/73
1. ♘d4

1506.
E. Backe 1985
1. c8♖

1507.
E. Backe 1969
1. ♗h4

1508.
E. Backe 1967
1. ♖a1

1509.
E. Backe 1967
1. ♕a7

1510.
Mrs. W. J. Baird 1924
1. ♔f6

1511.
Mrs. W. J. Baird 1924
1. ♔a2

1512.
Mrs. W. J. Baird 1907
1. ♔f8

1513.
Mrs. W. J. Baird 1907
1. ♘c6

1514.
Mrs. W. J. Baird 1907
1. ♔h6

1515.
Mrs. W. J. Baird 1907
1. ♘e6

1516.
Mrs. W. J. Baird 1907
1. ♘c8

1517.
Mrs. W. J. Baird 1907
1. ♗b8

1518.
Mrs. W. J. Baird 1907
1. ♘d7

1519.
Mrs. W. J. Baird 1907
1. ♘c5

1520.
Mrs. W. J. Baird 1907
1. ♗g8

1521.
Mrs. W. J. Baird 1907
1. ♔f6

1522.
Mrs. W. J. Baird 1907
1. ♘b5

1523.
Mrs. W. J. Baird 1907
1. ♗b8

1524.
Mrs. W. J. Baird 1907
1. ♗f6

1525.
Mrs. W. J. Baird 1907
1. ♘c6

1526.
Mrs. W. J. Baird 1907
1. ♕f8

1527.
Mrs. W. J. Baird 1907
1. c7

1528.
Mrs. W. J. Baird 1907
1. ♘c8

1529.
Mrs. W. J. Baird 1907
1. ♔b7

1530.
Mrs. W. J. Baird 1907
1. ♘b5

1531.
Mrs. W. J. Baird 1907
1. ♗g5

1532.
Mrs. W. J. Baird 1907
1. ♖b7

1533.
J. Bajtay 1926
1. ♔c3

1534.
J. Bajtay 1932
1. c8♖

1535.
J. Bajtay 1926
1. ♘e6

1536.
B. Bakay 1932
1. ♕a6

1537.
B. Bakay 1932
1. ♘e3

1538.
B. Bakay 1931
1. ♕g3

1539.
B. Bakay 1931
1. ♕d6

1540.
B. Bakay 1931
1. h4

1541.
B. Bakay 1931
1. ♘b8

1542.
B. Bakay 1931
1. ♘a5

1543.
B. Bakay – T. Feldmann 1932
1. ♘b1

1544.
Gy. Bakcsi 1981
1. ♕f5

1545.
Gy. Bakcsi 1973
1. ♕d3

1546.
Gy. Bakcsi 1990
1. d8♘

1547.
Gy. Bakcsi 1948
1. b6

1548.
Gy. Bakcsi – L. Zoltán 1993
1. ♘g3

1549.
Gy. Bakcsi 1981/82
1. ♖h5

1550.
Gy. Bakcsi 1982
1. ♖f6

1551.
Gy. Bakcsi 1982
1. ♖e5

1552.
Gy. Bakcsi 1981/82
1. ♖h4

1553.
Gy. Bakcsi – Á. Molnár 1961
1. ♘eg7

1554.
Gy. Bakcsi – L. Zoltán 1993
1. ♕a2

1555.
Gy. Bakcsi 1981
1. ♕c4

1556.
N. A. Bakke 1972
1. c8♕

1557.
N. A. Bakke 1972
1. b8♕

1558.
N. A. Bakke 1990
1. 0-0

1559.
N. A. Bakke 1972
1. b8♖

1560.
N. A. Bakke 1972
1. b8♘

1561.
N. A. Bakke 1972
1. b8♘

1562.
N. A. Bakke 1972
1. b8♗

1563.
N. A. Bakke 1972
1. b8♖

1564.
N. A. Bakke 1972
1. c8♗

1565.
N. A. Bakke 1972
1. c8♘

1566.
N. A. Bakke 1972
1. c8♖

1567.
Gy. Balló 1927
1. ♗a3

1568.
Gy. Balló 1927
1. a7

1569.
Gy. Balló 1926
1. a5

1570.
Gy. Balló 1926
1. gh4

1571.
Gy. Balló 1926
1. f3

1572.
Gy. Balló 1926
1. d3

1573.
Gy. Balló 1928
1. ♕f8

1574.
Gy. Balló 1928
1. ♗a5

1575.
Gy. Balló 1928
1. ♗b4

1576.
Gy. Balló 1925
1. ♕b7

1577.
Gy. Balló 1926
1. ♗g6

1578.
Gy. Balló 1926
1. c7

1579.
Gy. Balló 1926
1. ♕c5

1580.
Gy. Balló 1926
1. ♕a2

1581.
Gy. Balló 1926
1. c3

1582.
Gy. Balló 1926
1. b5

1583.
Gy. Balló 1926
1. ♔a3

1584.
Gy. Balló 1926
1. a3

1585.
Gy. Balló – R. Pikler 1926
1. ♕e6

1586.
Gy. Balló – L. Lindner – E. Szentgyörgyi 1936
1. ♕d1

1587.
Gy. Balló – E. Szentgyörgyi 1935
1. ♕a4

1588.
Gy. Balló – E. Szentgyörgyi 1927
1. e6

1589.
D. Banny 1969
1. ♗g2

1590.
O. Barda 1961
1. ♘e7

1591.
Y. Barsky 1936
1. ♗a7

1592.
E. Barthélemy 1935
1. ♕d3

1593.
E. Barthélemy 1935
1. ♖e1

1594.
E. Barthélemy 1935
1. ♗h1

1595.
E. Barthélemy 1935
1. ♕a1

1596.
H. Bartolovich 1993
1. ♕f3

1597.
H. Bartolovich 1993
1. ♕e6

1598.
L. Bata 1942
1. ♕g5

1599.
E. Battaglia 1992
1. ♕b7

1600.
E. Battaglia 1990
1. b7

1601.
H. van Beek 1926
1. ♘c4

1602.
P. Bekkelund 1947
1. ♔a5

1603.
V. Beloborodov 1974
1. ♘a3

1604.
N. Belchikov 1971
1. ♘d7

1605.
N. Belchikov 1971
1. ♗f1

1606.
A. Benedek 1968
1. ♘b6

1607.
A. Benedek 1968
1. ♗g8

1608.
A. Benedek 1968
1. d3

1609.
A. Benedek 1968
1. ♗g4

1610.
A. Benedek 1970
1. ♘c2

1611.
A. Benedek 1969
1. ♗g6

1612.
A. Benedek 1970
1. ♕c7

1613.
P. Benkõ Original
1. d8♕

1614.
P. Benkõ Original
1. e8♘

1615.
P. Benkõ Original
1. dc8♕

1616.
P. Benkõ Original
1. ef8♗

1617.
P. Benkõ Original
1. ♕a1

1618.
P. Benkõ Original
1. ♕h1

1619.
P. Benkõ Original
1. ♘g3

1620.
P. Benkõ Original
1. ♗b6

1621.
P. Benkõ Original
1. ♘d3

1622.
P. Benkõ 1980
1. ♕c4

1623.
P. Benkõ 1974
1. ♖e5

1624.
M. Benoit 1975
1. ♕a6

1625.
H. D'O. Bernard 1902/3
1. ♕h8

1626.
F. Bethge 1951
1. ♕b1

1627.
H. P. Bie 1967
1. ♖e5

1628.
O. Blumenthal 1907
1. ♔c3

1629.
O. Blumenthal 1901
1. ♕h7

1630.
O. Blumenthal 1902
1. ♕b8

1631.
E. Bogdanov 1973
1. ♕f2

1632.
E. Bogdanov 1973
1. ♘g5

1633.
E. Bogdanov 1973
1. ♗f2

1634.
E. Bogdanov 1973
1. ♗e7

1635.
E. Bogdanov 1977
1. ♔b8

1636.
E. Bogdanov 1972
1. ♖e1

1637.
E. Bogdanov 1972
1. ♖f2

1638.
E. Bogdanov 1972
1. ♔b2

1639.
E. Bogdanov 1972
1. ♖d5

1640.
E. Bogdanov 1971
1. ♖b2

1641.
L. Boissy 1951
1. ♖a5

1642.
S. Bondtke 1915
1. ♗e3

1643.
S. Boros 1929
1. ♘f6

1644.
S. Boros 1929
1. ♕g4

1645.
S. Boros 1925
1. ♖h5

1646.
S. Boros 1926
1. ♖h1

1647.
S. Boros 1926
1. ♘h6

1648.
S. Boros 1926
1. ♕g6

1649.
S. Boros 1926
1. f4

1650.
S. Boros 1926
1. ♔e1

1651.
S. Boros 1926
1. ♔e2

1652.
S. Boros 1926
1. ♘b5

1653.
S. Boros – F. Lazard 1926
1. ♘e2

1654.
E. Boswell 1929
1. ♘a4

1655.
E. Boswell 1953
1. d7

1656.
E. Boswell 1929
1. ♕a8

1657.
H. Brantberg 1972
1. ♕h3

1658.
J. Breuer 1934
1. ♕d6

1659.
V. Bron 1955
1. ♕c4

1660.
V. Bron 1955
1. ♕e1

1661.
H. H. Brouwer 1919
1. ♘e8

1662.
J. Brown 1902
1. ♕g1

1663.
E. Brunner 1927
1. ♔d3

1664.
E. Brunner 1926
1. ♔f4

1665.
J. Buglos 1990
1. ♕f8

1666.
R. Bukne 1946
1. d5

1667.
R. Bukne 1946
1. ♘bd1

1668.
R. Bukne 1955
1. ♕d4

1669.
J. J. Burbach 1948
1. ♕g7

1670.
J. J. Burbach 1948
1. ♔f6

1671.
A. Burmeister 1904
1. d4

1672.
T. H. Bvee 1976
1. ♖b5

1673.
T. H. Bvee 1976
1. ♖d7

1674.
T. H. Bvee 1976
1. ♖f5

1675.
T. H. Bvee 1976
1. ♖d3

1676.
W. E. Candy 1911
1. 0-0

1677.
G. E. Carpenter 1902
1. ♗d7

1678.
G. E. Carpenter 1873
1. ♕a6

1679.
G. E. Carpenter 1872
1. ♕h8

1680.
G. E. Carpenter 1873
1. ♕h3

1681.
P. T. Cate 1965
1. ♔b2

1682.
P. T. Cate 1972
1. ♕c7

1683.
P. T. Cate 1926
1. ♕h3

1684.
P. T. Cate 1926
1. ♕h3

1685.
P. T. Cate 1962
1. c7

1686.
V. Chayko 1987
1. 0-0-0

1687.
A. Charlick 1896
1. ♖a2

1688.
A. Charlick 1896
1. ♖b8

1689.
A. Chéron 1936
1. ♗f3

1690.
A. Chéron 1936
1. ♗g5

1691.
A. Chéron 1936
1. ♕e6

1692.
A. Chéron 1936
1. ♕b5

1693.
N. Chebanov 1982
1. ♘a1

1694.
N. Chebanov 1982
1. ♕d8

1695.
V. Chepizhny 1987
1. ♗c2

1696.
V. Chepizhny 1984
1. ♘b7

1697.
V. Chepizhny 1986
1. ♔h2

1698.
V. Chepizhny 1986/87
1. ♔b4

1699.
V. Chepizhny – M. Locker 1974
1. ♕h8

1700.
V. Chepizhny 1986
1. ♗g8

1701.
V. Chepizhny 1987
1. ♕b1

1702.
V. Chepizhny 1987
1. ♗h8

1703.
V. Chepizhny 1966
1. ♕d6

1704.
V. Chepizhny 1957
1. ♕c2

1705.
V. Chepizhny 1987
♘d2

1706.
V. Chepizhny 1987
1. ♖b5

1707.
V. Chepizhny 1986
1. ♖b6

1708.
V. Chepizhny – B. Pustovoy 1987
1. ♗h3

1709.
V. Chepizhny 1987
1. ♖h4

1710.
V. Chepizhny 1987
1. ♕b6

1711.
V. Chepizhny 1972
1. ♘d2

1712.
V. Chepizhny 1987
1. ♗f5

1713.
V. Chepizhny 1968
1. ♕h1

1714.
V. Chepizhny 1968
1. ♕f1

1715.
V. Chepizhny 1967
1. ♗c1

1716.
V. Chepizhny 1968
1. ♘d3

1717.
V. Chepizhny 1968
1. ♛e1

1718.
V. Chepizhny 1984
1. ♗a2

1719.
V. Chepizhny 1984
1. ♜b4

1720.
V. Chepizhny 1983/84
1. ♗d1

1721.
N. Cherniavsky 1976
1. ♘e8

1722.
E. B. Cook 1868
1. g8♜

1723.
N. N. after E. B. Cook 1913
1. ♗f1

1724.
E. B. Cook 1868
1. ♛a1

1725.
E. B. Cook 1868
1. ♜h3

1726.
C. H. Courtenay 1868
1. ♛e3

1727.
C. H. Courtenay 1870
1. d8♗

1728.
J. Cumpe 1908
1. ♔d6

1729.
S. Cirulik 1976
1. ♜d7

1730.
J. Csontos 1926
1. ♛f5

1731.
J. Csontos 1925
1. ♘c2

1732.
J. Csontos – E. Szentgyörgyi 1926
1. ♜b5

1733.
A. W. Daniel 1906
1. ♜c6

1734.
F. Davidenko 1983
1. ♗h2

1735.
T. R. Dawson 1922
1. ♗b5

1736.
T. R. Dawson 1920
1. b8♘

1737.
T. R. Dawson 1929
1. ♗a7

1738.
A. Decker 1903
1. ♔e2

1739.
O. Dehler 1925
1. ♘c6

1740.
O. Dehler 1919
1. ♜d7

1741.
O. Dehler 1928
1. ♛h7

1742.
O. Dehler 1923
1. ♗h2

1743.
E. Delaliau 1957
1. ♗h4

1744.
N. Derevenko 1936
1. ♜f7

1745.
N. Derevenko 1936
1. f4

1746.
N. G. G. van Dijk 1957
1. ♗c8

1747.
N. G. G. van Dijk 1959
1. ♔f2

1748.
N. G. G. van Dijk 1959
1. ♔f7

1749.
N. G. G. van Dijk 1972
1. e5

1750.
N. G. G. van Dijk 1972
1. c5

1751.
N. G. G. van Dijk 1948
1. ♗d1

1752.
N. G. G. van Dijk 1972
1. ♜h8

1753.
N. G. G. van Dijk 1968
1. ♜d5

1754.
N. Dimitrov 1960
1. ♜g3

1755.
N. Dimitrov 1971
1. ♘h5

1756.
R. Diot 1953
1. ♘b8

1757.
E. Dobrescu 1955
1. c4

1758.
A. Dombrovskis 1970
1. ♔a3

1759.
Y. Dorohov 1958
1. ♗g4

1760.
Y. Dorohov 1960
1. ♘c3

1761.
S. Dragostinov 1968
1. ♛b5

1762.
G. H. Drese 1947
1. ♔d7

1763.
D. D'Sylva 1929
1. ♜g1

1764.
A. Dzekcer 1955
1. ♔h6

1765.
A. Dzekcer 1968/69
1. ♗f5

1766.
A. Dzekcer 1968/69
1. ♘d2

1767.
A. Dzekcer 1968/69
1. ♜b4

1768.
A. Dzekcer 1970
1. e4

1769.
A. Dzekcer 1970
1. f4

1770.
A. Dzekcer 1970
1. b4

1771.
A. Dzekcer 1972
1. ♛e1

1772.
F. Dubbe 1902
1. ♘e6

1773.
S. H. East 1937
1. ♗c2

1774.
F. Eidem 1958
1. d8♘

1775.
E. A. Ekholm 1930
1. ♜b7

1776.
E. A. Ekholm 1931
1. a8♘

1777.
D. Elekes 1926
1. ♛e7

1778.
D. Elekes 1927
1. ♔e4

1779.
B. Ellinghoven 1971
1. ♛h3

1780.
B. Elmgren 1950
1. ♛e3

1781.
I. Enroth 1938
1. ♘h1

1782.
V. Erohin 1986
1. ♘g4

1783.
V. Erohin 1986
1. ♘c4

1784.
K. Fabel 1937
1. ♔e7

1785.
N. Fadeyev 1982
1. ♛a1

1786.
L. Faivuzinsky 1938
1. ♛g3

1787.
V. Fedorin 1982
1. ♛a4

1788.
R. Fedorovich 1986
1. ♛f5

1789.
R. Fedorovich 1986
1. f6

1790.
K. Fedoseyev 1936
1. ♜d5

1791.
T. Feldmann (Flórián) 1933
1. ♕g5

1792.
T. Feldmann (Flórián) 1933
1. ♘g5

1793.
T. Feldmann (Flórián) 1933
1. ♘g5

1794.
T. Feldmann (Flórián) 1933
1. ♖d3

1795.
T. Feldmann (Flórián) 1933
1. ♖hh2

1796.
T. Feldmann (Flórián) 1932
1. ♘a5

1797.
A. Feoktistov 1970
1. ♕c6

1798.
A. Feoktistov 1970
1. ♕d2

1799.
F. Ferber 1903
1. c7

1800.
A. Fetmaks 1990
1. ♘e5

1801.
N. Firbás 1979
1. ♕h6

1802.
A. Fossum 1951
1. ♘e1

1803.
W. F. von Holzhausen 1902
1. ♖a8

1804.
V. Frigin 1985/86
1. ♕d3

1805.
G. Z. 1893
1. ♗g8

1806.
A. Galitzky 1905
1. ♖d2

1807.
A. Galitzky 1903
1. ♔e5

1808.
A. Galitzky 1890
1. ♔h2

1809.
N. Galiletzky 1967/68
1. ♕c5

1810.
K. Gavrilov 1926
1. ♕e5

1811.
K. Gavrilov 1931
1. g4

1812.
K. Gavrilov 1905
1. ♕h2

1813.
K. Gavrilov 1904
1. ♕c7

1814.
K. Gavrilov 1933
1. ♘d5

1815.
K. Gavrilov 1902
1. ♕c7

1816.
K. Gavrilov 1904
1. ♖b6

1817.
Z. Gavrilovski 1989
1. ♕b4

1818.
I. Gazimon 1969
1. ♗d5

1819.
I. Gazimon 1969
1. ♗d6

1820.
I. Gegelsky 1958
1. ♕g2

1821.
I. Gegelsky 1938
1. ♖h2

1822.
B. Genkin 1964
1. ♕e7

1823.
A. George 1984
1. ♕a8

1824.
I. Gersits 1930
1. ♘f6

1825.
N. M. Gibbins 1916
1. ♕e5

1826.
B. Giöbel 1927
1. ♖b3

1827.
B. Giöbel 1945
1. ♕e4

1828.
B. Giöbel 1923
1. ♕h3

1829.
G. Gnilomedov 1971
1. ♖g3

1830.
G. Gnilomedov 1971
1. ♕e7

1831.
G. Gnilomedov 1971
1. ♖g3

1832.
G. Gnilomedov 1971
1. ♕e7

1833.
G. Gnilomedov 1971
1. ♔c3

1834.
G. Gnilomedov 1974
1. ♕h1

1835.
G. Gnilomedov 1974
1. ♕f7

1836.
I. Godal 1965
1. ♘g1

1837.
V. Gofman 1900
1. ♕d8

1838.
S. Gold 1896
1. ♕d8

1839.
S. Gold 1902
1. ♔e3

1840.
S. Gold 1902
1. ♔c5

1841.
C. Goldschmeding 1953/54
1. ♘d5

1842.
E. Goldschmiedt 1926
1. g8♗

1843.
E. Goldschmiedt 1926
1. ♘c5

1844.
E. Goldschmiedt 1926
1. g7

1845.
E. Goldschmiedt 1927
1. ♘e4

1846.
E. Goldschmiedt 1927
1. ♗d1

1847.
H. von Gottschal 1892
1. ♗f5

1848.
H. von Gottschal 1926
1. ♗h1

1849.
H. von Gottschal 1926
1. ♕a7

1850.
H. von Gottschal 1907
1. ♕h7

1851.
M. Gordian 1923
1. ♔g7

1852.
M. Gorislavsky 1971
1. ♕b4

1853.
M. Gorislavsky 1971
1. ♖b1

1854.
M. Gorislavsky 1971
1. c8♘

1855.
M. Gorislavsky 1971
1. c8♗

1856.
M. Gorislavsky 1971
1. b8♕

1857.
M. Gorislavsky 1982
1. ♕f1

1858.
H. Le Grand 1953
1. ♘g7

1859.
H. Le Grand 1936
1. ♕g4

1860.
W. Greenwood 1880
1. ♗b1

1861.
R. Grewe – F. Karge – H. Voigt 1940
1. ♕e8

1862.
A. Grin 1972
1. c7

1863.
A. Grin 1985
1. ♖g6

1864.
A. Grin 1954
1. ♕a8

1865.
A. Grin 1987
1. ♕b2

1866.
A. Grin 1985/86
1. ♕c3

1867.
D. Grinchenko 1979
1. ♗c3

1868.
M. Grönroos 1980
1. ♕h5

1869.
A. Grunenwald 1950
1. ♖e8

1870.
C. Groeneveld 1935
1. ♕f7

1871.
G. Guidelli 1924
1. ♔g3

1872.
A. Gulyaev 1936
1. ♕g4

1873.
A. Gulyaev 1957
1. ♕f1

1874.
A. Gulyaev 1929
1. ♗e6

1875.
J. Gunst 1925
1. ♕d4

1876.
E. M. H. Gutmann 1930
1. ♘e8

1877.
A. H. 1868
1. ♕d8

1878.
S. Hachaturov 1972
1. ♖e5

1879.
E. Hällström 1903
1. ♘a4

1880.
E. Hällström 1925
1. ♕h7

1881.
E. Hällström 1903
1. ♔e4

1882.
J. Hannelius 1982
1. ♖d3

1883.
J. Hannelius 1982
1. ♕e3

1884.
J. Hannelius 1978
1. ♕g5

1885.
J. Hannelius 1981
1. ♗e3

1886.
J. Hannelius 1974
1. ♕b6

1887.
J. Hannelius 1957
1. ♘c4

1888.
J. Hannelius 1971
1. ♖g5

1889.
J. Hannelius 1945
1. g7

1890.
K. Hannemann 1924
1. ♔f7

1891.
K. Hannemann 1932
1. c8♖

1892.
K. Hannemann 1932
1. d8♕

1893.
K. Hannemann 1932
1. e8♗

1894.
K. Hannemann 1932
1. b8♘

1895.
K. Harczi 1926
1. ♘c6

1896.
B. Harley – C. G. Watney 1921
1. ♕c8

1897.
J. Haring 1965
1. ♖b8

1898.
J. Haring 1968
1. d8♕

1899.
J. Haring 1968
1. d8♘

1900.
J. Haring 1968
1. d8♖

1901.
J. Haring 1968
1. d8♗

1902.
J. Haring 1955
1. ♖f4

1903.
H. Hartogh 1881
1. ♕d2

1904.
J. Hartong 1949
1. ♕a7

1905.
J. Hartong 1965
1. ♕b4

1906.
J. Hartong 1942
1. ♕h4

1907.
J. Hartong 1948
1. ♕a1

1908.
J. Hartong – J. Albarda 1962
1. ♘g8

1909.
E. M. Hassberg 1945
1. ♕c2

1910.
M. Havel 1903
1. ♘b3

1911.
M. Havel 1916
1. ♕c7

1912.
M. Havel 1900
1. ♖g4

1913.
F. Healey 1903
1. ♖d8

1914.
H. Hermanson 1958
1. d5

1915.
H. Hermanson – O. Stocchi 1929
1. e4

1916.
H. Hermanson 1954
1. ♕e3

1917.
H. Hermanson 1930
1. ♕b1

1918.
H. Hermanson 1957
1. ♕e4

1919.
Z. Hernitz 1971
1. ♕b8

1920.
E. Hinkka 1961
1. ♕f7

1921.
E. Hinkka 1966
1. ♕e5

1922.
D. Hjelle 1956
1. a5

1923.
D. Hjelle 1957
1. ♔d3

1924.
N. Høeg 1937
1. ♘c7

1925.
N. Høeg 1907
1. ♕f2

1926.
N. Høeg 1926
1. c8♘

1927.
W. Hoek 1970
1. ♗a3

1928.
W. Hoek 1972
1. ♘c7

1929.
W. Hoek 1969
1. ♖a4

1930.
S. Hoffenreich 1925
1. ♗d1

1931.
S. Hoffenreich 1925
1. ♕d4

1932.
E. Holladay 1978
1. ♔g7

1933.
W. von Holzhausen 1899
1. ♖a8

1934.
W. von Holzhausen 1924
1. ♖e5

1935.
M. Hramzevich 1985/86
1. ♗h3

1936.
B. Hülsen 1915
1. ♕c5

1937.
H. Hurme 1962–64
1. ♕hb2

1938.
H. Hurme 1964
1. ♔h2

1939.
S. Inostroza 1986
1. ♗f3

1940.
S. Inostroza 1990
1. ♘c6

1941.
S. Inostroza 1986
1. ♕a7

1942.
S. Inostroza 1990
1. ♖f7

1943.
S. Inostroza 1986
1. ♖d7

1944.
S. Inostroza 1986
1. ♗a1

1945.
N. Ivanovsky 1936
1. ♖a3

1946.
N. Ivanovsky 1936
1. ♔f3

1947.
M. Ivanov 1982
1. ♖h2

1948.
L. Ivanova 1947
1. ♗c3

1949.
A. Ivanov 1966
1. ♕f6

1950.
L. Iskra 1970
1. ♖c6

1951.
L. Iskra 1970
1. ♖a4

1952.
I. Ignatev 1965
1. ♖b7

1953.
Y. Ishta 1983
1. ♕h3

1954.
C. E. M. Jago 1986
1. ♕a5

1955.
Á. Jakab 1922
1. ♖e7

1956.
Á. Jakab 1922
1. ♔c3

1957.
Á. Jakab 1963
1. ♘b6

1958.
Zh. Janevski 1970
1. ♕c7

1959.
A. Yaroslavtzev 1967
1. ♖ga8

1960.
A. Jenssen 1925
1. ♗h5

1961.
J. Jespersen 1880
1. ♕b7

1962.
C. Jones 1991
1. ♗e6

1963.
K. Junker 1956
1. ♘c6

1964.
K. Junker – W. Speckmann 1964
1. ♗f5

1965.
A. Jussila 1944
1. ♗d7

1966.
O. Justnes 1922
1. ♘d8

1967.
E. Yusupov 1985/86
1. ♕g3

1968.
G. Kaiser 1939
1. ♕e7

1969.
G. Kaiser 1950
1. ♗c3

1970.
G. Kakabadze 1973
1. ♕a6

1971.
A. Kakovich 1972
1. ♕f6

1972.
V. Kalina 1925
1. ♕e4

1973.
V. Kalina 1925
1. ♗e1

1974.
V. Kamenetzky 1978/81
1. ♖g5

1975.
H. H. Kamstra 1929
1. ♖ff7

1976.
H. H. Kamstra 1929
1. ♖a1

1977.
D. Kanonik 1974
1. ♗b6

1978.
D. N. Kapralos – P. Moutecidis 1961
1. ♘c6

1979.
T. Kardos 1969
1. ♕h1

1980.
T. Kardos 1970
1. ♖h1

1981.
T. Kardos 1962
1. e8♕

1982.
T. Kardos 1970
1. g8♕

1983.
T. Kardos 1970
1. ♕e1

1984.
T. Kardos 1978
1. c7

1985.
T. Kardos 1975
1. b8♕

1986.
T. Kardos 1981
1. ♕g7

1987.
T. Kardos 1980
1. ♕a1

1988.
T. Kardos 1983
1. ♕h6

1989.
T. Kardos 1982
1. ♕g8

1990.
T. Kardos 1985
1. ♘c4

1991.
T. Kardos 1981
1. 0-0-0

1992.
T. Kardos 1986
1. ♕c1

1993.
T. Kardos 1988
1. ♕b6

1994.
T. Kardos 1990
1. ♖e2

1995.
T. Kardos 1987
1. ♕a3

1996.
T. Kardos 1988
1. ♖e5

1997.
T. Kardos 1990
1. ♖e5

1998.
T. Kardos 1992
1. ♕f6

1999.
T. Kardos 1990
1. b8♕

2000.
T. Kardos 1971
1. f8♕

2001.
T. Kardos 1971
1. ♘g2

2002.
T. Kardos 1972
1. a8♕

2003.
T. Kardos 1962
1. ♕h7

2004.
T. Kardos 1962
1. ♕e2

2005.
T. Kardos 1966
1. ♔c7

2006.
T. Kardos 1962
1. e8♕

2007.
T. Kardos 1966
1. c8♘

2008.
T. Kardos 1963
1. ♕h4

2009.
T. Kardos 1970
1. ♔c7

2010.
T. Kardos 1987
1. ♖g8

2011.
T. Kardos 1987
1. ♖gb3

2012.
T. Kardos 1987
1. ♖c3

2013.
T. Kardos 1971
1. ♗d5

2014.
T. Kardos 1971
1. e8♖

2015.
T. Kardos 1984
1. 0-0

2016.
T. Kardos 1987
1. ♖e7

2017.
T. Kardos 1962
1. b6

2018.
T. Kardos 1984
1. 0-0-0

2019.
T. Kardos 1981
1. e8♕

2020.
T. Kardos 1970
1. ♖b5

2021.
T. Kardos 1970
1. ♖g8

2022.
T. Kardos 1970
1. ♖b2

2023.
T. Kardos 1970
1. ♖f2

2024.
T. Kardos 1969
1. ♕g2

2025.
T. Kardos 1969
1. ♕f5

2026.
T. Kardos 1968
1. ♕e4

2027.
T. Kardos 1971
1. ♘d2

2028.
T. Kardos – W. Speckmann 1971
1. c8♖

2029.
T. Kardos – W. Speckmann 1970
1. ♕h2

2030.
T. Kardos – W. Speckmann 1971
1. f8♕

2031.
A. Karlström 1946
1. ♔c2

2032.
A. Karlström 1935
1. ♖c4

2033.
V. Karpov 1970
1. ♕c7

2034.
N. Kashcheyev 1976
1. ♕e1

2035.
N. Kashcheyev 1955
1. ♖b1

2036.
N. Kashcheyev 1966
1. ♕f7

2037.
N. Kashcheyev 1964
1. ♘e2

2038.
N. Kashcheyev 1930
1. ♕d5

2039.
I. Katkó 1927
1. ♖h8

2040.
I. Katkó 1927
1. ♘d6

2041.
I. Katkó 1927
1. ♕a5

2042.
I. Katkó 1927
1. ♗c7

2043.
I. Katkó 1927
1. ♗f8

2044.
I. Katkó 1927
1. ♖f7

2045.
I. Katkó 1927
1. ♗e1

2046.
I. Katkó 1927
1. ♘c2

2047.
I. Katkó 1927
1. a5

2048.
I. Katkó 1927
1. ♗h6

2049.
I. Katkó 1927
1. ♕h8

2050.
H. Keidanski 1902
1. d4

2051.
A. Keirans 1936
1. ♕c5

2052.
P. Keres 1933
1. ♖d7

2053.
R. Kintzig 1919
1. ♖c4

2054.
R. Kintzig 1923
1. ♘d4

2055.
C. S. Kipping 1910
1. c8♗

2056.
C. S. Kipping 1947
1. ♕g6

2057.
A. Kirilenko 1971
1. ♔f4

2058.
S. Kirillov 1985
1. ♖a3

2059.
J. Kiss 1970
1. ♗d8

2060.
J. Kiss 1941
1. ♘e6

2061.
V. Kiselev 1965
1. d8♘

2062.
I. Kisis 1981
1. ♘c3

2063.
D. Klark 1876
1. ♕a4

2064.
D. Klark 1879
1. ♖d6

2065.
J. Klemensiewicz 1926
1. ♘e2

2066.
J. Kling 1849
1. ♕h2

2067.
J. Kloostra 1986
1. ♕h6

2068.
J. Kloostra 1985
1. ♕b3

2069.
L. Knotek 1922
1. ♘h3

2070.
H. Knuppert 1941
1. ♕d7

2071.
G. Koder 1974
1. ♘c8

2072.
R. Kofman 1968
1.♖a5

2073.
J. Kohtz – C. Kockelkorn 1876
1. f8♗

2074.
M. Kokkonen 1967
1. ♗b3

2075.
A. M. Koldijk 1938
1. ♗h7

2076.
Z. Kolodnas 1927
1. e8♗

2077.
J. W. Te Kolste 1891
1. ♔d2

2078.
N. Kondratyuk 1974
1. ♔e6

2079.
N. Kondratyuk 1974
1. ♔g6

2080.
N. Kondratyuk 1969
1. ♗f5

2081.
N. Kondratyuk 1969
1. ♘c5

2082.
J. de Koning 1909
1. ♘h5

2083.
V. Kopaev 1956
1. c4

2084.
V. Korenev 1973
1. ♘g5

2085.
A. Korepin 1937
1. ♖cc1

2086.
J. Korponai 1967
1. ♗a2

2087.
J. Korponai 1967
1. ♕c2

2088.
J. Korponai 1955/56
1. g8♖

2089.
N. Kosolapov 1968
1. ♕g5

2090.
N. Kosolapov 1964
1. ♕h3

2091.
I. Kostlan 1937
1. ♗f7

2092.
V. Kotlyar 1974
1. ♕h6

2093.
F. Kovács 1924
1. ♘e5

2094.
N. Kovács 1928
1. ♖g4

2095.
N. Kovács 1928
1. ♕b4

2096.
N. Kovács 1930
1. ♔b1

2097.
N. Kovács 1932
1. ♖c2

2098.
N. Kovács 1926
1. ♕f2

2099.
N. Kovács 1927
1. ♕e7

2100.
N. Kovács – Gy. Neukomm 1926
1. ♘c5

2101.
N. Kovács – D. Elekes 1924
1. ♕e5

2102.
V. Kovalenko 1968
1. ♖b6

2103.
V. Kovalenko 1968
1. ♕b5

2104.
V. Kozhakin 1985
1. ♗f3

2105.
V. Kozhakin 1985/86
1. ♖d5

2106.
A. Kraemer 1926
1. ♖c8

2107.
A. Kraemer 1925
1. ♔e4

2108.
S. Kreinin 1929
1. ♘c8

2109.
V. Krivenko 1976
1. ♖g6

2110.
L. Kubbel 1940
1. ♗f7

2111.
L. Kubbel 1908
1. ♕d4

2112.
L. Kubbel 1909
1. ♕h1

2113.
L. Kubbel 1933
1. ♕g7

2114.
L. Kubbel 1911
1. ♖f7

2115.
L. Kubbel 1940
1. ♕a7

2116.
L. Kubbel 1939
1. 0-0-0

2117.
L. Kubbel 1939
1. ♗c4

2118.
L. Kubbel 1941
1. ♕h1

2119.
L. Kubbel 1908
1. c4

2120.
L. Kubbel 1929
1. ♕d4

2121.
L. Kubbel 1909
1. ♕a2

2122.
L. Kubbel 1958
1.♕g3

2123.
L. Kubbel 1958
1. ♖g1

2124.
L. Kubbel 1940
1. ♕h6

2125.
L. Kubbel 1941
1. ♕e3

2126.
L. Kubbel 1941
1. ♔d4

2127.
G. Kunos 1937
1. ♗c4

2128.
V. V. Kuzmichev 1988
1. h4

2129.
V. V. Kuzmichev 1988
1. ♖d2

2130.
V. V. Kuzmichev 1987
1. ♖e5

2131.
An. Kuznetzov 1981
1. ♔e8

2132.
An. Kuznetzov – V. Chepizhny 1971
1. ♔c1

2133.
A. Kuzovkov 1982
1. ♖d2

2134.
V. Kviatovsky 1977
1. ♖a6

2135.
V. Kviatovsky 1977
1. ♗f8

2136.
V. Kviatovsky 1984
1. ♖a6

2137.
H. Laaksonen 1944
1. ♕e5

2138.
H. Laaksonen 1942
1. ♕b4

2139.
H. Laaksonen 1946
1. ♕e3

2140.
H. Laaksonen 1945
1. ♕h1

2141.
H. Laaksonen 1950
1. ♘d5

2142.
W. Langstaff – E. C. Mortimer 1922
1. ♕e3

2143.
B. Larsson 1952
1. ♕g3

2144.
B. Larsson 1962
1. ♖f5

2145.
E. Larson–Letzen 1921
1. ♔g8

2146.
K. A. K. Larsen 1921
1. ♗b1

2147.
G. Latzel 1956
1. ♘g5

2148.
F. Lazard 1933
1. c4

2149.
F. Lazard 1924
1. ♖b8

2150.
F. Lazard 1924
1. ♖g2

2151.
F. Lazard 1926
1. ♕h7

2152.
F. Lazard 1939
1. c8♕

2153.
F. Lazard 1926
1. ♗f3

2154.
F. Lazard 1926
1. ♕e4

2155.
F. Lazard 1926
1. ♘f6

2156.
F. Lazard 1926
1. ♘h5

2157.
F. Lazard 1935
1. ♕d2

2158.
F. Lazard 1935
1. ♕c5

2159.
F. Lazard 1935
1. d4

2160.
F. Lazard 1926
1. ♕a3

2161.
F. Lazard 1935
1. ♕d1

2162.
F. Lazard – T. Schönberger 1926
1. ♕f6

2163.
E. Lazdinsh 1958
1. ♗e2

2164.
A. Lebedev 1929
1. ♕e4

2165.
A. Lebedev 1936
1. ♕c8

2166.
A. Lebedev 1933
1. f7

2167.
A. Lebedev 1929
1. ♗f5

2168.
A. Lebedev 1929
1. ♗a7

2169.
A. Lebedev 1929
1. ♖e5

2170.
A. Lebedev 1929
1. ♕c3

2171.
A. Lebedev 1929
1. ♗e4

2172.
A. Lebedev 1935
1. ♕a1

2173.
A. Lebedev 1933
1. ♗d5

2174.
A. Lebedev 1934
1. ♔h3

2175.
A. Lebedev 1934
1. ♘b5

2176.
A. Lebedev 1934
1. ♔a5

2177.
A. Lebedev 1934
1. ♕g6

2178.
A. Lebedev 1935
1. ♕f1

2179.
A. and V. Lebedev 1933
1. ♕c2

2180.
A. and V. Lebedev 1930
1. ♕b4

2181.
A. and V. Lebedev 1936
1. ♔b1

2182.
V. Lebedev 1932
1. ♕a5

2183.
V. Lebedev 1932
1. ♕d4

2184.
V. Lebedev 1931
1. ♕f4

2185.
V. Lebedev 1934
1. ♕b2

2186.
M. Lebedinetz 1982
1. ♗f3

2187.
H. Lehner 1874
1. ♕e5

2188.
P. Leibovici 1933
1. ♕g8

2189.
G. Leon–Martin 1926
1. ♘f7

2190.
R. Leopold 1923
1. ♕h7

2191.
W. E. Lester 1924
1. ♕b1

2192.
W. E. Lester 1924
1. ♕h1

2193.
R. L'hermet 1923
1. ♕e5

2194.
S. Liberali 1881
1. ♘f6

2195.
V. Lider 1970
1. ♕e7

2196.
V. Lider 1971
1. ♕h3

2197.
R. T. Lewis 1985
1. ♕b6

2198.
R. T. Lewis 1985
1. ♘h2

2199.
R. T. Lewis 1985
1. ♕f2

2200.
R. T. Lewis 1988
1. ♖e2

2201.
R. T. Lewis 1993
1. ♕g3

2202.
R. T. Lewis 1993
1. ♘b4

2203.
R. T. Lewis 1990
1. ♖b1

2204.
R. T. Lewis 1990
1. d4

2205.
R. T. Lewis 1990
1. ♕e5

2206.
R. T. Lewis 1985
1. ♖g7

2207.
R. T. Lewis 1985
1. ♘h4

2208.
R. T. Lewis 1985
1. ♔g5

2209.
R. T. Lewis 1985
1. ♔b7

2210.
R. T. Lewis 1985
1. ♕f3

2211.
R. T. Lewis 1985
1. ♖ge2

2212.
R. T. Lewis 1986
1. ♗c6

2213.
R. T. Lewis 1986
1. ♔g6

2214.
S. Liljestrand 1957
1. ♘a5

2215.
S. Liljestrand 1962
1. ♘e8

2216.
R. Lincoln 1988
1. ♗f2

2217.
R. Lincoln 1992
1. ♗b2

2218.
R. Lincoln 1989
1. ♕d8

2219.
R. Lincoln 1993
1. ♘c4

2220.
R. Lincoln 1993
1. ♘c5

2221.
R. Lincoln 1990
1. ♖h8

2222.
R. Lincoln 1990
1. ♘h4

2223.
R. Lincoln 1993
1. ♖a5

2224.
R. Lincoln 1993
1. ♖e7

2225.
R. Lincoln 1993
1. ♕b1

2226.
R. Lincoln 1993
1. ♕c3

2227.
R. Lincoln 1992
1. ♕b4

2228.
R. Lincoln 1989
1. ♕f3

2229.
R. Lincoln 1989
1. ♕c3

2230.
R. Lincoln 1989
1. ♘e4

2231.
R. Lincoln 1990
1. ♕f1

2232.
R. Lincoln 1990
1. ♕b4

2233.
R. Lincoln 1992
1. ♕a1

2234.
R. Lincoln 1992
1. ♕h2

2235.
R. Lincoln 1992
1. ♗g4

2236.
R. Lincoln 1991
1. ♕b7

2237.
R. Lincoln 1991
1. ♕e6

2238.
R. Lincoln 1988
1. ♕d6

2239.
R. Lincoln 1988
1. ♕c7

2240.
R. Lincoln 1990
1. ♕e2

2241.
F. Lindgren 1928
1. ♕e3

2242.
L. Lindner 1933
1. b4

2243.
L. Lindner 1933
1. e4

2244.
L. Lindner 1933
1. ♔f5

2245.
L. Lindner 1933
1. c6

2246.
L. Lindner 1933
1. cb4

2247.
L. Lindner 1933
1. ♕e5

2248.
L. Lindner 1933
1. ♔c2

2249.
L. Lindner 1933
1. c3

2250.
L. Lindner – E. Szentgyörgyi 1933
1. ♕d1

2251.
L. Lindner – E. Szentgyörgyi 1933
1. d4

2252.
L. Lindner – E. Szentgyörgyi 1933
1. ♘h4

2253.
L. Lindner – E. Szentgyörgyi 1933
1. ♔h6

2254.
L. Lindner – E. Szentgyörgyi 1933
1. ♔b5

2255.
J. F. Ling 1953
1. ♕c6

2256.
M. Lipton 1956
1. ♗d6

2257.
M. Lipton 1960
1. ♗b6

2258.
M. Lipton 1952
1. ♕g6

2259.
E. Livshitz 1967
1. ♔g2

2260.
E. Livshitz 1967
1. ♔f2

2261.
E. Livshitz – L. Melnichenko 1967
1. ♗f4

2262.
I. Lyapunov – V. Fishman 1961
1. ♕d6

2263.
I. Lyapunov – V. Fishman 1961
1. ♖d6

2264.
E. Lobanov 1970
1. ♘e5

2265.
A. Lobusov 1975
1. ♗f2

2266.
M. Locker 1967
1. ♕g1

2267.
M. Locker 1966
1. ♘c6

2268.
M. Locker 1966
1. ♘b5

2269.
M. Locker 1970
1. ♗h6

2270.
M. Locker 1970
1. ♗f8

2271.
M. Locker 1970
1. ♘f4

2272.
M. Locker 1971
1. ♕e5

2273.
M. Locker 1968
1. ♘d5

2274.
M. Locker 1969
1. ♖f4

2275.
M. Locker 1969
1. ♔c3

2276.
M. Locker 1971
1. ♕f2

2277.
M. Locker 1969
1. ♔e2

2278.
M. Locker 1969
1. ♗c5

2279.
M. Locker 1969
1. ♘c2

2280.
M. Locker 1969
1. ♘c6

2281.
M. Locker 1969
1. ♘d5

2282.
M. Locker 1969
1. ♘4d3

2283.
M. Locker 1970
1. ♔g6

2284.
M. Locker 1972
1. h8♖

2285.
M. Locker 1970
1. ♗f7

2286.
M. Locker 1965
1. ♘d1

2287.
M. Locker 1971
1. a8♘

2288.
M. Locker 1972
1. ♘ce4

2289.
M. Locker 1973
1. ♕e2

2290.
M. Locker 1974
1. ♕e6

2291.
M. Locker 1965
1. ♔h4

2292.
M. Locker 1966
1. ♗b7

2293.
M. Locker 1967
1. ♘b2

2294.
M. Locker – V. Melnichenko 1968
1. ♗c3

2295.
M. Locker – V. Melnichenko 1968
1. ♗d2

2296.
M. Locker – V. Melnichenko 1968
1. ♗b4

2297.
M. Locker – V. Melnichenko 1968
1. ♗a5

2298.
M. Locker – A. Sidorenko 1969
1. ♗e6

2299.
M. Locker – A. Sidorenko 1969
1. ♕c3

2300.
M. Locker – A. Tuseev 1955
1. ♗f3

2301.
P. Loquin 1843
1. ♘b7

2302.
L. Loshinsky – A. Dombrovskis 1971
1. ♖h6

2303.
S. Love 1987
1. ♖h2

2304.
S. Loyd 1877
1. ♕a1

2305.
S. Loyd 1877
1. ♕a1

2306.
S. Loyd 1859
1. ♖e1

2307.
S. Loyd 1881
1. ♕a2

2308.
S. Loyd 1885
1. ♕a8

2309.
S. Loyd 1859
1. ♘c8

2310.
S. Loyd 1902
1. ♔f5

2311.
S. Loyd 1902
1. ♕g4

2312.
S. Loyd 1859
1. ♕a1

2313.
V. Lukyanov – S. Shedey 1975
1. ♖h1

2314.
V. Lukyanov 1967
1. ♕b7

2315.
V. Lukyanov 1966
1. ♗b5

2316.
J. H. Lund 1961
1. ♗c7

2317.
J. H. Lund 1958
1. ♕a4

2318.
Z. Mach 1905
1. ♕e3

2319.
Z. Mach 1907
1. ♕e5

2320.
Z. Mach 1924
1. ♕g8

2321.
Z. Mach 1904
1. ♕h3

2322.
Z. Mach 1899
1. ♖h6

2323.
Z. Mach 1910
1. ♖g5

2324.
V. Mach 1948
1. ♕h8

2325.
T. Maeder 1988
1. ♕c4

2326.
N. Meistrenko – V. Shevtzov 1971
1. ♖g6

2327.
L. Makaronetz 1972/73
1. ♕h1

2328.
L. Makaronetz 1974
1. ♕d1

2329.
I. Mäkihovi 1945
1. ♘a6

2330.
K. Makovsky 1881
1. ♕b8

2331.
N. Maksimov 1892
1. ♕g1

2332.
A. Maksimovskih 1972
1. ♗g5

2333.
A. Maksimovskih 1972
1. ♗d2

2334.
N. Maksimov 1903
1. ♕f5

2335.
N. Maksimov 1895
1. ♗e6

2336.
N. Maksimov 1903
1. ♕f1

2337.
G. Maleika 1982
1. ♕g4

2338.
C. Mansfield 1929
1. ♘f4

2339.
M. Marandyuk 1969
1. ♕h2

2340.
M. Marandyuk 1969
1. ♖a8

2341.
M. Marandyuk 1969
1. ♗g6

2342.
M. Marandyuk 1971
1. ♕a7

2343.
M. Marandyuk 1972
1. ♗c4

2344.
M. Marandyuk 1971
1. ♗g5

2345.
M. Marandyuk 1971
1. d3

2346.
M. Marandyuk 1971
1. ♕d3

2347.
M. Marandyuk 1973
1. ♕e4

2348.
M. Marandyuk 1976
1. ♕c4

2349.
M. Marandyuk 1969
1. ♕h2

2350.
M. Marble – H. W. Bettmann 1915
1. ♕e8

2351.
V. Marin 1895
1. ♖f5

2352.
G. Mariz 1989
1. ♗a3

2353.
G. Mariz 1986
1. ♔g3

2354.
G. Mariz 1985
1. ♕a1

2355.
G. Mariz 1986
1. ♘d7

2356.
P. J. Markkola 1960
1. ♕d4

2357.
G. Markovsky 1969
1. f8♘

2358.
G. Markovsky 1969
1. f8♗

2359.
G. Martin 1951
1. ♗g6

2360.
E. O. Martin 1934
1. ♖b7

2361.
W. Massmann 1947
1. ♔d1

2362.
W. Massmann 1962
1. ♖h3

2363.
O. Mathiassen 1960
1. ♔h3

2364.
N. Matyushin 1972/73
1. ♗h3

2365.
N. Matyushin 1973
1. ♖g4

2366.
E. Mazel 1902
1. ♖c8

2367.
E. Mazel 1903
1. ♕h3

2368.
O. Mazin 1986
1. ♖f3

2369.
M. McDowell 1991
1. ♔a2

2370.
M. McDowell 1989
1. ♕e3

2371.
M. McDowell 1992
1. ♘f6

2372.
B. McWilliam 1988
1. ♕d5

2373.
W. J. G. Mees 1959
1. d8♖

2374.
W. J. G. Mees 1959
1. d8♗

2375.
W. J. G. Mees 1942
1. ♔e5

2376.
W. J. G. Mees 1964
1. ♕g4

2377.
W. J. G. Mees 1953
1. ♕d6

2378.
F. H. von Meijenfeldt 1972
1. ♘b5

2379.
F. Meinärtz 1932
1. ♖b1

2380.
F. Meinärtz 1932
1. d7

2381.
F. Meinärtz 1932
1. g4

2382.
F. Meinärtz 1932
1. ♘a5

2383.
F. Meinärtz 1932
1. e7

2384.
F. Meinärtz 1932
1. ♕a1

2385.
F. Meinärtz 1932
1. ♕d7

2386.
F. Meinärtz 1932
1. ♘h3

2387.
F. Meinärtz 1933
1. ♕c3

2388.
F. Meinärtz 1933
1. ♖g3

2389.
F. Meinärtz 1933
1. ♕h4

2390.
F. Meinärtz 1933
1. ♔e6

2391.
F. Meinärtz 1933
1. ♖a1

2392.
H. Melehov 1979
1. ♕a1

2393.
V. Melnichenko 1992
1. ♔f2

2394.
V. Melnichenko 1992
1. ♘f8

2395.
V. Melnichenko 1988
1. ♕e5

2396.
V. Melnichenko 1986
1. ♕d7

2397.
V. Melnichenko 1978
1. ♕a1

2398.
V. Melnichenko 1975
1. 0-0

2399.
V. Melnichenko 1962
1. ♖h7

2400.
V. Melnichenko 1958
1. ♗h5

2401.
L. Merényi 1926
1. ♔b6

2402.
L. Merényi 1926
1. ♕b8

2403.
L. Merényi 1926
1. f7

2404.
L. Merényi 1926
1. ♕b7

2405.
V. Metlitzky 1969
1. ♔f7

2406.
V. Metlitzky 1972/73
1. ♕d2

2407.
V. Metlitzky 1967
1. ♕f3

2408.
F. M. Mihalek 1979
1. ♗f6

2409.
A. Mihailov 1983
1. ♕e5

2410.
J. Minckwitz 1875
1. ♕c3

2411.
A. Mironov 1982
1. c8♗

2412.
A. Mironov 1982
1. ♘b3

2413.
A. Mironov 1982
1. c8♘

2414.
A. Miskolczy 1911
1. g4

2415.
W. Moykin 1973
1. ♕h8

2416.
V. Moykin 1973
1. ♕h8

2417.
E. Montvid 1893
1. ♘h4

2418.
E. Montvid 1901
1. ♘f3

2419.
E. Montvid 1901
1. ♘a6

2420.
G. Mosiashvili 1982
1. ♕g6

2421.
J. Mostert 1960
1. ♘c5

2422.
G. Mott–Smith 1937
1. ♖d3

2423.
G. Mott–Smith 1942
1. ♖a6

2424.
G. Mott–Smith 1937
1. ♕e1

2425.
G. Mott–Smith 1939
1. ♔a2

2426.
G. Mott–Smith 1932
1. ♕e5

2427.
G. Mott–Smith 1937
1. ♘c4

2428.
G. Mott–Smith 1937
1. ♘e4

2429.
G. Mott–Smith 1934
1. ♘f2

2430
G. Mott–Smith 1937
1. ♕a2

2431.
J. Möller 1914
1. ♔e2

2432.
I. Murarasu 1988
1. ♖a1

2433.
M. Myllyniemi 1954
1. ♕f5

2434.
E. Myhre 1928
1. ♔g7

2435.
M. Myllyniemi 1981
1. ♕g7

2436.
M. Myllyniemi 1970/71
1. ♖a2

2437.
M. Myllyniemi 1965
1. ♕e1

2438.
M. Myllyniemi 1957
1. ♘b4

2439.
N. Nadezhdin 1969
1. ♕h8

2440.
E. Nogovitzin 1974
1. ♗c3

2441.
B. Nazarov 1982
1. ♔d6

2442.
B. Nazarov 1982
1. ♗b5

2443.
V. Negodayev 1971
1. ♔g5

2444.
O. Nemo 1929
1. ♕e3

2445.
V. Nestorescu 1959
1. ♕g1

2446.
Gy. Neukomm 1928
1. ♔d5

2447.
Gy. Neukomm 1929
1. b4

2448.
Gy. Neukomm 1930
1. ♘c5

2449.
Gy. Neukomm 1931
1. ♘f4

2450.
Gy. Neukomm 1931
1. ♕b8

2451.
Gy. Neukomm 1927
1. ♔f7

2452.
Gy. Neukomm 1931
1. ♗b4

2453.
Gy. Neukomm 1928
1. e8♖

2454.
Gy. Neukomm 1927
1. f8♖

2455.
Gy. Neukomm 1927
1. g8♖

2456.
Gy. Neukomm 1927
1. h8♖

2457.
Gy. Neukomm 1947
1. ♕a2

2458.
Gy. Neukomm 1924
1. ♕a7

2459.
Gy. Neukomm – V. Onitiu 1932
1. ♕e6

2460.
Gy. Neukomm – N. Kovács 1929
1. ♗f2

2461.
K. Nielsen 1957
1. ♘h3

2462.
K. Nielsen 1971
1. ♗e3

2463.
A. Nieman 1985
1. ♕c8

2464.
M. Niemeijer 1971
1. ♔b3

2465.
M. Niemeijer 1970
1. ♘df4

SOLUTIONS (2466 – 2541)

2466.
A. R. Niemelä 1976
1. ♗h1

2467.
K. Nikolov 1962
1. d6

2468.
R. Nobre – G. Maritz 1985
1. ♔c1

2469.
R. Notaro 1978
1. ♖g1

2470.
R. Notaro 1977
1. ♖f8

2471.
C. E. Nylund 1944
1. ♗c7

2472.
T. Nylund 1967
1. ♘f5

2473.
J. Oehquist 1902
1. ♖c4

2474.
J. Oehquist 1889
1. ♗g4

2475.
A. G. Ojanen 1942
1. ♕g2

2476.
A. G. Ojanen 1942
1. ♕f5

2477.
Y. Oksala 1954
1. ♔d1

2478.
A. S. van Ommeren 1915
1. f8♗

2479.
V. Onitiu 1927
1. ♖e8

2480.
V. Onitiu 1930
1. ♕e8

2481.
V. Onitiu 1930
1. ♖h3

2482.
V. Onitiu 1930
1. ♔c3

2483.
V. Onitiu 1930
1. ♘c4

2484.
V. Onitiu 1930
1. ♗f7

2485.
V. Onitiu 1930
1. ♔g7

2486.
V. Onitiu 1930
1. ♕h4

2487.
V. Onitiu 1931
1. ♘a6

2488.
V. Onitiu 1932
1. ♔b2

2489.
V. Onitiu 1932
1. ♕d6

2490.
V. Onitiu 1932
1. ♔e5

2491.
V. Onitiu 1932
1. ♔e6

2492.
V. Onitiu 1932
1. ♕d3

2493.
V. Onitiu 1932
1. ♔f2

2494.
V. Onitiu 1932
1. ♔b4

2495.
V. Onitiu 1932
1. ♗c6

2496.
V. Onitiu 1932
1. ♘e5

2497.
V. Onitiu 1926
1. ♕g2

2498.
V. Onitiu 1926
1. ♔c7

2499.
V. Onitiu 1926
1. g7

2500.
V. Onitiu 1926
1. ♕a8

2501.
V. Onitiu 1926
1. ♕f2

2502.
V. Onitiu 1931
1. ♘h5

2503.
V. Onitiu 1931
1. ♔h4

2504.
V. Onitiu 1931
1. ♕h6

2505.
V. Onitiu 1931
1. ♕c4

2506.
V. Onitiu 1927
1. ♕g4

2507.
V. Onitiu 1927
1. ♘d2

2508.
V. Onitiu – N. Kovács – Gy. Neukomm – L. Talabér 1932
1. ♕g5

2509.
V. Onitiu – Ö. Nagy 1926
1. ♕e5

2510.
V. Onitiu – Z. Zilahi 1926
1. ♔f2

2511.
J. Opdenoordt 1917
1. ♕f3

2512.
A. Oreshin 1936
1. ♖g3

2513.
E. Orlov 1969
1. ♕e5

2514.
E. Orlov 1969
1. ♕e1

2515.
G. Osipov 1985/86
1. ♕a8

2516.
E. Paalanen 1966
1. ♘c3

2517.
E. Paalanen 1969
1. ♕a1

2518.
E. Paalanen 1969
1. ♘f5

2519.
E. Paalanen 1969
1. ♘g7

2520.
F. Palatz 1936
1. ♕d3

2521.
F. Palatz 1936
1. g6

2522.
F. Palatz 1929
1. e8♘

2523.
F. Palatz 1929
1. ♘g3

2524.
F. Palatz 1929
1. ♕h4

2525.
F. Palatz 1929
1. ♗c8

2526.
F. Palatz 1917
1. ♖e1

2527.
E. Palkoska 1901
1. ♘a2

2528.
E. Palkoska 1925/26
1. ♕d3

2529.
E. Pape 1927
1. ♕g2

2530.
E. Pape 1935
1. ♕b1

2531.
E. Pape 1935
1. ♖h1

2532.
E. Pape 1935
1. ♕e3

2533.
J. Papp 1924
1. ♕b3

2534.
M. Papp 1931
1. ♕c2

2535.
M. Papp 1927
1. ♕f8

2536.
N. Parhomenko 1984
1. ♕f2

2537.
V. Pavlenko 1967
1. ♕a7

2538.
V. Pavlenko 1967
1. ♕c8

2539.
V. Pavlenko 1969
1. ♗c3

2540.
W. Pauly 1923
1. ♘c4

2541.
J. Peris 1946
1. ♔c8

2542.
P. A. Petkov 1962
1. ♘f7

2543.
A. Petrushenko 1974
1. ♗f3

2544.
A. Petrushenko 1974
1. ♖h7

2545.
A. Petrushenko 1982
1. ♕e5

2546.
A. Petrushenko 1978
1. ♔g4

2547.
A. Petrushenko 1975
1. ♗g3

2548.
O. Pettersen 1956
1. ♗f8

2549.
O. Pettersen 1959
1. ♖c2

2550.
O. Pettersen 1956
1. ♗c1

2551.
V. D. Petrosani 1967
1. ♕b2

2552.
N. Petrovic 1977
1. ♕a4

2553.
R. Pikler 1926
1. ♕c7

2554.
R. Pikler 1926
1. ♕b4

2555.
R. Pikler 1926
1. ♔f5

2556.
R. Pikler 1926
1. ♔c8

2557.
R. Pikler 1926
1. ♕g6

2558.
R. Pikler 1926
1. ♔h6

2559.
R. Pikler 1926
1. c7

2560.
R. Pikler 1926
1. ♗e7

2561.
R. Pikler 1926
1. ♔f6

2562.
R. Pikler 1926
1. ♕d7

2563
R. Pikler 1926
1. ♔f1

2564.
R. Pikler 1927
1. ♕b4

2565.
R. Pikler 1927
1. ♔h3

2566.
R. Pikler 1927
1. ♕d7

2567.
A. Pakulik 1961
1. ♕g1

2568.
V. Pilipenko 1973
1. ♗a5

2569.
V. Pilipenko 1973
1. ♘f3

2570.
V. Pilipenko 1971
1. ♘d2

2571.
V. Pilipenko 1971
1. ♘d6

2572.
V. Pilipenko 1970
1. ♖f4

2573.
V. Pilipenko 1970
1. ♖f4

2574.
V. Pilipenko 1969
1. ♔e6

2575.
V. Pilipenko – S. Podushkin 1963
1. ♗e5

2576.
V. Pilipenko – S. Podushkin 1963
1. ♗f4

2577.
V. and S. Pimenov 1929
1. ♗e6

2578.
N. Plechikova 1950
1. ♕g2

2579.
S. Podushkin 1970
1. ♕a6

2580.
S. Podushkin 1969
1. ♖d2

2581.
S. Podushkin 1969
1. ♗f7

2582.
S. Podushkin 1972
1. ♕b7

2583.
S. Podushkin 1979
1. b8♘

2584.
S. Podushkin 1982
1. ♖h7

2585.
S. Podushkin 1982
1. ♕g8

2586.
S. Podushkin 1982
1. ♖g6

2587.
Zsu. Polgár 1973
1. ♔d1

2588.
V. Posessor 1927
1. ♘f4

2589.
B. Postma 1947
1. ♔a8

2590.
V. Potemsky 1893
1. ♕e6

2591.
E. Pradignat 1895
1. ♕g3

2592.
W. Preiswerk 1908
1. ♔c6

2593.
R. Prytz 1924
1. ♔d6

2594.
S. Pugachev 1947
1. ♔c1

2595.
S. Pugachev 1947
1. ♕c4

2596.
E. Puig – Y. Puig 1906
1. ♖a4

2597.
C. Pushkin 1981
1. ♔b5

2598.
B. Pustovoy 1954
1. ♖f5

2599.
B. Pustovoy 1972/73
1. ♗c8

2600.
B. Pustovoy 1972
1. ♗f3

2601.
B. Pustovoy 1972
1. ♖a5

2602.
B. Pustovoy 1972
1. ♗a5

2603.
B. Pustovoy 1971
1. ♕a1

2604.
B. Pustovoy 1971
1. ♕h1

2605.
B. Pustovoy 1971
1. ♕d6

2606.
B. Pustovoy 1969
1. ♘f6

2607.
B. Pustovoy 1969
1. ♘f6

2608.
B. Pustovoy 1969
1. ♔b5

2609.
B. Pustovoy 1969
1. ♕a1

2610.
B. Pustovoy 1966
1. ♕f7

2611.
B. Pustovoy 1963
1. ♗a4

2612.
B. Pustovoy 1964
1. ♔f5

2613.
A. Ragainis 1934
1. ♗c7

2614.
B. Restad 1928
1. ♕a8

2615.
B. Restad 1929
1. ♕e5

2616.
J. Rice 1992
1. ♕f2

2617.
J. Rice 1992
1. ♔d2

2618.
J. Rice 1992
1. ♖d2

2619.
J. Rice 1989
1. ♗f7

2620.
J. Rice 1982
1. ♘e7

2621.
J. Rice 1982
1. ♖ee3

2622.
F. Richter 1933
1. ♗h1

2623.
L. Riczu 1969
1. ♕e7

2624.
L. Riczu 1968
1. ♗d1

2625.
M. Ritchings 1988
1. ♘f2

2626.
M. Ritchings 1988
1. ♗d5

2627.
J. Roosendaal 1990
1. ♔f8

2628.
J. Ropelt 1902
1. ♘f4

2629.
J. Rosenhouse 1992
1. ♖f1

2630.
M. Rosenthal 1967
1. ♕g2

2631.
S. Rothwell 1991
1. ♕g4

2632.
C. A. H. Russ 1949
1. ♗a6

2633.
F. Sackmann 1909
1. ♖a3

2634.
K. Sampakoski 1977
1. ♗h8

2635.
L. B. Salkind 1907
1. ♕g7

2636.
B. Sallay – T. Kardos 1986
1. ♖d5

2637.
B. Sallay – T. Kardos 1986
1. ♕g6

2638.
B. Sallay – T. Kardos 1986
1. ♕f5

2639.
B. Sallay – T. Kardos 1986
1. ♕h7

2640.
V. Samilo 1982
1. ♕e4

2641.
D. M. Saunders 1985
1. ♕a6

2642.
A. de Savignac 1925
1. ♕e7

2643.
V. Sazhin 1978/81
1. ♘c7

2644.
V. I. Solovyev 1952
1. ♗c4

2645.
J. Scheel 1937
1. ♖f2

2646.
J. Scheel 1932
1. ♖h7

2647.
J. Scheel 1926
1. ♔d5

2648.
J. Scheel 1926
1. ♖a6

2649.
J. Scheel 1936
1. d8♘

2650.
J. Scheel 1919
1. ♖h5

2651.
S. Schett 1881
1. ♕a6

2652.
Gy. Schiffert 1927
1. ♖a2

2653.
Gy. Schiffert 1927
1. ♖f8

2654.
Gy. Schiffert 1927
1. ♕c5

2655.
Gy. Schiffert 1927
1. ♕d1

2656.
Gy. Schiffert 1927
1. ♖b2

2657.
Gy. Schiffert 1927
1. ♗e7

2658.
Gy. Schiffert 1928
1. ♖g8

2659.
Gy. Schiffert 1928
1. ♕c6

2660.
Gy. Schiffert 1928
1. ♖a7

2661.
Gy. Schiffert 1931
1. ♖b7

2662.
Gy. Schiffert 1931
1. ♗c2

2663.
Gy. Schiffert 1931
1. ♔d3

2664.
Gy. Schiffert 1931
1. ♕c4

2665.
Gy. Schiffert 1931
1. ♔d7

2666.
Gy. Schiffert 1931
1. ♕d4

2667.
Gy. Schiffert 1931
1. b3

2668.
Gy. Schiffert 1925
1. ♕e8

2669.
Gy. Schiffert 1925
1. g7

2670.
Gy. Schiffert 1926
1. ♔c5

2671.
Gy. Schiffert 1926
1. ♔g3

2672.
Gy. Schiffert 1926
1. ♕a5

2673.
Gy. Schiffert 1926
1. ♕e2

2674.
Gy. Schiffert 1926
1. ♔f6

2675.
Gy. Schiffert 1926
1. ♘d6

2676.
Gy. Schiffert 1926
1. ♕b4

2677.
Gy. Schiffert 1926
1. ♔e4

2678.
Gy. Schiffert 1926
1. c3

2679.
Gy. Schiffert 1926
1. ♔c2

2680.
Gy. Schiffert 1926
1. ♕e4

2681.
Gy. Schiffert 1926
1. gh6

2682.
Gy. Schiffert 1926
1. d7

2683.
Gy. Schiffert 1926
1. ♔c2

2684.
Gy. Schiffert 1931
1. e4

2685.
Gy. Schiffert 1931
1. g3

2686.
Gy. Schiffert 1931
1. d3

2687.
Gy. Schiffert 1931
1. b5

2688.
Gy. Schiffert 1931
1. ♗e6

2689.
Gy. Schiffert 1931
1. ♔a8

2690.
Gy. Schiffert 1931
1. ♖b6

2691.
Gy. Schiffert 1931
1. ♕a2

2692.
Gy. Schiffert 1932
1. ♔f1

2693.
Gy. Schiffert 1931
1. ♖a2

2694.
Gy. Schiffert 1931
1. ♗b8

2695.
Gy. Schiffert 1931
1. ♗g2

2696.
Gy. Schiffert 1931
1. ♖hh2

2697.
Gy. Schiffert 1931
1. ♘c1

2698.
Gy. Schiffert 1931
1. b4

2699.
Gy. Schiffert 1927
1. ♖c2

2700.
Gy. Schiffert 1927
1. ♖a5

2701.
Gy. Schiffert 1927
1. ♖g6

2702.
Gy. Schiffert 1927
1. ♕h3

2703.
Gy. Schiffert 1927
1. ♗e1

2704.
Gy. Schiffert 1927
1. ♖d2

2705.
Gy. Schiffert 1927
1. ♔f5

2706.
Gy. Schiffert 1927
1. ♘f8

2707.
Gy. Schiffert 1927
1. b4

2708.
Gy. Schiffert 1927
1. h3

2709.
Gy. Schiffert 1927
1. ♖g5

2710.
Gy. Schiffert 1927
1. ♗b4

2711.
Gy. Schiffert 1927
1. f6

2712.
Gy. Schiffert 1927
1. ♖d1

2713.
Gy. Schiffert 1928
1. ♕h1

2714.
Gy. Schiffert 1931
1. ♖a2

2715.
Gy. Schiffert 1926
1. ♔b1

2716.
Gy. Schiffert – S. Boros 1926
1. ♔c4

2717.
Gy. Schiffert – L. Szász 1928
1. ♖d1

2718.
Gy. Schiffert– Gy. Balló 1926
1. ♕b5

2719.
Gy. Schiffert – Gy. Neukomm 1927
1. ♗e3

2720.
J. Schlarkó 1926
1. ♖b7

2721.
C. Schlechter 1908
1. ♕a6

2722.
M. Schneider 1943
1. ♖h7

2723.
M. Schneider 1927
1. b4

2724.
T. Schönberger 1925
1. ♗a4

2725.
T. Schönberger 1926
1. ♔f2

2726.
T. Schönberger 1926
1. ♗d2

2727.
T. Schönberger 1926
1. ♘a3

2728.
T. Schönberger 1926
1. ♗c4

2729.
T. Schönberger 1926
1. ♖e1

2730.
T. Schönberger 1926
1. ♗b1

2731.
T. Schönberger 1926
1. ♕e1

2732.
T. Schönberger 1935
1. ♖b1

2733.
T. Schönberger 1935
1. ♕b2

2734.
T. Schönberger 1935
1. ♖b1

2735.
T. Schönberger 1935
1. ♕d4

2736.
T. Schönberger 1935
1. ♖a3

2737.
T. Schönberger 1935
1. ♕d6

2738.
T. Schönberger 1935
1. ♕g6

2739.
T. Schönberger 1935
1. ♕e1

2740.
T. Schönberger 1935
1. ♕d2

2741.
T. Schönberger 1935
1. ♘b4

2742.
T. Schönberger 1935
1. ♕g2

2743.
T. Schönberger 1935
1. ♕h1

2744.
T. Schönberger 1935
1. ♕e6

2745.
T. Schönberger 1935
1. ♕b2

2746.
T. Schönberger 1935
1. ♘b8

2747.
T. Schönberger 1935
1. ♕e7

2748.
T. Schönberger 1935
1. ♘f4

2749.
T. Schönberger 1925
1. ♗f7

2750.
T. Schönberger 1925/26
1. ♕f7

2751.
T. Schönberger 1925
1. h8♗

2752.
T. Schönberger 1925
1. ♕g5

2753.
T. Schönberger 1925
1. ♕c4

2754.
T. Schönberger 1925
1. ♘e2

2755.
T. Schönberger 1925
1. ♖ag1

2756.
T. Schönberger 1925
1. ♕g2

2757.
T. Schönberger 1925
1. h8♘

2758.
T. Schönberger 1925
1. ♕h7

2759.
T. Schönberger 1925
1. ♔d3

2760.
T. Schönberger 1925
1. ♗g8

2761.
T. Schönberger 1925
1. ♕a5

2762.
T. Schönberger 1926
1. ♖b8

2763.
T. Schönberger 1926
1. ♕f3

2764.
T. Schönberger 1926
1. ♖a2

2765.
T. Schönberger 1926
1. c8♗

2766.
T. Schönberger 1926
1. ♕e3

2767.
T. Schönberger 1926
1. ♖d2

2768.
T. Schönberger 1926
1. ♕c7

2769.
T. Schönberger 1926
1. ♗a6

2770.
T. Schönberger 1926
1. ♕d1

2771.
T. Schönberger 1925
1. ♕b1

2772.
T. Schönberger 1925
1. ♖h7

2773.
T. Schönberger 1925
1. ♕g4

2774.
T. Schönberger 1925
1. ♕h8

2775.
T. Schönberger 1933
1. ♕b2

2776.
T. Schönberger 1925/26
1. ♕f8

2777.
T. Schönberger 1925
1. ♖fe1

2778.
T. Schönberger 1933
1. ♗g3

2779.
T. Schönberger 1925/26
1. ♖f6

2780.
T. Schönberger 1925
1. ♘d4

2781.
T. Schönberger 1925
1. ♗a5

2782.
T. Schönberger 1925
1. ♕e5

2783.
T. Schönberger 1935
1. ♕d2

2784.
T. Schönberger 1933
1. ♕d4

2785.
T. Schönberger 1936
1. ♖b3

2786.
T. Schönberger 1933
1. ♕a7

2787.
T. Schönberger 1925
1. f7

2788.
T. Schönberger 1925
1. h8♕

2789.
T. Schönberger 1925
1. ♕e4

2790.
T. Schönberger 1933
1. ♕f5

2791.
T. Schönberger– R. Pikler 1926
1. ♖b4

2792.
T. Schönberger – Gy. Schiffert 1926
1. ♗e7

2793.
T. Schönberger – E. Szentgyörgyi 1926
1. ♔e2

2794.
E. Schulz 1936
1. d8♘

2795.
J. Schumer 1906
1. ♗c8

2796.
M. Segers 1936
1. ♖g8

2797.
M. Segers 1935
1. ♕g5

2798.
J. J. P. Seilberger – J. J. Ebben 1948/49
1. ♗d6

2799.
C. Seneca 1936
1. ♖b8

2800.
V. Sevastyanov 1971
1. 0-0

2801.
O. Shaligin 1985
1. ♘f3

2802.
I. Shanahan 1993
1. ♕e3

2803.
S. Shedey 1969
1. ♗h4

2804.
S. Shedey 1969
1. ♔e6

2805.
S. Shedey 1969
1. ♔b3

2806.
S. Shedey 1979
1. ♕a4

2807.
A. Sheynin 1945
1. ♗e7

2808.
W. A. Shinkman 1902
1. ♗d2

2809.
W. A. Shinkman 1902
1. ♖d1

2810.
W. A. Shinkman 1915
1. ♕e2

2811.
W. A. Shinkman 1902
1. ♕e8

2812.
W. A. Shinkman 1903
1. ♗e8

2813.
W. A. Shinkman 1919
1. ♗d4

2814.
W. A. Shinkman 1874
1. ♕c7

2815.
W. A. Shinkman 1903
1. ♕d8

2816.
W. A. Shinkman 1874
1. ♖a7

2817.
D. J. Shire 1974
1. ♘h3

2818.
A. Shuryakov 1982
1. ♕c5

2819.
T. Siers 1930
1. ♕g4

2820.
P. Siklósi 1973
1. ♔g5

2821.
P. Siklósi 1973
1. ♔g3

2822.
P. Siklósi 1973
1. d6

2823.
P. Siklósi 1955/56
1. ♘f3

2824.
F. H. Singer 1985
1. ♖a1

2825.
F. H. Singer 1985
1. ♖ee5

2826.
G. J. Slater 1886
1. ♗g7

2827.
V. Sljuter 1926
1. 0-0

2828.
R. Smook 1985
1. ♖h3

2829.
Bonus Socius Thirteenth Century
1. ♖hg7

2830.
H. Sokka 1967
1. ♘e4

2831.
L. Sokolov 1955
1. c8♕

2832.
L. Sokolov 1973
1. f8♗

2833.
L. Sokolov 1973
1. f8♘

2834.
A. Sokolov 1938
1. e8♗

2835.
L. Sokolov 1937
1. ♖a1

2836.
L. Sokolov 1955
1. 0-0-0

2837.
P. Sola 1925
1. ♕e7

2838.
I. Solheim 1935
1. ♕b1

2839.
A. F. Solovjev 1929
1. ♕h5

2840.
W. Speckmann 1964
1. d8♖

2841.
W. Speckmann 1963
1. f8♗

2842.
W. Speckmann 1984
1. ♔g8

2843.
W. Speckmann 1986
1. c7

2844.
W. Speckmann 1985
1. f5

2845.
W. Speckmann 1984
1. ♗g4

2846.
W. Speckmann 1986
1. ♕a4

2847.
W. Speckmann 1984
1. ♖a3

2848.
W. Speckmann 1986
1. ♕a7

2849.
W. Speckmann 1986
1. ♗d7

2850.
W. Speckmann 1985
1. ♕e8

2851.
W. Speckmann 1963
1. ♕c6

2852.
W. Speckmann 1956
1. ♔b6

2853.
W. Speckmann 1963
1. ♗h7

2854.
W. Speckmann 1964
1. ♖e4

2855.
W. Speckmann 1985
1. ♕b6

2856.
W. Speckmann 1959
1. ♖h8

2857.
W. Speckmann 1963
1. f8♕

2858.
W. Speckmann 1985
1. ♕d4

2859.
W. Speckmann 1963
1. f8♖

2860.
W. Speckmann 1962
1. ♘f4

2861.
W. Speckmann 1938
1. ♖a8

2862.
W. Speckmann 1962
1. ♘d5

2863.
W. Speckmann 1964
1. ♕d8

2864.
W. Speckmann 1962
1. ♘b5

2865.
W. Speckmann 1962
1. ♕d2

2866.
W. Speckmann 1963
1. d8♗

2867.
W. Speckmann 1982
1. ♕b1

2868.
W. Speckmann 1941
1. ♔g4

2869.
W. Speckmann 1941
1. ♔g5

2870.
W. Speckmann 1941
1. f4

2871.
W. Speckmann 1963
1. ♘b1

2872.
W. Speckmann 1964
1. ♕f5

2873.
W. Speckmann 1963
1. ♘e6

2874.
W. Speckmann 1970
1. ♕h3

2875.
W. Speckmann 1985
1. ♕c2

2876.
W. Speckmann 1970
1. ♕c1

2877.
W. Speckmann 1964
1. ♕ff2

2878.
W. Speckmann 1964
1. a8♗

2879.
W. Speckmann 1964
1. a8♖

2880.
W. Speckmann 1964
1. a8♕

2881.
W. Speckmann 1968
1. ♕f1

2882.
W. Speckmann 1988
1. ♕d6

2883.
W. Speckmann 1967
1. ♗c2

2884.
W. Speckmann 1963
1. d8♖

2885.
W. Speckmann 1961
1. ♔g3

2886.
W. Speckmann 1963
1. d8♘

2887.
W. Speckmann 1961
1. ♕d6

2888.
W. Speckmann 1964
1. ♕h1

2889.
W. Speckmann 1964
1. c8♕

2890.
W. Speckmann 1964
1. c8♖

2891.
W. Speckmann 1964
1. ♖h4

2892.
W. Speckmann 1947
1. d8♕

2893.
W. Speckmann 1963
1. c8♘

2894.
W. Speckmann 1962
1. ♕e5

2895.
W. Speckmann 1982
1. ♕b1

2896.
W. Speckmann 1964
1. c8♗

2897.
W. Speckmann 1960
1. ♗d1

2898.
W. Speckmann 1947
1. ♕e2

2899.
W. Speckmann 1963
1. f8♘

2900.
W. Speckmann 1962
1. ♖b3

2901.
W. Speckmann 1961
1. ♗f7

2902.
W. Speckmann 1962
1. f4

2903.
S. Stambuk 1952
1. ♔e4

2904.
J. Stanton 1986
1. ♕g8

2905.
H. Staudte 1963
1. f8♕

2906.
H. Staudte 1963
1. f8♘

2907.
H. Staudte 1963
1. f8♖

2908.
H. Staudte 1963
1. c8♗

2909.
H. Staudte 1963
1. f8♗

2910.
H. Staudte 1963
1. c8♖

2911.
H. Staudte 1963
1. c8♕

2912.
H. Staudte 1963
1. b8♘

2913.
H. Staudte 1973
1. ♘c6

2914.
H. Staudte 1964
1. e8♗

2915.
H. Staudte 1964
1. c8♖

2916.
H. Staudte 1964
1. d8♕

2917.
H. Staudte 1964
1. b8♘

2918.
A. Steif 1902
1. ♕a8

2919.
P. Steiner 1989
1. ♗h1

2920.
P. Steiner 1988
1. ♕b4

2921.
R. Steinmann 1902
1. g3

2922.
R. Steinweg 1903
1. ♕b7

2923.
O. Stocchi 1929
1. d4

2924.
O. Stocchi 1929
1. ♕g3

2925.
O. Stocchi 1934
1. ♕d4

2926.
V. Stolyarov 1981
1. ♔b3

2927.
P. Strigunov 1958
1. ♗g5

2928.
A. Studenetzky 1958
1. ♕h6

2929.
V. Sushkov 1973
1. ♖h5

2930.
V. Sushkov 1981
1. ♗g5

2931.
I. Suratkai 1985
1. ♕b5

2932.
I. Suratkai 1985
1. ♕h2

2933.
Y. Sushkov 1986
1. ♗e1

2934.
Y. Sushkov 1986
1. ♗c5

2935.
Y. Sushkov 1986
1. ♕f3

2936.
A. Sutter 1952
1. ♘h5

2937.
R. Svoboda 1948
1. ♕c7

2938.
L. Szabó 1978
1. ♕d8

2939.
L. Szász 1926
1. ♖e1

2940.
L. Szász 1927
1. ♘e7

2941.
L. Szász 1927
1. ♖h1

2942.
L. Szász 1927
1. ♔f3

2943.
L. Szász 1927
1. c3

2944.
L. Szász 1927
1. ♔f1

2945.
L. Szász 1926
1. ♔d1

2946.
L. Szász 1927
1. ♕e2

2947.
L. Szász – N. Kovács 1926
1. d8♘

2948.
L. Szász – Gy. Schiffert – E. Szentgyörgyi 1928
1. f8♘

2949.
L. Szász – N. Kovács 1927
1. ♗c5

2950.
E. Szentgyörgyi 1926
1. b8♕

2951.
E. Szentgyörgyi 1928
1. ♔e1

2952.
E. Szentgyörgyi 1928
1. d8♘

2953.
E. Szentgyörgyi 1928
1. d8♘

2954.
E. Szentgyörgyi 1928
1. ♘c5

2955.
E. Szentgyörgyi 1928
1. g8♘

2956.
E. Szentgyörgyi 1928
1. ♘8e7

2957.
E. Szentgyörgyi 1928
1. ♔e4

2958.
E. Szentgyörgyi 1928
1. ♕f1

2959.
E. Szentgyörgyi 1928
1. g8♘

2960.
E. Szentgyörgyi 1928
1. ♕e2

2961.
E. Szentgyörgyi 1928
1. ♕a4

2962.
E. Szentgyörgyi 1928
1. ♗g4

2963.
E. Szentgyörgyi 1929
1. ♕e4

2964.
E. Szentgyörgyi 1929
1. ♗c7

2965.
E. Szentgyörgyi 1929
1. ♕d8

2966.
E. Szentgyörgyi 1929
1. ♕f2

2967.
E. Szentgyörgyi 1929
1. ♖5c4

2968.
E. Szentgyörgyi 1929
1. ♕a3

2969.
E. Szentgyörgyi 1929
1. ♕h8

2970.
E. Szentgyörgyi 1929
1. ♔h4

2971.
E. Szentgyörgyi 1929
1. ♖e1

2972.
E. Szentgyörgyi 1929
1. ♖h7

2973.
E. Szentgyörgyi 1929
1. ♘h2

2974.
E. Szentgyörgyi 1929
1. ♕a6

2975.
E. Szentgyörgyi 1929
1. ♔f7

2976.
E. Szentgyörgyi 1929
1. ♘b7

2977.
E. Szentgyörgyi 1929
1. ♔h7

2978.
E. Szentgyörgyi 1929
1. ♖e1

2979.
E. Szentgyörgyi 1929
1. ♔e6

2980.
E. Szentgyörgyi 1929
1. ♕h5

2981.
E. Szentgyörgyi 1929
1. ♖aa4

2982.
E. Szentgyörgyi 1929
1. c4

2983.
E. Szentgyörgyi 1930
1. g5

2984.
E. Szentgyörgyi 1930
1. ♖a7

2985.
E. Szentgyörgyi 1930
1. ♖a2

2986.
E. Szentgyörgyi 1930
1. ♖h4

2987.
E. Szentgyörgyi 1930
1. d4

2988.
E. Szentgyörgyi 1930
1. ♖a7

2989.
E. Szentgyörgyi 1930
1. ♕h1

2990.
E. Szentgyörgyi 1930
1. ♕g4

2991.
E. Szentgyörgyi 1930
1. ♗d7

2992.
E. Szentgyörgyi 1930
1. d7

2993.
E. Szentgyörgyi 1930
1. ♕f5

2994.
E. Szentgyörgyi 1930
1. ♖d1

2995.
E. Szentgyörgyi 1930
1. ♕g2

2996.
E. Szentgyörgyi 1930
1. ♖2a4

2997.
E. Szentgyörgyi 1930
1. ♕e7

2998.
E. Szentgyörgyi 1930
1. c4

2999.
E. Szentgyörgyi 1930
1. ♘g5

3000.
E. Szentgyörgyi 1930
1. ♕d7

3001.
E. Szentgyörgyi 1930
1. ♕c5

3002.
E. Szentgyörgyi 1930
1. ♔f3

3003.
E. Szentgyörgyi 1930
1. e4

3004.
E. Szentgyörgyi 1930
1. ♔c3

3005.
E. Szentgyörgyi 1930
1. ♕d1

3006.
E. Szentgyörgyi 1930
1. ♘e3

3007.
E. Szentgyörgyi 1930
1. ♔f2

3008.
E. Szentgyörgyi 1930
1. ♕f2

3009.
E. Szentgyörgyi 1930
1. ♖h1

3010.
E. Szentgyörgyi 1930
1. ♔h7

3011.
E. Szentgyörgyi 1931
1. ♕e5

3012.
E. Szentgyörgyi 1931
1. ♕a5

3013.
E. Szentgyörgyi 1931
1. ♕c4

3014.
E. Szentgyörgyi 1931
1. ♗g2

3015.
E. Szentgyörgyi 1931
1. f3

3016.
E. Szentgyörgyi 1931
1. ♘f3

3017.
E. Szentgyörgyi 1931
1. a4

3018.
E. Szentgyörgyi 1931
1. ♔e1

3019.
E. Szentgyörgyi 1931
1. ♕e1

3020.
E. Szentgyörgyi 1931
1. ♔e2

3021.
E. Szentgyörgyi 1931
1. d7

3022.
E. Szentgyörgyi 1925
1. ♘f6

3023.
E. Szentgyörgyi 1925
1. ♔g7

3024.
E. Szentgyörgyi 1925
1. ♖h3

3025.
E. Szentgyörgyi 1925
1. ♕b8

3026.
E. Szentgyörgyi 1925
1. b3

3027.
E. Szentgyörgyi 1925
1. ♘a6

3028.
E. Szentgyörgyi 1925
1. ♗h6

3029.
E. Szentgyörgyi 1925
1. ♘f3

3030.
E. Szentgyörgyi 1926
1. ♖b1

3031.
E. Szentgyörgyi 1929
1. ♕h6

3032.
E. Szentgyörgyi 1926
1. ♘f7

3033.
E. Szentgyörgyi 1935
1. ♔a6

3034.
E. Szentgyörgyi 1926
1. ♔d1

3035.
E. Szentgyörgyi 1935
1. ♕a3

3036.
E. Szentgyörgyi 1935
1. ♘c3

3037.
E. Szentgyörgyi 1935
1. ♕e3

3038.
E. Szentgyörgyi 1935
1. d3

3039.
E. Szentgyörgyi 1935
1. ♗g2

3040.
E. Szentgyörgyi 1935
1. ♘c6

3041.
E. Szentgyörgyi 1935
1. ♔f1

3042.
E. Szentgyörgyi 1935
1. ♕g4

3043.
E. Szentgyörgyi 1935
1. ♗d2

3044.
E. Szentgyörgyi 1926
1. e6

3045.
E. Szentgyörgyi 1926
1. ♖b5

3046.
E. Szentgyörgyi 1926
1. ♗a8

3047.
E. Szentgyörgyi 1926
1. f8♖

3048.
E. Szentgyörgyi 1926
1. ♖h7

3049.
E. Szentgyörgyi 1926
1. ♖e8

3050.
E. Szentgyörgyi 1926
1. d8♘

3051.
E. Szentgyörgyi 1926
1. e8♘

3052.
E. Szentgyörgyi 1926
1. ♗c8

3053.
E. Szentgyörgyi 1925
1. ♗f2

3054.
E. Szentgyörgyi 1925
1. ♕h2

3055.
E. Szentgyörgyi 1925
1. ♕c2

3056.
E. Szentgyörgyi 1925
1. ♕a2

3057.
E. Szentgyörgyi 1925
1. c8♖

3058.
E. Szentgyörgyi 1925
1. ♔b2

3059.
E. Szentgyörgyi 1925
1. f7

3060.
E. Szentgyörgyi 1925
1. ♔b6

3061.
E. Szentgyörgyi 1925
1. ♕c6

3062.
E. Szentgyörgyi 1925
1. ♘f4

3063.
E. Szentgyörgyi 1929
1. ♘a5

3064.
E. Szentgyörgyi 1929
1. ♕a5

3065.
E. Szentgyörgyi 1929
1. ♔h5

3066.
E. Szentgyörgyi 1929
1. ♖h8

3067.
E. Szentgyörgyi 1929
1. ♖f6

3068.
E. Szentgyörgyi 1929
1. ♗b2

3069.
E. Szentgyörgyi 1931
1. ♖f5

3070.
E. Szentgyörgyi 1931
1. ♘c2

3071.
E. Szentgyörgyi 1931
1. ♗a6

3072. *E. Szentgyörgyi 1931* 1. ♔d3

3073. *E. Szentgyörgyi 1931* 1. ♗d7

3074. *E. Szentgyörgyi 1931* 1. ♘f3

3075. *E. Szentgyörgyi 1931* 1. ♕e7

3076. *E. Szentgyörgyi 1931* 1. g8♘

3077. *E. Szentgyörgyi 1931* 1. ♘a5

3078. *E. Szentgyörgyi 1931* 1. ♕d1

3079. *E. Szentgyörgyi 1931* 1. g8♘

3080. *E. Szentgyörgyi 1931* 1. d3

3081. *E. Szentgyörgyi 1932* 1. ♘d5

3082. *E. Szentgyörgyi 1932* 1. d7

3083. *E. Szentgyörgyi 1932* 1. ♔f4

3084. *E. Szentgyörgyi 1932* 1. ♕e8

3085. *E. Szentgyörgyi 1932* 1. ♗d4

3086. *E. Szentgyörgyi 1932* 1. ♘c1

3087. *E. Szentgyörgyi 1932* 1. ♕g2

3088. *E. Szentgyörgyi 1932* 1. f3

3089. *E. Szentgyörgyi 1932* 1. ♘a3

3090. *E. Szentgyörgyi 1932* 1. ♕h1

3091. *E. Szentgyörgyi 1932* 1. ♗f5

3092. *E. Szentgyörgyi 1932* 1. h4

3093. *E. Szentgyörgyi 1932* 1. ♘c3

3094. *E. Szentgyörgyi 1932* 1. ♘c2

3095. *E. Szentgyörgyi 1932* 1. ♖b7

3096. *E. Szentgyörgyi 1932* 1. ♘c2

3097. *E. Szentgyörgyi 1932* 1. ♔g4

3098. *E. Szentgyörgyi 1932* 1. d7

3099. *E. Szentgyörgyi 1933* 1. f4

3100. *E. Szentgyörgyi 1933* 1. ♗a3

3101. *E. Szentgyörgyi 1933* 1. ♕d3

3102. *E. Szentgyörgyi 1933* 1. ♕d7

3103. *E. Szentgyörgyi 1933* 1. ♘h3

3104. *E. Szentgyörgyi 1933* 1. ♕d1

3105. *E. Szentgyörgyi 1933* 1. ♕c6

3106. *E. Szentgyörgyi 1933* 1. ♕h2

3107. *E. Szentgyörgyi 1933* 1. ♕f5

3108. *E. Szentgyörgyi 1933* 1. ♕b8

3109. *E. Szentgyörgyi 1933* 1. ♘c4

3110. *E. Szentgyörgyi 1933* 1. ♗d5

3111. *E. Szentgyörgyi 1933* 1. ♕g2

3112. *E. Szentgyörgyi 1933* 1. ♕g7

3113. *E. Szentgyörgyi 1933* 1. ♕b7

3114. *E. Szentgyörgyi 1933* 1. ♘d2

3115. *E. Szentgyörgyi 1933* 1. ♘g4

3116. *E. Szentgyörgyi 1933* 1. ♕a7

3117. *E. Szentgyörgyi 1933* 1. ♕e1

3118. *E. Szentgyörgyi 1933* 1. ♕a2

3119. *E. Szentgyörgyi 1933* 1. ♕g7

3120. *E. Szentgyörgyi 1933* 1. ♕g2

3121. *E. Szentgyörgyi 1933* 1. ♗f3

3122. *E. Szentgyörgyi 1933* 1. ♕h1

3123. *E. Szentgyörgyi 1933* 1. ♖b2

3124. *E. Szentgyörgyi 1933* 1. ♕f7

3125. *E. Szentgyörgyi 1933* 1. b8♕

3126. *E. Szentgyörgyi 1927* 1. ♗d2

3127. *E. Szentgyörgyi 1927* 1. b8♕

3128. *E. Szentgyörgyi 1927* 1. ♔g8

3129. *E. Szentgyörgyi 1927* 1. h8♘

3130. *E. Szentgyörgyi 1927* 1. b4

3131. *E. Szentgyörgyi 1927* 1. ♖g2

3132. *E. Szentgyörgyi 1927* 1. ♔g3

3133. *E. Szentgyörgyi 1927* 1. ♘g6

3134. *E. Szentgyörgyi 1927* 1. ♗e5

3135. *E. Szentgyörgyi 1927* 1. ♕g4

3136. *E. Szentgyörgyi 1927* 1. ♕h3

3137. *E. Szentgyörgyi 1927* 1. ♕d1

3138. *E. Szentgyörgyi 1927* 1. ♘dc5

3139. *E. Szentgyörgyi 1927* 1. ♔d6

3140. *E. Szentgyörgyi 1927* 1. ♔f5

3141. *E. Szentgyörgyi 1927* 1. ♖c7

3142. *E. Szentgyörgyi 1927* 1. g6

3143. *E. Szentgyörgyi 1927* 1. ♖h4

3144. *E. Szentgyörgyi 1927* 1. ♖b8

3145. *E. Szentgyörgyi 1927* 1. ♗h7

3146. *E. Szentgyörgyi 1927* 1. ♗a2

3147. *E. Szentgyörgyi 1927* 1. ♘c6

3148.
E. Szentgyörgyi 1927
1. ♘g6

3149.
E. Szentgyörgyi 1927
1. ♔f2

3150.
E. Szentgyörgyi 1927
1. ♔d6

3151.
E. Szentgyörgyi 1927
1. ♖b8

3152.
E. Szentgyörgyi 1927
1. ♕h3

3153.
E. Szentgyörgyi 1927
1. e6

3154.
E. Szentgyörgyi 1927
1. ♘d1

3155.
E. Szentgyörgyi 1927
1. ♔d2

3156.
E. Szentgyörgyi 1926
1. g8♘

3157.
E. Szentgyörgyi 1926
1. c8♘

3158.
E. Szentgyörgyi 1926
1. ♘h3

3159.
E. Szentgyörgyi 1926
1. ♔e8

3160.
E. Szentgyörgyi 1926
1. ♖ba8

3161.
E. Szentgyörgyi 1926
1. ♔d5

3162.
E. Szentgyörgyi 1926
1. e4

3163.
E. Szentgyörgyi 1926
1. ♖b1

3164.
E. Szentgyörgyi 1926
1. ♖b7

3165.
E. Szentgyörgyi 1926
1. ♔a6

3166.
E. Szentgyörgyi 1926
1. ♗c3

3167.
E. Szentgyörgyi 1926
1. ♗g3

3168.
E. Szentgyörgyi 1926
1. ♔b3

3169.
E. Szentgyörgyi 1926
1. ♗a7

3170.
E. Szentgyörgyi 1928
1. ♕e7

3171.
E. Szentgyörgyi 1927
1. f8♗

3172.
E. Szentgyörgyi 1926
1. ♔b6

3173.
E. Szentgyörgyi 1926
1. ♔h3

3174.
E. Szentgyörgyi 1926
1. ♔d2

3175.
E. Szentgyörgyi 1936
1. ♕h7

3176.
E. Szentgyörgyi 1936
1. ♕f1

3177.
E. Szentgyörgyi 1925
1. ♕c6

3178.
E. Szentgyörgyi 1932
1. ♖d2

3179.
E. Szentgyörgyi 1928
1. ♔f5

3180.
E. Szentgyörgyi 1928
1. ♕f6

3181.
E. Szentgyörgyi 1927
1. ♕d4

3182.
E. Szentgyörgyi 1927
1. ♘c5

3183.
E. Szentgyörgyi 1933
1. ♕a7

3184.
E. Szentgyörgyi 1928
1. ♕a3

3185.
E. Szentgyörgyi 1928
1. ♔a3

3186.
E. Szentgyörgyi 1928
1. ♕e2

3187.
E. Szentgyörgyi 1928
1. ♘d3

3188.
E. Szentgyörgyi 1928
1. ♗d3

3189.
E. Szentgyörgyi 1928
1. ♕b3

3190.
E. Szentgyörgyi 1928
1. ♕f7

3191.
E. Szentgyörgyi 1928
1. ♔c3

3192.
E. Szentgyörgyi 1936
1. ♔b4

3193.
E. Szentgyörgyi 1933
1. ♕b7

3194.
E. Szentgyörgyi 1933
1. b4

3195.
E. Szentgyörgyi 1933
1. ♔c2

3196.
E. Szentgyörgyi 1933
1. d4

3197.
E. Szentgyörgyi 1933
1. c3

3198.
E. Szentgyörgyi 1933
1. ♔f5

3199.
E. Szentgyörgyi 1933
1. c6

3200.
E. Szentgyörgyi 1933
1. ♗b1

3201.
E. Szentgyörgyi 1933
1. ♔b5

3202.
E. Szentgyörgyi 1933
1. ♕e5

3203.
E. Szentgyörgyi 1933
1. ♘c7

3204.
E. Szentgyörgyi 1933
1. ♕d3

3205.
E. Szentgyörgyi 1933
1. ♘f7

3206.
E. Szentgyörgyi 1933
1. ♕c4

3207.
E. Szentgyörgyi 1933
1. ♘h3

3208.
E. Szentgyörgyi 1933
1. ♘b4

3209.
E. Szentgyörgyi 1933
1. ♕a4

3210.
E. Szentgyörgyi 1933
1. ♖b5

3211.
E. Szentgyörgyi 1933
1. ♖h4

3212.
E. Szentgyörgyi 1933
1. ♔f7

3213.
E. Szentgyörgyi 1933
1. ♖g1

3214.
E. Szentgyörgyi 1933
1. ♘b6

3215.
E. Szentgyörgyi 1933
1. ♖e8

3216.
E. Szentgyörgyi 1933
1. ♔g2

3217.
E. Szentgyörgyi 1933
1. ♖d5

3218.
E. Szentgyörgyi 1933
1. ♕f3

3219.
E. Szentgyörgyi 1933
1. ♕f5

3220.
E. Szentgyörgyi 1933
1. ♕b4

3221.
E. Szentgyörgyi 1933
1. d3

3222.
E. Szentgyörgyi 1933
1. ♖a1

3223.
E. Szentgyörgyi 1933
1. ♘d3

3224.
E. Szentgyörgyi 1933
1. ♕d6

3225.
E. Szentgyörgyi 1933
1. ♕g2

3226.
E. Szentgyörgyi 1933
1. ♗h8

3227.
E. Szentgyörgyi 1933
1. ♕f4

3228.
E. Szentgyörgyi 1933
1. ♔e7

3229.
E. Szentgyörgyi 1933
1. f4

3230.
E. Szentgyörgyi 1933
1. ♘e8

3231.
E. Szentgyörgyi 1933
1. ♖g2

3232.
E. Szentgyörgyi 1933
1. f6

3233.
E. Szentgyörgyi 1933
1. ♕e3

3234.
E. Szentgyörgyi 1933
1. ♕a2

3235.
E. Szentgyörgyi 1933
1. g5

3236.
E. Szentgyörgyi 1933
1. c4

3237.
E. Szentgyörgyi 1933
1. ♔g5

3238.
E. Szentgyörgyi 1933
1. ♖1h3

3239.
E. Szentgyörgyi 1933
1. ♖g7

3240.
E. Szentgyörgyi 1933
1. g7

3241.
E. Szentgyörgyi 1933
1. ♘e6

3242.
E. Szentgyörgyi 1933
1. ♕e8

3243.
E. Szentgyörgyi 1933
1. ♕a5

3244.
E. Szentgyörgyi 1933
1. ♗b2

3245.
E. Szentgyörgyi 1933
1. ♕e6

3246.
E. Szentgyörgyi 1933
1. ♗g1

3247.
E. Szentgyörgyi 1933
1. ♖c4

3248.
E. Szentgyörgyi 1933
1. ♘e8

3249.
E. Szentgyörgyi 1933
1. ♔g6

3250.
E. Szentgyörgyi 1933
1. ♗g8

3251.
E. Szentgyörgyi 1936
1. d7

3252.
E. Szentgyörgyi 1928
1. ♖ab4

3253.
E. Szentgyörgyi 1936
1. ♔c4

3254.
E. Szentgyörgyi 1936
1. ♘h4

3255.
E. Szentgyörgyi 1936
1. ♔c6

3256.
E. Szentgyörgyi 1936
1. ♔d5

3257.
E. Szentgyörgyi 1936
1. ♕c7

3258.
E. Szentgyörgyi 1936
1. ♕g7

3259.
E. Szentgyörgyi 1936
1. g8♗

3260.
E. Szentgyörgyi 1936
1. ♔f4

3261.
E. Szentgyörgyi 1936
1. ♗g3

3262.
E. Szentgyörgyi 1936
1. ♖b3

3263.
E. Szentgyörgyi 1936
1. ♖f7

3264.
E. Szentgyörgyi 1936
1. ♔g5

3265.
E. Szentgyörgyi 1936
1. h7

3266.
E. Szentgyörgyi 1936
1. ♘d2

3267.
E. Szentgyörgyi 1936
1. ♔d7

3268.
E. Szentgyörgyi 1936
1. ♗f2

3269.
E. Szentgyörgyi 1936
1. d3

3270.
E. Szentgyörgyi 1936
1. ♗f7

3271.
E. Szentgyörgyi 1936
1. ♘g6

3272.
E. Szentgyörgyi 1936
1. c4

3273.
E. Szentgyörgyi 1936
1. ♔f3

3274.
E. Szentgyörgyi 1936
1. g7

3275.
E. Szentgyörgyi 1936
1. ♕f3

3276.
E. Szentgyörgyi 1936
1. ♘a5

3277.
E. Szentgyörgyi 1936
1. e7

3278.
E. Szentgyörgyi 1936
1. ♖g1

3279.
E. Szentgyörgyi 1936
1. g3

3280.
E. Szentgyörgyi 1936
1. f4

3281.
E. Szentgyörgyi 1936
1. ♔f5

3282.
E. Szentgyörgyi 1936
1. ♗g1

3283.
E. Szentgyörgyi 1936
1. h7

3284.
E. Szentgyörgyi 1936
1. ♗a3

3285.
E. Szentgyörgyi 1936
1. f7

3286.
E. Szentgyörgyi 1936
1. ♔c2

3287.
E. Szentgyörgyi 1936
1. ♖a5

3288.
E. Szentgyörgyi 1936
1. ♕f3

3289.
E. Szentgyörgyi 1936
1. ♔h3

3290.
E. Szentgyörgyi 1936
1. ♗f2

3291.
E. Szentgyörgyi 1936
1. ♔d7

3292.
E. Szentgyörgyi 1936
1. ♕c6

3293.
E. Szentgyörgyi 1936
1. ♖a5

3294.
E. Szentgyörgyi 1936
1. ♔e4

3295.
E. Szentgyörgyi 1936
1. ♕h8

3296.
E. Szentgyörgyi 1936
1. ♗d6

3297.
E. Szentgyörgyi 1936
1. ♔c6

3298.
E. Szentgyörgyi 1936
1. ♗f6

3299.
E. Szentgyörgyi 1936
1. ♘b7

3300.
E. Szentgyörgyi 1936
1. b4

3301.
E. Szentgyörgyi 1936
1. ♗g7

3302.
E. Szentgyörgyi 1936
1. ♔h6

3303.
E. Szentgyörgyi 1936
1. ♗h2

3304.
E. Szentgyörgyi 1936
1. ♕d7

3305.
E. Szentgyörgyi 1936
1. h3

3306.
E. Szentgyörgyi 1936
1. ♔b3

3307.
E. Szentgyörgyi 1936
1. ♕b2

3308.
E. Szentgyörgyi 1936
1. ♔b5

3309.
E. Szentgyörgyi 1936
1. ♔b7

3310.
E. Szentgyörgyi 1936
1. ♕c6

3311.
E. Szentgyörgyi 1936
1. ♘g1

3312.
E. Szentgyörgyi 1936
1. ♕c8

3313.
E. Szentgyörgyi 1936
1. ♖a6

3314.
E. Szentgyörgyi 1936
1. ♗f3

3315.
E. Szentgyörgyi 1936
1. ♔f7

3316.
E. Szentgyörgyi 1936
1. ♕b5

3317.
E. Szentgyörgyi 1936
1. e5

3318.
E. Szentgyörgyi 1936
1. h4

3319.
E. Szentgyörgyi 1936
1. ♕b5

3320.
E. Szentgyörgyi 1936
1. ♘a2

3321.
E. Szentgyörgyi 1936
1. ♖e6

3322.
E. Szentgyörgyi 1936
1. ♕b1

3323.
E. Szentgyörgyi 1936
1. ♔b3

3324.
E. Szentgyörgyi 1936
1. ♘c2

3325.
E. Szentgyörgyi 1936
1. ♘f3

3326.
E. Szentgyörgyi 1936
1. ♘f6

3327.
E. Szentgyörgyi 1933
1. ♕d7

3328.
E. Szentgyörgyi 1927
1. h7

3329.
E. Szentgyörgyi 1927
1. ♘f6

3330.
E. Szentgyörgyi 1927
1. ♕h2

3331.
E. Szentgyörgyi 1927
1. ♕b8

3332.
E. Szentgyörgyi 1927
1. ♖h8

3333.
E. Szentgyörgyi 1927
1. b8♘

3334.
E. Szentgyörgyi 1927
1. ♕h7

3335.
E. Szentgyörgyi 1927
1. ♕g3

3336.
E. Szentgyörgyi 1927
1. ♕a6

3337.
E. Szentgyörgyi 1928
1. ♗c5

3338.
E. Szentgyörgyi 1928
1. ♔c4

3339.
E. Szentgyörgyi 1928
1. ♕e3

3340.
E. Szentgyörgyi 1928
1. ♕g8

3341.
E. Szentgyörgyi 1928
1. ♗a4

3342.
E. Szentgyörgyi 1928
1. ♘e3

3343.
E. Szentgyörgyi 1928
1. c6

3344.
E. Szentgyörgyi 1928
1. d5

3345.
E. Szentgyörgyi 1928
1. ♖aa1

3346.
E. Szentgyörgyi 1928
1. e6

3347.
E. Szentgyörgyi 1928
1. ♔f7

3348.
E. Szentgyörgyi 1928
1. ♕e7

3349.
E. Szentgyörgyi 1928
1. ♔d3

3350.
E. Szentgyörgyi 1928
1. d4

3351.
E. Szentgyörgyi 1928
1. ♕c6

3352.
E. Szentgyörgyi 1928
1. b3

3353.
E. Szentgyörgyi 1928
1. ♗a6

3354.
E. Szentgyörgyi 1928
1. ♕d4

3355.
E. Szentgyörgyi 1928
1. ♕e7

3356.
E. Szentgyörgyi 1928
1. ♘f4

3357.
E. Szentgyörgyi 1928
1. ♘c4

3358.
E. Szentgyörgyi 1928
1. ♘c7

3359.
E. Szentgyörgyi 1928
1. ♔c5

3360.
E. Szentgyörgyi 1928
1. ♔b6

3361.
E. Szentgyörgyi 1928
1. ♔f5

3362.
E. Szentgyörgyi 1928
1. ♖h8

3363.
E. Szentgyörgyi 1928
1. ♕a5

3364.
E. Szentgyörgyi 1928
1. ♕b4

3365.
E. Szentgyörgyi 1925
1. ♔h4

3366.
E. Szentgyörgyi 1928
1. ♗a3

3367.
E. Szentgyörgyi 1928
1. h3

3368.
E. Szentgyörgyi 1928
1. ♕b5

3369.
E. Szentgyörgyi 1936
1. ♔a6

3370.
L. Talabér 1932
1. ♖a8

3371.
L. Talabér 1932
1. ♗g4

3372.
L. Talabér 1932
1. ♖d3

3373.
L. Talabér 1932
1. ♔c3

3374.
L. Talabér 1932
1. ♕c4

3375.
L. Talabér 1932
1. ♗e7

SOLUTIONS (3376 - 3450)

3376.
L. Talabér 1932
1. f3

3377.
L. Talabér 1931
1. ♘b4

3378.
L. Talabér 1931
1. ♕h5

3379.
L. Talabér 1931
1. ♗g3

3380.
L. Talabér 1931
1. ♕g3

3381.
L. Talabér 1931
1. ♕f2

3382.
L. Talabér 1931
1. ♗h4

3383.
L. Talabér 1931
1. ♔a4

3384.
L. Talabér 1931
1. ♖e4

3385.
L. Talabér 1931
1. ♔c2

3386.
L. Talabér 1931
1. ♖b3

3387.
L. Talabér 1931
1. ♕f6

3388.
L. Talabér 1930
1. ♕f3

3389.
L. Talabér 1933
1. ♕c1

3390.
L. Talabér 1931
1. ♕c2

3391.
L. Talabér 1971
1. ♗a4

3392.
L. Talabér 1971
1. ♖a4

3393.
L. Talabér 1972
1. ♕f6

3394.
L. Talabér 1970
1. ♖h3

3395.
I. Telkes 1924
1. ♕e3

3396.
N. Tereshchenko 1893
1. ♕e2

3397.
T. Tikkanen 1985
1. ♘h2

3398.
C. J. Timenes 1956
1. ♖c4

3399.
I. Tkeshelashvili 1893
1. ♕g1

3400.
I. Tokar 1975
1. ♗b3

3401.
S. Tolstoy 1971
1. ♕a2

3402.
C. Tomlinson 1845
1. c8♖

3403.
Gy. Tóth 1928
1. ♕c2

3404.
Gy. Tóth 1928
1. ♕b4

3405.
I. Tóth 1955/56
1. ♖ce2

3406.
P. Törngren 1928
1. ♕c4

3407.
P. Törngren 1928
1. ♔f8

3408.
P. Törngren 1929
1. ♔c8

3409.
P. Törngren 1929
1. ♔f1

3410.
St. Trcala – before 1901
1. ♕b2

3411.
A. Trilling 1932
1. ♗b4

3412.
A. Trilling 1932
1. ♗h3

3413.
A. Trilling 1933
1. ♘d2

3414.
A. Trilling 1933
1. ♕c3

3415.
L Tryssesoone 1984
1. ♗c8

3416.
R. Turnbull 1986
1. ♕b5

3417.
R. Turnbull 1986
1. ♘a4

3418.
V. Turuntayev 1976
1. ♕f8

3419.
H. V. Tuxen 1959
1. ♕e7

3420.
H. V. Tuxen 1962
1. c5

3421.
H. V. Tuxen 1962
1. ♗e7

3422.
H. V. Tuxen 1961
1. ♕c7

3423.
H. V. Tuxen 1953
1. ♗c6

3424.
H. V. Tuxen 1962
1. ♕g3

3425.
H. V. Tuxen 1963
1. ♖a5

3426.
H. V. Tuxen 1962
1. ♖d3

3427.
H. V. Tuxen 1963
1. ♗d3

3428.
G. Umnov 1978
1. ♕c8

3429.
A. Ursic 1905
1. ♗e3

3430.
R. Usmanov 1986
1. ♕f5

3431.
E. Uzunov 1964
1. ♗b7

3432.
L. Valve 1943
1. f8♗

3433.
L. Valve 1943
1. g8♕

3434.
L. Valve 1943
1. f8♖

3435.
L. Valve 1943
1. g8♘

3436.
A. Vaulin – M. Marandyuk 1969
1. ♕a4

3437.
A. Vaulin – M. Marandyuk 1969
1. ♕g4

3438.
V. Veders 1971
1. ♗b1

3439.
S. Veselenchuk 1985
1. ♕d7

3440.
K. Virtanen 1955
1. ♕e5

3441.
V. Vishnevsky 1969
1. ♔b2

3442.
V. Vladimirov 1971
1. ♖h6

3443.
J. Vladimirov 1974
1. Vc3

3444.
V. Voinov 1983
1. ♔e1

3445.
I. Vulfovich 1969
1. ♕a3

3446.
R. H. de Waard 1914
1. ♕f5

3447.
F. von Wardener 1903
1. c7

3448.
F. von Wardener 1925/26
1. ♕f4

3449.
H. Weenink 1917
1. ♕c4

3450.
H. Weenink 1919
1. ♘d5

3451.
H. Weenink 1918
1. ♔c7

3452.
H. Weenink 1920
1. c8♗

3453.
H. Weenink 1926
1. ♗d3

3454.
H. Weenink 1917
1. ♘f6

3455.
R. Weinheimer 1891
1. ♘b4

3456.
A. Weisz 1933
1. ♕c5

3457.
A. Weisz 1933
1. ♕e1

3458.
J. G. West 1882
1. ♕g7

3459.
J. R. R. White 1989
1. ♗d2

3460.
K. Widlert 1968
1. ♕b1

3461.
P. H. Williams 1897
1. ♖d2

3462.
P. H. Williams 1917
1. ♕e5

3463.
P. H. Williams 1890
1. ♕h3

3464.
L. C. Willemsens 1954
1. ♘a6

3465.
A. Willmott 1989
1. ♕h3

3466.
A. Willmott 1990
1. ♕h5

3467.
E. A. Wirtanen 1942
1. ♗e6

3468.
E. A. Wirtanen 1966
1. ♔g4

3469.
E. A. Wirtanen 1944
1. ♗c6

3470.
E. A. Wirtanen 1967
1. ♕c7

3471.
E. A. Wirtanen 1971
1. ♕a8

3472.
E. Wolf 1933
1. ♕c4

3473.
E. Wolf 1932
1. ♖g3

3474.
E. Wolf 1932
1. ♘h3

3475.
E. Wolf 1932
1. ♕d5

3476.
E. Wolf 1932
1. ♘e6

3477.
E. Wolf 1932
1. ♔b3

3478.
M. Wróbel 1953
1. ♕c5

3479.
O. Würzburg 1932
1. ♖e2

3480.
V. Zagoruyko 1971
1. ♔e1

3481.
G. Zahodyakin 1980/81
1. ♘b4

3482.
G. Zahodyakin 1980/81
1. ♕a6

3483.
G. Zahodyakin 1974
1. h8♘

3484.
G. Zahodyakin 1974
1. f8♖

3485.
G. Zahodyakin 1974
1. f8♕

3486.
G. Zahodyakin 1974
1. f8♗

3487.
G. Zahodyakin 1968
1. ♔g7

3488.
G. Zahodyakin 1967
1. ♕c3

3489.
G. Zahodyakin 1967
1. ♗a4

3490.
G. Zahodyakin 1967
1. ♔a1

3491.
G. Zahodyakin 1967
1. ♖c3

3492.
G. Zahodyakin 1966
1. ♕b7

3493.
G. Zahodyakin 1966
1. ♔d6

3494.
G. Zahodyakin 1966
1. ♕g4

3495.
G. Zahodyakin 1966
1. ♕c3

3496.
G. Zahodyakin 1962
1. ♖g7

3497.
G. Zahodyakin 1962
1. ♖a7

3498.
G. Zahodyakin 1949
1. ♕e1

3499.
G. Zahodyakin 1930
1. g8♘

3500.
L. Zalkind 1907
1. ♕g7

3501.
R. Zalokotzky 1963
1. ♘h3

3502.
Z. Zilahi 1929
1. ♔f4

3503.
Z. Zilahi 1931
1. ♘f6

3504.
Z. Zilahi 1931
1. g5

3505.
Z. Zilahi 1931
1. ♕c8

3506.
Z. Zilahi 1926
1. ♔c2

3507.
Z. Zilahi 1926
1. ♖h3

3508.
Z. Zilahi 1926
1. f6

3509.
Z. Zilahi 1926
1. ♖g2

3510.
Z. Zilahi 1926
1. ♔d4

3511.
Z. Zilahi 1926
1. ♕g7

3512.
S. Zimmermann 1908
1. ♕d7

3513.
S. Zlatik 1954
1. ♗a8

3514.
L. Zoltán 1966
1. ♖e8

3515.
1. ♘d2

3516.
1. ♗e6

3517.
1. ♔e7

3518.
1. ♘d5

3519.
1. d5

3520.
1. 0-0

3521.
1. ♕g6

3522.
1. ♕h5

3523.
1. ♕b4

3524.
1. ♕d5

3525.
1. ♘f4

3526.
1. ♘f3

3527.
1. ♖f5

3528.
1. ♖f4

3529.
1. ♕c6

3530.
1. ♕d3

3531.
1. ♕e2

3532.
1. ♕d4
3533.
1. ♕c6
3534.
1. ♕e5
3535.
1. ♖e6
3536.
1. ♘d4
3537.
1. ♕e6
3538.
1. ♕f6
3539.
1. ♕d5
3540.
1. ♕e4
3541.
1. ♕g8
3542.
1. ♕h5
3543.
1. ♕f7
3544.
1. ♘d4
3545.
1. ♕d5
3546.
1. ♕f4
3547.
1. ♕e5
3548.
1. ♕f4
3549.
1. ♕d4
3550.
1. ♕e4
3551.
1. ♕e4
3552.
1. ♕g5
3553.
1. ♕d5
3554.
1. ♕c4
3555.
1. ♕e4
3556.
1. ♕d6
3557.
1. ♕e5
3558.
1. ♕b5
3559.
1. ♕e5
3560.
1. ♕d4
3561.
1. ♕d4
3562.
1. ♕d5
3563.
1. ♕c6
3564.
1. ♕f4
3565.
1. ♕f6
3566.
1. ♕c6
3567.
1. ♕f6
3568.
1. ♕e4
3569.
1. ♕h7
3570.
1. ♕g7
3571.
1. ♕f5
3572.
1. ♘cb2
3573.
1. ♗d2
3574.
1. ♖g1
3575.
1. ♗c5
3576.
1. ♖h4
3577.
1. ♖d5
3578.
1. ♘3f5
3579.
1. ♖a1
3580.
1. ♕d2
3581.
1. ♕e3
3582.
1. ♘c5
3583.
1. ♘d5
3584.
1. ♗h6
3585.
1. ♖e5
3586.
1. e4
3587.
1. ♖c5
3588.
1. ♗d5
3589.
1. ♘f5
3590.
1. ♘f5
3591.
1. ♘g4
3592.
1. ♖e1
3593.
1. ♖e4
3594.
1. ♖a1
3595.
1. ♕d4
3596.
1. ♖c4
3597.
1. ♘b2
3598.
1. ♘e5
3599.
1. ♕c4
3600.
1. ♘f6
3601.
1. ♕e5
3602.
1. ♕d4
3603.
1. ♖e5
3604.
1. ♔a2
3605.
1. ♕h5
3606.
1. ♕e4
3607.
1. ♖d6
3608.
1. ♘c5
3609.
1. ♖f5
3610.
1. ♘g5
3611.
1. ♘e5
3612.
1. ♖e5
3613.
1. ♘d6
3614.
1. ♘d6
3615.
1. ♖d4
3616.
1. ♘c5
3617.
1. ♘b6
3618.
1. ♗d6
3619.
1. ♖a4
3620.
1. ♘d4
3621.
1. ♘d6
3622.
1. ♖f2
3623.
1. ♖e5
3624.
1. ♖cd6
3625.
1. ♘c5
3626.
1. ♗f5
3627.
1. ♘b6
3628.
1. ♕e5
3629.
1. d4
3630.
1. ♘e4
3631.
1. ♕a7
3632.
1. ♕e4
3633.
1. ♕d4
3634.
1. ♕g1
3635.
1. ♕f6
3636.
1. ♕e5
3637.
1. ♕e5
3638.
1. ♕f5
3639.
1. ♕e4
3640.
1. ♕c4
3641.
1. ♕d5
3642.
1. ♕c4
3643.
1. ♕e5

3644.
1. ♕d5
3645.
1. ♕h6
3646.
1. ♘fe4
3647.
1. ♗f7
3648.
1. ♖c4
3649.
1. ♖a4
3650.
1. ♘f4
3651.
1. ♖e8
3652.
1. ♘c4
3653.
1. ♘c8
3654.
1. ♘d5
3655.
1. ♖c8
3656.
1. ♕d2
3657.
1. g5
3658.
1. ♖h2
3659.
1. ♗b8
3660.
1. d8♘
3661.
1. ♕d3
3662.
1. ♗h7
3663.
1. ♖e4
3664.
1. ♘e6
3665.
1. ♕d6
3666.
1. ♕e4
3667.
1. ♕e3
3668.
1. ♕c1
3669.
1. ♕a4
3670.
1. ♕f3
3671.
1. ♕f5
3672.
1. ♕d5
3673.
1. ♕a1
3674.
1. ♕f3
3675.
1. ♕e3
3676.
1. ♕e4
3677.
1. ♕f3
3678.
1. ♕d1
3679.
1. ♕g4
3680.
1. ♕b4
3681.
1. ♕c3
3682.
1. ♕c4
3683.
1. ♕g3
3684.
1. ♕d4
3685.
1. ♕d4
3686.
1. ♕e6
3687.
1. ♕f4
3688.
1. ♕c5
3689.
1. ♕h4
3690.
1. ♕e3
3691.
1. ♕c3
3692.
1. ♕f6
3693.
1. ♕a3
3694.
1. ♕h6
3695.
1. ♕c4
3696.
1. ♕d4
3697.
1. ♖c4
3698.
1. ♕e4
3699.
1. ♕d3
3700.
1. ♖e1
3701.
1. ♕c6
3702.
1. ♕e3
3703.
1. ♕f5
3704.
1. ♕e2
3705.
1. ♕d3
3706.
1. ♕f5
3707.
1. ♕d6
3708.
1. ♕e4
3709.
1. ♕d4
3710.
1. ♕g4
3711.
1. ♕e4
3712.
1. ♕c3
3713.
1. ♕d4
3714.
1. ♕e5
3715.
1. ♕e5
3716.
1. ♖c3
3717.
1. ♕g5
3718.
1. ♕d5
3719.
1. ♘c5 ♕c5 2. ♘c7 ♕c7 3. ♕e4#
3720.
1. ♘c7 ♕c7 2. ♘c5 ♕c5 3. ♕e1#
3721.
1. ♘g6 hg6 2. ♘d5 ed5 3. ♕e5#
3722.
1. ♕f7 ♔f7 2. ♗c4 ♕d5 3. ♗d5#
3723.
1. ♕f7 ♔h6 2. ♗g5 ♔g5 3. ♕f4#
3724.
1. ♕d8 ♔d8 2. ♗g5 ♔c7 3. ♗d8#
3725.
1. ♘f7 ♔g8 2. ♖h8 ♘h8 3. ♘h6#
3726.
1. ♘e7 ♔h8 2. ♖h7 ♔h7 3. ♖h5#
3727.
♘e7 ♔h8 2. ♘g6 hg6 3. hg3#
3728.
1. ♖h6 gh6 2. ♖g8 ♔f5 3. e4#
3729.
1. ♗g5 ♔g5 2. ♕g7 ♔h5 3. g4#
3730.
1. ♘e6 fe6 2. ♕g5 ♔h7 3. ♖h6#
3731.
1. ♖h8 ♘h8 2. ♕h7 ♔f8 3. ♕h8#
3732.
1. ♖f8 ♔f8 2. ♕a8 ♔e7 3. ♕e8#
3733.
1. ♖g8 ♔g8 2. ♖g1 ♔h8 3. ♗f6#
3734.
1. ♖c8 ♕c8 2. ♕g7 ♖g7 3. ♖g7#
3735.
1. g7 ♖g7 2. ♕h6 ♖h7 3. ♕h7#
3736.
1. ♗g6 ♔g6 2. ♖g4 ♔f5 3. ♕h5#
3737.
1. ♗f7 ♘f7 2. ♕c6 ♔d8 3. ♘f7#
3738.
1. ♖f5 gf5 2. ♕f5 ♔h6 3. ♕g5#
3739.
1. ♕e6 fe6 2. ♗e6 ♘e6 3. ♘e5#
3740.
1. ♕b8 ♔b8 2. ♖d8 ♔b7 3. ♖b8#
3741.
1. ♕g6 hg6 2. ♗g6 ♔e7 3. ♗c5#
3742.
1. ♗f8 ♗h5 2. ♕h5 gh5 3. ♖h6#

3743. 1. ♖a6 ba6 2. ♗g2 ♖c6 3. ♗c6#

3744. 1. ♗f6 gf6 2. ♔f8 f5 3. ♘f7#

3745. 1. ♕g8 ♖g8 2. ♘g6 hg6 3. ♖h4#

3746. 1. ♕g8 ♔g8 2. d8♕ ♕e8 3. ♕e8#

3747. 1. ♘e6 fe6 2. ♖d8 ♔f7 3. 0-0#

3748. 1. ♖a6 ba6 2. ♕c6 ♗d6 3. ♕d6#

3749. 1. ♕h6 ♔h6 2. ♖h3 ♗h4 3. ♖h4#

3750. 1. ♘a6 ♖d8 2. ♕b8 ♖b8 3. ♘c7#

3751. 1. ♕h7 ♔h7 2. ♘g5 ♔h8 3. ♘f7#

3752. 1. ♖a8 ♔a8 2. ♘d7 h3 3. ♖a4#

3753. 1. ♕c4 ♔c4 2. e4 ♖a4 3. ♘d6#

3754. 1. ♘f6 ♗f6 2. ♕d5 ♗e6 3. ♕e6#

3755. 1. ♕g7 ♖g7 2. hg7 ♔g8 3. ♖h8#

3756. 1. ♕d6 ♘ge7 2. ♕d8 ♘d8 3. ♖d8#

3757. 1. ♘g4 ♖g4 2. ♖f5 ♔f5 3. ♖d5#

3758. 1. ♘e5 ♖e5 2. ♖f4 ♔f4 3. ♖h4#

3759. 1. ♕d7 ♖d7 2. ♘c7 ♖c7 3. ♖d8#

3760. 1. ♕h8 ♔h8 2. ♘g6 ♔g8 3. ♖h8#

3761. 1. ♕g8 ♔g8 2. ♖e8 ♖e8 3. ♖e8#

3762. 1. ♘7e6 ♔e8 2. ♕f8 ♘f8 3. ♘g7#

3763. 1. ♕g8 ♘g8 2. ♖h7 ♘h7 3. g7#

3764. 1. ♘e6 de6 2. ♗h6 ♔g8 3. ♘f6#

3765. 1. ♖h5 gh5 2. ♕h5 ♗h6 3. ♕h6#

3766. 1. ♕d7 ♖d7 2. ♖c8 ♖d8 3. ♗b5#

3767. 1. ♖a3 ♘a3 2. b4 ♔a4 3. ♘c5#

3768. 1. ♘e6 ♔e8 2. ♕e7 ♘e7 3. ♖d8#

3769. 1. ♕h8 ♔h8 2. ♗f6 ♔g8 3. ♖e8#

3770. 1. ♗e7 ♕e7 2. ♕a8 ♘a8 3. ♖c8#

3771. 1. ♖a8 ♔a8 2. ♕c8 ♔a7 3. ♕b7#

3772. 1. ♕g6 ♘g6 2. hg6 ♔g8 3. ♖h8#

3773. 1. ♖a8 ♔a8 2. ♕c8 ♔a7 3. ♕b7#

3774. 1. ♖e4 ♘e5 2. ♖e5 ♔d6 3. ♕e6#

3775. 1. ♘h6 ♔h8 2. ♕g8 ♖g8 3. ♘f7#

3776. 1. ♖h6 ♔h6 2. ♕g7 ♘g7 3. ♖h3#

3777. 1. ♘a7 ♗a7 2. ♕c6 bc6 3. ♗a6#

3778. 1. ♘g6 ♘g6 2. ♕h7 ♔h7 3. ♖h5#

3779. 1. ♕h7 ♔h7 2. ♖h3 ♖h6 3. ♖h6#

3780. 1. ♘h7 ♖h7 2. ♕h6 ♖h6 3. ♗h6#

3781. 1. ♕g6 ♘g6 2. ♖h4 ♘h4 3. g4#

3782. 1. ♘g7 ♗g7 2. ♖e6 fe6 3. ♕g6#

3783. 1. ♕g6 fg6 2. ♖g7 ♔f8 3. ♘g6#

3784. 1. ♕c7 ♔a6 2. ♕c8 ♔b5 3. ♕c4#

3785. 1. ♕h7 ♘h7 2. ♘g6 ♔g8 3. ♗d5#

3786. 1. ♕g3 hg3 2. ♔g1 h4 3. ♘f4#

3787. 1. ♗f2 ♘f2 2. ♘d4 ♘c4 3. ♘f3#

3788. 1. ♕g7 ♔g7 2. ♖g4 ♔h8 3. ♗f6#

3789. 1. ♕g6 fg6 2. f7 ♕f7 3. ♖h8#

3790. 1. ♕c7 ♖c7 2. ♖d8 ♖c8 3. ♖1c8#

3791. 1. ♘b6 ab6 2. ♕c6 bc6 3. ♗a6#

3792. 1. ♕g4 ♗g4 2. ♖h6 gh6 3. ♗f7#

3793. 1. ♗g7 ♔g7 2. ♕g5 ♔h8 3. ♕f6#

3794. 1. ♕c5 dc5 2. ♘c4 ♔b5 3. ♖b6#

3795. 1. ♖g5 ♔f7 2. ♕h7 ♔f8 3. ♖g8#

3796. 1. ♖b7 ♗b7 2. ♘c2 ♘c2 3. ♖b5#

3797. 1. ♖b7 ♔a8 2. ♖a7 ♔a7 3. ♕b7#

3798. 1. ♕f7 ♘f7 2. ♗f7 ♔d8 3. ♘e6#

3799. 1. ♕h8 ♔h8 2. ♘f7 ♔g8 3. ♘h6#

3800. 1. ♕h7 ♘h7 2. ♗h7 ♔f8 3. ♘g6#

3801. 1. ♕g8 ♔g8 2. ♖h8 ♔h8 3. ♗f7#

3802. 1. ♕g6 fg6 2. hg6 ♔h8 3. ♘f7#

3803. 1. ♖d1 ♖a4 2. ♖a1 ♗a1 3. ♘c1#

3804. 1. ♗a5 ♔a5 2. c3 ♕h1 3. b4#

3805. 1. ♖g1 ♕g1 2. ♕f5 ♔h4 3. ♕h5#

3806. 1. ♕e2 ♔c2 2. d3 ♔c1 3. 0-0#

3807. 1. ♕d8 ♖d8 2. ♖d8 ♘d8 3. ♖e8#

3808. 1. ♘g5 ♖g5 2. ♖f6 ♔f6 3. ♖d6#

3809. 1. ♘g5 hg5 2. ♖h1 ♕h1 3. ♖h1#

3810. 1. ♕g7 ♖g7 2. hg7 ♔g8 3. ♘h6#

3811. 1. ♕f8 ♗g8 2. ♕f6 ♗f6 3. ♗f6#

3812. 1. ♖f8 ♕f8 2. ♖f8 ♖f8 3. ♕g6#

3813. 1. ♕a6 ba6 2. ♘b5 ♔a8 3. ♖a7#

3814. 1. ♕g5 ♖g8 2. ♕h6 gh6 3. ♖g8#

3815.
1. ♕f8 ♔f8 2. ♖d8 ♔e7 3. ♖e8#

3816.
1. ♕a6 ♘a6 2. ♗b7 ♔a7 3. ♘c6#

3817.
1. ♕g5 ♘g5 2. ♘f5 ♔g6 3. h5#

3818.
1. ♕f8 ♔f8 2. ♗h6 ♔g8 3. ♖e8#

3819.
1. ♘g5 fg5 2. ♖e7 ♔f6 3. ♕g7#

3820.
1. ♕f8 ♔f8 2. ♖c8 ♔e7 3. d6#

3821.
1. ♕g6 fg6 2. ♘e7 ♔h8 3. ♘g6#

3822.
1. ♘g6 hg6 2. ♖h1 ♔g7 3. ♖h7#

3823.
1. ♕g8 ♕g8 2. ♔d3 ♕g6 3. c4#

3824.
1. ♕g7 ♔g7 2. ♘f5 ♔g8 3. ♘h6#

3825.
1. ♕f8 ♖f8 2. ♖h7 ♔g8 3. ♘h6#

3826.
1. ♘e5 ♘e5 2. ♗e8 ♔f8 3. ♗g6#

3827.
1. ♖h1 ♖h2 2. ♖h7 ♖h7 3. g7#

3828.
1. ♕f7 ♘f7 2. ♖g8 ♖g8 3. ♘f7#

3829.
1. ♕g7 ♔g7 2. ♗f6 ♔g8 3. ♘h6#

3830.
1. ♕a5 ♔a5 2. ab7 ♔b6 3. b8♕#

3831.
1. ♗f7 ♖f7 2. ♕g6 ♖g7 3. ♕g7#

3832.
1. ♘c6 ♔b7 2. a6 ♔c6 3. b5#

3833.
1. ♕f6 ♖f6 2. ef6 ♔g6 3. ♖g8#

3834.
1. ♕e7 ♕c7 2. ♕f8 ♖f8 3. ♖f8#

3835.
1. ♕e5 de5 2. ♖h8 ♔g7 3. ♖1h7#

3836.
1. ♕e6 fe6 2. ♗h5 ♗g6 3. ♗g6#

3837.
1. ♕g5 ♔g5 2. h4 ♔h6 3. g5#

3838.
1. ♘g6 ♔h7 2. ♖g7 ♔g7 3. ♖c7#

3839.
1. ♕h7 ♔h7 2. ♖h3 ♔g7 3. ♗e7#

3840.
1. ♕e2 ♗e2 2. ♖e4 de4 3. d4#

3841.
1. ♕g7 ♔g7 2. ♗h6 ♔f6 3. g5#

3842.
1. ♕c8 ♖c8 2. ♖d8 ♖d8 3. ♖d8#

3843.
1. ♖h5 gh5 2. ♕f5 ♔h4 3. ♕h5#

3844.
1. ♖h8 ♔h8 2. ♖f8 ♖f8 3. ef8♕#

3845.
1. h4 ♔h5 2. ♖f5 gf5 3. ♗f7#

3846.
1. d8♘ ♖b6 2. ♕b7 ♖b7 3. ♘c6#

3847.
1. ♕h7 ♔h7 2. hg6 Lg6 3. ♗e4#

3848.
1. ♕g7 ♘g7 2. ♖f8 ♔h7 3. ♘f6#

3849.
1. ♕d8 ♔d8 2. ♖f8 ♔c7 3. ♖c8#

3850.
1. ♕e8 ♘e8 2. d8♘ ♔f6 3. ♘e4#

3851.
1. ♕g6 hg6 2. ♖g7 ♔h8 3. ♖h4#

3852.
1. ♕d8 ♖d8 2. ♖e8 ♔e8 3. ♘f6#

3853.
1. ♕g6 ♘g6 2. ♗f7 ♔f8 3. ♘g6#

3854.
1. ♕h7 ♔h7 2. ♖h5 ♔g8 3. ♖h8#

3855.
1. ♕g7 ♖g7 2. ♗g7 ♔g8 3. ♗e5#

3856.
1. ♕c8 ♔d5 2. ♕c4 ♔c4 3. ♘b6#

3857.
1. ♗b7 ♘b7 2. ♕c6 dc6 3. d7#

3858.
1. ♕d8 ♔d8 2. ♘c6 ♔e8 3. ♖d8#

3859.
1. ♕g7 ♗g7 2. ♗g7 ♔g8 3. ♗f6#

3860.
1. ♕h7 ♔h7 2. hg6 ♔g6 3. ♖h6#

3861.
1. ♕c8 ♘b8 2. ♕b8 ♖b8 3. ♖b8#

3862.
1. ♕g6 hg6 2. ♗g6 ♔e7 3. ♘f5#

3863.
1. ♗b5 ab5 2. ♘c7 ♖c7 3. ♖d8#

3864.
1. ♖g7 ♗g7 2. ♖a8 ♗f8 3. ♕g5#

3865.
1. ♕h8 ♗h8 2. ♖g8 ♔e7 3. ♖e8#

3866.
1. ♕d6 ♗d6 2. ♗d6 ♔g7 3. f6#

3867.
1. ♕d6 ♗d6 2. ♗d6 ♔g8 3. ♘h6#

3868.
1. ♕d8 ♖d8 2. ♖f7 ♔e8 3. ♘f6#

3869.
1. ♕d8 ♗d8 2. ♖e8 ♔g7 3. ♖g8#

3870.
1. ♕f8 ♔f8 2. ♖d8 ♗e8 3. ♖ee8#

3871.
1. ♕h7 ♔h7 2. ♖h1 ♔g6 3. f5#

3872.
1. ♖d4 ♖e2 2. ♖e4 ♖e4 3. f3#

3873.
1. ♘g6 fg6 2. ♗g6 b5 3. ♖e8#

3874.
1. ♕d8 ♘d8 2. ♖e8 ♔g7 3. f6#

3875.
1. ♕h6 ♗h6 2. ♖h6 ♔g7 3. ♘f5#

3876.
1. ♘g4 hg4 2. ♗e5 ♔e5 3. ♕d4#

3877.
1. ♘c5 bc5 2. ♗b5 ♘b5 3. ♕e6#

3878.
1. ♘g6 ♔g8 2. ♕g7 ♖g7 3. ♘h6#

3879.
1. ♕h6 ♕e5 2. ♕h7 ♔h7 3. ♔g2#

3880.
1. gf8♘ ♔h8 2. ♗c3 ♖c3 3. ♖h6#

3881.
1. ♕e7 ♔h6 2. ♕f8 ♔g5 3. ♕f4#

3882.
1. ♕f4 ♖f4 2. ♖e5 ♘e5 3. ♘g5#

3883.
1. ♕f6 gf6 2. ♗h6 ♔g8 3. ♘f6#

3884.
1. ♕f7 ♔d8 2. ♕d7 ♘d7 3. ♘f7#

3885.
1. ♖g3 ♗e6 2. ♖h3 ♗h3 3. g3#

3886.
1. ♖d8 ♘d8 2. ♘d6 ♔f8 3. ♕e8#

3887.
1. ♕h6 ♔g8 2. ♕h8 ♔f7 3. ♕h7#

3888.
1. ♕f7 ♔f7 2. ♘e5 ♔e6 3. ♗c4#

3889.
1. h4 ♖h8 2. ♕h7 ♖h7 3. ♖h7#

3890.
1. ♕h6 ♔h6 2. ♘f5 ♔h5 3. ♗f7#

3891.
1. ♖b7 ♔c8 2. ♖b8 ♔b8 3. ♖d8#

3892.
1. ♖g5 ♔h6 2. ♕f8 ♖f8 3. ♗g7#

3893.
1. ♕d5 ♗d5 2. ♗d5 ♔b8 3. ♘a6#

3894.
1. g3 hg3 2. ♖f3 ♖f3 3. ♘g2#

3895.
1. g4 ♖g4 2. ♖h3 ♖h4 3. g4#

3896.
1. g4 hg4 2. ♖f4 ♗f4 3. e4#

3897.
1. g4 ♔h4 2. ♔g2 ♖b2 3. ♘f5#

3898.
1. ♕h5 h6 2. ♕h6 ♖fd8 3. ♕h8#

3899.
1. ♕h5 ♔h5 2. ♖h8 ♔g6 3. h5#

3900.
1. ♕f6 ♗f6 2. gf6 ♔f8 3. ♖h8#

3901.
1. ♖d7 ♔c8 2. ♕c6 bc6 3. ♗a6#

3902.
1. ♖d5 ♕d5 2. ♕f7 ♕f7 3. ♘d7#

3903.
1. ♕f6 ♗f6 2. ♗f6 ♔g8 3. ♘h6#

3904.
1. ♕h6 ♖g8 2. ♕h7 ♔h7 3. ♖h5#

3905.
1. ♕d6 ♔b5 2. ♖a5 ♔a5 3. ♕c5#

3906.
1. ♖a7 ♕a7 2. ♖a5 ♕a6 3. ♖a6#

3907.
1. ♖a7 ♘a7 2. ♕b6 ♔a8 3. ♘c7#

3908.
1. ♖a7 ♘a7 2. ♕a7 ♔d6 3. ♕c5#

3909.
1. ♘h3 gh3 2. ♔f2 h2 3. ♘g3#

3910.
1. ♕e7 ♕e7 2. ♖d8 ♗e8 3. ♖e8#

3911.
1. ♖a6 ba6 2. b7 ♔a7 3. b6#

3912.
1. ♖f8 ♗f8 2. ♕f7 ♔d8 3. ♕d7#

3913.
1. ♖d7 ♔e5 2. ♖d5 ♖d5 3. ♘c4#

3914.
1. ♗h7 ♔h7 2. ♕h5 ♔g8 3. ♕h8#

3915.
1. ♘d5 ♔a8 2. ♘b6 ab6 3. ♖a1#

3916.
1. ♕h5 ♕h5 2. d7 ♘g7 3. d8♘#

3917.
1. ♖h5 ♗h5 2. ♖f5 ♖f5 3. ♘e6#

3918.
1. ♖c8 ♔f7 2. ♖f8 ♘f8 3. ♘d8#

3919.
1. ♕h7 ♔h7 2. ♖h4 ♕h6 3. ♖h6#

3920.
1. ♕h7 ♔h7 2. ♖g7 ♔h8 3. ♘g6#

3921.
1. ♕h7 ♔h7 2. ♖h1 ♔g8 3. ♖h8#

3922.
1. ♕h7 ♘h7 2. ♗e5 ♔h6 3. ♗g7#

3923.
1. ♖g5 ♔g5 2. ♘f7 ♔h5 3. g4#

3924.
1. ♖c6 ♗c6 2. ♘c5 ♔a5 3. ♗c7#

3925.
1. ♕c6 bc6 2. ♗a6 ♔d7 3. ♘c5#

3926.
1. ♖e8 ♗f8 2. ♕g5 ♕g5 3. ♖f8#

3927.
1. ♘f5 ♔h5 2. ♕h7 ♘h7 3. g4#

3928.
1. ♕h6 gh6 2. ♖g8 ♖g8 3. ♖g8#

3929.
1. ♖e8 ♔e8 2. ♖g8 ♔e7 3. ♘f5#

3930.
1. ♕h6 gh6 2. ♗e5 ♕f6 3. ♗f6#

3931.
1. ♕c6 bc6 2. ♗a6 ♔b8 3. ♘d7#

3932.
1. ♕h6 ♗f6 2. ♘f6 ♔h8 3. ♕h7#

3933.
1. ♘f7 ♔g8 2. ♖h8 ♘h8 3. ♘h6#

3934.
1. ♖b8 ♕b8 2. ♕c6 ♔d8 3. ♕d7#

3935.
1. ♖e8 ♗f8 2. ♗h6 ♕d5 3. ♖f8#

3936.
1. ♖e8 ♖e8 2. ♕g4 ♕g4 3. ♘f6#

3937.
1. ♕c6 ♗c6 2. ♖d8 ♕c8 3. ♖c8#

3938.
1. ♖c8 ♖c8 2. ♕b8 ♖b8 3. ♘c7#

3939.
1. ♖c6 dc6 2. ♗e6 ♔c7 3. e8♘#

3940.
1. ♖a4 ♕a4 2. ♘a3 ♕a3 3. ♖c2#

3941.
1. ♕c4 ♗c4 2. ♗c4 ♔d6 3. ♘b5#

3942.
1. ♗d7 ♘d7 2. ♕b8 ♘b8 3. ♖d8#

3943.
1. ♕b5 ♕b5 2. ♖c3 ♔a4 3. ♖a3#

3944.
1. ♕b4 ♔b4 2. ♗a3 ♔a3 3. ♘c2#

3945.
1. ♕b3 ♔a5 2. ♕b5 ab5 3. ♖a8#

3946.
1. ♕a8 ♖a8 2. ♘c5 ♔a5 3. ♖b5#

3947.
1. ♕a8 ♕b8 2. ♕c6 dc6 3. d7#

3948.
1. ♕c7 ♖c8 2. ♕b8 ♖b8 3. ♘c7#

3949.
1. ♕c6 ♕c6 2. ♘d4 ♘d4 3. ♖e7#

3950.
1. ♘f7 ♖f7 2. ♖d8 ♖f8 3. ♖f8#

3951.
1. ♕h6 gh6 2. ♖h6 ♖h7 3. ♗f6#

3952.
1. ♖e8 ♕e8 2. ♕f6 ♖g7 3. ♕g7#

3953.
1. ♕h7 ♔g5 2. ♕h4 ♔f5 3. ♕f4#

3954.
1. ♖g7 ♘g7 2. ♖h6 ♔g8 3. ♘e7#

3955.
1. ♕b6 ab6 2. ♘c6 ♔a8 3. ♘b6#

3956.
1. ♖c5 ♗c5 2. ♗e4 ♘e4 3. d5#

3957.
1. ♕h6 ♔h6 2. ♗f8 ♔h5 3. ♗e2#

3958.
1. ♖e8 ♕e8 2. ♕g5 ♔h8 3. ♕g7#

3959.
1. ♘f7 ♕f7 2. ♕h4 ♔g6 3. ♕h5#

3960.
1. ♖b7 e2 2. ♖b8 ♖b8 3. ♘c7#

3961.
1. ♖g7 ♔h4 2. ♕h3 gh3 3. g3#

3962.
1. cd7 ♔f8 2. ♘g6 ♔f7 3. h8♘#

3963.
1. b4 ♗b4 2. ♗b6 ab6 3. ♕a8#

3964.
1. b3 ♘b3 2. ♖b4 ♗b4 3. ♘b6#

3965.
1. ♕b8 ♖b8 2. ♖a7 ♗a7 3. ♘c7#

3966.
1. ♖g8 ♕g8 2. ♕h6 ♕h7 3. ♕h7#

3967.
1. ♖b8 ♔b8 2. a7 ♔a8 3. ♘c7#

3968.
1. ♕h6 ♗h6 2. ♘g5 ♔h8 3. ♖h7#

3969.
1. ♘f8 ♔h8 2. ♕h7 ♖h7 3. ♖h7#

3970.
1. ♕h8 ♔h8 2. ♖f8 ♔g7 3. ♘e6#

3971.
1. ♘f6 ♔h8 2. ♕h6 gh6 3. ♖h7#

3972.
1. ♖b8 ♗c8 2. ♕d7 ♘d7 3. ♘e6#

3973.
1. ♗e4 ♕d4 2. ♖a6 ♗a6 3. b6#

3974.
1. ♘f6 ♔h8 2. ♖h6 gh6 3. ♖h7#

3975.
1. c5 ♔g6 2. f5 ♔h7 3. ♖h2#

3976.
1. ♕d6 ♖d6 2. ♖f7 ♔g8 3. ♖e8#

3977.
1. ♘f6 gf6 2. ♕g4 ♔h8 3. ♖h1#

3978.
1. ♗f5 ♖f5 2. ♘g7 ♖h1 3. gf5#

3979.
1. ♗h6 ♔h6 2. ♕f6 ♔h5 3. g4#

3980.
1. ♗h7 ♔h7 2. ♖h3 ♔g8 3. ♖h8#

3981.
1. ♗e4 ♗e4 2. h3 ♔g3 3. ♗e1#

3982.
1. ♕h7 ♘h7 2. ♗h7 ♔h8 3. ♘g6#

3983.
1. ♗e6 de6 2. ♕g4 ♘g6 3. ♕e6#

3984.
1. ♗e5 f6 2. ♗f6 ♖f6 3. ♖g8#

3985.
1. ♗e8 ♔e8 2. ♕h5 ♔d7 3. ♕b5#

3986.
1. ♕h8 ♔e7 2. ♘f5 ef5 3. ♗c5#

3987.
1. ♕c4 ♔d2 2. ♕c1 ♔c1 3. ♘b3#

3988.
1. ♕h7 ♔h7 2. ♘f6 gf6 3. ♖h4#

3989.
1. ♗c6 ♕c8 2. ♕a7 ♔a7 3. ♖a1#

3990.
1. ♕h7 ♔h7 2. ♘g5 fg5 3. ♖h3#

3991.
1. ♕h7 ♔h7 2. ♖h3 ♔g6 3. ♘e7#

3992.
1. ♘f6 gf6 2. ♕e6 fe6 3. ♗h5#

3993.
1. ♖a8 ♘cb8 2. ♖b8 ♘b8 3. ♖c7#

3994.
1. ♘5f6 ♖c8 2. ♘e5 f1♕ 3. ♘g6#

3995.
1. ♕h6 ♗h6 2. ♖h6 ♕b1 3. ♖h8#

3996.
1. ♕h8 ♔e7 2. ♘g6 fg6 3. ♕g7#

3997.
1. ♖a8 ♕a8 2. ♗b4 ♕e8 3. b3#

3998.
1. ♖h8 ♗h8 2. ♘e7 ♔f8 3. ♖g8#

3999.
1. ♖h8 ♔h8 2. ♕h6 ♔g8 3. ♕g7#

4000.
1. ♖h8 ♔h8 2. ♕h4 ♔g8 3. ♕h7#

4001.
1. ♖g8 ♖g8 2. ♗f6 ♖g7 3. ♗g7#

4002.
1. ♖h4 ♔h4 2. g3 ♔h3 3. ♘f4#

4003.
1. ♗d7 ♘d7 2. ♘c7 ♔e7 3. ♘c8#

4004.
1. ♗d4 cd4 2. cd4 ♔f4 3. ♘e6#

4005.
1. ♘f6 ♕c4 2. ♖e8 ♖e8 3. ♖e8#

4006.
1. ♕h8 ♔h8 2. ♖f8 ♔h7 3. ♖h4#

4007.
1. ♘f6 ef6 2. ef6 ♖d1 3. ♕g7#

4008.
1. ♖h7 ♖h7 2. ♘g8 ♔g5 3. ♗f6#

4009.
1. ♖h7 ♔h7 2. ♕h5 ♔g7 3. ♕h8#

4010.
1. ♖h7 ♔h7 2. ♕f7 ♔h8 3. ♘g6#

4011.
1. ♖h7 ♔h7 2. ♗f8 ♔g8 3. ♕g7#

4012.
1. ♖h8 ♗h8 2. ♘e6 ♔g8 3. ♗h7#

4013.
1. ♖h8 ♕g3 2. ♔g3 c3 3. ♖h7#

4014.
1. ♖h8 ♔h8 2. ♘e7 ♖c2 3. ♖h1#

4015.
1. ♖c8 ♘c8 2. ♕b7 ♔b7 3. ♗d5#

4016.
1. ♕b8 ♖b8 2. ♖a7 ♗a7 3. ♘c7#

4017.
1. ♕b7 ♔b7 2. ♖a7 ♔c8 3. ♖c7#

4018.
1. ♕b6 ab6 2. ♘c7 ♔a7 3. ♖a8#

4019.
1. ♕b6 ♘b6 2. ♗c3 ♘b4 3. ♗b4#

4020.
1. ♕b6 ♔d6 2. ♖d5 ♔d5 3. ♕d4#

4021.
1. ♕d8 ♕e8 2. ♕e8 ♗f8 3. ♕f8#

4022.
1. ♕g6 ♗g6 2. hg6 ♔g6 3. ♗d3#

4023.
1. ♕a8 ♔h7 2. ♕h8 ♘h8 3. ♖g7#

4024.
1. ♕h7 ♖h7 2. ♖g8 ♖g8 3. ♖g8#

4025.
1. ♗g3 hg3 2. ♔g1 h4 3. ♘f4#

4026.
1. ♕c8 ♖c8 2. ♖c8 ♔e7 3. ♘c6#

4027.
1. ♕h7 ♔b6 2. ♕c7 ♘c7 3. ♗d4#

4028.
1. ♘g6 hg6 2. hg3 ♕h4 3. ♖h4#

4029.
1. ♕c8 ♔a7 2. ♗b6 ♔b6 3. ♕b8#

4030.
1. ♖f8 ♔h7 2. ♗f5 g6 3. ♖h8#

SOLUTIONS (4031 – 4102)

4031.
1. ♘e7 ♘e7 2. ♕f8 ♔f8 3. ♖d8#

4032.
1. ♘f6 ♔f8 2. ♗d6 ♕d6 3. ♖e8#

4033.
1. ♗a8 ♗a8 2. ♔f1 ♗e4 3. ♘f2#

4034.
1. ♘g3 ♔h2 2. ♖h1 ♗h1 3. ♘f1#

4035.
1. ♗a4 ♔a4 2. b3 ♔b5 3. c4#

4036.
1. ♕h7 ♔h7 2. ♖h6 ♔g8 3. ♖h8#

4037.
1. ♕a4 ♔a4 2. ♖a1 ♔b5 3. ♗d7#

4038.
1. ♕h6 ♖g8 2. ♕h7 ♔h7 3. ♖h3#

4039.
1. ♕h8 ♗h8 2. ♖g8 ♔e7 3. ♖e8#

4040.
1. ♘c6 bc6 2. ♕b3 ♗b4 3. ♕b4#

4041.
1. ♖d6 ♔e5 2. ♘f7 ♔f5 3. g4#

4042.
1. ♖d6 ♔d6 2. e5 ♔c6 3. d5#

4043.
1. ♖e7 ♔f8 2. ♖e8 ♔e8 3. ♕e7#

4044.
1. ♖b1 ♔a7 2. ♕d4 ♕d4 3. ♘c6#

4045.
1. ♕h7 ♔h7 2. ♖d7 ♔h8 3. ♖h7#

4046.
1. ♘g6 fg6 2. ♕h7 ♔h7 3. ♖h3#

4047.
1. ♖h8 ♔h8 2. ♕h1 ♔g8 3. ♕h7#

4048.
1. ♖h8 ♔f7 2. ♖f8 ♕f8 3. d6#

4049.
1. ♖h8 ♗h8 2. ♘e6 ♔g8 3. ♘h6#

4050.
1. ♖h8 ♗h8 2. ♕h7 ♔f8 3. ♕f7#

4051.
1. ♕h6 ♗f6 2. gf6 g6 3. ♕g7#

4052.
1. ♖e8 ♘e8 2. ♗h7 ♔h8 3. ♕f8#

4053.
1. ♖c7 ♔b6 2. ♕a5 ♔a5 3. ♖c4#

4054.
1. ♘f5 ef5 2. ♗e5 ♔e6 3. ♘c5#

4055.
1. ♖e8 ♖e8 2. ♖e8 ♕e8 3. ♕g7#

4056.
1. ♕f7 ♘f7 2. ♘g6 ♔e8 3. ♖e7#

4057.
1. ♗c5 ♗c5 2. gf6 ♔f8 3. ♖h8#

4058.
1. ♖g7 ♔g7 2. ♗h6 ♔h6 3. ♕g5#

4059.
1. ♖g8 dc4 2. ♖h5 ♗h5 3. g5#

4060.
1. ♖f6 ♔g5 2. g3 d6 3. h4#

4061.
1. ♖f6 gf6 2. ♕h6 ♖h6 3. ♗h6#

4062.
1. ♖g8 ♔g8 2. h7 ♔f8 3. h8♕#

4063.
1. ♖g8 ♔g8 2. ♘f6 ♔f8 3. ♖g8#

4064.
1. ♕h6 ♔f6 2. ♖f5 ♔f5 3. ♕f4#

4065.
1. ♘f7 ♔c8 2. ♖e8 ♗e8 3. ♖d8#

4066.
1. ♕h6 ♘h6 2. ♗h6 ♔g8 3. f7#

4067.
1. ♕c5 ♔b7 2. ♕c8 ♔c8 3. ♘d6#

4068.
1. ♕h7 ♔h7 2. ♘f6 ♔h8 3. ♘g6#

4069.
1. ♕g6 ♗g6 2. ♘g5 hg5 3. hg6#

4070.
1. ♘e7 ♔h8 2. ♖h7 ♔h7 3. ♖h1#

4071.
1. ♕c5 bc5 2. bc5 ♔d5 3. ♖d7#

4072.
1. ♗h5 ♖h5 2. ♕h5 ♘h5 3. ♖g8#

4073.
1. ♘h6 ♗d6 2. ♕g8 ♖g8 3. ♘f7#

4074.
1. ♘e3 ♔a4 2. ♘c2 e3 3. b3#

4075.
1. ♘d7 ♖d7 2. ♖e8 ♔e8 3. ♖g8#

4076.
1. ♘e5 ♘f7 2. ♗e2 ♗g4 3. ♗g4#

4077.
1. ♘e8 ♖f7 2. ♖f6 ♖f6 3. ♘g7#

4078.
1. ♘5f6 gf6 2. ♗h6 hg4 3. ♘f6#

4079.
1. ♗b6 ♔b6 2. c8♘ ♔a5 3. b4#

4080.
1. ♗b5 ♘b5 2. ♖e7 ♘e7 3. ♘f6#

4081.
1. ♗b7 ♖b7 2. ♖c8 ♖b8 3. ♘c7#

4082.
1. ♘f6 ♕f6 2. ♗a4 ♘a4 3. ♕d7#

4083.
1. ♘f6 ♔f8 2. ♘gh7 ♖h7 3. ♕g8#

4084.
1. ♖a4 ♔a4 2. ♘c5 ♔a5 3. b4#

4085.
1.♕d7 ♕d7 2. ♖b8 ♕d8 3. ♗b5#

4086.
1. ♕g8 ♔g8 2. ♘e7 ♔f8 3. ♘g6#

4087.
1. ♘h6 ♕c3 2. ♖g8 ♖g8 3. ♘f7#

4088.
1. ♖a5 ♔a5 2. ♕a8 ♔b5 3. ♕a4#

4089.
1. ♕g6 hg6 2. ♖f3 ♔h7 3. ♖h3#

4090.
1. ♕d5 ed5 2. ♗b6 ab6 3. ♖e8#

4091.
1. ♘e6 ♕e6 2. ♕h6 ♔h6 3. ♗f8#

4092.
1. ♖f4 ♘f4 2. g3 ♔h3 3. ♘f4#

4093.
1. ♗h6 ♘h6 2. ♕g5 ♔h8 3. ♕h6#

4094.
1. ♖h7 ♘h7 2. ♕h7 ♔f8 3. ♗h6#

4095.
1. ♕h6 ♗h6 2. f8♕ ♔h7 3. ♖h6#

4096.
1. ♗f6 ♗f6 2. ♕h7 ♔h7 3. ♖h5#

4097.
1. ♕h6 gh6 2. ♗e5 ♖f6 3. ♗f6#

4098.
1. ♘f6 ♗f6 2. ♘g6 ♕e7 3. ♕f8#

4099.
1. ♕g7 ♔g7 2. ♗e5 ♔h6 3. ♗g7#

4100.
1. ♗h7 ♘h7 2. ♕f7 ♔h8 3. ♘g6#

4101.
1. ♖a7 ♘a7 2. ♕b6 ♔a8 3. ♘c7#

4102.
1. ♕g6 ♘g6 2. hg6 ♗h1 3. ♖h1#

4103.
1. ♘c4 bc4 2. b4 cb4 3. ab4#

4104.
1. ♕f8 ♗f8 2. ♖h7 ♔g8 3. ♗e6#

4105.
1. ♖h6 gh6 2. d5 ♘e5 3. ♕e5#

4106.
1. ♕e6 ♗e6 2. ♗e6 ♔d6 3. ♘b5#

4107.
1. ♕g7 ♘g7 2. ♘h6 ♔h8 3. fg7#

4108.
1. h6 fg5 2. h7 ♔h8 3. ♘g6#

4109.
1.♘e7 ♘e7 2. ♗h7 ♕h7 3. ♕h7#

4110.
1. ♕d7 ♖d7 2. ♖c8 ♖d8 3. ♗b5#

4111.
1. ♗g7 ♔g7 2. ♖h7 ♔f8 3. ♖f7#

4112.
1. ♕g6 hg6 2. ♗g6 ♔e7 3. ♗c5#

4113.
1. ♖g7 ♔g7 2. ♕f6 ♔g8 3. ♕g7#

4114.
1. ♘e3 ♘e3 2. ♖c2 ♘c2 3. ♘d2#

4115.
1. ♗c6 ♖ac6 2. ♘c7 ♖c7 3. ♘d6#

4116.
1. ♗c2 ♖c2 2. ♘f8 ♖f2 3. ♘g6#

4117.
1. ♘h2 gh2 2. ♖f2 ♔g1 3. ♖f1#

4118.
1. ♗e5 de5 2. b4 ♔c4 3. d6#

4119.
1. ♕d6 ♔d6 2. ♖d5 ♔d5 3. ♗f3#

4120.
1. ♖d5 ♗d5 2. ♘g6 ♕g6 3. f4#

4121.
1. ♖d4 ♔d4 2. ♘c6 ♔d3 3. ♘c5#

4122.
1. ♘e4 ♘e5 2. ♖g5 ♘g5 3. ♘f6#

4123.
1. ♗d3 ♕d3 2. ♘f3 ♔e2 3. ♖e1#

4124.
1. ♘f6 ♗f6 2. ♗c4 c1♕ 3. ♗f7#

4125.
1. ♕d5 ♗d5 2. ♘f3 ♗f3 3. ♗d6#

4126.
1. ♗b6 ♖bb6 2. ♘c6 ♖c6 3. ♘b5#

4127.
1. ♘f5 ef5 2. ♖cd3 ♔c5 3. ♖d5#

4128.
1. ♖d5 ♔d5 2. ♕e5 ♔c6 3. ♕d6#

4129.
1. ♖h6 ♔h6 2. ♗e8 g4 3. ♗f4#

4130.
1. ♕d5 ♘d5 2. ♗c2 ♔f4 3. ♗g3#

4131.
1. ♕e5 ♔e5 2. ♖f5 ♔e4 3. f3#

4132.
1. ♕h6 gh6 2. ♘f6 ♔h8 3. ♖g8#

4133.
1. ♕e5 ♘e5 2. ♘d2 ♔d4 3. ♘f5#

4134.
1. ♖d4 ♔d4 2. ♗b2 ♔c2 3. d3#

4135.
1. ♖d4 ♖d4 2. ♘d2 ♖d2 3. ♖c3#

4136.
1. ♘g5 fg5 2. ♕e4 ♔f6 3. ♕e5#

4137.
1. ♘a5 ♕a5 2. ♖c3 bc3 3. ♘a3#

4138.
1. ♖b7 ♔b7 2. ♗d5 ♔b6 3. c5#

4139.
1. – ♗b2 2. ♔b2 ♕c2 3. ♔a3 ♕c3#

4140.
1. – ♕h2 2. ♔h2 ♘f3 3. gf3 ♖h4#

4141.
1. – ♕h3 2. gh3 ♖h2 3. ♔g1 ♘e2#

4142.
1. – ♖h2 2. ♔h2 ♕g3 3. ♔h1 ♕h2#

4143.
1. – ♕b2 2. ♔b2 ♖b8 3. ♔a1 ♗c3#

4144.
1. – h5 2. ♔h3 ♕f5 3. g4 ♕g4#

4145.
1. – b5 2. ab5 ♘e5 3. ♔c5 ♗b4#

4146.
1. – ♕a2 2. ♖a2 ♖a2 3. ♔g1 ♖d1#

4147.
1. – ♖h3 2. gh3 ♗f3 3. ♔h2 ♗g1#

4148.
1. – ♘d2 2. ♖d2 ♖e1 3. ♔e1 ♖g1#

4149.
1. – ♖h2 2. ♔h2 ♖h4 3. ♔g1 ♕g3#

4150.
1. – ♖h4 2. ♗h4 g4 3. ♔g3 f4#

4151.
1. – ♕f3 2. ♔h3 ♖h4 3. ♔h4 ♕g4#

4152.
1. – ♕g5 2. ♔g5 ♗f6 3. ♔h6 ♖h8#

4153.
1. – ♖h1 2. ♔h1 ♕h3 3. ♔g1 ♕h2#

4154.
1. – ♖g2 2. ♔g2 ♕f3 3. ♔g1 ♕g4#

4155.
1. – ♕h2 2. ♔h2 ♘g3 3. ♔g3 f4#

4156.
1. – f2 2. ♕e4 ♕g1 3. ♖g1 fg1♕#

4157.
1. – ♕g2 2. ♔g2 ♗f3 3. ♔f1 ♘h2#

4158.
1. – ♕g2 2. ♖g2 ♖a1 3. ♖g1 ♖g1#

4159.
1. – ♕h3 2. ♘h3 ♖h3 3. ♔g1 ♖h1#

4160.
1. – ♕h3 2. ♘h3 ♖h3 3. any ♖h1#

4161.
1. – ♖g2 2. ♔g2 ♕g3 3. ♔h1 ♖h7#

4162.
1. – ♕g1 2. ♔g1 ♖dg8 3. ♔f1 ♖h1#

4163.
1. – ♕d1 2. ♗d1 ♖d1 3. ♕e1 ♖e1#

4164.
1. – ♕f2 2. ♔f2 ♖g2 3. ♔e3 ♖e2#

4165.
1. – ♕c1 2. ♔c1 ♗f4 3. ♔d1 ♖b1#

4166.
1. – ♕f1 2. ♘f1 ♖f1 3. ♔f1 ♖d1#

4167.
1. – ♕e1 2. ♔e1 ♗g3 3. ♔d1 ♖e1#

4168.
1. – ♕g1 2. ♘g1 ♗h4 3. ♔f1 ♖e1#

4169.
1. – ♕h3 2. ♔h3 ♖h6 3. ♔g4 ♖h4#

4170.
1. – ♘df3 2. gf3 ♖d1 3. ♔g2 ♗h3#

4171.
1. – ♕b2 2. ♔b2 ♖a2 3. ♔c1 ♖c2#

4172.
1. – ♖a2 2. ♔a2 ♖a3 3. ♔a3 ♕a1#

4173.
1. – ♕b2 2. ♔b2 ♖8a2 3. ♔c1 ♗a3#

4174.
1. – ♘c3 2. ♖e1 ♘e2 3. ♖e2 ♖f1#

4175.
1. – ♗h4 2. ♔h2 ♗f3 3. any ♖h1#

4176.
1. – ♘f3 2. gf3 ♕g3 3. ♔h1 ♕g2#

4177.
1. – ♖e3 2. fe3 ♕f3 3. ♕f2 ♕f2#

4178.
1. – ♖h2 2. ♔h2 ♘g4 3. ♔g1 ♗h2#

4179.
1. – ♕h2 2. ♔h2 ♖h8 3. ♔g1 ♖h1#

4180.
1. – ♕f1 2. ♗f1 ♘f3 3. ♔e2 ♗c4#

4181.
1. – ♕f2 2. ♖f2 ♖c1 3. ♖f1 ♖f1#

4182.
1. – ♕d2 2. ♕d2 ♖b1 3. ♕e1 ♖e1#

4183.
1. – ♕f2 2. ♗f2 ♗f2 3. ♔h1 ♘g3#

4184.
1. – ♘f3 2. gf3 ♖g6 3. ♔h1 ♗f3#

4185.
1. – ♘h3 2. gh3 ♖g6 3. ♔f2 ♖e2#

4186.
1. – ♕f4 2. ♔f4 g5 3. ♔g3 f4#

4187.
1. – ♕g2 2. ♔g2 ♖g6 3. ♔h1 ♘f2#

4188.
1. – ♖h3 2. gh3 ♗f3 3. ♔h2 ♘g4#

4189.
1. – ♘f2 2. ♖f2 ♕g2 3. ♗g2 hg2#

4190.
1. – ♘e5 2. fe5 ♕g4 3. ♖g4 fg4#

4191.
1. – ♕h2 2. ♔h2 ♘g3 3. ♔g3 f4#

4192.
1. – ♖c1 2. ♔a2 ♗b1 3. ♔a1 ♘b3#

4193.
1. – ♕g2 2. ♖g2 ♖g2 3. ♔h1 ♖h3#

4194.
1. – ♕b2 2. ♖b2 f1♕ 3. ♖b1 ♕b1#

4195.
1. – ♘f3 2. h3 ♕h3 3. gh3 ♖h2#

4196.
1. – ♕b1 2. ♘d1 ♕d1 3. ♔d1 ♖c1#

4197.
1. – ♖e1 2. ♖e1 ♕g4 3. ♕g2 ♕g2#

4198.
1. – ♘e4 2. fe4 ♗h4 3. ♔f3 ♕e4#

4199.
1. – ♖d2 2. ♕d2 ♖b1 3. ♕c1 ♕c2#

4200.
1. – ♘g4 2. fg4 ♕e4 3. ♗f3 ♕f3#

4201.
1. – ♕g1 2. ♔g1 ♘e2 3. ♔h1 ♖c1#

4202.
1. – ♕g1 2. ♔g1 ♗d4 3. ♔h1 ♖g1#

4203.
1. – ♖h2 2. ♔h2 ♕g3 3. ♔h1 ♕g1#

4204.
1. – ♖h2 2. ♔h2 ♕h4 3. ♔g1 ♕h1#

4205.
1. – ♗g4 2. ♖g4 ♕f2 3. ♔e4 ♕f5#

4206.
1. – ♖c1 2. ♔c1 ♖a1 3. ♔d2 ♗b4#

4207.
1. – ♕f1 2. ♖f1 ♘h3 3. ♔h1 ♖f1#

4208.
1. – ♖f2 2. ♕f2 ♕h5 3. ♔g1 ♕h1#

4209.
1. – ♖h1 2. ♔h1 ♗f2 3. any ♖h8#

4210.
1. – ♕d2 2. ♔d2 ♖8e2 3. ♔c3 ♖c2#

4211.
1. – ♖h3 2. gh3 ♕g1 3. ♔h4 ♕g5#

4212.
1. – ♕g2 2. ♗g2 ♖d1 3. ♗f1 ♖f1#

4213.
1. – h1♕ 2. ♔h1 ♔h3 3. any g2#

4214.
1. – ♗h3 2. ♖a1 ♕g2 3. ♔e1 ♕g1#

4215.
1. – ♕f3 2. ♘f3 ef3 3. ♔f1 ♖d1#

4216.
1. – ♕c1 2. ♗c1 ♖c1 3. ♔e2 d3#

4217.
1. – ♕g3 2. fg3 hg3 3. ♔h1 ♘f2#

4218.
1. – ♕h3 2. gh3 ♖h3 3. ♔g2 f3#

4219.
1. – ♕f2 2. ♖f2 gf2 3. ♔f1 ♘g3#

4220.
1. – ♕e1 2. ♕f1 ♕f1 3. ♘g1 ♕g1#

4221.
1. – ♖e1 2. ♖e1 ♕g1 3. ♔g1 ♖e1#

4222.
1. – ♖e4 2. ♕e4 ♕c1 3. ♖d1 d2#

4223.
1. – ♕e1 2. ♖e1 ♖e1 3. ♔g2 ♖g1#

4224.
1. – ♕g1 2. ♖g1 ♖g1 3. ♔h3 ♘f2#

4225.
1. – ♖h2 2. ♔h2 ♕h3 3. ♔g1 ♕g2#

4226.
1. – g5 2. ♔g5 ♖g8 3. ♔h6 ♕h3#

4227.
1. – ♖b1 2. ♔b1 ♕a2 3. ♔c1 ♕a1#

4228.
1. – ♕f2 2. ♖f2 ♖g1 3. ♖f1 ♖f1#

4229.
1. – ♘f1 2. ♔h2 ♕h2 3. ♘h2 ♘fg3#

4230.
1. – ♘f3 2. gf3 ♖g5 3. ♔h1 ♕f1#

4231.
1. – ♖g2 2. ♗g2 ♕f2 3. ♔h2 ♕g2#

4232.
1. – ♖f1 2. ♔f1 ♕h1 3. ♔f2 ♘g4#

4233.
1. – ♕g1 2. ♔g1 ♖c1 3. ♘e1 ♖e1#

4234.
1. – ♖f5 2. ♔h1 ♘g3 3. hg3 ♖h5#

4235.
1. – ♕f2 2. ♖f2 ♖f2 3. ♔h1 ♘g3#

4236.
1. – ♕h3 2. ♔h3 ♗f1 3. ♔g4 h5#

4237.
1. – ♘f2 2. ♗f2 ♕f1 3. ♗g1 ♕f3#

4238.
1. – ♕h3 2. gh3 ♗c6 3. ♔h2 ♗g3#

4239.
1. – ♘f2 2. ♔g1 ♖h1 3. ♘h1 ♘h3#

4240.
1. – ♕h5 2. ♔h5 g6 3. ♔g4 h5#

4241.
1. – ♕g5 2. ♔g5 gh6 3. ♔h5 ♖g5#

4242.
1. – ♕h3 2. ♔h3 ♖h1 3. ♔g4 h5#

4243.
1. – ♕h3 2. ♔h3 hg6 3. ♔g4 gf5#

4244.
1. – ♕h4 2. gh4 ♖g1 3. ♔h5 g6#

4245.
1. – ♖d4 2. ed4 ♕b5 3. ♔c3 ♕c4#

4246.
1. – ♘e5 2. fe5 ♖d2 3. ♗d2 ♕d2#

4247.
1. – ♛g1 2. ♗g1 ♜g1 3. ♔h2 ♘f3#

4248.
1. – ♜a1 2. ♔a1 ♛a3 3. ♔b1 ♜b8#

4249.
1. – ♛e3 2. ♔e3 ♜d3 3. ed3 f4#

4250.
1. – ♛c1 2. ♔c1 ♘e2 3. ♔b1 ♜d1#

4251.
1. – ♛a2 2. ♔a2 ♘c3 3. bc3 ♜a8#

4252.
1. – ♛c1 2. ♗c1 ♜e1 3. ♜g1 ♜g1#

4253.
1. – ♛h1 2. ♜h1 ♜h1 3. ♔f2 ♗h4#

4254.
1. – ♜a1 2. ♔a1 ♛a6 3. ♔b1 ♛f1#

4255.
1. – ♛g2 2. ♗g2 ♜d1 3. ♗f1 ♜f1#

4256.
1. – ♜a2 2. ♔a2 ♛c4 3. ♔a1 ♛a4#

4257.
1. – ♜e1 2. ♗f1 ♜f1 3. ♛f1 ♘f3#

4258.
1. – ♛h3 2. ♔h3 ♜h6 3. ♔g4 ♗d7#

4259.
1. – ♘e3 2. ♗e3 ♗f1 3. ♔g1 ♗h3#

4260.
1. – ♛f2 2. ♔f2 ♗c5 3. ♗d4 ♗d4#

4261.
1. – ♘e2 2. ♔b1 ♛g6 3. ♜d3 ♛d3#

4262.
1. – ♛h3 2. gh3 ♜e2 3. ♔g1 ♜d1#

4263.
1. – ♗c4 2. ♔c4 ♛f7 3. ♔d3 c4#

4264.
1. – ♜d1 2. ♗d1 ♘g3 3. hg3 ♛e1#

4265.
1. – ♛c1 2. ♗c1 ♜c1 3. ♛d1 ♜d1#

4266.
1. – ♜c1 2. ♔c1 ♜e1 3. ♘e1 ♛e1#

4267.
1. – ♛c1 2. ♔c1 ♜e1 3. ♔d2 ♗b4#

4268.
1. – ♘g4 2. ♔h1 ♜h3 3. gh3 ♜h2#

4269.
1. – ♜b1 2. ♔b1 ♗a2 3. ♔c1 ♗b2#

4270.
1. – ♜h3 2. ♔h3 ♛g4 3. ♔h2 ♛h4#

4271.
1. – ♘a2 2. ♗a2 ♜c3 3. bc3 ♗a3#

4272.
1. – ♗g2 2. ♛g2 ♜h2 3. ♔h2 ♛h4#

4273.
1. – ♗g2 2. ♘g2 ♛g4 3. ♔h2 ♛g2#

4274.
1. – ♘g4 2. hg4 ♛h6 3. ♗h4 ♛h4#

4275.
1. – ♜e1 2. ♘e1 ♛e1 3. ♔e1 ♜d1#

4276.
1. – ♛d2 2. ♗d2 ♜f2 3. ♛g2 ♜gg2#

4277.
1. – ♘e3 2. ♔h2 ♛g3 3. ♔g3 ♗e5#

4278.
1. – ♛d4 2. ♔d4 ♗c5 3. ♔d3 ♘e5#

4279.
1. – ♘g4 2. hg4 ♜h6 3. ♔g1 ♜d1#

4280.
1. – ♜e1 2. ♘e1 ♛h2 3. ♔f1 ♛f2#

4281.
1. – ♛a5 2. ♔a5 ♜a2 3. ♔b4 a5#

4282.
1. – ♛e1 2. ♜e1 ♜d2 3. ♛d2 ♜d2#

4283.
1. – ♛h2 2. ♔f1 ♛h1 3. ♘h1 ♜h1#

4284.
1. – ♛f3 2. ♜f3 ♜e1 3. ♜f1 ♜f1#

4285.
1. – ♜e3 2. fe3 ♛g3 3. hg3 ♗g3#

4286.
1. – g2 2. ♔g2 ♛f2 3. ♔h3 d5#

4287.
1. – ♛h1 2. ♔h1 ♜h3 3. ♔g1 ♜h1#

4288.
1. – ♛d2 2. ♔b1 ♛d1 3. ♜d1 ♜d1#

4289.
1. – ♜h1 2. ♔h1 ♔g3 3. any ♜e1#

4290.
1. – ♜h1 2. ♔g3 ♛h4 3. ♜h4 gh4#

4291.
1. – ♛e4 2. ♔e4 d5 3. ♔d3 ♗f5#

4292.
1. – ♛h2 2. ♔h2 ♘f3 3. ♔h1 ♘f2#

4293.
1. – ♘d2 2. ♛d2 ♜e1 3. ♛e1 fe1♛#

4294.
1. – ♜b3 2. cb3 ♛d3 3. ♔a2 ♛b3#

4295.
1. – ♘g4 2. ♗g4 ♜8f2 3. ♔h3 ♜h1#

4296.
1. – ♘h5 2. ♔h3 ♛g3 3. ♜g3 ♘f4#

4297.
1. – ♛a4 2. ♔a4 ♘c3 3. ♔a5 ♘b3#

4298.
1. – ♛e1 2. ♔e1 ♘f3 3. ♔f1 ♘d2#

4299.
1. – ♘g4 2. hg4 ♗g3 3. ♔g3 ♛h4#

4300.
1. – ♜d2 2. ♘d2 ♘d4 3. ♔e1 ♘c2#

4301.
1. – ♛h2 2. ♔h2 ♜h8 3. ♛h5 ♜h5#

4302.
1. – ♛g2 2. ♛g2 ♜e1 3. ♛f1 ♜f1#

4303.
1. – ♛f1 2. ♜g1 ♘g3 3. hg3 ♛h3#

4304.
1. – ♛g4 2. ♔h2 ♛h5 3. ♔g2 ♜g8#

4305.
1. – ♗f3 2. ♔g1 ♗e3 3. ♔f1 g2#

4306.
1. – ♗c5 2. ♔c5 ♛b6 3. ♔d5 ♛d6#

4307.
1. – ♗d4 2. ♜d4 ♛e3 3. ♔g2 h3#

4308.
1. – ♗d4 2. ♔g2 ♛e4 3. ♔g3 h4#

4309.
1. – ♜h2 2. ♔h2 ♘f3 3. ♔h1 ♜g1#

4310.
1. – ♜h2 2. ♔h2 ♛g3 3. ♔h1 ♛h3#

4311.
1. – ♛h3 2. gh3 ♜f1 3. ♔g2 ♜8f2#

4312.
1. – ♗c3 2. bc3 ♛f2 3. ♔d1 ♘c3#

4313.
1. – ♛c4 2. ♗c4 ♜h2 3. ♔g1 ♜h1#

4314.
1. – ♗h3 2. ♘h3 ♛f3 3. ♔g1 ♛h1#

4315.
1. – ♗h3 2. ♜h3 ♛e1 3. ♔g2 ♛g1#

4316.
1. – ♛h3 2. ♔h3 ♘e3 3. ♔h2 ♜h8#

4317.
1. – ♛h3 2. gh3 ♗f3 3. ♜f3 ♜g1#

4318.
1. – ♛b1 2. ♗c1 ♜e1 3. ♔e1 ♛c1#

4319.
1. – ♕b2 2. ♔b2 ♘c4 3. ♔c2 ♘a3#

4320.
1. – ♖eb8 2. ♔a4 ♖b4 3. ♔a3 ♗b2#

4321.
1. – ♘f3 2. gf3 ♖g6 3. ♔h1 ♘f2#

4322.
1. – ♘f4 2. ♔e3 ♖e2 3. ♔f4 ♗e5#

4323.
1. – ♖e2 2. ♔g1 ♖e1 3. ♔f2 ♖f1#

4324.
1. – ♘h3 2. gh3 ♖g4 3. hg4 ♕h2#

4325.
1. – ♘g3 2. ♖g3 ♖g1 3. ♔g1 ♖e1#

4326.
1. – ♕d3 2. ♔c5 b6 3. ♔c6 ♕d7#

4327.
1. – h4 2. ♔g4 f5 3. ♖f5 ♖g2#

4328.
1. – g5 2. ♔h5 ♕e2 3. g4 ♕e8#

4329.
1. – ♕h3 2. gh3 ♗f3 3. any ♘h3#

4330.
1. – ♕a3 2. ba3 ♘c2 3. ♖c2 ♖b1#

4331.
1. – ♘f3 2. ♔h1 ♖h3 3. gh3 ♖h2#

4332.
1. – ♕g2 2. ♗g2 ♖d1 3. ♗f1 ♖f1#

4333.
1. – ♕e3 2. ♕e3 ♖d1 3. ♔f2 ♖f1#

4334.
1. – ♕g4 2. ♔h1 ♕f3 3. ♔g1 ♕f1#

4335.
1. – ♖h3 2. gh3 ♕h3 3. ♕h2 ♘f2#

4336.
1. – ♗g2 2. ♔g2 ♕g4 3. ♔h1 ♕f3#

4337.
1. – ♖h3 2. gh3 ♗d5 3. ♔h2 ♗e5#

4338.
1. – ♘4g3 2. fg3 ♖h2 3. ♔h2 ♕h7#

4339.
1. – ♘g3 2. hg3 ♕h6 3. ♗h3 ♕h3#

4340.
1. – ♘g1 2. ♔h4 ♖h7 3. ♕h6 ♖hh6#

4341.
1. – ♕h2 2. ♔h2 ♘f3 3. ♔h3 ♖h8#

4342.
1. – ♕d1 2. ♖d1 ♘e2 3. ♗e2 ♘b3#

4343.
1. – ♕c2 2. ♔c2 ♗e4 3. ♔d2 ♖c2#

4344.
1. – ♕h3 2. ♕g3 ♕g3 3. any ♕f2#

4345.
1. – ♘e2 2. ♔h1 ♘g3 3. ♔g1 ♖f1#

4346.
1. – ♖a3 2. ♔a3 ♕a1 3. ♔b3 a4#

4347.
1. – ♘e3 2. ♔h1 ♕f1 3. ♔h2 ♕g2#

4348.
1. – ♕c3 2. ♕c3 bc3 3. any ♖a1#

4349.
1. – g2 2. ♗g2 ♕h3 3. ♗h3 ♖g1#

4350.
1. – ♕g3 2. hg3 ♖h1 3. ♘h1 ♘h3#

4351.
1. – ♕f2 2. ♔f2 ♗e3 3. ♔f1 ♘g3#

4352.
1. – ♗f5 2. ♔f5 ♕f6 3. ♔e4 ♕g6#

4353.
1. – ♖g3 2. hg3 ♕f2 3. ♔h1 ♕h2#

4354.
1. – ♕e3 2. ♔g2 ♕e2 3. ♔g1 ♗d4#

4355.
1. – ♘h5 2. ♔e4 ♗b7 3. ♔d3 ♗f2#

4356.
1. – ♖g2 2. ♖g2 f5 3. ♔h5 ♕h3#

4357.
1. – ♗g2 2. ♖g2 ♖e1 3. ♖g1 ♖gg1#

4358.
1. – ♖8d2 2. ♔e3 ♗d4 3. cd4 cd4#

4359.
1. – ♖e5 2. de5 ♗e7 3. ♔h5 ♕f5#

4360.
1. – ♕g3 2. ♖g3 fg3 3. ♔g1 ♖e1#

4361.
1. – ♖g3 2. ♗h3 ♕f3 3. ♗g2 ♕g2#

4362.
1. – ♕f1 2. ♕f2 ♕f2 3. ♔g4 ♕f5#

4363.
1. – ♖d4 2. ♘d4 ♕d4 3. ♔c1 ♕a1#

4364.
1. – ♕d1 2. ♖d1 ♖d1 3. ♔c2 ♖8d2#

4365.
1. – ♕a3 2. ♔a3 ab4 3. ♔b2 ♖a2#

4366.
1. – ♕h2 2. ♔h2 ♘g4 3. ♔g1 ♘h3#

4367.
1. – ♕f1 2. ♔f1 ♘g3 3. ♔e1 ♖c1#

4368.
1. – ♕f1 2. ♔f1 ♖c1 3. ♔e2 ♖d2#

4369.
1. – ♕f1 2. ♔f1 ♖d1 3. ♔e2 ♘c3#

4370.
1. – ♕h2 2. ♔h2 ♘f3 3. ♔g3 f4#

4371.
1. – ♕h3 2. ♔h3 ♖h5 3. ♔g3 ♘e4#

4372.
1. – ♕h2 2. ♔h2 ♘f3 3. ♔g3 ♘h5#

4373.
1. – ♕g2 2. ♘g2 ♘f3 3. ♔h1 ♖h2#

4374.
1. – g1♕ 2. ♗g1 ♕f3 3. ♔h4 ♕f4#

4375.
1. – ♗b5 2. cb5 ♖d4 3. ♗c4 ♖cc4#

4376.
1. – ♘h3 2. ♔h1 ♘g5 3. ♔g1 ♘f3#

4377.
1. – ♖f3 2. gf3 ♗f1 3. ♖g2 ♗g2#

4378.
1. – ♕e3 2. ♔e3 ♗h6 3. ♔d3 ♖d2#

4379.
1. – ♘f2 2. ♔g2 ♕h3 3. ♔f2 ♕f1#

4380.
1. – ♖g2 2. ♔g2 ♕f3 3. ♔g1 ♕g4#

4381.
1. – ♕f1 2. ♕f1 ♗f3 3. ♕f3 ♖g1#

4382.
1. – ♕e7 2. ♔e5 ♘c6 3. ♔e4 f5#

4383.
1. – ♕h2 2. ♔h2 hg3 3. ♔g1 ♖h1#

4384.
1. – ♖h5 2. gh5 ♖h3 3. ♔g5 ♖h5#

4385.
1. – ♕h3 2. gh3 ♖h2 3. ♔g1 ♘e2#

4386.
1. – ♖h1 2. ♔h1 ♕b1 3. ♔h2 ♕g1#

4387.
1. – ♕h2 2. ♔h2 hg3 3. ♔g2 ♖h2#

4388.
1. – ♕h2 2. ♔h2 hg2 3. ♔g1 ♖h1#

4389.
1. – ♘d2 2. ♖d2 ♖e1 3. ♔e1 ♖g1#

4390.
1. – ♗g3 2. ♔g1 ♕h2 3. ♔f1 ♕f2#

4391.
1. – g5 2. ♔h5 ♕g4 3. ♕g4 ♗f7#

4392.
1. – ♕e4 2. ♔g3 ♕g4 3. ♔h2 ♖h4#

4393.
1. – ♕h2 2. ♔h2 ♖h8 3. ♖h4 ♖h4#

4394.
1. – ♕h2 2. ♔h2 ♘g3 3. ♔g3 f4#

4395.
1. – ♕h1 2. ♔h1 ♖f1 3. ♗g1 ♗f3#

4396.
1. – ♘d4 2. ♔d1 ♘e3 3. ♔c1 ♘e2#

4397.
1. – ♘d3 2. ♔d1 ♕e1 3. ♖e1 ♘f2#

4398.
1. – ♘d3 2. ♔d1 ♕e1 3. ♘e1 ♘f2#

4399.
1. – ♕h4 2. g3 ♕g3 3. hg3 ♗g3#

4400.
1. – f5 2. gf6 ♕f5 3. ♔h4 ♕h5#

4401.
1. – ♕f2 2. ♔f2 ♖d1 3. ♗e3 ♗e3#

4402.
1. – f2 2. ♔f2 ♕g2 3. ♔e3 ♕f3#

4403.
1. – ♘d4 2. ♔b2 ♕c2 3. ♔a3 ♕b3#

4404.
1. – ♕h4 2. g3 ♕g3 3. hg3 ♗g3#

4405.
1. – f6 2. ♔h4 ♗f2 3. g3 ♗g3#

4406.
1. – ♖g3 2. hg3 ♕h1 3. ♘h1 ♖g2#

4407.
1. – ♖g1 2. ♖g1 ♕f3 3. ♖g2 ♖d1#

4408.
1. – ♕g2 2. ♗g2 hg2 3. ♔g1 ♖h1#

4409.
1. – ♖h1 2. ♗h1 ♕f2 3. ♗g2 ♕g1#

4410.
1. – ♕f2 2. ♖f2 gf2 3. ♔f1 ♘g3#

4411.
1. – ♘a4 2. ♖a4 ♖b3 3. ♔b3 ♖d3#

4412.
1. – ♖h3 2. ♔f4 ♖f3 3. ♕f3 ♕e5#

4413.
1. – ♖e1 2. ♕e1 ♕f3 3. ♖g2 ♕g2#

4414.
1. –h6 2. ♔f4 g5 3. ♔e5 ♕e6#

4415.
1. – ♖d5 2. ♘d5 g6 3. ♔h6 ♘g4#

4416.
1. – ♕f1 2. ♔f1 ♘e3 3. ♔g1 ♖f1#

4417.
1. – ♕f1 2. ♔f1 ♖c1 3. ♔e2 ♖b2#

4418.
1. – ♕f2 2. ♕f2 ♖h5 3. ♗h5 g5#

4419.
1. – g5 2. ♔h5 ♘f4 3. ef4 ♖h3#

4420.
1. – ♕g3 2. hg3 ♗f3 3. any ♖h1#

4421.
1. – ♕e1 2. ♗e1 ♖e1 3. ♖f1 ♖f1#

4422.
1. – ♘g3 2. hg3 hg3 3. ♔g1 ♗d4#

4423.
1. – ♕f4 2. ♔f4 ♗h6 3. ♔f5 ♘e5#

4424.
1. – ♖h1 2. ♗h1 ♖h1 3. ♔g4 h5#

4425.
1. – ♕d5 2. ♕e4 ♕e4 3. ♔g1 ♕g2#

4426.
1. – ♖h2 2. ♔h2 ♕g3 3. ♔h1 ♕h3#

4427.
1. – ♕g3 2. fg3 ♖f1 3. ♖f1 ♖f1#

4428.
1. – ♖h2 2. ♔h2 ♕h4 3. ♔g2 ♕g3#

4429.
1. – ♕f2 2. ♕f2 ♖d1 3. ♕f1 ♖f1#

4430.
1. – ♕f3 2. ♗f3 ♗f3 3. ♔g1 h2#

4431.
1. – ♖h4 2. ♗h4 ♕h4 3. gh4 ♖h4#

4432.
1. – ♘g4 2. hg4 ♕f2 3. ♔h3 ♖h1#

4433.
1. – ♖h1 2. ♔h1 ♕h6 3. ♔g1 ♕h2#

4434.
1. – ♖h3 2. ♔h3 ♕h1 3. ♔g3 ♕h4#

4435.
1. – ♖f5 2. ♖f5 ♕g3 3. ♔h5 ♖h8#

4436.
1. – ♘d4 2. ♔g4 ♗c8 3. ♔h4 ♘f3#

4437.
1. – ♖h4 2. ♔h4 ♕h2 3. ♔g4 ♕h3#

4438.
1. – ♕g5 2. ♔f2 ♕h4 3. ♔g1 ♘h3#

4439.
1. – ♕c1 2. ♖d1 ♖e1 3. ♖e1 ♕e1#

4440.
1. – ♕f1 2. ♖f1 ♖f1 3. ♗f1 ♖f1#

4441.
1. – ♗h3 2. ♔e1 ♕g2 3. any ♕g1#

4442.
1. – ♖c1 2. ♔c1 ♕b2 3. ♔d1 ♕d2#

4443.
1. – ♕f1 2. ♔f1 ♗d3 3. ♔g1 ♖f1#

4444.
1. – ♕g1 2. ♔g1 ♖d1 3. ♕e1 ♖e1#

4445.
1. – ♘f3 2. ♔h1 ♘g3 3. ♘g3 ♖h2#

4446.
1. – ♕e1 2. ♔e1 ♗f2 3. ♔f1 ♖d1#

4447.
1. – ♗f2 2. ♖f2 ♕f2 3. ♔h1 ♕e1#

4448.
1. – ♗g1 2. ♔g3 ♕f2 3. ♔h3 ♕h2#

4449.
1. – ♖d5 2. any b5 3. ♔c3 ♘a2#

4450.
1. – ♘f5 2. ♔d2 ♖d1 3. ♔c2 ♘e3#

4451.
1. – ♖f5 2. ♔g4 h5 3. ♔h3 ♖f3#

4452.
1. – ♖f1 2. ♔h2 ♗g1 3. ♔h1 ♘g3#

4453.
1. – ♖c2 2. ♔b1 ♖e2 3. ♔c1 ♖e1#

4454.
1. – ♖c4 2. ♔b3 ♖c3 3. ♔a4 ♖a3#

4455.
1. – ♕g2 2. ♕g2 ♘g3 3. hg3 ♖h8#

4456.
1. – ♖f4 2. gf4 ♖h4 3. ♔g5 ♕f4#

4457.
1. – ♖a1 2. ♔a1 ♕a3 3. ♖a2 ♕c1#

4458.
1. – ♖h3 2. ♘h3 ♖e1 3. ♔h2 ♕h1#

4459.
1. – ♕h2 2. ♔h2 ♘f3 3. ♔h1 ♖g1#

4460.
1. – ♕f1 2. ♖f1 ♖f1 3. ♔g2 ♘e3#

4461.
1. – ♖h5 2. gh5 ♗g2 3. ♔h4 ♗g3#

4462.
1. – ♗h3 2. ♘h3 ♕g3 3. ♔h1 ♕h3#

5063.
1. ♔e6 =

SOLUTIONS (5064 – 5150)

5064.
1. ♔f2 ♔f4 2. ♔f1 ♔g3 3. ♔g1 f2 4. ♔f1 ♔f3 =

5065.
1. ♔b1 =

5066.
1. ♔c3 ♔b1 2. ♔b4 =

5067.
1. ♔c1 ♔a2 2. ♔c2 ♔a1 3. ♔c1 =

5068.
1. ♔d4 a3 2. ♔c3 a2 3. ♔b2 =

5069.
1. ♔g5 =

5070.
1. ♔c6 =

5071.
1. g4 hg4 =

5072.
1. ♔h8 ♛f7 =

5073.
1. ♔g7 ♛g4 2. ♔h8 ♛g6 =

5074.
1. ♔b7 =

5075.
1. ♔a8 ♛c7 =

5076.
1. ♜c5 ♛c5 =

5077.
1. ♜d2 =

5078.
1. ♔h8 ♜g7 =

5079.
1. e8♘ =

5080.
1. ♔g2 =

5081.
1. ♜f8 c1♛ 2. ♜d8 ♔c2 3. ♜c8 =

5082.
1. ♜b5 c1♛ 2. ♜c5 ♛c5 =

5083.
1. ♜g3 ♔e2 2. ♜g2 =

5084.
1. ♜b3 ♔c2 2. ♔c4 ♔d2 3. ♜b8 =

5085.
1. ♔b4 =

5086.
1. c8♛ ♔c8 2. e8♛ ♜e8 3. ♔e8 =

5087.
1. ♜g8 ♔f4 2. ♜g3 ♔g3 =
1. – ♔h4 2. ♜h8 =

5088.
1. ♜f3 ♔g4 2. ♔g2 h1♛ 3. ♔h1 ♔f3 =

5089.
1. ♜e2 de2 =

5090.
1. ♗e8 d1♛ 2. ♗h5 =

5091.
1. ♔h1 =

5092.
1. ♔c3 ♔b6 2. ♔b4 =
1. – ♗g8 2. ♔c2 ♔b6 3. ♔b1 =

5093.
1. ♔b3 =

5094.
1. ♔d5 ♔b1 2. ♔c4 =

5095.
1. ♘e5 f2 2. ♘g4 f1♛ 3. ♘e3 =

5096.
1. ♘e4 f1♛ 2. ♘g3 =

5097.
1. ♔g5 ♔b7 2. ♔h4 ♔c6 3. ♔g5 =

5098.
1. ♔d4 ♔g2 2. ♔e3 ♔h1 3. ♔f2 =

5099.
1. ♔f2 ♘h3 2. ♔f1 ♘f4 3. ♔f2 =

5100.
1. ♗d4 ♜e1 2. ♗g1 =

5101.
1. ♘c2 ♗c2 =

5102.
1. ♜d5 h1♛ 2. ♜c5 ♔b6 3. ♗h1 =

5103.
1. ♔e3 ♘g4 2. ♔f3 ♔c6 3. ♔g3 h2 4. ♔g2 ♔d5 5. ♔h1 =

5104.
1. ♔d4 ♘c6 2. ♔e3 ♘b8 3. ♔f3 =

5105.
1. ♔d6 +–

5106.
1. ♔e6 ♔d8 2. d7 ♔c7 3. ♔e7 +–

5107.
1. h7 +–

5108.
1. b6 +–

5109.
1. h8♛ ♔h8 2. ♔f6 ♔g8 3. g7 +–

5110.
1. h7 a2 2. h8♛ ♔c2 3. ♛a1 +–

5111.
1. h7 a1♛ 2. h8♛ +–

5112.
1. a7 h2 2. a8♛ h1♛ 3. ♔b6 +–

5113.
1. e8♛ e1♛ 2. ♔f6 +–

5114.
1. b6 ♔d5 2. b7 +–

5115.
1. ♔c7 ba6 2. b7 +–

5116.
1. ♔b7 +–

5117.
1. ♔b8 ♔d6 2. ♔b7 +–

5118.
1. ♔g7 ♔c6 2. ♔f8 ♔d6 3. ♔f7 +–

5119.
1. ♔c4 b3 2. ab3#

5120.
1. ♛b4 ♔a2 2. ♔e6 ♔a1 3. ♛a3 +–

5121.
1. ♛c2 h2 2. ♛c1#

5122.
1. ♔g3 +–

5123.
1. ♔g3 h1♛ 2. ♛f2#

5124.
1. f8♘ +–

5125.
1. ♔f6 +–

5126.
1. ♜a1 ♔b3 2. ♜b1 +–

5127.
1. ♔b3 +–

5128.
1. ♔f7 ♜f1 2. ♔e6 ♜e1 3. ♜e5 +–

5129.
1. f7 ♔g7 2. ♜g1 +–
1. – ♜a7 2. ♜d7 +–

5130.
1. ♜h7 ♔g8 2. f7 ♔f8 3. ♜h8 +–

5131.
1. ♜h8 ♜a7 2. ♜h7 +–

5132.
1. ♜c8 ♜a7 2. ♔b6 +–

5133.
1. ♔e6 ♔f8 2. ♔f6 ♔e8 3. ♜b8 +–

5134.
1. ♜a7 ♔f8 2. e7 ♔f7 3. e8♛ ♔e8 4. ♜a8 +–

5135.
1. d8♛ ♔d8 2. e7 ♔e8 3. ♜a8 +–

5136.
1. ♜f8 ♜f8 2. gf8♛ ♔f8 3. ♔h7 +–

5137.
1. ♜b7 ♔a8 2. ♜cb6 ♜hg8 3. ♜a7#

5138.
1. ♜1e7 ♜a8 2. ♜h7 ♔g8 3. ♜dg7 ♔f8 4. ♜h8#

5139.
1. ♔c2 +–

5140.
1. ♜h1 ♔c2 2. ♔e2 +–

5141.
1. ♜e8 ♔d1 2. ♔f2 +–

5142.
1. ♔c6 +–

5143.
1. ♔f6 e3 2. ♔f5 ♔d3 3. ♔f4 e2 4. ♔f3 +–

5144.
1. ♔d7 ♔e4 2. ♔c6 +–
1. – ♔c4 2. ♔e6 +–

5145.
1. ♔b3 ♔b1 2. ♜h8 a1♘ 3. ♔c3 ♔a2 4. ♜b8 +–

5146.
(Réti, 1928) 1. ♜d2 d4 2. ♜d1 ♔d5 3. ♔d7 +–

5147.
1. ♜f1 ♔g2 2. ♔e2 +–

5148.
1. ♔b3 a1♘ 2. ♔c3 ♔a2 3. ♜b7 +–

5149.
1. ♔e6 e3 2. ♔f5 ♔d3 3. ♔f4 e2 4. ♔f3 +–

5150.
1. ♔e5 ♔c3 2. ♔e4 d2 3. ♔e3 +–

5151.
1. ♔b5 ♔b7 2. a8♕ ♔a8 3. ♔c6 ♔a7 4. c8♖ +–

5152.
1. ♖e1 ♔e3 2. ♔c2 ♔f2 3. ♔d2 +–

5153.
1. b8♕ ♔b8 2. ♔d6 ♔c8 3. c7 +–

5154.
1. f7 +–

5155.
1. c7 ♖c1 2. d6 ♔e3 3. d7 +–

5156.
1. c7 ♖c1 2. b7 +–

5157.
1. ♔d3 ♔f3 2. ♖f1 ♔g2 3. ♔e2 +–

5158.
1. ♖h8 ♔f2 2. ♖h2 ♔f1 3. ♖h1 ♔f2 4. ♔d2 +–

5159.
1. ♖h1 ♔f2 2. ♔d2 +–

5160.
1. f8♖ ♖h4 2. ♔g3 +–

5161.
1. e8♕ ♖e8 2. d7#

5162.
1. d6 ♖d6 2. b8♕ +–

5163.
1. ♘a6 ♔a8 2. ♗e4#

5164.
1. ♗g4 ♔a8 2. ♔a6 ♔b8 3. ♗f4 ♔a8 4. ♗f3#

5165.
1. ♔g4 d1♕ 2. ♗f3 ♔e1 3. ♗d1 ♔d1 4. b6 +–

5166.
1. ♗f8 ♔f5 2. ♗g7 +–

5167.
1. ♗g4 ♔b7 2. ♗d7 +–

5168.
1. ♗g6 ♗g8 2. ♗e8 +–

5169.
1. d7 +–

5170.
1. a7 +–

5171.
1. f7 ♘e7 2. ♔g5 +–

5172.
1. ♘e8 ♘e8 2. d7 +–
1. – ♔f7 2. ♘f6 ♔f6 3. d7 ♔e7 4. ♔c7 +–

5173.
1. ♖f8 ♘f8 2. e7 +–

5174.
1. ♔f7 ♔d6 2. ♔g8 ♔e6 3. g6 +–

5175.
1. ♔c6 ♔e4 2. ♔b7 ♔d5 3. ♔a8 ♔c6 4. ♔b8 +–

5176.
1. ♘a6 ♔d7 2. ♔b7 +–

5177.
1. ♘c1 a2 2. ♘b3#

5178.
1. e6 ♘e2 2. ♔h2 +–

5179.
1. ♘c6 bc6 2. b7 +–

5180.
1. ♘b4 a2 2. ♘d5 a1♕ 3. ♘c7#

5181.
1. ♘a3 c2 2. ♘b5 c1♕ 3. ♘c7#

5182.
1. ♖g7 +–

5183.
1. ♖g8 ♗c8 2. ♖h8 +–

5184.
1. ♔f3 +–

5185.
1. ♔b3 +–

5186.
1. ♔f3 ♔h5 2. ♔f4 +–

5187.
1. ♔f3 +–

5188.
1. ♔f6 +–

5189.
1. ♖a7 ♔b8 2. ♖d7 +–

5190.
1. ♖f3 ♔f3 2. a8♕ +–

5191.
1. g4 +–

5192.
1. ♗f7 +–

5193.
1. ♔e7 ♔d4 2. ♗g2 +–

5194.
1. ♗b1 ♗d1 2. ♘e3 +–
1. – ♗e2 (♗f3) 2. ♘d4 +–

5195.
1. ♘e7 ♗f7 2. ♘c6 ♔a6 3. ♗d3 ♔b7 4. ♘d8 +–

5196.
1. ♔b8 ♗d7 2. ♔c7 ♗e8 3. ♔d8 ♗f7 4. ♔e7 ♗g8 5. ♔f8 ♗h7 6. ♔g7 +–

5197.
1. ♔d7 ♗f6 2. ♘e4 +–
1. – ♗h4 2. ♘f3 +–
1. – ♗a5 (♗b6) 2. ♘c4 +–

5198.
1. ♔b1 ♗c3 2. ♘d6 ♔c7 3. ♘b5 +–

5199.
1. ♔c1 ♗d3 2. d7 ♔c7 3. ♘e6 ♔d7 4. ♘c5 +–

5200.
1. ♔h1 +–

5201.
1. ♗e4 +–

5202.
1. ♘e5 +–

5203.
1. ♘d4 +–

5204.
1. ♘f3 ♗f2 2. ♘d4 ♗g3 3. a6 ♗b8 4. ♘c6 +–

5205.
1. ♘g3 ♔f4 2. ♔d6 ♔g3 3. ♔e6 ♔g4 4. ♔f6 +–

5206.
1. d5 ♔a5 2. ♔d4 ♔a6 3. ♔e5 ♖e8 4. ♔d6 ♖d8 5. ♔e6 ♖e8 6. ♔d7 ♖e2 7. d6 ♖d2 8. ♔c7 ♖c2 9. ♔d8 ♖d2 10. d7 ♖e2 11. ♖b4 ♖e1 12. ♔c7 ♖c1 13. ♔d6 ♖d1 14. ♔c6 ♖c1 15. ♔d5 ♖d1 16. ♖d4 +–

5207.
Zsu. Polgár – Z. Endrõdy, Budapest 1977
1. ♖f6 ♘d5 2. ♖f7 ♔f7 3. ed5 ♕d4 4. ♕d4 ed4 5. ♖d1 +–

5208.
Zsu. Polgár – J. Németh, Ceske Budejovice 1980
1. ♕c4 ♗f3 2. ♕c8 ♖c8 3. ♘d5 ♕c5 4. ♘e7 ♔f8 5. ♘c8 ♕c1 6. ♖c1 ♗e2 7. ♘a7 +–

5209.
Zsó. Polgár – G. Hajdú, Debrecen 1980
1. ♕g7 ♖g7 2. hg7#

5210.
Zsu. Polgár – S. Boicu, Panonija 1981
1. ♗e8 ♖bc7 2. ♘c6 g6 3. ♗g6 hg6 4. ♖g6 ♘g6 5. ♘e7 ♖e7 6. ♕g6 ♔f8 7. ♗e1 d4 8. ♗h4 de3 9. ♗g5 ♗f7 10. ♕f6 ♔e8 11. ♖d1 ♕e6 12. ♕e6 ♗e6 13. ♗e7 ♗e7 14. ♖e1 ♔d7 15. ♖e3 ♔c6 16. ♖h3 ♗f8 17. ♖h8 ♗g7 18. ♖a8 ♗h6 19. ♖a6 ♔d7 20. ♖d6 ♔e7 21. a6 ♗c8 22. ♖h6 +–

5211.
Zsu. Polgár – J. Cirakov, Targoviste 1981
1. ♘h7 fe4 2. ♘f8 ♔f8 3. ♖a1 ♗c4 4. ♗g5 ♘f6 5. ♕d4 +–

5212.
Zsu. Polgár – A. Pollák, Budapest 1982
1. ♕h6 ♘h6 2. ♖h6 ♔g7 3. ♘f7 ♔f7 4. ♖h7 ♔f6 5. e5 de5 6. de5#

5213.
Zsó. Polgár – G. Nikolov, Targoviste 1982
1. ♖g7 ♘bd7 2. ♖d7 ♕g7 3. ♕f6 ♕f6 4. ♗f6 ♔g8 5. ♖d1 ♖fc8 6. ♖g1 ♔f8 7. ♖g7 b5 8. ♖h7 ♔e8 9. ♗b5 +–

5214.
Zsó. Polgár – J. Kontra, Budapest 1982
1. ♗b6 ♘e7 2. ♘c7 ♕c6 3. ♘e8 ♕b6 4. ♘d6 ♔b8 5. ♘c4 ♖d2 6. ♘b6 ♖d1 7. ♖d1 +–

5215.
Zsu. Polgár – V. Basagic, Varna 1984
1. e5 fg5 2. ♕e4 ♗f5 3. ♖f5 gf5 4. ♕f5 ♘e7 5. de7 ♖e7 6. ♕c8 ♔f7 7. ♖f1 ♔g6 8. ♗d3 +–

5216.
Zsu. Polgár – Kr. Dimitrov, Targoviste 1984
1. ♕g6 ♖g7 2. ♖h7 +–

5217.
Zsu. Polgár – Vl. Dimitrov, Ivajlograd 1984
1. ♗e6 fe6 2. ♗f6 ♕d4 3. ♕h5 +–

5218.
Zsó. Polgár – L. Gáti, Budapest 1984
1. ♖b6 ab6 2. ♕b4 +–

5219.
Zsu. Polgár – J. Remlinger, New York 1985
1. ♗g7 ♔g7 2. ♕h7 +–

5220.
Zsu. Polgár – J. Federowicz, New York 1985
1. ♖a8 +–

5221.
Zsu. Polgár – Gy. Rajna, Budapest 1985
1. ♗b5 cb5 2. ♘b5 ♖aa8 3. ♘d6 ♖c7 4. ♖c1 ♗e8 5. b5 ♖b8 6. ♖cc2 ♔d8 7. b6 ♖c6 8. ♘b5 ♗f8 9. ♘d4 ♗d7 10. ♘c6 ♗c6 11. ♖a2 gf4 12. gf4 ♔d7 13. ♖a7 ♖b7 14. ♖a8 ♗e7 15. ♖h8 ♗b5 16. ♖h7 ♗c4 17. ♔f3 ♔c6 18. ♖b2 ♗c5 19. ♖b7 ♔b7 20. h4 ♗d3 21. h5 ♗f8 22. ♖a2 ♔b6 23. ♖a8 ♗g7 24. ♖g8 ♗c2 25. ♖g7 +–

5222.
Zsu. Polgár – Shoengold, New York 1985
1. ♘f7 ♔f7 2. ♗e6 ♔e6 3. ♕b3 ♔d6 4. ♗f4 ♘e5 5. de5 ♔c7 6. ef6 ♗d6 7. ♕f7 ♗d7 8. ♗d6 +–

5223.
Zsu. Polgár – P. Hardicsay, Budapest 1985
1. ♗b5 ab5 2. ♖e1 ♔f8 3. ♗h6 ♔g8 4. ♖e7 ♗d7 5. ♕b8 ♕b8 6. ♘e4 +–

5224.
Zsó. Polgár – A. Pollák, Dresden 1985
1. ♖h6 ♔g8 2. ♖e6 ♕f4 3. ♗e3 ♕b4 4. ♗f2 ♘e5 5. ♗d5 ♘f7 6. ♕f7 ♔f7 7. ♖ef6#

5225.
Zsu. Polgár – W. Browne, New York 1986
1. ♖b4 +–

5226.
Zsó. Polgár – E. Mortensen, Copenhagen 1986
1. ♘d7 ♕d7 2. ♗f6 ♕d5 3. ♗h7 ♔h7 4. ♖d5 ed5 5. ♖e8 +–

5227.
Zsó. Polgár – G. Canfell, Rio Gallegos 1986
1. ♘e7 ♕e7 2. ♖e7 ♗e7 3. ♖f8 ♖f8 4. ♗b5 ab5 5. ♕d3 +–

5228.
Zsó. Polgár – NOVAG Constellation Forte, Sydney 1986
1. ♗d4 ♕b1 2. ♔f2 ♕h1 3. ♕e5 +–

5229.
Zsó. Polgár – L. Félegyházi, Tapolca 1986
1. ♘e6 fe6 2. ♕h6 ♔e8 3. fe6 d5 4. ♕h5 ♔e7 5. ♕f7 ♔d6 6. ♖f6 ♖e8 7. ♕d7#

5230.
J. Polgár – B. Dorosiev, Teteven 1986
1. ♕h6 +–

5231.
J. Polgár – V. Metodiev, Albena 1986
1. ♖h7 ♖h7 2. ♕g6 ♔h8 3. ♕e8 +–

5232.
Zsu. Polgár – J. Costa, San Bernardino 1987
1. ♘h7 ♔h7 2. ♕h5 ♔g8 3. ♗g6 fe5 4. ♗e8 ♖e8 5. ♕g5 +–

5233.
Zsó. Polgár – U. Dresen, Biel 1987
1. ♕h6 ♔h6 2. hg6 ♔g7 3. ♖h7 ♔g8 4. gf7 +–

5234.
Zsu. Polgár – Liu She–Lan, Thessaloniki 1988
1. ♖d5 ♖d5 2. ♕g5 +–

5235.
Zsó. Polgár – MEPHISTO Roma, Budapest 1988
1. ♖f7 ♖fe8 2. ♕d8 ♖d8 3. ♖f8 ♖f8 4. ♗d5 ♖f7 5. ♗f7 ♔h8 6. ♗b3 +–

5236.
Zsó. Polgár – K. Commons, Mazatlan 1988
1. ♕h7 +–

5237.
Zsó. Polgár – R. Akesson, Reykjavik 1988
1. e5 ♗f3 2. ♘f6 ♔h8 3. ♕f3 g6 4. ♕a8 de5 5. ♘d7 ♖c8 6. ♘b8 +–

5238.
J. Polgár – P. Angelova, Thessaloniki 1988
1. ♕f8 +–

5239.
J. Polgár – S. Singh, Hastings 1988
1. ♗d5 ♕d5 2. ♕c8 ♘e8 3. ♕d7 +–

5240.
J. Polgár – A. Muir, Haifa 1989
1. ♗g6 ♖e3 2. ♗e3 ♘c3 3. ♕f6 ♗g7 4. ♗h7 ♔h8 5. ♗f5 +–

5241.
Zsu. Polgár – L. Schandorff, Abenraa 1989
1. ♗d6 ♕d6 2. e5 fe5 3. ♗h7 ♔f7 4. ♖f3 ♔e7 5. ♕e5 ♕e5 6. ♖e5 ♗e6 7. ♖f8 ♔f8 8. de6 ♘c6 9. ♖f5 ♔e7 10. ♖g5 ♖d8 11. ♗e4 ♖d1 12. ♔f2 ♘d8 13. ♖g7 ♔e6 14. ♗d5 ♔f6 15. ♖d7 +–

5242.
Zsu. Polgár – C. Hansen, Abenraa 1989
1. ♘d5 ♖a7 2. ♖ad1 c4 3. ♘d4 ♘d4 4. ♖d4 ♗c5 5. ♖d2 0-0 6. ♖fd1 ed5 7. ♖d5 ♖e8 8. e5 ♕b6 9. ♗d7 ♖e7 10. ♗h3 ♖a8 11. ♗g3 ♖ee8 12. ♔h2 ♖ad8 13. f4 ♗d3 14. ♖d8 ♗e2 15. ♖e8 ♔h7 16. ♗f5 g6 17. ♖dd8 +–

5243.
Zsó. Polgár – A. Chernin, Rome 1989
1. ♘d5 ♘d5 2. ♘e6 g6 3. ♘d8 ♕d8 4. ed5 ♖c2 5. ♖ab1 +–

5244.
J. Polgár – H. Ree, Amsterdam 1989
1. ♘e6 fe6 2. ♗e6 ♔h8 3. ♘d5 ♕b8 4. ♗f7 ♖f8 5. ♗d4 ♘de5 6. ♘e7 ♘e7 7. fe5 de5 8. ♗c5 ♔g7 9. ♗e7 ♕a7 10. ♔h1 ♗h3 11. ♕f3 +–

5245.
J. Polgár – K. Hulak, Amsterdam 1989
1. ♖g5 ♕f2 2. ♖f1 ♕f1 3. ♗f1 hg5 4. ♗d3 f5 5. ♗f5 ♗e5 6. ♕h7 ♔f7 7. ♕g6 ♔e7 8. ♕e6 ♔d8 9. ♕e5 +–

5246.
J. Polgár – L. B. Hansen, Vejstrup 1989
1. ♕g7 +–

5247.
Zsó. Polgár – R. Gocheva, Novi Sad 1990
1. ♕f6 +–

5248.
Zsu. Polgár – Z. Hracek, Stara Zagora 1990
1. ♘e6 fe6 2. fe6 ♘f8 3. ♕a4 ♘6d7 4. c5 ♗c8 5. ed7 ♗d7 6. c6 +–

5249.
Zsu. Polgár – Z. Azmajparasvili, Dortmund 1990
1. ♘b5 cb5 2. ♖b5 ♔c7 3. ♖a6 ♖d1 4. ♖ba5 ♘d8 5. ♔c2 ♖h1 6. b5 ♖h3 7. ♘d2 h5 8. gh5 gh5 9. ♘c4 ♘b7 10. ♖a1 ♗f8 11. ♖c6 ♔b8 12. ♖a8 ♔a8 13. ♖c8 ♔a7 14. c6 ♖e3 15. fe3 +–

5250.
Zsu. Polgár – V. Kotronias, Corfu 1990
1. gh5 gh4 2. h6 ♘h5 3. ♖f5 ♘f4 4. ♖f4 ef4 5. e5 f5 6. e6 ♕e7 7. ♘g4 f3 8. ♗b2 ♕g5 9. h3 ♕d2 10. ♗b1 ♕g2 11. ef7 ♖f7 12. ♖e1 f2 13. ♗f5 ♖f5 14. ♖e7 ♔g8 15. h7 +–

5251.
Zsu. Polgár – T. Károlyi, Miskolc 1990
1. ♘f5 gf5 2. ♖d8 ♕c5 3. ♖c1 ♕e7 4. ♕g3 ♔h6 5. ♖g8 +–

5252.
Zsu. Polgár – M. Todorcevic, Pamplona 1990
1. ♕f5 ♖f5 2. ♖e8 ♖f8 3. ♗f8 ♗e5 4. ♔h1 +–

5253.
Zsó. Polgár – L. van Wely, Wijk aan Zee 1990
1. f5 ♘g3 2. ♕e1 ♘h1 3. g4 gf5 4. gf5 ♕d8 5. ♗c4 ♔h8 6. ♘g5 ♗h6 7. ♕h4 ♗g5 8. ♗g5 f6 9. ♗h6 ♘d7 10. ♖g1 ♕e7 11. ♕g4 +–

5254.
J. Polgár – R. Knaak, Cologne 1990
1. f4 ♘e5 2. fe5 ♕e5 3. ♔f2 ♕g7 4. ♖g1 ♕b2 5. ♗b4 f6 6. ♖e1 0-0-0 7. ♖e6 ♔b8 8. ♕f6 +–

5255.
Zsu. Polgár – A. Grosar, Portoroz 1991
1. ♘d7 ♗d7 2. ♕f7 ♖d8 3. ♖e6 ♗h6 4. ♖e7 +–

5256.
Zsu. Polgár – R. Lau, Polanica Zdroj 1991
1. ♗g6 ♘g6 2. ♘h5 ♔h8 3. ♘f6 ♖e4 4. ♘ce4 de4 5. ♕d1 +–

5257.
Zsu. Polgár – L. Yudasin, Munich 1991
1. ef6 gf6 2. ♕f6 c3 3. ♘e4 c2 4. ♕h6 ♗g6 5. ♘f6 ♔h8 6. ♘e8 +–

5258.
Zsu. Polgár – P. Schaffart, San Bernardino 1991
1. ♘f7 ♕b6 2. ♗g6 ♖f8 3. ♘3g5 +–

5259.
J. Polgár – J. Molnár, Budapest 1991
1. dc5 ♗b2 2. ♔b2 ♕d1 3. ♗h6 g6 4. ♕e5 f6 5. ♕e6 ♖f7 6. ♘c3 ♕d7 7. ♕d7 ♖d7 8. c6 ♗c6 9. ♗a6 +–

5260.
J. Polgár – Jagerflung, Gothenburg (simul) 1991
1. ♘f6 gf6 2. gf6 ♔h8 3. ♖g1 ♘d7 4. ♖g7 ♕d6 5. ♖h7 +–

5261.
J. Polgár – P. Sinkovics, Budapest 1991
1. ♖g7 ♔g7 2. ♖h6 ♖h8 3. ♖f6 ♔g8 4. ♖f4 h5 5. ♖f6 +–

5262.
Zsu. Polgár – M. Voiska, Budapest 1992
1. ♖f6 gf6 2. ♕h7 ♔f8 3. ♕h8 ♔e7 4. ♖e6 fe6 5. ♕g7 ♔e8 6. ♕g8 ♔e7 7. ♕f7 +–

5263.
J. Polgár – L. Pliester, Aruba 1992
1. ♖f7 ♘f7 2. ♖f7 ♔f7 3. ♘b6 ♔g6 4. ♗h5 ♔h5 5. ♕f7 +–

5264.
J. Polgár – E. Bareev, Hastings 1992
1. ♖g7 ♔g7 2. ♖g1 ♔h8 3. ♘f7 ♔h7 4. ♘h6 +–

5265.
J. Polgár – P. Benkõ, Aruba 1992
1. ♗g6 ♕g7 2. ♖g3 fe5 3. de5 +–

5266.
J. Polgár – J. Rubinetti, Buenos Aires 1992
1. ♘b5 ♖a1 2. ♗a1 ♕a1 3. ♘d6 ♖b8 4. e5 ♗f3 5. ♗h7 ♔h7 6. ♖a1 ♗g5 7. ♕c3 ♗d5 8. dc5 ♘d7 9. ♖a4 +–

5267.
Zsu. Polgár – V. Smyslov, Vienna 1993
1. ♕a2 ♘a2 2. ♖c8 ♔h7 3. ♖h8 ♔h8 4. ♘g6 ♔h7 5. ♘h4 +–

5268.
J. Polgár – B. Spassky, Budapest 1993
1. ♗e5 ♘e5 2. ♕f5 ♔g7 3. ♕e5 ♔g6 4. ♕f5 ♔h6 5. ♕f6 ♔h7 6. ♕f7 +–

5269.
J. Polgár – J. Gdanski, Budapest 1993
1. ♖a4 ♕a4 2. h6 g6 3. ♗f6 ♖f6 4. ef6 ♔g8 5. ♕c7 ♘b4 6. cb4 ♕b4 7. ♔e3 g5 8. ♘g5 +–

5270.
J. Polgár – D. Barua, Biel 1993
1. ♖f6 ♖f6 2. ♖e2 ♖e7 3. ♖e6 ♖f8 4. ♔d2 ♖e6 5. de6 ♔e7 6. ♔e3 ♔f6 7. ♔f4 b5 8. ♘f3 ♖g8 9. ♘d2 ♔e7 10. ♘f1 d5 11. ♘d2 ♔d6 12. ♘f3 d4 13. ♘d2 c4 14. ♘e4 ♔e7 15. ♔e5 +–

5271.
J. Polgár – J. L. Fernandez, Dosttermanas 1993
1. ♕h7 ♔h7 2. ♖f7 ♔h6 3. ♖h8 +–

5272.
J. Polgár – J. Rogers, Biel 1993
1. ♗g6 ♖d3 2. ♗h7 +–

5273.
J. Polgár – F. Hellers, Biel 1993
1. ♖d5 ♖g6 2. ♖d7 ♔f8 3. ♖f7 ♔f7 4. ♕d7 +–

5274.
J. Polgár – L. Polugaevsky, Aruba 1991
1. ♗f6 ♖c1 2. ♖c1 ♖f6 3. ♖cg1 ♖f1 4. ♖f1 de4 5. ♖fg1 ♕d3 6. ♕d3 ed3 7. ♖e1 ♗c2 8. ♔c1 +–

5275.
Zsó. Polgár – M. Chiburdanidze, Budapest 1992
1. ♘hf5 gf5 2. ♘f5 ♔f8 3. ♖g3 f6 4. ♕g4 +–

5276.
Zsó. Polgár – A. Galyamova, Budapest 1992
1. ♖d7 ♕d7 2. ♘h6 ♔h8 3. ♘f7 +–

5277.
Zsó. Polgár – Garcia Nunez, Guarapuava 1991
1. ♖h6 gh6 2. ♕h6 ♔g8 3. ♕g6 ♔f8 4. ♖f5 ♗f6 5. ♘f6 ♖f7 6. ♘h7 +–

5278.
Zsó. Polgár – J. Banas, Pila 1992
1. ♖e7 ♔e7 2. ♖g7 ♔f8 3. ♖d7 ♖g8 4. ♔f2 ♖g7 5. ♕d6 ♔g8 6. ♕e6 ♔h7 7. ♘d5 ♖f8 8. ♕f5 ♔h6 9. ♘e7 ♖ff7 10. ♘g8 +–

5279.
Dr. Pataky – Zsu. Polgár, Corr. 1978
1. – ♕h2 2. ♖h2 gh2 3. ♘c6 h1♕ 4. ♔e2 ♕h5 5. ♔f2 ♕f5 6. ♔g1 ♗d7 7. ♕d4 ♗c6 8. ♕h8 0-0-0 9. ♕g7 ♗e5 10. g4 ♗h2 –+

5280.
A. Smirnov – Zsu. Polgár, Teteven 1981
1. – ♕h2 2. ♔h2 ♖h3 –+

5281.
T. Needham – Zsu. Polgár, Westergate 1981
1. – ♗h3 –+

5282.
M. Nedelkov – Zsu. Polgár, Targoviste 1982
1. – ♘h3 2. ♕h5 f4 3. ♘g2 ♘f2 –+

5283.
T. Kántor – Zsó. Polgár, Zánka 1982
1. – ♕f5 2. ♕g3 ♕f1 3. ♕g1 ♖g1#

5284.
T. Balogh – Zsó. Polgár, Budapest 1982
1. – ♕d4 2. h3 ♘f2 3. ♔h2 ♕f4 4. g3 ♘g4 +–

5285.
Torma – Zsó. Polgár, Budapest 1982
1. – ♖g2 –+

5286.
I. Hausner – Zsu. Polgár, Kecskemét 1983
1. – ♘f5 2. ♗a3 ♕d4 3. ♔h1 ♘e3 4. ♕e6 ♔h7 5. ♗f8 ♖f8 6. ♖fe1 ♕e4 7. ♕h3 ♕d4 8. ♖ac1 ♘g4 –+

5287.
I. Varasdy – Zsu. Polgár, Fonyód 1983
1. – h6 2. ♗f6 ♗f6 3. ♖a2 ♗c3 4. ♕c3 ab4 5. ♕d2 ♖a2 6. ♕a2 ♕b6 –+

5288.
J. Orsó – Zsu. Polgár, Kecskemét 1983
1. – ♕b4 2. ♕d3 e4 3. ♕a6 e3 4. ♔f1 f3 5. ♗g3 e2 –+

5289.
R. Herendi – Zsu. Polgár, Kecskemét 1984
1. – ♕g6 –+

5290.
V. Pavlov – Zsu. Polgár, Targoviste 1984
1. – a4 2. ab4 ab3 3. ♘a3 b2 4. b5 ♔d5 5. b6 ♔c6 6. ♔g2 ♘e3 7. ♔g3 ♔b6 8. f5 ♔c5 9. ♔f4 ♘f5 –+

5291.
P. Dimitrov – Zsu. Polgár, Teteven 1984
1. – ♗h3 2. gh3 ♕h3 3. ♗d1 ♘d2 4. ♖c2 ♘f3 5. ♗f3 ♗h2 6. ♔h1 ♕f3 7. ♔h2 ♖e4 –+

5292.
S. Stefanov – Zsó. Polgár, Teteven 1984
1. – ♕f2 2. ♖g1 ♖d1 3. ♗e3 ♕g2#

5293.
Nief – Zsu. Polgár, Hamburg 1985
1. – ♘g3 2. hg3 hg3 3. ♔g1 ♕d8 4. ♗f2 ♖h1 –+

5294.
S. Lewine – Zsu. Polgár, Las Vegas 1986
1. – ♘d5 –+

5295.
L. Lengyel – Zsu. Polgár, Budapest 1986
1. – ♘c4 2. ♘d1 ♕d1 3. ♕d1 ♘e3 4. ♔f2 ♘d1 5. ♖d1 ♗b5 –+

5296.
R. Rowley – J. Polgár, Hastings 1988
1. – ♗g3 2. hg3 ♖d8 3. a5 ♕f2 4. ♕b7 ♔f6 5. ♕b2 ♔g6 6. ♗e4 f5 7. ♗f5 ef5 8. ♖e6 ♔h7 9. ♕c1 ♖d2 –+

5297.
T. Pfeifer – J. Polgár, Budapest 1988
1. – ♗g4 2. fg4 ♖c3 3. bc3 ♘e4 4. ♕d3 ♘c3 5. ♔c1 ♕a3 6. ♔d2 ♘d1 7. ♔d1 ♕a2 8. hg6 hg6 9. ♖h2 ♕a4 10. g5 ♖c3 11. ♕d2 a5 12. ♕f2 ♕e4 13. ♖h3 a4 14. ♘c1 a3 15. ♖f3 a2 –+

5298.
E. Klimova – Zsu. Polgár, Thessaloniki 1988
1. – ♗g3 2. ♗d3 ♕h4 3. h3 ♕g5 4. ♘c6 ♘h4 5. ♔h1 ♘g2 6. ♕g2 ♕h4 7. ♔g1 ♗h3 8. ♕h1 ♖f6 9. ♗d8 ♖d8 10. ♘d8 ♖h6 11. ♘e6 ♕e7 –+

5299.
G. Gislasson – Zsó. Polgár, Reykjavik 1988
1. – ♖e2 –+

5300.
S. Bergsson – Zsó. Polgár, Reykjavik 1988
1. – ♘e5 2. fe5 ♕e5 3. ♗f2 ♖d4 –+

5301.
Xie Jun – J. Polgár, Thessaloniki 1988
1. – ♗g2 2. ♕g5 ♖4h7 3. ♔g2 ♖g8 4. ♕g8 ♔g8 –+

5302.
D. Rodriguez – Zsu. Polgár, Tunja 1989
1. – ♕c3 –+

5303.
S. Mohr – J. Polgár, Amsterdam 1989
1. – ♖f4 2. gf4 ♔d5 3. ♔g2 ♔e4 4. b4 ♖d1 5. ♖d1 ♗b4 –+

5304.
A. Greenfeld – J. Polgár, Haifa 1989
1. – ♖d2 2. ♖e1 ♖d4 –+

5305.
S. Drazic – Zsó. Polgár, Novi Sad 1990
1. – ♗f3 2. ♖ef3 ♖b3 3. ♗b3 ♖g7 4. ♘c5 ♕f5 –+

5306.
M. Chiburdanidze – Zsu. Polgár, Novi Sad 1990
1. – ♘h3 2. d4 ♘f4 3. ♗e1 ♕g5 4. d5 ♗d7 5. ♗b1 ♘e2 6. ♘g4 e4 7. ♘h2 ♗e5 8. ♖f2 ♕g3 9. f4 ♗f4 10. ♖f4 ♕f4 –+

5307.
R. Bertholee – J. Polgár, Amsterdam 1990
1. – ♖b4 2. ab4 b2 –+

5308.
A. Chernin – J. Polgár, New Delhi 1990
1. – ♖g2 2. ♖g2 ♗h3 3. ♘e4 ♘e5 4. ♘e5 ♗e5 5. ♘g5 ♗g2 6. ♔g2 ♕g5 7. ♔f3 ♖g8 8. ♔e2 f3 –+

5309.
L. Portisch – Zsu. Polgár, Budapest 1991
1. – ♖d2 2. ♖a8 ♘f2 3. ♔g2 ♘e1 4. ♔g1 ♘h3 5. ♔h1 ♕f1 6. ♗f1 ♘f2 7. ♔g1 ♘f3 –+

5310.
A. Grosar – Zsu. Polgár, Portoroz 1991
1. – ♗g3 2. ♖g2 ♗h2 3. ♔f1 ♕h3 4. ♕d2 ♖e3 –+

5311.
K. Mokry – Zsu. Polgár, Brno 1991
1. – ♖d6 2. ♕e2 ♖d5 3. ♖c1 g6 4. b3 ♗f4 5. ♖d1 ♖h5 6. ♖d8 ♔g7 7. ♘d6 ♕b3 –+

5312.
D. Sahovic – Zsó. Polgár, Royan 1988
1. – ♖c4 2. bc4 ♖a4 3. ♖h1 ♔g7 4. ♖h5 ♔g6 5. ♖h1 ♖c4 6. ♔f3 ♖c2 7. ♖b3 c4 8. ♖b7 ♖c3 9. ♔g2 g4 10. ♖h4 f5 11. ♖e7 ♗e5 12. ef5 ♔f5 13. ♖h5 ♔e4 14. ♖he5 de5 15. d6 ♖d3 16. ♖c7 ♖d4 –+

5313.
I. Balogh – J. Polgár, Budapest 1984
1. – ♗h3 2. ♗h3 ♖g3 3. ♗g2 ♖e3 4. ♔f1 ♕g4 –+

5314.
J. Costa – J. Polgár, Biel 1987
1. – ♘f2 2. ♔f2 fe3 3. ♔e1 ♖f2 4. ♖g1 ♕f8 5. ♕d3 ♘a6 6. a3 ♗f5 7. ♗e4 ♗e4 8. ♕e4 ♗d4 9. ♗e3 ♘c5 10. ♗d4 ed4 11. ♕d4 ♖e8 12. ♔d1 ♕f3 –+

5315.
J. Tisdall – J. Polgár, Reykjavik 1988
1. – ♕a4 –+

5316.
Gy. Horváth – J. Polgár, Zalaegerszeg 1990
1. – ♖f2 2. ♘e7 ♖g2 –+

5317.
N. Gaprindashvili – Zsu. Polgár, Shanghai 1992
1. – ♘f2 2. ♔f2 ♔e7 3. ♔e2 ♕g3 –+

5318.
J. Timman – J. Polgár, Paris 1992
1. – g4 2. fg4 ♗f4 3. g3 hg3 4. ♔g2 ♖f2 5. ♔g1 ♖a2 –+

5319.
B. Gulko – J. Polgár, Aruba 1992
1. – ♘fe4 2. fe4 ♗e4 3. ♘e4 ♘e4 4. ♗e7 ♖fe8 5. ♖d7 ♘f2 6. ♗f3 e4 7. ♗g2 ♖ac8 8. c5 ♗f8 9. ♗d6 ♖cd8 10. ♖d8 ♖d8 11. ♘h3 ♘h1 12. ♗e4 ♗d6 13. cd6 ♖d6 14. ♔c2 ♖d4 15. ♗h1 ♖g4 16. ♗b7 ♖c4 17. ♔d3 ♖b4 18. ♗d5 ♖b2 19. ♘g5 ♖h2 20. ♘f7 ♔g7 21. ♘e5 h5 22. ♘f3 ♖b2 23. ♔e3 ♔f6 24. ♔f4 ♖g2 25. ♔e3 g5 26. ♘d2 ♖g3 27. ♗f3 ♔f5 28. ♘e4 ♖g1 29. ♔f2 ♖a1 30. ♘g3 ♔e5 31. ♗h5 ♖a2 32. ♔f3 a3 33. ♗f7 ♖b2 34. ♔g4 ♔f6 –+

5320.
L. Rentero – J. Polgár, Oviedo 1992
1. – ♖b3 2. ab3 ♕b3 3. ♖c1 ♕a3#

5321.
D. Bronstein – J. Polgár, Oviedo 1992
1. – e6 2. de6 ♖c6 3. ed7 ♖c2 4. ♔h1 ♘f2 5. ♔g2 ♘h3 6. ♖e2 ♖e2 7. ♔f1 ♖f2 8. ♔e1 ♘d7 –+

5322.
D. Garcia Ilundain – J. Polgár, Oviedo 1992
1. – ♘f3 2. ♖f3 hg4 3. hg4 ♗g4 4. ♖g2 ♗f3 5. ♕f3 g4 6. ♕e2 ♖g7 7. ♕e1 g3 8. ♗e2 ♘h7 9. ♘d2 ♕h4 10. ♘f3 ♕h3 11. ♗f1 ♘f6 12. ♖d2 ♕h5 –+

5323.
M. Adams – J. Polgár, Monaco 1992
1. – ♗g2 2. ♗b8 ♗f3 3. ♗e5 ♘e4 4. c4 f6 5. cb5 fe5 6. ba6 ♖a8 7. ♖c2 ♖a6 8. ♘c4 ♗d4 9. h4 ♔f7 10. ♘e5 ♗e5 11. ♖e3 ♗g4 12. ♖e4 ♗f5 13. ♖ec4 ♗c2 14. ♖c2 ♔f6 –+

5324.
A. Chernin – Zsu. Polgár, Budapest 1993
1. – ♖gg3 2. fg3 ♕g3 3. ♔h1 ♕h3 4. ♔g1 ♖g3 5. ♔f2 ♕h4 6. ♖h1 ♖h3 –+

5325.
A. Kolev – J. Polgár, Budapest 1993
1. – ♖a3 2. ♖d1 e4 3. ♘ce4 ♖a1 4. ♗a1 ♕a3 –+

5326.
Y. Seirawan – J. Polgár, Monte–Carlo 1993
1. – ♕h2 –+

5327.
A. Karpov – J. Polgár, Monte–Carlo 1993
1. – ♘c4 2. ♖aa1 ♘d2 3. ♕b7 ♘e4 4. ♖dc1 ♕e2 5. ♕e4 ♕c4 6. ♗g3 a5 7. ♕b1 a4 –+

5328.
O. Renet – Zsó. Polgár, Oviedo 1992
1. – ♘e4 2. ♗h3 ♘d2 –+

5329.
Favaro – Zsó. Polgár, Guarapuava 1991
1. – ♘h3 2. gh3 ♗h3 3. ♗e4 ♕g3 4. ♔h1 cd5 5. ♗f3 ♖ac8 6. ♕d3 ♖c6 7. ♗d5 ♗f1 8. ♖f1 ♖g6 –+

5330.
J. Mestel – J. Polgár, Oviedo 1993
1. – ♕f5 2. ♗f5 ♖h2 –+

5331.
A. Sznapik – Zsu. Polgár, Budapest 1993
1. – ♖ac8 2. ♖c8 ♖c8 3. ♖c8 ♕c8 4. ♘e2 ♗b2 –+

5332.
A. Ker – Zsu. Polgár, Wellington 1988
1. – ♖f2 2. ♔f2 e3 3. ♕e3 ♖f8 4. ♕f3 ♗c6 5. ♕f8 ♔f8 6. ♖hf1 ♔e8 7. ♖c3 ♕f5 8. ♔e1 ♕b1 9. ♔f2 ♕f5 10. ♔e1 ♕h3 11. ♘d2 ♗g2 12. ♖f2 ♕h1 13. ♘f1 h5 14. e3 ♗f1 15. ♖f1 ♕g2 16.♖b3 ♕g3 17. ♔e2 h4 18. ♖b8 ♔d7 19. ♖h8 g5 20. ♖ff8 ♕g4 21. ♔d3 ♕d1 22. ♔c3 ♕c1 23. ♔d3 ♕b2 24. ♖d8 ♔e6 25. ♖h6 ♔f5 –+

5333.
B. Ruban – J. Polgár, Groningen 1993
1. – ♗f3 2. gf3 ♖g8 3. ♘h2 ♗f8 4. ♔h1 ♗c5 5. ♖e2 ♖d7 6. ♘a4 ♗f2 –+

5333+1.
Zsó. Polgár 1981
1. ♕b2

BIOGRAPHY OF LÁSZLÓ POLGÁR

László Polgár was born in 1946 in Hungary, where he studied philosophy and wrote a doctoral thesis on chess pedagogy. After working as a teacher for fifteen years, he became full-time coach, manager and public relations officer to his daughters, who have won eleven Olympic gold medals and several world championships. Winners of numerous Hungarian and international chess awards, the Polgár girls have topped the world rankings fourteen times, and two of them are currently first and second in the list. László Polgár holds the title of master coach, and has published widely on chess tutoring. Among his books are *Nevelj zsenit!* (Budapest, Interart, 1989) and *Le Phénomène Polgár, ou l'Art de Former des Génies* (Bruxelles, La Renaissance du Livre, 1991).

Bibliography

Abrahams, G.: *Test Your Chess*, New York, 1964.

Aguilera, R.: *El error en la apertura*, Madrid, 1964.

Aguilera, R.: *500 Celadas*, Madrid, 1966.

Alexander, A.: *Praktische Sammlung bester und höchst interessanter Schachspiel-Probleme*, Zürich, 1979.

Alster, L.: *Šachy Hra Kralovska*, Praha, 1987.

Alster, L.: *Miniaturni Šachove Partie*, Praha, 1978.

Andrić, D.: *Matni Udar*, Zagreb, 1981.

Anton, R.: *Opferkombinationen im Schach*, Leipzig, 1952.

Aptekar, L.: *Wisdom in Chess*, New Zealand, 1987.

Averbach, Y.: *Chess Tactics for Advanced Players*, Berlin, 1984.

Averbach, Y.: *Schachtaktik für Fortgeschrittene*, Berlin, 1983.

Ayal, S. – Hajadi, L.: *Menang Cepat And Indah Pilihan 66 Partai Mini*, Jakarta, 1974.

Bachmann, L.: *Schach-Spiele*, Berlin, 1917.

Bachmann, L.: *Schachmentor*, Ansbach, 1909.

Bagnoli, P.: *Scacchi Matti*, Milano, 1974.

Baird, W. J.: *The Twentieth Century Retractor and Chess Novelties*, London, 1907.

Bakcsi, Gy.: *Kevésbábos magyar feladványok*, Budapest, 1982.

Bán, J.: *Végjáték-iskola*, Budapest, 1965.

Barden, L. – Heidenfeld, W.: *Modern Chess Miniatures*, New York, 1960.

Barden, L.: *Chess Puzzle Book*, London, 1977.

Baswedan, A.: *Gerakan Taktis Diatas Papan Catur*, Surabaya, 1982.

Benzinger, J.: *Lehrreiche Kurzpartien*, Hamburg, 1964.

Bertolo, R. – Risacher, L.: *Objectif mat*, Paris, 1976.

Bhend, E.: *Kombinieren und Angreifen I-II-III*, Basel, 1986.

Bjelica, D.: *Čudesni Šahovski Svijet*, Beograd, 1978.

Blumenthal, O.: *Schachminiaturen*, Leipzig, 1902.

Bondarevski, I.: *Kombinacije u Sredisnici*, Beograd, 1975.

Böök, E. E.: *120 Lyhyttä Loistopeliä*, Helsinki, 1949.

Caputto, Z. R.: *El arte del estudio de ajedrez*, Madrid, 1992.

Chernev, I.: *Combinations – The Heart of Chess*, New York, 1960.

Chernev, I.: *1000 Best Short Games of Chess*, London, 1957.

Chernev, I. – Reinfeld, F.: *Winning Chess*, London, 1947.

Chéron, A.: *Lehrbuch und Handbuch der Endspiele*, Berlin, 1957.

Chéron, A.: *Miniatures stratégiques françaises*, Nancy–Strasbourg–Paris, 1936.

Clarke, P. H.: *Cien Miniatures Rusas,* London, 1963.

Czarnecki, T.: *Szachowe Klejnoty*, Warszawa, 1980.

Czarnecki, T.: *Pulapki Szachowe*, Warszawa, 1956.

Diemer, E. J.: *Vom ersten Zug an auf Matt!*, Amsterdam, 1957.

Djamal, D.: *Perangkap Dan Muslihat*, Jakarta, 1987. IV.

Djamal, D.: *Perangkap Dan Muslihat*, Jakarta, 1987. I.

Du Mont, J.: *The Basics of Combination in Chess*, London, 1938.

Du Mont, J.: *200 Miniature Games of Chess*, London, 1941.

Dyckhoff, E.: *Fernschach Kurzschlüsse*, Bad Tölz, 1948.

Edwards, R.: *Chess Tactics and Attacking Techniques*, London–Boston, 1978.

Eeden, E. van: *De Polgár-zusters, Of: De creatie van drie schaakgenieën*, Amsterdam, 1990.

Euwe, M.: *Strategy and Tactics In Chess*, London, 1937.

Forbes, C.: *The Polgar Sisters, Training or Genius*, London, 1992.

Geisler, W.: *Kombinationen und Opfer*, Nürnberg, 1910?

Gelenczei, E.: *200 megnyitási sakkcsapda*, Budapest, 1958.

Gelenczei, E.: *175 új megnyitási sakkcsapda*, Budapest, 1972.

Gillam, A. J.: *Starting Chess*, London, 1977.

Gillam, A. J.: *Simple Checkmates*, London, 1978.

Gillam, A. J.: *Simple Chess Tactics*, London, 1978.

Godal, I.: *154 Norske Miniatyrproblemer*, København, 1973.

Golz, W. – Keres, P.: *Schönheit der Kombinationen*, Berlin, 1974.

Gutmayer, F.: *Der fertige Schach-Praktiker*, Leipzig, 1923.

Gutmayer, F.: *Turnierpraxis*, Berlin–Leipzig, 1922.

Gutmayer, F.: *Die Schachpartie*, Leipzig, 1913.

Gutmayer, F.: *Die Geheimnisse der Kombinationkunst*, Leipzig, 1914.

Gutmayer, F.: *Der Weg zur Meisterschaft*, Berlin–Leipzig, 1923.

Hairabedian, K.: *Shakhmatni Zadachi-Miniaturi*, Sofia, 1980.

Hannelius, J.: *100 År Finländska Miniatyrer*, København, 1984.

Хенкин, В.: *Последний шах*, Москва, 1979.

Hodgson, J.: *Chess Traveller's Quiz Book*, London, 1993.

Hooper, D. – Cafferty, B.: *Play For Mate!*, London, 1977.

Horowitz, I. A.: *Winning Chess Tactics Illustrated*, New York, 1972.

Hütler, C.: *Schnellmatt, 333 kurze, brillante Schachpartien*, München, 1913.

Hütler, C.: *Schnellmatt, 700 kurze, brillante Schachpartien*, Leipzig, 1924.

Jørgensen, W.: *Albino Og Pickaninny*, København, 1970.

Kagan, B.: *300 kurze Glanzpartien in 6 Heft*, Berlin, 1921?

Карпов, А.: *Миниатюры чемпионов мира*, Москва, 1986.

Kasparyan, G. M.: *Domination in 2545 Endgame Studies*, Moskva, 1980.

Kautsky, V. – Prokeš, L.: *Sbornik Partii*, Praha, 1917.

Korn, W.: *The Brilliant Touch in Chess*, New York, 1965.

Krogius, N. – Livsic, A. – Parma, B. – Taimanov, M.: *Encyclopedia of Chess Middlegames*, Beograd, 1980.

Larsen, B.: *Du kan kombinera*, Stockholm, 1975.

Link, O.: *Vidi Vici auf dem Schachbrett*, Berlin, 1921.

Livshitz, A.: *Test Your Chess IQ, I-II*, Oxford–New York–Toronto–Sydney–Paris–Frankfurt, 1981.

Lossa, G.: *Matt dem König, aber wie?*, Stuttgart, 1987.

Marič, R.: *Šahovske Minijature*, Osijek, 1993.

Matanovic, A.: *Encyclopedia Of Chess Endings I-V*, Beograd, 1986–1991.

Mazukewich, A.: *Verflixte Fehler, 500 lehrreiche Minipartien*, Berlin, 1986.

Mingrelie, D.: *Fins De Partie*, Kiev, 1903.

Müller, G.: *Springertaktik von A–Z*, Berlin, 1986.

Müller, H.: *Lerne kombinieren*, Hamburg, 1969.

Надареишвили, Г. – Акобия, Я.: *Мат в этюдах, Тбилиси*, 1990.

Négyesy, Gy. – Hegyi, J.: *Így kombináljunk*, Budapest, 1965.

Négyesy, Gy. – Hegyi, J.: *Sakk-kombinációk*, Budapest–Pécs, 1955.

Нейштадт, Я.: *Шахматный практикум*, Москва, 1980

Neistadt, Y.: *Eröffnungsfehler und lehrreiche Kombinationen*, Heidelberg, 1978.

Нейштадт, Я.: *Жертва ферзя, Москва*, 1989.

Nesis, G.: *Tactical Chess Exchanges*, London, 1981.

Niemeijer, M.: *Nederland in miniatuur*, Wassenaar, 1972.

Orbán, L.: *Schach-Taktik. Kunstgriffe und Kombinationen*, München, 1979.

Orbán, L.: *Schach als Denkspiel*, München, 1974.

Ovadia, J. M.: *Brillante Kombinationen*, Belgrad, 1925.

Pachman, L.: *Moderne Schachtaktik, I-II*, Berlin, 1966.

Palau, L.: *Ejercicios de Combination con Finales Brillantes*, Buenos Aires, 1958.

Paoli, E.: *L'Arte della combinazione scacchistica*, Milano, 1976.

Pohla, H.: *Ettevaatust-Loks!*, Tallinn, 1975.

Polgár, Zsu.: *Schachmatt in 2 Zügen*, Düsseldorf, 1986.

Polgár, Zsu.: *Schachmatt in 2 Zügen. 200 neue Mattkombinationen*, Düsseldorf, 1987.

Polgár, Zsu.-Zsó.-J.: *Polgar Tactics, 77 Chess Combinations*, Budapest, 1991.

Polgár, L.: *Nevelj zsenit!*, Budapest, 1989.

Polgár, L.: *Le Phénomène Polgár, ou l'Art de Former des Génies*, Bruxelles, 1991.

Pongó, I.: *A sakktaktika titkai*, Budapest, 1986.

Postma, S.: *De Polgár-zusjes*, Venlo, 1989.

Pötzsch, A.: *Spaß am kombinieren*, Berlin, 1986.

Randviir, J.: *Matt*, Tallinn, 1975.

Reinfeld, F.: *Chess Secrets Revealed*, Hollywood, 1959.

Reinfeld, F.: *1001 Winning Chess Sacrifices and Combinations*, Hollywood, 1955.

Reinfeld, F.: *The Way to Better Chess*, London, 1959.

Reinfeld, F.: *Great Short Games of the Chess Masters*, New York, 1961.

Reinfeld, F.: *Improving Your Chess*, London, 1959.

Reinfeld, F.: *How to Win Chess Games Quickly*, New York, 1956.

Renaud, G. – Kahn, V.: *Der erfolgreiche Mattangriff*, Hamburg, 1969.

Richter, K.: *Schachmatt*, Berlin, 1958.

Richter, K.: *Combinaciones en Ajedrez*, Barcelona, 1972.

Richter, K.: *Das Matt*, Berlin, 1942.

Richter, K.: *Der Weg zum Matt*, Berlin, 1941.

Richter, K.: *666 Kurzpartien*, Berlin, 1966.

Ройзман, А.: *300 Партий-миниатюр*, Минск, 1972.

Ройзман, А.: *444 сраженных короля*, Минск, 1987.

Roisman, A.: *400 Kurzpartien*, Berlin, 1980.

Ройзман, А.: *Шахматные дуэли, Минск*, 1976

Russ, C.: *Miniature Chess Problems*, New York, 1982.

Ruttkay, A.: *420 miniatür – Középcsel, Falkbeer ellencsel, Királycsel*, Budapest, 1978.

Ruttkay, A.: *1000 szicíliai miniatür*, Budapest, 1981.

Ruttkay, A.: *204 miniatür, Francia Védelem*, Budapest, 1971?

Salvioli, C.: *Il gioco degli scacchi*, Livorno?

Савин, П.: *В мире шахматных комбинаций, Кишинев*, 1981.

Schulte, W. – Limbeck, W.: *Typische Mattstellungen*, Solingen, 1907.

Shamkovich, L.: *The Tactical World of Chess*, New York, 1981?

Shosin, J. W.: *Kombinationen und Fallen*, Leipzig, 1915?

Шумилин, Н.: *Шахматный задачник*, Москва, 1964.

Snosko-Borowsky, E.: *Eröffnungsfallen am Schachbrett*, Berlin, 1965.

Сокольский, А.: *Внимание, ловушка,* Москва, 1970.

Speckmann, W.: *Schachminiatüren. Zweizüger*, Berlin, 1965.

Speckmann, W.: *Kleinste Schachaufgaben*, Berlin, 1970.

Суэтин, А.: *Дебют и миттельшпиль, Минск*, 1980.

Суэтин, А.: *Как играть дебыт, Москва*, 1981.

Szilágyi, S.: *Debrecena Bulteno*, Debrecen, 1983.

Teschner, R. – Miles, A. J.: *It's Your Move*, London, 1972.

Тишков, А. – Чепижный, В.: *Шахматные задачи–миниатюры, Москва*, 1987.

Tranquille, H.: *Voir clair aux echecs*, Ottawa, 1972.

Vaisman, V.: *O Idee Strabate Deschiderile, București*, 1983.

Вайнштейн, Б.: *Комбинации и ловушки в дебюте,* Москва, 1965.

Вольчок, А.: *Уроки шахматной тактики*, Киев, 1977.

Vuković, V.: *Das Buch vom Opfer*, Berlin, 1964.

Vuković, V.: *Skola Kombiniranja*, Zagreb, 1978.

Deutsche Schachzeitung, 1846–1986.

Fide-Album: 1914–1944 I-III, 1945–1955, 1956–1958, 1959–1961, 1962–1964, 1965–1967, 1968–1970, 1971–1973, 1974–1976, 1977–1979, 1980–1982, 1983–1985.

Magyar Sakkélet – Sakkélet, 1951–1993.

Magyar Sakkvilág, 1922–1950.

Šahovski Informator, 1–57, 1972–1993.

Шахматы в СССР, 1931–1990.

The Problemist, 1926–1993.